D0152485

DIGITAL LOGIC

DIGITAL LOGIC

Applications and Design

by

John M. Yarbrough

Oregon Institute of Technology

I(T)P™ *An International Thomson Publishing Company*

Boston • Albany • Bonn • Cincinnati • Detroit • London • Madrid • Melbourne • Mexico City
New York • Pacific Grove • Paris • San Francisco • Singapore • Tokyo • Toronto • Washington

PWS Publishing Company
20 Park Plaza
Boston, MA 02116-4324

Reprinted in 1997 by PWS Publishing Company, a division of International Thomson Publishing Inc.

I(T)P™
International Thomson Publishing
The trademark ITP is used under license.

Library of Congress Cataloging-in-Publication Data

Yarbrough, John M.
Digital logic : applications and design / John M. Yarbrough.
p. cm.
Includes index.
ISBN 0-314-06675-6 (alk. paper)
1. Logic circuits—Design and construction. I. Title.
TK7868.L6Y37 1997
621.39'5—dc20 96-1561
CIP

Sponsoring Editor: Bill Barter
Marketing Manager: Nathan Wilbur
Manufacturing Buyer: Andrew Christensen
Printing & Binding: West Publishing

Printed and bound in the United States of America.
97 98 99 00 01 02 — 10 9 8 7 6 5 4 3 2

This text is printed on recycled, acid-free paper.

For more information, contact:
PWS Publishing Company
20 Park Plaza
Boston, MA 02116

International Thomson Publishing Europe
Berkshire House
168–173 High Holborn
London WC1V 7AA
England

Thomas Nelson Australia
102 Dodds Street
South Melbourne, 3205
Victoria, Australia

Nelson Canada
1120 Birchmont Road
Scarborough, Ontario
Canada M1K 5G4

International Thomson Editores
Campos Eliseos 385, Piso 7
Col. Polanco
11560 Mexico D.F., Mexico

International Thomson Publishing GmbH
Königswinterer Strasse 418
53227 Bonn, Germany

International Thomson Publishing Asia
221 Henderson Road
#05-10 Henderson Building
Singapore 0315

International Thomson Publishing Japan
Hirakawacho Kyowa Building, 31
2-2-1 Hirakawacho
Chiyoda-ku, Tokyo 102
Japan

This book is dedicated to my wife Carol and my daughter Adelle
for their unwavering love, patience and support.

PREFACE

The purpose of *Digital Logic: Applications and Design* is to provide electrical, electronic, and computer engineering students, and engineering technology students with material fundamental to the design and analysis of digital circuits. It can also be used by practicing engineers, technologists, and technicians or anyone else, for that matter, with a basic science background, for self-study or as a review.

The typical audience might consist of college sophomores taking the first or second of a series of quarter or semester course(s) in digital logic and circuits. The material is comprehensive; that is, it covers classical combinational and sequential logic. It also covers material on PAL, PLA, GAL, and EPLD realization of logic circuits and has a chapter on circuit analysis of TTL, ECL, and CMOS logic families.

There are 10 chapters in the book, starting with basic principles and number systems, continuing on through combinational logic topics, sequential circuit analysis and design, asynchronous sequential circuit material, and ending with coverage of digital switching circuits. The organization of the book is as follows:

- Chapter 1 introduces digital concepts, number systems, binary codes, and arithmetic.
- Chapter 2 develops logic symbols and Boolean algebra.
- Chapter 3 defines combinational logic and the generation of switching equations from truth tables. It continues in the simplification of switching equations using Karnaugh maps, Quine-McClusky, and map-entered variable algorithms.
- Chapter 4 continues the treatment of combinational logic with design problems, multiplexers, decoders, adders, subtractors, and arithmetic logic units.
- Chapter 5 treats flip-flops, including timing specifications. The chapter continues with simple counters, MSI integrated circuit counters, and registers.
- Chapter 6 introduces synchronous sequential circuit design. Included are Mealy and Moore sequential circuit models, state diagram notation, development of transition and excitation tables, and derivation of excitation and output equations. Analysis of synchronous circuits and the design of counters is included.
- Chapter 7 continues with such sequential circuit design topics as state equivalence and reduction, state assignment, algorithm state machines, and linked sequential machines.

Chapter 8 develops asynchronous sequential circuit analysis and design. Fundamental and pulse mode circuits are discussed, analysis of asynchronous circuits, derivation of flow tables, state assignment problems, and design examples are included. Some discussion of data synchronizers and mixed-operating mode logic is given.

Chapter 9 introduces memory chips as a means of realizing combinational and sequential circuits. PLD design is discussed and examples given. The chapter also introduces two field programmable gate array (FPGA) architectures; the Xilinx and Actel FPGA devices. Chapter 9 covers digital switching circuits, starting with simple switches and progressing through the various TTL, ECL, and CMOS logic family circuits.

Chapter 10 concludes the text by presenting a fairly comprehensive treatment of digital integrated circuit logic families. Included are discussions of TTL, ECL, and CMOS circuits. In addition, tristate, open collector, and mixed logic subfamily interfacing is presented.

The electronics aspect of digital design and analysis is absent from many texts on digital logic. Combinational and sequential logic design from a strictly logic perspective is, in my view, incomplete. Digital integrated circuits are electronic devices. Factors such as fan-out considerations, power supply current, propagation delay, and timing analysis, all affect the design of working systems. To address the electronics as well as the logic aspects of digital design, the entire book uses actual IC devices. Becoming familiar with the wide array of ICs available to the logic designer is as important as knowing Boolean algebra and state diagram notation. I have attempted to show the internal logic as well as the logic symbols for many of the SSI and MSI devices commonly used.

Even though PLDs, FPGAs, and ASIC devices are the choices of industry logic designers, an understanding of fundamental principles imparted by using standard ICs prepares the student for dealing with ASIC libraries. Once the logic design principles are learned, any FPGA or ASIC design system can be mastered and used. It would be difficult to move directly from Boolean algebra, combinational, and sequential logic design principles into designing circuits using FPGAs and ASIC devices. FPGA and ASIC design platforms and software are still too expensive to find their way into undergraduate curricula. Writing a text using a given hardware platform and design software would necessarily exclude those not using that particular hardware and software combination.

I know that *almost* anyone can connect a series of digital integrated circuits to provide some useful result under ideal conditions. But what about nonideal conditions? The answer is found in Chapter 10 on "Digital Integrated Circuits," whose purpose is to develop the connection between the "logic" and the electronics that implement that logic. For those who have a good transistor circuits background, much of the early part of Chapter 10 can be eliminated by going straight to the specifications and characteristics of the IC themselves. For students more concerned with digital logic as a prelude to computer architecture courses and software, Chapter 10 can be skipped altogether.

COURSE PARTITIONING

Several partitions are possible with the material presented in the book, depending on relative emphasis, length of lecture and laboratory course (quarter or semester), and the amount of credit given. Some examples are shown here.

1. Quarter calendar, three-course sequence with emphasis on circuits as logic. Three lectures and three laboratory courses.
   a. First quarter: 3 credit lecture, 2 credit laboratory.
      Chapter 1. "Digital Concepts and Number Systems"
      Chapter 2. "Boolean Switching Algebra"
      Chapter 3. "Principles of Combinational Logic"
      Chapter 10. "Digital Integrated Circuits" (portions of Chapter 10 may be treated lightly or omitted depending on the emphasis of IC families and transistor circuits)
   b. Second quarter: 3–4 credit lecture, 1–2 credit laboratory.
      Chapter 4. "Analysis and Design of Combinational Logic"
      Chapter 5. "Flip-Flops, Simple Counters, and Registers"
      Chapter 6. "Introduction to Sequential Circuits"
   c. Third quarter: 3–4 credit lecture, 1–2 credit laboratory.
      Chapter 7. "Sequential Circuit Design"
      Chapter 8. "Asynchronous Sequential Circuits"
      Chapter 9. "Programmable Logic and Memory"
2. Semester calendar with the same coverage as given in the three-term course sequence.
   a. First semester.
      Chapter 1. "Digital Concepts and Number Systems"
      Chapter 2. "Boolean Switching Algebra"
      Chapter 10. "Digital Integrated Circuits"
      Chapter 3. "Principles of Combinational Logic"
      Chapter 4. "Analysis and Design of Combinational Logic"
   b. Second semester.
      Chapter 5. "Flip-Flops, Simple Counters, and Registers"
      Chapter 6. "Introduction to Sequential Circuits"
      Chapter 7. "Sequential Circuit Design"
      Chapter 8. "Asynchronous Sequential Circuits"
      Chapter 9. "Programmable Logic and Memory"

Laboratory assignments can be made from selected problems given at the end of each chapter. For example, Chapter 5, "Flip-Flops, Simple Counters, and Registers," might yield the following laboratory assignments:

1. Latches and flip-flops (problems 3, 4, 5, 6, 7, 8, 9)
   a. Construct an $S'R'$ latch from NAND gates. Verify the characteristic table and derive the characteristic equation.
   b. Add additional NAND gates to convert the $S'R'$ latch to a level triggered J–K flip-flop. Verify the operation by constructing a characteristic table. Identify the operation when $J = K = 1$. Suggest a solution for the problem encountered (convert the level trigger to a pulse trigger).
   c. Convert the pulse triggered J–K flip-flop to a D flip-flop and a T flip-flop. Verify the operation of each by creating characteristic tables. Derive the characteristic equations.
   d. Predict and measure power consumption and propagation delay.
2. Counters (problem 25)
   a. Design a modulo-8 counter using standard IC flip-flops.
      1. Construct the state diagram.

2. Determine the number of flip-flops needed.
3. Make a binary state assignment (000, 001, 010, 111).
4. Create a state table and a transition table.
5. Using *D* flip-flops create an excitation table (compare the transition and excitation tables).
6. Generate the simplified excitation equations.
7. Construct the circuit using TTL integrated circuit packages (*D* flip-flops, AND, NAND, etc. gates).

b. Predict the timing diagram for the modulo-8 counter.

c. Test the circuit operation.

1. Single step through each state and verify state transitions.
2. Connect a function generator to the clock inputs. Set the function generator to produce a square wave between 0 and 5 V at a frequency of 10 KHz. Measure the *Q* outputs of the flip-flops using an oscilloscope and produce a timing diagram.

3. Johnson counter (problem 26)

4. Special purpose pulse generator (problem 33)

5. Register system (problem 43)

SUPPLEMENTAL PACKAGE

- A solutions manual to all text problems (only odd answers are given in the text)
- For selected text illustrations, a set of transparency masters to use as overheads
- A lab manual is being developed, which is unique because other texts at this level do not have one (ask your West sales representative when it will be available)

ACKNOWLEDGMENTS

No author can complete a text without the assistance of many people. I would like to thank Jack Stewart, a student at Oregon Institute of Technology; the publishing professionals at West, and the following reviewers and accuracy checkers who deserve particular mention:

Gail Allwine
Gonzaga University (WA)

Kenneth J. Ayala
Western Carolina University (NC)

Dee Bailey
Mt. Hood Community College (OR)

William E. Barnes
New Jersey Institute of Technology (NJ)

Robert Belge
Syracuse University (NY)

John Brannon
Embry–Riddle Aeronautical University (FL)

Thomas Casavant
University of Iowa (IA)

Aaron Collins
Clemson University (SC)

Darrow Dawson
University of Missouri–Rolla (MO)

Richard Eason
University of Maine (ME)

Joe Enekes
Humber College (Ontario, Canada)

Milos Ercegovac
University of California–Los Angeles (CA)

Barry S. Fagin
Dartmouth College (NH)

Verne Hansen
Weber State University (UT)

Ralph Horvath
Michigan Technological University (MI)

Michael Horn
University of Hartford (CT)

Ernest Joerg
Westchester Community College (NY)

Harold Klock
Ohio University (OH)

Willard Korfhage
Polytechnic University (NY)

Sura Lekhakul
St. Cloud State University (MN)

Albert L. McHenry
Arizona State University (AZ)

James Migel
Red River Community College (Manitoba, Canada)

John Nagi
Hudson Valley Community College (NY)

John R. Pavlat
Iowa State University (IA)

James Rankin
St. Cloud State University (MN)

Dennis Suchecki
San Diego Mesa College (CA)

CONTENTS

CHAPTER 1

DIGITAL CONCEPTS AND NUMBER SYSTEMS

INTRODUCTION

Digital circuits, processing discrete information, are found in an astonishingly wide range of electronic systems. Consumer microwave ovens and television, advanced industrial controls, space communications, traffic control radar and hospital critical care units are all designed using digital techniques. Advances in digital circuit manufacturing have provided systems designers with more functions contained in less space, thereby improving system reliability and increasing speed. Digital systems operate on numbers that represent some real logical or arithmetic function. It is the designer's task to establish the relationship between the numbers and the task being performed. This chapter develops the use of numbers and how they represent real world conditions in digital systems.

The use of the words *digit* and *digital* was probably in association with counting; it can apply to decimal numbers (0, 1, 2, 3, . . . , 9) or numbers in another base such as binary (0, 1). Counting in decimal numbers and the number of fingers or toes are probably directly related in human history. More appropriate for students of electronics and computers, digital systems are systems that process discrete information. A digital system is simply one that receives input, processes or controls activity, and outputs information in a discrete or noncontinuous manner. The information may be encoded or represented in a variety of ways; it can be in the familiar base 10 number system or in some other number system, such as base 2 (binary).

Digital design is the *application of a set of rules and techniques for developing digital circuits and subsystems to create a solution for some problem.*

1.1 DIGITAL AND ANALOG: BASIC CONCEPTS

The difference between analog and digital lies in the method of encoding information. Digital electronics uses discrete quantities to represent information while analog electronics deals with continuous signals. **Discrete** means distinct or separated as opposed to **continuous** or connected. Analog systems process information that varies continuously. Examples of analog represented variables include temperature as measured with a mercury thermometer, clocks with hands, sine wave voltages indicated on a galva-

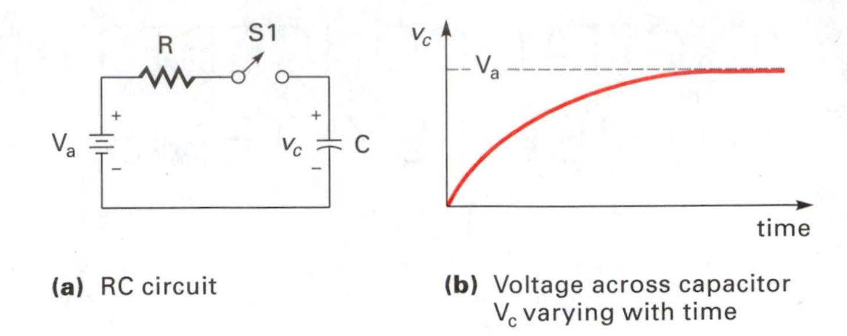

Figure 1.1 RC circuit and continuous plot of v_c

(a) RC circuit **(b)** Voltage across capacitor V_c varying with time

nometer or a needle speedometer in a car. The same variables can also be represented digitally in discrete instead of continuous quantities.

Consider the simple RC circuit in Figure 1.1(a). Perhaps you have encountered such a circuit in a physics or introductory electronics course. Let the initial charge of capacitor C equal zero; when switch S1 is closed, the voltage across C will vary continuously from 0 volts to the maximum of the battery voltage, V_a, as described by the formula

$$\begin{aligned} v_c(0) &= 0 \text{ V}, \qquad \text{Initial voltage at } t_0 \text{ is 0 V} \\ v_c &= V_a\,(1 - e^{-t/RC}) \end{aligned} \tag{1.1}$$

The voltage developed across C, Figure 1.1(b), illustrates an analog function; C increases exponentially until it reaches the same value as V_a at $t = \infty$. The "continuous" nature of the information is what causes it to be classified as **analog**.

Electronic systems tend to be classified as either analog (continuous) or digital (discrete). The mathematics of analog systems relates input to output in a continuous fashion; an amplifier gain equation describes how the input voltage will be amplified as a function of frequency, for example. Digital systems also have the ability to transform inputs into outputs. However, data are represented discretely, usually in a binary (base-2) number system, instead of in the more familiar decimal number system. Binary numbers are used, instead of decimal (base-10), due to the ease with which they can be implemented using electronic devices. It is easier to make a transistor operate between two discrete values than ten, as would be necessary if decimal numbers were used. Number systems—decimal, binary, and others—will be discussed later in this chapter.

A sine wave can be represented in either a continuous (analog) fashion or in a discrete (digital) fashion. Figure 1.2 illustrates the two methods of representing the same function.

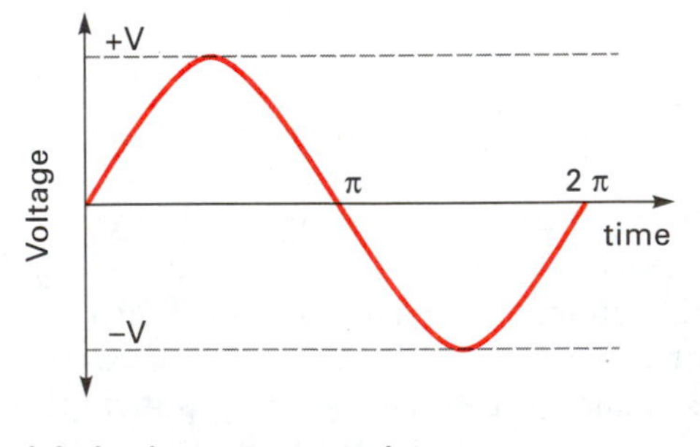

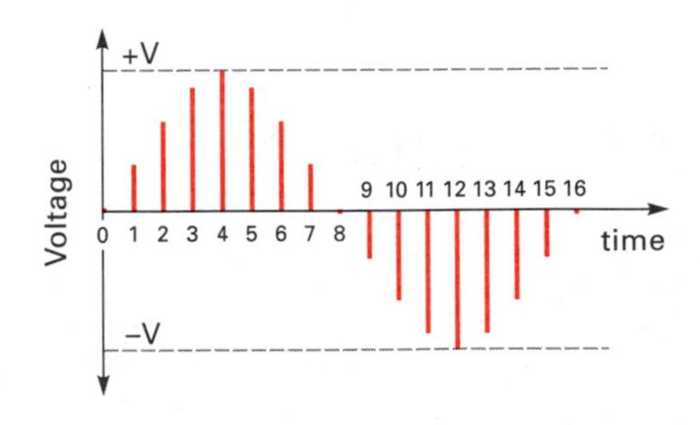

Figure 1.2 A sine wave voltage represented in continuous form and discrete form

(a) Analog representation **(b)** Discrete representation

TABLE 1.1
Discrete voltage values for time intervals

Degrees	v_c volts	Time interval
00.0	0.00	0
22.5	3.83	1
45.0	7.07	2
67.5	9.23	3
90.0	10.0	4
112.5	9.23	5
135.0	7.07	6
157.5	3.83	7
180.0	0.00	8
202.5	−3.83	9
225.0	−7.07	10
247.5	−9.23	11
270.0	−10.0	12
292.5	−9.23	13
315.0	−7.07	14
337.5	−3.83	15
360.0	0.00	16

Figure 1.2(a) illustrates a sine wave function in a continuous or analog form, and Figure 1.2(b) represents the same sine wave in a discrete or digital form. Each point of the sine wave shown in Figure 1.2(b) is represented by a discrete voltage at some time (t). Digital representation of a discrete voltage may be in a decimal number system or some other number system, such as binary. It is not the number system used that makes the representation digital but rather that the measurements are taken at separate intervals of time.

Let each discrete value for the sine wave shown in Figure 1.2(b) be expressed in a three-digit decimal number as presented in Table 1.1. Let a cycle of the sine wave be divided into 16 equal time intervals, each representing 22.5 degrees. The time division is arbitrary for this example; it certainly could be divided differently. The number of horizontal divisions determines time resolution, and the number of digits for each interval determines the voltage resolution of the function. Table 1.1 could also be represented in base-2 numbers, or any other number system.

Discrete representation of a function provides useful information only at the time it was generated. If, for instance, a measurement was taken at time interval 6, then $v_c = 7.07$ V. Voltage v_c is updated at time interval 7 to 3.83 V; in between time intervals 6 and 7 the value of v_c is unknown.

Figure 1.3 shows a block diagram of a typical analog meter for measuring voltage. The input voltage is analog. A range select switch allows the user to choose the setting for making a reading. An output meter continuously indicates the resulting voltage values. Figure 1.4 illustrates a digital volt meter. Notice that the input section is the same as the analog meter. An analog-to-digital converter is used to transform the continuous analog voltage to a discrete decimal value prior to display.

An example of a larger digital system is illustrated in Figure 1.5. Analog and digital inputs, representing certain automobile functions, are brought into subsystems. The analog input data are converted to digital signals and sent to a digital control system for processing. Digital signals between the control system and the various subsystems provide the necessary communication to control data flow. Outputs are sent from the subsystems to displays or to the main control system. For instance, engine RPM is required to control fuel flow. The engine RPM is sensed by a transducer (which converts physical RPM to an electrical signal), converted from an analog to a digital quantity, and then sent to the control system as part of the information it needs to control fuel flow. Other

Figure 1.3
Analog volt meter

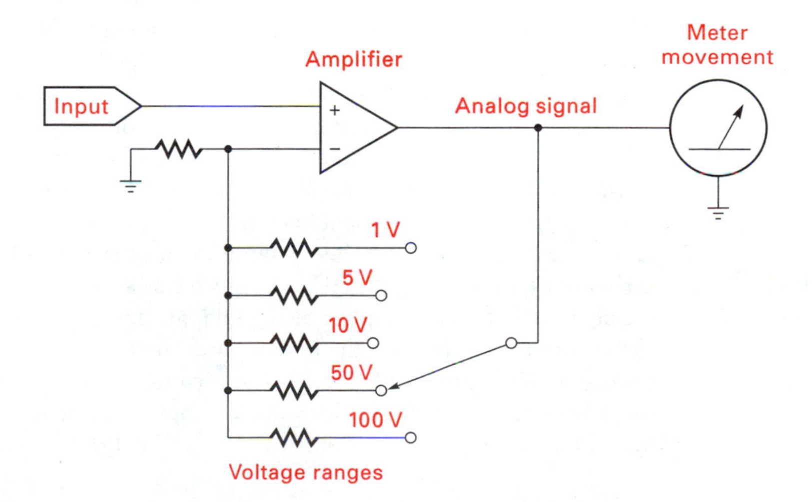

Figure 1.4
Digital volt meter

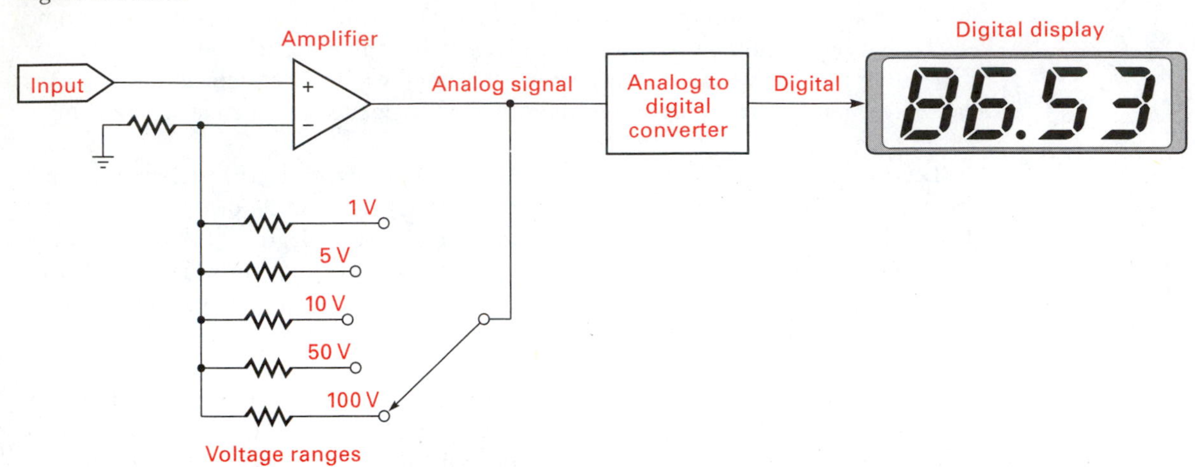

interactions occur in a similar manner. The digital control system may in fact be a microprocessor. A step-by-step procedure for performing an operation is called an **algorithm.** Control algorithms define a process of measuring needed information (input data), determining the desired responses (sequence of operations), and sending necessary output information to displays or output subunits.

1.2 SOME HISTORY OF DIGITAL SYSTEMS

The mathematical developments of Pascal, Boole, DeMorgan, Babbage, and others laid the foundations for modern digital systems. Blaise Pascal designed a mechanical adder in 1642; and in 1671, the German mathematician Gottfried developed a machine to multiply and divide. Charles Babbage, an English mathematician living in the 1800s, set out to develop an automatic calculating machine, used to compute values for navigation tables, that has been recognized as the forerunner to modern computers. His machine had reliability problems because of the technological limitations of that time. George Boole, another English mathematician, developed a special algebra that we now call *Boolean algebra* that has formed the core of modern logic design.

The Frenchman Jacquard developed an automatic loom for weaving cloth. He applied mathematical logic in the machine design for commercial use in the 19th century. In the early 20th century people were looking for ways to create practical adding machines, typewriters, and bookkeeping machines. Claude Shannon, working for Bell Laboratories during the 1930s, wanted to automate telephone switching. He took the earlier work of Boole and, in what is now a classical paper, outlined the modern switching algebra that is now applied to digital logic design. Development of electronics, from the vacuum tube through the invention of the transistor by Walter Brattain, John Bardeen, and William Shockley in 1947 to the integrated circuits of the 1960s, all furthered the advancement of digital logic and computer systems. Von Neumann, Elbert, and Mauchly, working at Princeton in 1946, outlined the structure that was to become the

Figure 1.5
Automobile digital control system

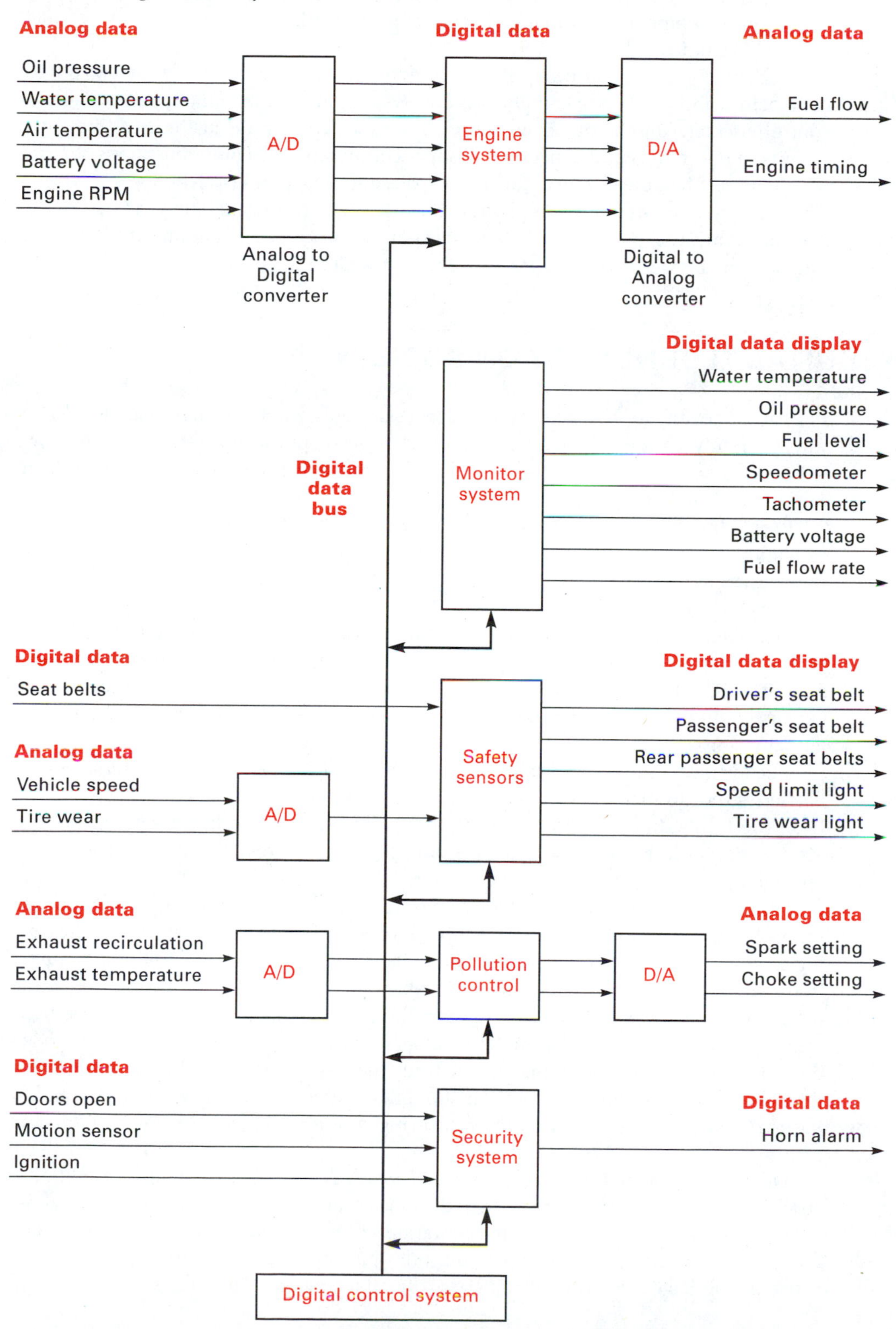

classical architecture for designing computers. In the early 1970s the first microprocessor was designed by Intel. This first "computer on a chip" was designed to provide a reduction in component count for a large desk top calculator (this was before the small handheld calculators hit the market).

Modern digital system designers are standing on the shoulders of those who have gone before and producing components and systems at a astonishing rate, from microcontrollers metering the fuel in an automobile to the large-scale customized integrated circuits found in the latest generation of super computers. Telecommunications circuits, which enable data to be transmitted from spacecraft to ground control, are available to the digital systems architect. Based on such work, the current generation can design and develop an array of products to help people control processes, communicate, entertain, calculate the mysteries of deep space, or forecast weather.

1.3 IMPACT OF DIGITAL TECHNOLOGY ON SOCIETY

Electronics, in any form, analog or digital, is useful only when it benefits people in some way. Certainly nothing about an electronic device, except for the materials from which it was made, makes it intrinsically valuable. The true value in such devices comes from the human benefit derived.

People buy television sets and stereo systems for entertainment, informational, and educational purposes, not as an investment. Those who fish buy consumer electronic sonar units to improve their chances of finding fish. Truck drivers use their CB radios to inform and entertain.

Manufacturing companies, ranging from automobile makers to lumber mills, tire manufacturers to toy makers, all use electronics for measuring, counting, controlling, and informing. Products from food processors to machine tools are made better and less expensive due to the impact of electronics, as automated control reduces the human tendency for error in the manufacturing process.

The average accounting firm uses telephones, desktop personal computers, and fax machines, not to mention copy machines and the microwave oven in the lunch room. Even the local auto parts store probably has a computer controlled inventory system. Many grocery stores have laser scanners inputting prices into a digitally controlled cash register.

Digital technology has affected computing from large, multimillion dollar mainframe computers to small personal computers and microprocessor controllers measuring the fuel to be injected in a new automobile engine. Many of us think of computers in the sense of medium to large systems in a bank or insurance company or for scientific research or military use. In the last 10 years the growth of personal computing has been tremendous. But most of the microprocessors and microcontrollers being sold today do not go into personal computers. Instead, they are being used by the millions in dishwashers, automobiles, aircraft, oscilloscopes, stereo systems, television sets, and in industrial control systems used for measuring lumber thickness or controlling steel making.

Digital technology is finding its way into areas of electronics normally thought of as strictly analog. Commercial two-way radios have microcontrollers setting the local oscillator frequency. Electronic instrumentation such as oscilloscopes, logic analyzers, function generators, power supplies, and multimeters measuring current, voltage, and resistance all have made widespread use of digital technology. In fact it is difficult to find a recently designed electronic instrument that does not use digital technology in some way. **Digital signal processing** (DSP) is a methodology that involves the conversion of analog information into digital form, processes it in some rather sophisticated manner, and produces an output, either analog or digital. Modern submarines use DSP to make sense of the acoustic noises that occur under the ocean, for example.

Digital technology is especially useful because it is less susceptible to electrical noise (spikes or fluxations in voltage or current), temperature effects, component aging, and other factors that affect analog systems. This does not mean that analog technology is not important or that students of electronics (and even computers) should not study analog circuits and systems. In fact, just the contrary is true. The concepts of radio propagation, antenna design, amplifier analysis and active filters, to mention a brief few, are more important than ever. To appropriately apply digital design principles to these areas certainly requires an in-depth understanding of analog principles. And, as the frequencies of digital devices continue to increase, digital systems designers are faced with solving analog-related problems. A 50-Mhz square-wave clock signal for a microcomputer system necessitated printed circuit layout practices that, until just recently, only microwave engineers worried about.

The application of digital technology will continue to grow in a wide range of consumer product, industrial control, military/defense, and service industries. Look for digital system speeds to increase, new devices with added complexity, integrated systems comprising both digital and analog functions embedded into a single integrated circuit, and even possibly the use of organic material for constructing memory devices.

1.4 DEFINING THE PROBLEM, AN INTRODUCTION TO ALGORITHMS

Designing digital systems often requires analysis of a task so that it may be partitioned into easily described functions. Each function interacts with the others in the system in some prescribed manner. Designers who attempt to solve logic problems without clearly defining requirements and constraints often find an unworkable or clumsy result. This section introduces a way of thinking that encourages defining the problem before attempting a solution.

Algorithms are normally associated with mathematics and computer programming, but the basic principle of problem definition is equally valid for logic hardware design and software development. Algorithms are systematic processes for solving a problem or controlling a process. The problem may be simple, requiring only a pencil and paper solution, or so complex a super computer is needed to find a solution.

The process may be making bread, measuring and displaying automotive variables such as speed and RPM, or controlling flight surfaces of a jet aircraft. An algorithm is a list of instructions specifying a sequence of operations that will generate a correct result to any problem of a particular type. A solution to a logic problem may find its final realization in hardware (what this book is about) or in software. An approach to problem definition that can result in designing digital systems involves

1. Determination of processing tasks (development of task performance).
   a. Determination of inputs to each task.
   b. Determination of outputs from each task.
2. Specification of component functions to accomplish the tasks.

Figure 1.6
Horn alarm flowchart

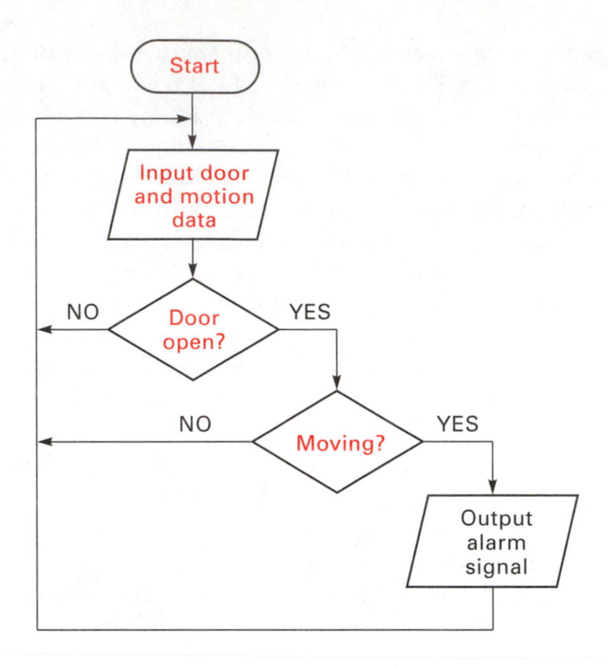

Figure 1.7
Fuel flow flowchart

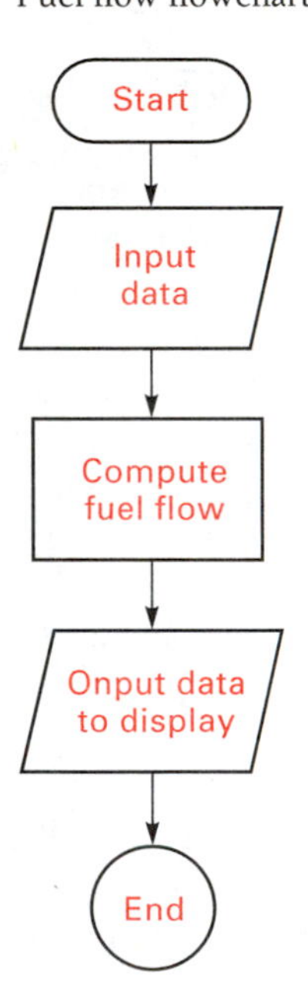

Any activity can be broken down into individual tasks, and the steps involved in accomplishing each task can be defined. Once the task has been defined, the inputs and outputs can be determined, as well as the relationships between tasks.

Designing digital hardware often requires the two basic steps just mentioned. We cannot design a digital circuit without defining its inputs and outputs as well as the circuit function. Many times it is useful to describe a problem using the aid of a **flowchart**, which is a diagram illustrating the "flow" of tasks. Later in the text we will study a design technique called *algorithm state machines,* ASM. ASM borrowed the flowchart symbols from computer programming and applied it to digital hardware design. Whether writing a program or designing digital system hardware, problem definitions may be accomplished by generating a "flowchart." Simple flowcharts are illustrated in Figures 1.6 and 1.7. Other flowcharts illustrating the same problem could also be designed. The flowchart "ellipse" symbol establishes the starting point in the problem solution. A trapezoid is used to indicate input to or output from the system. A rectangle indicates a function, which can be very simple or quite complex, at the discretion of the designer. A diamond indicates a decision or a branch in the process flow, usually determined by an input or an internal operation condition.

The automobile system block diagram presented in Figure 1.5 shows input and output signal transmission between the subunits and the main control. The digital control block in the diagram indicates where the functional processes occur. For instance, if the car door is ajar and the car in motion, then the horn alarm will sound. The requirement for turning on the horn would be represented by an output trapezoid in a flowchart. The decision to invoke the alarm is made by monitoring the door ajar and the car motion inputs, as represented by decision diamonds. Sounding the horn requires sending a signal to activate the horn solenoid.

Problem definitions can be described in general or in detail. Simple problems often do not need any graphical aids to visualize. Figure 1.6 illustrates the decision process for the horn alarm problem. Figure 1.7 shows a simple flowchart to indicate a complex function (calculate fuel flow) in a single function box. The fuel flow calculation function can, in turn, be represented by its own flowchart. Algorithms can be described several ways, only one of which is a flowchart; others include mathematical equations and graphical block diagrams.

1.5 DIGITAL SYSTEMS OVERVIEW

Digital design can be considered as a hierarchy. The hierarchy levels start at the bottom with basic electronic circuits, and progress upward, with each level indicating more complex functional units. The basic circuit is composed of individual components and designed to perform specified functions. Components, such as resistors, capacitors, inductors, and transistors, behave in specific ways of interest to the circuit designer but are not of immediate interest to the system designer. Digital circuits are designed and constructed to provide functions that become building blocks for logic designers in the development of more complex functions. The basic circuit building blocks are, in turn, considered to be components for even larger functional units, and so on. The basic functional units that logic designers consider are called *gates*. Gates perform specified logic functions needed to design larger, more complex functions.

Once an understanding of basic logic gates is accomplished, we need not be overly concerned with the internal electronic circuit detail; we are then free to concentrate on their use in designing higher level functions. A digital adder (adds two binary numbers) is a standardized logic function constructed from gates. Many other logic functions have been standardized and are available as integrated circuit packages. Many times a designer's task is to identify the necessary functions and choose the appropriate integrated circuits to accomplish the task. Digital integrated circuits, constructed from electronic components, are designed to provide specified logic functions. Simple integrated circuits contain a few logic gates and is called *small-scale integration* (SSI contains 0 to 9 gates). More complex functions require a larger number of gates and perform what is called *medium-scale integration* (MSI contains 10 to 99 gates). The process continues from MSI to *large-scale integration* (LSI contains 100 or more gates) and *very large-scale integration* (VLSI does not have a gate count but is usually thought of as more than 1000 gates). Integrated circuit manufacturers have identified common logic functions and provided these functions in integrated circuit packages.

At each level of digital design, additional concepts and principles are introduced. The functional complexity increases at each higher level. As gates become the building blocks for adders, so adders become a building block for microprocessors (VLSI component), and microprocessors become functional units for process control systems. This is presented graphically in Figure 1.8.

1.6 INTRODUCTION TO NUMBER SYSTEMS

In the previous sections, a distinction was made between analog and digital systems, by noting that digital systems represented information in a discrete form. Number systems of a given *radix* or *base* provide the means of quantifying information for processing by

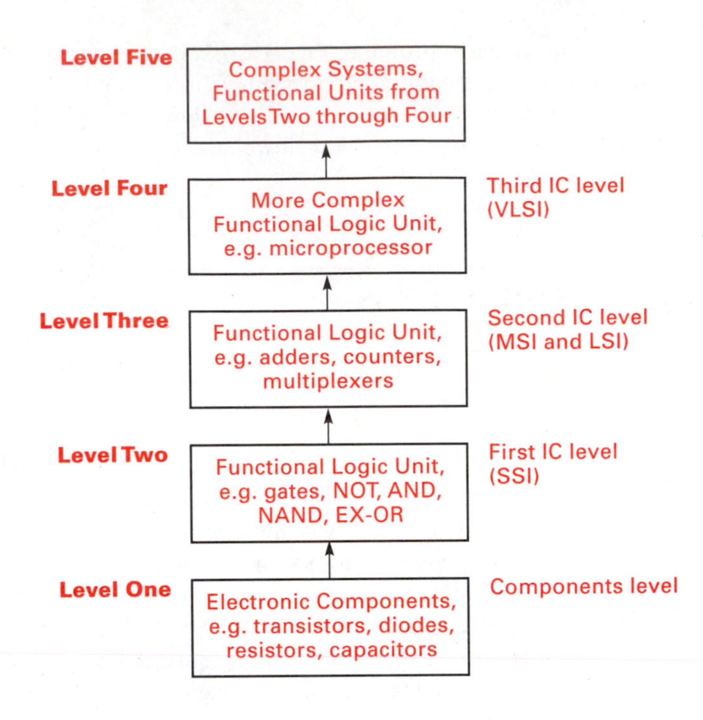

Figure 1.8
Simple digital hierarchy

digital systems. We are familiar with the base-10 number system; we have been using it since childhood for counting and arithmetic. The remaining part of this chapter deals with numbers (represented using several radices), the process of conversion from one base to another and some simple arithmetic using various bases.

Digital circuits using a binary radix are the most common, because the electronics necessary to process the information is simpler than the electronics needed to process decimal numbers. Binary numbers are not the only ones used in digital systems. Base-3 and base-4 number systems are being used by people doing research and development work in a field called multivalued logic. Integrated circuits have been designed and built in research laboratories using base-3 and base-4 numbers. Several different "formats" of binary numbers are in use, providing a wide application range.

1.7 POSITIONAL NUMBER SYSTEMS

Number systems use **positional notation** to represent value. The position of the character (numeral) in a character string (collection of possible numerals) indicates value as well as the character itself. For example, the base-10 number system contains 10 characters (0, 1, 2, 3, 4, 5, 6, 7, 8, 9); no character other than those in the set is used to represent a value no matter how large. Larger values, those greater than the largest character in the set, are indicated by position.

1.7.1 Decimal Numbers

Consider the decimal number 536. Notice that all the characters in the number are contained within the character set for **decimal** numbers (0 through 9). The decimal number 536, is also written 536_{10}, where the 10 subscript indicates the radix or base.

The position of each of the characters also represents its value:

$$(5 \times 10^2) + (3 \times 10^1) + (6 \times 10^0) = 536_{10}$$

Each position has a value determined by the radix or base and by its location with respect to a reference, in this case the decimal point. Fractional decimal numbers also use positional notation. The numbers to the left of the reference or decimal point have positive exponents and those to the right have negative exponents. The mixed decimal number 536.159 would be written

$$(5 \times 10^2) + (3 \times 10^1) + (6 \times 10^0) + (1 \times 10^{-1}) + (5 \times 10^{-2}) + (9 \times 10^{-3})$$

A general relationship from which we can describe any number in any base is the following one:

Let r = radix or base of a number system;

Let c = character from the character set of the radix;

Let N = number to be represented in radix r;

Let n = the number of digits in the integer portion of N;

Let m = the number of digits in the fractional portion of N;

Then the number N would be represented in radix r as

$$(N)_r = (c_{n-1}r^{n-1} + c_{n-2}r^{n-2} + \cdots + c_1r^1 + c_0r^0 + c_{-1}r^{-1} + c_{-2}r^{-2} + \cdots + c_{-m}r^{-m}) \quad (1.2)$$

The most significant digit is c_{n-1} and the least significant digit is c_{-m}. A more concise format for writing the expression is

$$(N)_r = \sum_{i=-m}^{n-1} c_i r^i \quad (1.3)$$

where i represents any digit position of number N.

1.7.2 Binary Numbers

Only two characters exist in the **binary** character set (0, 1). If a 1 occurs, the value associated with its position is noted and added to the value for other positions where a 1 occurs. Positions holding a 0 are not added to the total.

❖ **EXAMPLE 1.1** Let 11001.011_2 represent a binary number; find its decimal equivalent.

SOLUTION Writing the number in positional notation gives,

$$N = (1 \times 2^4) + (1 \times 2^3) + (0 \times 2^2) + (0 \times 2^1) + (1 \times 2^0)$$
$$+ (0 \times 2^{-1}) + (1 \times 2^{-2}) + (1 \times 2^{-3})$$

Performing the arithmetic on each term in the polynomial,

$$N = 16 + 8 + 0 + 0 + 1 + 0 + .25 + .125 = 25.375_{10}$$

Evaluation of any number, in any radix, is accomplished in a similar fashion. ❖

1.7.3 Octal Numbers

The octal number system has eight characters (0, 1, 2, 3, 4, 5, 6, 7). Counting in octal is similar to counting in base-10 (decimal) numbers, in that each position character represents value. When a position digit exceeds the value represented by the numerals for that base, a carry into the next higher position is produced. Consider the number 231.25_8. Each numeral has a value that is determined by its position and the fact that the numbers are octal.

❖ **EXAMPLE 1.2** Write the octal number 231.25_8 in position notation.

SOLUTION

$$N_8 = (2 \times 8^2) + (3 \times 8^1) + (1 \times 8^0) + (2 \times 8^{-1}) + (5 \times 8^{-2})$$

Each character occupies a power of eight position that is indicated by the value and sign of an exponent. Every character in the number is multiplied by 8 raised to a power, including fractional values that have negative exponents. The decimal equivalent is determined by evaluation of each term in the polynomial:

$$\begin{aligned} N_{10} &= 2(8^2) + 3(8^1) + 1(8^0) + 2(8^{-1}) + 5(8^{-2}) \\ &= 2(64) + 3(8) + 1(1) + 2(.125) + 5(.015625) \\ &= 128 + 24 + 1 + .25 + .078125 \\ &= 153.328125_{10} \end{aligned}$$

❖

1.7.4 Hexadecimal Numbers

Radix-16 is also called *hexadecimal*. The term **hexadecimal** comes from the combination of the words decimal (10) and hex (6). The number system has widespread use in digital systems because it is easily converted to and from binary. Sixteen characters exist in the hexadecimal **character set** (0, 1, 2, 3, . . . , 8, 9, A, B, C, D, E, F). Both upper- and lowercase letters can be used for hex numerals. Each character represents value and, like other positional notation number systems, when we run out of characters to represent the value in a given position, a carry into the next higher order position is produced. Characters A through F represent numerals, not letters of the alphabet. Any other universally agreed upon characters could have been used; the first six letters of the English alphabet were simply chosen to represent numerals in the hexadecimal number system.

❖ **EXAMPLE 1.3** Consider the number $A59C.3A_{16}$.

SOLUTION

$$N_{16} = (A \times 16^3) + (5 \times 16^2) + (9 \times 16^1) + (C \times 16^0) + (3 \times 16^{-1}) + (A \times 16^{-2})$$

Each position has a positive or negative exponent like the other positional number systems. The hexadecimal character set may appear strange at first but the letters are simply numerals that have a given value. The base-16 numeral is multiplied by 16 raised to a power indicated by the sign and value of its exponent. That value is multiplied by the character holding the position:

$$N_{10} = (A \times 4096) + (5 \times 256) + (9 \times 16) + (C) + (3/16) + (A/256)$$

Before we can complete the indicated arithmetic we must know the values assigned the hexadecimal numerals A through F. Table 1.2 gives the decimal equivalent of the letter hexadecimal characters.

TABLE 1.2
Decimal equivalent of hexadecimal letter characters

Hexadecimal	Decimal
A	10
B	11
C	12
D	13
E	14
F	15

Substituting the decimal equivalent for the letter hex characters the arithmetic continues:

$$N_{10} = 40{,}960 + 1{,}280 + 144 + 12 + 3/16 + 10/256$$
$$N_{10} = 42{,}396.2265625_{10}$$

The greater value range of the hexadecimal characters allows larger values to be represented with fewer digits than with decimal numbers. ❖

1.7.5 Counting in Base *r*

A number represented in radix *r*, where *r* can be any value, has *r* characters in its set.

Consider Table 1.3, which illustrates several bases and the numerals used in their respective sets.

Positional notation numbers for any radix, *r*, requires a set of characters. A positional place holder is required for indicating values greater than that of a single digit. Counting in any radix involves sequencing through the characters in the set, carrying over to the next position when the set is exhausted for that digit and repeating the sequencing of characters for the next digit.

Decimal counting involves sequencing through the radix-10 character set (0, 1, 2, 3, 4, 5, 6, 7, 8, 9) until character 9 is reached. Then a carry from the 10^0 position to the 10^1 position occurs and the sequence from 0 to 9 in the 10^1 position repeats. The process continues for each position until the terminal count is reached. The terminal count is the maximum value that can be represented in a single digit for a given radix.

TABLE 1.3
Bases and characters

Radix (base)	Characters in set
2 (binary)	(0,1)
3	(0,1,2)
4	(0,1,2,3)
5	(0,1,2,3,4)
⋮	⋮
8 (octal)	(0,1,2,3,4,5,6,7)
⋮	⋮
10 (decimal)	(0,1,2,3,4,5,6,7,8,9)
⋮	⋮
16 (hexadecimal)	(0, 1, 2, 3, 4, 5, 6, 7, 8, 9, A, B, C, D, E, F)

For example in binary, a count from 0 to 3, written in binary is 00, 01, 10, 11. A binary digit is called a **bit** (short for *b*inary dig*it*). Each increment in value changes the least significant bit, producing a carry to the next higher order bit when the position value exceeds 1. Because each binary digit represents only one of two values (0, 1), more positions are necessary than for higher order radix numbers to represent a given quantity.

❖ **EXAMPLE 1.4** Count from 0 to 9_{10} in radix 3.

SOLUTION Table 1.3 indicates that radix 3 has three characters. A count sequence from 0 decimal to 9 decimal is

000, 001, 002, 010, 011, 012, 020, 021, 022, 100 ❖

❖ **EXAMPLE 1.5** Count from 0 to 11_{10} in radix 5.

SOLUTION Table 1.3 indicates that radix 5 has five characters. A count from 0 to 11 decimal is

00, 01, 02, 03, 04, 10, 11, 12, 13, 14, 20, 21 ❖

❖ **EXAMPLE 1.6** Count from 0 to 25_{10} in hexadecimal.

SOLUTION 00, 01, 02, 03, 04, 05, 06, 07, 08, 09, 0A, 0B, 0C, 0D, OE, 0F, 10, 11, 12, 13, 14, 15, 16, 17, 18, 19 ❖

1.8 NUMBER SYSTEM CONVERSION

Digital systems often convert from one number system to another during information processing. Digital electronics use binary numbers, yet people function best using decimal numbers. Transforming binary numbers used by digital systems to decimal numbers used by people is necessary.

Binary numbers are used by digital electronics hardware (for the majority of applications) due to circuit simplicity. Octal (base-8) and hexadecimal (base-16) number systems were developed to facilitate human understanding of long binary numbers. The conversion from binary to octal and hexadecimal and octal or hexadecimal to binary is very straightforward.

Consider the binary number 111001010011_2. For people to keep track of large binary numbers presents a problem. By converting the binary value to hexadecimal, the number of digits to be remembered is reduced. One hex digit contains the same information content as four binary digits. Table 1.4 illustrates the number of digits in several number systems.

TABLE 1.4
Decimal, binary, octal, and hexadecimal numbers

Decimal	Binary	Octal	Hexadecimal
0	0000	00	0
1	0001	01	1
2	0010	02	2
3	0011	03	3
4	0100	04	4
5	0101	05	5
6	0110	06	6
7	0111	07	7
8	1000	10	8
9	1001	11	9
10	1010	12	A
11	1011	13	B
12	1100	14	C
13	1101	15	D
14	1110	16	E
15	1111	17	F

1.8.1 Binary to Hexadecimal Conversion

❖ **EXAMPLE 1.7** Convert the binary number 010011110111.110101010_2 to hexadecimal.

Solution

1. Partition the binary number into groups of four, starting at the radix point and going left and right.

   $0100, 1111, 0111.1101, 0101,0000_2$

2. Each group of four corresponds to a single hexadecimal digit. Using Table 1.4 find the hexadecimal character for each group.

   $0100,1111,0111.1101,0101,0000_2 = 4F7.D50_{16}$ ❖

❖ **EXAMPLE 1.8** Convert the binary number 111111001010.01111111_2 to hexadecimal.

Solution

1. Partition the binary number into groups of four, starting at the radix point and going left and right.

   $1111,1100,1010.0111,1111_2$

2. Each group of bits represents a single hex character. From Table 1.4 find the hex character that corresponds to each group of four bits.

   $1111,1100,1010.0111,1111_2 = FCA.7F_{16}$ ❖

1.8.2 Hexadecimal and Octal to Binary Conversion

❖ **EXAMPLE 1.9** Convert the hexadecimal number $F37A.B2_{16}$ to binary.

SOLUTION Each hexadecimal digit corresponds to four binary digits. Using Table 1.4, find the hexadecimal character that corresponds to each four-bit binary group.

$$F37A.B2_{16} = 1111,0011,0111,1010.1011,0010_2$$ ❖

❖ **EXAMPLE 1.10** Convert the hexadecimal number $2DE5.6A_{16}$ to binary.

SOLUTION Using Table 1.4, find the four bits corresponding to each base-16 digit and replace the digit with the four bits.

$$2DE5.6A_{16} = 0010,1101,1110,0101.0110,1010_2$$ ❖

❖ **EXAMPLE 1.11** Convert the octal number 735.5_8 to binary.

SOLUTION Each octal digit represents three binary digits. Using Table 1.4 find the three bits that correspond to each octal digit and replace it with the binary digits.

$$735.5_8 = 111,011,101.101_2$$ ❖

PRACTICE PROBLEMS

1. Convert 1101100010101001.1101_2 to octal, to hexadecimal.
2. Convert $FECA_{16}$ to binary then to octal.
3. Convert 367.236_8 to binary then hexadecimal.
4. What is the maximum value for a 16-bit binary number:
   a. In decimal?
   b. In octal?
   c. In hexadecimal?
5. A 4-digit octal number requires how many binary digits (bits)?
6. Write the expanded positional notation for
   a. 325.78_{10}.
   b. 325.78_{16}.
   c. 1001011.110_2.

1.8.3 Binary to Decimal Conversion

Position in number systems conveys value information. Equation (1.2) works for any radix. Exponents indicate position value for a binary digit. If the digit is 1 then the

TABLE 1.5
Decimal equivalent values for some binary positions

Binary position	Decimal value
2^{-4}	.0625
2^{-3}	.125
2^{-2}	.25
2^{-1}	.5
2^0	1
2^1	2
2^2	4
2^3	8
2^4	16
2^5	32
2^6	64
2^7	128
2^8	256
2^9	512
2^{10}	1024

decimal equivalent of 2^x, where x is the bit position, is added to the total; if the binary digit is 0, no value is added:

$$b_{10}\, 2^{10} + b_9\, 2^9 + b_8\, 2^8 + b_7\, 2^7 + \cdots + b_0\, 2^0 + b_{-1}\, 2^{-1} + b_{-2}\, 2^{-2} + b_{-3}\, 2^{-3}$$

Bit b_x can be a 0 or a 1. If $b_x = 1$, then the value of the position is added to the total; if $b_x = 0$, then nothing is added. Table 1.5 lists the decimal values for some binary positions.

1.8.4 Successive Division Radix Conversion

Radix conversion can be accomplished by positional notation, but a more succinct method is available through the use of an algorithm called **successive division**. Conversion is accomplished by repeated division of N (the number being converted) by r (the new radix). The remainder of each division becomes the numeral in the new radix. The process is repeated until the least significant numeral is generated. MSB means most significant bit and LSB means least significant bit.

❖ **EXAMPLE 1.12** Convert 119_{10} to binary.

SOLUTION Number Being Converted.

1. Divide 119 by 2.

Radix of new number

$$\begin{array}{r} 59 \\ 2\,\overline{)\,119} \\ \underline{118} \\ 1 \end{array} \longrightarrow 1 \quad \text{LSB (Integer)}$$

Remainder

2. Divide 59 by 2.

$$\begin{array}{r} 29 \\ 2\,\overline{)\,59} \\ \underline{58} \\ 1 \end{array} \longrightarrow 1$$

3. Divide 29 by 2.

$$\begin{array}{r} 14 \\ 2\,\overline{)\,29} \\ \underline{28} \\ 1 \end{array} \longrightarrow 1$$

4. Divide 14 by 2.

$$\begin{array}{r} 7 \\ 2\,\overline{)\,14} \\ \underline{14} \\ 0 \end{array} \longrightarrow 0$$

5. Divide 7 by 2.

$$2 \overline{)\,7\,} = 3,\quad 7 - 6 = 1 \longrightarrow 1$$

6. Divide 3 by 2.

$$2 \overline{)\,3\,} = 1,\quad 3 - 2 = 1 \longrightarrow 1$$

7. Divide 1 by 2.

$$2 \overline{)\,1\,} = 0,\quad 1 - 0 = 1 \longrightarrow 1 \quad \text{MSB}$$

Starting with the most significant bit (MSB) and collecting the remainders, the resulting binary number is

$$119_{10} = 1110111_2$$

Checking the results, we find

$$119_{10} = 1 \cdot 2^6 + 1 \cdot 2^5 + 1 \cdot 2^4 + 0 \cdot 2^3 + 1 \cdot 2^2 + 1 \cdot 2^1 + 1 \cdot 2^0$$

$$119_{10} = 64 + 32 + 16 + 0 + 4 + 2 + 1$$

❖

1.8.5 Fractional Radix Conversion, Successive Multiplication

Conversion of fractional numbers from one radix to another is accomplished using a successive multiplication algorithm. The number to be converted is multiplied by the radix of the new number, producing a product that has an integer and a fractional portion. The integer part of the result becomes a numeral in the new radix number. The fraction is again multiplied by the radix, producing a product, and so on until the fractional portion of the product reaches 0 or until the number is carried out to sufficient digits for the application and truncation occurs.

❖ **EXAMPLE 1.13** Convert $.75_{10}$ to binary.

SOLUTION

Integers

Multiply .75 by 2.	$(.75)2 = 1.5$	$\longrightarrow$ 1	(MSB)
Multiply .5 by 2.	$(.5)2 = 1.0$	$\longrightarrow$ 1	
Multiply 0 by 2.	$(0)2 = 0.0$	$\longrightarrow$ 0	(LSB)

Checking the result, we find

$$.75_{10} = (1)2^{-1} + (1)2^{-2} = 1/2 + 1/4 = 3/4$$

❖

EXAMPLE 1.14 Convert 95.0625_{10} to binary.

SOLUTION The conversion is accomplished in two parts. First, convert the integer part of the decimal number by **successive division.**

1. Divide 95 by 2. Number Being Converted

   Radix of New Number

$$2\overline{)95} = 47,\quad 94,\quad \text{remainder } 1 \longrightarrow 1 \text{ (LSB) (integer)}$$

2. Divide 47 by 2.

$$2\overline{)47} = 23,\quad 46,\quad \text{remainder } 1 \longrightarrow 1$$

3. Divide 23 by 2.

$$2\overline{)23} = 11,\quad 22,\quad \text{remainder } 1 \longrightarrow 1$$

4. Divide 11 by 2.

$$2\overline{)11} = 5,\quad 10,\quad \text{remainder } 1 \longrightarrow 1$$

5. Divide 5 by 2.

$$2\overline{)5} = 2,\quad 4,\quad \text{remainder } 1 \longrightarrow 1$$

6. Divide 2 by 2.

$$2\overline{)2} = 1,\quad 2,\quad \text{remainder } 0 \longrightarrow 0$$

7. Divide 1 by 2.

$$2\overline{)1} = 0,\quad 0,\quad \text{remainder } 1 \longrightarrow 1 \text{ (MSB)}$$

$$1011111_2 = 95_{10}$$

Checking the results, we find

$$95_{10} = 1 \cdot 2^6 + 0 \cdot 2^5 + 1 \cdot 2^4 + 1 \cdot 2^3 + 1 \cdot 2^2 + 1 \cdot 2^1 + 1 \cdot 2^0$$

$$95_{10} = 64 + 0 + 16 + 8 + 4 + 2 + 1$$

Second, convert the fraction.

1. $(.0625)2 = 0.125 \longrightarrow 0$ (MSB)
2. $(.125)2 = 0.25 \longrightarrow 0$

3. $(.25)2 = 0.5 \longrightarrow 0$
4. $(.5)2 = 1.0 \longrightarrow 1$ (LSB)

Because the fractional portion of step 4 is 0, any further multiplication will yield 0, so the process ends.

Checking the results, we find $.0001_2 = .0625_{10}$.

$$0 \cdot 2^{-1} + 0 \cdot 2^{-2} + 0 \cdot 2^{-3} + 1 \cdot 2^{-4} = 2^{-4} = 1/16 = .0625_{10}$$ ❖

1.8.6 Radix Conversion Algorithm

Having applied successive division for the conversion of decimal integers to binary integers, we can see a definite pattern. By defining the variables and steps involved in a systematic procedure, we can develop an algorithm and illustrate it in a flowchart. The same is true for successive multiplication in the conversion of fractional numbers to another radix. The successive division variables are

N = number to be converted (original number).
r = radix of new number.
R = remainder of divide operation.
Q = quotient of divide operation.

The flowchart of Figure 1.9 receives as inputs, the original integer number to be converted and the radix of the new number. Number N is divided by the radix of the new number, r, producing a quotient, Q, and a remainder, R. The first division produces a remainder, R, that becomes the least significant digit of the new number. Dividing each successive quotient, Q, by the radix, r, and saving the resulting remainders, R, as

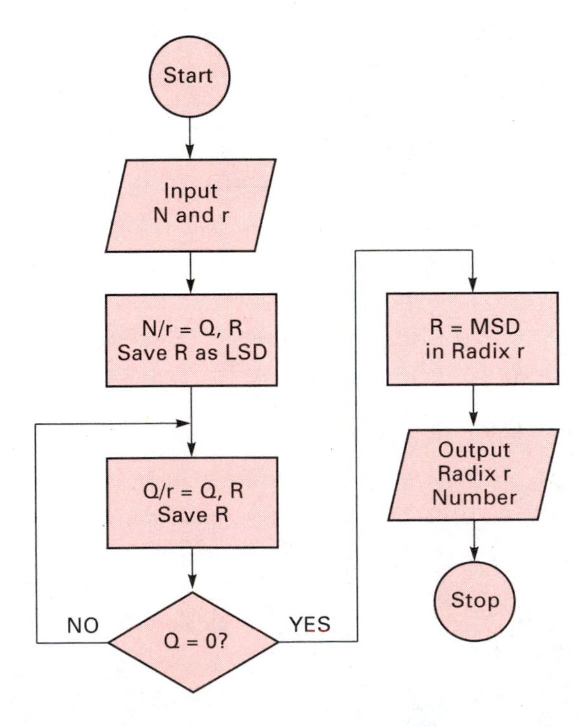

Figure 1.9
Flowchart illustrating successive division radix conversion

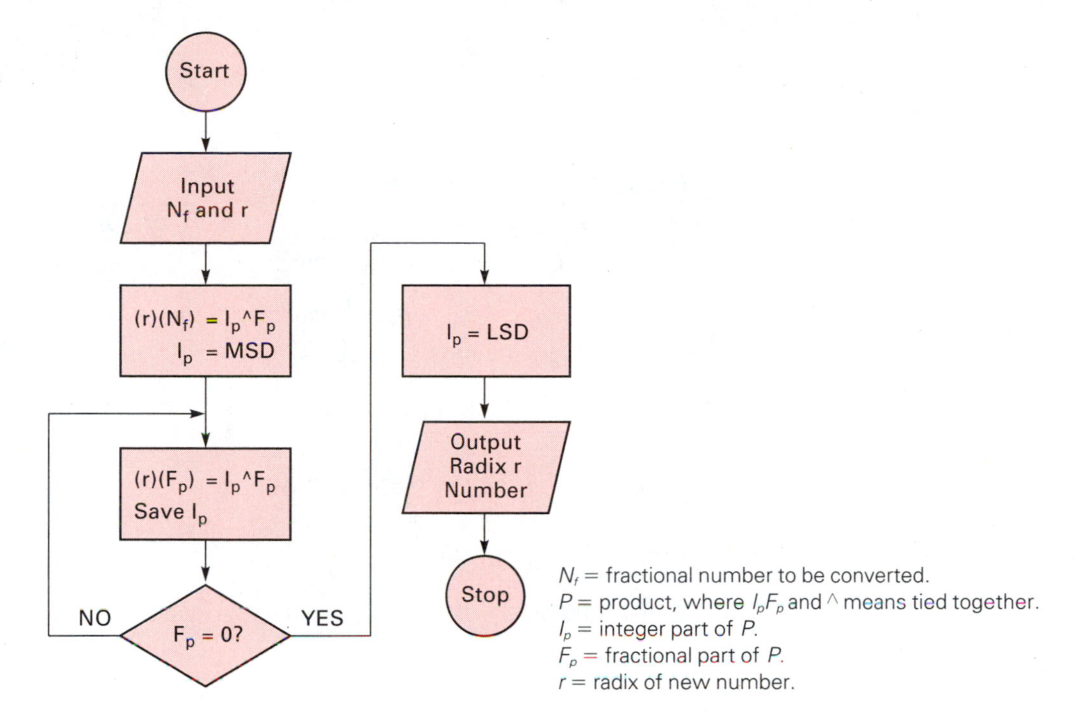

Figure 1.10
Fractional radix conversion flowchart

digits in the new number produces the converted number in radix, r. The process is terminated when the quotient reaches 0 or some previously determined number of digits.

The reason for terminating at some previously determined number of digits is because of limitations of a given digital system to represent a number. For instance, if 16 binary digits are available in a digital system then it makes no sense to convert more than 16 bits. The end of the algorithm is determined by testing the value of Q or when the maximum number of digits is reached. If Q is 0 or the maximum number of conversions has occurred, then output the R values and stop; if Q is not 0 or the maximum number of conversions has not occurred, then continue the successive division. Fractional radix conversion is illustrated in Figure 1.10.

1.8.7 Decimal to Any Radix

The conversion of decimal numbers to any other radix applies the successive division and successive multiplication algorithms.

EXAMPLE 1.15 Convert 23.625_{10} to octal (base 8).

SOLUTION Convert the integer portion by successive division:

Divide 23 by 8.

$$\begin{array}{r} 2 \\ 8\overline{)23} \\ \underline{16} \\ 7 \end{array} \longrightarrow 7 \quad \text{(LSD)}$$

Divide 2 by 8.
$$\begin{array}{r|l} & 0 \\ \hline 8 & 2 \\ & \underline{0} \\ & 2 \longrightarrow 2 \text{ (MSD)} \end{array}$$

Check:

$$23_{10} = 27_8$$
$$2 \cdot 8^1 + 7 \cdot 8^0 = 16 + 7 = 23_{10}$$

Convert the fractional portion by successive multiplication:

$$8(.625) = 5.00 \longrightarrow 5 \quad \text{(MSD)}$$

Because the fractional portion, F_p, is 0, the process is ended, and 5_8 is the octal fractional digit.

Check: $5(8^{-1}) = 5(1/8) = .625_{10}$. ❖

❖ **EXAMPLE 1.16** Convert 235.2_{10} to radix 4.

SOLUTION Convert the integer 235_{10} to radix 4 by successive division.

Divide 235_{10} by 4.
$$\begin{array}{r|l} & 58 \\ \hline 4 & 235 \\ & \underline{232} \\ & 3 \longrightarrow 3 \text{ (LSD)} \end{array}$$

Divide 58 by 4.
$$\begin{array}{r|l} & 14 \\ \hline 4 & 58 \\ & \underline{56} \\ & 2 \longrightarrow 2 \end{array}$$

Divide 14 by 4.
$$\begin{array}{r|l} & 3 \\ \hline 4 & 14 \\ & \underline{12} \\ & 2 \longrightarrow 2 \end{array}$$

Divide 3 by 4.
$$\begin{array}{r|l} & 0 \\ \hline 4 & 3 \\ & \underline{0} \\ & 3 \longrightarrow 3 \text{ (MSD)} \end{array}$$

Therefore, $3223_4 = 235_{10}$.

Check: $3 \cdot 4^3 + 2 \cdot 4^2 + 2 \cdot 4^1 + 3 \cdot 4^0 = 3(64) + 2(16) + 2(4) + 3 = 235_{10} = 3223_4$, which is the integer part of the number.

The fractional part is found by successive multiplication.

$$4(.2) = 0.8 \longrightarrow 0 \quad \text{(MSD)}$$
$$4(.8) = 3.2 \longrightarrow 3$$
$$4(.2) = 0.8 \longrightarrow 0$$
$$4(.8) = 3.2 \longrightarrow 3 \quad \text{Repeating 0303 . . .}$$
$$235.2_{10} = 3223.\dot{0}303\ldots_4$$
❖

1.8.8 Any Radix to Decimal

Equation (1.3) describes positional notation number systems. Conversion from any radix to decimal is accomplished by applying the equation to the radix being converted.

❖ **EXAMPLE 1.17** Convert 324.2_5 to decimal.

SOLUTION

$$3 \cdot 5^2 + 2 \cdot 5^1 + 4 \cdot 5^0 + 2 \cdot 5^{-1} = 3(25) + 2(5) + 4(1) + 2(1/5)$$
$$75 + 10 + 4 + 2/5 = 89.4_{10}$$ ❖

❖ **EXAMPLE 1.18** Convert 65.32_7 to decimal.

SOLUTION

$$6 \cdot 7^1 + 5 \cdot 7^0 + 3 \cdot 7^{-1} + 2 \cdot 7^{-2} = 6(7) + 5(1) + 3(1/7) + 2(1/49)$$
$$65.32_7 = 47\ 23/49 = 47.469\ldots$$ ❖

PRACTICE PROBLEMS

1. Convert 110011.1101_2 to decimal.
2. Convert 23.75_{10} to binary.
3. Convert 392.84_{10} to octal.
4. Convert 274.623_8 to decimal.
5. Convert 345.2_6 to decimal.
6. Convert 34.8_{10} to base 5.

1.9 BINARY CODES

A **code** is a symbol or group of symbols that stand for something. Binary bits are often used in groups to stand for things such as decimal or alphabetic characters. **Binary coded decimal**, BCD, uses a binary representation for decimal numbers and is useful in driving visual displays. Certain binary codes are suited for arithmetic operations. Other codes facilitate the creation of digital transducers for entering information into a system. Binary codes are available for representing signed numbers. This section covers a variety of binary codes.

TABLE 1.6
Decimal to natural BCD

Decimal	BCD
0	0000
1	0001
2	0010
3	0011
4	0100
5	0101
6	0110
7	0111
8	1000
9	1001

1.9.1 Natural Binary Coded Decimal

Many applications in which people interact with digital systems require binary information to be encoded in such a way as to be easily translated into decimal information. People can read decimal displays more readily than a row of lights indicating information in straight binary form. For this reason, a code was developed where groups of four binary digits (bits) represent a single decimal digit. The information being processed in a digital system is in binary; however, the results can be displayed in decimal form by the use of BCD codes.

Table 1.6 illustrates the relationship between decimal and natural BCD. BCD is a coding scheme, not a number system. Hexadecimal and BCD both use four bits to represent a value but they are not the same thing. Notice that the 10 characters in the decimal character set requires four BCD bits. Also notice that some combinations of binary codes do not exist in BCD; binary numbers 1010 to 1111 are not used in BCD.

Decimal numbers can be encoded directly into BCD numbers because each decimal digit translates into a four-bit binary code.

EXAMPLE 1.19 Convert 9275.6_{10} into BCD.

SOLUTION Each decimal digit requires a four-bit BCD value; from Table 1.6 we obtain

$$9 = 1001,\quad 2 = 0010,\quad 7 = 0111,\quad 5 = 0101,\quad 6 = 0110$$

$$9275.6_{10} = 1001{,}0010{,}0111{,}0101.0110 \text{ in BCD}$$

Notice that the BCD code needs more bits to represent a decimal number than straight binary. This is because each BCD digit takes four bits to encode.

1.9.2 Binary Codes (Weighted)

Many binary coded decimal, BCD, codes exist. All use four bits to represent a single decimal digit, and as a result, all have six unassigned values. **Weighted codes** are simply codes that have assigned weights or values for each bit position. To the degree that different weights can be assigned to the bit positions, different codes can be created. Table 1.7 illustrates several weighted BCD codes.

The 8, 4, 2, 1 natural BCD code is the most widely used and considered to be the "normal" BCD code. **Excess-3 code** values are derived from adding 3 to the natural BCD code value. The excess-3 code is also a self-complementing code. Other self-complementing codes exist, and we will treat them as a group in Section 1.9.3. Another

TABLE 1.7
Examples of BCD weighted codes

Decimal	8421 Code	Ex-3 Code	7421 Code	5311 Code	84^-2^-1* Code	5421 Code
0	0000	0011	0000	0000	0000	0000
1	0001	0100	0001	0001	0111	0001
2	0010	0101	0010	0011	0110	0010
3	0011	0110	0011	0100	0101	0011
4	0100	0111	0100	0101	0100	0100
5	0101	1000	0101	1000	1011	1000
6	0110	1001	0110	1001	1010	1001
7	0111	1010	1000	1011	1001	1010
8	1000	1011	1001	1100	1000	1011
9	1001	1100	1010	1101	1111	1100

*Indicates negative value, so $^-2$ and $^-1$ are negative numbers.

TABLE 1.8
Self-complementing BCD codes

Decimal	Ex-3	631-1	2421
0	0011	0011	0000
1	0100	0010	0001
2	0101	0101	0010
3	0110	0111	0011
4	0111	0110	0100
5	1000	1001	1011
6	1001	1000	1100
7	1010	1010	1101
8	1011	1101	1110
9	1100	1100	1111

valuable characteristic of certain BCD codes is the reflective property. That is the case with the 8, 4, −2, −1, 5, 3, 1, 1, and 5, 4, 2, 1 codes. Notice that in each of these codes the bottom half of the code is reflective with the top half. Study the 8, 4, −2, −1 code. The first number, 0000, is the reflection of the last number 1111; also the second number, 0111, reflects with the next to the last value, 1000. In the 8, 4, −2, −1 code the reflection is the inverse; that is, the first code is the inverse of the last, the second with the next to last, and so on.

The reflection is different in the 5, 3, 1, 1 code. Notice that the first half of the code starts at 0000 and proceeds to 0101; whereas the second half starts with 1000 and proceeds to 1101. The difference between the top half and bottom half of the code, in Table 1.7, is the most significant bit position. It is 0 from 0000 to 0101 and 1 from 1000 to 1101. Certain BCD codes use less logic than others when used in specific digital applications. This is particularly true for the self-complementing codes. Table 1.8 illustrates three complementing codes.

1.9.3 BCD Self-Complementing Codes

Self-complementing codes are codes whose arithmetic and logic complements are the same. Two arithmetic **complements** are commonly used, radix and diminished radix. The arithmetic radix complement of x is written x':

$$x' = (b) - a,$$

where x' is the complement of numeral a, b is the base (radix) of the numeral (a) being complemented.

The arithmetic **diminished radix** complement is written

$$x'_{-1} = (b - 1) - a,$$

where x'_{-1} is the diminished radix complement, b is the base of the numeral (a) being complemented.

The arithmetic radix (10s) complement of 6_{10} is $10 - 6 = 4_{10}$. The arithmetic diminished radix (9s) complement of 6_{10} is $(10 - 1) - 6 = 3_{10}$. The logical complement of a binary digit is its opposite value. For instance, the logical complement of 0 is 1 and the logical complement of 1 is 0. BCD codes are often used in digital systems that perform decimal arithmetic, where self-complementing codes reduce the logic needed to design arithmetic units. We will study more about BCD adders in Chapter 4, Section 4.6.6. For now we are concerned with only defining self-complementing BCD codes. Self-complementing BCD codes are designed so the *arithmetic diminished complement* can be found by taking the logical complement, a bit-by-bit inversion of the BCD code. We will investigate logical inverter functions in Chapter 2 ($1 \rightarrow 0, 0 \rightarrow 1$).

In excess-3 (ex-3) BCD, a self-complementing code, the representation for 5_{10} is 1000_2. The arithmetic diminished (9s) complement of 5_{10} is 4_{10}, and the ex-3 BCD representation of 4_{10} is 0111. So, in ex-3 BCD code, the arithmetic 9s complement of 1000 is 0111, which is also the logical complement. To perform arithmetic 9s complement of any ex-3 BCD code requires finding the opposite value for each bit in the number. Contrast the ex-3 BCD code with natural BCD. The natural BCD code for 5_{10} is 0101, and 4_{10} is 0100. Note the logical complement of 0101 is not 0100; therefore the logical complement of the code is not the numeric 9s complement.

The 2, 4, 2, 1 code is also a self-complementing code. The 2, 4, 2, 1 code representation for 5_{10} is 1011, and 4_{10} is represented as 0100. After evaluation of the 2, 4, 2, 1 code we see that 1011 is the logical complement of 0100, illustrating the self-complementing action of the code.

1.9.4 Unit Distance Code

Other binary codes are used to input data from transducers measuring physical parameters into digital systems. Often it is desirable to have these codes behave so that only one bit can change between successive values. This particular property is desirable to reduce measurement error that may occur if two bits were to change at the same time. The ability to discriminate between different values in a code is essential. If slight time delays between bit changes were to occur the digital system might receive incorrect data. Figure 1.11 illustrates this concept, where a logical 0 is represented by a low voltage and a logical 1 is represented by a high voltage. Changes in voltage, representing binary logical values, occur over time.

As seen in Figure 1.11, a delay between changes in D_0 and D_1 creates a different code for a brief period of time. Assume that D_1D_0 made a transition from 00 to 11 in some

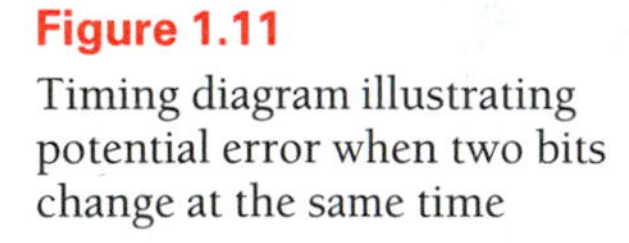

Figure 1.11
Timing diagram illustrating potential error when two bits change at the same time

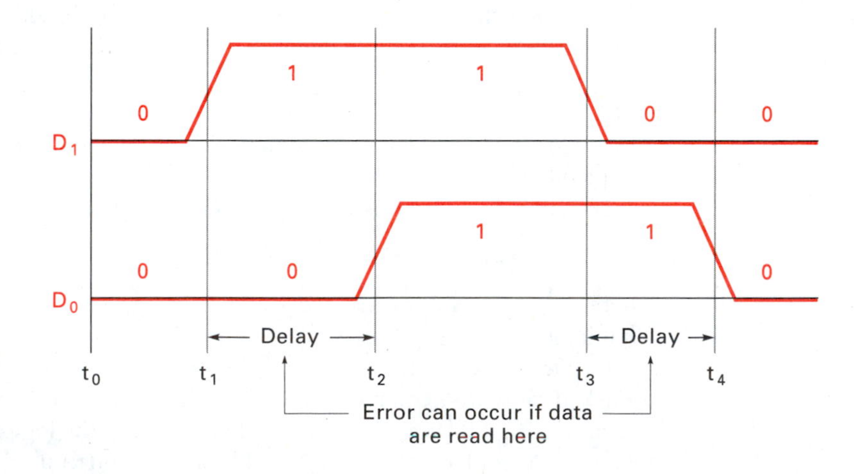

TABLE 1.9
Three-bit and Four-bit Gray codes

Decimal	3-bit	4-bit
	Gray	Gray
0	000	0000
1	001	0001
2	011	0011
3	010	0010
4	110	0110
5	111	0111
6	101	0101
7	100	0100
8		1100
9		1101
10		1111
11		1110
12		1010
13		1011
14		1001
15		1000

digital transducer providing input information to a digital system. However, due to the electronic circuitry, a time delay occurred so that D_0 did not go high at t_1 but was delayed until t_2. During the time interval between t_2 and t_1 the output is not 11, as it should be, but 10, which is incorrect. If a digital system were to capture the information during the change, incorrect data would be read. To eliminate this problem requires that a code be developed where only one bit change occurs between successive values in the code. Such a code is called a **unit distance code**. The most common unit distance code is the **Gray code**, which is also a reflective code.

Table 1.9 illustrates three-bit and four-bit Gray codes. Gray codes of any length can be constructed.

Gray codes are often generated from rotational position transducers as a means of encoding the shaft or angular position as digital information. Consider a circular disk mounted to a shaft that is free to rotate. If holes or slits were cut into the disk so that light could be sensed through the holes and if a light detecting circuit were positioned to sense the presence of light, then shaft position could be determined. The pattern into which the holes are cut in the disk generates the Gray code. The number of Gray-code bits determines the shaft position resolution.

Figure 1.12
Gray-code shaft position encoder

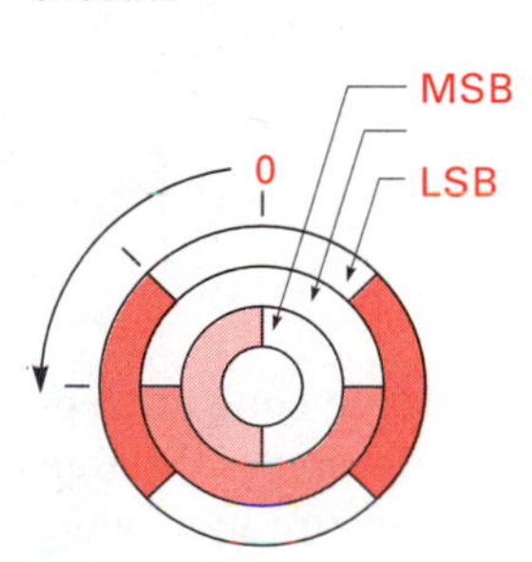

Use of the Gray code so that only a one-bit change can occur between adjacent positions eliminates the possibility of position error occurring during transitions from one code value to another. A three-bit code would determine the shaft position to 1 part in 8, or every 45 degrees; a four-bit code would resolve 1 part in 16.

The three-bit Gray-code encoder illustrated in Figure 1.12 indicates a logic 0 by the light-colored areas and a logic 1 by the dark-colored areas. A light source positioned on one side of the disk passes light through the light-colored areas (slits) onto a light sensitive detector on the other side. When the detector senses light it transmits a logical 0 and when no light is detected a 1. Because a three-bit detector is indicated, three such light sensors, one for each bit, would be needed.

1.9.5 Alphanumeric Codes

Digital systems, especially computers, need a means of encoding data other than numbers, such as the alphabet and punctuation. The binary information (representing numbers, letters, and punctuation) can be transmitted to a digital system, stored, manipulated, and then sent to the outside world via a printer, plotter, or a display. The most common **alphanumeric codes** are the seven- and eight-bit **ASCII** (American standard code for information interchange) **codes**.

The seven-bit ASCII code encodes all upper- and lowercase alphabet letters, numbers, punctuation, and control information for typewriters and printers. Control information sent to printers and typewriters includes carriage returns, tabs, and paper line feeds. The eight-bit ASCII code is extended to include some additional graphic characters. Another alphanumeric code is the EBCDIC (extended BCD interchange code). It encodes the upper- and lowercase alphabet, numbers, punctuation, and control codes.

Alphanumeric codes provide the means for sending data from a keyboard to a computer, and after processing, to be transmitted to a printer, cathode ray tube screen, plotter, or some other output device. Consider the standard typewriter keyboard; it is typical of the keyboard entry unit into a personal computer. When a key is struck, digital electronics determine the alphanumeric code for the struck key and send the information to the computer to be processed. Data are stored in the computer's memory in a binary code, not letters or numbers or punctuation marks. The alphanumeric code provides a translation of the struck key, in binary form, suitable for processing by the computer. For example, when you type your laboratory report on a personal computer, using a word processing software package, each key stroke is translated into an alphanumeric code prior to being sent to the computer for processing. When the computer prints the letters, numbers, and punctuation marks on the terminal screen, the reverse is true: alphanumeric coded binary is converted to letters, numbers and punctuation marks. Table 1.10 shows a partial listing of seven-bit ASCII and eight-bit EBCDIC codes.

1.9.6 Signed Number Binary Codes

Signed binary numbers provide the means by which both positive and negative numbers may be represented. We have been dealing with only positive binary numbers, represented in one of several codes. This section and the next 2 subsections develop the ideas necessary to represent negative as well as positive binary numbers.

Binary signed magnitude number convention uses the most significant bit position to indicate sign and the remaining lesser significant bits to represent magnitude. Three main signed number binary codes are used: sign magnitude, 2s complement, and 1s complement. Table 1.11 illustrates the **sign-magnitude**, 1s and 2s complement codes for the decimal numbers -15 to $+15$.

1.9.7 Signed Magnitude Codes

The sign magnitude column in Table 1.11 illustrates the **sign magnitude code** for representing decimal numbers $+15$ to -15. The most significant bit position is 0 for all positive values and 1 for negative values. Magnitude bits represent the absolute value.

TABLE 1.10
Partial listing of seven-bit ASCII and eight-bit EBCDIC codes in hexadecimal

Character	ASCII code	EBCDIC code	Character	ASCII code	EBCDIC code
A	41	C1	0	30	F0
B	42	C2	1	31	F1
C	43	C3	2	32	F2
D	44	C4	3	33	F3
E	45	C5	4	34	F4
F	46	C6	5	35	F5
G	47	C7	6	36	F6
H	48	C8	7	37	F7
I	49	C9	8	38	F8
J	4A	D1	9	39	F9
K	4B	D2	blank	20	40
L	4C	D3	!	21	5A
M	4D	D4	"	22	7F
N	4E	D5	#	23	7B
O	4F	D6	‡	24	5B
P	50	D7	%	25	6C
Q	51	D8	&	26	50
R	52	D9	'	27	70
S	53	E2	(	28	4D
T	54	E3	)	29	5D
U	55	E4	*	2A	5C
V	56	E5	+	2B	4E
W	57	E6	,	2C	6B
X	58	E7	–	2D	60
Y	59	E8	.	2E	4B
Z	5A	E9	/	2F	61

Note that two values for zero exist: +0 and −0. Sign magnitude codes are seldom used; complement codes are much more common.

1.9.8 Complement Codes

The 2s and 1s **complement codes** (binary radix and diminished radix) codes are also represented in Table 1.11 for the same range of decimal numbers. The most significant bit position is used to represent sign for these two codes. The codes are identical for positive numbers; the difference occurs in representing 0 and negative values. The **2s complement code** has only one value for 0 whereas the **1s complement code** has two. The 2s complement code is the same as the 1s complement code plus 1. Table 1.12 illustrates the radix and radix-1 complements for single digit decimal, octal, and binary characters.

TABLE 1.11
Sign magnitude and complement codes

Decimal	Sign magnitude	2s complement	1s complement
+15	01111	01111	01111
+14	01110	01110	01110
+13	01101	01101	01101
+12	01100	01100	01100
+11	01011	01011	01011
+10	01010	01010	01010
+9	01001	01001	01001
+8	01000	01000	01000
+7	00111	00111	00111
+6	00110	00110	00110
+5	00101	00101	00101
+4	00100	00100	00100
+3	00011	00011	00011
+2	00010	00010	00010
+1	00001	00001	00001
+0	00000	00000	00000
−0	10000	00000	11111
−1	10001	11111	11110
−2	10010	11110	11101
−3	10011	11101	11100
−4	10100	11100	11011
−5	10101	11011	11010
−6	10110	11010	11001
−7	10111	11001	11000
−8	11000	11000	10111
−9	11001	10111	10110
−10	11010	10110	10101
−11	11011	10101	10100
−12	11100	10100	10011
−13	11101	10011	10010
−14	11110	10010	10001
−15	11111	10001	10000
−16		10000	

❖ **EXAMPLE 1.20** Find the 10s (radix) complement for the decimal number 412.

SOLUTION $10^3 - 412_{10} = 1000 - 412 = 588$ ❖

❖ **EXAMPLE 1.21** Find the 10s complement of the decimal number 9864.

SOLUTION $10^4 - 9864_{10} = 10{,}000 - 9864 = 0136$ ❖

TABLE 1.12
Radix and radix-1 complements for decimal, octal, and binary characters

Decimal	10s	9s	Octal	8s	7s	Binary	2s	1s
0	(1)*0	9	0	(1)*0	7	0	(1)*0	1
1	9	8	1	7	6	1	1	0
2	8	7	2	6	5			
3	7	6	3	5	4			
4	6	5	4	4	3			
5	5	4	5	3	2			
6	4	3	6	2	1			
7	3	2	7	1	0			
8	2	1						
9	1	0						

*Indicates a carryover to the next higher digit.

❖ **EXAMPLE 1.22** Find the 2s complement of the binary number 110011.

SOLUTION

$$2^6 - 110011_2 = 1000000 - 110011 = 001101$$ ❖

❖ **EXAMPLE 1.23** Find the 1s complement code for -12_{10}.

SOLUTION Write the 1s complement binary number for +12. Find the 1s complement code for +12: $+12_{10} = 0\wedge1100$. The ^ sign separates the sign bit from the magnitude bits.

A 0 sign bit indicates a positive number, and a 1 sign bit indicates a negative number.

Complement each bit. The bit-by-bit complement of 0^1100 is 1^0011, which is the 1s complement code for -12_{10}. ❖

❖ **EXAMPLE 1.24** Find the 2s complement code for -18_{10}.

SOLUTION Write the binary number for +18. Recall that the positive numbers in both 2s and 1s complement codes are the same as straight binary with the addition of a leading sign bit:

$$+18_{10} = 0\wedge10010$$

Complement each bit, including the sign bit.

The complement of 0^10010 becomes 1^01101. Add 1 to the complement:

```
  1^01101
+ 0^00001
  1^01110
```

which is the 2s complement of -18_{10}. ❖

The complement code value for positive numbers is equivalent to straight binary numbers with a "0" sign bit added in the most significant position.

The 1s complement code for all negative numbers can be found by first writing the binary value for the absolute value of the negative number, with a 0 in the sign bit position. Then, the positive value is complemented.

Negative 2s complement codes are found by first finding the 1s complement code and then adding 1 to the result.

1.10 ARITHMETIC

Simple addition and subtraction of binary complement codes, as well as addition, subtraction, multiplication, and division of binary and hexadecimal codes are considered in this section.

1.10.1 Binary Arithmetic

Binary numbers can be added, subtracted, multiplied, and divided just like any other radix number. Each numeral position has a value, and when the addition of two numbers produces a result that exceeds the value that can be represented by a character in the set, a carry is generated to the next higher position. Subtraction generates borrows when a larger number in a digit position is subtracted from a smaller. The multiplication table for binary is quite simple: 0 multiplied by either 0 or 1 = 0, and 1 multiplied by 1 = 1. Binary division produces quotients with remainders like decimal division.

Table 1.13 illustrates binary addition with three inputs (two operands, the addend and augend, and a carry-in) and two outputs (the sum and the carry-out). The table represents any bit position of a binary number. In the case of the least significant bit of a binary number, the carry-in would never be a 1 because no carry-out from a preceding position would be possible.

The binary character set only has two characters (0,1), so any time the sum is greater than 1, a carry-out is produced. For instance, 1 + 1 = 2, but no 2 exists as a single digit, so a carry to the next higher bit position is necessary. If a carry-in were a 1 (a carry from a less significant bit position) and two 1s were added, the result would be 3, producing sum = 1 and carry-out = 1. The carry-out of one position becomes the carry-in to the next higher order position.

TABLE 1.13
Binary addition

Inputs			Outputs	
Augend	Addend	Carry-in	Sum	Carry-out
0	0	0	0	0
0	0	1	1	0
0	1	0	1	0
0	1	1	0	1
1	0	0	1	0
1	0	1	0	1
1	1	0	0	1
1	1	1	1	1

❖ **EXAMPLE 1.25** Find the binary sum.

SOLUTION

Carry	0 0011		1
Augend	10011_2	=	19_{10}
Addend	01011_2	=	11_{10}
Sum	0 11110_2		30_{10}

❖

❖ **EXAMPLE 1.26** Find the binary sum.

SOLUTION

Carry	1 11111		
Augend	01110.10_2	=	14.50_{10}
Addend	11011.11_2	=	27.75_{10}
Sum	1 01010.01_2	=	42.25_{10}

❖

In Example 1.26 the fractional portion of the binary number generated a carry into the integer portion; in this case the least significant bit is fractional. A binary carry has a power of 2 weight, transferred to the next higher order digit, similar to a decimal carry producing a power of 10 weight.

Table 1.14 illustrates binary subtraction, the inverse operation of addition. The minuend, subtrahend, and borrow-in constitute inputs and the difference and borrow-out, the outputs of a binary subtracter. A borrow-out of one bit position becomes the borrow-in of the next higher order bit position. The least significant bit position's borrow-in is always a 0. Borrows from a bit position occur when the value of that position is less than the amount being subtracted from it.

When insufficient value is present, from which a larger value is to be subtracted, then additional value is "borrowed" from the next higher position. A borrow from a column is the equivalent of subtracting 1 from the column. In the case of binary, subtracting 01 from 10 requires that a value of 2 (10_2) must be borrowed from the second bit and lent to the first bit position.

TABLE 1.14
Binary subtraction

Inputs			Outputs	
Minuend	Subtrahend	Borrow-in	Difference	Borrow-out
0	0	0	0	0
0	0	1	1	1
0	1	0	1	1
0	1	1	0	1
1	0	0	1	0
1	0	1	0	0
1	1	0	0	0
1	1	1	1	1

EXAMPLE 1.27

Find the following binary difference.

SOLUTION

```
 1  0
 0  1
-----
 0 10
 1  0.
 0  1
-----
 0  1
```

$0 - 1$ produces a negative value. A borrow value of 2 (10_2) is obtained from the higher position.

A borrow-out is generated by the LSB that becomes the borrow-in for the second bit position.

EXAMPLE 1.28

Find the following binary difference.

SOLUTION

Minuend	$1\ 0\ 1\ 1\ 0\ 1\ 0_2$	$= 90_{10}$
Subtrahend	$0\ 1\ 0\ 1\ 1\ 1\ 0_2$	$= 46_{10}$
Difference	$0\ 1\ 0\ 1\ 1\ 0\ 0_2$	$= 44_{10}$
Column	7 6 5 4 3 2 1	

No borrow occurred in the first two columns ($0 - 0 = 0$) and ($1 - 1 = 0$). The third column subtrahend is larger than the minuend so a borrow-out is required ($10 - 1 = 1$). The 1 in the fourth column minuend became a 0 with a borrow-in from the third column, necessitating a borrow-out from the fourth column. The new fourth column minuend of 0 became 10 with a borrow-out from the fifth column ($10 - 1 = 1$). The fifth column does not need a borrow out because $0 - 0 = 0$. The sixth column subtrahend is larger than the minuend requiring a borrow-out to the seventh column ($10 - 1 = 1$). The seventh column does not need a borrow-out ($0 - 0 = 0$).

Binary multiplication is accomplished using an add and shift process we learned in grade school. A partial product is produced for each significant bit of the multiplier when that bit value is multiplied with the multiplicand. Each partial product is "shifted" one bit position to the left from the preceding partial product. The "shifted" partial products are added, producing the product.

EXAMPLE 1.29

How is binary multiplication performed?

SOLUTION

```
          1 0 1 1 0.1 =  22.5
        × 0 1 0 0 1.1 =   9.5
          -----------
          1 0 1 1 0 1
        1 0 1 1 0 1
      0 0 0 0 0 0
    0 0 0 0 0 0
  1 0 1 1 0 1
0 0 0 0 0 0
---------------------
0 1 1 0 1 0 1 0 1.1 1 = 213.75₁₀
```

Binary multiplication table:

$$0 \times 0 = 0$$
$$0 \times 1 = 0$$
$$1 \times 0 = 0$$
$$1 \times 1 = 1$$

❖

Binary division is the arithmetic inverse of multiplication. It uses a subtract and shift technique. Consider Example 1.30. A seven-bit value 11011.10 is divided by a three-bit value 101. Because 1 is not divisible by 101, a leading 0 is the first quotient bit. Nor is 11 divisible by 101, so the second quotient bit is also 0. But 110 is divisible by 101, so the third quotient bit is a 1, and so on until the process is complete. In many cases the final quotient may be endless or end in a repeating value. The division process is terminated when the number of quotient bits available are all used.

❖ **EXAMPLE 1.30** How is binary division performed?

SOLUTION Binary:

```
        00101.10
   101 |11011.10
        101
          111
          101
           10 1
           10 1
              0
```

Decimal:

```
     5.5
  5 |27.5
     25
      2 5
```

❖

1.10.2 Binary Arithmetic Using Complement Codes

Addition and subtraction of signed numbers can generate eight distinct operations. Each operand (X and Y) can be either positive or negative and the operation can be either addition or subtraction. Table 1.15 illustrates the permutations of adding and subtracting two signed numbers. The operation column indicates the sign of the operands and the operation. Two main operation columns, Add and Subtract, are indicated. The Subtract column has three subcolumns indicating the relative magnitude of the operands.

For example, in the first row two positive operands are added together producing $+(X + Y)$ under the Add column. The second row indicates adding $+X$ to $-Y$. The

TABLE 1.15
Addition and subtraction of signed numbers

Operation	Add	Subtract		
		$X > Y$	$X < Y$	$X = Y$
$(+X)+(+Y)$	$+(X+Y)$			
$(+X)+(-Y)$		$+(X-Y)$	$-(Y-X)$	$+(X-Y)=0$
$(-X)+(+Y)$		$-(X-Y)$	$+(Y-X)$	$+(Y-X)=0$
$(-X)+(-Y)$	$-(X+Y)$			
$(+X)-(+Y)$		$+(X-Y)$	$-(Y-X)$	$+(X-Y)=0$
$(+X)-(-Y)$	$+(X+Y)$			
$(-X)-(+Y)$	$-(X+Y)$			
$(-X)-(-Y)$		$-(X-Y)$	$+(Y-X)$	$+(Y-X)=0$

final form in the second row depends on the relative magnitude of X and Y. If $X > Y$ then the result is $+(X - Y)$, if $X < Y$ the result is $-(Y - X)$ and if $X = Y$ the result is $+(X - Y)$.

❖ EXAMPLE 1.31

Add $(+X) + (+Y)$.

SOLUTION Let $X = 0\wedge01101.1_2$ and $Y = 0\wedge01011.0_2$:

$$\begin{array}{r} 0\wedge01101.1 \\ +\ \underline{0\wedge01011.0} \\ 0\wedge11000.1 \end{array}$$

❖

Notice that both operands X and Y were written using six magnitude bits, yet only five magnitude bits were necessary to express the value of each operand. If the operands were not written using six bits then the carry from the fifth bit position would have entered into the sign bit causing the result to be negative instead of positive. The result must always have enough bit positions to hold any possible answer. Consider what would happen if insufficient bit positions were available to store the result as in Example 1.32.

❖ EXAMPLE 1.32

Let X and Y be the same operands as in Example 1.31 but represented in only five bits each.

SOLUTION

$$\begin{array}{r} 0\wedge1101.1 \\ +\ \underline{0\wedge1011.0} \\ 1\wedge1000.1 \end{array}$$

which is incorrect, because a positive X added to a positive Y must yield a positive result. The error occurred due to insufficient bit positions to hold the answer. ❖

❖ **EXAMPLE 1.33** Add X and Y. Let $X = +12_{10}$ and $Y = -8_{10}$, both in 2s complement code.

SOLUTION Find the 2s complement code values for X and Y, where $X = 0$^1100 and $Y = 1$^1000.

$$\begin{array}{r} 0\wedge 1100 \\ +\ \underline{1\wedge 1000} \\ (1)\ 0\wedge 0100,\ +12_{10} + (-8_{10}) = +4_{10}. \end{array}$$

The apparent overflow is ignored because there is a -100000 implied in the -8_{10} term. ❖

❖ **EXAMPLE 1.34** Add X and Y. Let $X = -9_{10}$ and $Y = +5_{10}$, both in 2s complement.

SOLUTION Let $X = 1$^0111 and $Y = 0$^0101.

$$\begin{array}{l} \ \ 1\wedge 0111 \\ +\underline{0\wedge 0101} \\ \ \ 1\wedge 1100,\ -9_{10} + (+5_{10}) = -4_{10};\ 1\wedge 1100 = -4 \text{ in 2s complement.} \end{array}$$ ❖

❖ **EXAMPLE 1.35** Add X and Y. Let $X = -8_{10}$ and $Y = -10_{10}$, in 2s complement.

SOLUTION

$$\begin{array}{r} 1\wedge 1000 \\ +\ \underline{1\wedge 0110} \\ (1)\ 0\wedge 1110.\ -8_{10} + (-10_{10}) = -18_{10} \text{ and } 0\wedge 1110 \text{ is not } -18_{10}. \end{array}$$

Overflow is what happens when insufficient magnitude bit positions are available for the answer. ❖

The example result is positive, but the addition of two negative numbers should have been negative. The problem is that overflow occurred because not enough bits were available to store the answer. Reworking the problem so that the result is contained in a five-bit magnitude,

$$\begin{array}{r} 1\wedge 11000 \\ +\ \underline{1\wedge 10110} \\ (1)\ 1\wedge 01110 \end{array}$$

No overflow occurred.

Notice that both operands are negative and the most significant bit of the magnitude is a 1. Table 1.11 shows that $-8 = 1$^1000 and $-10 = 1$^0110, but using these values created the overflow problem, resulting in an incorrect answer.

An overflow occurs when addition of two positive numbers gives a negative answer or addition of two negative numbers gives a positive result.

❖ **EXAMPLE 1.36** Find the 2s complement of -10_{10} with five magnitude bits.

SOLUTION $+10_{10} = 0$^01010

The 1s complement of -10_{10} is 1^10101.
The 2s complement is the 1s complement + 1.
The 2s complement of $-10_{10} = 1$^10101 + 1 = 1^10110. ❖

EXAMPLE 1.37 Subtract Y from X. Let $X = +12_{10} = 0\text{^}1100$ and $Y = +8_{10} = 0\text{^}1000$.

SOLUTION

$$\begin{array}{r} 0\text{^}1100 \\ -\ \underline{0\text{^}1000} \\ 0\text{^}0100 \end{array} \quad (+12_{10}) - (+8_{10}) = +4_{10}$$

Subtraction can be accomplished by finding the radix complement of the minuend and adding it to the subtrahend. For example, Let $X = 23_{10}$, and $Y = 18_{10}$. Find $Z = X + (-Y)$. By subtraction,

$$\begin{array}{r} 23_{10} \\ -\ \underline{18_{10}} \\ 05_{10} \end{array}$$

Because 8 was greater than 3 we "borrowed" 10 from 20 and subtracted 8 from 13.

By complementing the minuend and adding, find the 10s complement of 18_{10} and then add the complement to 23_{10}. The 10s complement of -18_{10} is $100_{10} - 18_{10} - 100_{10} = 82_{10} - 100_{10}$. Add $82_{10} - 100_{10}$ to 23_{10}.

$$\begin{array}{l} \ 23_{10} \\ \underline{+\ 82_{10} - 100_{10}} \\ 1\ 05_{10} - 100_{10} = 05_{10} \end{array}$$

EXAMPLE 1.38 Subtract Y from X. Let $X = +8_{10} = 0\text{^}1000$ and $Y = -6_{10} = 1\text{^}1010$, $X - Y = X + (-Y)$.

SOLUTION Use the complement and add technique.

$$0\text{^}1000 = +8_{10}$$

The 2s complement of -6_{10} $(1\text{^}1010)$ is $0\text{^}0110$.

$$\begin{array}{r} 0\text{^}1000 = +8_{10} \\ +\ \underline{0\text{^}0110} = \text{(radix complement of } -6 \text{ in 2s complement)} \\ 0\text{^}1110 = +14_{10} \end{array}$$

EXAMPLE 1.39 Subtract Y from X. Let $X = -10_{10} = 1\text{^}0110$ and $Y = -5_{10} = 1\text{^}1011$, $(-10_{10}) - (-5_{10}) = -5_{10}$.

SOLUTION The radix complement of $1\text{^}1011$ (-5_{10}) is $0\text{^}0101$.

$$\begin{array}{r} 1\text{^}0110 = -10_{10} \\ +\ \underline{0\text{^}0101} = \text{(radix complement of } -5 \text{ in 2s complement).} \\ 1\text{^}1011 = -5_{10} \end{array}$$

1.10.3 Hexadecimal Arithmetic

Hexadecimal arithmetic is performed using the same principles as binary and decimal. Tables 1.16 and 1.17 illustrate hexadecimal addition and subtraction.

TABLE 1.16
Hexadecimal addition

(A + B)	Addend (A)															
Augend (B)	0	1	2	3	4	5	6	7	8	9	A	B	C	D	E	F
0	0	1	2	3	4	5	6	7	8	9	A	B	C	D	E	F
1	1	2	3	4	5	6	7	8	9	A	B	C	D	E	F	10
2	2	3	4	5	6	7	8	9	A	B	C	D	E	F	10	11
3	3	4	5	6	7	8	9	A	B	C	D	E	F	10	11	12
4	4	5	6	7	8	9	A	B	C	D	E	F	10	11	12	13
5	5	6	7	8	9	A	B	C	D	E	F	10	11	12	13	14
6	6	7	8	9	A	B	C	D	E	F	10	11	12	13	14	15
7	7	8	9	A	B	C	D	E	F	10	11	12	13	14	15	16
8	8	9	A	B	C	D	E	F	10	11	12	13	14	15	16	17
9	9	A	B	C	D	E	F	10	11	12	13	14	15	16	17	18
A	A	B	C	D	E	F	10	11	12	13	14	15	16	17	18	19
B	B	C	D	E	F	10	11	12	13	14	15	16	17	18	19	1A
C	C	D	E	F	10	11	12	13	14	15	16	17	18	19	1A	1B
D	D	E	F	10	11	12	13	14	15	16	17	18	19	1A	1B	1C
E	E	F	10	11	12	13	14	15	16	17	18	19	1A	1B	1C	1D
F	F	10	11	12	13	14	15	16	17	18	19	1A	1B	1C	1D	1E

Hexadecimal addition of two digits is shown in Table 1.16. The addend digits (0 through F) are listed across the top and the augend digits (0 through F) are listed vertically. The sum digit(s) are contained in the matrix. For instance, $F_{16} + A_{16} = 19_{16}$; the sum is located at the junction of the addend digit F_{16} and the augend digit A_{16}.

TABLE 1.17
Hexadecimal subtraction

(A − B)	(A) Minuend																								
Subtrahend (B)	0	1	2	3	4	5	6	7	8	9	A	B	C	D	E	F	10	11	12	13	14	15	16	17	18
0	0	1	2	3	4	5	6	7	8	9	A	B	C	D	E	F	10	11	12	13	14	15	16	17	18
1	−1	0	1	2	3	4	5	6	7	8	9	A	B	C	D	E	F	10	11	12	13	14	15	16	17
2	−2	−1	0	1	2	3	4	5	6	7	8	9	A	B	C	D	E	F	10	11	12	13	14	15	16
3	−3	−2	−1	0	1	2	3	4	5	6	7	8	9	A	B	C	D	E	F	10	11	12	13	14	15
4	−4	−3	−2	−1	0	1	2	3	4	5	6	7	8	9	A	B	C	D	E	F	10	11	12	13	14
5	−5	−4	−3	−2	−1	0	1	2	3	4	5	6	7	8	9	A	B	C	D	E	F	10	11	12	13
6	−6	−5	−4	−3	−2	−1	0	1	2	3	4	5	6	7	8	9	A	B	C	D	E	F	10	11	12
7	−7	−6	−5	−4	−3	−2	−1	0	1	2	3	4	5	6	7	8	9	A	B	C	D	E	F	10	11
8	−8	−7	−6	−5	−4	−3	−2	−1	0	1	2	3	4	5	6	7	8	9	A	B	C	D	E	F	10
9	−9	−8	−7	−6	−5	−4	−3	−2	−1	0	1	2	3	4	5	6	7	8	9	A	B	C	D	E	F
A	−A	−9	−8	−7	−6	−5	−4	−3	−2	−1	0	1	2	3	4	5	6	7	8	9	A	B	C	D	E
B	−B	−A	−9	−8	−7	−6	−5	−4	−3	−2	−1	0	1	2	3	4	5	6	7	8	9	A	B	C	D
C	−C	−B	−A	−9	−8	−7	−6	−5	−4	−3	−2	−1	0	1	2	3	4	5	6	7	8	9	A	B	C
D	−D	−C	−B	−A	−9	−8	−7	−6	−5	−4	−3	−2	−1	0	1	2	3	4	5	6	7	8	9	A	B
E	−E	−D	−C	−B	−A	−9	−8	−7	−6	−5	−4	−3	−2	−1	0	1	2	3	4	5	6	7	8	9	A
F	−F	−E	−D	−C	−B	−A	−9	−8	−7	−6	−5	−4	−3	−2	−1	0	1	2	3	4	5	6	7	8	9

❖ **EXAMPLE 1.40** Two examples of how hexadecimal addition is performed are demonstrated next.

SOLUTION First example,

$$\begin{array}{rcccc} & & 1 & & \\ & A & 5 & C & 4_{16} \\ + & 3 & 9 & A & 5_{16} \\ \hline & D & F & 6 & 9_{16} \end{array} \qquad \begin{array}{rcl} 4 + 5 &=& 9 \\ C + A &=& 16 \\ 1 + 5 + 9 &=& F \\ A + 3 &=& D \end{array}$$

Second example,

$$\begin{array}{rccccc} & & 1 & 1 & & \\ & & 9 & A & F & 2 & 7_{16} \\ + & & A & B & C & D & 4_{16} \\ \hline & 1 & 4 & 6 & B & F & B_{16} \end{array} \qquad \begin{array}{rcl} 7 + 4 &=& B \\ 2 + D &=& F \\ F + C &=& 1B \\ 1 + A + B &=& B + B = 16 \\ 1 + 9 + A &=& A + A = 14 \end{array}$$

❖

Hexadecimal subtraction of two digits is shown in Table 1.17. The minuend digits (0 through 18) are shown across the top and the subtrahend digits (0 through F) are listed vertically. The difference digit(s) are contained in the matrix. For instance $C_{16} - 9_{16} = 3_{16}$; the difference is located at the junction of C_{16} (top) and 9_{16} (side).

❖ **EXAMPLE 1.41** An example of how hexadecimal subtraction is performed is shown below.

SOLUTION

$$\begin{array}{rcccc} & 9 & 19 & 5 & 1B \\ & A & 9 & 6 & B_{16} \\ - & 9 & F & 2 & C_{16} \\ \hline & 0 & A & 3 & F_{16} \end{array}$$

B − C produces a negative result, so a borrow is generated, 1B − C = F:

$$5 - 2 = 3$$

$$19 - F = A$$

$$9 - 9 = 0$$

❖

The hexadecimal multiplication table (Table 1.18) shows the multiplicand across the top and the multiplier along the side. The product digits are contained in the matrix. For example, the product of $D_{16} \times F_{16} = C3_{16}$. The intersection of the D_{16} multiplicand column with the F_{16} multiplier row contains the $C3_{16}$ product.

❖ **EXAMPLE 1.42** An example of how hexadecimal multiplication is performed follows below.

SOLUTION

$$\begin{array}{rcccccc} & & & & A & F & C & 4_{16} \\ \times & & & & & B & 9 & C_{16} \\ \hline & & & 8 & 3 & D & 3 & 0 \\ & & 6 & 2 & D & E & 4 & \\ & 7 & 8 & D & 6 & C & & \\ \hline & 7 & F & 8 & 8 & 7 & 7 & 0_{16} \end{array}$$

TABLE 1.18
Hexadecimal multiplication table

($A \times B$)		Multiplicand (A)															
		0	1	2	3	4	5	6	7	8	9	A	B	C	D	E	F
M	0	0	0	0	0	0	0	0	0	0	0	0	0	0	0	0	0
u	1	0	1	2	3	4	5	6	7	8	9	A	B	C	D	E	F
l	2	0	2	4	6	8	A	C	E	10	12	14	16	18	1A	1C	1E
t	3	0	3	6	9	C	F	12	15	18	1B	1E	21	24	27	2A	2D
i	4	0	4	8	C	10	14	18	1C	20	24	28	2C	30	34	38	3C
p	5	0	5	A	F	14	19	1E	23	28	2D	32	37	3C	41	46	4B
l	6	0	6	C	12	18	1E	24	2A	30	36	3C	42	48	4E	54	5A
i	7	0	7	E	15	1C	23	2A	31	38	3F	46	4D	54	5B	62	69
e	8	0	8	10	18	20	28	30	38	40	48	50	58	60	68	70	78
r	9	0	9	12	1B	24	2D	36	3F	48	51	5A	63	6C	75	7E	87
	A	0	A	14	1E	28	32	3C	46	50	5A	64	6E	78	82	8C	96
(B)	B	0	B	16	21	2C	37	42	4D	58	63	6E	79	84	8F	9A	A5
	C	0	C	18	24	30	3C	48	54	60	6C	78	84	90	9C	A8	B4
	D	0	D	1A	27	34	41	4E	5B	68	75	82	8F	9C	A9	B6	C3
	E	0	E	1C	2A	38	46	54	62	70	7E	8C	9A	A8	B6	C4	D2
	F	0	F	1E	2D	3C	4B	5A	69	78	87	96	A5	B4	C3	D2	E1

The first partial product row was found:

a. $C \times 4 = 3(0) \longrightarrow 0$
Carry $= 3$

b. $C \times C = 90$
$90 + 3 = 9(3) \longrightarrow 3\ 0$
Carry $= 9$

c. $C \times F = B4$
$B4 + 9 = B(D) \longrightarrow D\ 3\ 0$
Carry $= B$

d. $C \times A = 78$
$78 + B = (83) \longrightarrow 8\ 3\ D\ 3\ 0$

The second partial product row was found:

a. $9 \times 4 = 2(4) \longrightarrow 4$
Carry $= 2$

b. $9 \times C = 6C$
$6C + 2 = 6(E) \longrightarrow E\ 4$
Carry $= 6$

c. $9 \times F = 87$
$87 + 6 = 8(D) \longrightarrow D\ E\ 4$
Carry $= 8$

d. $9 \times A = 5A$
$5A + 8 = (62) \longrightarrow 6\ 2\ D\ E\ 4$

The third partial product row was found:

a. $B \times 4 = 2(C) \longrightarrow C$
Carry = 2

b. $B \times C = 84$
$84 + 2 = 8(6) \longrightarrow 6\ C$
Carry = 8

c. $B \times F = A5$
$A5 + 8 = A(D) \longrightarrow D\ 6\ C$
Carry = A

d. $B \times A = 6E$
$6E + A = (78) \longrightarrow 7\ 8\ D\ 6\ C$

The sum of the partial products was found using Tables 1.16 and 1.18 (addition and multiplication):

```
     8 3 D 3 0     first partial product
   6 2 D E 4       second partial product
 + 7 8 D 6 C       third partial product
 -------------
   7 F 8 8 7 7 0   Sum of partial products
```

❖

❖ **EXAMPLE 1.43** Shown next is an example of how hexadecimal division is performed.

SOLUTION

```
     0 C 5 C. A 2
  F | B 9 6 D. 8 0
      0
      -
      B 9
      B 4
      ---
        5 6
        4 B
        ---
          B D
          B 4
          ---
            9 8
            9 6
            ---
              2 0
              1 E
              ---
                2 = Remainder
```

❖

SUMMARY

A. The chapter started with a definition of digital design. Digital design is the science of organizing information, defining operations, accepting inputs, and generating outputs using discrete quantities.

B. The chapter went on to develop these concepts:
1. Discrete information is separate, distinct.
2. Continuous information is connected, not separate.

3. Analog representation deals with the continuous representation of data.
4. Digital representation deals with the discrete representation of data.

C. Some historical notes were used to illustrate that many of the basic principles of digital design have existed for a long time.

D. The future holds a very bright promise for digital design in the development of new devices and products.

E. Algorithms are step-by-step procedures for accomplishing a task.

F. Digital quantities are represented using binary, octal, or hexadecimal numbers.
1. We must know how to use numbers of different bases to represent a quantity.
2. We must be able to convert numbers in one base to numbers of another base, especially in binary, octal, decimal, and hexadecimal bases.

G. Binary codes used for different purposes include binary coded decimal (BCD) codes, weighted binary codes, unit distance binary codes, alphanumeric codes, and signed binary codes.

H. Arithmetic (addition, subtraction, multiplication, and division) in binary and hexadecimal is a useful part of a digital designer's toolbox.

REFERENCES

Texts that present material on basic digital concepts, number systems, radix conversion, codes, and code conversion as well as arithmetic include these:

Fundamentals of Computer Engineering (Logic Design and Microprocessors) by Herman Lam and John O'Malley (New York: John Wiley and Sons, 1990). Chapter 1 covers number systems and digital arithmetic.

Digital Circuits by Kenneth Muchow, Anthony Zeppa, and Bill Deem (Englewood Cliffs, N.J.: Prentice-Hall, 1987). Chapter 1 is a survey of basic digital computer concepts; Chapter 2 covers number systems and codes; and Chapter 3 deals with basic arithmetic.

Digital Concepts and Applications by Amin R. Ismail and Victor M. Rooney (Orlando, Fla.: Holt, Rinehart and Winston, 1990). Chapter 1 treats number systems and basic digital concepts.

Digital Logic Design by Gideon Langholz, Abraham Kandel, and Joe L. Mott (Dubuque, Iowa: William Brown, 1988). Chapter 1 deals with introductory digital concepts; Chapter 2 covers number systems, codes, and arithmetic.

Digital Design Principles and Practices by John F. Wakerly (Englewood Cliffs, N.J.: Prentice-Hall, 1990). Chapter 1 presents number systems, basic arithmetic, and digital codes.

Fundamentals of Logic Design, 4th Edition by Charles H. Roth, Jr. (Minneapolis–St. Paul: West, 1992). Chapter 1 discusses an introduction to numbers and radix conversion, arithmetic and codes.

Introduction to Logic Design by Sajjan G. Shiva (Glenview, Ill.: Scott, Foresman and Little/Brown, 1988). Chapter 1 covers number systems, radix conversion codes, and digital arithmetic.

Digital Design, A Pragmatic Approach by E. L. Johnson and M. A. Karim (Boston: PWS-Kent, 1987). Chapter 1 deals with an introduction to number systems and binary codes.

Digital Systems, Principles and Applications by Ronald J. Tocci (Englewood Cliffs, N.J.: Prentice-Hall, 1988). Chapter 1 provides some introductory concepts of digital systems; Chapter 2 covers number systems and digital codes.

Digital Fundamentals by Thomas L. Floyd (Columbus, Ohio: Merrill Publishing, 1990). Chapter 1 discusses basic digital concepts; Chapter 2 introduces number systems, binary codes, and digital arithmetic.

GLOSSARY

Algorithm A step-by-step procedure for accomplishing a given task or solving a problem. Usually partitioned into simple steps that do not require detailed knowledge of the process.

Alphanumeric code A binary code used to represent the alphabet, numbers, and punctuation marks as well as control characters for controlling a printer or display.

Analog In electronics, the term usually refers to a means of representing information in a continuous fashion. Amplifiers produce analog outputs to drive a speaker, for example.

ASCII code One of several alphanumeric codes for representing the alphabet, numbers, and punctuation in a digital format.

Base The number of characters in the character set of a positional number system. Decimal numbers have a base of 10, binary numbers a base of 2. *Base* and *radix* are used interchangeably.

Binary The name given the base-2 number system. *Binary* means two.

Binary coded decimal (BCD) A special binary code used to directly represent decimal characters. Each four-bit value in BCD represents a single decimal character.

Bit A single binary digit. Each position in the binary number system is represented by a bit. Bit is short for *bi*nary digi*t*.

Character set The group of characters used to represent value in any positional number system. The decimal number system contains a set of 10 characters (0, 1, 2, 3, 4, 5, 6, 7, 8, 9).

Code A unique way to represent a value or a character in the alphabet. Digital codes use bits configured in a given way to represent numbers or alphabet characters.

Complement The complement of a single character represented in a positional number system is the difference between the total number of characters in the character set and the value of the character for which a complement is being sought. The 10s complement of 4 is 6 because $10 - 4 = 6$. The 2s complement of a binary $1 = 1$, since $2 - 1 = 1$.

Complement codes Binary codes are used to represent signed numbers. Two such codes are common: the 1s complement and the 2s complement. The codes represent negative numbers as the 1s or 2s complement of the positive value with the same magnitude.

Continuous Changing without interruption, no breaks in the manner in which a value changes. Analog data are continuous.

Decimal Refers to base-10 number system.

Digital signal processing (DSP) Techniques used to convert analog data into a digital format process the information in some manner with a digital system, usually a specialized computer, and then convert it back to analog for output.

Diminished radix A term used when referring to the radix -1 complement. The diminished radix of the 10s complement is the 9s complement. The diminished radix of the binary 2s complement is the 1s complement, which can be obtained by changing all 1s to 0s and all 0s to 1s.

Discrete Separate, not connected. Discrete data are represented in a series of numbers.

Excess-3 code A self-complementing code used to represent BCD numbers. It is self-complementing because the 1s complement (bit by bit inversion) is also the 9s complement of the BCD number. It is widely used in BCD arithmetic circuits.

Flowchart A structured notational technique for graphically representing data flow and operational control of some process. Algorithms are often represented using flowcharts.

Gray code A reflective, unit distance code where only one bit position changes for each adjacent change in value.

Hexadecimal A base-16 number system. The system has 16 characters in its character set (0 through 9 plus A, B, C, D, E, F). It is widely used in digital systems due to its ease of conversion with binary numbers. Four bits are used to represent a single hexadecimal numeral.

Inverter A basic electronic circuit that changes logical "0s" to "1s" and "1s" to "0s." It performs the radix minus-1 complement of a bit. It is one of the primitive gates used in digital systems.

Logic The term was borrowed from the philosophers. It originally had to do with the process of determining the "truth" or "falsity" of a statement or proposition.

When used in conjunction with digital systems and design, it refers to the process and rules for designing digital systems.

Logic design The application of a set of rules and techniques for developing digital circuits and sub-systems to create a solution to some problem.

1s complement code A digital code used to represent signed binary numbers. The negative numbers are represented as the 1s complement of the positive value.

Positional number system A number system of any radix or base that assigns value to a character based on its relative position. In decimal the units, tens, hundreds, and so forth positions indicate value not indicated by the character (0 through 9).

Radix See Base.

Sign magnitude code A binary code used to represent signed binary numbers. The most significant bit is used to indicate sign; a 1 represents a negative number and a 0 a positive number.

Successive division A technique for converting from one base or radix to another. The number being converted is successively divided by the radix of the new number, with the remainder from each division operation representing a digit in the new radix.

2s complement code A code used to represent signed numbers. Negative 2s complement code numbers are found by inverting the absolute value of the binary number and adding 1 to the result.

Unit distance code A binary code in which only one bit changes between successive values. The Gray code is a unit distance code.

Weighted code A code in which each bit position has a certain numeric value assigned. Several such codes exist, such as 8, 4, 2, 1; 7, 4, 2, 1; 6, 3, 1, −1.

QUESTIONS AND PROBLEMS

Section 1.1 to 1.5

1. Describe each of the following in your own words:
   a. Discrete
   b. Continuous
   c. Digital
   d. Analog
2. Which English mathematician developed the algebra that bears his name?
3. What company first developed the "computer on a chip" and in what year?
4. List some of the common consumer products that use digital concepts in some way.
5. Define an *algorithm*; explain why they are useful in digital design.
6. What is a flowchart? Explain.
7. Define in your own words the concept of a hierarchy. How does the term apply to the field of digital systems?

Section 1.6 to 1.7

8. Write the following numbers in expanded positional notation.
   a. 639.58_{10}
   b. 110010.110_2
   c. 234.2_8
   d. $1{,}239.53_{10}$
   e. 456.234_7
   f. $FE2.C_{16}$
9. How many characters are in the character set for numbers in the following bases:
   a. octal
   b. binary
   c. decimal
   d. base 5
   e. base 3
   f. hexadecimal
   g. base 12

10. What are the hexadecimal characters for numbers greater than 9?

11. Count to 30_{10} in the following bases:

a. binary

b. octal

c. base 5

d. hexadecimal

12. Equations (1.2) and (1.3) say the same thing. Explain the equations in your own words.

Section 1.8

13. Convert the following binary numbers to octal:

a. 100110_2

b. 100101101.110_2

c. 10000111001.100101_2

14. Convert the following binary numbers to hexadecimal:

a. 110100111100.0101_2

b. 111111011010.10100111_2

15. Convert the following decimal numbers to binary:

a. 12_{10}

b. 34.25_{10}

c. $1{,}024.5_{10}$

d. 255.75_{10}

16. Explain the "successive division" base conversion algorithm in your own words.

17. Convert the following binary numbers to decimal:

a. 11001.1_2

b. 100111.11_2

c. 11001101.111_2

18. Convert the following octal numbers to binary:

a. 234.6_8

b. 656.25_8

c. 2365.123_8

19. Convert the following hexadecimal numbers to binary:

a. $FAC.B_{16}$

b. $27AD.9B_{16}$

c. $CDE2.F5_{16}$

20. Convert the following to decimal:

a. 234_5

b. 453.6_9

c. $CF.5_{16}$

d. 12.2_3

Section 1.9

21. Draw a table comparing decimal numbers 0 through 9 and their binary coded decimal equivalents.

22. Convert the following decimal numbers to BCD:

a. 325.6_{10}

b. 1985.67_{10}

c. 2954.13_{10}

23. Convert the following BCD numbers into decimal:

a. $100100111000.0111_{(BCD)}$

b. $1000011000100011.10010111_{(BCD)}$

24. Write the equivalent in the following BCD weighted codes for the given decimal numbers:

a. 82_{10} in 8, 4, 2, 1; 7, 4, 2, 1; 8, 4, −2, −1

b. 32.1_{10} in 8, 4, 2, 1; 5, 4, 2, 1; 5, 3, 1, 1

c. 295.47_{10} in 8, 4, −2, −1; ex-3; 7, 4, 2, 1

25. Describe in your own words the purpose of a "self-complementing" code.

26. Construct a table showing the ex-3 self-complementing code and its decimal equivalent.

27. Describe a "unit distance" code in your own words.

28. Write your name using the seven-bit ASCII code.

29. Sketch a wheel that can be fastened to a shaft that is free to rotate, with "slots" cut so that light can shine through to indicate relative position for a three-bit digital optical encoder that uses the unit distance Gray code.

30. Write the sign magnitude, 1s complement, and 2s complement codes for the following decimal numbers:

a. +12

b. −12

c. +9.5

d. −22.5

e. +19.75

f. −17.25

31. Find the radix and radix-1 complements for the following numbers:

a. 1011.11_2

b. 23.412_5

c. 92.1_{10}

d. 327.4_8

e. $AF2.4_{16}$

f. 21.3_4

Section 1.10

32. Perform the indicated operations:

a. $10110.1_2 + 01111.1_2$

b. $1000.0_2 - 0111.1_2$

c. $312_8 + 123_8$

d. $241_8 - 176_8$

33. Perform the indicated operations:

a. $FDC_{16} + A29_{16}$

b. $FE6_{16} - EFC_{16}$

c. $34_5 + 12_5$

d. $24_6 - 15_6$

34. Perform the indicated operations:

a. $10110_2 \times 011_2$

b. $1110.11_2 \times 11.1_2$

c. $11001.11_2 \times 101_2$

d. $010110.101_2 \times 11.1_2$

35. Perform the indicated operations:

a. $234_8 \times 24_8$

b. $FC2_{16} \times DE_{16}$

c. $FDE_{16} \times F_{16}$

d. $CD.35_{16} \times 4.C_{16}$

36. Perform the following subtractions by taking the 2s complement of the subtrahend and adding it to the minuend:

a. $101101_2 - 01111_2$

b. $11101000.11_2 - 0110101.1_2$

37. Perform the following operations by finding the radix complement of the subtrahend and adding the result to the minuend:

a. $\begin{array}{r} 23.4_{10} \\ -\ \underline{19.8_{10}} \end{array}$

b. $\begin{array}{r} 135.7_8 \\ -\ \underline{67.7_8} \end{array}$

c. $\begin{array}{r} 321.2_4 \\ -\ \underline{33.3_4} \end{array}$

d. $\begin{array}{r} FA.3_{16} \\ -\ \underline{0F.F_{16}} \end{array}$

e. $\begin{array}{r} 10011.11_2 \\ -\ \underline{1111.01_2} \end{array}$

38. Perform the indicated operations. Convert each decimal number into its corresponding 2s complement code prior to performing the indicated operation. Refer to Table 1.15.

a. $\begin{array}{r} +32 \\ (-)\ \underline{-19} \end{array}$

b. $\begin{array}{r} -29 \\ (-)\ \underline{-17} \end{array}$

c. $\begin{array}{r} -17 \\ (+)\ \underline{-12} \end{array}$

d. $\begin{array}{r} +12 \\ (-)\ \underline{+14} \end{array}$

e. $\begin{array}{r} -29 \\ (-)\ \underline{+8} \end{array}$

f. $\begin{array}{r} +21 \\ (+)\ \underline{+14} \end{array}$

g. $\begin{array}{r} +19 \\ (+)\ \underline{-20} \end{array}$

h. $\begin{array}{r} -13 \\ (+)\ \underline{+23} \end{array}$

39. Construct an addition table for

a. base-4 numbers

b. base-9 numbers

40. Construct a multiplication table for

a. base-3 numbers

b. base-12 numbers

41. Explain the procedure showing how numbers can be subtracted by finding the radix complement of the subtrahend and adding it to the minuend.

CHAPTER 2

BOOLEAN SWITCHING ALGEBRA

INTRODUCTION

Boolean algebra is a mathematical system that defines a series of logical operations (AND, OR, NOT) performed on sets of variables (a, b, c, . . .). When stated in this form, the expression is called a **Boolean equation or switching equation**. In our discussion in this text variables will be restricted to two values (0 and 1). George Boole introduced the mathematical system of logic that bears his name in the 19th century. Huntington, in 1904, formulated a series of mathematical statements that gave Boole's work additional structure. Digital systems are built using electronics that perform the functions that Boole and Huntington formulated. In the late 1930s Claude Shannon presented a paper on the use of Boolean algebra in telephone switching. Shannon is considered to be the "father" of modern digital design.

In this chapter we will study the basic "rules" governing logic analysis and design. First we need to develop an understanding of the logic functions and symbols. Then we can proceed to Boolean algebra. The topics covered in this chapter lay a foundation for the actual analysis and design of working digital circuits.

2.1 BINARY LOGIC FUNCTIONS

A binary **variable** is like a variable in regular algebra, except it can only have the values of zero (0) or one (1). Binary variables are designated by names, symbols, letters, numbers, or their combinations. Usually the names or symbols associated with a binary variable are related to the original logic problem being solved. The designer is free to assign any name or letter or combinations of letters, numbers, or names to a variable.

Three logic **functions** (AND, OR, and NOT or complement) provide the foundation for all digital systems analysis and design. A function is a term used in mathematics and logic to denote a relationship between input and output variables. Each variable is restricted to binary (0,1) values. A tabular representation of the combinations that a group of binary input and output variables can assume is called a **truth table**. The *truth* in the name comes from its original background in philosophy where "true" and "false" were assigned to logic statements. True is most often represented by a 1 and false by a 0.

The three primitive logic functions are

And An AND is represented by operator symbols ( ), *, ·, or no space, which are the same operator symbols used for multiplication in regular algebra. The AND function operating on binary variables x and y is shown in equations (2.1). If x AND y are true, then s is true.

$$s = xy \tag{2.1a}$$

$$s = x*y \tag{2.1b}$$

$$s = (x)(y) \tag{2.1c}$$

The expression is read as $s = x$ AND y.

Let x, y, and s represent binary variables, where x and y are inputs and s is an output of the logic AND function.

A two-input AND truth table is:

Inputs		Output
x	y	s
0	0	0
0	1	0
1	0	0
1	1	1

All four possible combinations of x and y are represented. The output column represents the value of s for each combination of the values of x and y.

Gates are typically realized by digital circuits that perform the logic functions discussed in this section. An AND gate (Figure 2.1) is a circuit that performs the AND logic function. **Distinctive shape logic symbols** are used to represent AND, OR, and NOT functions. Logic functions are not limited to just two input variables. As many input variables as desired can be ANDed to form a function. A three- and four-variable expression would be written

$$p = xyz \tag{2.2a}$$

$$t = wxyz \tag{2.2b}$$

The truth tables for three- and four-input variable AND functions would contain $2^3 = 8$ and $2^4 = 16$ rows, respectively. Practical conditions limit the number of input variables an AND gate (or other logic gate) circuit may have.

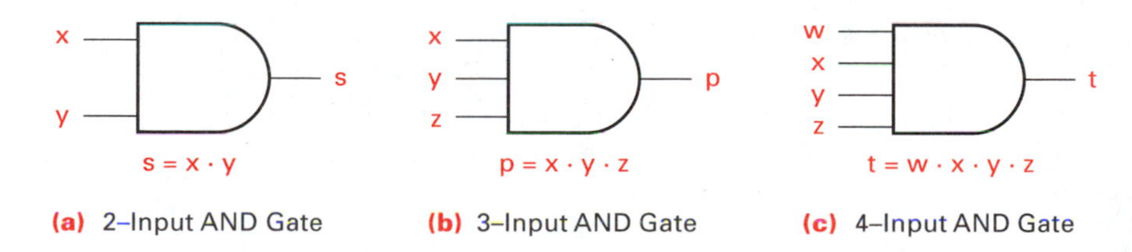

Figure 2.1
Distinctive shape symbols for AND gates

Equations (2.2a) and (2.2b) are read: $p = x$ AND y AND z; $t = w$ AND x AND y AND z. This means that *all* the input variables must be true (1) for the output to be true (1); otherwise the output is false.

Or An OR is represented by the operator symbol +. The OR function operating on binary variables x and y is shown in equation (2.3). If either x or y, or both, is true, then the output is true.

$$s = x + y \tag{2.3}$$

The expression is read as $s = x$ OR y. A two-input OR truth table is

Inputs		Output
x	y	s
0	0	0
0	1	1
1	0	1
1	1	1

As with the AND function, no logical limit is imposed on the number of possible input variables that can be formed into a single OR function. The number of rows in a multiple input OR function truth table is found similar to that for an AND truth table. Figure 2.2 shows the symbol used to represent an OR gate.

Not A NOT is represented by the operator bar symbol, —, or an apostrophe, ′. The NOT operation inverts a variable: if the input variable, x, is 0, the output, x', is 1; if $x = 1$ then $x' = 0$. NOT is also called a *complement* or *invert function:*

$$s = x' \qquad s = \overline{x} \tag{2.4}$$

The expression is read as $s = x$ NOT. The NOT truth table is

x	s
0	1
1	0

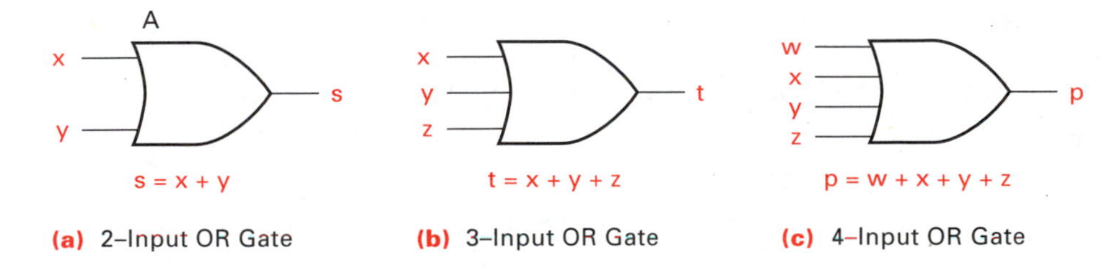

Figure 2.2 Distinctive shape for an OR gate symbol

Figure 2.3
Distinctive shape of the NOT (inverter) symbol

If x is false (0), then s is true (1). If x is true (1), then s is false (0). Figure 2.3 shows the NOT symbol.

Other commonly used logic functions are derived from the three primitives. The NAND, NOR, EX-OR (exclusive OR), and EX-NOR (exclusive NOR or equivalence), are all based on combinations of AND, OR, and NOT functions.

Nand A NAND function is derived from the NOT and the AND, and it is a contraction of NOT AND. The function is defined as an AND operation $(x\,y)$ followed by a NOT operation $(x\,y)'$, also written as $\overline{xy}$.

$$s = (xy)' \tag{2.5a}$$

$$s = \overline{xy} \tag{2.5b}$$

The expression is read as: $s =$ NOT (xy). A two-input NAND truth table is

Inputs		Output
x	y	s
0	0	1
0	1	1
1	0	1
1	1	0

The logic symbols for the NAND function are derived from the AND and NOT functions. Figure 2.4 illustrates the separate AND–NOT realization of NAND.

The NAND symbol combines the AND gate symbol and the circle of the inverter gate as shown in Figure 2.5.

Figure 2.4
NAND function using separate AND and NOT symbols

$s = (xy)'$ or $s = \overline{xy}$

Nor A NOR function is derived from the NOT and the OR and it is a contraction of NOT OR. The function is defined as an OR operation $(x + y)$ followed by a NOT function $(x + y)'$ or $\overline{x + y}$.

$$s = (x + y)' \tag{2.6a}$$

$$s = \overline{x + y} \tag{2.6b}$$

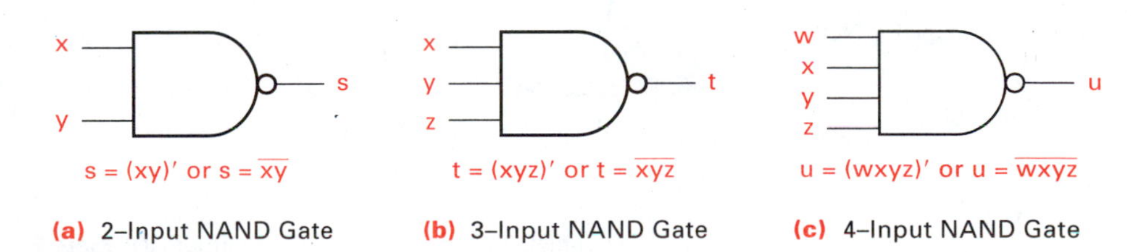

Figure 2.5
Distinctive shape for NAND gate symbols

The expression is read as $s = \text{NOT } (x \text{ OR } y)$. A two-input NOR truth table is

Inputs		Output
x	y	s
0	0	1
0	1	0
1	0	0
1	1	0

Figure 2.6 illustrates the combination of the OR and NOT symbols to form the NOR function. Figure 2.7 illustrates the contraction of the OR and NOT symbols to form the NOR symbol.

Two additional, commonly used logic functions are also derived from the AND, OR, and NOT functions. They are Exclusive OR and Exclusive NOR, also called EX-OR and EX-NOR.

Ex-Or An exclusive OR function produces a true output when an odd number of input variables are true. The exclusive part of the name is taken from the fact that the function "excludes" the case where even numbers of input variables are true.

A two input EX-OR function is written

$$s = x \oplus y \tag{2.7}$$

Three- and four-variable EX-OR functions are written

$$s = x \oplus y \oplus z$$
$$p = w \oplus x \oplus y \oplus z$$

A two-input EX-OR truth table is

Inputs		Output
x	y	s
0	0	0
0	1	1
1	0	1
1	1	0

Notice that the input combination $x = y = 1$ causes the output $s = 0$.

Figure 2.6
NOR function using separate OR and NOT symbols

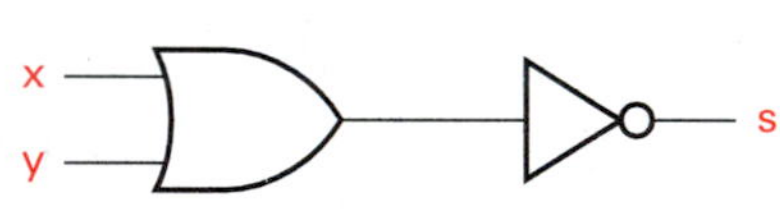

$s = (x + y)'$ or $s = \overline{x + y}$

Figure 2.7
Distinctive shape for NOR gate symbol

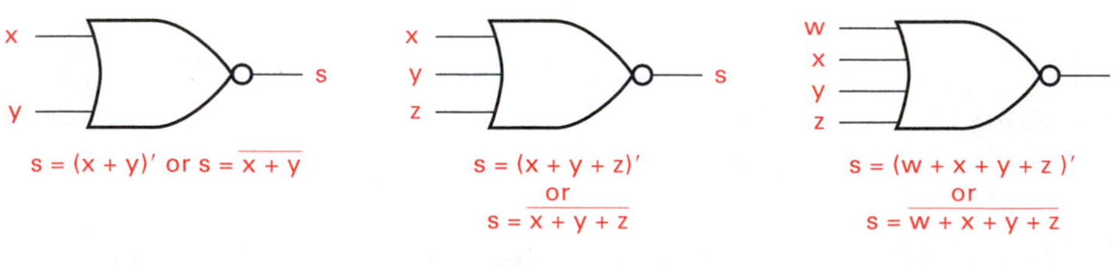

$s = (x + y)'$ or $s = \overline{x + y}$

$s = (x + y + z)'$ or $s = \overline{x + y + z}$

$s = (w + x + y + z)'$ or $s = \overline{w + x + y + z}$

(a) 2–Input NOR Gate **(b)** 3–Input NOR Gate **(c)** 4–Input NOR Gate

A two-input exclusive OR function, written $s = x \oplus y$, can also be written $s = xy' + x'y$. A two-input exclusive NOR function, written $s = \overline{x \oplus y}$, can also be written $s = x'y' + xy$.

A three-input EX-OR truth table is

Inputs			Output
x	*y*	*z*	*p*
0	0	0	0
0	0	1	1
0	1	0	1
0	1	1	0
1	0	0	1
1	0	1	0
1	1	0	0
1	1	1	1

Any odd combination of 1 inputs causes the output to be a 1.

Exclusive OR and exclusive NOR functions are, in integrated circuit packages, available as two-input gates. Exclusive OR and exclusive NOR functions with three or more inputs are derived from two-input gates as shown in Figures 2.8 and 2.9.

Ex-Nor EX-NOR is also called an *equivalence gate*. The output is true when an even number of inputs are equal. An EX-NOR is the inverse of an EX-OR; that is, an EX-OR combined with a NOT function. The result is that the EX-NOR excludes the condition where odd numbers of input variables are true. Instead it includes the cases where even numbers of inputs are true.

Figure 2.8
Two- and three-input EX-OR functions

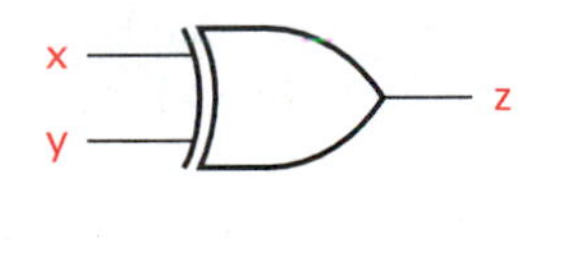

(a) 2–Input EX-OR Gate

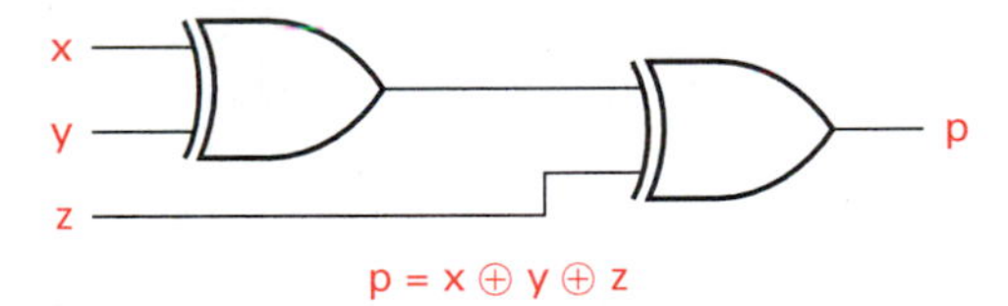

(b) 3–Input EX-OR function derived from Two 2-input-EX-OR gates

Figure 2.9
Two- and three-input EX-NOR functions

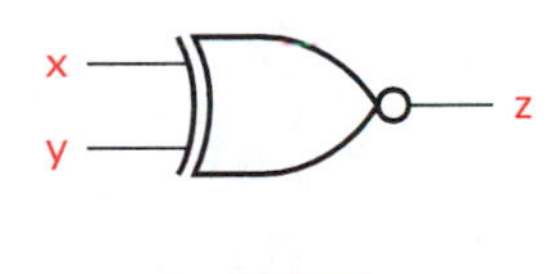

(a) 2–Input EX-NOR Gate

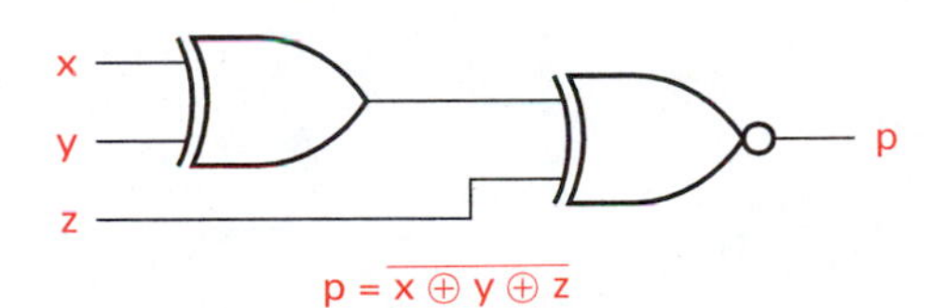

(b) 3–Input EX-NOR function derived from Two 2-input EX-NOR gates

Two-, three-, and four-variable EX-NOR functions are expressed as

$$z = \overline{x \oplus y}, \qquad z = x \odot y$$
$$p = \overline{x \oplus y \oplus z}, \qquad p = x \odot y \odot z$$
$$t = \overline{w \oplus x \oplus y \oplus z}, \qquad t = w \odot x \odot y \odot z$$

Two- and three-input EX-NOR truth tables are

Inputs		Output
x	y	z
0	0	1
0	1	0
1	0	0
1	1	1

Inputs			Output
x	y	z	p
0	0	0	1
0	0	1	0
0	1	0	0
0	1	1	1
1	0	0	0
1	0	1	1
1	1	0	1
1	1	1	0

Compare the two- and three-variable truth tables for the EX-OR and EX-NOR. Notice that the output columns of the EX-NOR are the logical inverse of the EX-OR. EX-NOR functions requiring more than two variables are derivied from two-input gates like the EX-OR.

2.1.1 IEEE Logic Symbols

It is necessary to graphically represent logic functions just like symbols are required to represent resistors, capacitors, diodes, transistors, and other electronic and mechanical components or functions. The distinctive shape symbols, shown in the previous section, are the traditional method of representing logic functions. The **IEEE logic symbols**, illustrated in this section, are gaining acceptance. We show both because new designs will increasingly use the IEEE symbols; however, the distinctive shape symbols are still widely used.

The American National Standards Institute (ANSI) and the Institute of Electrical and Electronic Engineers (IEEE) developed a standard for logic symbols. Revision of the original standard by ANSI/IEEE in 1984 resulted in the symbols shown in Figure 2.10.

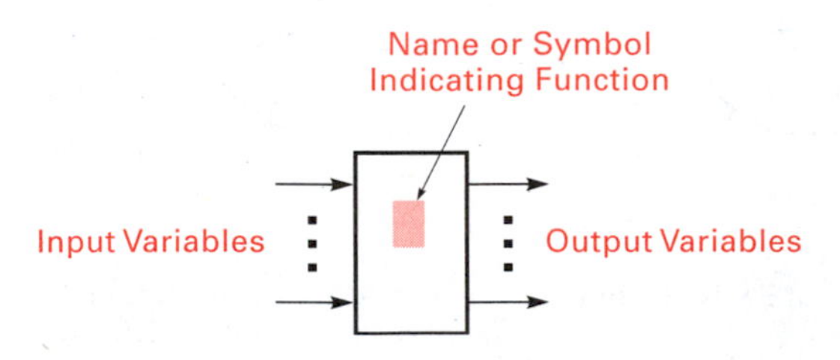

Figure 2.10
IEEE general logic symbols

A rectangle forms the basic shape of all logic functions, with some information inside the rectangle to indicate a specific function. Inputs to the function are on the left of the rectangle with outputs on the right.

Some of the symbols used to designate a function are

& AND

>1 OR (+ sometimes used)

=1 EX-OR ($\oplus$ sometimes used)

As shown both inputs and outputs can be inverted. The reason for this will become apparent later when we study mixed logic.

Figures 2.11 through 2.17 show the IEEE symbols for AND, OR, NOT, NAND, NOR, EX-OR and EX-NOR. Each symbol has a rectangle with function indicator located inside.

2.1.2 Functions, Symbols, and Truth Tables

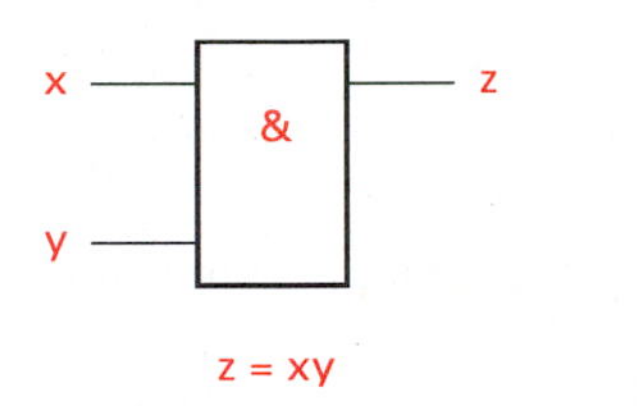

Figure 2.11
IEEE symbol for the AND gate

Each logic symbol represents a function (AND, OR, NOT, NAND, NOR, EX-OR, and EX-NOR) that can be described in terms of input variables or their complements. Truth tables illustrate all of the input variable combination values and the output variable(s) value(s). By combining the basic functions (AND, OR, and NOT), other functions were created (NAND, NOR, EX-OR, and EX-NOR). Designers use the functions described to solve any combinational (logic composed of gates without feedback loops) logic problem.

Typically a problem statement is derived from the problem to be solved; a truth table that describes the relationship between the input and output variables is constructed;

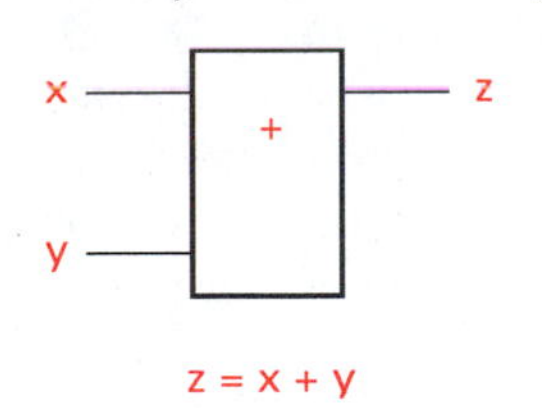

Figure 2.12
IEEE symbol for the OR gate

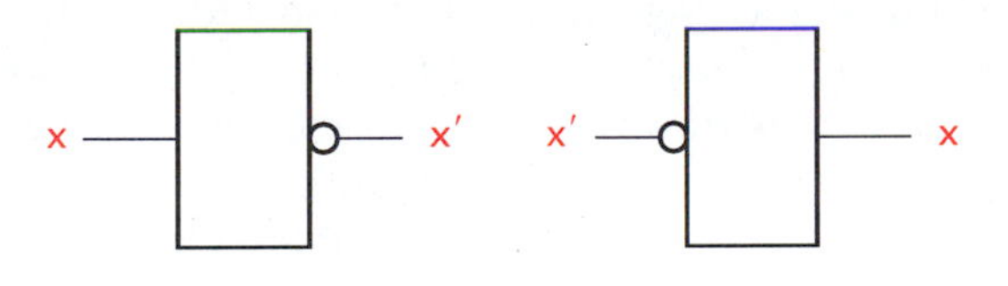

Figure 2.13
IEEE symbols for the NOT gate

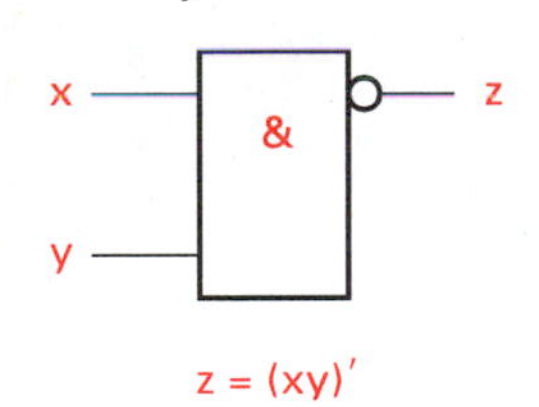

Figure 2.14
IEEE symbol for the NAND gate

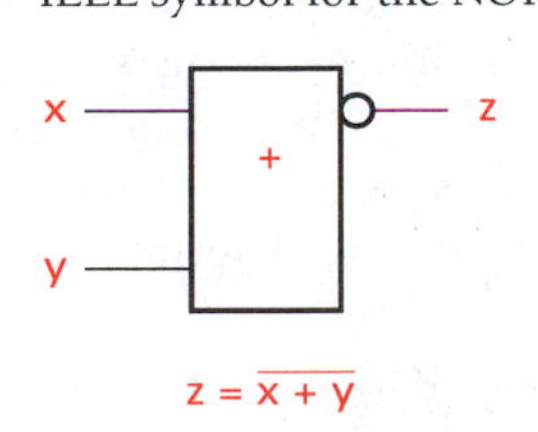

Figure 2.15
IEEE symbol for the NOR gate

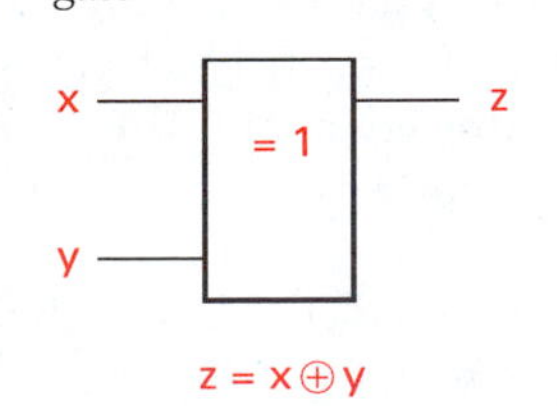

Figure 2.16
IEEE symbol for the EX-OR gate

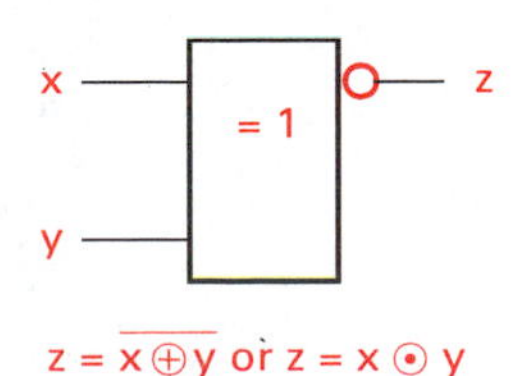

Figure 2.17
IEEE symbol for the EX-NOR gate

then the logic expressions are derived from the truth table and finally a diagram is drawn using logic symbols. This is an oversimplification of the many details necessary to accomplish combinational logic design, but it provides an outline of the process.

A switching equation defines the relationship between an output variable and a set of input variables. Switching equations must follow a prescribed set of rules called a switching algebra. What follows in the next several pages are examples of switching equations.

Consider the following switching expression:

$$T = xy'z + xy'z' + x'yz \tag{2.8}$$

T is the symbol used for the output variable; x, y, and z are the symbols used for the input variables. Each of the three input variables can have the values of either zero or one (0, 1). The total number of input variable combinations is eight and is found by the equation

$$R^n = P \tag{2.9}$$

Let R represent the radix of the number system, in this case 2.

Let n represent the number of binary variables and P the number of combinations.

In our example of three variables, we write

$$2^3 = 8 \tag{2.10}$$

The output T is true (1) when the conditions established by the switching expression are met. Table 2.1 is the truth table for these expressions.

TABLE 2.1
Truth table for equations (2.8)–(2.10)

x	y	z	T
0	0	0	0
0	0	1	0
0	1	0	0
0	1	1	1
1	0	0	1
1	0	1	1
1	1	0	0
1	1	1	0

The output variable, T, is found by evaluating each of the input terms in equation (2.8). Three terms exist in the original equation ($xy'z$, $xy'z'$, and $x'yz$), creating a (1) in three places in the output column. The equation (2.8) output terms represent three three-input AND gates (one for each term). The three-input AND gate outputs become inputs to a three-input OR gate, which produces output T. Each term in the expression causes the output variable to be true (1). The first term, $xy'z$, is 1 when the input variable x is 1, y is false (0), and z is 1; $xy'z = 101$. The second term, $xy'z'$, is true (1) when x is a 1, y is a 0, and z is a 0; $xy'z' = 100$. The third term, $x'yz$, is 1 when $x'yz = 011$. No other conditions exist where the output T is true, so the remaining possible combinations of the input variables generate a 0 for T. We see from equation (2.8) that four variables, three input and one output, are involved. The truth table, shown in Table 2.1, has eight input combinations, as described in equation (2.10), each of which represents a possible three-variable output term. Truth table input variable combinations are generated by counting, in binary, from 000 to 111.

Three approaches were used to describe the same function: a **switching equation**, a truth table, and a logic diagram (Figure 2.18). By combining logic symbols, whether distinctive shape or IEEE, **logic diagrams** describing any combination of inputs and outputs can be effectively designed and constructed. Given the logic diagram, designers can derive the truth table or switching equation(s). The design process usually involves the development of a truth table from which the switching equations are determined and finally the logic diagram drawn.

Troubleshooting digital systems often requires analysis of logic diagrams to determine how the circuit is supposed to work.

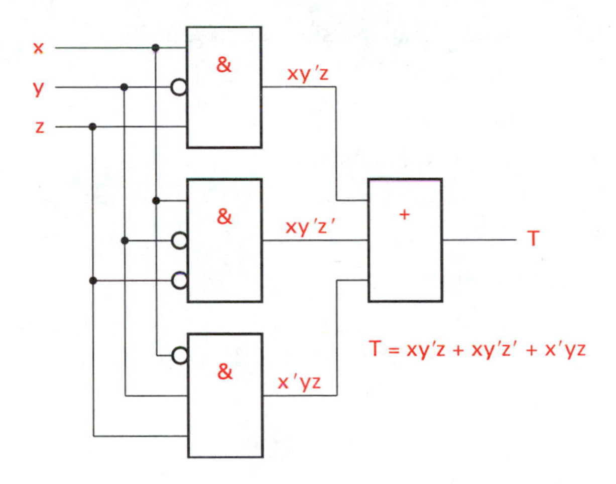

Figure 2.18
Logic diagram using IEEE symbols for equation (2.8)

2.2 SWITCHING ALGEBRA

In this section we will explore the basic definitions of a binary switching algebra to lay a foundation for combinational logic design. A switching algebra is a set of rules that govern the behavior of switching equations. Like ordinary algebra has a set of rules, so does switching algebra. This material may appear, in some cases, as simple common sense, in other cases it may appear to be rather esoteric. However, it provides a foundation that cannot be ignored; to do so leaves the logic designer with a disconnected set of procedures that cannot be transferred beyond the task at hand. Several necessary definitions follow.

A **set** of elements is a collection of items that have something in common. Let

$$D = \{0,1,2,3,4,5,6,7,8,9\}$$

D is a set, whose elements consist of the numerals 0 through 9. Each number is called an **element**. To denote that an element is a part of a set, we write

$$3 \in D$$

which means that the element 3 is a member of set D. Nonmembership of an element in a set is written

$$P \notin D$$

which means that P is not a member of set D.

Operations on members of a set must define a manipulation of two or more members of the set producing a result that must also be a member of the same set. Operations that manipulate two or more elements of a set producing a result that is not a member of the set are not allowed because one or more algebraic properties would be violated. We will see this when we look at the switching algebra properties.

For example, let the + symbol represent addition on set D so that

$$\begin{array}{rl} 2 \in D & \text{2 is an element of set } D. \\ +\,3 \in D & \text{3 is an element of set } D. \\ \hline 5 \in D & \text{Result 5 is an element of } D. \end{array}$$

Other operations are possible. For arithmetic,

1. + addition.
2. − subtraction.
3. ( ) or * or · multiplication
4. ÷ or / division.
5. = equality.
6. ≠ inequality.

Operations of AND, OR, and NOT are the basic functions from which all logic design can proceed. For switching algebra, three operations exist:

1. ( ) or * or · AND
2. + OR
3. − or ′ NOT

Groups of statements that establish procedures for a mathematical system are called **theorems.** A theorem is simply the establishment or definition of a mathematical idea. We have a need to establish such ideas to provide a foundation for combinational logic. The theorems presented here are for two-valued switching algebra and have their foundation in the propositional mathematics of Boolean algebra.

2.2.1 Equivalence

Let $\{0,1\} \in B$. B is a set containing the elements 0 and 1. If two variables, x and y, have the same value they are said to be *equivalent*. For example, let $x = 0$ and $y = 0$, then $x = y$; let $x = 1$ and $y = 1$, then $x = y$.

2.2.2 Closure

A set is closed with respect to a binary operator (· , +) if, when the operation is applied to members of the set, the result is a member of the set. The property of closure would not hold if an operation could produce a result not in the initial set.

B Is Closed with Respect to AND ( )

Let a, b, and c be binary variables. Each variable is restricted to the binary character set (0, 1). If, when the AND operation is applied to a and b, the output variable c produces a result that is a binary value, then the operation is said to have *closure*. We can illustrate closure by constructing a truth table (Table 2.2) for the AND operation.

The "truth table" provides a "proof" by a process called **perfect induction.** Because all of the input combinations of two binary variables exist in the truth table, each possible output condition exists, and all have values contained within the binary character set. This is perfect induction.

TABLE 2.2
Truth table illustrating closure for AND

a	b	c
0	0	0
0	1	0
1	0	0
1	1	1

B Is Closed with Respect to +

Let a, b, and c be binary variables. Each variable is restricted to the binary character set. If, when the OR operation is applied to the input variables a and b, the output variable

TABLE 2.3
Truth table illustrating closure for OR

a	b	c
0	0	0
0	1	1
1	0	1
1	1	1

c produces a result that is binary, then the operation is said to have closure. We can show closure for the OR operation by creating a truth table (Table 2.3).

In each example the results of the binary operators ($\cdot$ and $+$) produce output variable results whose value is contained within the binary character set, so the operations are said to have closure.

2.2.3 Identity

A binary operation ($\cdot$, $+$) has an identity element we will call I_e. I_e must be contained in the binary number set $\{0, 1\}$. When I_e is ANDed with a variable x the result is x; when I_e is ORed with the variable x the result is x.

$$xI_e = x \tag{2.11a}$$

$$x + I_e = x \tag{2.11b}$$

The identity element does not change the value of the element x. Let the value of $I_e = 1$; then $x1 = x$. The binary constant 1 is the identity element for the AND operation (see Table 2.4).

TABLE 2.4
Truth tables illustrating identity for AND and OR

AND			OR		
x	I_e	xI_e	x	I_e	$x + I_e$
0	1	0	0	0	0
1	1	1	1	0	1

Let the value of $I_e = 0$; then $x + 0 = x$. The binary constant 0 is the identity element for the OR operation.

Notice that the x and xI_e columns are identical for the AND table, and the x and $x + I_e$ columns for the OR table are identical. Figure 2.19 illustrates the identity element of the AND and OR functions.

2.2.4 Associative Properties

Binary operations ($\cdot$ and $+$) performed on a set, B, are associative if

$$(xy)z = x(yz) \tag{2.12a}$$

$$(x + y) + z = x + (y + z) \tag{2.12b}$$

Where the parentheses are placed is incidental, as can be seen in equations (2.12). The group of variables can be rearranged or "reassociated" for AND and OR operations.

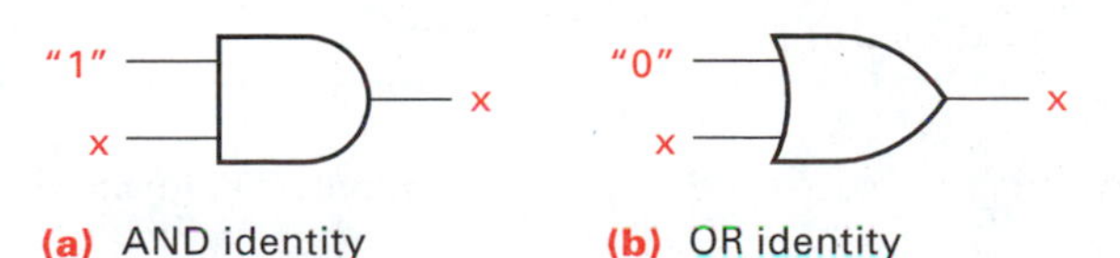

Figure 2.19
Identity elements for the AND and OR functions

TABLE 2.5
Truth table illustrating associative properties

x	y	z	xy	$(xy)z$	yz	$x(yz)$	$x + y$	$y + z$	$(x + y) + z$	$(y + z) + x$
0	0	0	0	0	0	0	0	0	0	0
0	0	1	0	0	0	0	0	1	1	1
0	1	0	0	0	0	0	1	1	1	1
0	1	1	0	0	1	0	1	1	1	1
1	0	0	0	0	0	0	1	0	1	1
1	0	1	0	0	0	0	1	1	1	1
1	1	0	1	0	0	0	1	1	1	1
1	1	1	1	1	1	1	1	1	1	1
1	2	3	4	5	6	7	8	9	10	11

The associative properties of both AND and OR can be shown through construction of a truth table and comparison of output columns. Table 2.5 illustrates the associative property.

The truth table illustrates what the two associative equations state algebraically. Columns 5 and 7 are identical (the numbers are at the bottom of each column), confirming the AND associative theorem. Likewise the OR associative theorem is confirmed by the fact that columns 10 and 11 are equal. Figures 2.20 and 2.21 show the use of a logic diagram to illustrate the association of variables by using two two-input gates to form a three-input function.

Figure 2.20
Association of a three-variable AND function

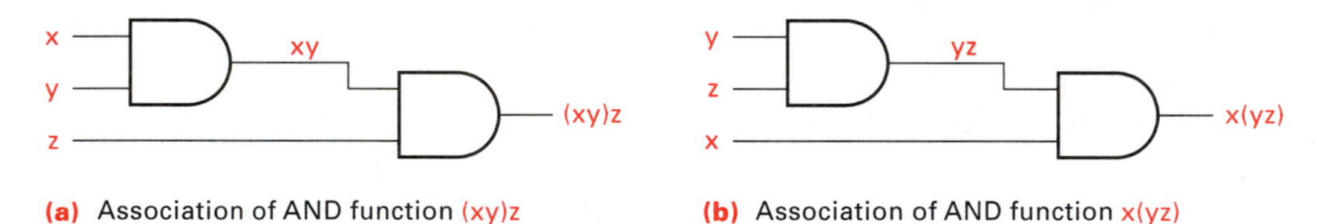

(a) Association of AND function (xy)z

(b) Association of AND function x(yz)

Figure 2.21
Association of a three-variable OR function

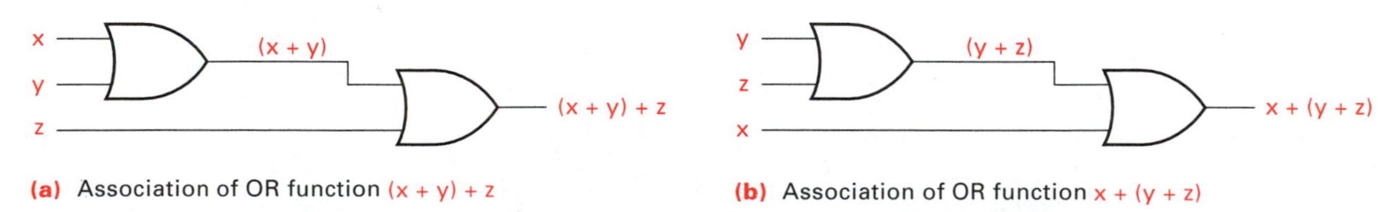

(a) Association of OR function (x + y) + z

(b) Association of OR function x + (y + z)

TABLE 2.6
Truth table proving distributive properties

x	y	z	yz	$x + (yz)$	$x + y$	$x + z$	$(x + y)(x + z)$
0	0	0	0	0	0	0	0
0	0	1	0	0	0	1	0
0	1	0	0	0	1	0	0
0	1	1	1	1	1	1	1
1	0	0	0	1	1	1	1
1	0	1	0	1	1	1	1
1	1	0	0	1	1	1	1
1	1	1	1	1	1	1	1
1	2	3	4	5	6	7	8

2.2.5 Distributive Properties

Binary operations ($\cdot$ and $+$) performed on a set, B, are distributive if

$$x(y + z) = xy + xz \tag{2.13a}$$

$$x + (yz) = (x + y)(x + z) \tag{2.13b}$$

Equation (2.13a) shows the distribution of AND over OR. Variable x is a factor of both y and z. The AND operation is distributed over the OR for both variables y and z, resulting in $xy + xz$.

Equation (2.13b) shows the distribution of OR over AND. Variable x can be redistributed over y and z resulting in $(x + y)(x + z)$. Equation (2.13b) is not like an ordinary algebra expression. Note in the first case that the variables contained inside the parenthesis are ORed; the OR connects the two terms in the redistribution. In the second case the variables contained within the parentheses are ANDed; the AND connects the two terms in the redistribution. The distributive properties can be illustrated by a truth table (Table 2.6).

Table 2.6, column 5 represents the left side of equation (2.13b), and column 8 represents the right side of equation (2.13b). Columns 5 and 8 are the same, proving equation (2.13b). The arrangements of equations (2.13a) and (2.13b) can also be illustrated by the logic diagrams in Figures 2.22 and 2.23.

Figure 2.22
Logic diagrams illustration distribution of $x(y + z)$ to $xy + xz$

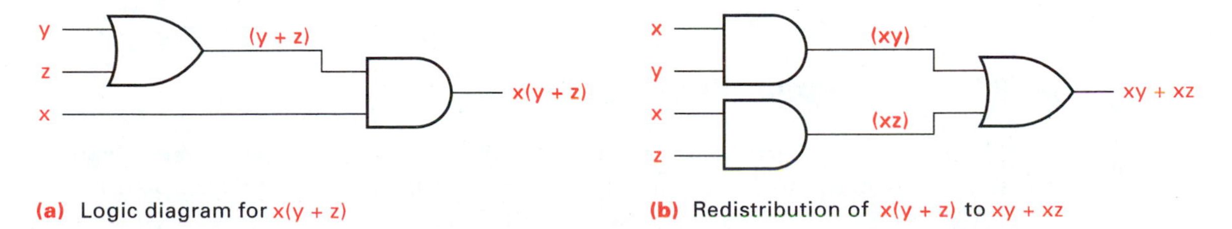

(a) Logic diagram for x(y + z)

(b) Redistribution of x(y + z) to xy + xz

Figure 2.23
Logic diagram illustrating distribution of $x + (yz)$ to $(x + y)(x + z)$

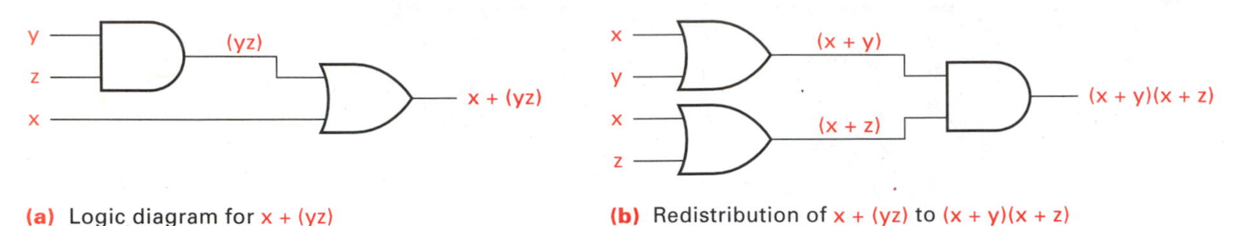

(a) Logic diagram for x + (yz)

(b) Redistribution of x + (yz) to (x + y)(x + z)

2.2.6 Commutative Properties

Binary operators (· and +) are commutative on set, B, if

$$xy = yx \tag{2.14a}$$
$$x + y = y + x \tag{2.14b}$$

The commutative property simply lets us rearrange the order of the variables for a logic operation. The logic to realize the equations (2.14a) and (2.14b) would be a single gate in each case. The order of connecting a variable to a gate input is immaterial.

2.2.7 Complement Property

A binary operator called **complement** or negation exists on binary variables such that

$$xx' = 0 \tag{2.15a}$$
$$x + x' = 1 \tag{2.15b}$$

where x' is the complement of x.

The complement of 0 is 1. The complement of 1 is 0. The complement of a variable x is written x'. A simple truth table (Table 2.7) illustrates equations (2.15).

TABLE 2.7
Truth table illustrating equations (2.15)

x	x'	xx'	$x + x'$
0	1	0	1
1	0	0	1

2.2.8 Duality Property

Duals are opposites or mirror images of original operators or constants. Switching logic using binary operators exhibits duality between the AND and OR operators. The dual of AND is OR and the dual of OR is AND. This is shown in Table 2.8.

The principles of duality and complementing provide features that will be used later in the development of DeMorgan's theorems.

TABLE 2.8
Duals for binary operations

Operator	Dual
AND	OR
OR	AND
0	1
1	0

2.2.9 Absorption Property

The absorption property provides a useful tool for reducing switching algebra expressions. Sometimes switching expressions are generated from initial problem statements

TABLE 2.9
Truth table illustrating absorption

x	y	xy	$x + xy$	$x + y$	$x(x + y)$
0	0	0	0	0	0
0	1	0	0	1	0
1	0	0	1	1	1
1	1	1	1	1	1
1	2	3	4	5	6

that contain redundancies. Eliminating (absorbing) these redundancies lowers the cost of the final electronic circuit. For example,

$$x + xy = x \tag{2.16a}$$

$$x(x + y) = x \tag{2.16b}$$

In equation (2.16a), x is ORed with xy; for an output to be true, x must be true, therefore, $x + xy = x$. We can further illustrate this and equation (2.16b) with a truth table (Table 2.9).

Equation (2.16a) stated that $x + xy = x$; from Table 2.9 we see that columns 1 and 4 are identical. Also equation (2.16b) stated that $x(x + y) = x$; from Table 2.9 we see that columns 1 and 6 are identical. Therefore equations (2.16a) and (2.16b) are true. A single piece of wire, necessary to transmit binary variable x is much cheaper than the two gates necessary to realize $x + xy$. If you draw the logic diagram for $x + xy$, this becomes apparent.

2.2.10 Idempotency Property

Idempotency refers to the property of sameness:

$$x + x = x \tag{2.17a}$$

$$xx = x \tag{2.17b}$$

Any variable ANDed or ORed with itself results in the original variable.

2.2.11 Binary Variables and Constants

A binary variable, x, can have the value of $\{0,1\}$ and can be associated with a binary constant $\{0,1\}$ by ANDing and ORing such that

$$x + 0 = x \tag{2.18a}$$

$$x + 1 = 1 \tag{2.18b}$$

$$x\,0 = 0 \tag{2.18c}$$

$$x\,1 = x \tag{2.18d}$$

Equation (2.18a) ($x + 0 = x$) is illustrated by column 2 in Table 2.10. Equations (2.18a) and (2.18b) illustrate that variable x can be either a 0 or a 1; therefore, the pos-

TABLE 2.10
Truth table illustrating equations (2.18)

x	$x + 0$	$x + 1$	$x \cdot 0$	$x \cdot 1$
0	0	1	0	0
1	1	1	0	1
1	2	3	4	5

sible values for Equation (2.18) are $0 + 0 = 0$, $0 + 1 = 1$, $1 + 0 = 1$, and $1 + 1 = 1$. Equations (2.18b), (2.18c), and (2.18d) are illustrated by columns 3, 4, and 5, respectively.

2.2.12 DeMorgan's Theorems

DeMorgan, an English mathematician, was a contemporary of Boole. He developed a pair of logic theorems that provide a very useful tool for converting logic switching expressions. Both of DeMorgan's theorems use the principle of duality.

1. The complement of an AND function is the OR of the complemented input variables.

a. Let $z = \overline{x \cdot y}$, then by DeMorgan's theorem $z = \overline{x} + \overline{y}$; therefore,

$$\overline{x \cdot y} = \overline{x} + \overline{y} \tag{2.19}$$

b. Let $p = \overline{x \cdot y \cdot z}$, then by DeMorgan's theorem $p = \overline{x} + \overline{y} + \overline{z}$; therefore

$$\overline{x \cdot y \cdot z} = \overline{x} + \overline{y} + \overline{z} \tag{2.20}$$

2. The complement of an OR function is the AND of the complemented input variables.

a. Let $\overline{x + y} = z$, then by DeMorgan's theorem $\overline{x} \cdot \overline{y} = z$; therefore,

$$\overline{x + y} = \overline{x} \cdot \overline{y} \tag{2.21}$$

b. Let $\overline{x + \overline{y} + z} = p$, then by DeMorgan's theorem $\overline{x} \cdot y \cdot \overline{z} = p$; therefore,

$$\overline{x + \overline{y} + z} = \overline{x}\, y\, \overline{z} \tag{2.22}$$

In general,

$$\overline{x_1 x_2 x_3 \ldots x_n} = \overline{x}_1 + \overline{x}_2 + \overline{x}_3 + \cdots + \overline{x}_n \tag{2.23}$$

$$\overline{x_1 + x_2 + x_3 + \cdots + x_n} = \overline{x}_1 \cdot \overline{x}_2 \cdot \overline{x}_3 \ldots \overline{x}_n \tag{2.24}$$

The two theorems can be combined in a single statement. The complement of any switching function can be found by replacing each variable with its complement (x is replaced by x'), each AND with OR, and each OR with AND; constants are replaced by their complements (0 is replaced by 1 and 1 by 0).

❖ **EXAMPLE 2.1** Let

$$D = (x \cdot y)' \tag{2.25}$$

Find the equivalent of equation (2.25) by applying DeMorgan's theorems.

TABLE 2.11
Truth table showing that $(xy)' = x' + y'$

x	x'	y	y'	$(xy)'$	$(x' + y')$
0	1	0	1	1	1
0	1	1	0	1	1
1	0	0	1	1	1
1	0	1	0	0	0

SOLUTION

1. Find the complement of x, which is x'
2. Find the dual of AND (the dual of AND is OR), giving $x' +$
3. Find the complement of y, which is y'

$$(x' + y')'' = x' + y'$$

A truth table (Table 2.11) shows that $(xy)' = x' + y'$. ❖

❖ EXAMPLE 2.2

Use DeMorgan's Theorem to find an equivalent to $B = x + y$. Let

$$B = x + y \tag{2.26}$$

SOLUTION $B = (x' \cdot y')'$. Complement each input variable, replace the + with ·, and then complement the entire expression. ❖

❖ EXAMPLE 2.3

Use DeMorgan's Theorem to find an equivalent to $G = xy + xz$. Let

$$G = xy + xz \tag{2.27}$$

SOLUTION Two main steps are required to find the equivalent of equation (2.27) applying DeMorgan's theorems. The logical OR function connects the two terms xy and xz. When finding the dual of the OR function we must keep the two terms intact.

1. Start with the complete function:
   a. Let $A = xy$ and $B = xz$.
   b. Then $G = A + B$.
   c. Complementing step b, $G = (A' B')'$.
2. Having converted the OR to AND, we can now proceed to convert terms xy and xz:
   a. Converting, $xy = (x' + y')'$.
   b. Converting, $xz = (x' + z')'$.
   c. Therefore,

$$G = [(x' + y')(x' + z')]' \tag{2.28}$$

The original expression, $G = xy + xz$, implies two two-input AND gates and one two-input OR gate, as illustrated in Figure 2.24.

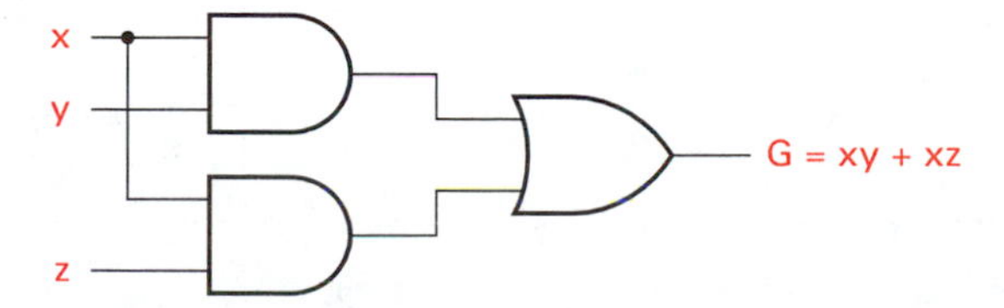

Figure 2.24
Distinctive shape logic diagram for the expression $g = xy + xz$

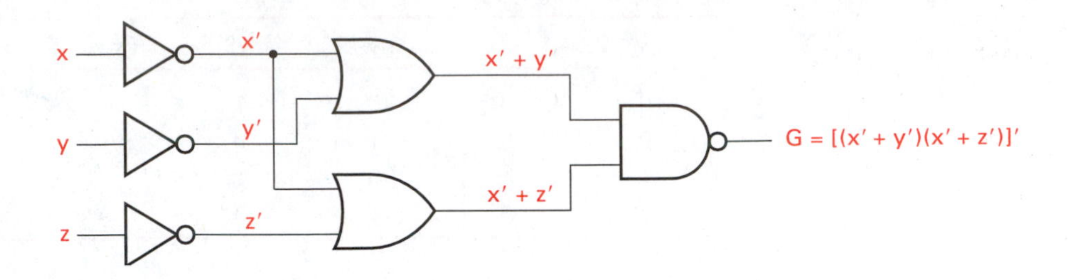

Figure 2.25
Logic diagram for equation (2.28)

The conversion of equation (2.27) by application of DeMorgan's theorems results in the logic illustrated in Figure 2.25. ❖

❖ EXAMPLE 2.4

Find the DeMorgan equivalent to equation (2.29). Let

$$F = x(y + z')' \tag{2.29}$$

1. Term $(x + z')'$ is inverted, implying that it be treated as a unit when converting the AND to an OR. Figure 2.26 illustrates the two ANDed terms.

Figure 2.26
x ANDed with $(y + z')'$

2. By treating $(y + z')'$ as a unit the conversion yields

$$F = [x' + (y + z')'']' = (x' + y + z')' \tag{2.30}$$

Figure 2.27 illustrates the resulting three-input NOR logic diagram.

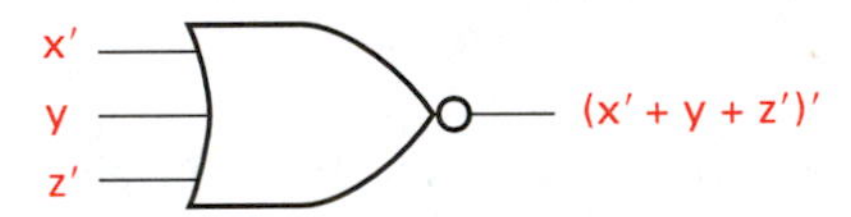

Figure 2.27
Three-input NOR equivalent to equation (2.30)

Equation (2.30) is an equivalent of equation (2.29); however, by applying DeMorgan's theorem once more, a single three-input AND gate equivalent can also be found.

Figure 2.28
Three-input AND equivalent to equation (2.31)

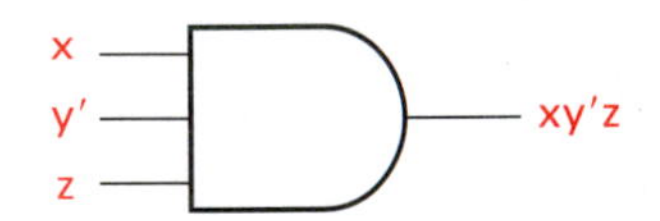

3. $$F = (x' + y + z')' = x\, y'\, z \tag{2.31}$$

Figure 2.28 illustrates equation (2.31) and Table 2.12 illustrates the equivalencies of equations (2.29)–(2.31).

TABLE 2.12
Truth table illustrating equality among equations (2.29)–(2.31)

x	y	z	$y + z'$	$(y + z')'$	$x(y + z')'$	$xy'z$	$(x' + y + z')'$
0	0	0	1	0	0	0	0
0	0	1	0	1	0	0	0
0	1	0	1	0	0	0	0
0	1	1	1	0	0	0	0
1	0	0	1	0	0	0	0
1	0	1	0	1	1	1	1
1	1	0	1	0	0	0	0
1	1	1	1	0	0	0	0
1	2	3	4	5	6	7	8

❖

❖ **EXAMPLE 2.5** Find the equivalent of equation (2.32). Let

$$G = u + vw' + x(y' + z)' \tag{2.32}$$

SOLUTION

1. Let $A = u + vw'$ and $B = x(y' + z)'$. Then $G = A + B$. By DeMorgan's theorem, $G = (A'B')'$. This process converted the OR, connecting the two terms represented by A and B, to an AND. Notice that the terms that A and B represent were not changed.
2. Because $A = u + vw'$, by DeMorgan's theorem we get

$$A = [u'(vw')']' = [u'(v' + w)]'$$
$$A' = u'(v' + w)$$
$$B = x(y' + z)'$$

and by DeMorgan's theorem,

$$B = x(y' + z)' = [x' + (y' + z)]' = [x' + (yz')']'$$
$$B' = x' + (yz')'.$$

3. Substituting for A' and B' into $G = (A'B')'$ gives

$$G = \{u'(v' + w)\,[x' + (yz')']\}'$$

Table 2.13 shows some of the more important Boolean identities and theorems for convenient referral.

TABLE 2.13
Boolean identities and theorems

Expression	Postulate or theorem
$x + 1 = 1$	
$x0 = 0$	
$x1 = x$	Identity
$x + 0 = x$	Identity
$xx' = 0$	Complement
$x + x' = 1$	Complement
$x'' = x$	Involution or double complement
$x + x = x$	Idempotence
$xx = x$	Idempotence
$xy = yx$	Commutative
$x + y = y + x$	Commutative
$x(y + z) = xy + xz$	Distributive
$x + (yz) = (x + y)(x + z)$	Distributive
$x(yz) = (xy)z$	Associative
$x + (y + z) = (x + y) + z$	Associative
$x + xy = x$	Absorption
$x(x + y) = x$	Absorption
$x(x' + y) = xy$	Absorption
$x'y' = (x + y)'$	DeMorgan
$(xy)' = x' + y'$	DeMorgan
$xy + xy' = x$	Adjacency

❖

2.3 FUNCTIONALLY COMPLETE OPERATION SETS

A **functionally complete operation set** is a set of logic functions from which any combinational logic expression can be realized. Set FC_1 consists of logic operations AND, OR, and NOT; set FC_2 consists only of the logic operation NOR; and set FC_3 consists only of the logic operation NAND. An additional set contains EX-OR and AND gates.

$$FC_1 = \{\text{AND, OR, NOT}\}$$

$$FC_2 = \{\text{NOR}\}$$

$$FC_3 = \{\text{NAND}\}$$

$$FC_4 = \{\text{EXOR, AND}\}$$

The three Boolean algebra operations of AND, OR, and NOT can be realized using NOR or NAND gates. The NOT function is realized by NAND and NOR gates, as indicated in Figure 2.29. AND and OR functions can be realized using NAND and NOR gates, as indicated in Figures 2.30 and 2.31. It is often desirable to realize a logic function with all NAND or all NOR gates. Some integrated circuit technologies (e.g., TTL, CMOS, and ECL) implement functions more cheaply than others. For instance, TTL realizes a NAND function more cheaply than a NOR but ECL realizes a NOR more cheaply than a NAND. It is also sometimes desirable to minimize inventory of IC types. NAND or NOR equivalent expressions can be found by application of one or more Boolean identities or theorems.

Figure 2.29(a) illustrates "NANDing" x with a constant 1, created by the pull-up resistor connecting the gate input to V_{CC}, resulting in an output of $(x \cdot 1)^1 = x^1$. Figure 2.29b shows the NOR of x with a constant 0, by connecting the input to ground, resulting in an output of $(x + 0)^1 = x^1$. Using NAND and NOR gates as inverters permits generation of AND and OR functions using only NAND or NOR gates, as shown in Figures 2.30 and 2.31. Example 2.6 illustrates the realization of the same logic expression using all three functionally complete logic sets.

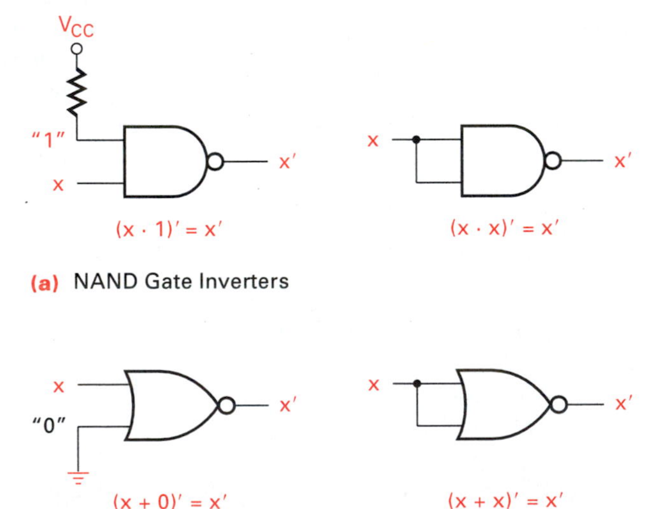

Figure 2.29
Using NAND and NOR gates as inverters

Figure 2.30
AND and OR functions using NAND gates

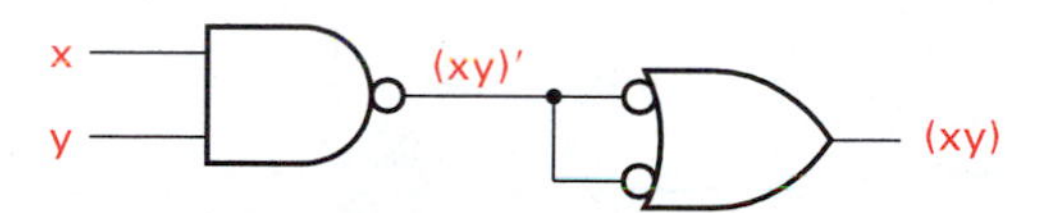

(a) AND Function Using NAND Gates

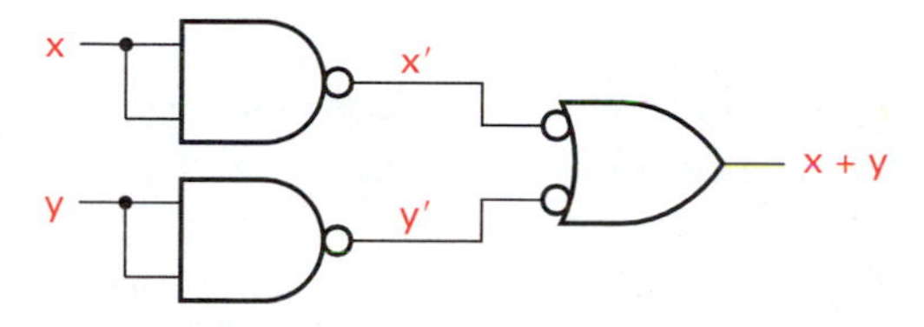

(b) OR Function Using NAND Gates

Figure 2.31
AND and OR functions using NOR gates

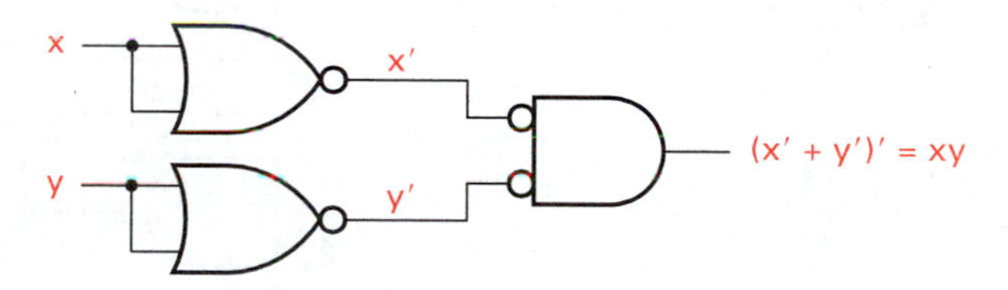

(a) AND Function Using NOR Gates

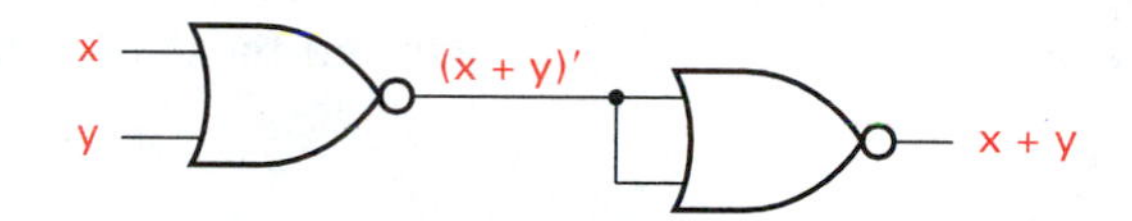

(b) OR Function Using NOR Gates

❖ EXAMPLE 2.6

Let

$$R = x'y + xy' \tag{2.33}$$

We want to realize equation (2.33) using AND, OR, NOT, NOR, and NAND gates.

SOLUTION

1. Figure 2.32 shows the realization of equation (2.33) using FC_1 with AND, OR, and NOT gates.
2. Figure 2.33 shows the realization of equation (2.33) using NOR gates. By application of DeMorgan's theorems we can convert equation (2.33) to all NORs and inverters.

$$x'y = (x + y')'$$
$$xy' = (x' + y)'$$
$$x'y + xy' = (x + y')' + (x' + y)'$$

Figure 2.32
Realization of equation (2.33)

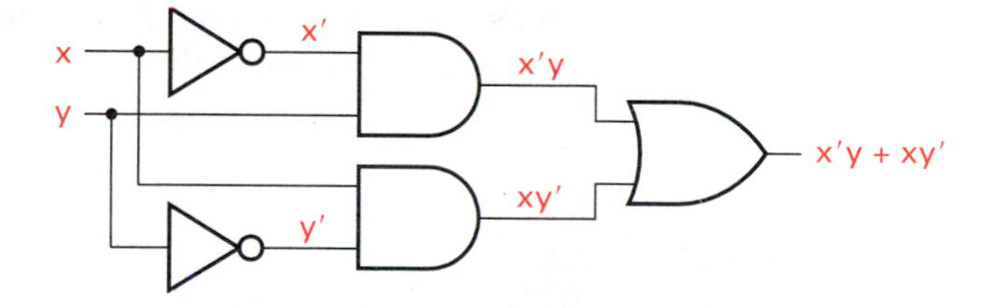

Figure 2.33
Realization of equation (2.33) using NOR gates

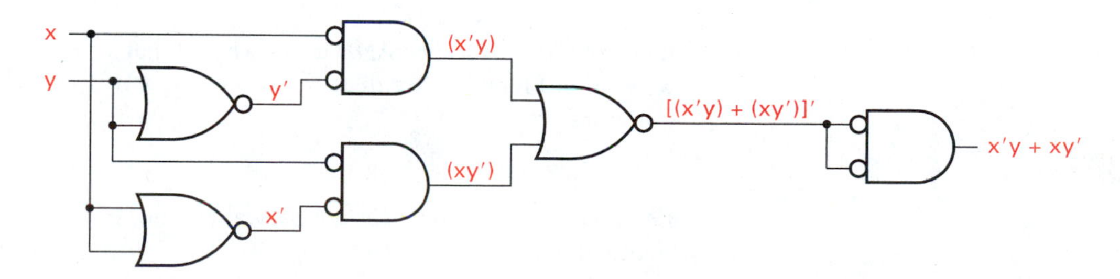

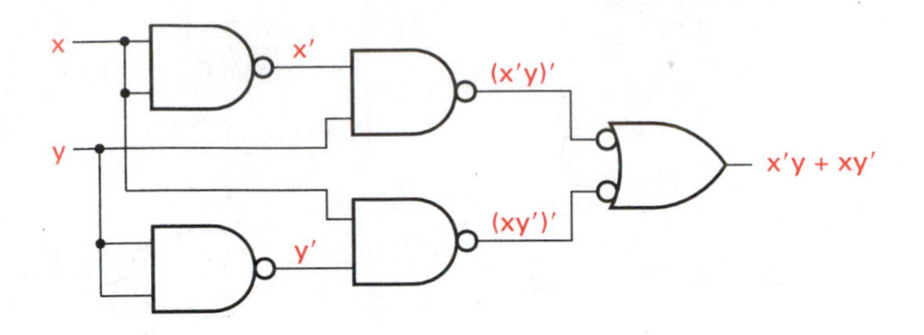

Figure 2.34
NAND realization of equation (2.33)

3. Figure 2.34 shows the realization of equation (2.33) using NAND gates. We must convert the OR function into a NAND by using DeMorgan's theorem:

$$R = x'y + xy'$$

Let $x'y = A$ and $xy' = B$, then

$$R = A + B$$

By DeMorgan's theorem, $R = (A'B')'$. Substituting A and B we obtain

$$R = [(x'y)'(xy')']'$$ ❖

The ability to restructure any combinational logic equation by application of DeMorgan's theorems allows the designer to configure the result using any functionally complete operation sets. In other words, logic designs can be implemented using only NAND gates, only NOR gates, or AND, OR, and NOT gates.

2.4 REDUCTION OF SWITCHING EQUATIONS USING BOOLEAN ALGEBRA

Any variables or terms in a switching equation that are redundant should be eliminated, because they add cost to the circuit. Consider the switching equation in Example 2.7.

❖ EXAMPLE 2.7

$$F = xy'z + xyz \tag{2.34}$$

We want to eliminate the redundant terms.

SOLUTION By inspection of the two AND terms we find that xz is common to both terms; therefore, we do the following. By distribution, $xz(y + y') = xy'z + xyz$, also, $y + y' = 1$; therefore,

$$F = xz \tag{2.35}$$

Equations (2.34) and (2.35) are equivalent; they produce the same result in a truth table, as illustrated by Table 2.14.

TABLE 2.14
Truth table illustrating equality between equations (2.34) and (2.35)

x	y	z	$xy'z$	xyz	$xy'z + xyz$	xz
0	0	0	0	0	0	0
0	0	1	0	0	0	0
0	1	0	0	0	0	0
0	1	1	0	0	0	0
1	0	0	0	0	0	0
1	0	1	1	0	1	1
1	1	0	0	0	0	0
1	1	1	0	1	1	1

❖

The reduction of equation (2.34) was accomplished by the application of the property of distribution, $xy'z + xyz = xz(y + y')$, and the complement $(y + y' = 1)$. In fact, as we will see later when Karnaugh maps are used to simplify switching equations, the complement property is a significant key to reduction in switching equations.

❖ EXAMPLE 2.8

Let

$$P = x'yz' + x'yz + xyz' + xyz \tag{2.36}$$

Again, we want to eliminate the redundant terms.

SOLUTION By comparison of each term in equation (2.36) with every other term, we can identify any groups of variables that can be isolated by the property of distribution. We are looking for terms with one variable change, in the form of $(x + x')$ with the remaining variables remaining constant. Because each term in our example consists of three variables, we need to find common groups of two variables.

1. Compare term 1 with term 2. We find that the variable group xy' is common; thus,

   $$x'y(z' + z) = x'y(1) = x'y$$

2. Compare term 1 with 3. We find that yz' is common to both; thus,

   $$yz'(x' + x) = yz'(1) = yz'$$

3. Compare term 1 with term 4. Only the variable y is common, so no single variable in the form of $(x + x')$ is available; therefore, no reduction is possible.
4. Compare terms 2 and 3. Again no reduction is possible.
5. Compare term 2 with 4. We find that yz is common; thus,

   $$yz(x' + x) = yz$$

6. Compare term 3 with 4. We find xy common; thus,

   $$xy(z + z') = xy$$

7. Putting the results together for steps 1 through 6 gives

   $$P = x'y + yz' + yz + xy \tag{2.37}$$

8. An evaluation of equation (2.37) indicates the necessity of repeating the term-by-term comparison of steps 1 through 6.
   a. A comparison of equation (2.37) terms 1 and 4 results in

      $y(x' + x) = y$

   b. The results of a comparison of terms 2 and 3 gives

      $y(z' + z) = y$

   c. Combining results from 8a and 8b gives

      $y + y = y$

      Therefore, equation (2.37) becomes

      $$P = y \tag{2.38}$$

❖

❖ **EXAMPLE 2.9** Let

$$S = A(B' + C)'(BC)' \tag{2.39}$$

SOLUTION We notice that, in equation (2.39), groups of variables, as well as individual variables, are complemented. Prior to application of the distributive and complement properties we must rearrange the expression to clear the complements from groups and entire terms.

1. Clearing the complement from $(BC)'$ gives

   $(BC)' = B' + C'$ DeMorgan's theorem

2. And clearing the complement from $(B' + C')'$

   $(B' + C)' = BC'$ DeMorgan's theorem

3. Substituting, $S = A(BC')(B' + C')$.
4. Then, by distribution

   $S = (ABC')(B' + C')$

5. Again by distribution,

   $S = ABB'C' + ABC'C'$

6. By idempotence (the property of sameness) and identity,

   $BB' = 0, \qquad C'C' = C'$

   therefore, $S = A0C + ABC'$.
7. But anything ANDed with 0 is 0; therefore, $A0C = 0$.
8. Finally,

   $$S = ABC' \tag{2.40}$$

Equations (2.39) and (2.40) are equivalent; that equality can be verified through construction of a truth table. ❖

Any or all of the Boolean algebraic relationships illustrated in Table 2.13 can be applied to the simplification of switching equations. In fact, all switching expression reduction algorithms are derived from the algebraic postulates and theorems.

Practice Problems

1. $F = a'b + ab$ (answer $F = b$).
2. $W = abc' + abc$ (answer $W = ab$).
3. $Q = a'bc + a'b'c + a'b'c' + abc$ (answer $Q = a'b' + bc$).
4. $P = x(y + z')'(x'y)'$ (answer $P = xy'z$).
5. $R = (x' + y')'(x + y')$ (answer $R = xy$).
6. $S = ab + a'b + ab' + a'b'$ (answer $S = 1$).

2.5 REALIZATION OF SWITCHING FUNCTIONS

As has been previously stated, switching functions can be described in three ways: switching equations, truth tables, and logic diagrams. This section develops more fully the relationship between the switching equations and the logic diagrams.

2.5.1 Conversion of Switching Functions to Logic Diagrams

After having developed the switching equations necessary to realize the combinational logic functions needed to solve a problem, the logic designer must convert those switching equations to a logic diagram. At the same time, a wiring diagram is created to facilitate the physical layout and construction of the circuit board. Documentation, in the form of logic diagrams, parts lists, and a circuit layout, is still needed after the design of the logic is completed, tested, and under production. Test engineers, production and manufacturing engineers and technicians, as well as field service people must have correct documentation. This section covers the translation of combinational switching functions to a logic diagram. Converting from switching functions to logic diagrams involves using the symbols assigned to logic functions and replacing the terms of a switching equation with the appropriate logic symbol. (This process is best illustrated through examples.)

❖ EXAMPLE 2.10 Convert the following switching function to a logic diagram, use AND, OR, and NOT symbols.

$$G = x'y'z' + xy'z + x'yz' + xyz = xz + x'z' \quad (2.41)$$

Solution Notice that four three-variable terms are present. Each of the terms is an AND function with some individual variables inverted. The four AND terms are ORed to produce the output variable G.

Converting a switching equation to a logic diagram requires substituting the implied symbols for the terms in the equation. Figure 2.35 illustrates the logic diagram, drawn using distinctive shape symbols.

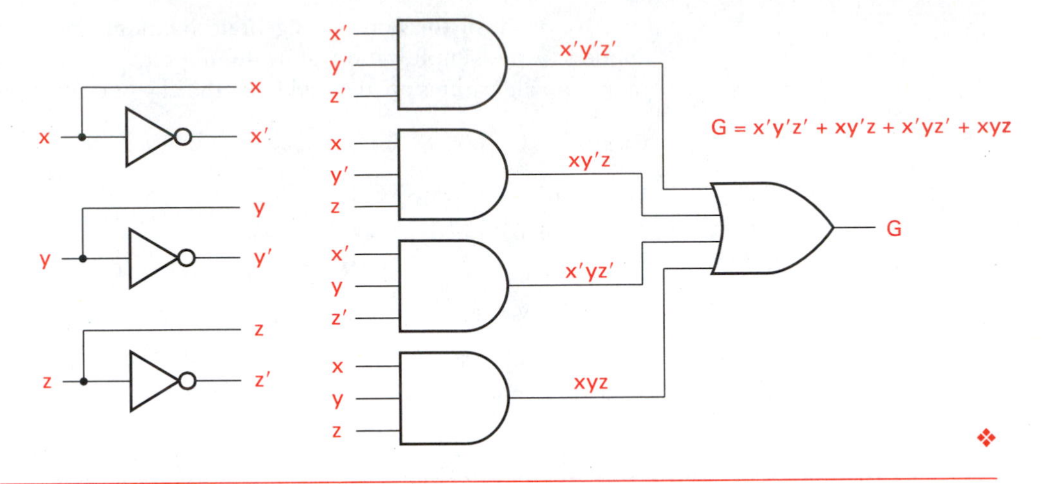

Figure 2.35 Distinctive shape logic diagram for equation (2.41)

❖

❖ **EXAMPLE 2.11** Convert the following switching equation to a logic diagram, use IEEE standard symbols, and realize the equation using AND, OR, and NOT gates.

$$H = (x + y' + z)(x' + y + z) \tag{2.42}$$

SOLUTION Equation (2.42) is an OR-AND; the logic diagram is illustrated in Figure 2.36.

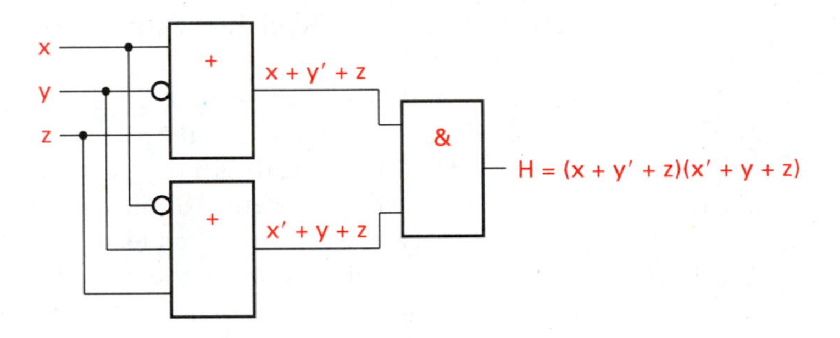

Figure 2.36 IEEE symbol logic diagram for equation (2.42)

❖

❖ **EXAMPLE 2.12** Draw the logic diagram for the following switching equation:

$$T = a'b + ab' \tag{2.43}$$

SOLUTION Equation (2.43) is an AND-OR form of the EX-OR function. The AND-OR diagram is shown in Figure 2.37 (on the left), and the EX-OR symbol is shown on the right. They are functionally equivalent.

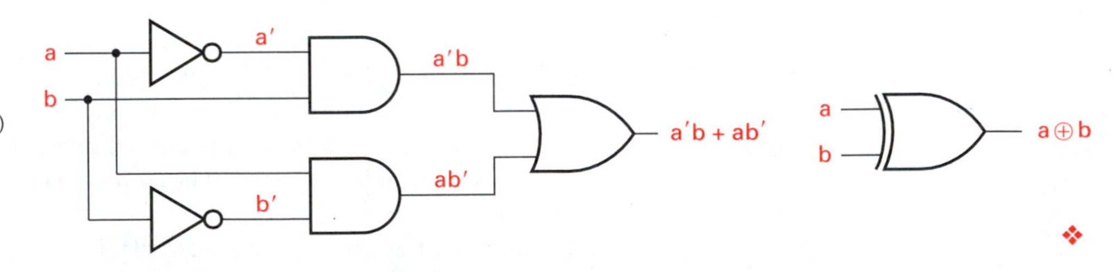

Figure 2.37 EX-OR function shown in an AND-OR form (left) and in its single symbol equivalent (right)

❖

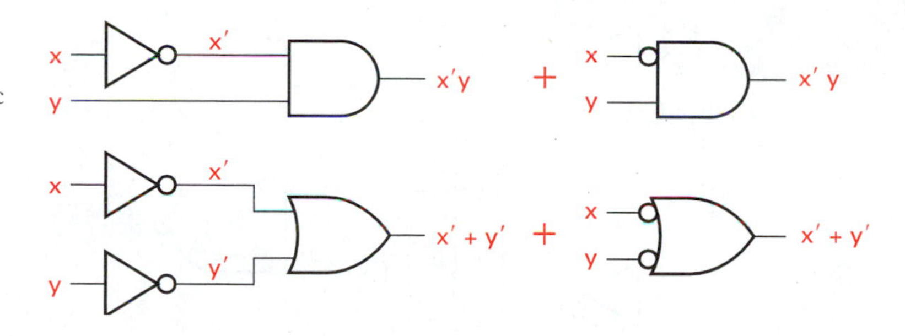

Figure 2.38
Using inverting bubbles on the inputs of distinctive shape logic gate symbols

Inverting bubbles on the inputs of logic gates indicate the inversion of the input variable. This is a drawing notation, the actual implementation using integrated circuit gates still requires the inverter. Figure 2.38 illustrates this with the distinctive shape logic symbols. We have already noted that the IEEE symbol notation accommodates this idea. We have also noted two notations for indicating complementing or inversion: the overbar and the apostrophe. Figure 2.38 uses the ′ notation to indicate logical inversion.

❖ EXAMPLE 2.13

Combinational logic is not always represented in a single switching equation. Many useful logic functions are represented with multiple outputs, requiring multiple output equations. Draw the logic diagram for the following switching equations using only NAND gates and distinctive shape symbols. Use the bubble notation illustrated in Figure 2.38 for inputs requiring complementing instead of drawing the individual inverters. Keep in mind that the realization of the circuit using integrated circuit components could use NAND gates as inverters, as illustrated in Figure 2.29. Also use the apostrophe symbol to indicate inversion.

$$F_1 = x'y'z + x'yz' \tag{2.44}$$

$$F_2 = x'yz' + xy' \tag{2.45}$$

SOLUTION Both equations (2.44) and (2.45) are AND-OR equations. To realize the equations using only NAND gates, they must be converted using DeMorgan's theorem:

$$F_1 = [(x'y'z)'(x'yz')']'$$
$$F_2 = [(x'yz')'(xy')']'$$

Inversion of an entire term or group of variables is indicated by the ( )′ notation. The ′ symbol to indicate inversion is used to the ease of entering Boolean expressions into a computer. The overbar symbol cannot be entered into a text string for computer entry.

Notice that the term $(x'yz')'$ occurs in both output functions F_1 and F_2. The logic diagram in Figure 2.39 shows the term being created twice, once for F_1 and again for F_2. Once a term has been generated, it can be used more than once. The term $(x'yz')'$ should be shared between output functions as illustrated in Figure 2.40.

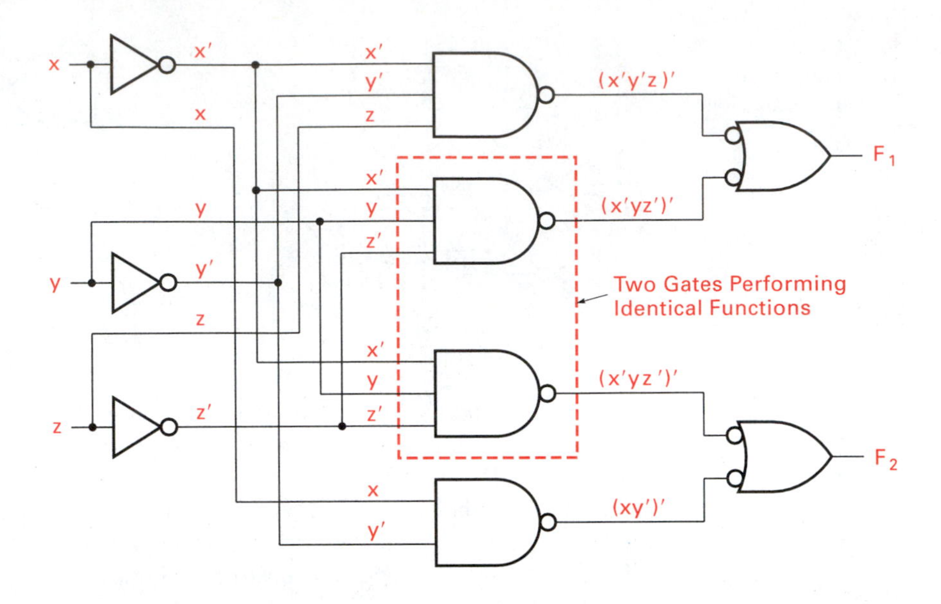

Figure 2.39
Logic diagram for equations (2.44) and (2.45)

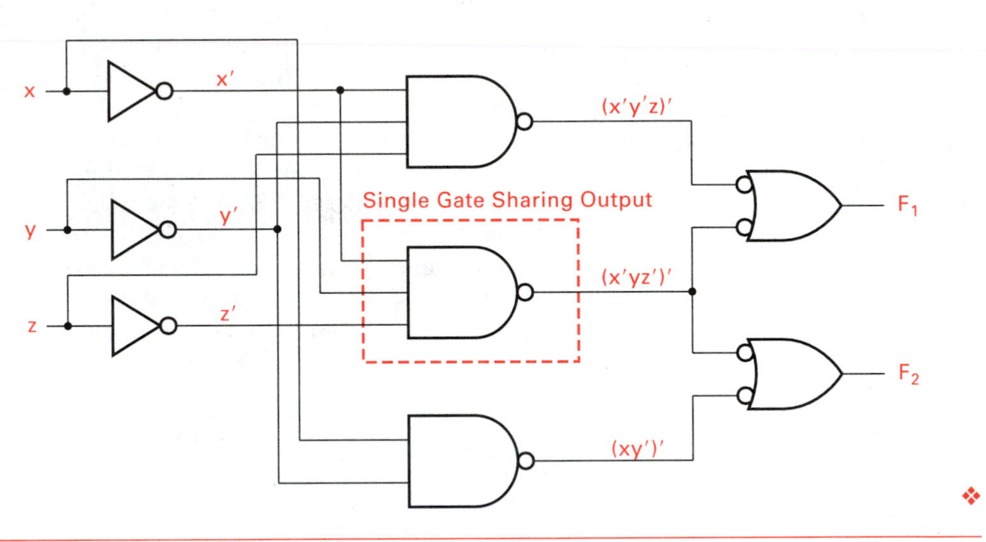

Figure 2.40
Logic diagram showing the shared $(x'yz')'$ term between F_1 and F_2

❖

❖ EXAMPLE 2.14

Draw the logic diagram for the following switching equations using only NOR gates. Use the IEEE symbols.

SOLUTION

$$K_1 = w'x + xz + yz' \tag{2.46}$$

$$K_2 = x'y' + xz + w'x' \tag{2.47}$$

$$K_3 = yz' + x'y \tag{2.48}$$

Each equation must be converted to NORs.

K_1: $w'x = (w + x')'$, $xz = (x' + z')'$, $yz' = (y' + z)'$

K_2: $x'y' = (x + y)'$, $xz = (x' + z')'$, $w'x' = (w + x)'$

K_3: $yz' = (y' + z)'$, $x'y = (x + y')'$

all by DeMorgan's theorems.

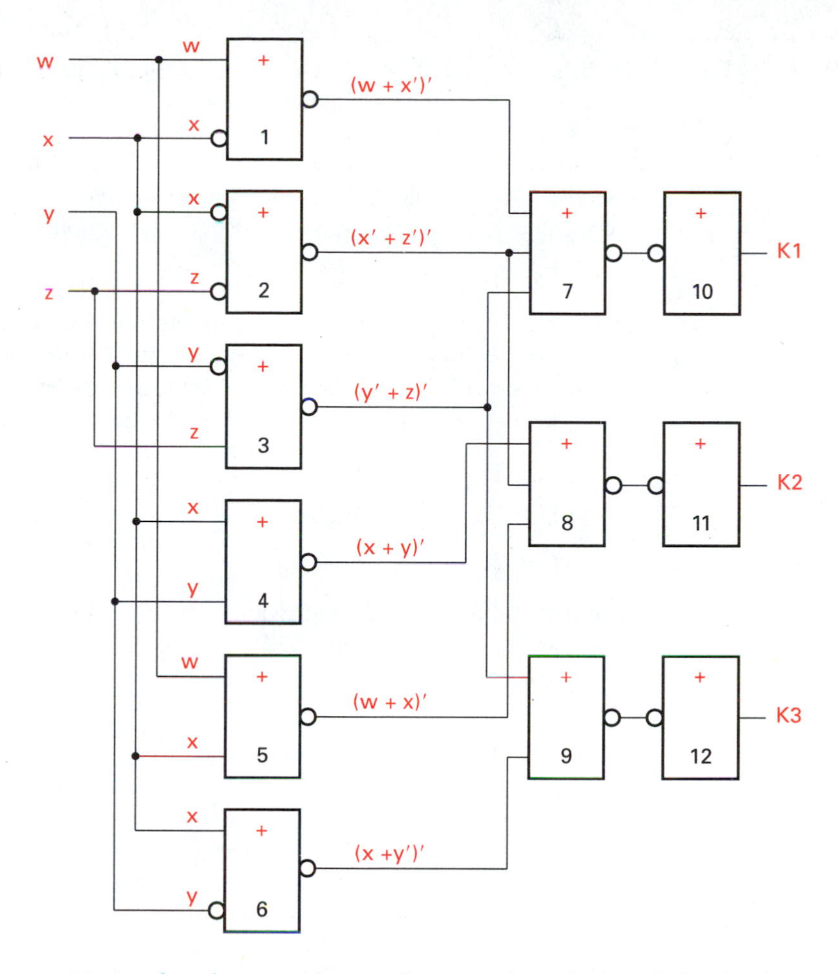

Figure 2.41
Logic diagram for Example 2.14 using IEEE symbols

Notice that the second terms for K_1 and K_2 (xz) and the third term of K_1 and the first term of K_3 (yz') are identical. Therefore, they need be generated only once and shared between the required output functions. Figure 2.41 shows a logic diagram for this example. ❖

2.5.2 Converting Logic Diagrams to Switching Equations

Converting logic diagrams into switching equations is the opposite process from that described in the preceding section. Here we will analyze logic functions to determine their purpose when troubleshooting or modifying existing logic designs. Often, when troubleshooting digital systems, the only documentation available is a logic diagram, no equations or truth tables are given. It can be difficult to determine what the circuit is supposed to do based only on the logic diagram. When that happens a conversion from the diagram to equation(s) and truth table(s) will clarify the circuit's purpose. Then, by knowing what the circuit was designed to do, we can determine if it is performing correctly. An introduction to analysis is presented here. The ability to read logic diagrams and extract the Boolean equations is introduced in this section. More will follow in later chapters.

❖ EXAMPLE 2.15

Consider the logic diagram in Figure 2.42. Write the switching equation, convert it to readable form, and construct a truth table from the result.

SOLUTION By inspection we can write the gate output terms. The AND gate has an output of $x'y$, the OR gate of $x' + z$, and these two terms are NANDed together to form $[(x'y)(x' + z)]'$. So we can write the output function, F, as

$$F = [(x'y)(x' + z)]' \tag{2.49}$$

Equation 2.49 describes the logic output but it is not in an easily understandable form. It would be helpful to remove the complement from the equation by application of DeMorgan's theorems and then to convert the equation to a truth table. Once the truth table is known we can always determine whether the circuit is performing as originally intended.

Let $x'y = A$, and $x' + z = B$. Then $F = (AB)'$ and, by DeMorgan's theorem, $F = A' + B'$. Substituting for A and B gives

$$F = (x'y)' + (x' + z)' \tag{2.50}$$

Further application of DeMorgan's theorems will remove the complement from each of the two terms in Equation (2.50).

$$\begin{aligned} (x'y)' &= x + y' \\ (x' + z)' &= xz' \\ F &= x + y' + xz' \end{aligned} \tag{2.51}$$

Output, F, is a 1 except when inputs $xyz = 010$ or 011, and then it is a 0. Troubleshooting the circuit is now much easier because we know exactly the value of F in all input conditions. By applying Boolean algebra principles to equation (2.51) we see that it simplifies to $x + y'$, which is the same thing that Table 2.15 gives.

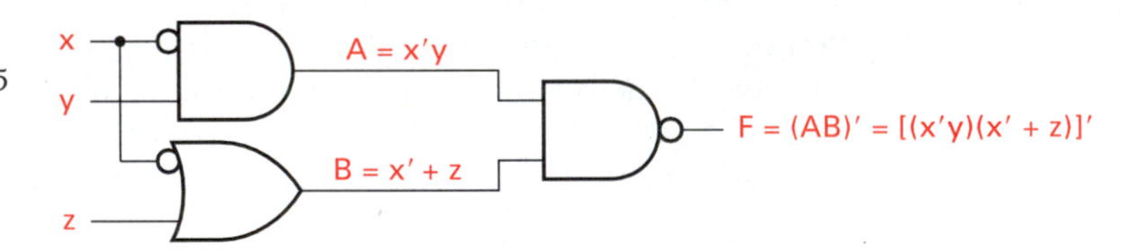

Figure 2.42
Logic diagram for Example 2.15

TABLE 2.15
Truth table for the logic circuit in Figure 2.38

x	y	z	xz'	$x + y'$	F
0	0	0	0	1	1
0	0	1	0	1	1
0	1	0	0	0	0
0	1	1	0	0	0
1	0	0	1	1	1
1	0	1	0	1	1
1	1	0	1	1	1
1	1	1	0	1	1

❖

❖ EXAMPLE 2.16

Write the output equations and construct a truth table given the logic diagram in Figure 2.43. Each gate is numbered in the logic diagram to facilitate discussion. Individual gate outputs are labeled G_1 to G_5. Each gate output is derived by inspection and then manipulated using DeMorgan's theorems to a more convenient form.

SOLUTION The gate outputs are

$$G_1 = (AB)'$$

$$G_2 = (CD)'$$

$$G_3 = [(AB)'C]' = AB + C'$$

$$G_4 = [(AB)'(CD)']' = AB + CD$$

$$G_5 = [(AB + C')(CD)']' = A'C + B'C + CD$$

The outputs are $O_1 = AB + CD$ and $O_2 = A'C + B'C + CD$.

From Table 2.16 we can determine exactly which input variable conditions produce output variable 1s.

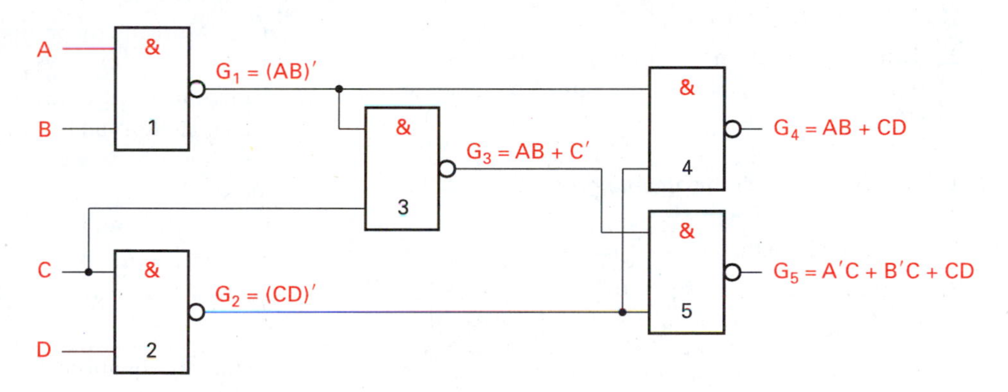

Figure 2.43 IEEE symbol logic diagram for Example 2.16

TABLE 2.16 Truth table for logic circuit in Figure 2.43

A	B	C	D	AB	CD	$A'C$	$B'C$	O_1	O_2
0	0	0	0	0	0	0	0	0	0
0	0	0	1	0	0	0	0	0	0
0	0	1	0	0	0	1	1	0	1
0	0	1	1	0	1	1	1	1	1
0	1	0	0	0	0	0	0	0	0
0	1	0	1	0	0	0	0	0	0
0	1	1	0	0	0	1	0	0	1
0	1	1	1	0	1	1	0	1	1
1	0	0	0	0	0	0	0	0	0
1	0	0	1	0	0	0	0	0	0
1	0	1	0	0	0	0	1	0	1
1	0	1	1	0	1	0	1	1	1
1	1	0	0	1	0	0	0	1	0
1	1	0	1	1	0	0	0	1	0
1	1	1	0	1	0	0	0	1	0
1	1	1	1	1	1	0	0	1	1

The ability to translate switching equations into logic diagrams and to translate logic diagrams into switching equations and, if necessary into truth tables, is a very useful and powerful analysis tool for debugging logic designs or troubleshooting existing designs. In later chapters, we shall develop more design and analysis tools that will add to the "toolbox" of the logic designer. ❖

SUMMARY

A. Chapter 2 introduces switching algebra; it distinguishes between Boolean algebra and two-valued switching algebra, which is a subset of Boolean algebra.

B. Logic functions were expressed as
   1. Switching variables
   2. Switching functions, including expressions, truth tables, and symbols, both distinctive shape and IEEE symbols

C. The following switching algebra properties and theorems are presented:
   1. Sets and elements
   2. Arithmetic and logic operators
   3. Properties:
      a. Identity
      b. Equivalence
      c. Closure
      d. Associative
      e. Distributive
      f. Commutative
      g. Complement
      h. Duality
      i. Absorption
      j. Idempotency
   4. DeMorgan's theorems

D. Functionally complete operation sets are developed. *Functionally complete* means that any switching expression can be realized using the functions within the set.
   1. AND, OR, and NOT
   2. NAND
   3. NOR

E. Switching equations are reduced using the postulates and theorems of Boolean algebra.

F. Switching equations are converted to logic diagrams.

G. Switching equations are derived from logic diagrams.

REFERENCES

Texts that cover material on Boolean or switching algebra include the following.

Digital Design by M. Morris Mano (Englewood Cliffs, N.J.: Prentice-Hall, 1984), Chapter 2.

Logic Design Principles, with Emphasis on Testable Semicustom Circuits by Edward J. McCluskey (Englewood Cliffs, N.J.: Prentice-Hall, 1986). Chapter 2 provides a comprehensive treatment of switching algebra.

Digital Design Principles and Practices by John F. Wakerly (Englewood Cliffs, N.J.: Prentice-Hall, 1990). Part of Chapter 3 deals with switching algebra theorems.

Fundamentals of Logic Design by Charles H. Roth, Jr. (Minneapolis–St. Paul: West, 1985). Roth uses short chapters called *units* in his text. Units 2 and 3 deal with Boolean algebra, including simplification.

Digital Technology by Gerald E. Williams (Chicago: Science Research Associates, 1986). Chapter 1 presents numbers systems and Boolean algebra.

Digital Logic Design by B. Holdsworth (Stoneham, Mass.: Butterworth, 1987). Chapter 1 covers Boolean algebra.

Digital Design Fundamentals by Kenneth J. Breeding (Englewood Cliffs, N.J.: Prentice-Hall, 1989). Chapter 3 covers Boolean and switching algebra along with simplification techniques.

Digital Electronics, A Practical Approach by William Kleitz (Englewood Cliffs, N.J.: Prentice-Hall, 1990). A practical approach to Boolean algebra theorems and simplification techniques is covered in Chapter 5.

Digital Fundamentals by Thomas L. Floyd (Columbus, Ohio: Merrill, 1990). Both Boolean algebra and simplification techniques are covered in Chapter 4.

Digital Logic and State Machine Design by David J. Comer (Philadelphia: Saunders, 1990). Boolean algebra and mapping methods are treated in Chapter 2.

Modern Digital Systems Design by John Y. Cheung and Jon G. Bredeson (Minneapolis–St. Paul: West, 1990. Chapter 2 covers fundamentals of Boolean algebra and minimization procedures.

GLOSSARY

AND One of several combinational logic functions. The AND function provides a 1 output when all of the input variables are 1.

Boolean algebra The algebra developed by George Boole, an English mathematician, "to give expression to the fundamental laws of reasoning in the symbolic laws of a calculus." It provided the foundation for the two-valued switching algebra used in logic design and analysis.

Complement The term means "to make complete." In two-valued switching algebra, the complement is the opposite case. For example, the complement of A is written A', the complement of 1 is 0. The terms *complement* and *inverse* are often used interchangeably. The complement function is provided by the NOT gate.

Distinctive shape logic symbols The set of symbols, each with a different shape, used to designate logic functions: AND, OR, NOT, NAND, NOR, EX-OR, and EX-NOR. See Figures 2.1 through 2.9.

Element An item, a single variable, or part of a group. A single number 5 is an element of a set of numbers. A single person in a room full of people can be considered to be an element in the set of people in the room.

EX-OR A combinational logic function. When either one or an odd number of inputs is "true," the output is "true." All other conditions generate a "false" output.

EX-NOR The inverse function of the EX-OR.

Function A mathematical and logic term used to denote a relationship between input and output variables. An output variable x is said to be a function of (related to) a set of input variables, a, b, and c. We would write $x = f(a, b, c)$; x is a function of input variables a, b, and c. AND, OR, NOT, NAND, NOR, EX-OR, and EX-NOR are logical functions.

Functionally complete operation set A set of logical functions from which any combinational logic expression can be realized.

Logic diagram A drawing made up of logic symbols, distinctive shape or IEEE, showing input and output connections between various logic functions. A direct relationship exists between a logic diagram and the logic expression(s) that it represents.

Logic or switching equation A set of input variables linked together using logical operatives (AND, OR, etc.) that is equal to an output variable. For example, $x = a'bc + (a + b')'$.

IEEE logic symbols A group of symbols established by the Institute of Electrical and Electronic Engineers as standards for representing logic functions. See Figures 2.10 through 2.17.

NAND A combinational logic function composed of the AND and NOT functions; NAND is a contraction for NOT AND.

NOR A combinational logic function composed of the OR and NOT functions; NOR is the contraction for NOT OR.

NOT A combinational logic function. The NOT function complements a variable; A' is read as A NOT, or the complement of A.

OR A combinational logic function. The OR function provides a 1 output when *any* of the input variables is a 1.

Perfect induction A method of proof by examination of all possible input variable combinations, therefore showing the output variable value in every possible condition. It is an exhaustive proof of a postulate or theorem.

Set A group of items (elements) brought together by some defined commonality. A set consists of elements; a set A {0, 1, 2, 3, 4, 5, 6, 7, 8, 9} consists of all the positive integers from 0 to 9.

Switching algebra Switching algebra is a subset of Boolean algebra. Two-valued switching algebra deals only with binary valued variables. Other switching algebras exist that operate on number systems other than binary.

Theorem A postulate or group of postulates that have been "proven." If a single postulate or a set of postulates is assumed to be true, then it may be used to "prove" a theorem. Perfect induction, proof by trying all of the possible combinations, may be used to prove a postulate.

Truth table A graphical two-dimensional matrix showing the "truth" relationship between a set of input and output variables. *Truth* may be defined as when a switching variable is a 1 (positive logic) or when it is a 0 (negative logic).

Variable A changeable quantity. In binary switching logic a variable can have the value of 0 or 1. Usually a symbol, letter, number, or short name (mnemonic) is assigned to distinguish one variable from another.

QUESTIONS AND PROBLEMS

Section 2.0 to 2.1

1. When is Boolean algebra called *switching algebra*?
2. What are the properties of Boolean algebra?
3. List the three primitive logic functions; write the switching equation; draw the distinctive shape and IEEE symbols, and construct the truth table for each.
4. Write the switching equation, draw the distinctive shape and IEEE logic symbols, and construct a truth table for the NAND and NOR logic functions. What two primitive functions make up the NAND and NOR functions?
5. Repeat Problem 4 for the EX-OR and EX-NOR (equivalence) functions.
6. How many permutations do the following number of two-valued variables have?
   - **a.** 1
   - **b.** 2
   - **c.** 3
   - **d.** 4
   - **e.** 5
7. What three methods can describe a logic function?

Section 2.2

8. Define a *set* in your own words.
9. How many elements are in the set of students taking your digital course? Are these the same elements taking a mathematics or English class?
10. What is an operator in arithmetic? In logic?
11. Define *closure*. Give an example using AND and OR functions.
12. Define *identity*. What is the identity element for
    - **a.** AND
    - **b.** OR
13. Define *equivalence*. Give an example.
14. Recite the associative postulate. Draw the logic diagram illustrating association for
    - **a.** AND
    - **b.** OR
15. Prove the AND and OR postulate of association using a truth table.
16. Show the principle of distribution for AND and OR by constructing a truth table.
17. Recite in your own words the commutative postulate.
18. What is the dual or complement of
    - **a.** A
    - **b.** B'
    - **c.** 0
    - **d.** 1
    - **e.** AND
    - **f.** OR
19. Is $xy = yx$? By what property?
20. Show that $xx' = 0$, $x + x' = 1$. Illustrate this using a truth table.
21. Show that $x + xy = x$. (Construct a truth table.)
22. Complete the following:
    - **a.** $xx =$
    - **b.** $xx' =$

c. $x + x' =$
d. $x + 0 =$
e. $x + 1 =$
f. $xy + x =$
g. $x(y + z) =$
h. $x + (yz) =$
i. $x(x' + y) =$

23. State DeMorgan's theorem in words.

24. Apply DeMorgan's theorem to the following:
a. $x + y =$
b. $(xy)' =$
c. $xy + xz =$
d. $(A'B')' =$
e. $A'(B + C) =$
f. $(A'B')'(CD')' =$
g. $W + Q =$
h. $(A + B + C)D$
i. $(AB' + C'D + EF)' =$
j. $[(A + B)' + C')]' =$

Section 2.3 to 2.4

25. Explain, in your own words, a functionally complete operation set.

26. Show how the function $AB' + CD$ can be realized:
a. Using AND, OR, and Inverter gates.
b. Using NAND gates.
c. Using NOR gates.

27. Reduce the following using the Boolean algebra postulates (including DeMorgan's theorems):
a. $xy'z + xyz'$
b. $AB + A'BC' + BC$
c. $A'B'C' + AB'C' + AB'C$
d. $ab(cd + c'd)$
e. $[x(xy)'][y(xy)']$
f. $x'(y + z)' + xy$
g. $(a + b)'(a' + b')'$
h. $a'b'c' + a'b'c + ab'c' + abc$

Section 2.5

28. Realize (draw the logic diagram) for the following:
a. $x'y' + xy$
b. $AC + AB'$
c. $(x' + y)'$
d. $(x + y + z')(x' + y' + z')$
e. $(AB'C'D)' + AB'C'$

29. Realize the following (draw the logic diagram) using only NAND gates and inverters.
a. $AB' + CD$
b. $xyz + x'y'z'$
c. $(A + B' + C)(A' + B' + C)$
d. $x'y'z' + xy'z + x'yz' + xyz$

30. Write the switching expressions for the following logic circuits.

31. Construct truth tables for each of the logic circuits illustrated in Problem 30.

Section 2.5

Problem 30

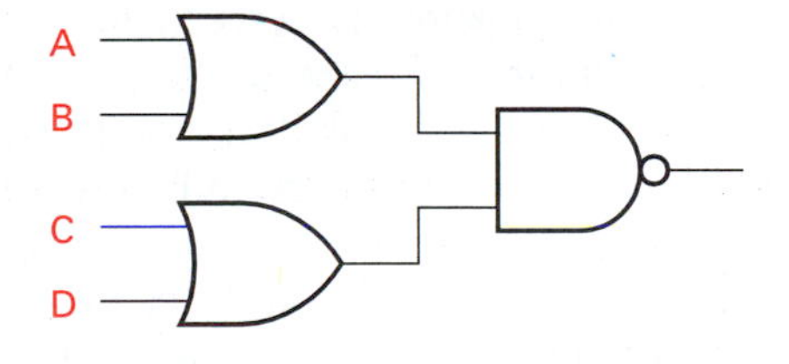

(a)

A
B
C
D
(b)

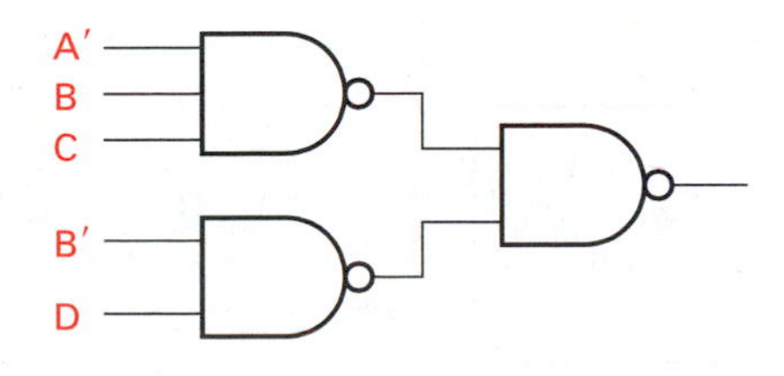

(c)

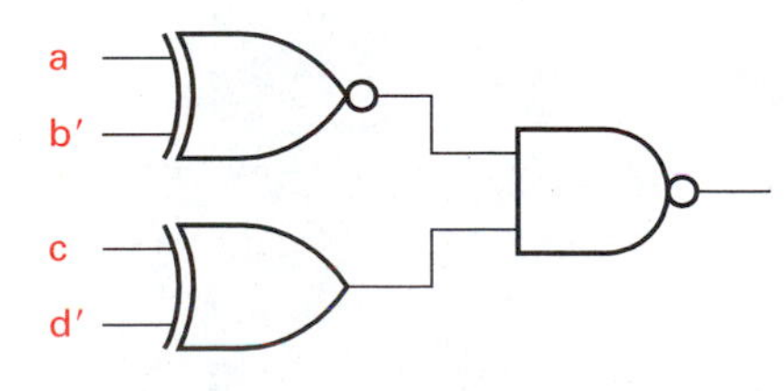

(d)

CHAPTER

3 PRINCIPLES OF COMBINATIONAL LOGIC

INTRODUCTION

Chapter 1 covered number systems and Chapter 2 developed the fundamentals of switching algebra. Now we are ready to put the skills developed in the previous chapters to work in the analysis and design of combinational logic. The primitives of combinational logic are the gates AND, OR, NOT, NAND, and NOR plus their derivatives EX-OR and EX-NOR (equivalence).

3.1 DEFINITION OF COMBINATIONAL LOGIC

Combinational logic deals with the techniques of "combining" the basic gates, mentioned previously, into circuits that perform some desired function. Examples of useful combinational logic functions are adders, subtractors, decoders, encoders, multipliers, dividers, display drivers, and keyboard encoders.

Logic circuits without feedback from output to the input, constructed from a functionally complete gate set, are said to be *combinational*. Logic circuits that contain no memory (ability to store information) are combinational; those that contain memory, including flip-flops, are said to be *sequential*. We will study sequential logic in later chapters; for now our task is to develop skills in dealing with combinational logic. Combinational logic can be modeled as illustrated in Figure 3.1.

Let X be the set of all input variables $\{x_0, x_1, \ldots, x_n\}$, and Y be the set of all output variables $\{y_0, y_1, \ldots, y_m\}$. The combinational function, F, operates on the input variable set X, to produce the output variable set, Y. Notice that the output variables y_0 through y_m are not fed back to the input. The output is related to the input as

$$Y = F(X)$$

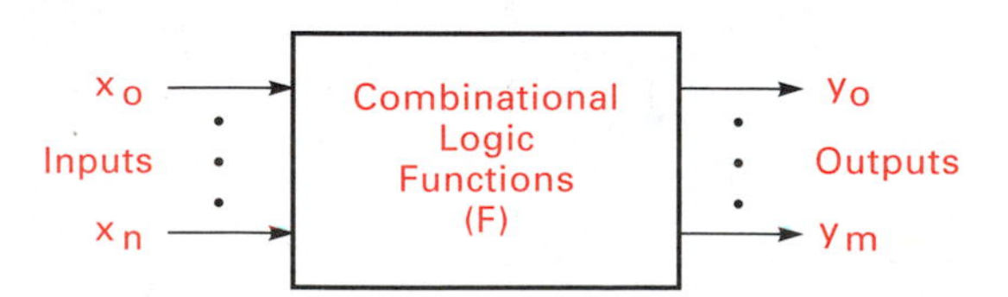

Figure 3.1
Combinational logic model

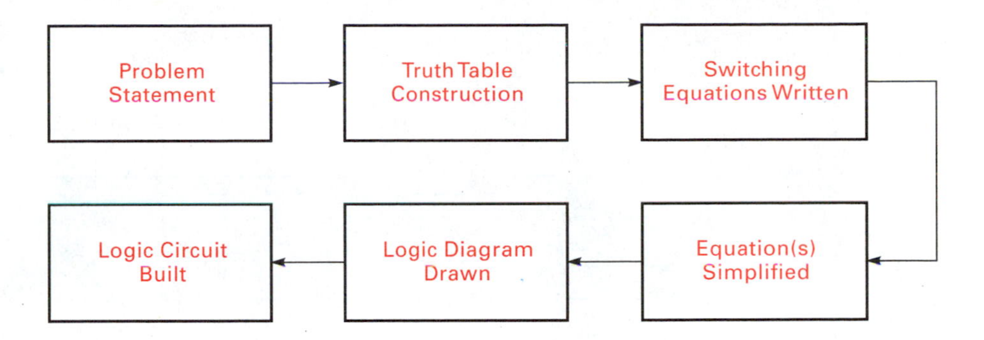

Figure 3.2
General logic design sequence

The logic circuits developed in Chapter 2, illustrating various switching properties and theorems, were all combinational. The relationship between the input and output variables can be expressed in equations, logic diagrams, or truth tables. A truth table specifies the input conditions under which the outputs are true or false (1 or 0). Switching equations are then derived from the truth tables and realized (constructed) using gates. This chapter deals with developing truth tables and graphically simplifying switching equations using several techniques.

3.1.1 Problem Statements to Truth Tables

Before any combinational logic system can be designed it must be defined. Proper statement of a problem is the most important part of any digital design task. Once correctly and clearly stated, any problem can be converted to the necessary logic for implementation. Figure 3.2 illustrates the sequence of design tasks in a general way.

Notice that the first task is to define the problem to be solved. Nothing can occur until that is correctly accomplished. The problem is then "rewritten" in the form of a truth table. From the truth table, the switching equations can be written and simplified and the logic diagram drawn. The logic diagram can be realized using any one of the three main digital integrated circuit families: transistor-transistor logic (TTL), emitter-coupled logic (ECL), or complementary metal-oxide silicon (CMOS). Practical applications rarely come in a prepackaged "truth table," ready for logic design. Truth tables must be constructed from verbal problem descriptions. Examples are given here to develop a sense of constructing truth tables from verbal problem statements.

EXAMPLE 3.1

An electric motor powering a conveyor used to move material is to be turned on when one of two operators is in position, if material is present to be moved and if the protective interlock switch is not open. Input and output variables are to be expressed in binary; that is, if operator 1 is in position then the associated variable is a logical 1. The motor is running (on) if its output control variable is a 1, and the motor is off if the variable is a 0. Figure 3.3 shows a simple diagram of the conveyor system.

The first task is to identify the input and output variables and to assign names to them.

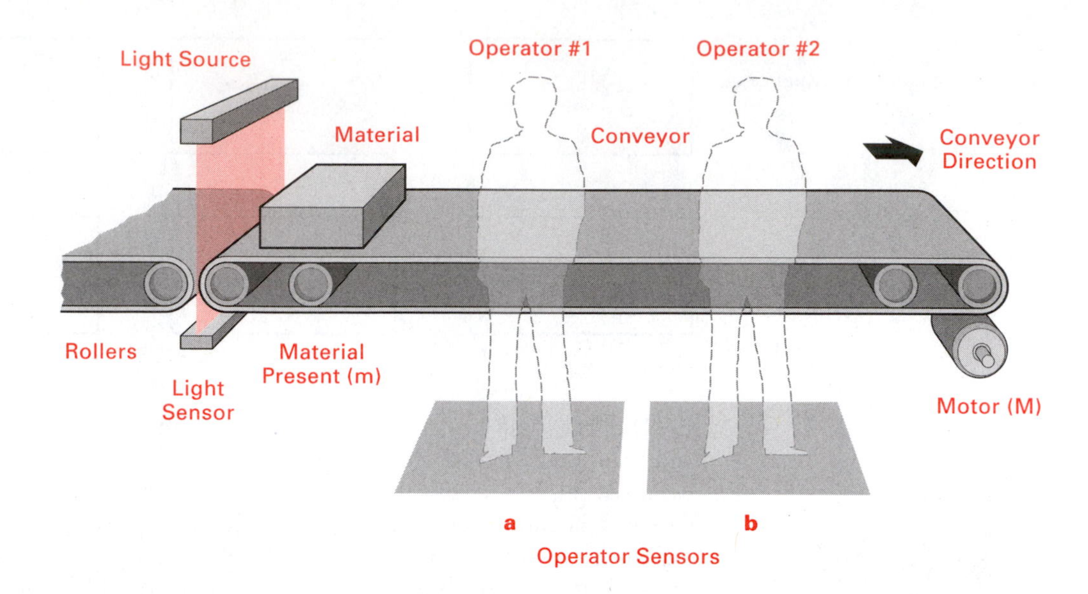

Figure 3.3
Conveyor system for Example 3.1

SOLUTION Variables a and b signify that the two operators are in position:

a, operator 1 is in position

b, operator 2 is in position

Let the motor variable be signified by M and the interlock switch be specified by s. If the interlock switch is closed, then s is true, 1. Let the material present variable be m.

M is the signal to turn the motor on and off

s means the interlock switch is closed

m means material is present

a, b, s, and m are input variables; M is an output variable

Because four input variables exist we know that 16 combinations can occur; $2^4 = 16$. The truth table describing the input/output relationship is illustrated in Table 3.1.

To turn on the motor one of the two operators must be present, material must be present, and the interlock switch must be closed. Such a system may be found in any number of industrial environments. The actual control may be implemented in any number of ways. The actual means of implementation is not important, for now; what is needed is to develop a mental picture of the process of converting a written problem statement to a truth table.

Evaluate the truth table. Notice that the following set of switching terms causes the output M to be a logical 1: $\{a'bms, ab'ms, abms\}$. Each element in the set of conditions that causes the output to be a logical 1 is a term in the output equation, which is written

$$M = a'bms + ab'ms + abms$$

When either one or both of the two operators, represented by variables a and b, is present, material is present ($m = 1$), and the interlock switch is closed ($s = 1$), then the motor, represented by the output variable M, is started. ❖

TABLE 3.1
Truth table for control problem Example 3.1

a	b	m	s	M
0	0	0	0	0
0	0	0	1	0
0	0	1	0	0
0	0	1	1	0
0	1	0	0	0
0	1	0	1	0
0	1	1	0	0
0	1	1	1	1
1	0	0	0	0
1	0	0	1	0
1	0	1	0	0
1	0	1	1	1
1	1	0	0	0
1	1	0	1	0
1	1	1	0	0
1	1	1	1	1

❖ EXAMPLE 3.2

A NASA system consists of three computers, two of which are on-line (connected to the system) at any given time. The system uses three computers to ensure safety in spacecraft operation through redundancy. If one computer experiences a problem it is taken off-line and another computer is brought on-line. Self-checking diagnostics determine each computer's operating status and generate an output in the event of failure. When one computer fails it must be switched off-line. No more than two computers are to be on-line at any given time. Design the control logic to connect or disconnect the computers. In the event that two computers are unavailable, generate a warning and allow the third computer to come on-line. If all three computers are unavailable, generate a second warning signal that invokes emergency procedures.

SOLUTION Determine input and output variables and assign names.

Let variables represent the operation status of the three computers: C_1 is computer 1, C_2 is computer 2, and C_3 is computer 3. When $C_1 = 1$, for example, then that computer has failed. Let O_1, O_2, and O_3 be the computer disconnect control signal outputs. If $O_x = 1$, for example, then that computer is connected. Let W_1 and W_2 be the two warning output signals. If $W_x = 1$, then the warning is activated.

Construct a truth table that generates the proper output responses based on input variable values. Because three input variables exist an eight row truth table is needed (one row for each input combination). Each connect/disconnect control signal is generated based on the values of the computer operation status signals and the requirement for two computers to be available. For instance, if $C_1 = C_2 = C_3 = 0$, then all three computers are operational; which two are brought on-line is the designer's choice. In this example computers 2 and 3 are connected. The second case, when $C_1 = C_2 = 0$ and $C_3 = 1$, computer 3 has failed, so computers 1 and 2 are switched on-line. The truth table (Table 3.2) illustrates all of the output values for each of the input combinations. A separate equation must be written for each output variable.

TABLE 3.2
Truth table for Example 3.2

C_1	C_2	C_3	O_1	O_2	O_3	W_1	W_2
0	0	0	0	1	1	0	0
0	0	1	1	1	0	0	0
0	1	0	1	0	1	0	0
0	1	1	1	0	0	1	0
1	0	0	0	1	1	0	0
1	0	1	0	1	0	1	0
1	1	0	0	0	1	1	0
1	1	1	0	0	0	0	1

❖

❖ EXAMPLE 3.3

Design a combinational logic truth table so that an output is generated indicating when a majority of four inputs is true.

SOLUTION The circuit has four inputs and a single output. Let I_1, I_2, I_3, and I_4 represent the input variables, and O_1 be the output variable.

A majority occurs any time three or more of the input variables are true. The truth table (see Table 3.3) must contain 16 rows, 1 for each input permutation.

TABLE 3.3
Truth table for Example 3.3

I_4	I_3	I_2	I_1	O_1
0	0	0	0	0
0	0	0	1	0
0	0	1	0	0
0	0	1	1	0
0	1	0	0	0
0	1	0	1	0
0	1	1	0	0
0	1	1	1	1
1	0	0	0	0
1	0	0	1	0
1	0	1	0	0
1	0	1	1	1
1	1	0	0	0
1	1	0	1	1
1	1	1	0	1
1	1	1	1	1

❖

❖ EXAMPLE 3.4

A conveyor system brings raw material in from three different sources. The three sources converge into a single output conveyor. Sensors mounted adjacent to each source conveyor indicate the presence of raw material. All four conveyors have separate motors so they can be individually controlled. Each source conveyor can have a different speed. The output product flow rate is fixed; it can be turned only on or off. The output product rate must match the source flow rates. To accomplish this the following conditions must be met. If source 1 has product, then sources 2 and 3 must be turned off; if source 1 is empty, then either 2 or 3 or both can be turned on. In the event that no product is available from the three sources, the output conveyor must be turned off. If no product is available, the respective source conveyor must be turned off.

SOLUTION A diagram illustrating the conveyor system with the product switches and motors helps us to visualize the problem and define the variables. This diagram is Figure 3.4. Table 3.4 is a truth table for this example.

TABLE 3.4
Truth table for Example 3.4

S3	S2	S1	M_4	M_3	M_2	M_1
0	0	0	0	0	0	0
0	0	1	1	0	0	1
0	1	0	1	0	1	0
0	1	1	1	0	0	1
1	0	0	1	1	0	0
1	0	1	1	0	0	1
1	1	0	1	1	1	0
1	1	1	1	0	0	1

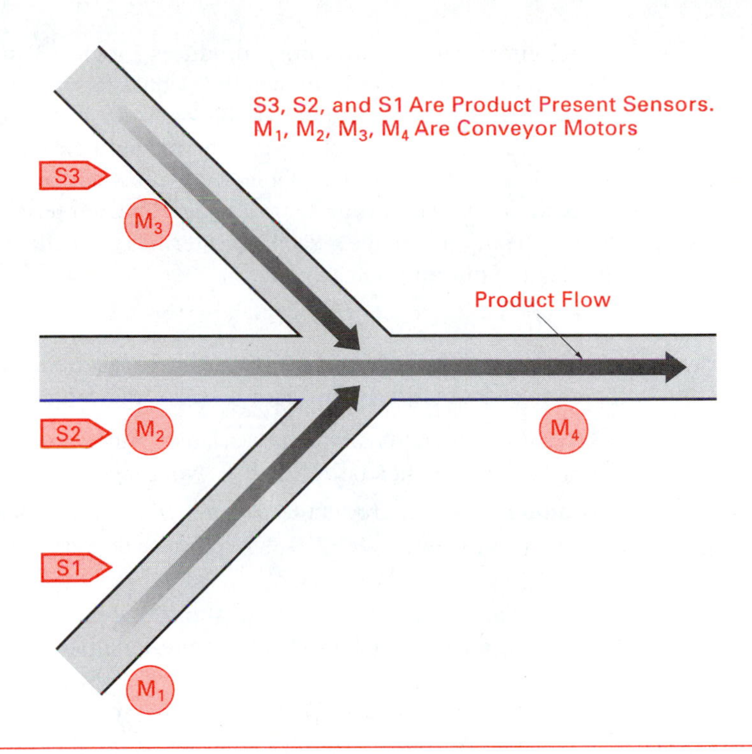

Figure 3.4
Conveyor system for Example 3.4

❖

Once the truth table is complete, "think" your way through it by careful comparison with the original problem description. In many real world applications, the original problem statement may be incompletely defined. When that occurs, additional discussion or investigation into the objectives is necessary. In some cases several "passes" or trials may be necessary to accurately define the problem.

In summary, the process of converting a verbal problem statement into a truth table involves the following steps:

1. Determine the input variables and output variables that are involved.
2. Assign mnemonic or letter or number symbols to each variable.
3. Determine the size of the truth table; how many input combinations exist:

   $2^x = y$

   where x = number of input variables and y = number of combinations.
4. Construct a truth table containing all of the input variable combinations.
5. By careful reading of the problem statement determine the combinations of inputs that cause a given output to be true.

3.1.2 Deriving Switching Equations

We stated earlier that logic can be described in several ways. One way is in a truth table, another is in logic diagrams, and a third is by Boolean equation(s). Boolean equations can be derived directly from a truth table or from the logic diagram. Likewise a truth table or logic diagram can be constructed from the Boolean equations(s).

Each input variable group that produces a logical 1 in a truth table output column can form a term in a Boolean switching equation. For instance, in the truth table for Example 3.4, Table 3.4, output variable M_1 is a 1 in four cases: {S3′, S2′, S1}, {S3′, S2, S1}, {S3, S2′, S1}, and {S3, S2, S1}. The remaining input variable combinations cause output M_1 to be a logical 0. A Boolean equation can be written that defines all the conditions under which output M_1 is a logical 1. Each term in the equation is formed by ANDing the input variables. Each AND term is then ORed with the other AND terms to complete the output Boolean equation. For M_1 the Boolean equation would be written

$$M_1 = S3'S2'S1 + S3'S2S1 + S3S2'S1 + S3S2S1$$

Notice that each AND term (also called a *product term*) identifies one input condition where the output is a 1. Before continuing, some definitions are necessary:

Literal A literal is a Boolean variable or its complement. For instance, let X be a binary variable, then both X and X' would be literals.

Product term A product term is a literal or the logical product (AND) of multiple literals. For instance, let X, Y, and Z be binary variables. Then a representative product term could be X, XY, or $X'YZ$.

Sum term A sum term is a literal or the logical OR of multiple literals. Let X, Y, and Z be binary variables. Then a representative sum term could be X, $X' + Y$, or $X + Y' + Z'$.

Sum of products A sum of products (SOP) is the logical OR of multiple product terms. Each product term is the AND of binary literals. For example, $XY' + X' + YZ + XY'Z'$ is a SOP expression.

Product of sums A product of sums (POS) is the logical AND of multiple OR terms. Each sum term is the OR of binary literals. For example, $(X + Y')(X + Y + Z')(Y' + Z')$ is a POS expression.

Minterm A minterm is a special case product (AND) term. A minterm is a product term that contains all of the input variables (each literal no more than once) that make up a Boolean expression.

Maxterm A maxterm is a special case sum (OR) term. A maxterm is a sum term that contains all of the input variables (each literal no more than once) that make up a Boolean expression.

Canonical sum of products A canonical sum of products is a complete set of minterms that defines when an output variable is a logical 1. Each minterm corresponds to the row in the truth table where the output function is 1; that is, the SOP for the output M in Table 3.1 is $M = a'bms + ab'ms + abms$.

Canonical product of sums A canonical product of sums is a complete set of maxterms that defines when an output is a logical 0. Each maxterm corresponds to a row in the truth table where the output is a 0; that is, the POS for the output O_1 in Table 3.2 is $O_1' = (C_1 + C_2 + C_3)(C_1' + C_2 + C_3)(C_1' + C_2 + C_3') \times (C_1' + C_2' + C_3)(C_1' + C_2' + C_3')$.

Table 3.5 shows the complement nature of minterms and maxterms. Note that an input variable is complemented when it has a value of 0 if we are writing minterms. The input variables are complemented when they have a value of 1, if we are writing maxterms. Minterms represent output variable 1s and maxterms represent output variable 0s. Lower case m is used to denote a minterm and upper case M is used to denote a maxterm. The number subscript indicates the decimal value of the term.

TABLE 3.5
Minterm and maxterm designations for three variables

Input variables			Minterm		Maxterm	
a	b	c	Term	Designation	Term	Designation
0	0	0	$a'b'c'$	m_0	$a + b + c$	M_0
0	0	1	$a'b'c$	m_1	$a + b + c'$	M_1
0	1	0	$a'bc'$	m_2	$a + b' + c$	M_2
0	1	1	$a'bc$	m_3	$a + b' + c'$	M_3
1	0	0	$ab'c'$	m_4	$a' + b + c$	M_4
1	0	1	$ab'c$	m_5	$a' + b + c'$	M_5
1	1	0	abc'	m_6	$a' + b' + c$	M_6
1	1	1	abc	m_7	$a' + b' + c'$	M_7

Output equations can be written directly from the truth table using either minterms or maxterms. When an output equation is written in minterms or maxterms, it is a canonical expression.

3.2 CANONICAL FORMS

Canonical is a word used to describe a condition of a switching equation. In normal use the word means "conforming to a general rule." The "rule," for switching logic, is that each term used in a switching equation must contain all of the available input variables. Two formats generally exist for expressing switching equations in a canonical form: sums of minterms and products of maxterms. We looked at examples of both in the previous section when generating minterms and maxterms from truth tables.

In the previous chapter we studied Boolean postulates and theorems and applied them to simplifying Boolean equations. Postulates are a set of basic definitions we assume to be true. The simplification rationale was to eliminate redundancy and lower the cost of the final logic circuits. Canonical expressions are not simplified. Often a canonical expression is the opposite of simplification; that is, it contains redundancies. Situations occur when logic designers must manipulate Boolean equations for purposes other than simplification. For instance, logic problems called *hazards* (combinational logic hazards are described in Chapter 4) require that certain logic redundancies be designed into a circuit.

Designers may also be faced with converting logic realized in one form to another form (TTL NAND gates to ECL NOR gates) by changing an expression into a canonical form before conversion. Later in this chapter (Section 3.4) we will study a graphical technique for simplifying Boolean equations called the *Karnaugh map*. Each square in the Karnaugh map represents either a minterm or maxterm. A noncanonical POS or SOP expression can be placed into a canonical form before entry into the Karnaugh map.

Often switching equations are written in a sum-of-products (SOP) form, or in a product-of-sums form (POS) that is not canonical. By this we mean that each term may not contain all of the original variables used to express the output variable.

To place a SOP equation into canonical form using Boolean algebra, we do the following:

1. Identify the missing variable(s) in each AND term.
2. AND the missing term and its complement with the original AND term,

   $xy\,(z + z')$.

   Because $(z + z') = 1$, the original AND term value is not changed.
3. Expand the term by application of the property of distribution, $xyz + xyz'$.

To place a POS equation into canonical form using Boolean algebra, we do this:

1. Identify the missing variable(s) in each OR term.
2. OR the missing term(s) and its complement with the original OR term, $x + y' + zz'$. Because $zz' = 0$, the original OR term value is not changed.
3. Expand the term by application of distributive property,

   $(x + y' + z)(x + y' + z')$.

❖ **EXAMPLE 3.5** Place the following equations into the proper canonical form:

a. $P = f(a, b, c) = ab' + ac' + bc$ (SOP)

SOLUTION

1. First term, ab', is missing the variable c. So we AND $(c + c')$ with ab':

   $ab' = ab'(c + c') = ab'c + ab'c'$

2. Second term, ac', is missing the variable b. So we AND $(b + b')$ with ac'.

   $ac' = ac'(b + b') = abc' + ab'c'$

3. Third term is missing the variable a. So we AND $(a + a')$ with bc.

   $bc = bc(a + a') = abc + a'bc$

4. The final canonical SOP form is

   $P = ab'c + ab'c' + abc' + ab'c' + abc + a'bc$

Note that two terms, the second and the fourth, are identical. Only one is needed, because any variable or group of variables ORed with itself is redundant: $x + x = x$ (property of idempotency or sameness).

The final equation becomes

$P = ab'c + ab'c' + abc' + abc + a'bc$

b. $G = f(w, x, y, z) = w'x + yz'$ (SOP)

SOLUTION

1. In this example the terms are missing two variables. The first term is missing variables y and z.

   $w'x = w'x(y + y')(z + z')$

   Expand in two stages:

   $(w'xy + w'xy')(z + z')$
   $w'xyz + w'xyz' + w'xy'z + w'xy'z'$

2. The second term is also missing two variables.

$$yz'(w + w')(x + x')$$

Expand in two stages:

$$(yz'w + yz'w')(x + x')$$
$$yz'wx + yz'wx' + yz'w'x + yz'w'x'$$

Rearrange to reflect the original binary order and eliminate redundant terms.

$$G = f(w, x, y, z)$$
$$= wxyz' + wx'yz' + w'xyz + w'xyz' + w'xy'z + w'xy'z' + w'x'yz'$$

c. $T = f(a, b, c) = (a + b')(b' + c)$ (POS)

Solution

1. The variable c is missing from the first term.

$$a + b' + cc' = (a + b' + c)(a + b' + c')$$

2. The variable a is missing from the second term.

$$b' + c + aa' = (a + b' + c)(a' + b' + c)$$

3. The complete equation is

$$T = f(a, b, c) = (a + b' + c)(a + b' + c')(a' + b' + c)$$

d. $J = f(A, B, C, D) = (A + B' + C)(A' + D)$ (POS)

Solution

1. $A + B' + C + DD' = (A + B' + C + D)(A + B' + C + D')$

2. $A' + D + CC' + BB'$

Distribute in two steps:

$$(A' + D + C)(A' + D + C') + BB'$$
$$(A' + D + C + B)(A' + D + C + B')(A' + D + C' + B)$$
$$(A' + D + C' + B')$$

3. The final canonical equation is

$$J = f(A, B, C, D) = (A + B' + C + D)(A + B' + C + D')$$
$$(A' + B + C + D)(A' + B' + C + D)(A' + B + C' + D)(A' + B' + C' + D)$$ ❖

3.3 GENERATION OF SWITCHING EQUATIONS FROM TRUTH TABLES

Switching equations can be written more conveniently by using the minterm or maxterm numerical designation, as shown in Table 3.5, instead of writing the variable names or their complements. In fact, the m and M designation can be dropped and the decimal equivalent value for the term can be written directly. For example, consider the canonical SOP equation $P = (ab'c + ab'c' + abc' + abc + a'bc)$. If we were to decode each of the minterms based on the binary weighing of each variable and produce a list of decimal decoded minterms, the result would be $P = \Sigma(5, 4, 6, 7, 3)$. To keep the input variable notation from being lost in a minterm list the relationship $P = f(a, b, c)$ is used. This means that output variable P is a function of the set of input variables $\{a, b, c\}$, with input variable a being the most significant one.

The sign Σ indicates summation and stands for the sum of products canonical form. When numbers in a decimal decoded set are preceded by a Σ sign, a SOP expression is indicated. Each number represents a minterm.

Canonical product of sums expressions can be written in a fashion similar to that just developed for sum of products expressions. The π (pi) sign is used to indicate product of sums canonical form. The decimal numbers listed in a POS set indicate maxterms. The input variable names are indicated in the same manner as in SOP equations, $X = f(A, B, C)$. Each maxterm is decoded as indicated in Table 3.5. Keep in mind that maxterms are the complement of minterms.

❖ EXAMPLE 3.6

Write the canonical minterm and maxterm expressions for Tables 3.1 through 3.2.

SOLUTION This can be solved using two different approaches. The first one is from Example 3.1, Table 3.1. Only one output variable (M) exists so only one equation is needed. Notice where each 1 occurs in the M column. Then determine the value of the input variables (a, b, m, s). The complete output equation contains a set of minterms ORed together.

The minterm expression for the output variable, M, is

$$M = f(a, b, m, s) = a'bms + ab'ms + abms = \sum(7, 11, 15)$$

A maxterm expression for the same output variable can be found by describing the places in the M column where 0 occurs and ORing together the set of input variables. Each maxterm is ANDed with the other maxterms to complete the output equation. $M = 0$ in Table 3.1 in the first row, where input variables a, b, c, and d are all 0. The maxterm is written $(a + b + c + d)$. Note that the complement of each input variable is written in the maxterm. If the input x is a 0, then x is written in the maxterm; if x is a 1, then x' is written in the maxterm.

The maxterm expression for the output variable, M, is

$$\begin{aligned} M &= f(a, b, m, s) \\ &= (a + b + m + s)(a + b + m + s') \\ &\quad (a + b + m' + s)(a + b + m' + s') \\ &\quad (a + b' + m + s)(a + b' + m + s')(a + b' + m' + s) \\ &\quad (a' + b + m + s)(a' + b + m + s')(a' + b + m' + s) \\ &\quad (a' + b' + m + s)(a' + b' + m + s')(a' + b' + m' + s) \\ M &= \pi(0, 1, 2, 3, 4, 5, 6, 8, 9, 10, 12, 13, 14) \end{aligned}$$

It is easy to see that in this case the minterm expression contains fewer terms.

The second approach is from Example 3.2. Five output variables exist: O_1, O_2, O_3, and W_1, W_2. Each output variable requires a separate switching expression.

Writing the output equations, we get

$$\begin{aligned} O_1 &= f(C_1, C_2, C_3) = C_1'C_2'C_3 + C_1'C_2C_3' + C_1'C_2C_3 \\ O_2 &= f(C_1, C_2, C_3) = C_1'C_2'C_3' + C_1'C_2'C_3 + C_1C_2'C_3' + C_1C_2'C_3 \\ O_3 &= f(C_1, C_2, C_3) = C_1'C_2'C_3' + C_1'C_2C_3' + C_1C_2'C_3' + C_1C_2C_3' \\ W_1 &= f(C_1, C_2, C_3) = C_1'C_2C_3 + C_1C_2'C_3 + C_1C_2C_3' \\ W_2 &= f(C_1, C_2, C_3) = C_1C_2C_3 \end{aligned}$$

Writing the same equations using decimal list shorthand notation, we get

$$O_1 = f(C_1, C_2, C_3) = \sum(1, 2, 3)$$
$$O_2 = f(C_1, C_2, C_3) = \sum(0, 1, 4, 5)$$
$$O_3 = f(C_1, C_2, C_3) = \sum(0, 2, 4, 6)$$
$$W_1 = f(C_1, C_2, C_3) = \sum(3, 5, 6)$$
$$W_2 = f(C_1, C_2, C_3) = \sum(7)$$

The maxterm expressions written in decimal form are

$$O_1 = f(C_1, C_2, C_3) = \pi(0, 4, 5, 6, 7)$$
$$O_2 = f(C_1, C_2, C_3) = \pi(2, 3, 6, 7)$$
$$O_3 = f(C_1, C_2, C_3) = \pi(1, 3, 5, 7)$$
$$W_1 = f(C_1, C_2, C_3) = \pi(0, 1, 2, 4, 7)$$
$$W_2 = f(C_1, C_2, C_3) = \pi(0, 1, 2, 3, 4, 5, 6)$$

❖

❖ EXAMPLE 3.7 Express the following SOP equations in a minterm list (shorthand decimal notation) form:

a. $H = f(A, B, C) = A'BC + A'B'C + ABC$

SOLUTION

1. $A'BC = 011_2 = 3_{10}$.
2. $A'B'C = 001_2 = 1_{10}$.
3. $ABC = 111_2 = 7_{10}$.

Rearrange the decoded decimal values in numerical order:

$$H = f(A, B, C) = \sum(1, 3, 7)$$

b. $G = f(w, x, y, z) = wxyz' + wx'yz' + w'xyz' + w'x'yz'$

SOLUTION

1. $wxyz' = 1110_2 = 14_{10}$.
2. $wx'yz' = 1010_2 = 10_{10}$.
3. $w'xyz' = 0110_2 = 6_{10}$.
4. $w'x'yz' = 0010_2 = 2_{10}$.

Rearrange the decimal equivalent minterms in numerical order:

$$G = f(w, x, y, z) = \sum(2, 6, 10, 14)$$

c. From Example 3.3, a single output variable, O_1, exists requiring one minterm equation:

$$O_1 = f(I_4, I_3, I_2, I_1) = \sum(7, 11, 13, 14, 15)$$

SOLUTION Expressed as maxterms, we get

$$O_1 = f(I_4, I_3, I_2, I_1) = \pi(0, 1, 2, 3, 4, 5, 6, 8, 9, 10, 12)$$

❖

❖ **EXAMPLE 3.8** Express the following POS equations in a maxterm list (decimal notation) form:

a. $T = f(a, b, c) = (a + b' + c)(a + b' + c')(a' + b' + c)$

SOLUTION

1. $a + b' + c = 010_2 = 2_{10}$.
2. $a + b' + c' = 011_2 = 3_{10}$.
3. $a' + b' + c = 110_2 = 6_{10}$.

$$T = f(a, b, c) = \pi(2, 3, 6)$$

b. $J = f(A, B, C, D) = (A + B' + C + D)(A + B' + C + D')(A' + B + C + D)$
$(A' + B' + C + D)(A' + B + C' + D)(A' + B' + C' + D)$

SOLUTION $J = f(A, B, C, D) = \pi(4, 5, 8, 10, 12, 14)$ ❖

TABLE 3.6
Truth table for example canonical equations

Inputs			Output	
x	y	z	A	A'
0	0	0	0	1
0	0	1	0	1
0	1	0	1	0
0	1	1	0	1
1	0	0	1	0
1	0	1	1	0
1	1	0	0	1
1	1	1	0	1

It is not necessary to first write switching equations in a variable name format and then convert to a minterm or maxterm list (decimal format). It is easier to write the canonical output equations directly from the truth table in a minterm or maxterm numerical list than to write the equations using input variable names. For example, consider the canonical equations in Table 3.6.

1. Write the SOP equation for A:

   $A = f(x, y, z) = \sum(2, 4, 5)$

   Collect minterms for A output 1s.

2. Write the SOP equation for A':

   $A' = f(x, y, z) = \sum(0, 1, 3, 6, 7)$

   Collect minterms for A output 0s or A' output 1s.

3. Write the POS equation for A:

   $A = f(x, y, z) = \pi(0, 1, 3, 6, 7)$

   Collect maxterms for A output 0s.

4. Write the POS equation for A':

   $A' = f(x, y, z) = \pi(2, 4, 5)$

   Collect maxterms for A' output 0s.

Notice that equations in steps 1 and 4 are the complements of the equations in steps 2 and 3.

3.4 KARNAUGH MAPS

In Chapter 2 we investigated some algebraic techniques for simplifying switching equations. In Section 3.2 we investigated canonical forms. It may have occurred to you that rendering an equation into canonical form and simplification were opposites; they are.

Figure 3.5
Two-variable Karnaugh map

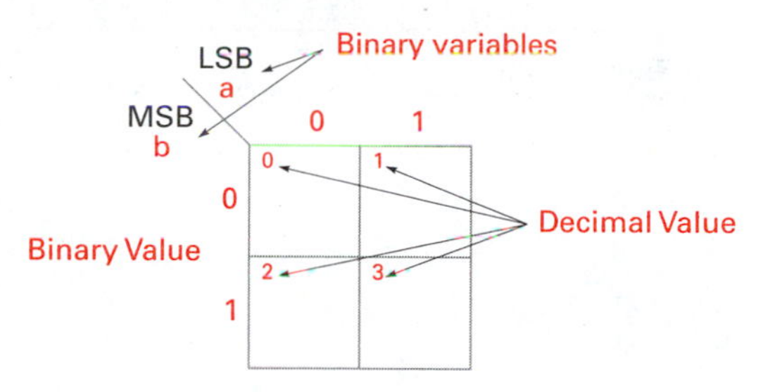

Simplification of switching equations reduces the amount of hardware needed to realize a given function. Reduction of gates and gate inputs may result in fewer integrated circuits, which in turn decreases cost and improves reliability.

Boolean algebra can be used to simplify equations but the process is lengthy and error prone. What is needed is a more systematic method for finding and eliminating any redundancies in an equation.

A better approach is the use of the Karnaugh map. The **Karnaugh map** is a matrix of squares. Each square represents a minterm or maxterm from a Boolean equation. The arrangement of the matrix squares permits identification of input variable redundancies, which helps reduce the output equation.

The Karnaugh map identifies all of the cases for a given set of input variables where groups of minterms may contain redundant variables of the form of $x + x' = 1$. When these groups are identified, the redundant variables can be eliminated, resulting in a simplified output function ($abc + abc' = ab$).

If a given switching equation contains a minterm, then a 1 is entered into the square that represents that term. A maxterm is represented by a 0.

A simple two-variable Karnaugh map is shown in Figure 3.5. Note that all four possible combinations of input variables are represented.

3.4.1 Three- and Four-Variable Karnaugh Maps

Each square in the map shown in Figure 3.6 represents a possible minterm or maxterm of a three-variable function. The upper left square represents binary 000, minterm ($A'B'C'$) or maxterm ($A + B + C$). Because three binary variables can represent eight unique combinations, we find eight squares are needed. If minterm $A'B'C'$ (binary 000, decimal 0) occurred in a switching equation, then a 1 would be inserted into the upper left square. Assignment of a square to each of the eight minterms resulting from three input variables results in Table 3.7.

Figure 3.6
Three-variable Karnaugh map

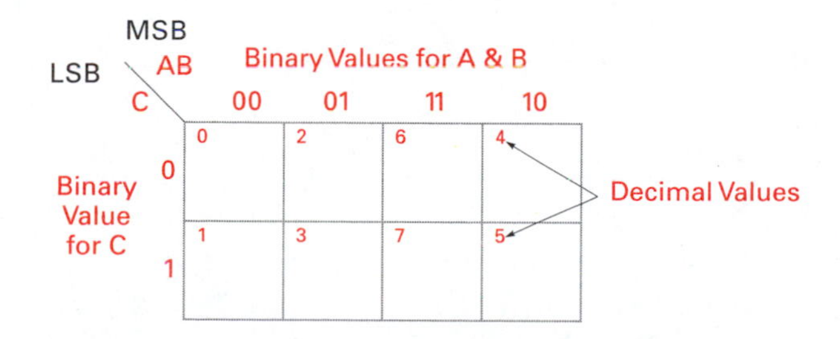

TABLE 3.7
Three-variable minterms and maxterms

Inputs			Minterm	Maxterm
A	*B*	*C*	*m*	*M*
0	0	0	0	0
0	0	1	1	1
0	1	0	2	2
0	1	1	3	3
1	0	0	4	4
1	0	1	5	5
1	1	0	6	6
1	1	1	7	7

We may label the decimal value for each square in a Karnaugh map by decoding the binary numbers as indicated in Figure 3.6. This means that a direct connection can be made between the minterm or maxterm equation list and the appropriate square in the Karnaugh map.

Notice that across the top and down the side of the three- and four-variable Karnaugh maps only one-bit changes occur between adjacent squares for each column and row. Building maps for a greater number of input variables requires increasing the number of squares to match the number of variables and making sure that each adjacent row or column differs by only one variable.

Notice that the four-variable map (as in Figure 3.7) is the same length both horizontally and vertically. Also notice that the variable representation is the same in both directions; two variables across (00, 01, 11, 10) and two down.

The input variable binary weighting is

MSB			LSB
W	*X*	*Y*	*Z*
$2^3 = 8$	$2^2 = 4$	$2^1 = 1$	$2^0 = 1$

where MSB is the most significant bit, and LSB is the least significant bit.

Loading Karnaugh maps involves writing a 1 in the squares corresponding to a minterm or a 0 in the square corresponding to a maxterm. Consider the simplification of three-variable switching equations in the following examples.

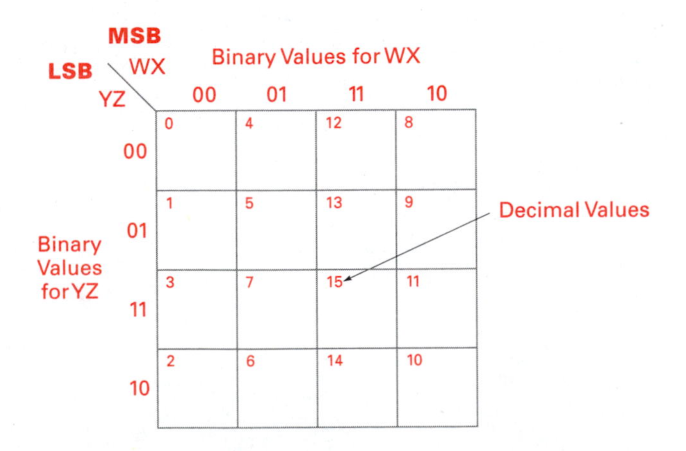

Figure 3.7
Four-variable Karnaugh map

❖ EXAMPLE 3.9

$$Y = f(a, b, c) = \sum(0, 1, 4, 5)$$

SOLUTION The decimal numbers in this canonical SOP equation are indicated in the three-variable Karnaugh map in Figure 3.8 by 1 in each appropriate square.

Figure 3.8
Three-variable Karnaugh map for $Y = f(a, b, c) = \Sigma(0, 1, 4, 5)$

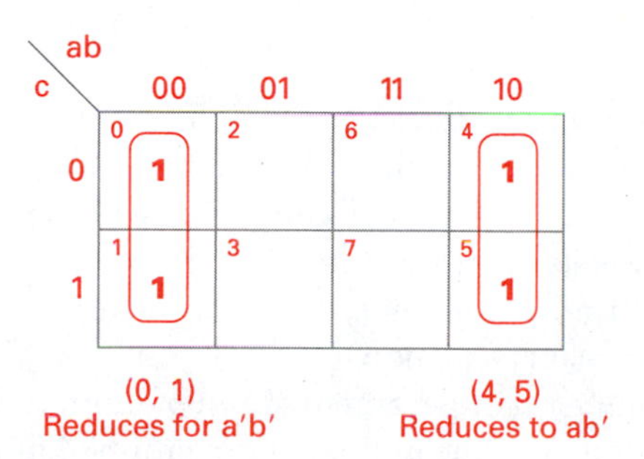

(a) Two groups of two minterms

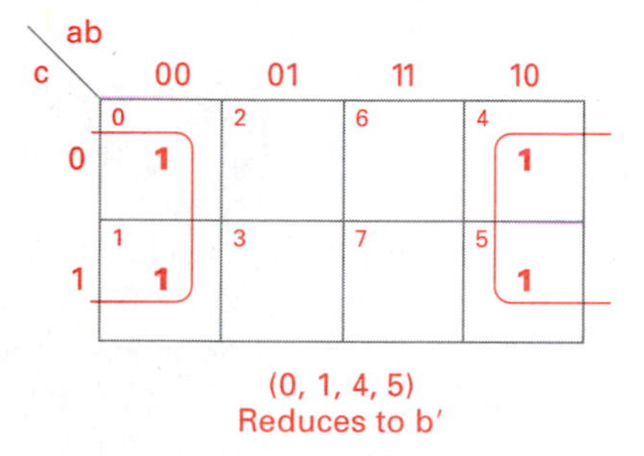

(b) One group of four minterms

Each minterm is indicated by a 1 in the appropriate square. Once the minterms are loaded, the next step is to determine any logical adjacencies. *Logical adjacency* has been defined as the case where only one variable changes between minterms, or groups of minterms {0, 1} are logically adjacent because only one variable change separates the two minterms.

Minterms {0, 1} = $a'b'c' + a'b'c$, which can be reduced to $a'b'$, because $a'b'(c + c') = a'b'(1) = a'b'$. Thus {0, 1} simplifies to $a'b'$. The group {4, 5} also can be reduced; $ab'c' + ab'c = ab'$, by the same reasoning. The two groups {0, 1} and {4, 5} are also logically adjacent (one variable change between groups), allowing even further reduction, because $a'b' + ab' = b'$. A group of four is indicated, in the three-variable map in Figure 3.8(b), by drawing circles around the four minterms. ❖

We have noted that a Karnaugh map, or K-map, aids in identifying variable redundancies by systematically grouping sets of logically adjacent minterms. This was the case in Figure 3.8 between {0, 1} and {4, 5}, where {0, 1, 4, 5} forms a single group of four minterms where two variables are permitted to change, resulting in $Y = b'$.

From the example illustrated in Figure 3.8, we note that a group of two logically adjacent minterms eliminated one variable from the final expression. A group of four logically adjacent minterms eliminated two variables from the result. All minterm groups must occur in a power of two as follows: $2^n = M$, where M is the maximum number of minterms in a group, and n is the number of variables to be eliminated by the group.

$$2^1 = 2, \qquad 2^2 = 4, \qquad 2^3 = 8, \qquad \text{and so on}$$

This means that groups of 3, 5, 6, 7, and so forth minterms are not possible combinations, because they are not an integer power of 2. The number of squares or cells containing minterms in a Karnaugh map that may be combined is defined as follows.

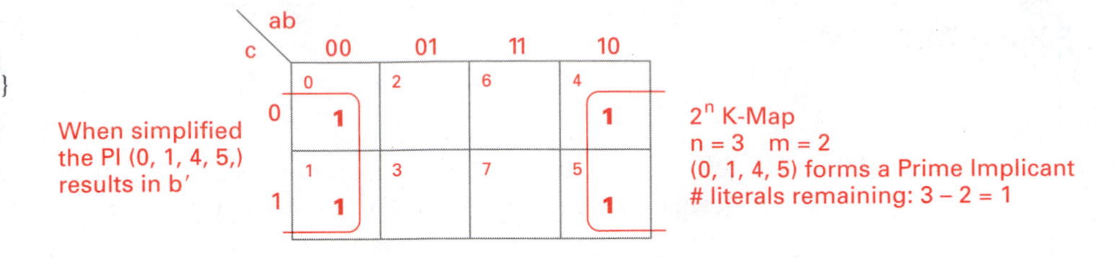

Figure 3.9
Three-variable K-map showing prime implicant (PI) {0, 1, 4, 5}

DEFINITION

A set of 2^n K-map squares (cells) are combined to form a *prime implicant*, if *n* variables of the equation being simplified have 2^n permutations within the set and the remaining *m–n* variables have the same value within the set. The resulting product term has *m–n* literals. In other words 2^n K-map squares may be combined to form a group of minterms containing *m–n* literals, where *m* represents the number of variables in the original equation.

Any single minterm or permitted group of minterms is called an *implicant* of an output function. A **prime implicant** is a group of minterms that cannot be combined with any other minterms or groups.

Minterms {0, 1, 4, 5} are each implicants of output *Y*, from the previous example. The prime implicant consists of the four minterms {0, 1, 4, 5} grouped together (see Figure 3.9).

❖ EXAMPLE 3.10

$$G = f(x, y, z) = \sum(0, 2, 3, 7)$$

SOLUTION Expression *G* is loaded into the K-map shown in Figure 3.10. Three groups of two minterms are found: {0, 2}, {2, 3}, and {3, 7}. None of these can be combined into groups of four, because more than one variable change occurs between the groups of two. Each of the three groups forms a prime implicant. Prime implicant {0, 2} reduces to $x'z'$. Notice that the variable *y* was a 0 in minterm {0} and a 1 in minterm {2}. For this reason it was eliminated. Similarly prime implicant {2, 3} eliminated variable *z* and prime implicant {3, 7} eliminated variable *x*.

Groups {0, 2} and {3, 7} both contain a minterm that is not in any other group. Minterm {0} is unique to prime implicant {0, 2} and minterm {7} is unique to prime implicant {3, 7}. Having one or more unique minterms creates what is called an *essential prime implicant* (EPI).

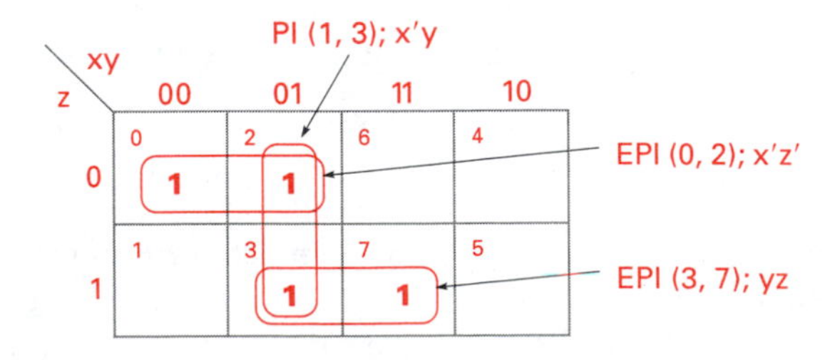

Figure 3.10
Three-variable K-map for the expression $G = \Sigma\ (0, 2, 3, 7)$

An **essential prime implicant** is a prime implicant in which one or more minterms are unique; that is, it contains at least one minterm not contained in any other prime implicant.

The simplified expression for G includes only the essential prime implicants. The non-essential prime implicant {2, 3} term is redundant because all of its minterms are covered by the two essential prime implicants.

The simplified expression is $G = x'z' + yz$.

A procedure for finding essential prime implicants is this:

1. Find the prime implicants by finding all permitted (integer power of 2) maximum-sized groups of minterms.
2. Find essential prime implicants by identifying those prime implicants that contain at least one minterm not found in any other prime implicant. ❖

❖ EXAMPLE 3.11

Simplify the following expression (also in Figure 3.11) using a three-variable Karnaugh map.

$$D = f(x, y, z) = \sum(0, 2, 4, 6)$$

SOLUTION Minterms {0, 2, 4, 6} form an essential prime implicant. Every variable is either (1) a 0 in every square, in which case it appears complemented in the final product term; or (2) a 1 in every square, in which case it appears as is in the final product term; or (3) a 0 in half and a 1 in the other half of the squares, in which case it does not appear at all in the final product term. This idea of input variable symmetry (equal number of 0 and 1) for that variable to be eliminated holds for any size group.

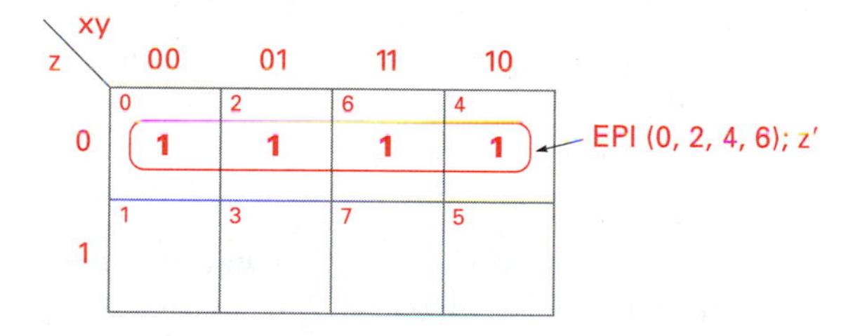

Figure 3.11
Three-variable K-map for $D = \Sigma\ (0, 2, 4, 6)$

❖

❖ EXAMPLE 3.12

Simplify the following expression (also in Figure 3.12) using a three-variable Karnaugh map.

$$Q = f(a, b, c) = \sum(1, 2, 3, 6, 7)$$

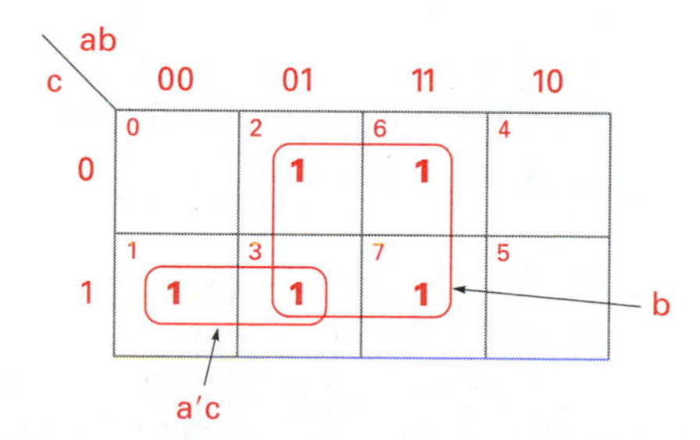

Figure 3.12
Three-variable K-map for $Q = \Sigma\ (1, 2, 3, 6, 7)$

SOLUTION Two prime implicants exist: {1, 3} and {2, 3, 6, 7}. Both are essential because each has a unique minterm. Essential prime implicant {2, 3, 6, 7} eliminates two variables, a and c, leaving b. Essential prime implicant {1, 3} eliminates one variable, b, leaving $a'c$.

The simplified expression is $Q = b + a'c$. ❖

A group of eight minterms in a three-variable K-map reduces to 1; for example, let $G = f(a, b, c) = \Sigma(0, 1, 2, 3, 4, 5, 6, 7)$, then $G = 1$. Before we continue on to higher order K-maps, we need to systematically evaluate the steps involved in loading and determining the essential prime implicants.

1. Load the minterms into the K-map by placing a 1 in the appropriate square.
2. Look for groups of minterms (prime implicants).
   a. The group size must be a power of 2:

      $2^0 = 0$

      $2^1 = 2$

      $2^2 = 4$

      $2^3 = 8$

      $2^4 = 16$

      $\vdots$

      2^n

      The number 2 is used because we are dealing with the binary number system. Each exponent represents the number of variables. The numbers 2, 4, 8, 16, . . . , 2^n represent the permissible group size.

   b. Prime implicants are formed from groups of minterms whose size is a power of 2. This means that minterm group sizes that are not an integer power of 2 are not permitted. Find the largest groups of minterms first, then progressively evaluate smaller collections of minterms until all groups are found.
3. Once all of the possible prime implicants have been identified, we determine whether any have minterms that are unique. If so that prime implicant is an essential prime implicant.
4. Select all essential prime implicants and a minimal set of remaining prime implicants that cover all remaining 1s in the K-map.
5. More than one equally simplified result is possible when more than one set of remaining prime implicants contain the same number of minterms or maxterms.

❖ **EXAMPLE 3.13** Simplify the following three-variable equation (also in Figure 3.13):

$$J = f(x, y, z) = \sum(0, 2, 3, 4, 5, 7)$$

SOLUTION No groups of four minterms exist; however, six prime implicants containing two minterms exist. Note that none of the prime implicants is essential. The final simplified equation is not unique; that is, more than one equally simplified result is possible. The product terms in the simplified result must cover all of the minterms. By *cover,* we mean

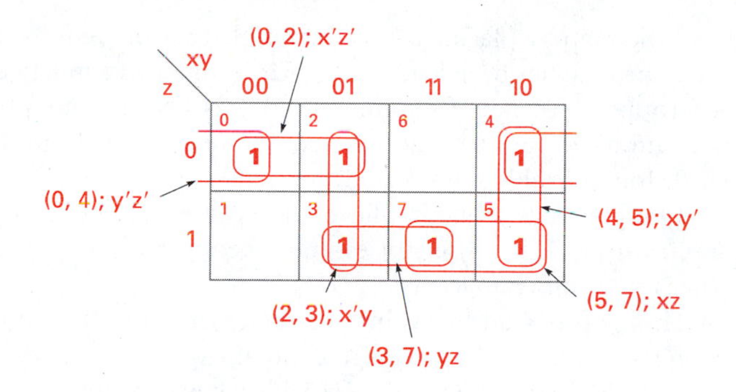

Figure 3.13
Three-variable K-map for $J = \Sigma\,(0, 2, 3, 4, 5, 7)$

that all of the minterms in the original equation must be contained by a prime implicant or essential prime implicant. The prime implicants from Figure 3.13 are

1. $\{0, 2\} \rightarrow x'z'$
2. $\{0, 4\} \rightarrow y'z'$
3. $\{2, 3\} \rightarrow x'y$
4. $\{3, 7\} \rightarrow yz$
5. $\{4, 5\} \rightarrow xy'$
6. $\{5, 7\} \rightarrow xz$

For the final equation to cover the original, each minterm must be implied by a product term in the answer. For instance, we must pick a prime implicant $\{0, 2\}$ or $\{0, 4\}$ to cover the minterm $\{0\}$. Two equally simplified equations are

$$J = x'z' + xy' + yz$$

which covers minterms $\{0, 2, 3, 4, 5, 7\}$.

$$J = y'z' + x'y + xz$$

which covers minterms $\{0, 2, 3, 7, 4, 5\}$. ❖

❖ **EXAMPLE 3.14** Simplify the following four-variable equations (Figure 3.14):

a. $K = f(w, x, y, z) = \sum(0, 1, 4, 5, 9, 11, 13, 15)$

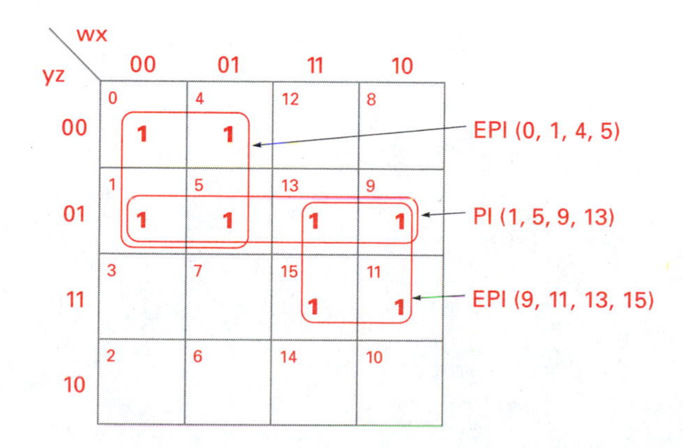

Figure 3.14
Karnaugh map for $K = f(w, x, y, z)$ $= \Sigma\,(0, 1, 4, 5, 9, 11, 13, 15)$

SOLUTION Load the map by placing a 1 in the numbered square for each number in the equation. Once the map has been loaded, the prime implicants must be identified. In this example, no groups of 16 (all of the minterms) exist. No essential prime implicants containing eight minterms are present because symmetry (equal number of 0s and 1s) is lacking. Three four-variable prime implicants exist: {0, 1, 4, 5}, {1, 5, 9, 13}, and {9, 11, 13, 15}.

Two essential prime implicants are present: {0, 1, 4, 5} and {9, 11, 13, 15}. Prime implicant {1, 5, 9, 13} is not essential because all of its minterms are contained in the other two prime implicants.

The map is read by evaluating the essential prime implicants and then any prime implicants covering minterms not contained in the essential prime implicants. The first, {0, 1, 4, 5}, reduces to $w'y'$. Variables x and z are eliminated because they contain an equal number of 0s and 1s (symmetry). The second essential prime implicant, {9, 11, 13, 15}, reduces to wz.

The final simplified equation is $K = w'y' + wz$.

b. Reduce the following equation (Figure 3.15) using a four-variable K-map.

$$L = f(a, b, c, d) = \sum(0, 2, 5, 7, 8, 10, 13, 15)$$

SOLUTION Prime implicant {5, 7, 13, 15} is easy to identify because the minterms are both logically and physically adjacent. Prime implicant {0, 2, 8, 10}, is more difficult to see at first. It is, however, a prime implicant because the adjacency requirements are met. The K-map wraps around on itself so that the four corners are logically adjacent. Minterms {0, 2, 8, 10} also form a prime implicant. It may be difficult to see the symmetry of group {0, 2, 8, 10}. Notice that variable a is a 0 in column 00 and a 1 in column 10. Also notice that c is a 0 in the 00 row and a 1 in the 10 row. Because both a and c can be eliminated, leaving variables b and d intact, prime implicant {0, 2, 8, 10} forms an essential prime implicant. The EPI (essential prime implicant) {0, 2, 8, 10}, reduces to $b'd'$, and EPI {5, 7, 13, 15} reduces to bd.

The simplified equation from the K-map in Figure 3.15 is $L = b'd' + bd$, which is also the equation for EX-NOR $L = (b \oplus d)'$. The EX-NOR is not a further simplification but rather a different way of realizing the same function. EX-OR medium-scale integrated circuits (MSI) are available.

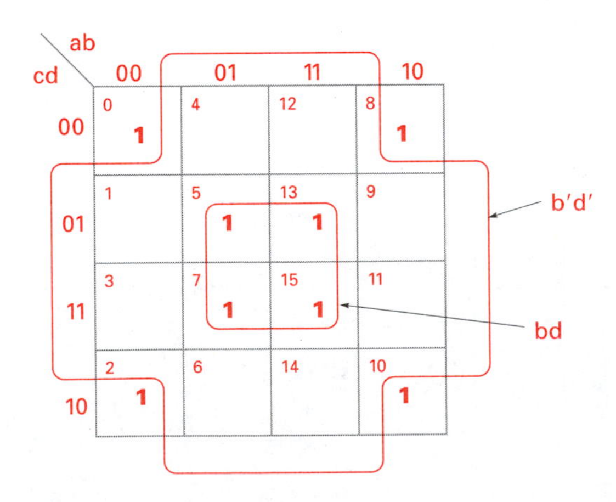

Figure 3.15
Four-variable K-map for
$L = f(a, b, c, d)$
$= \Sigma\ (0, 2, 5, 7, 8, 10, 13, 15)$

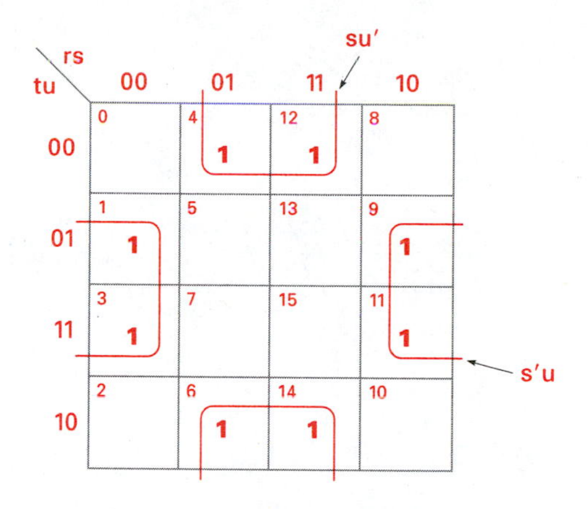

Figure 3.16

Four-variable K-map for $P = f(r, s, t, u)$ $= \Sigma(1, 3, 4, 6, 9, 11, 12, 14)$

c. Simplify the following equation (Figure 3.16):

$$P = f(r, s, t, u) = \sum(1, 3, 4, 6, 9, 11, 12, 14)$$

Solution The two prime implicants illustrated in Figure 3.16 are both essential. They reduce to $\{1, 3, 9, 11\} = s'u$ and $\{4, 6, 12, 14\} = su'$.

The simplified equation is $P = s'u + su'$. This is the form for an EX-OR. So $P = (s \oplus u)$.

d. Consider the following switching equation (Figure 3.17):

$$D = f(w, x, y, z) = \sum(5, 7, 8, 9, 13)$$

Solution Groups {5, 7} and {8, 9} form EPIs. Minterm {13} must be included in some group, and two choices are available; either one produces an equally simplified result. EPI {5, 7} reduces to $w'xz$, and EPI {8, 9} reduces to $wx'y'$. Prime implicant {5, 13} gives $xy'z$, and prime implicant {9, 13} produces $wy'z$. The complete simplified equation is

$$D = w'xz + wx'y + xy'z \quad \text{or} \quad w'xz + wx'y + wy'z$$

Either of the two expressions is equally simplified and will produce the correct value for the output function, D. The result is a nonunique simplification. Other interesting four-variable simplification problems follow (Figures 3.18–3.20).

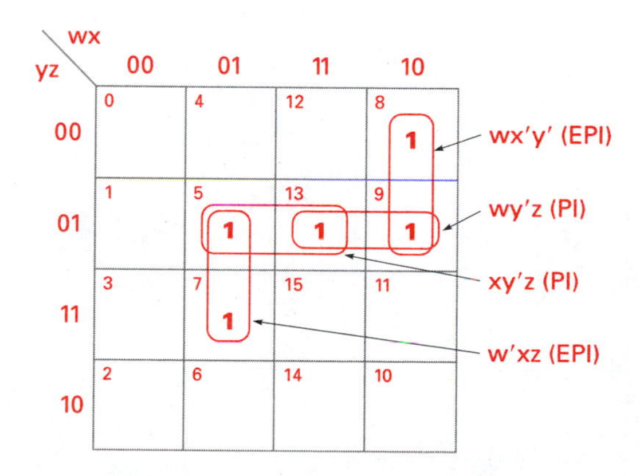

Figure 3.17

Four-variable K-map illustrating nonunique simplification

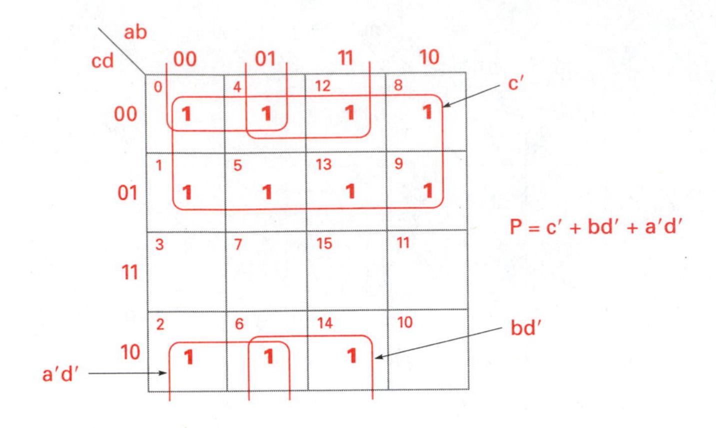

Figure 3.18
Four-variable K-map for
$P = f(a, b, c, d)$
$= \Sigma\ (0, 1, 2, 4, 5, 6, 8, 9, 12, 13, 14)$

e. Simplify

$$P = f(a, b, c, d) = \sum(0, 1, 2, 4, 5, 6, 8, 9, 12, 13, 14)$$

SOLUTION A group of eight minterms form EPI {0, 1, 4, 5, 8, 9, 12, 13}, which reduces to c'. Two groups of four minterms, {0, 2, 4, 6} and {4, 6, 12, 14}, reduce to $a'd'$ and bd' respectively and are also EPIs.

f. Let

$$S = f(a, b, c, d) = \sum(1, 3, 4, 5, 7, 8, 9, 11, 15)$$

Simplify using a Karnaugh map (Figure 3.19).

SOLUTION The EPIs and the reduced terms are

$$\{1, 3, 5, 7\} \rightarrow a'd, \qquad \{3, 7, 11, 15\} \rightarrow cd,$$
$$\{4, 5\} \rightarrow a'bc', \qquad \{8, 9\} \rightarrow ab'c'$$

Prime implicants are formed when any maximally sized group of minterms occurs. Essential prime implicants are prime implicants that contain at least one minterm not occurring in any other prime implicant. Later in our study of static and dynamic hazards we will find prime implicants of interest in designing hazard-free logic circuits.

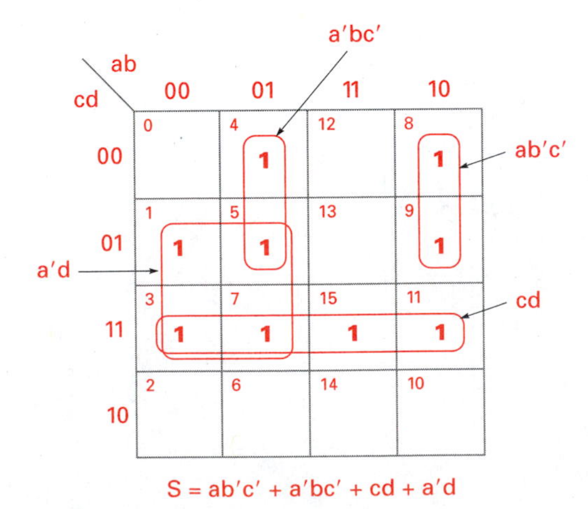

Figure 3.19
Four-variable K-map for
$S = f(a, b, c, d)$
$= \Sigma\ (1, 3, 4, 5, 7, 8, 9, 11, 15)$

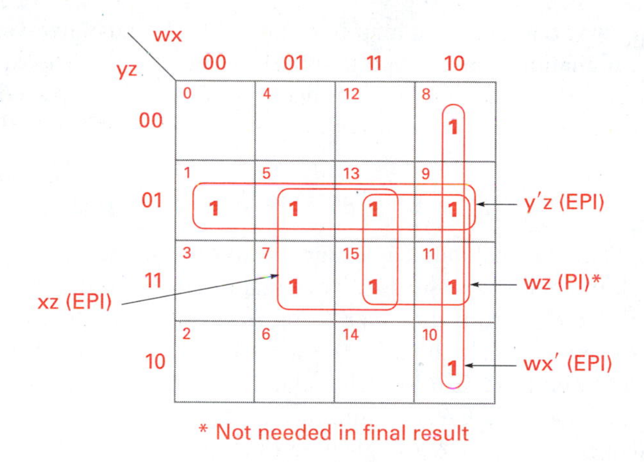

Figure 3.20
Four-variable K-map for illustrating prime implicant and essential prime implicants

g. Determine the prime implicants and essential prime implicants for the following (Figure 3.20):

$$U = f(w, x, y, z) = \sum(1, 5, 7, 8, 9, 10, 11, 13, 15)$$

SOLUTION The EPIs are {1, 5, 9, 13}, {8, 9, 10, 11}, and {5, 7, 13, 15}. Notice that each group contains one or more minterms that are unique. Minterm {1} is unique to EPI {1, 5, 9, 13}, minterm {7} is unique to EPI {5, 7, 13, 15}, and minterms {8, 10} are unique to EPI {8, 9, 10, 11}. The prime implicant {9, 11, 13, 15} is a maximally sized group, yet it contains no unique minterms; hence, it is a prime implicant but not an essential prime implicant. ❖

3.4.2 Five- and Six-Variable Karnaugh Maps

Five-variable Karnaugh maps can be constructed so that three variables are laid out horizontally and two vertically, as illustrated in Figure 3.21. *A*, *B*, *C*, *D*, and *E* are the

Figure 3.21
A mirror image version five-variable K-map

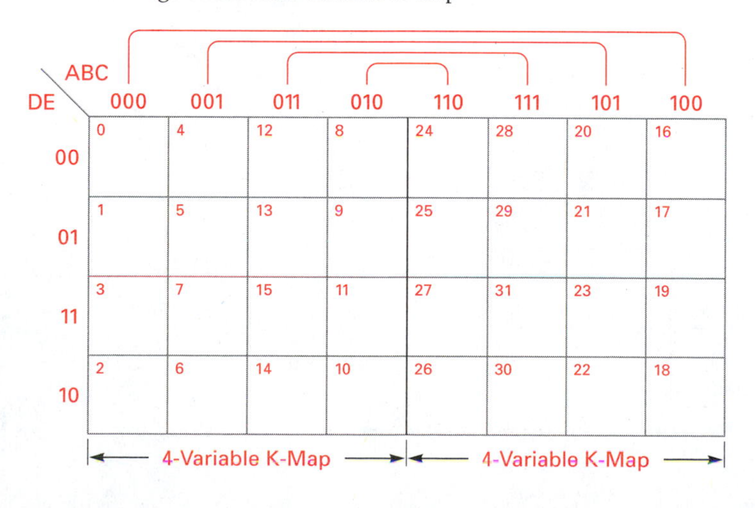

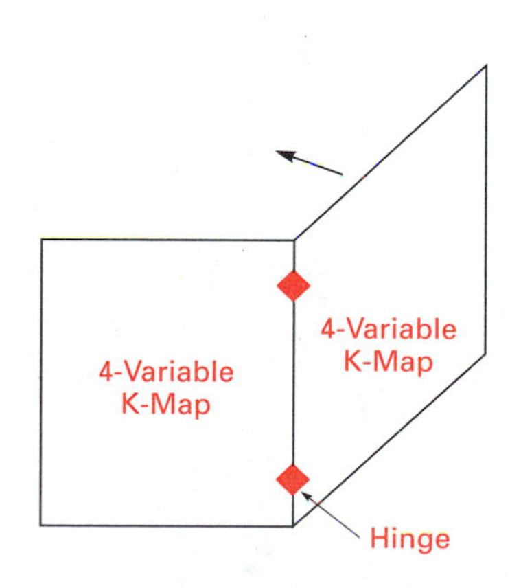

input variables to the map in Figure 3.21. Because five variables are used 2^5, possible combinations exist, ranging from 00000 to 11111_2. The square in the upper left corner will be the starting point and therefore contain the input combination 00000. A weighting of each variable *A* through *E* will aid in showing how the squares are numbered.

A	*B*	*C*	*D*	*E*	Variables
$2^4 = 16$	$2^3 = 8$	$2^2 = 4$	$2^1 = 2$	$2^0 = 1$	Power of 2

The decimal value associated with each square in the map is found by adding the column and row values. The decimal numbers, in the upper left corner of each square, are decoded from the binary numbers. Notice that the column logical adjacencies are determined by the binary code written across the top of the map. Each physically adjacent column is also logically adjacent. Using the numbering system shown in Figure 3.21 logically adjacent columns occur as mirror images. Each half of the five-variable map corresponds to a single four-variable K-map.

Figure 3.22 illustrates a stack version of a five-variable Karnaugh map. The same number of squares exist in both types of five-variable K-maps. The difference is in how the two halves of the map are viewed. In the stacked version, the right half of the map is viewed as sliding over the left half of the map. The column adjacencies are as indicated in Figure 3.22.

Any size K-map can be constructed in a similar manner. The numbering of the squares in the stacked version differs from that in the mirrored version.

Six-variable maps are constructed by extending the vertical length by one additional variable. Notice that a four-variable map had 16 squares; a five-variable map, 32; and now the six-variable map needs 64 squares. The map size increases as an integer power of 2; that is, for each variable increase the size of the map doubles. A four-variable map is constructed from two three-variable maps, a five-variable map is the equivalent of two four-variable maps, and so on. The original requirement of a single variable change between adjacent squares is true for any K-map.

Figure 3.23 illustrates a stacked six-variable map. Notice the two heavy lines dividing the map into four large pieces. Each quadrant is the equivalent of a single four-

Figure 3.22
A stacked numbering version of the five-variable K-map

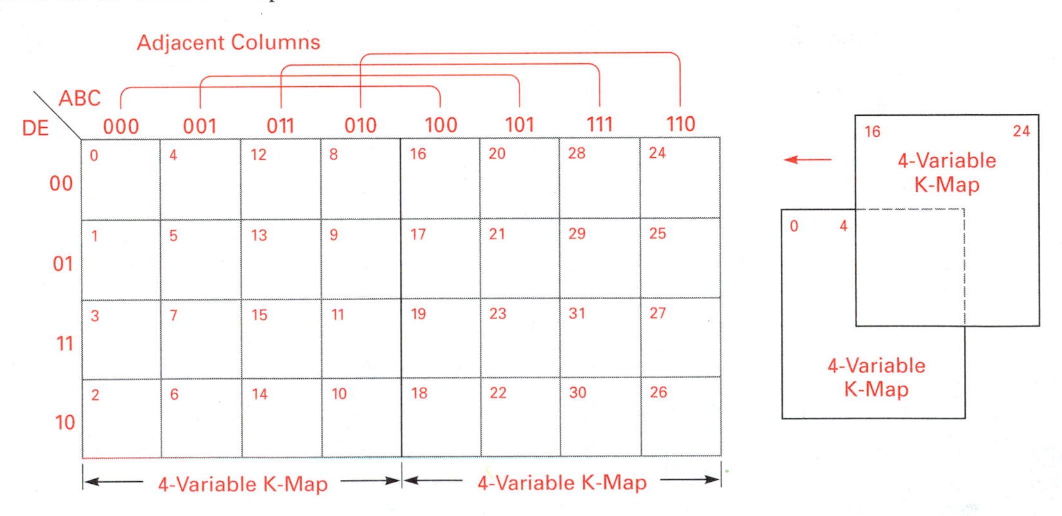

Figure 3.23
Six-variable Karnaugh map

def \ abc	(0) 000	(8) 001	(24) 011	(16) 010	(32) 100	(40) 101	(56) 111	(48) 110
(0) 000	0	8	24	16	32	40	56	48
(1) 001	1	9	25	17	33	41	57	49
(3) 011	3	11	27	19	35	43	59	51
(2) 010	2	10	26	18	34	42	58	50
(4) 100	4	12	28	20	36	44	60	52
(5) 101	5	13	29	21	37	45	61	53
(7) 111	7	15	31	23	39	47	63	55
(6) 110	6	14	30	22	38	46	62	54

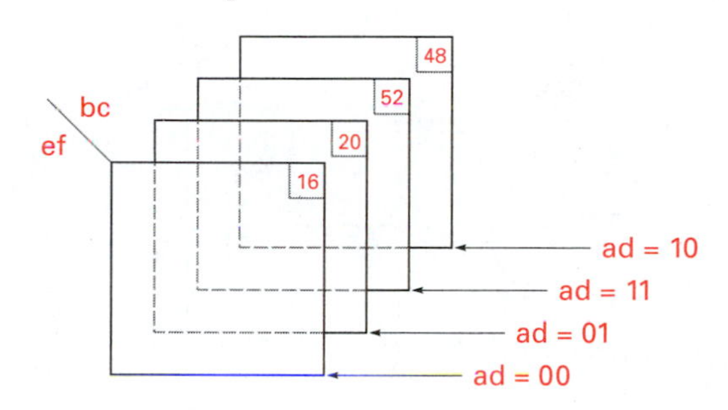

Figure 3.24
Four four-variable K-maps stacked to form a six-variable K-map

variable map. If we think of the quadrants as being movable, then each quadrant can be overlaid on top of other quadrants to form a stack of four four-variable maps, as in Figure 3.24.

Each square in the six-variable map is numbered, using the stack technique applied to the five-variable map in Figure 3.22. Note the darkened squares in each of the "stacked" four-variable maps. They represent squares 16, 20, 48, and 52 in the six-variable map. If each of these squares were loaded with a 1, they would be logically adjacent. Any square in one of the four-variable maps has a counterpart in the other four-variable maps. For example, squares 0, 4, 32, and 36 are aligned in a vertical path, one from the top left corner of each quadrant.

3.4.3 Simplification Using Five-Variable Karnaugh Maps

Five-variable maps are formed using two connected four-variable maps. A five-variable K-map contains 32 squares, one for each possible minterm or maxterm. The same rules and principles that applied to three- and four-variable maps apply here as well. The added "trick" is to see the logical adjacencies with more squares in the map.

For example, simplify the following five-variable expression using a Karnaugh map as in Figure 3.25:

$$T = f(a, b, c, d, e) = \sum(0, 2, 8, 10, 16, 18, 24, 26)$$

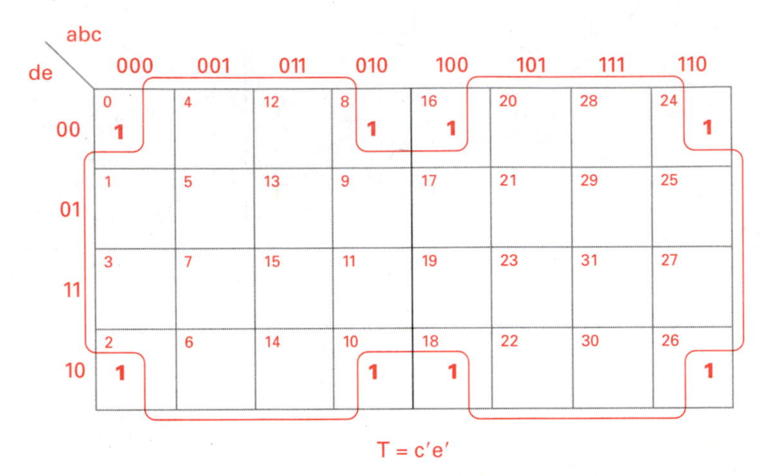

Figure 3.25
Five-variable K-map for $T = f(a, b, c, d, e) = \Sigma\ (0, 2, 8, 10, 16, 18, 24, 26)$

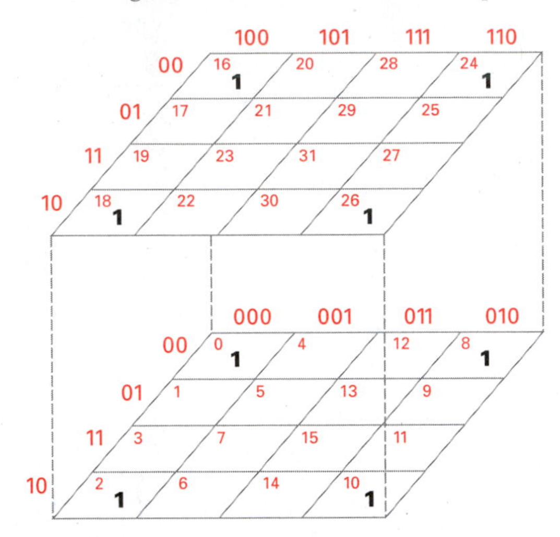

Figure 3.26
Vertical alignment of two four-variable K-maps

We have already noted that the corners in a four-variable map are logically adjacent. If we mentally cut the five-variable map in the center and slide the right half over on top of the left we see a picture of two four-variable maps aligned vertically. The square of one four-variable map is logically adjacent to the square in the same relative position on the other four-variable map. Figure 3.26 illustrates this idea. Minterm {0} is vertically aligned with minterm {16}; similarly minterms {2, 18}, {8, 24}, and {10, 26} are vertically aligned. The vertical alignment produces logical adjacency.

Minterms located in all of the corners of each four-variable map produce an EPI that simplifies to $T = c'e'$. The centers of the two four-variable maps are also logically adjacent by the same reasoning.

Simplify the Boolean function (Figure 3.27):

$$R = f(v,\ w,\ x,\ y,\ z) = \sum(5,\ 7,\ 13,\ 15,\ 21,\ 23,\ 29,\ 31)$$

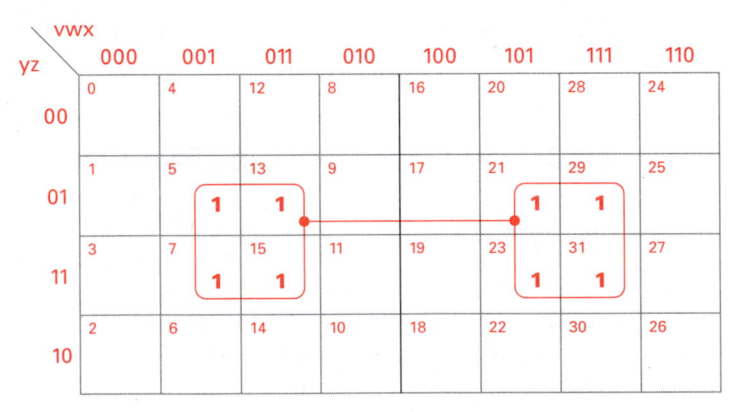

Figure 3.27
Five-variable K-map for $R = f(v, w, x, y, z) = \Sigma\ (5, 7, 13, 15, 21, 23, 29, 31)$

Figure 3.28

Karnaugh map for the equation $W = f(a, b, c, d, e) = \Sigma\,(1, 3, 4, 6, 9, 11, 12, 14, 17, 19, 20, 22, 25, 27, 28, 30)$

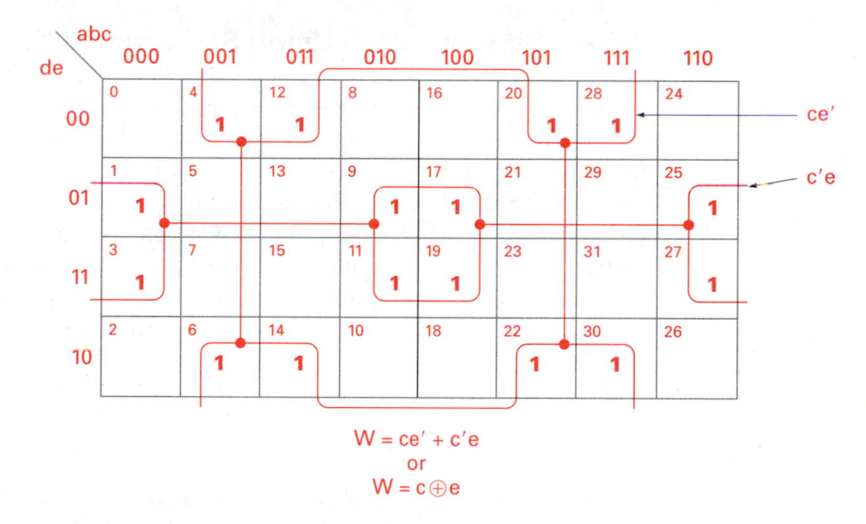

Each half of the five-variable K-map has a group of four minterms that are vertically aligned (when stacked) with each other. This forms an EPI of {5, 7, 13, 15, 21, 23, 29, 31} that simplifies to xz.

Simplify the Boolean function (Figure 3.28):

$$W = f(a, b, c, d, e)$$
$$= \sum(1, 3, 4, 6, 9, 11, 12, 14, 17, 19, 20, 22, 25, 27, 28, 30)$$

Two EPIs occur: {1, 3, 9, 11, 17, 19, 25, 27}, which reduces to $c'e$, and {4, 6, 12, 14, 20, 22, 28, 30}, which simplifies to ce'.

Find the prime implicants and essential prime implicants for (Figure 3.29):

$$J = f(v, w, x, y, z) = \sum(4, 5, 6, 7, 9, 11, 13, 15, 25, 27, 29, 31)$$

The two EPIs are {4, 5, 6, 7}, which simplifies to $v'w'x$, and {9, 11, 13, 15, 25, 27, 29, 31}, which reduces to wz. The non-essential prime implicant, {5, 7, 13, 15}, contains minterms already covered in the two EPI groups.

The final equation reduces to $J = v'w'x + wz$.

Figure 3.29

Five-variable K-map for $J = f(v, w, x, y, z) = \Sigma\,(4, 5, 6, 7, 9, 11, 13, 15, 25, 27, 29, 31)$

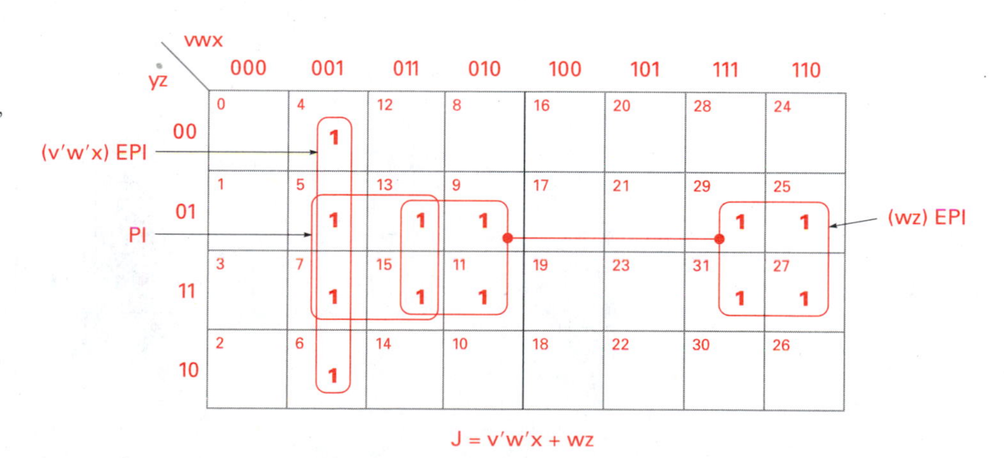

3.4.4 Simplification Using Six-Variable Karnaugh Maps

Six-variable Karnaugh maps are two connected five-variable maps. Or one can think of the six-variable map as four vertically stacked four-variable maps as indicated in Figure 3.24. For example, the minterms {9, 13, 41, 45}, from Figure 3.23, are logically adjacent. Squares found in the same relative position in each of the four variable maps are logically adjacent. Loading the six-variable map is an extension of the process previously discussed; simply find the K-map square that corresponds to a minterm number and put a 1 in it. Finding the essential prime implicants is a bit tricky at first; however, with a little practice it becomes a straightforward procedure.

Simplify

$$K = f(a, b, c, d, e, f)$$
$$= \sum(9, 11, 13, 15, 25, 27, 29, 31, 41, 43, 45, 47, 57, 59, 61, 63)$$

This particular group of minterms produces a subset of four minterms in the same relative position for each of the four-variable "sub" maps. As a result, each "sub" group is logically adjacent to the others producing a single EPI of 16 minterms.

The six-variable problem shown in Figure 3.30 gives $K = cf$.

Simplify

$$I = f(a, b, c, d, e, f)$$
$$= \sum(0, 2, 4, 6, 8, 10, 12, 14, 16, 18, 20, 22, 24, 32, 40, 48, 56)$$

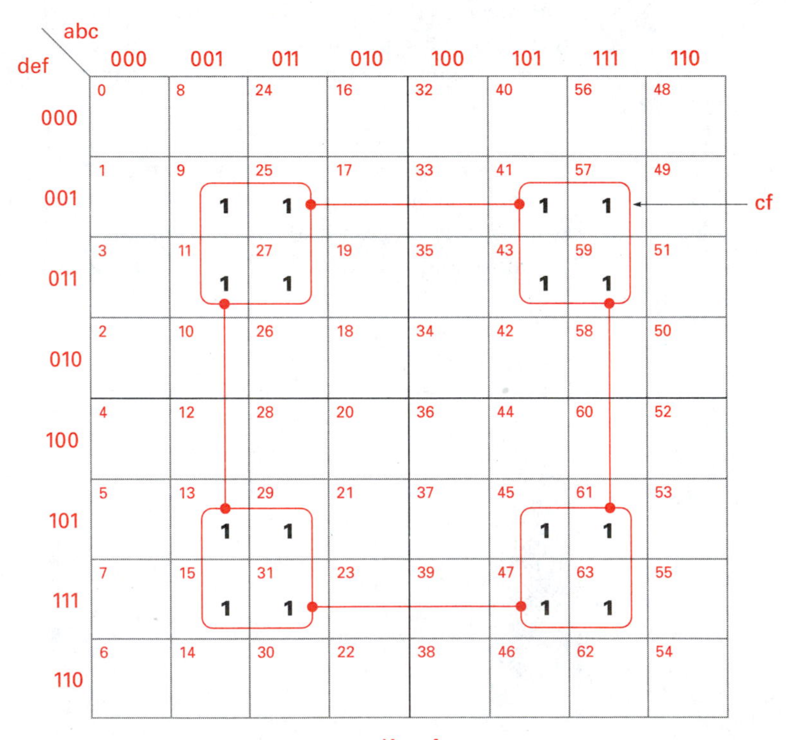

Figure 3.30

Six-variable K-map for equation $K = f(a, b, c, d, e, f) = \Sigma$ (9, 11, 13, 15, 25, 27, 29, 31, 41, 43, 45, 47, 57, 59, 61, 63)

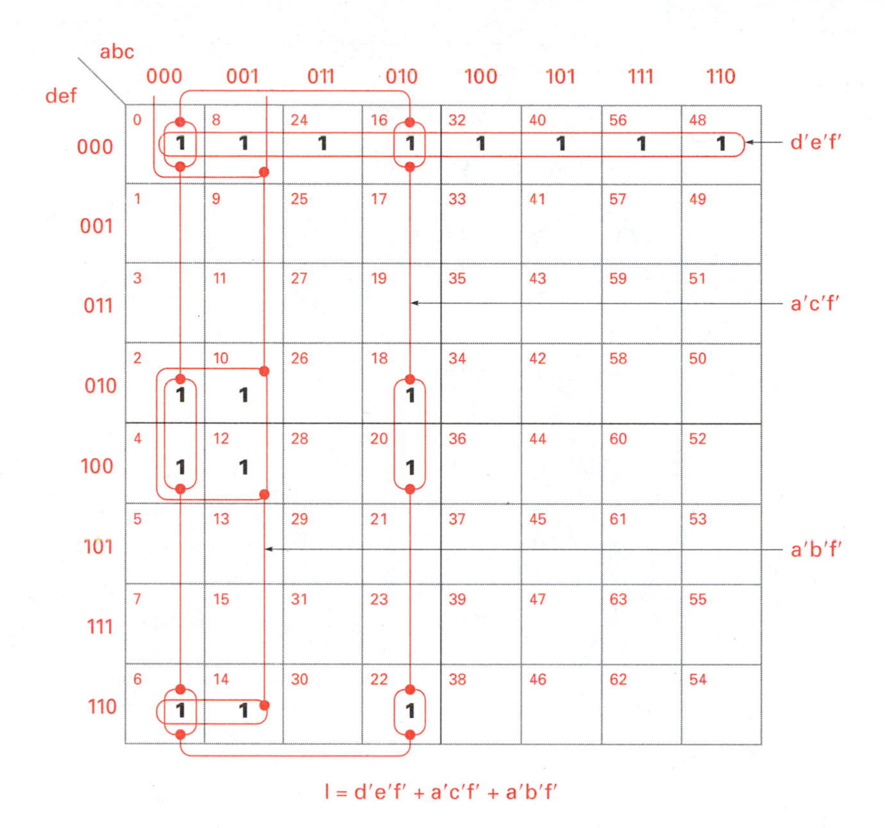

Figure 3.31
Six-variable K-map for the equation $I = f(a, b, c, d, e, f) = \Sigma\,(0, 2, 4, 6, 8, 10, 12, 14, 16, 18, 20, 22, 24, 32, 40, 48, 56)$

Determination of EPIs is a bit more difficult in this problem (see Figure 3.31). Look for the largest groups first. The top row forms a EPI {0, 8, 16, 24, 32, 40, 48, 56} that simplifies to $d'e'f'$. Another EPI consists of minterms {0, 2, 4, 6, 8, 10, 12, 14) located in the left half of the map, which reduces to $a'b'f'$. The last EPI consists of minterms {0, 2, 4, 6, 16, 18, 20, 22}, which simplifies to $a'c'f'$.

3.4.5 Incompletely Specified Functions (Don't Care Terms)

When an output value is known for every possible combination of input variables, the function is said to be *completely specified*. However, when an output value is not known for every combination of input variables, usually because all combinations cannot occur, the function is said to be *incompletely specified*. This means that the truth table does not generate an output value for every possible combination of input variables. The minterms or maxterms that are not used as part of the output function are called **don't care terms**. For example, the truth table for conversion of binary to EX-3 BCD (see Table 1.8 for the EX-3 BCD code) is shown in Table 3.8.

Binary coded decimal (BCD) codes, including EX-3, can represent only a single digit decimal character (0 through 9). The decimal number 9 takes four bits in EX-3 BCD code. Input combinations 1010, 1011, 1100, 1101, 1110, and 1111 are not used to specify any output variable value. Outputs A, B, C, and D are incompletely specified for those input code values that are "don't care" terms.

TABLE 3.8
Truth table for binary to EX-3 BCD code conversion

Binary				EX-3 BCD			
W	X	Y	Z	A	B	C	D
0	0	0	0	0	0	1	1
0	0	0	1	0	1	0	0
0	0	1	0	0	1	0	1
0	0	1	1	0	1	1	0
0	1	0	0	0	1	1	1
0	1	0	1	1	0	0	0
0	1	1	0	1	0	0	1
0	1	1	1	1	0	1	0
1	0	0	0	1	0	1	1
1	0	0	1	1	1	0	0
1	0	1	0	Don't care			
1	0	1	1	Don't care			
1	1	0	0	Don't care			
1	1	0	1	Don't care			
1	1	1	0	Don't care			
1	1	1	1	Don't care			

Writing the equations for output variables A, B, C, and D, including the don't care terms, we get

$$A = f(W, X, Y, Z) = \sum(5, 6, 7, 8, 9) + \sum d(10, 11, 12, 13, 14, 15)$$
$$B = f(W, X, Y, Z) = \sum(1, 2, 3, 4, 9) + \sum d(10, 11, 12, 13, 14, 15)$$
$$C = f(W, X, Y, Z) = \sum(0, 3, 4, 7, 8) + \sum d(10, 11, 12, 13, 14, 15)$$
$$D = f(W, X, Y, Z) = \sum(0, 2, 4, 6, 8) + \sum d(10, 11, 12, 13, 14, 15)$$

The don't care terms are the same for each output variable in the problem, because the same set of input variable combinations are used. Don't care terms are distinguished from regular minterms in that it does not matter whether we assign them a value of 0 or 1, because these combinations of input variables never occur. This becomes a distinct advantage when simplifying the output equations using Karnaugh maps. If a don't care term can be used to create a larger group of minterms then we can assign it (the don't care term) a 1. If it does not help in creating a larger group then we assign it a 0 value. This means that we can use don't care terms to our advantage in creating the largest group possible, but it is not necessary that we cover them when selecting a minimal set of prime implicants.

To illustrate this principle, consider simplification of the four output variables for the binary to EX-3 converter. From outputs A, B, C, and D we load the K-maps in Figure 3.32, using the don't care terms to create larger prime implicants.

Don't care terms are indicated by a d in each of the K-maps. The resulting K-map for A, with EPI {8, 9, 10, 11, 12, 13, 14, 15}, consisting of both minterms and don't care terms, simplified to W. Variable A K-map EPIs {5, 7, 13, 15} and {7, 6, 14, 15} include two don't care terms each; they reduce to XZ and XY, respectively. Function $A = W +$

Figure 3.32
Binary to EX-3 BCD K-maps

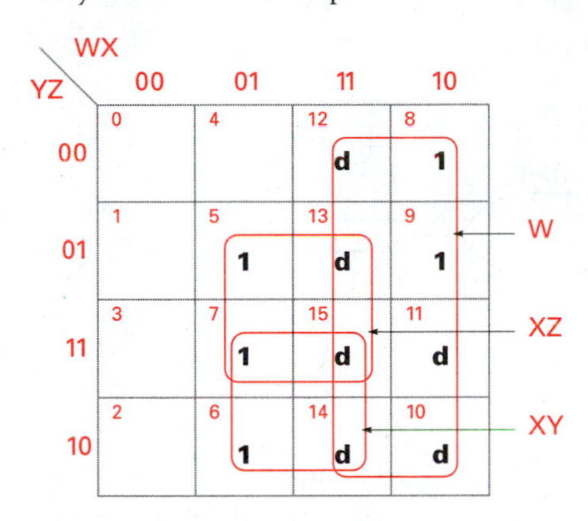

(a) K-map for output A

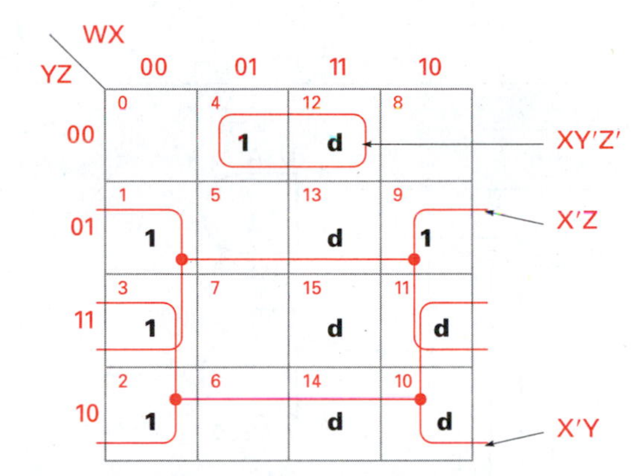

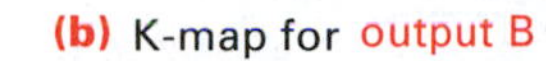

(b) K-map for output B

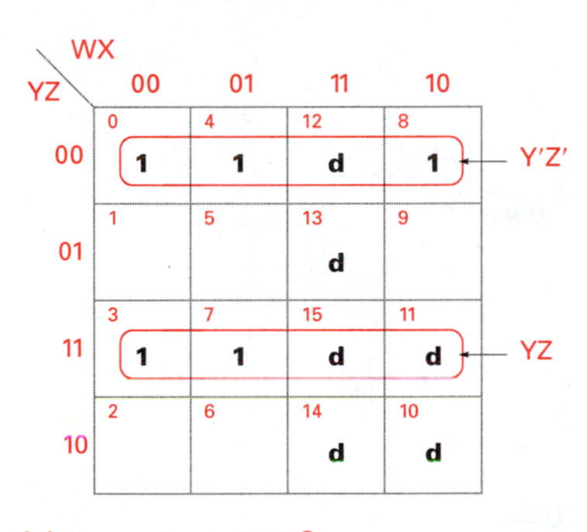

(c) K-map for output C

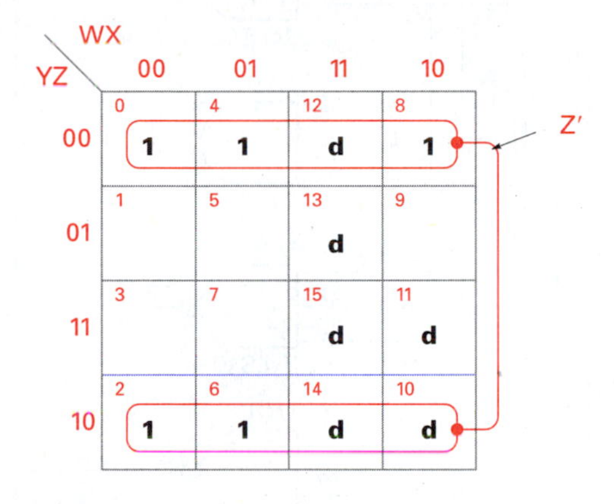

(d) K-map for output D

$XZ + XY$ using don't care terms, and $A = WX'Y' + W'XZ + W'XY$ without using don't care terms.

In the map for output variable B, don't care terms {10, 11, 12} were used. Don't care terms {13, 14, 15} were not used because they did not combine with minterms to create a larger grouped EPI. Notice that don't care terms are never grouped by themselves.

Output C map used don't care term {12} with EPI {0, 2, 8, 12}, which reduced to $Y'Z'$, and used don't care terms {11, 15} with EPI {3, 7, 11, 15} producing the result YZ. Output variable D used don't care terms {10, 12, 14} to produce EPIs {0, 2, 4, 6, 8, 10, 12, 14}, which reduced to Z'.

The final set of output equations are (Figure 3.33):

$$A = W + XZ + XY$$
$$B = X'Y + X'Z + XY'Z'$$
$$C = Y'Z' + YZ$$
$$D = Z'$$

Figure 3.33
Binary to EX-3 code converter

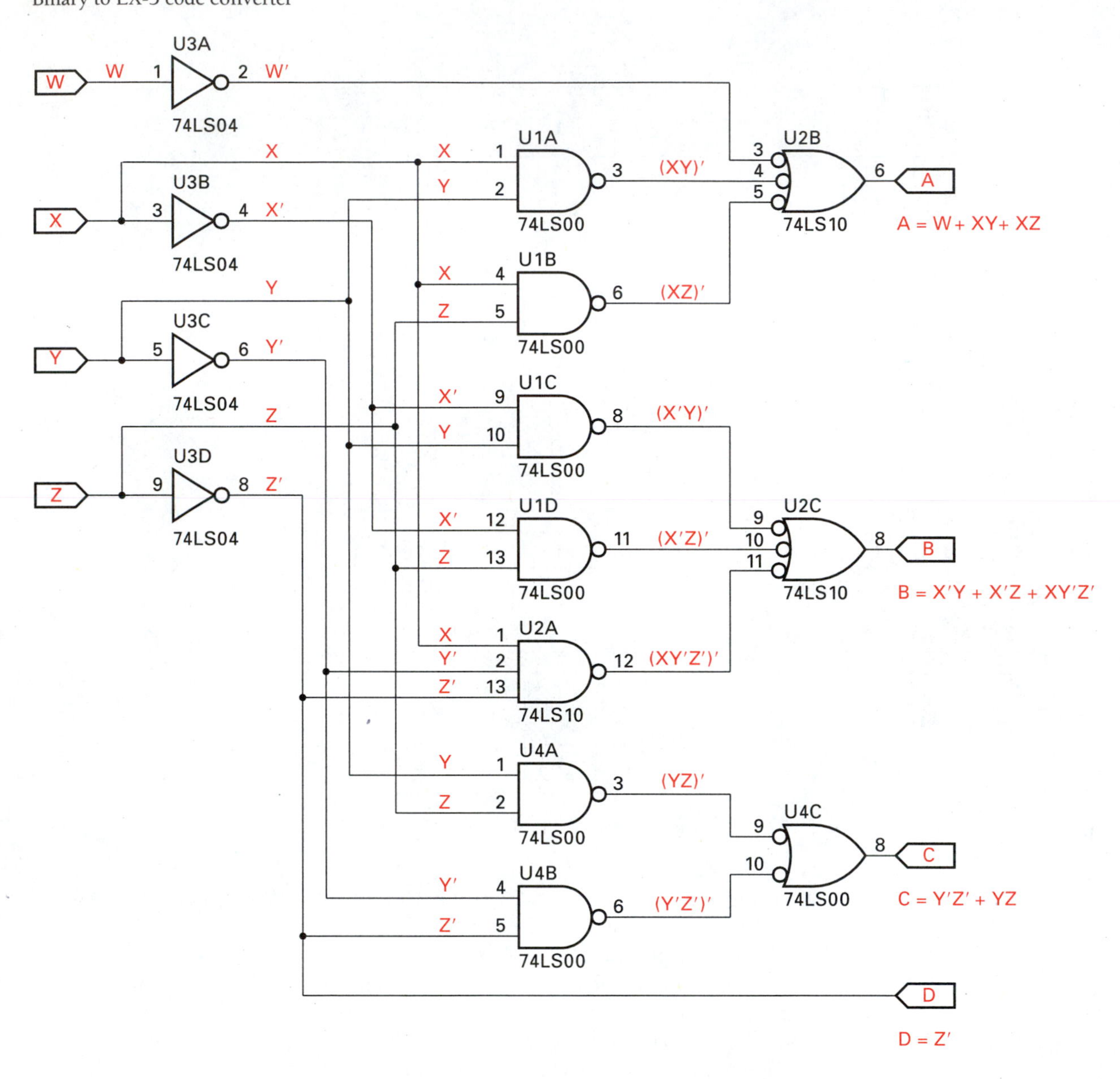

❖ **EXAMPLE 3.15** Map the same binary to EX-3 code converter output functions without using the don't care terms (assume that they are all 0s) and compare the results. Which produces fewer gates or inputs to gates or both?

The procedures for the determination and use of don't care terms in combinational logic are these:

1. Develop the truth table that describes the input/output relationship.
2. Determine if all of the input combinations are used to generate the output(s).
   a. If so, then no don't care terms exist.
   b. If not, then those combinations of input variables not used to determine output values are don't care terms.
3. Once the don't care terms have been identified, use a separate symbol, in the K-map squares, so they will not be confused with normal minterms or maxterms.
4. Create as large an EPI grouping as possible, including don't care terms that have been combined with normal minterms.
5. Do not group don't care terms by themselves. A don't care prime implicant is meaningless, because the don't care terms are not used to generate the output function.

3.4.6 Simplifying Maxterm Equations

Loading and grouping maxterms is exactly the same as loading and grouping minterms, except that 0s are loaded into the map and grouped to form PI and EPI groups. Reading the map is, however, different.

EXAMPLE 3.16 Consider the following maxterm simplifications (Figure 3.34):

a. $J = f(x, y, z) = \pi(0, 3, 4, 7)$

SOLUTION Two EPIs are formed. EPI {0, 4} reduces to $y + z$ and EPI {3, 7} reduces to $y' + z'$. Notice that the reduced terms are OR expressions. The simplification of maxterms produces ORed variables ANDed together with other ORed variables. The procedure for reading maxterms is as follows:

1. Variables that are contained in an essential prime implicant, which are not eliminated, are complemented. If input n is a 1 for all maxterms in an EPI, then it would be read as n'. If n were a 0 for all maxterms in an EPI, then it would be read as n.

   For instance, in EPI {0, 4} in Figure 3.34, variable x is a 1 in square 4 and a 0 in square 0; therefore, it is eliminated. Input variable y is a 0 in both squares 0 and 4; therefore, it is read as y. Input z is a 0 in both maxterms, so it is read as z. Input

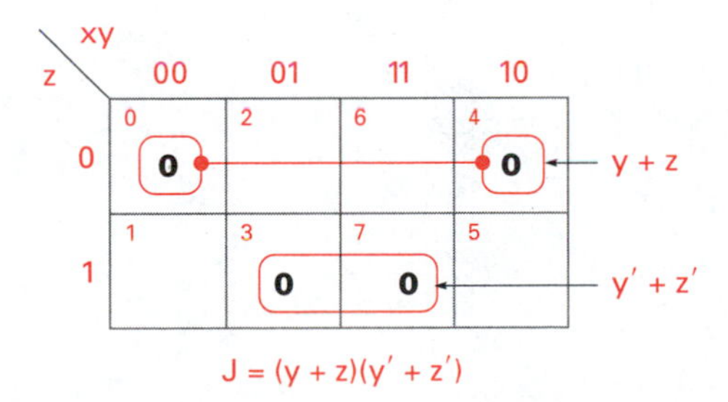

Figure 3.34
Three-variable K-map using maxterms

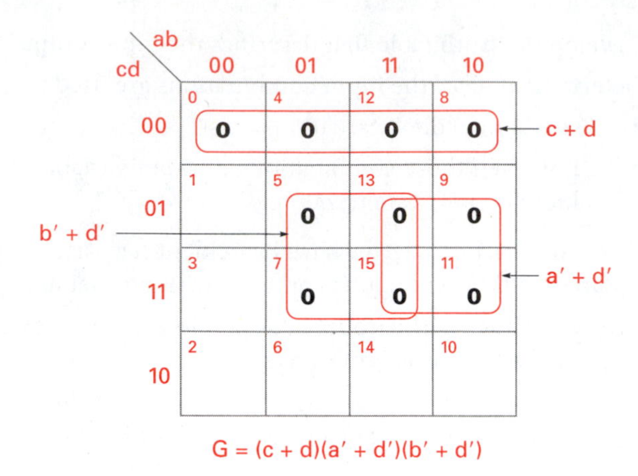

Figure 3.35
Four-variable K-map simplifying $G = f(a, b, c, d) = \pi(0, 4, 5, 7, 8, 9, 11, 12, 13, 15)$

variable x is eliminated from maxterm EPI {3, 7} in Figure 3.34. Input y is a 1 for both maxterms in the EPI and is read as y'. Likewise input z is a 1 for both maxterms and is read as z'.

2. Each remaining variable, n, is ORed with the other remaining variables, n_m, so that a term $n_0 + n_1 + \cdots + n_m$ is formed.
3. Each EPI contains a set of maxterms that when reduced form a constant (0 or 1), a variable (n), or a set of variables ($n_0 + n_1 + n_2 + \cdots + n_m$) ORed together.
4. Each reduced EPI (ORed term) is ANDed with any other OR terms

$$(x_0 + \cdots + x_n)\,(x_b + \cdots + x_m)$$

b. Simplify the function (Figure 3.35):

$$G = f(a, b, c, d) = \pi(0, 4, 5, 7, 8, 9, 11, 12, 13, 15)$$

Solution Three four-variable maxterm EPIs exist. The first {0, 4, 8, 12} simplifies to $c + d$; the second {5, 7, 13, 15} reduces to $b' + d'$; and the third {9, 11, 13, 15} produces $a' + d'$. The simplified result is $G = (c + d)(b' + d')(a' + d')$.

c. Simplify the function (Figure 3.36):

$$T = f(w, x, y, z) = \pi(1, 3, 8, 10, 12, 13, 14, 15)$$

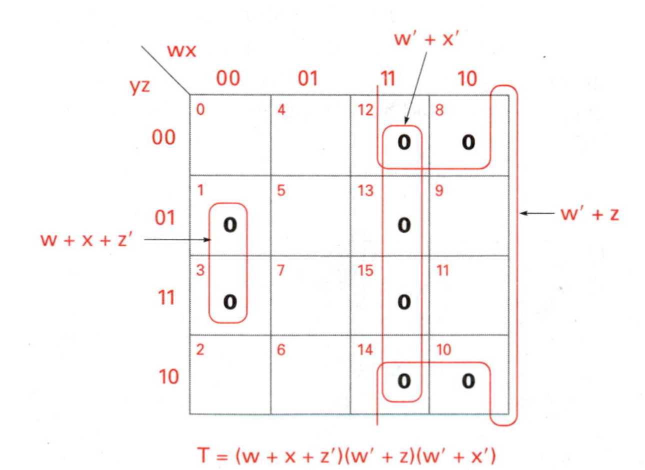

Figure 3.36
K-map for the equation $T = f(w, x, y, z) = \pi(1, 3, 8, 10, 12, 13, 14, 15)$

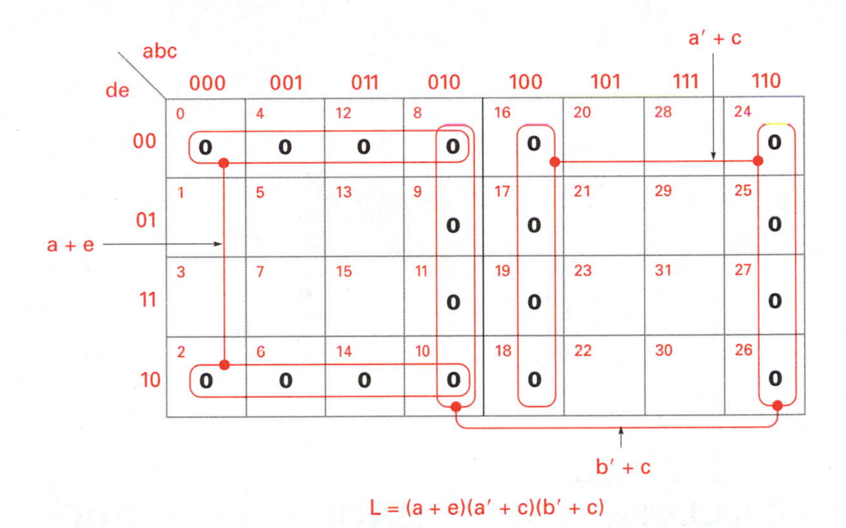

Figure 3.37
K-map for the equation $L = f(a, b, c, d, e) = \pi(0, 2, 4, 6, 8, 9, 10, 11, 12, 14, 16, 17, 18, 19, 24, 25, 26, 27)$

SOLUTION Load the maxterms by placing a 0 in the square with the same number as the maxterm. Group the maxterms PIs and EPIs in the same fashion as minterms were grouped. Identify the redundant input variables and eliminate them.

Three maxterm EPIs are found: {1, 3}, {8, 10, 12, 14}, and {12, 13, 14, 15}. Each EPI produces a sum (ORed) term:

$$\{1, 3\} \rightarrow (w + x + z'\}$$
$$\{8, 10, 12, 14\} \rightarrow (w' + z)$$
$$\{12, 13, 14, 15\} \rightarrow (w' + x')$$

d. Consider the following five-variable problem (Figure 3.37):

$$L = f(a, b, c, d, e) = \pi(0, 2, 4, 6, 8, 9, 10, 11, 12, 14, 16, 17, 18, 19, 24, 25, 26, 27)$$

Three maxterm EPIs exist: {0, 2, 4, 6, 8, 10, 12, 14}, which reduces to $(a + e)$; {8, 9, 10, 11, 24, 25, 26, 27}, which reduces to $(b' + c)$; and {16, 17, 18, 19, 24, 25, 26, 27}, which reduces to $(a' + c)$. The simplified equation is $L = (a + e)(a' + c)(b' + c)$. ❖

Another way to view simplification of maxterms is to consider them as minterms (i.e., as if they were 1s), write the simplified expression, which gives the complement of the desired function, and then apply DeMorgan's theorem to complement the result. Consider the following simplification problem (Figure 3.38):

$$A = f(w, x, y, z) = \pi(2, 3, 8, 9, 10, 11, 12, 13, 14, 15)$$

Load a four-variable map with the maxterms. Treat the maxterms as though they were minterms when generating the simplified results. The equation is $A' = w + x'y$.

Function A is complemented. The DeMorgan equivalent of the equation $A' = w + x'y$ is $A' = [(w')(x + y')]'$. Complementing both sides gives $A = (w')(x + y')$, which is the same result that would have been obtained if the EPIs in the map had been read as maxterms.

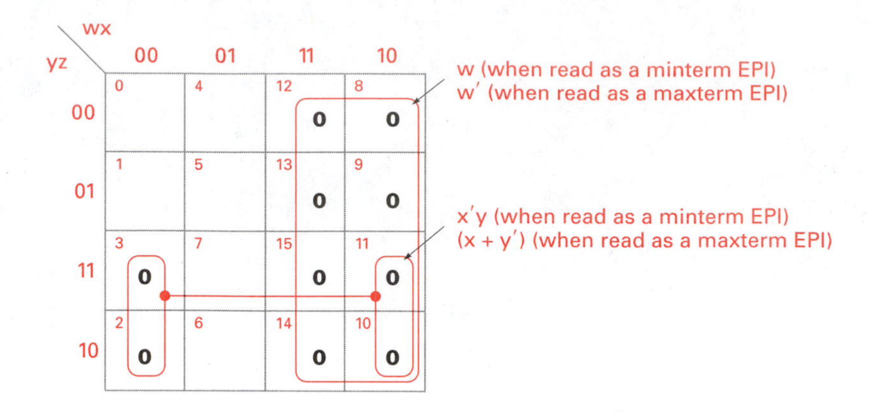

Figure 3.38
K-map for the equation $A = f(w, x, y, z) = \pi\,(2, 3, 8, 9, 10, 11, 12, 13, 14, 15)$

3.5 QUINE-McCLUSKEY MINIMIZATION TECHNIQUE

For many applications the number of variables in a problem is too large to simplify manually, using Karnaugh maps. Karnaugh maps, although capable of handling any number of variables, become unwieldy beyond six or seven variables. Contemporary design techniques for realizing combinational circuits include the use of programmable logic arrays, programmable logic devices, and other large-scale integrated circuits that can be configured by the end user. Simplification typically means that a logic designer can obtain more functional use from a given component. Therefore, some "automatic" or computer-driven simplification routine is desirable.

The **Quine-McCluskey minimization technique** is an algorithm that uses the same Boolean algebra postulates that were used with Karnaugh maps but in a form suitable for a computer solution. Large Karnaugh maps require recognition of groups of terms that may form essential prime implicants. The larger the map, the more difficult this pattern recognition becomes. The Quine-McCluskey approach eliminates the need for such pattern recognition.

The way to develop an understanding of the Quine-McCluskey approach is to follow an example.

❖ **EXAMPLE 3.17** Simplify the following using the Quine-McClusky minimization technique:

$$D = f(a, b, c, d) = \sum(0, 1, 2, 3, 6, 7, 8, 9, 14, 15)$$

SOLUTION

1. Arrange all of the minterms, in a list of increasing order, so that groups of terms contain the same number of 1s. Each minterm in the original expression, is contained in Table 3.9, arranged in increasing order according to the number of 1s contained in each term. Group 0 contains no 1s; group 1 contains only those minterms that have a single 1 {1, 2, 8}; group 2 contains minterms with two 1s {3, 6, 9}; group 3 contains minterms with three 1s {7, 14}; and group 4 contains minterms with four 1s {15}.

The relation $x + x' = 1$ can be applied to pairs of minterms where only one variable changes value. Candidates for this relationship can come from only adjacent groups of minterms.

TABLE 3.9
Grouping minterms according to the number of 1s

Group	Minterm	Variables a	b	c	d
0	0	0	0	0	0√
1	1	0	0	0	1√
	2	0	0	1	0√
	8	1	0	0	0√
2	3	0	0	1	1√
	6	0	1	1	0√
	9	1	0	0	1√
3	7	0	1	1	1√
	14	1	1	1	0√
4	15	1	1	1	1√

2. Create a new table showing the minterms in group n that matched with those from group $n + 1$ such that they differ in only one position. This is the equivalent to $x + x' = 1$; that is, $ab' + ab = a$.

Eliminated variable bit positions are indicated by the dash (—). This corresponds to $a'b'c'd' + a'b'c'd = a'b'c'(d + d') = a'b'c'-$.

Compare minterm {0}, in group 0 of Table 3.9, with each of the minterms in group 1. Minterm {0} (group 0) and minterm {1} (group 1) differ by only one variable, giving $a'b'c'$—. In a similar fashion, minterm {0}, group 0, can combine with minterms {2, 8} in group 1 producing $a'b'$—d' (minterms 0, 2), —$b'c'd'$ (minterms 0, 8). The combination of minterms 0, 1; 0, 2; and 0, 8, forms group 0 in Table 3.10.

When all of the minterms in group 0 have been compared with those in group 1, we compare the minterms in group 1 with those in group 2. This process is repeated

TABLE 3.10
Creation of minterm groups of two

Group	Minterms	Variables a	b	c	d
0	0, 1	0	0	0	— √
0	0, 2	0	0	—	0 √
0	0, 8	—	0	0	0 √
1	1, 3	0	0	—	1 √
1	1, 9	—	0	0	1 √
1	2, 3	0	0	1	— √
1	2, 6	0	—	1	0 √
1	8, 9	1	0	0	— √
2	3, 7	0	—	1	1 √
2	6, 7	0	1	1	— √
2	6, 14	—	1	1	0 √
3	7, 15	—	1	1	1 √
3	14, 15	1	1	1	— √

Note: A dash (—) indicates a bit position where a variable is 0 in one group and 1 in the other.

TABLE 3.11
Creation of minterm groups of four

Group	Minterm	Variable			
		a	b	c	d
0	0, 1, 2, 3	0	0	—	—
0	0, 1, 8, 9	—	0	0	—
1	2, 6, 3, 7	0	—	1	—
2	6, 7, 14, 15	—	1	1	—

until all of the minterms in each group have been compared to those in the next higher group.

When a minterm in a group is combined with a minterm in an adjacent group, a dash (—) is used to indicate an eliminated variable. The combined minterms are grouped together in Table 3.10. Each combination results in a new minterm group. As each minterm, from a group in Table 3.9, combines with a minterm in the next higher group it is checked (√), indicating that it is now part of a larger group. If a minterm did not combine with another, then no check would be made. If a term does not simplify, it is a prime implicant.

3. All of the adjacent minterm groups contained in Table 3.10 are compared to see if groups of four can be made. The criteria for forming groups of four are as follows: The dashes in the groups of two must be in the same bit position and only one variable change (0 in one group and 1 in the other) is allowed.

   A comparison is made of each minterm in group n with each minterm in group $n + 1$. Those that meet the criteria are combined in a larger group.

   Minterms {0, 1} in group 0 combine with minterms {2, 3} in group 1, Table 3.10, to form a group {0, 1, 2, 3}. Each minterm in group n is compared with each minterm group in $n + 1$. If the dashes (—) are in the same position and only one other variable changes, then a new, larger group is created and entered in Table 3.11. Those minterm groups that are used in the creation of larger groups are checked. Any unchecked groups are prime implicants.

4. Repeat the process outlined in step 3. In this case both dashes (—) must be in the same bit position with only one other variable allowed to change. The creation of a new table further groups the sets of minterms. This same process is repeated until no further combination of minterm groups is possible.
5. All nonchecked minterm groups are now considered to be prime implicants.
6. All of the prime implicants are formed into a prime-implicant table as shown in Table 3.12.

   The prime implicant table lists each of the minterms contained in the original switching equation across the top of the table. Each prime implicant is listed verti-

TABLE 3.12
Prime implicant table

PI terms	Decimal	Minterms									
		0	1	2	3	6	7	8	9	14	15
$a'b'$	0, 1, 2, 3	x	x	x	x						
$b'c'$	0, 1, 8, 9	x	x					ⓧ	ⓧ		
$a'c$	2, 6, 3, 7			x	x	x	x				
bc	6, 7, 14, 15					x	x			ⓧ	ⓧ

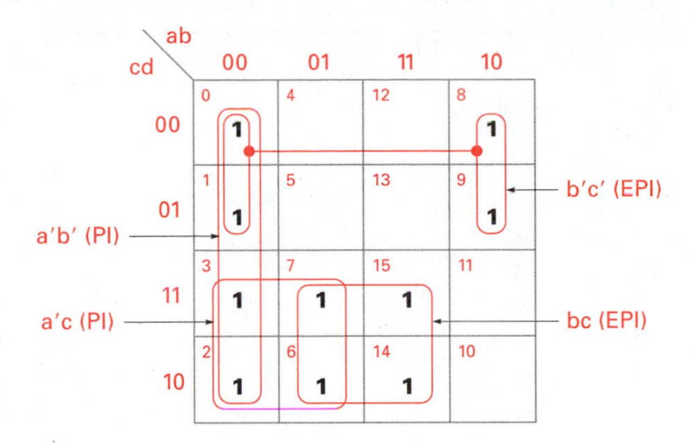

Figure 3.39
K-map for the equation $D = f(a, b, c, d) = \Sigma\,(0, 1, 2, 3, 6, 7, 8, 9, 14, 15)$

cally in two forms, PI terms and the decimal list of minterms that make up the PI. Note in Table 3.11 that PI {0, 1, 2, 3}, $ab = 00$, is read as $a'b'$. Each of the prime implicants consists of the largest possible group of minterms.

7. Evaluate the prime implicants by circling those minterms that are contained in only one prime implicant (only one x in a column).

Note that minterms {8, 9, 14, 15} meet this condition. Circled minterms represent essential prime implicants (EPI). Minterms, {0, 1, 8, 9} and {6, 7, 14, 15} are essential prime implicants. Minterms {2, 3} are contained in two prime implicants, {0, 1, 2, 3} and {2, 3, 6, 7}. We need one or the other of these prime implicants to cover minterms in the equation but not both. This implies that two equally simplified results are possible. We can take our pick:

$$D = b'c' + bc + a'b' \quad \text{or} \quad D = b'c' + bc + a'c$$

To compare the results found by application of Quine-McClusky (QM) to that of a Karnaugh map, consider Figure 3.39.

We find that four prime implicants exist: {0, 1, 2, 3}, {2, 3, 6, 7}, {0, 1, 8, 9}, and {6, 7, 14, 15}. Of these, two have unique terms: {0, 1, 8, 9} and {6, 7, 14, 15}. The other two do not contain unique terms, minterms {2, 3} are contained in both: therefore, one or the other, but not both is necessary in the final result. The QM simplification technique requires that we list minterms according to the number of 1s in a table. A second table is created that lists all implicants covering 2 minterms. A third table is created listing all implicants covering 4 minterms, and so on until no further grouping is possible. All implicant groups not contained in a larger group become prime implicants. The PI table is created that lists all prime implicants, and the essential prime implicants are identified. ❖

3.5.1 Quine-McClusky Using Don't Care Terms

The same rules that applied to using don't care terms with the Karnaugh map are appropriate for Quine-McClusky. Consider the following problem:

$$S = f(w, x, y, z) = \sum(1, 3, 13, 15) + \sum d(8, 9, 10, 11)$$

1. Construct a list of minterms and don't care terms classified according to the number of 1s. Indicate the don't care terms by using a * symbol. Don't care terms are never included as prime implicants by themselves.

TABLE 3.13
Quine-McClusky for
$S = f(w, x, y, z)$
$= \Sigma(1, 3, 13, 15)$
$+ \Sigma d(8, 9, 10, 11)$

Group	Minterms	Variables			
		w	x	y	z
Minterm/*Don't care term list					
1	1	0	0	0	1 √
1	8*	1	0	0	0 √
2	3	0	0	1	1 √
2	9*	1	0	0	1 √
2	10*	1	0	1	0 √
3	11*	1	0	1	1 √
3	13	1	1	0	1 √
4	15	1	1	1	1 √
Minterm/*Don't care term sets					
1	1, 3	0	0	—	1 √
1	1, 9*	—	0	0	1 √
1	8*, 9*	1	0	0	— √
1	8*, 10*	1	0	—	0 √
2	3, 11*	—	0	1	1 √
2	9*, 11*	1	0	—	1 √
2	9*, 13	1	—	0	1 √
2	10*, 11*	1	0	1	— √
3	11*, 15	1	—	1	1 √
3	13, 15	1	1	—	1 √

2. Compare terms in group n, including don't care terms, with terms in group $n + 1$, looking for a single variable change. Treat don't care terms as a 1 in finding prime implicants.
3. Repeat step 2, creating an additional table indicating groups of four minterm/don't care term groups. Repeat step 3 until no further grouping can occur.
4. Construct a prime implicant chart and determine essential prime implicants. Treat any don't care terms not part of a group containing a minterm as 0s. They do not need to be covered.

Each group is checked (√) in Table 3.13. Table 3.14 represents groups of four minterms/don't care terms. Each is a candidate for becoming an EPI. However, set {8*, 9*, 10*, 11*} contains only don't care terms and is, therefore, not a PI.

TABLE 3.14
Sets of four minterms/don't care terms

Group	Minterms	Variables			
		w	x	y	z
1	1, 3, 9*, 11*	—	0	—	1
1	8*, 9*, 10*, 11*	1	0	—	—
2	9*, 13, 11*, 15	1	—	—	1

*Don't care terms.

TABLE 3.15
Prime implicant table

PI terms	Decimal	Minterms 1	3	13	15
$x'z$	1, 3, 9*, 11*	⊗	⊗		
wz	9*, 11*, 13, 15			⊗	⊗

* Don't care terms.

Table 3.15 indicates that both prime implicants {1, 3, 9*, 11*} and {9*, 11*, 13, 15} contain unique terms and are essential prime implicants. The resulting simplified equations is

$$S = x'z + wz$$

In this example, all of the PIs were also EPIs. Many switching equations are not so neatly simplified; that is, the essential prime implicants alone do not necessarily satisfy the original function. The first example illustrated this principle. In Table 3.12 we noted that two PIs existed but only one was needed to OR with the two EPI terms. Often the interpretation of the prime implicant table is not so straightforward. In such cases reduced prime implicant tables are constructed.

3.5.2 Reduced Prime Implicant Tables

The following five-variable example illustrates the need for a modified prime implicant table.

EXAMPLE 3.18 Reduce

$$Q = f(a, b, c, d, e) = \sum(1, 3, 4, 5, 6, 7, 10, 11, 12, 13, 14, 15, 18, 19, 20, 21, 22, 23, 26, 27)$$

SOLUTION Arranging minterms into increasing order groups according to the number of 1s in each term results in Table 3.16.

Adjacent groups are compared to determine if they can be combined, producing Table 3.17. The combined minterms, listed in Table 3.17, are compared to form larger groups in Table 3.18. Each minterm group from Table 3.17 is covered in Table 3.18; all are checked (√). Adjacent groups, in Table 3.18, are compared, producing Table 3.19. Any groups from Tables 3.18 and 3.19 that cannot be combined any further are prime implicants.

In Tables 3.18 and 3.19 the prime implicants are

{1, 3, 5, 7}, $a'b'e$; {3, 7, 11, 15}, $a'de$; {3, 7, 19, 23}, $b'de$;
{3, 11, 19, 27}, $c'de$; {10, 11, 26, 27}, $bc'd$;
{10, 11, 14, 15}, $a'bd$; {18, 19, 22, 23}, $ab'd$;
{18, 19, 26, 27}, $ac'd$; {4, 5, 6, 7, 20, 21, 22, 23}, $b'c$;
{4, 5, 6, 7, 12, 13, 14, 15}, $a'c$

TABLE 3.16
Minterms grouped according to number of 1 bits in the term

Group	Minterms	Variables *a*	*b*	*c*	*d*	*e*
1	1	0	0	0	0	1√
1	4	0	0	1	0	0√
2	3	0	0	0	1	1√
2	5	0	0	1	0	1√
2	6	0	0	1	1	0√
2	10	0	1	0	1	0√
2	12	0	1	1	0	0√
2	18	1	0	0	1	0√
2	20	1	0	1	0	0√
3	7	0	0	1	1	1√
3	11	0	1	0	1	1√
3	13	0	1	1	0	1√
3	14	0	1	1	1	0√
3	19	1	0	0	1	1√
3	21	1	0	1	0	1√
3	22	1	0	1	1	0√
3	26	1	1	0	1	0√
4	15	0	1	1	1	1√
4	23	1	0	1	1	1√
4	27	1	1	0	1	1√

Having identified all of the prime implicants, the next task is to create a prime implicant table and as many reduced prime implicant tables as needed to determine EPIs and the final equation. Table 3.20 is the first prime implicant table.

Each original minterm is listed across the top of Table 3.20. Prime implicants are listed vertically. Prime implicants containing unique terms are indicated by circles, which are drawn when only one *x* exists in a column.

A reduced PI table is constructed by removing all of the EPIs. A set of minterms with the EPI minterms removed are listed across the top of the reduced table (Table 3.21); {10, 11, 18, 19, 26, 27} remain. The remaining prime implicants are listed vertically: {3, 7, 11, 15}, *a′de*; {3, 7, 19, 23}, *b′de*; {3, 11, 19, 27}, *c′de*; {10, 11, 26, 27}, *bc′d*; {10, 11, 14, 15}, *a′bd*; {18, 19, 22, 23}, *ab′d*; {18, 19, 26, 27}, *ac′d*.

The reduced PI is analyzed for dominant prime implicants. A prime implicant, P_1, is dominant over another prime implicant, P_2, when

1. P_1 contains all of the minterms found in P_2.
2. The number of literals in P_1 is less than or equal to the number of literals in P_2. A literal is a variable or the variable complement; both *x* and *x′* are literals.

A dominant row (PI) replaces a dominated row. For example, *a′de* contains only minterm {11} in the reduced table. It is dominated by several other PIs and can be eliminated. The minterms are actually being absorbed into the EPIs or the dominant PIs. Refer to the absorption property in Chapter 2. Analysis of dominant prime implicants

TABLE 3.17
Combined minterms

Group	Minterms	Variables				
		a	*b*	*c*	*d*	*e*
1	1, 3	0	0	0	—	1 √
1	1, 5	0	0	—	0	1 √
1	4, 5	0	0	1	0	— √
1	4, 6	0	0	1	—	0 √
1	4, 12	0	—	1	0	0 √
1	4, 20	—	0	1	0	0 √
2	3, 7	0	0	—	1	1 √
2	3, 11	0	—	0	1	1 √
2	3, 19	—	0	0	1	1 √
2	5, 7	0	0	1	—	1 √
2	5, 13	0	—	1	0	1 √
2	5, 21	—	0	1	0	1 √
2	6, 7	0	0	1	1	— √
2	6, 14	0	—	1	1	0 √
2	6, 22	—	0	1	1	0 √
2	10, 11	0	1	0	1	— √
2	10, 14	0	1	—	1	0 √
2	10, 26	—	1	0	1	0 √
2	12, 13	0	1	1	0	— √
2	12, 14	0	1	1	—	0 √
2	18, 19	1	0	0	1	— √
2	18, 22	1	0	—	1	0 √
2	18, 26	1	—	0	1	0 √
2	20, 21	1	0	1	0	— √
2	20, 22	1	0	1	—	0 √
3	7, 15	0	—	1	1	1 √
3	7, 23	—	0	1	1	1 √
3	11, 15	0	1	—	1	1 √
3	11, 27	—	1	0	1	1 √
3	13, 15	0	1	1	—	1 √
3	14, 15	0	1	1	1	— √
3	19, 23	1	0	—	1	1 √
3	19, 27	1	—	0	1	1 √
3	21, 23	1	0	1	—	1 √
3	22, 23	1	0	1	1	— √
3	26, 27	1	1	0	1	— √

reduces Table 3.21 to Table 3.22. We find that PI $a'de$ is dominated by PIs $c'de$, $bc'd$, and $a'bd$, because all contain minterm {11} and all have the same number of literals. PI $a'bd$ is dominated by PI $bc'd$ for the same reason. Likewise PI $ab'd$ is dominated by PI $ac'd$ and $ac'd$ dominates $b'de$. Elimination of the dominated PI rows produces Table 3.22.

TABLE 3.18
Groups of four minterms

Group	Minterms	Variables					
		a	*b*	*c*	*d*	*e*	
1	1, 3, 5, 7	0	0	—	—	1	EPI
1	4, 5, 6, 7	0	0	1	—	—	√
1	4, 5, 12, 13	0	—	1	0	—	√
1	4, 5, 20, 21	—	0	1	0	—	√
1	4, 6, 12, 14	0	—	1	—	0	√
1	4, 6, 20, 22	—	0	1	—	0	√
2	3, 7, 11, 15	0	—	—	1	1	PI
2	3, 7, 19, 23	—	0	—	1	1	PI
2	3, 11, 19, 27	—	—	0	1	1	PI
2	5, 7, 13, 15	0	—	1	—	1	√
2	5, 7, 21, 23	—	0	1	—	1	√
2	6, 7, 14, 15	0	—	1	1	—	√
2	6, 7, 22, 23	—	0	1	1	—	√
2	10, 11, 14, 15	0	1	—	1	—	PI
2	10, 11, 26, 27	—	1	0	1	—	PI
2	12, 13, 14, 15	0	1	1	—	—	√
2	18, 19, 22, 23	1	0	—	1	—	PI
2	18, 19, 26, 27	1	—	0	1	—	PI
2	20, 21, 22, 23	1	0	1	—	—	√

TABLE 3.19
Groups of eight minterms

Group	Minterms	Variables					
		a	*b*	*c*	*d*	*e*	
1	4, 5, 6, 7, 12, 13, 14, 15	0	—	1	—	—	EPI
1	4, 5, 6, 7, 20, 21, 22, 23	—	0	1	—	—	EPI

TABLE 3.20
Prime implicant table

PI	Decimal	1	3	4	5	6	7	10	11	12	13	14	15	18	19	20	21	22	23	26	27
$a'b'e$	1, 3, 5, 7	⊗	*x*		*x*		*x*														
$a'de$	3, 7, 11, 15		*x*				*x*		*x*				*x*								
$b'de$	3, 7, 19, 23		*x*				*x*								*x*				*x*		
$c'de$	3, 11, 19, 27		*x*						*x*						*x*						*x*
$bc'd$	10, 11, 26, 27							*x*	*x*											*x*	*x*
$a'bd$	10, 11, 14, 15							*x*	*x*			*x*	*x*								
$ab'd$	18, 19, 22, 23													*x*	*x*			*x*	*x*		
$ac'd$	18, 19, 26, 27													*x*	*x*					*x*	*x*
$a'c$	4, 5, 6, 7, 12, 13, 14, 15			*x*	*x*	*x*	*x*			⊗	⊗	*x*	*x*								
$b'c$	4, 5, 6, 7, 20, 21, 22, 23			*x*	*x*	*x*	*x*									⊗	⊗	*x*	*x*		

TABLE 3.21
Reduced prime implicant table

PI	Decimal	10	11	18	19	26	27
** $a'de$	3, 7, 11, 15		x				
$c'de$	3, 11, 19, 27		x		x		x
$bc'd$	10, 11, 26, 27	x	x			x	x
** $a'bd$	10, 11, 14, 15	x	x				
** $ab'd$	18, 19, 22, 23			x	x		
$ac'd$	18, 19, 26, 27			x	x	x	x
** $b'de$	3, 7, 19, 23				x		

** Note: Dominated rows and columns are eliminated.

TABLE 3.22
Second reduced prime implicant table

PI	Decimal	10	11	18	19	26	27
$c'de$	3, 11, 19, 27		x		x		x
$bc'd$	10, 11, 26, 27	⊗	x			x	x
$ac'd$	18, 19, 26, 27			⊗	x	x	x

Note that minterm {10} is present only in PI $bc'd$ and therefore must be part of the final simplified equation. PI $ac'd$ is the only one containing minterm {18}, and so it also becomes part of the final equation. All of the minterms contained in PI $c'de$ are present in the two PI terms $bc'd$ and $ac'd$, therefore $c'de$ is absorbed. The final simplified equation is a collection of all the remaining PI terms:

$$Q = a'c + b'c + a'b'e + bc'd + ac'd$$

The reduced prime implicant tables provides a systematic approach for finding the PI terms necessary to fully cover the original equation, by determination of which terms can be absorbed by others. ❖

3.6 MAP-ENTERED VARIABLES

Karnaugh mapping is the best manual technique for Boolean equation simplification, yet when the map sizes exceed five or six variables it becomes unwieldy. The technique called **map entered variables** (MEVs) increases the effective size of a Karnaugh map, allowing a smaller map to handle a greater number of variables. Normally the map dimensions and the number of problem variables are related by $2^n = m$, where n is the number of problem variables and m is the number of squares in a K-map. Using map entered variables permits the map dimensions to be compressed.

In the normal Karnaugh map each square represents a minterm, a maxterm, or a don't care term. MEV K-maps permit a cell to contain a single variable (x) or a complete switching expression, ($x + y'z$) for instance, in addition to the 1s, 0s, and don't care terms. Consider the following two-variable K-map simplification of a three-variable problem.

Let

$$P = f(a, b, c) = \sum(0, 1, 4, 5, 7)$$

TABLE 3.23
Truth table when c is a MEV

Decimal minterm		Binary variables			Output
MEV	Standard	a	b	c (MEV)	P
0	0	0	0	0	1
0	1	0	0	1	1
1	2	0	1	0	0
1	3	0	1	1	0
2	4	1	0	0	1
2	5	1	0	1	1
3	6	1	1	0	0
3	7	1	1	1	1

Let variable c become the map entered variable. By constructing a special MEV truth table we can see how reducing the 2^n map creates MEV terms that can be loaded into a reduced size K-map.

Table 3.23 shows two minterm columns, one for a MEV K-map and one for the standard map. Note that two standard K-map squares are covered by one MEV K-map square. The reduction of the number of minterm and maxterm squares in the MEV K-map occurred because one variable, c, was designated a MEV. A four-variable standard K-map has 16 minterm and maxterm squares. If one of the four variables was considered a MEV, then the resulting MEV K-map would contain eight squares. If two variables were considered as MEVs, then the resulting MEV K-map would contain four squares. The reason for designating one or more variables a MEV is to reduce the K-map size. The numbering of the MEV minterm column is accomplished by decoding variables a and b, because c has been designated a MEV. Loading maps with map entered variables requires a decision as to which input variable(s) is to become a MEV. There is no standardized approach to this task. Typically the MEVs are chosen from the least significant variables or the variables that appear the least often in the unsimplified expression.

The procedure for loading a K-map using MEVs is as follows:

1. If the output variable is a 0 for both standard minterms covered by a MEV map square, then a 0 is written in that MEV map square.
2. If the output variable is a 1 for both standard minterms covered by a MEV map square, then a 1 is written in that MEV square.
3. If, for minterms covered by a MEV map square, the output variable has the same value as the MEV, then the MEV is written into MEV map square:

MEV	Output
0	0
1	1

4. If, for standard minterms covered by a MEV map square, the output and MEV variables are complements, write the MEV complement into the MEV map square:

MEV	Output
0	1
1	0

5. If, for standard minterms covered by a MEV map square, the output variable is a don't care term, write a don't care symbol (x) into the MEV map square.
6. If, for standard minterms covered by a MEV map square, the output variable is a don't care term in one case and a 0 in the other, write 0 in the appropriate square:

MEV	Output	MEV	Output
0	x	0	0
1	0	1	x

If the output variable is a don't care term in one case and a 1 in the other, write a 1 in the appropriate square:

MEV	Output	MEV	Output
0	x	0	1
1	1	1	x

Figure 3.40 compares a two-variable K-map with a map entered variable to a standard three-variable K-map.

Truth Table 3.23 shows output function P to be a 1 for standard minterms 0 and 1. Therefore a 1 is written into MEV map square 0. Output P is a 0 for standard minterm locations 2 and 3, so a 0 is written into MEV map square 1. P is a 1 for standard minterm locations 4 and 5, requiring a 1 to be written in MEV square 2. Finally, standard minterm locations 6 and 7 contain 0 and 1, respectively, requiring a c MEV to be written in MEV map square 3.

To read the simplified function from a MEV K-map, follow these steps:

1. Determine the EPIs consisting of only 1s along with any don't care terms that may exist (i.e., cover the 1s in the K-map).
2. Consider the 1s as don't care terms once step 1 is completed, because all of the 1s have been previously covered.
3. Group all identical MEV terms with 1s or don't care terms to maximize the MEV EPI size. Any minterms that are not contained in the MEV EPI are considered to be 0s (i.e., cover all the MEVs in the K-map).
4. Determine the MEV EPIs by reading the K-map in the normal fashion. Then AND the MEV variable or expression with the remaining map variables.

From Figure 3.40 we find the simplified expression

$$T = b' + a(c)$$

Figure 3.40
Two-variable MEV K-map and three-variable standard K-map

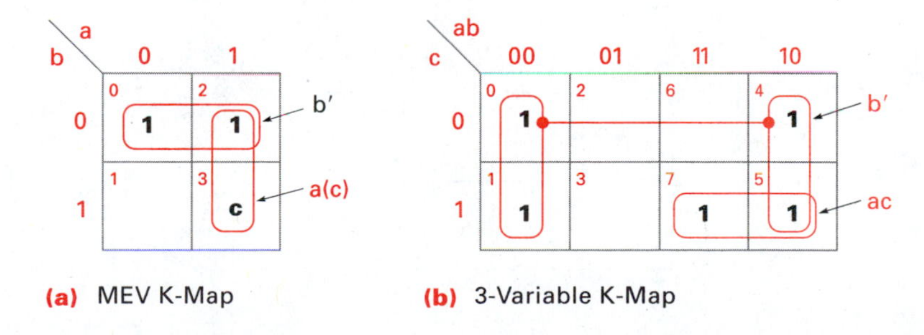

(a) MEV K-Map **(b)** 3-Variable K-Map

TABLE 3.24
MEV truth table for
$T = f(w, x, y, z)$
$= \Sigma(2, 4, 5, 10, 11, 14)$
$+ \Sigma' d(7, 8, 9, 12, 13, 15)$

Decimal minterm		Binary variables				Output
MEV	Standard	w	x	y	z (MEV)	T
0	0	0	0	0	0	0
0	1	0	0	0	1	0
1	2	0	0	1	0	1
1	3	0	0	1	1	0
2	4	0	1	0	0	1
2	5	0	1	0	1	1
3	6	0	1	1	0	0
3	7	0	1	1	1	d
4	8	1	0	0	0	d
4	9	1	0	0	1	d
5	10	1	0	1	0	1
5	11	1	0	1	1	1
6	12	1	1	0	0	d
6	13	1	1	0	1	d
7	14	1	1	1	0	1
7	15	1	1	1	1	d

Term b' is found by grouping squares {0, 2} and term $a(c)$ resulted from grouping squares {2, 3}. If squares {2, 3} contained minterms the result would simplify to a; however, square {3} contains MEV (c), which must be ANDed with variable (a), giving ac.

Consider the following Boolean equation:

$$T = f(w, x, y, z) = \sum(2, 4, 5, 10, 11, 14) + \sum d(7, 8, 9, 12, 13, 15)$$

Simplify T using a three-variable MEV K-map. Assign variable z to be the MEV.

First construct the MEV truth table (Table 3.24) to identify the MEV minterms and don't care terms.

Construct a three-variable MEV K-map as in Figure 3.41 and load terms and group according to the four steps given previously. Reading the map gives

$$\text{EPI } \{4, 5, 6, 7\} \rightarrow w; \qquad \text{EPI } \{2, 6\} \rightarrow xy'; \qquad \text{EPI } \{1, 5\} \rightarrow x'yz'$$

Notice that EPI {4, 5, 6, 7} and {2, 6} contained only 1s and don't care terms and EPI {1, 5} contained MEV z' and a 1.

Consider the function $M = f(w, x, y, z) = (2, 9, 10, 11, 13, 14, 15)$. The MEV

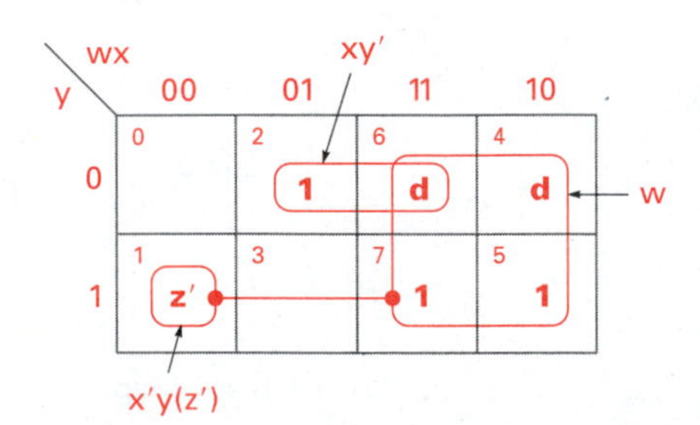

Figure 3.41
Three-variable MEV K-map resulting from Table 3.24

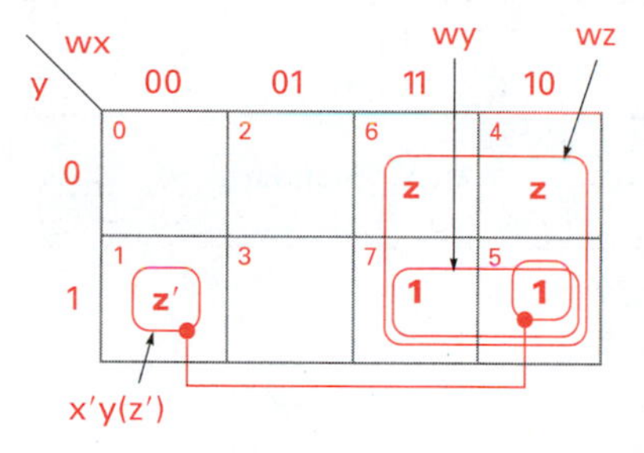

Figure 3.42

MEV K-map for $M = f(w, x, y, z) = \Sigma$ (2, 9, 10, 11, 13, 14, 15)

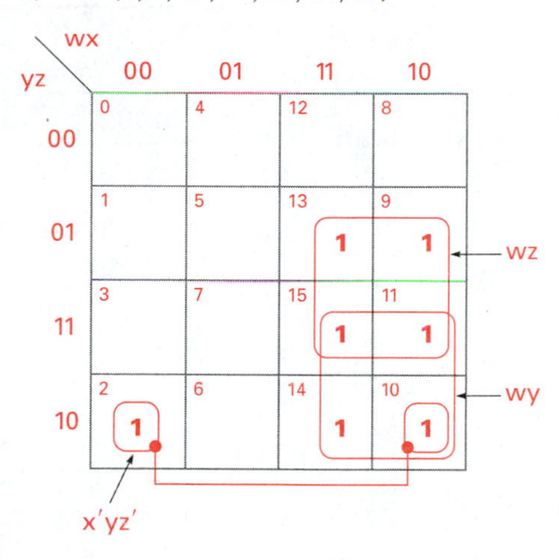

Figure 3.43

Four-variable MEV K-map for $M = f(w, x, y, z) = \Sigma$ (2, 9, 10, 11, 13, 14, 15)

K-map is shown in Figure 3.42. The 1s must be grouped to cover the function. Compare the MEV K-map in Figure 3.42 with the four-variable K-map in Figure 3.43.

However, always covering the 1s by themselves or in combination with don't care terms may lead to a nonoptimum solution. Consider the function

$$A = f(x, y, z) = \sum(0, 4, 5, 7)$$

The MEV K-map is shown in Figure 3.44. Compare the MEV K-map with the three-variable K-map shown in Figure 3.45.

The use of MEVs is not restricted to single variables; logic expressions can be included, as indicated in the next example. Developing MEV K-maps from standard combinational logic truth tables usually does not produce complex Boolean equations as MEV entries, but when we study synchronous sequential machines in later chapters we will encounter algorithm state machine design techniques that will produce complex MEV entries. Our purpose in this section is to develop a procedure for reading a MEV K-map where complex function entries are included.

Consider the MEV K-map shown in Figure 3.46. In this example squares 2 and 3

Figure 3.44

MEV K-map for the equation $A = f(x, y, z) = \Sigma$ (0, 4, 5, 7)

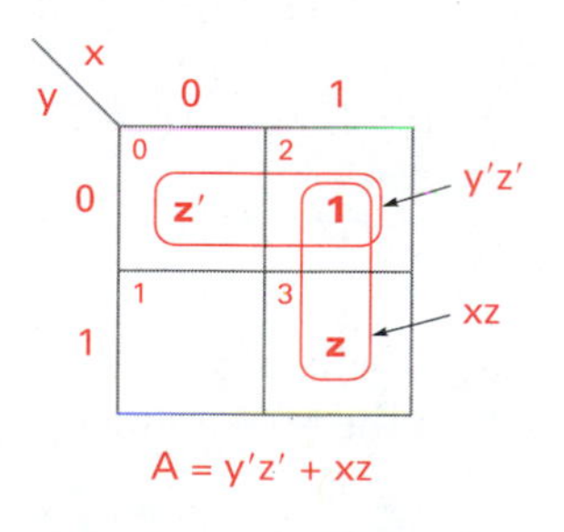

Figure 3.45

Three-variable K-map for the equation $A = f(x, y, z) = \Sigma$ (0, 4, 5, 7)

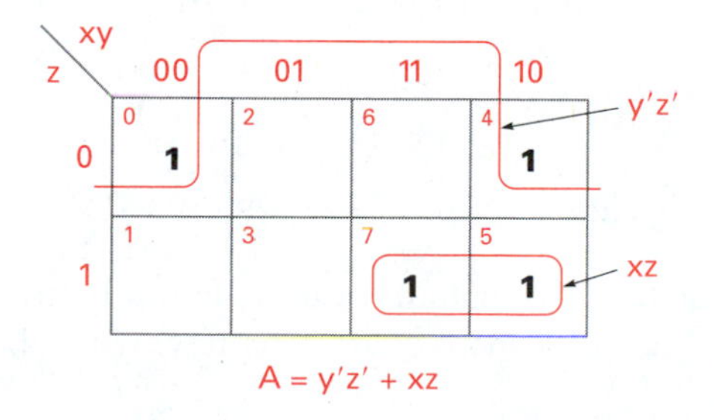

Figure 3.46

Three-variable MEV K-map using a complex MEV entry

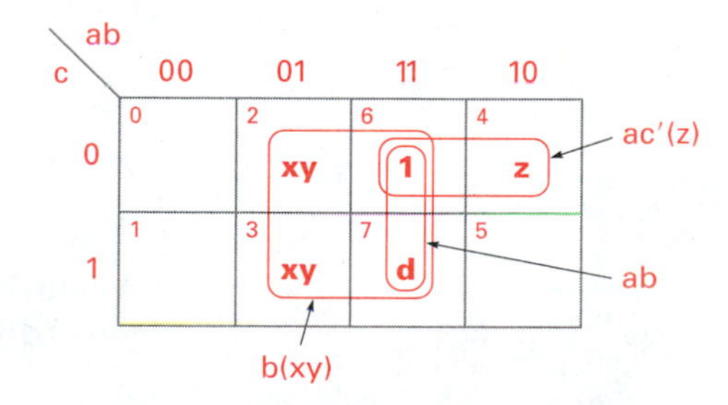

TABLE 3.25

Truth table for five-variable MEV problem

Decimal minterm		Variables						Decimal minterm		Variables					
		MAP				MEV	F			MAP				MEV	F
MEV 4	Standard 5	*a*	*b*	*c*	*d*	*e*		MEV 4	Standard 5	*a*	*b*	*c*	*d*	*e*	
0	0	0	0	0	0	0	0	8	16	1	0	0	0	0	1
0	1	0	0	0	0	1	0	8	17	1	0	0	0	1	0
1	2	0	0	0	1	0	0	9	18	1	0	0	1	0	1
1	3	0	0	0	1	1	1	9	19	1	0	0	1	1	0
2	4	0	0	1	0	0	0	10	20	1	0	1	0	0	0
2	5	0	0	1	0	1	0	10	21	1	0	1	0	1	0
3	6	0	0	1	1	0	0	11	22	1	0	1	1	0	0
3	7	0	0	1	1	1	1	11	23	1	0	1	1	1	0
4	8	0	1	0	0	0	0	12	24	1	1	0	0	0	*x*
4	9	0	1	0	0	1	0	12	25	1	1	0	0	1	*x*
5	10	0	1	0	1	0	0	13	26	1	1	0	1	0	*x*
5	11	0	1	0	1	1	1	13	27	1	1	0	1	1	*x*
6	12	0	1	1	0	0	1	14	28	1	1	1	0	0	*x*
6	13	0	1	1	0	1	1	14	29	1	1	1	0	1	*x*
7	14	0	1	1	1	0	1	15	30	1	1	1	1	0	*x*
7	15	0	1	1	1	1	1	15	31	1	1	1	1	1	*x*

contain the MEV xy, and square 4 contains the MEV z. Minterm {6} and the don't care term {7}, form the PI ab. MEV xy, in squares 2 and 3 form an EPI with minterm {6} and don't care term {7}, to give $b(xy)$. EPI $ac'(z)$ was formed from minterm {6} and MEV z. Note that only identical MEVs can be grouped with minterms and don't care terms to form a PI.

The simplified expression from the K-map in Figure 3.46 is

$$f = ab + b(xy) + ac'(z)$$

A five-variable problem further reinforces MEV K-maps, so consider the problem in Table 3.25, where x represents don't care terms for this problem.

From Table 3.25 we can write the expression for output function F (Figure 3.47):

$$\begin{aligned} F &= f(a, b, c, d, e) \\ &= \sum(3, 7, 11, 12, 13, 14, 15, 16, 18) \\ &\quad + \sum x(24, 25, 26, 27, 28, 29, 30, 31) \end{aligned}$$

A four-variable K-map solution can be found by using input variable e as a MEV (Figure 3.48). Each MEV map square is represented in Table 3.25 in the MEV minterm

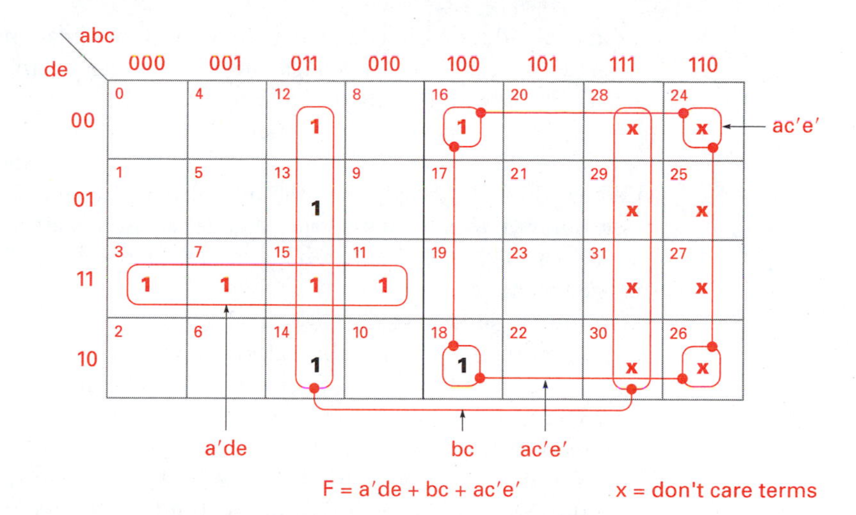

Figure 3.47 Five-variable K-map solution for example problem

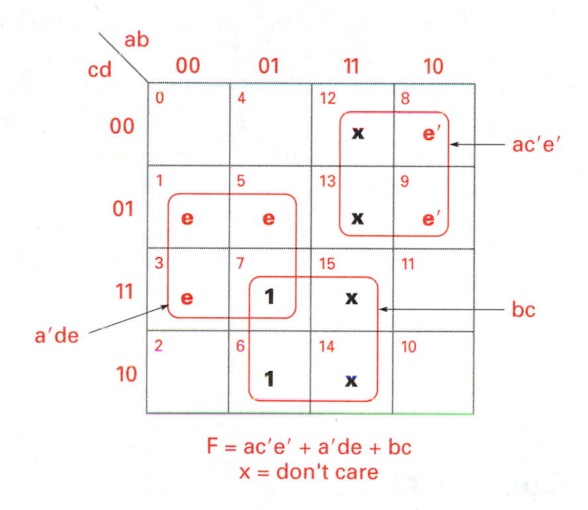

Figure 3.48 MEV K-map simplification of Table 3.25

column. Each five-variable minterm is represented in the standard minterm column. Note the partitions: two five-variable minterms are contained in one MEV row. Input variables *a*, *b*, *c*, and *d* are decoded to produce the MEV column values. The MEV *e* is included when decoding the values for the five-variable minterm column.

3.7 MIXED LOGIC COMBINATIONAL CIRCUITS

The logic presented so far in this book has defined "true" as a high voltage or a logical 1 and "false" as a low voltage or a logical 0. An idea is developed in this section that adds versatility in defining the "truth" of a Boolean variable. Each variable can have an *active* level associated with "truth." By this we mean that a variable can define *truth* as either a 1 or a 0. A variable is said to be **active high** when its "true" value is a logical 1. The same variable could have been defined as an **active low** signal, meaning that it is "true" when the signal output is 0. *Active high* and 1 indicating "truth" is a positive logic notation. *Active low* and 0 indicating "truth" is a negative logic notation.

Logic designers speak of an **asserted** signal, which is another term for "truth." A signal that is asserted when its value is a logical 1 is an active high signal (positive logic). A signal that is asserted when the output is a 0, active low, is a negative logic signal. Assertion is associated with "true" and deassertion is associated with "false." A variable can be asserted as a 1 (active high) or a 0 (active low). **Deasserted** is a synonym for negated or "false."

In **mixed logic** systems (most real logic designs use mixed logic), some means of keeping track of which variable is active high and which is active low is necessary. This is typically done by the assignment of the variable name. The signal name (which indicates its assertion value) and its status in a logic circuit are not the same thing. For example, the variable $A.H$, is the name for a Boolean variable, A, and it is asserted high (active high) as indicated by $.H$. The variable $A.H$ can be inverted through a logical operation and thus be primed or negated, $A.H'$. This simply means that the active high signal $A.H$ has been complemented. A variable $b.L$ is an active low signal, b is "true" (asserted) when the logic level is 0. Variable $b.L$ can also be inverted, $b.L'$, which means that the signal is deasserted (false).

We have been using single letters or letters with subscripts as variable names for reasons of simplicity. Most practical logic designs assign variable names that indicate a function in the system. For example, a power switch closure can generate a Boolean variable used in a logic circuit. Instead of giving the variable an arbitrary name, X, we could name it *PWR*. In this way the variable name is related to its function in the circuit.

Table 3.26 lists several binary functions and some examples of "useful" names. Variable names that indicate the assertion level are not the same thing as logical operations. For example, *PWR.L* means that the signal *PWR* is asserted when the signal is a 0; it does not mean that *PWR* has been complemented. The advantage of mixed logic notation will soon become apparent.

3.7.1 Logic Symbols

Each combinational logic function can be realized using mixed logic notation. Consider Figure 3.49, where i is the input variable and o the output variable, each expressed as active high or active low.

TABLE 3.26
Active high and active low variable name notation

Function	Arbitrary name	Useful name	
		Active high	Active low
Power on	X	PWR.H	PWR.L
Run	Y	RUN.H	RUN.L
Reset	Z	RST.H	RST.L
Interlock	W	ITL.H	ITL.L
Data ready	P	DRDY.H	DRDY.L

Figure 3.49
Single variable mixed logic

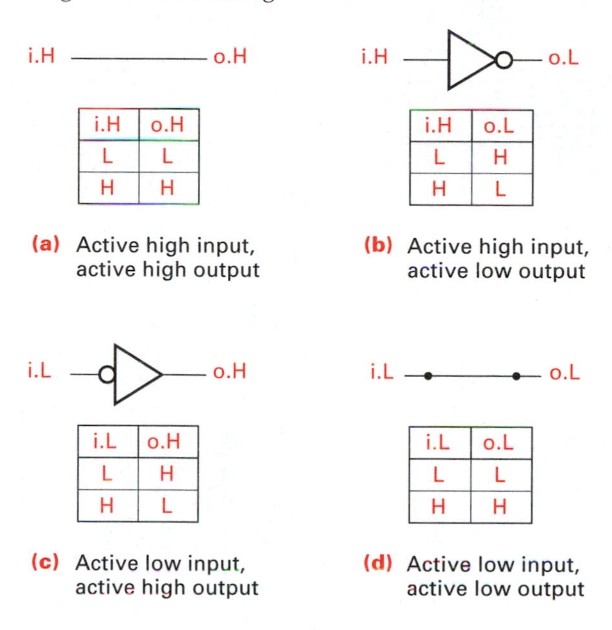

i.H	o.H
L	L
H	H

(a) Active high input, active high output

i.H	o.L
L	H
H	L

(b) Active high input, active low output

i.L	o.H
L	H
H	L

(c) Active low input, active high output

i.L	o.L
L	L
H	H

(d) Active low input, active low output

We can change the assertion level from the input to the output as indicated in Figure 3.49. If the assertion levels remain the same then a piece of wire is sufficient to transmit the input to the output. Inverter bubbles are shown on the side with the active low signal. Multiple variable logic also uses mixed logic notation.

Consider the AND function. Figure 3.50 illustrates the eight combinations possible with two input and one output variables, where each can be asserted high or asserted low. An active high output, $O.H$, occurs only once in each $O.H$ truth table. An active low output, $O.L$, occurs only once in each $O.L$ truth table. For example, consider Figure 3.50(e). The output is active low as indicated by $O.L$ (the output is a 0 when the input conditions $X.H = 1$ and $Y.L = 0$).

To read mixed logic we need to know the active state of the variables. In the case of the AND function, we are ANDing input variable active states, not 1s. For instance, we can AND an active high state with an active low state and produce an active low output. Each of the eight truth tables and logic symbols indicates different active state AND combinations.

Figure 3.51 shows the truth tables and logic for the eight combinations of mixed logic input and output variables for the OR function. Notice that we are ORing the active status of the variables and producing an active output as a result. Variable notation tells us the active status and the symbol indicates the logic function. The original assignment of the active state to a logic variable is generally driven by the application. Usually the logic designer must meet existing requirements that indicate the active status to be assigned to a variable. Now we will combine the assignment of variable names and active status to a logic circuit to illustrate the ideas.

Figure 3.50
AND: mixed logic symbols and tables

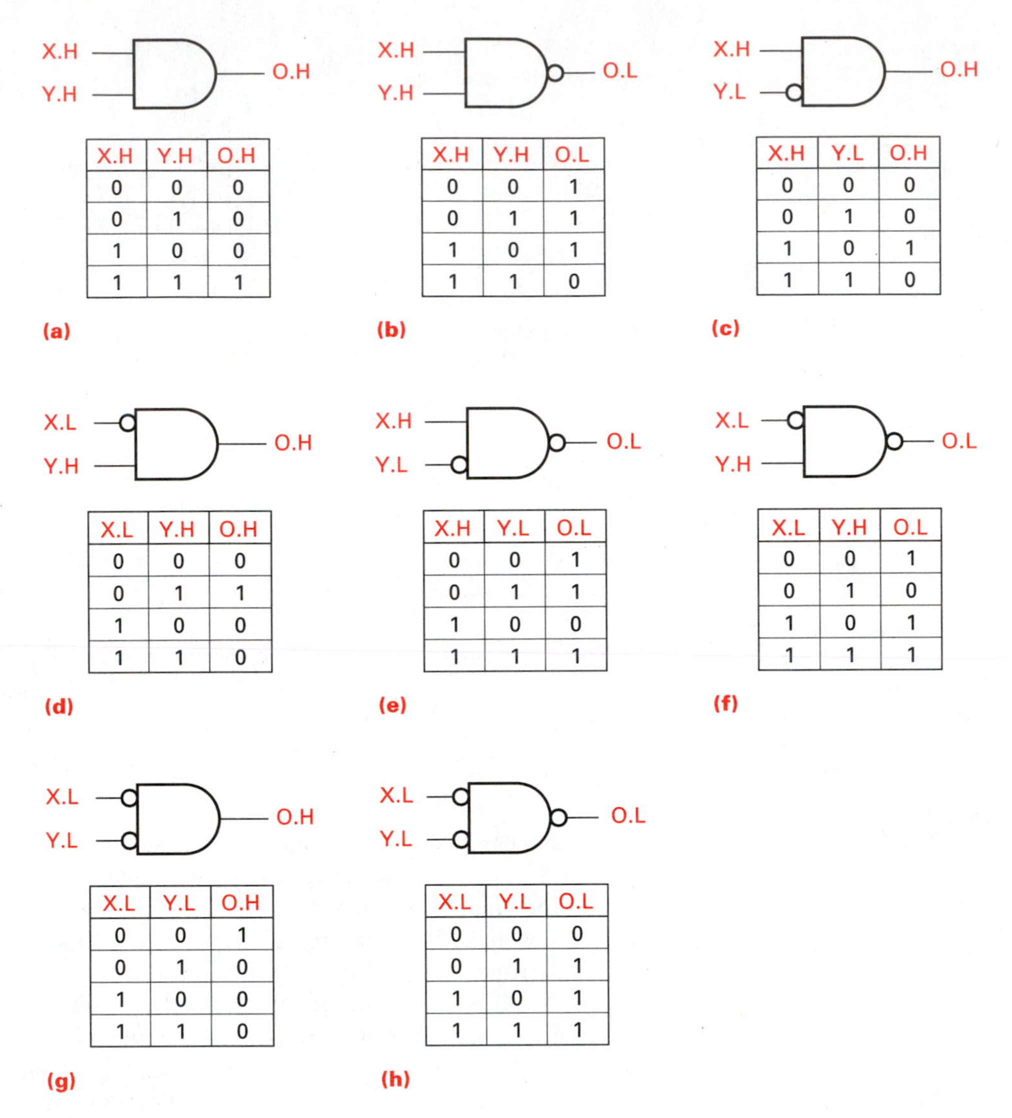

X.H	Y.H	O.H
0	0	0
0	1	0
1	0	0
1	1	1

(a)

X.H	Y.H	O.L
0	0	1
0	1	1
1	0	1
1	1	0

(b)

X.H	Y.L	O.H
0	0	0
0	1	0
1	0	1
1	1	0

(c)

X.L	Y.H	O.H
0	0	0
0	1	1
1	0	0
1	1	0

(d)

X.H	Y.L	O.L
0	0	1
0	1	1
1	0	0
1	1	1

(e)

X.L	Y.H	O.L
0	0	1
0	1	0
1	0	1
1	1	1

(f)

X.L	Y.L	O.H
0	0	1
0	1	0
1	0	0
1	1	0

(g)

X.L	Y.L	O.L
0	0	0
0	1	1
1	0	1
1	1	1

(h)

❖ EXAMPLE 3.19

Design the logic so an active low output is generated when power is on (an active high signal), the system is not reset (an active low signal), an interlock is closed (an active low signal), a run signal is present (active low), and data are ready (active high).

SOLUTION **1.** Determine the signal names and indicate active status:

Input variables

Power on = PWR.H
Reset = RST.L
Interlock = ITL.L
Run = RUN.L
Data Ready = DRDY.H

Output variable

Out = OUT.L

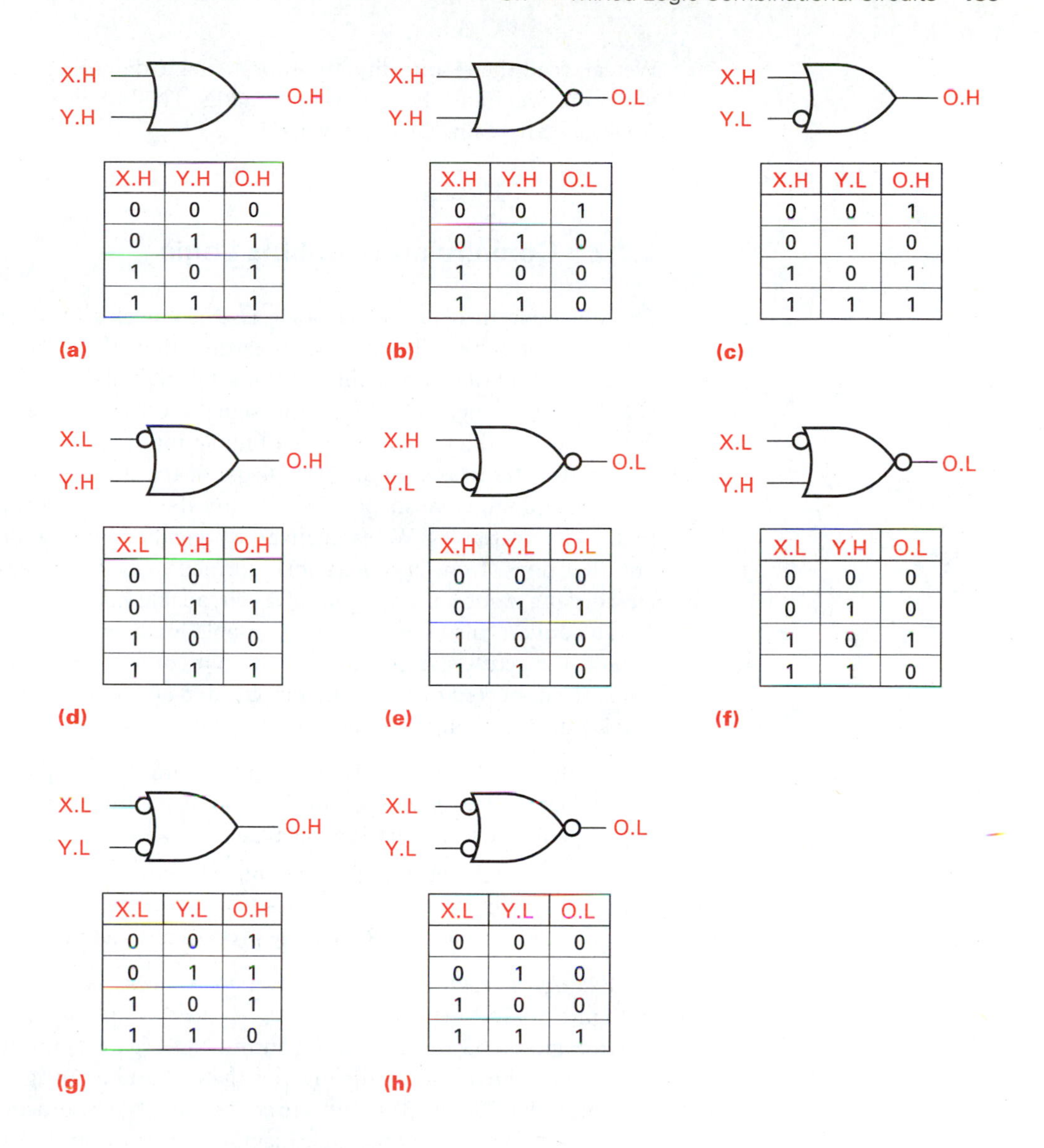

(a)

X.H	Y.H	O.H
0	0	0
0	1	1
1	0	1
1	1	1

(b)

X.H	Y.H	O.L
0	0	1
0	1	0
1	0	0
1	1	0

(c)

X.H	Y.L	O.H
0	0	1
0	1	0
1	0	1
1	1	1

(d)

X.L	Y.H	O.H
0	0	1
0	1	1
1	0	0
1	1	1

(e)

X.H	Y.L	O.L
0	0	0
0	1	1
1	0	0
1	1	0

(f)

X.L	Y.H	O.L
0	0	0
0	1	0
1	0	1
1	1	0

(g)

X.L	Y.L	O.H
0	0	1
0	1	1
1	0	1
1	1	0

(h)

X.L	Y.L	O.L
0	0	0
0	1	0
1	0	0
1	1	1

Figure 3.51
OR: mixed logic symbols and tables

2. Each of the input variables' active status is ANDed to produce the active low output signal:

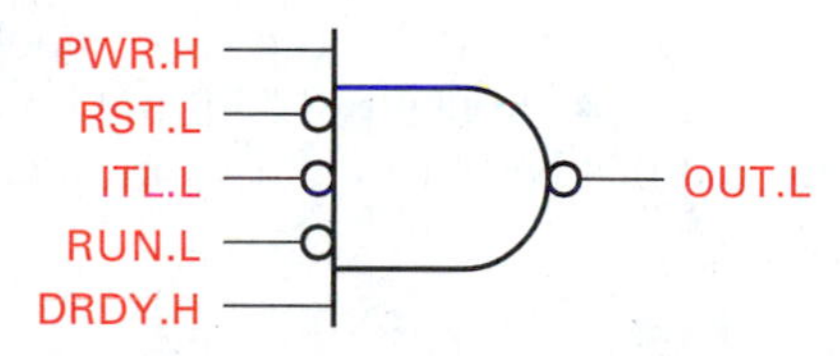

All of the input variables are ANDed, independent of the active status level, to produce an active output. The mix in mixed logic comes from mixing active status levels; the logic function remains the same.

In reading the AND logic function we note that the output OUT.L is going to be a 0 (active low true) when *PWR.H* = 1, *RST.L* = 0, *ITL.L* = 0, *RUN.L* = 0, and *DRDY.H* = 1.

We can read the results directly from the logic diagram and need not construct a truth table to understand the output logic status. The bubbles on the inputs of the five-input AND gate are realized using inverters. ❖

3.7.2 Conversion to Bubble Logic

The objective in using mixed logic is to connect a bubble output to a bubble input whenever possible. This permits determination of the logic output function directly from the logic diagram without having to mentally keep track of any complemented variables. This means that an input signal level is matched to the input status (active low input variables connected to a bubble input and active high input variables connected directly to the input). The shape of the logic symbol indicates an AND or OR operation (distinctive shape symbols) and the use of inverting bubbles indicates active high or low variables. We can convert a logic diagram that does not use **bubble logic** to one that does; from cryptic, poorly formed logic to a readable diagram, by applying DeMorgan's theorem. The result gives a readable bubble-to-bubble logic diagram.

To convert mismatched logic to bubble logic, we do the following. The objective is to have the active status of input to output connections match; that is, active high outputs are connected to active high inputs and active low outputs are connected to active low input. To accomplish this,

1. Start at the output gate(s). Convert any output gate that has a logic level mismatch. An active low output to an active high input or an active high output to an active low input is a mismatch.
2. Redraw the output gate to conform to the conversion. Convert the next level of mismatch so that logic levels between outputs and inputs are matched. Redraw the logic diagram. Repeat the process until all mismatches are converted.

Examples best illustrate the conversion process, so consider the logic diagram shown in Figure 3.52.

In Figure 3.52 we see that the output NOR (gate 3) input and output logic levels do not match. The output of the inverter (gate 1) and NOR (gate 2) are active low whereas the output NOR (gate 3) inputs are active high. It is confusing to determine the purpose of the circuit by reading the logic diagram drawn in this fashion. To do so would require that we mentally apply DeMorgan's theorems to the circuit. Under what input conditions is output z asserted (true)?

Convert the output NOR logic symbol to an AND symbol with inverting input bubbles as shown in Figure 3.53. This is done by application of DeMorgan's theorems.

The logic function shown in Figure 3.52 is not immediately obvious. We could construct a truth table, but that becomes clumsy for functions containing many input vari-

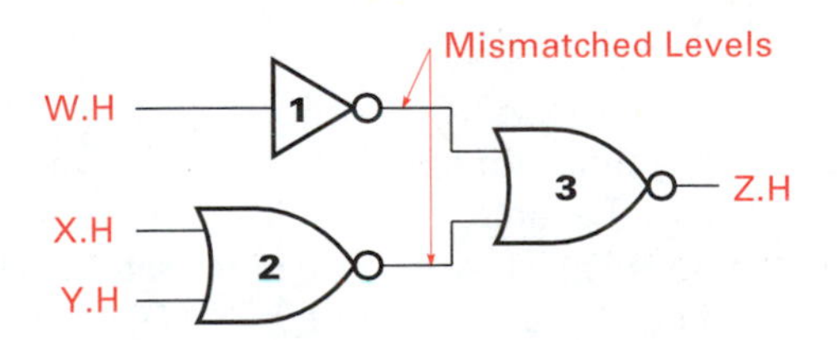

Figure 3.52
Logic mismatch example

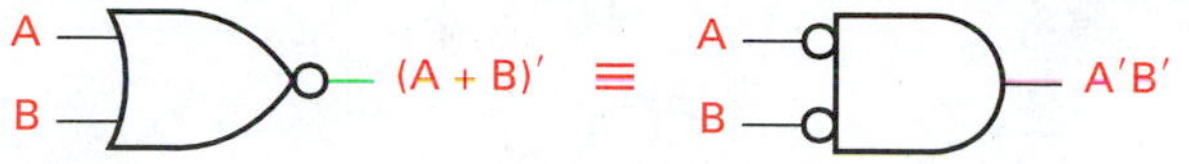

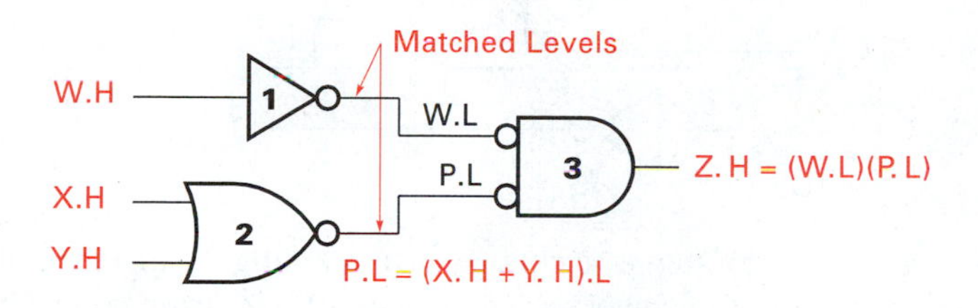

Figure 3.53
Conversion of a NOR with active high inputs to an AND with active low inputs

Figure 3.54
Matched logic levels

ables. Now, consider the matched output logic diagram shown in Figure 3.54. The bubbles cancel, so we can ignore them in our circuit function analysis. The output function, $Z.H = W.H(X.H + Y.H)$, which is easier to determine than constructing a complex switching equation or truth table based on the original logic diagram. Trouble shooting the logic, based on the drawing in Figure 3.54, can be done directly by inspection of the diagram.

Consider the circuit illustrated in Figure 3.55. Two levels of logic are mismatched in this circuit. Starting with the output gate we convert the active high input NAND to an active low input OR, as shown in Figure 3.56.

The output gate is changed from a NAND with active high inputs and an active low output to an OR with active low inputs to match the levels of the input signals. Redrawing the original logic to include the output gate conversion gives Figure 3.57.

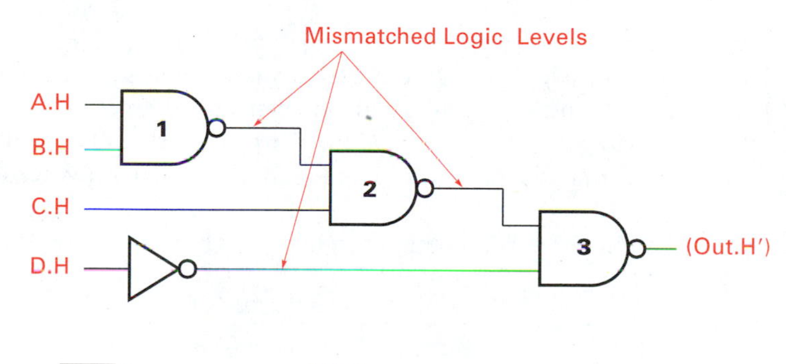

Figure 3.55
Mismatched logic level diagram

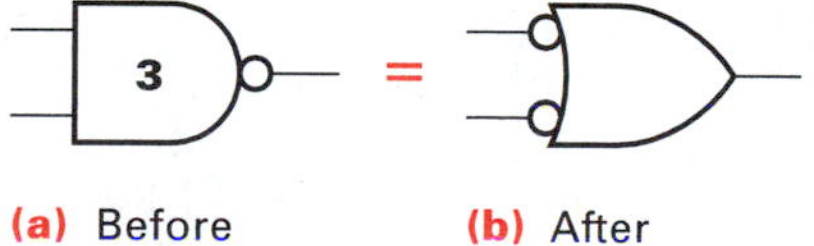

Figure 3.56
Gate 3 before and after conversion

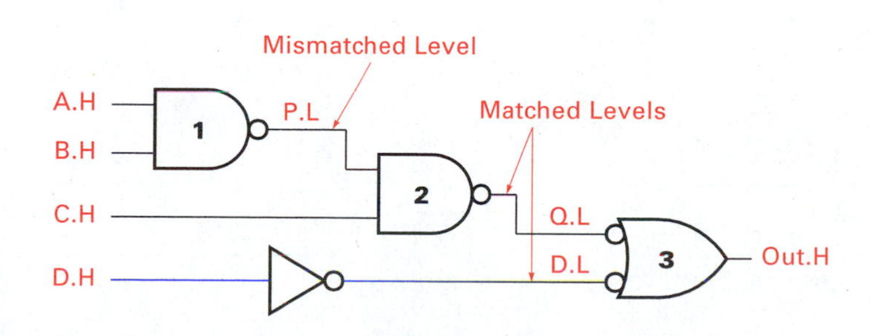

Figure 3.57
Step 1 in converting the logic from Figure 3.55 to mixed notation

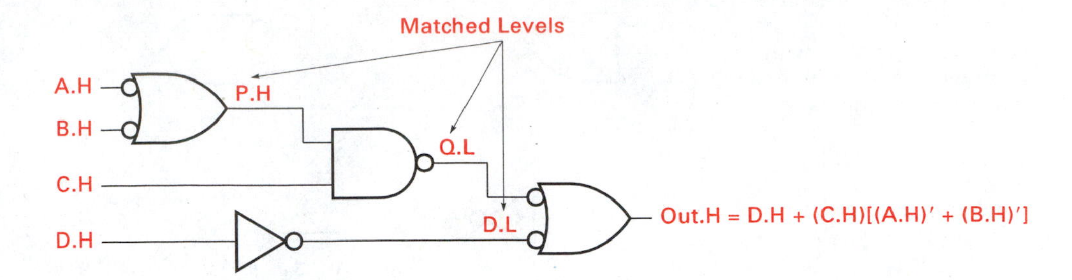

Figure 3.58
Final mixed logic diagram

The next step is to change the remaining mismatch. Output P, gate 1, is mismatched to the input of the NAND gate (gate 2), requiring conversion of gate 1 to an invert OR gate. Figure 3.58 shows the final mixed logic diagram. From the logic diagram we can determine that the output function X is a 1 if either or both of the inputs (Q or D) is low. Mixed logic permits troubleshooting at the gate level without knowing exactly what the final function is to accomplish.

3.7.3 Synthesizing Switching Functions Using Bubble Notation

Developing bubble-to-bubble logic notation can be done algebraically or by drawing the logic diagram directly from the switching equation and then applying the techniques previously developed to the circuit. Designing a mixed logic circuit directly from the original simplified switching functions is illustrated in the following example.

❖ **EXAMPLE 3.20** Assume a logic design was partially implemented using minterm equations and part was done using maxterm equations, then a third function was added later that used the two previously generated functions. This sort of thing happens all the time as designs evolve over time and often with more than one designer involved in the process.

SOLUTION Let the minterm function (Figure 3.59) be

$$z = f(a,\ b,\ c) = \sum(0,\ 1,\ 2,\ 3,\ 4,\ 6)$$

Let the maxterm function (Figure 3.60) be

$$p = f(a,\ b,\ c) = \pi(2,\ 3,\ 5,\ 7)$$

Figure 3.59
K-map for $z = f(a, b, c) = \Sigma\ (0, 1, 2, 3, 4, 6)$

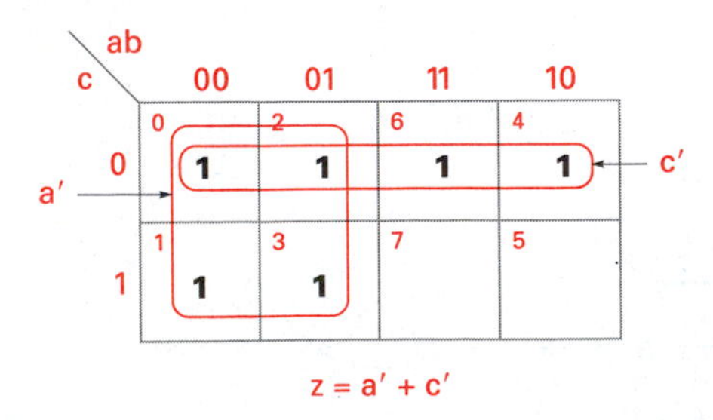

Figure 3.60
K-map for $p = f(a, b, c) = \pi\ (2, 3, 5, 7)$

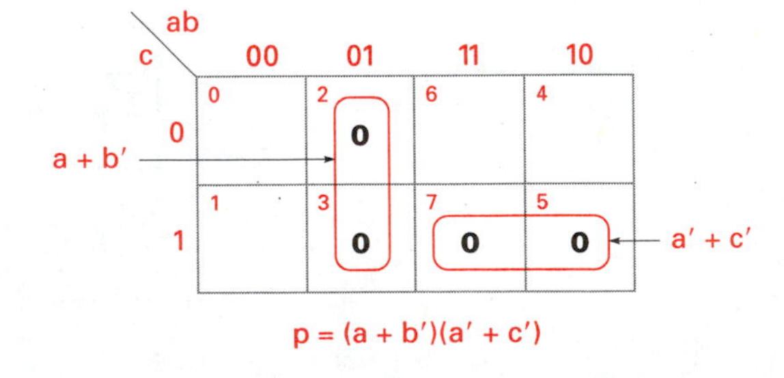

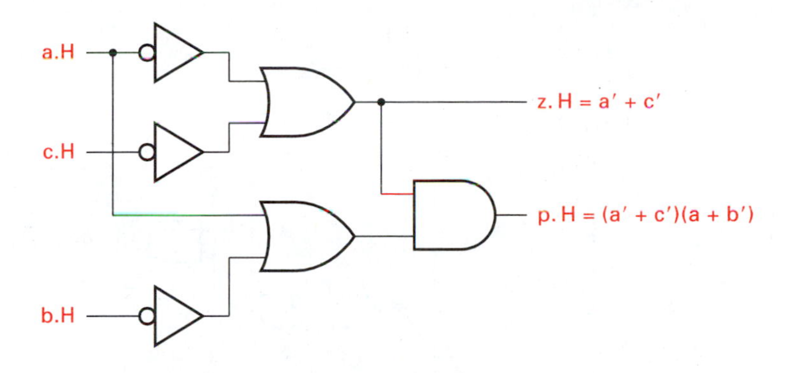

Figure 3.61
Resulting logic for *z* and *p*

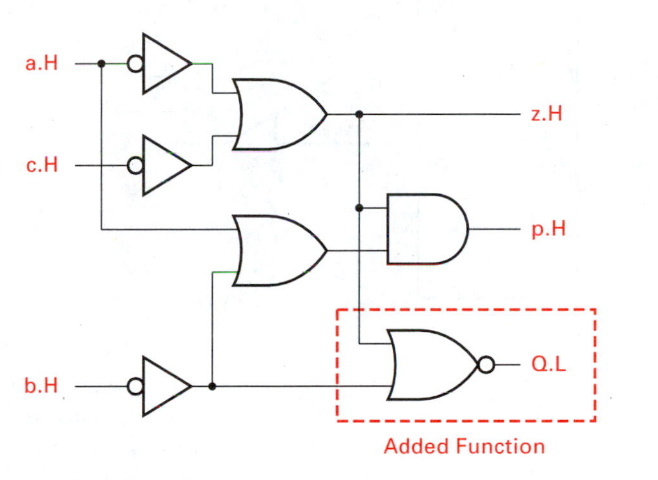

Figure 3.62
Combination of functions *z.H* and *p.H* and the inclusion of function *Q.L*

The resulting logic for the minterm and maxterm functions is shown in Figure 3.61. To this we are going to add a third output function expressed as

$$Q.L = f(a, b, c) = \sum(0, 1, 2, 3, 4, 5, 6)$$

which simplifies to

$$Q.L = a' + b' + c'$$

Combining the two circuits and adding a third output function, *Q.L*, results in Figure 3.62. ❖

Situations occur where a logic designer is required to use existing variable definitions and logic functions to create additional output functions. In the preceding example the additional output function *Q.L* was active low and produced from existing terms.

❖ **EXAMPLE 3.21** Design the mixed logic necessary to realize the following function. A motor can be turned on and off from any one of three separate locations without changing the switch setting at the other two locations. The ON variables are active high. The three-way motor switches must operate so that a change in position will cause a corresponding change in the motor status.

Figure 3.63
Block diagram of example logic system

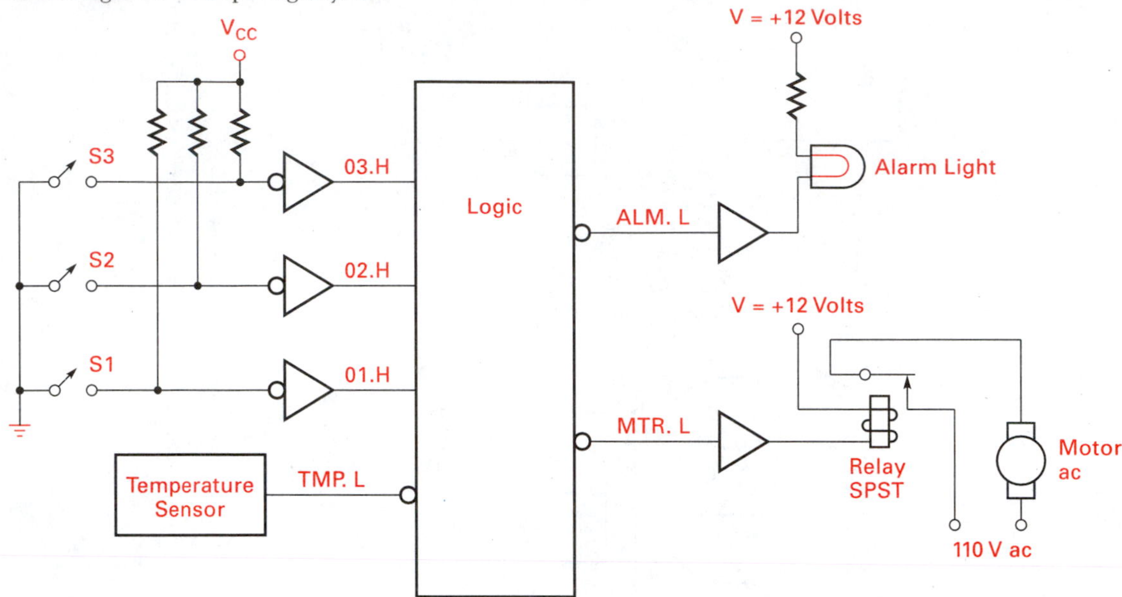

For example, if no switches were active, then any one switch being asserted will cause the motor to start. Likewise a change in any switch position must produce a change in the motor status. If any two switch positions are asserted the motor is to be turned off; yet if three are asserted, the motor is to be turned on. An active low signal is necessary to close the motor control relay.

In addition the motor must be shut off if the temperature exceeds a given set point. The temperature sensor provides an active low output. If the temperature sensor has been activated then each of the three turn-on points are to be deactivated and a temperature alarm light is turned on. The alarm light is to be an active low output variable. The logic system is illustrated in Figure 3.63.

SOLUTION The first step is to designate input and output variables.

Let $O1.H$, $O2.H$, and $O3.H$ be the three ON-OFF variables.

Let $TMP.L$ be the temperature sensor variable.

Let $MTR.L$ be the motor ON output variable.

Let $ALM.L$ be the alarm ON output variable.

A truth table consisting of four input and two output variables is illustrated in Table 3.27.

$TMP.L = 0$ is an active low input signal and the outputs $MTR.L$ and $ALM.L$ are active low signals. When writing the equation for $MTR.L$, we note the conditions that make it a logical 0. The best way to treat the mixed input variables is to separate them. If we consider only the active high variables $O3.H$, $O2.H$, and $O1.H$ then the equation for $MTR.L$ becomes

$$MTR.L = f(O3.H, O2.H, O1.H) = (1, 2, 4, 7)$$

TABLE 3.27
Truth table for mixed logic example problem

Input variables				Output variables	
O3.H	*O2.H*	*O1.H*	*TMP.L*	*MTR.L*	*ALM.L*
0	0	0	0	1	0
0	0	0	1	1	1
0	0	1	0	1	0
0	0	1	1	0	1
0	1	0	0	1	0
0	1	0	1	0	1
0	1	1	0	1	0
0	1	1	1	1	1
1	0	0	0	1	0
1	0	0	1	0	1
1	0	1	0	1	0
1	0	1	1	1	1
1	1	0	0	1	0
1	1	0	1	1	1
1	1	1	0	1	0
1	1	1	1	0	1

Note: For *X.L* signals, 1 = OFF and 0 = ON.

which is a three-input EX-OR function. Note that when the equation is mapped, as in Figure 3.64, it does not reduce.

The output of the EX-OR is gated with the active low *TMP.L*, so that the *MTR.L* function can be true (active low) only when *TMP.L* is high (inactive). The logic diagram for *MTR.L* is shown in Figure 3.65.

Input *TMP.L* is active low; a 0 means that the temperature has exceeded the set point value. We want to turn the motor ON (*MTR.L* = 0) only when *TMP.L* is inactive or high. We can rename the *TMP.L* signal as *TOK.H* (temperature is OK or not above the set point).

Study the truth table in Table 3.27, and compare the *ALM.L* output with the input variables. Notice that *ALM.L* follows the *TMP.L* signal exactly. The two inverters, shown

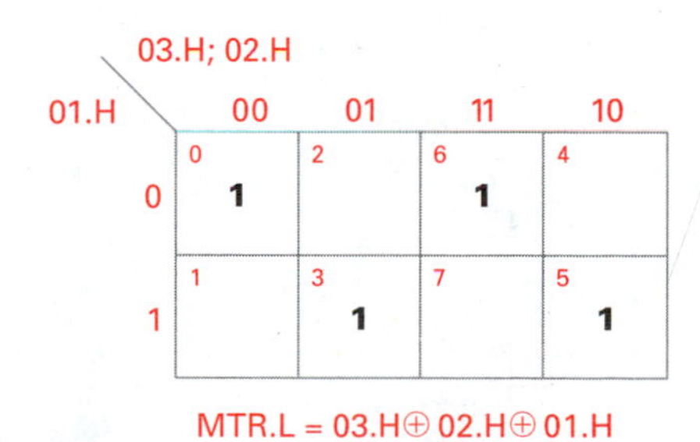

Figure 3.64
K-map for *MTR.L*

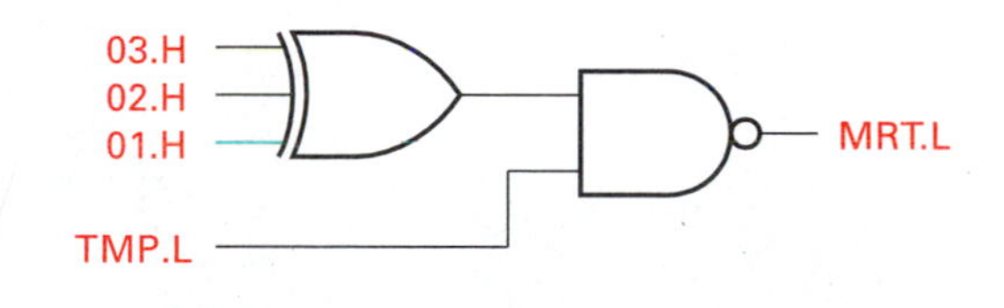

Figure 3.65
Logic for *MTR.L*

Figure 3.66
Isolation buffering for *ALM.L*

TMP.L ALM.L

in Figure 3.66, are used to isolate the temperature sensor from the alarm driver and keep the logic level notation straight. ❖

Designs using medium- and large-scale integrated circuits use mixed logic notation extensively. Often the interconnection of MSI and LSI components requires mixed logic notation. We will cover selected MSI components relative to combinational logic design in Chapter 4.

3.8 MULTIPLE OUTPUT FUNCTIONS

Many digital applications require multiple outputs derived from the same input variables. We could simplify each output function separately by the use of Karnaugh maps or the Quine-McClusky algorithm. Sometimes, however, terms used by one output function can be shared by other output functions, thereby reducing the total number of gates.

The objective is to find a minimal covering (simplified result) for all of a system's output functions. Consider the following three-variable multiple-output system:

$$F_3 = f(a, b, c) = \sum(2, 4, 5, 6)$$
$$F_2 = f(a, b, c) = \sum(2, 3, 6, 7)$$
$$F_1 = f(a, b, c) = \sum(2, 5, 6, 7)$$

Create a K-map for each of the output functions (Figure 3.67).

Figure 3.67
K-maps for multiple output functions F_3, F_2, and F_1

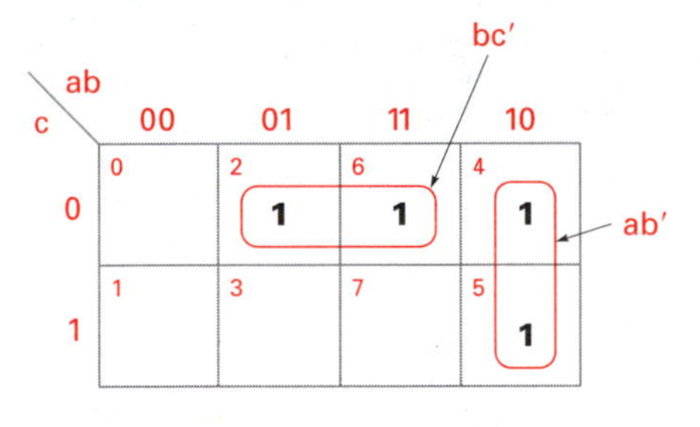

(a) K-map for F_3

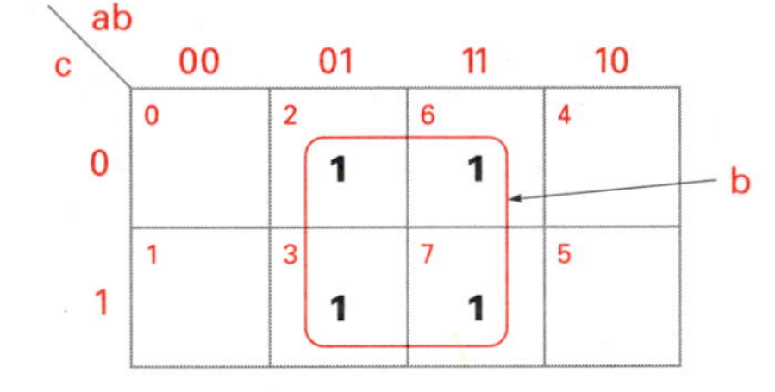

(b) K-map for F_2

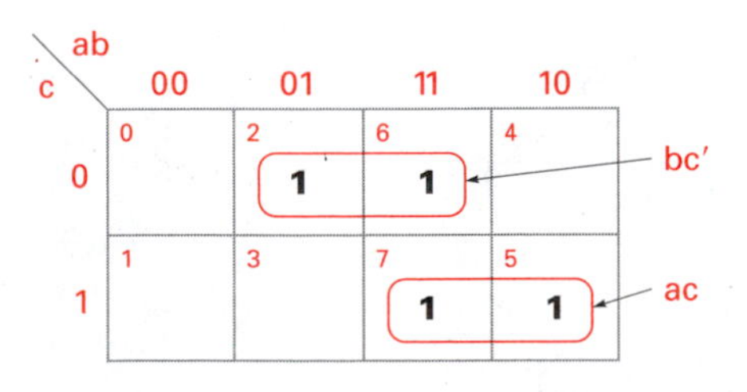

(c) K-map for F_1

Figure 3.68
Logic diagram for multiple output functions when each is treated separately

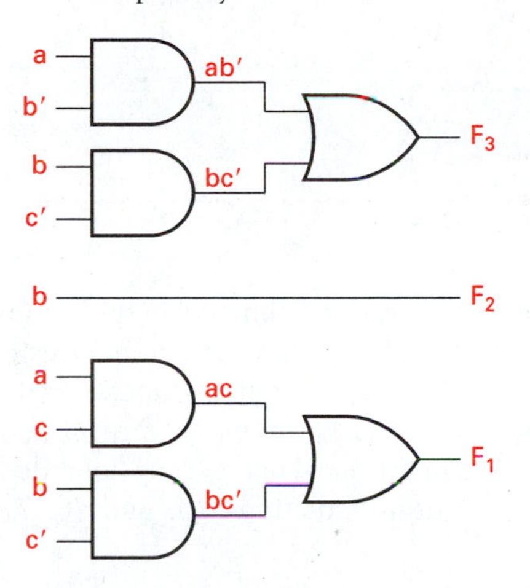

Figure 3.69
Logic for functions F_3 and F_1 when sharing term bc'

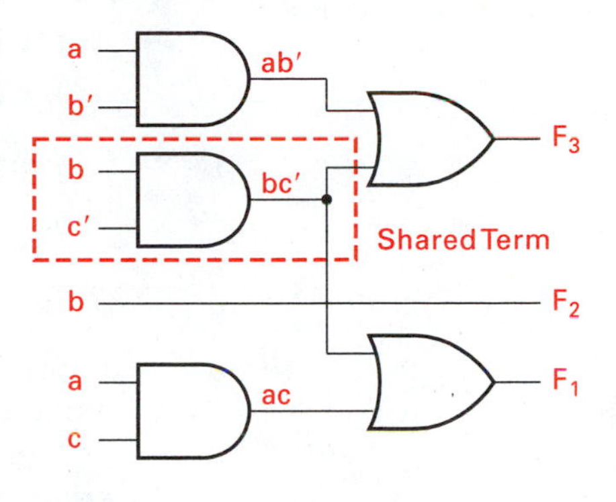

The reduced equations, when treated separately (Figure 3.68), are as follows:

$$F_3 = ab' + bc'$$
$$F_2 = b$$
$$F_1 = ac + bc'$$

Term bc' occurs in both F_3 and F_1 output equations. Figure 3.69 shows the logic when term bc' is shared by both F_3 and F_1.

The previous example did not require any special techniques to recognize the common term. Now, consider the following problem:

$$F_2 = f(a, b, c) = \sum(1, 3, 7)$$
$$F_1 = \sum(2, 6, 7)$$

The K-maps for F_2 and F_1 are shown in Figure 3.70. The reduced equations for F_2 and F_1 are

$$F_2 = a'c + bc$$
$$F_1 = bc' + ab$$

Figure 3.70
K-maps for F_2 and F_1

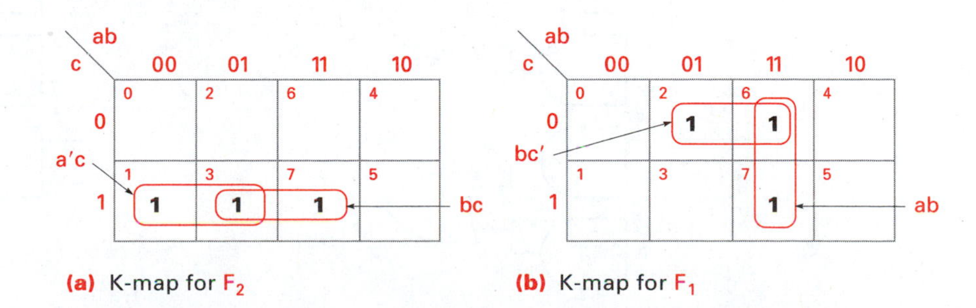

(a) K-map for F_2 **(b)** K-map for F_1

TABLE 3.28

Prime implicant chart for determination of multiple output common terms

		Minterms			
Prime implicants	1	2	3	6	7
1, 3 (F_2)	x		x		
3, 7 (F_2)			x		x
2, 6 (F_1)		x		x	
6, 7 (F_1)				x	x

In this example no common term is immediately obvious. The construction of a prime implicant table (Table 3.28) helps show common terms more readily.

Minterm {7} is common to both output functions. If we create a gate realizing minterm {7}, *abc*, then both F_2 and F_1 can use it. Any minterms not shared between output functions are simplified in the usual manner. By not reducing the groups associated with minterm {7}, {3, 7} (output function F_2), and {6, 7} (output F_1), we obtain the following:

$$F_2 = a'c + abc$$

$$F_1 = bc' + abc$$

Minterm *abc* is common to both output functions and can, therefore, be shared. The logic required when there is no consideration for multiple outputs is illustrated in Figure 3.71. The resulting logic diagram when multiple outputs are considered is shown in Figure 3.72.

Consideration of multiple outputs together does not guarantee a minimum number of ICs. That also depends on how the gates in an IC are packaged (how many gates and what type). The process does guarantee a minimum number of product terms. Sometimes that is also the minimum number of packages. Consideration of multiple outputs is especially helpful when realizing a design using PALs (programmable array logic) or PLDs (programmable logic devices). We will cover these devices in a later chapter.

Finding the terms shared between two or more output functions requires finding the AND of the prime implicants involved. By ANDing the prime implicants we create what

Figure 3.71

Logic for F_2 and F_1 when multiple outputs are considered separately

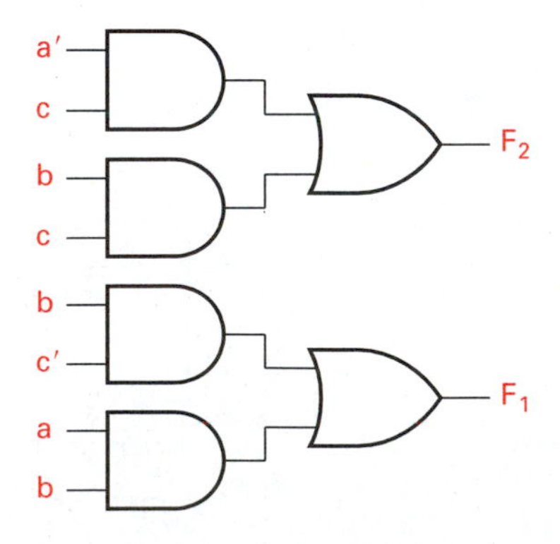

Figure 3.72

Logic for functions F_2 and F_1 when multiple outputs are considered together

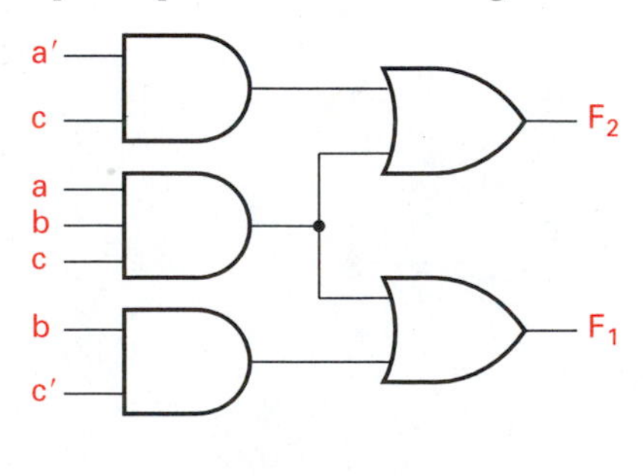

is called a *product function*. When the prime implicants for each output function are ANDed, we obtain a common prime implicant that covers all of the individual output prime implicants. In other words, we have found a term that can be shared with each of the involved outputs.

More formally, a *product function* is defined as follows.

DEFINITION

The product of two functions $F_2(a_1, a_2, a_3, \ldots, a_n)$ and $F_1(a_1, a_2, a_3, \ldots, a_n) = F_2F_1(a_1, a_2, a_3, \ldots, a_n)$. When F_2 or F_1 or both $= 0$, then $F_2F_1 = 0$. When either F_2 or F_1 but not both $= 1$, then $F_2F_1 = 0$. When F_2 and $F_1 = 1$, then $F_2F_1 = 1$.

The product function for F_2 and F_1 prime implicants, ab and bc, is abc.

Techniques exist for finding the optimum solution to multiple output functions that involve finding the prime implicants of each output function and each product function. Once this is done the multiple output prime implicants are placed in a prime implicant chart similar to the ones used in the Quine-McClusky simplification algorithm. A text that covers this approach in some detail is *Digital Logic Design* by Langholz, Kandel, and Mott (Dubuque, Iowa: Wm. C. Brown).

SUMMARY

A. Chapter 3 defines combinational logic and introduces a model.

1. Combinational logic can be described by Boolean equations, truth tables, or logic diagrams.
2. Boolean equations are derived from truth tables.
   a. Minterms are defined as where output variable logical 1s occur in a truth table.
   b. Maxterms are defined as where output variable logical 0s occur in a truth table.
3. Logic diagrams use symbols to represent logical functions such as AND, OR, NOT, NAND, NOR, EX-OR, and EX-NOR.
4. Canonical Boolean equations contain minterms or maxterms (each term includes all the input variables or their complements).

B. Karnaugh maps are developed as a means of eliminating input variable redundancies in a Boolean equation.

1. Maps of three, four, five, and six variables are illustrated.
2. Incompletely specified outputs create don't care terms, which are used to further reduce Boolean equations.
3. Prime implicants and essential prime implicants are defined. A prime implicant is a maximally sized group of minterms or maxterms. An essential prime implicant is a prime implicant that contains unique minterms or maxterms.

C. The Quine-McClusky algorithm is developed as a means of reducing Boolean functions.

1. Don't care terms are used with the Quine-McClusky algorithm.
2. Prime implicant and reduced prime implicant tables are developed.

D. Map entered variable Karnaugh maps are developed as a technique for reducing Boolean equations.

E. Mixed logic (positive and negative logic notation) is introduced.

1. Mixed logic variable and symbol notation are presented.
2. Synthesizing mixed logic annotated circuits are presented.

F. Multiple output function simplification is introduced.

REFERENCES

Texts that cover combinational logic modeling, truth table development, and simplification techniques include the following:

Digital Design Principles and Practices by John F. Wakerly (Englewood Cliffs, N.J.: Prentice-Hall, 1994). Chapters 4 and 5 cover combinational logic.

Digital Logic: Analysis, Application and Design by Susan A. R. Garrod and Robert Borns (Philadelphia: Saunders, 1991). Chapter 3 covers simplification techniques (K-maps).

Introduction to Logic Design by Sajjan G. Shiva (Glenview, Ill.: Scott, Foresman and Company, 1988). Multiple output simplification is covered in Chapter 3.

The Art of Digital Design, an Introduction to Top-Down Design by Franklin P. Prosser and David E. Winkel (Englewood Cliffs, N.J.: Prentice-Hall, 1987). Chapter 1 covers Karnaugh mapping, and Chapter 2 provides extensive coverage of mixed logic notation.

Digital Logic Design by Gideon Langholz, Abraham Kandel, and Joe L. Mott (Dubuque, Iowa: Wm. C. Brown, 1988). Chapter 4 provides extended coverage of Karnaugh maps and Quine-McClusky simplification techniques.

Fundamentals of Logic Design, 4th ed., by Charles H. Roth, Jr. (Minneapolis–St. Paul: West, 1992). Unit 5 covers minterm and maxterm development, Unit 6 covers Karnaugh maps, and Unit 7 covers Quine-McClusky techniques.

Fundamentals of Digital Electronics by Robert K. Dueck (Minneapolis–St. Paul: West, 1994). Chapter 5 covers number systems and codes and Chapter 9 covers digital arithmetic.

GLOSSARY

Active high The "true" status of a switching variable occurs when the variable is a 1, as represented by a positive voltage. Opposite of active low.

Active low The "true" status of a switching variable occurs when the variable is a 0, as represented by a low voltage. Opposite of active high.

Asserted The "true" condition of a variable. A variable can be asserted as an active high or an active low signal.

Bubble logic Another name for mixed logic, where variables are represented by both active high and active low notations in the same logic circuit.

Canonical The basic definition means "conforming to a general rule." The "rule" for switching logic equations is that each minterm or maxterm must contain all of the input variables (or their complements) used in the expression.

Canonical Sum of Products A sum (OR function) of products (AND function) is a set of product terms ORed together. For the sum of products to be canonical all of the input variables used in the expression must be contained in each individual product term. For example, if X, Y, and Z are Boolean variables then a canonical sum of products could be $T = X'YZ + XY'Z + XYZ$.

Combinational logic Combinational logic consists of circuits whose outputs are determined by the present combination of inputs, independent of previous input combinations. Combinational logic circuits produce outputs that are specified by a set of Boolean equations.

Deasserted The "false" condition of a variable. A deasserted variable can be represented by a high or low voltage signal, depending on its active status. See also active high and active low.

Don't care terms Minterms or maxterms that are not used in specifying a logical output function. This occurs when the truth table for an output function is incompletely specified. The leftover terms become don't care terms. Because they are don't care terms, they can be assigned a value of 0 or 1.

Essential prime implicant A prime implicant that contains one or more minterms or maxterms that are unique; that is, terms not contained in any other prime implicant. See also prime implicant.

Karnaugh map A Karnaugh map is a diagram made up of squares that is used to simplify Boolean equations. Each square represents a minterm or maxterm from an equation. The squares are arranged so that only one binary digit changes between adjacent squares, providing visualization of variable redundancies. The number of squares in a Karnaugh map is related to the number of variables in an equation by $2^x = y$, where x is the number of problem variables and y is the number of map squares.

Literal A literal is a Boolean variable or its complement. For instance, if X is a Boolean variable then both X and X' are literal.

Map entered variables A variable or expression that is entered into a Karnaugh map square that is not a minterm or maxterm. The use of map entered variables (or expressions) allows the size of a Karnaugh map to be reduced for a given number of input variables. See Karnaugh map.

Maxterm A term in a Boolean equation that represents a condition where an output variable is a logical 0 in the output function truth table. A maxterm is the complement of a minterm. Maxterms are expressed as a set of input variables ORed together and ANDed with other maxterms. An example maxterm expression would be $F = (x + y' + z)(x + y' + z')$.

Minterm A term in a Boolean equation that represents a condition where an output variable is a logical 1 in the output function truth table. A minterm is the complement of a maxterm. Minterms are expressed as a set of input variables ANDed together and ORed with other minterms. An example minterm expression would be $F = x'yz' + x'yz$.

Mixed logic Mixed logic contains both active high and active low signals in one logic circuit. *Bubble logic* is another name for mixed logic. Mixed logic diagrams match logic levels (active high or low) between inputs and outputs so that an active high output is never connected to an active low input and so on.

Multiple output functions Multiple outputs derived from the same set of input variables are said to be *multiple output functions*. The term is used in conjunction with simplification of multiple outputs, so that common terms can be shared between outputs.

POS (product of sums) Maxterms are used to form POS. The arithmetic equivalent of the logical AND is a product and the equivalent of the logical OR is a sum. A POS term is a logical ANDing of ORed terms. The difference between a POS term and a maxterm is that the POS need not be canonical and the maxterm is canonical.

Prime implicant A prime implicant is a permitted group of minterms or maxterms. The grouping of terms requires that the group size be an integer power of 2 (2, 4, 8, 16 . . .) and that symmetry exist for each variable in the group. Variable symmetry means that a variable has an equal number of 0s and 1s for all terms in the group. See also *essential prime implicant*.

Product term A product term is a literal or the logical product (AND) of multiple literals. For example, if X, Y, and Z are Boolean variables then a representative product term could be X, or XY, or $X'YZ'$.

Quine-McClusky minimization technique A Boolean minimization technique that uses a series of ordered tables to determine prime and essential prime implicants instead of a single table, as used in Karnaugh maps. The Q-M algorithm is more structured and lends itself to realization using a computer.

SOP (sum of products) Minterms are used to form SOP equations. A sum is the arithmetic equivalent of the logical operation OR. A product is the arithmetic equivalent of the logical operation AND. A sum of products is the logical ORing of ANDed terms. The difference between a SOP term and a minterm is that the SOP term is not canonical and the minterm is canonical.

Sum term A sum term is a literal or the logical OR of multiple literals. For example, if X, Y, and Z are Boolean variables then a representative sum term could be X, or $X + Y'$ or, $X + Y' + Z'$.

QUESTIONS AND PROBLEMS

Section 3.1 to 3.3

1. Draw a model representing combinational circuits. Label the input and output variables. Write a general expression showing the input and output relationship.
2. Describe what is meant by *combinational logic* in your own words.
3. How does a "truth table" express a combinational circuit?
4. Construct a truth table and write the Boolean output

equations for the following verbal problem statements:

a. A single output variable, Z, is to be true when the input variables a and b are true and when b is false but a and c are true.

b. An output is to be true (logical 1) when the value of the inputs exceeds 3. The weighting for each input variable is as follows:

$$w = 3, \qquad x = 3, \qquad y = 2, \qquad z = -1$$

c. Two motors, M_2 and M_1, are controlled by three sensors, S_3, S_2, and S_1. One motor, M_2, is to run any time all three sensors are on (true). The other motor is to run whenever sensors S_2 or S_1, but not both, are on and S_3 is off. For all sensor combinations where M_1 is on, M_2 is to be off, except when all three sensors are off and then both motors must remain off.

d. A four-bit binary character is presented to a circuit that must detect whether the input is a legitimate BCD code. If a non-BCD code is entered, the output is to be true (logical 1).

e. An automobile system is to be designed that will sound an alarm under certain conditions. The alarm is to sound if the seat belt is not fastened and the engine is running, or if the lights are left on when the key is not in the ignition, or if the key is in the ignition, the engine is not running, and the driver's door is open. Determine the number of inputs and outputs, assign names to variables, and construct a truth table that describes the system.

5. Write the minterm Boolean output equations for the following truth tables using variable names ($a'bc$, etc.) and in decimal format (0, 1, 2, . . . , 7).

a.

Inputs			Outputs	
a	b	c	W	X
0	0	0	0	1
0	0	1	1	0
0	1	0	0	1
0	1	1	1	1
1	0	0	1	0
1	0	1	0	1
1	1	0	0	0
1	1	1	1	1

b.

Inputs				Outputs			
w	x	y	z	A	B	C	D
0	0	0	0	0	0	0	0
0	0	0	1	0	0	0	0
0	0	1	0	0	0	0	0
0	0	1	1	0	0	0	0
0	1	0	0	0	0	0	0
0	1	0	1	0	0	0	1
0	1	1	0	0	0	1	0
0	1	1	1	0	0	1	1
1	0	0	0	0	0	0	0
1	0	0	1	0	0	1	0
1	0	1	0	0	1	0	0
1	0	1	1	0	1	1	0
1	1	0	0	0	0	0	0
1	1	0	1	0	0	1	1
1	1	1	0	0	1	1	0
1	1	1	1	1	0	0	1

6. Write the decimal format maxterm Boolean output equations for the truth tables in Problem 5.

7. Convert the following equations into their requested canonical forms:

a. (SOP) $X = a'b + bc$

b. (POS) $P = (w' + x)(y + z')$

c. (SOP) $T = p(q' + s)$

d. (SOP) $R = L + M'(N'M + M'L)$

e. (POS) $U = r' + s(t + r) + s't$

8. Write the minterm output variable Boolean equations for each of the truth tables developed in Problem 4a through 4e.

9. Write the maxterm output variable Boolean equations for each of the truth tables developed in Problem 4a through 4e.

Section 3.4 to 3.5

10. Simplify the following using Karnaugh maps:

a. $X = a'bc + ab'c' + abc$

b. $y = f(a, b, c) = \Sigma(1, 3, 5, 6, 7)$

c. $T = w'xy + wz' + xyz'$

d. $P = f(w, x, y, z) = \Sigma(0, 2, 8, 10)$

e. $R = f(w, x, y, z)$
$= \Sigma(1, 3, 4, 5, 6, 9, 11, 12, 13, 14)$

11. Simplify the following using Karnaugh maps:

a. $V = f(a, b, c, d)$
$= \Sigma(2, 3, 4, 5, 13, 15)$
$+ \Sigma\, d(8, 9, 10, 11)$

b. $Y = f(u, v, w, x)$
$= \Sigma(1, 5, 7, 9, 13, 15)$
$+ \Sigma\, d(8, 10, 11, 14)$

c. $P = f(r, s, t, u)$
$= \Sigma(0, 2, 4, 8, 10, 14) + \Sigma\, d(5, 6, 7, 12)$

d. $F = f(u, v, w, x, y)$
$= \Sigma(0, 2, 8, 10, 16, 18, 24, 26)$

e. $H = f(a, b, c, d, e)$
$= \Sigma(5, 7, 9, 12, 13, 14, 15, 20, 21, 22, 23, 25, 29, 31)$

f. $M = f(v, w, x, y, z)$
$= \Sigma(1, 3, 4, 6, 9, 11, 12, 14, 17, 19, 20, 22, 25, 27, 28, 30)$
$+ \Sigma\, d(8, 10, 24, 26)$

g. $J = f(a, b, c, d, e, f)$
$= \Sigma(7, 12, 22, 23, 28, 34, 37, 38, 40, 42, 44, 46, 56, 58, 60, 62)$

h. $K = f(r, s, t, u, v, w)$
$= \Sigma(9, 11, 13, 15, 25, 27, 29, 31, 41, 43, 45, 47, 57, 59, 61, 63)$

12. Simplify the following noncanonical expressions using Karnaugh maps:

a. $T = a'b'c'de + a'bc'de + abcde + ab'c'de$

b. $P = v'w' + v'wy' + vw'z$

c. $G = y'z + w'xy' + w'xy + xy'z$

13. Identify the prime and essential prime implicants for the following expressions:

a. $S = f(a, b, c, d) = \Sigma(1, 5, 7, 8, 9, 10, 11, 13, 15)$

b. $T = f(a, b, c, d, e)$
$= \Sigma(0, 4, 8, 9, 10, 11, 12, 13, 14, 15, 16, 20, 24, 28)$

14. Simplify the following maxterm equations using Karnaugh maps:

a. $h = f(a, b, c, d) = \pi(2, 3, 4, 6, 7, 10, 11, 12)$

b. $I = f(d, e, f, g, h)$
$= \pi(5, 7, 8, 21, 23, 26, 30)$
$+ \pi\, d(10, 14, 24, 28)$

15. Show that $G = f(a, b, c, d) = \Sigma(0, 2, 5, 7, 8, 10, 13, 15)$ is the complement of $G = f(a, b, c, d) = \pi(1, 3, 4, 6, 9, 11, 12, 14)$. Use a Karnaugh map to illustrate the complement nature of the two equations.

16. Analyze the illustrated logic circuit. Is it implemented as simply as possible? Redraw the circuit with as few gates as possible.

Section 3.4 to 3.5
Problem 16

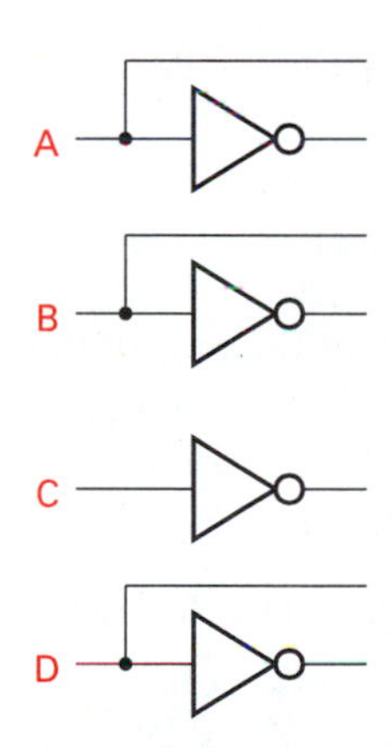

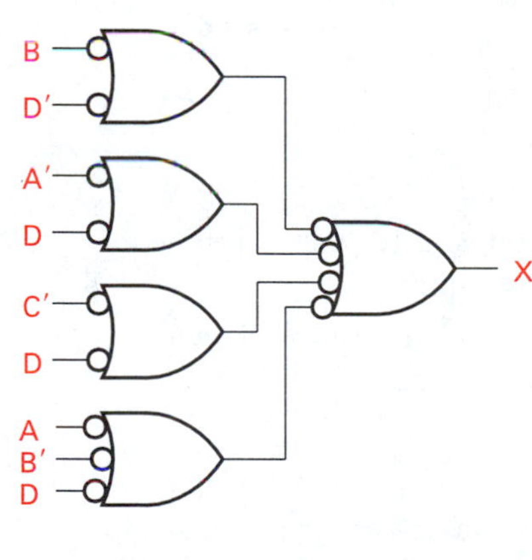

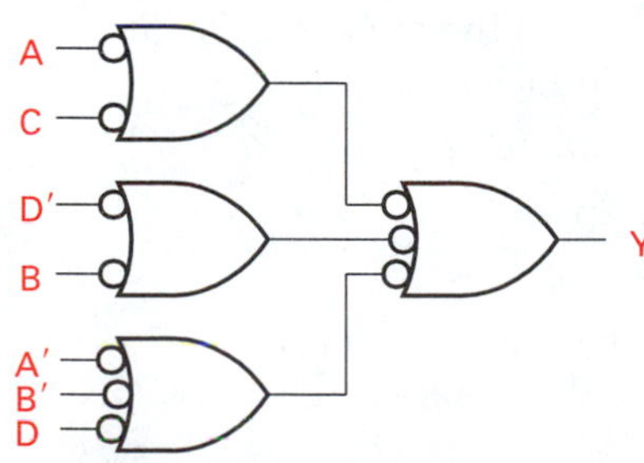

Section 3.6

Problem 20

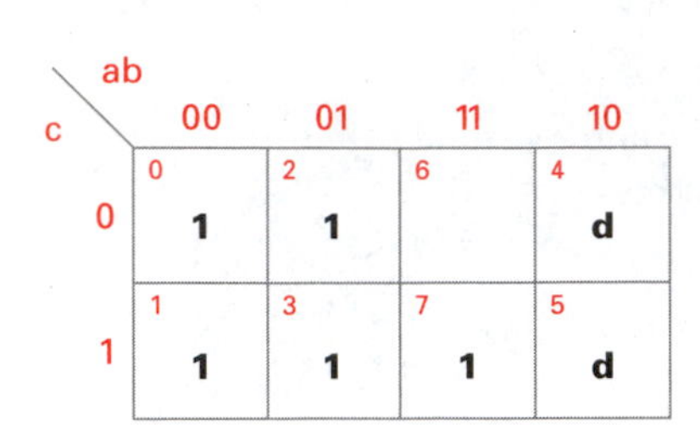

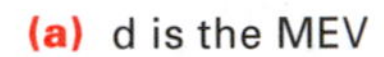
(a) d is the MEV

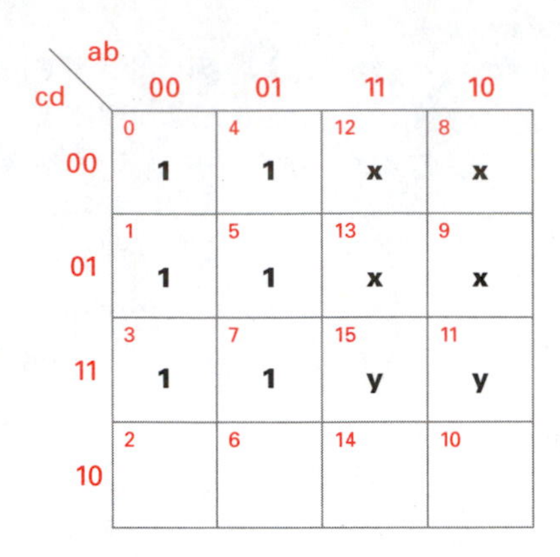

(b) x and y are MEVs

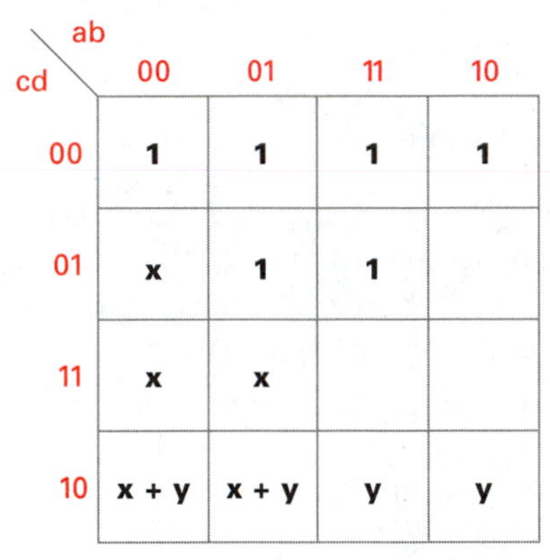

(c) x and y are MEVs

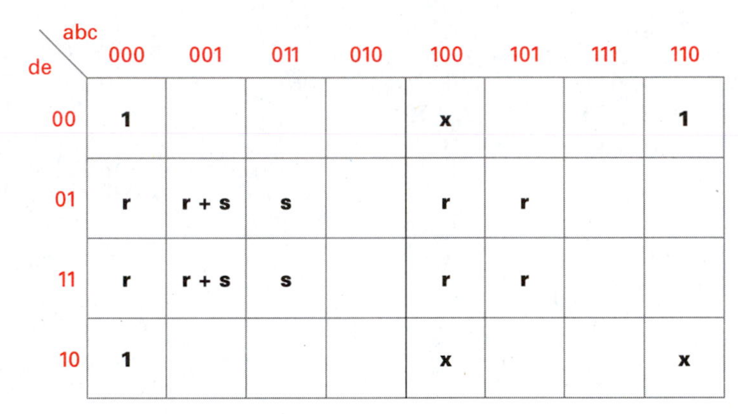

(d) x is a don't care term, r and s are MEVs

17. Repeat Problem 10b, 10d, and 10e using the Quine-McClusky tabulation algorithm.

18. Repeat Problem 11a through 11h using the Quine-McClusky tabulation algorithm.

Section 3.6

19. Simplify the SOP equations given in Problem 11a through 11h using MEVs. Let the MEV term be the least significant variable in each expression.

- **a.** Construct the MEV truth table.
- **b.** Create the MEV K-map.
- **c.** Write the simplified equations.
- **d.** Is the final expression optimal (compare it to a regular K-map simplified expression).

20. Find the output function for each of the illustrated MEV K-maps.

Section 3.7

21. Define the following in your own words:

- **a.** Asserted
- **b.** Active high, active low
- **c.** Bubble logic
- **d.** Mixed logic

22. Show eight ways that the AND function can be realized.

23. Show eight ways that the OR function can be realized.

Section 3.7

Problem 24

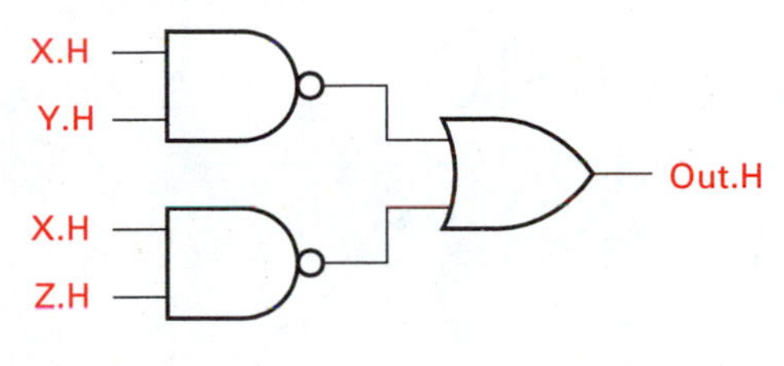

(a)

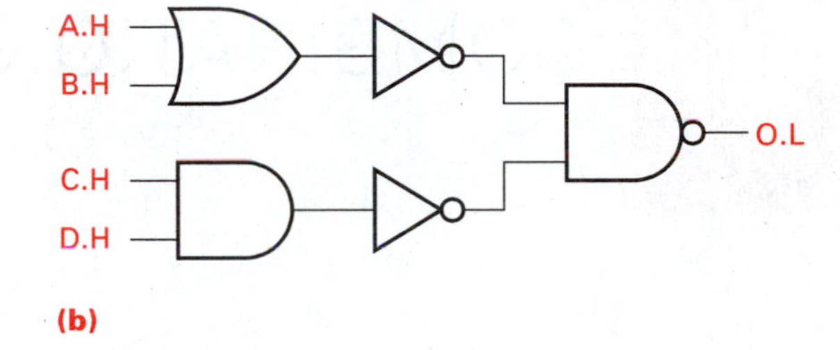

(b)

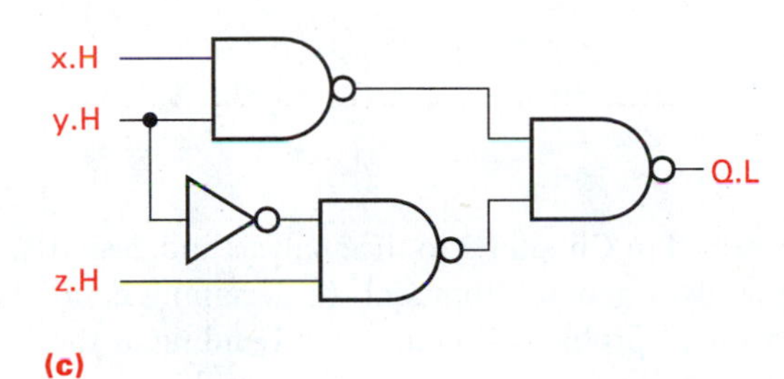

(c)

x.H
y.H
z.L
(A.H)′

(d)

24. Convert the illustrated logic diagrams to mixed "bubble" logic.

Section 3.8

25. Simplify the following multiple output functions:

a. $X = f(a, b, c) = \Sigma(1, 3, 7)$
$Y = f(a, b, c) = \Sigma(2, 6, 7)$

b. $X = f(a, b, c) = \Sigma(3, 4, 5, 7)$
$Y = f(a, b, c) = \Sigma(3, 4, 6, 7)$

26. Find the product functions for the following multiple output functions.

$X = f(a, b, c) = \Sigma(1, 2, 3, 7)$
$Y = f(a, b, c) = \Sigma(1, 2, 3, 6)$
$Z = f(a, b, c) = \Sigma(2, 4, 6)$

CHAPTER 4

ANALYSIS AND DESIGN OF COMBINATIONAL LOGIC

INTRODUCTION

This chapter will apply the principles developed in Chapter 3 to the analysis and design of combinational logic circuits. We will study a general approach to designing combinational logic circuits, starting with an initial problem statement and ending at the logic diagram. Combinational logic circuits such as decoders, encoders, multiplexers, adders, subtractors, multipliers, comparators, parity generators, and tristate buffers will be analyzed.

Because combinational logic is realized using integrated circuits, some of the problems associated with the devices themselves will be discussed. The chapter also provides an introduction to static hazards encountered with combinational logic circuits.

4.1 GENERAL APPROACH TO COMBINATIONAL LOGIC DESIGN

The synthesis of combinational logic starts with a problem statement and progresses through a series of steps, terminating in the final circuit design. The combinational logic design steps are these:

1. Develop a statement describing the problem to be solved.
2. Based on the problem statement, construct a truth table that clearly establishes the relationship between the input and output variables. Pay particular attention to the assignment of variable assertion levels.
3. Use Karnaugh maps or Quine-McClusky techniques to simplify the functions in deriving the output equations. This may require that the output equations be expressed as either SOP or POS. The best solution will require the fewest gates and gate inputs.
4. Arrange the simplified equations to suit the logic primitive type to be used in realizing the circuit. If at all possible design logic applying bubble-to-bubble techniques, using NANDS, NORs, or AND-OR logic as required.
5. Draw the final logic diagram.
6. Document the design by identifying variables' names that indicate assertion levels, and if possible provide a truth table.

❖ **EXAMPLE 4.1** Design a combinational circuit that will multiply two two-bit binary values.

SOLUTION

1. Four input variables (A_1, A_0, B_1, B_0) and four output variables (P_3, P_2, P_1, P_0) are needed. The four output variables are necessary because the maximum product of two two-bit values ($3_{10} \times 3_{10} = 9_{10}$) requires four bits. Let each variable be active high.
2. Construct a truth table like Table 4.1.

TABLE 4.1
2 × 2 multiplier truth table

Inputs				Outputs			
A_1	A_0	B_1	B_0	P_3	P_2	P_1	P_0
0	0	0	0	0	0	0	0
0	0	0	1	0	0	0	0
0	0	1	0	0	0	0	0
0	0	1	1	0	0	0	0
0	1	0	0	0	0	0	0
0	1	0	1	0	0	0	1
0	1	1	0	0	0	1	0
0	1	1	1	0	0	1	1
1	0	0	0	0	0	0	0
1	0	0	1	0	0	1	0
1	0	1	0	0	1	0	0
1	0	1	1	0	1	1	0
1	1	0	0	0	0	0	0
1	1	0	1	0	0	1	1
1	1	1	0	0	1	1	0
1	1	1	1	1	0	0	1

The output SOP equations are
$P_3 = f(A_1, A_0, B_1, B_0) = \Sigma\,(15)$
$P_2 = f(A_1, A_0, B_1, B_0) = \Sigma\,(10, 11, 14)$
$P_1 = f(A_1, A_0, B_1, B_0) = \Sigma\,(6, 7, 9, 11, 13, 14)$
$P_0 = f(A_1, A_0, B_1, B_0) = \Sigma\,(5, 7, 13, 15)$

3. The individually simplified equations are

$$P_3 = A_1A_0B_1B_0, \qquad P_2 = A_1A_0'B_1 + A_1B_1B_0'$$
$$P_1 = A_1'A_0B_1 + A_0B_1B_0' + A_1B_1'B_0 + A_1A_0'B_0$$
$$P_0 = A_0B_0$$

4. The equations from step 3 can be realized using a two-level AND-OR network. If realization in NANDs or NORs were necessary then bubble-to-bubble conversions would be required.
5. Draw the mixed logic diagram that satisfies the equations. The logic diagram is illustrated in Figure 4.1.

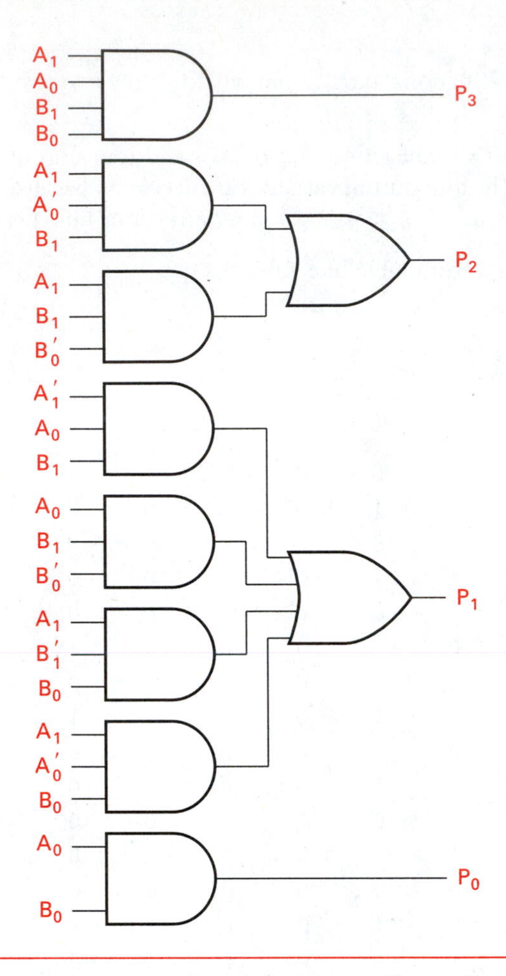

Figure 4.1
Logic diagram for a 2 × 2-bit multiplier

❖

❖ **EXAMPLE 4.2** Design a combinational circuit that will accept a 2421 BCD code and drive a TIL-312 seven-segment display.

SOLUTION The 2421 code is shown in Table 4.2. The TIL-312 is a common-anode, red light-emitting-diode display package. Each LED is indicated by a letter as shown in Figure 4.2.

TABLE 4.2
2421 code

Decimal	*w*	*x*	*y*	*z*
0	0	0	0	0
1	0	0	0	1
2	1	0	0	0
3	1	0	0	1
4	1	0	1	0
5	1	0	1	1
6	1	1	0	0
7	1	1	0	1
8	1	1	1	0
9	1	1	1	1

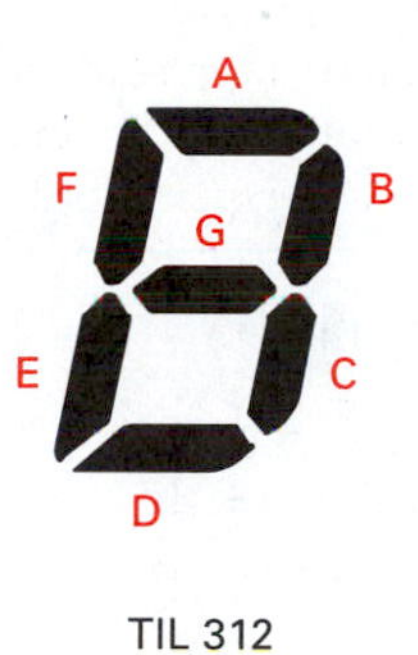

TIL 312
TIL 314
TIL 315

Pin 1 Cathode A
Pin 2 Cathode F
Pin 3 Anode Digit and Decimal
Pin 4 Omitted
Pin 5 Omitted
Pin 6 Cathode Left Decimal
Pin 7 Cathode E
Pin 8 Cathode D
Pin 9 Cathode Right Decimal
Pin 10 Cathode C
Pin 11 Cathode G
Pin 12 Omitted
Pin 13 Cathode B
Pin 14 Anode Digit and Decimal

Pin 3 is internally connected to Pin 14

Figure 4.2
LED diagram and package pin assignments

Figure 4.3
Digit display format

A combinational logic output is needed to drive the cathodes of the seven LEDs in the display package. The LED common anodes are connected to a +5 V dc supply. The LED is lit when the control signal connected to its cathode is a logic 0. The display format is illustrated in Figure 4.3.

Generate a truth table (like that in Table 4.3) illustrating active low outputs to turn on each of the seven-segment LEDs. Write the output equations and simplify them using either K-maps or Quine-McClusky techniques. You may assume that both levels of the four inputs are present (x and x'). Realize the final logic diagram using NAND gates. Draw the diagram using mixed logic convention.

The 2421 BCD code value and the minterm/maxterm value are not the same, as indicated in the truth table. The decimal values in the K-map squares are decoded from

TABLE 4.3
2421 BCD to seven-segment display truth table

2421	Min-max	Inputs				Outputs						
Decimal	Term Decimal	w	x	y	z	A	B	C	D	E	F	G
0	0	0	0	0	0	0	0	0	0	0	0	1
1	1	0	0	0	1	1	0	0	1	1	1	1
2	8	1	0	0	0	0	0	1	0	0	1	0
3	9	1	0	0	1	0	0	0	0	1	1	0
4	10	1	0	1	0	1	0	0	1	1	0	0
5	11	1	0	1	1	0	1	0	0	1	0	0
6	12	1	1	0	0	0	1	0	0	0	0	0
7	13	1	1	0	1	0	0	0	1	1	1	1
8	14	1	1	1	0	0	0	0	0	0	0	0
9	15	1	1	1	1	0	0	0	0	1	0	0

the binary code values (8, 4, 2, 1), thus giving a different minterm or maxterm than the 2421 code value.

The active low output functions can be expressed either as maxterms or as minterms. To illustrate the need to view combinational logic from each perspective we will do both.

The active-low product-of-sums (POS) equations are

$$
\begin{aligned}
A.L &= \pi(0,\ 8,\ 9,\ 11,\ 12,\ 13,\ 14,\ 15)\\
B.L &= \pi(0,\ 1,\ 8,\ 9,\ 10,\ 13,\ 14,\ 15)\\
C.L &= \pi(0,\ 1,\ 9,\ 10,\ 11,\ 12,\ 13,\ 14,\ 15)\\
D.L &= \pi(0,\ 8,\ 9,\ 11,\ 12,\ 14,\ 15)\\
E.L &= \pi(0,\ 8,\ 12,\ 14)\\
F.L &= \pi(0,\ 10,\ 11,\ 12,\ 14,\ 15)\\
G.L &= \pi(8,\ 9,\ 10,\ 11,\ 12,\ 14,\ 15)
\end{aligned}
$$

Figure 4.4
Maxterm K-maps for 2421 to seven-segment decoder

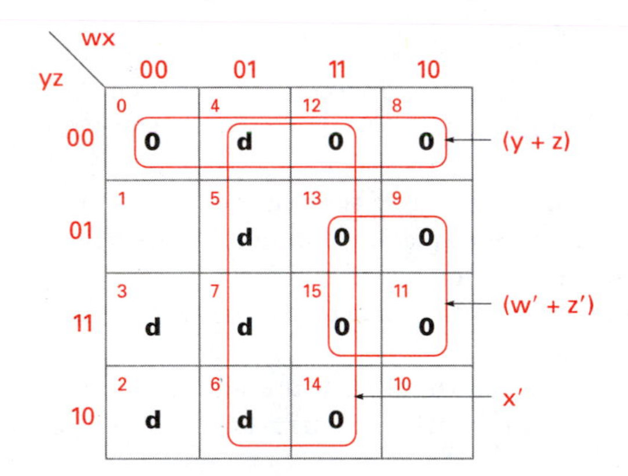

(a) A.L output K-map
A.L = (x')(y + z)(w' + z')

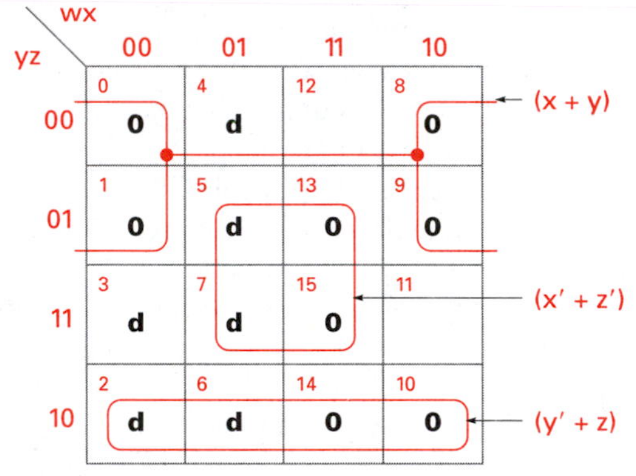

(b) B.L output K-map
B.L = (x + y)(x' + z')(y' + z)

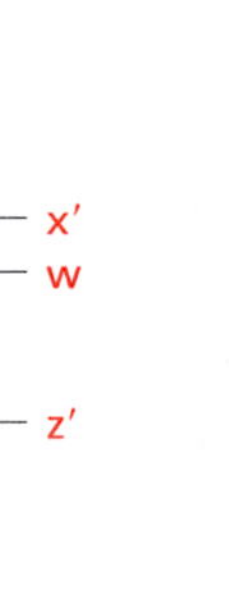

(c) C.L output K-map
C.L = wx'y'z'

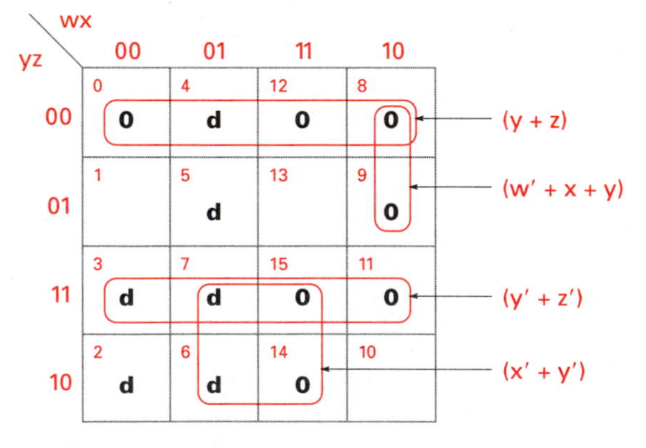

(d) D.L output K-map

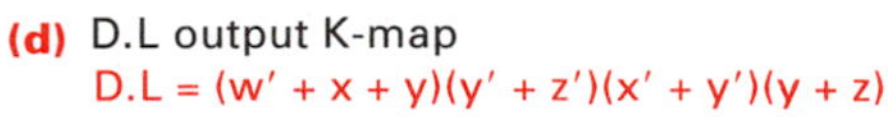
D.L = (w' + x + y)(y' + z')(x' + y')(y + z)

The active low sum-of-products (SOP) equations are

$$A.L = \sum(1,\ 10)$$
$$B.L = \sum(11,\ 12)$$
$$C.L = \sum(8)$$
$$D.L = \sum(1,\ 10,\ 13)$$
$$E.L = \sum(1,\ 9,\ 10,\ 11,\ 13,\ 15)$$
$$F.L = \sum(1,\ 8,\ 9,\ 13)$$
$$G.L = \sum(0,\ 1,\ 13)$$

Both POS and SOP don't care terms are $\Sigma d(2, 3, 4, 5, 6, 7)$, because these terms do not occur.

Simplifying maxterms produces an OR-AND realization and simplifying the minterms produces an AND-NOR realization.

The K-maps for the POS equations are illustrated in Figure 4.4.

Figure 4.4
Continued

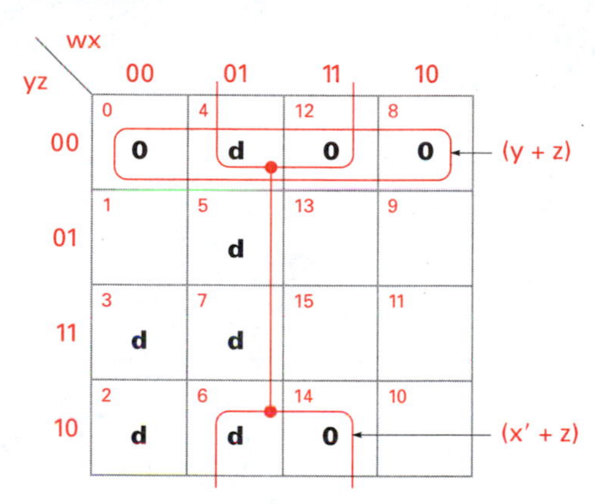

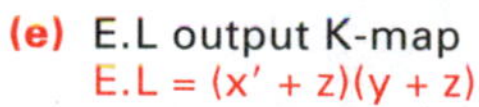

(e) E.L output K-map
E.L = (x' + z)(y + z)

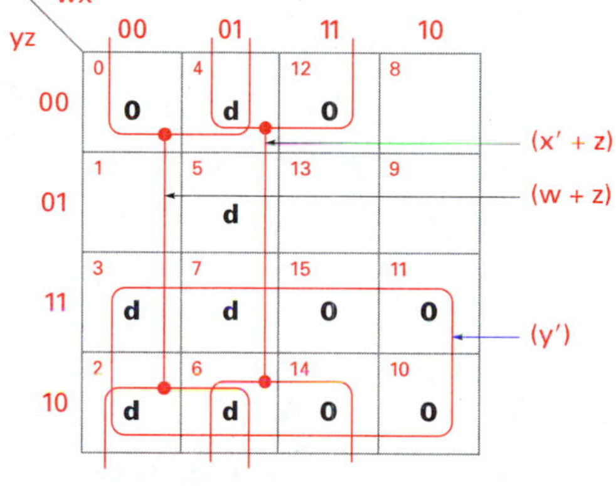

(f) F.L output K-map
F.L = y'(x' + z)(w + z)

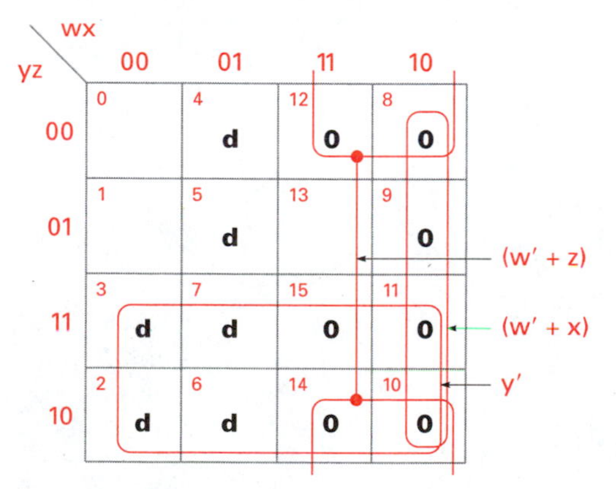

(g) G.L output K-map
G.L = (y')(w' + z)(w' + x)

The simplified POS equations are

$$A.L = (x')(y + z)(w' + z')$$
$$B.L = (x + y)(x' + z')(y' + z)$$
$$C.L = wx'y'z'$$
$$D.L = (w' + x + y)(y' + z')(y + z)(x' + y')$$
$$E.L = (y + z)(x' + z)$$
$$F.L = (y')(w + z)(x' + z)$$
$$G.L = (y')(w' + x)(w' + z)$$

The Karnaugh maps to simplify minterms from Table 4.3 are shown in Figure 4.5.

The following SOP output simplified equations:

$$A.L = w'z + x'yz'$$
$$B.L = xy'z' + x'yz$$
$$C.L = wx'y'z'$$

Figure 4.5

K-maps simplifying minterms

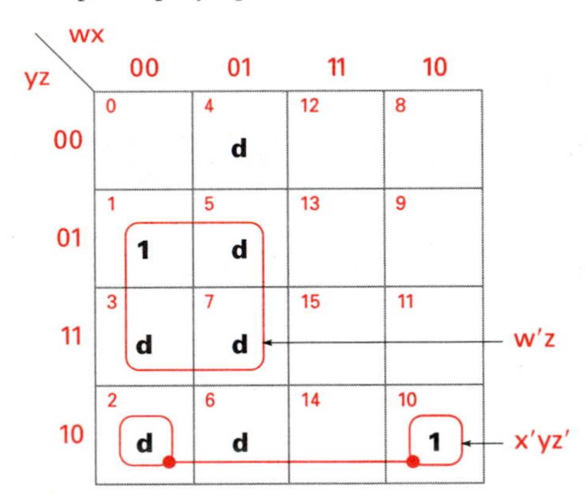

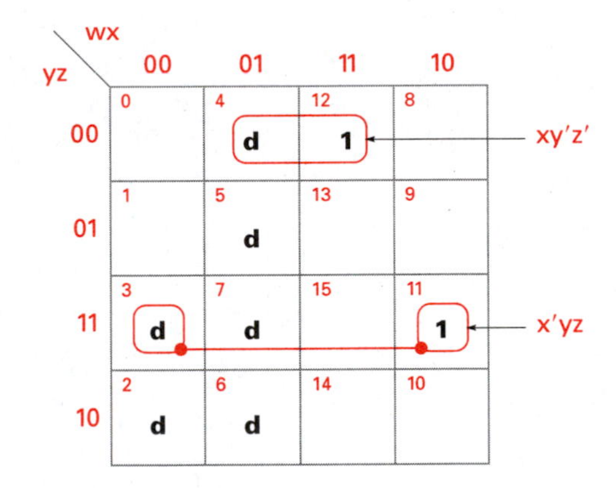

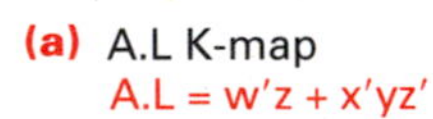
(a) A.L K-map
A.L = w'z + x'yz'

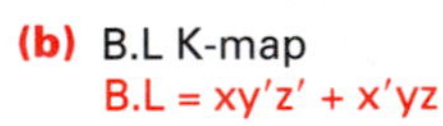
(b) B.L K-map
B.L = xy'z' + x'yz

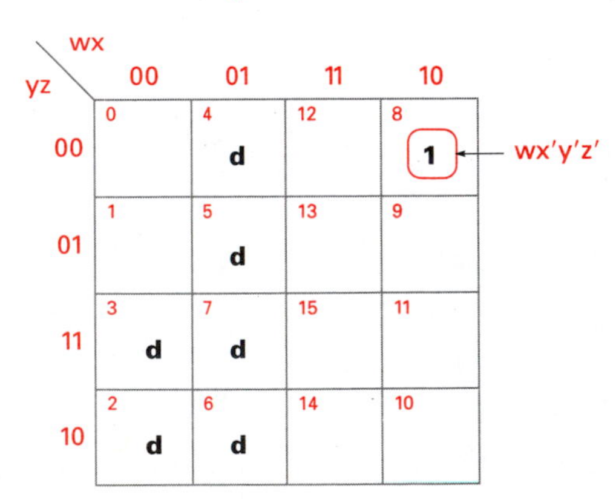

(c) C.L K-map
C.L = wx'y'z'

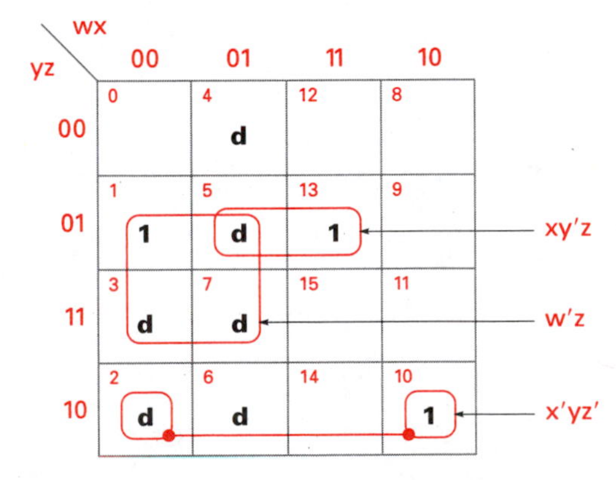

(d) D.L K-map
D.L = xy'z + w'z + x'yz'

$$D.L = xy'z + x'yz' + w'z$$
$$E.L = x'y + z$$
$$F.L = wx'y' + y'z$$
$$G.L = w' + xy'z$$

The initial requirement to realize the circuit using NAND gates can better be met by using the SOP equations. Using DeMorgan's theorems the conversion of the preceding equations yields a set of NAND functions. Figure 4.6 shows the logic diagram.

$$A.L = [(w'z)'(x'yz')']'$$
$$B.L = [(xy'z')'(x'yz)']'$$
$$C.L = (wx'y'z')'$$
$$D.L = [(xy'z)'(x'yz')'(w'z)']'$$
$$E.L = [(x'y)'(z)']'$$
$$F.L = [(wx'y')'(y'z)']'$$
$$G.L = [(w)(xy'z)']'$$

Figure 4.5
Continued

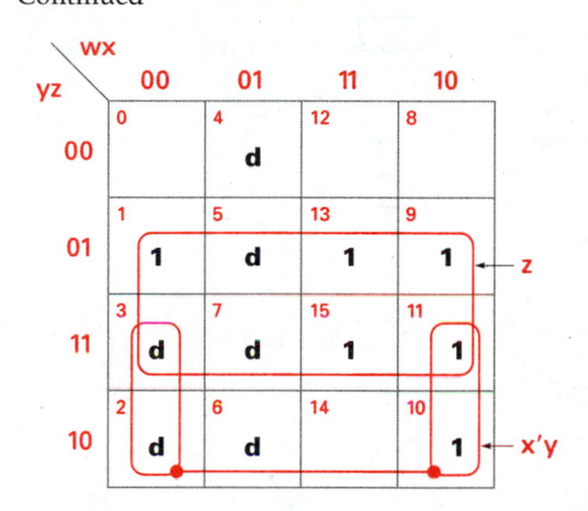

(e) E.L K-map
E.L = z + x'y

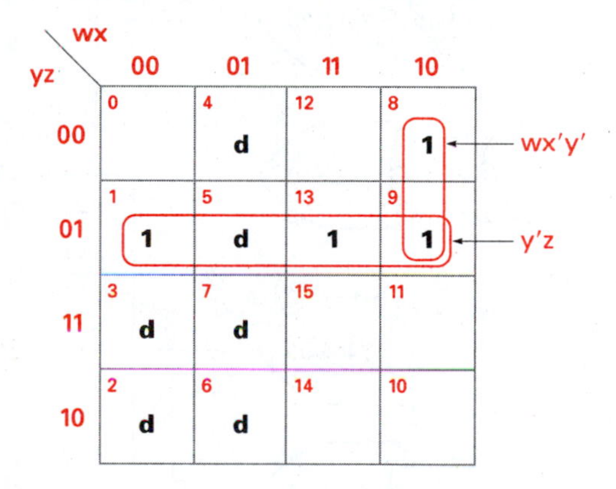

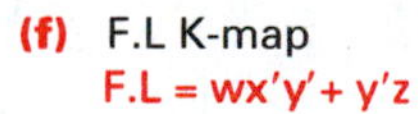

(f) F.L K-map
F.L = wx'y' + y'z

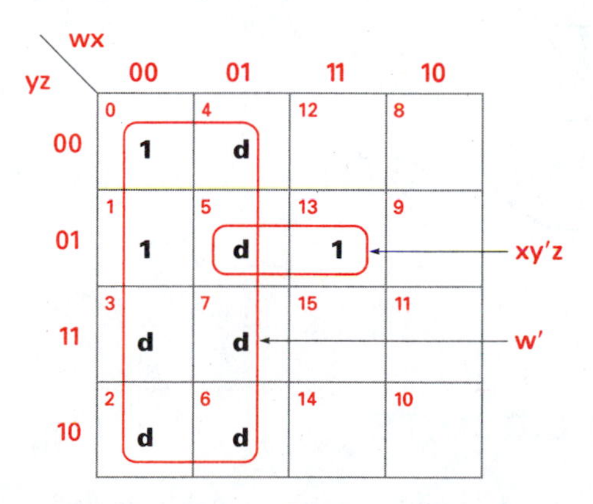

(g) G.L K-map
G.L = xy'z + w'

Figure 4.6
Logic diagram for 2421 BCD to seven-segment decoder

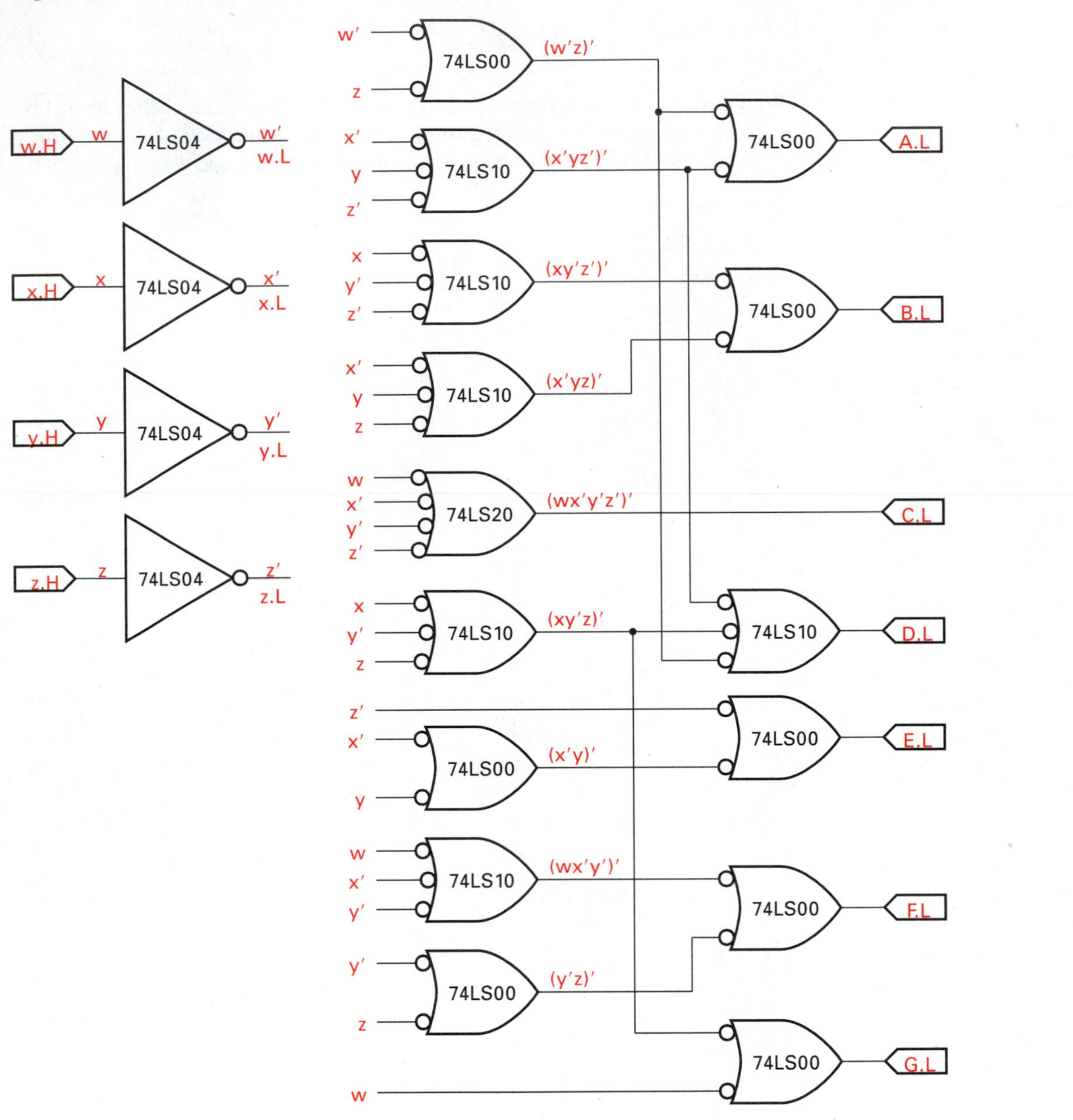

4.2 INTRODUCTION TO DIGITAL INTEGRATED CIRCUITS

In the preceding examples we have been concerned with combinational logic design. Nothing has been said about the actual circuits used to implement the logic. This section introduces small-scale integrated circuits (ICs) that can be interconnected to realize combinational logic designs. In Chapter 1 we introduced a hierarchy of integrated circuits based on the number of equivalent gates. Small-scale integrated circuits contain up to 10 equivalent two-input gates. Medium-scale ICs contain up to 100 equivalent gates, and large-scale ICs contain over 100 gates. Very large-scale ICs may contain thousands of equivalent gates. In this section, we introduce the fundamentals of small-scale digital ICs. We will not cover the internal circuit operation; that is left to Chapter 10.

Several digital integrated circuit logic families exist; the most common are transistor-transistor logic (TTL), emitter-coupled logic (ECL), and complementary metal-oxide semiconductor logic (CMOS). Each of these main logic families is made up of numerous subfamilies. For example, the TTL families include low power (L), high speed (H), low power Schottky (LS), Schottky (S), advanced low power Schottky (ALS), and advanced Schottky (AS). The same sort of subfamily divisions exist in the ECL and CMOS main families. Each type of logic device occupies a niche in the logic design structure. Some consume less power and therefore are used in power-sensitive designs; others switch much faster and are used in high-speed designs. As new subfamilies become available they will often supersede older families. The Schottky TTL subfamilies have made the older subfamilies (low power and high speed) obsolete. Chapter 10 covers the three main logic familes in more detail.

Continuing the design example started in the preceding section we will use actual TTL ICs in implementing a logic design.

The numbering system used to indicate the various subfamilies is partially indicated at the top of Figure 4.7. SN is a prefix for Texas Instruments (the prefix indicates the manufacturer). The number 54 indicates a military operating temperature range (-55 to 125°C) and 74 indicates a commercial temperature range (0 to 70°C). Both temperature ranges are available for most functions and subfamilies. The letter-number combination following the SN74/54 indicates the subfamily and the logical function of the IC. For example, the logical function for the IC indicated in Figure 4.7 is NAND. Two subfamilies are represented, advanced low-power Schottky (ALS) and advanced Schottky (AS).

A commercial temperature range advanced low-power Schottky NAND gate IC is indicated by the name, SN74ALS00. The package type is represented by the suffix. A variety of package types with different pin-out configurations are available as indicated in Figure 4.7.

Four two-input NAND gates are available in the SN74ALS00 integrated circuit. The pin-out diagrams and truth tables for the 74ALS10 and 74ALS20 are shown in Figure 4.8. The 74ALS10 and 74ALS20 are triple three-input and dual four-input NAND gate IC packages.

By using all three NAND gate packages we can realize the 2421 BCD to seven-segment decoder designed previously. Figure 4.9 shows the IC interconnection for the 2421 to seven-segment decoder.

Definitions of TTL circuit parameters, input voltages and currents, output voltages and currents, power supply voltages and currents, fan-out, power consumption, and switching speed follow and are demonstrated in Table 4.4.

V_{CC}, power supply voltage This is the voltage supplied to the integrated circuit. The range of allowable supply voltages is dependent on the temperature range (military 54 series or commercial 74 series). The nominal V_{CC} value is $+5$ volts for TTL circuits.

I_{CCH}, power supply current This is the supply current required to operate the IC when the output voltage is a logical 1. The value is dependent on the TTL subfamily.

Figure 4.7
Two-input NAND gates (reprinted by permission of Texas Instruments)

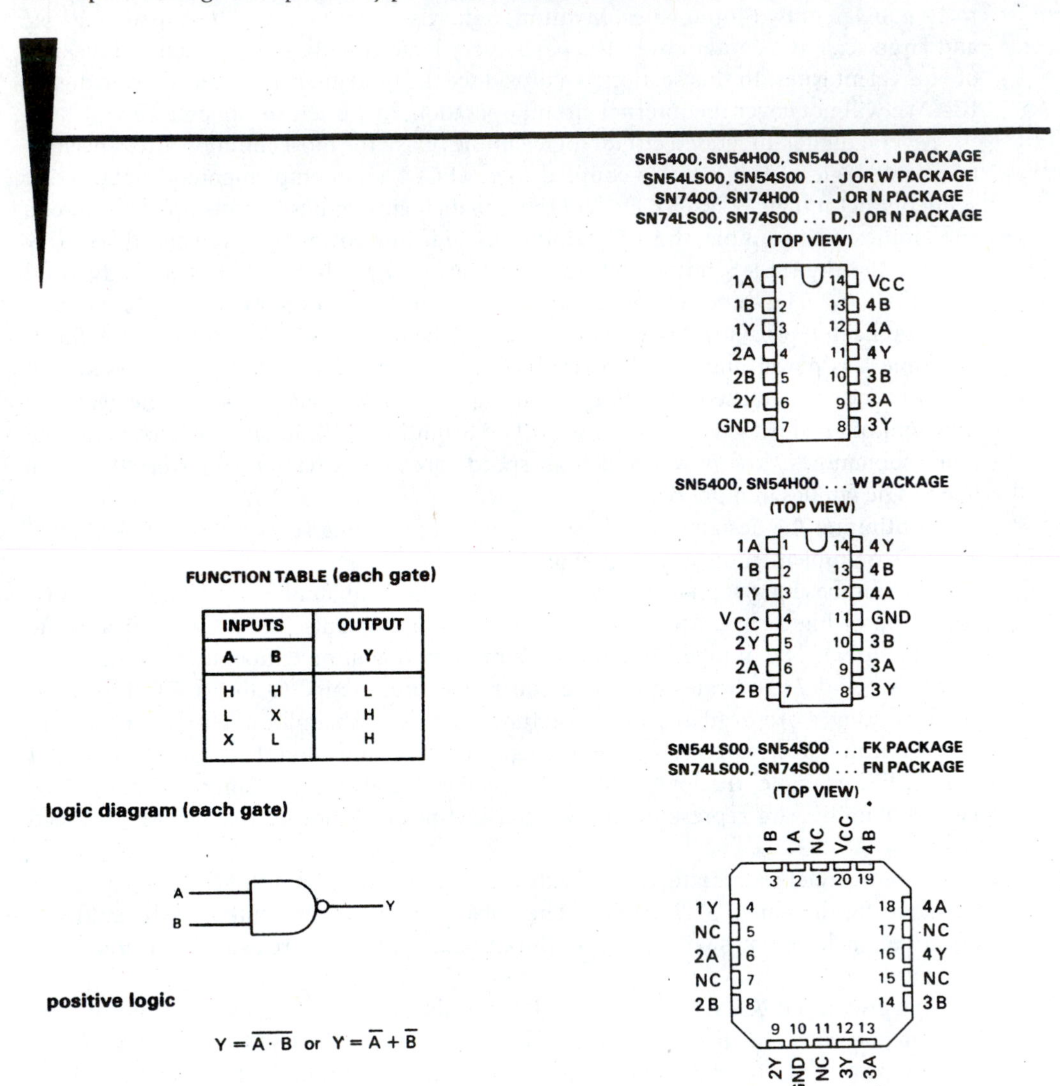

FUNCTION TABLE (each gate)

INPUTS		OUTPUT
A	B	Y
H	H	L
L	X	H
X	L	H

logic diagram (each gate)

positive logic

$Y = \overline{A \cdot B}$ or $Y = \overline{A} + \overline{B}$

Figure 4.8
74ALS10 and 74ALS20 IC package pin-out diagrams. (reprinted by permission of Texas Instruments)

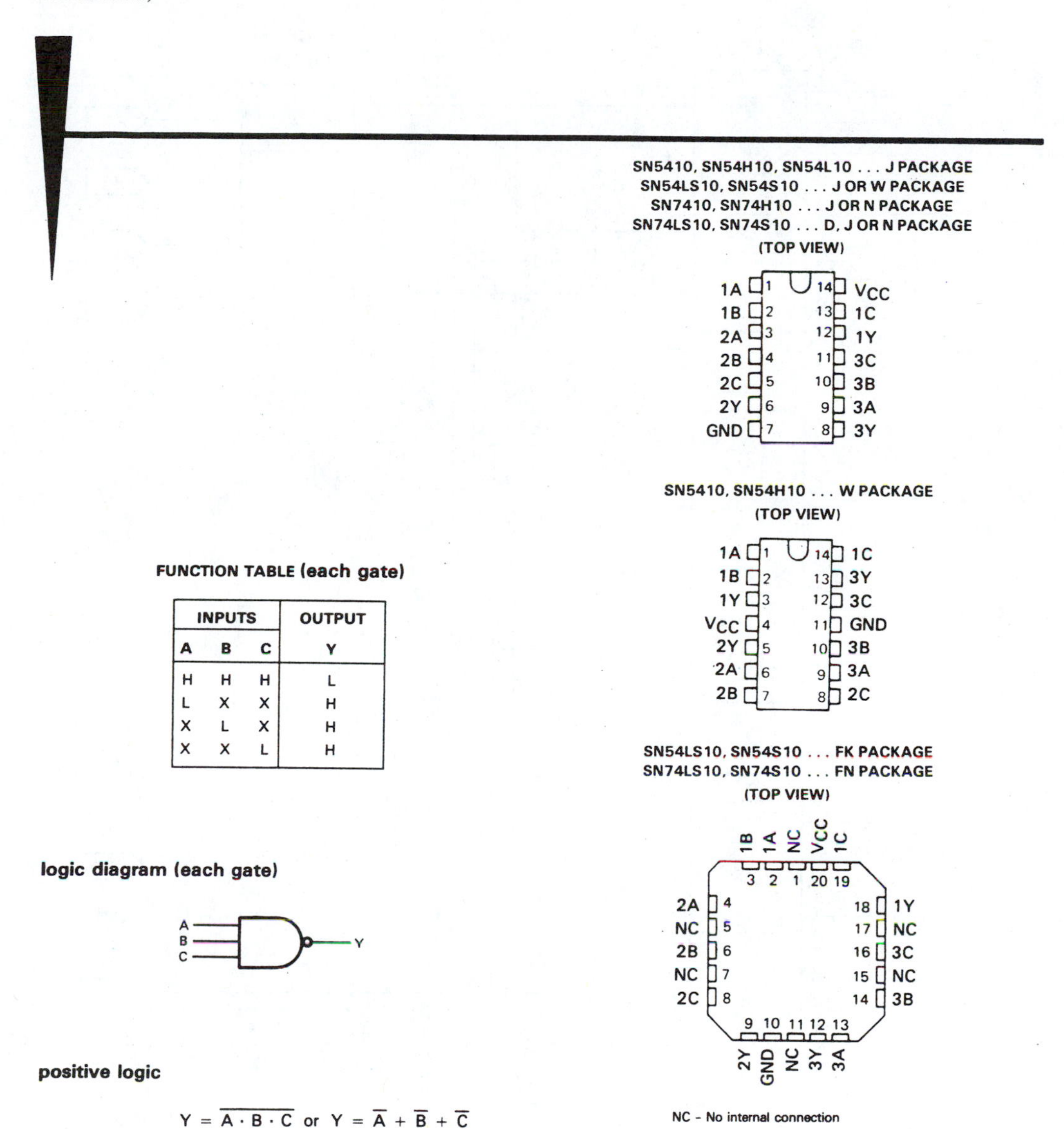

INPUTS			OUTPUT
A	B	C	Y
H	H	H	L
L	X	X	H
X	L	X	H
X	X	L	H

Figure 4.9
IC interconnection diagram for 2421 to seven-segment decoder

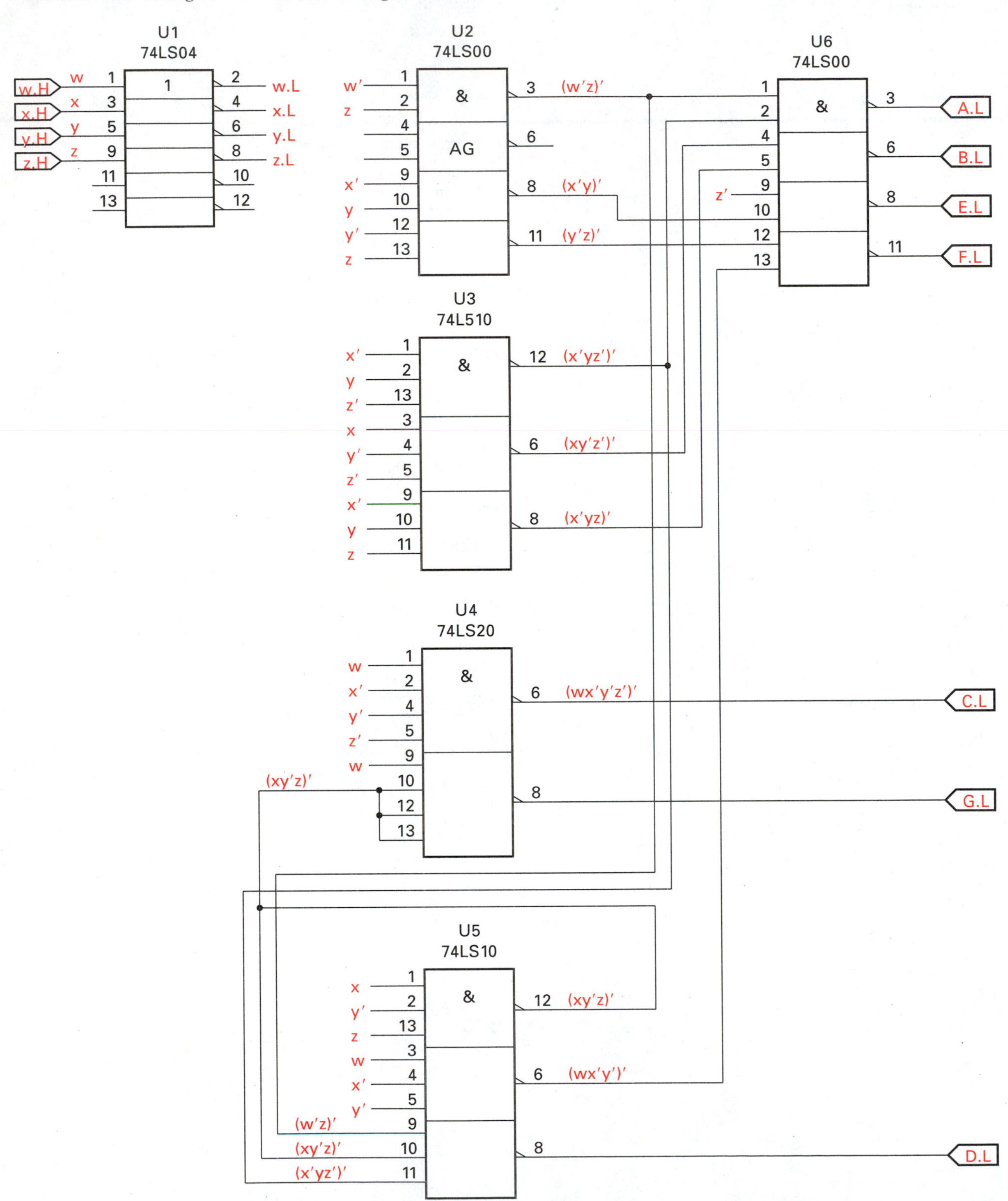

TABLE 4.4

Two-input NAND (74xxx00) TTL subfamily parameter values

Specification		TTL subfamily 7400	74LS00	47S00	74ALS00	74AS00
V_{CC}	4.75 V − 5.0 V →					
I_{CCH}	(max)	8 mA	1.6 mA	16 mA	.85 mA	3.2 mA
I_{CCL}	(max)	22 mA	4.4 mA	36 mA	3 mA	17.4 mA
I_{OS}	(max)	55 mA	100 mA	100 mA	70 mA	112 mA
I_{OL}	(max)	16 mA	20 mA	20 mA	8 mA	20 mA
I_{OH}	(max)	.4 mA	.4 mA	1 mA	.4 mA	2 mA
I_{IH}	(max)	50 μA	20 μA	50 μA	20 μA	20 μA
I_{IL}	(max)	1.6 mA	.4 mA	2 mA	.1 mA	.5 mA
V_{OL}	(max)	.4 V	.5 V	.5 V	.5 V	.5 V
V_{OH}	(min)	2.4 V	2.4 V	2.7 V	V_{CC} − 2V	V_{CC} − 2V
V_{IH}	(min)	2.0 V	2.0 V	2.0 V	2.0 V	2.0 V
V_{IL}	(max)	.8 V	.8 V	.8 V	.8 V	.8 V
t_{pLH}	(max)	22 ns	15 ns	4.5 ns	11 ns	4.5 ns
t_{pHL}	(max)	15 ns	15 ns	5 ns	8 ns	4 ns

I_{CCL}, power supply current This is the supply current required to operate the IC when the output voltage is a logical 0. The value is dependent on the TTL subfamily.

I_{OS}, short circuit output current This is the amount of current the TTL output circuit can deliver into a short circuit. The input logic levels are at the logical values necessary to drive the output to a logical 1 if it were not shorted to ground.

I_{OL}, output current low This is the maximum output terminal sink current, when the output voltage is a logical 0, that will guarantee a logical 0 output. The value depends on the logic family.

I_{OH}, output current high This is the maximum output terminal source current, when the output voltage is a logical 1, that will guarantee a logical 1 output. The value depends on the logic family.

I_{IH}, current input high This is the maximum current at the input terminal when the input voltage represents a logical 1. The value of this parameter is dependent on the TTL subfamily.

I_{IL}, current input low This is the maximum input current when the input voltage represents a logical 0. The value of this parameter is dependent upon the TTL subfamily.

V_{OL}, voltage output low This is the maximum allowable output voltage that can represent a logical 0. The value of this parameter is dependent on the TTL subfamily.

V_{OH}, voltage output high This is the minimum output voltage that can represent a logical 1. The value of this parameter is dependent on the TTL subfamily.

V_{IH}, voltage input high This is the minimum logical 1 input voltage. For all TTL families $V_{IH} = 2.0$ V.

V_{IL}, voltage input low This is the maximum logical 0 input voltage. For all TTL families V_{IL} is a maximum of .8 V.

t_{pLH}, propagation delay time from output low to high This is the delay between the input and output voltage levels as the output changes (propagates) from a logical 0 to a logical 1 (see Figure 4.10).

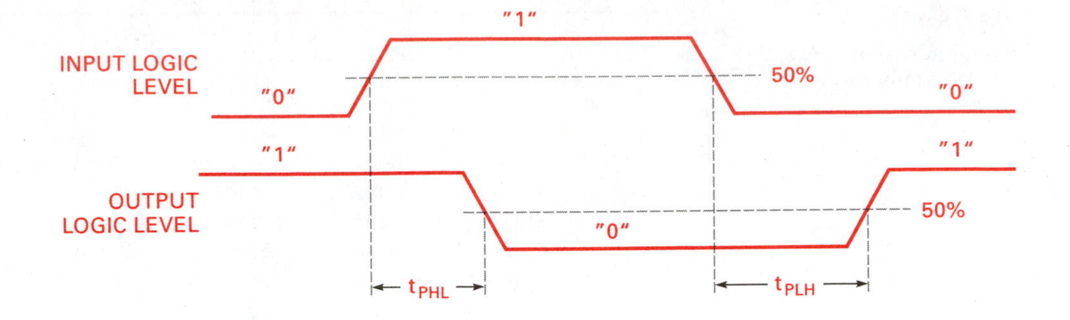

Figure 4.10 Input and output voltage waveforms illustrating t_{pHL} and t_{pLH}

t_{pHL}, propagation delay time from output high to low This is the delay time between the input and output voltage levels as the output changes (propagates) from a logical 1 to a logical 0 (see Figure 4.10).

Fan-out *Fan-out* is defined as the number of gate inputs that a single gate output can drive and still maintain voltage and current specifications. A fan-out of 10 is permitted, within the same TTL subfamily.

The two parameters that govern the choice of TTL subfamily for a given logic application are power supply currents (power consumption) and switching speed (propagation delays). Applications requiring high speed use fast integrated circuits, and applications requiring low power consumption use ICs that draw minimum supply currents.

For example, consider the power consumption and switching speed for the 2421 to seven-segment decoder, using 74ALS small-scale ICs presented in the first section. The power dissipated by the circuit shown in Figure 4.7 is the product of the average current and the power supply voltage, $P = (I_{CC}\text{ave})(V_{CC})$. Assume $V_{CC} = 5.0$ V. The average current drawn by the circuit assumes that each device output is 0 for half of the time and 1 the other half. We assume an average current because there is no way to know what the device output levels are without specifying a given input code value. The supply current values, for each gate in the IC, for the 74LS00 (two-input NAND), 74LS10 (three-input NAND), 74LS20 (four-input NAND), and 74LS04 (Hex inverters) are found in a TTL data book and rewritten here:

	LS00	LS10	LS20	LS04
I_{CCH}	.8 mA	.6 mA	.4 mA	1.2 mA
I_{CCL}	2.4 mA	1.8 mA	1.2 mA	3.6 mA
I_{CCave}	1.6 mA	1.2 mA	.8 mA	2.4 mA

The total average supply current is found to be

$$I_{CCave} = 8(1.6 \text{ mA}) + 6(1.2 \text{ mA}) + 2(.8 \text{ mA}) + 6(2.4 \text{ mA}) = 36 \text{ mA}$$

Dissipated power is found to be

$$P = 5.0 \text{ V}(36 \text{ mA}) = 180 \text{ mW}$$

Compare the dissipated power using LS devices with the standard TTL and the ALS TTL realization of the same circuit. Look up the values for $I_{CC}H$ and $I_{CC}L$ in a TTL data book.

Calculating the total propagation delay requires finding the longest delay path and then adding up the worst case (maximum values) propagation delays. The longest path is chosen because new data cannot be presented to the inputs until the outputs have

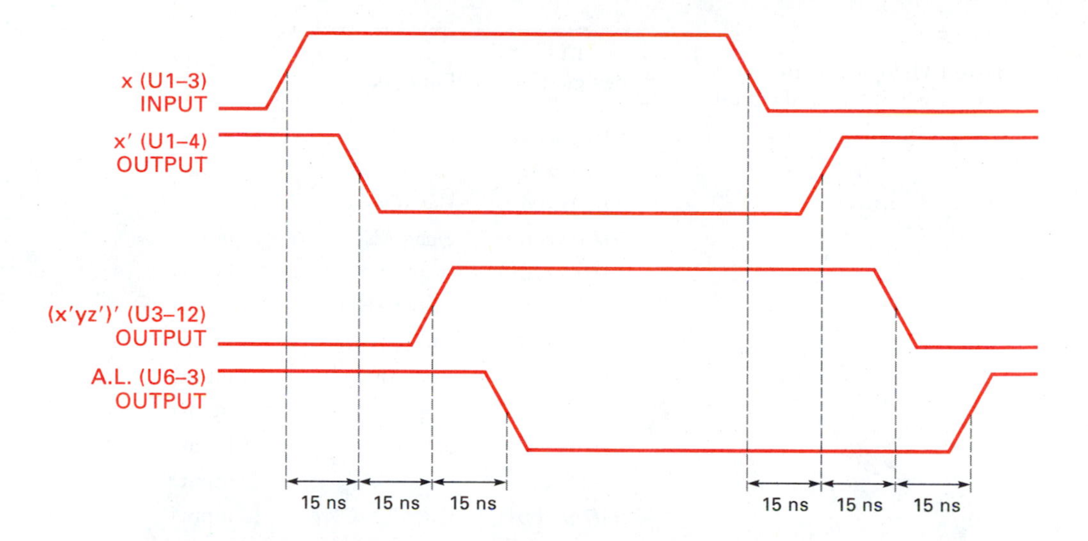

Figure 4.11
Propagation delay

responded to the prior data. By using worst case delay values, we can ensure that new input data will not be presented until the circuit has properly passed the prior data to its output. Propagation delays for the four ICs used are rewritten from a TTL data book:

	LS00	LS04	LS10	LS20
t_{pHL}	15 ns	15 ns	15 ns	15 ns
t_{pLH}	15 ns	15 ns	15 ns	15 ns

The longest delay path is through one LS04 (inverter) and two LS NAND gates. Figure 4.11 illustrates the delay path through two LS10 and one LS00 gates.

To illustrate propagation delay calculations, assume that the x input to U–1 (pin 3) changes as shown in Figure 4.11. The x' output at U1–4 is connected to a three-input NAND gate at U3–1. Assume that the other inputs to U3A are such that the U3–12 output changes as indicated. The U3–12 output is connected to an input at U6–2. U6A is acting as an inverter, producing the change in its output at U6-3. The result is the timing diagram in Figure 4.11. The total delay is the sum of the individual delays:

$$T_{\text{pdtotal}} = 15 \text{ ns} + 15 \text{ ns} + 15 \text{ ns} = 45 \text{ ns}$$

Compare the propagation delays using the same input transition and delay path using standard 74xx and advanced Schottky (AS) logic. Look up the delay times in a TTL data book.

A wide variety of combinational logic devices are available to the designer. Table 4.5 illustrates the TTL designation for the basic logic functions. Other special purpose devices such as open collector, Schmidt triggers, high-current buffers, and combinations of logic functions are available and will be covered in Chapter 10.

4.3 DECODERS

Decoders are a class of combinational logic circuits that convert a set of input variables representing a code into a set of output variables representing a different code. The relationship between the input and output codes can be expressed in a truth table. En-

TABLE 4.5
Basic TTL logic functions available in integrated circuits

TTL designation	Function	
54/74xx00	Quad NAND	(2 input)
54/74xx02	Quad NOR	(2 input)
54/74xx04	Hex inverter	
54/74xx08	Quad AND	(2 input)
54/74xx10	Triple NAND	(3 input)
54/74xx11	Triple AND	(3 input)
54/74xx20	Dual NAND	(4 input)
54/74xx21	Dual AND	(4 input)
54/74xx27	Triple NOR	(3 input)
54/74xx30	NAND	(8 input)
54/74xx32	Quad OR	(2 input)
54/74xx86	Quad EX-OR	(2 input)

coded information is presented as n inputs producing 2^n outputs. The 2^n output values can range from 0 to $2^n - 1$. Some decoders generate outputs over a truncated portion of possible values. For example a 4-line to 10-line decoder has 4 inputs ($n = 4$) and only 10 outputs. Decoders also usually have inputs to activate or enable decoded outputs based on data inputs. Figure 4.12 is one model of a typical decoder.

A wide range of decoders are available as integrated circuits, so the logic designer is rarely required to develop decoders from scratch. Figure 4.13 shows a 2-to-4 decoder. This particular decoder function is integrated into a 74xx139, a dual 2-to-4 decoder IC.

The 74xx139 contains two, 2-to-4 decoders, qualifying it for medium-scale integrated circuit status (over 10 equivalent gates). Because $n = 2$, four outputs are possible. One, and only one, of the outputs, Y_0 to Y_3, is active for a given input. Y_0 is asserted when inputs $A = B = 0$, and the active low enable input, $G' = 0$. Refer to the function table in Figure 4.13(b), where the relationship between the enable input G', the select inputs A and B, and the outputs are given. For the select inputs to be decoded and one of the Y outputs to be asserted, the enable must be asserted.

Note the buffers in Figure 4.13(a) for each of the inputs. The enable input has an inverting bubble drawn on the input side, indicating active low status. The two data inputs are active high; the inverting bubble is on the pointed side of the buffer. The inputs and input complements (A, A', and B, B') are presented to each of the four output NAND gates, along with the enable input, G'.

Decoders can be used as minterm or maxterm generators. For example, the 2-to-4 decoder can be used to generate minterms/maxterms (0, 1, 2, 3). A two-variable Boolean

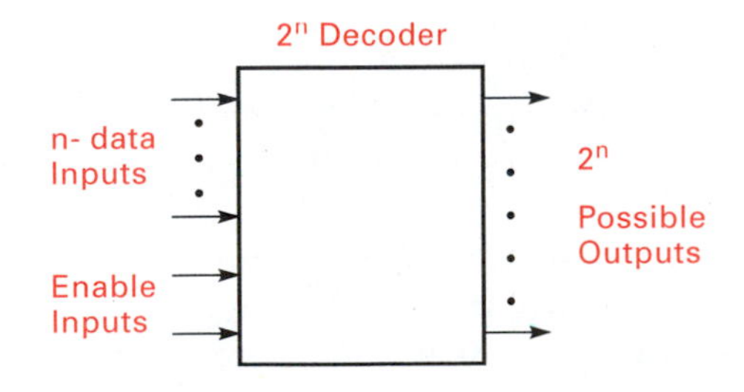

Figure 4.12
Typical decoder

Figure 4.13
TTL 74xx139 IC 2-to-4 decoder

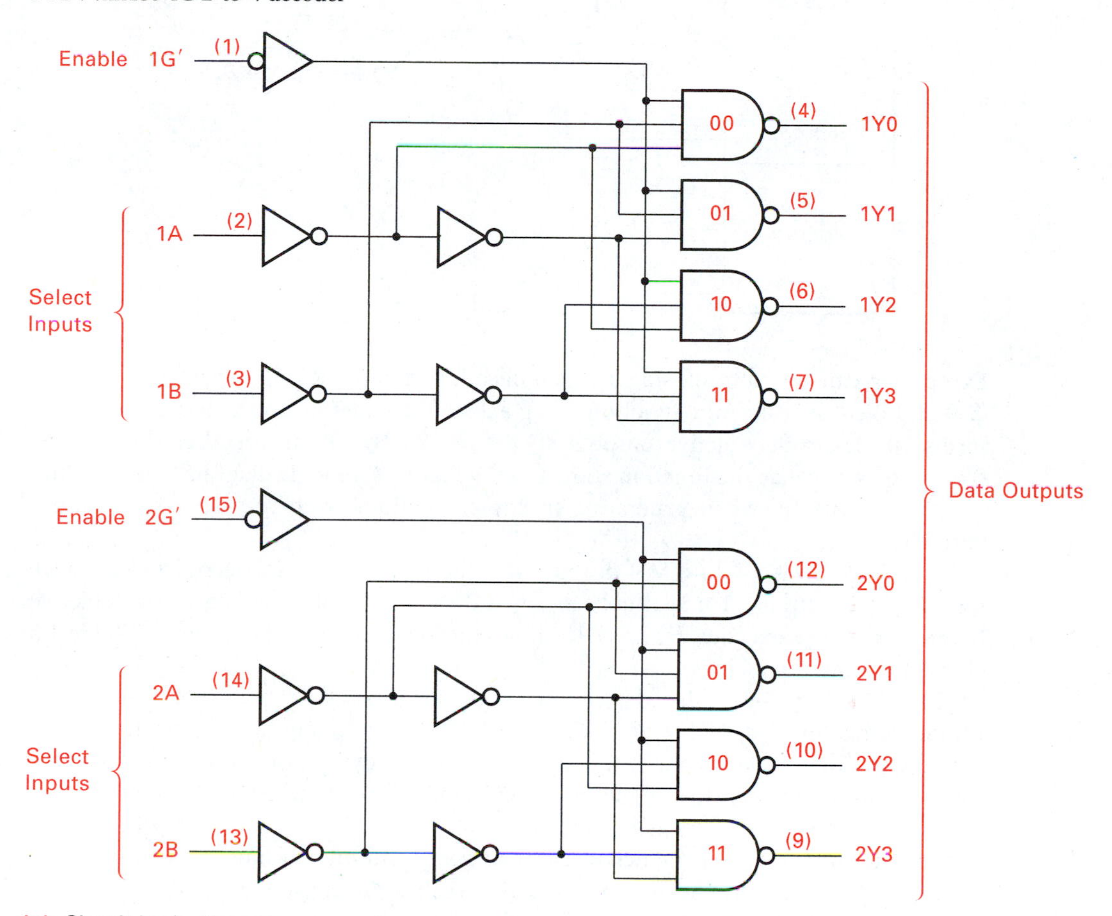

(a) Circuit logic diagram

Inputs			Outputs			
Enable G′	Select B	A	Y_0	Y_1	Y_2	Y_3
H	X	X	H	H	H	H
L	L	L	L	H	H	H
L	L	H	H	L	H	H
L	H	L	H	H	L	H
L	H	H	H	H	H	L

(b) Function table

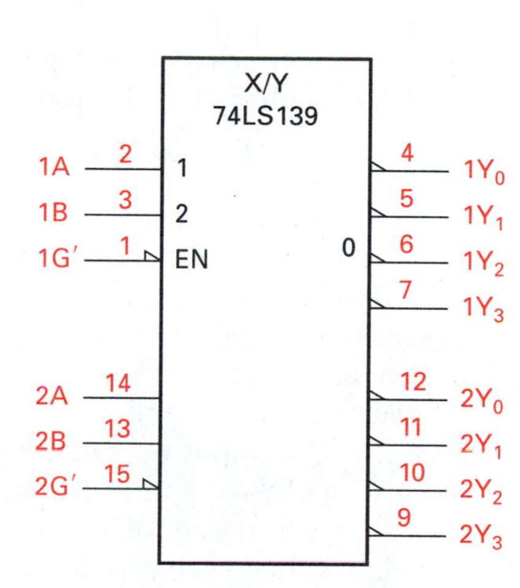

(c) IEEE logic symbol

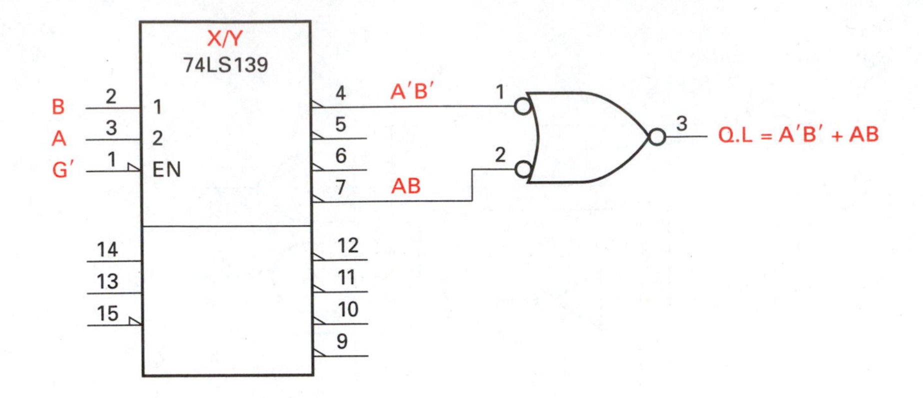

Figure 4.14
Using a 2-to-4 decoder to realize a two-variable Boolean function

expression can be realized using a 2-to-4 decoder and an OR gate. Let $Q = f(A, B) = \Sigma(0, 3)$. We can realize the function using a 2-to-4 decoder and an external gate. Minterms (0, 3) produce active outputs at Y_0 and Y_3. By ORing together these active 0 outputs, we realize the function shown in the Figure 4.14 logic diagram. This method is especially useful when generating multiple functions from the same set of input variables.

The 74xx138 is a 3-to-8 MSI decoder IC. The three inputs are decoded to produce one of eight outputs. Three enable inputs are provided, all of which must be active before decoding can occur. Figure 4.15 shows the logic circuit and the symbol. Table 4.6 shows the function table.

The 74xx138, 3-to-8 decoder can be used as a three-variable minterm generator. Minterm/maxterms 0, 1, 2, 3, 4, 5, 6, 7 potentially exist in any three-variable Boolean expression. The 3-to-8 decoder generates the minterms by decoding the input variables. The advantage of using the decoder to realize combinational functions is that the number of integrated circuits needed is reduced.

To determine whether the decoder is a simpler solution you must first simplify the function and determine the gates necessary to realize the expression. If the number of ICs is greater than 2, the solution is better rendered by using a decoder and external gates.

Consider the example in Figure 4.16. Let $X = \text{f}(\text{a, b, c}) = \Sigma\,(0, 3, 5, 6, 7)$.

After simplification, the resulting equation is

$$X = a'\,b'\,c' + ab + bc + ac$$

which results in the conventionally realized logic diagram of Figure 4.17. Figure 4.18 shows a decoder realized logic diagram.

Multiple 3-to-8 decoders can be cascaded to form 4-to-16 or even larger decoder networks by using an enable as a data input. Consider the two 3-to-8 decoders interconnected in Figure 4.19 to form a 4-to-16 decoder.

Data inputs x, y, and z are connected in parallel to the select inputs of both 3-to-8 decoders. Data input w, the most significant one, is connected to the active low enable input G_2A of the top IC and to the active high enable input G_1 on the bottom IC. This allows input w to control which 3-to-8 decoder is to be enabled. For all minterms (0, 1, 2, 3, 4, 5, 6, 7) the top decoder is enabled, and for all minterms (8, 9, 10, 11, 12, 13, 14, 15) the bottom decoder is enabled. This provides decoding for all 16 minterms possible from four input variables. The 4-to-16 decoder is controlled by a single active low enable connected to the G_2B enable input of the top and G_2A enable input of the bottom IC.

Figure 4.15
74xx138 decoder

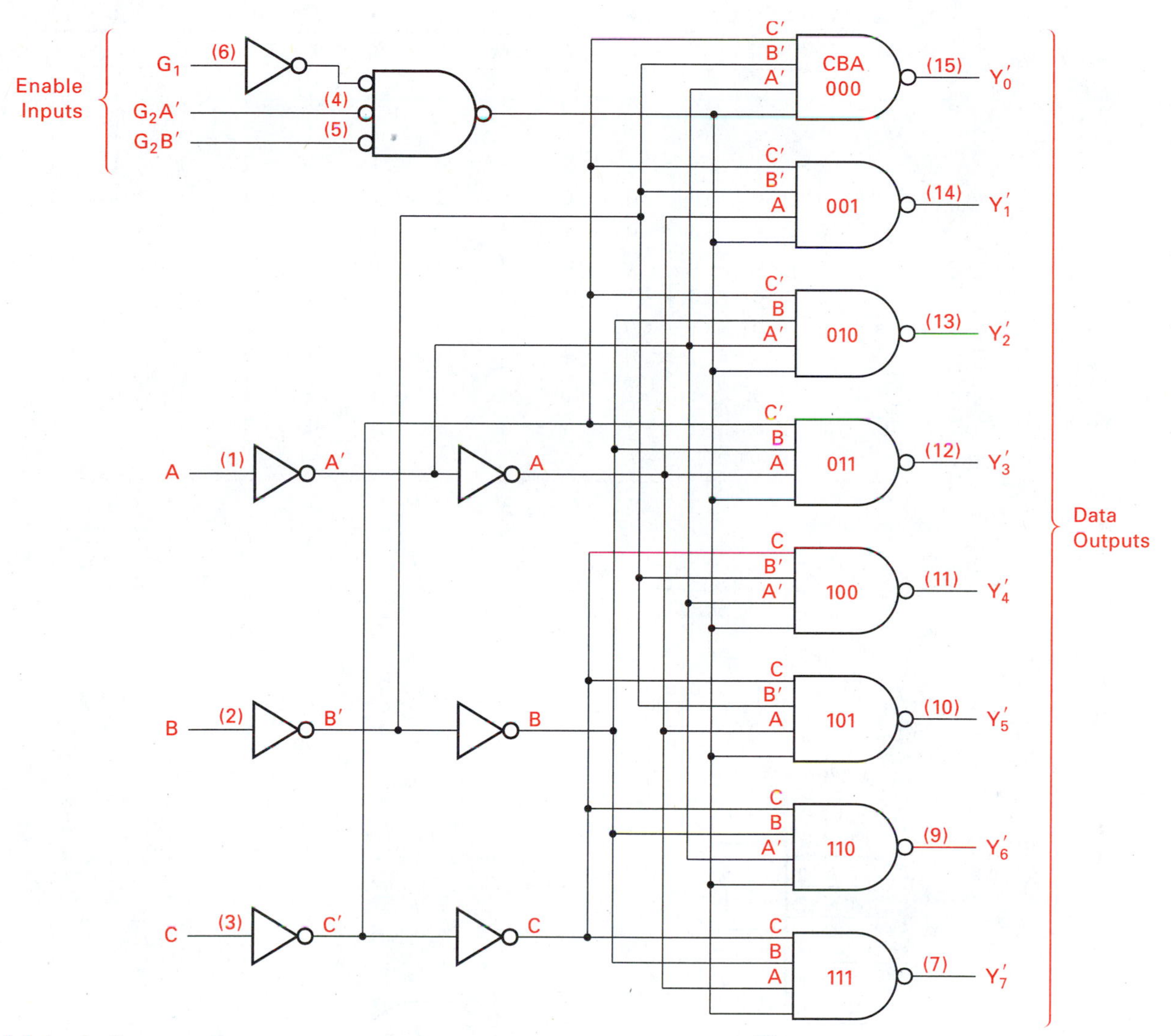

(a) Logic diagram

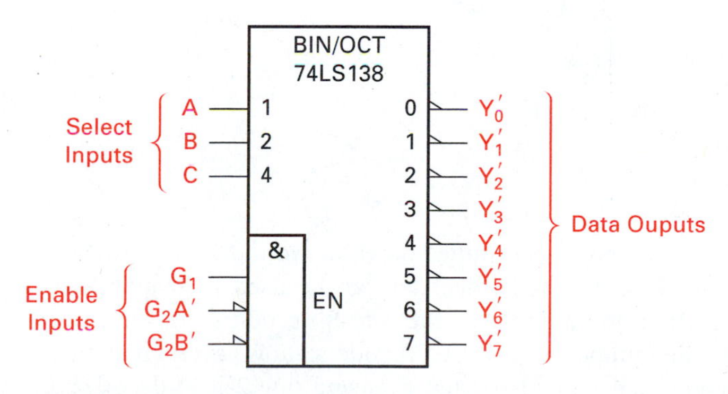

(b) IEEE logic symbol

TABLE 4.6
74xxx138 decoder function table

Inputs						Outputs							
Enable			Select										
G_1	G_2A'	G_2B'	C	B	A	Y_0	Y_1	Y_2	Y_3	Y_4	Y_5	Y_6	Y_7
0	x	x	x	x	x	1	1	1	1	1	1	1	1
x	1	x	x	x	x	1	1	1	1	1	1	1	1
x	x	1	x	x	x	1	1	1	1	1	1	1	1
1	0	0	0	0	0	0	1	1	1	1	1	1	1
1	0	0	0	0	1	1	0	1	1	1	1	1	1
1	0	0	0	1	0	1	1	0	1	1	1	1	1
1	0	0	0	1	1	1	1	1	0	1	1	1	1
1	0	0	1	0	0	1	1	1	1	0	1	1	1
1	0	0	1	0	1	1	1	1	1	1	0	1	1
1	0	0	1	1	0	1	1	1	1	1	1	0	1
1	0	0	1	1	1	1	1	1	1	1	1	1	0

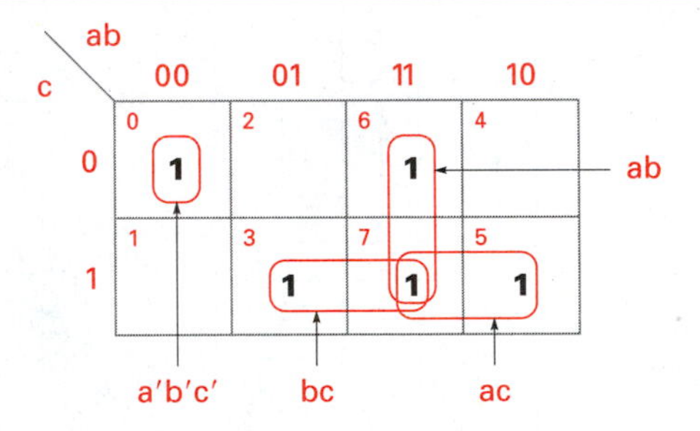

Figure 4.16
K-map for $X = \Sigma\ (0, 3, 5, 6, 7)$

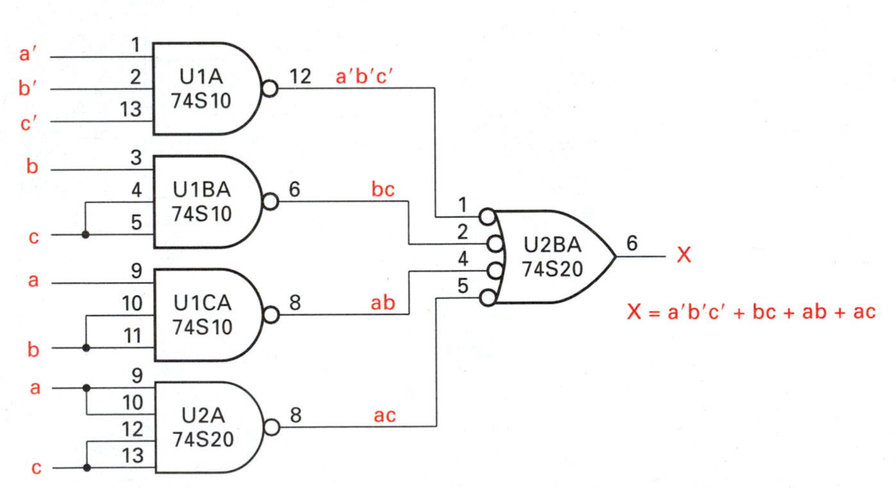

Figure 4.17
Conventionally realized logic circuit

G_1 is pulled up to V_{CC} on the top IC and G_2B is pulled down to ground, completing the enable inputs for both decoder ICs. A 4-to-16 decoder can be used to realize four-variable Boolean functions in the same manner as the 3-to-8 decoders did for three-variable problems. Increasing the number of cascaded decoders allows even greater decoding levels. A 5-to-32 decoder can be constructed by using one 2-to-4 decoder to

Figure 4.18
Decoder realized logic circuit

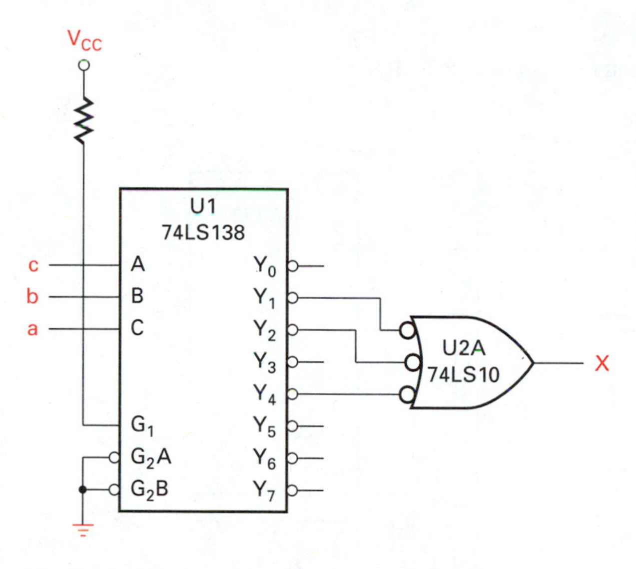

The Y_X outputs from the decoder are active low, e.g. (a′b′c′)′, (a′b′c)′ etc. By using a 3-input NAND (converted to show active low input OR), we get the same results if we NANDed active high inputs.
$X = ((a'b'c')'(a'bc')'(ab'c')')' = (a'b'c + a'bc' + ab'c')'$

Figure 4.19
Two 74xx138 decoders forming a single 4-to-16 decoder

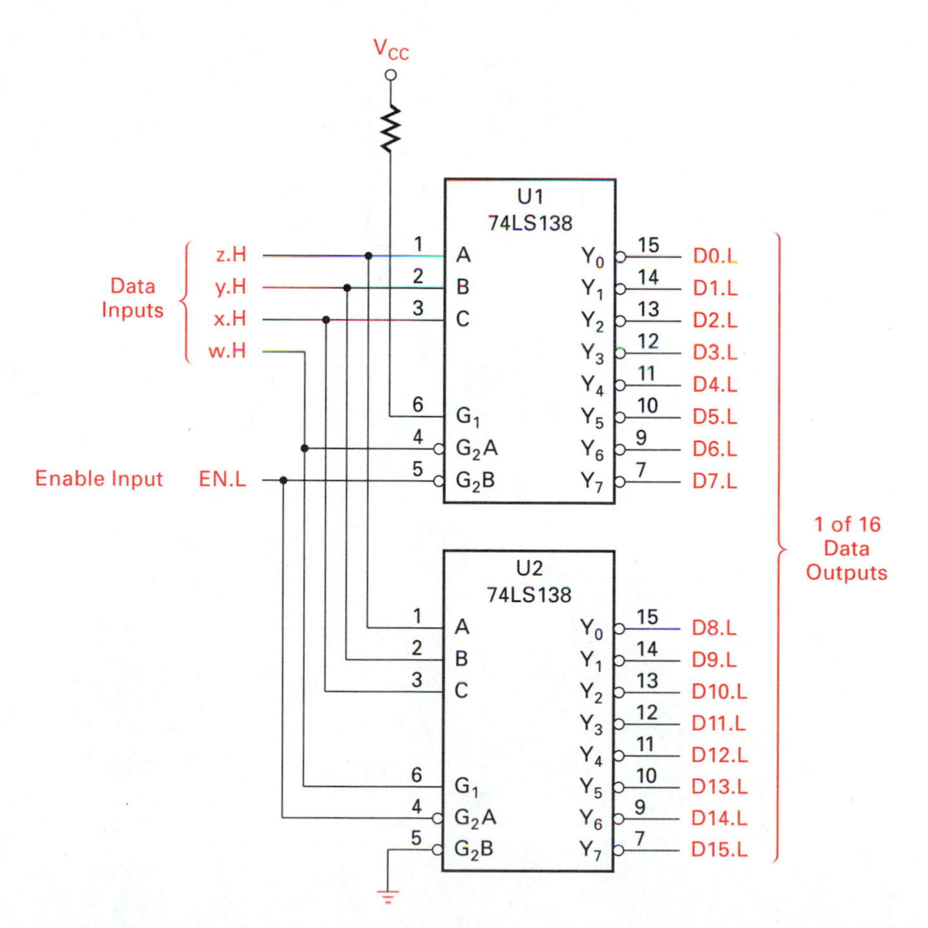

Figure 4.20
A 5-to-32 decoder using one 2-to-4 and four 3-to-8 decoder ICs

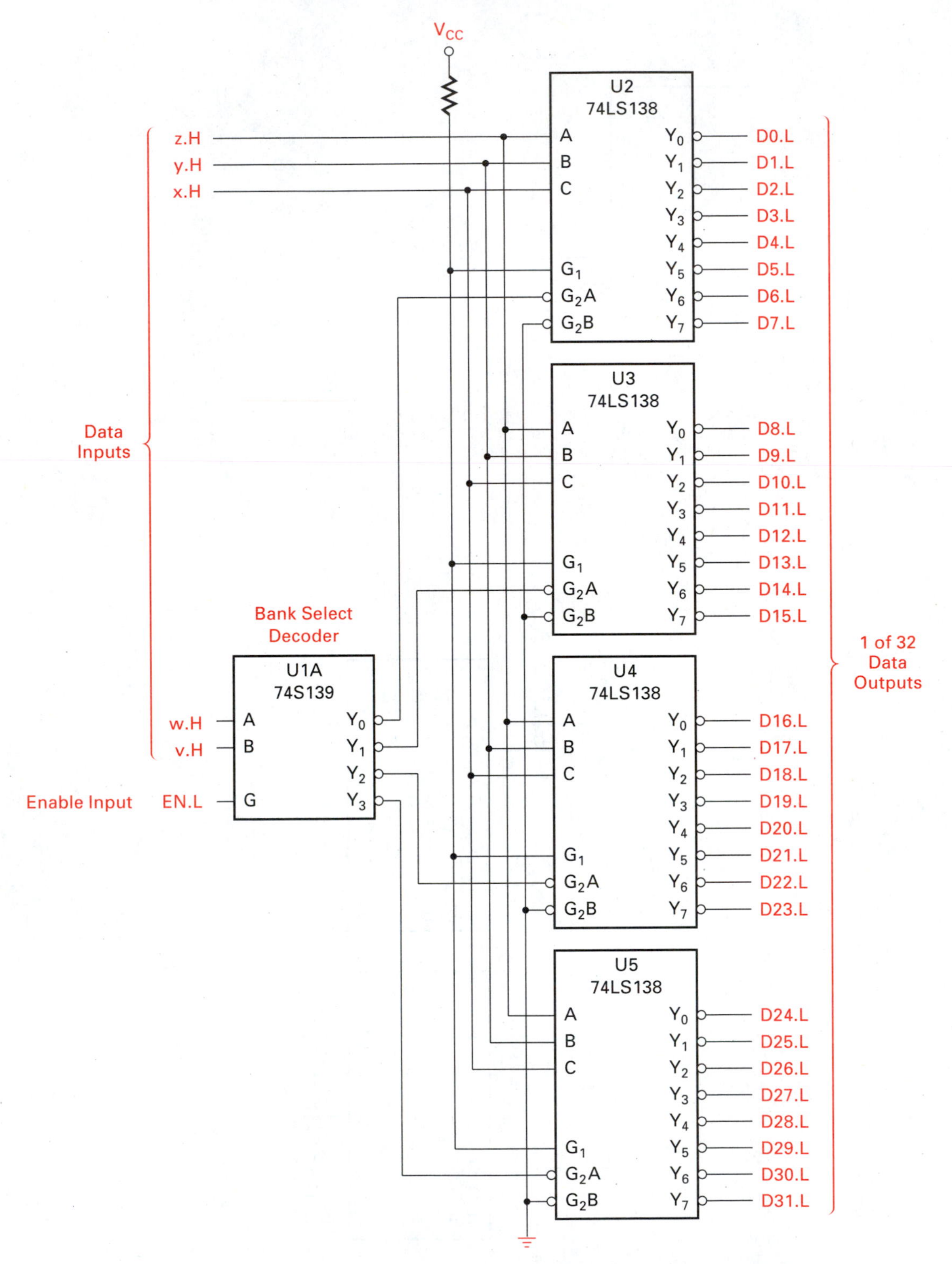

select one of four 3-to-8 decoders, as illustrated in Figure 4.20. The 2-to-4 IC decodes the two higher order inputs *v*.H and *w*.H, enabling one of the four 3-to-8 decoders. Notice that the three lower order inputs, *x*.H, *y*.H, and *z*.H, are connected in parallel to each of the four 3-to-8 decoders. This means that the same output pin of each of the four 3-to-8 decoders is selected but only one is enabled.

Consider the realization of the following multiple output four-variable problem using a 4-to-16 decoder.

Let $P = f(w, x, y, z) = \Sigma\,(1, 4, 8, 13)$ and $Q = g(w, x, y, z) = \Sigma\,(2, 7, 13, 14)$. Simplification of the two equations is not possible. A single 4-to-16 decoder can realize all 16 minterms possible in a four-variable Boolean equation. Because all of the minterms are available at the decoder output we can realize as many output functions as needed. The resulting decoder realization for functions *P* and *Q* is illustrated in Figure 4.21.

Sometimes the number of variables can be reduced through the simplification process, thus allowing the logic realization using a smaller decoder than the unsimplified minterm list would initially indicate. For example, consider the following four-variable problem, which can be simplified to three variables and then realized using a single 3-to-8 decoder. Let

$$R = f(w, x, y, z) = \Sigma\,(1, 5, 8, 9, 12, 13)$$

Figure 4.21
Using two 74xx138 decoders to realize a four-variable multiple output function

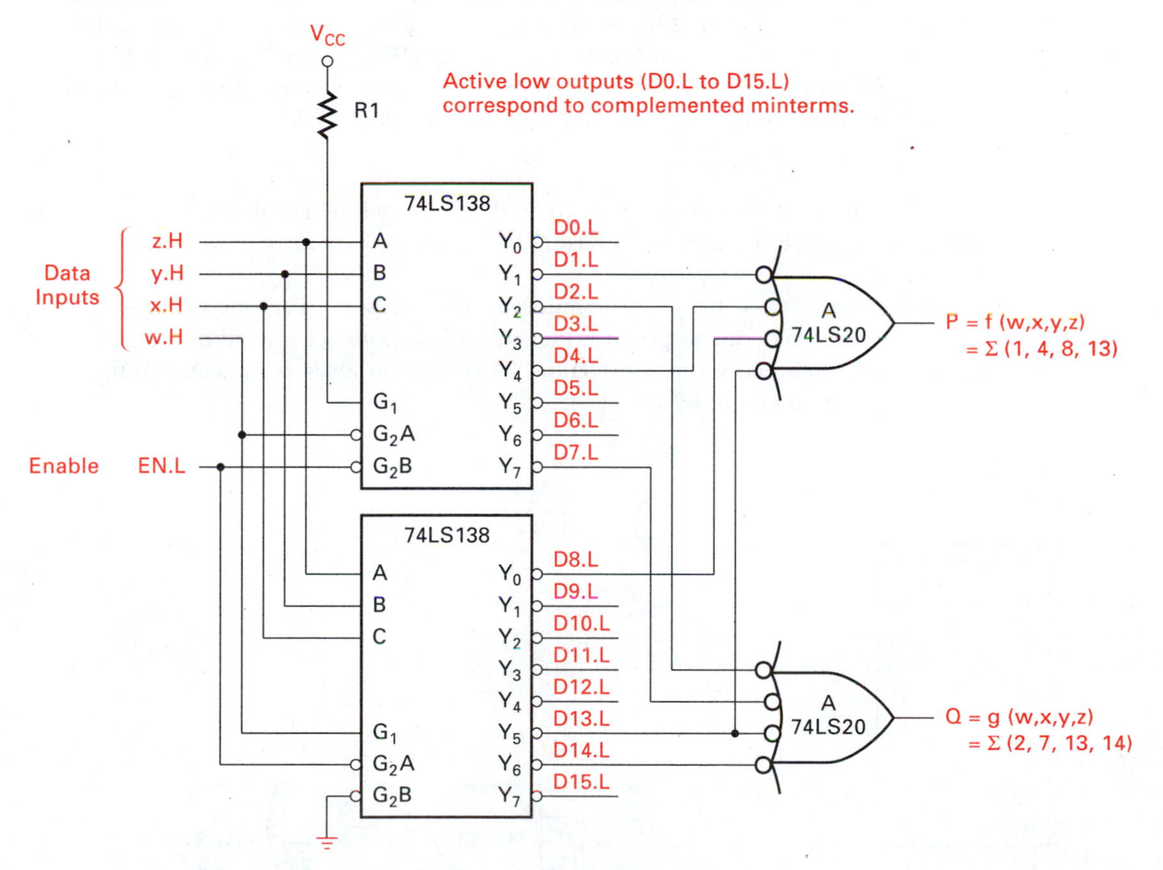

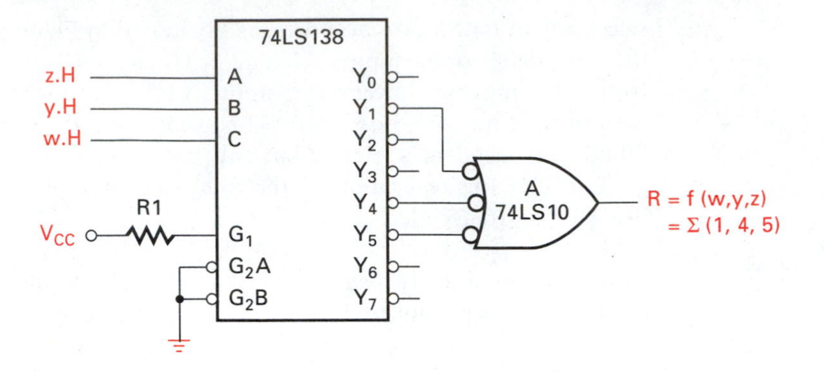

Figure 4.22
3-to-8 decoder realization of equation (4.1)

which simplifies to

$$R = wy' + y'z \tag{4.1}$$

The original four-variable problem is now reduced to three variables, x has been eliminated through simplification. The equation is not canonical, however; so it must be converted to canonical form to identify the new minterm set for three variables prior to realization. Converting to canonical form gives

$$R = wy'z' + wy'z + w'y'z + wy'z$$

which produces the new three-variable minterms (1, 4, 5) as shown in Figure 4.22.

Most, if not all, TTL MSI IC decoders have active low outputs. When minterms are collected to produce an output function, we end up ORing active low terms as indicated in the previous examples. When maxterms are collected, we must AND the active low decoder outputs. For instance, consider the POS equation

$$A = f(x, y, z) = \pi(0, 1, 3, 5)$$

Maxterms (0, 1, 3, 5) are collected to produce output function A. We can accomplish this using a 3-to-8 decoder and ANDing the resulting four active low maxterms (see Figure 4.23).

Maxterm (0) $= (x + y + z) =$ minterm (0) $= x'y'z'$. Each maxterm is ANDed with the other maxterms to produce the output function. Because all of the maxterms are asserted when low, we must AND the active low variables as indicated in the logic diagram shown in Figure 4.23.

Figure 4.23
Decoder realization of POS π (0, 1, 3, 5)

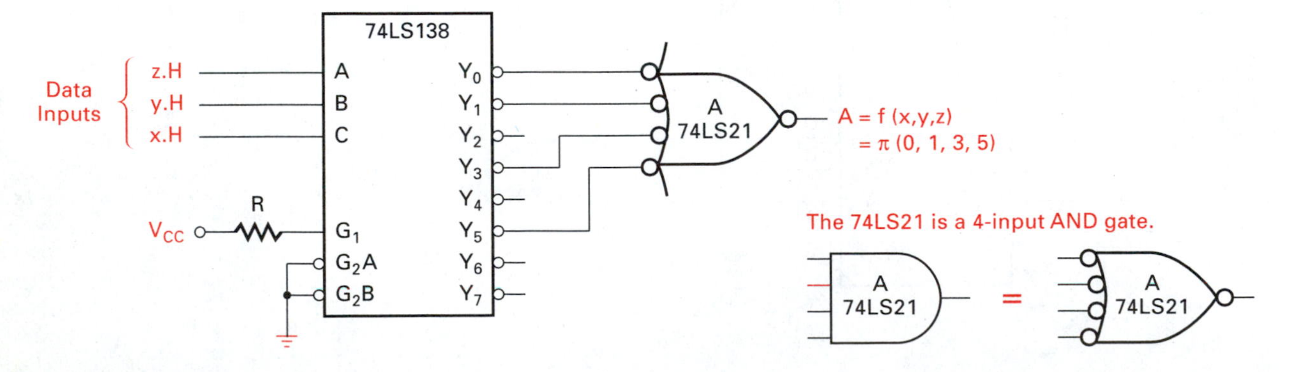

Figure 4.24
POS and SOP realizations to the same equation

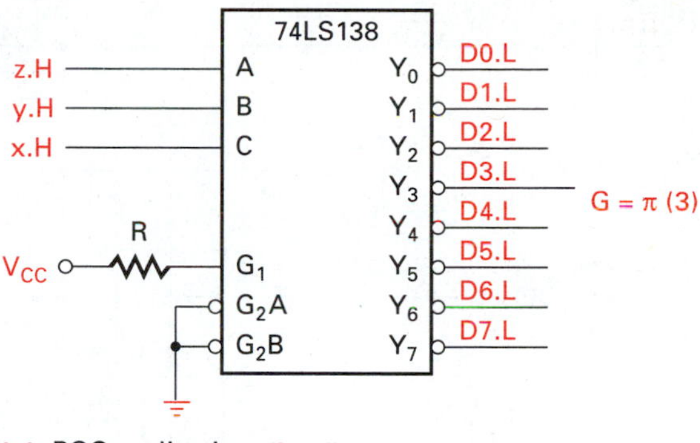

(a) POS realization. $G = f(x,y,z) = \pi\ (3)$

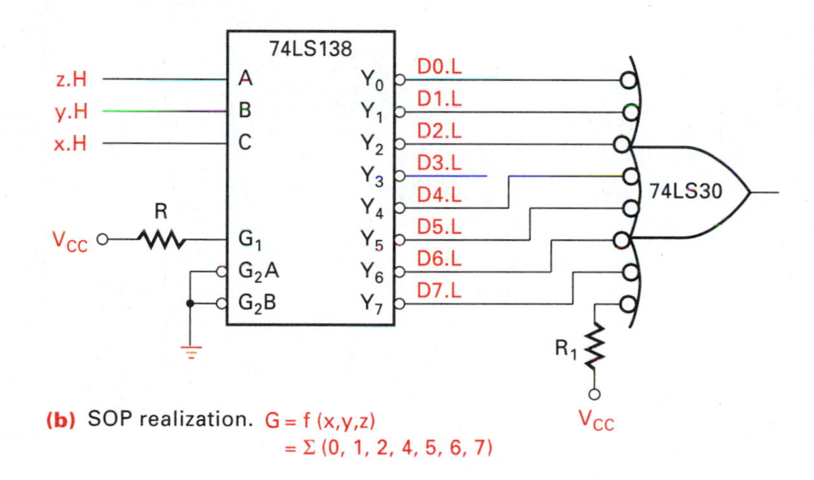

(b) SOP realization. $G = f\ (x,y,z) = \Sigma\ (0, 1, 2, 4, 5, 6, 7)$

Whether a minterm (SOP) or maxterm (POS) realization is used depends on which produces the simpler result. For example, we realize the expression $G = f(x, y, z) = \Sigma\ (0, 1, 2, 4, 5, 6, 7)$ using a decoder and external gates as necessary (see Figure 4.24). The SOP (minterm) realization would require a seven-input gate plus the decoder. The POS solution would require only the decoder because only one maxterm (3) is needed.

4.3.1 BCD Decoders

BCD decoders have four inputs and 10 outputs. The four-bit BCD input is decoded to one of ten outputs. The SN74xxx42 is a BCD to decimal decoder MSI integrated circuit. Figure 4.25 shows the logic diagram for the 74xxx42 BCD decoder.

The 74xxx47 is a BCD (8421 code) to seven-segment decoder driver MSI circuit. An 8421 BCD input is decoded, selecting which of the seven outputs are to be low, directly driving one of the seven-segment display LEDs. Inputs are provided to permit display blanking and testing. The logic circuit is shown in Figure 4.26.

The active low outputs can be connected, through current-limiting resistors, to the appropriate common anode seven-segment LED display IC pins. An active-low ripple blanking input (RBI′) is available (pin 5) that forces all outputs to a logical 1, when inputs A, B, C, and D are all 0, turning off the display digit. The ripple blanking input is used as an enable to control activating digit displays. Applications exist where leading

Figure 4.25
4-Line BCD to 10-line decimal decoder

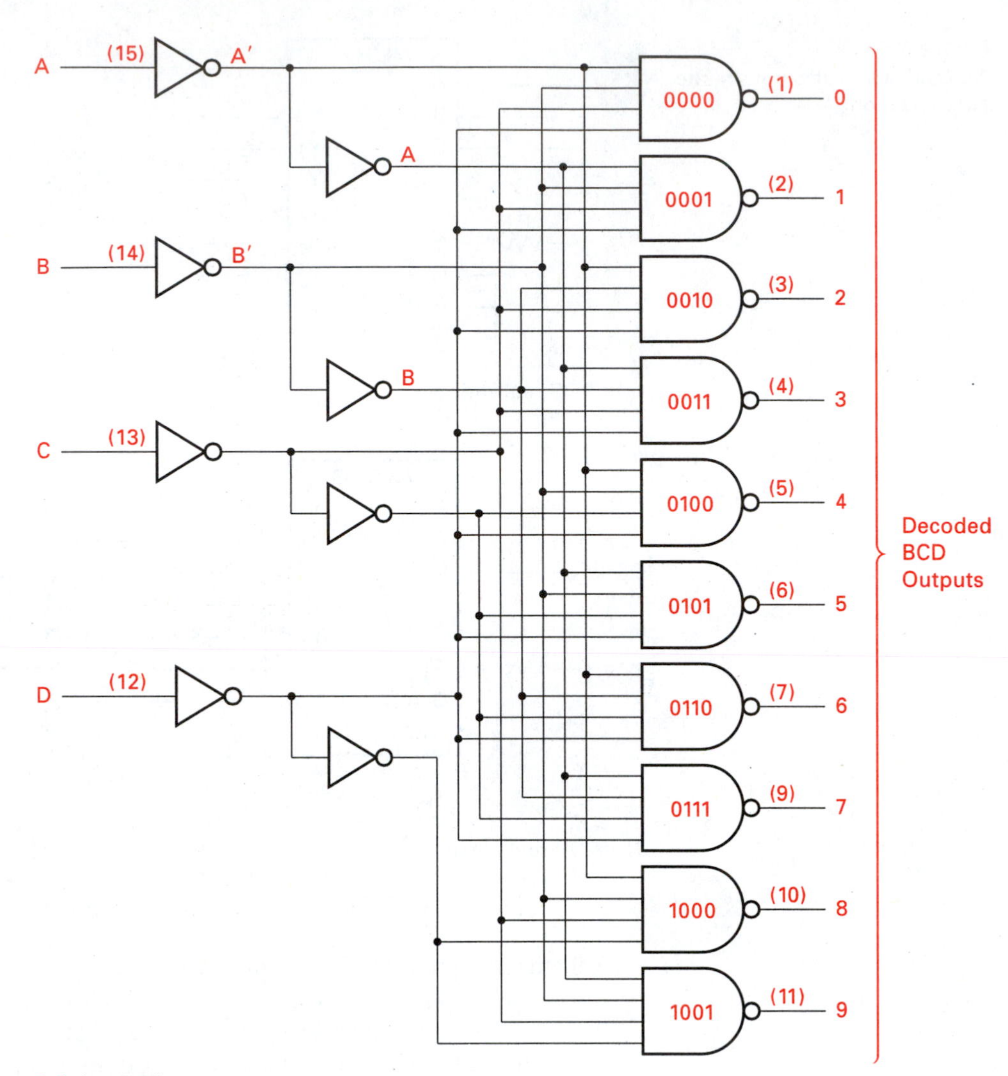

(a) Logic diagram

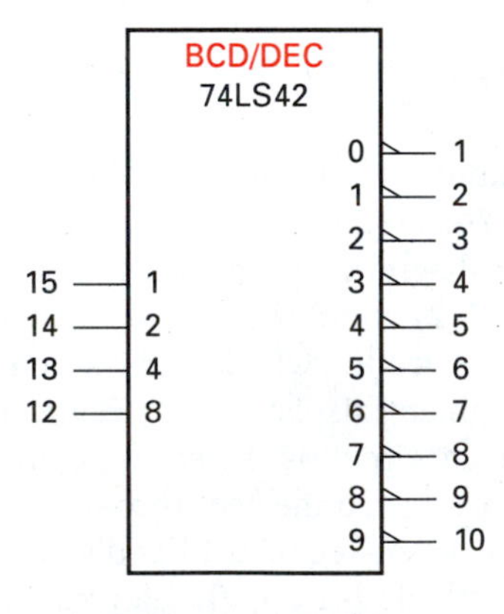

(b) IEEE logic symbol

Figure 4.26
74xx47 BCD to seven-segment decoder/driver

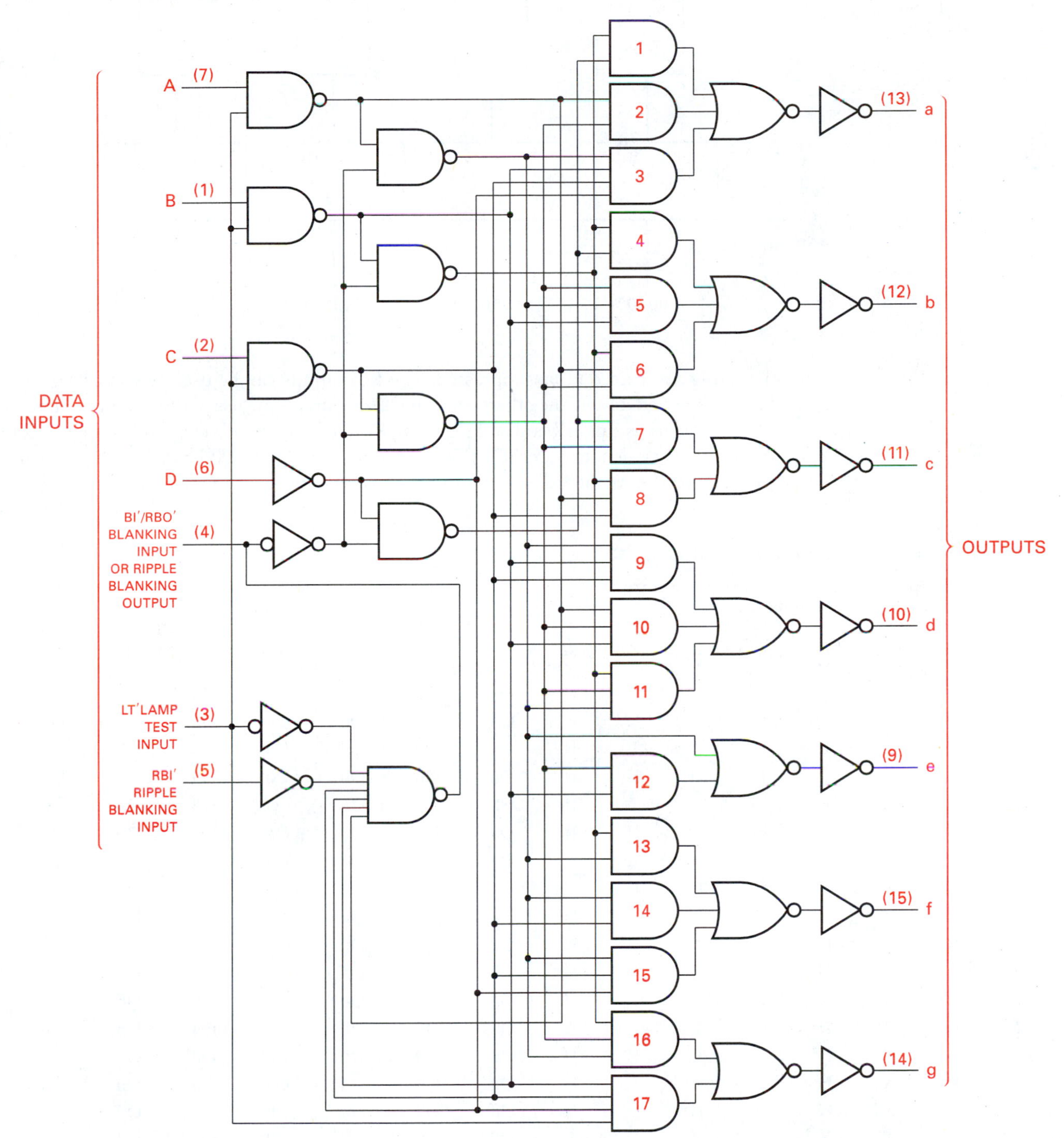

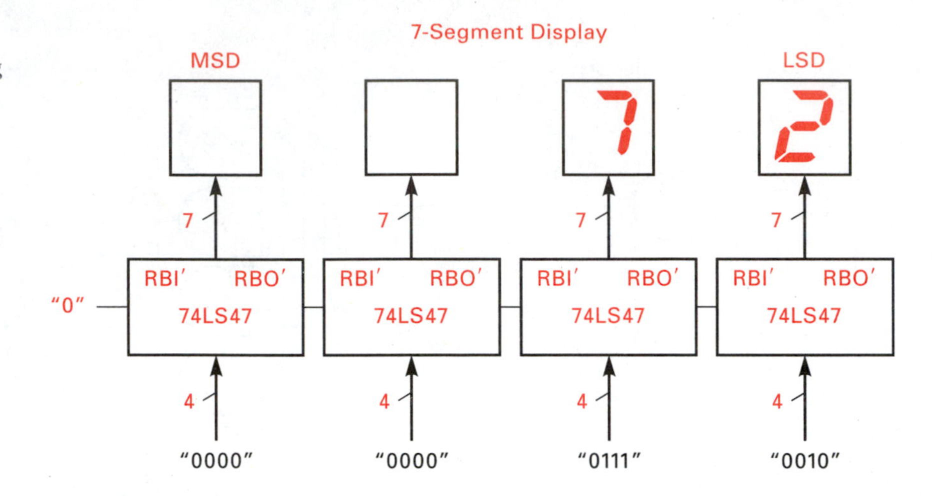

Figure 4.27
Using RBI′ to control blanking leading zeros of a multiple seven-segment display

or trailing zeros need to be suppressed. The RBI′ inputs can be used to enable a given display digit and disable the others. Figure 4.27 illustrates the use of RBI′ in this manner.

Trailing zeros can be blanked by forcing RBI′ low on the least significant digit BCD to the seven-segment decoder IC. When a leading or trailing digit has a BCD input of all zeros and RBI′ = 0, then all seven outputs are forced high, turning off that digit's LEDs.

TABLE 4.7
74xxx47 BCD to seven-segment decoder truth table

Decimal or function	Inputs							Outputs						
	$\overline{LT}$	$\overline{RBI}$	*D*	*C*	*B*	*A*	*BI′/RBO′*[9]	*a*	*b*	*c*	*d*	*e*	*f*	*g*
0	H	H	L	L	L	L	H	On	On	On	On	On	On	Off
1	H	X	L	L	L	H	H	Off	On	On	Off	Off	Off	Off
2	H	X	L	L	H	L	H	On	On	Off	On	On	Off	On
3	H	X	L	L	H	H	H	On	On	On	On	Off	Off	On
4	H	X	L	H	L	L	H	Off	On	On	Off	Off	On	On
5	H	X	L	H	L	H	H	On	Off	On	On	Off	On	On
6	H	X	L	H	H	L	H	Off	Off	On	On	On	On	On
7	H	X	L	H	H	H	H	On	On	On	Off	Off	Off	Off
8	H	X	H	L	L	L	H	On	On	On	On	On	On	On
9	H	X	H	L	L	H	H	On	On	On	Off	Off	On	On
10	H	X	H	L	H	L	H	Off	Off	Off	On	On	Off	On
11	H	X	H	L	H	H	H	Off	Off	On	On	Off	Off	On
12	H	X	H	H	L	L	H	Off	On	Off	Off	Off	On	On
13	H	X	H	H	L	H	H	On	Off	Off	On	Off	On	On
14	H	X	H	H	H	L	H	Off	Off	Off	On	On	On	On
15	H	X	H	H	H	H	H	Off	Off	Off	Off	Off	Off	Off
BI	X	X	X	X	X	X	L	Off	Off	Off	Off	Off	Off	Off
RBI	H	L	L	L	L	L	L	Off	Off	Off	Off	Off	Off	Off
LT	L	X	X	X	X	X	H	On	On	On	On	On	On	On

RBO′, a ripple blanking output signal, is forced low by the previously mentioned conditions. RBO′, when connected to the next digit's RBI′ input, allows interconnecting all digits in the display for either leading or trailing zero suppression. Table 4.7 illustrates the input/output/control signal relationships for the 74xxx47 IC.

4.4 ENCODERS

Encoders perform a function that is the inverse of decoders. Encoders have multiple inputs and outputs, like decoders. Encoders, however, have more input than output variables. The decoder produced 2^n outputs from n inputs. An encoder produces n outputs from 2^n inputs. A typical encoder function can be modeled as shown in Figure 4.28.

The logic diagram for the 74xx147, a 10-line to BCD encoder medium-scale integrated circuit, is shown in Figure 4.29.

The nine input lines to the encoder shown in Figure 4.29 each represent a unique BCD code. Both input and output lines are asserted active low for the device. If, for example, line 1 (representing a decimal 1) were 0, then output A (the LSB of the BCD digit) would be a 0. Notice that no input line for decimal zero is present; when this condition occurs all output lines are 1, indicating no valid BCD output. The function table for the encoder is shown in Table 4.8.

Encoding decimal keypad or switch panel output data into BCD prior to transmission to a digital system is a typical application of the 74xx147. Consider a 10-key keypad used to input set point values (values that specify when a control activity is to occur, such as, if the temperature = 100°C, then turn off the heater) into a digital control system. When a key is pressed it forces its corresponding digit line to a logical 0. That value is presented to the encoder producing a BCD output that is in turn sent to the digital control system for processing. Figure 4.30 shows the keypad encoder structure.

Figure 4.31 shows the logic diagram and symbol of another encoder. This priority encoder accepts a three-bit input and produces a 1 out of 2^3 coded output. It can be expanded by using multiple devices to form even larger value encoding systems. The 1 out of 2^n code output structure is used to encode information for transmission to a digital system similar to that described for a 10-line-to-BCD encoder.

Consider a situation where several possible events may occur in an industrial system, and you want to identify an event and assign and transmit a code to the control unit based on some priority. Such events may be assigned a priority based on their function in the system. By connecting the highest priority event to the highest value input line, the next highest priority event to the next highest input line and so on, down to the lowest priority event being assigned the lowest input line, the device can transmit priority information to the control system. The control system determines a specific course of action to be taken for each event occurrence.

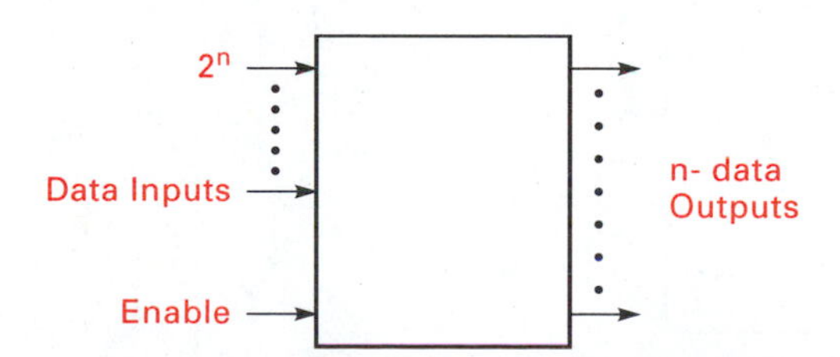

Figure 4.28
Typical encoder

Figure 4.29
10-line to BCD encoder

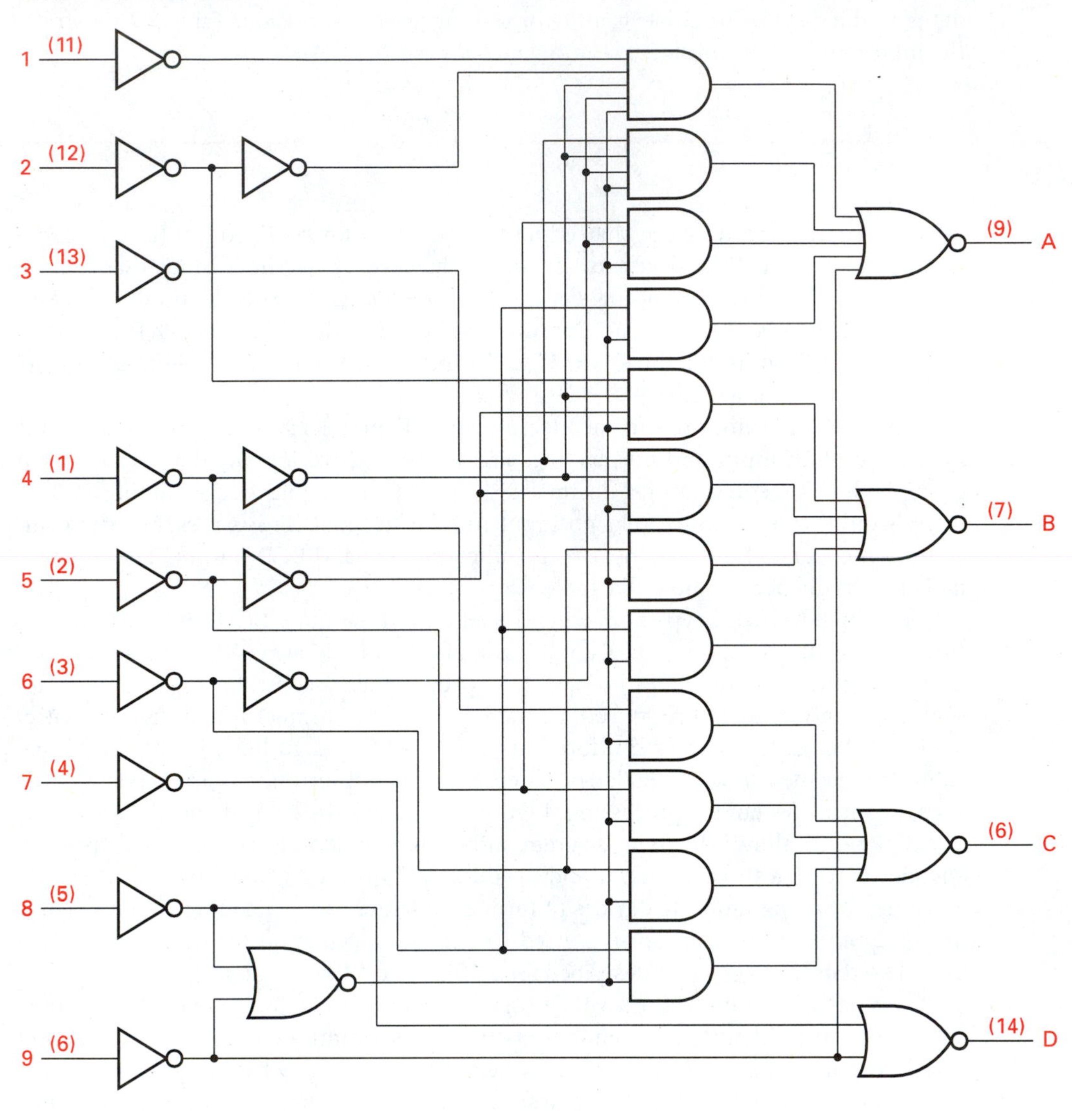

(a) Logic Diagram

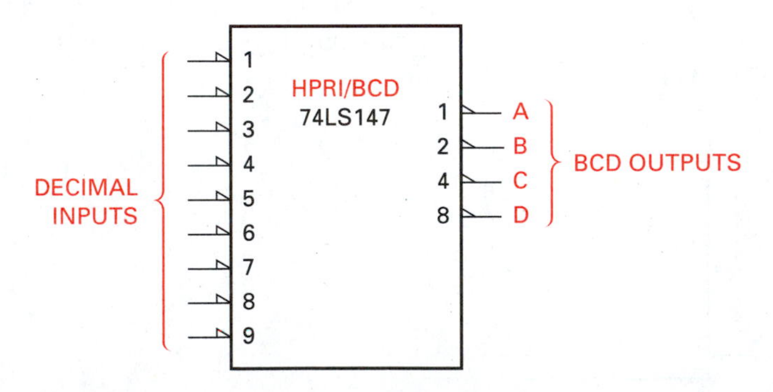

(b) IEEE logic Symbol

TABLE 4.8
Truth table for a 10-line to BCD encoder

Decimal value	Inputs									Outputs			
	1	2	3	4	5	6	7	8	9	*D*	*C*	*B*	*A*
0	1	1	1	1	1	1	1	1	1	1	1	1	1
1	0	1	1	1	1	1	1	1	1	1	1	1	0
2	x	0	1	1	1	1	1	1	1	1	1	0	1
3	x	x	0	1	1	1	1	1	1	1	1	0	0
4	x	x	x	0	1	1	1	1	1	1	0	1	1
5	x	x	x	x	0	1	1	1	1	1	0	1	0
6	x	x	x	x	x	0	1	1	1	1	0	0	1
7	x	x	x	x	x	x	0	1	1	1	0	0	0
8	x	x	x	x	x	x	x	0	1	0	1	1	1
9	x	x	x	x	x	x	x	x	0	0	1	1	0

x indicates a don't-care condition.

Notice that output *GS*′, at the top of the diagram, goes low when any of the inputs to the encoder is active. The purpose of the signal is to communicate to the control system logic that an event has occurred and priority encoded data are present. *EO* is an active low signal that can be used to cascade several devices to form a larger priority

Figure 4.30
Key pad interface to a digital system using a 74xx147 10-line to BCD encoder

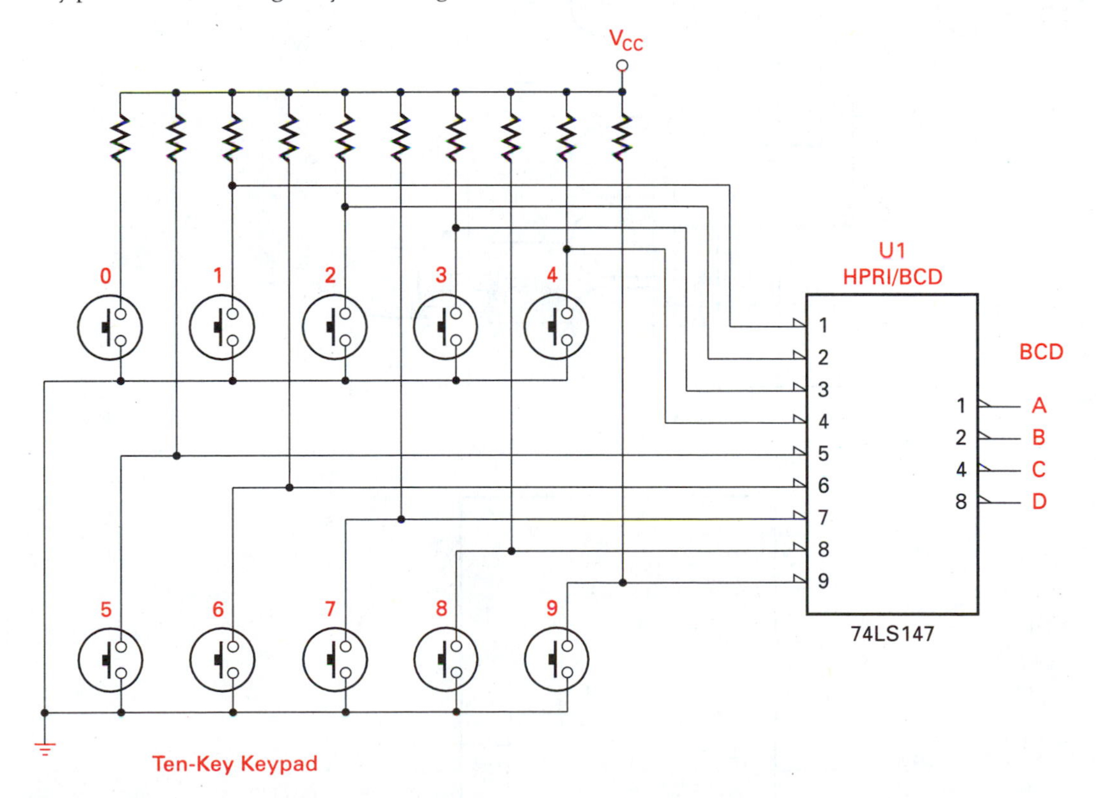

Figure 4.31
8-line to 3-line encoder

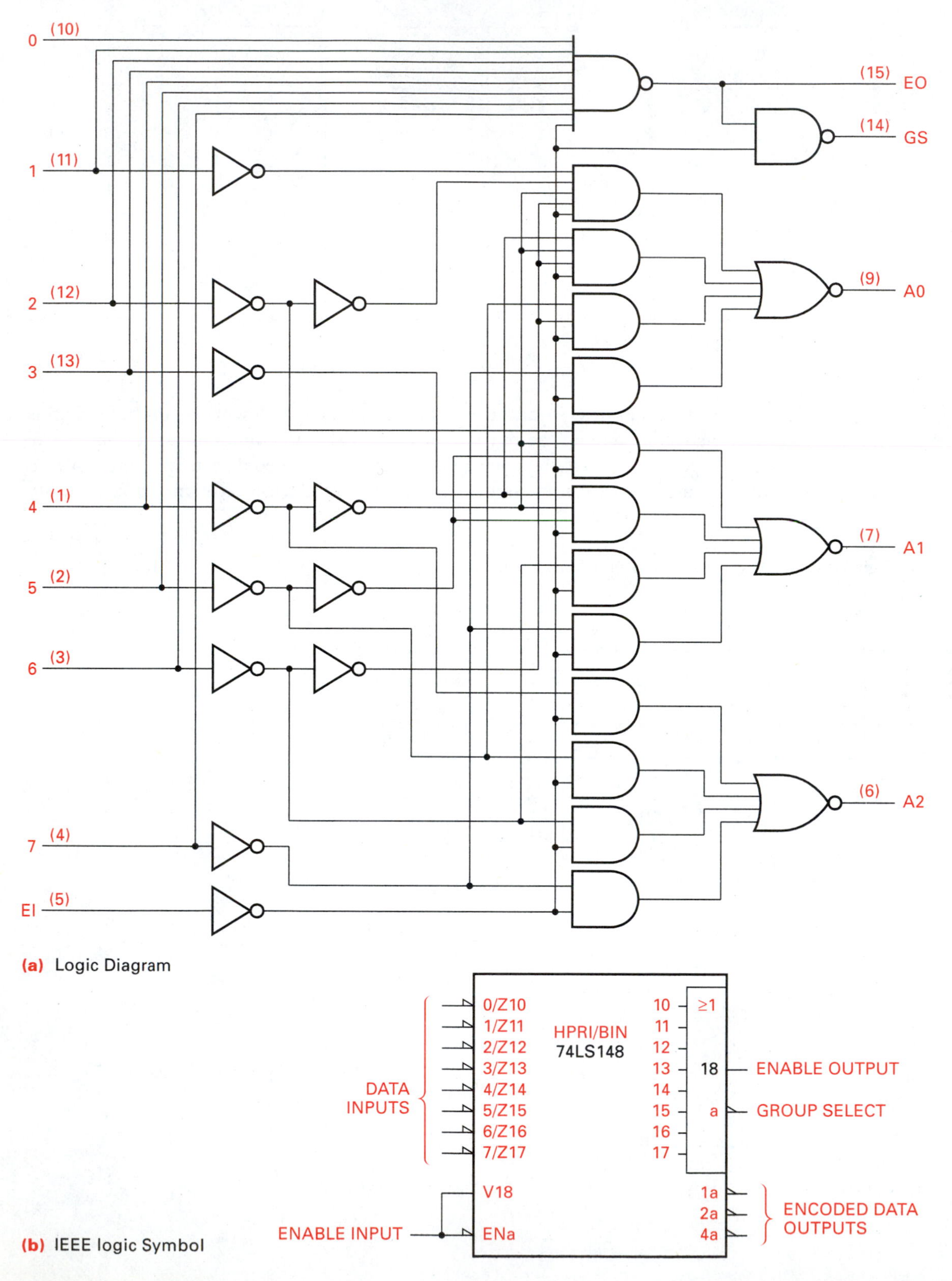

(a) Logic Diagram

(b) IEEE logic Symbol

TABLE 4.9
Truth table for the 74xx148 8-line to 3-line priority encoder

Inputs									Outputs				
EI	0	1	2	3	4	5	6	7	*A2*	*A1*	*A0*	*GS*	*EO*
1	x	x	x	x	x	x	x	x	1	1	1	1	1
0	1	1	1	1	1	1	1	1	1	1	1	1	0
0	0	1	1	1	1	1	1	1	1	1	1	0	1
0	x	0	1	1	1	1	1	1	1	1	0	0	1
0	x	x	0	1	1	1	1	1	1	0	1	0	1
0	x	x	x	0	1	1	1	1	1	0	0	0	1
0	x	x	x	x	0	1	1	1	0	1	1	0	1
0	x	x	x	x	x	0	1	1	0	1	1	0	1
0	x	x	x	x	x	x	0	1	0	0	1	0	1
0	x	x	x	x	x	x	x	0	0	0	0	0	1

encoding system. When *EO* is active no priority event connected to the IC is present. *EO* is used as the enable input to the next lower priority encoder. The priority encoder truth table is shown in Table 4.9.

The following priority list is assigned for the industrial process control system in Figure 4.32:

E7 Highest priority—fire alarm. If input *E7* is active, a fire has been detected. Invoke a fire evacuation procedure by calling the fire department and sounding fire alarms.

E6 Next highest priority—main power failure. If *E6* is active, then the main power system has failed. Invoke emergency power procedures. Go to a machine fail-safe condition to prevent injury and product loss.

E5 System safety interlock 1. If *E5* is active, the system safety interlock 1 is open. A dangerous condition exists. Shut down appropriate machine stations to prevent injury. Set alarm conditions; automatically notify system maintenance people.

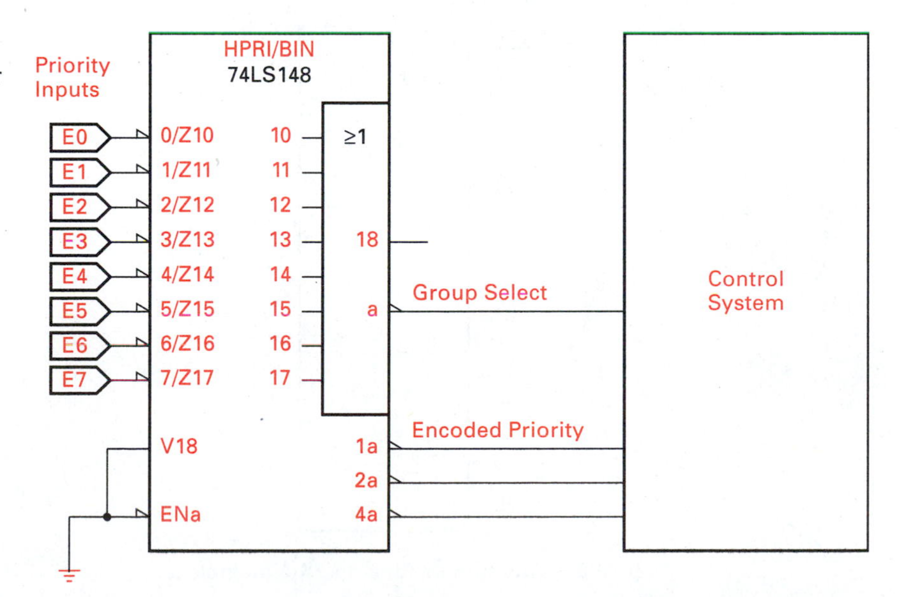

Figure 4.32
Control system priority encoder

$E4$ System interlock 2. Same procedure as $E4$, except a less dangerous condition exists, so a lower priority is required.

$E3$ Machine station 1 operator interlock. If $E3$ is active then the operator for machine station 1 is not ready. Automatically notify appropriate machine stations that a potential product flow bottleneck exists. If the signal is not rectified in the proper time, invoke appropriate machine station shutdowns.

$E2$ Machine station 2 operator interlock. Same sort of response as for machine station 1 except different response algorithm invoked as appropriate.

$E1$ Machine station 3 operator interlock. Same type of response as for $E2$.

$E0$ Machine station 4 operator interlock. Same type of response as for $E2$.

4.5 DIGITAL MULTIPLEXERS

Digital multiplexers provide the digital equivalent of an analog selector switch. A digital **multiplexer** connects one of n inputs to a single output line, so that the logical value of the input is transferred to the output. The one of n input selection is determined by m select inputs, where $n = 2^m$. A 4-to-1 multiplexer requires 2 select inputs, an 8-to-1 multiplexer requires 3 select inputs. Figure 4.33 illustrates the analog select function and the digital functional block for a 4-to-1 multiplexer.

Figure 4.33
Multiplexer model and IEEE symbol

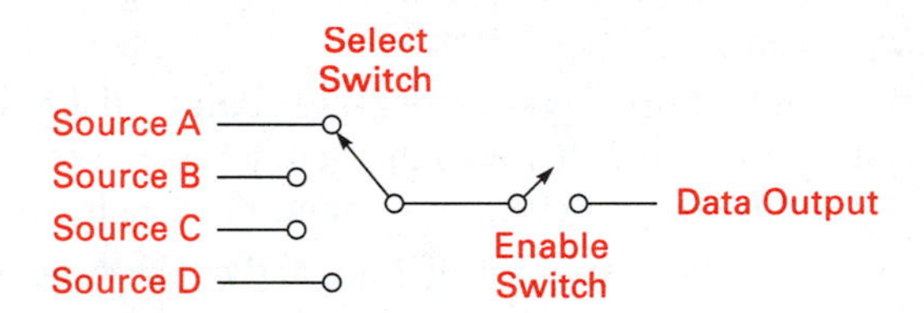

(a) Analog select switch

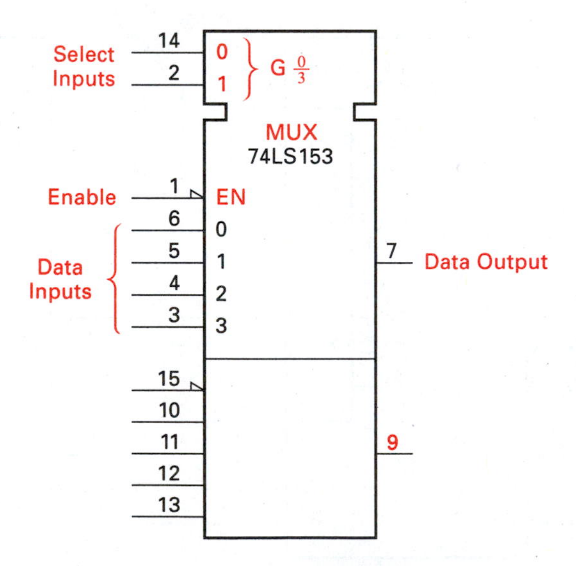

(b) IEEE logic symbol for a dual 4-1 multiplexer

When the analog selector, shown in Figure 4.33(a), is in position *a*, data are routed from source *A* to the output, through the select switch. An enable switch is in series with the data output line, which must be closed for data transmission. The channel selector knobs on older television sets illustrate an analog multiplexer.

The digital multiplexer, shown in Figure 4.33(b), routes digital data from source *A* to the output when the *BA* select input lines are 00. Other data input to output connections are made based on the code presented to the select inputs. If $BA = 01$, source *B* is selected, and so on. The data output of a digital multiplexer is dependent on the value of the data at a given input at the time the select lines switch the input through to the output.

Several digital multiplexer MSI devices are available. The 74xx153 is a dual 4-to-1 multiplexer, the 74xx151 is a single 8-to-1 and the 74xx150 is a single 16-to-1 multiplexer.

The 74xx151, 8-to-1 digital multiplexer IC logic diagram is shown in Figure 4.34. Two outputs are provided; one is active high, the other is active low. A truth table for the device is given in Table 4.10. Notice that 1s and 0s are not present in the truth table output columns. Because the multiplexer acts as a data switch it does not generate any data of its own, but passes external input data from the selected input to the output. For this reason the two output columns represent data by D*n* and D*n*′. When multiple digital signals need to be multiplexed at the same time, we use several multiplexers in parallel. Figure 4.35 illustrates switching four-bit inputs from several sources to a single output.

The logical value at one of eight input lines, D0 to D7, is transferred to the output when selected by three select inputs *C*, *B*, and *A*, and when the strobe or enable input, *G*′, is low. Two output levels exist, active high output *Y* and active low output *W*′. Each of the three select lines is buffered by inverters. Input buffering is done to present a constant load to the driving source. Both high and low levels of the select inputs are used to drive the data select AND gates.

The multiplexer shown in Figure 4.35 is not available in an off-the-shelf MSI circuit. Four data sources, each four-bits, are to be switched to a single four-bit data destination.

The slash mark (/) (used in Figure 4.35) with a number indicates the number of wires (one wire for each bit). The notation simplifies drawing all four data wires. Figure 4.36 shows the use of two 74S153 ICs, each containing a dual 4-to-1 multiplexer, to form the system four-bit multiplexer illustrated in Figure 4.35.

Notice that the enable and select inputs are connected in parallel. This means that the chips will be enabled and the inputs selected together. All position 3 bits from each data source provide inputs to the top multiplexer; all position 2 bits provide inputs to the next multiplexer, and so on. Data source *A* is selected when select inputs $BA = 11$; then all four individual multiplexers switch the appropriate input bit to its output, creating a four-bit destination.

Expansion of multiplexers beyond that available in integrated circuits can be accomplished by interconnecting several multiplexers. For example, two 74xx150, 16-to-1 multiplexers can be used together to form a 32-to-1 multiplexer. Four 74xx150 and a single 4-to-1 (one-half of a 74xx153 4-to-1) multiplexer can combine to create a 64-to-1 multiplexer. Figure 4.37 shows a 32-to-1 multiplexer using two 74xx150 devices.

Each 16-to-1 multiplexer has four select inputs and a single enable. Because 5 select inputs are needed to perform a 32-to-1 multiplexing function, we can use the enable input of each IC as the fifth select bit. For example, select input *V*, the most significant select bit position, is connected directly to the active low enable of the top IC. When

Figure 4.34
8-to-1 digital multiplexer

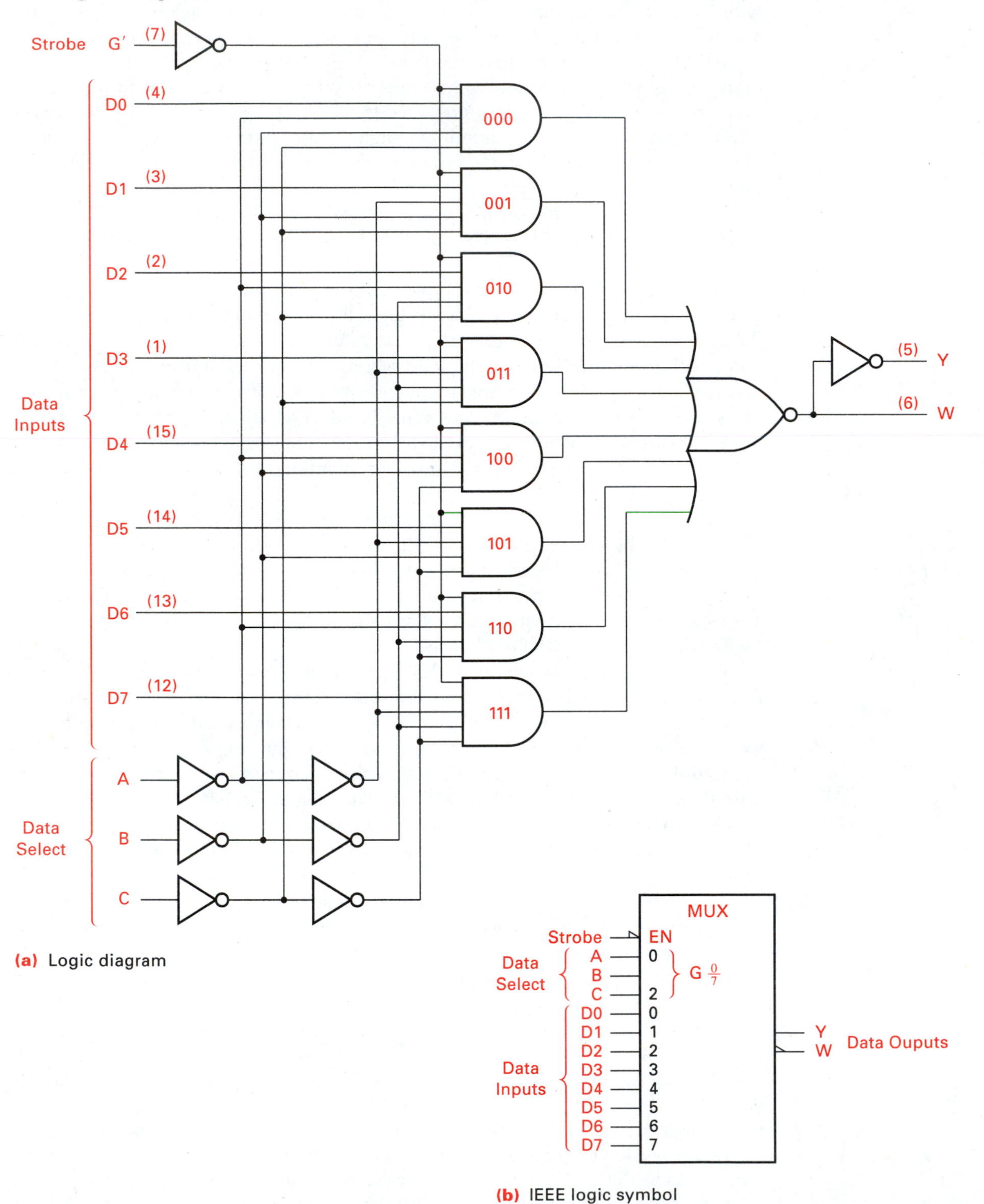

(a) Logic diagram

(b) IEEE logic symbol

TABLE 4.10
8-to-1 multiplexer truth table

Inputs				Outputs	
Select			Strobe		
C	B	A	G'	Y	W
x	x	x	1	0	1
0	0	0	0	D0	D0′
0	0	1	0	D1	D1′
0	1	0	0	D2	D2′
0	1	1	0	D3	D3′
1	0	0	0	D4	D4′
1	0	1	0	D5	D5′
1	1	0	0	D6	D6′
1	1	1	0	D7	D7′

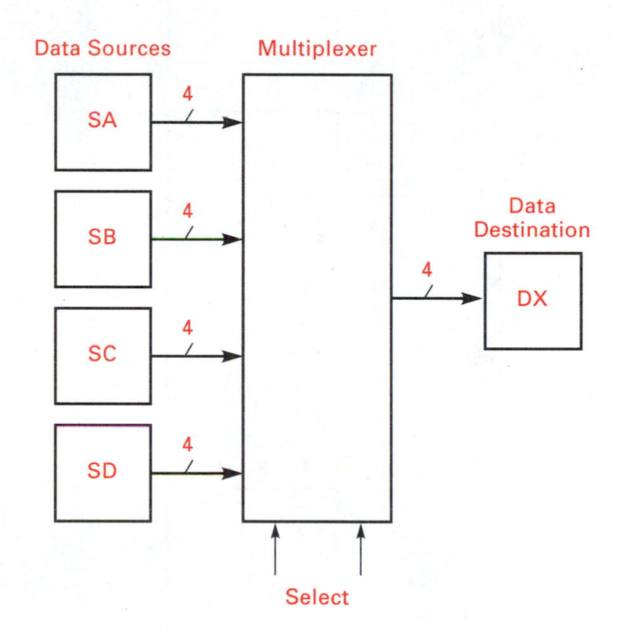

Figure 4.35
Four four-bit data sources multiplexed to a single four-bit destination

select input V is low, the top IC is enabled and the remaining select lines W through Z determine which of the 16 data inputs will be connected to the output. Select input V is inverted and presented to the enable input of the bottom IC. When $V = 1$ the bottom IC is enabled and the select inputs determine which of 16 data lines (D16 through D31) will be connected to its output. The top multiplexer switches data lines D0 to D15 and the bottom IC switches data lines D16 to D31.

A 32-to-1 multiplexer can also be realized by using four 74xx151 (8-to-1 multiplexer) and a 74xx139 (dual 2-to-4 decoder) as illustrated in Figure 4.38.

V through Z are the select inputs to the multiplexer system. V and W are decoded by the 74xx139 to determine which of the four 8-to-1 multiplexers will be enabled. Because only one of the four outputs from the decoder can be active (low), only one of the 8-to-1 multiplexers can be enabled at a time. Select inputs X through Z are connected in parallel to the four multiplexer ICs. The decoded select input switches the

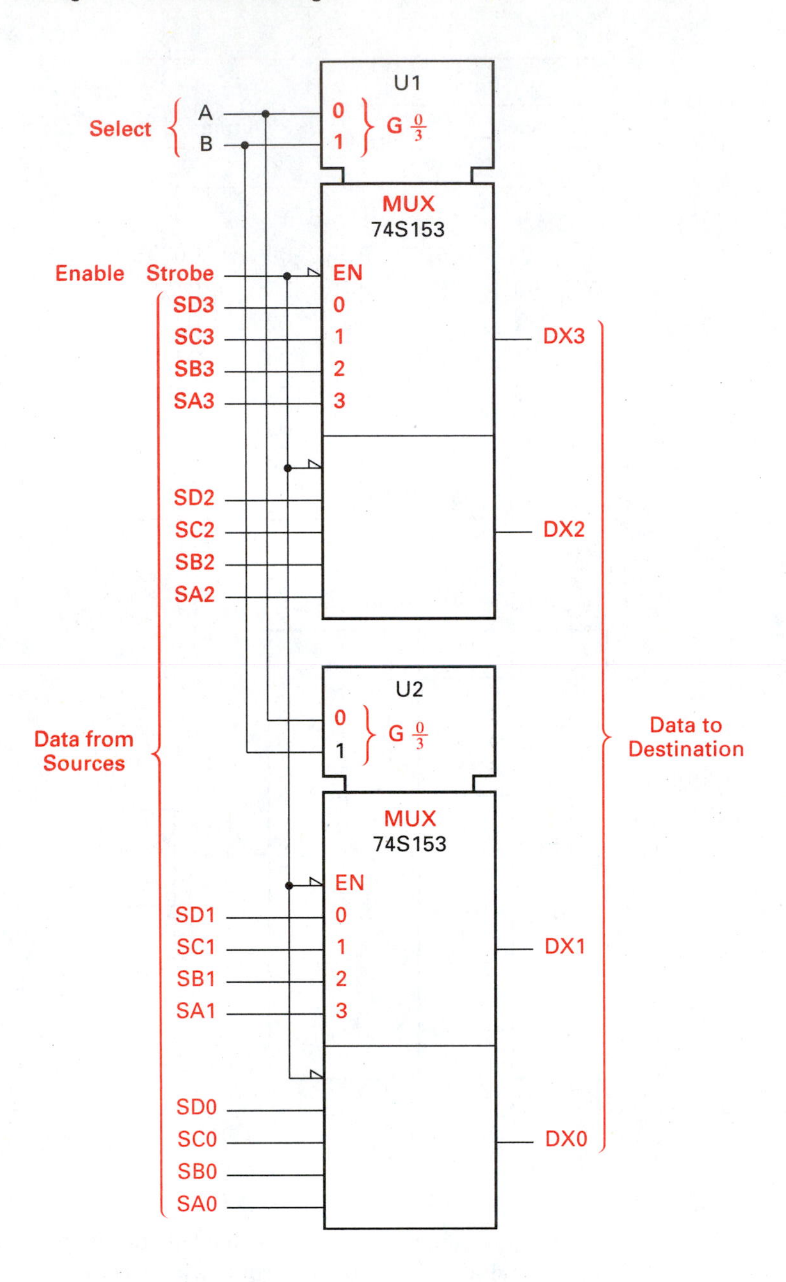

Figure 4.36
A four-bit multiplexer system using two 74S153 integrated circuits

data input to the output of the enabled 8-to-1 multiplexer. The top multiplexer in Figure 4.38 switches inputs D0 to D7 to the output. The next multiplexer switches outputs D8 to D15 and so on. The four multiplexer active low outputs are ORed together to provide a system 1-of-32 output.

4.5.1 Using Multiplexers as Boolean Function Generators

A multiplexer consists of a set of AND gates feeding a single output (N)OR gate. Because any Boolean function can be realized using AND-OR and NOT primitives, everything

Figure 4.37
A 32-to-1 multiplexer using two 74xx150 ICs

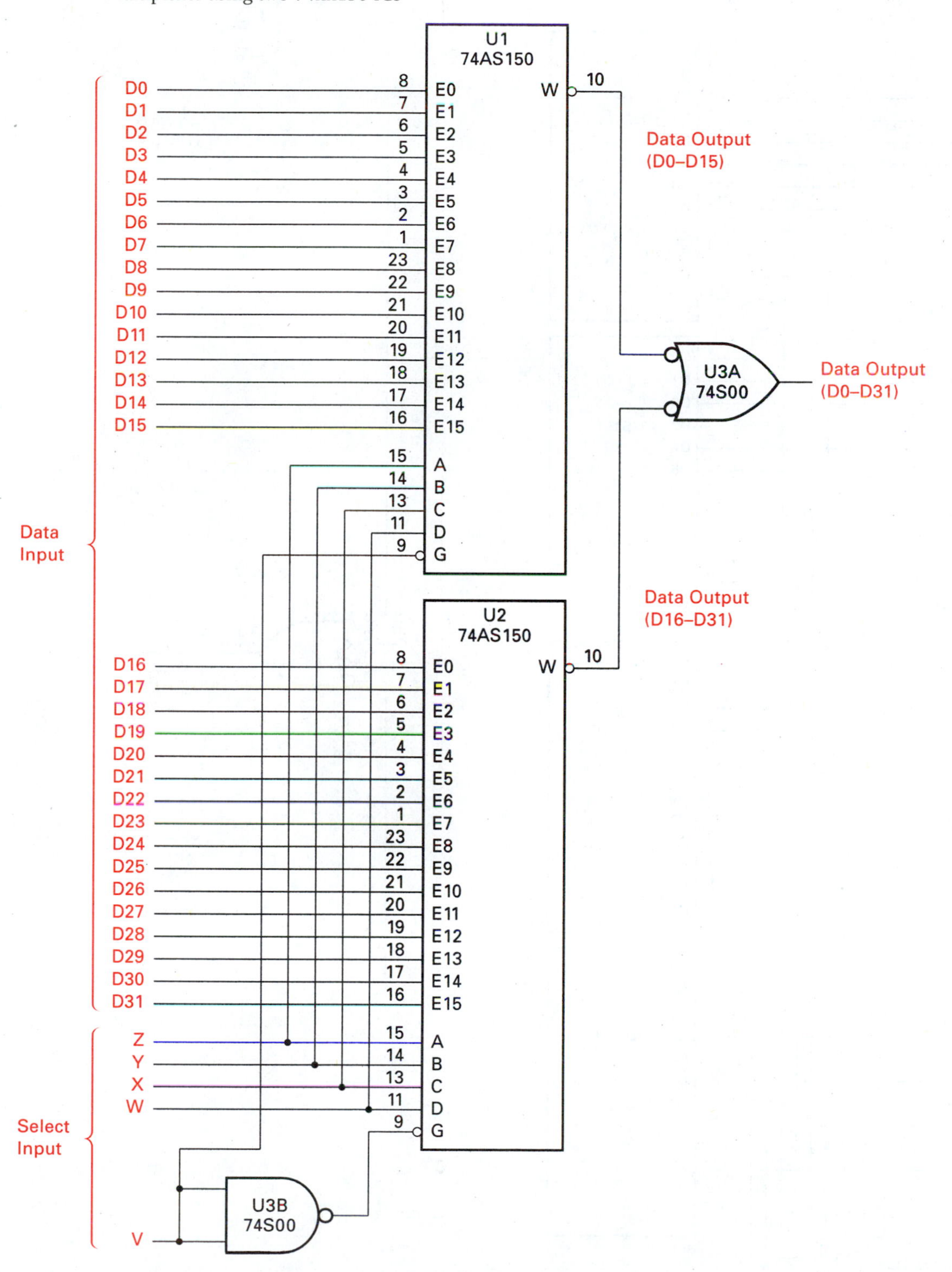

Figure 4.38
A 32-to-1 multiplexer using four 8-to-1 multiplexers and a 2-to-4 decoder

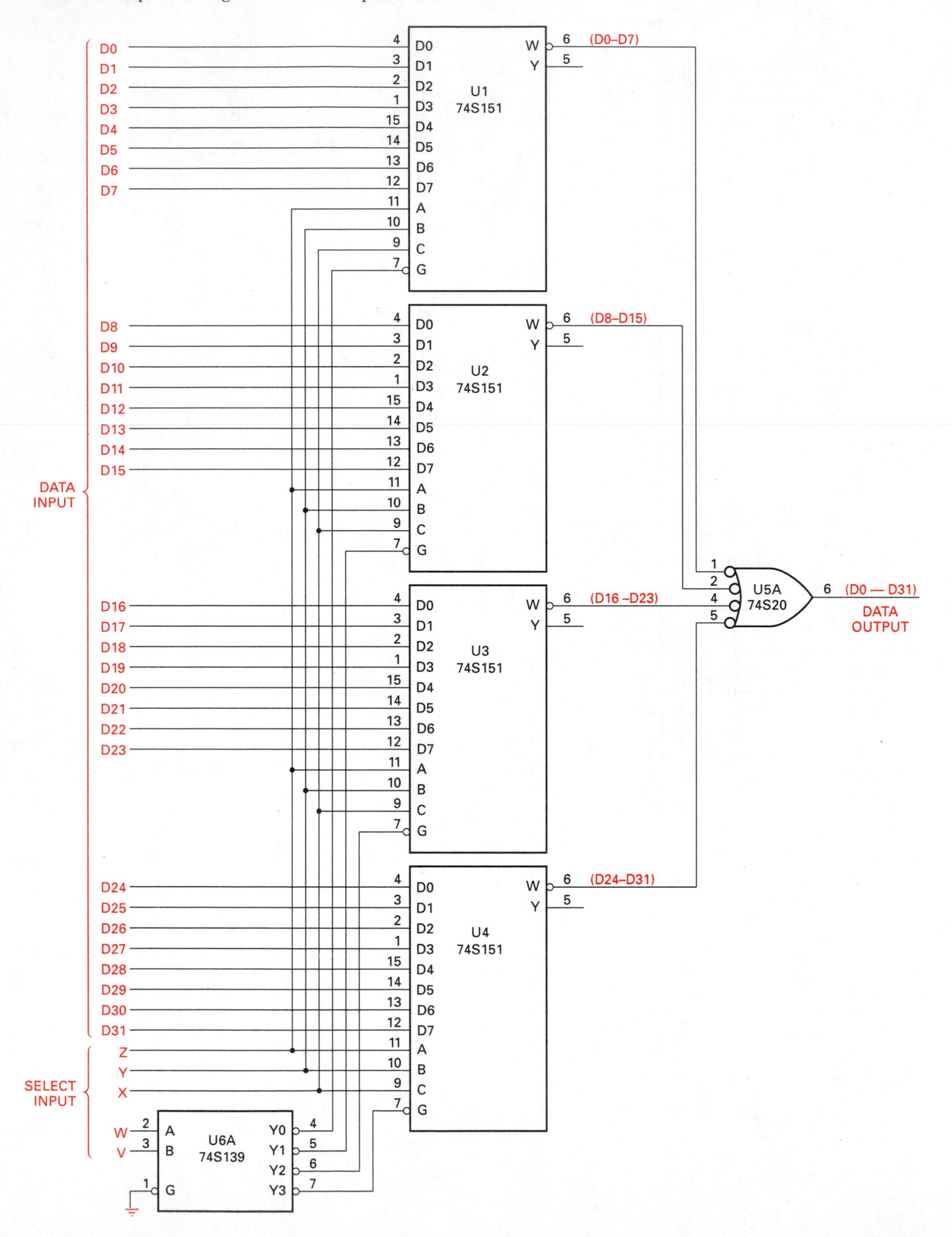

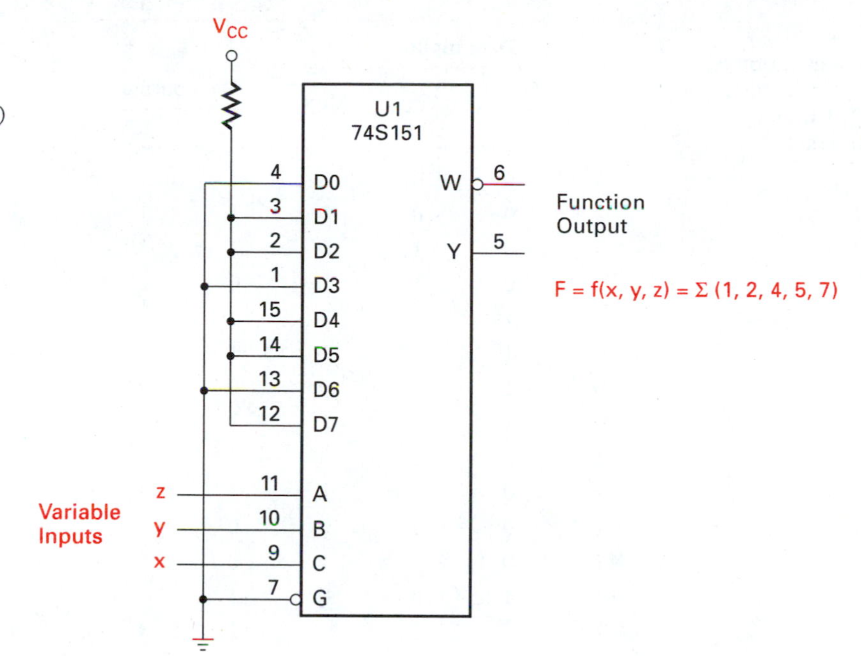

Figure 4.39
Using an 8-to-1 multiplexer to realize the Boolean function $F = f(x, y, z) = \Sigma\ (1, 2, 4, 5, 7)$

that is needed to realize a logic equation is found in a multiplexer. Each AND gate in a multiplexer can be used to generate a minterm when the number of variables in the minterm is equal to the number of select lines on the multiplexer. For example, an 8-to-1 multiplexer has three select lines, so it can be used to generate up to 2^3 minterms. By connecting the function variables directly to the select inputs, a multiplexer can be made to select the AND gate that corresponds to the minterm in the function. If a minterm exists in a function, we connect the AND gate data input to a 1. If it does not exist we connect it to a 0. Then, when a minterm occurs, the select lines switch the AND input 1 to the output. Figure 4.39 illustrates the use of an 8-to-1 multiplexer to realize the expression $F = f(x, y, z) = \Sigma\ (1, 2, 4, 5, 7)$.

Boolean expressions of any size can be realized by using multiplexer systems configured from individual multiplexer ICs as indicated in Figures 4.37 and 4.38. A four-variable problem can be realized directly using a 74xx150. Five-variable functions would require a 1-of-32 multiplexer.

We can increase the number of variables that a given multiplexer can realize by connecting the multiplexers data inputs to 0, 1, variable, or complemented variable. For example, a four-variable Boolean function $T = f(w, x, y, z) = \Sigma\ (0, 1, 2, 4, 5, 7, 8, 9, 12, 13)$ can be realized using an 8-to-1 multiplexer. Three variables are used as select inputs and the fourth is connected as needed to the multiplexer (mux) data inputs.

A truth table is constructed to aid in determination of the value to be connected to the multiplexer data inputs. Table 4.11 is that table. Figure 4.40 is the MEV K-map for this problem. The multiplexer data inputs are listed across the top of the table. The least significant variable, z, is separated from the three higher order inputs (w, x, y) to distinguish between the multiplexer input variable and the multiplexer select inputs. As a result the MEV map uses variable z as the MEV, shown in Figure 4.40. Notice that variables w, x, and y are connected to the select inputs of the multiplexer and define the MEV K-map squares and z, the MEV, is the data input.

TABLE 4.11
Truth table for realization of $T = f(w, x, y, z) = \Sigma\,(0, 1, 2, 4, 5, 7, 8, 9, 12, 13)$ using an 8-to-1 multiplexer

Data inputs				Data output	
Select			Mux		
w	x	y	z	T	MEV
0	0	0	0	1	
0	0	0	1	1	1
0	0	1	0	1	
0	0	1	1	0	z'
0	1	0	0	1	
0	1	0	1	1	1
0	1	1	0	0	
0	1	1	1	1	z
1	0	0	0	1	
1	0	0	1	1	1
1	0	1	0	0	
1	0	1	1	0	0
1	1	0	0	1	
1	1	0	1	1	1
1	1	1	0	0	
1	1	1	1	0	0

The completed MEV K-map shows that multiplexer inputs 0, 2, 4, and 6 need to be connected to V_{CC}. It also shows that inputs 5 and 7 need to be connected to ground. Input 1 is connected to z' and 3 is connected to z.

An inverter is used to negate input variable z; if both asserted and complemented input variables are available, the inverter is not needed.

See Figure 4.41. The number of variables in the Boolean function used in this example cannot be reduced. After simplification we still have four variables, so the multiplexer solution uses the least number of ICs. This is not always the case, however, so it is a good practice to see if the number of variables in an expression can be reduced before assuming a multiplexer solution. If, for example, a function specified for five variables is reduced to only one variable, then no logic gates are necessary, only a piece of wire is needed. Another example further illustrates the idea. Let $S = f(a, b, c, d, e) = \Sigma\,(5, 7, 13, 15, 21, 23, 29, 31)$. The realization of the five-variable function would need a 32-to-1 multiplexer if we did not simplify it first. After simplification we find that $S = ce$, which is realized using a two-input AND gate.

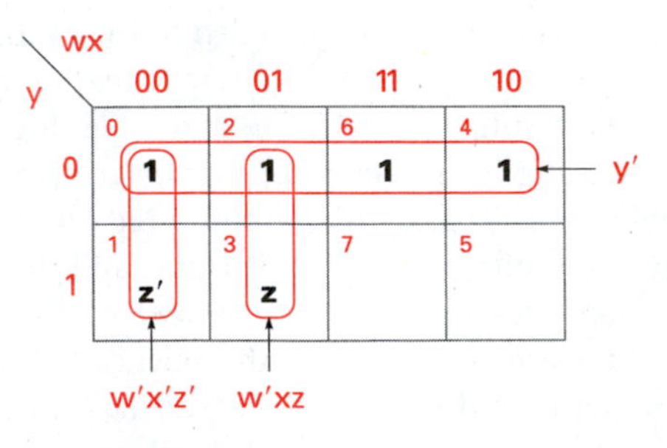

Figure 4.40
MEV K-map for determination of 8-to-1 multiplexer input values

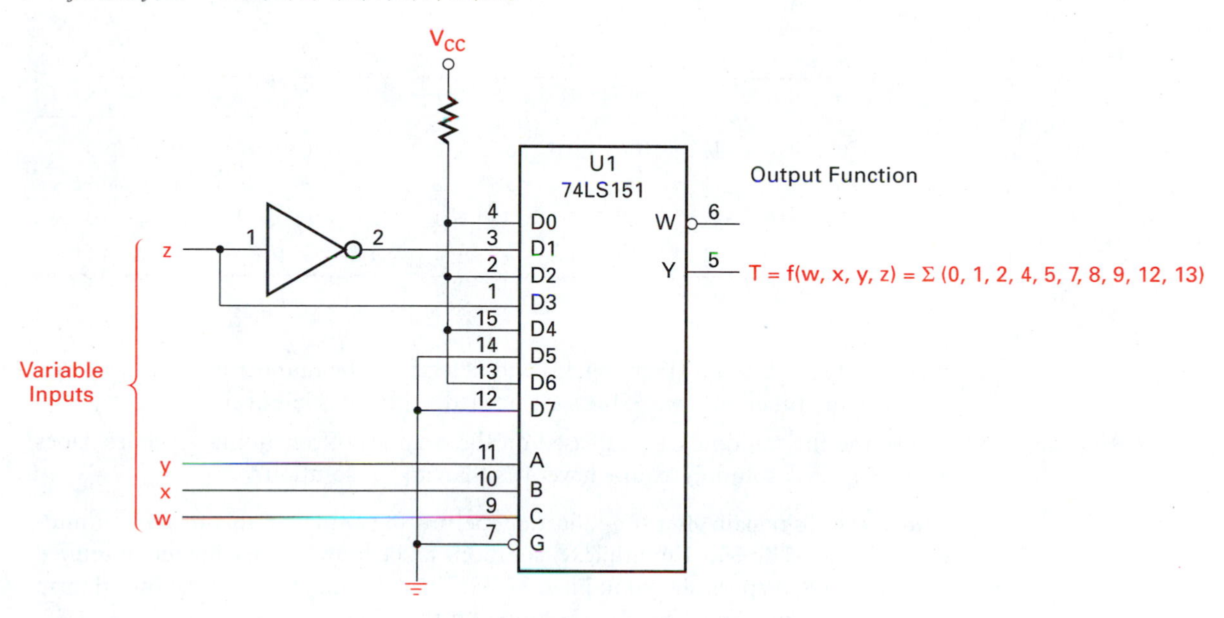

Figure 4.41
8-to-1 multiplexer realization of Boolean function
$T = f(w, x, y, z) = \Sigma\,(0, 1, 2, 4, 5, 7, 8, 9, 12, 13)$

❖ EXAMPLE 4.3

Realize the following Boolean function using the least number of ICs:

$$S = f(a, b, c, d, e) = \sum(8, 9, 10, 11, 13, 15, 17, 19, 21, 23, 24, 25, 26, 27, 29, 31)$$

SOLUTION Assume that both asserted and complemented input variables are available.

1. Simplify the expression (Figure 4.42).
2. Write the simplified expression and determine the number of remaining variables:

$$S = ae + bc' + be$$

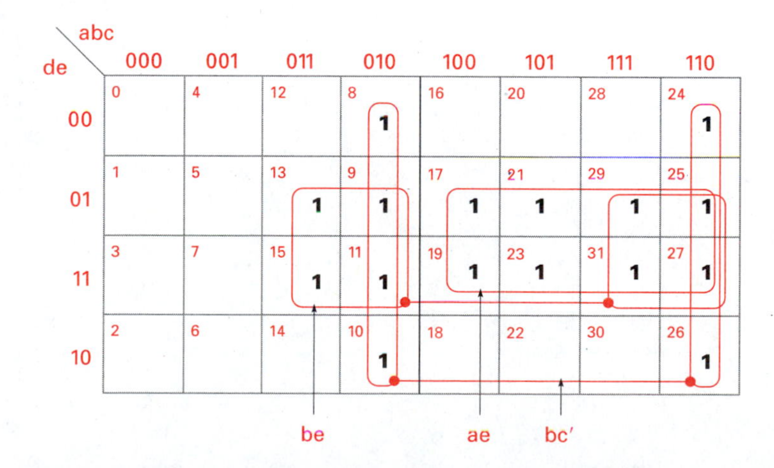

Figure 4.42
Five-variable Karnaugh map

TABLE 4.12
Tables showing different choices for select inputs to a 4-to-1 multiplexer solution to the equation $S = ae + bc' + be$

A			B			C			D			E			F		
a	b	S	a	c	S	a	e	S	b	c	S	b	e	S	c	e	S
0	0	0	0	0	b	0	0	bc'	0	0	ae	0	0	0	0	0	b
0	1	$c'+e$	0	1	be	0	1	b	0	1	ae	0	1	a	0	1	$a+b$
1	0	$c'+e$	1	0	$b+e$	1	0	bc'	1	0	1	1	0	c'	1	0	0
1	1	e	1	1	e	1	1	1	1	1	e	1	1	1	1	1	$a+b$

3. Draw the logic diagram (Figure 4.43) and determine the number of SSI ICs needed to realize the function. Two SSI ICs are needed, with gates left over.
4. Determine the multiplexer size based on the number of remaining variables. Does the multiplexer solution require fewer ICs than an SSI solution?

Four variables remain after simplification, permitting a single 8-to-1 or a 4-to-1 multiplexer solution. The 4-to-1 multiplexer solution can be found by placing the simplified equation into a K-map, as shown in Figure 4.44. This also identifies the canonical form of the simplified equation. The 8-to-1 solution for the equation is shown in Figure 4.45. The K-map in Figure 4.44 permitted identification of minterms (4, 5, 7, 9, 11, 12, 13, 15), which are used to obtain the 8-to-1 multiplexer solution.

The K-map shown in Figure 4.44 also permitted identification of the select inputs to a 4-to-1 multiplexer and the variables that will provide multiplexer data inputs. The two variables used for multiplexer select inputs form a "patch" of four cells in the four-variable K-map. Each "patch" is evaluated and the resulting value is placed in the S column for that combination. For example, consider using variables a and b for the select inputs. The ab = 00 location on the K-map in Figure 4.44 corresponds to cells (0, 1, 2, 3), all of which contain 0s, so a 0 is written in the S column of the A group in

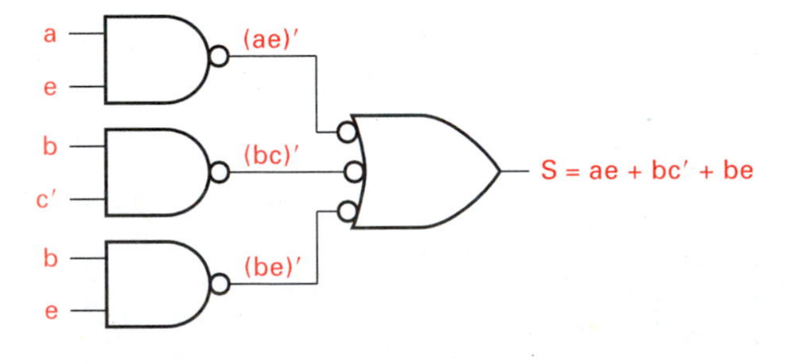

Figure 4.43
Logic diagram

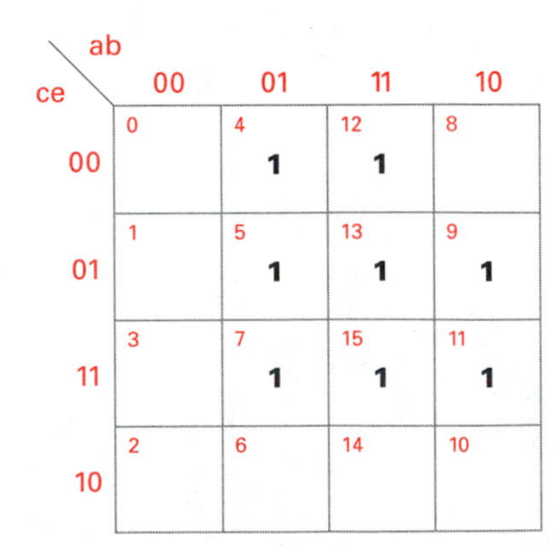

Figure 4.44
K-map identifying minterms for $S = ae + bc' + be$

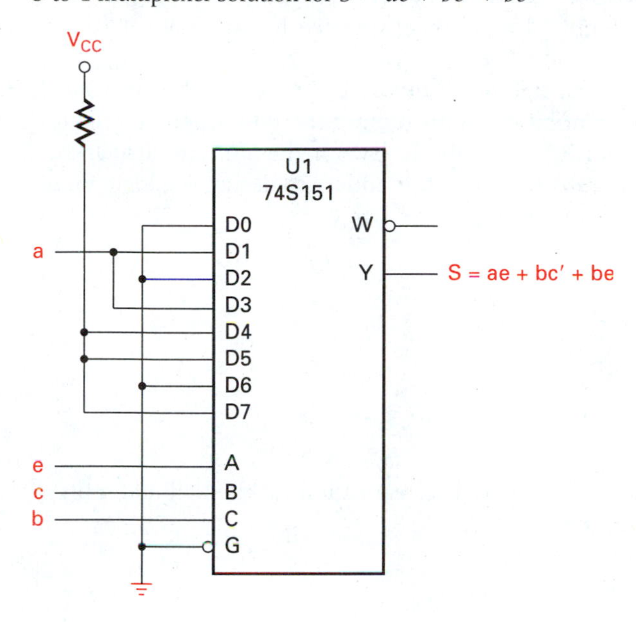

Figure 4.45
8-to-1 multiplexer solution for $S = ae + bc' + be$

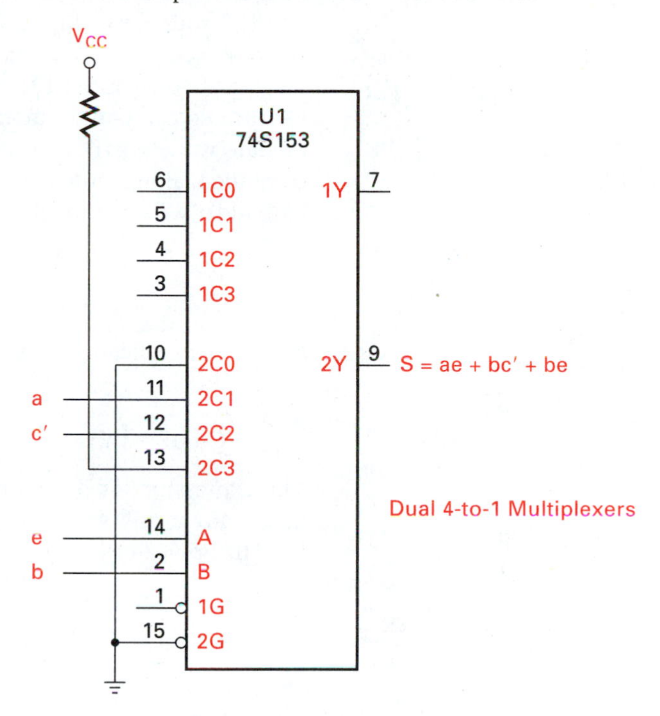

Figure 4.46
A 4-to-1 multiplexer solution to $S = ae + bc' + be$

Table 4.12. The ab = 01 corresponds to the second column representing cells (4, 5, 6, 7). Cells 4, 5, and 7 contain minterms. Finding the simplified result of the ab = 01 column produces $c' + e$, which is written in the S column of the A group in Table 4.12. Input ab = 11 produces $c' + e$ and ab = 10 yields e, which are written in the S column of the A group in Table 4.12.

The trick in finding the best solution is to choose the correct variables to use for the select inputs. The groups in Table 4.12 show all of the combinations of possible select inputs and the resulting inputs to the 4-to-1 multiplexer. We will select the best one.

The best choice is group E, which uses variables b and e as the multiplexer select inputs. In this example, choosing variables b and e requires no additional gates; all of the other combinations of select inputs require gating external to the multiplexer. Figure 4.46 shows the 4-to-1 multiplexer solution. ❖

4.6 ADDERS AND SUBTRACTORS

Section 10 of Chapter 1 introduced binary arithmetic. In that section we investigated binary addition and subtraction operations. Now we will apply those operations in the logic design of binary adders and subtractors.

Adding two single-bit binary values produces a sum and a carry output. This operation is called a *half-add* and the circuit that realizes the function is called a **half-adder**.

TABLE 4.13
Binary adder truth tables

Inputs			Outputs	
X	Y	C-in	S	C-out
Half Adder:				
0	0		0	0
0	1		1	0
1	0		1	0
1	1		0	1
Full Adder:				
0	0	0	0	0
0	0	1	1	0
0	1	0	1	0
0	1	1	0	1
1	0	0	1	0
1	0	1	0	1
1	1	0	0	1
1	1	1	1	1

Adding two single-bit binary values with the inclusion of a carry input produces two outputs, a sum and a carry; this circuit is called a **full-adder**. The truth table for the half- and full-adders was first given in Table 1.13 and is reproduced here as Table 4.13 for convenience.

A half-adder is used to add the least significant binary bits of a n-bit binary value, because no carry-in occurs. Subsequent bit positions require carry-in inputs to accommodate value overflow from lower order bit positions. Two half-adder circuits can be combined, along with an additional gate, to form a full-adder. To design the logic for a half-adder we write the Boolean equations for the two outputs.

$$S = f(X, Y) = \sum(1, 2)$$
$$\text{C-out} = f(X, Y) = \sum(3)$$

Because simplification is not possible,

$$S = X'Y + XY'$$
$$\text{C-out} = XY$$

The expression for S is an EX-OR: $S = X \oplus Y$. The realization for the half-adder may take many forms, as we can see from Figure 4.47.

The Boolean equations for the full-adder are

$$S = f(X, Y, \text{C-in}) = \sum(1, 2, 4, 7)$$
$$\text{C-out} = f(X, Y, \text{C-in}) = \sum(3, 5, 6, 7)$$

Figure 4.47
Half-adder logic diagrams

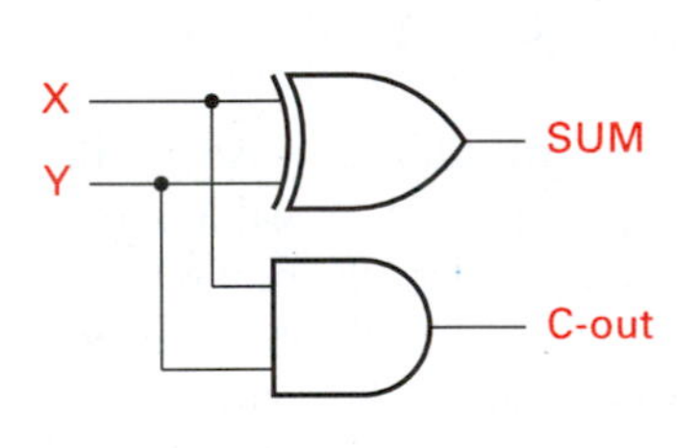

(a)

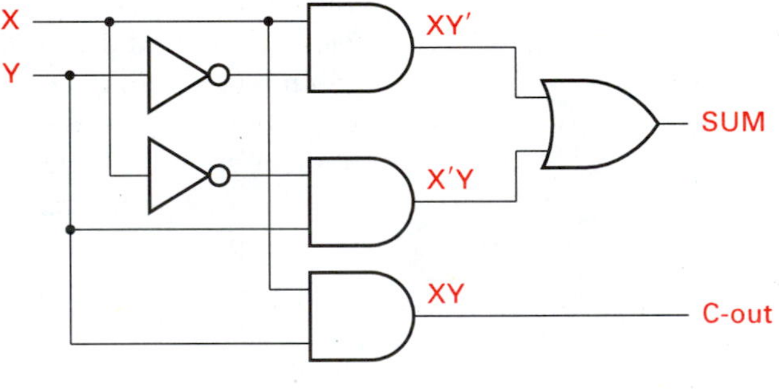

(b)

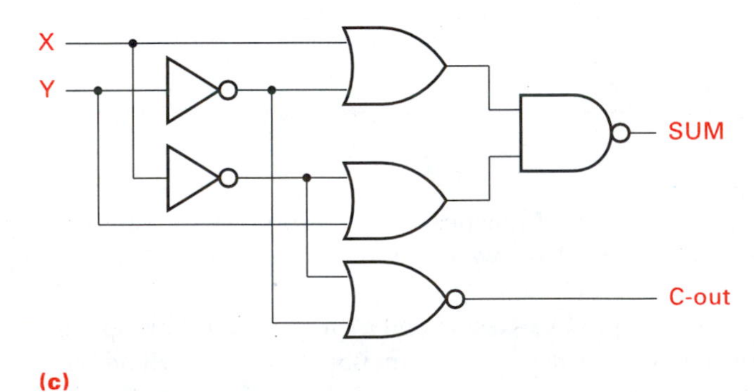

(c)

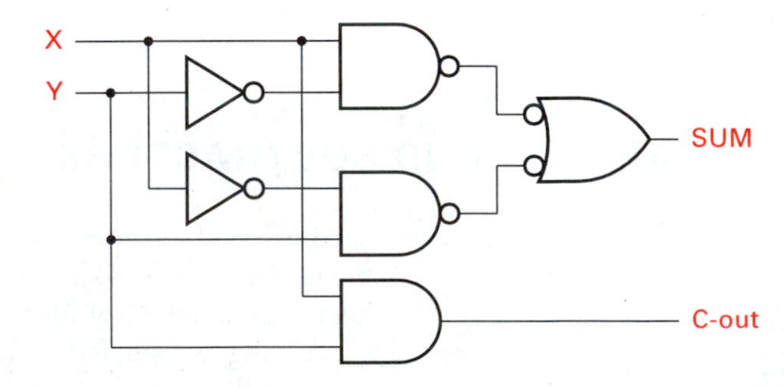

(d)

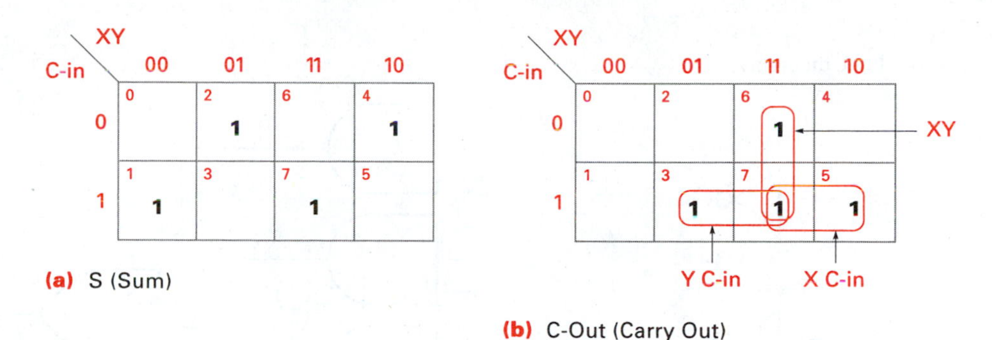

Figure 4.48
Full-adder K-maps

The K-maps for the two output functions are shown in Figure 4.48. The sum K-map shows no possible simplification. However, the equation for S is that of a three-input EX-OR function (see Figure 4.49). The final equations are

$$S = X'Y'\text{ (C-in)} + X'Y\text{ (C-in)}' + XY'\text{ (C-in)}' + XY\text{ (C-in)}$$
$$S = X \oplus Y \oplus \text{(C-in)}$$
$$\text{C-out} = XY + X\text{ (C-in)} + Y\text{ (C-in)}$$

Two half-adders and an OR gate form a full adder as shown in Figure 4.50.

The sum output is the same as that shown in Figure 4.49(b). C-out is a different logic configuration. Derive the C-out equation from Figure 4.50 and compare it to the simplified C-out equation for verification.

Binary Subtractors

The logic for binary subtractors can be developed in the same fashion as was done for adders. A half-subtractor does not have a borrow input. A full-subtractor has a borrow input. The truth tables for half- and full-subtractors, where the Y-bit is subtracted from the X-bit ($X - Y$), are given in Table 4.14.

The output equations for the half-subtractor are

$$D = f(X, Y) = \sum(1, 2)$$
$$\text{B-out} = f(X, Y) = \sum(1)$$

The results cannot be simplified, so

$$D = X'Y + XY'$$

which is the same EX-OR function we found for the half-adder,

$$\text{B-out} = X'Y$$

The half-subtractor logic diagram is shown in Figure 4.51.

The output equations for the full-subtractor are

$$D = f(X, Y, \text{B-in}) = \sum(1, 2, 4, 7)$$

which is the same three-input EX-OR function we found for the adder.

$$\text{B-out} = f(X, Y, \text{B-in}) = \sum(1, 2, 3, 7)$$

Figure 4.49
Full-adder logic diagrams

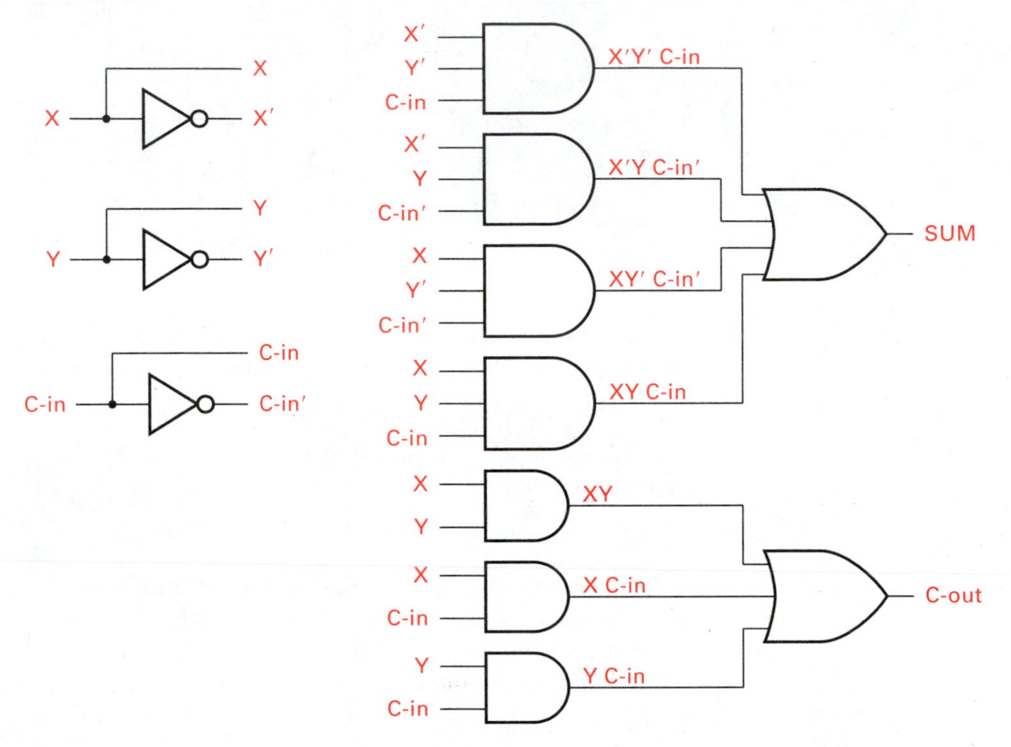

(a) Sum-of-Products

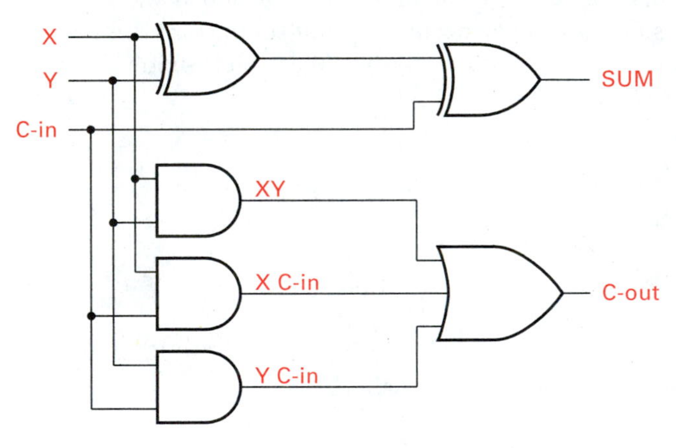

(b) EX-OR

which simplifies to

$$\text{B-out} = X'Y + X'(\text{B-in}) + Y(\text{B-in})$$

The full-subtractor logic diagram is given in Figure 4.52.

Compare the logic diagram for the full-adder with that for the full subtractor. Notice that the sum and difference outputs are realized identically. Also the carry-out and bor-

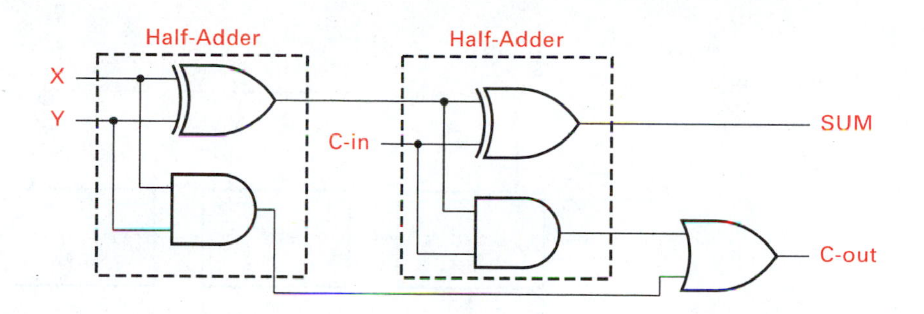

Figure 4.50
Two half-adders and a carry OR-gate form a full-adder

TABLE 4.14
Subtractor truth tables

Inputs			Outputs	
X	*Y*	B-in	*D*	B-out
Half-Subtractor:				
0	0		0	0
0	1		1	1
1	0		1	0
1	1		0	0
Full-Subtractor:				
0	0	0	0	0
0	0	1	1	1
0	1	0	1	1
0	1	1	0	1
1	0	0	1	0
1	0	1	0	0
1	1	0	0	0
1	1	1	1	1

row-out differ only in that the input *X* is inverted when subtracting. This implies that subtraction can be realized using an adder by complementing and adding.

4.6.1 Cascading Full-Adders

We showed that two half-adders can form a full-adder. An *n*-bit adder can be built by cascading (connecting in series) *n* full-adders. Each full-adder represents a bit position. The least significant bit position can be a half-adder, because a carry from a less significant position does not exist. Each carry-out from a full-adder becomes the carry-in to the next higher order adder. Figure 4.53 illustrates cascading four full adders to form a four-bit adder.

The carry-in input of the least significant full-adder is grounded. EX-OR gates are available in 74xx86 TTL ICs; AND gates, in 74xx08; and the OR gate, in 74xx802 TTL ICs. The two binary inputs (*X* and *Y*) are presented to the adder from some external source. The logic designer must know how much time the logic requires for the addition of two binary numbers to know when new data inputs can be presented. The add time required is determined by calculating the **propagation delays** through the full-adders. (See Figure 4.11 where propagation delays are illustrated.) Notice that the carries propagate from one full-adder to the next higher order full-adder. This type of addition is called *ripple-carry propagation.*

Total add delay time is the product of the sum of the number of stages in the adder, and the carry-in to carry-out propagation delay time. The carry-in to carry-out propa-

Figure 4.51
Half-subtractor logic diagram

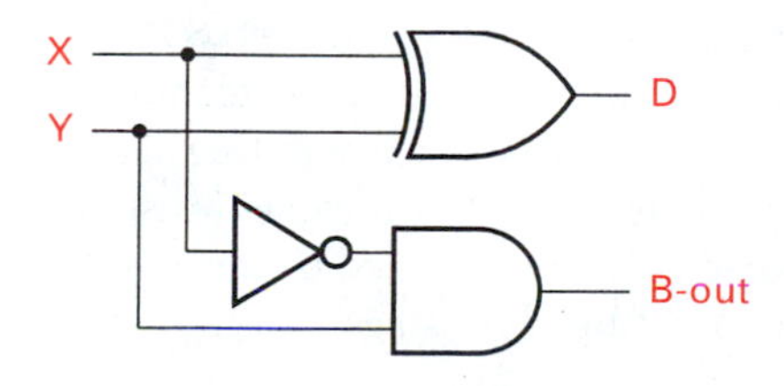

Figure 4.52
Full-subtractor logic diagram

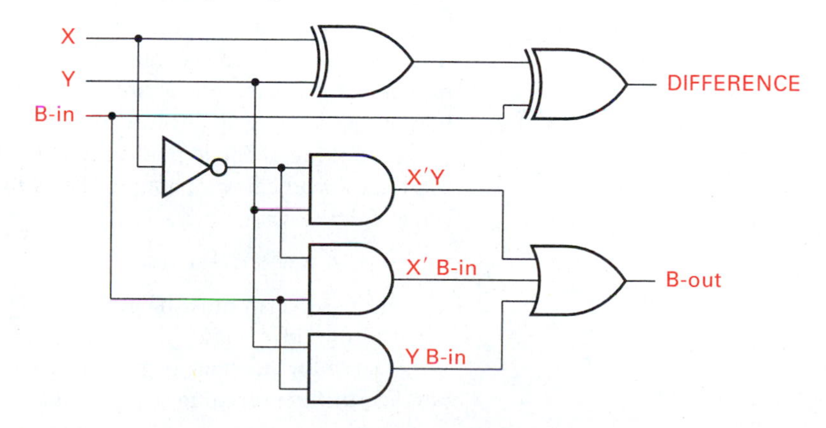

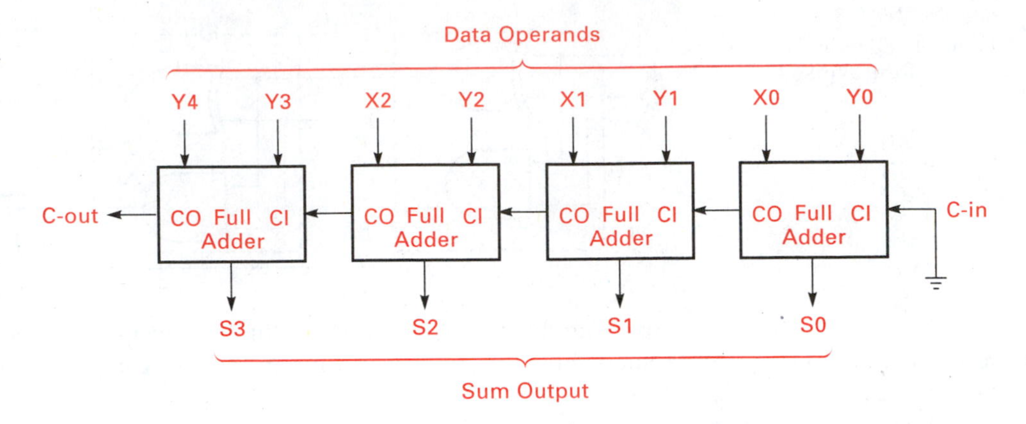

Figure 4.53
Four full-adders forming a four-bit adder

gation delay is used instead of the input-to-sum output delay because it is the signal that ripples from one full-adder to another.

Assume that an adder is realized using an 74ALS86 (EX-OR), 74AS08 (two-input AND), and 74AS802 (four-input OR). The TTL data book tells us the delay times for each type of IC:

Parameter	AS08	AS802	ALS86
t_{pLH}	6 ns	4 ns	8 ns
t_{pHL}	6 ns	5 ns	7 ns

The propagation time from any data input (X, Y, C-in) to the sum output is

$$t_{pd} = 8 \text{ ns} + 8 \text{ ns} = 16 \text{ ns} \qquad \text{(2 EX-OR gates)}$$

Propagation time from any data input (X, Y, C-in) to the C-out of a full-adder (refer to Figure 4.50), using two half-adders, is

$$t_{pd} = 8 \text{ ns} + 6 \text{ ns} + 5 \text{ ns} = 19 \text{ ns} \qquad \text{(EX-OR, AND, and OR gates)}$$

The total propagation delay for the four-bit adder (using 2 half adders per adder) is

$$t_{pd} = 19 \text{ ns} + 3(11 \text{ ns}) = 52 \text{ ns} \qquad \text{(19 ns for LSB adder, 11 ns each for the other three adders).}$$

4.6.2 Look-Ahead Carry

Logic designers rarely need to realize binary adders using SSI devices. The 74xx83A is a four-bit adder with a fast carry feature. The **look-ahead carry**, or fast carry, technique reduces propagation times through the adder. This is accomplished by generating all the individual carry terms needed by each full-adder in as few levels of logic as possible. Before looking at the logic diagrams for the IC four-bit adders, we need to investigate the look-ahead carry technique. The sum output of any full-adder stage in a multiple-stage adder is written

$$S_j = X_j \oplus Y_j \oplus C\text{-in}_j$$

The $C\text{-in}_j$ term must be generated based on all the values of X and Y that precede the jth full-adder. Figure 4.54 illustrates the idea.

A carry for jth stage is generated when inputs X_j and Y_j produce $\text{C-out}_j = 1$, independent of any preceding stage inputs, X_{j-1}, $Y_{j-1} \ldots$, X_0, Y_0.

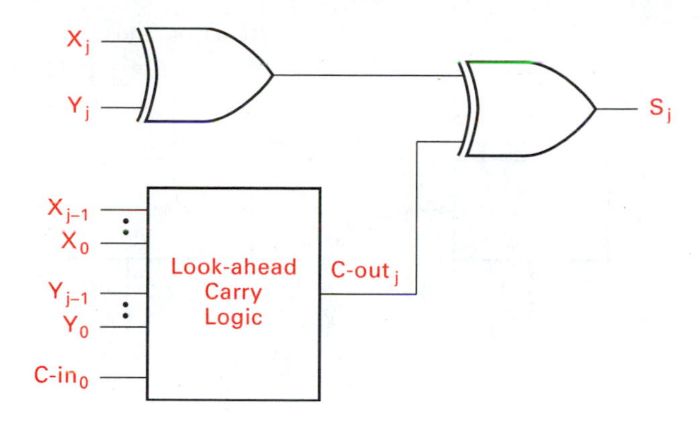

Figure 4.54
Generation of $C\text{-}in_j$ for a look-ahead carry

A carry for the jth stage is propagated when $C\text{-}out_j = 1$, based on receiving a $C\text{-}in_j$ from $C\text{-}out_{j-1}$.

Carry generate, Cg, and carry propagate, Cp, signals can be formed for each stage of a look-ahead carry adder. Let $Cg_j = X_j Y_j$, and $Cp_j = X_j + Y_j$. Each stage automatically generates a C-out if both of its data inputs (X, Y) are a 1, and it sends a carry if either of its inputs is a 1. The carry output, $C\text{-}out_{j+1}$, can be written in terms of the carry generate and carry propagate signals:

$$C\text{-}out_j = Cg_j + Cp_j\, C\text{-}in_j$$
$$C\text{-}out_j = C\text{-}in_{j+1}$$

The carries for a four-stage look-ahead adder are

$$\begin{aligned}
C\text{-}out_0 &= Cg_0 + Cp_0(C\text{-}in_0) \\
C\text{-}out_1 &= Cg_1 + Cp_1(C\text{-}out_0) = Cg_1 + Cp_1(Cg_0 + Cp_0(C\text{-}in_0)) \\
&= Cg_1 + Cp_1Cg_0 + Cp_1Cp_0C\text{-}in_0 \\
C\text{-}out_2 &= Cg_2 + Cp_2(C\text{-}out_1) \\
&= Cg_2 + Cp_2(Cg_1 + Cp_1Cg_0 + Cp_1Cp_0C\text{-}in_0) \\
&= Cg_2 + Cp_2Cg_1 + Cp_2Cp_1Cg_0 + Cp_2Cp_1Cp_0C\text{-}in_0 \\
C\text{-}out_3 &= Cg_3 + Cp_3(C\text{-}out_2) \\
&= Cg_3 + Cp_3(Cg_2 + Cp_2Cg_1 + Cp_2Cp_1Cg_0 + Cp_2Cp_1Cp_0C\text{-}in_0) \\
&= Cg_3 + Cp_3Cg_2 + Cp_3Cp_2Cg_1 + Cp_3Cp_2Cp_1Cg_0 + Cp_3Cp_2Cp_1Cp_0C\text{-}in_0
\end{aligned}$$

Each of the four look-ahead carry equations requires three levels of logic to realize. The first level of logic is needed to produce the Cp_j and Cg_j terms. Then two additional levels are needed to realize the $C\text{-}out_j$ functions. This means that the adder propagation delay is constant, not dependent on the ripple carry.

4.6.3 MSI Adders

Cascading multiple four-bit adders allow the logic designer to design adders of any length that are a multiple of four. Figure 4.55 illustrates cascading four four-bit adders to add two 16-bit values.

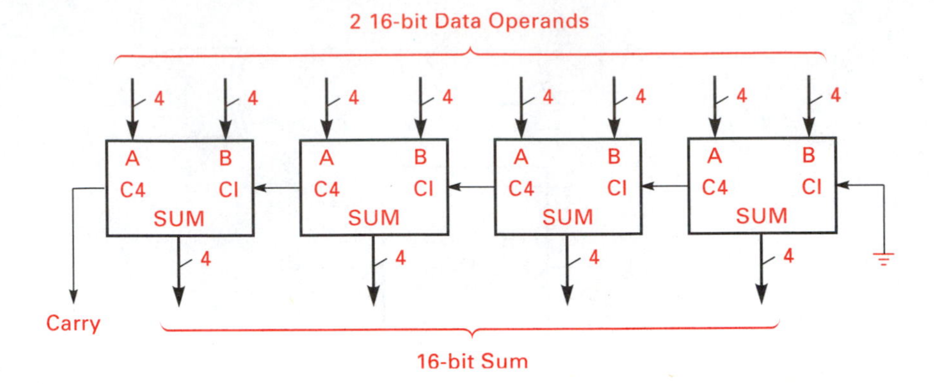

Figure 4.55
Cascading four 74LS83A four-bit adders to form a 16-bit adder

C_0 of the least significant four-bit adder is grounded to produce a no-carry input. The carry-out, C4, of the most significant adder stage represents the most significant bit of the result. Two 16-bit inputs produce a 17-bit result.

Figure 4.56 illustrates the internal logic of the 74LS83A device. Note that the C4 output of the IC uses a partial look-ahead carry to reduce the carry-out propagation delay times. The TTL data book lists the worst case delay times from carry-in to carry-out as 22 ns. A worst case delay time for the 16-bit adder illustrated in Figure 4.55 is 88 ns.

The logic shown in Figure 4.56 can be evaluated by applying the logic implied in Figure 4.54 to each of the full-adder stages. A half-sum is produced for each stage and gated with the carry-out ($C\text{-out}_j$) to produce the sum output. Each output EX-OR receives the half-sum (hs_j) and $C\text{-out}_j$ to generate the S_j output. The Cp_j and Cg_j terms for each of the four full-adder stages are generated by the first level of logic. The carry-out of the four-bit adder, $C\text{-out}_4$, is generated without any stage to stage ripple, by the top NOR gate.

4.6.4 Using MSI Adders as Subtractors

In Chapter 1, Section 10.2 we discussed the binary arithmetic operation of subtraction by finding the radix complement of the subtrahend and adding it to the minuend to produce the difference. Recall that the radix complement of a number is the difference between the largest value that n digits of radix r can express plus 1 and the number for which a complement is sought.

For a number N, the radix complement is found as

$$r^n - N = \text{radix complement of } N$$

Let $r = 10$, $n = 2$, and $N = 36$. Find the radix complement:

$$10^2 - 36 = 100 - 36 = 64$$

The radix complement of a binary number can be found by inverting each bit and adding 1 to the result. For example, let us find the 2s complement of 1101. The inversion of each bit gives 0010, adding 1 gives 0011 = 2s complement of 1101.

By finding the 2s complement of the subtrahend and adding the result to the minuend, we can realize subtraction. Let $A = 0{,}11011$ and $B = 0{,}01110$. Find $A - B$ by adding the 2s complement of B to A. A leading 0 indicates a positive number and a

Figure 4.56
74LS83A four-bit adder

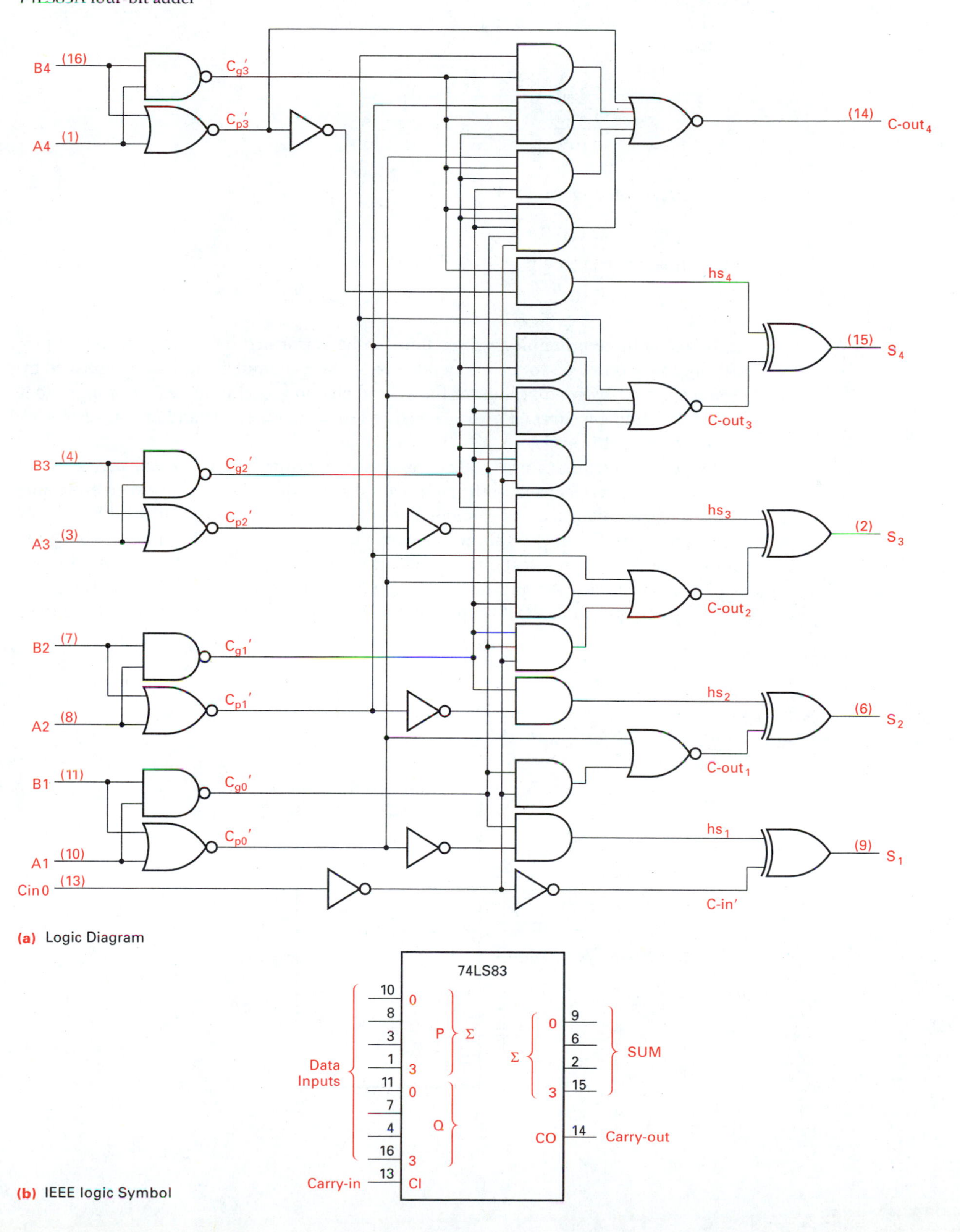

(a) Logic Diagram

(b) IEEE logic Symbol

leading 1 a negative number; the comma separates the sign from the magnitude portion of the number.

The 2s complement of $B = B' + 1 = 1{,}10010$.

By adding,

$$\begin{array}{rl} A & = 0{,}11011 \\ B' + 1 & = \underline{1{,}10010} \\ & \ 0{,}01101 \end{array}$$

By direct subtraction,

$$\begin{array}{rl} A = & 0{,}11011 \\ B = & -\underline{0{,}01110} \\ & 0{,}01101 \end{array}$$

By using these principles, subtraction can be performed by adding the 2s complement of the subtrahend to the minuend. Taking the 2s complement is accomplished by producing a bit-by-bit inversion of the subtrahend and forcing the carry-in input to a logical 1. A four-bit subtractor can be realized using one 74xx83 and four inverters, as illustrated in Figure 4.57.

Larger subtractors can be realized by cascading four-bit adders and inverters. By replacing the inverters used in the subtractor circuit with EX-OR gates, with one input from each gate connected to a common control line, the system can realize both addition and subtraction. Figure 4.58 shows such an adder/subtractor circuit. Input, A'/S, provides mode control for selection of an add or subtract operation. When the A'/S input is a 1 the EX-OR gates invert the input subtrahend bits prior to addition. It also forces the carry-in input to be a 1, completing the 2s complement operation for subtraction.

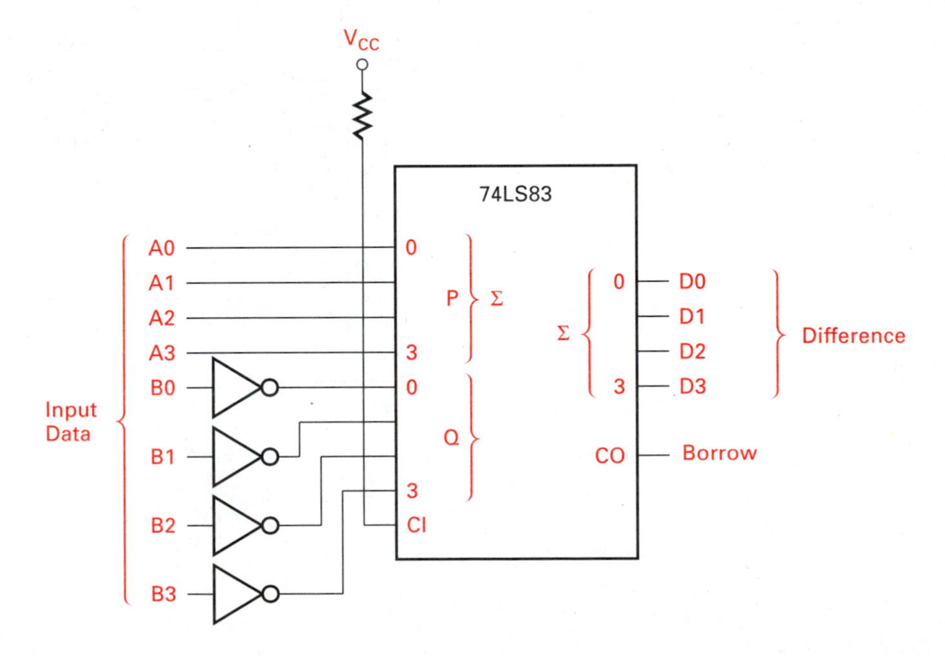

Figure 4.57
A four-bit subtractor using a four-bit adder and inverters

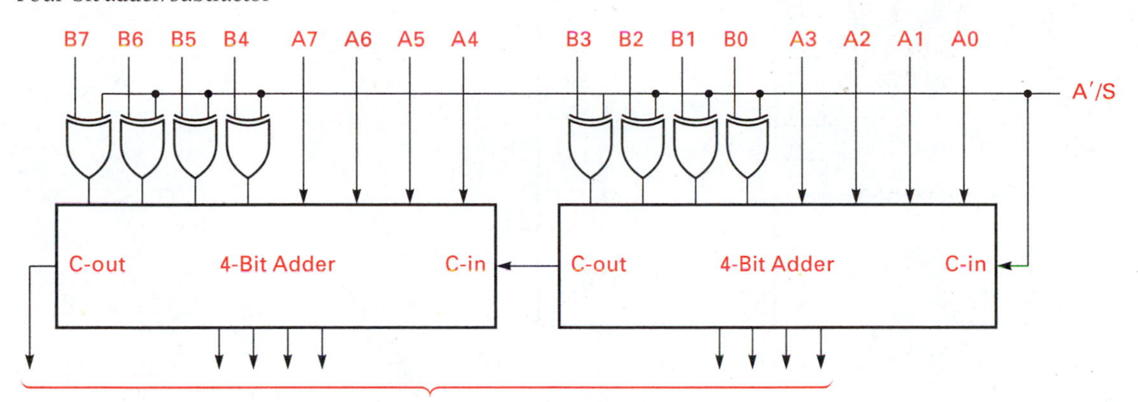

Figure 4.58
Four-bit adder/subtractor

When the A'/S mode control input is a 0 the EX-OR gates do not invert, so the input data are added unchanged. The carry-in bit is forced to logical 0.

4.6.5 Using an MSI Adder as a BCD to Excess-3 Code Converter

The Excess-3 code is a special self-complementing BCD code (refer to Table 1.8) used in decimal arithmetic units. Standard BCD can be easily converted to ex-3 BCD by using a four-bit adder for each BCD digit to be converted. Table 4.15 illustrates the truth table for a BCD to Excess-3 code conversion.

Evaluation of this truth table shows that the output code value is 3 greater than the value of the input code. An MSI four-bit adder can be used to add the input BCD code to a constant of 3, producing an Excess-3 BCD code output.

TABLE 4.15
BCD to Ex-3 code converter truth table

BCD Input				Ex-3 BCD Output			
A	*B*	*C*	*D*	*W*	*X*	*Y*	*Z*
0	0	0	0	0	0	1	1
0	0	0	1	0	1	0	0
0	0	1	0	0	1	0	1
0	0	1	1	0	1	1	0
0	1	0	0	0	1	1	1
0	1	0	1	1	0	0	0
0	1	1	0	1	0	0	1
0	1	1	1	1	0	1	0
1	0	0	0	1	0	1	1
1	0	0	1	1	1	0	0

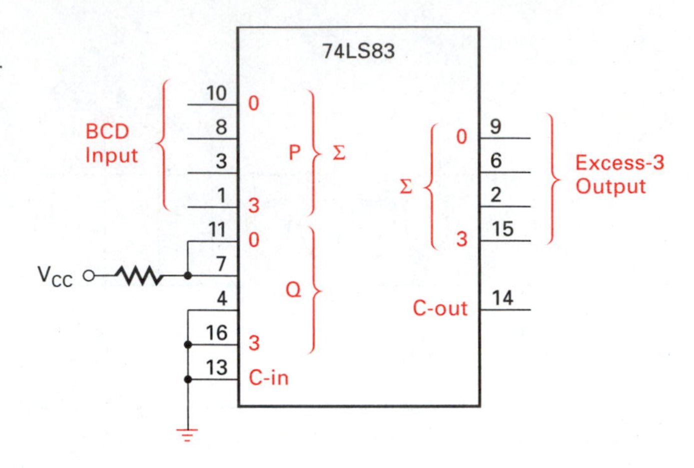

Figure 4.59
BCD to Excess-3 code converter using a four-bit adder

Adders can be used to add or subtract two binary operands or a single operand with a constant, as was the case with the BCD to Excess-3 code converter. Figure 4.59 shows the realization of the BCD to Excess-3 code converter.

4.6.6 BCD Adder

BCD adders require two four-bit adders for each BCD digit that is to be added. The reason for this becomes apparent when we analyze the process of adding together two BCD digits. Consider the following.

Let $A = 1000$ and $B = 0111$, both legitimate BCD codes. When the two operandi are added directly together, the result is not a legitimate BCD code:

$$\begin{array}{r} 1000 \\ +\ \underline{0111} \\ 1111 \end{array}$$

The correct response of the BCD adder should be

$$\begin{aligned} 1000 + 0111 &= (1)\ 0101 \\ 8 + 7 &= (1)\ 5 \end{aligned}$$

The digit in parentheses represents the carry output from the adder.

Other BCD input operand combinations also produce incorrect BCD output results. In fact, any input operand combination that results in a sum that is greater than nine or that produces a carry-out generates an incorrect BCD output if correction is not taken. Consider the preceding example. The result from direct addition was 1111, which is not a BCD code value; it should have been 0101. If we construct a table of all possible binary and BCD sums we can determine the correction needed. Such a table would have 20 rows because adding two BCD operandi can produce a result from 0_{10} to 19_{10} ($0_{10} + 0_{10}$ to $9_{10} + 9_{10}$ with a carry-in).

Consider Table 4.16. Below the middle line we see that all the BCD result rows produce a carry output. Above the middle line both binary and BCD results are the same. From this analysis we can conclude that two conditions require correction: first, any binary result greater than 9_{10}; and second, any binary add carry produced. Knowing

TABLE 4.16
Binary and BCD sum table

Binary result					BCD result				
C	E3	E2	E1	E0	C	E3	E2	E1	E0
0	0	0	0	0	0	0	0	0	0
0	0	0	0	1	0	0	0	0	0
0	0	0	1	0	0	0	0	1	0
0	0	0	1	1	0	0	0	1	1
0	0	1	0	0	0	0	1	0	0
0	0	1	0	1	0	0	1	0	1
0	0	1	1	0	0	0	1	1	0
0	0	1	1	1	0	0	1	1	1
0	1	0	0	0	0	1	0	0	0
0	1	0	0	1	0	1	0	0	1
0	1	0	1	0	1	0	0	0	0
0	1	0	1	1	1	0	0	0	1
0	1	1	0	0	1	0	0	1	0
0	1	1	0	1	1	0	0	1	1
0	1	1	1	0	1	0	1	0	0
0	1	1	1	1	1	0	1	0	1
1	0	0	0	0	1	0	1	1	0
1	0	0	0	1	1	0	1	1	1
1	0	0	1	0	1	1	0	0	0
1	0	0	1	1	1	1	0	0	1

when a correction is needed is only part of the solution; we must also know what correction is needed. Again, from studying Table 4.16, we see that if the binary sum is greater than 9, then it is 10_{10} greater than the BCD sum. To correct the BCD result we would need to subtract 10_{10} from the binary sum. Also, when the binary add produces a carry output its sum is a value of 6_{10} lower than the BCD sum.

If the binary sum is greater than 9, 10_{10} must be subtracted to produce the BCD sum; and if a binary carry occurs, 6 must be added to produce the correct BCD sum. Consider subtracting 10_{10} from the binary sum to produce the correct BCD sum. We know that subtraction can be accomplished by adding the minuend complement to the subtrahend. In this case the complement is the difference between the highest value contained in the four-bit sum and the constant 10_{10}.

The highest value contained in a four-bit binary number value is 15_{10}. Let X be the binary number. Then $X - 10_{10}$ is the corrected result:

$$X - 10_{10} = X + (16_{10}\text{s complement of } 10_{10})$$

$$X + [(15_{10}\text{s complement of } 10_{10}) + 1)]$$

And the 15_{10}s complement of $10_{10} = 5_{10}$. So the 16_{10}s complement of $10_{10} = 6_{10}$.

Adding 6_{10} to the binary sum corrects the BCD result for any case where the binary sum is greater than 9. We also add 6 any time a binary carry is generated. This means that both conditions can be satisfied by adding 6_{10} to the binary sum.

The correction logic needs to detect two conditions. The carry is straightforward, because a binary carry output is available from a four-bit adder. The second condition requires that the binary sum outputs ranging from 1010 to 1111 must be detected (10_{10}, 11_{10}, 12_{10}, 13_{10}, 14_{10}, 15_{10}). By treating these as minterms of a four-variable Boolean expression, we can simplify the logic and determine the gates required.

A simplification of the above minterms gives

$$\text{Result} = \text{E3E2} + \text{E3E1}$$

E3, E2, E1, and E0 are the sum outputs from the adder.

Combining the binary carry-out and the greater than nine conditions, the BCD correction logic produces the logic shown in Figure 4.60.

A second four-bit binary adder adds the constant six to the output of the first binary adder, when necessary. Either the constant 6_{10} or 0 is added to the binary sum of the first adder. Figure 4.61 shows the complete logic for a single BCD digit adder.

Multiple-digit BCD adders can be constructed by cascading as many single-digit units as needed. The BCD carry-out from each stage would be connected to the carry-in of the next higher order stage.

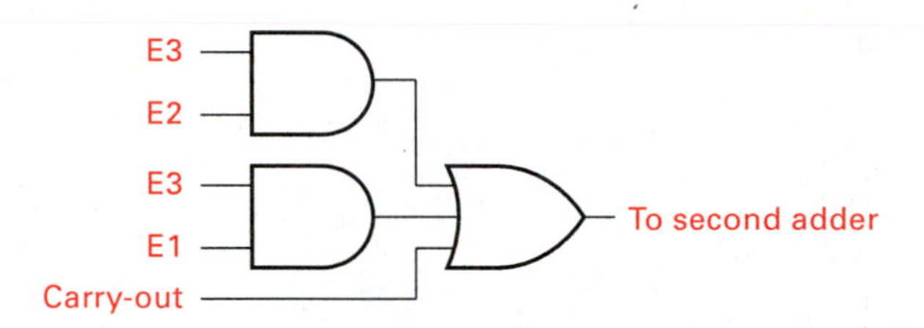

Figure 4.60
Binary to BCD correction logic

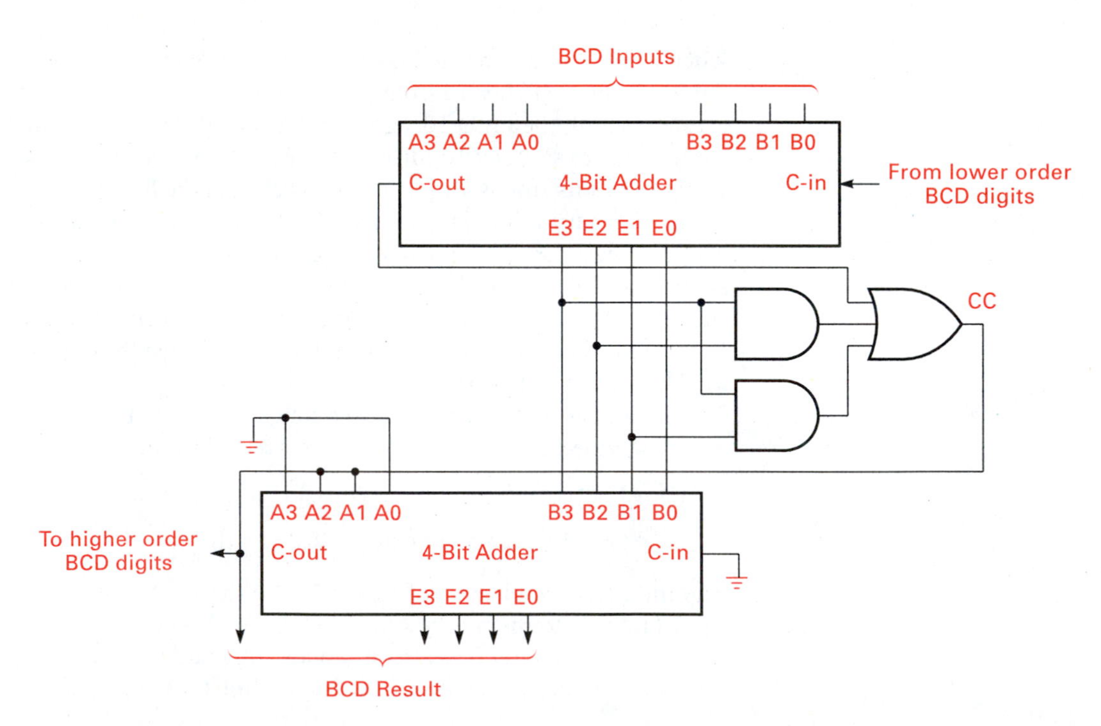

Figure 4.61
Single-digit BCD adder

TABLE 4.17
Truth table for a one-bit comparator

A	B	A EQ B	$A > B$	$A < B$
0	0	1	0	0
0	1	0	0	1
1	0	0	1	0
1	1	1	0	0

4.7 BINARY COMPARATORS

A binary magnitude **comparator** is a logic circuit that provides output information indicating the relative magnitude of two input operandi. Three output conditions exist as a result of comparison of two operands: A is equal to B, A is greater than B, or A is less than B.

Consider the case where both operands A and B are single bit binary numbers. The construction of a truth table for such a comparator is Table 4.17.

The equations for each output are

$$A \text{ EQ } B = A'B' + AB = (A \oplus B)'$$
$$A > B = AB'$$
$$A < B = A'B$$

The resulting logic is shown in Figure 4.62. Increasing the size of the operand to two bits results in Table 4.18, where

$$(A \text{ EQ } B) = f(a1, a0, b1, b0) = \sum(0, 5, 10, 15)$$
$$(A > B) = f(a1, a0, b1, b0) = \sum(4, 8, 9, 12, 13, 14)$$
$$(A < B) = f(a1, a0, b1, b0) = \sum(1, 2, 3, 6, 7, 11)$$

The K-maps for all three are given in Figure 4.63. The resulting equations are

$$(A \text{ EQ } B) = a1'a0'b1'b0' + a1'a0b1'b0 + a1a0b1b0 + a1a0'b1b0'$$
$$(A > B) = a1b1' + a0b1'b0' + a1a0b0'$$
$$(A < B) = a1'b1 + a1'a0'b0 + a0'b1b0$$

Increasing the size of a comparator beyond two bits would produce an unwieldy truth table. Another approach is necessary. Recall that, once the design for a full-adder

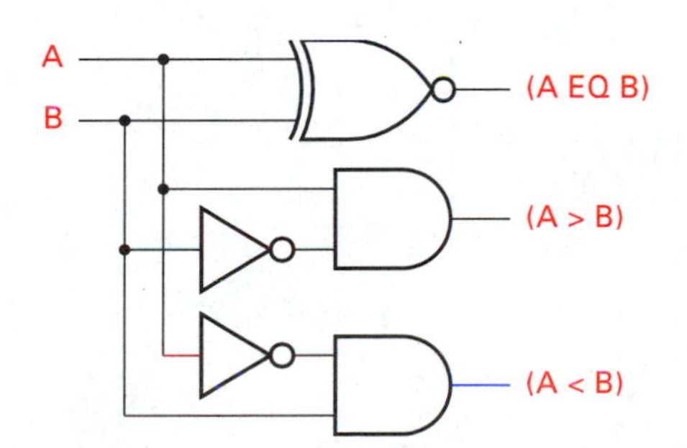

Figure 4.62
Logic for a single-bit comparator

Table 4.18
Truth table for a two-bit magnitude comparator

Inputs				Outputs		
A		B				
$a1$	$a0$	$b1$	$b0$	$A = B$	$A > B$	$A < B$
0	0	0	0	1	0	0
0	0	0	1	0	0	1
0	0	1	0	0	0	1
0	0	1	1	0	0	1
0	1	0	0	0	1	0
0	1	0	1	1	0	0
0	1	1	0	0	0	1
0	1	1	1	0	0	1
1	0	0	0	0	1	0
1	0	0	1	0	1	0
1	0	1	0	1	0	0
1	0	1	1	0	0	1
1	1	0	0	0	1	0
1	1	0	1	0	1	0
1	1	1	0	0	1	0
1	1	1	1	1	0	0

Figure 4.63
K-maps for a two-bit binary comparator

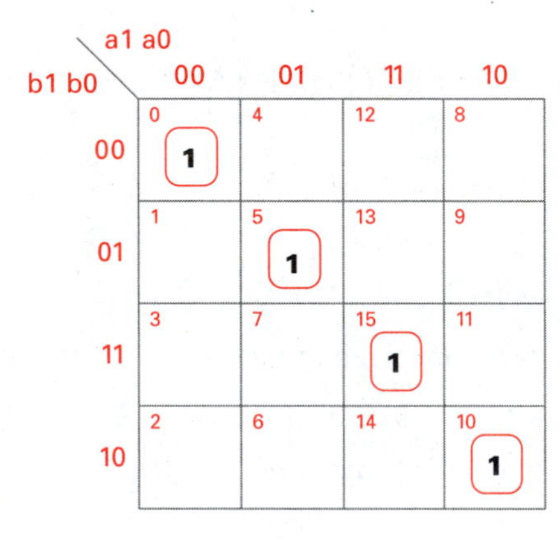

(a) (A EQ B)

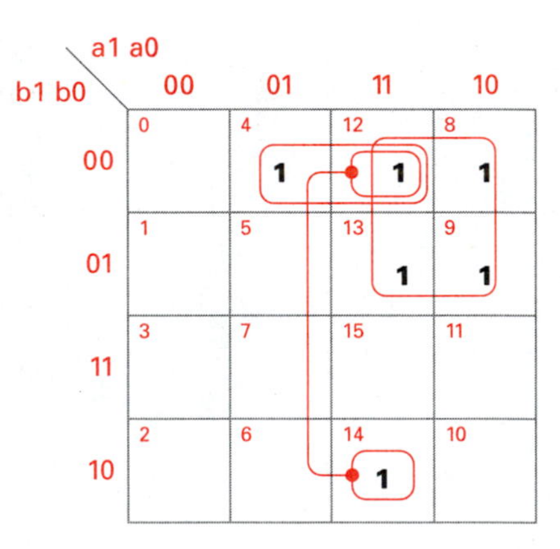

(b) (A > B)

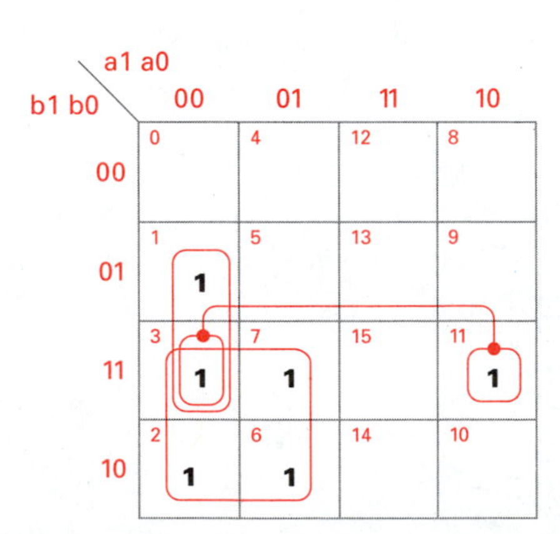

(c) (A < B)

Figure 4.64
Basic iterative circuit model

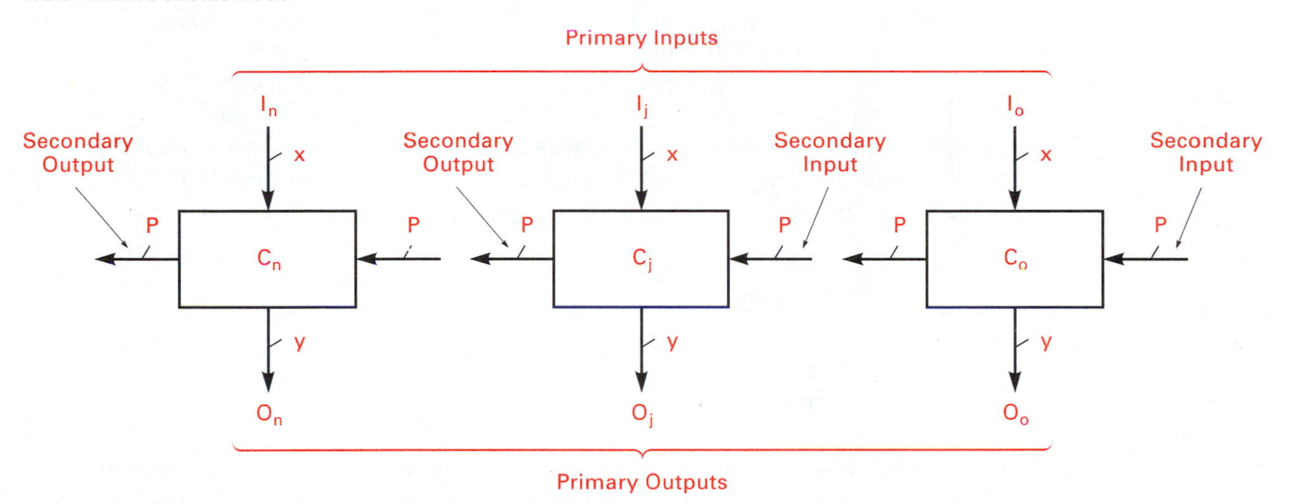

was completed, we could cascade them to form multiple-bit adders. The principle of designing a single cell or stage and then repeating the design also applies to comparators.

The technique of designing a single cell, each with a set of inputs and outputs, and then cascading the cells to form larger circuits is called *iterative design*. Two classes of inputs and outputs exist: primary inputs(outputs) enter(exit) the cell from(to) external data sources(destinations) and secondary inputs(outputs) enter(exit) the cell from(to) neighboring cells.

The secondary I/Os (inputs/outputs) communicate information between cells. Consider the full-adder, where the primary inputs were data from an external source and the primary output is the sum. The secondary input is the carry-in and the secondary output is the carry-out. Primary inputs to a comparator cell, C_j, are provided from external data sources (A_j and B_j). Comparator secondary output, SO_j, conveys to cell C_{j+1} the results of the comparison up to that cell position. No primary outputs are necessary for individual comparator cells, because all cells must respond before the final result can be known. The basic model of an **iterative circuit** intercell relationship is illustrated in Figure 4.64.

Inputs that enter the entire iterative circuit and outputs that exit the same circuit (primary I/O) are sometimes called *boundary I/O*. In the case of the comparator (see Figure 4.65), the intercell secondary inputs and outputs are encoded, using two bits, so

Figure 4.65
Four-bit iterative comparator block diagram

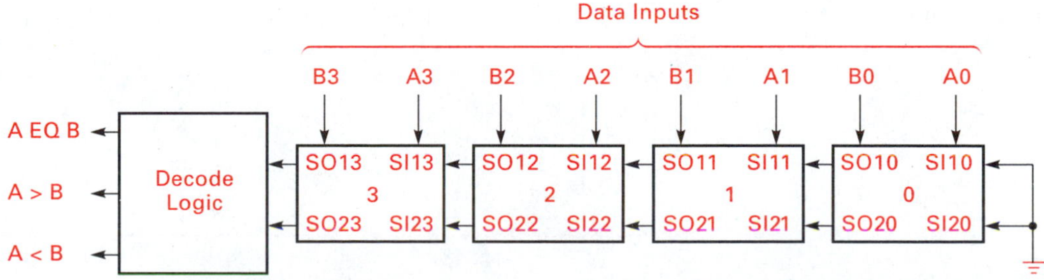

that the three pieces of information (A EQ B, $A > B$, $A < B$) can be economically conveyed.

All of the comparator cells are identical. The least significant cell has its secondary inputs grounded. Here, this indicates that A is equal to B. Each cell compares the external data it receives, checks the result from previous cells, and then transmits the result to the next cell.

A code must be assigned to the two secondary input/output signals from each cell. Let S_{O2} be defined as secondary output 2, S_{I1} be defined as secondary input 1, and so on. The secondary inputs and outputs are connected between cells:

I/O		Function
$S_{I/O2}$	$S_{I/O1}$	
0	0	$A_j = B_j$
0	1	$A_j > B_j$
1	1	$A_j < B_j$

A modified truth table can now be constructed that illustrates the primary and secondary inputs and the secondary outputs to be generated. The primary inputs and secondary inputs define the input part of the table, and the two secondary outputs form the table outputs. We must keep in mind the definitions of the secondary input/output signals to determine the values to write into the truth table. The modified truth table, shown in Table 4.19, is partitioned into two sections. The left side of the table indicates the comparison state (present condition) being sent to cell j from cell $j - 1$. The right side of the table conveys two pieces of information. The first is the condition of the primary data inputs (A_jB_j) to that cell. A_jB_j requires four columns to cover all possible combinations of the two data bits. The second is the value of the two encoded secondary outputs (S_{O2}, S_{O1}), indicating the status of the comparison to that point. The value of S_{O2} and S_{O1} convey relative comparison information on to the next higher order cell $j + 1$.

The table was constructed as indicated, the values for S_{O2} and S_{O1} filled in as the present condition ($A = B$, $A > B$, $A < B$), and the data inputs were evaluated. For example, assume that S_{I2} and S_{I1} indicated that previous comparisons yielded $A = B$, and the primary inputs to cell C_j, $A_jB_j = 10$, then the secondary outputs of C_j, $S_{O2}S_{O1} = 01$, indicating to cell C_{j+1} that $A > B$. The secondary signals propagate from a less to a more significant bit position. Boolean equations can now be derived from Table 4.19, simplified, and the logic diagram drawn.

TABLE 4.19 Modified truth table for a single cell of a binary comparator

Present condition			Next condition—Primary inputs A_j B_j							
	Secondary inputs		00		01		10		11	
	S_{I2}	S_{I1}	S_{O2}	S_{O1}	S_{O2}	S_{O1}	S_{O2}	S_{O1}	S_{O2}	S_{O1}
$A = B$	0	0	0	0	1	1	0	1	0	0
$A > B$	0	1	0	1	1	1	0	1	0	1
$A < B$	1	1	1	1	1	1	0	1	1	1
Don't care			x	x	x	x	x	x	x	x

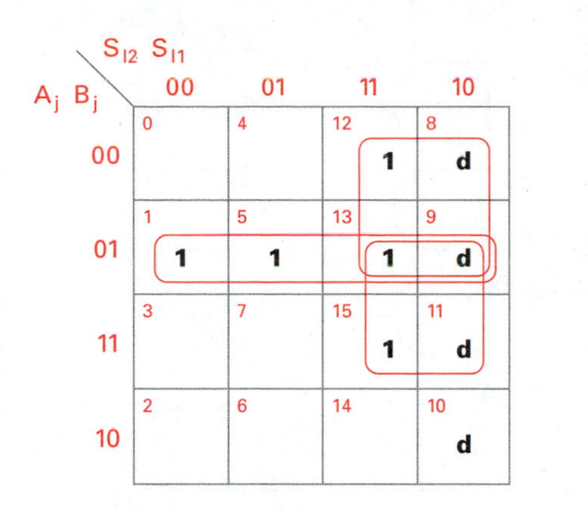

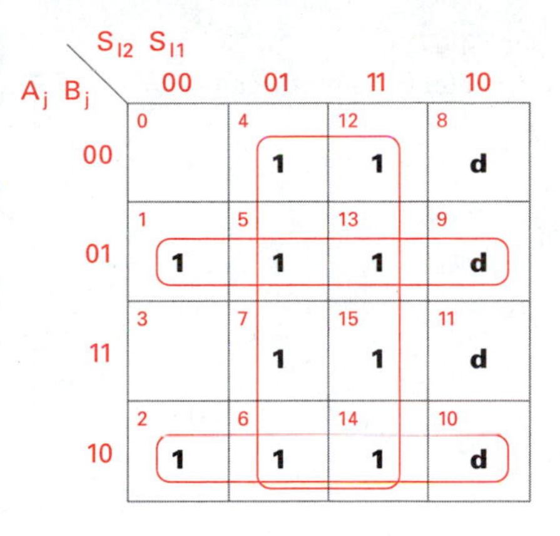

(a) SO2 **(b)** SO1

Figure 4.66
K-maps for S_{O2} and S_{O1}

The minterms are found as follows:

1. Assign a binary weight to each input variable

   S_{I2}, S_{I1}, A_j, B_j
   8 4 2 1

   Note that two most significant variables are located in the present condition half of the table (S_{I2}, S_{I1}) and two least significant variables are in the next condition half of the table (A_j, B_j).

2. Locate all positions where the secondary output variable is a 1, determine the input variable combination, and record the decimal minterm value.

3. Any incompletely specified input conditions become don't care terms:

$$S_{O2} = f(S_{I2}, S_{I1}, A_j, B_j) = \sum(1, 5, 12, 13, 15) + \sum d\,(8, 9, 10, 11)$$
$$S_{O1} = f(S_{I2}, S_{I1}, A_j, B_j) = \sum(1, 2, 4, 5, 6, 7, 12, 13, 14, 15) + \sum d\,(8, 9, 10, 11)$$

The two four-variable K-maps in Figure 4.66 simplify the secondary output equations, from which we can determine the logic diagram. The simplified equations are

$$S_{O2} = A_j'B_j + S_{I2}A_j' + S_{I2}B_j$$
$$S_{O1} = S_{I1} + A_j'B_j + A_jB_j' = S_{I1} + (A_j \oplus B_j)$$

The logic diagram for the cell is shown in Figure 4.67.

The secondary outputs' connecting cells are encoded. To determine the result of an N-bit comparison, the final secondary outputs must be decoded to produce $A = B$, $A > B$, and $A < B$ boundary outputs. A table decoding S_{O2} and S_{O1} to produce boundary outputs $A = B$, $A > B$, and $A < B$ is shown in Table 4.20.

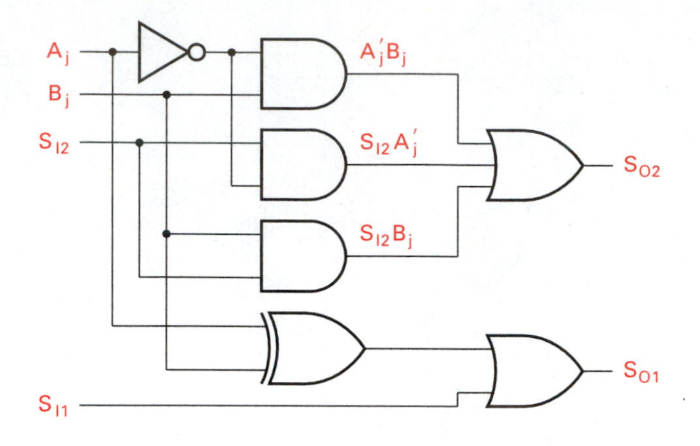

Figure 4.67
Comparator cell logic diagram

TABLE 4.20
Truth table for boundary outputs

S_{O2}	S_{O1}	$A = B$	$A > B$	$A < B$
0	0	1	0	0
0	1	0	1	0
1	1	0	0	1
1	0		don't care	

The three boundary output variable equations are

$$(A \text{ EQ } B) = f(S_{O2}, S_{O1}) = \sum(0) + \sum d(2)$$
$$(A > B) = f(S_{O2}, S_{O1}) = \sum(1) + \sum d(2)$$
$$(A < B) = f(S_{O2}, S_{O1}) = \sum(3) + \sum d(2)$$

which reduces to

$$(A \text{ EQ } B) = S'_{O1}$$
$$(A > B) = S'_{O2}S_{O1}$$
$$(A < B) = S_{O2}$$

The logic diagram is shown in Figure 4.68.

The use of iterative techniques, that is, partitioning a design problem into identical cells, can be used to design comparators, adders/subtractors, and other combinational logic circuits. MSI IC comparators are also available. The TTL 74xx85 is a four-bit comparator and the 74xx682 is an eight-bit comparator. The logic diagram for the 74LS85 is shown in Figure 4.69. Note that the 74LS85 has boundary inputs $A = B$, $A > B$, and

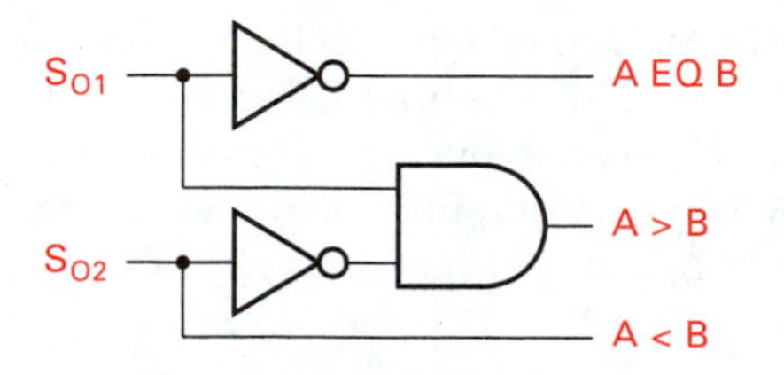

Figure 4.68
Logic diagram for boundary outputs

Figure 4.69
Logic diagram and symbol for the 74LS85 four-bit integrated circuit comparator

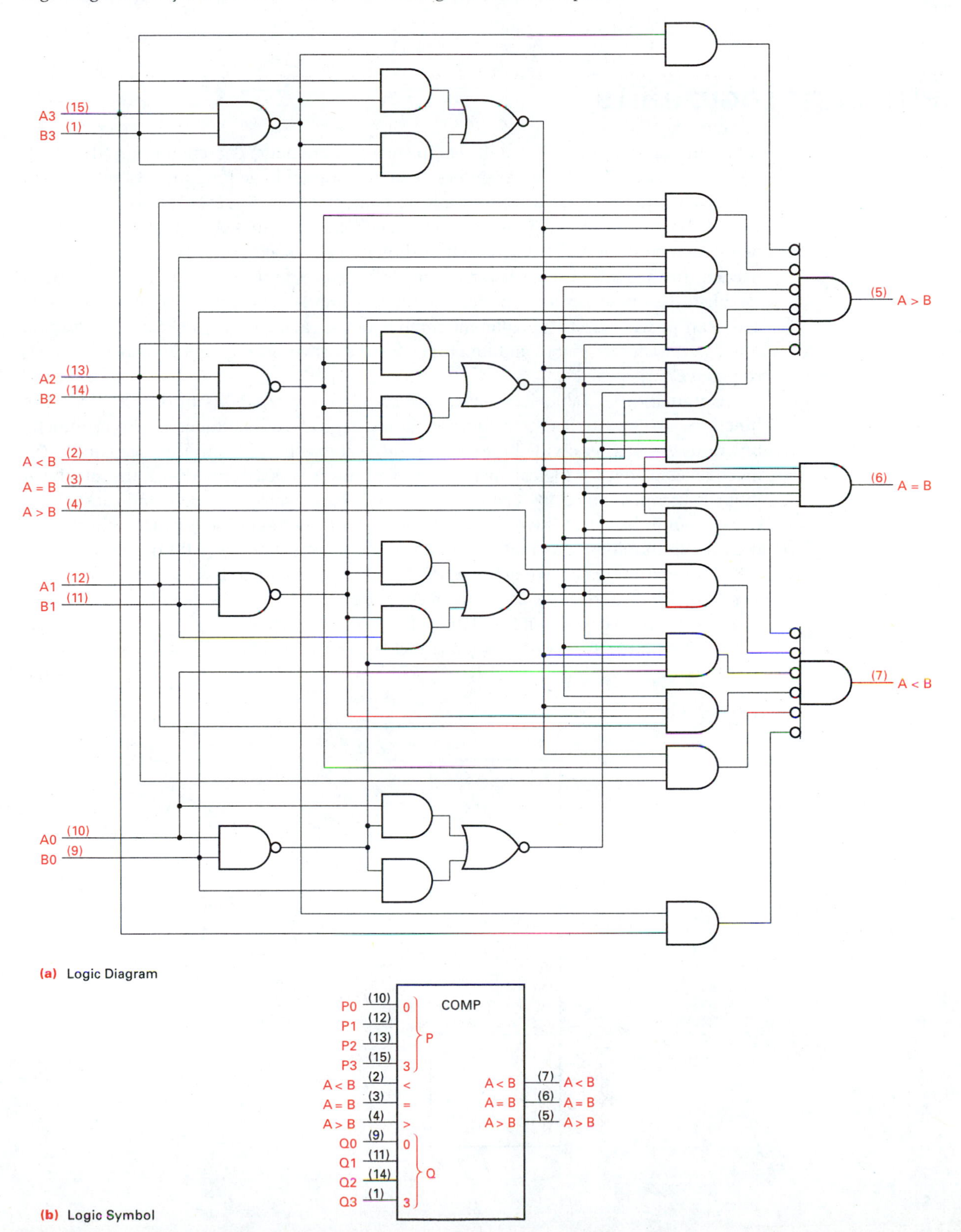

(a) Logic Diagram

(b) Logic Symbol

$A < B$ as well as the primary data inputs. This allows the device to be cascaded with other devices to form N-bit comparators.

4.8 ARITHMETIC LOGIC UNITS

Addition, subtraction, AND, OR, complement, increment, decrement, NAND, NOR, EX-OR, and EX-NOR are all examples of combinational logic functions. Combining all of these functions into a single medium-scale integrated circuit provides logic designers with a very powerful tool. One example of such an integrated circuit is the 74xx **arithmetic logic unit, or ALU**. When multiple functions are used, some means of selection is needed. In the case of the 74xx181, a mode bit and a four-bit select code are used to establish the operation to be performed on data inputs. The five-bit (mode and select) code can provide up to 32 different combinations, each causing a separate operation. The 74xx181 logic symbol and function table are shown in Figure 4.70 and Table 4.21, respectively.

Examination of Table 4.21 reveals the considerable capability of the 74xx181. The functions are partitioned into two categories: logical and arithmetic. The arithmetic functions are further divided into two groups: carry active (low in this case) and carry inactive. Consider the logical operations; when select = 0000 and M = 1, the four-bit A input is complemented. By changing M to 0, keeping select = 0000, and causing the carry input to be active (low), the output is $A + 1$ (increment A). Further inspection of the function table reveals that a NOR $(A + B)'$ occurs when S = 0001 and M = 1. A NAND operation results when S = 0100, M = 1. Adding A and B occurs when S = 1001 and M = 0. To perform a 2s complement subtraction, we let S = 0110, M = 0, and C–in = 0 (no carry). When S = 1010, M = 0, $C_n = 0$, the result is (A OR B') plus (A AND B). Whether such an operation is useful is questionable; it simply exists.

Designing a simplified ALU from scratch proves helpful. It further illustrates the

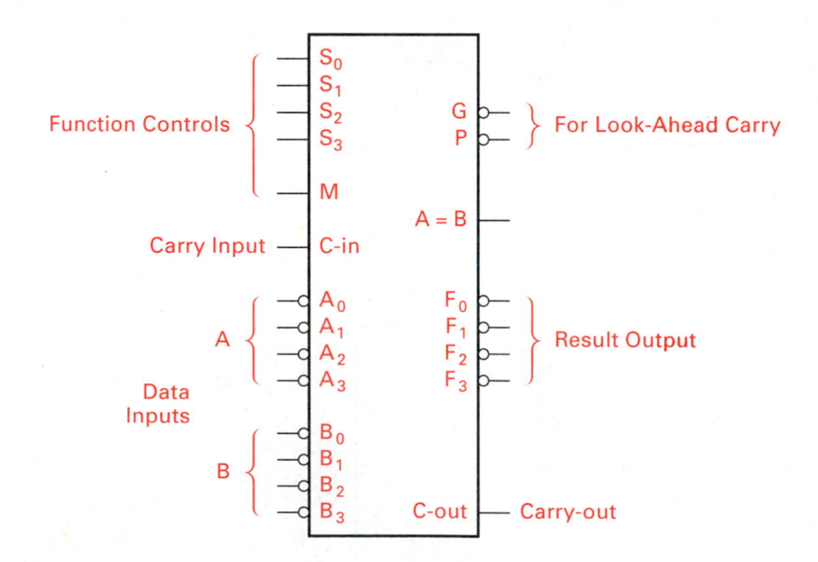

Figure 4.70
ALU logic symbol

TABLE 4.21
74xx181, ALU, function table (courtesy of Texas Instruments Inc.)

Selection				Active high data		
				$M = H$ logic functions	$M = L$: Arithmetic operations	
					$C_n' = \text{H}$ (no carry)	$C_n' = \text{L}$ (with carry)
S_3	S_2	S_1	S_0			
L	L	L	L	$F = \overline{A}$	$F = A$	$F = A$ plus 1
L	L	L	H	$F = \overline{A + B}$	$F = A + B$	$F = (A + B)$ plus 1
L	L	H	L	$F = \overline{A}B$	$F = A + \overline{B}$	$F = (A + B)$ plus 1
L	L	H	H	$F = 0$	$F =$ minus 1 (2s compl)	$F =$ zero
L	H	L	L	$F = \overline{AB}$	$F = A$ plus $A\overline{B}$	$F = A$ plus $A\overline{B}$ plus 1
L	H	L	H	$F = \overline{B}$	$F = (A + B)$ plus $A\overline{B}$	$F = (A + B)$ plus $A\overline{B}$ plus 1
L	H	H	L	$F = A \oplus B$	$F = A$ minus B minus 1	$F = A$ minus B
L	H	H	H	$F = A\overline{B}$	$F = A\overline{B}$ minus 1	$F = A\overline{B}$
H	L	L	L	$F = \overline{A} + B$	$F = A$ plus AB	$F = A$ plus AB plus 1
H	L	L	H	$F = \overline{A \oplus B}$	$F = A$ plus B	$F = A$ plus B plus 1
H	L	H	L	$F = B$	$F = (A + \overline{B})$ plus AB	$F = (A + B)$ plus AB plus 1
H	L	H	H	$F = AB$	$F = AB$ minus 1	$F = AB$
H	H	L	L	$F = 1$	$F = A$ plus A'	$F = A$ plus A plus 1
H	H	L	H	$F = A + \overline{B}$	$F = (A + B)$ plus A	$F = (A + B)$ plus A plus 1
H	H	H	L	$F = A + B$	$F = (A + B)$ plus A	$F = (A + \overline{B})$ plus A plus 1
H	H	H	H	$F = A$	$F = A$ minus 1	$F = A$

iterative nature of many combinational logic circuits, both in function and size. If we can design a one-bit ALU, then repeating the basic design n times provides an n-bit ALU. Likewise we can partition the functions (add, subtract, AND, etc.) and combine them to form the complete ALU.

Consider the design of a combinational ALU that will perform the following functions: add (with and without carry), subtract (with and without borrow, by complementing and adding), transfer an operation to the output, AND, NAND, OR, complement, and EX-OR.

We have already designed a full-adder in Section 6 of this chapter. We can use the same full-adder to realize subtraction (see Figure 4.71) by providing a minuend operand complement function (this was accomplished using an EX-OR gate to selectively invert the minuend). Because we need to complement the minuend to subtract, we already have the logic in place to provide a complement function ($F = B'$).

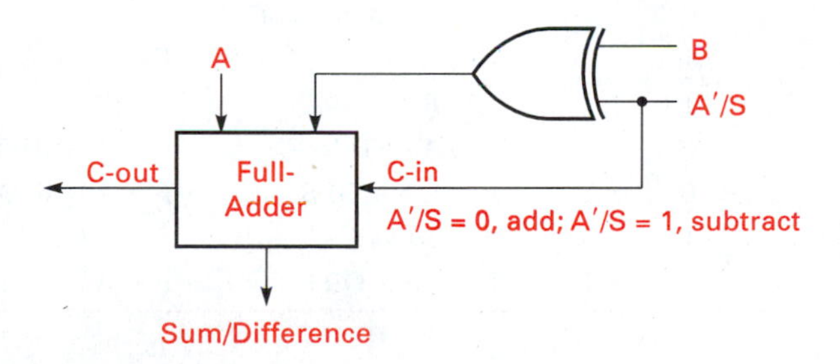

Figure 4.71
Single-cell Adder/Subtractor for an ALU

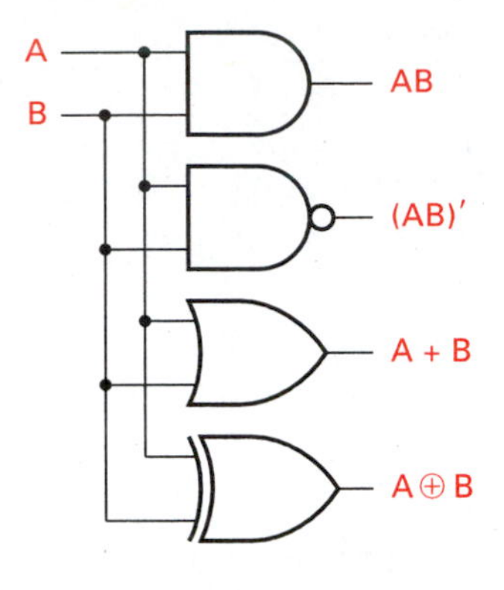

Figure 4.72
AND, NAND, OR, and EX-OR functions

Logic functions such as AND, OR, NAND, NOR, and EX-OR can be realized by performing the required bit-by-bit logic operation on the input data bits. The logical operations of AND, NAND, OR, and EX-OR are accomplished by using logic gates as illustrated in Figure 4.72.

Each ALU function produces an intermediate output that must be selected and sent to the final output. In our example ALU, we have eight functions (addition/subtraction, transfer, complement, AND, NAND, OR, and EX-OR); so an 8-to-1 multiplexer can be used to provide function select logic. Additional logic is used to control the EX-OR gates, permitting subtraction and complement functions. Additional functions could be added by increasing the multiplexer size and providing the logic required to realize the functions.

The output of a two-input AND gate controls the B_n inverting EX-OR gate. The two conditions that require inversion are these: when select = 001 subtract, and when select = 011 $F = B'$. Simplifying the equations from the two conditions produces $S_2'S_0$. The function table for the ALU is shown in Table 4.22. The complete diagram for a single cell ALU is shown in Figure 4.73.

A 74xx181 medium-scale integrated circuit ALU, shown in Figure 4.74, provides inputs and outputs so that a look-ahead carry generator can be connected to improve add times. Without the look-ahead generator, a 74LS181 device can add or subtract two four-bit values in a maximum of 41 ns. Adding a 74S182 look-ahead carry generator improves the add time to 24 ns. Figure 4.75 shows the connection of the ALU and the look-ahead carry generator.

The block diagram shown in Figure 4.75 illustrates how four four-bit ALU devices can be interconnected to form a 16-bit ALU with look-ahead carry capability. The A and B inputs are each 16-bits ($A_{15}, A_{14}, \ldots, A_1, A_0$; $B_{15}, B_{14}, \ldots, B_1, B_0$). A bus is a collection of signal wires. The least significant four-bit ALU operates on the least significant four bits of the A and B bus inputs and so on. The select and mode bus is five bits wide. It presents the same function code to each of the four ALUs. The G and P look-ahead carry outputs originate from each ALU IC. They are presented to the look-ahead carry generator so that the C-in inputs to each four-bit ALU can be generated more quickly. Each ALU 4-bit result (for a total of 16 ouput bits) and the carry-out form a 17-bit output result bus.

TABLE 4.22
Design example ALU function table

Select inputs			Function	
S_2	S_1	S_0	$C = 0$	$C = 1$
0	0	0	A plus B plus 1	A plus B
0	0	1	A minus B minus 1	A minus B
0	1	0	$F = A$	$F = A$
0	1	1	$F = B'$	$F = B'$
1	0	0	$F = A$ AND B	$F = A$ AND B
1	0	1	$F = A$ NAND B	$F = A$ NAND B
1	1	0	$F = A$ OR B	$F = A$ OR B
1	1	1	$F = A$ EX-OR B	$F = A$ EX-OR B

Figure 4.73
Single-cell eight-function ALU

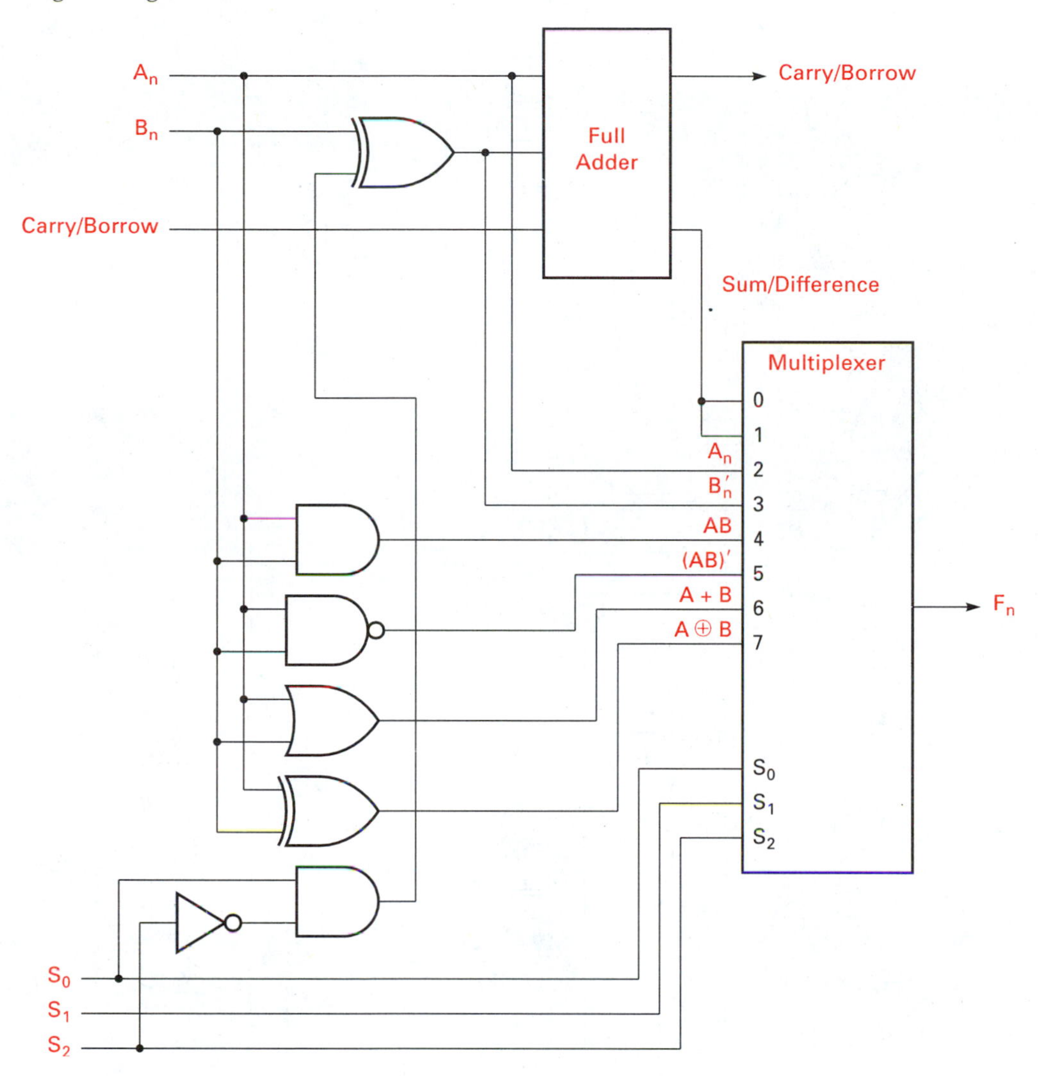

4.9 ARRAY MULTIPLIERS

Binary multiplication can be accomplished by several approaches. The approach presented here is realized entirely with combinational circuits. Such a circuit is called an **array multiplier**. The term *array* is used to describe the multiplier because the multiplier is organized as an array structure. Consider the following multiplication of two four-bit operands. It is the same method of doing longhand multiplication that we learned in grade school.

Each row, called a *partial product*, is formed by a bit-by-bit multiplication of each

Figure 4.74

74LS181 ALU logic diagram, with provision for look-ahead carry to reduce addition propagation delay

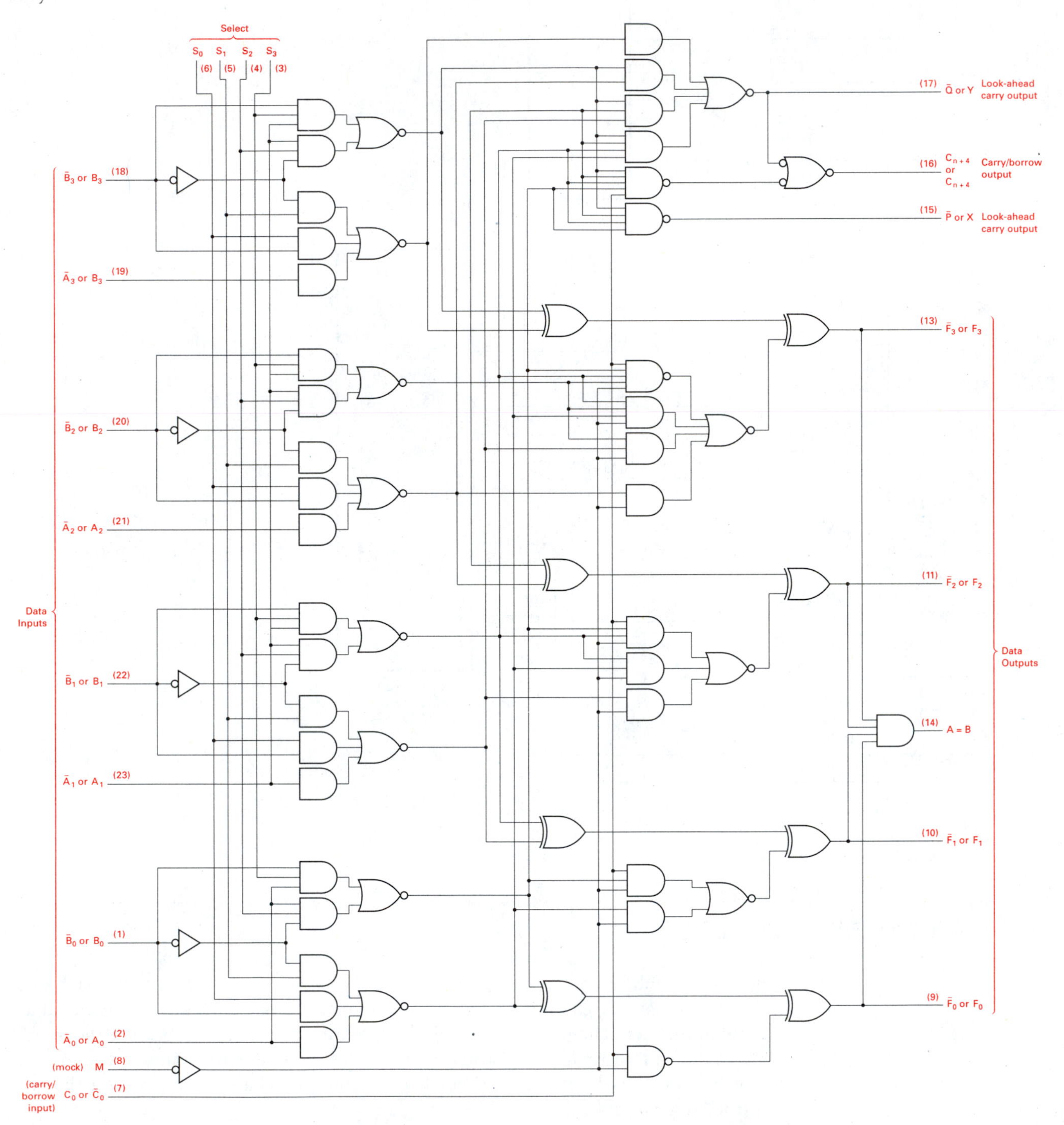

Figure 4.75
16-bit ALU constructed from four 74xx181 and one 74xx182 integrated circuits

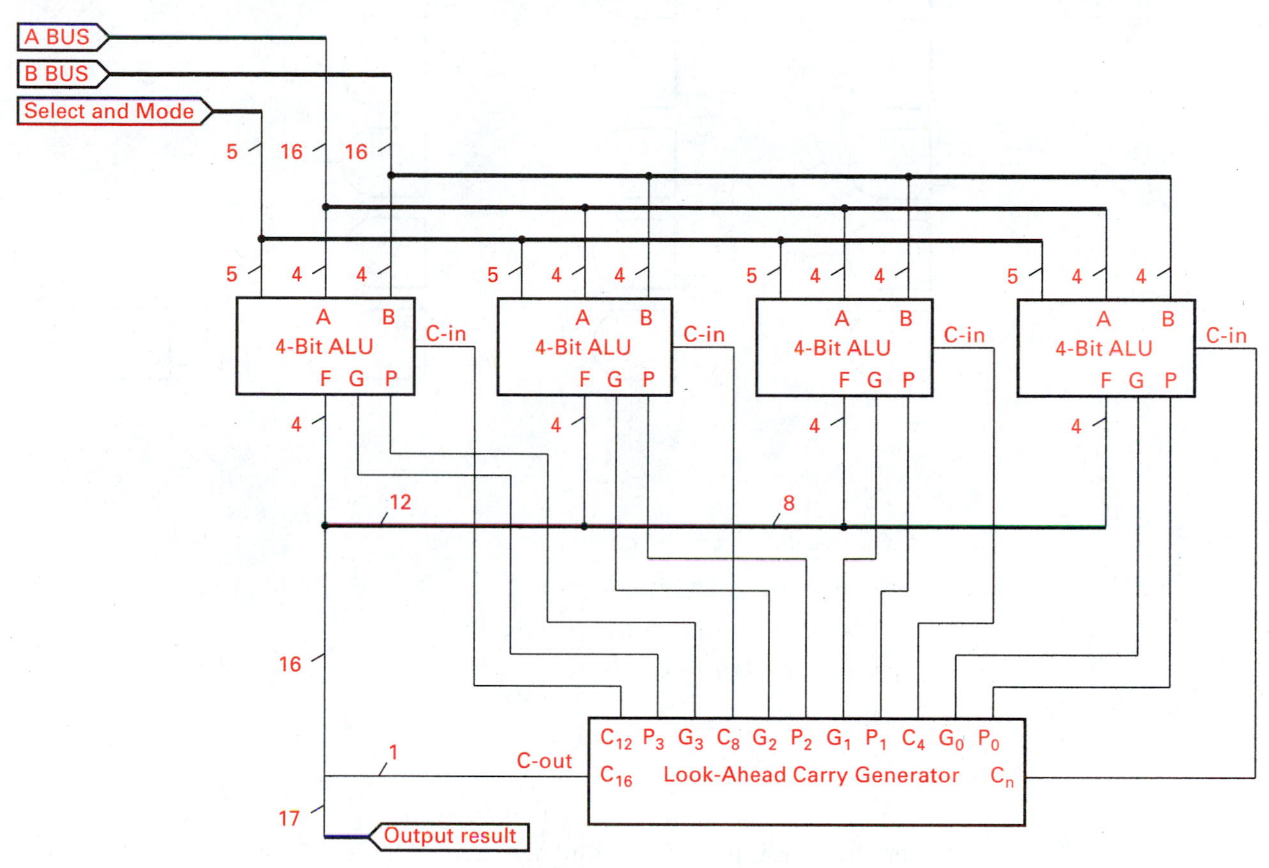

operand. For example, a partial product is formed when each bit of operand a is multiplied by b_0, resulting in a_3b_0, a_2b_0, a_1b_0, a_0b_0:

$$
\begin{array}{cccccccc}
 & & & & a_3 & a_2 & a_1 & a_0 \\
 & & & \times & b_3 & b_2 & b_1 & b_0 \\
\hline
 & & & & a_3b_0 & a_2b_0 & a_1b_0 & a_0b_0 \\
 & & & a_3b_1 & a_2b_1 & a_1b_1 & a_0b_1 & \\
 & & a_3b_2 & a_2b_2 & a_1b_2 & a_0b_2 & & \\
 & a_3b_3 & a_2b_3 & a_1b_3 & a_0b_3 & & & \\
O_7 & O_6 & O_5 & O_4 & O_3 & O_2 & O_1 & O_0
\end{array}
$$

TABLE 4.23
AND truth and binary multiplication tables

AND			Multiply		
A	B	C	A	B	A × B
0	0	0	0	0	0
0	1	0	0	1	0
1	0	0	1	0	0
1	1	1	1	1	1

The binary multiplication table is identical to the AND truth table (see Table 4.23).

Each product bit, O_x, is formed by adding partial product columns. The product equations, including the carry-in, c_x, from column c_{x-1}, are (the plus sign indicates addition not OR)

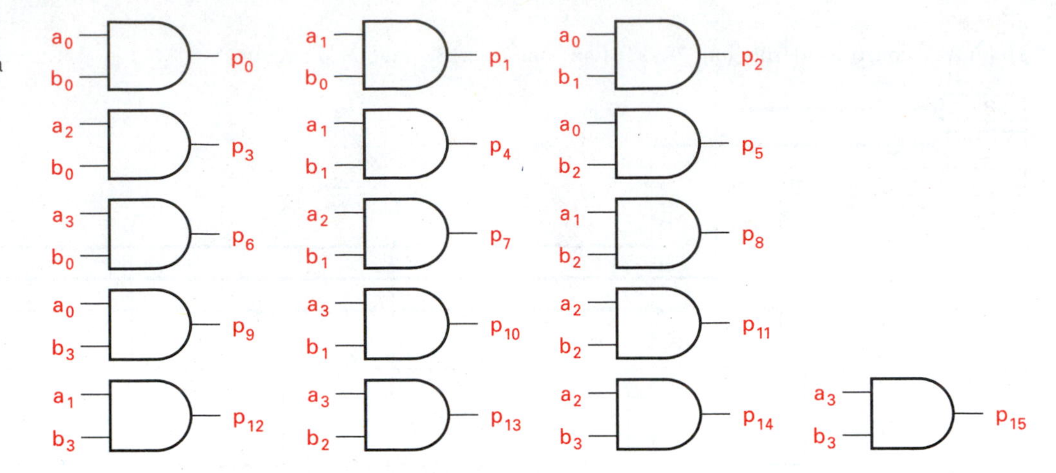

Figure 4.76
Product terms generated by a collection of AND gates

$$O_0 = a_0b_0$$
$$O_1 = a_1b_0 + a_0b_1 + c_0$$
$$O_2 = a_2b_0 + a_1b_1 + a_0b_2 + c_1$$
$$O_3 = a_3b_0 + a_2b_1 + a_1b_2 + a_0b_3 + c_2$$
$$O_4 = a_3b_1 + a_2b_2 + a_1b_3 + c_3$$
$$O_5 = a_3b_2 + a_2b_3 + c_4$$
$$O_6 = a_3b_3 + c_5$$
$$O_7 = c_6$$

Each product term, p_x, is formed by AND gates. Figure 4.76 shows the collection of product terms needed for the multiplier.

By adding appropriate p term outputs, the multiplier output equations are realized, as shown in Figure 4.77.

4.10 TRISTATE BUFFERS

To this point we have been concerned with only logical 0 and 1 values for input and output variables. A very useful combinational logic function differs from the normal logic we have been discussing thus far. In this section we will introduce logic that acts like a switch that is in series with the data signal wire. Such logic is built into an integrated circuit function called *tristate*. The name *tristate* implies that three distinct states for a data line are possible. We are not talking about a base-3 number system but rather the ability to open (high impedance state) or close a switch (normal) that is in series with a binary data line.

Consider the circuit diagram in Figure 4.78. The circuit behaves functionally like a tristate logic device. A relay or solenoid contact is in series with the data line that carries the electrical signal used to represent logic values (0 and 1). When the enable signal is active, the relay is activated and the contact switch is closed, allowing passage of the

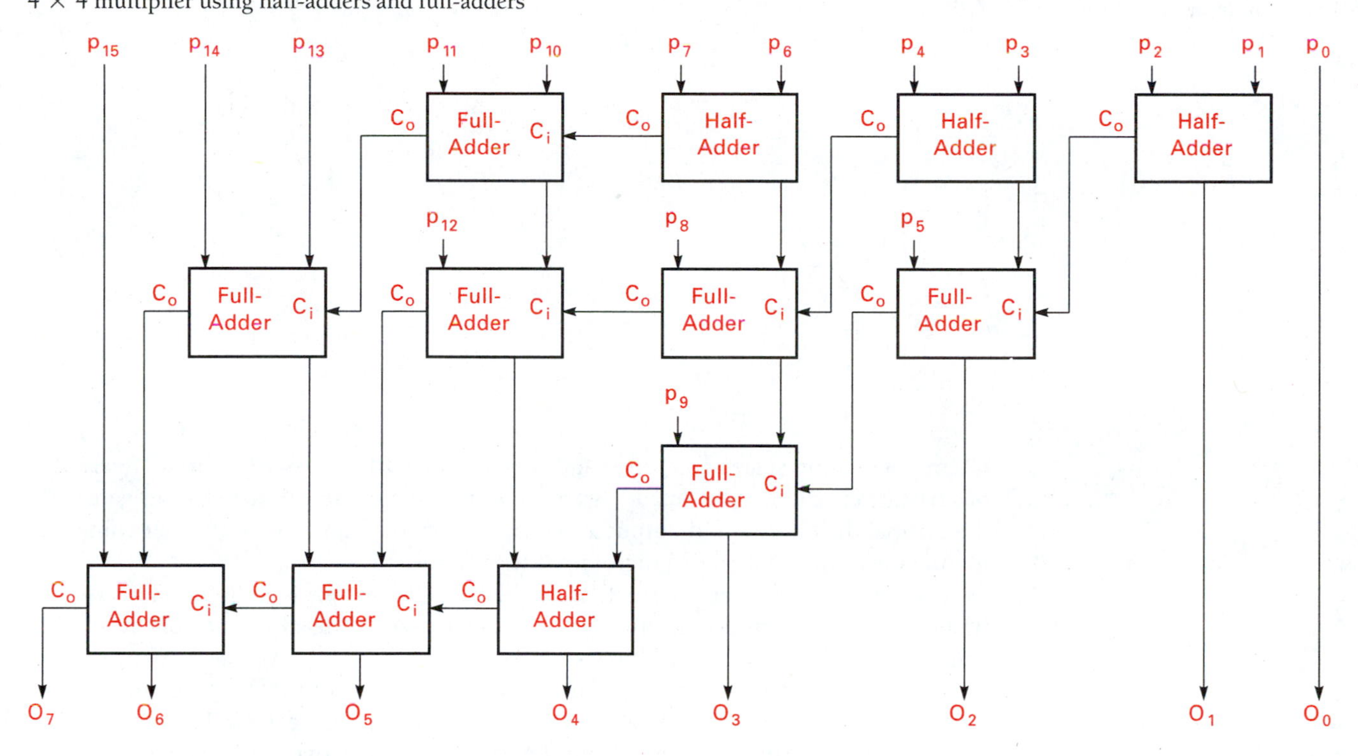

Figure 4.77
4 × 4 multiplier using half-adders and full-adders

logic signal over the conductor. When the enable is inactive, the data line appears as an open circuit.

A data signal representing logical conditions, and an enable signal controlling the closure of the series switch work together to create three conditions: logical 0 and 1 and open (high impedance). Tristate integrated circuits accomplish the same without a physical switch.

The output stage of a transistor-transistor logic (TTL) IC consists of two transistors arranged as shown in Figure 4.79. Such an arrangement is called a *totem pole*, because one transistor appears stacked on top of the other in the circuit schematic. Normal binary operation causes either Q1 or Q2 to be conducting while the other transistor is not conducting (cut off). Either of the conducting transistors presents a low-impedance path from the output directly to ground or through V_{CC} to ground. To achieve a high-impedance path from the output to ground requires forcing both transistors in the TTL

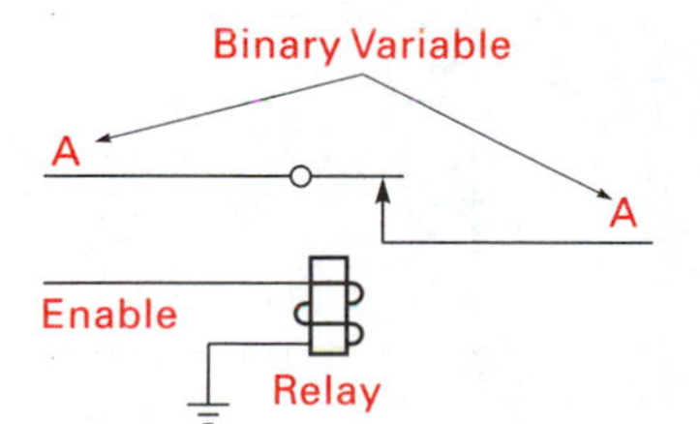

When the relay contacts are closed variable A can be either a logical 0 or a 1. When the relay contacts are open the output is in a high impedance state.

Figure 4.78
Relay model of a tristate switch

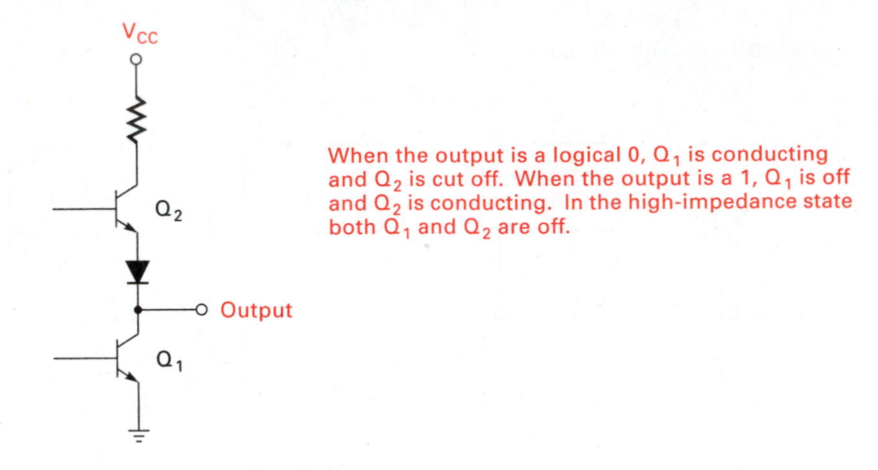

Figure 4.79
TTL totem-pole output stage

totem pole output stage to an "off" condition. Chapter 10 discusses the circuit operation of a tristate circuit in more detail. Figure 4.80 shows a complete tristate NAND gate with the normal data inputs and output as well as the enable input to control switching the totem pole output state into a high-impedance condition.

Many tristate integrated circuits are available. The two we will discuss here are the tristate buffer and tristate inverter. Tristate devices are all used to control access to a single digital data line from multiple sources.

Figure 4.81 illustrates the idea of multiple data sources connected to a common data line. The logic symbols for **tristate buffers** are shown in Figure 4.82. The data may be buffered (noninverted) or inverted and the enable may be active low or active high.

The enable is activated with a high or low input that conforms to the same voltage level definition as binary data. An active high enable (no bubble) closes the series data switch when a 1 is present, and an active low enable closes the switch when a 0 is present.

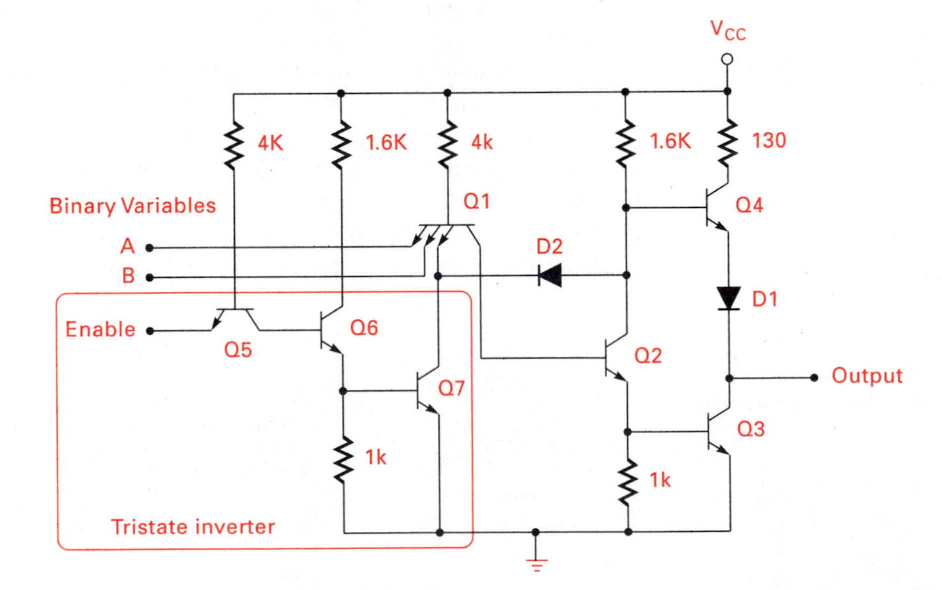

Figure 4.80
Tristate TTL two-input NAND gate

Figure 4.81
Multiple data sources connected to a common data line

Multiple Data Sources

W, X, Y, Z

S_1, S_2, S_3, S_4

Output

The logical value of the output depends upon which relay contact is closed. Only one contact may be closed at any one time.

Relay contacts S_1, S_2, S_3, and S_4

Figure 4.82
Logic symbols for tristate buffers

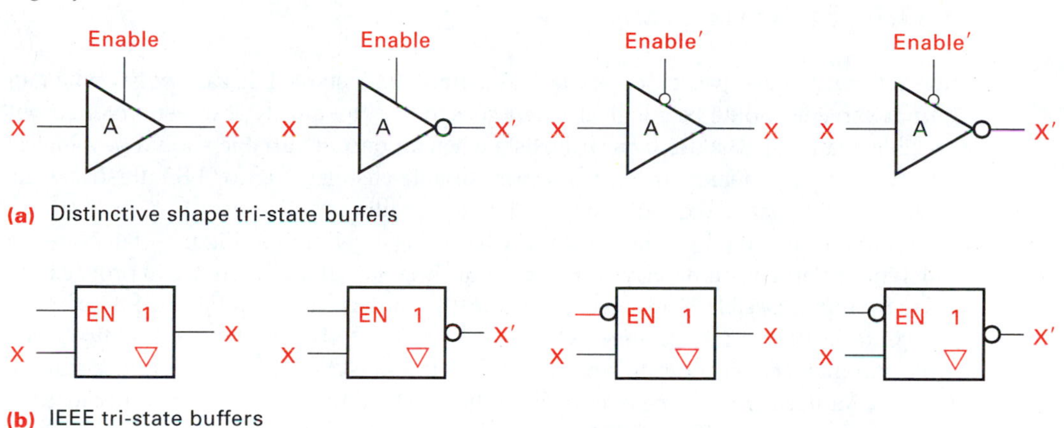

(a) Distinctive shape tri-state buffers

(b) IEEE tri-state buffers

Tristate buffers are often used in switching multiple data sources over a common line. Chapter 5 discusses the use of multiple data registers (binary storage devices) connected to a common data bus by tristate devices.

4.11 COMBINATIONAL LOGIC HAZARDS

Boolean algebra assumes that variable values are constant during evaluation. However, in actual circuits, variable values are free to change. The changing of variable values brings about the potential for incorrect results if corrective measures are not taken. Boolean algebra also does not account for propagation delays through signal paths of actual circuits. These delays can cause glitches to occur. A glitch is an unwanted signal, usually a short pulse. Glitches are caused by the transient behavior of signal paths that have different delays. A hazard exists any time the potential for glitches is present.

Glitches may not be a problem or they may create havoc. It depends on the specific application. Later we will study topics in sequential circuit analysis and design where a synchronizing clock is used to control variable changes. If we delay evaluation of a combinational logic output until all propagation delays have transpired, then hazards are no problem. This approach works well when circuit speed is not an issue.

Hazards begin to present problems when logic circuits are pushed close to the limits of the propagation delay paths. Assume that the worst case propagation path for a combination logic function is 25 ns. That means that data inputs cannot be changed any faster than every 25 ns. From the time that data are presented to the input of the circuit until the data are stable at the output takes 25 ns. If the output data are evaluated every 25 ns, and if the delay were at the critical edge of the integrated circuit specifications, then all that would be required to create a glitch is for the delay to increase by some amount. This would happen if an IC were replaced in the circuit, if the temperature were to rise, if a connector were not mated properly, or for a host of other reasons. The point is that any increase in delay would create a hazard. Two types of hazards are static and dynamic.

4.11.1 Static Hazards

Static hazards have two cases: static 1 and static 0. A static-1 hazard exists when an output variable should be a logical 1, but goes to 0 momentarily as a result of an input variable changing. A static-0 hazard exists when an output variable should be a logical 0 but goes to a 1 momentarily as an input variable changes. Figure 4.83 illustrates the waveforms for a static-0 condition and a static-1 condition.

A circuit that could produce a static-1 hazard is illustrated in Figure 4.84. Note the difference in propagation delay paths for input variable y. This difference in propagation delay for input variable y produces the static-1 hazard as shown in Figure 4.83(a).

Minterm generating circuits (AND-OR, NAND-NAND) can produce static-1 hazards, and maxterm generating circuits (OR-AND and NOR-NOR) can produce static-0 hazards. Minterm generating circuits have the potential for static-1 hazards because the true output is a 1, so a transient to a 1 produces a hazard. Maxterm generating circuits produce a 0 true output, so a transient to a 0 produces a hazard.

The Boolean equation describing the circuit of Figure 4.84 is $P = xy' + yz$. The timing of a static-1 hazard, produced as a result of the difference in propagation delay, is shown in Figure 4.85.

When y went from a 1 to a 0 at time t_0, y' followed with a 0 to a 1 at time t_1. Time $t_1 - t_0$ corresponds to the propagation delay of the inverter. The output of G2, xy', occurs at time t_2. The output of G3 changes from a 1 to a 0 at approximately t_1 (assuming the delays are equal for all the gates). Output variable P is a 1 when either G2 or G3 outputs are 1. Notice that between t_2 and t_1 neither output was a 1. Accounting for the delay through G4 the momentary 0 time is $t_3 - t_2$.

We have stated that minterm-generating circuits have the potential for static-1 hazards. Consider the three-variable K-map shown in Figure 4.86. It illustrates two EPI

Figure 4.83
Static hazards

1 1 0

(a) Static "1" hazard

1 0 0

(b) Static "0" hazard

Figure 4.84
Circuit producing a static-1 hazard

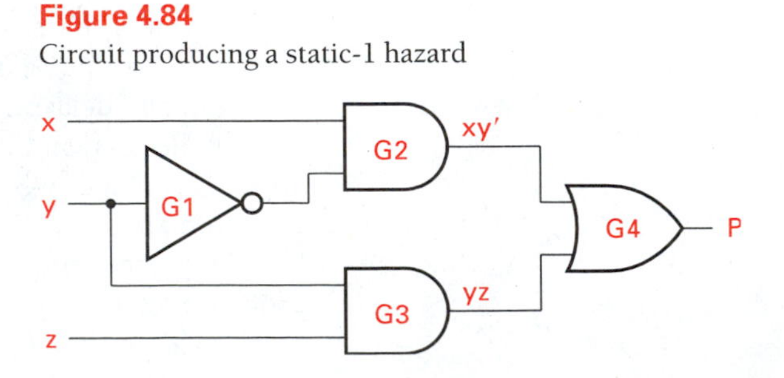

Figure 4.85
Static-1 timing diagram for the circuit shown in Figure 4.84

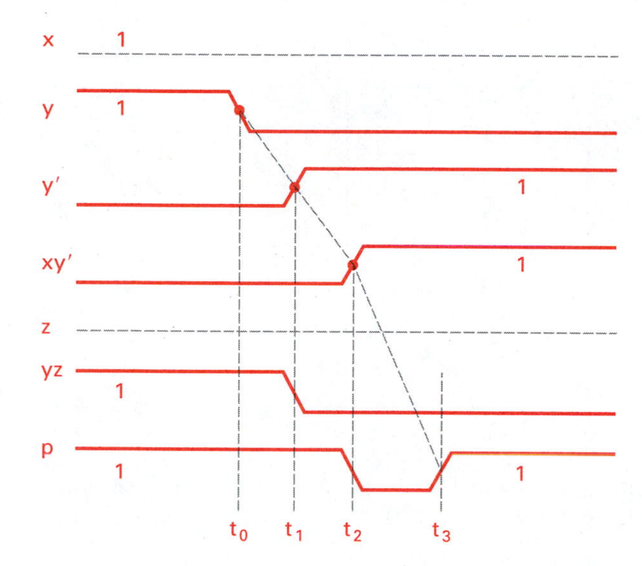

Figure 4.86
Three-variable K-map illustrating static-1 hazard

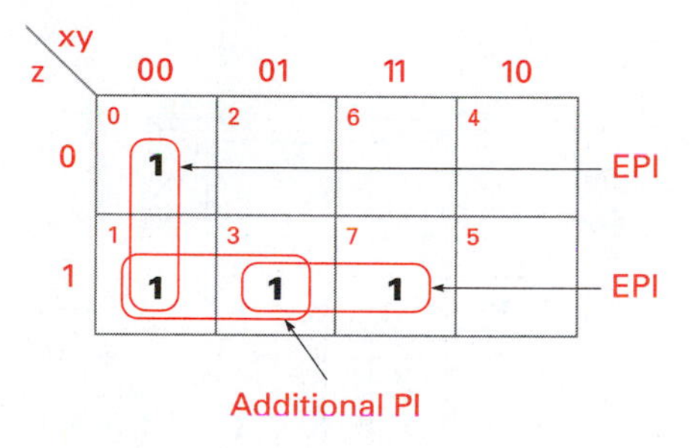

terms that realize a logic function. Each minterm produces a 1 output. If for example, minterm (1) were present ($xyz = 001$) and then if y were to change from a 0 to a 1, EPI {0, 1} would no longer cover the new minterm. It would be covered by EPI {3, 7}. If no propagation delays existed in making the transition from EPI {0, 1} to EPI {3, 7} then no hazard would exist. However, making transitions from one EPI to another, when delays do exist, may create a static-1 hazard. For example, when $xyz = 001$ and changes to $xyz = 011$ no hazard occurs. However, when $xyz = 011$ and changes to $xyz = 001$ a static-1 hazard occurs.

To protect against the static-1 hazard shown in Figure 4.86 we must generate a new term consisting of PI {1, 3}. Prime implicant {1, 3} is chosen because it covers both EPIs. The resulting logic (shown in Figure 4.87) is no longer simplified but is hazard free.

Figure 4.87
Hazard and hazard-free realizations

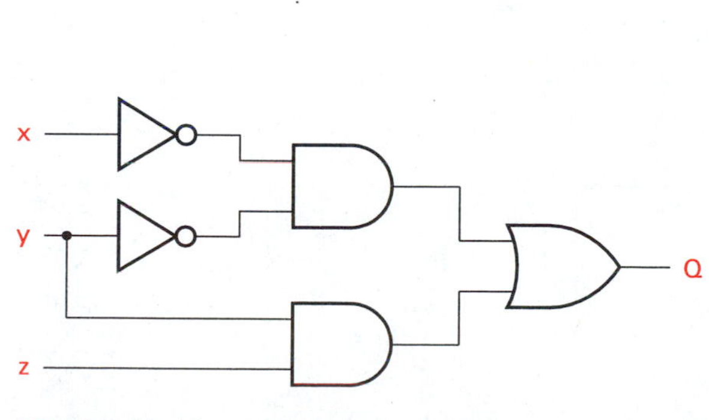

(a) Circuit with a static-1 hazard

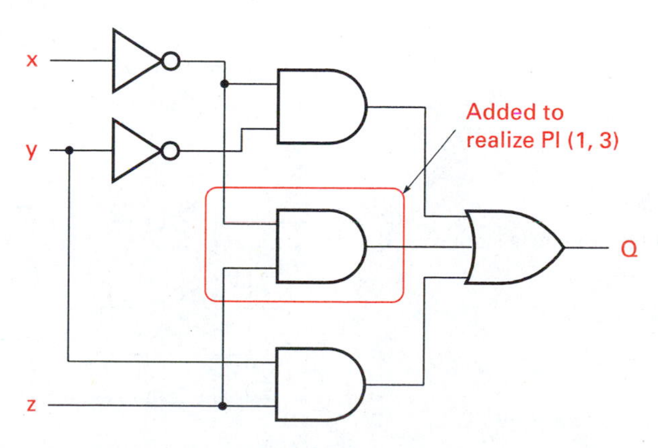

(b) Static-1 hazard free circuit

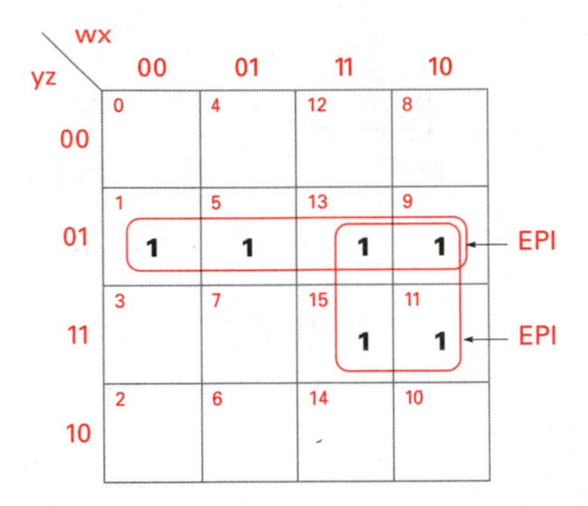

Figure 4.88

K-map illustrating a circuit that is static-1 hazard free

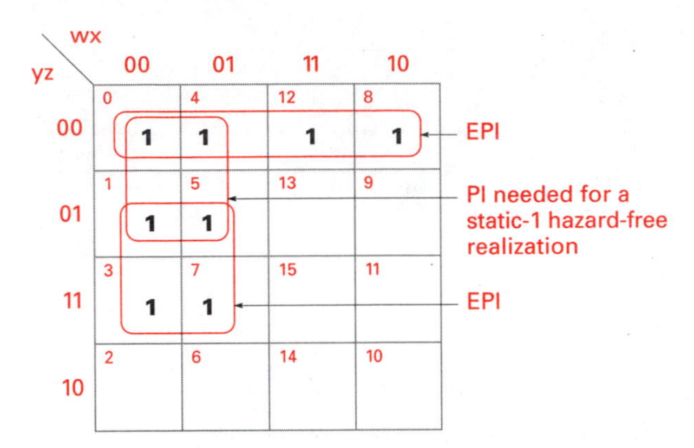

Figure 4.89

K-map illustrating a static-1 hazard and the product term that eliminates the problem

Not all AND-OR realized circuits have static-1 hazards. Consider the realization of the Boolean equation $A = f(w, x, y, z) = \Sigma\ (1, 5, 9, 11, 13, 15)$. By mapping the equation we find two EPIs result: {1, 5, 9, 13} and {9, 11, 13, 15}. Because the two EPIs overlap (that is, they share common minterms) any single input variable change that creates an output 1 is always covered (contained in) a product term. This is illustrated in Figure 4.88.

Design a hazard-free circuit to realize the Boolean equation $G = f(a, b, c, d) = \Sigma\ (0, 1, 3, 4, 5, 7, 8, 12)$. When we map the expression, as in Figure 4.89, we find two EPIs, {0, 4, 8, 12) and {1, 3, 5, 7}, that are not overlapping. Any single input variable change that would result in a change in product-term coverage may cause a static-1 hazard. For example, when a variable changes such that coverage moves from EPI (0, 4, 8, 12) to EPI (1, 3, 5, 7), no glitch (hazard) occurs, but changing from EPI (1, 3, 5, 7) to EPI (0, 4, 8, 12) creates a hazard.

A static-0 hazard is illustrated by the circuit and timing diagram in Figures 4.90 and 4.91. Consider the circuit of Figure 4.90 and the timing diagram of Figure 4.91. When input x changed from 0 to 1, it produced an output 1 pulse static-0 hazard, as illustrated in Figure 4.91. The time from t_0 to t_1 represents the x input inverter propagation delay. Time $t_2 - t_1$ represents the propagation delay through gate G1. Gate G2 $(x + z')$ changes from a 0 to a 1 at t_1. At t_2 the output goes to a 1 and remains a 1 for one gate propagation delay time.

Designing static-0-free OR-AND circuits requires adding maxterm prime implicants to bridge between adjacent maxterm EPIs. Consider the equation $L = f(w, x, y, z) =$

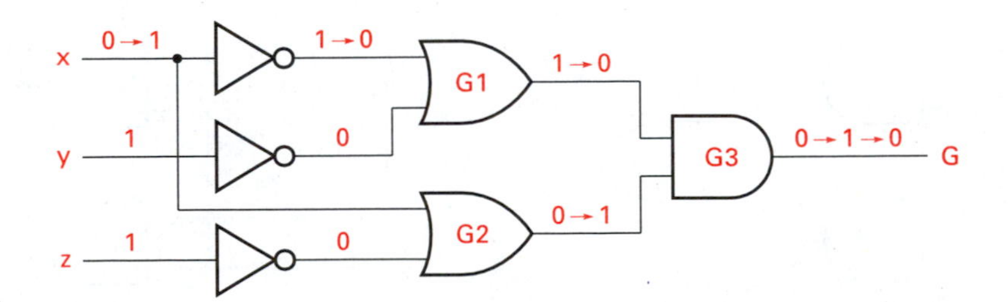

Figure 4.90

A circuit producing a static-0 hazard

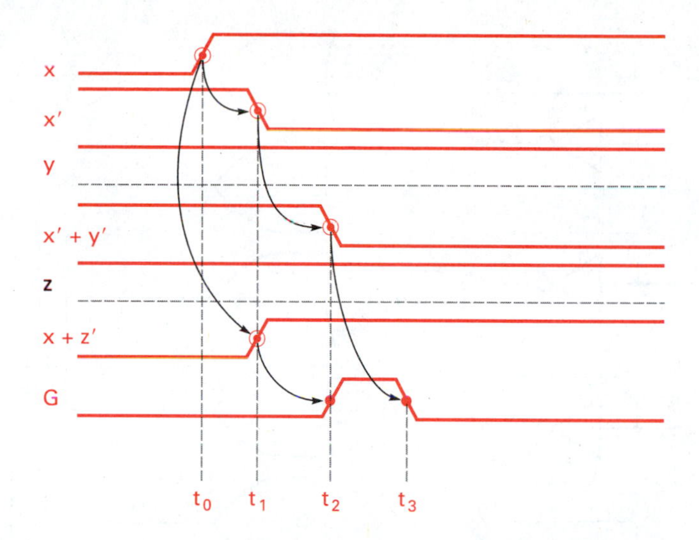

Figure 4.91
Timing diagram for the circuit in Figure 4.90, illustrating a static-0 hazard

π (1, 3, 4, 5, 7, 10, 11, 12, 14, 15). The K-map for the expression is shown in Figure 4.92. Two maxterm EPIs exist: {1, 3, 5, 7} = $w + z'$, {10, 11, 14, 15} = $w' + y'$; and PI {4, 12} = $x' + y + z$. A static-0-free circuit must provide bridging between each sum term (OR). Each sum term produces a logical 0 output. If an input variable changes so that a shift from one sum term to another occurs, then a static-0 hazard will occur. Notice that none of the three sum terms in Figure 4.92 overlap. Maxterm (14), $w' + x' + y' + z = 0 + 0 + 0 + 1$, resides in EPI {10, 11, 14, 15} and maxterm (12), $w' + x' + y + z = 0 + 0 + 1 + 1$, occurs in PI {4, 12}. A 0 to 1 change in variable y changes which EPI is producing the output. A bridging sumterm PI, {12, 14} $w' + x' + z$, between the two EPIs eliminates this particular static-0 hazard. Likewise bridging PIs are needed between sum term EPIs {1, 3, 5, 7} and {10, 11, 14, 15}, and between EPI {1, 3, 5, 7} and PI {4, 12}. Generating PIs {3, 7, 11, 15} = $y' + z'$ and {4, 5} = $w + x' + y$ completes the elimination of static-0 hazards. To summarize,

Maxterm (14) ⟶ maxterm (12), no hazard
Maxterm (12) ⟶ maxterm (14), hazard

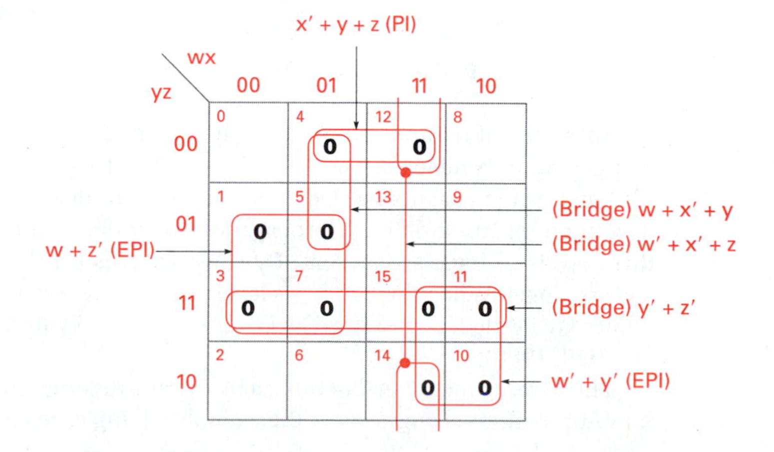

Figure 4.92
K-map illustrating static-0 hazards

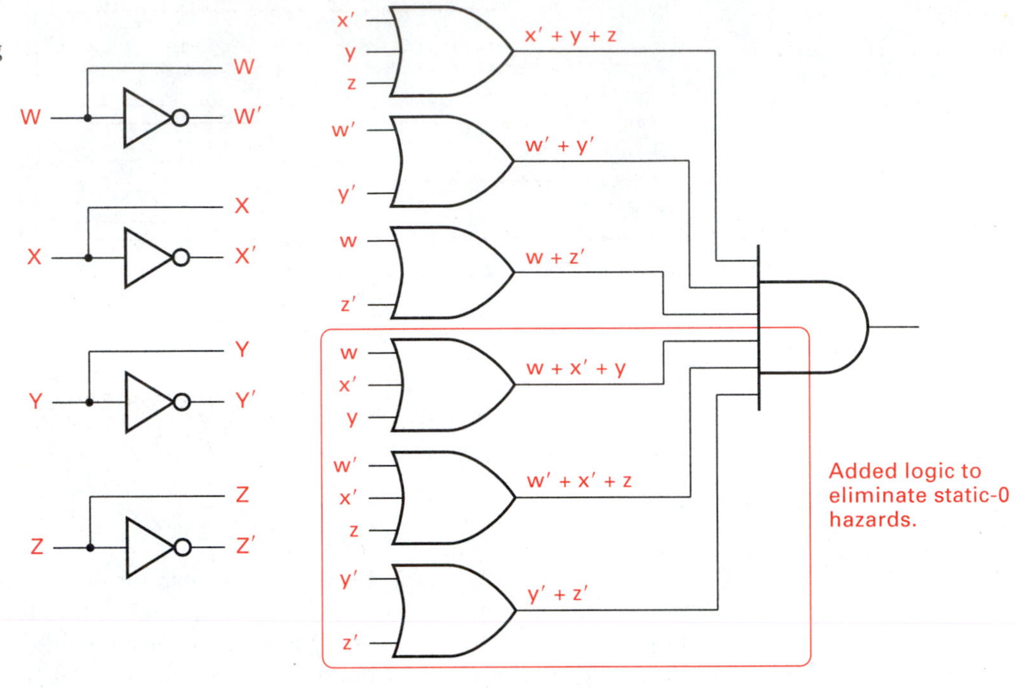

Figure 4.93
Complete circuit illustrating original simplified and static-0 hazard-free logic

Maxterm (7) ⟶ maxterm (15), hazard
Maxterm (15) ⟶ maxterm (7), no hazard
Maxterm (4) ⟶ maxterm (5), hazard
Maxterm (5) ⟶ maxterm (4), no hazard
Maxterm (3) ⟶ maxterm (11), hazard
Maxterm (11) ⟶ maxterm (3), no hazard

The original simplified circuit with hazards and the addition of sum terms to eliminate the static-0 hazards is shown in Figure 4.93. The color line in Figure 4.93 separates the logic resulting from the PIs and EPIs (top half) and those added to eliminate the static-0 hazards (bottom half).

4.11.2 Dynamic Hazards

Another class of hazards exists where the output changes more than once as a result of a single input variable change. A 0-1-0-1 or a 1-0-1-0 change is called a *dynamic hazard.* Although static hazards needed only two different delay paths, dynamic hazards must have three or more different propagation delay data paths. This occurs when at least three levels of logic are present. Dynamic hazards can be avoided by using only two levels of logic. When three or more levels of logic are required, more sophisticated techniques are needed, as described by Edward J. McClusky in *Logic Design Principles* (Prentice-Hall, 1986).

The hazards mentioned in this section deal with only single-input variables changes. If multiple-input changes occur then simply adding terms to cover transitions are insufficient. Fletcher in *An Engineering Approach to Digital Design* describes the use of holding registers to avoid glitches.

SUMMARY

A. A general approach to combination logic design was presented that included

1. Problem statements converted to truth tables.
2. Output equations generated and simplified.
3. Logic diagram drawn using bubble-to-bubble notation.

B. Digital integrated circuit specifications introduced included

1. Package types, pin-out diagrams, and symbols.
2. Specification definitions.

C. Medium-scale devices discussed included

1. Decoders, a device with n inputs and 2^n outputs: 2-to-4 decoder ICs, 3-to-8 decoder ICs, generation of Boolean functions using decoders, and BCD decoders.
2. Encoders, a device with 2^n inputs and n outputs: decimal to BCD, 8-to-3 line, and priority system.
3. Multiplexers: basic model, 8-to-1 MUX, cascading smaller multiplexers to create a larger multiplexer, and generation of Boolean functions using a multiplexer.
4. Adders and subtractors: basic half-adders, full-adders, and subtractors; cascading full-adders and full-subtractors; look-ahead carry generation; and integrated circuit adders—cascading MSI four-bit adders, using adders to subtract, using adders as BCD-to-Ex-3 code converters, and analysis of a BCD adder.
5. Comparators: iterative design techniques, comparator design, and an MSI 4-bit comparator IC.
6. Arithmetic logic units (ALUs): description and design of a simple ALU, MSI four-bit ALU, and cascading MSI ALUs.

D. Additional combinational logic topics included

1. Array multipliers
2. Tristate buffers.
3. Combinational logic hazards, static 1, static 0, and designing static-hazard-free combinational circuits.

REFERENCES

Texts and references that cover combinational logic design material presented in this chapter include:

Digital Design Principals and Practices, by John F. Wakerly (Englewood Cliffs, N.J.: Prentice-Hall, 1994. Chapters 4 and 5 cover combinational logic.

Fundamentals of Logic Design, 4th ed., by Charles H. Roth Jr. (Minneapolis–St. Paul: West, 1992). Unit 8 covers multilevel networks, Unit 9 covers decoders and multiplexers, and Unit 10 covers combinational network 'design'.

An Engineering Approach to Digital Design, by William I. Fletcher (Englewood Cliffs, N.J.: Prentice-Hall, 1980). Chapter 4 discusses MSI and LSI circuits and their applications.

Digital Logic: Analysis, Applications and Design, by Susan A. R. Garrod and Robert Borns (Sanders, 1991). Chapters 4, 5, and 6 cover MSI applications.

Switching and Finite Automata Theory, 2d ed., by Zvi Kohavi (New York: McGraw-Hill, 1978). Chapter 5 covers combinational design, and Chapter 8 treats reliability and fault tolerance in detail.

The Art of Digital Design, an Introduction to Top-Down Design, by Franklin P. Processer and David E. Winkel

(Englewood Cliffs, N.J.: Prentice-Hall, 1987). Chapter 3 covers MSI combinational logic circuits.

Analysis and Synthesis of Logic Systems, by Daniel Mange (Dedham, Mass.: Artech House, 1986). Chapter 2 treats synthesis and analysis of combinational circuits.

Digital System Design, by Barry Wilkinson (Englewood Cliffs, N.J.: Prentice-Hall International, 1987). Chapter 3 covers MSI circuits, hazards, adders/subractors, and multipliers.

Electronic Logic Systems, by A. E. A. Almaoni (Englewood Cliffs, N.J.: Prentice-Hall, 1989). Chapter 5 covers hazards, and Chapter 6 covers arithmetic circuits.

GLOSSARY

Arithmetic logic unit (ALU) A combinational logic circuit that performs arithmetic and logical operations such as add, subtract, AND, OR, complement, and increment. Medium-scale integrated circuit devices such as the TTL 74xx181 provide these and other functions in a single device.

Array multiplier A combinational logic circuit that performs binary multiplication. Binary multiplication can be performed sequentially as well, using data registers, an adder, and a sequential circuit control. Array multipliers use a matrix or "array" of product terms added together to produce a result.

BCD adder A combinational logic circuit that adds two binary coded decimal numbers. The adder is more complex than a simple binary adder due to the logic needed to correct the result under certain input conditions.

Comparator A combinational logic circuit that compares two binary input values and produces results that specify the relative value of one input with respect to the other. Outputs of some comparators specify $A = B$, others may specify $A = B$, $A > B$, $A < B$.

Decoder A combinational logic circuit that accepts binary inputs, determines the input value, and then activates an output that corresponds to the decoded input value. A decoder accepts n data inputs and produces up to 2^n outputs. A three-input circuit has eight outputs.

Encoder A combinational logic circuit that accepts 2^n binary inputs and produces n data encoded output values. An 8-to-3 encoder has eight input lines and three output lines. When one of the data inputs is active, the output code that represents that value is generated.

Full-adder A combinational logic circuit that adds two one-bit values plus a carry and produces a sum and a carry output. The *full* part of the name comes from the carry input bit. Contrast that with the half-adder, which has no carry input bit. An n-bit adder requires $n - 1$ full-adders and one half-adder.

Half-adder A combinational logic circuit that adds together two one-bit values and produces a sum and a carry output. The *half* part of the name comes from the lack of a carry input. Contrast the half-adder with the full-adder.

Iterative circuits *Iterative* means to repeat. An iterative circuit is one in which a single cell can be identified and then repeated as often as needed to form a larger circuit. For example, adders and subtractors can be designed using iterative approaches. The basic adder cell is the full-adder. Addition of two n-bit operands would require n full-adders, with the carry-out of one cell feeding the carry-in to the next.

Look-ahead carry Adders and subtractors are iterative circuits in which the carry or borrow output from one cell provides the carry-in or borrow-in to the next cell. This means that, in an n-cell circuit, the carry and borrow signals must ripple from one cell to the next before the result is valid. This ripple effect takes time. To increase circuit speed a technique is used that looks at all of the inputs together to determine the carry and borrow signals. The look-ahead carry and borrow requires additional logic but greatly increases operation speed.

Multiplexer A digital multiplexer selects one of 2^n inputs to send to a single data output. It is the digital equivalent of a multiple position selector switch. One of 2^n inputs is selected by n select inputs.

Propagation delay The time it takes a signal to propagate from the input of a logic circuit to the output of the circuit. All logic circuits exhibit propagation delays. Each logic family (TTL, CMOS, ECL) has different propagation delays, with ECL being the fastest.

Static hazards Two types of static hazards can exist in combinational logic circuits; static 0 and static 1. A static 1 hazard occurs when a combinational logic input, with two delay paths to the output, changes, causing a momentary 1-0-1 pulse at the output. A static-0 hazard occurs when a momentary 0-1-0 pulse occurs at the output.

Tristate buffers Inverting or noninverting buffers whose output can also be placed into a high-impedance or hi-Z condition. The three "states" are logical 1, logical 0, and high impedance. The ability to switch a logic circuit into a high-impedance state permits the connection of several data output lines to a single data transfer line.

QUESTIONS AND PROBLEMS

Section 4.1 to 4.2

1. A logic circuit is needed that multiplies a two-bit binary value by the constant 3_{10}, as indicated by Figure 4.94.

Figure 4.94

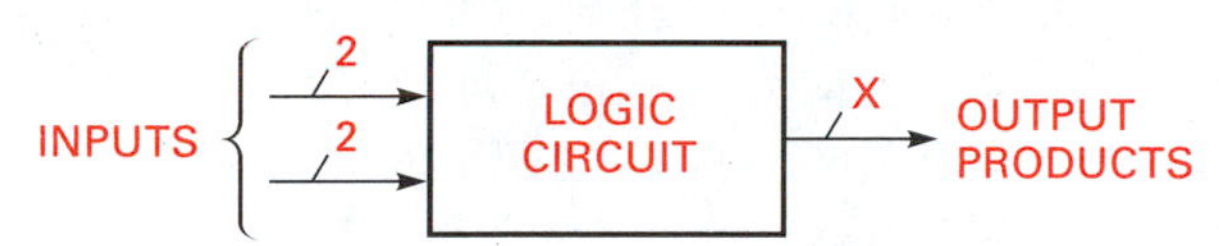

   a. Determine the number of input variables. How many rows does the truth table contain?
   b. How many output variables are needed?
   c. Construct a truth table, indicating the input and output variables.
   d. Write the minterm list equation for each output variable.
   e. Simplify each output function using K-maps, and write the reduced equations.

2. Design a combinational logic circuit to find the 9s complement of a BCD number.
   a. Construct the truth table for the logic circuit. How many output variables are needed?
   b. Write the minterm list equation for each output variable.
   c. Simplify each output function using K-maps, and write the reduced equations.
   d. Draw the resulting logic diagram.

3. Design a combinational logic circuit that will convert a straight BCD digit to an excess-3 BCD digit.
   a. Construct the truth table.
   b. Write the minterm list equation for each output function.
   c. Simplify each output function, and write the reduced logic equations.
   d. Draw the resulting logic diagram.

4. Design a circuit that will find the 2s complement of a four-bit binary number.
   a. Construct the truth table.
   b. Write the minterm list equations.
   c. Simplify the equations.
   d. Write the reduced equations, and draw the logic diagram.

5. Design the circuit necessary to drive a seven-segment display when driven with an excess-3 code BCD input. Assume that the display requires an active low input to turn on the LED segment. The segments are indicated in Figure 4.95.

Figure 4.95

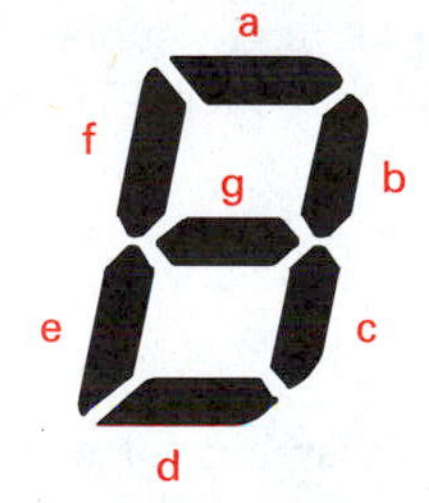

   a. Construct the truth table and simplify the output equations.
   b. Draw the logic diagram.

6. Determine the standard SSI TTL logic ICs required to realize the logic of Problem 1. Draw the logic diagram and identify the IC type number for each gate. How many ICs are needed for your design.

7. Assume that the logic designed in Problem 2 were realized using standard SSI integrated circuits. Determine the longest propagation path from any input to any output. Calculate the propagation delay.

8. What would be the speed advantage of converting the design of Problem 7 from standard SSI ICs to Schottky ICs? Support your answer with calculations.

9. Determine the SSI low power Schottky ICs necessary to realize the design in Problem 3. Determine how many LS ICs are needed. Calculate the average power dissipated by the circuit.

10. Determine the longest propagation delay path for the circuit designed in Problem 4. Assume that Schottky TTL SSI ICs are used to realize the circuit. Sketch a timing diagram for the longest delay path. Show a logic level transition for each gate in the path, and show the effects of propagation delay times. What is the propagation delay when the input variable changes from a 0 to a 1? What is the propagation delay when the input variable changes from a 1 to a 0?

11. Define the following in your own words:

a. Fan-out

b. Noise margin

c. $I_{CC}H$

d. V_{OH}

e. T_{pLH}

f. Current source

12. Using the information presented in Chapter 4 determine which TTL logic subfamily has the fastest propagation delay. Which TTL subfamily consumes the least power?

Section 4.3 to 4.4

13. Draw the logic diagram for a 2-to-4 logic decoder with an active low encode enable and active high data outputs. Construct a truth table and identify the data inputs, the enable input, and the outputs. Describe the circuit's function. What does it do? Draw the logic symbol for the decoder.

14. Using the logic symbol for the decoder described in Problem 13, describe how the circuit can be used as a minterm generator.

15. Draw the logic symbol for a 3-to-8 logic decoder that has active low data inputs, an active high enable, and active low data outputs. Use such a decoder to realize the Boolean equation $X = f(a, b, c) = (0, 3, 5, 6)$.

16. The 74LS138 is an MSI IC 3-to-8 logic decoder. Sketch a diagram showing how the device could be used to create a 4-to-16 decoder. Identify all the data inputs, enable signals, and data outputs. Identify the minterm data output association (which minterm is associated with which output?).

17. Realize the following functions with a decoder. Determine the size decoder necessary for each Boolean function.

a. $A = f(x, y) = \Sigma\,(0, 3)$; $B = f(x, y) = \Sigma\,(1)$

b. $X = f(a, b, c) = \Sigma\,(1, 3, 5, 7)$

c. $X = f(a, b, c) = \Sigma\,(2, 5)$;
$Y = f(a, b, c) = \Sigma\,(3, 5, 6)$

d. $P = f(w, x, y, z) = \Sigma\,(0, 5, 11)$

e. $Q = f(w, x, y, z) = \Sigma\,(4, 7, 8)$;
$R = f(w, x, y, z) = \Sigma\,(3, 9, 15)$

f. $T = f(a, b, c, d, e) = \Sigma\,(15, 17, 31)$

g. $D = f(a, b, c) = \pi\,(3, 5, 7)$

h. $F = f(w, x, y, z) = \pi\,(1, 7, 13, 15)$

18. A logic system must identify a single event from one of 64 events. A decoder will do the job; however, no such IC is available off the shelf. Design a 1-of-64 decoder using available TTL ICs. Determine which ICs you will use. Show all the inputs, enables, and outputs. Identify the minterm list produced by each output decoder.

19. Sketch the logic symbol for a 10-line to BCD encoder. Show how 10 events can be encoded into a four-bit data bus.

20. Show how a 74LS148 8-line to 3-line encoder can be used in conjunction with a 74LS138 3-to-8 decoder to save wires in transmitting information from point A to point B in a digital system.

Section 4.5

21. Sketch a diagram illustrating how a digital multiplexer is like a channel selector switch on an old TV. Explain in your own words what digital *multiplexing* means.

22. Sketch the logic symbol for a 8-to-1 digital multiplexer. Identify the data, select, and strobe inputs as well as the data output. Show how several data sources can be connected to a common data destination using a multiplexer.

23. Sketch a diagram showing how several 74S151 ICs can be connected to form a 64-to-1 multiplexer.

24. Sketch a diagram illustrating eight four-bit data sources connected to a single eight-bit destination using multiplexers.

25. Realize the following Boolean functions using the appropriate multiplexer, whose data inputs are connected directly to logical 1 and 0 levels.
   - **a.** $x = f(a, b, c) = \Sigma\,(0, 1, 3, 5, 7)$
   - **b.** $y = f(a, b, c, d) = \Sigma\,(1, 4, 5, 7, 8, 12, 13, 15)$
   - **c.** $z = (a, b, c, d, e)$
     $= \Sigma\,(0, 1, 3, 4, 6, 7, 15, 21, 25)$

26. Realize the following Boolean functions using the appropriate multiplexer when a single variable MEV is permitted.
   - **a.** $x = (a, b, c) = \Sigma\,(0, 1, 4, 5, 7)$
   - **b.** $y = (a, b, c, d) = \Sigma\,(0, 3, 4, 5, 7, 9, 13, 15)$
   - **c.** $z = (a, b, c, d, e) = \Sigma\,(0, 2, 3, 4, 6, 9, 12, 13, 15, 19, 23, 25, 26, 31)$

27. Realize the following Boolean equations with as few ICs as possible. Assume that input variables and their complements are available.
   - **a.** $T = (w, x, y, z) = \Sigma\,(0, 2, 4, 6, 8, 10, 12, 14)$
   - **b.** $S = (w, x, y, z) = \Sigma\,(0, 1, 3, 9, 10, 12, 14)$
   - **c.** $U = (a, b, c, d, e) = \Sigma\,(0, 2, 5, 7, 8, 10, 13, 15, 16, 18, 21, 23, 24, 26, 29, 31)$
   - **d.** $A = (v, w, x, y, z) = \Sigma\,(1, 3, 5, 7, 12, 13, 14, 15, 17, 19, 21, 23)$

Section 4.6 to 4.7

28. Design a full-adder.
   - **a.** Construct the truth table, and simplify the output equations.
   - **b.** Draw the resulting logic diagram.
   - **c.** Realize the adder using a decoder.
   - **d.** Which design takes fewer ICs?
   - **e.** Determine the propagation delays from data inputs to the sum and carry outputs if you use low power Schottky SSI ICs.
   - **f.** Determine the propagation delays from the data inputs to sum and carry outputs if a low power Schottky decoder is used.
   - **g.** Calculate the power consumption for each design. Which requires a lower supply current?

29. Design a full-subtractor.
   - **a.** Construct the truth table, and simplify the output equations.
   - **b.** Draw the resulting logic diagram.
   - **c.** Realize the subtractor using a decoder.

30. Show how a full adder can be made to subtract.

31. Using the design in Problem 28, determine the ripple-carry propagation delay for a four-bit adder. Determine the total average current requirement.

32. Draw the block diagram for a single-cell look-ahead carry adder. Label all inputs and outputs. Describe the function of the look-ahead carry generator.

33. Write the logic equations for a two-stage look-ahead carry adder in terms of data and carry inputs.

34. Show how the 74xx83 four-bit adder can be cascaded to add 24 bits. Determine the carry propagation delay for such an adder when the 74LS83 IC is used without look-ahead carry.

35. Draw the logic diagram showing the 74LS83 in an eight-bit adder/subtractor. Identify all input and output signals.

36. Design a two-digit BCD adder using the 74LS83.

37. Design a four-bit comparator that propagates its secondary signals from left to right.
   - **a.** Sketch a block diagram showing the primary inputs, secondary inputs, and outputs.
   - **b.** Construct a modified table (see Table 4.19).
   - **c.** Write the minterm list equations for the secondary outputs and the boundary outputs.
   - **d.** Simplify the output equations, and draw the resulting logic diagram.

38. Sketch a diagram showing how the 74LS85 can be used to create a 16-bit comparator.

Section 4.8 to 4.9

39. Determine the select and mode bit and the carry-in values for the following ALU functions when using a 74xx181 integrated circuit: $MS_3S_2S_1S_0$.

a. AND =

b. ADD (with carry) =

c. SUB (with borrow) =

d. NOR =

e. Transfer A to output =

f. Decrement A =

40. Given the following value sequences for mode, select, and carry-in, determine the the function sequence, such as ADD, SUB, and AND.

a. M, $S_3S_2S_1S_0$, C_n = 1,0000,1; 0,1001,0

b. M, $S_3S_2S_1S_0$, C_n = 0,1111,1; 1,1001,0; 0,1100,0

c. M, $S_3S_2S_1S_0$, C_n = 1,1011,0; 1,0000,1; 0,1001,1; 0,0000,0

41. Design a simple ALU using SSI ICs that will add, subtract, increment, decrement, and transfer the A operand to the output. Design a single cell and show how several can be connected to realize a four-bit ALU.

42. Design a 12-bit ALU using the 74181 MSI IC and a look-ahead carry logic generator. Show how all the inputs, outputs, carries, mode, select, and look-ahead carry signals are connected.

43. Assume that the array multiplier shown in Figure 4.77 is realized using full-adders designed with advanced Schottky SSI ICs. Also realize the product terms using AS ICs. Calculate the propagation delay for the multiplier.

44. Determine how may product terms are needed for a six-bit array multiplier.

Section 4.10 to 4.11

45. Explain why a tristate device is not using base-3 arithmetic. What is the purpose of the device? Sketch a diagram showing how a tristate device can be thought of as a solenoid.

46. Sketch a diagram showing the output transistor arrangement of a TTL gate. Explain how the gate output can be put into a high-impedance state.

47. Sketch a diagram illustrating several data sources being connected to several data destinations. Each source and destination must have separate enable signals. Describe how source and destination select data could be generated using decoder ICs.

48. Write out the definitions of static-0 and static-1 hazards.

49. Explain why a maxterm circuit cannot generate a static-1 hazard.

50. Why are two levels of logic necessary to generate static hazards?

51. Determine any static-0 or static-1 hazards in the following Boolean equations. Identify where the hazards are and what must be done to eliminate them.

a. $A = f(w, x, y, z) = \Sigma\,(5, 7, 13, 15)$

b. $X = f(a, b, c, d) = \Sigma\,(5, 7, 8, 9, 10, 11, 13, 15)$

c. $T = f(a, d, c, d) = \Sigma\,(0, 2, 4, 6, 8, 10, 12, 14)$

d. $T = f(a, b, c, d) = \Sigma\,(0, 2, 4, 6, 12, 13, 14, 15)$

CHAPTER 5

FLIP-FLOPS, SIMPLE COUNTERS, AND REGISTERS

INTRODUCTION

Chapter 5 begins the study of sequential circuits. Recall that digital logic was partitioned into two major sections: combinational and sequential. Combinational logic excluded memory or any feedback from logic outputs back to the input. Sequential logic expands combinational logic functions to include the ability to store and later retrieve binary information.

Sequential implies that events are ordered in time, that one event then another occurs, separated by time. Time separation of events, in digital logic, requires the use of memory. Two types of sequential logic exist: synchronous and asynchronous. Asynchronous logic will be discussed in Chapter 8.

The purpose of Chapter 5 is to develop an understanding of the functional components that are used in sequential logic as a storage medium. These basic storage building blocks are called *flip-flops*. *Flip-flop* was coined as a name descriptive of the circuit's function. It flipped and flopped; that is, it had two stable states, either a 0 or a 1.

The difference between the states of a flip-flop and those of a combinational logic device (NAND, NOR, etc.), is that the flip-flop stays at a given logic level after the input signal level changes, whereas the combinational gate provides a valid output only as long as the input remains steady. A gate has no memory capacity but the flip-flop does.

First, we will develop the basic modeling structure for sequential logic, so that we can see how flip-flops are incorporated in sequential circuits. Then we will explore some details of the devices themselves, identify four different types of flip-flops, and use them to form storage registers and simple counters. Once we have mastered flip-flop basics and used them in actual applications, we can move on to additional topics in sequential circuit analysis and design.

5.1 SEQUENTIAL CIRCUIT MODELS

Models provide a means to represent an idea, concept, or process, mathematically or graphically, as an aid in enhancing our understanding. The two main components of sequential circuits are combinational logic and storage or memory. The two are related as illustrated in Figure 5.1. Binary input variables, I_0 to I_n, are presented to the system,

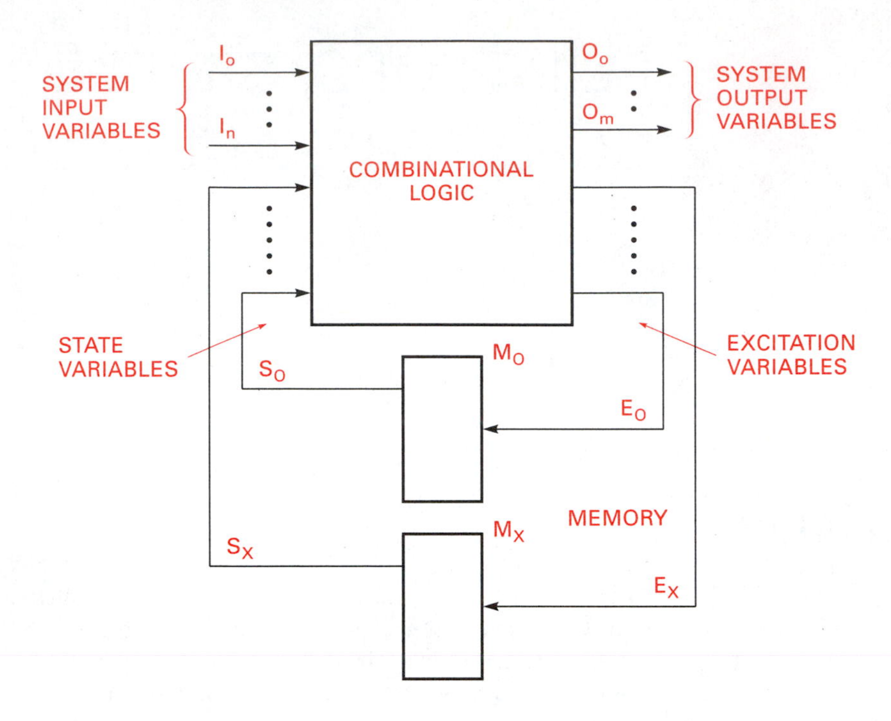

Figure 5.1
Universal sequential circuit model

and binary output variables, O_0 to O_m, are generated. Internally, the system generates excitation variables, E_0 to E_x, and state variables, S_0 to S_x. Note that the combinational logic unit has both system input and state variable inputs and generates both system output and excitation output variables. We can see how the basic combinational logic model, introduced in Chapter 3, is expanded to include memory.

The model illustrated in Figure 5.1 contains all of the elements of a sequential circuit. However, the combinational logic function is represented by a single block. It is convenient to separate the combinational function into two parts for reasons that will become apparent as the discussion progresses.

The model illustrated in Figure 5.2 separates the combinational logic function into two parts, an input transform (f) and an output transform (g). The two logic blocks "transform" the input logic values into output logic values. The transform is defined by Boolean equations. The input combinational block receives external inputs (I) and state variables (S), from which the input excitation to the memory is generated. The memory

Figure 5.2
A general sequential circuit model with the combinational logic functions separated into two blocks

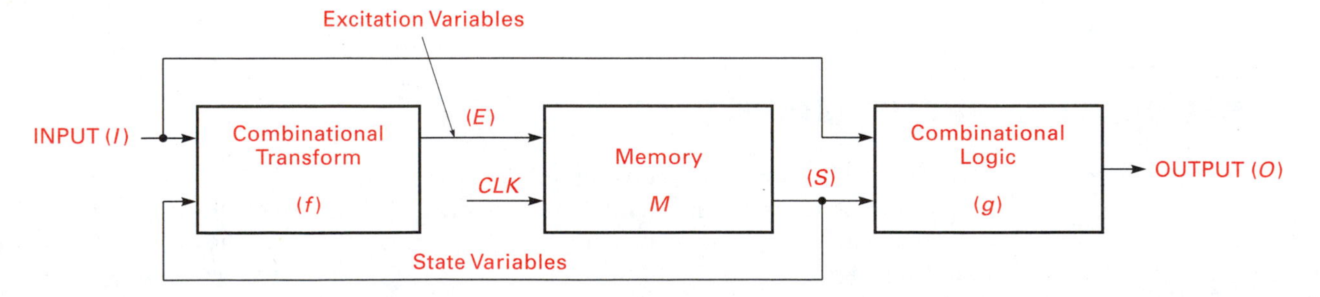

excitation input relationship can be written as $E = f(I, S)$, which means that the input to the memory is a result of the combinational operation, f, on input variables I and state variables S. Excitation inputs (E) force the memory elements, usually flip-flops, to the desired state variable output value. We must make it clear that, when we speak of "memory," we are referring to the ability to store binary data. Memory, in a sequential circuits context, takes on a slightly different meaning than memory internal to a personal computer, even though both store binary data. The real difference is in the amount and the technology used to realize a given memory function. We will study more about integrated circuit memory for larger scale applications later. For now, the "memory" we speak of is realized with a few flip-flops.

Output, O, in Figure 5.2, is generated by the combinational logic transform, (g), operating on input variables I and state variables S. The relationship is written, $O = g(I, S)$.

The value of the next-state variables (new memory output), S_{t+1}, is determined by the present value of the state variables (old memory output), S_t, and the input, I. S_t and S_{t+1} represent the binary value of state variables, S, at different times, t. A key principle in understanding sequential circuits is that memory data are free to change with time. It can be one value at time t_1 and another at time t_2. The memory function in Figure 5.2 has a synchronizing clock input, C_{lk}. The clock input is not a binary value representing the time of day, but rather a "synchronizing" train of pulses. Synchronous memory changes its data only at certain time intervals.

The present state is considered to be the flip-flop's value prior to the arrival of a clock pulse of interest. Often it is necessary to know what changes may occur in memory data if the present state is known and a certain input condition exists. A distinction is made between the present state, S_t, and the next-state, S_{t+1}. We can write

$$S_{t+1} = f(I, S_t)$$

which is read as the next state is dependent on the present state (value in memory) as well as the input and will change to S_{t+1} on the next clock pulse.

Consider the timing diagram shown in Figure 5.3. A series of pulses called the *clock* produces a falling or negative edge at times labeled t, $t + 1$, $t + 2$, and so on. Variables Q_1 and Q_2 are free to change value only on the falling or negative clock edges. Let Q_1 and Q_2 be **state variables** that change as indicated in Figure 5.3. Because two variables can have four permutations, we can identify four "states": $Q_2\,Q_1 = 00, 01, 11, 10$. Each state becomes a "present" state after a clock pulse and will go to a "next-state" on the next clock pulse. When we write S_t we mean the state at some clock time t. Then "next" state, S_{t+1}, will occur on the next clock pulse.

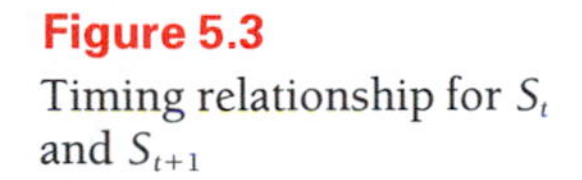

Figure 5.3

Timing relationship for S_t and S_{t+1}

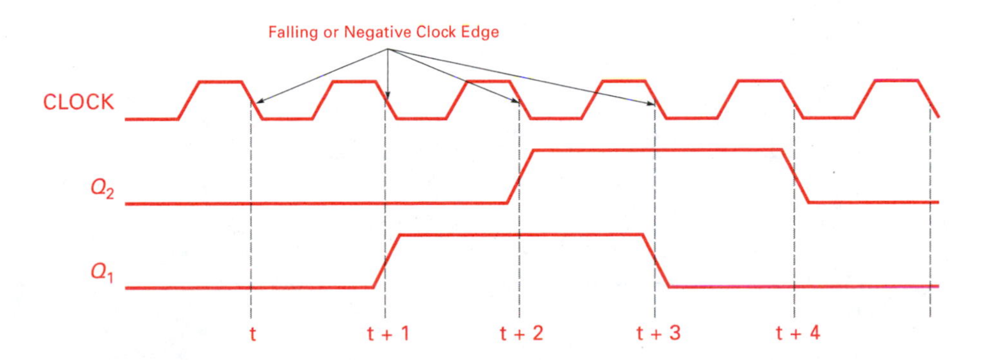

Figure 5.4
Combinational logic case of the general sequential model

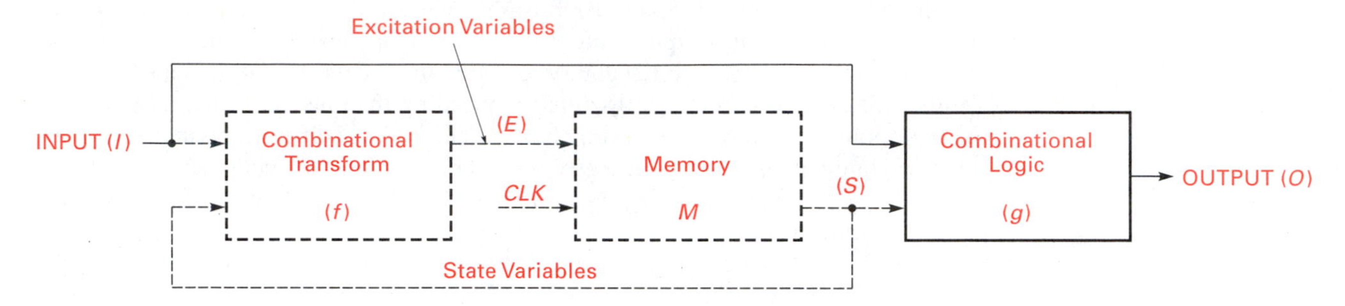

The general sequential circuit model can be simplified to show several subsets of the general condition. Figure 5.4 illustrates the application of the general model for a combinational circuit.

In this case output O is written as a function of input I. This is the same form as the combinational logic model illustrated in Figure 3.1, Chapter 3.

$$O = g(I) \text{ Combinational input to output transform}$$

The simplest sequential circuit is one where the output changes are based only on the input changes but delayed by time, t. Sequential delay circuits have no feedback path from memory to the input combinational logic. Input variables provide the excitation to memory, which generates the delay, and the input to the output logic. Figure 5.5 illustrates a sequential delay model.

Excitation inputs are $E = f(I)$, and the outputs are $O = g(S_{t+1})$ for sequential delay circuits. The next state, S_{t+1}, is dependent on the present state, S_t; the excitation input, E; and the clock. Output, O, is dependent on the next-state value, S_{t+1}.

Figure 5.6 illustrates a simple sequential counter that changes on every clock pulse. Note that no external input is present in this case. The input logic translates the state variables, S, to provide the excitation input to memory. The present-state values are converted to output variables by the output logic.

$$E = f(S_t), \quad \text{excitation}$$
$$S_{t+1} = f(S_t, E), \quad \text{next state}$$
$$O = g(S_t), \quad \text{outputs}$$

The next classification of sequential circuits shows the generation of the memory **excitation variable** inputs as a function of both the present state, S_t, and the inputs, I.

Figure 5.5
Sequential delay model

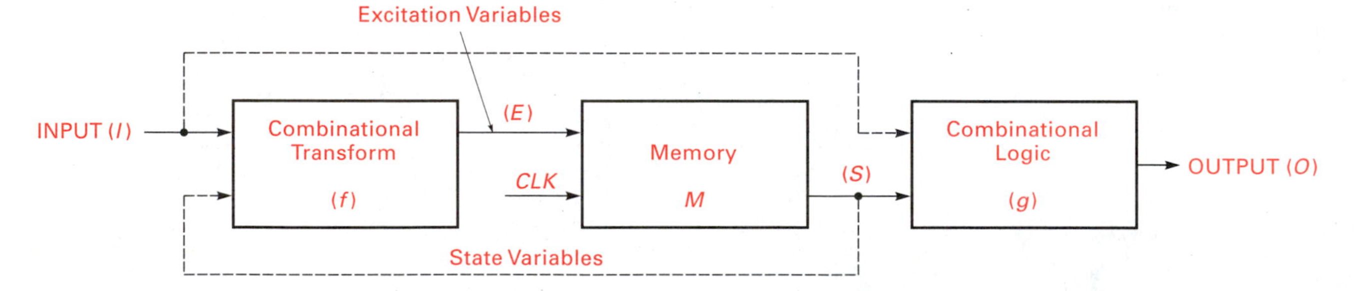

Figure 5.6
Simple sequential counter model

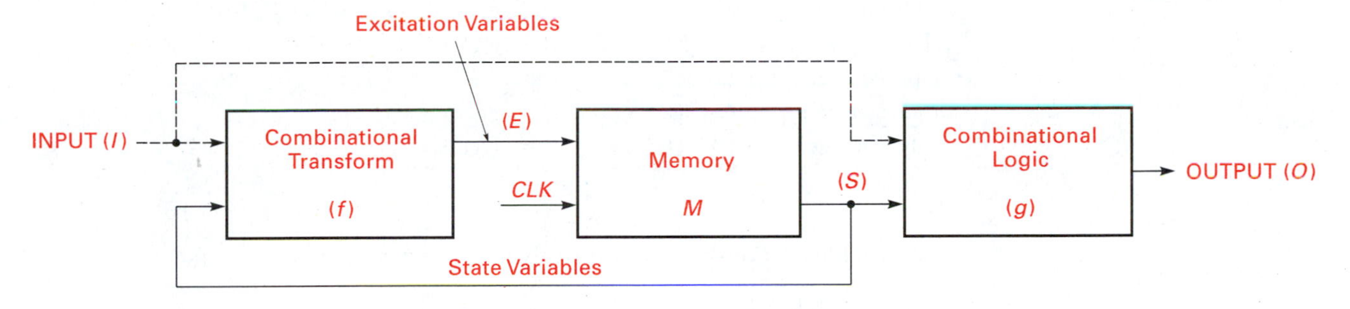

Figure 5.7
Moore machine sequential circuit model

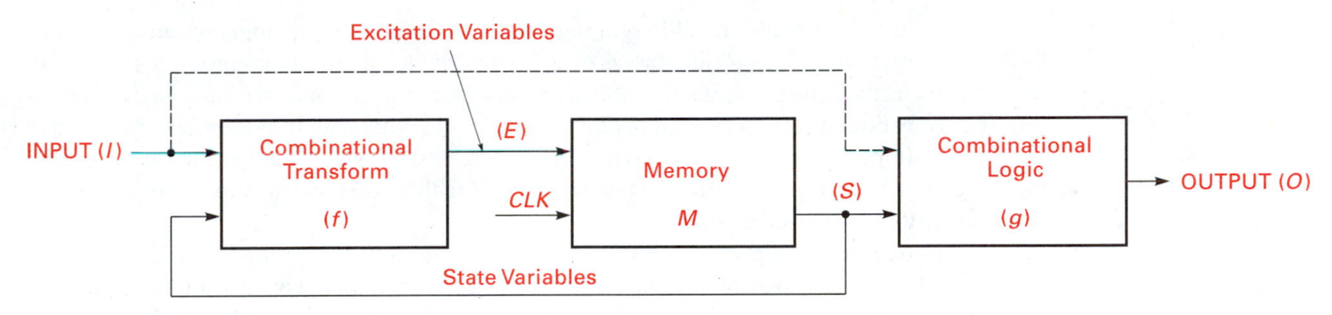

The output, O, is a function of only the present state, S_t. The model is shown in Figure 5.7. It is commonly called a **Moore** machine model.

The variable relationships for the Moore model are

$$E = f(I, S_t)$$
$$S_{t+1} = f(S_t, E)$$
$$O = g(S_t)$$

The last sequential circuit model is the **Mealy** machine. It differs from the Moore model in that the outputs are dependent on both the present state, S_t, and the input variables, I. The model diagram is shown in Figure 5.8.

Figure 5.8
Mealy machine sequential circuit model

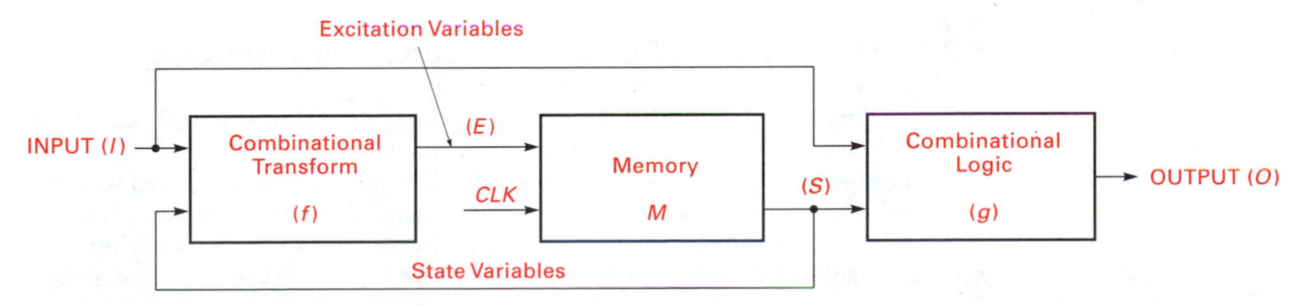

TABLE 5.1
Comparison of different logic classes

Circuit class type	Excitation	Output
Combinational	None	$O = (g)(I)$
Delay	$E = f(I)$	$O = (g)(S_t)$
Simple counter	$E = f(S_t)$	$O = (g)(S_t)$
Moore machine	$E = f(I, S_t)$	$O = (g)(S_t)$
Mealy machine	$E = f(I, S_t)$	$O = (g)(I, S_t)$

The variable relationships for the Mealy machine model are

$$E = f(I, S_t)$$
$$S_{t+1} = f(S_t, E)$$
$$O = (g)(I, S_t)$$

Table 5.1 compares the different logic classes. Combinational logic requires no storage or memory. The various classes of sequential logic all require memory. A sequential delay circuit is like combinational logic in that the output is affected only by the input, but only after a time delay that is represented by clocking data into memory.

A simple counter has no external input variables; the state changes on the clock and the output is decoded from the state variables. Mealy and Moore machines differ in how the outputs are generated.

Two subcategories of sequential circuits are commonly recognized: synchronous and asynchronous. **Synchronous** sequential logic is clocked, and **asynchronous** sequential logic is not clocked. We will study clocked or synchronous sequential logic first.

Prior to developing sequential logic design techniques, we need to introduce flip-flops and analyze some IC counters and registers. Then, in Chapter 6, we will develop synchronous circuit analysis and design techniques. Chapter 7 will continue coverage of synchronous circuits. Chapter 8 will cover asynchronous sequential circuits.

5.2 FLIP-FLOPS

Flip-flops are digital electronic devices that have the ability to store binary information after the excitation input has changed. For this reason, they are considered to be the basic memory cell for the majority of electronic binary data storage applications. This section explores the flip-flop from a functional perspective.

5.2.1 Flip-Flop Logic Symbols, Function, and Triggering

Before dealing with the internal workings of flip-flops, we are going to look at them in terms of function. Four types of flip-flops are commonly considered: *JK*, *SR*, *D*, and *T*. Each stores binary data but has a unique set of input variables. Each flip-flop type can be described in terms of its input and output characteristics by writing a special truth table, called an excitation table. A characteristic equation can be generated from the excitation tables. The names *flip-flops* and *latches* are sometimes used interchangeably;

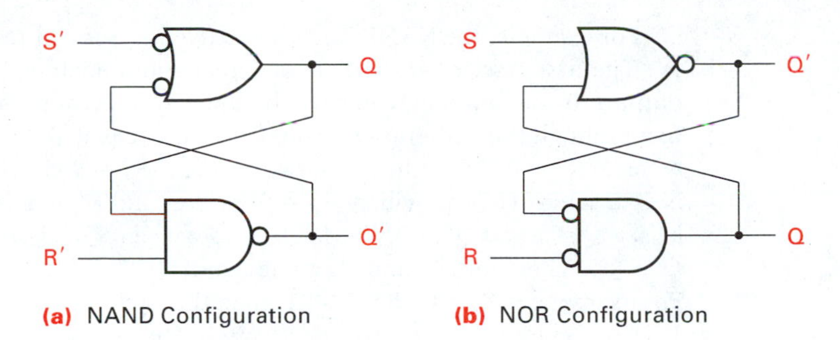

Figure 5.9
Set-reset latches with truth tables

S'	R'	Q	Q'	S	R	Q	Q'
0	0	1*	1*	0	0	last Q	last Q'
0	1	1	0	0	1	0	1
1	0	0	1	1	0	1	0
1	1	last Q	last Q'	1	1	0*	0*

* Q and Q' cannot both be the same value; therefore, the input conditions that cause $Q = Q'$ is not allowed.

however, the term *flip-flop* is more appropriately associated with devices that change state only on a clock edge or pulse, whereas latches change state without being clocked. Flip-flops are clocked and latches are not; both are bistable (two stable states) and both are used to store binary data. Some latches are allowed to change state only when gated with an external "enable" signal.

Latches

The simplest binary storage circuit is the **latch**. By cross-connecting the outputs of a multiple-input inverting gate (NAND, NOR) back to an input, a latch can be constructed. Figure 5.9 illustrates two simple latches.

The NAND configured latch remembers which input was the last to be a 0; the NOR latch remembers the last input to be a 1. The logic symbols for the two latches are shown in Figure 5.10.

Note that the NAND latch is drawn with active low inputs connected into an inverting bubble OR gate. Recall the reasons for this from the mixed logic treatment in Section 3.7 of Chapter 3. The inputs to the latch are called S (set) and R (reset). A flip-flop or latch is said to be "**set**" when the Q output is a logical 1 and "**reset**" when the same Q output is a logical 0. To set the flip-flop, the two inputs, S and R, must be the values indicated in the appropriate truth table.

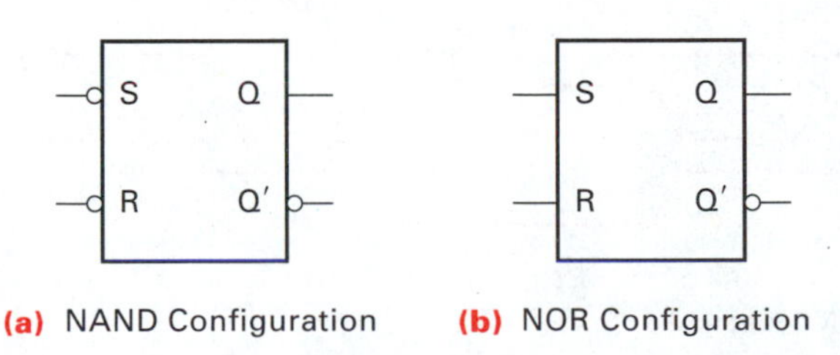

(a) NAND Configuration (b) NOR Configuration

Figure 5.10
Logic symbols for latches

For example, the NAND latch has active low set and reset inputs, labeled as S' and R'. The first case, when $S'R' = 00$, creates an undefined condition for the Q and Q' outputs. By definition, Q can never be equal to Q' because one is the inverse of the other. Even though the definition says that Q cannot equal Q', the circuit will cause $Q = Q'$ when $S'R' = 00$. Therefore, the case where $S'R' = 00$ is illegal and not allowed for the NAND latch. The case where $S' = 0$ (active) and $R' = 1$ (inactive) causes the latch to be set; $Q = 1$ and $Q' = 0$. The latch is reset when $S' = 1$ and $R' = 0$, causing $Q = 0$ and $Q' = 1$. When both S' and R' are inactive, $S'R' = 11$, the Q output does not change. For example, if $S' = 0$ and $R' = 1$, then the latch output $Q = 1$. The output remains a 1 when the inputs go inactive, $S'R' = 11$. The reverse is true when R' is active and S' is inactive; the latch is reset to $Q = 0$. After both inputs go inactive, $S'R' = 11$, the Q output remains a 0.

A NOR latch operates in the same manner except that the two inputs, S and R, are active high. The truth table includes set and reset input conditions, an illegal or undefined condition, and a no change condition just like the NAND latch configuration.

NAND and NOR latch timing diagrams, in Figure 5.11, show the relationship of latch inputs and outputs as they might be viewed with an oscilloscope or logic analyzer. The undefined conditions were not shown in the timing diagram because they are not used. Note that the Q and Q' outputs are complements of one another. Also note the memory capability—the S input for the NOR latch sets the flip-flop, so that $Q = 1$, $Q' = 0$, and the outputs remain in that condition after the excitation input go inactive. The latch stored the active set input by causing $Q = 1$. It stored the active reset input by causing $Q = 0$, $Q' = 1$.

Input pulse widths, shown in Figure 5.11, are of arbitrary length. The only requirement is that they be wide enough to allow the latch to change state. We will discuss input pulse widths and other timing considerations associated with latches and flip-flops later in this chapter.

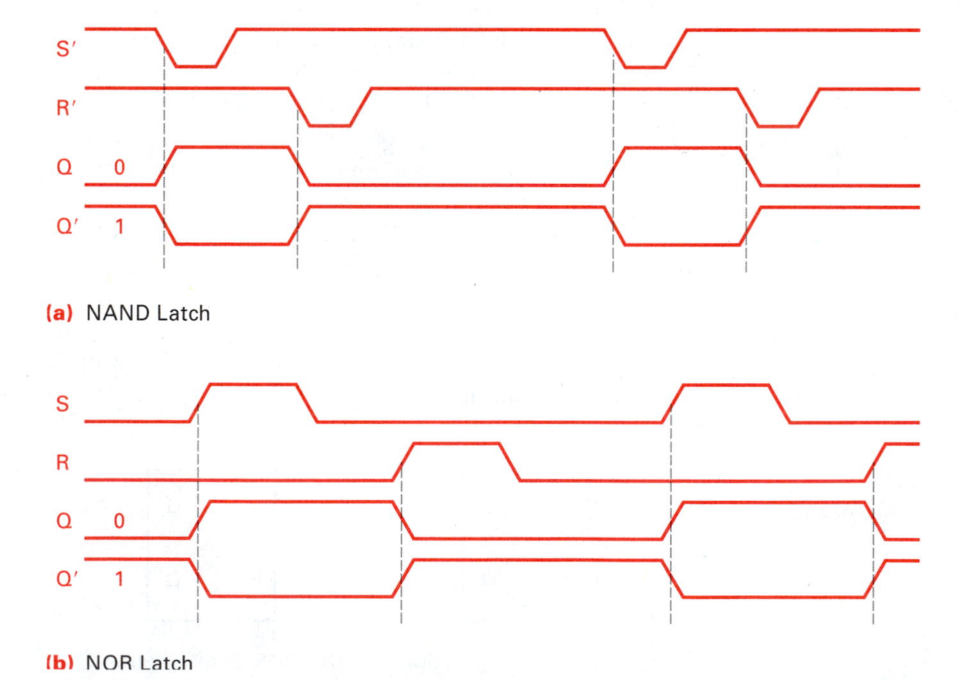

Figure 5.11
NAND S'–R' and NOR S'–R' latch timing diagram

Figure 5.12

Gated S–R latch, excitation table and characteristic equation $Q_{t+1} = S + R' Q_t$

Excitation table

Q_t	S	R	Q_{t+1}
0	0	0	0
0	0	1	0
0	1	0	1
0	1	1	Indeterminate
1	0	0	1
1	0	1	0
1	1	0	1
1	1	1	Indeterminate

Characteristic Equation

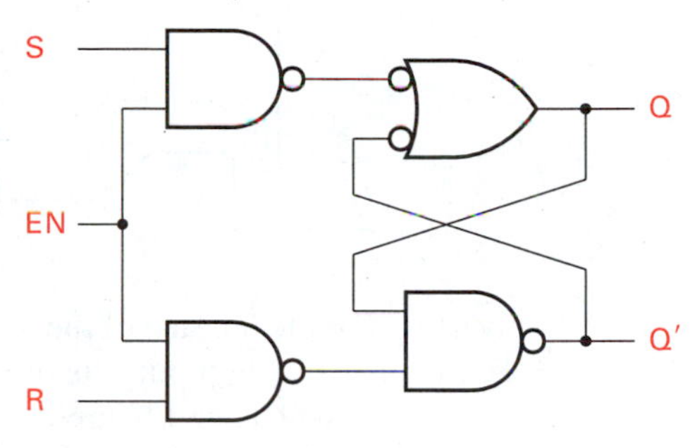

The characteristic equation is derived from the excitation table. Both indeterminate conditions create don't-care terms.

In Figure 5.9 no clock is present to gate the excitation inputs. Because no synchronizing clock is present, the latches are said to be asynchronous. A gated latch that uses an ENABLE input, EN, is shown in Figure 5.12.

The EN input, in Figure 5.12, is present at both input NAND gates; therefore, EN must be active (1) before any changes in the output can occur. In other words, if EN is not a 1, the circuit Q output does not change when the S and R inputs change.

The SR characteristic equation is derived by mapping all of the minterms for Q_{t+1} including both indeterminate conditions used as don't-care terms (see Figure 5.13).

A gated S–R latch (Figure 5.12) is the simplest type of flip-flop. The enable input level may be active high, as shown in Figure 5.12, or active low. Figure 5.14 shows a timing diagram for a gated R-S latch. An active low enable would result from placing an

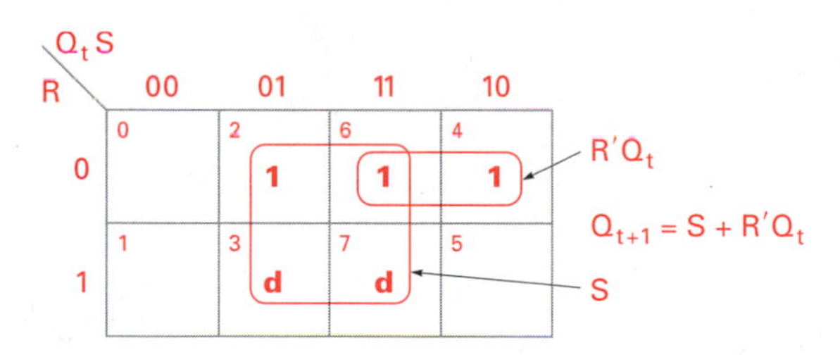

Figure 5.13

SR NAND latch characteristic equation K-map

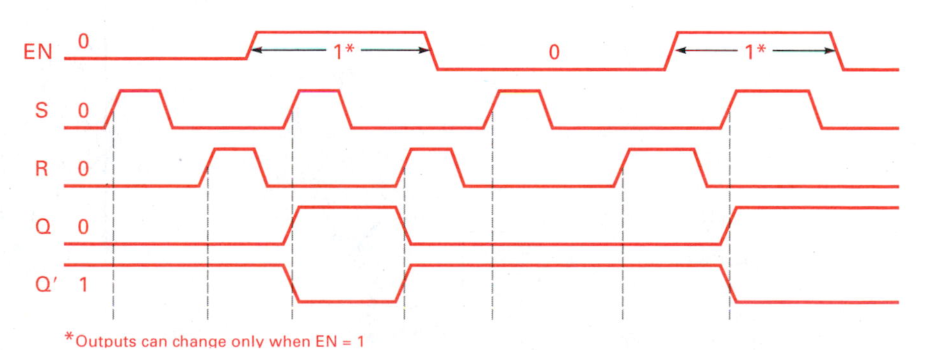

Figure 5.14

Timing diagram for the gated S–R latch

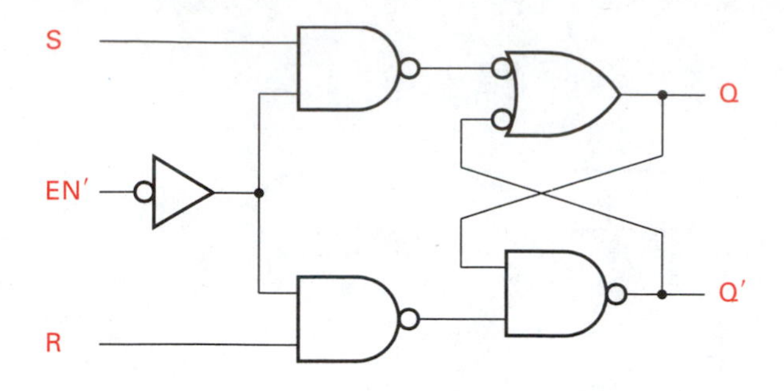

Figure 5.15
Negative level enable S–R latch

inverting bubble on each of the input gates as shown in Figure 5.15. In each case, active high or low enable input, the output is free to change according to the S and R input values only when the EN input is active. This type of enable control is sometimes referred to as either a *positive* or *negative* **level trigger.**

Both NOR and NAND latches, whether gated or not, have an indeterminate or undefined condition that is a not-allowed input combination. In the case of the NAND latch, the undefined condition is $S'R' = 00$. In the case of the NOR latch, it is $SR = 11$.

Several integrated circuit latches are commercially available. The SN74LS279 is a quadruple $S'R'$ latch. The logic diagram, pin assignment, and truth table for a single latch are shown in Figure 5.16.

Another type of latch can be formed from the SR latch by connecting the S and R inputs, as shown in Figure 5.17. This single input latch is called a *data* or D latch. The D (data) input is only an excitation input and the input C is an enable. Connecting the S and R inputs to the latch through an inverter provides a reduction of one input. The D input causes the Q output to change states; whenever $D = 1$, $Q = 1$, when $D = 0$, then $Q = 0$. The Q output follows the D input as shown in the truth table.

Figure 5.16
Logic diagram, truth table, and pin assignment for the SN74LS279 IC $S'R'$ latch

Truth table for each latch

Inputs		Output
S'	R'	Q
1	1	Remains same
0	1	1
1	0	0
0	0	Indeterminate

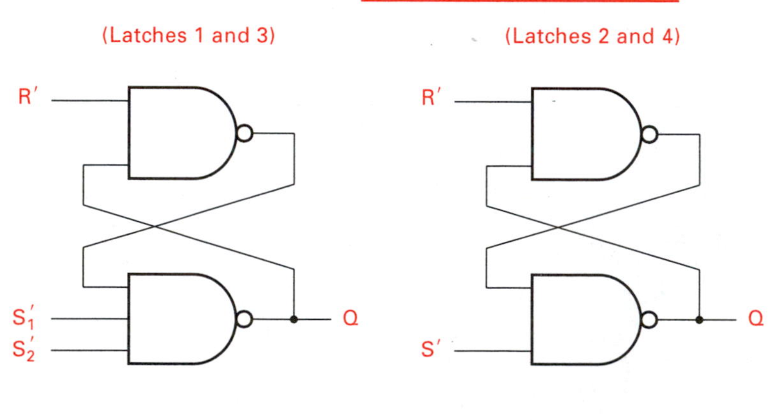

Pin assignment

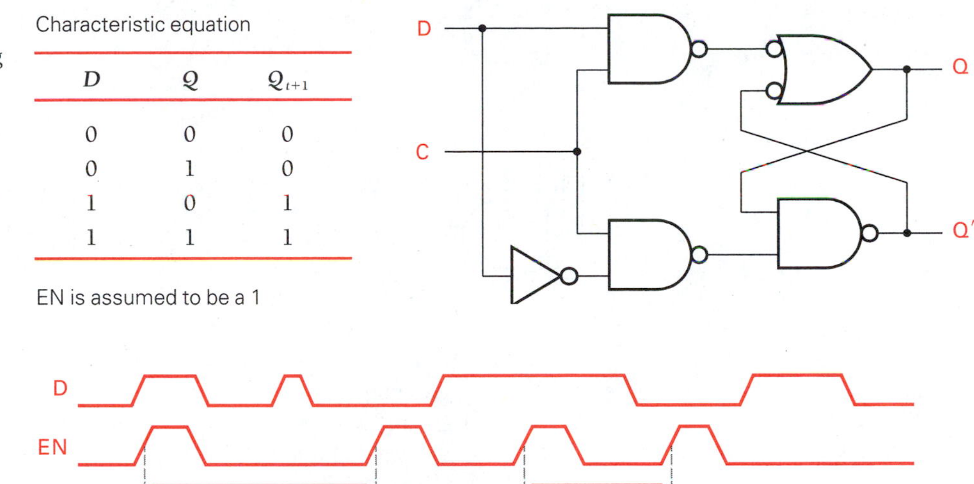

Characteristic equation

D	Q	Q_{t+1}
0	0	0
0	1	0
1	0	1
1	1	1

EN is assumed to be a 1

Figure 5.17
D latch formed from connecting together the S and R inputs through an inverter

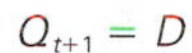

$Q_{t+1} = D$

Figure 5.18
Timing diagram for the D latch

A representative timing diagram for the D latch is given in Figure 5.18.

Notice that the Q output follows the D input when the enable input is active. Anytime EN is inactive the Q output remains at the previous level. The enable input, C, can be configured as active high or low as was the case with the gated RS latch.

The SN74LS75 is a quad D latch integrated circuit, whose logic diagram and pin assignment are given in Figure 5.19.

J–K Clocked Flip-Flops

A simple latch forms the basis for the J–K flip-flop. We have looked briefly at the enabled S–R and D latches. Now we want to investigate edge and pulse triggered flip-flops. Flip-flops differ from latches due to the triggering method used. Latches are not pulse or clock edge triggered and flip-flops are.

The J–K gated latch does not have an indeterminate input combination as did the S–R gated latch. The J input is analogous to the S (set) input and the K input to the R

Figure 5.19
Logic diagram and pin assignment for the SN74LS75 quad D-latch IC

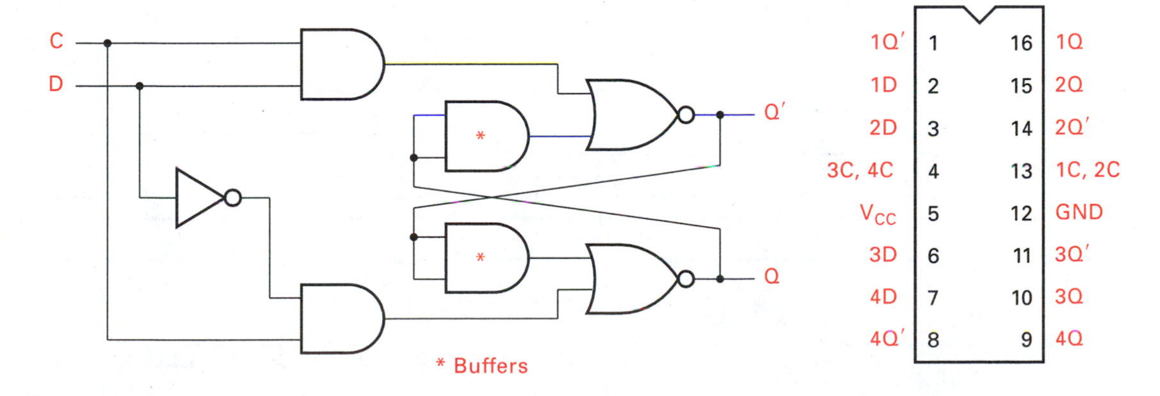

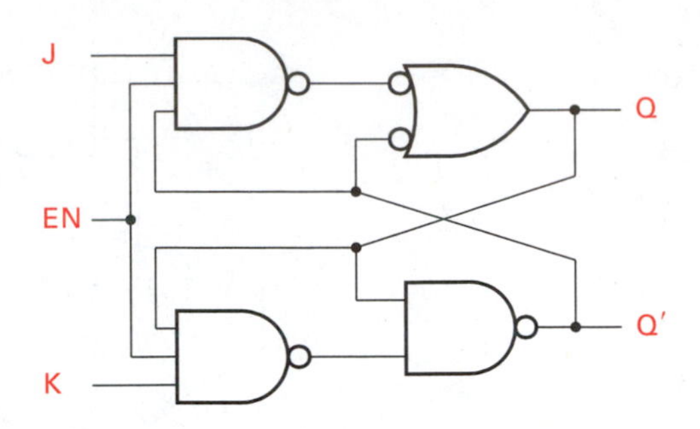

Figure 5.20
Simple *J–K* gated latch logic diagram

TABLE 5.2
Truth table for a *J–K* gated latch

Inputs			Output
J	K	Q_t	Q_{t+1}
0	0	0	0
0	0	1	1
0	1	0	0
0	1	1	0
1	0	0	1
1	0	1	1
1	1	0	1
1	1	1	0

Characteristic equation for the *J–K* gated latch:

$$Q_{t+1} = JQ_t' + K'Q_t$$

(reset). The indeterminate ($SR = 11$) input condition of the *RS* latch causes the *J–K* gated latch to toggle; when $J = K = 1$ then $Q_{t+1} = Q_t'$. By connecting the Q' and Q outputs back to the J and K excitation inputs, the undefined condition that existed with the *S–R* latch is no longer undefined. Figure 5.20 shows the simplest form of a *J–K* enabled latch. The circuit shown in Figure 5.20 is not practical. It is shown here to illustrate the evolution from the *RS* latch to the *J–K* gated latch and finally the edge triggered *J–K* flip-flop.

J, K, and Q_t are inputs and Q_{t+1} is an output in Table 5.2. Implied is the fact that the C input must be active.

The following discussion explains the impracticality of the *J–K* gated latch shown in Figure 5.20. If the C input were to remain 1 longer than the propagation delay from the output back to the *J–K* inputs, the latch would oscillate at some frequency.

Trace the output back to the input gates under the condition when $C = 1$ and $JK = 11$. As long as the C input remained a 1 when $J = K = 1$, the Q and Q' outputs will oscillate. Consider the timing diagram shown in Figure 5.21. Notice that, when $J = K = 0$, no change occurs in the Q and Q' outputs. When $J = 1$ and $K = 0$, the Q output changes to a 1 when $C = 1$. When $J = K = 1$, however, the Q and Q' outputs oscillate during the time that $C = 1$. For this reason the pulse width of C must be less than the delay time from the J or K input gate to the Q or Q' output gate.

The solution to the problem of an oscillating *J–K* gated latch is to create an edge triggered *J–K* flip-flop. Figure 5.22 shows the logic diagram of a positive edge triggered *J–K* flip-flop, the 74xx109. We will discuss the design of edge triggered *J–K* flip-flops in the chapter on asynchronous sequential logic, for edge triggered flip-flops are really asynchronous machines internally.

The 74xx109 is a dual JK' positive edge triggered flip-flop. An example timing diagram is shown in Figure 5.23. Notice that K is active low, so that when $J = 1$ and $K' = 0$ the flip-flop toggles on every positive edge of the clock input. Two asynchronous

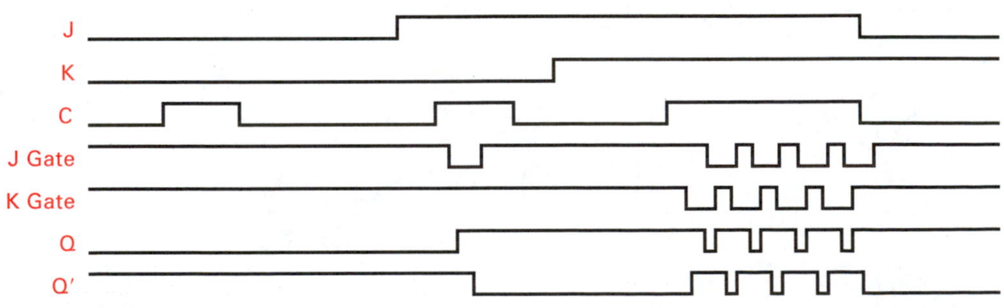

Figure 5.21
Timing diagram showing the oscillations that occur in a gated *J–K* latch when $J = K = 1$ and $EN = 1$

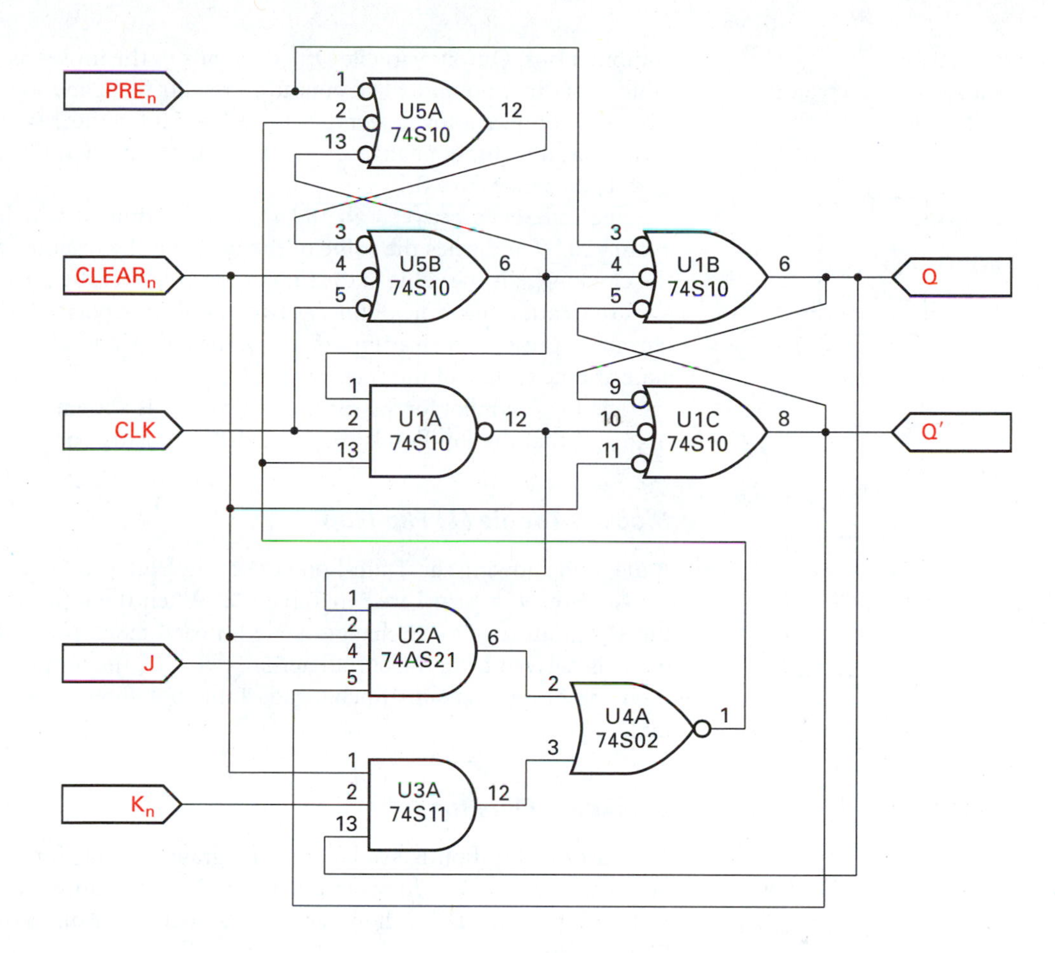

Figure 5.22
74xx109, positive edge triggered J–K' flip-flop

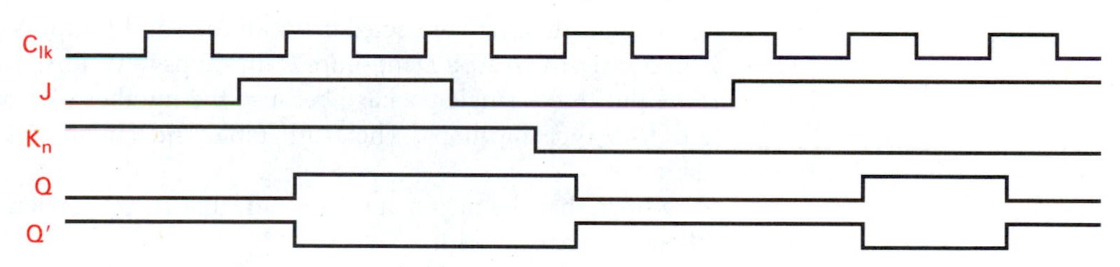

Figure 5.23
Example timing diagram for the 74xx109 J–K' flip-flop

inputs are also present on the device. **PRE*n*** and **CLEAR*n*** are active low signals that cause the flip-flop to set and reset independent of the clock. When PRE*n* is low, the flip-flop sets; that is, $Q = 1$ and $Q' = 0$. When CLEAR*n* is low, the flip-flop resets; that is, $Q = 0$ and $Q' = 1$. The timing diagram shows how the synchronous inputs J and K cause the Q output to change on the positive clock edge.

Table 5.2 is appropriate for both the J–K gated latch (which does not exist in IC form) and the clocked flip-flop. Because the gated J–K latch is useful only in showing the progression from the RS latch to the clocked flip-flop and only pulse or edge triggered J–K flip-flops are practical devices, we will limit future discussion to the clocked J–K flip-flops.

The three input variables, shown in Table 5.2 (J, K, and Q_t), produce eight combinations. Output variable Q_{t+1} responds to the inputs as indicated in the J–K truth table. Note that no undefined condition results from any input combination. The $JK = 11$ case produces a toggle condition ($JK' = 10$ for the 74xx109 device). By *toggle,* we mean that the output changes from one state to another; if a $Q = 0$, it changes to a 1; if $Q = 1$, it changes to a 0.

The difference between Q_t and Q_{t+1} is the time at which the Q (or Q') output is evaluated. Q_t indicates the value of the Q output before the arrival of the active edge of the clock. Q_{t+1} is the value of the same Q output after the arrival of the active clock edge. We are treating the same output as two variables separated in time. In the case of asynchronous preset and clear inputs, the values of Q_t and Q_{t+1} simply indicate the present state and next state of the Q output.

The logic symbol for a typical J–K flip-flop is shown in Figure 5.24. The > symbol indicates that the flip-flop is triggered on the positive or rising edge of the clock pulse.

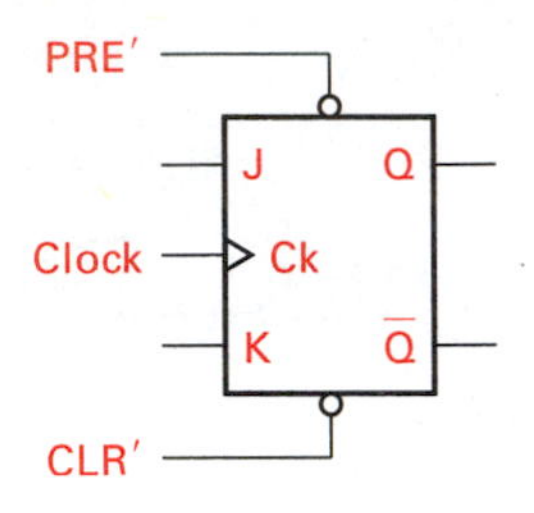

Figure 5.24
Logic symbol for a typical *JK* flip-flop

Clocked Toggle (T) Flip-Flop

A diagram showing the T flip-flop derivation from the J–K is illustrated in Figure 5.25.

Note the $J = K = 1$ case in Table 5.2. When the input condition $J = K = 1$ exists, the Q output toggles or changes state on each clock pulse. The T or toggle flip-flop's name is derived from the output action: if $T = 1$, the output will change state; if $T = 0$, then the output remains unchanged. Table 5.3 shows the T flip-flop excitation truth table.

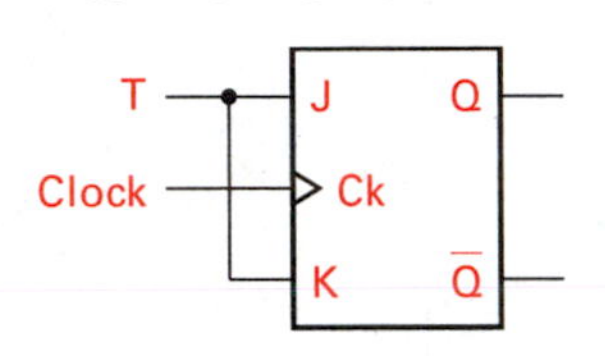

Figure 5.25
Toggle flip-flop logic diagram

TABLE 5.3
Truth table for a *T* flip-flop

Input		Output
T	Q_t	Q_{t+1}
0	0	0
0	1	1
1	0	1
1	1	0

Characteristic equation for the T flip-flop:

$$Q_{t+1} = T'Q_t + TQ_t'$$

Clocked D Flip-Flop

A data or D flip-flop is available in integrated circuit form in TTL, ECL, and CMOS devices. The 74xx374, for example, is a TTL IC containing eight positive edge triggered D flip-flops. The D flip-flop can also be derived from a J–K flip-flop as shown in Figure 5.26.

A single inverter connected between the J and K inputs of the J–K flip-flop is all that is necessary to create a D flip-flop. D flip-flops in IC form usually permit more flip-flops to be put into a single package because the number of inputs is reduced over the J–K and R–S type flip-flops. The truth table and characteristic equations are shown in Table 5.4.

Notice that the next-state Q output, that is, Q_{t+1}, changes to what the D input was when the clock arrived.

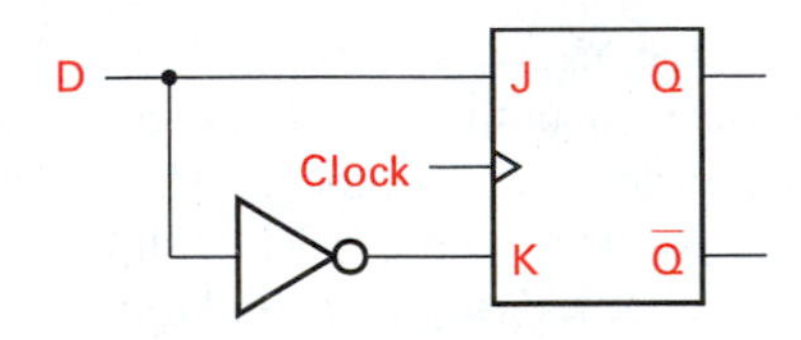

Figure 5.26
A *D* flip-flop derived from a *J*–*K* flip-flop

TABLE 5.4
Truth table for a *D* flip-flop

D	Q_t	Q_{t+1}
0	0	0
0	1	0
1	0	1
1	1	1

Characteristic equation: $Q_{t+1} = D$

TABLE 5.5
Summary of flip-flop excitation equations

Flip-flop type	Characteristic equations
RS	$Q_{t+1} = S + R'Q_t$
JK	$Q_{t+1} = JQ'_t + K'Q_t$
D	$Q_{t+1} = D$
T	$Q_{t+1} = Q_tT' + Q_t'T$

A summary of the characteristic equations for the *J–K*, *R–S*, *T*, and *D* flip-flops is given in Table 5.5.

Master-Slave Flip-Flops

Master-slave flip-flops are constructed by connecting two flip-flops in a cascade, operating from a common clock input. The purpose of master-slave flip-flops is to protect a flip-flop's output from inadvertent changes caused by glitches on the input. Master-slave flip-flops are used in applications where glitches may be prevalent on inputs.

A master flip-flop passes its output to the slave flip-flop. Each flip-flop is driven by a common clock with the level at the second flip-flop inverted from that presented to the first.

Master-slave flip-flops can be constructed to behave as a *J–K*, *R–S*, *T*, or *D* flip-flop. The basic core is a *RS* latch, one for the master and another for the slave. Figure 5.27 shows a basic *J–K* master-slave flip-flop.

Figure 5.27
J–K master-slave flip-flop

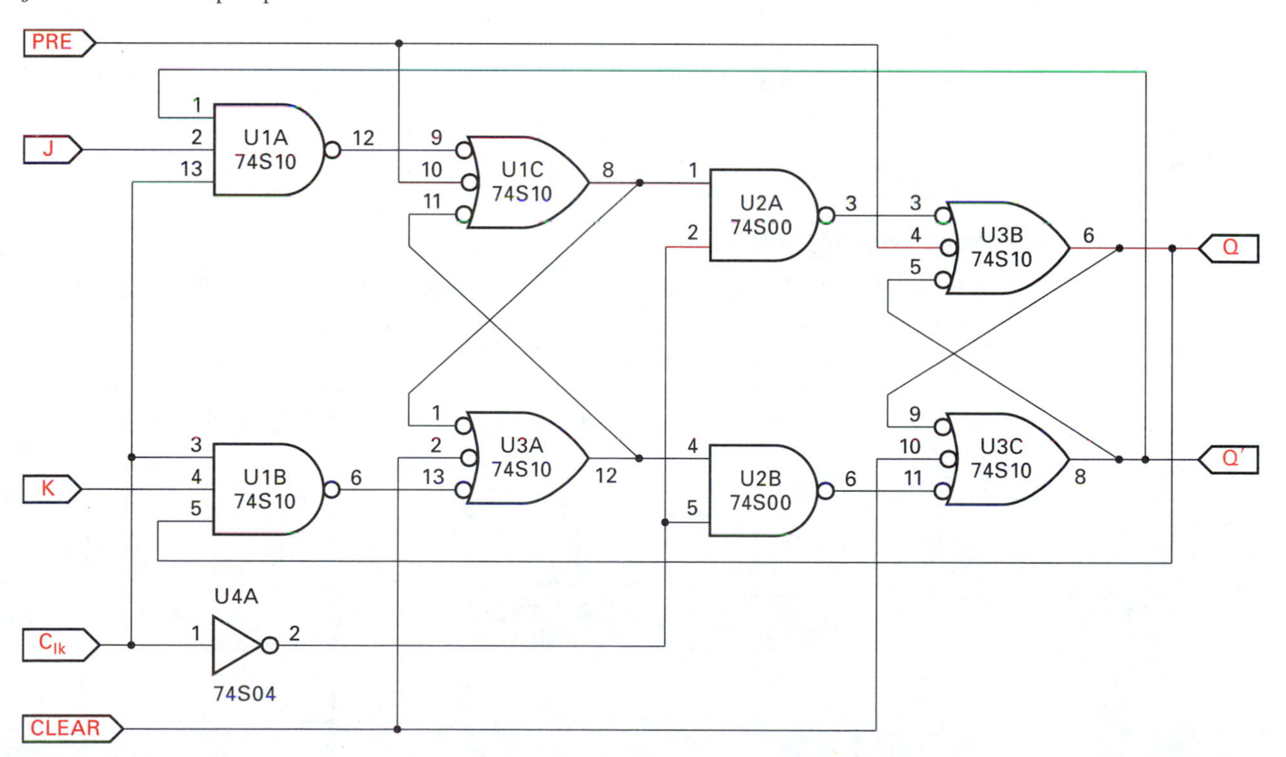

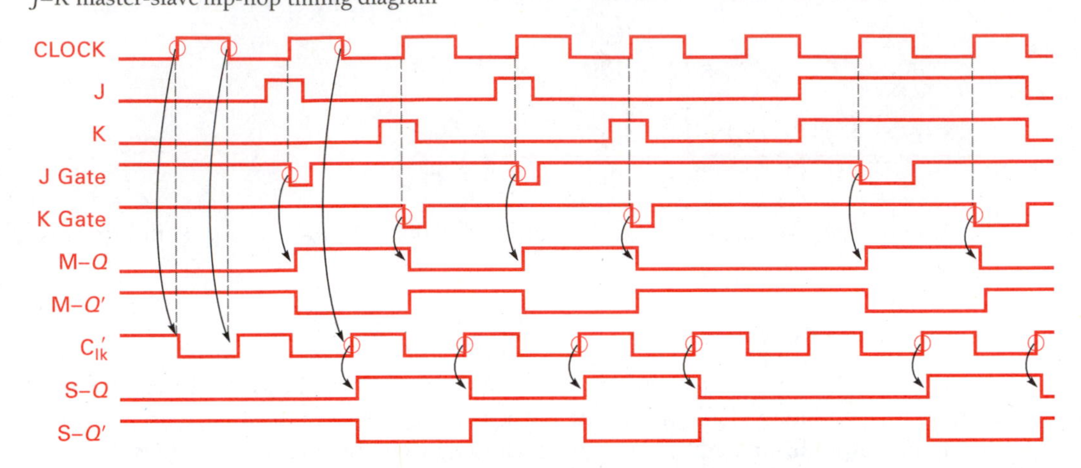

Figure 5.28
J–K master-slave flip-flop timing diagram

Figure 5.28 shows the timing diagram for the *J–K* master-slave flip-flop shown in Figure 5.27. The master flip-flop changes state when the clock input is positive, and the slave flip-flop changes state when the clock input is negative.

The isolation of the master from the slave is accomplished by triggering the two flip-flops at different levels of the input clock. Note that the master flip-flop changes when the input clock is positive, and because the slave clock is inverted, the slave flip-flop changes when the clock input is negative.

The *J* and *K* gate waveform shows the output of U1A and U1B. The arrows indicate the cause and result relationship between waveforms. For instance, arrows originate from the falling edge of the *J*-gate pulses that causes the M-*Q* (master Q) output to go high. Notice the delay between the M-*Q* and S-*Q* outputs. Data are loaded when the clock is positive and transferred to the slave when the clock is negative.

Master-slave flip-flops are not as widely used as they used to be; edge triggered flip-flops are replacing the master-slave in most new designs.

Edge Triggered Flip-Flops

Edge triggered flip-flops initiate *Q* output changes on the rising or falling edge of the clock input. By causing state changes to occur on the clock edges, many of the problems attributed to improper triggering of latches or gated latches are eliminated. Edge triggering requires an edge detector circuit that produces an output when a clock edge occurs. An edge detector can be created from an AND gate and an inverter, as shown in Figure 5.29.

Figure 5.29
Edge detector

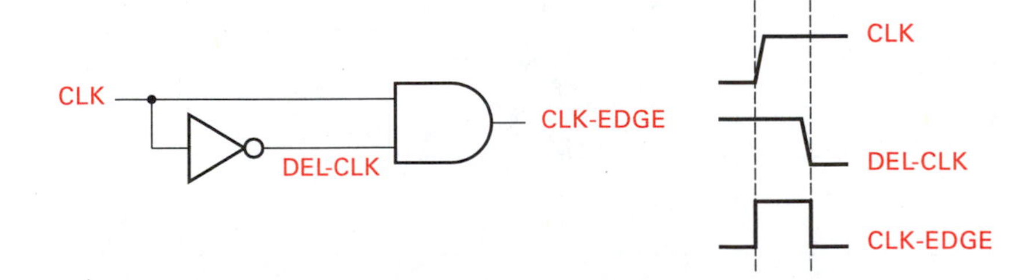

Figure 5.30
Positive edge triggered *D* flip-flop

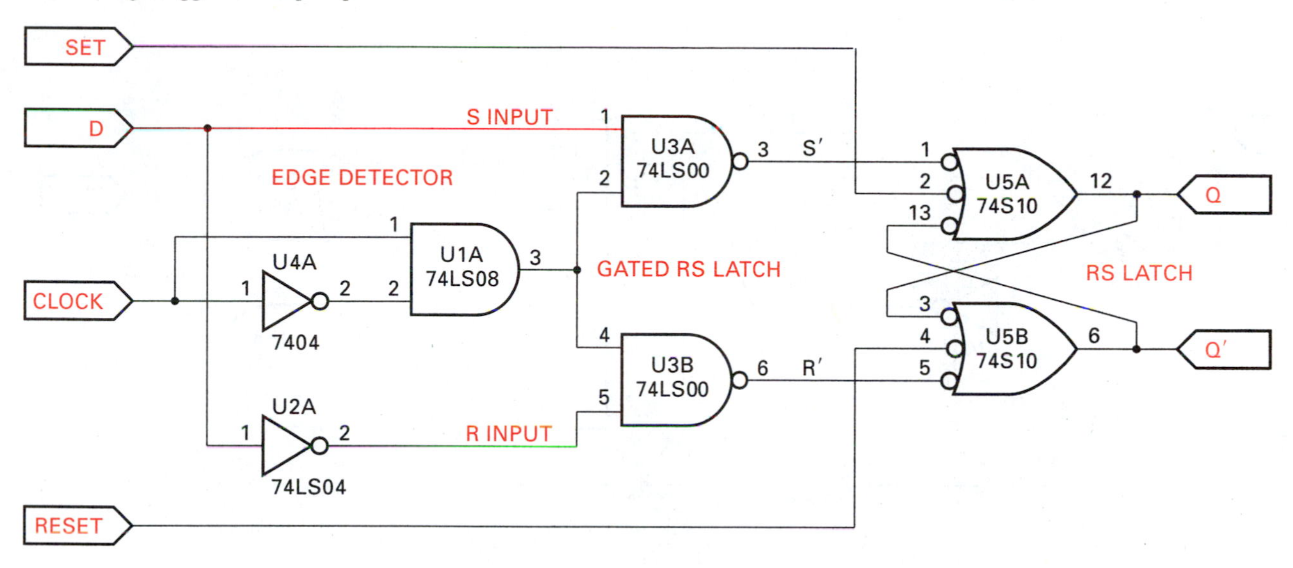

Figure 5.30 shows a positive edge triggered *D* flip-flop. The addition of the edge detector between the clock input and U3A and U3B converts the gated *RS* latch to a positive edge triggered flip-flop.

The timing diagram for the edge triggered flip-flop is shown in Figure 5.31. Note the CLK-DEL signal timing with respect to the clock. It is delayed by the propagation delay of U4A (inverter) in Figure 5.30. The CLK-EDGE signal is a pulse whose width is a function of the propagation delay of the inverter. The CLK-EDGE pulse is presented to the gated *RS* latch along with the *S* Input and *R* Input signals. S' and R' are produced as inputs to the output latch, causing *Q* to change states.

Edge triggered *J–K* flip-flops can be created from a basic gated *RS* latch and an edge detector when the *Q* and Q' outputs are fed back to the input gates as indicated in Figure 5.32.

The edge detector for the *J–K* flip-flop shown in Figure 5.32 produces pulses on the negative edge of the input clock. The negative edge triggered flip-flop timing diagram is shown in Figure 5.33. Notice when $JK = 10$ the flip-flop sets ($Q = 1$) on the negative

Figure 5.31
Timing diagram for positive edge triggered *D* flip-flop

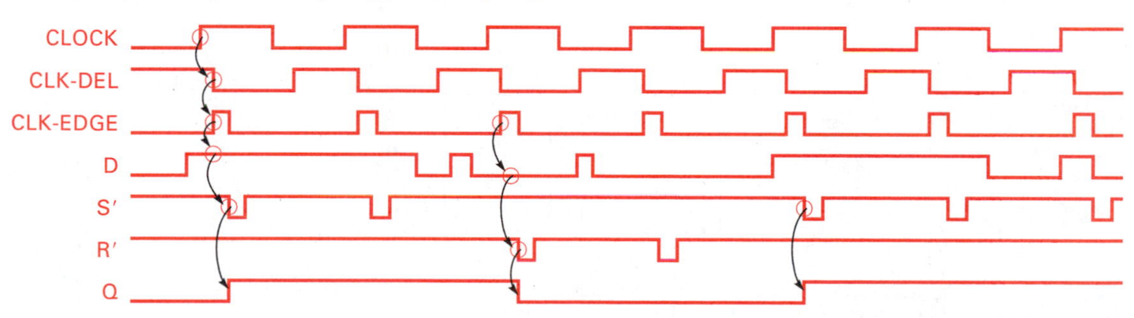

Figure 5.32
Negative edge triggered *J–K* flip-flop

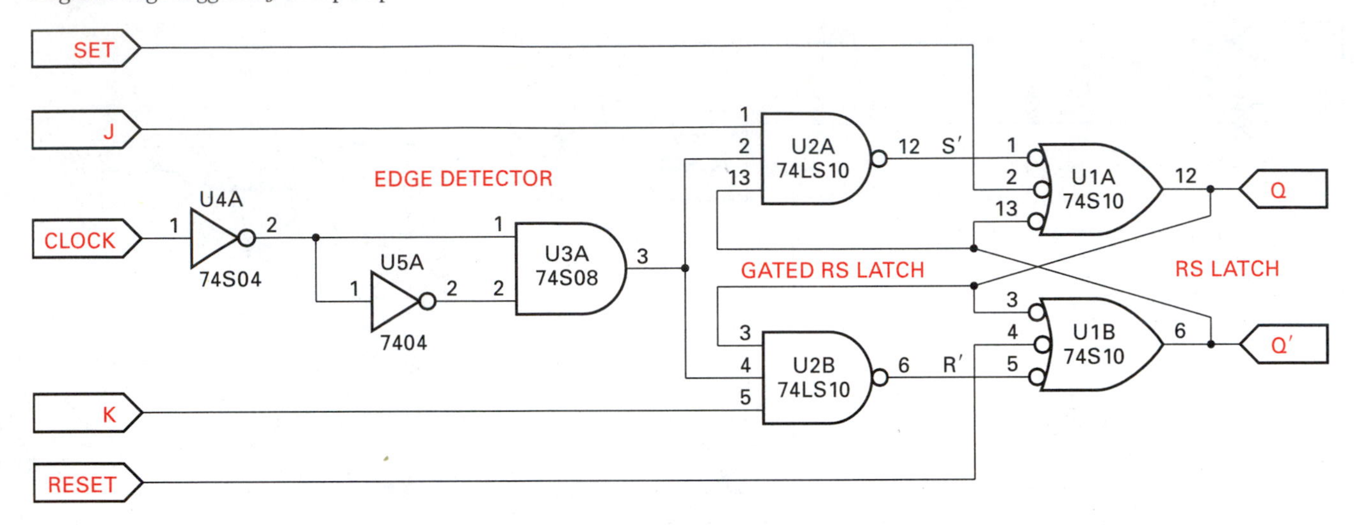

Figure 5.33
Negative edge triggered *J–K* flip-flop timing diagram

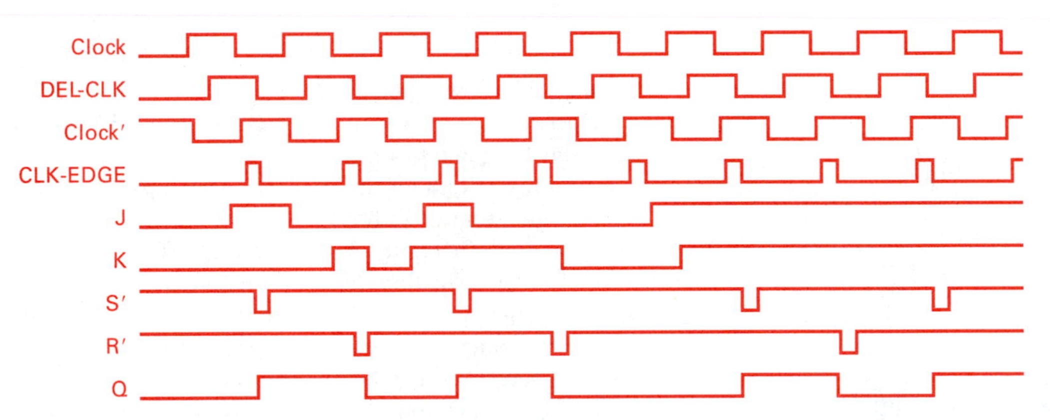

edge of the input clock pulse. When $JK = 01$ the flip-flop resets, and when $JK = 11$ the output toggles.

Pulse triggered master-slave flip-flops with data lockout can be constructed using the edge detector and two gated latches as shown in Figure 5.34. The data lockout version of the master-slave flip-flop includes the edge detector to prevent spurious changes in the output.

Figure 5.34
Pulse triggered master-slave flip-flop with data lockout

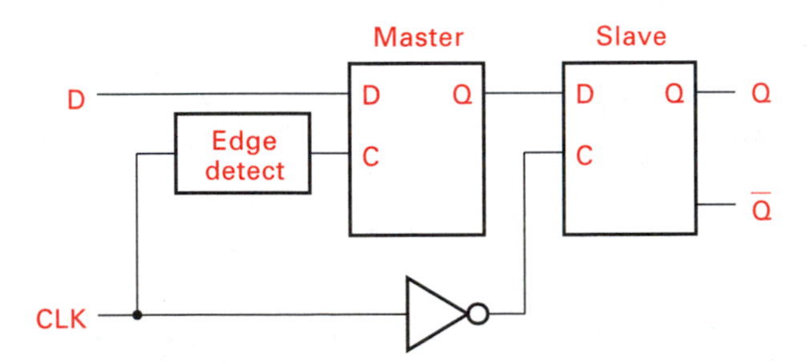

Figure 5.35
Individual flip-flop clock input symbols

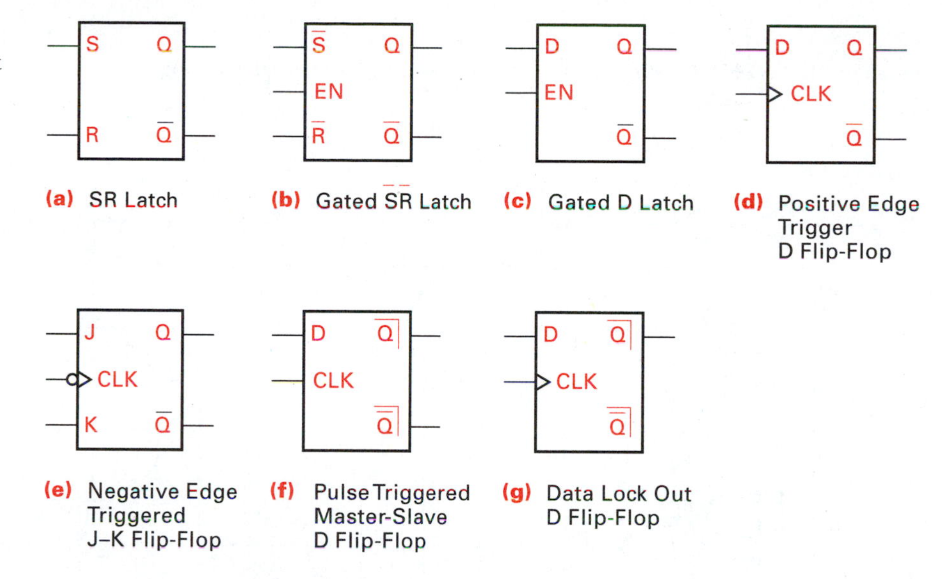

(a) SR Latch (b) Gated S̄R̄ Latch (c) Gated D Latch (d) Positive Edge Trigger D Flip-Flop

(e) Negative Edge Triggered J–K Flip-Flop (f) Pulse Triggered Master-Slave D Flip-Flop (g) Data Lock Out D Flip-Flop

Flip-Flop Clock Triggering Symbols

All four flip-flop types can have a positive or negative level or positive or negative edge triggering. Often a logic designer must evaluate the function of an integrated circuit that contains many flip-flops interconnected internally. The recognition of triggering modes is essential for that evaluation. Individual flip-flop clock triggering symbols are illustrated in Figure 5.35.

The IEEE standard symbols for illustrating flip-flops are somewhat different. The inverting bubble is replaced by an inverting triangle symbol on the input or output of the logic symbol. In either symbol, the > indicates a dynamic or edge triggered clock input. The absence of such a > is an indication of a level enabled device. The inverting bubble or triangle without the dynamic > indicates an active low level enable. Figure 5.36 shows several IEEE symbols for commercially available flip-flops.

Figure 5.37 shows the SN74AS95, a four-bit parallel-access shift register consisting of four *RS* flip-flops internally connected to form a register. The important issue here is

Figure 5.36
IEEE symbols for some commercial flip-flops

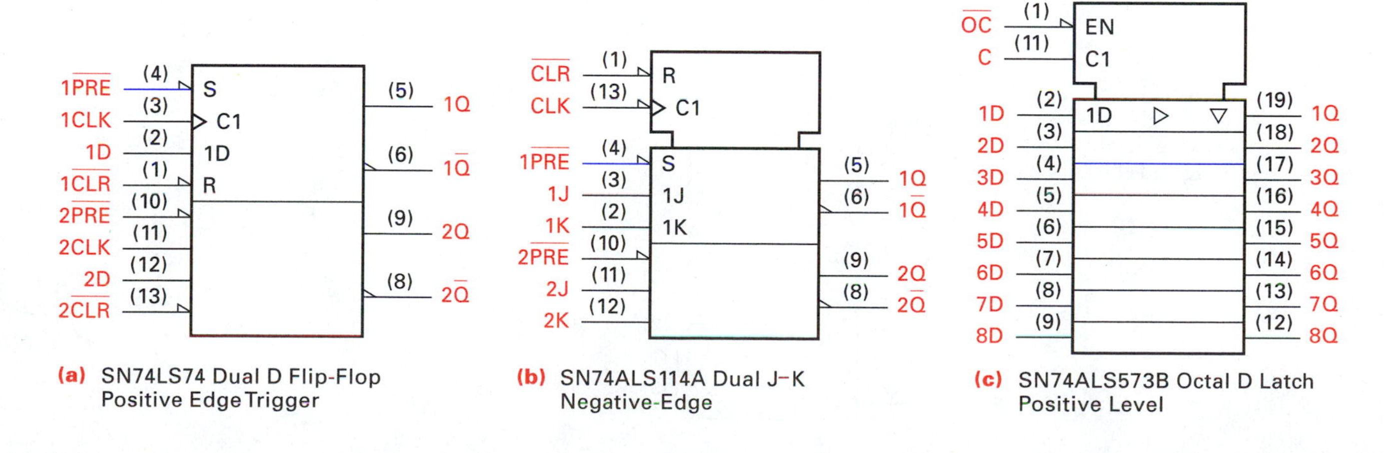

(a) SN74LS74 Dual D Flip-Flop Positive Edge Trigger (b) SN74ALS114A Dual J–K Negative-Edge (c) SN74ALS573B Octal D Latch Positive Level

Figure 5.37
Logic diagram for the SN74AS95 four-bit shift register, showing the internal clock logic

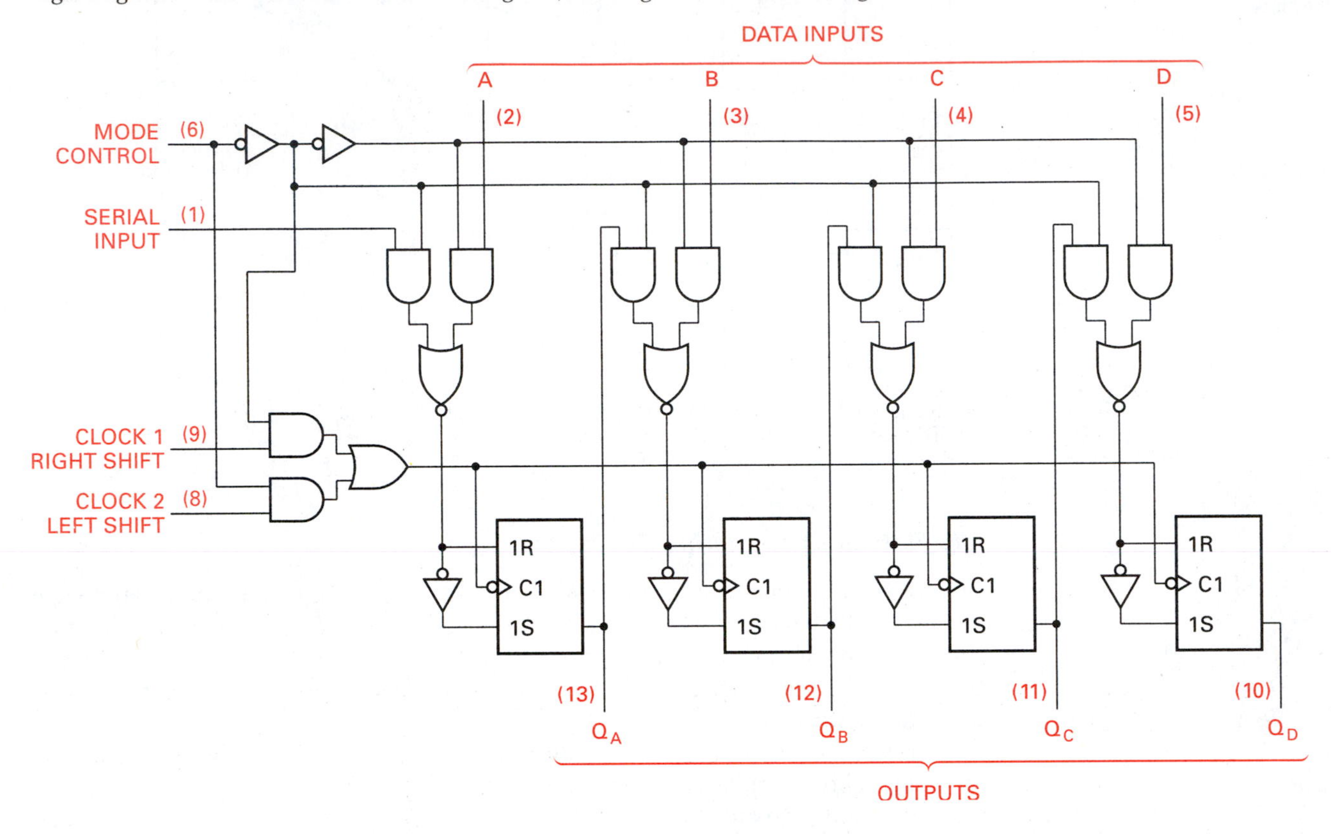

the recognition of clocking techniques rather than the operation of the shift register. Note that either of two external clocks can be presented to the flip-flops, depending on the mode control level. Also note that the flip-flops are clocked on the negative edge of either of the two input clocks. The logic diagram is useful when analysis of internal logic is required. However, it takes too much space to draw all the details when documenting a large digital system. The logic symbol for the same integrated circuit is shown in Figure 5.38. Notice both clock inputs show negative edge triggering.

Figure 5.39(a) illustrates the SN74ALS161, a four-bit binary counter. Again the pur-

Figure 5.38
Logic symbol for the SN74AS95 shift register showing the dual negative edge triggered clocks

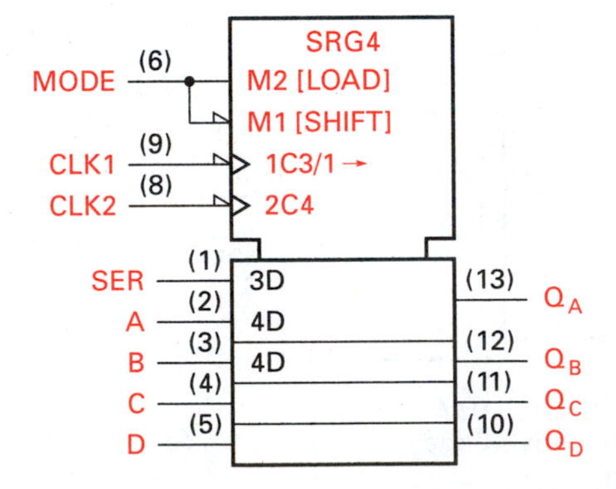

Figure 5.39
Logic diagram for the SN74ALS161 counter

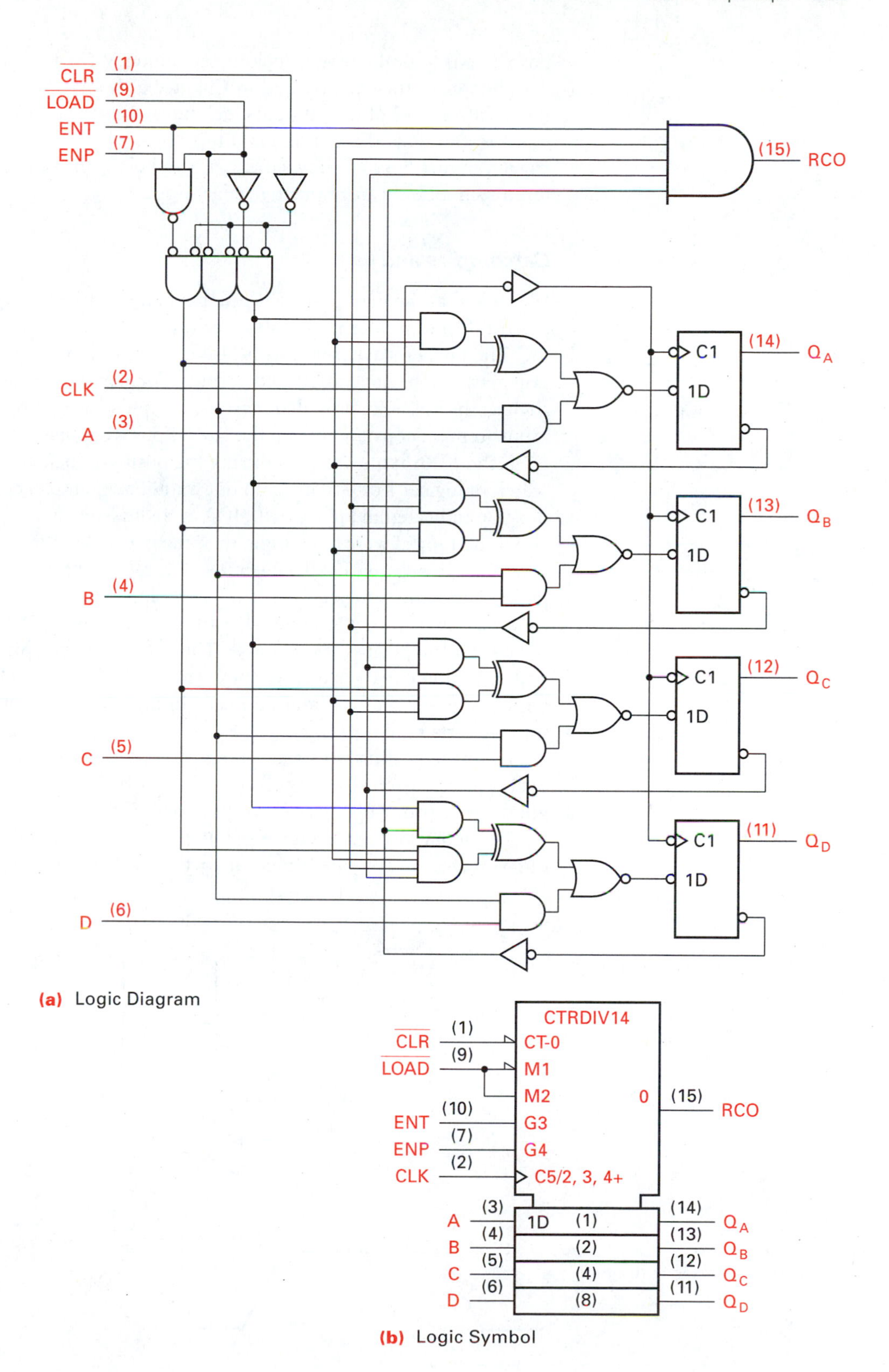

pose here is to evaluate the clocking technique. We will investigate counter analysis and design later in this chapter and in Chapter 6. Notice that each flip-flop is triggered on the negative edge of the internal clock pulse. However, an inverter is between the clock input to the IC and the individual flip-flops, which results in a positive edge external trigger. Figure 5.39(b) shows the logic symbol for the same counter. Note that the clock input symbol indicates a positive edge trigger.

Catching 1s and 0s

We looked at the master-slave flip-flop when we analyzed the operation of an edge triggered *D* flip-flop. The master-slave technique is commonly used to provide edge triggering. The master-slave principle is also useful for designing flip-flops that provide isolation between the data inputs and outputs. Some of the earlier master-slave *J–K* flip-flops, such as the SN7476, provided master-slave pulse triggering but stipulated that the data inputs remain constant during the clock's positive pulse width.

If the *J–K* inputs changed during the positive clock pulse width, the flip-flop could "catch" a logical 1 or 0 that we do not want. For example, consider the *J–K* master-slave flip-flop indicated in Figure 5.40. If Q_s were low during the positive portion of the clock and if the *J* input were to go high, then the master would set ($Q_m = 1$), causing the slave to set on the negative clock edge ($Q_s = 1$), catching a 1 on the end of the positive clock pulse. The same thing can happen for catching a 0. If $Q_s = 1$ and *K* became high after the clock pulse went high then Q_m would reset, causing Q_s to reset, thus catching a 0 on the end of the positive clock pulse. The solution to catching 1s and 0s is to keep the *J* and *K* inputs constant during the positive clock pulse. This, however, is not always possible, as glitches can and do occur in digital circuits in spite of the designer's best initial efforts to the contrary.

Figure 5.41 illustrates the timing of the master and slave flip-flops internal to the overall *J–K* master-slave. Figure 5.42 illustrates the problem of catching 1s and 0s when glitches occur during the positive clock pulse.

The actual voltage levels for points 1 through 4 in Figure 5.41 depend on the threshold voltages of the gates internal to the master-slave integrated circuit; ideally, the

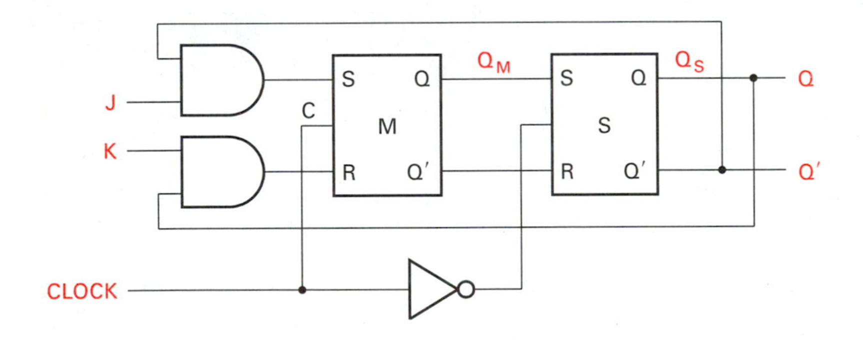

Figure 5.40
J–K master-slave flip-flop logic diagram

1. Slave is isolated from master.
2. Master data inputs are enabled.
3. Master data inputs are disabled.
4. Data are transferred from master to slave.

Figure 5.41
J–K master-slave clock timing

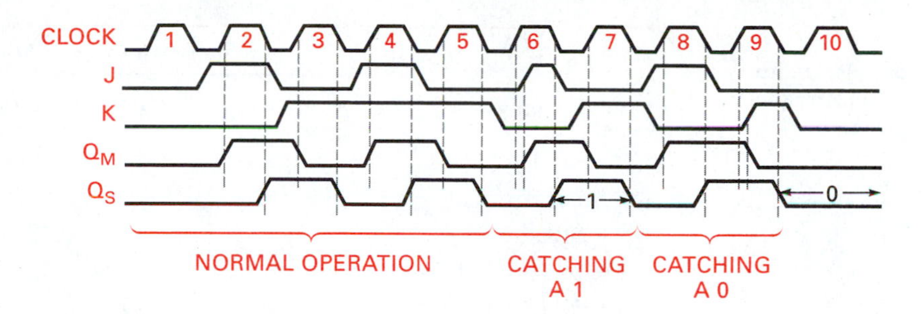

Figure 5.42
Master-slave problem of catching 1s and 0s

threshold voltages at points 1 and 2 will be the same and the threshold voltages for points 3 and 4 will be the same. The time interval between points 2 and 3 are considered to be the **pulse width.** During this time the master-slave flip-flop is vulnerable to changes in the *J* and *K* inputs.

Companies like Texas Instruments, aware of the problem of catching 1s and 0s, designed master-slave flip-flops with data lockout. The SN74xx111 is a *J–K* master-slave flip-flop with data lockout. The *J* and *K* inputs are enabled only during a 20 ns maximum setup time plus a 5 ns hold time. During this 25 ns time the inputs must remain stable, then they can change during the positive clock without presenting catching difficulties. The master-slave data lockout feature provides a flip-flop much less sensitive to input glitches that cause unwanted state transitions. For this reason these flip-flop types are used in circumstances where clock **skew** (variations in delay between clocks presented to one flip-flop and another) is considerable.

5.3 FLIP-FLOP TIMING SPECIFICATIONS

In Section 4.2 of Chapter 4, we introduced TTL integrated circuit specifications, such as the 0 and 1 voltage and current levels, propagation delay, rise time, fall time, and other typical logic parameters. A quick referral back to those basic definitions may be helpful. We will now investigate the timing properties of flip-flops. The integrated circuit flip-flops and MSI/LSI functions containing flip-flops specify timing parameters that the logic designer must obey to develop a useful working digital system.

5.3.1 Clock Parameters, Pulse Width, and Skew

Figure 5.43 shows a typical flip-flop clock waveform. Although most timing diagrams illustrate a flip-flop clock waveform as a square wave, any sort of periodic or aperiodic rectangular waveform that meets the logic level and timing requirements may be used.

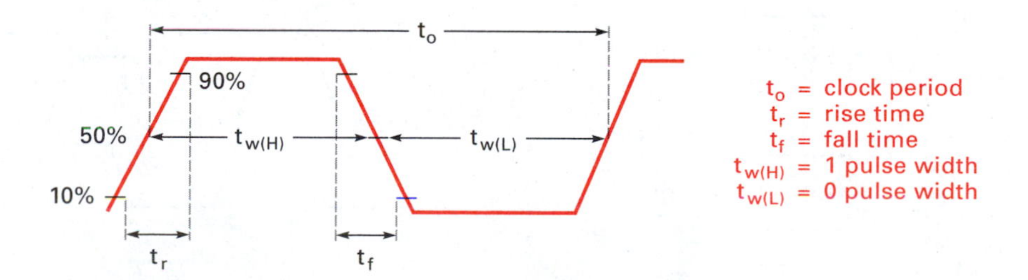

Figure 5.43
Flip-flop clock waveform

TABLE 5.6
Comparison of pulse width and maximum clock frequency for several logic families

Logic family	Min. pulse width Low (nS)	Min. pulse width High (nS)	Min.-max. clock (M Hz)
74AS109 (TTL)	5.5	4	105
74ALS109 (TTL)	16.5	16.5	34
74LS109 (TTL)	25	25	25
MC1670 (MECL III)	None given		350
74HC11109 (CMOS)	5	5	100
74HCT11109 (CMOS)	5	5	100

Rise and fall times are not typically listed in the data manuals for digital integrated circuits. The assumption is that the output of a given logic device will be producing the clock for a device in the same family. If the clock is generated by something other than a logic device from the same family, consideration of rise and fall times may be required. If clock rise and fall times are too slow then the use of a Schmitt trigger gate may be necessary. Chapter 10 discusses the Schmitt trigger circuit in detail. The faster the logic family, the more critical clock rise and fall times become.

Pulse width and maximum frequency are specified clock parameters independent of logic family. For example, the TTL SN74AS109, a dual J–K' positive edge triggered flip-flop, establishes the maximum clock frequency (from a given edge to the next edge of a waveform) at a minimum of 105 MHz. The same device has a specified clock pulse width of 4 ns high and 5.5 ns low. Table 5.6 compares several TTL, CMOS, and ECL family clock frequencies and pulse width specifications.

Skew refers to the phase shift between rectangular clock waveforms presented in parallel to multiple synchronous sequential elements that require the same timing. Consider Figure 5.44, where a single clock generator is producing a waveform for distribution to two parts of a digital system. Both clock waveforms should be occurring at the same time; however, due to the different logic paths, one clock lags behind the other. The clock distribution logic paths are represented by 1 and 2. The time delay between the two clock pulses is called *clock skew.*

At low clock speeds skew rarely is troublesome. It is when the flip-flops are operating at or close to maximum clock frequency that skew presents problems.

For example, if a sequential circuit is constructed using two 74AS109 IC flip-flops and one clock input is derived from a different delay path than the other, then the delay could prevent the circuit from working. Consider the circuit illustrated in Figure 5.45

Figure 5.44
Clock skew

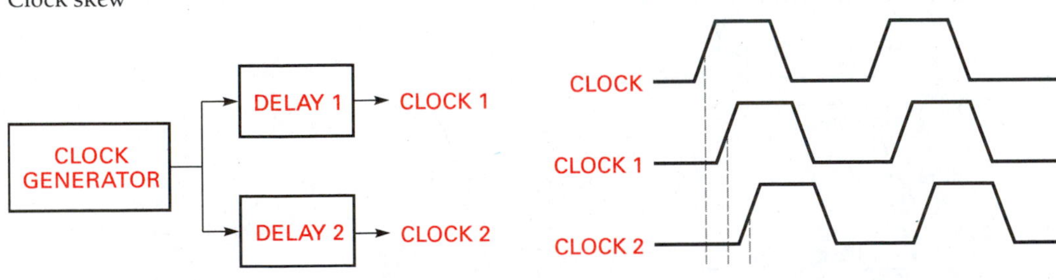

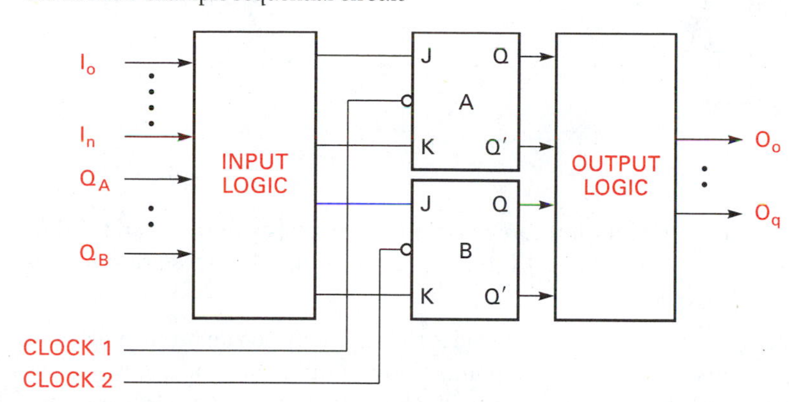

Figure 5.45
Clock skew example sequential circuit

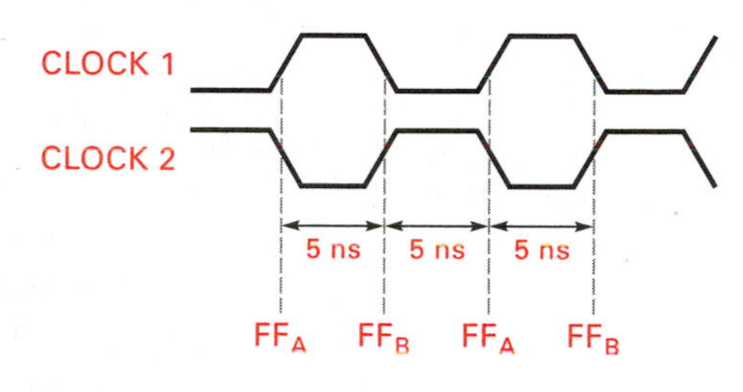

Figure 5.46
Skewed clock inputs for the circuit in Figure 5.45

and the timing diagram in Figure 5.46. Assume that the delay between clocks 1 and 2 is 5 ns.

At a frequency of 100 MHz a positive clock edge arrives at the SN74AS109s every 10 ns. The 5 ns delay between clock 1 and clock 2 is shown in Figure 5.46. Because both flip-flops should be triggered at the same time, the circuit is rendered inoperative by the clock skew. It does not matter if we are discussing two flip-flops or two complete digital modules, clock skew can still create timing problems. In this example both flip-flops A and B are required to trigger and therefore change state at the same time. If they do not, then unwanted state transitions will occur. We will investigate state transitions in more detail later, but for now we need to understand that, for a sequential circuit or sequential modules to work, they must conform to trigger timing specifications.

5.3.2 Flip-Flop Timing, Setup, Hold, and Delay

Three additional timing parameters relative to flip-flops are of importance to our understanding of flip-flop operation: setup time, hold time, and propagation delay time.

Setup time Setup time refers to the time that the excitation inputs to a flip-flop device must remain stable prior to the arrival of the clock edge that triggers the device.

Hold time Hold time is the time that the excitation inputs must remain stable after the clock edge has occurred.

Figure 5.47 illustrates the setup and hold time definitions relative to the clock trigger pulse. In this case we are considering a negative edge trigger device.

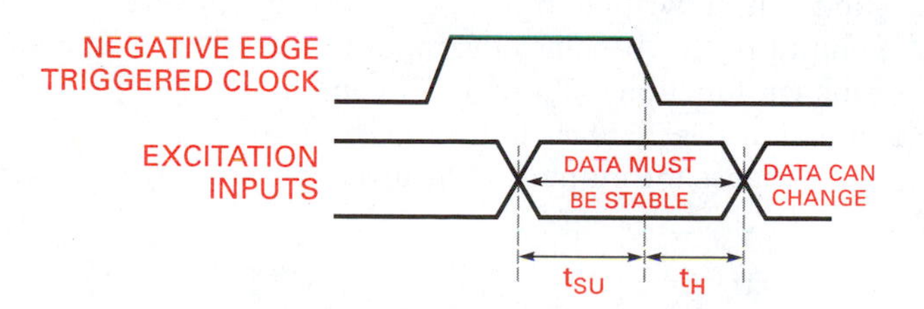

Figure 5.47
Setup and hold time definitions

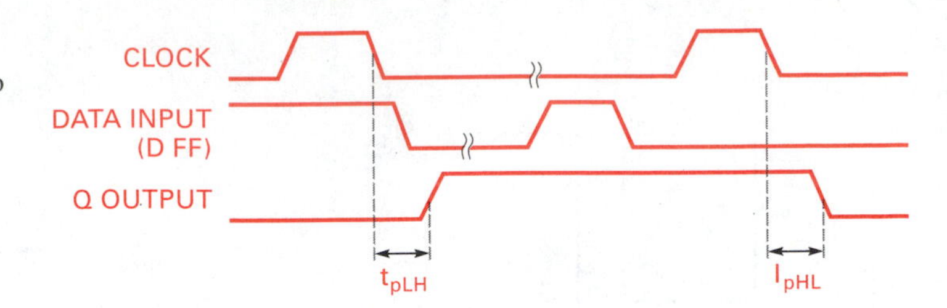

Figure 5.48
Propagation delay from clock to *Q* output

The excitation input to the flip-flop can be a single variable, as in the *D* and *T* flip-flops, or two variables, as in the *SR* and *JK* devices. Setup and hold times are synchronous input timing parameters.

Propagation delay Propagation delay time is the time between an input signal and the output signal measured at the 50% points. Several different propagation delays may be specified, depending on the device and which propagation path is being considered. Typically, propagation delays from the clock to the *Q* and *Q*′ outputs and the asynchronous inputs, CLEAR, and PRE to *Q* and *Q*′ are specified. Propagation delay times for multiple clock devices or where additional data or control inputs are used are included. Figure 5.48 illustrates a typical propagation delay timing relationship.

Two types of propagation delay times are the delay as the output switches from low to high, t_{pLH}; and the delay as the output switches from high to low, t_{pHL}. When considering timing relationships the designer must use the worst-case delays, usually the longest. For example, t_{pHL} from the clock to the *Q*-output for the SN74LS78, a dual *J–K* flip-flop IC, is a maximum of 20 ns. The longest delay time would be considered when calculating propagation delay paths and the flip-flop was to be included in the path.

5.3.3 Flip-Flop Metastability

Under normal conditions a flip-flop has two states. When the flip-flop setup and hold timing parameters are met, the *Q* output can change from a 0 to a 1 and back without causing problems. If, however, the setup and hold timing parameters are not met, a third condition may exist, a metastable state. A **metastable** state exists halfway between a logical 0 and a logical 1. It is an undefined state, because the voltage levels do not conform to either a 0 or a 1. This undefined condition (in terms of logic levels) causes problems in the operation of a digital system.

Figure 5.49
Inverter

V_i — $V_o = f(V_i)$

Flip-flops contain feedback as we have discussed earlier. This feedback is what sets the stage for a metastable state. Digital circuits exhibit voltage gain. The characteristic of voltage gain is what allows the circuit to switch from one state to another. Consider the simple inverter in Figure 5.49. The output voltage is a function of the input voltage and can be written as $V_o = F(V_i)$. If two inverters were cross-connected so that the output of one becomes the input to the other, as shown in Figure 5.50, we have two voltage functions: $V_{o1} = F(V_{i1})$ and $V_{o2} = F(V_{i2})$. We assume that the voltage gain function *F* is identical in both inverters.

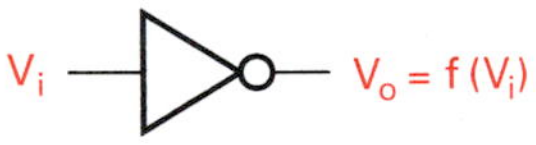

Figure 5.50
Cross-connected inverters

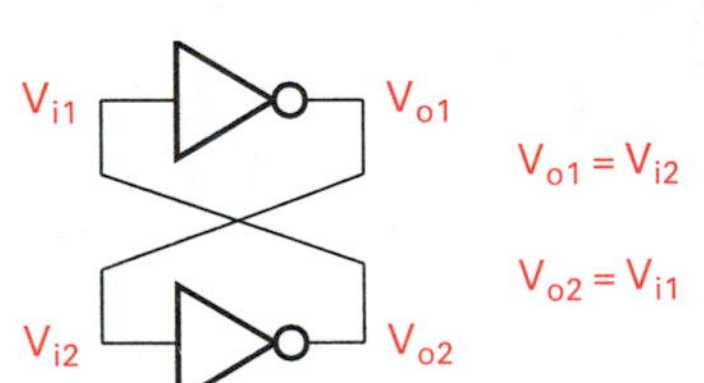

Because the outputs of the inverters are fed back to the input we can write

$$V_{o1} = V_{i2}$$
$$V_{o2} = V_{i1}$$

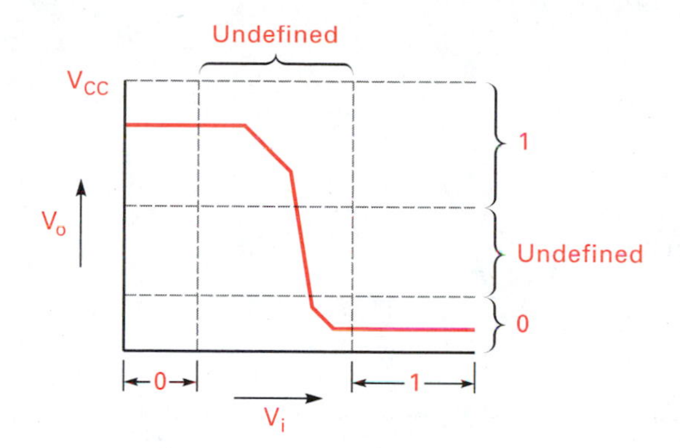

Figure 5.51
Voltage transfer characteristics of an inverting gate

As the input voltage switches from 0 to 1 the output switches from a 1 to a 0. There is a time when the output voltage is undefined as it makes the transition from either a 1 to a 0 or from a 0 to a 1.

Figure 5.51 is the result if we plot the input-output transfer behavior of a single inverter. Note that, as V_i increases from 0 volts toward V_{CC}, V_o stays within the range of permissible logical 1 voltages until about the middle of the graph. V_o then drops toward a logical 0 voltage level and remains there. This, called a *voltage transfer curve,* describes the switching behavior of the inverter. The same sort of transfer curve behavior is seen when both inputs of a NAND gate switch simultaneously.

The behavior of two inverters that are cross-connected as shown in Figure 5.50 can also be plotted as shown in Figure 5.52. The vertical axis shows V_{o1} and V_{i2}, and the horizontal axis shows V_{i1} and V_{o2}. The transfer characteristics of both inverters are plotted on the same graph, with one axis rotated by 90°.

When the two inverters (or any inverting gate) are cross-connected the result is a third metastable state. We see from the graph in Figure 5.52 that the two voltage transfer curves intersect. The metastable state exists at this intersection. If the right input conditions were to occur, the flip-flop would enter the metastable state. The length of time a flip-flop will stay in a metastable state is indeterminate, depending on the input conditions. A stray glitch occurring at the wrong time can put a flip-flop into a metastable state, which is in between a logical 0 and a logical 1.

The principle of metastability can be modeled as shown in Figure 5.53. Either side of the "hill" is considered to be a stable state. We can assign a logical 0 to the left side and a logical 1 to the right side. The top of the hill is considered to be the metastable state. To cause a flip-flop to change from a 0 to a 1, a force is applied (excitation inputs) that causes the output to switch. If the excitation inputs are of insufficient strength (which usually applies to the pulse width), then a complete output transition is not made and the flip-flop ends up in the metastable condition for a period of time.

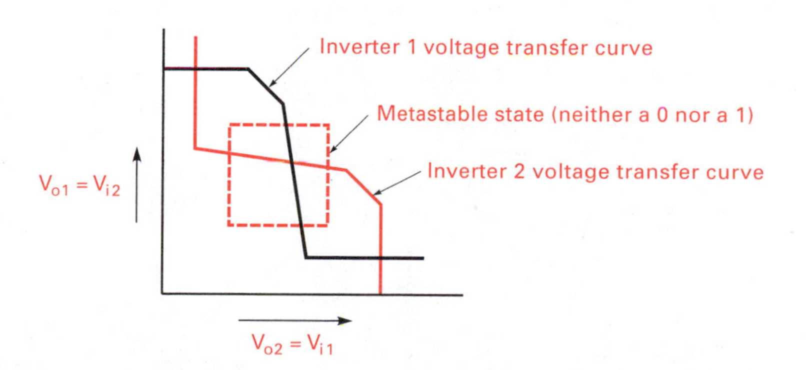

Figure 5.52
Voltage transfer characteristics of cross coupled inverting gates

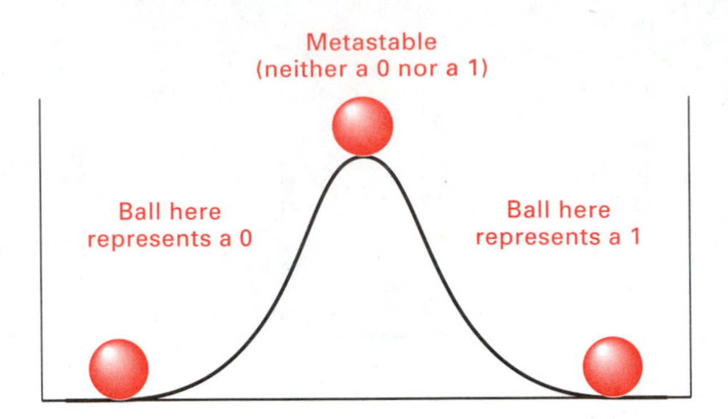

Figure 5.53
Metastability modeled with a hill and ball

The following are some characteristics, causes, and effects of flip-flop metastability:

1. Metastable states exist halfway between logical 0 and 1 states.
2. Flip-flops can go into a metastable state when the setup and hold times and clock pulse width timing requirements are violated.
3. External inputs to a digital system containing flip-flops often occur asynchronously with respect to system signals (including clock pulses). This means that external signals can occur in a random fashion. Because of this, the setup and hold times cannot always be guaranteed.
4. Metastable states are to be avoided.

Consider the circuit shown in Figure 5.54. A combinational gate is providing the input to a D flip-flop. The inputs to the gate can occur at any time. The flip-flop is triggered according to the setup and hold time constrains. If the inputs to the flip-flop occur as indicated in Figure 5.55, a short or "runt" pulse is created. This runt pulse may have sufficient energy to cause the flip-flop to try to set (leave the 0 state) but not enough to finish the job, leaving the flip-flop in a metastable state.

The runt pulse is created by the random behavior of inputs A and B. The pulse attempted to set the flip-flop, but lacked sufficient amplitude or pulse width and put the flip-flop into the metastable state instead.

Reducing the probability of flip-flop metastability is accomplished by the circuit shown in Figure 5.56. The first flip-flop is interfaced directly to the asynchronous excitation inputs. The output of the first flip-flop is connected directly to the second flip-

Figure 5.54
Logic circuit illustrating metastability

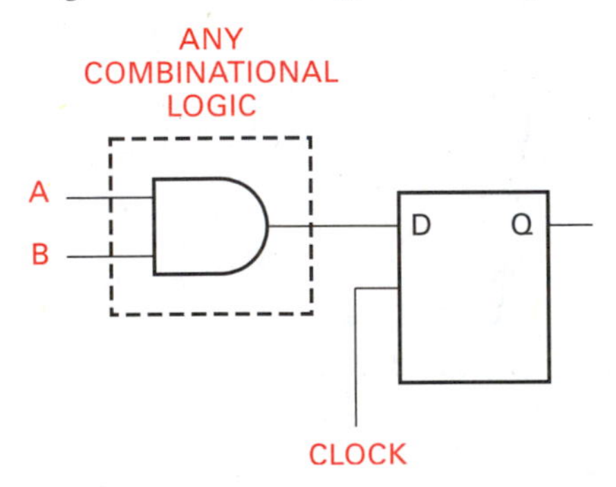

Figure 5.55
Timing diagram for Figure 5.54

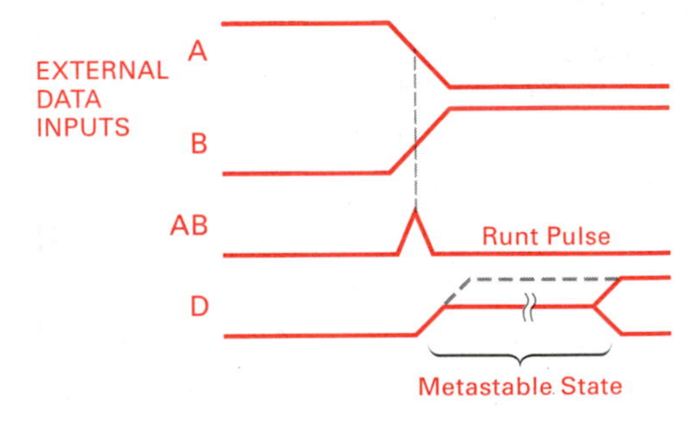

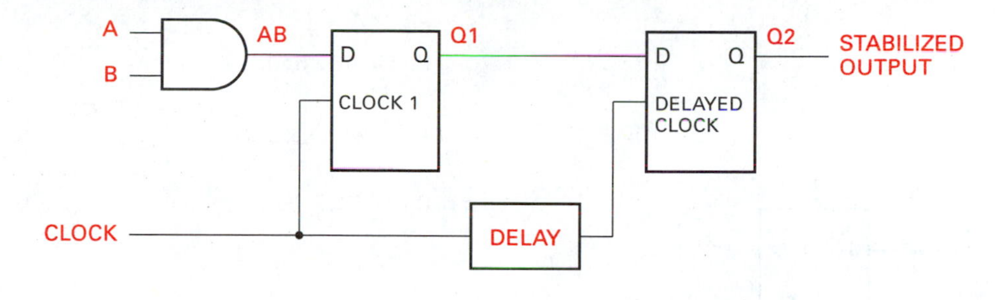

Figure 5.56
Circuit to reduce metastability problem

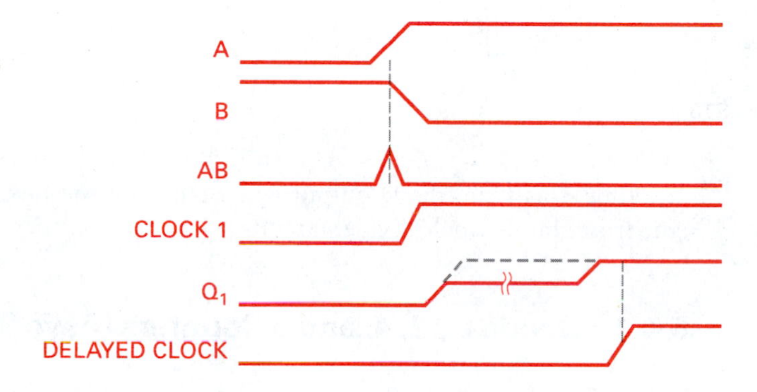

Figure 5.57
Timing diagram for Figure 5.56

flop. The clock of the second flip-flop is delayed from the first. The reason for the delay is to allow the first flip-flop to leave the metastable state before clocking the second. The second flip-flop output is now synchronized with the rest of the system. Figure 5.57 shows a typical timing diagram where the delayed clocking of flip-flop 2 eliminated the metastability state.

The propagation delay (clock to Q output) through flip-flop 1 may be sufficient to eliminate the need for a clock delay. When the clock is applied to both flip-flops simultaneously, Q_1 may go into a metastable state, but because Q_1 is not available at D_2 on the clock edge, Q_2 should not become metastable. The assumption is that Q_1 has left the metastable state prior to the arrival of the next clock edge. When setup and hold time requirements are met, metastability is not a problem. If they are not met, then the flip-flop may enter the metastable state for an indeterminate amount of time. Because of the unpredictability of the metastable state duration, no universal solution is currently available. We shall visit metastability again in Chapter 8, when we discuss asynchronous circuit design.

5.4 SIMPLE COUNTERS

Counting is a fundamental function of digital circuits. A digital counter consists of a collection of flip-flops that change state (set or reset) in a prescribed sequence. We will study several types of counters in this section. Flip-flops are commonly used to design counters, and the most straightforward counter is a ripple divider. A J–K flip-flop can be converted to a T or toggle flip-flop by connecting together the J and K inputs. The truth table for the T flip-flops was given in Table 5.3. From the truth table we note that,

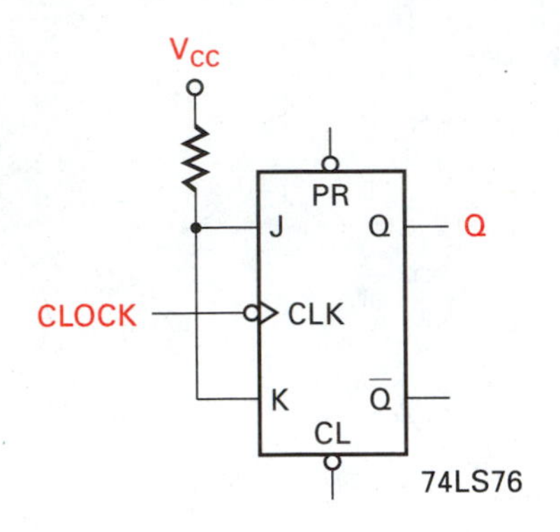

Figure 5.58
Divide-by-2 counter

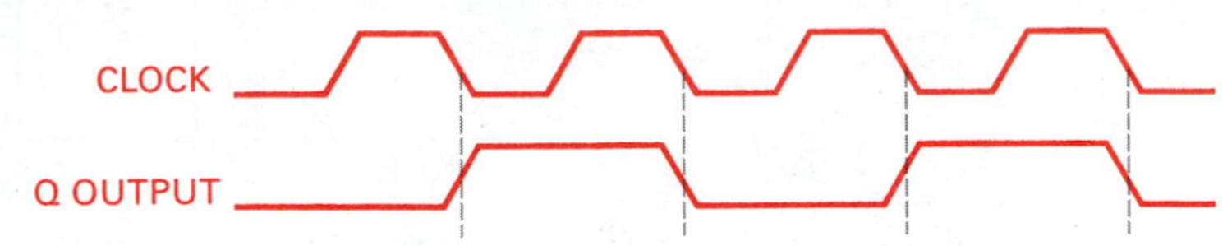

Figure 5.59
Divide-by-2 timing waveforms

any time T is high, the Q output will change on the next clock. We can use this to construct simple ripple divider circuits.

5.4.1 Divide by 2, 4, and 8 Counters (Asynchronous)

The division by 2, 4, and 8 counters indicated in this section are said to be asynchronous because no common synchronizing clock input is applied to all the flip-flops. The clock input is actually used as a data input in these applications. A divide-by-2 circuit divides the input clock frequency by 2 and requires a single flip-flop connected as shown in Figure 5.58. Note that the J and K inputs are connected together and pulled-up to V_{CC}. This forces both excitation inputs high, causing the flip-flop to toggle on every clock pulse. If a negative edge trigger IC were used, then the resulting timing diagram would be as shown in Figure 5.59. Note that the input frequency is divided by 2, or in other words, the period of the output is twice that of the clock input.

A divide-by-4 counter requires two similarly connected toggle flip-flops, as shown in Figure 5.60 with the corresponding timing diagram illustrated in Figure 5.61. The divide-by-4 counter divides the input clock frequency by 4.

Figure 5.60
Divide-by-4 counter

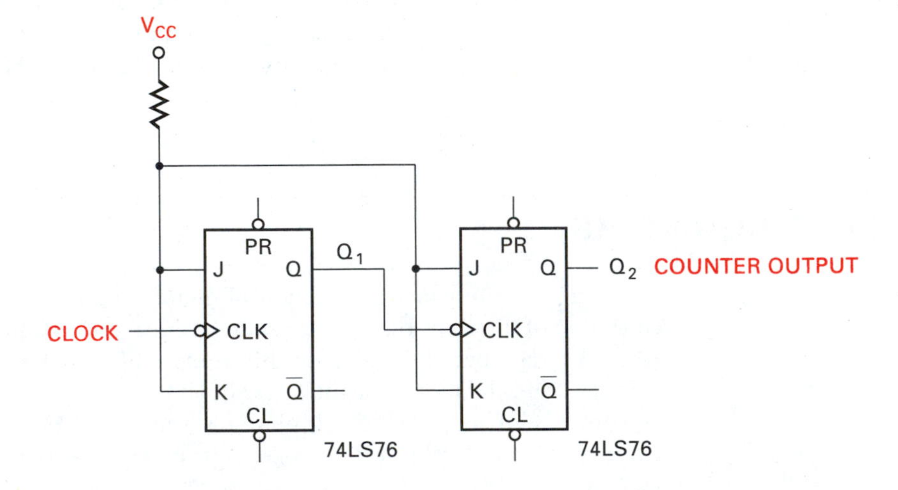

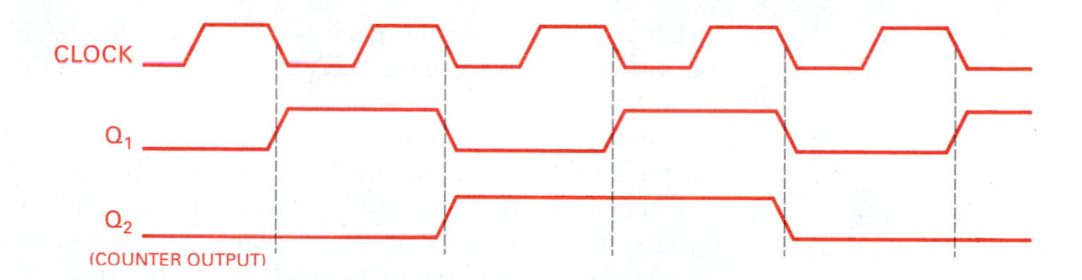

Figure 5.61
Divide-by-4 counter timing diagram

Both *J–K* flip-flops are connected to form a toggle flip-flop, with the *Q* output of the first flip-flop providing the input to the second. The first flip-flop divides the input frequency by 2. The second flip-flop divides the *Q* output from the first flip-flop by 2, thereby dividing the input frequency by 4.

A divide-by-8 ripple counter requires three flip-flops, each connected as a toggle flip-flop. The *Q* output of the first becomes the clock input to the second, and the *Q* output of the second becomes the clock input to the third. Each flip-flop divides its input by 2 resulting in an overall division by 8.

This process may be continued to include as many flip-flops as needed. Notice that each flip-flop added to such a counting scheme provides an additional divide-by-2 factor. This is the power-of-2 progression we saw earlier when dealing with base-2 numbers. The logic diagram for a divide-by-8 counter is shown in Figure 5.62, with its corresponding timing diagram shown in Figure 5.63.

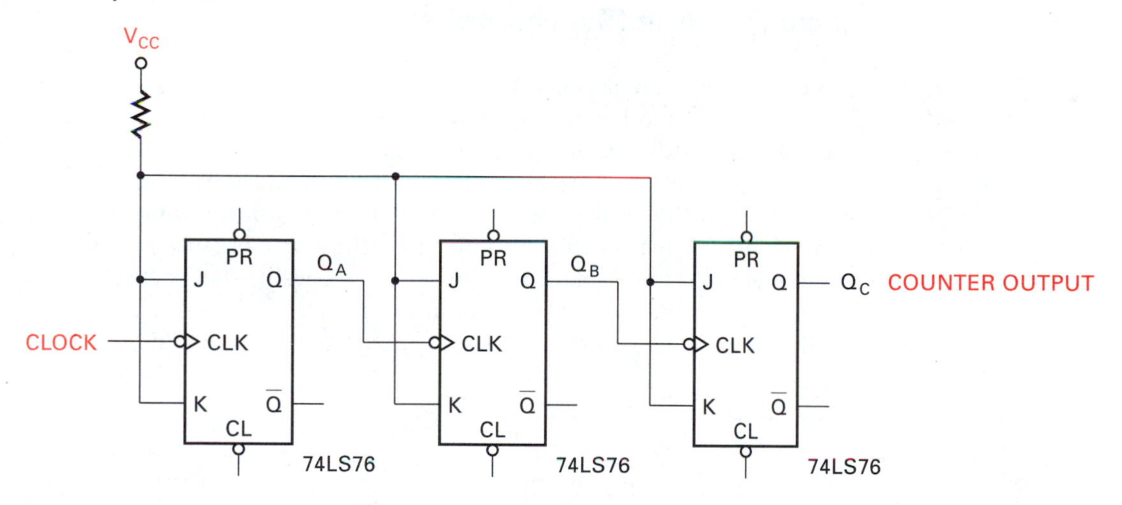

Figure 5.62
Divide-by-8 counter

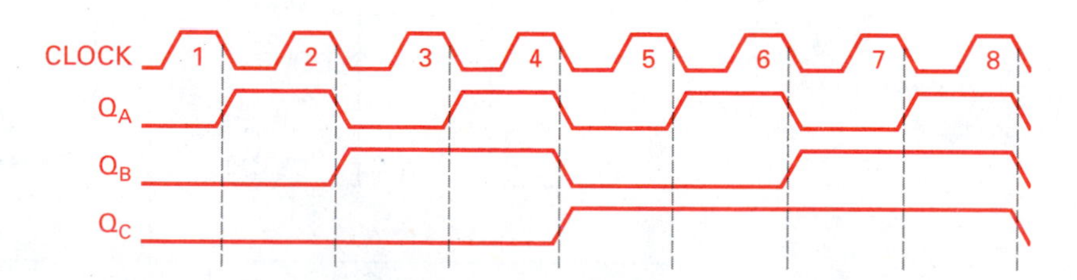

Figure 5.63
Divide-by-8 counter timing diagram

Each of the three simple divider counters presented is called an *asynchronous counter* because the output of one flip-flop provides the clock for the next flip-flop. The counters are said to be *rippled* because the input ripples from one flip-flop to another. A divide-by-2 counter consists of one flip-flop and has two states: $Q = 0$, $Q = 1$. The divide-by-4 counter has four states, and the divide-by-8 counter has eight states.

Counters are often used to keep track of a number of events. When all of the flip-flop outputs are available, these are called **modulo-*n* counters** where *n* represents the terminal count. For example, a modulo-4 counter would have two outputs representing all four states. A modulo-8 counter has eight states encoded by three flip-flop outputs. Let the flip-flops of a modulo-8 counter be labeled Q_A, Q_B, Q_C, then the states are $Q_CQ_BQ_A = 000, 001, 010, 011, 100, 101, 110, 111$. A divide-by-8 and modulo-8 counter have the same number of flip-flops connected in identical ways but different outputs. The divide-by-8 counter has a single output that is active on the eighth input pulse, and the modulo-8 counter has three outputs, one from each flip-flop.

The relationship between the number of flip-flops (state variables) and the number of states is expressed as

$$n = 2^{\text{number of flip-flops}}$$

The modulo-8 counter (divide by 8) timing diagram shown in Figure 5.63 has eight states. Each state is identified by a unique counter value (combination of Q output values from flip-flops C through A; C is the MSB). Each flip-flop output (state variable) can be considered a Boolean variable. A timing diagram, such as in Figure 5.63, is a time-dependent truth table. Each of the variables (flip-flop Q outputs) changes with time and at a given time has a unique value.

5.4.2 Johnson Counter (Synchronous)

Another type of counter that can be constructed directly from connecting the output of one flip-flop to the input of another is the Johnson counter. The **Johnson counter** can produce a series of outputs from each flip-flop that is offset by a clock pulse from the preceding flip-flop output. Figure 5.64 illustrates a four-bit Johnson counter using D flip-flops. Figure 5.65 shows the timing diagram for the four-bit Johnson counter.

The Johnson counter shown in this example is synchronous, because a common synchronizing clock is applied to all of the flip-flops simultaneously; so the state changes for all affected flip-flops occur at the same clock time.

Notice that the Q_D' output is fed back to the D_A input. Assuming that all four flip-

Figure 5.64
Four-bit Johnson counter

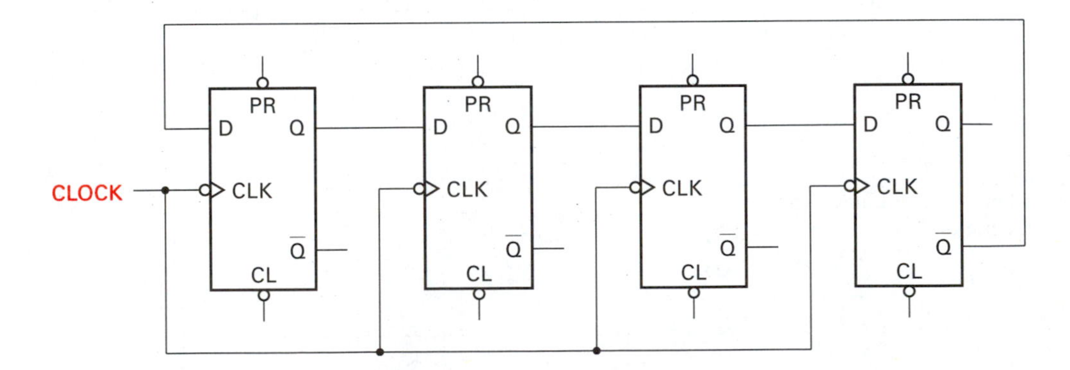

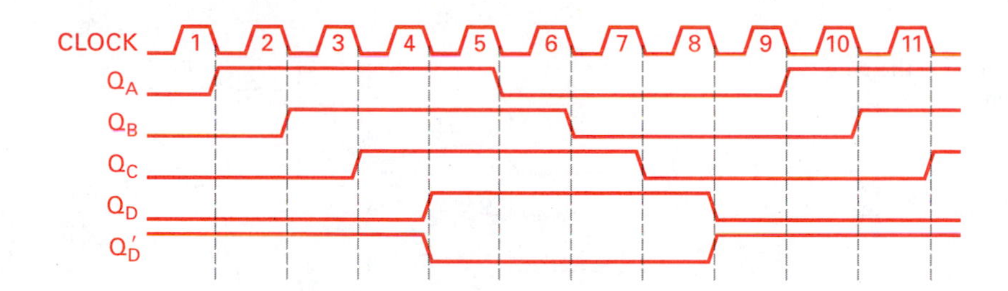

Figure 5.65
Four-bit Johnson counter timing diagram

flops are initially reset, the first negative clock edge will cause flip-flop A to set. Flip-flops B through D do not set on the first clock because the Q outputs for flip-flops A through C are still at a logical 0. The second clock causes flip-flop B to set ($Q_B = 1$), the third clock pulse causes flip-flop C to set ($Q_C = 1$), and the fourth causes flip-flop D to set ($Q_D = 1$). Once flip-flop D has set the Q_D' output goes low causing F_A to reset on the next clock negative edge. The process keeps repeating as long as power is applied and the clock input is present. The Johnson counter can be used to produce time delays; the output of each flip-flop is delayed by one clock pulse from the previous output.

5.4.3 Ring Counter (Synchronous)

The **ring counter** is another counter where the output of one flip-flop connects directly into the input of another to produce a particular output pattern. The ring counter produces a continuous pattern of pulses from the flip-flop outputs. The ability to load the flip-flops to a particular state permits a repeatable output pattern. The counter is initialized to the appropriate data pattern, using the flip-flops' asynchronous inputs. Consider the eight-bit ring counter illustrated in Figure 5.66. Each of the eight flip-flops has separate preset and clear inputs so that any input bit pattern can be preloaded. For example, the counter could be initialized to 00101101 by first reseting all eight flip-flops and then bringing the active low PRE′ inputs to flip-flops 3, 5, 6, and 8 to logical 0s.

Data are shifted from one flip-flop to the next on each negative edge of the clock pulse. Because the Q output of flip-flop 8 is fed back to the D input of flip-flop 1, the preloaded pattern repeats every eight clock pulses.

Figure 5.66
Ring counter

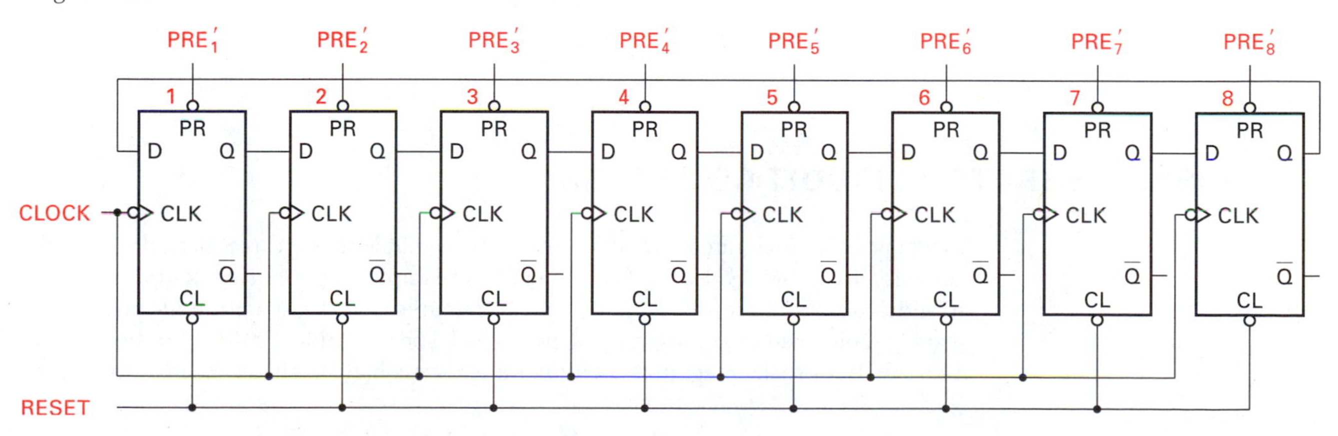

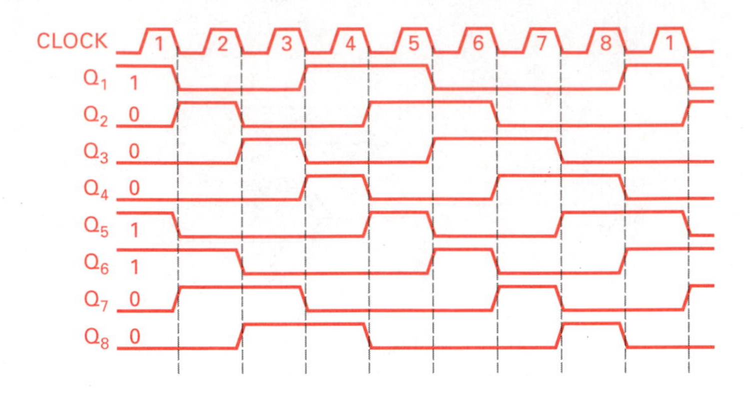

Figure 5.67 Timing diagram for the ring counter when preloaded to 10001100

TABLE 5.7 Illustration of repeating bit pattern for an eight-bit ring counter when preloaded to 10001100

Bit position								
1	2	3	4	5	6	7	8	
1	0	0	0	1	1	0	0	Initial condition
0	1	0	0	0	1	1	0	1st clock pulse
0	0	1	0	0	0	1	1	2nd clock pulse
1	0	0	1	0	0	0	1	3rd clock pulse
1	1	0	0	1	0	0	0	4th clock pulse
0	1	1	0	0	1	0	0	5th clock pulse
0	0	1	1	0	0	1	0	6th clock pulse
0	0	0	1	1	0	0	1	7th clock pulse
1	0	0	0	1	1	0	0	8th clock pulse
								Return to initial condition

Assume that a bit pattern of 10001100 were preloaded into flip-flops 1 through 8. The application of a succession of clock pulses would produce the waveforms illustrated in Figure 5.67. Table 5.7 shows the same information somewhat differently.

Ring counters can be constructed to generate a repeating *n*-bit pattern by connecting a series of flip-flops together and feeding back the most significant output to the least significant input. Such bit patterns can be very useful in the generation of timing sequences for control applications.

5.5 MSI INTEGRATED CIRCUIT COUNTERS

A wide array of integrated circuit counters are available in the three main digital technologies (TTL, CMOS, ECL). They range from simple ripple binary counters to synchronous up-down decade counters. Various features include asynchronous presets and clears, parallel load capability, and different clock speeds. Tables 5.8–5.10 indicate some of the medium-scale integrated circuit counters available. Detailed information on pin-

TABLE 5.8
TTL medium-scale integrated circuit counters

Synchronous		Ripple	
Function	Type (74)	Function	Type (74)
4-bit binary	161/163/561/669/691/693	4-bit binary	69/93/177/197/293
4-bit binary	169/191/193/569/	Dual 4-bit	393
Up-down	697/699	Binary	
Decade	160/162/560/668/690/692	Decade	68/90/176/196/290
Decade	168/190/192/568/696	Dual decade	390/490
Up-down	698	Divide by 12	92
8-bit up-down	867/869		
50:1 divider	56		
60:1 divider	57		
Rate multiplier	97/167		

TABLE 5.9
ACT CMOS medium-scale integrated circuit counters

Function (4-Bit Synchronous)	Type 74ACT
Decade	160/162
Decade up-down	168/190/192/568
Binary	161/163
Binary up-down	169/191/193/569

TABLE 5.10
ECL medium-scale integrated circuit counters

Function	Type MC
Binary counter	10H016/1654/10154
Bi-quinary	10138
Prog. modulo-n	4018
Divide-by-4 GigaHz counter	1699

outs, propagation delay, and features are contained in the digital data books published by the companies that manufacture the devices.

5.5.1 MSI Asynchronous Counters

Several integrated circuit counters are listed in manufacturer's data books as asynchronous, because the flip-flops are not clocked with the same clock input. The SN74176 is a TTL MSI decade (modulo-10) or BCD (binary coded decimal) counter with two clock inputs, and it is asynchronous. The first clock input is presented only to the first flip-flop. This flip-flop can be used as a high-speed prescaler; that is, a divide-by-2 counter that operates at a higher speed than the other flip flops in the package. Figure 5.68 shows the logic internal to the IC.

Figure 5.68
Integrated circuit decade counter

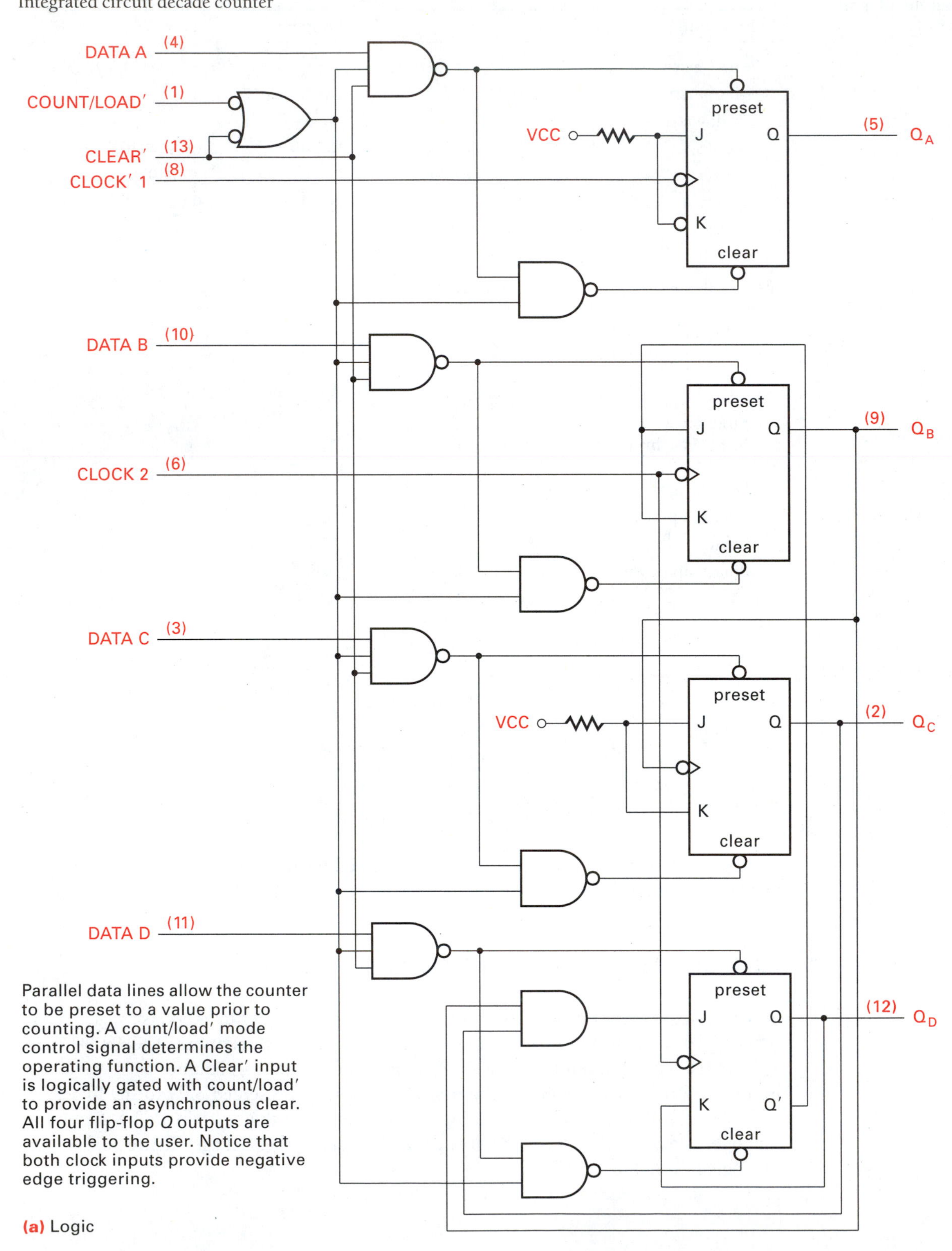

Parallel data lines allow the counter to be preset to a value prior to counting. A count/load′ mode control signal determines the operating function. A Clear′ input is logically gated with count/load′ to provide an asynchronous clear. All four flip-flop *Q* outputs are available to the user. Notice that both clock inputs provide negative edge triggering.

(a) Logic

Figure 5.68
Continued

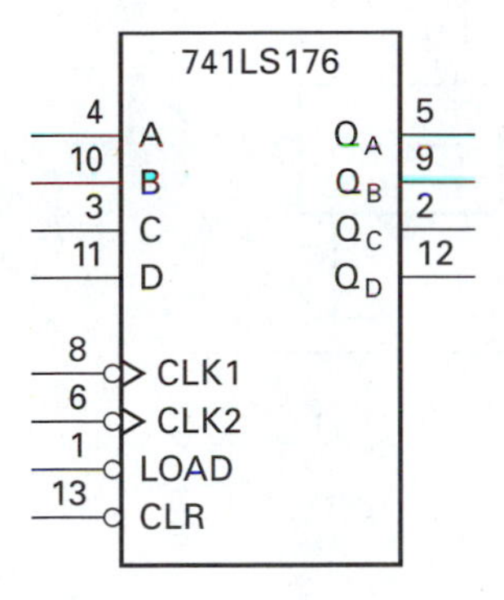

(b) Logic Symbol for 74LS176

TABLE 5.11
BCD count sequence

	Output			
Count	Q_D	Q_C	Q_B	Q_A
0	0	0	0	0
1	0	0	0	1
2	0	0	1	0
3	0	0	1	1
4	0	1	0	0
5	0	1	0	1
6	0	1	1	0
7	0	1	1	1
8	1	0	0	0
9	1	0	0	1

TABLE 5.12
Bi-quinary count sequence

	Output			
Count	Q_A	Q_B	Q_C	Q_D
0	0	0	0	0
1	0	0	0	1
2	0	0	1	0
3	0	0	1	1
4	0	1	0	0
5	1	0	0	0
6	1	0	0	1
7	1	0	1	0
8	1	0	1	1
9	1	1	0	0

Note that the Q output of flip-flop A is not connected to the rest of the internal flip-flops. This allows the user to connect it to the second clock input or to use the single flip-flop separately from the remaining three. Also note that flip-flops B and D are triggered by the clock 2 input, and that flip-flop C is triggered by Q_B. This is why the data books list the device as asynchronous.

The SN74xx390 and 393 are TTL dual four-bit counters. Two four-bit counters reside in the same package; the 390 is configurable as a decade (BCD), binary, or bi-quinary counter, depending on how it is connected. Table 5.11 shows the BCD count sequence. Bi-quinary is a reflective BCD code as shown in Table 5.12. The 393 is a binary counter.

Figure 5.69(a) shows the configurable counter (390) and Figure 5.69(b) shows a binary counter (393). By connecting the Q_A output to the B input, the logic circuit shown in Figure 5.69(a) is configured as a BCD counter. Each of the counters has its own active high clear input. The T flip-flops are negative edge triggered. Table 5.11 illustrates a BCD count sequence. To connect the configurable circuit as a bi-quinary counter, the Q_D output is connected to the A input. The bi-quinary count sequence is illustrated in Table 5.12.

5.5.2 MSI Synchronous Counters

Synchronous counters have all of the clock inputs to the flip-flops connected together, providing for state changes to occur at the same time. The SN74XX163 is a synchronous four-bit binary counter with a parallel load capability and an active low clear. Figure 5.70 shows the internal logic diagram for the integrated circuit.

Each of the four flip-flops has a two-level AND-OR circuit that provides a multiplexing function. Each flip-flop has two possible data sources to provide excitation inputs: one is the external parallel load data and the other is the new count. All the external

Figure 5.69
TTL dual four-bit counters

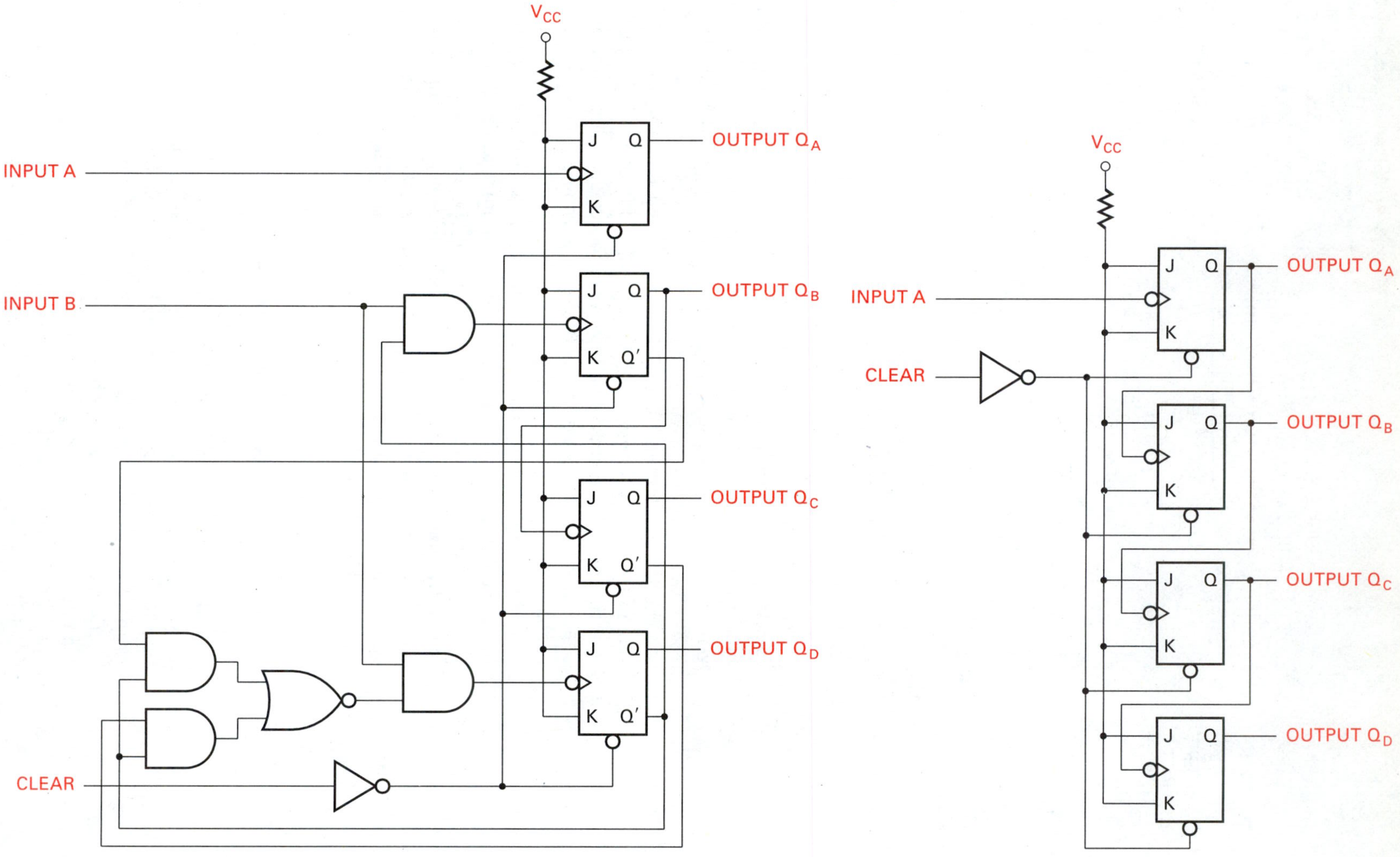

(a) Configurable Counter (390)

(b) Binary Counter (393)

Figure 5.70
Synchronous four-bit binary counter SN74S162

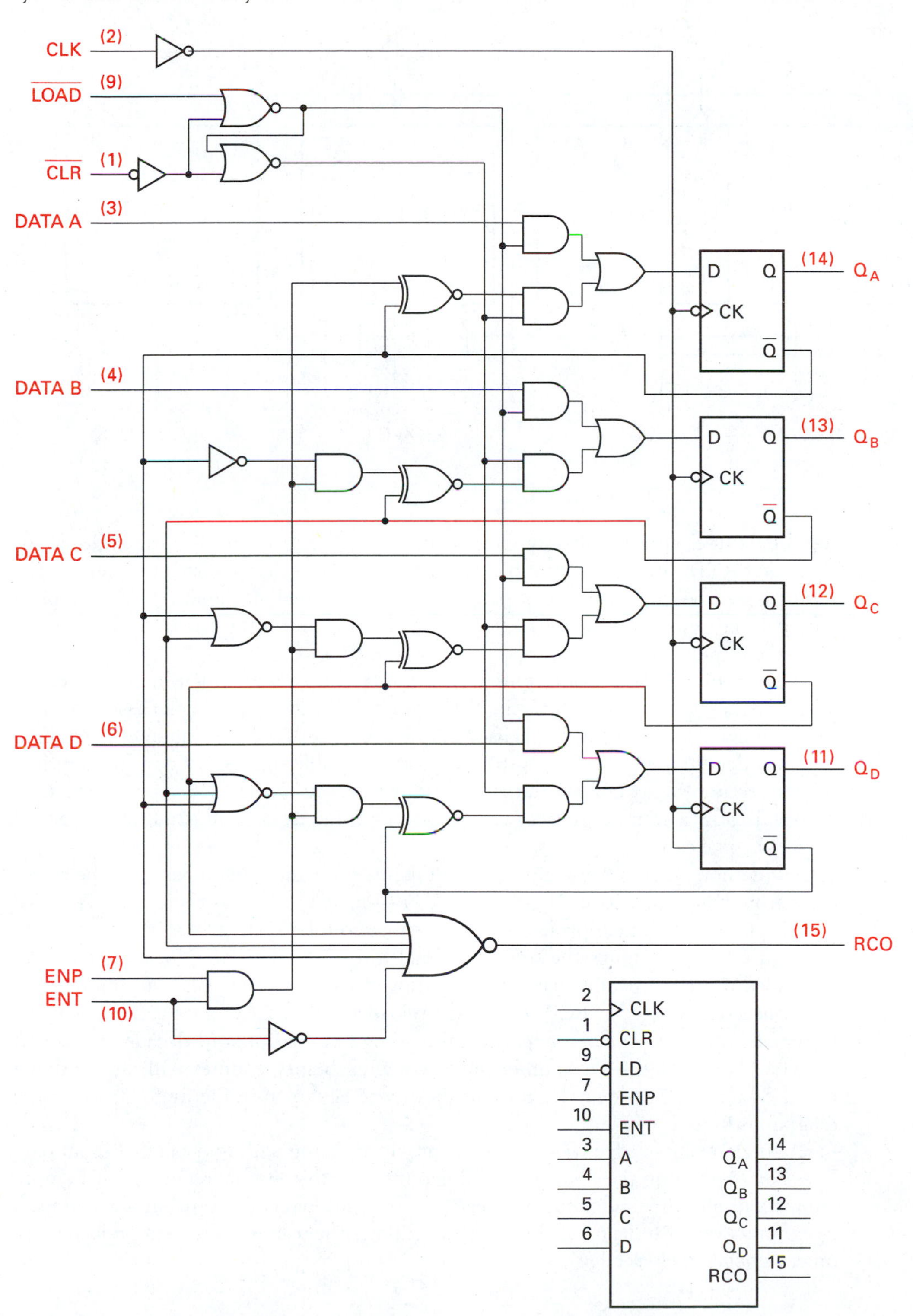

Figure 5.71
Cascaded binary counters using SN74S163 ICs

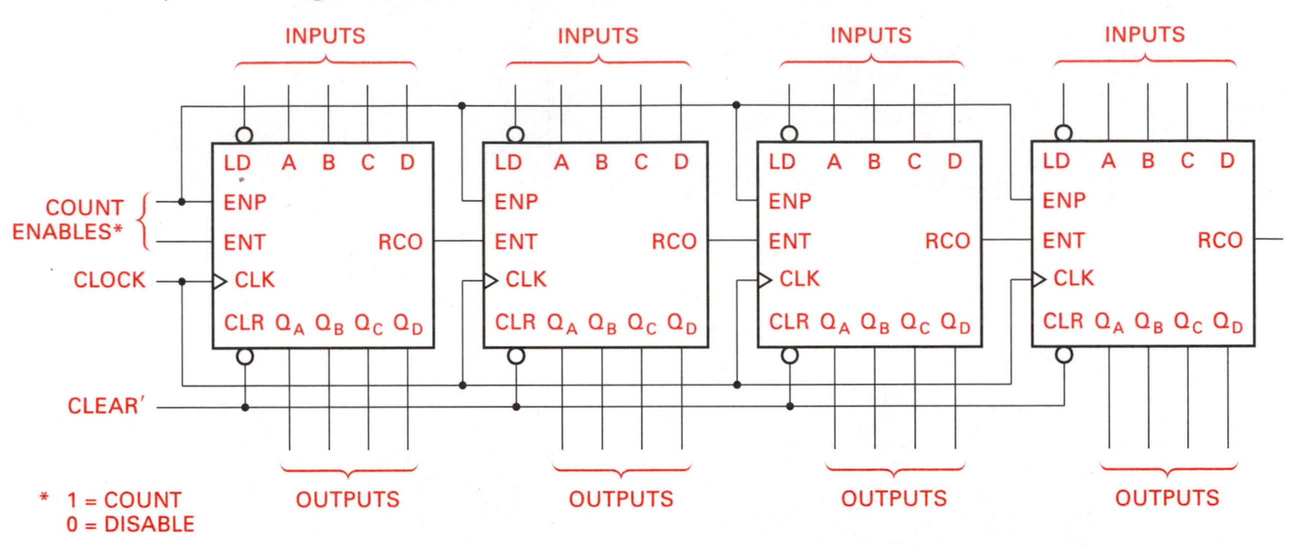

input data lines connect directly into one of the AND gates that form the multiplexer. The other AND gates in the multiplexers have data inputs derived from the count logic. A single positive edge trigger clock input is presented to all of the internal flip-flops in parallel. Active low clear and load signals are used to control the function, either count or load data.

The circuit can be cascaded with other counter chips to provide even longer count sequences, using the ENP, ENT, and RCO lines. RCO is an output signal asserted when the terminal count is reached. It is used to link a lower order decade counter to the next higher order counter. The ENP and ENT are input enable signals that control counting. By connecting the RCO from the output of one decade to the ENT input of the next higher decade, a series of devices can be cascaded. Figure 5.71 illustrates how several devices can be cascaded.

A common clock is presented to all of the integrated circuits in parallel. The ripple carry from one decade to another is provided by the RCO-to-ENT connection. The ENP signal is used in this example to control the entire counter system.

Figure 5.72 illustrates the timing internal to each IC and how the device can be preloaded using the parallel data inputs and the active LOAD signal. The counter then can proceed to count, in BCD, from the preloaded value. RCO goes active when the terminal count of the device is reached, providing a carry from one decade to another.

The SN74xx169 is yet another, more compex, binary counter with an up-down count capability. The logic diagram for the device is shown in Figure 5.73. Its timing diagram is Figure 5.74.

A single positive edge trigger clock is provided, along with four parallel input data lines for preloading the counter. The logic internal to the integrated circuit is more complex due to the up-down count feature. The up-down count capability, when used in conjunction with the parallel load feature, allows the user to increment or decrement on each positive clock edge.

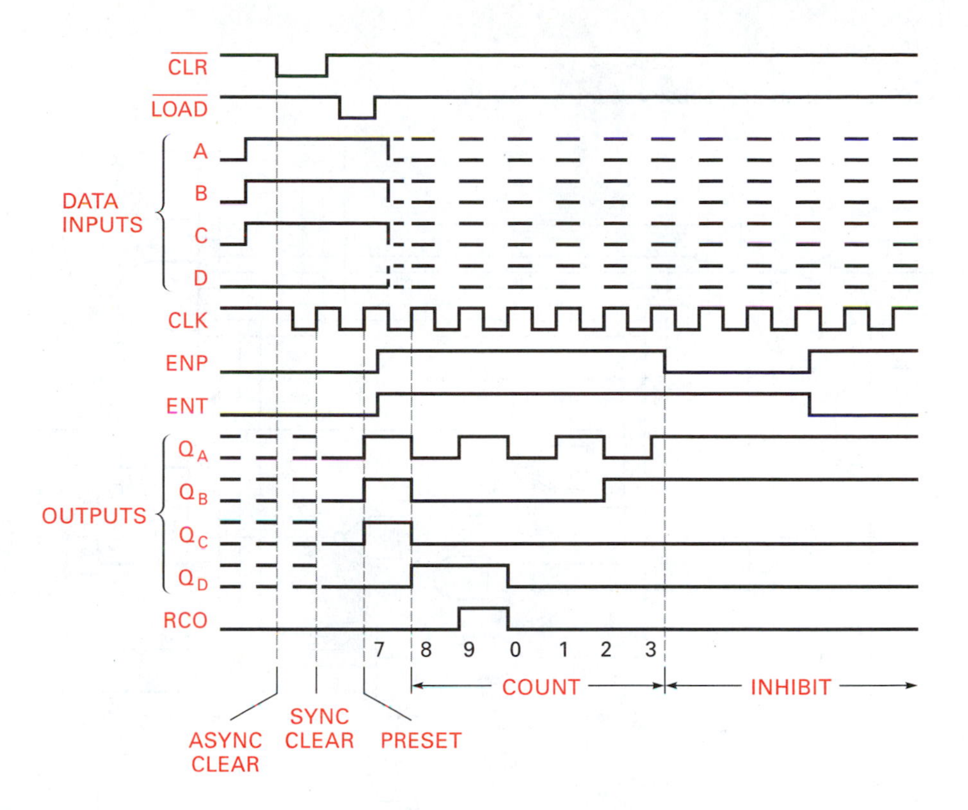

Figure 5.72
Timing diagram for the SN74S162 decade counter

5.5.3 Control Signal Generation by Decoding Counter Outputs

We have seen the logic and timing diagrams for several commercially available integrated circuit counters. Counting is the primary function of these devices. However, timing signals for controlling external events may be generated from the counter's output signals. Each counter, whether binary, BCD, or some other count sequence, generates a sequence of changing outputs. In a number of control applications a series of timing pulses is desired to cause some event to occur relative to other events. Figure 5.6 with some changes is repeated in Figure 5.75. The figure models a sequential counter in which no external input is available; the counter simply moves from one output value to the next on each clock pulse.

If a series of repeating control pulses are needed to control some external functions, a simple counter can do the job. For example, assume that a series of timing pulses were necessary to control three activities in some manufacturing process and must occur in the following sequence:

1. Event 1 must be active, inactive, and then active again.
2. Event 2 must not occur until event 1 has gone active and then inactive for the first time. Event 2 occurs only after the first time event 1 goes inactive.
3. Event 3 must go active during the second time event 1 is active and then only after event 2 is inactive. Event 3 must also return inactive prior to event 1 going inactive.

Such a timing sequence could look like the diagram in Figure 5.76.

Figure 5.73
SN74xx169 four-bit up-down binary counter

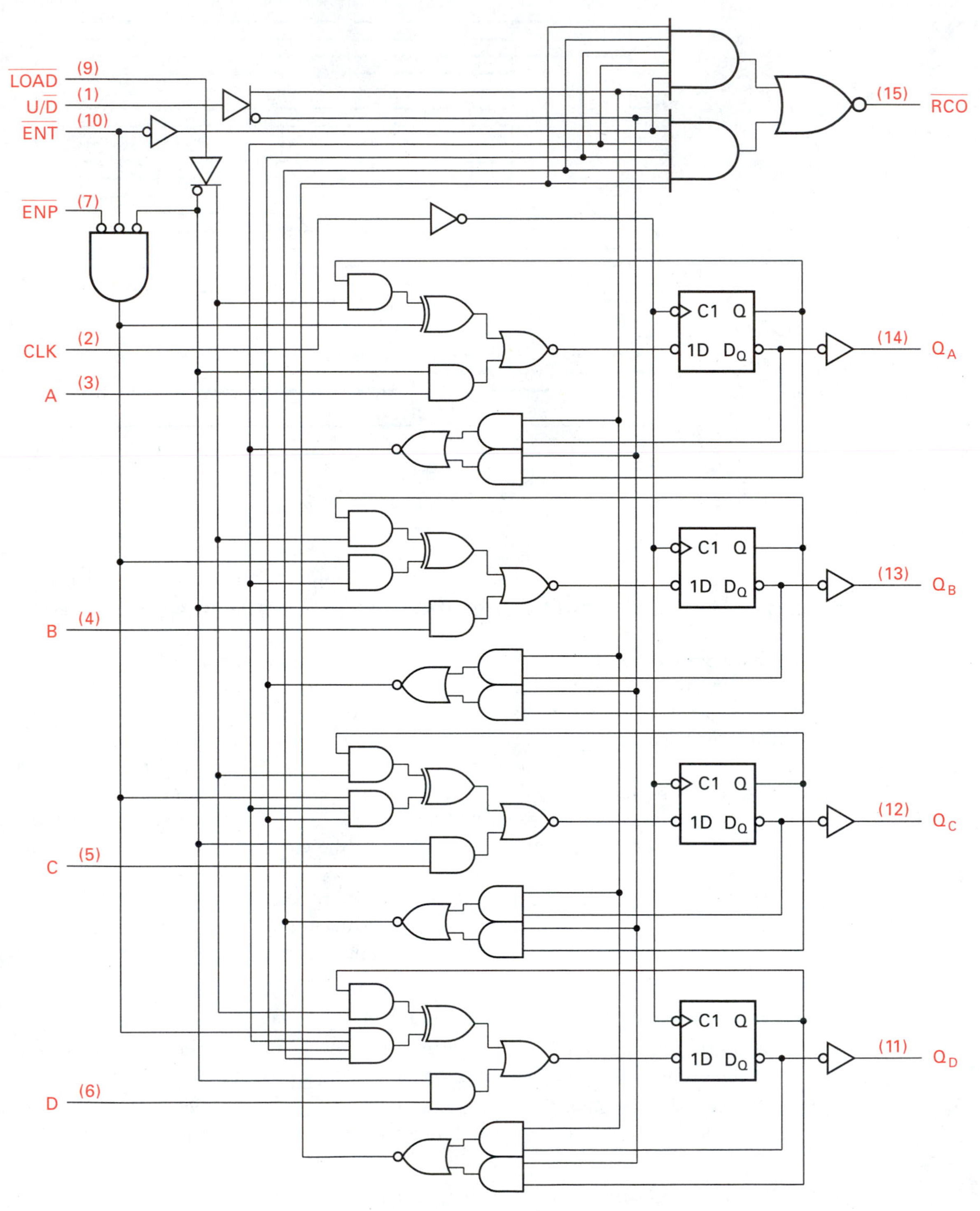

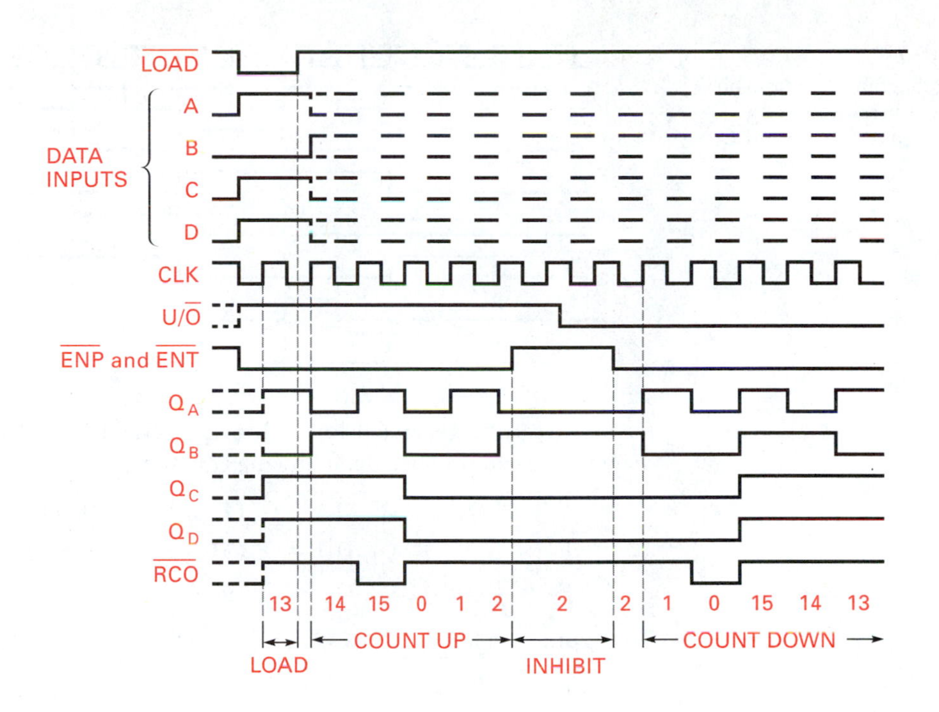

Figure 5.74
Timing diagram for the SN74xx169 up-down binary counter

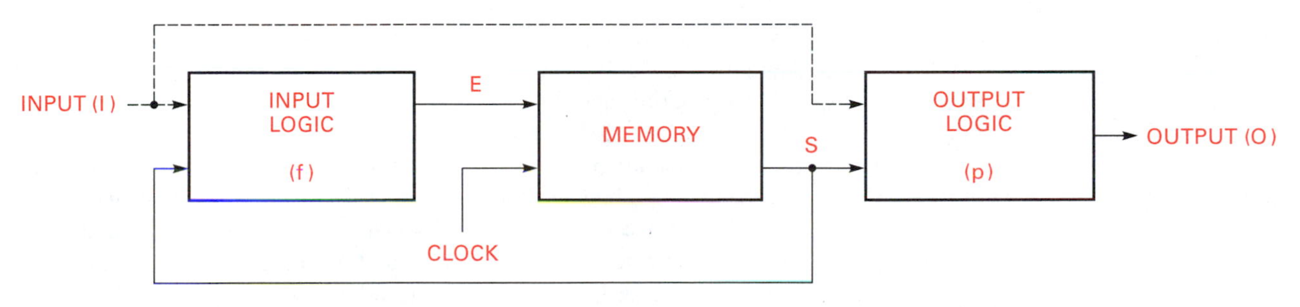

Figure 5.75
Sequential delay model

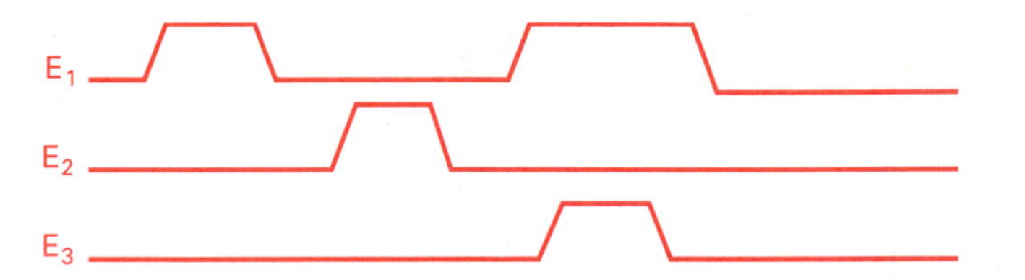

Figure 5.76
Relative occurrences of events 1, 2, and 3

The logic necessary to produce the timing pulses of Figure 5.76 can be realized using a counter. Figure 5.77 shows the timing diagram of a four-bit binary counter using a SN74xx393. A four-bit binary counter has 16 states, 0000 to 1111. The flip-flops in the counter provides us with Boolean variables Q_0 through Q_3.

Each clock pulse input causes the present counter value to increment by 1. The counter continues through the sequence of state transitions as long as power is applied and the clock is running. Each count consists of a unique set of flip-flop output values that we can write as minterms 0 through 15. A Boolean equation, consisting of a min-

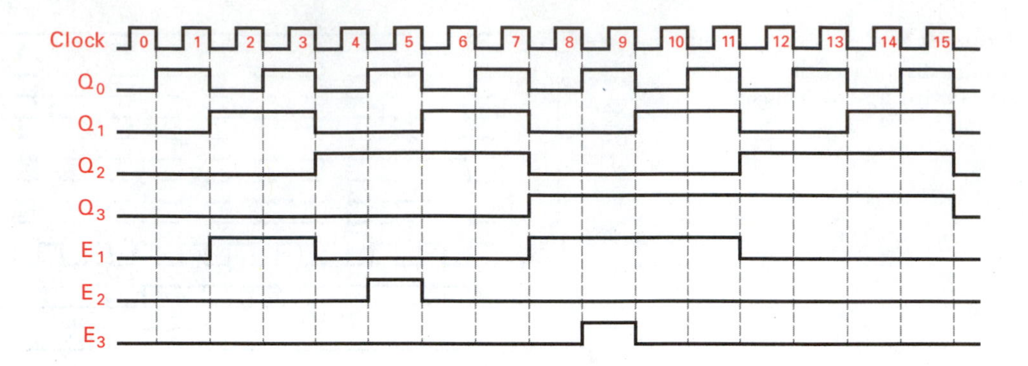

Figure 5.77
Timing diagram for a four-bit binary counter and timing control signals E_1, E_2, and E_3

term set, can be generated for each event variable, E_1, E_2, and E_3. From the timing diagram we can produce the following equations:

$$E_1 = f(Q_3, Q_2, Q_1, Q_0) = \Sigma\ (2, 3, 8, 9, 10, 11)$$
$$E_2 = f(Q_3, Q_2, Q_1, Q_0) = \Sigma\ (5)$$
$$E_3 = f(Q_3, Q_2, Q_1, Q_0) = \Sigma\ (9)$$

The expression for E_1 reduces to $Q_3Q_2' + Q_2'Q_1$

$$E_2 = Q_3'Q_2Q_1'Q_0$$
$$E_3 = Q_3Q_2'Q_1'Q_0.$$

The final circuit for the event timing controller is shown in Figure 5.78.

The timing diagram shown in Figure 5.77 is idealized; that is, shown without regard to different flip-flop propagation delays that may produce glitches when the state variables are decoded. In this example, all of the output function prime implicants were covered, as discussed in Chapter 3. However, consider the expanded timing diagram shown in Figure 5.79 where Q_0 changes at 428 ns and Q_1 changes at 436 ns. Because Q_3 and Q_2 were both at 0, the difference in timing between when Q_0 and Q_1 changed produced an 8 ns period of time when the counter switched from a count of 0001 to 0000 before assuming a value of 0010. If an output function contained minterms {1, 2} but not {0}, a glitch would have occurred. For this reason care should be taken when decoding counter outputs.

Figure 5.78
Event controller logic diagram

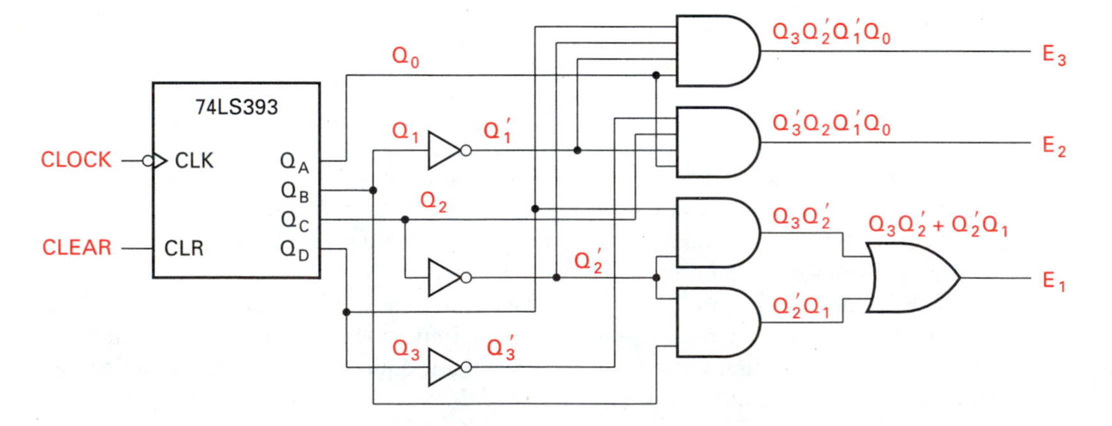

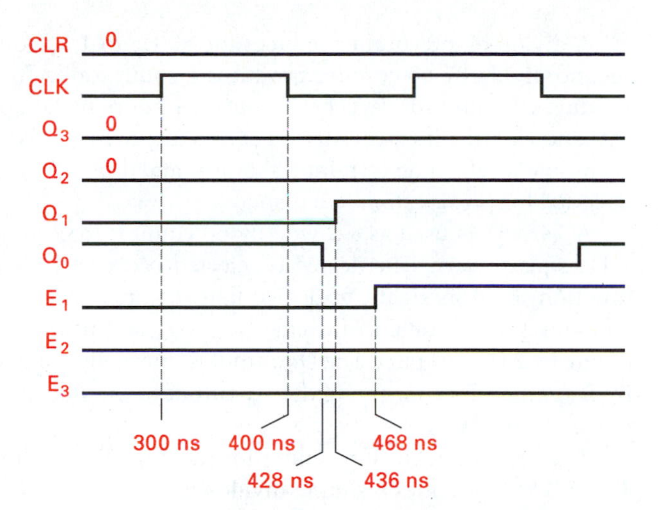

Figure 5.79
Expanded partial view of counter decoder timing

5.5.4 A Counter Application: Digital Clock

Counters have numerous applications in digital systems. A common application, one that illustrates the use of different modulo-*n* counters, is a 12 hour digital clock. Starting with a 60 Hz sine wave from the input power line and using a series of wave shaping circuits, counters, and decoders, a clock system can be produced. Consider the system block diagram shown in Figure 5.80. The clock timing frequency is derived from the 60 Hz sine wave shaped into a square wave (by clipping the sine wave and presenting the result to a Schmitt trigger gate). The output from the wave shaping circuit is a 60 Hz square wave suitable for a timing reference input to some counters.

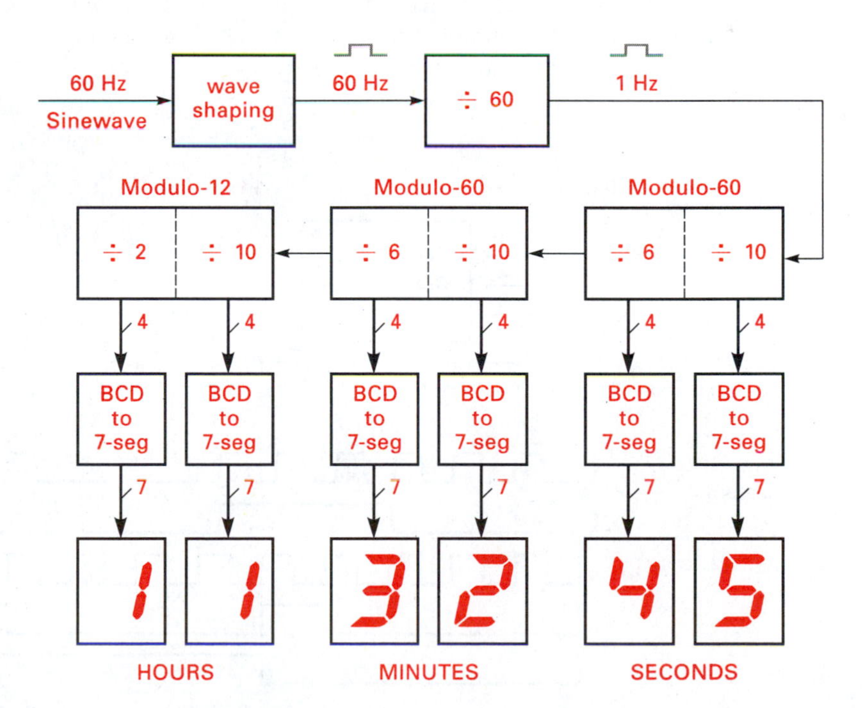

Figure 5.80
Digital clock block diagram

A divide-by-60 counter reduces the 60 Hz to 1 Hz, one count for each second. The second divide-by-60 counter produces a count for each minute, and the third counter produces a count for each hour. Since 24 hours make up a day, broken into two parts (AM and PM), the final counter must be a divide by 12. We need two decade counters for each time level: seconds, minutes, hours, and days. Each decade counter output must be decoded for presentation to a display.

A 74LS57 is used as a divide-by-60 counter to convert the 60 Hz square wave to a 1 Hz square wave. The device is selected because it matches exactly with the needed function. It is an eight-pin device that requires an I_{CC} of 17 mA. Internally, the chip contains three counters: a divide by 6 (Q_A output), a divide by 5 (Q_B output), and a divide by 2 (Q_C output). The Q_B output is internally connected to the input of the divide-by-2 counter. By connecting the Q_A output to the B clock input we get a divide by 60 in one eight-pin IC.

The detailed logic and timing for the 74LS57 are shown in Figures 5.81 and 5.82. The 74LS57 provides a simple divide-by-60 function. The timing diagram shows how the Q_A output divides the A clock by 6, and when the Q_A output is connected to the B clock input, the C counter output divides the input by 60.

The Q_A output of the 75LS57 IC goes low once for every 6 clock inputs. Output Q_C goes high once every 5 times that Q_A goes low and then Q_C goes low once every 5 times that Q_A goes low. The result is that Q_C goes low once for every 60 input clock pulses.

The seconds and minutes counters must use something other than the 74LS57. Access to the individual flip-flop outputs, for the purpose of decoding, is necessary. The SN74LS162 contains a decade counter that performs a divide-by-10 function. By using gates external to the counter any terminal count (less than 10) can be obtained. In the example we need a divide-by-6 function. By using the 74162, configured as shown in Figure 5.83, we can obtain a divide-by-6 function. The counter outputs are decoded by a single two-input NAND gate to detect when the terminal count of 5 (six events, 0–5) is reached. If the terminal count (TC) signal is coupled back to the active-low clear input, the counter is reset every six clock pulses. Figure 5.84 illustrates the timing diagram for the divide-by-6 counter.

Figure 5.81
SN74LS57 divide-by-60 counter

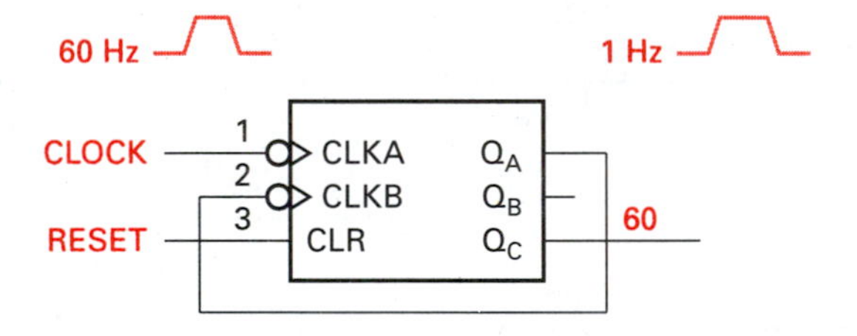

Figure 5.82
Divide-by-60 timing diagram

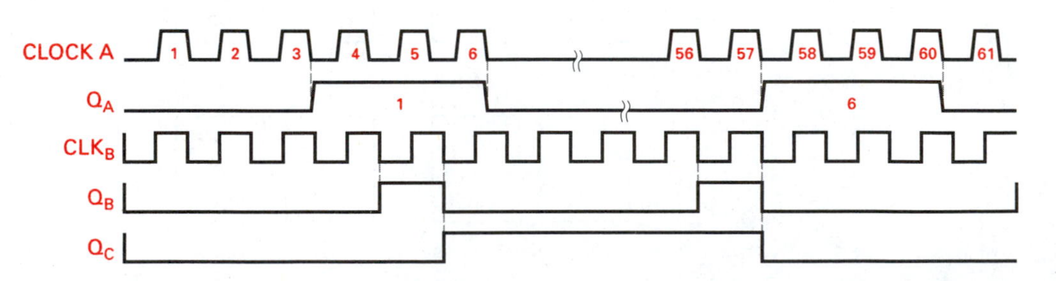

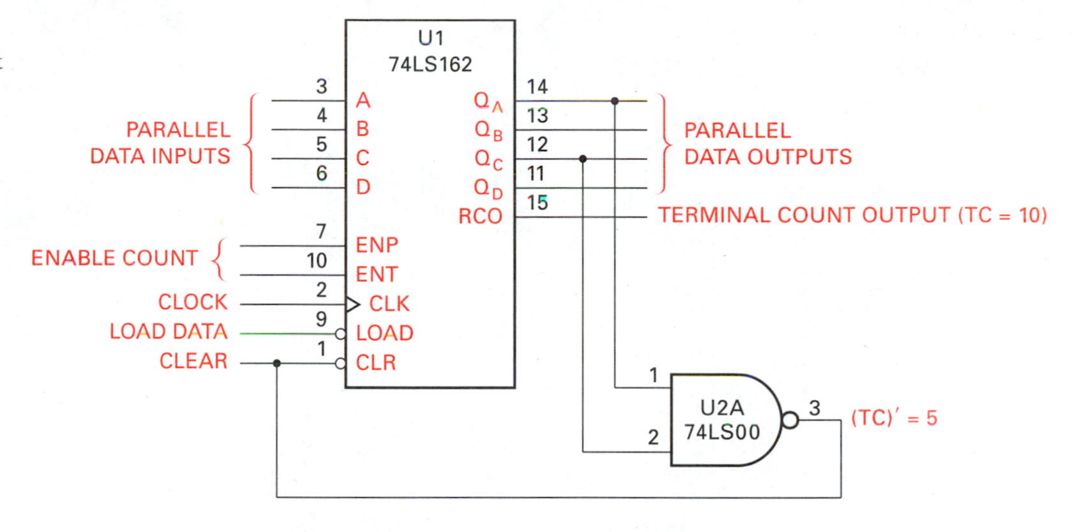

Figure 5.83
SN74LS162 integrated circuit configured as a divide-by-6 counter

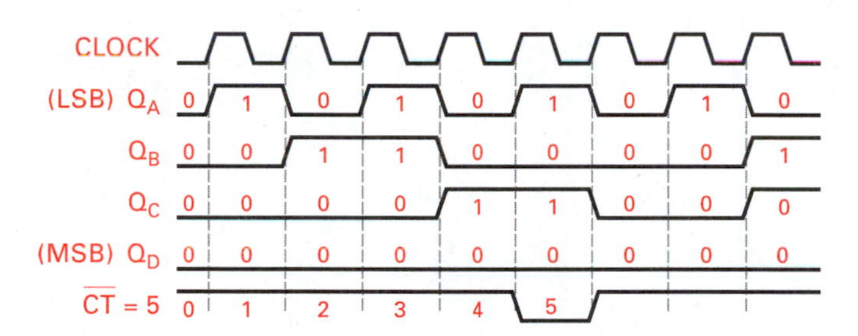

Figure 5.84
Timing diagram for a divide-by-6 counter

Using two SN74LS162 chips provides the divide-by-10 and divide-by-6 counters needed for the seconds and minutes.

A digital clock logic diagram is shown in Figure 5.85. The decade and divide-by-6 counters are used to count time: seconds, minutes, and hours. External logic is required to initialize the correct time of day. Once that is done the "set time" button would be pushed to load the correct values into the respective counters. SN74247 integrated circuits are used to translate the BCD and partial BCD data from the counters to seven-segment codes for distribution to a series of seven-segment display modules.

A binary counter with feedback could be used to produce a modulo-12 counter. However, the hours display consists of two BCD decoders and seven-segment displays. A single four-bit modulo-12 counter could not drive both hour display digits. Instead we use a modulo-12 hours counter that has a decade counter for the first digit (LSD). The second digit, however, is either 0 or 1, so a flip-flop can be used. A modulo-2 (flip-flop) and a modulo-10 (decade) counter forms a modulo-20 counter when no feedback is provided. By providing feedback from the flip-flop to the decade counter, we can cause the decade counter to reset at the correct count. The hours decade counter resets to 0001 each time the flip-flop changes state from 1 to 0. Table 5.13 illustrates the hours counter state changes.

The flip-flop sets the terminal count output, RCO, of the decade counter and remains set for two more hours; then the flip-flop and the decade counter reset to start the modulo-12 count all over. The logic needed for the feedback to clear the decade counter and generate the D input to the MSD flip-flop is shown in Figure 5.85.

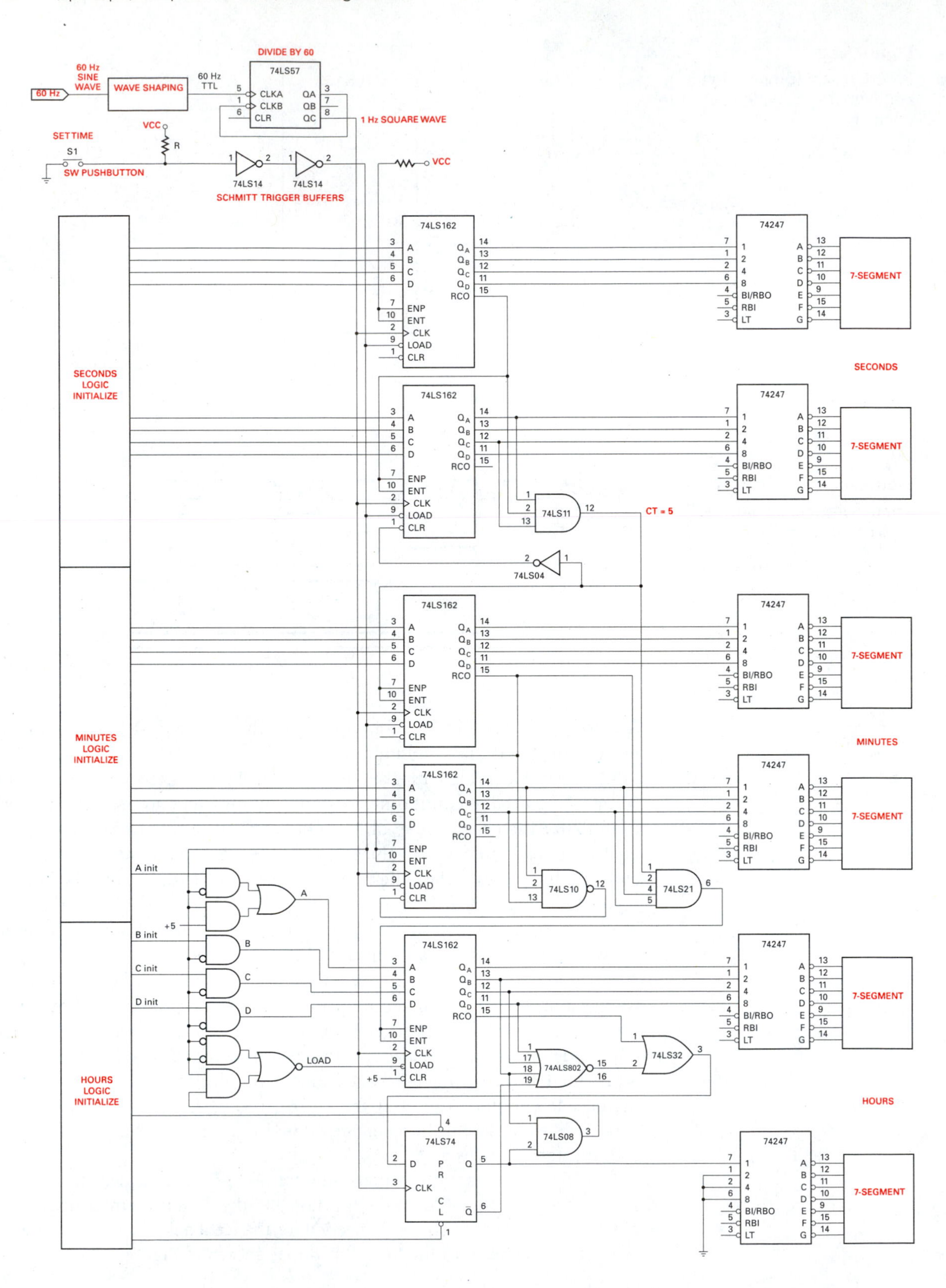

Figure 5.85
Digital clock

TABLE 5.13
Modulo-12 counter state changes

Hours Counter, MSD					
Flip-flop	Decade counter				
Q	D	C	B	A	
0	0	0	0	1	
0	0	0	1	0	
0	0	0	1	1	
0	0	1	0	0	
0	0	1	0	1	
0	0	1	1	0	
0	0	1	1	1	
0	1	0	0	0	
0	1	0	0	1	← RCO is active here
1	0	0	0	0	
1	0	0	0	1	
1	0	0	1	0	
0	0	0	0	1	← Decade counter reloads

5.5.5 IEEE Standard Symbols for MSI Counters

IEEE standard symbols divide counters into two major parts: a control block and a sequential block. These symbols reflect the modeling principles discussed earlier, where the control consists of combinational logic and the sequential block is constructed of flip-flops. The top portion of the logic symbol indicates the control block and the bottom portion the flip-flops (see Figure 5.86). The control receives such inputs as load, count enable, count up-down, clear, and the clock and generates the terminal count, carry, or overflow signals. The sequential section receives the data inputs and generates the counter state variable outputs.

Notation for input and output dependency is shown in Table 5.14. The word *dependency* is used to establish the relationship between inputs and outputs in the IEEE symbol: which output is dependent on which inputs. Inverted inputs or outputs use either a triangle or a bubble to indicate active-low status. Inputs always enter on the left and outputs exit on the right of the symbol. The > symbol indicates dynamic status such as edge triggering.

Table 5.14 shows the dependency notation for various functions. For instance, the

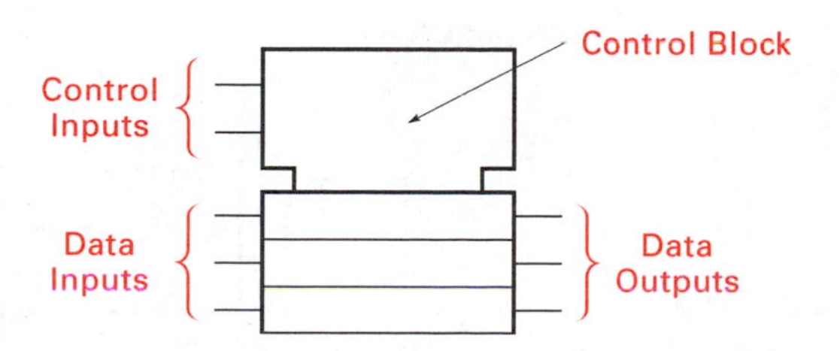

Figure 5.86
IEEE standard symbol composition

TABLE 5.14
IEEE dependency functions

Dependency notation	Function
G	AND
V	OR
N	Negate (EX-OR)
Z	Interconnection
C	Control
S	Set
R	Reset
EN	Enable
M	Mode
A	Address

letter G implies the AND function. If two inputs were labeled G_1 and G_2, then output $Y = G_1G_2$. The AND, OR, and exclusive OR dependencies are indicated as shown in Figures 5.87 through 5.91.

The AND dependency is indicated by using the rectangles with the & symbol. The variable y is active-high for the top AND box and active-low for the bottom one, as shown in Figure 5.87. Figure 5.88(a) shows multiple inputs being ANDed. An OR-AND configuration is shown in Figure 5.88(b), where variables x and y are ORed, then the output of the OR function is ANDed with variable z.

The dotted lines shown in Figures 5.89 through 5.91 indicate only that part of the device symbol associated with the function being discussed is shown.

An AND function can be indicated between input and output functions of a particular control box. Keep in mind that the control boxes we are discussing are part of a larger system. So the input to a control box can be ANDed with an output that provides an input to another control box within the system.

The letter V indicates the OR dependency function. In Figure 5.90, both output variables x and y are available. Note that variable x is not affected by the OR dependency. The variable not affected is denoted by the letter V_1. Other output variables having the same number as V_1 are ORed with the variable associated with V_1.

Figure 5.87
IEEE standard logical dependencies and distinctive shape equivalent

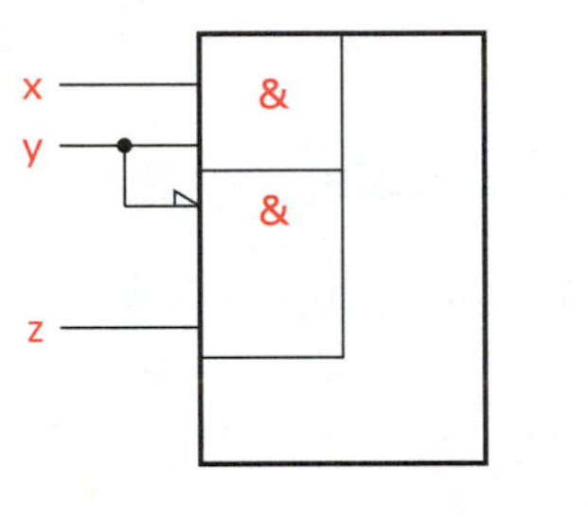

(a) IEEE Multiple AND

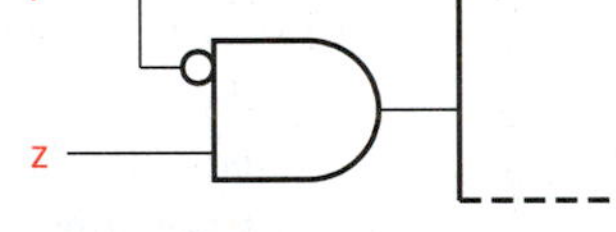

(b) Multiple AND distinctive shape

Figure 5.88
IEEE standard logical dependencies for multiple AND and OR AND

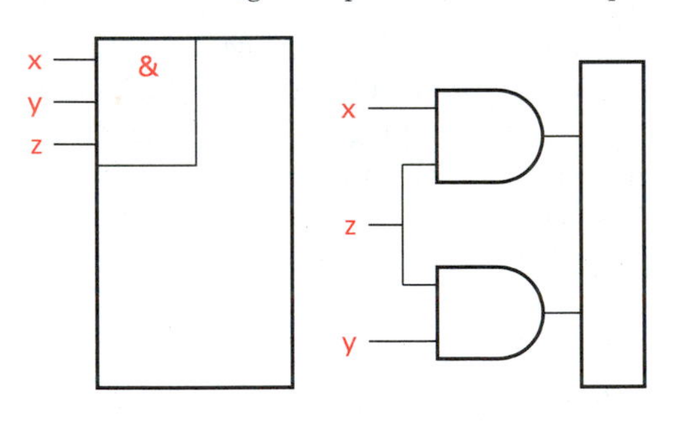

(a) Multiple AND

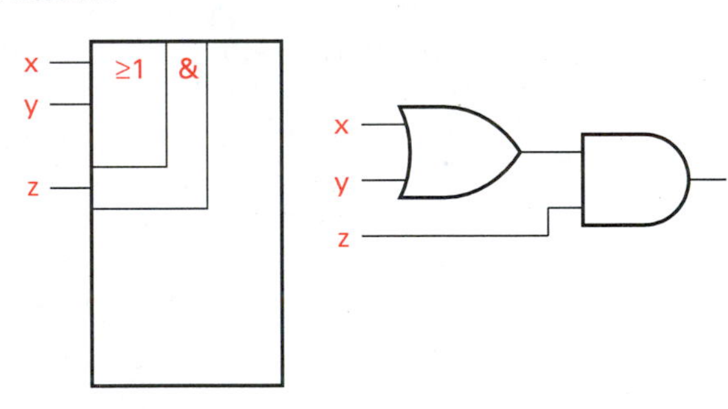

(b) Input OR AND

Figure 5.89
ANDing input with an output

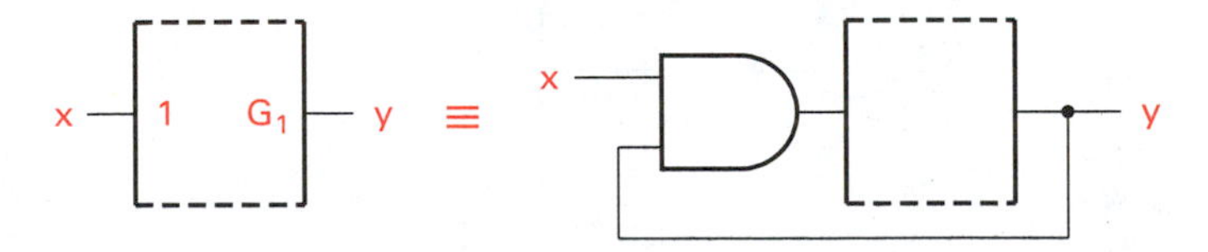

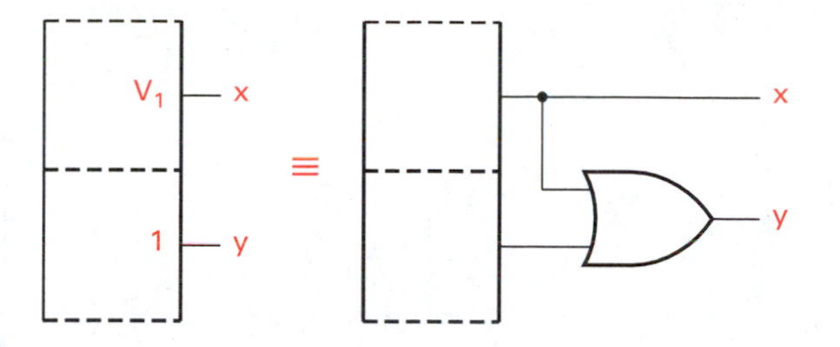

Figure 5.90
IEEE output OR dependency notation

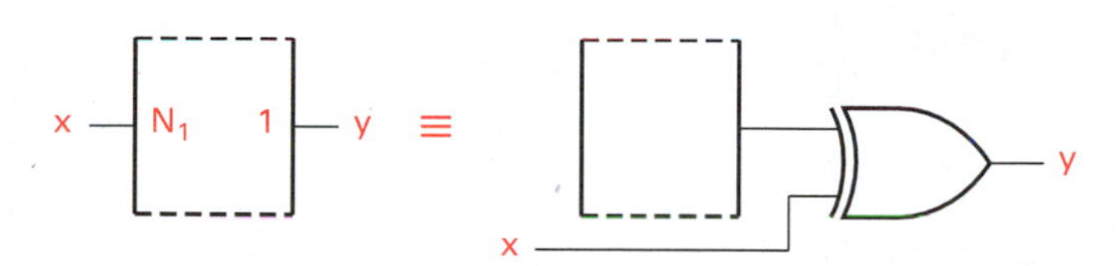

Figure 5.91
IEEE output EXOR dependency notation

Exclusive OR dependency is denoted by the letter *N*. Figure 5.91 illustrates that output variable *y* is generated by the EX-OR of input variable *x* with an intermediate system output to produce variable *y*.

Control dependency notation specifies the control input functions. Control inputs associated with counters and other MSI and LSI devices enable or disable data inputs or outputs. The *EN* notation is an enable input. It enables a function such as loading data or enabling a tristate output. Figure 5.92 illustrates the *EN* function for controlling a multiplexer.

Mode inputs (*M*) can specify logical function dependencies. Mode dependencies may consist of several encoded inputs that specify a function the device is to perform. These inputs are represented as *M* 0/3 in Figure 5.93, for example, with a bracket around the mode inputs, indicating that two mode inputs are used to decode one of four functions (mode 0 to mode 3). Numbers associated with the mode inputs are used to specify significance (the highest number is the most significant and 0 is the least significant bit positions), not dependency numbers. Output functions are often dependent on which mode the unit is in. These dependencies are specified by mode numbers. For instance, output *z*, Figure 5.93, is dependent on modes 1 and 2.

Figure 5.92
IEEE EN dependency function

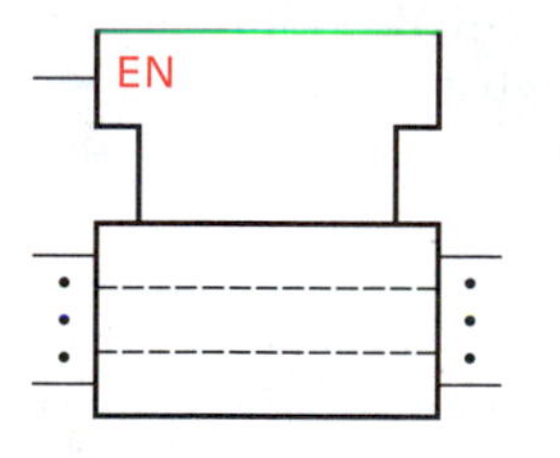

The 74ALS169 is an up-down binary counter that can be loaded with a preset count value and then count either up or down, depending on the mode inputs. In Figure 5.94, four mode inputs M_1, M_2, M_3, M_4 are indicated on the logic symbol of the device. M_1 is the logical inverse of M_2; they are connected together so that one input signal can cause the device to load (parallel data) or count. When the single input mode signal connected to M_1, M_2 is a 0 and the device loads data and when it is a 1, it counts. M_4 is the logical inverse of M_3. These two modes operate like M_1 and M_2, except that when the single input signal is a 0 the device counts down and when it is a 1 it counts up. Control inputs G_5 and G_6 are active-low. Both (AND) must be a 0 before any counting can occur.

Figure 5.93
IEEE notation mode dependency

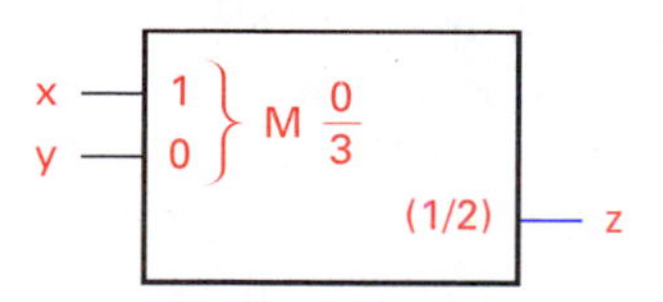

CT inputs or outputs indicate the "content" value of the device. For instance, on the 74193, CT = 0 on an input variable means to clear the counter or register. The content of the flip-flops is 0. CT = 15 on an input means to preset the flip-flops to all 1s. A CT = 10 output pin means that the pin will be active when the internal content or count is equal to 10. For instance, $1'CT = 15$ specifies that when 1 (the count down clock) is low and the internal count CT = 15, then the output signal CT will be asserted. The 74ALS193 binary counter shown in Figure 5.94(a) has a clear input, CT = 0. Counting

Figure 5.94
IEEE logic symbols

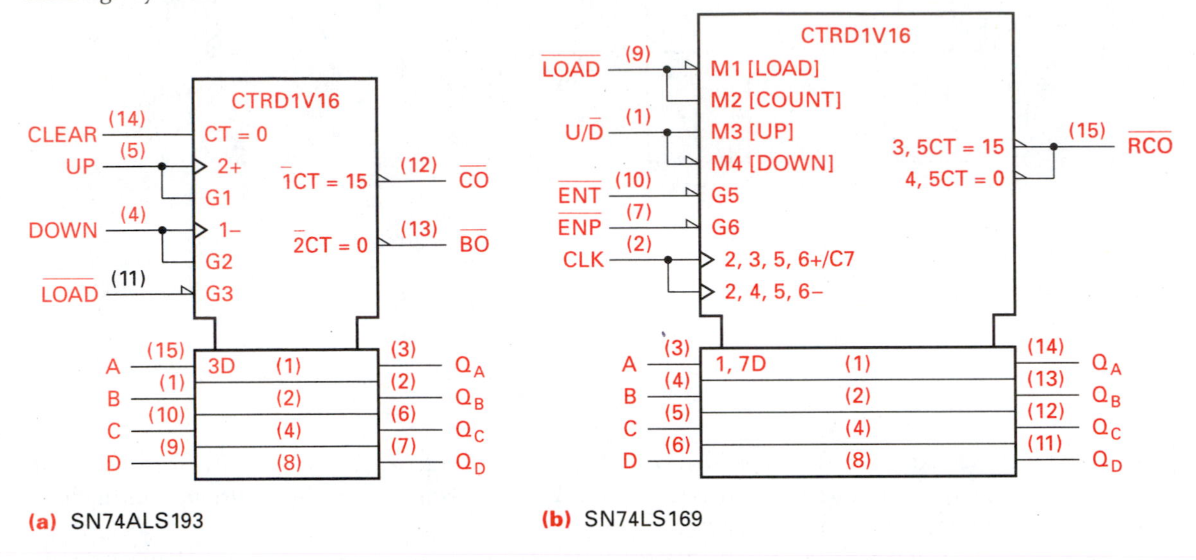

(a) SN74ALS193 **(b)** SN74LS169

up is controlled by input G_1 and the positive edge trigger clock, 2+; counting down occurs when G_2 and its associated positive edge clock 1− are both asserted.

Parallel loading data occurs when LOAD′ is low at G_3. Two outputs are associated with the 74AS193 device: CO′ occurs when the terminal count of 15 occurs and G_1 is low, likewise BO′ is low when the terminal count of 0 occurs and G_2 is low. The CO′ signal is asserted when the counter increments (count up) and can be likened to a carry out. BO′ occurs when the counter is decrementing (count down) and can be likened to a borrow out.

Data inputs, *D*-in, for the four flip-flops are affected when G_3 is low (LOAD′). The numbers in parentheses for the data section of the symbol represent the binary weighting of the flip-flops.

The clock input to the 74ALS169 is positive edge triggered as indicated by the > symbol. Note the numbers written adjacent to the clock input symbol in Figure 5.94(b). The numbers 2, 3, 5, and 6 refer to mode inputs M_2, M_3, G_5, and G_6. The + after the number 6 means that an up-count will occur on each positive clock edge when the indicated mode and control inputs are active. The positive edge trigger clock is designated C_7 and separated from the other control inputs by the / symbol. Count-up occurs when mode bits 2, 3, 5, and 6 are true and the positive edge of the input occurs. Count-down occurs when mode bits 2, 4, 5, and 6 are true on the positive edge of the clock. Because the same signal is applied to both, C_7 is listed only once.

The data flip-flops are changed when M_1 is high or on the positive edge of the input clock (for both count-up and -down). Two OR dependency outputs are present: 3, 5CT = 15 and 4, 5CT = 0. If M_3 and G_5 are true and the terminal count = 15, then RCO′ is low OR; if M_4 and G_5 are true and the terminal count = 0, then RCO is low.

At the top of the control block we find the term CTRDIV16. The term indicates the function of the counter, in this case it is a counter-register (CTR) that divides (DIV) by 16. A decade counter such as the SN74ALS192 has a title block of CTRDIV10, indicating a counter that divides by 10.

5.6 REGISTERS

Registers, like counters, are formed from a collection of flip-flops. Data storage is the main function of a register. Storage and counting functions are available in some integrated circuits, so separation into neat functional categories is not completely possible.

5.6.1 Register Data Input and Output

Registers, formed from a collection of flip-flops, are used to store or manipulate data or both. Input and output functions associated with registers include: Parallel input/parallel output, serial input/serial output, parallel input/serial output, and serial input/parallel output. Input data are presented to registers in either a parallel or a serial format. To input parallel data to a register requires that all the flip-flops be affected (set or reset) at the same time. To output parallel data requires that the flip-flop Q outputs be accessible. Serial input data loading requires that one data bit at a time be presented to either the most or least significant flip-flop. Data are shifted from the flip-flop initially loaded to the next one in series. Serial output data are taken from a single flip-flop, one bit at a time.

Serial data input or output operations require multiple clock pulses. Parallel data operations only take one clock pulse. Data can be loaded in one format and removed in another. Two functional parts are required by all shift registers: (1) data storage flip-flops, and (2) logic to load, unload, and shift the stored information. Generic logic diagrams of the four data input and output types are shown in Figure 5.95.

Parallel Input/Parallel Output

To load parallel data, each input data bit is presented to its respective storage flip-flop. An enable input is used in association with a clock in the case of synchronous registers.

Figure 5.95
Generic logic registers

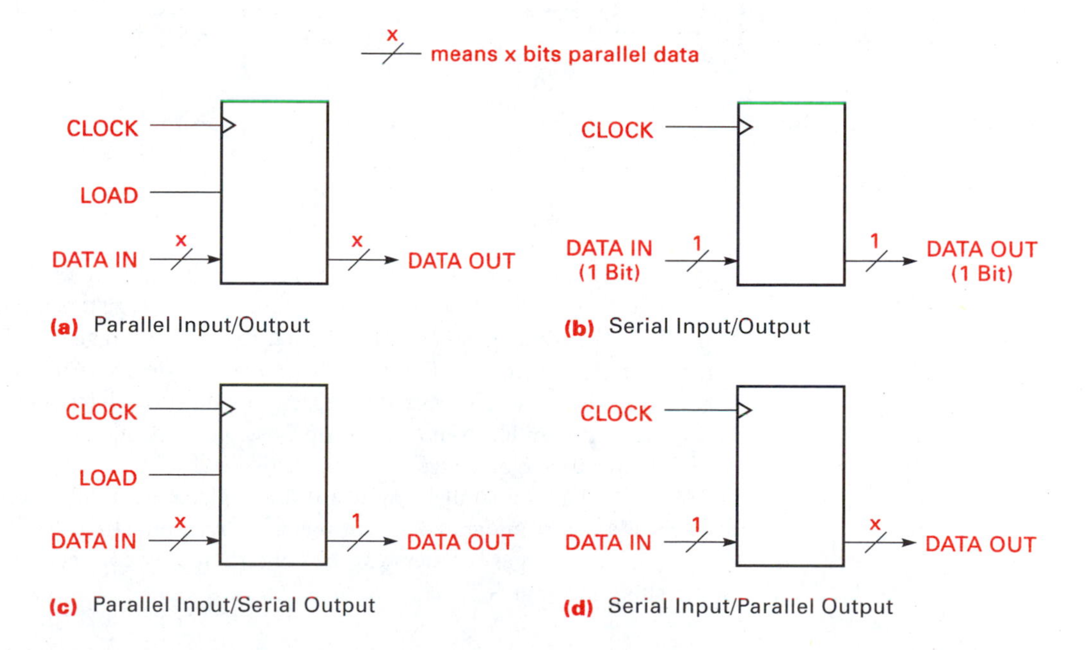

Figure 5.96
Parallel input-output register

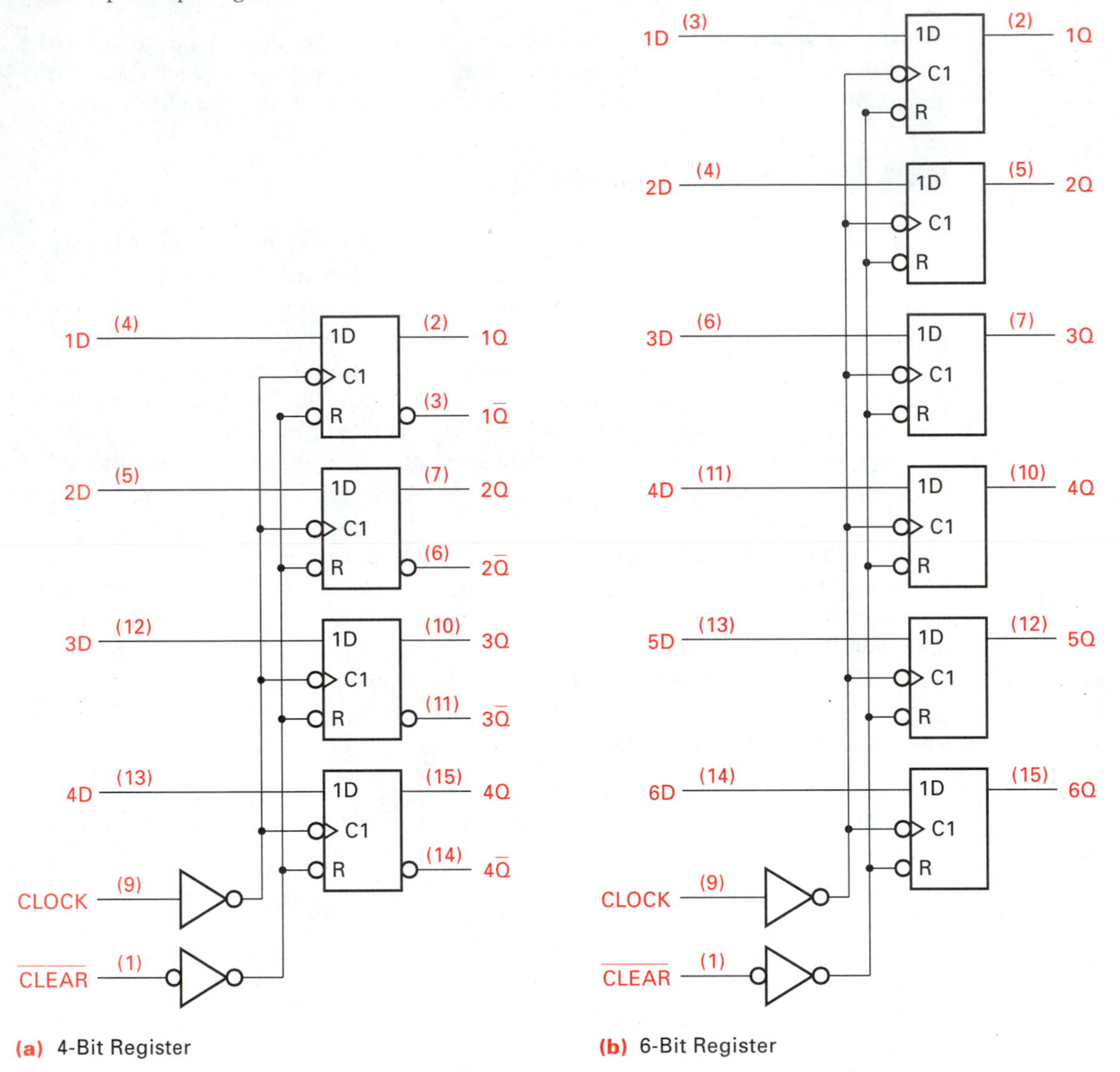

Asynchronous parallel load registers usually operate on level activated enable inputs. Simple parallel input/output integrated circuit registers are found in the SN74xx174/5 ICs. Figure 5.96 shows the internal logic of the registers. Figure 5.96(a) illustrates a four-bit device, and Figure 5.96(b) a six-bit device. An asynchronous active-low clear signal is present on both devices. All the flip-flops have separate input and output lines, making parallel in/out operation possible. A single clock input loads the flip-flops on the positive clock edge.

A 12-bit parallel input/output register is easily constructed from two 6-bit SN74xx174 integrated circuit registers by connecting the clock and clear inputs of the two ICs. Larger registers can be constructed by connecting several smaller registers as shown in Figure 5.97.

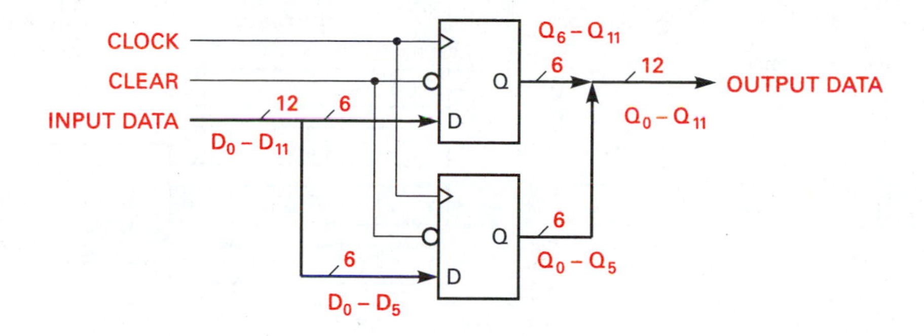

Figure 5.97
A 12-bit parallel input-output register formed from two 6-bit registers

Serial Input/Serial Output Shift Registers

Serial input-output registers receive the input data stream one bit at a time on succeeding clock pulses. Data are shifted between register flip-flops, with the direction dependent on internal output-input connections. Each clock pulse causes data stored in one flip-flop to be transferred into the next. The serial output is obtained from a single flip-flop at the register end. Figure 5.98 illustrates the internal logic for a SN7491A serial shift-right register, assuming that Q_0 is in the most significant bit position. This particular integrated circuit **shift register** can provide either shift-right or shift-left operation depending on the definition of the most significant bit position. Notice that the output of each flip-flop is projected to the next rightmost device. The final output is taken from the rightmost flip-flop.

The eight-bit serial shift register shown in Figure 5.98 takes eight clock transitions to fill. Consider the sequence of clock transitions in Figure 5.99 and the resulting data positions in the register. Assume that the register was cleared prior to serial loading of data. Let the input data stream start at the least significant bit and proceed to the most significant bit.

Figure 5.99 illustrates a shift-right sequence, with the internal data shifted from the most significant bit toward the least significant bit. External data are loaded into the register, one bit at a time, least significant bit first. The same integrated circuit, SN7491A, can also be used as a shift-left register. When starting with the most significant bit and proceeding to the least significant one, the process is said to be a *shift left*. Figure 5.100 illustrates the definitions of shift right and left.

Other combinations of parallel input/serial output or serial input/parallel output are possible. The SN74xx166, shown in Figure 5.101, is an eight-bit parallel load register with serial input/serial output capability. It can be used as a serial input/output or a

Figure 5.98
Right-shift serial input-output register

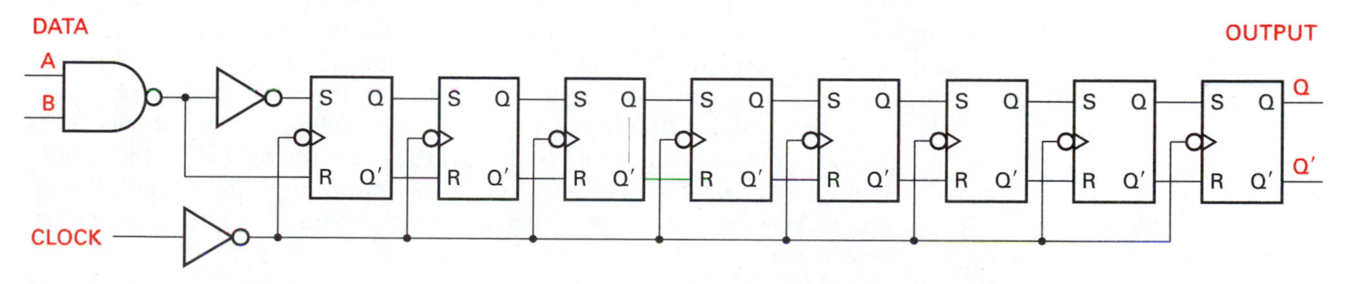

Figure 5.99
Example serial data stream being loaded into a shift register

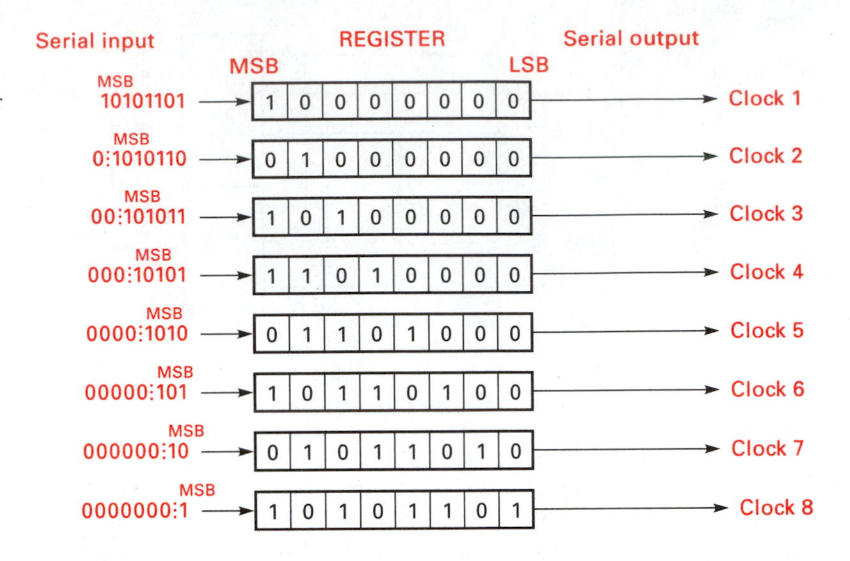

Figure 5.100
Shift-right, shift-left definitions

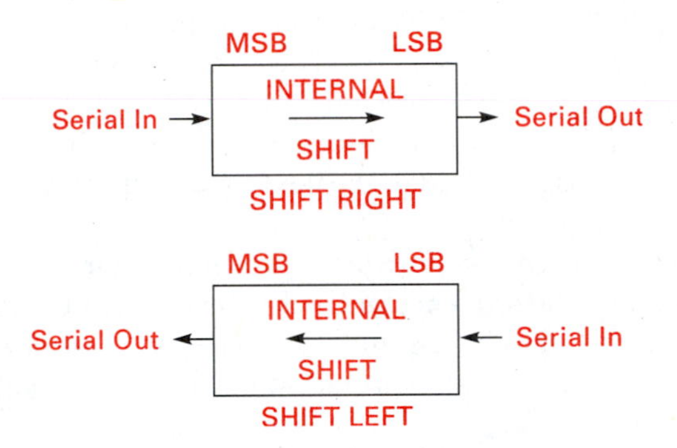

parallel input/serial output register. No parallel output capability exists; notice that the only Q output is from flip-flop H.

A timing diagram for positive edge triggered SN74xx166 shift register is shown in Figure 5.102. A separate clock inhibit controls shifting. A shift/load' input controls the register function allowing for parallel loading and then serial shifting to either the right or left, depending on the definition of bit order.

Universal Shift Registers

Several universal shift registers are available. By *universal,* we mean that they can be used in any possible input-output combination. Because of the additional features, requiring more device pins, the registers are limited to four-bit, medium-scale integrated circuits. Additional internal logic is also required to handle the added functions.

The logic diagram for one such device, the SN74194 integrated circuit, is shown in Figure 5.103. Four parallel data inputs permit parallel loading of external data. Two serial inputs are available, one for left-shift, and the other for right-shift operations. Serial output data are taken from either Q_A or Q_D depending on shift direction. A two-bit mode control provides feature selection. An asynchronous clear and positive edge trigger clock are used.

Figure 5.101

Logic diagram for the SN74xx166 shift register

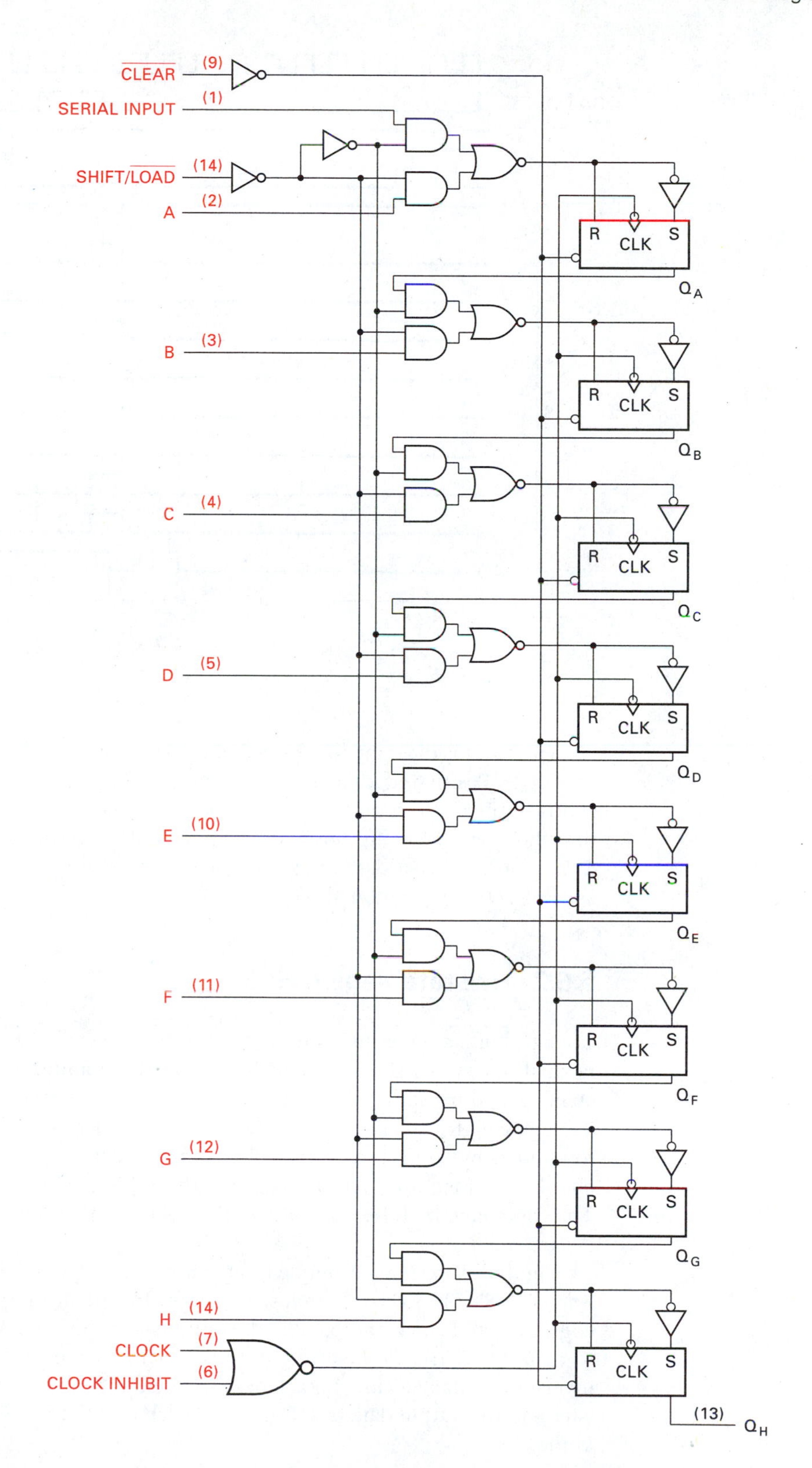

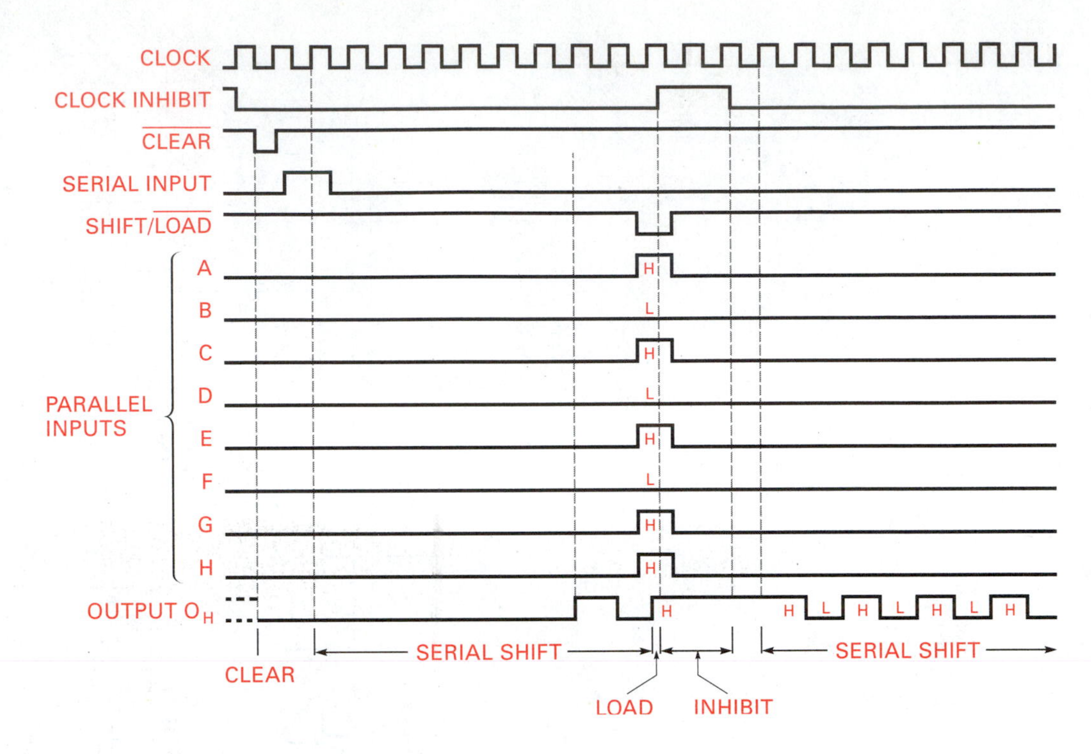

Figure 5.102
Timing diagram for the SN74xx166 shift register

The timing diagram for the SN74xx194 universal shift register is shown in Figure 5.104. Positive edge clocks synchronize loading or shifting of data depending on the status of the mode control inputs S_0 and S_1. When $S_1S_0 = 11$, external data are loaded into their respective flip-flops. When $S_1S_0 = 01$, internal data are shifted right. When $S_1S_0 = 10$, internal data are shifted left, and finally when $S_1S_0 = 00$, all loading or shifting activity is inhibited.

5.6.2 Tristate Registers

Tristate buffers were introduced in Chapter 4. Recall that tristate devices operate as normal binary switches when enabled and exhibit a high impedance (open) when disabled. The third state, a high impedance state, can be modeled as an open switch. Registers having tristate outputs can easily be connected to transfer data from one register to another by controlling which of several register outputs is connected to a common data bus. An eight-bit **tristate register**, the SN74AS534, is illustrated in Figure 5.105. The complement of the register's contents is loaded onto the bus when the tristate buffers are enabled.

Figure 5.106 shows a simple tristate inverter and an equivalent switch in the output line. The SN74AS534 is composed of eight *D*-type flip-flops with a common positive edge triggered clock and parallel data inputs and outputs. A single input, OC', is used to control the tristate inverting output buffers. When OC' is low each of the tristate inverters is enabled (low impedance), providing a logical connection between the register and the output data bus. Data are loaded into the SN74AS534 on the positive edge of the clock.

Figure 5.103
SN74xx194 universal shift register

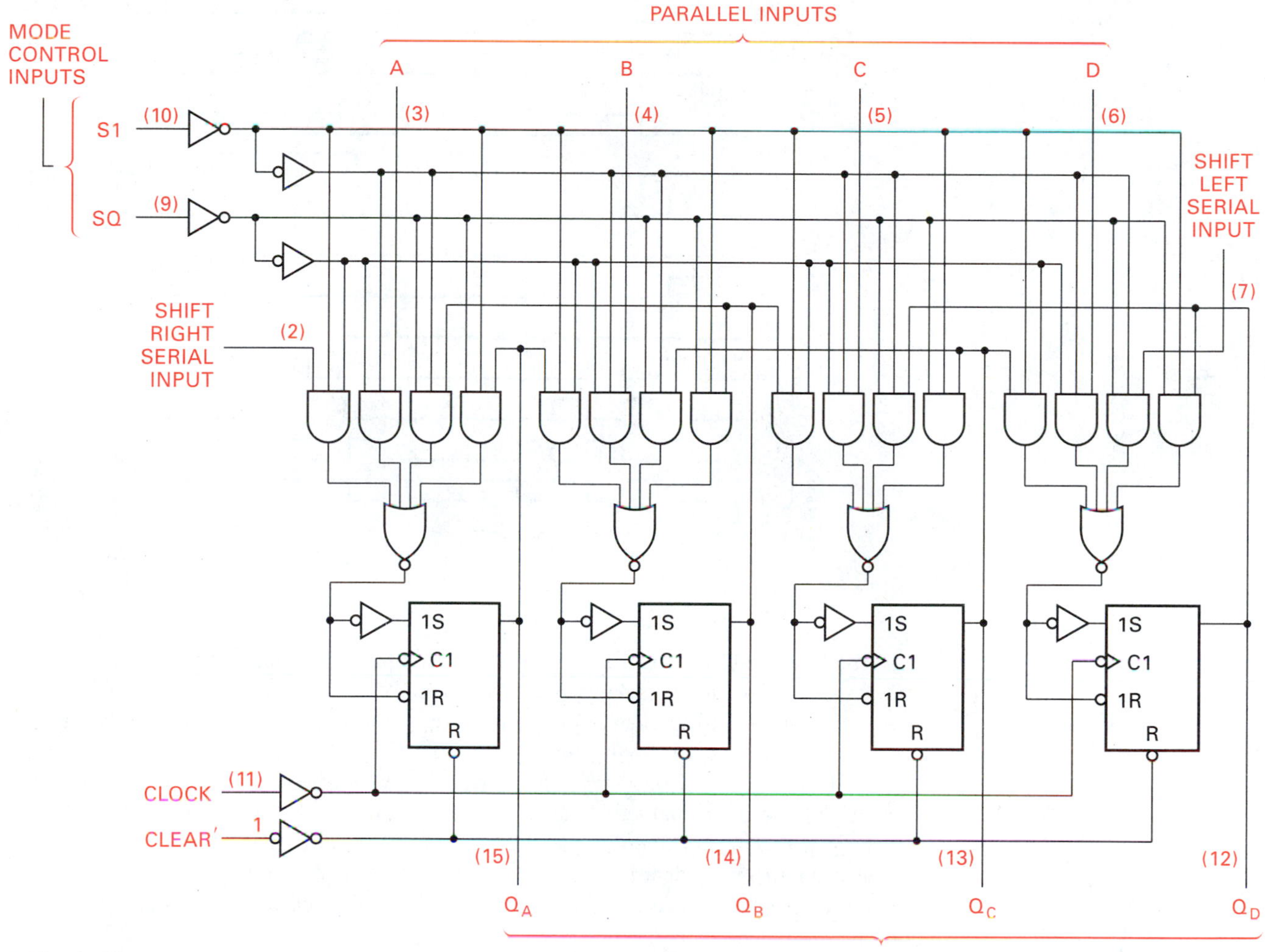

(a) Logic Diagram

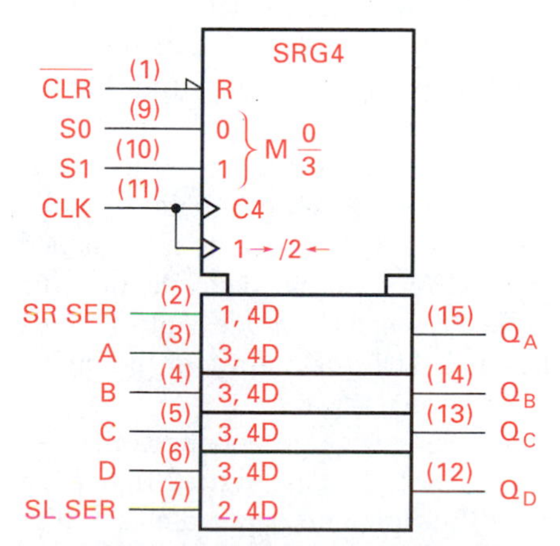

(b) IEEE Symbol for a 4-bit Shift Register

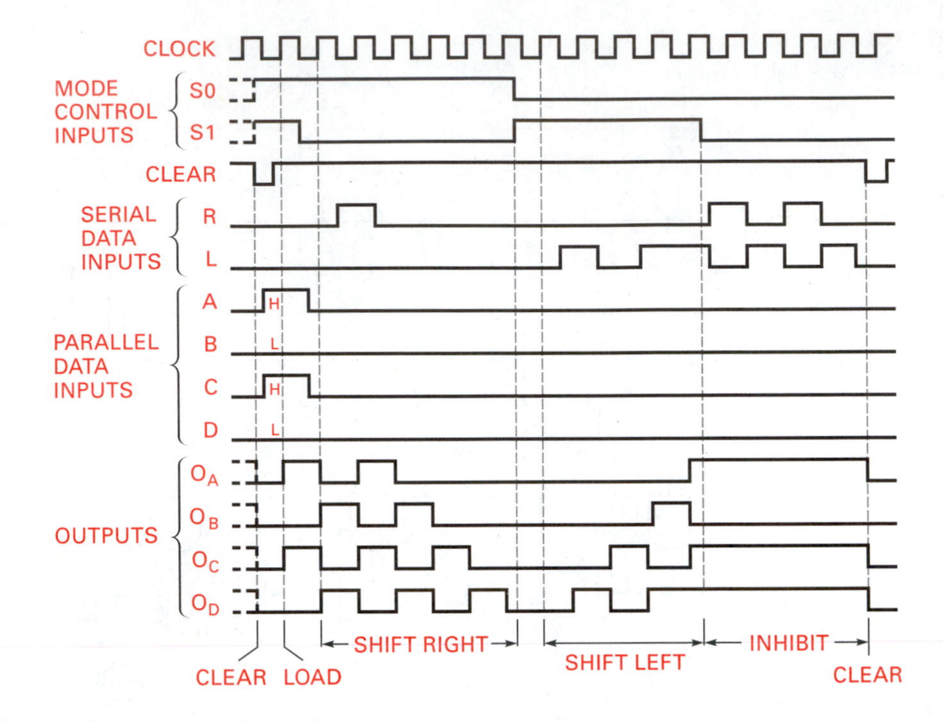

Figure 5.104
Timing diagram for the SN74xx194, universal shift register

The SN74LS173, shown in Figure 5.107, is a four-bit tristate register. Parallel data inputs are under the control of two active low input controls, G_1' and G_2', which control the quad 2-to-1 multiplexers. One multiplexer input is external data, and the other is the present contents of the flip-flop. When both G_1' and G_2' are low (G_1' AND G_2') external data are loaded into each of the four flip-flops on the positive clock edge. The tristate inverters are enabled by active-low control inputs M and N. When M and N are active low, the output data are passed through the tristate inverters to the data bus. A single asynchronous clear input can be used to reset the data register.

Figure 5.108 shows a more complicated tristate register, one that stores, shifts right or left, and has multiple tristate control inputs.

The IEEE logic symbol for the SN74ALS299 is shown in Figure 5.109. The control inputs are on the top left portion of the symbol; data inputs enter the left and the data outputs exit from the right of the bottom part of the diagram. Notice that inputs A through H can also be outputs Q_A through Q_H. Multiplexing the input-output lines is accomplished by tristate output controls G_1' and G_2'.

Notice, in the Figure 5.108 logic diagram, that a 4:1 multiplexer controls the D input to each of the storage flip-flops. One input to this multiplexer originates from the tristate buffer output associated with the flip-flop. When the tristate is not enabled ($G_1'G_2'$ are not both active low), then the X/Q_X line is free to input data. Data can be loaded into the flip-flops, through the 4-to-1 multiplexer, when the mode control inputs S_0 and S_1 are both asserted and the tristate is disabled. The dual nature of the multiplexed parallel input data lines are indicated by bidirectional arrows.

Shifting either right or left is also possible with this device, under the control of mode inputs S_1 and S_0. When $S_1S_0 = 01$ and $G_1'G_2' = 00$, data are shifted right. The shift-right serial input data are loaded into leftmost flip-flop, and data from each flip-flop is shifted right, through the 4-to-1 multiplexer. A shift left occurs when $S_1S_0 = 10$.

Figure 5.105
SN74AS534 eight-bit tristate data register

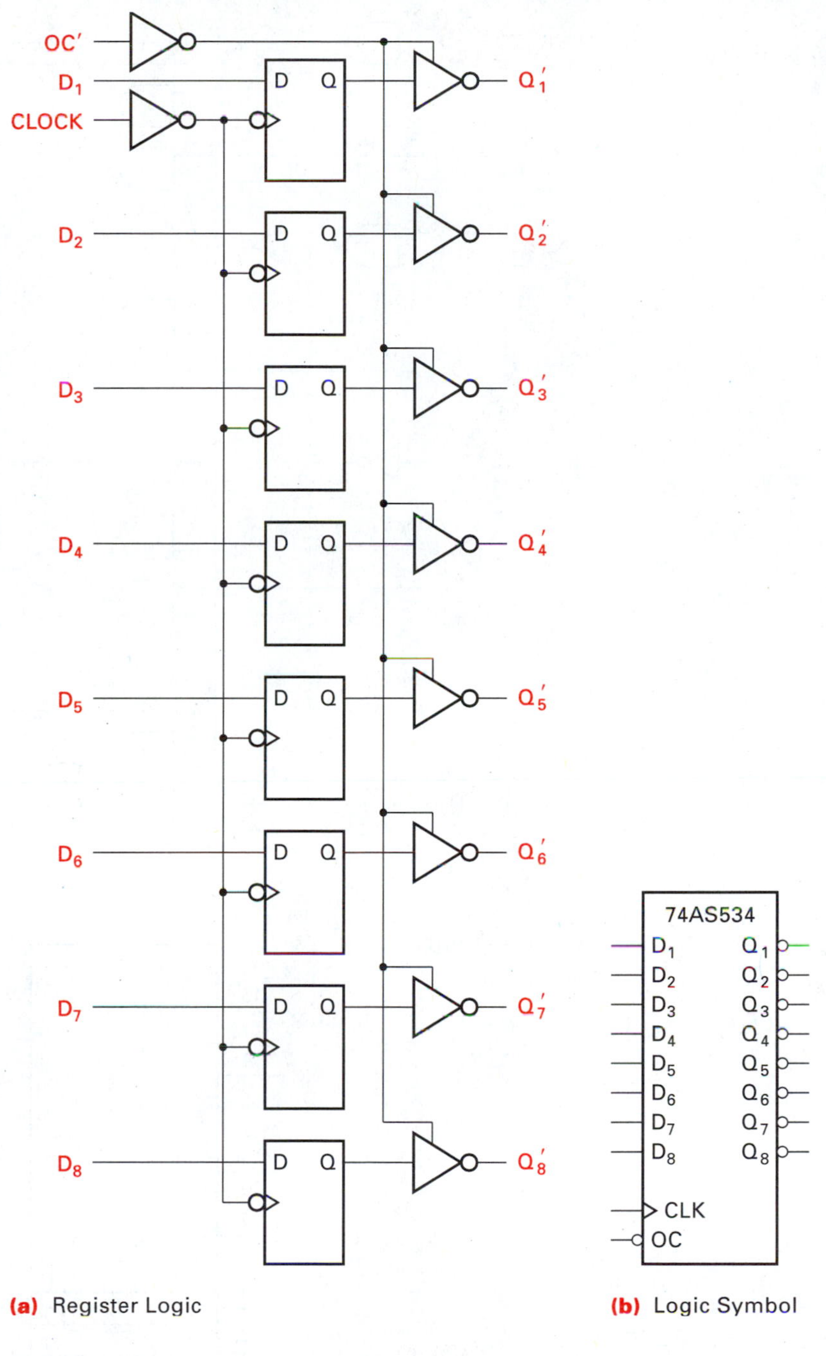

(a) Register Logic

(b) Logic Symbol

Figure 5.106
Tristate inverter

EN

A X

X = A′ WHEN EN = 1

EN

A X

X = A′ WHEN EN = 1

Figure 5.107
SN74LS173 four-bit tristate register

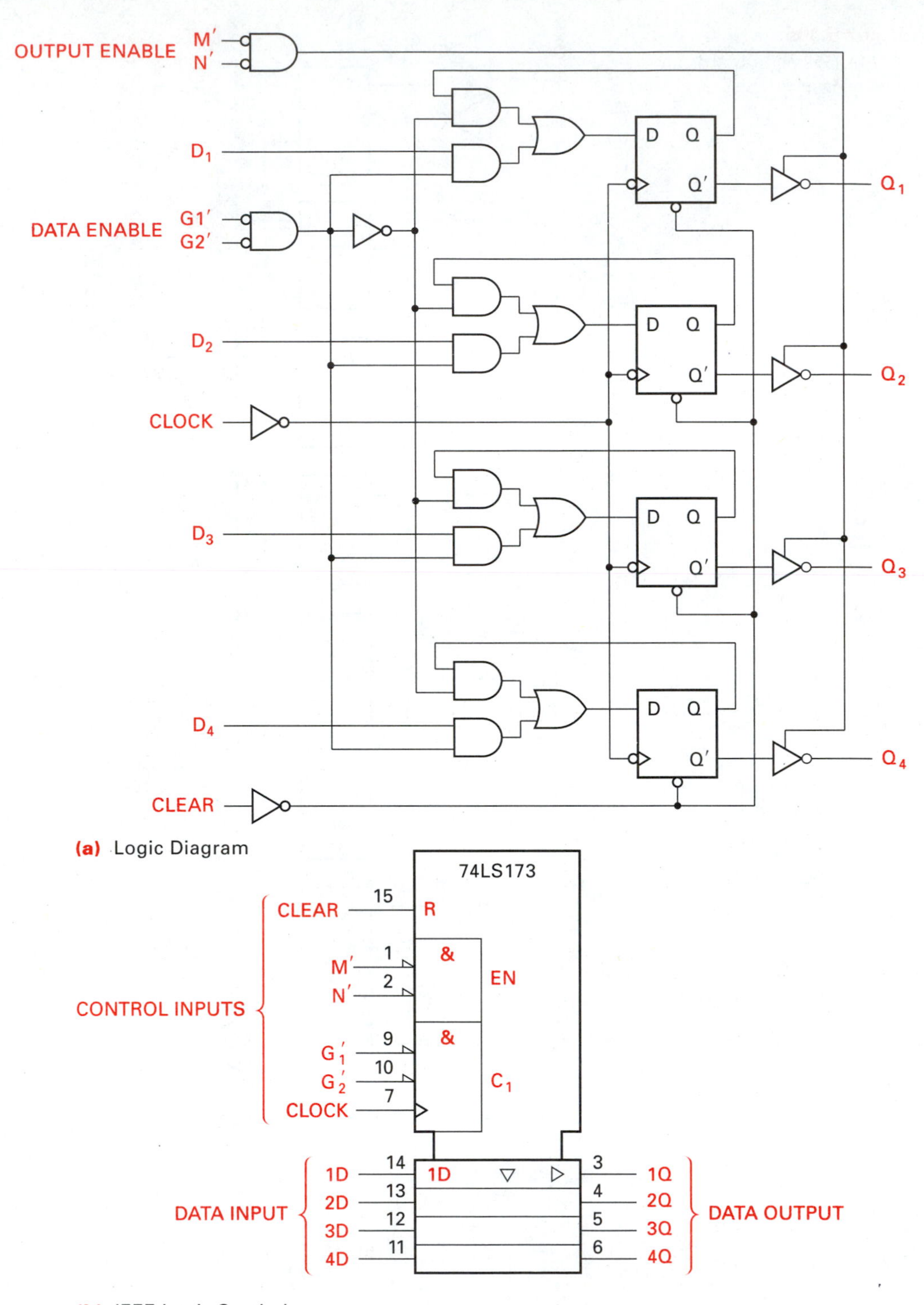

(a) Logic Diagram

(b) IEEE Logic Symbol

Figure 5.108
SN74ALS299 8-bit tristate parallel or serial load shift register

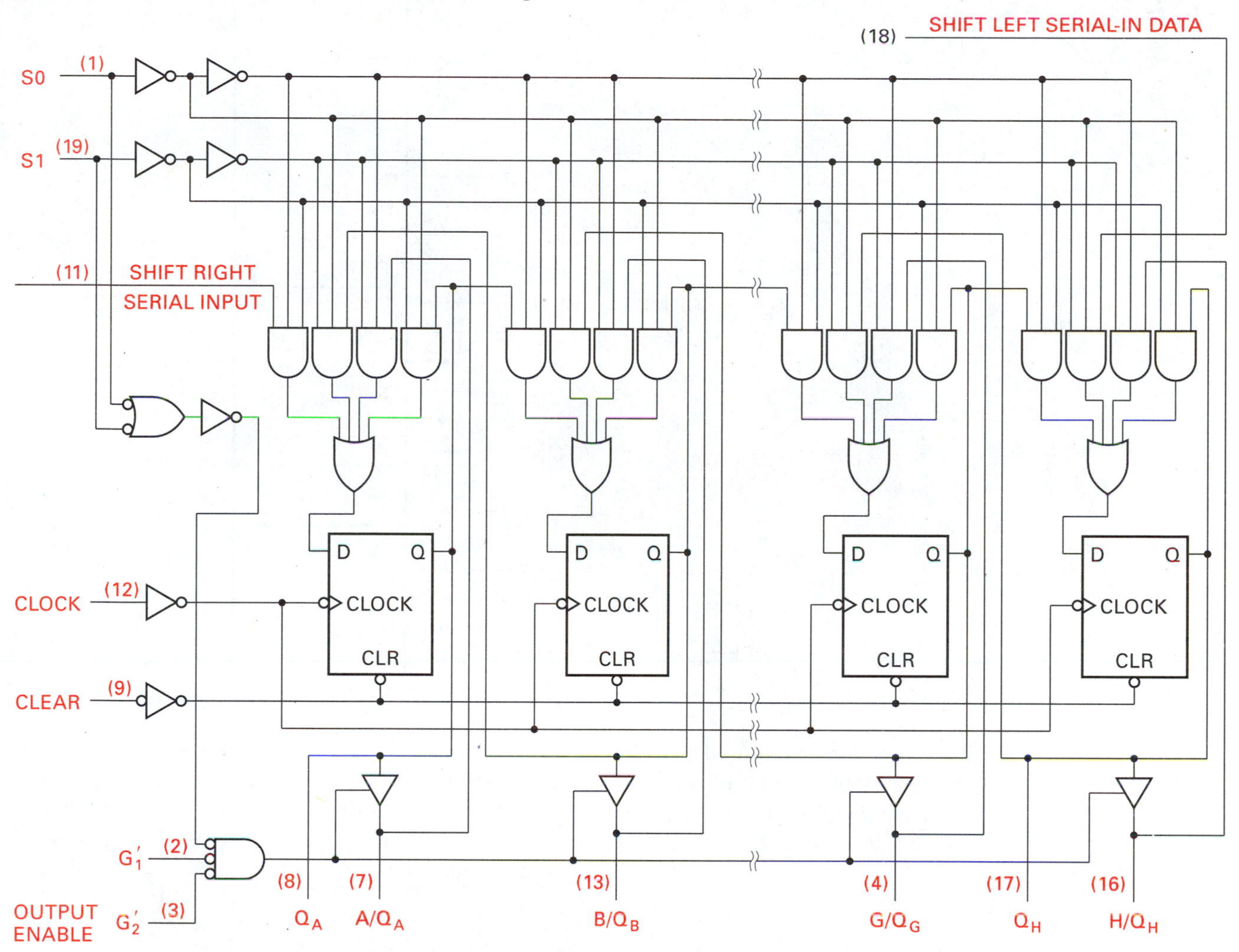

In this case, the rightmost flip-flop is loaded with data presented on the shift-left serial input.

Parallel loading of data occurs when $S_1S_0 = 11$. Storing previously loaded data values occurs when $S_1S_0 = 00$. All data changes, whether parallel loading or serial shifting, take place on the positive clock edge.

5.6.3 Registers Connected to a Common Data Bus

In Chapter 4 we discussed the use of tristate devices as an approach to interconnect several data sources to a common bus. Tristate, with its high-impedance condition, provides a means to switch data sources on- or off-line. Prior to the development of tristate circuits, transfer of data between registers was typically accomplished using multiplex-

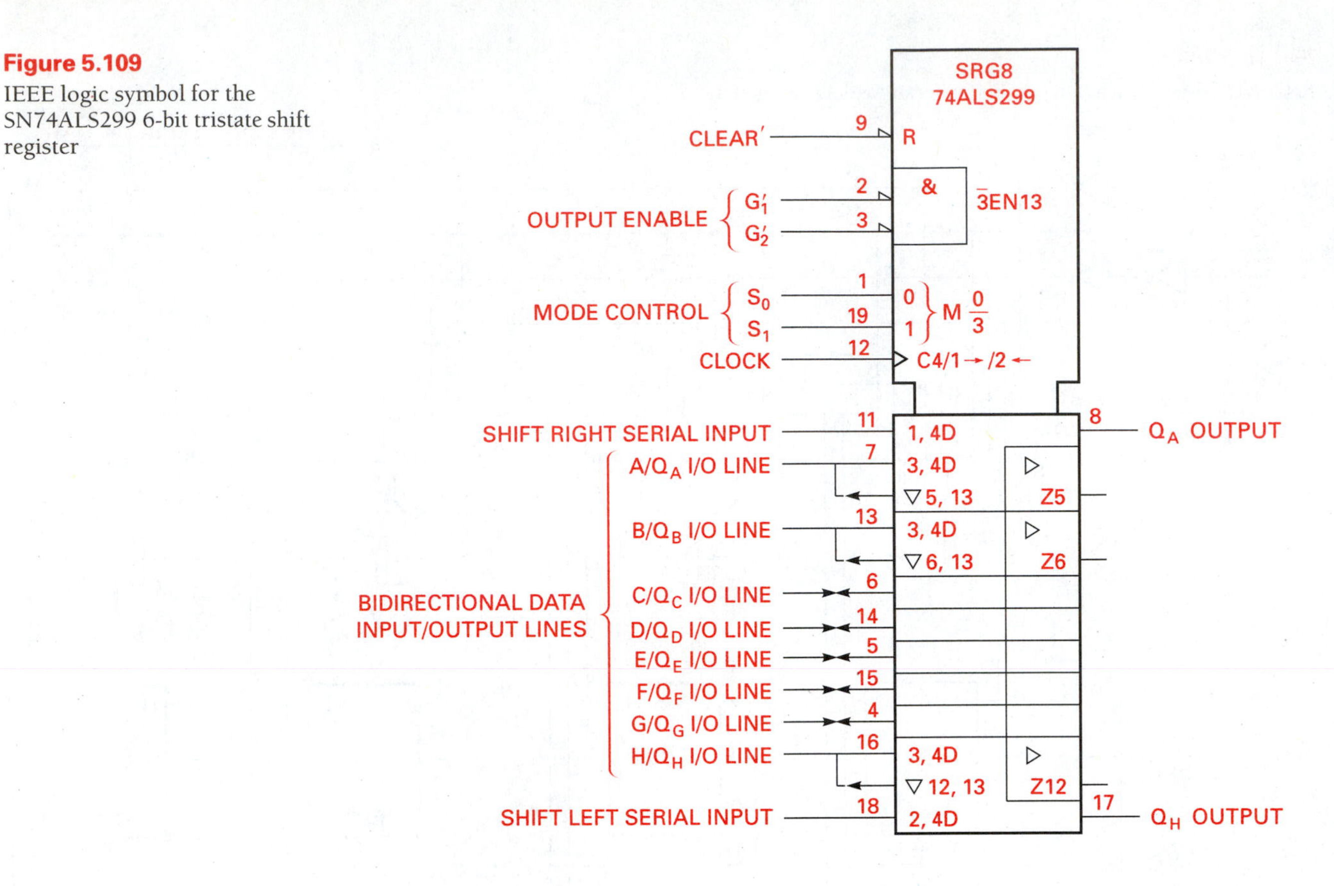

Figure 5.109
IEEE logic symbol for the SN74ALS299 6-bit tristate shift register

ers, which required considerably more logic than tristate registers. Consider the register system diagram illustrated in Figure 5.110. Each eight-bit register has tristate outputs. The eight-bit data bus provides a transmission path for sending and receiving digital information. Each bit position is common on all registers and the inputs and outputs for each register are connected. Separate input load and output enable commands are provided.

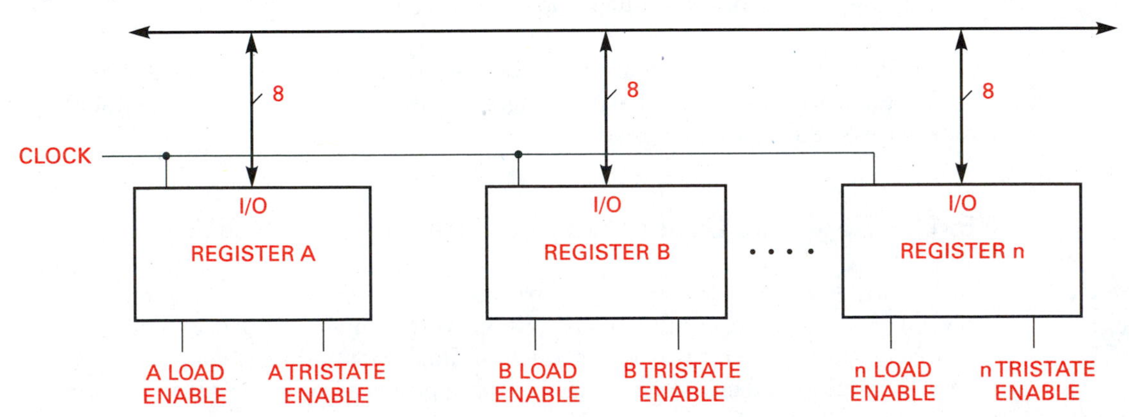

Figure 5.110
Register system illustrating a common data bus

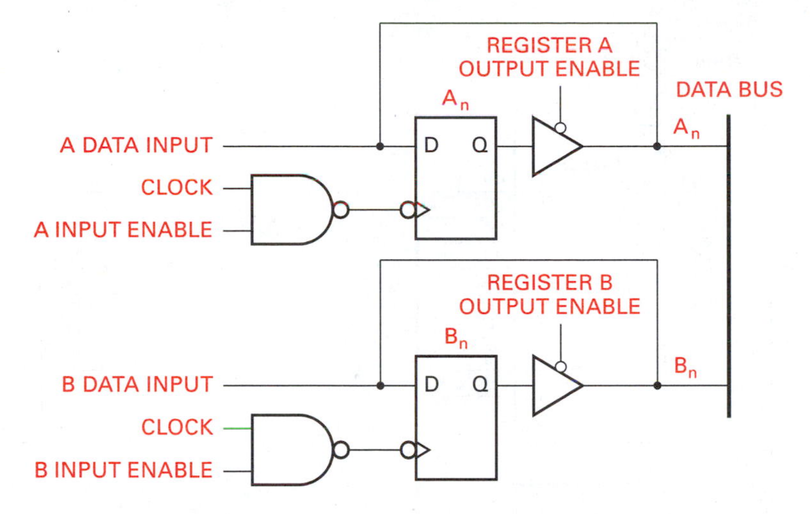

Figure 5.111
Partial logic diagram illustrating n-bit position flip-flops in a common bus register system

To transfer data stored in register *A* to register *B*, for example, the *A* output enable must be asserted and, after some time, the input enable for register *B* must be asserted. The source register output must be activated prior to the assertion of the destination register input enable to accommodate the register setup and hold timing requirements.

A partial logic diagram showing a single data line from each register is illustrated in Figure 5.111. Each flip-flop represents the same data bit position for each register. All are connected to the common data bus via the tristate devices. Only one tristate enable can be active at a time. To enable more than one register would cause a bus contention. Data bus contention occurs when more than one data source is connected to the bus at the same time. This results in data errors because a logical 0 from one register bit position will pull down a logical 1 from another (two different logic levels cannot exist on the same wire at the same time).

A register system, such as the one shown in Figure 5.112, can be realized using any number of integrated circuit registers. One good candidate is the SN74xx299, an eight-bit parallel I/O register with tristate outputs. The internal logic diagram for the SN74xx299 was shown in Figure 5.108. The device has separate load and tristate enable inputs, a clear and a positive edge trigger clock. The serial input-output feature is not used in this example.

By controlling the generation of individual register input load and output enable controls, digital data can be transferred from any one register in the system to any other register. Input and output enable control signals are generated such that only one source register is enabled at any time. Combinational logic decoders can provide the control needed to generate source and destination enable signals. For example, an eight-register system would need two 3-to-8 decoders, one for the source register and another for the destination register decoding. Both source and destination decoders receive register-code inputs, representing a unique binary value, identifying each of the eight registers. The decoder enable signals synchronize the timing for data transfer.

A logic diagram, showing the data bus and control functions, is illustrated in Figure 5.113. The source and destination codes are presented in Table 5.15. The timing parameters for the registers and decoders are given in Table 5.16. Figure 5.113 also shows the source and destination decoder logic to control data transfer among the registers shown in Figure 5.112.

Figure 5.112
Register system logic diagram

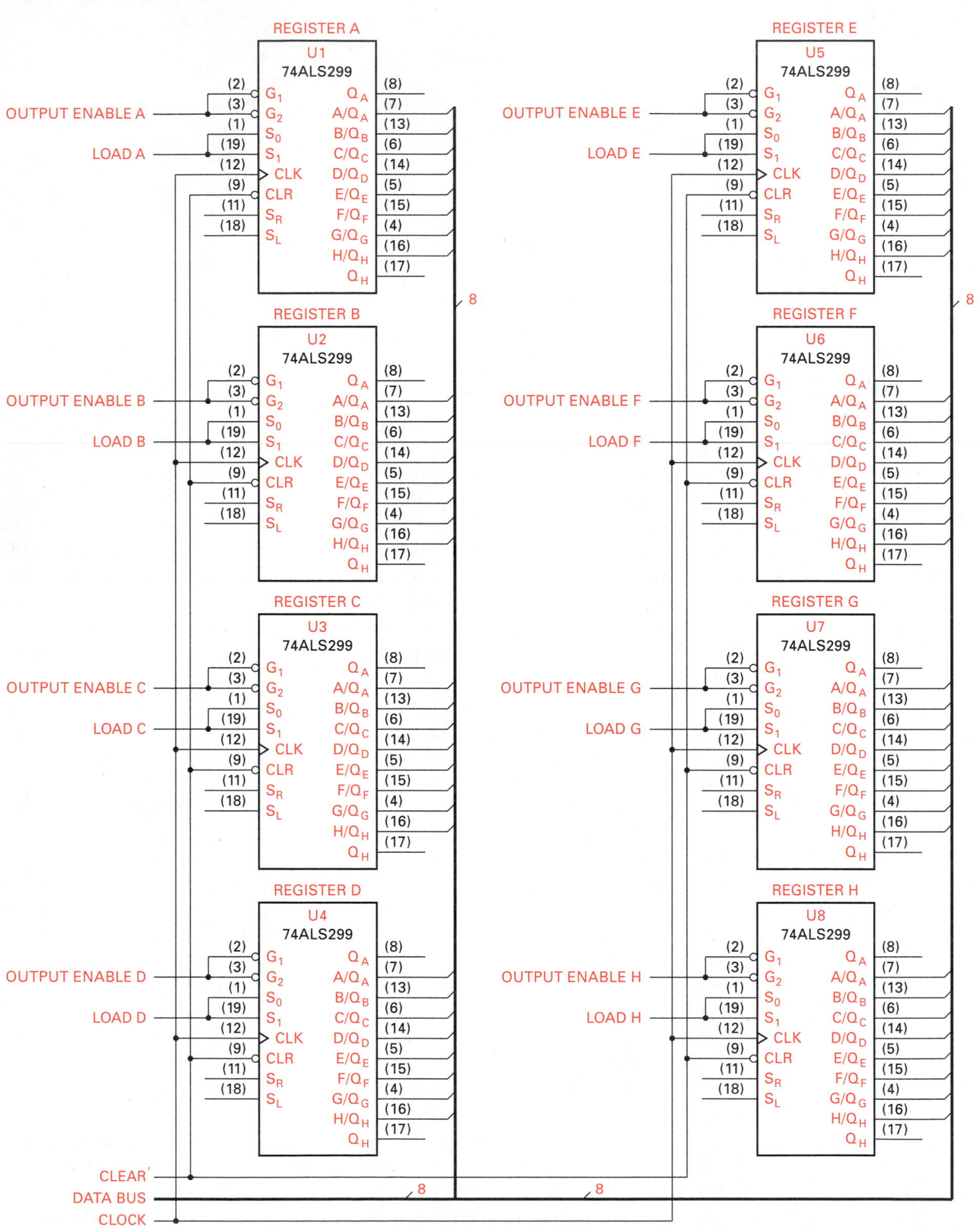

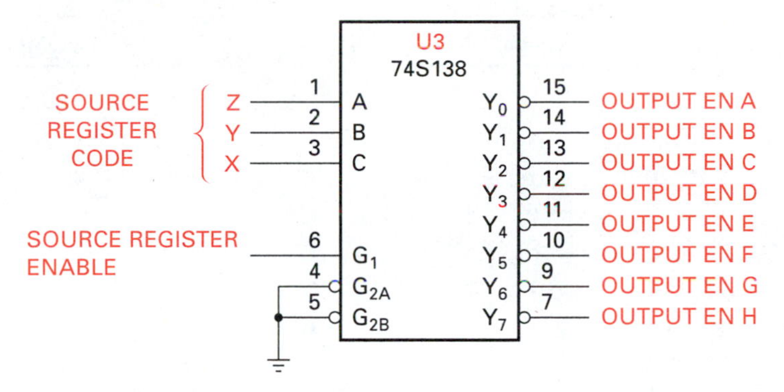

(a) Source Register Decoder

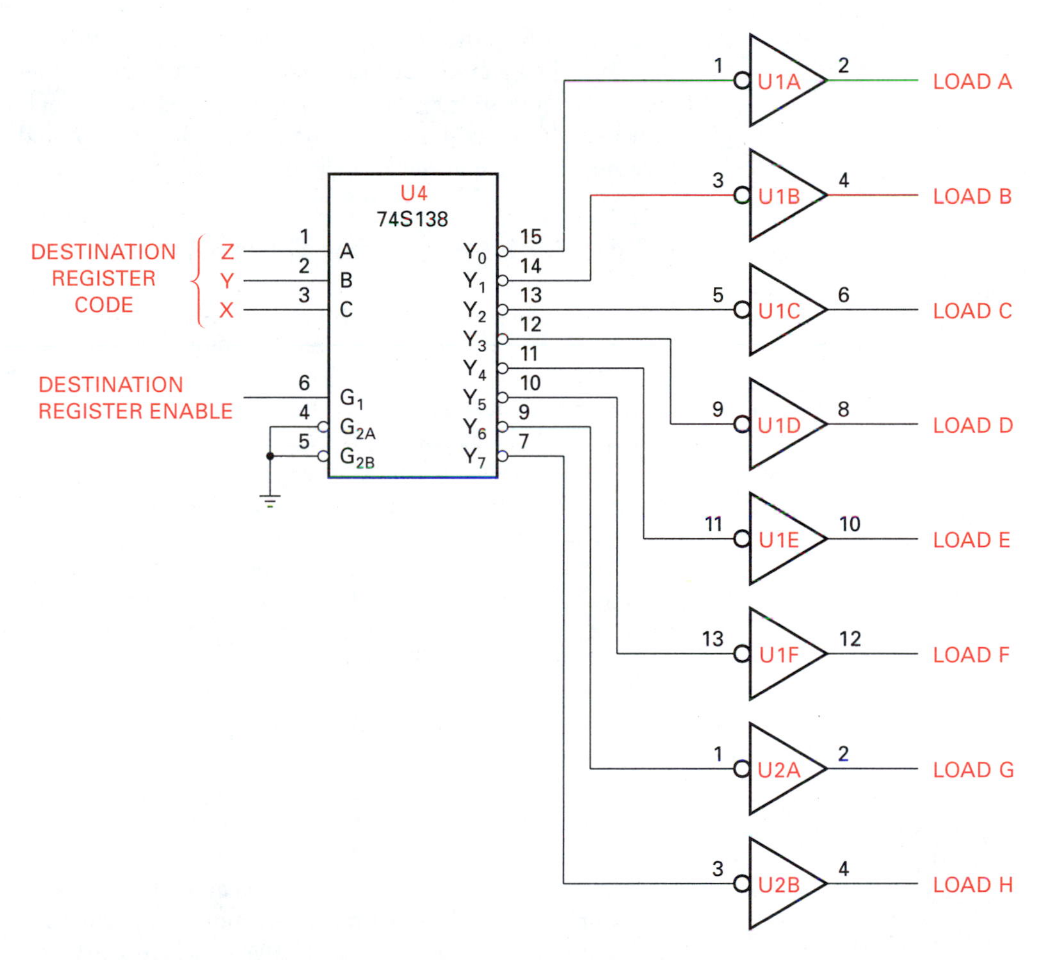

(b) Destination Register Decoder

Figure 5.113
Register system source and destination decoder logic

Register codes identify which register is the data source and which is the data destination. For example, if the source code is 101 and the destination code is 010, then the contents of register R_F would be transferred to register R_C. The register mode function table shows that a 00 on the mode input causes data to be held or stored. As long as S_1

TABLE 5.15
Source and destination register codes and register mode functions

Decoder			Source	Destination	Mode		
X	Y	Z	Reg	Reg	S_1	S_0	Function
0	0	0	*A*	*A*	0	0	Hold
0	0	1	*B*	*B*	0	1	Shift right
0	1	0	*C*	*C*	1	0	Shift right
0	1	1	*D*	*D*	1	1	Load data
1	0	0	*E*	*E*			
1	0	1	*F*	*F*			
1	1	0	*G*	*G*			
1	1	1	*H*	*H*			

and S_0 remain 00 the flip-flops will not change state. When mode control inputs $S_1S_0 =$ 11, parallel data are loaded on the next clock pulse. In this example, no shifting is required. Because shifting data is not required, the control of holding and loading data can be accomplished with a single data input. Therefore, S_1 and S_0 of each register are tied together and driven from a single inverter.

TABLE 5.16
Timing parameters for integrated circuits used in register system

Parameter	ALS299
t_{pZH}*	21 ns
t_{pZL}	30 ns
t_{pHZ}	20 ns
t_{pLZ}	15 ns
t_{pHL} (Ck to out)	25 ns
t_{pLH} "	39 ns
t_h	0 ns
t_{su} (select)	35 ns
t_{su} (data)	20 ns
Pulse width 0	10 ns
Pulse width 1	30 ns
f_{max}	30 MHz
	ALS138
t_{pLH} (ABC to Y)	22 ns
t_{pHL} (ABC to Y)	18 ns
t_{pLH} (EN to Y)	17 ns
	S04
t_{pHL}	4.5 ns
t_{pLH}	5 ns

*Z as in t_{pZH} indicates high impedance.

5.6.4 Register Transfer Timing Considerations

Timing specifications for the integrated circuits used in the register system are given in Table 5.16. Figure 5.114 is the logic diagram. The source and destination enable signals presented to each of the decoders are generated from external control logic. Decoder enable timing must conform to the register timing requirements. The source and destination enable signal timing is determined by analyzing the propagation delay, setup, hold, and high impedance to data times of the various integrated circuits. Because the register loads data on the positive edge of the clock, we can use that as a reference point to begin analysis.

From the timing data tables, we find that data must be present at least 20 ns, and the input enable (load RX) must be present at least 35 ns, prior to the clock. A positive load RX at the S_1S_0 inputs of the destination register, occuring 35 ns before the positive clock edge, loads the register from the data bus.

The destination decoder produces Load RX′, which is inverted by the S04 inverter. The low-to-high propagation delay of the inverter is 5 ns. The enable-to-Y output of the destination decoder, SNALS138, is 17 ns. This means that the enable input to the destination decoder must occur 22 ns prior to the positive clock edge. If the destination enable were present at the ALS 138 decoder sooner, the setup time for the select input to the register would be violated.

Data must be stable on the data bus at least 20 ns prior to the positive clock edge. The source decoder enable time can be found by calculating the delays from the decoder enable input to the positive clock edge. The source register's tristate outputs must be active to place data on the bus. A worst case time is used because eight data lines are changing and we have no way of knowing which is going from high to low or low to high. The t_{pZL} parameter, 30 ns, is the worst case. Output enable RY' is asserted 30 ns prior to when the data bus is stable. Source enable, S_E, is delayed 17 ns by the ALS138 decoder, placing the signal 67 ns ahead of the positive clock edge.

Figure 5.115 illustrates the timing requirements for transferring data from a source

Figure 5.114
Simplified source and destination register logic diagram

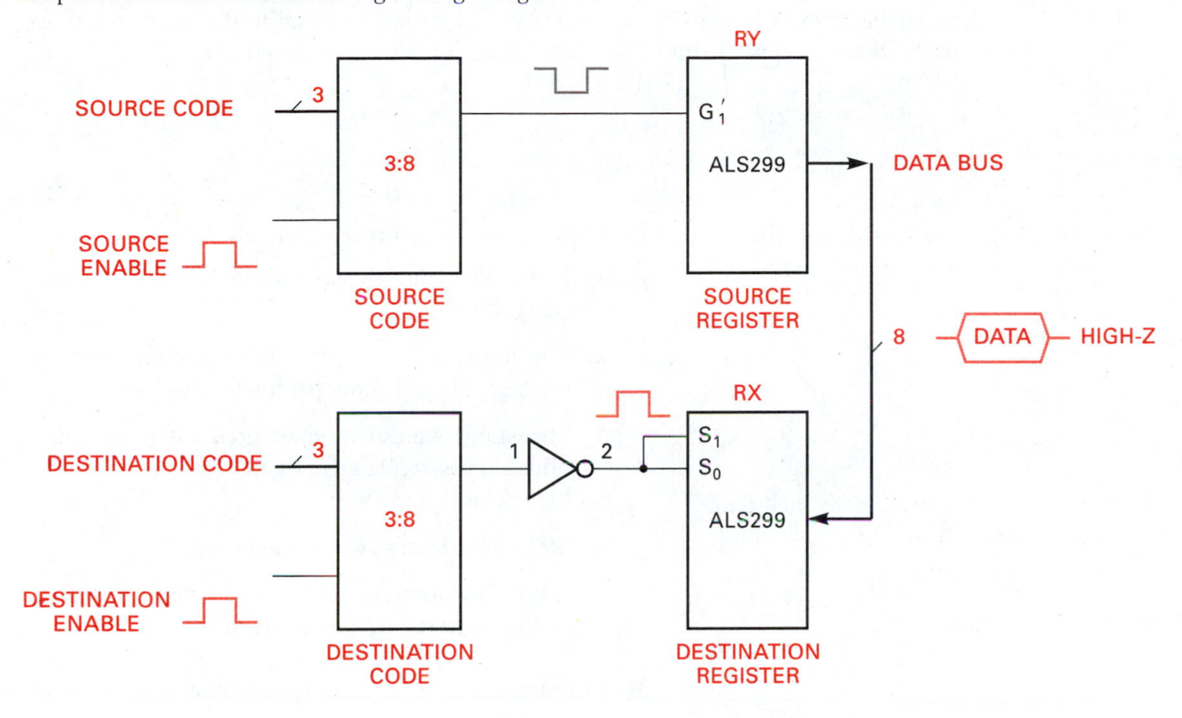

Figure 5.115
Register system timing diagram

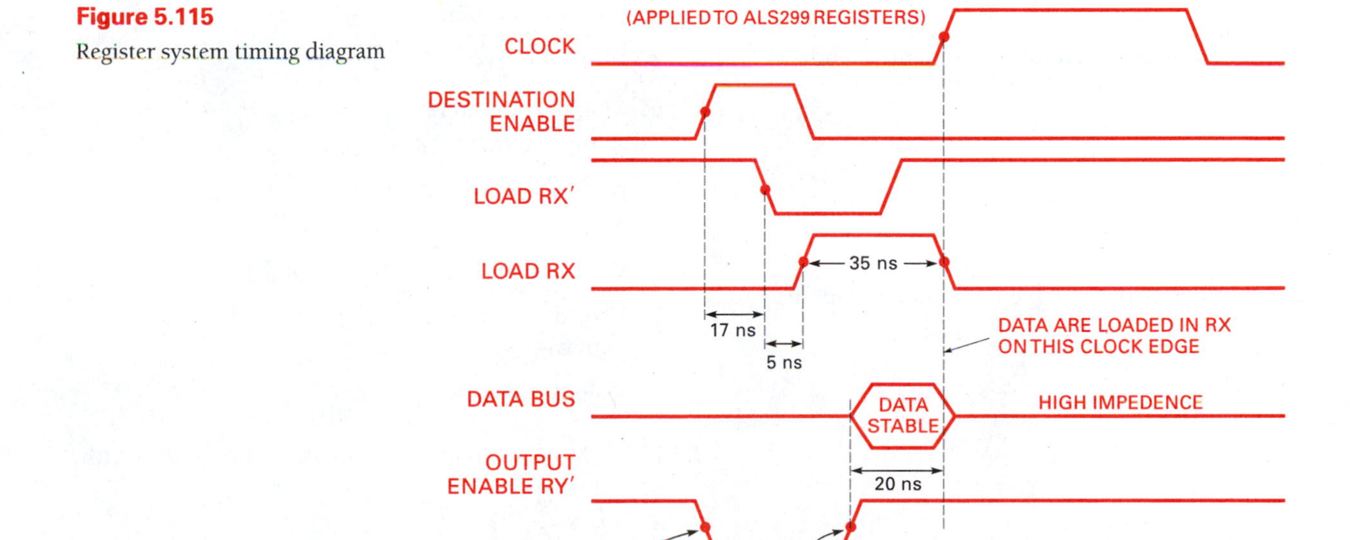

to a destination register. The times given are the minimum; that is, the system can be operated at a slower rate but no faster. The actual generation of the source and destination enable signals S_E and D_E is accomplished by logic external to the register system, and the timing requirements of that logic have not been considered here.

SUMMARY

A. Sequential circuit models lay the foundation for the relationship between combinational and sequential circuits:

1. Simple delay
2. Mealy
3. Moore

B. Four basic flip-flop types are introduced, including logic diagrams, excitation tables, and logic symbols:

1. *R–S* latch and clocked *R–S* flip-flops
2. *D* latch and clocked flip-flops
3. *J–K* flip-flops
4. *T* or toggle flip-flops

C. Flip-flop triggering is introduced, including

1. Positive and negative level triggering
2. Positive and negative edge triggering

D. Flip-flop timing parameters are presented and explained:

1. Setup time
2. Hold time
3. Propagation delay time

E. Several commercially available integrated circuit flip-flop packages are analyzed:

1. SN74LS74
2. SNALS114A
3. SN74ALS109A

F. Clock schemes are discussed.

1. Edge triggering
2. Master-slave

G. Flip-flop metastability was defined:

1. Metastable state exists halfway in between a logical 0 and 1.
2. Flip-flops can enter the metastable state when the setup and hold timing parameters are not met.
3. Metastable states can be entered when asynchronous inputs create signal glitches (short duration logical level transitions).
4. Metastable states are to be avoided.
5. A circuit example to reduce the probability of entering metastable states is given.

H. Counters, including logic diagrams and timing diagrams, are discussed:

1. Divide by 2 counters
2. Divide by 4 counters
3. Divide by 8 counters
4. Johnson counters
5. Ring counters
6. MSI TTL counters.
   - **a.** Binary (SN74163) counters
   - **b.** BCD (SN74390) counters
   - **c.** Binary (SN74393) counters
   - **d.** Binary up-down counter (SN74169) counters
7. Use of counters to generate pulse waveform patterns
8. Digital clock using counters

I. IEEE MSI symbols and notation are used with counters:

1. AND
2. OR
3. EX-OR

4. Enable
5. Control modes

J. Registers of different types are examined:
  1. Input-output registers
     a. Parallel input/parallel output
     b. Parallel input/serial output
     c. Serial input/parallel output
     d. Serial input/serial output
  2. Universal shift registers
  3. Tristate registers

K. Tristate registers are connected to a common data bus:
  1. Data source registers
  2. Data destination registers
  3. Control logic for identifying source and destination registers
  4. Timing considerations for data transfer between registers

REFERENCES

TTL Data Books, by Texas Instruments. These books include a wide range of digital integrated circuit functions and types. The integrated circuit flip-flops, counters, registers, and miscellaneous devices used in this chapter are Texas Instrument devices. The data books provide logic diagrams, functional descriptions, and truth tables as well as device parameters. Three such books are examples. *TTL Data Book,* vol. 2, by Texas Instruments, covers standard, high-speed, low-power, Schottky and low-power Schottky small- and medium-scale digital integrated circuit devices. *ALS/AS Logic Data Book,* by Texas Instruments, covers advanced low-power Schottky and advanced Schottky digital small- and medium-scale integrated circuit devices. *Advanced CMOS Data Book,* by Texas Instruments, covers advanced CMOS digital integrated circuit small- and medium-scale devices.

MECL Device Data Book, by Motorola, covers small-scale and some medium-scale ECL digital integrated circuit devices.

Digital Design Principles and Practices, by John F. Wakerly (Englewood Cliffs, N.J.: Prentice-Hall, 1994). Chapter 7 covers flip-flops, sequential circuit definitions, and models. Chapter 9 covers counters and registers. Wakerly makes liberal use of logic and timing diagrams in circuit analysis.

Fundamentals of Logic Design, by Charles H. Roth, Jr. (Minneapolis–St. Paul: West, 1992). Roth covers flip-flop types in Unit 11. Basic counters are developed in Unit 12.

Digital Logic and State Machine Design, by David J. Comer (Philadelphia: Saunders, 1990). Flip-flops, counters, and registers are treated in Chapter 5.

Digital Logic Design, by G. Langholz, A. Kandek, and J. Mott (Dubuque, Iowa: Wm. C. Brown, 1988). Treatment of flip-flops, modeling of sequential circuits, counters, and registers are covered in Chapter 7. The treatment is complete but does not cover actual integrated circuit devices.

Digital Design, by M. Morris Mano (Englewood Cliffs, N.J.: Prentice-Hall, 1991). Flip-flop types and simple counters are developed in Chapter 6. Chapter 7 covers additional counters and registers. The treatment is complete but lacks coverage of actual integrated-circuit devices.

Introduction to Logic Design, by Sajjan G. Shiva (Glencoe, Ill.: Scott, Foresman, and Company, 1988). Shiva treats flip-flops and sequential circuit modeling in Chapter 8. Registers, counters, and some integrated circuit examples including timing analysis are covered in Chapter 9.

Designing Logic Systems Using State Machines, by Christopher R. Clare (New York: McGraw-Hill, 1973). Although dated, this paperback treatment of sequential machines provides some excellent coverage of sequential machine modeling.

Computer Design and Architecture, by L. Howard Pollard (Englewood Cliffs, N.J.: Prentice-Hall, 1990). This is a text on computer architecture rather than logic design. It is included here due to the treatment of register transfer using actual integrated circuits complete with timing analysis.

GLOSSARY

Asynchronous Without synchronization. In a digital system a series of pulses called *clock pulses* are used to synchronize changes in state variables. In an asynchronous digital system no such synchronizing clock is used. The state variables change at a rate determined by the logic propagation delay.

CLEAR An input to a flip-flop that is not controlled by the clock; therefore, called an *asynchronous input*. A flip-flop with a clear input also has synchronous excitation inputs. When the CLEAR input is asserted the flip-flop Q output is forced to a logic 0. A clear input to an MSI device may cause some or all of the internal flip-flops to reset ($Q = 0$).

Counter A digital counter is a collection of flip-flops arranged in such a manner as to change state in a prescribed sequence. For example, binary counter state variables change state in the binary code. A BCD counter counts in BCD code. Counters can be either synchronous or asynchronous.

Edge trigger Flip-flops change state when commanded to do so by a synchronizing clock pulse. When the flip-flop changes state on the clock edge it is said to be *edge triggered*. Both positive and negative edge triggered flip-flops are commercially available.

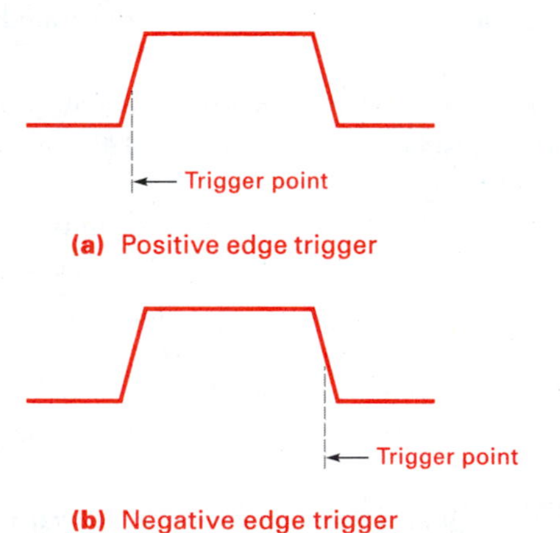

(a) Positive edge trigger

(b) Negative edge trigger

Excitation variable Inputs to flip-flops that excite the device to change state. Four basic types of synchronous flip-flops are commonly used: *R–S*, *J–K*, *T*, and *D*. The synchronous inputs cause the flip-flop to change state with respect to the clock pulse. Synchronous flip-flops can also have asynchronous excitation inputs such as preset and clear.

Flip-flop A bistable digital device that changes states based on its excitation inputs. A flip-flop stores data: it "remembers" what the last excitation input was, even after the input is changed. The synchronous flip-flops commonly used are *SR*, *JK*, *D*, and *T*.

Hold Time (t_h) The time that synchronous excitation input data must be held after the clock pulse is gone. For an edge triggered flip-flop, the hold time is the length of time the input data must be present after the clock edge is past.

Johnson counter A digital counter in which the Q output of one flip-flop is directly connected to the excitation input of the next in such a fashion that each Q value is transferred to the next flip-flop. The Q output of the least significant flip-flop is connected back to the input of the most significant flip-flop. The feedback connection allows the counter to produce a series of pulses from each flip-flop output that is delayed from the preceding flip-flop by the period time of the clock.

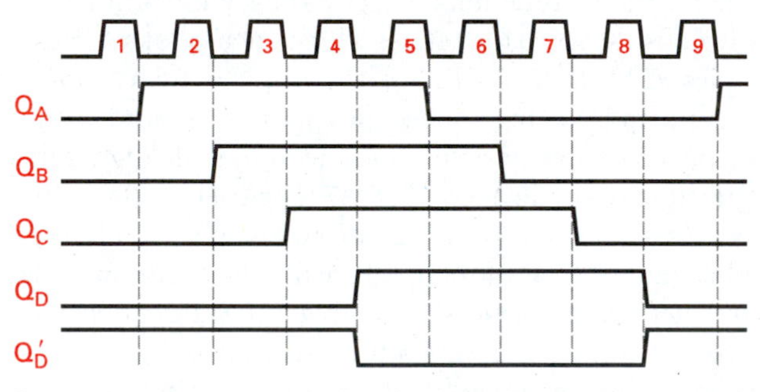

Four-bit Johnson counter timing of state variables

Latch A pair of inverting gates (NAND or NOR) arranged such that the output of one gate is connected to the input of another. This connection scheme permits the latch to remember (by changing state) what the excitation input was. A latch is not considered the same as a flip-flop, although flip-flops contain latches. A flip-flop is clocked and a latch is not.

Level trigger A trigger input value to a flip-flop that initiates a state change on an input voltage level instead of a clock edge. Both negative and positive level flip-flops are available. Level flip-flops are more sensitive to spurious state changes than edge triggered devices.

Mealy A class of sequential circuits in which the output variables (O) are a function of both the state variables

(S_t) and the input variables (I). The output relationship can be written as $O = f(S_t, I)$. The next-state (S_{t+1}) variable is a function of the present state (S_t) and the input (I), written as $(S_{t+1}) = f(S_t, I)$.

Metastable Flip-flops have two stable states. A third state (a metastable state, not a true stable state) can be entered under certain conditions. The metastable state is undesirable and great lengths are undertaken to keep high-speed sequential logic from entering metastability. Slow clock speeds prevent metastability problems; however, fast clock speeds greatly improve system performance so the metastability difficulties are a concern to logic designers.

Modulo-*n* counter A modulo counter produces a count that is evenly divisible by 2. Counters that count to 2, 4, 8, 16, . . . , are called *modulo counters.* A modulo-*n* counter is one that has a terminal count value (n) evenly divisible by 2.

Moore A class of sequential machines whose output variables are a function of the present state (S_t). The next-state variables (S_{t+1}) are a function of the present state (S_t) and the input (I).

Propagation delay Propagation delay time is the time interval between the input signal and the output signal measured at the 50% points.

Pulse width A voltage or current pulse has a minimum and maximum value. The time measured at the 50% points between the minimum and maximum values is called the *pulse width.*

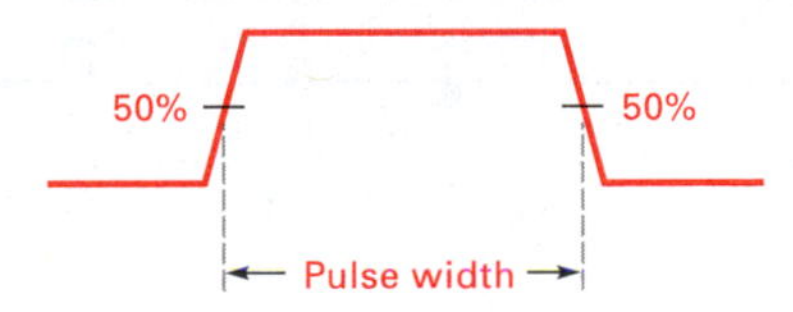

PRE (preset) An asynchronous excitation input to a flip-flop that causes the flip-flop Q output to be a 1. A preset input to an MSI device would cause some or all of the internal flip-flops to be set ($Q = 1$).

Register A collection of flip-flops arranged in such a manner to store binary input data. A register can be loaded in parallel (all internal flip-flops set or reset on one clock pulse) or in serial (data are presented one bit at a time and loaded into the register using a number of clock pulses).

Reset An asynchronous input to a flip-flop or latch that causes the Q output to be a 0. *Reset* is synonymous with clear.

Ring counter A ring counter is a collection of synchronously clocked flip-flops, usually D flip-flops. Each flip-flop output is connected to the synchronous excitation input of the next flip-flop so that the data are transferred from a left flip-flop to a right flip-flop on each clock pulse, much like a shift register. The rightmost flip-flop output is fed back into the synchronous input of the leftmost flip-flop, permitting the contents of the counter to recirculate. Separate parallel data inputs permit the parallel loading of all the flip-flops at the same time. Usually the parallel load inputs are asynchronous. Once a data pattern has been loaded into the counter, it will shift the counter contents right on each clock pulse to produce a repeating bit pattern output.

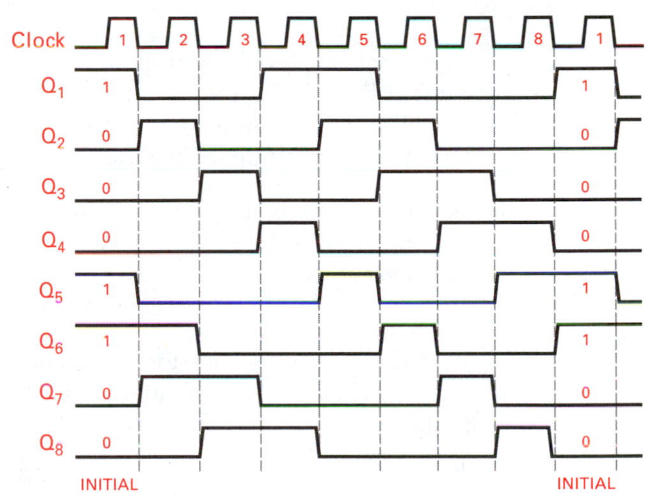

Four-bit Johnson counter timing of state variables

Set An input to a latch or flip-flop. When set is asserted, $Q = 1$. *Set* is synonymous with preset.

Setup time (t_{su}) Setup time is the time that an input to a flip-flop, counter, or register must be stable prior to the arrival of the clock. With edge triggered devices the data inputs must be stable prior to the clock edge that triggers the flip-flops.

Shift register A shift register is a collection of flip-flops whose inputs and outputs are connected in such a fashion to be able to (1) parallel load and output data, (2) parallel load and serial output data, (3) serial load and parallel output data, and (4) serial load and serial output data. Once data have been loaded, they can be shifted either right or left, depending on the register logic and mode control signal values.

Skew When a system clock pulse train is distributed to flip-flops or devices containing flip-flops through multiple distribution paths, a difference in the clock arrival times can exist. This difference in clock arrival times is called *clock skew*. Clock skew can cause problems in a digital system if flip-flop triggering must be simultaneous.

State variable Sequential circuits typically use flip-flops as the storage or memory devices. In a sequential circuit the flip-flop outputs are used to encode the circuit "state"; therefore, the flip-flop outputs are called *state variables*.

Synchronous To be synchronized with a common clocking scheme. Sequential circuits are classified as synchronous (with a clock) or asynchronous (without a clock).

Tristate register A register with tristate buffers connected between the flip-flop outputs and the rest of the system. Controlling the tristate buffers permits the contents of the register to be presented to a common data bus without bus contention.

QUESTION AND PROBLEMS

Section 5.1 to 5.2

1. Sketch the "universal" model block diagram for sequential circuits. Label all the system inputs, outputs, state variables, and excitation variables. Define each variable and its relationship to the other.

2. Sketch the model diagrams for a Mealey and a Moore sequential circuit. Identify all variables and describe the difference between a Mealey model and a Moore model.

3. Draw the logic diagram, construct the excitation table, and write the characteristic equations for the following flip-flops:

   a. *R–S* latch

   b. Gated *R–S* latch

   c. *JK*

   d. *T* or toggle

   e. *D* or data

4. In the figure for this problem, complete the timing diagram for a NAND latch.

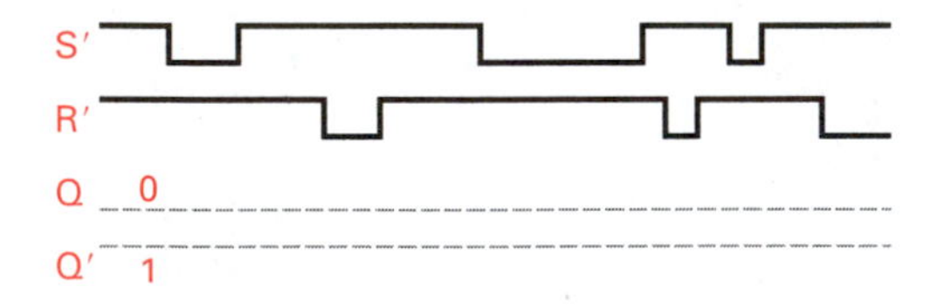

5. In the figure for this problem, complete the timing diagram for a gated *R–S* latch.

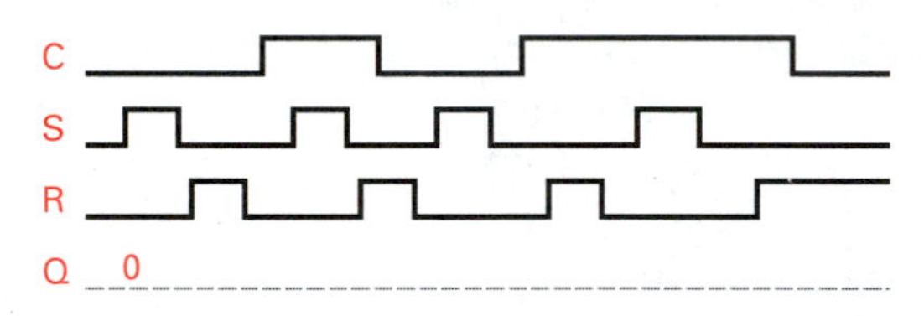

6. In the figure for this problem, complete the timing diagram for a positive edge triggered *J–K* flip-flop.

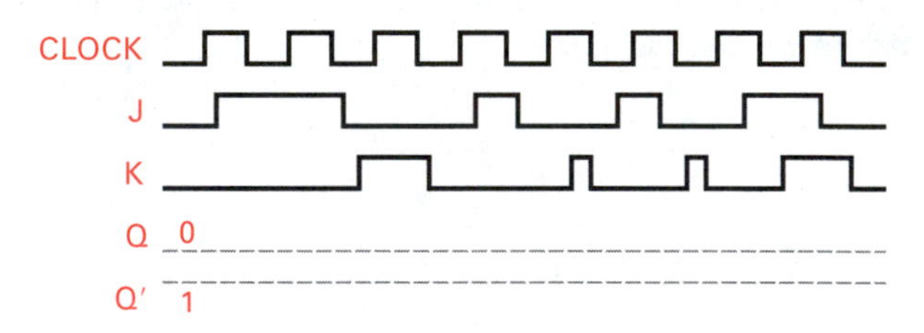

7. In the figure for this problem, complete the timing diagrams for negative edge triggered *T* and *D* flip-flops.

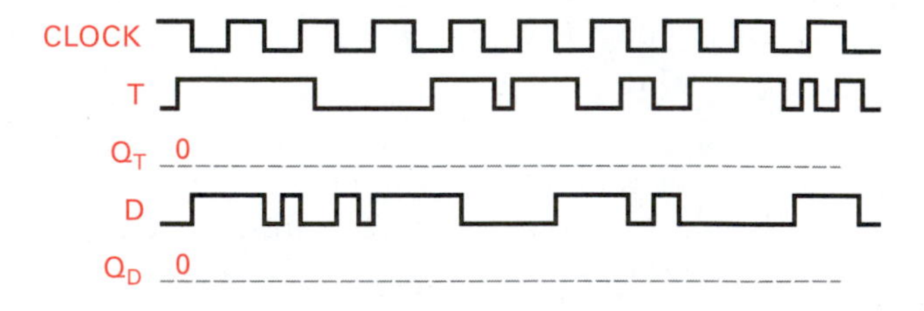

8. A diagram of a *J–K* flip-flop is shown as the figure for this problem. Explain why this is not a practical flip-flop.

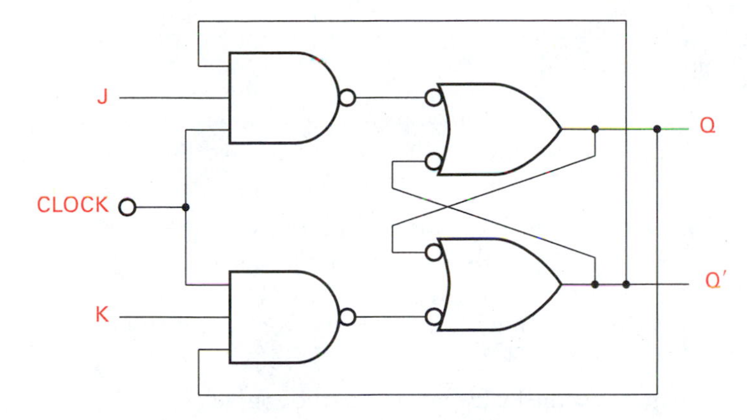

9. What are the advantages of an edge triggered flip-flop over a level triggered device?

10. Sketch the block diagram for a master-slave flip-flop. Explain its operation and features.

11. Sketch the standard distinctive shape logic symbols for the following:

- **a.** Negative edge triggered *D* flip-flop with active-low clear
- **b.** Positive edge triggered *J–K* flip-flop with active-low clear and active-high preset

12. Given the *J–K′* flip-flop shown in the figure for this problem, complete the timing diagram.

Sections 5.1 and 5.2

Problem 12

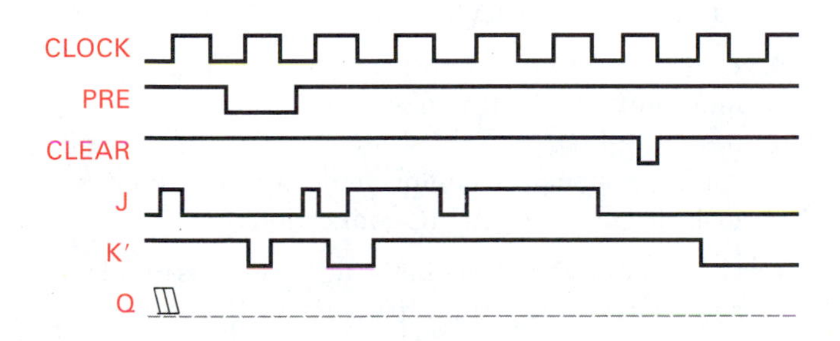

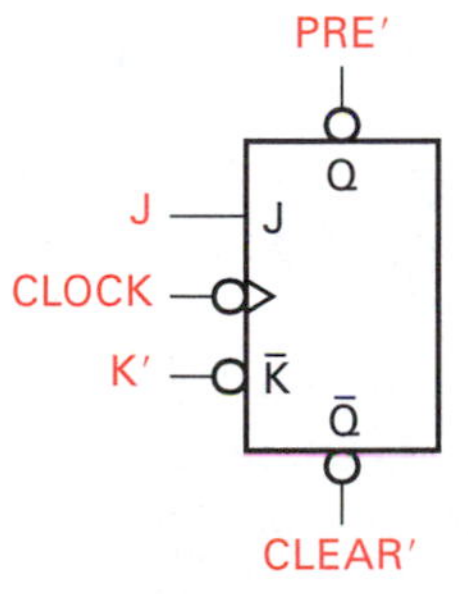

13. Sketch the IEEE symbols for the following:

- **a.** Positive level triggered *T* flip-flop with active-low preset and clear
- **b.** Negative edge triggered *D* flip-flop with active-high present and active-low clear

14. Draw a block diagram of a master-slave *J–K* flip-flop. Sketch a clock waveform and indicate on the waveform when key circuit activity occurs.

15. Describe the problem of catching 0 and 1 associated with a level triggered master-slave flip-flop. Draw a timing diagram to support your explanation.

16. Sketch a full period of a clock waveform. On the sketch, identify the following:

- **a.** Rise time
- **b.** Fall time
- **c.** Period time
- **d.** 1 pulse width

17. Sketch a diagram illustrating clock "skew." Describe the effects of clock skew in high-speed systems.

18. Given the logic diagram for this problem, complete the partial timing diagram.

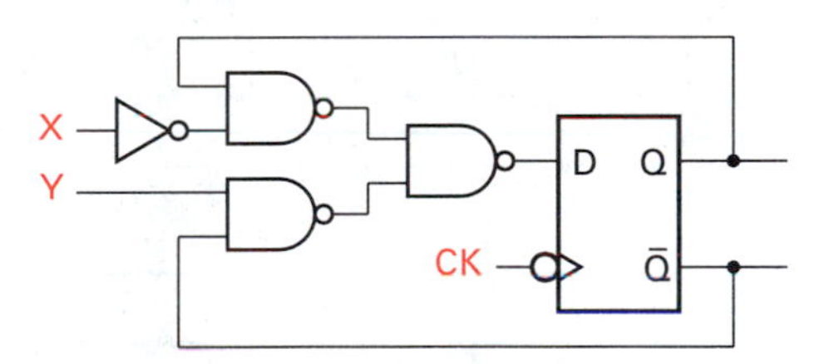

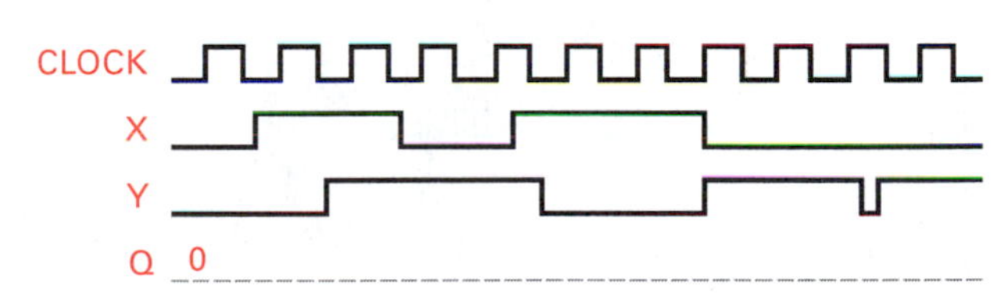

19. Given the logic diagram for this problem, complete the timing diagram.

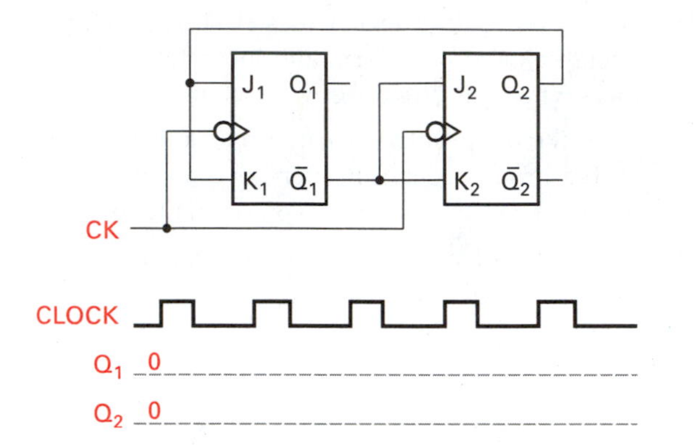

20. Given the figure for this problem, complete the timing diagram.

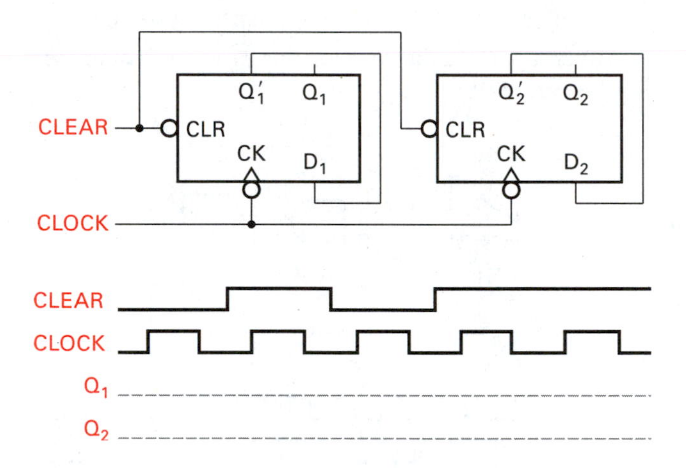

Section 5.3

21. A negative edge triggered flip-flop has the following timing parameters: t_{su} = 15 ns, t_{hold} = 5 ns, minimum positive clock pulse width = 20 ns. Sketch the clock, data input, and data output waveforms showing these timing relationships.

22. Define *metastability*. What is a metastable state? What causes a flip-flop to go into a metastable state? Sketch a voltage transfer diagram for a pair of cross-connected inverters to illustrate your response.

23. What general approach can be taken to reduce the possibility of a sequential circuit entering metastable states?

24. Sketch two timing diagrams, one showing a flip-flop entering a metastable state and the other showing how the metastable state may be avoided.

Section 5.4 to 5.5

25. Draw the logic diagram for a modulo-8 binary counter. Assume negative edge triggered flip-flops are used. Sketch the resulting state variable outputs.

26. Design a four-bit Johnson counter using a 74LS109 *J–K* flip-flop with the following parameters: t_{pLH} = 25 ns (maximum), t_{pHL} = 40 ns (maximum), t_{su} = 35 ns (minimum), t_h = 5 ns, t_w = 25 ns, and $f_{(max)}$ = 25 MHz.

a. Draw the logic diagram for the counter.

b. Draw the timing diagram showing the clock input and all the flip-flop outputs.

c. Determine the maximum operating speed of the counter.

27. Assume that a 1010 input data pattern is loaded into a four-bit ring counter. Sketch the resulting flip-flop *Q* output waveforms. Assume a positive edge trigger.

28. Determine the MSI integrated-circuits that accomplish the following functions:

a. TTL compatible CMOS decade up-down counter

b. A very fast modulo-4 binary counter

c. A TTL compatible divide-by-60 counter

d. A dual four-bit ripple binary counter

29. Show how the 74163 MSI counter IC can be configured as a bi-quinary counter.

30. A 74S162 decade counter is to be used to count an input pulse train. It is desirable to count the pulse train to three orders of magnitude. Design the counter system to accomplish the task (i.e., draw the logic diagram showing IC connections).

31. Given the BCD counter in the figure for this problem and the following circuit parameters, determine the counter's maximum clock frequency.

32. A 74168 up-down decade counter is used as a data register that must be incremented or decremented upon demand. Assume the data to be stored in the

Sections 5.4 and 5.5

Problem 31

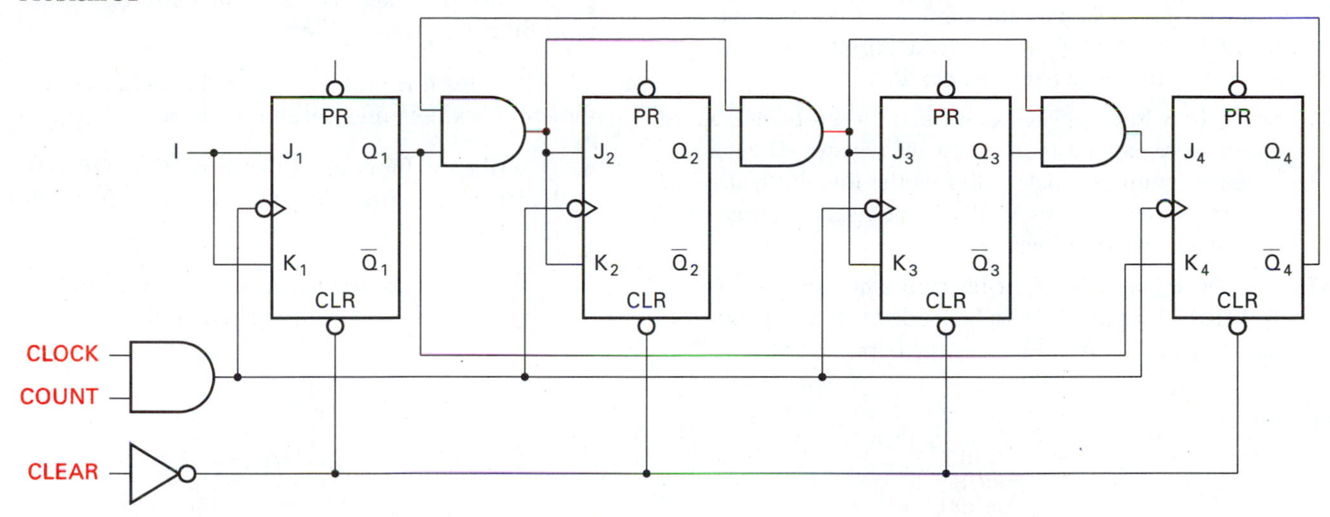

device is 0101 (0 is MSB). Determine the control sequence necessary to load the data, increment twice, and decrement once. What final data values are stored in the counter? Sketch a timing diagram showing the control signal sequences.

33. A 74LS393 binary counter is used to provide the state variables for a simple sequence controller. Three control pulses are derived from the state variables. They must conform to the following pattern:

a. O_1 must go high and remain high for 1 microsecond, then go low and remain low for .5 microsecond.

b. O_2 must go high 300 ns after O_1 goes high and remain high until 200 ns after O_1 goes low.

c. O_3 goes high 100 ns after O_2 goes high, remains high for 200 ns then goes low, remains low for 300 ns, then goes high again for 200 ns.

1. Draw a timing diagram of the three control outputs.
2. Determine the clock frequency needed to trigger the counter.
3. Design the output logic for the three outputs.

34. Given the IEEE symbol for the 74xx194 MSI integrated circuit shown in the figure for this problem, answer the following:

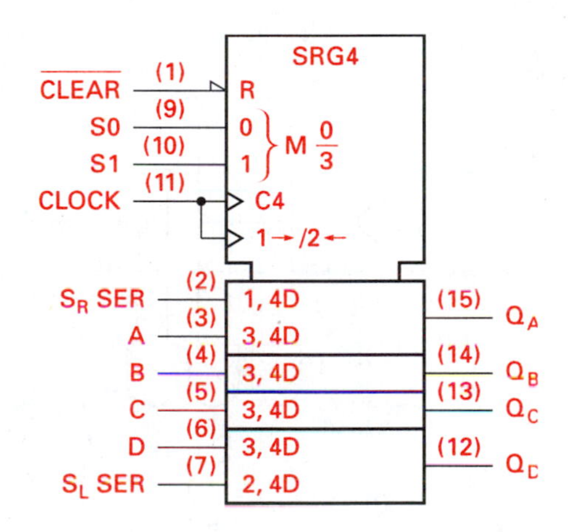

a. The two inputs on pins 9 and 10 are labeled S_0 and S_1. Inside the logic symbol input the S_0 and S_1 are labeled with a bracket and M 0/3. Explain the meaning of M 0/3.

b. Is the device edge or level triggered? What polarity?

c. What are the two signals labeled S_R SER and S_L SER?

d. Inputs A, B, C, and D have notation labels 3,4D inside the symbol box. What does the notation 3,4D mean?

Section 5.6

35. Draw the logic diagram for a four-bit parallel input/parallel output register. Indicate inputs, outputs, and a negative edge triggered clock.

36. Show how the register described in Problem 35 could be expanded to an eight-bit register. Draw a block diagram (a block indicates the four-bit register) and show data inputs, data outputs, and a negative edge triggered clock.

37. The 7491 is a TTL eight-bit right-shift register. A 11001010 bit pattern is to be loaded into the previously cleared register (assume the leftmost one-bit is the MSB). What is the bit pattern after five clock pulses have been applied? Assume 0s in the rightmost bit position are shifted.

38. The figure for this problem shows the logic diagram for a SN74xx194 universal shift register.

    a. Describe the function of the three AND-OR gate logic sections providing the excitation for each flip-flop.

    b. Is the clock (as applied to clock input pin 11) edge triggered or level triggered, and what is its polarity?

Problem 38

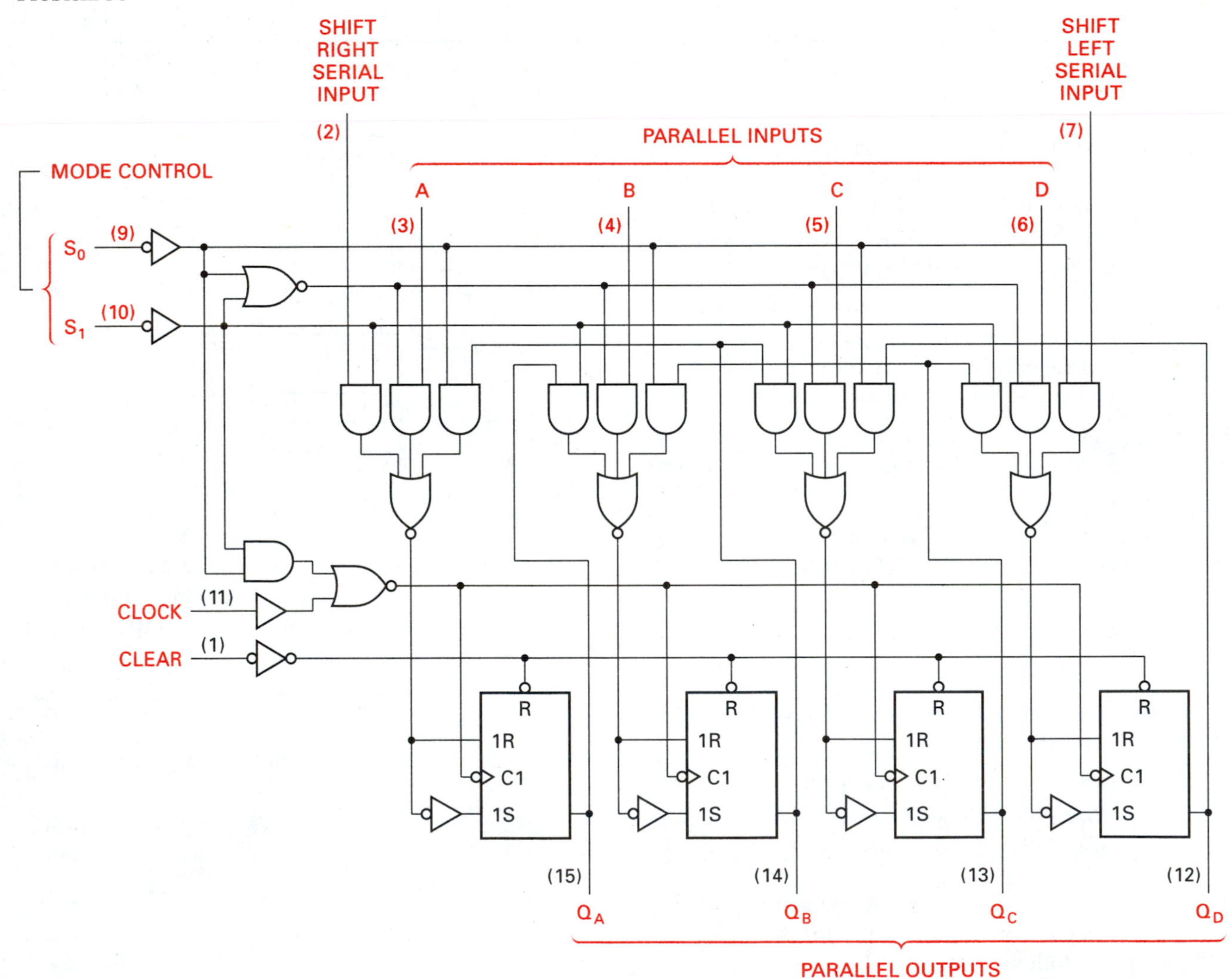

c. What is the code for the mode control signals S_1 and S_0 to perform a parallel load operation?

d. What is the mode control code for a shift-left?

39. Consider the logic diagram for Problem 38. If pin 12 were connected to pin 2, inputs $S_1,S_0 = 11$ for one clock pulse, and then inputs $S_1 S_0 = 01$ for two clock pulses, when the data at pins 3, 4, 5, 6 = 0110, respectively, determine the logic levels at pins 15, 14, 13, and 12.

40. Sketch a diagram of a single flip-flop so that the Q output is driving a tristate buffer. Connect the tristate buffer output back to the flip-flop excitation input. Explain how this single cell can be used to form a tristate data register.

41. Draw a diagram, using the tristate storage cell of Problem 40 as the building block, showing four single-bit data sources and destinations all connected to a common data bus. Label each one-bit cell *A* through *D*; explain how the logic value contained in cell *A* can be transferred to cell *D*.

42. Using the 74LS173, a four-bit tristate data register, as the basic building block, draw a block diagram showing four four-bit registers connected to a common data bus.

43. Design a register system consisting of eight eight-bit data registers that will permit the transfer of data over a common parallel data bus between any two of the registers.

a. Specify the integrated circuits used.

b. Draw the block diagram indicating all necessary control signals, including their assertion levels.

c. Determine a source and destination register selection codes and determine the source and destination select logic.

d. Sketch a timing diagram, include the system clock, source and destination enable control signals, and any control signals necessary to operate the select logic.

e. Determine the maximum data transfer rate between registers for the ICs you choose.

CHAPTER

6 INTRODUCTION TO SEQUENTIAL CIRCUITS

INTRODUCTION

In Chapter 5 we learned about flip-flops, counters, and registers. We also looked briefly at models for sequential circuits. In Chapter 6 we continue our investigation of sequential circuit material; we will develop notation and techniques for designing synchronous sequential machines. In Chapter 5 we introduced the idea that a series of clock pulses can synchronize state changes. In Chapters 6 and 7 we will cover clocked or synchronous sequential circuits. In Chapter 8 we will discuss asynchronous or nonclocked sequential machines.

6.1 MEALY AND MOORE MODELS

The two most prevalent sequential circuit models are the Mealy and Moore models, introduced in Chapter 5 and repeated here for convenience. As indicated in Figures 6.1 and 6.2 sequential circuit models include combinational and memory subunits.

Synchronous sequential circuit memory, usually consisting of flip-flops, updates circuit status information. The logic functions shown in the model diagrams translate input and flip-flop output information. The input logic produces the excitation inputs to the flip-flops, and the output logic converts input and flip-flop data to satisfy the output variable requirements.

Figure 6.1
Mealy sequential circuit model

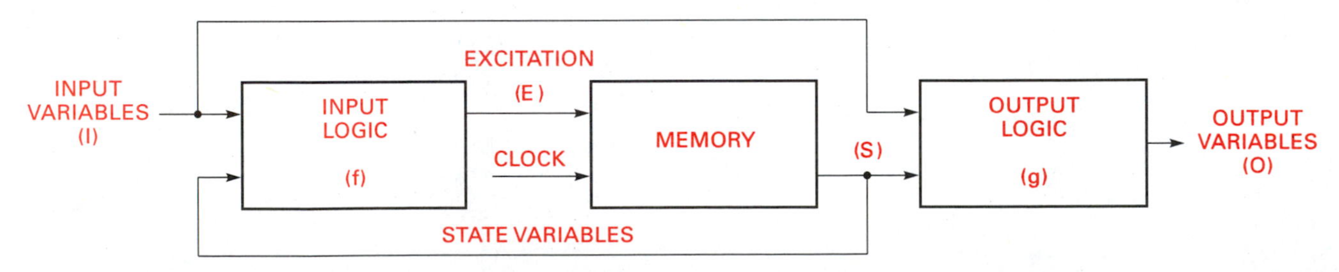

Figure 6.2
Moore sequential circuit model

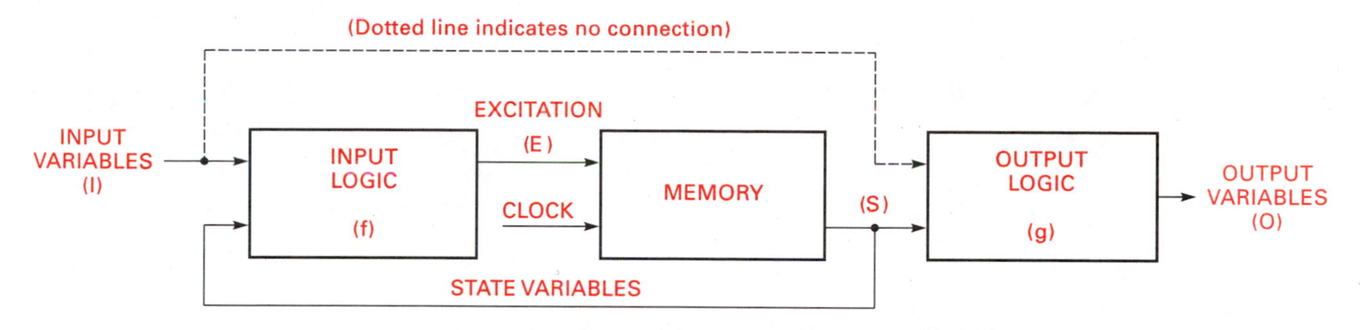

A Mealy sequential circuit differs from a Moore circuit in the variables used to generate output functions. In a Mealy machine the external input variables (I) and the state variables (S) are applied to the output logic function (g) to generate the output (O). A Moore circuit output is dependent only on the state variables (S).

6.2 STATE MACHINE NOTATION

Several different names are given to Boolean variables in **state machine**, or sequential circuit, nomenclature; the differences in name arise from where in the sequential model the variables are generated.

Input variable All variables that originate outside the sequential machine are said to be *input variables.*

Output variable All variables that exit the sequential machine are said to be *output variables.*

State variable The output of memory (flip-flops) defines the state of a sequential machine. Decoded state variables (along with the input variables for a Mealy machine) produce the output variables. A good working definition is that the state variables are the flip-flop outputs.

Excitation variable Excitation variables are the inputs to memory (flip-flops). The name *excitation* is used because the variable "excites" the memory to change. When flip-flops are used for the system memory, the excitation variables are the inputs (J–K, R–S, D, T) to the flip-flops. State variables are a function of the excitation variables. Excitation variables are generated by the input combinational logic operating on the state variables and input variables.

State The state of a sequential machine is defined by the content of memory. When memory is realized with flip-flops, the machine state is defined by the Q outputs. Each state of a sequential machine must be unique and unambiguous.

State variables and states are related by the expression

$$2^x = y$$

where x = number of state variables (e.g., flip-flops) and y = maximum number of states possible (e.g., 4 state variables can represent a maximum of 16 states).

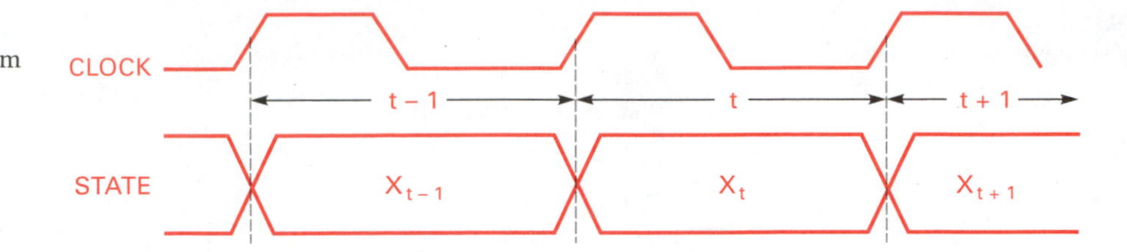

Figure 6.3
Present state/next state diagram

Sequential machine states are unique combinations of state variables; each state must be distinct from all other states.

6.2.1 Present State, Next State

We have discussed several methods used in clocking flip-flops in Chapter 5 and determined that edge triggered clocking was the preferred scheme. Because flip-flops can change state only on a clock pulse edge, we need a way to distinguish state variables before and after the clock pulse. State variable X can be represented as X_t before the arrival of a clock and as X_{t+1} after the arrival of a synchronizing clock pulse. Figure 6.3 illustrates the idea of present state/next state.

Present State

The status of all state variables, at some time, t, before the next clock edge, represents a condition called **present state.** The present state or status of sequential circuit memory is a reference point with respect to time.

Next State

The status of all state variables, at some time, $t + 1$, represents a condition called **next state.** The next state of a sequential machine is represented by the memory status after a particular clock, t. For example, let a state machine be realized using two flip-flops F_2 and F_1. Two state variables can have four states as illustrated in Table 6.1.

For instance, the present state variables $F_2F_1 = 00$, shown in Table 6.1 will change to 01 after the arrival of a clock pulse; the present state 11 will become next state 00 after the arrival of a clock pulse. Every possible state can be both a present state or a next state depending on time. For instance, if the present values of state variables $F_2F_1 = 01$ then the next state, at $t + 1$, is 11. A state table illustrates all of the possible present state conditions and the new state variable values (next state) after clocking.

TABLE 6.1
Simple state table

Present		Next State	
$F_{2(t)}$	$F_{1(t)}$	$F_{2(t+1)}$	$F_{1(t+1)}$
0	0	0	1
0	1	1	1
1	0	1	0
1	1	0	0

The same two state variables are listed in both the present and the next state columns. $F_{2(t)}$ and $F_{1(t)}$ represent the Q outputs of the two flip-flops before a clock pulse. $F_{2(t+1)}$ and $F_{1(t+1)}$ represent the same two flip-flop outputs after the arrival of a clock pulse.

6.2.2 State Diagram

In Table 6.1 we represented states by actual binary values of the state variables. Abstract symbols can also be used to represent states and must be distinguishable from all other symbols. A **state diagram** is a graphical rendition of a sequential circuit, where individual states are represented by a circle with an identifying symbol located inside. Changes from state to state are indicated by directed arcs. Input conditions that cause state changes to occur and the resulting output signals are written adjacent to the di-

Figure 6.4
Example Mealy state notation

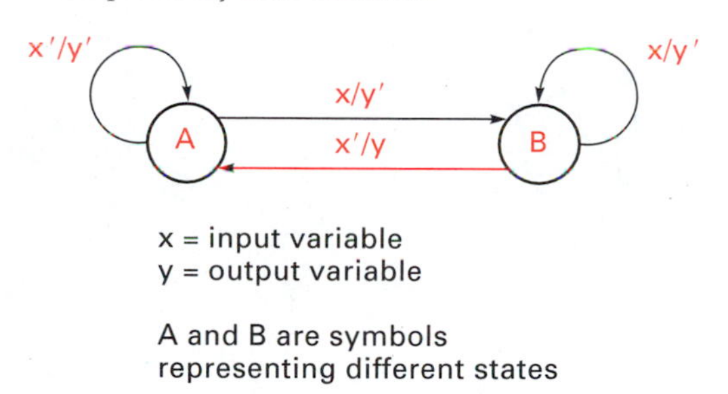

Figure 6.5
Moore circuit notation of a *J–K* flip-flop

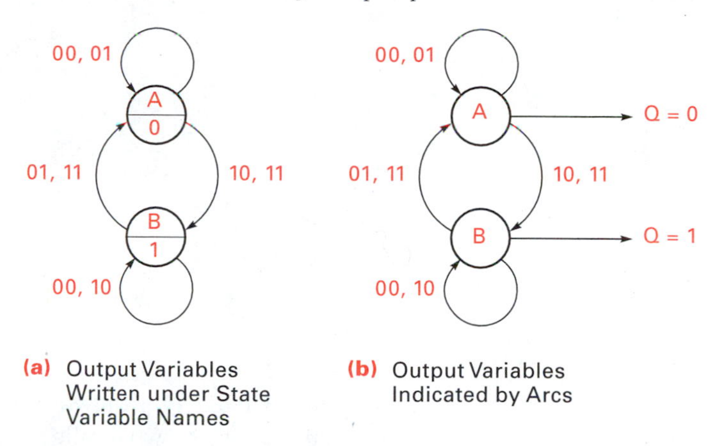

(a) Output Variables Written under State Variable Names

(b) Output Variables Indicated by Arcs

rected arc. Figure 6.4 illustrates the symbolic notation for a one input, one output Mealy sequential circuit.

State *A* is represented with a circle with the letter A written inside. A directed arc connects state *A* with state *B*. In this example, the next state can be *A* or it could change to *B*, depending on input variable, *x*. If input variable $x = 0$, then the machine remains in state *A*; if $x = 1$, then a transition to state *B* is made. An input-output statement is written adjacent to each directed arc. The notation places input variables to the left and the output variables to the right of the separating slash. Note that the output in this scheme is a function of both the present state and the input variables, making it a Mealy machine. Moore machine notation, where the output is dependent on only the present state, represents the output variable inside the state circle or with an arc terminating in the output variable value. Figure 6.5 demonstrates the two methods of output variable notation for a Moore circuit.

A *J–K* flip-flop can be modeled by a state diagram as shown in Figure 6.5. Each state in the state diagram represents a state of the flip-flop, either set or reset. For this example, *J* and *K* are both input and excitation variables and *Q* is both the output and state variable. State *A* represents the case where $Q = 0$ and state *B*, the case where $Q = 1$. Note that, if $JK = 00$ when the machine is in state *A*, then it remains in state *A* and $Q = 0$. If *A* is the present state and $JK = 10$, a transition to state *B* is made and the *Q* output changes to 1. A transition from state *A* to state *B* is made when $JK = 10$ or 11; the other two input combinations leave the machine in state *A*. If *B* is the present state ($Q = 1$), the two input conditions that cause a transition to state *A* are $JK = 01$ or $JK = 11$; when $JK = 00$ or $JK = 10$, the machine remains in state *B*.

In Figure 6.5(a), the *Q* output is written inside the circle and not in direct association with the *J–K* input variables; Figure 6.5(b) shows an arrow leaving the state, with the output variable value written at the end. Two input conditions are part of each trans state directed arc.

Sequential state machines can be represented by a mix of both Mealy and Moore notation. An example state diagram showing mixed Mealy and Moore notation is shown in Figure 6.6.

State diagrams graphically represent the relationships between states. They show the input-output variable combinations as well as the transitions from one state to the next.

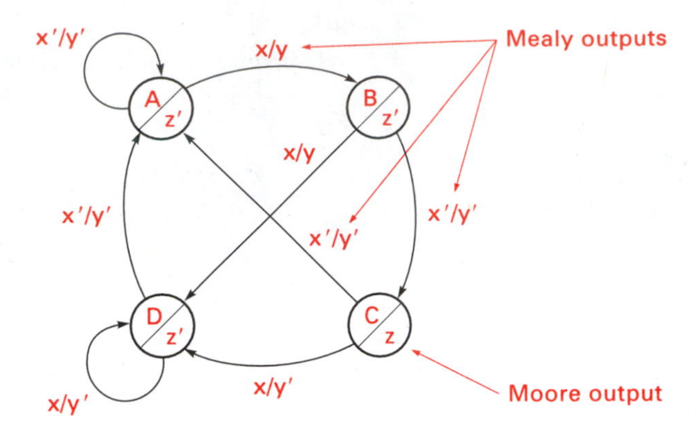

Figure 6.6
Mealy and Moore mixed notation state diagram

The state diagram shown in Figure 6.6 has four states A, B, C, and D, requiring two state variables (two flip-flops). Note that one input variable, x, and two output variables, y and z, are present; output y is represented in Mealy notation and output z is represented in Moore notation. Output y is true when the machine is in state A or B and input variable x is asserted. Output variable z is asserted only when the machine is in state C, independent of the input variable.

The operation of the state machine is as follows:

1. Starting in state A, if x is deasserted ($x = 0$), then the machine remains in state A. This is indicated by the directed arc that starts and stops with state A. Both y and z outputs are 0.
2. On the next clock edge, occurring after x is asserted, a transition from state A to B occurs and output y is asserted. Note that y is true only between the time that x is asserted and the clock edge that causes the state transition from A to B.
3. Two directed arcs leave state B, indicating two possible next states: to C if $x = 0$ and to D if $x = 1$. If $x = 0$, then, on the next clock, the machine will go to state C and produce an asserted z output. Note that output z is derived only from state C and is independent of x. If x is true, then output y is asserted (logical 1) and a transition from B to D is made on the next clock pulse.
4. Two arcs emanate from state C. When $x = 1$, an arc starts at state C and ends in state D; the other arc, when $x = 0$, starts at state C and ends in state A. When the machine is in state C, output z is asserted. As soon as the next clock edge occurs, a state transition takes place and output z is deasserted.
5. If state D is the present state, then the possible next states are D or A. The next state will be D if $x = 1$ and state A if $x = 0$. In either case neither output variable is asserted.

States represent different events. They can be steps in a process or an algorithm that is being controlled. State diagrams are not limited to realizations using electronics; hydraulics, pneumatics, relays, magnetics, or optics could be used as implementing technology. This text deals only with digital electronic implementation. State diagrams are a graphical representation of steps in a process. We identify each step with a symbol and indicate the next step, if certain conditions, as indicated by the input variables, are true. The process can be a traffic light control built with relays or the control unit of a computer. The method of describing the sequence of steps can be represented by a state diagram.

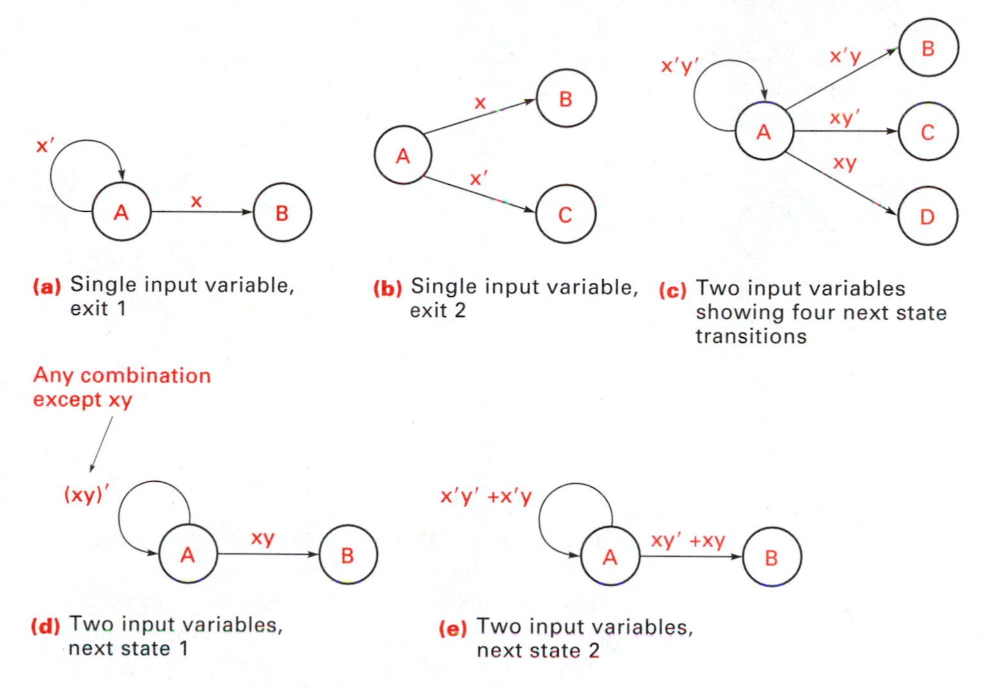

Figure 6.7
Partial state diagram illustrating different transitions and variable combinations

The addition of one input variable doubles the number of possible next state transitions. One input variable creates two exit path conditions from a given state: two input variables create four exit conditions, and three input variables cause eight exit conditions. From this, we can visualize the complexity of dealing with a large number of input variables.

Figure 6.7 illustrates several partial state diagrams with different state transitions under different input variable combinations. State A is the starting point. Each arc indicates the next state when the specified input conditions are present. Figure 6.7(a) shows a single input variable, x, with two possible next states. Figure 6.7(b) shows that the next states are both different than the present state. Figure 6.7(c) illustrates the present state/next state transitions that may occur when two input variables are present. With two input variables four possible next state combinations exist. Figure 6.7(d) and 6.7(e) illustrate that more than one input combination may result in the same next state transition.

Notice, in Figures 6.7(d) and 6.7(e), that the Boolean expressions indicating state transitions are not simplified. Often it is better not to reduce expressions at the state diagram stage of design. Simplification may cause loss of insight into the machine operation. Methods of state reduction will be covered later in the chapter.

A state diagram is somewhat analogous to the programmer's flowchart. The states can be viewed as tasks or operations and the directed arcs, indicating input conditions, as conditional statements. For example, consider Figure 6.7(c). Assume the system is in a condition defined as operation (state) A; if inputs $x'y' = 1$, then no operational change occurs; if inputs $x'y = 1$, the system performs operation B; if inputs $xy' = 1$, the system performs operation C; if the remaining input combination, $xy = 1$ occurs, the system performs operation D.

With large numbers of input variables, it is convenient to indicate only the asserted conditions that cause state or output variable changes. The unspecified input variable

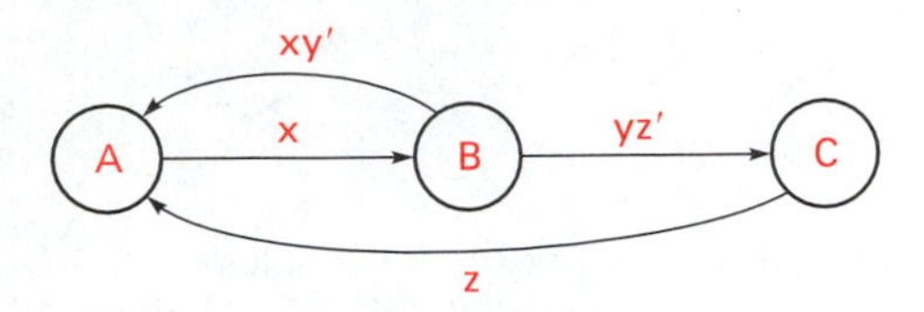

Figure 6.8
A three-input variable partial state diagram showing only input assertions producing state changes

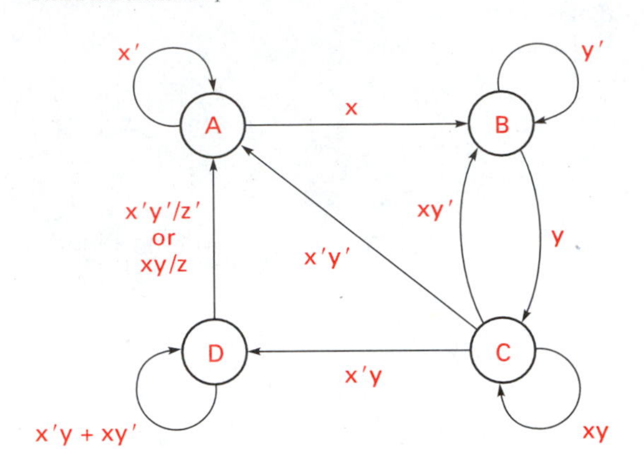

Figure 6.9
State machine M_1

combinations are assumed to leave the state machine in its present state. Consider the state diagram illustrated in Figure 6.8.

The change from state A to state B occurs only when $x = 1$. If $x = 0$, the machine remains in state A, independent of variable y and z. When B is the present state, the machine will return to state A when xy' is true or go to state C when $y'z$ is true. All other combinations produce no state changes. This approach greatly reduces the complexity of the state diagram without sacrificing understanding of the machine operation.

6.2.3 State Table

State tables are tabular forms of state diagrams. The present state column lists all of the possible states in the machine. A series of next state columns exist, one for each input combination. The purpose of the state table is to indicate the state transitions. The four-state, two-input-variable state diagram shown in Figure 6.9 is converted to the state table shown in Table 6.2.

Four states, two input variables, and one output variable are implied by the state diagram. Each combination of the two input variables produces a separate next state column for the state table. The states are listed by their symbol name, rather than by the assignment of state variables. The present state column lists the four states. The next state depends on the value of the state variables and the four input variable combinations. The output variable z has a separate subcolumn for each of the input columns. In this way the conditions that cause the output to be true can easily be determined.

TABLE 6.2
State table for state diagram from Figure 6.9

	Next state xy/z							
Present state	00	z	01	z	11	z	10	z
A	A	0	A	0	B	0	B	0
B	B	0	C	0	C	0	B	0
C	A	0	D	0	C	0	B	0
D	A	0	D	0	A	1	D	0

If the present state is A, then only the input variable x is considered. If $x = 1$, then the transition from state A to state B is made. This happens under two input variable combinations, $xy = 11$ or 10. The remaining cases, when $xy = 00$ or 01, cause no state change. This process is much like converting a noncanonical expression to a canonical expression by using a K-map. Completing the remainder of the state table is accomplished by evaluating each state in the present state column. This is done by considering the next state possibilities for given input variable values. In state machine M_1, the only time output z is asserted if the present state is D and the input is xy, as reflected in the state table. Notice that only the asserted conditions were indicated in the M_1 state diagram of Figure 6.9. The state table indicates all of the conditions, asserted or not.

6.2.4 Transition Table

A **transition table** takes the state table one step further. The state diagram and state table represent states using symbols or names. Creation of a transition table requires that specific state variable values be assigned to each state. Assignment of values to state variables is called *making the state assignment*. The state assignment links the abstract state symbols to actual state variable binary values. We observed this in Table 6.1, where the four states were represented by unique combinations of the two state variables.

The purpose of this section is to describe the transition table, not to develop the state assignment, we will leave that for a later section. Consider Table 6.2. There are four states labeled A, B, C, and D, requiring two state variables, F_A and F_B. Let the state assignment be

F_A	F_B	State
0	0	A
0	1	B
1	1	C
1	0	D

Each state is assigned a unique code. State A is defined when both flip-flops are reset; state B is defined when $F_AF_B = 01$, and so on.

Table 6.3, a transition table, can now be constructed, based on the state assignment just given. A transition table illustrates the changes that occur in the state variables as a state machine sequences from one state to the next. Where the state table showed only the abstract symbols, the transition table shows the actual state variable changes. At each new clock pulse the input variables x and y are evaluated. If a change of state is required then the appropriate flip-flop output must change, indicating the state transition.

TABLE 6.3
Transition table for state machine M_1

Present state		Next state xy/z							
		00	z	01	z	11	z	10	z
F_A	F_B	F_A F_B		F_A F_B		F_A F_B		F_A F_B	
0	0	0 0	0	0 0	0	0 1	0	0 1	0
0	1	0 1	0	1 1	0	1 1	0	0 1	0
1	1	0 0	0	1 0	0	1 1	0	0 1	0
1	0	0 0	0	1 0	0	0 0	1	1 0	0

TABLE 6.4
D flip-flop characteristic table and equation

D	Present, Q_t	Next, Q_{t+1}
0	0	0
1	0	1
0	1	0
1	1	1

$Q_{t+1} = D$

6.2.5 Excitation Table and Equations

Once the changes in flip-flop outputs are known, the next step is to determine the excitation inputs needed to cause the desired flip-flop output changes. This requires deciding on the type of flip-flop to be used in realizing the state machine. Four types, as discussed in Chapter 5, are available, each having a different input-output characteristic table. Two of the flip-flop types, T and D, require only a single input, but JK and RS require two inputs. The excitation tables developed in Chapter 5 illustrate the input-output relationship for each type. The D flip-flop excitation table is shown in Table 6.4.

The **excitation table** for any state machine M is exactly the same as its transition table when D flip-flops are used to realize the circuit. Note that the next state value for a state variable using a D flip-flop is the same as the D input. This means that the Q_{t+1} (next state) values for F_AF_B are the same as the D values. If we were to replace the state variable names, F_AF_B, in each of the next state columns in Table 6.3, the result would be a D flip-flop excitation table, like Table 6.5.

Excitation equations are derived directly from the excitation table. An equation is written for each flip-flop D input and for each output variable. For this example, two excitation equations and one output equation are needed. The weighting for the input and state variables can be made arbitrarily; however, once made, consistency is necessary. We will assign the weighting so F_A is most and y is the least significant variable. A minterm exists for each logical 1 in every next state column.

Each term in the D_A equation is derived from the table as follows:

Minterm	F_A	F_B	x	y
5	0	1	0	1
13	1	1	0	1
9	1	0	0	1
7	0	1	1	1
15	1	1	1	1
10	1	0	1	0

Each term in the D_B equation is derived from the table as follows:

Minterm	F_A	F_B	x	y
4	0	1	0	0
5	0	1	0	1
3	0	0	1	1
7	0	1	1	1
15	1	1	1	1
2	0	0	1	0
6	0	1	1	0
14	1	1	1	0

TABLE 6.5
State machine M_1 excitation table using D flip-flops

Present state		Next state xy/z							
		00	z	01	z	11	z	10	z
F_A	F_B	$D_A\ D_B$		$D_A\ D_B$		$D_A\ D_B$		$D_A\ D_B$	
0	0	0 0	0	0 0	0	0 1	0	0 1	0
0	1	0 1	0	1 1	0	1 1	0	0 1	0
1	1	0 0	0	1 0	0	1 1	0	0 1	0
1	0	0 0	0	1 0	0	0 0	1	1 0	0

Figure 6.10
Excitation and output function K-maps

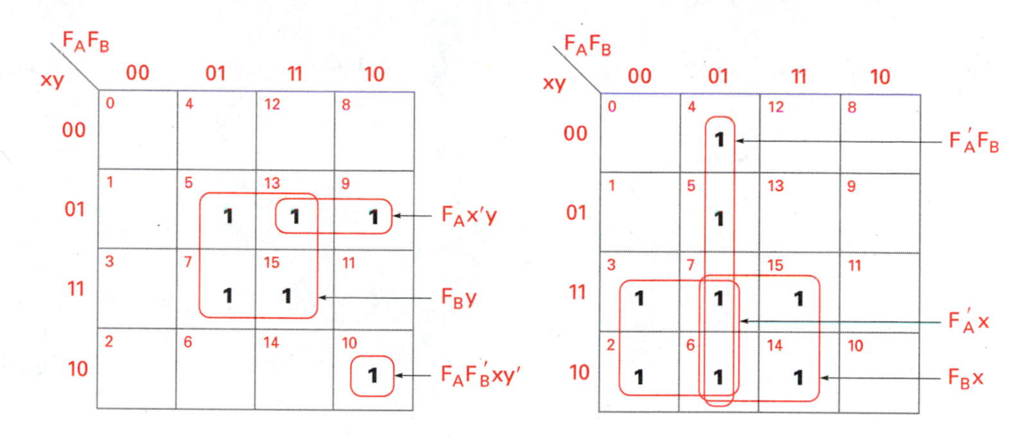

(a) D_A K-Map

(b) D_B K-Map

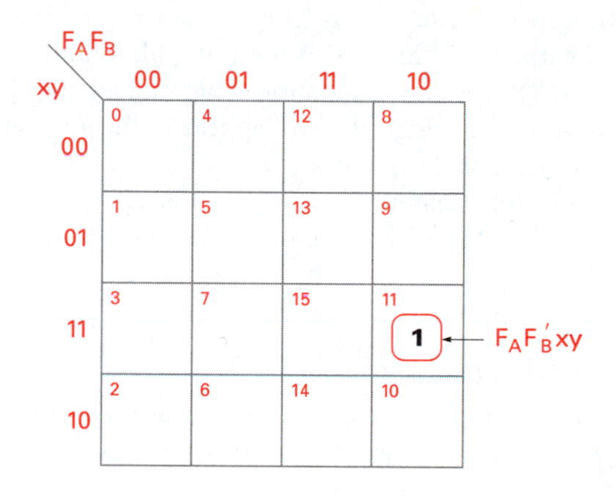

(c) Z K-Map

Writing the decimal notation equations for each excitation variable gives

$$D_A = f(F_A, F_B, x, y) = \Sigma\,(5, 7, 9, 10, 13, 15)$$

$$D_B = f(F_A, F_B, x, y) = \Sigma\,(2, 3, 4, 5, 6, 7, 14, 15)$$

Output variable z has a single term in its final equation:

$$z = f(F_A, F_B, x, y) = \Sigma\,(11)$$

After simplification, using Karnaugh maps (Figure 6.10), the equations become

$$D_A = F_A F_B' x y' + F_A x' y + F_B y$$

$$D_B = F_A' F_B + F_A' x + F_B x$$

and the output variable

$$Z = F_A F_B' x y$$

We will leave the D flip-flop realization of state machine M_1 at the equation level because we also want to evaluate the excitation tables for the other flip-flop types. After we have determined the excitation equations for all four flip-flop types, we will draw the logic diagram for the simplest realization.

TABLE 6.6
T flip-flop characteristic table and equation

	Present	Next
T	Q_t	Q_{t+1}
0	0	0
1	0	1
1	1	0
0	1	1

$Q_{t+1} = T \oplus Q_t$

TABLE 6.7
T flip-flop excitation table for state machine M_1

Present state		Next state xy/z							
		00	z	01	z	11	z	10	z
F_A	F_B	T_A T_B		T_A T_B		T_A T_B		T_A T_B	
0	0	0 0	0	0 0	0	0 1	0	0 1	0
0	1	0 0	0	1 0	0	1 0	0	0 0	0
1	1	1 1	0	0 1	0	0 0	0	1 0	0
1	0	1 0	0	0 0	0	1 0	1	0 0	0

The toggle or T flip-flop also has a single excitation input. Its excitation table is illustrated in Table 6.6.

To realize machine M_1 using T flip-flops, we combine data from the transition table (Table 6.3) with the T flip-flop excitation table (Table 6.6) to create a new excitation table (Table 6.7). The new excitation table merges the transition table and the T flip-flop excitation table. A logical 1 will appear in the appropriate T column each time the output of a state variable changes.

Excitation equations for T_A and T_B are derived in the same fashion as the excitation equations for the D flip-flop. Each 1 in the appropriate next state column produces a minterm. The excitation equations for T_A and T_B are

$$T_A = f(F_A, F_B, x, y) = \Sigma\ (5, 7, 8, 11, 12, 14)$$
$$T_B = f(F_A, F_B, x, y) = \Sigma\ (2, 3, 12, 13)$$

The equation for output variable z is the same as previously determined.

Simplification of the T excitation equations results in

$$T_A = F_A F_B' xy + F_A x' y' + F_A F_B y' + F_A' F_B y$$
$$T_B = F_A' F_B' x + F_A F_B x'$$

The K-maps for these equations are shown in Figure 6.11.

A brief comparison of the simplified D and T excitation equations indicates that the D solution is slightly simpler.

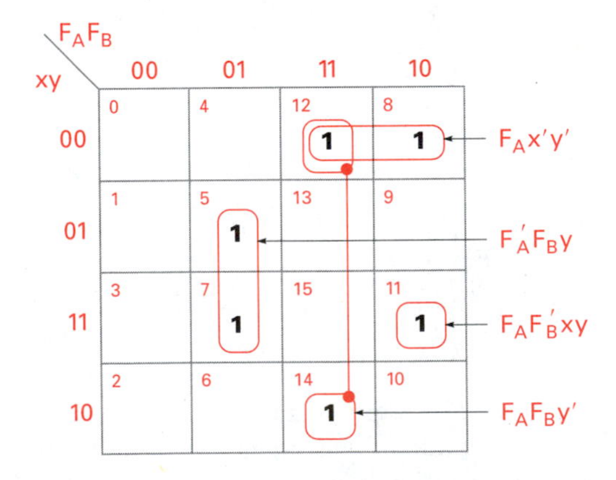

(a) T_A K-Map

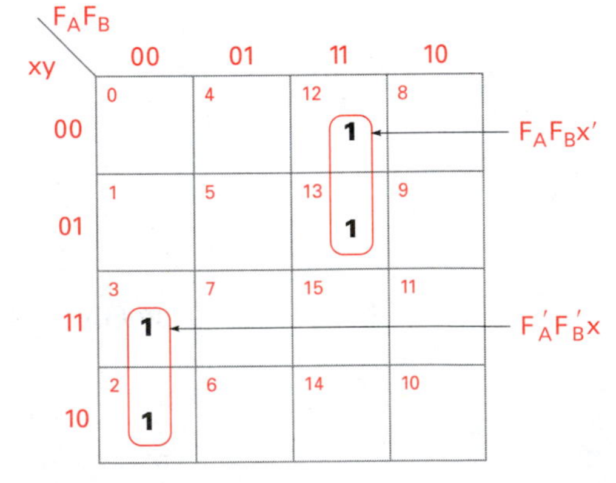

(b) T_B K-Map

Figure 6.11
T flip-flop excitation K-maps

TABLE 6.8
R–S flip-flop characteristic table and equation

Present			Next
S	R	Q_t	Q_{t+1}
0	0	0	0
0	0	1	1
0	1	0	0
0	1	1	0
1	0	0	1
1	0	1	1
1	1	0	Undefined
1	1	1	Undefined

$Q_{t+1} = S + R'Q_t$

S–R flip-flops have two excitation inputs to bring about a change in the Q output. The excitation table and characteristic equations are shown in Table 6.8.

A simplified characteristic table, more suitable for use with constructing problem excitation tables when using S–R flip-flops, is illustrated in Table 6.9.

The d in the S and R columns indicate don't-care conditions. Table 6.9 is more convenient for translating the problem transition table into an S–R excitation table than Table 6.8, because we can readily identify the state variable change and read the needed S and R inputs directly.

If flip-flops that have two excitation inputs are used to realize a state machine, we replace the state variable with two excitation variables. The S–R excitation table for state machine M_1 is generated by merging transition Table 6.3 with the S–R flip-flop excitation table (Table 6.9). For example, the transition table in Table 6.3 shows a present state value for F_A of 0 and a next state value of 0 under input column 00.

The state change of 0 to 0 requires that the S input be a 0 and that the K input be a d or don't-care. So a 0d is loaded into the excitation table for $S_A R_A$ under input column 00. When the present state is a 0 and changes to a 1, the $SR = 10$. If the present to next state change is $1 \rightarrow 0$, then $SR = 01$. The complete excitation table is created by evaluating the state transition and writing the necessary S–R values in the appropriate cells. This is done in Table 6.10.

TABLE 6.9
Simplified S–R flip-flop characteristic table

Input			
S	R	Q_t	Q_{t+1}
0	d	0	0
1	0	0	1
0	1	1	0
d	0	1	1

Equations for the S–R realization are

$$S_A = f(F_A, F_B, x, y) = \Sigma\,(5, 7) + \Sigma d(9, 10, 13, 15)$$
$$R_A = f(F_A, F_B, x, y) = \Sigma\,(8, 11, 12, 14) + \Sigma d(0, 1, 2, 3, 4, 6)$$
$$S_B = f(F_A, F_B, x, y) = \Sigma\,(2, 3) + \Sigma d(4, 5, 6, 7, 14, 15)$$
$$R_B = f(F_A, F_B, x, y) = \Sigma\,(12, 13) + \Sigma d(0, 1, 8, 9, 10, 11)$$

Don't-care terms are listed separately and were derived from the excitation table. They originated from the simplified S–R characteristic table. The simplified S–R excitation equations are

$$S_A = F_B y$$
$$R_A = F_B' xy + x'y' + F_B y'$$
$$S_B = F_A' x$$
$$R_B = F_A x'$$

Output variable z remains the same as in the previous realizations.

The Karnaugh maps for the S–R excitation equations are shown in Figure 6.12.

The J–K flip-flop is the last of the four flip-flop types for which we shall construct a state machine M_1 excitation table. The characteristic table and equation are shown in Tables 6.11 and 6.12.

TABLE 6.10
Problem excitation table using S–R flip-flops

Present state		Next state xy/z											
		00		z	01		z	11		z	10		z
F_A	F_B	$S_A R_A$	$S_B R_B$		$S_A R_A$	$S_B R_B$		$S_A R_A$	$S_B R_B$		$S_A R_A$	$S_B R_B$	
0	0	0d	0d	0	0d	0d	0	0d	10	0	0d	10	0
0	1	0d	d0	0	10	d0	0	10	d0	0	0d	d0	0
1	1	01	01	0	d0	01	0	d0	d0	0	01	d0	0
1	0	01	0d	0	d0	0d	0	01	0d	1	d0	0d	0

Figure 6.12
S–R excitation equation K-maps

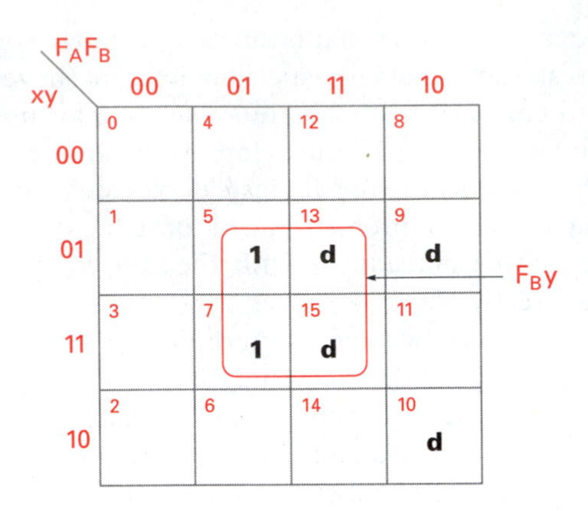

(a) S_A K-Map

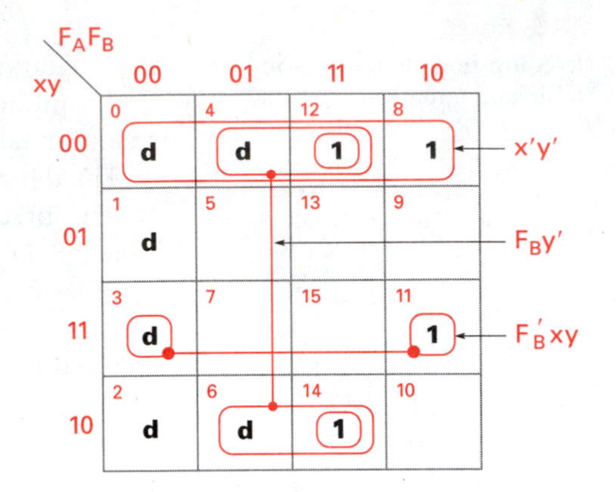

(b) R_A K-Map

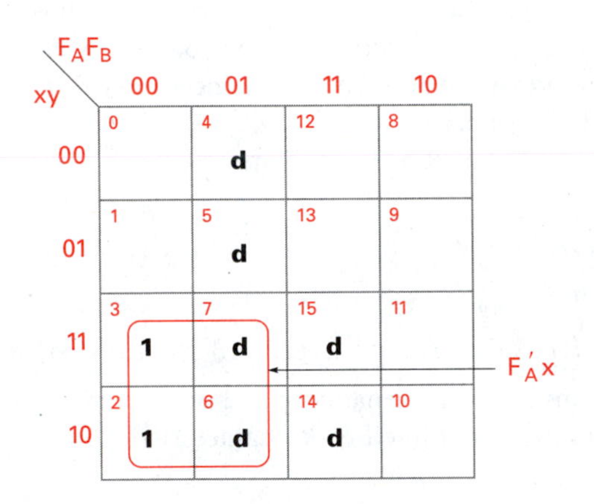

(c) S_B K-Map

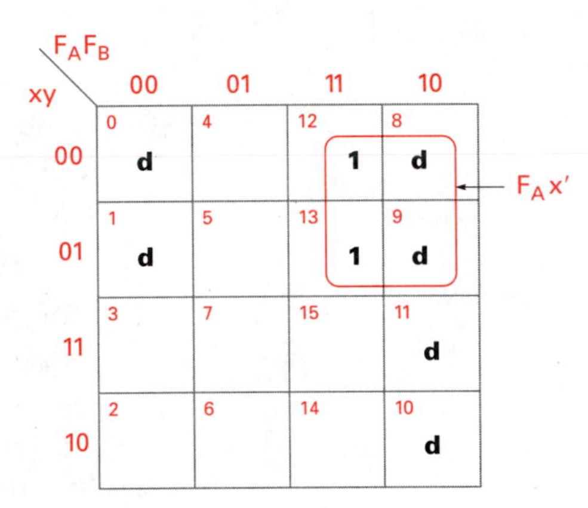

(d) R_B K-Map

TABLE 6.11
J–K flip-flop characteristic table and equation

Present			Next
J	K	Q_t	Q_{t+1}
0	0	0	0
0	0	1	1
0	1	0	0
0	1	1	0
1	0	0	1
1	0	1	1
1	1	0	1
1	1	1	0

$Q_{t+1} = JQ_t' + K'Q_t$

TABLE 6.12
Simplified *J–K* characteristic table

Input			
J	K	Q_t	Q_{t+1}
0	*d*	0	0
1	*d*	0	1
d	1	1	0
d	0	1	1

TABLE 6.13
State machine M_1 excitation table using J–K flip-flops

Present state		Next state xy/z											
		00		z	01		z	11		z	10		z
F_A	F_B	J_AK_A	J_BK_B		J_AK_A	J_BK_B		J_AK_A	J_BK_B		J_AK_A	J_BK_B	
0	0	$0d$	$0d$	0	$0d$	$0d$	0	$0d$	$1d$	0	$0d$	$1d$	0
0	1	$0d$	$d0$	0	$1d$	$d0$	0	$1d$	$d0$	0	$0d$	$d0$	0
1	1	$d1$	$d1$	0	$d0$	$d1$	0	$d0$	$d0$	0	$d1$	$d0$	0
1	0	$d1$	$0d$	0	$d0$	$0d$	0	$d1$	$0d$	1	$d0$	$0d$	0

As with the S–R flip-flop, d represents a don't-care condition. Two excitation variables exist for each state variable. Notice that one-half of the simplified characteristic table for the JK contains don't-care terms. This characteristic of J–K flip-flops is very useful when simplifying the excitation equations.

The J–K excitation table (Table 6.13) has the same appearance as the S–R table, except that SR has been replaced by JK.

A close inspection of Table 6.13 reveals that don't-care terms occupy one-half of the possible minterms for J and K. This results from the don't-care terms contained in the J–K characteristic table. The excitation equations for the J–K flip-flops are

$$J_A = f(F_A, F_B, x, y) = \Sigma\,(5, 7) + \Sigma d(8, 9, 10, 11, 12, 13, 14, 15)$$
$$K_A = f(F_A, F_B, x, y) = \Sigma\,(8, 11, 12, 14) + \Sigma d(0, 1, 2, 3, 4, 5, 6, 7)$$
$$J_B = f(F_A, F_B, x, y) = \Sigma\,(2, 3) + \Sigma d(4, 5, 6, 7, 12, 13, 14, 15)$$
$$K_B = f(F_A, F_B, x, y) = \Sigma\,(12, 13) + \Sigma d(0, 1, 2, 3, 8, 9, 10, 11)$$

The J and K minterms and the don't-care minterms are read directly from the excitation table. An easier, more straightforward method for determining don't-care terms in a J–K realization of a sequential machine is available. We already understand that one-half of the possible minterms for a given J or K excitation variable will be don't-care. Notice where the don't-care terms occur for a given excitation variable. For instance, J_A has don't-care terms for all minterms where F_A is a logical 1; K_A produces don't-care terms where F_A is a logical 0. This is easier to visualize when placed into a K-map, and for that reason we will include the K-maps for the J–K excitation variables.

Figure 6.13 provides the K-maps for the J–K excitation variables given previously. Compare the don't-care terms listed in the map with those in the original equations. Note that J don't-care terms exist where the state variable is a 1, and K don't-cares exist where the state variable is a 0.

Each K-map is half full of don't-care terms; which half depends on the excitation variable being simplified. Don't-care terms can be loaded directly into K-maps, instead of reading them from the table. To do this we note that the J maps must have don't-care terms where that state variable is 1, and the K maps must have don't-care terms where the state variable is 0. For example, J_A produces don't-care terms where F_A is a 1, which occurs as indicated in the J_A map in Figure 6.13. K_A produces don't-care terms when F_A is a 0, which occurs in the opposite half of the map from J_A. Because of the don't-care terms J–K excitation equations are usually simpler than the excitation equations of

Figure 6.13
K-maps for J–K excitation variables

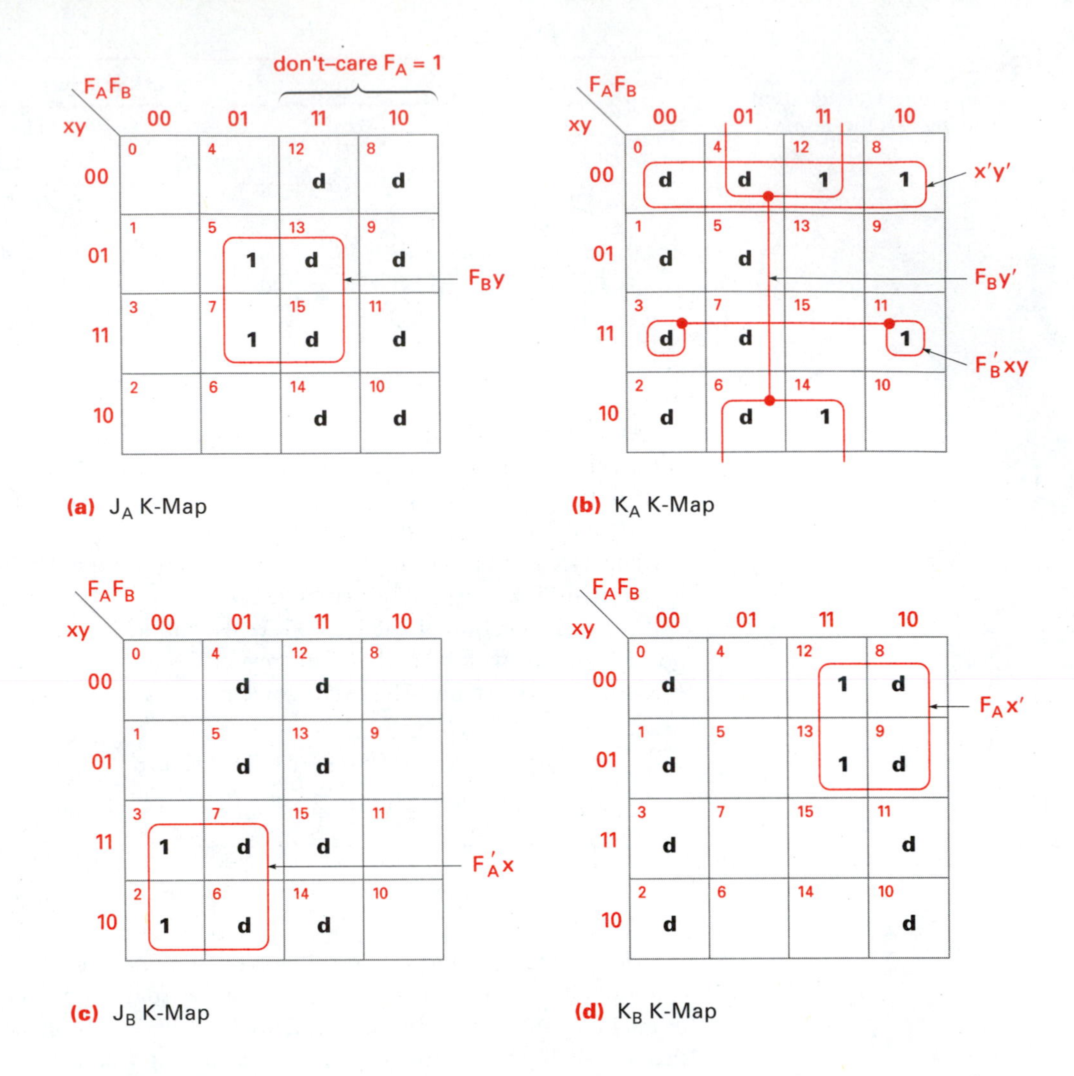

(a) J_A K-Map
(b) K_A K-Map
(c) J_B K-Map
(d) K_B K-Map

other flip-flop types. The resulting J–K excitation equations are given as part of Figure 6.13.

6.2.6 Excitation Realization Cost

Comparing the excitation equations for all four flip-flop types gives us some insight into the relative complexity of the logic necessary to implement state machine M_1. The actual logic realization for a given problem cannot be easily predetermined, but generally the J–K realization will be simpler. An S–R realization of a state machine is never simpler than the JK because of the two missing don't-care entries in the S–R excitation table. Even though the JK uses two inputs instead of the one needed by T and D the realization is usually simpler because of the don't-care terms.

1. J–K excitation equations (Figure 6.13):

$$J_A = F_B y$$
$$K_A = x'y' + F_B y' + F_B' xy$$
$$J_B = F_A' x$$
$$K_B = F_A x'$$

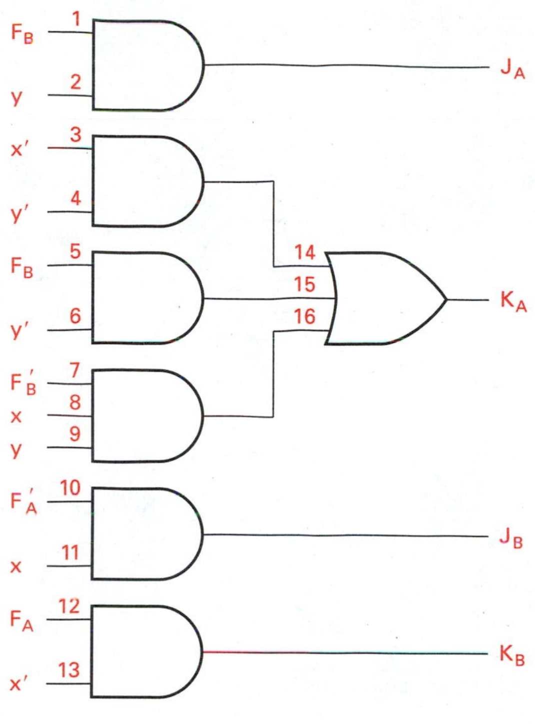

Figure 6.14
Logic for J–K excitation realization

2. S–R excitation equations (Figure 6.15):

$$S_A = F_B y$$
$$R_A = F'_B xy + x'y' + F_B y'$$
$$S_B = F'_A x$$
$$R_B = F_A x'$$

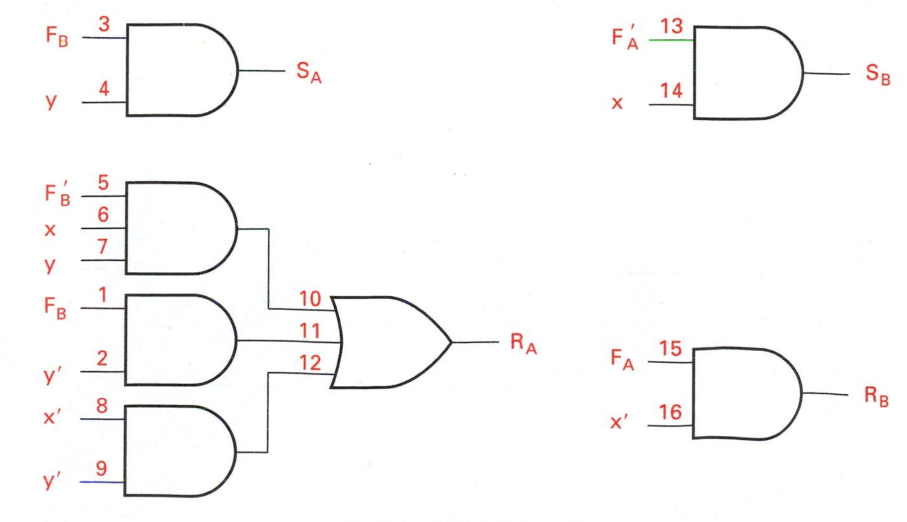

Figure 6.15
Logic for S–R realization

Figure 6.16
Logic for T excitation realization

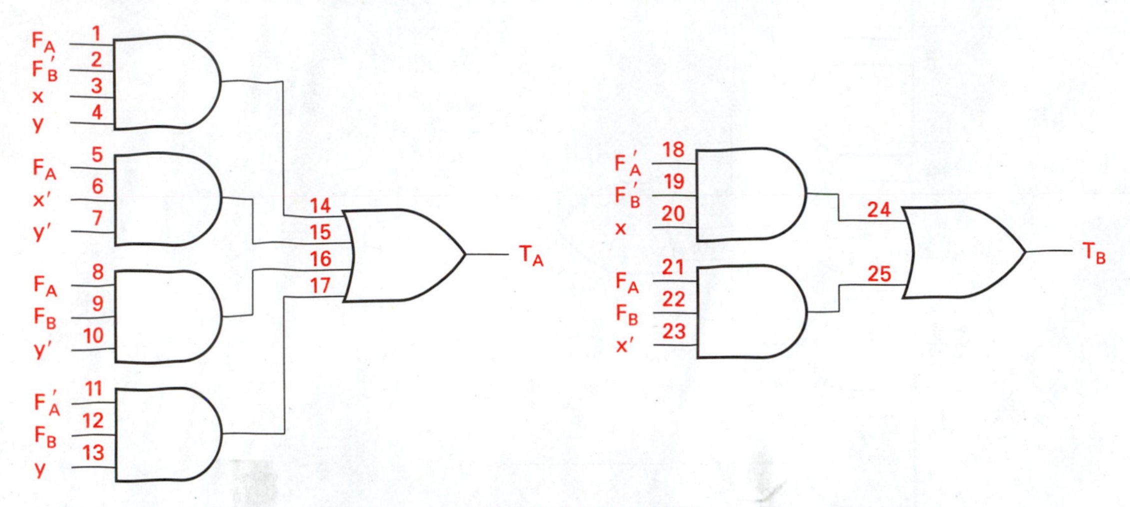

3. T excitation equations (Figure 6.16):

$$T_A = F_AF_B'xy + F_Ax'y' + F_AF_By' + F_A'F_By$$
$$T_B = F_A'F_B'x + F_AF_Bx'$$

4. D excitation equations (Figure 6.17):

$$D_A = F_AF_B'xy' + F_Ax'y + F_By$$
$$D_B = F_A'F_B + F_A'x + F_Bx$$

Relative cost, in terms of gate inputs, for each of the realizations can be seen from the excitation logic diagram. The output logic is not included because it is the same for each.

Comparing logic realization costs we find that the J–K realization and SR are the winners and the D and T flip-flop realizations considerably greater. In this example, the

Figure 6.17
Logic for D excitation realization

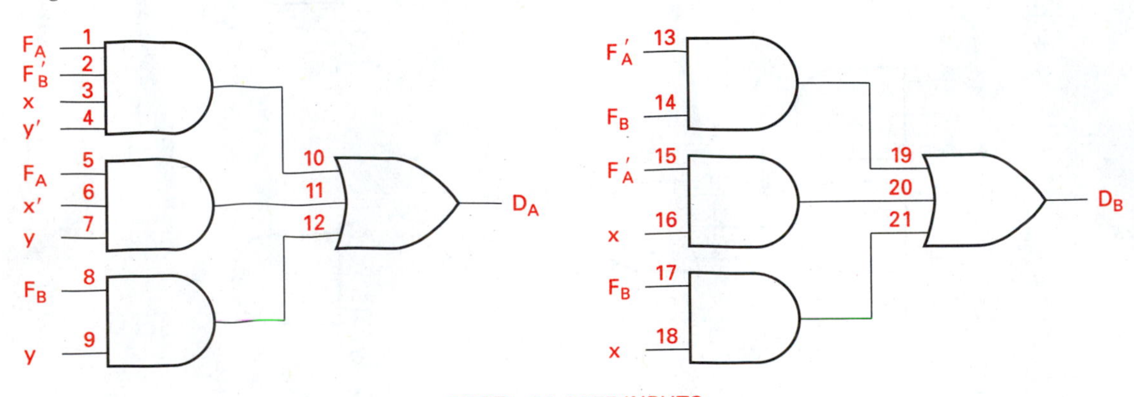

Figure 6.18
Logic diagram for *J–K* realization of example sequential circuit (M_1)

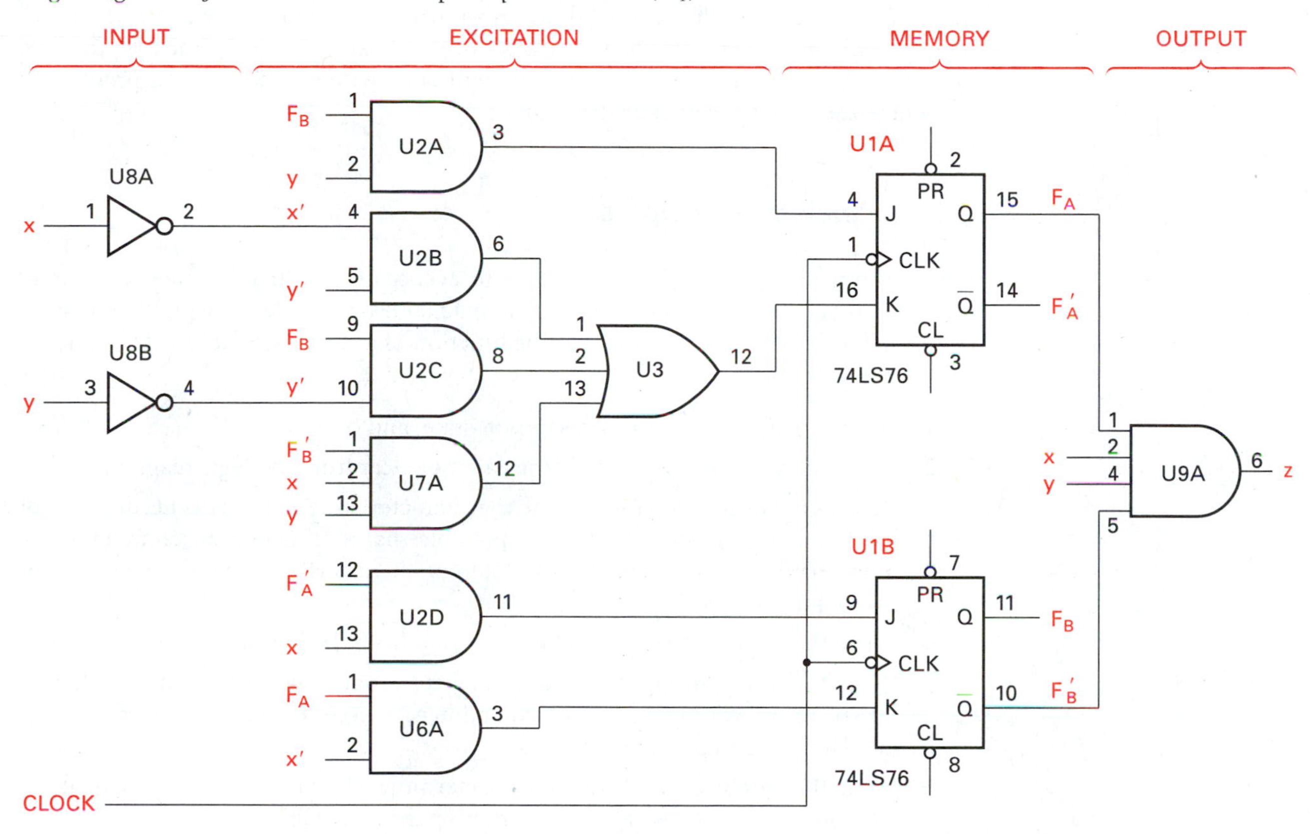

realization of state machine M_1, we compare gate inputs—not integrated circuits. Usually, but not always, the lower the gate input cost, the lower the number of ICs needed. The gate input cost is especially helpful if a programmable gate array is used to realize the system logic.

A complete logic diagram showing the system inputs, excitation variables, state variables (flip-flop outputs), and system outputs for a *J–K* flip-flop realization is shown in Figure 6.18.

6.3 SYNCHRONOUS SEQUENTIAL CIRCUIT ANALYSIS

In the preceding sections we discussed models, state diagram notation, state tables, transition tables, and excitation tables. The impact of flip-flop types on the final excitation equations was also discussed. In this section, we will analyze existing circuits to determine their function. Often logic designers must evaluate someone else's design or develop documentation so that another person can troubleshoot a circuit.

Analysis and design of digital circuits require the application of the same principles, only in opposite directions. That is to say, design is a **synthesis** process; to take pieces and put them together to form a functional whole. **Analysis**, on the other hand, requires breaking the whole into its component pieces. Often the synthesis process is easier than analysis, because the designer's reasons for making a particular decision are clear. Attempting to determine the original designer's purpose for a circuit, with no accompanying information, can be difficult. This section applies the same principles that we investigated in previous sections. Here we will start with the circuit and proceed to determine the state table or state diagram.

6.3.1 Analysis Principles

Working from a logic diagram, the goal is to eventually construct a state table, state diagram, and possibly a timing diagram. The state table or state diagram provides insight not available from a logic diagram into the function of a sequential circuit. The analysis steps are as follows:

1. Determine the system variables: input, state, and output.
2. Assign names to the variables if they are not clear from the logic diagram.
3. Determine the flip-flop type. Write the characteristic equations as needed for the flip-flops used in the circuit. (It is possible that a sequential machine may use more than one type of flip-flop, so different characteristic equations may be necessary for the same circuit.)
4. Write the excitation equations by inspection from the logic diagram.
5. Write the next state equations for each state variable, using the flip-flop characteristic equations and the circuit excitation equations, or construct the excitation table from the excitation equations.
6. Write the output variable equations. Determine whether the circuit is a Mealy or a Moore machine (this aids in constructing the state table and diagram).
7. Construct a transition table. Identify all of the states possible for a given number of state variables. Some states may be don't-care ones. We cannot tell the specified states from the unspecified (don't-care) states at this point. The transition table is determined by evaluation of the next state equations. Minterms, generated by the next state equations, are represented by logical 1s in the next state portion of the table. Or, as an alternative, construct the transition table from the present state and the excitation equations.
8. Assign symbols to the states and construct a state table or state diagram.
9. When possible, construct a timing diagram. Sometimes the number of input and state variables makes this task impractical.

6.3.2 Analysis Examples

We will present two synchronous sequential circuit logic diagrams for analysis.

EXAMPLE 6.1 Consider Figure 6.19, which has two *D*-type flip-flops with positive edge clock inputs, a single input variable *x*, and a single output variable *z*. Gate 1 provides the excitation *D* input to F_2 and gate 4 does the same for F_1.

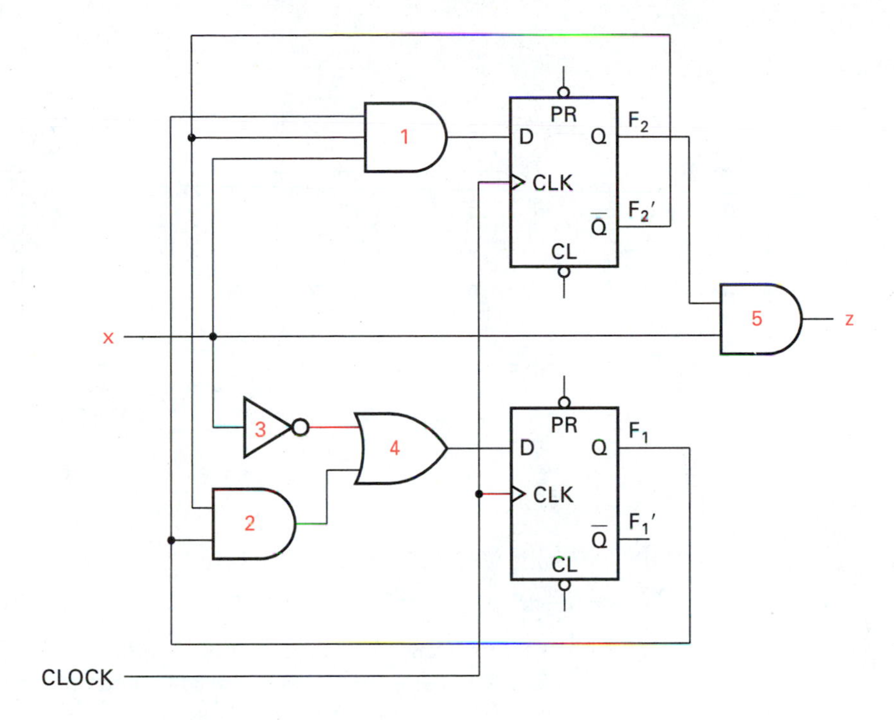

Figure 6.19
Sequential circuit analysis example

SOLUTION

1. Determine system variables:
   a. Input = x
   b. Output = z
   c. State variables = F_2 and F_1
2. Assign variable names. In this example, they are clear; we need do nothing further.
3. Determine flip-flop type. They are both D type. The characteristic equation is

   $$Q_{t+1} = D_t$$

4. Write the excitation equations by inspection:
   a. $D_2 = F_2' F_1 x$
   b. $D_1 = x' + F_2' F_1$
5. Write the next state equations:
   a. $F_2^+ = F_2' F_1 x$
   b. $F_1^+ = x' + F_2' F_1$

   or construct the excitation tables, as in Figure 6.20.

Figure 6.20
Excitation K-maps for D_2 and D_1

x \ F_2F_1	00	01	11	10
0	(0) **0**	(2) **0**	(6) **0**	(4) **0**
1	(1) **0**	(3) **1**	(7) **0**	(5) **0**

(a) D2

x \ F_2F_1	00	01	11	10
0	(0) **1**	(2) **1**	(6) **1**	(4) **1**
1	(1) **0**	(3) **1**	(7) **0**	(5) **0**

(b) D1

Figure 6.21
K-maps used to determine next state terms

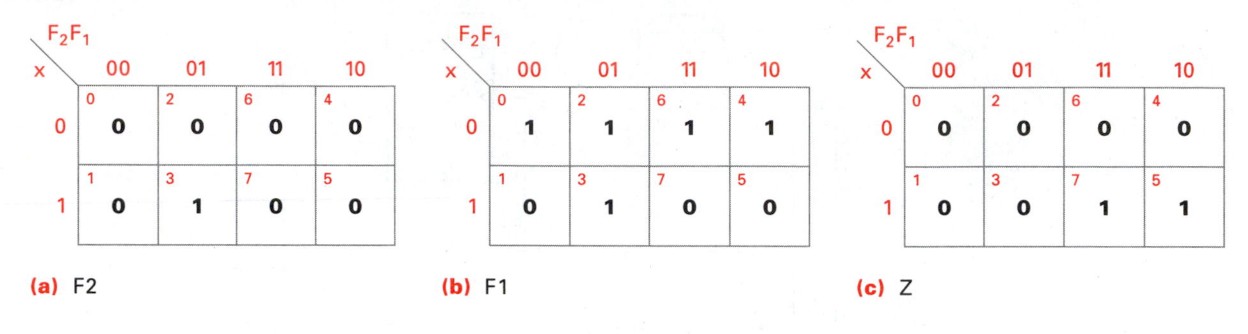

6. Write the output variable equation:

$$z = F_2x$$

7. Construct the transition table. We do this either from the excitation maps or by first creating a K-map for the two state variables and one for the output variable. The two state-variable K-maps in Figure 6.21 are used to generate the canonical expressions for the excitation variables.

   Notice that the excitation maps and the next state maps are the same for the *D* flip-flop. With *J–K*, *R–S*, and *T* flip-flops, the excitation and next state maps are different. The list of minterms (including any unknown don't-care terms) from the next state K-maps are

$$F_2^+ = f(F_2, F_1, x) = \Sigma\ (3)$$
$$F_1^+ = f(F_2, F_1, x) = \Sigma\ (0, 2, 3, 4, 6)$$
$$z = f(F_2, F_1, x) = \Sigma\ (5, 7)$$

   From the canonical next state equations we construct the transition table (Table 6.14). In the case of *D* flip-flops, it is also the excitation table. Let the most significant variable be F_2 so that F_2F_1x is the weighting order.

TABLE 6.14
Transition and excitation table

Present state		Next state xy/z					
		0			1		
F_2	F_1	F_2	F_1	z	F_2	F_1	z
0	0	0	1	0	0	0	0
0	1	0	1	0	1	1	0
1	1	0	1	0	0	0	1
1	0	0	1	0	0	0	1

TABLE 6.15
State table

Present state	Next state x/z			
	0	z	1	z
A	B	0	A	0
B	B	0	D	0
C	B	0	A	1
D	B	0	A	1

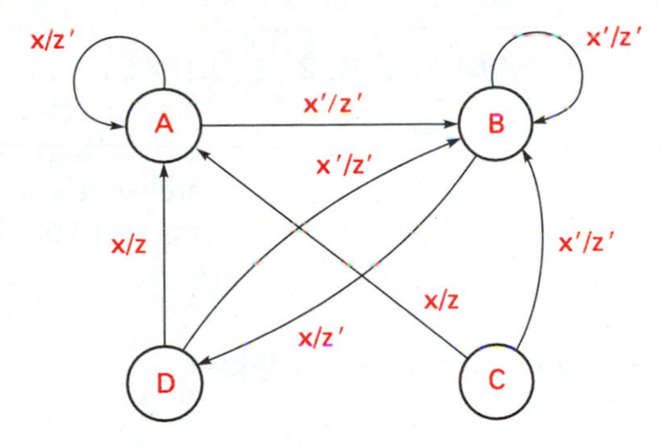

Figure 6.22
State diagram for analysis example

8. Assign state symbols and construct a state table or state diagram. Let the state symbols be

State	F_2	F_1
A	0	0
B	0	1
C	1	0
D	1	1

A state diagram is constructed from Table 6.15. The state diagram, shown in Figure 6.22, indicates that state C is a don't-care state, because no transition path exists from any state into state C. A potential problem exists, if the circuit were to initialize in state C and x were a 1. This would cause output z to be true when it was not supposed to be true. It would appear, because no present state has C as a next state, that such a condition was not intended.

The problem originated when the designer used state C as don't-care terms in the simplification map for z. By removing minterm 5 from the defining equation for z, the problem can be resolved. By removing minterm 5 from the K-map for output z, state C is no longer a factor in determination of when z is true. Removal of minterm 5 adds variable F_1 to the z output equation. Adding variable F_1 to the z output equation effectively removes minterm 5 from the output equation. Then no matter where the circuit is initialized it can not give an incorrect output. Another solution is to initally asynchronously reset the sequential circuit. Doing so disallows the possibility of the state machine starting in state C. The timing diagram shown in Figure 6.23 illustrates a correct state transition sequence, starting in state A.

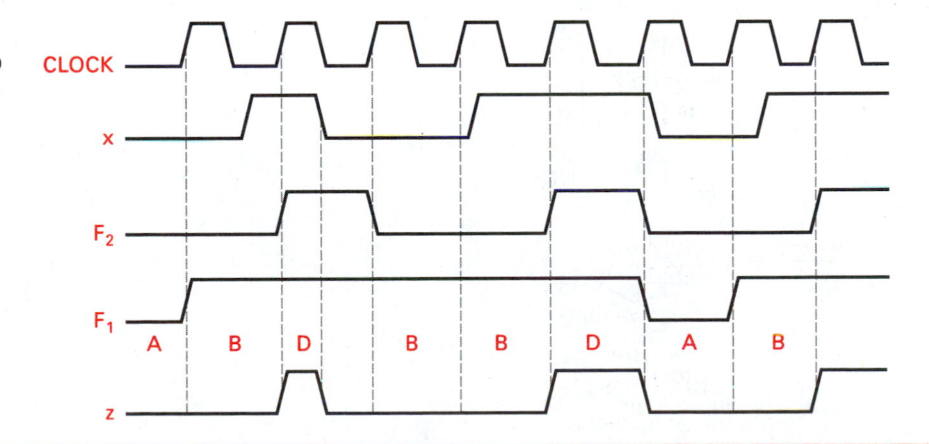

Figure 6.23
Timing diagram for Figure 6.19

❖ **EXAMPLE 6.2** The circuit diagram illustrated in Figure 6.24 is that of a portable, handheld, battery-powered chronograph. A chronograph is a device for measuring the velocity of a projectile, such as a rifle or pistol bullet. Such devices are used by competition shooters and archers for determining equipment performance or by baseball pitchers to determine the speed of a fast ball.

Figure 6.24
Ballistic chronograph circuit diagram

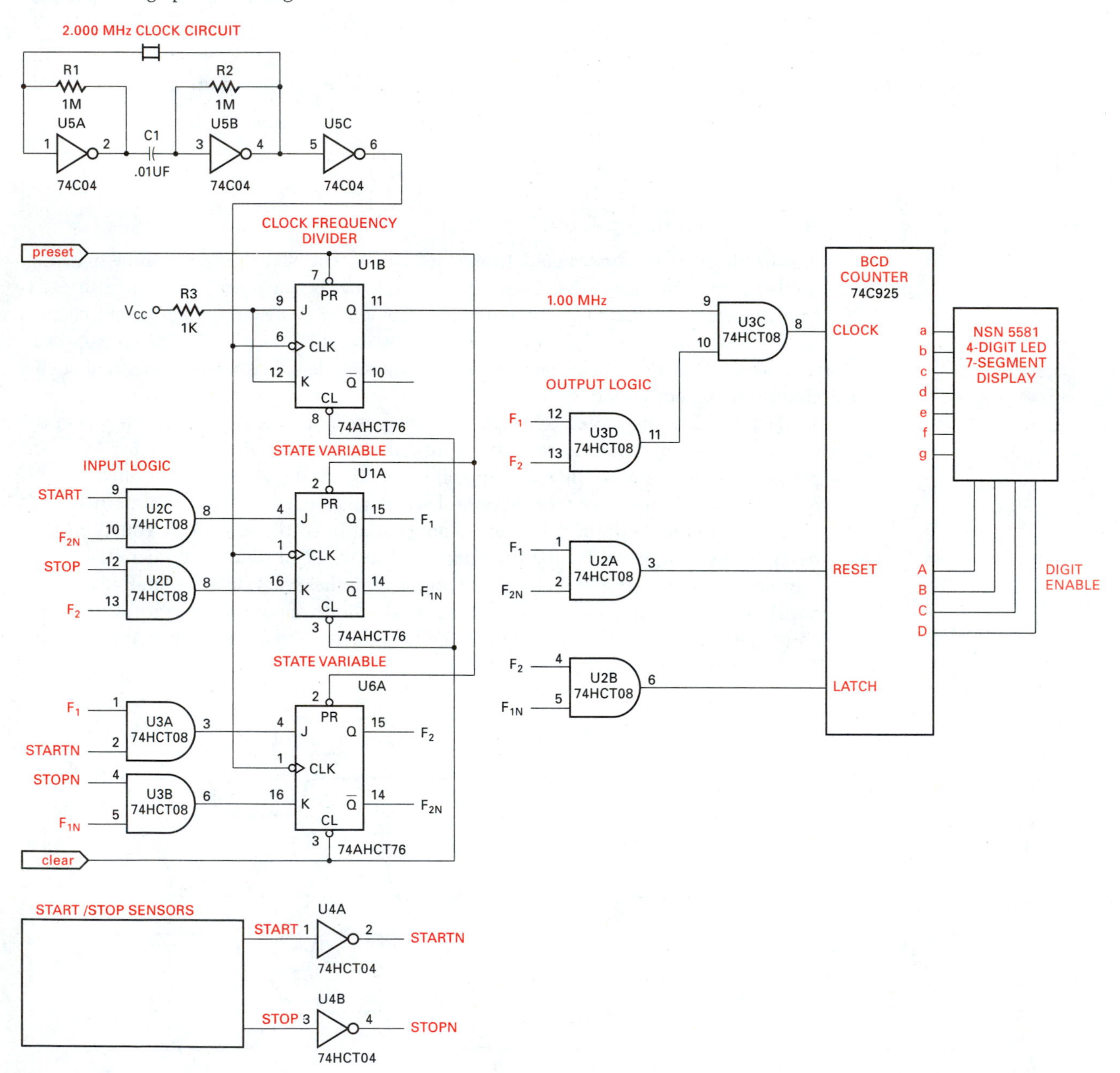

Basic operation requires that the shooter fire a projectile over two light sensitive transducers that generate the start and stop measurement pulses. The light sensors (shadow sensors in this case) are placed a known, precise distance apart. The time of flight between the sensors is directly proportional to the velocity of the projectile that created the shadow as it passed over the sensor. A crystal-controlled square-wave generator produces an accurate frequency that can be measured by a digital counter. Flip-flop U1B divides the 2 MHz clock frequency by 2.

The resulting 1 MHz square wave is gated through U3C to the clock input of the counter. Each square wave has a 1 microsecond period; therefore, the number of pulses measured by the counter is a direct measurement of time. Controlling the counter is the job of the sequential circuit we wish to analyze.

SOLUTION The 74C925 integrated circuit shown in the chronograph logic diagram requires a count input, a reset, and a latch signal to produce a seven-segment display of the number of clock pulses counted. The count output is displayed as the elapsed time of the projectile. From knowing the elapsed time and distance, velocity can be easily calculated.

U1A and U6A, *J–K* flip-flops, form the state variables for the sequential circuit controller. U2C, U2D, U3A, and U3B provide the input logic. Output logic functions are produced by gates U3D, U2A, and U2B. Three functions are needed to control the 74C925 IC counter: count enable, count reset, and count latch. Two input variables, start and stop along with two state variables, F_2 and F_1, make up the remainder of the system.

The analysis of the sequential controller is accomplished by the following steps:

1. Make a list of input, state, excitation, and output variables:

a. Input variables: Start = SRT, Stop = STP.

b. State variables: F_1, F_2.

c. Output variables: Count reset = CRST, Count enable = CEN, Count latch = CLTCH.

2. Write the excitation and output equations from inspection:

a. Excitation equations:

$$J_2 = \text{SRT}'F_1$$
$$K_2 = \text{STP}'F_1'$$
$$J_1 = \text{SRT}F_2'$$
$$K_1 = \text{STP}F_2$$

b. Output equations:

$$\text{CRST} = F_2'F_1$$
$$\text{CEN} = F_2F_1$$
$$\text{CLTCH} = F_2F_1'$$

3. Write the next state equations for each state variable, using the characteristic equation for the *J–K* flip-flop and the excitation equations from step 2. The *J–K* characteristic equation is

$$Q_{t+1} = JQ_t' + K'Q_t$$

The state variable next state equations are

$$F_{2,(t+1)} = J_2F'_{2,t} + K'_2F_{2,t}$$
$$F_{2,(t+1)} = \text{SRT}'F_1F'_2 + (\text{STP}'F'_1)'F_2$$
$$F_{2,(t+1)} = \text{SRT}'F'_2F_1 + (\text{STP} + F_1)F_2$$
$$F_{2,(t+1)} = \text{SRT}'F'_2F_1 + \text{STP}F_2 + F_2F_1 \quad \text{Final } F_2 \text{ equation}$$
$$F_{1,(t+1)} = J_1F'_{1,t} + K'_1F_{1,t}$$
$$F_{1,(t+1)} = \text{SRT}F'_2F'_1 + (\text{STP}F_2)'F_1$$
$$F_{1,(t+1)} = \text{SRT}F'_2F'_1 + (\text{STP}' + F'_2)F_1$$
$$F_{1,(t+1)} = \text{SRT}F'_2F'_1 + \text{STP}'F_1 + F'_2F_1 \quad \text{Final } F_1 \text{ equation}$$

The next state equations define the conditions that cause the state variables to be logical 1s. At all other times, they are 0s. Using these data we can proceed to complete a transition table for the circuit under analysis.

4. Construct a transition table. Assume a starting point of $F_2,F_1 = 00$. This is reasonable because an asynchronous power-up reset establishes 00 as an initial state. It is helpful to establish a weighting priority for the input and state variables when determining next state values. For this example, the following are used. Variable weighting assignment is

F_2, F_1, SRT, STP

8 4 2 1

The first term in the F_2 next state equation produces (d indicates a don't-care variable)

$$(F_2, F_1, \text{SRT}, \text{STP}) = 010d = 0100 + 0101, \qquad (4, 5)$$

The F_2 second term produces

$$(F_2, F_1, SRT, STP) = 1dd1 = 1001 + 1011 + 1101 + 1111, \qquad (9, 11, 13, 15)$$

The F_2 third term produces

$$(F_2, F_1, \text{SRT}, \text{STP}) = 11dd = 1100 + 1101 + 1110 + 1111, \qquad (12, 13, 14, 15)$$

The complete minterm list for F_2 is

$$F_{2,(t+1)} = f(F_2, F_1, \text{SRT}, \text{STP}) = \Sigma(4, 5, 9, 11, 12, 13, 14, 15)$$

In a like manner the minterm list for $F_{1,(t+1)}$ is

$$F_{1,(t+1)} = f(F_2, F_1, \text{SRT}, \text{STP}) = \Sigma(2, 3, 4, 5, 6, 7, 12, 14)$$

From the two minterm expressions we can construct the transition table (Table 6.16).

The next state equations produce a logical 1 for every minterm in the state variable function. The output variables are derived only from the state variables; because they do not include the input variables, we have a Moore machine.

TABLE 6.16
Transition table for example analysis problem

	Present state		Next state, SRT/STP								Output variables		
			00		01		11		10				
	F_2	F_1	F_2	F_1	F_2	F_1	F_2	F_1	F_2	F_1	CRST	CEN	CLTCH
(A)	0	0	0	0	0	0	0	1	0	1	0	0	0
(B)	0	1	1	1	1	1	0	1	0	1	1	0	0
(C)	1	1	1	1	1	0	1	0	1	1	0	1	0
(D)	1	0	0	0	1	0	1	0	0	0	0	0	1

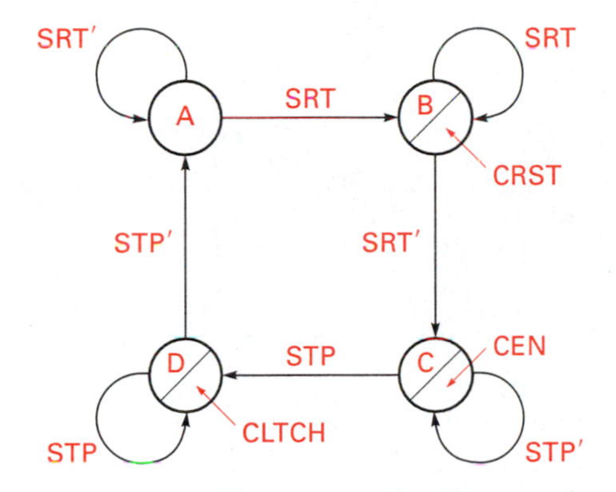

Figure 6.25
State diagram derived from the sequential controller in Figure 6.24

A state diagram (Figure 6.25) can now be constructed from the transition table by assigning each combination of state variables some symbol. These symbols are listed in the transition table.

From the state diagram and the timing diagram in Figure 6.26, we can determine the circuit function. Figure 6.27 illustrates the generation of the start and stop inputs to the sequential circuit controller.

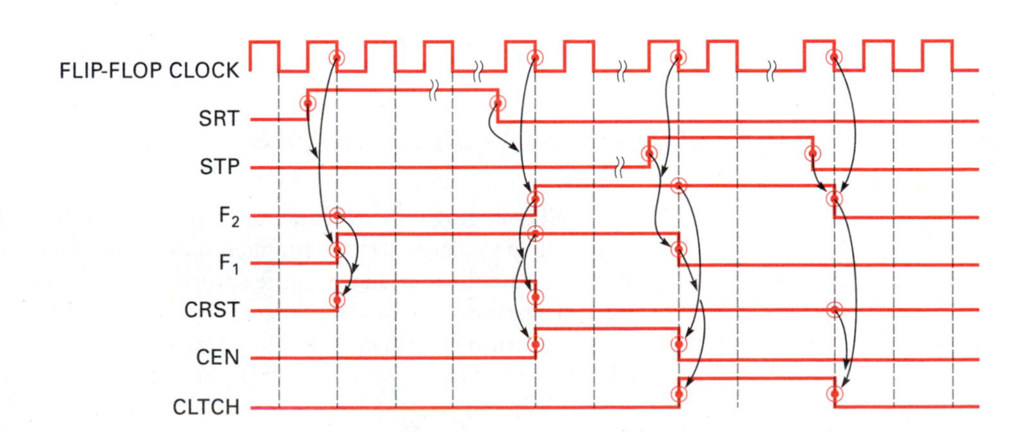

Figure 6.26
Timing diagram for chronograph sequential controller in Figure 6.24

Figure 6.27
Cross-sectional diagram of a bullet passing over the photosensitive start and stop sensors

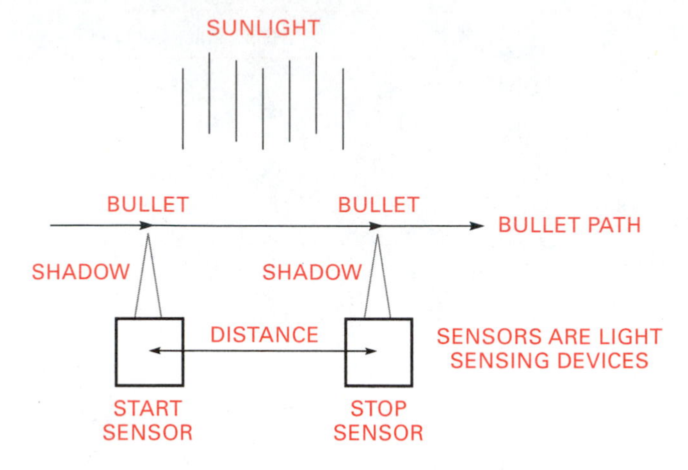

As the bullet passes over the start sensor, the SRT signal is generated. When the same bullet passes over the stop sensor, the STP signal is generated. The distance, s, between sensors is fixed. By measuring the elapsed time between the sensors, the velocity, v, can be found. v = s/t ❖

Having the state and timing diagrams will aid in any future circuit modification or troubleshooting. The process of determining the state variable equations can be accomplished by using K-maps for the state variables, as in the first example, or by converting the next state equations to minterms directly as in the second example.

It is far easier to troubleshoot a sequential circuit from a transition or excitation table than by simply placing probes on various points in the circuit. Often troubleshooting documentation, other than circuit schematics, is missing; and the person responsible for maintenance is left in the dark. Logic designers should, as a matter of course, provide additional documentation in the form of state diagrams, state tables, or excitation tables in system maintenance manuals. Field service and maintenance engineers should generate such documentation for repair personnel, if it was not included by the original designers. In any regard, analysis skills are needed by both designers and field maintenance people.

6.4 CONSTRUCTION OF STATE DIAGRAMS

The most challenging part of sequential circuit synthesis is the development of a state diagram from a verbal or written problem statement. The translation of a problem statement into a state diagram may require several attempts before the final state diagram is reached. Construction of a state diagram in sequential design is analogous to the construction of a truth table in combinational logic design. We will present several problems for which we must construct a state diagram prior to continuing the design of a complete circuit.

6.4.1 Up-Down Decade Counter

❖ **EXAMPLE 6.3**

Create the state diagram for a synchronous decade counter. The counter is to count up or down, in binary, depending on the value of a mode control input signal. When the mode control input $M = 0$, the counter is to count up; when $M = 1$, the counter is to count down. The counter is to repeat or cycle; that is, to count from 0 to 9 and then start over if $M = 0$ or count down from 9 to 0 if $M = 1$. The counter output y is to be a 1 if counting up and the terminal count is reached. Another output z is to be a 1 at the terminal count if counting down.

Should the mode control value change during a count sequence, outputs y and z will be asserted when the terminal count is reached.

SOLUTION We know from the description of the problem that 10 states are needed. We also know that a single mode control determines the count direction. Because the problem stated that the counter must be synchronous, the circuit is to count an input external event when it is coincident with a clock. The state diagram is developed as follows:

1. Draw 10 state symbols and label them 0 through 9 (10 events).
2. Determine the mode control value: Let M be 0 = count-up and 1 = count-down.
3. Connect each state symbol with directed arcs, indicating the next state for the two values of M.

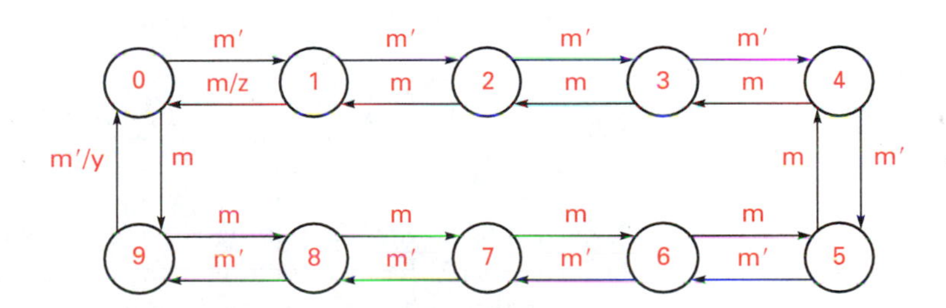

Figure 6.28 Mealy state diagram for up-down decade counter

The state diagram is a Mealy machine (Figure 6.28) because the two output variables are dependent on both the mode input and the present state. ❖

6.4.2 Sequence Detectors

❖ **EXAMPLE 6.4**

Construct a Mealy state diagram that will detect a serial input sequence of 10110. The detection of the required bit pattern can occur in a longer data string and the correct pattern can overlap with another pattern. When the input pattern has been detected, cause an output z to be asserted high.

For example, let the input string be

$$x = 1\ 0\ 1\ 1\ 0\ 1\ 1\ 0\ 1\ 1\ 0$$
$$z = 0\ 0\ 0\ 0\ 1\ 0\ 0\ 1\ 0\ 0\ 1$$

SOLUTION The input pattern requires a single input variable. A single output variable is used to indicate detection of a correct input pattern. If an incorrect pattern is presented, no output occurs.

1. Develop the state diagram one state at a time. The LSB to be detected is a 1. State *A* is the initial state; it checks for a 1. If a 1 occurs, the machine changes from state *A* to state *B*. If a 0 occurs, the machine remains in state *A* waiting for the first 1.

Checking for the initial 1

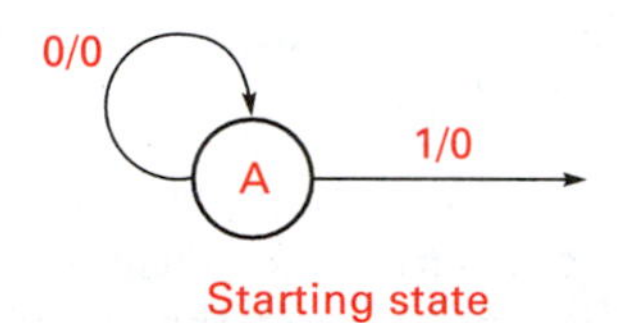

2. The correct second bit in the string is a 0. If it is detected, the machine goes to state *C*. If a second 1 is detected, it is to be counted as the first correct bit in the string. State *B* represents the fact that a 1 has been detected. If multiple 1s occur, the last 1 is counted as the first correct bit in the desired input string.

State transition for initial 1 and second 0

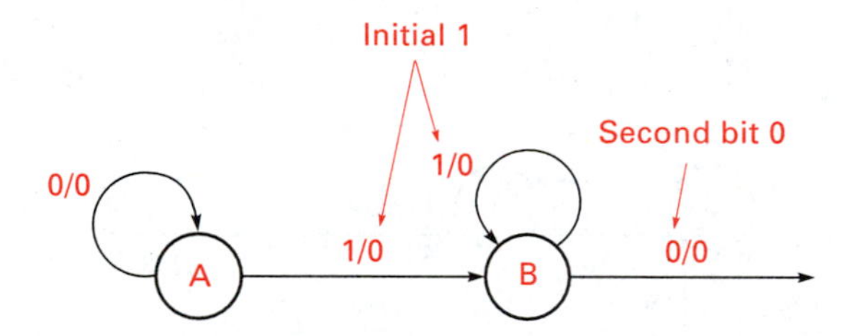

3. The third correct bit in the input string is a 1. A transition from state *C* to state *D* occurs if a string of 101 has been detected (the rightmost bit is the LSB). If an input string of 100 is detected, the machine goes back to state *A* because a 100 string cannot be part of the correct input string.

Checking the third bit

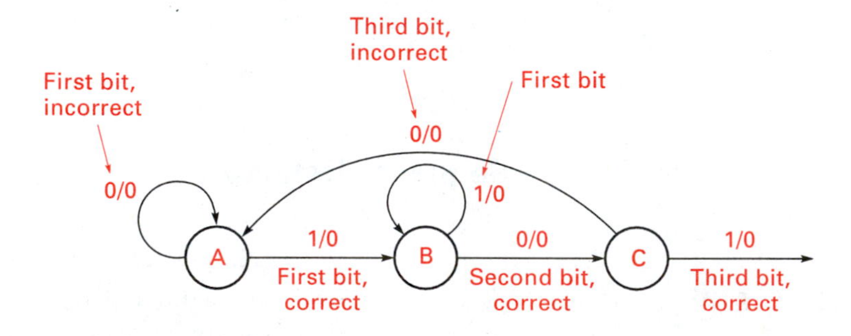

4. The fourth correct input bit in the serial string is checked by the transition from state *D* to either state *E* or back to state *C*. State *D* indicates that three correct input values have been received. The first three correct inputs did not necessarily have to occur

on the first three clock pulses. If $x = 0$, the system returns to state C, indicating two correct inputs (overlap is possible); if $x = 1$ the next state is E (indicating four correct inputs).

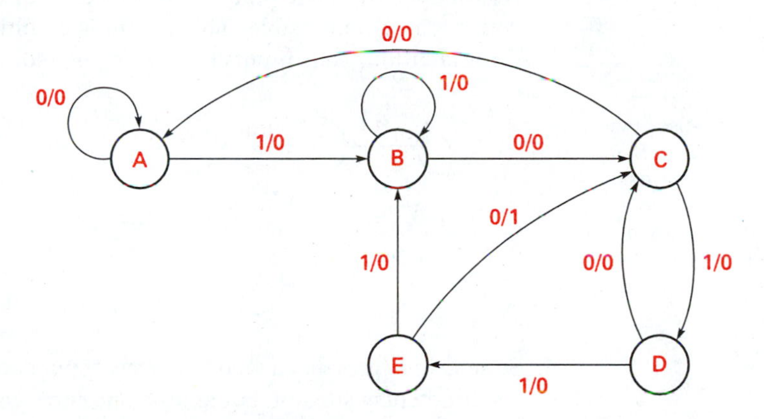

5. The fifth correct input bit in the serial string is checked by the transition from state E to either state B or state C. If $x = 0$, the fifth correct input, the system outputs a 1 and changes to state C. The reason for changing from state E to state C is that the strings can overlap and first a 1 and then a 0 of a new correct string has been detected. If $x = 1$, the machine changes from state E to state B, indicating that the first 1 of a new string has been detected. ❖

❖ EXAMPLE 6.5

Sequence detectors are not restricted to a single input variable. Consider the design of a Mealy state machine that can detect a serial string of four inputs where each input is a four-bit code. If the string of four four-bit codes is correctly received, then an output is to be generated. An incorrect input code pattern is to generate a second output. The second output is to be asserted only after receiving the sequence of four four-bit codes.

A pattern detector like this can be used as an electronic combination lock that opens when the correct combination is received and turns on an alarm if the incorrect pattern is tried.

SOLUTION Let the input combination be 0110, 1010, 0100, 1000 (input the leftmost code first).

1. Start in some initial state A. The first pattern is either correct or incorrect. If it is correct, then change to state B; if the code is incorrect, change to a new state C. Note that only 1 of the 16 possible combinations (0110) causes an A to B transition, the remaining 15 combinations cause an A to C transition.

First four-bit code is checked

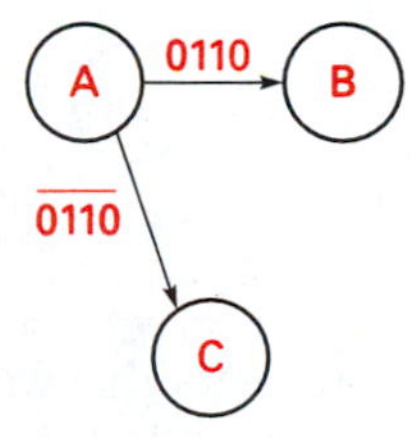

2. State B indicates the first correct code input. State C indicates an incorrect first code input. The second four-bit input code is entered next. A correct code input causes a

transition from state B to state D. The state changes from B to E after one correct and one incorrect code inputs have been entered. Two incorrect code inputs leaves the system in state E (A to C and C to E). Note that the C to E transition occurs regardless of the code input. This is true because the correct combination of four-bit codes is now impossible. The system is counting attempts in order to indicate an error after four code inputs have been entered.

Second correct code causes the state machine to change to *D*. The first incorrect code leaves the machine in state *C*. Two incorrect codes or one correct code and one incorrect code puts the machine in state *E*.

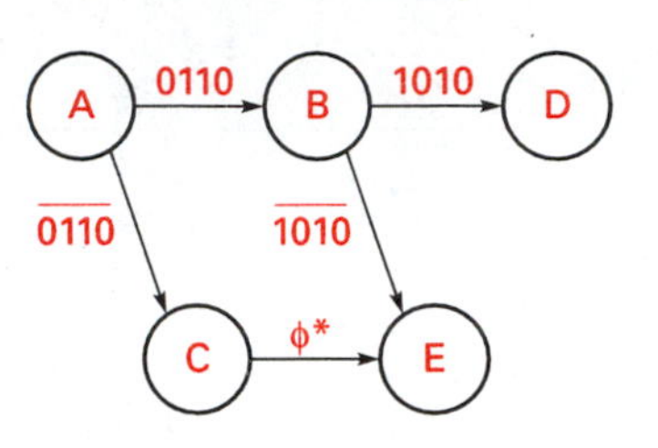

ϕ* indicates that all input combinations cause a transition

3. State D indicates that a second correct input code has been entered. State E indicates a second entry attempt, but at least one entry was incorrect. Again notice that only 1 of 16 input combinations enables the B to D state transition. The third correct input causes a state transition from D to F. If, on the third attempt, any input is incorrect, the machine changes to state G.

State *F* indicates three correct code entries. State *G* indicates at least one incorrect code entry in three tries

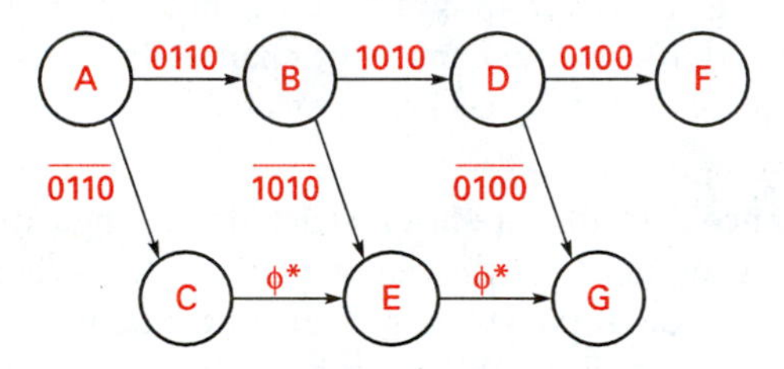

4. The fourth correct input causes a state transition from F back to A, while generating an "open" output. An incorrect input attempt after three correct entries also causes an F-to-A state transition; however, "alarm" is asserted instead of "open." The fourth input code when starting from state G causes a G-to-A transition and an "alarm" output.

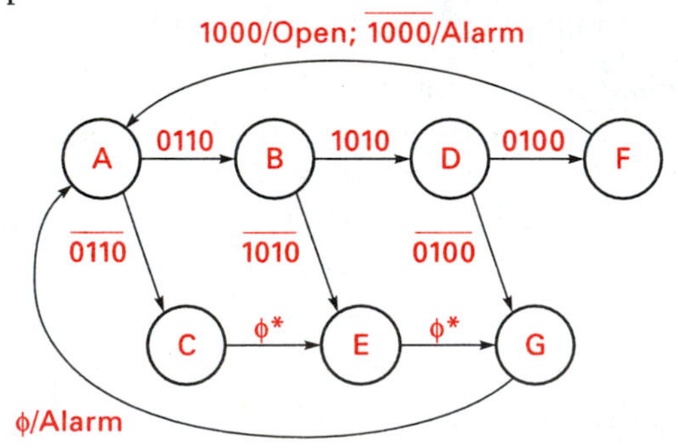

ϕ* indicates that all input combinations cause a transition

❖

❖ **EXAMPLE 6.6**

Design a Mealy serial bit-pattern detector that will detect the input sequence 01010 in a longer bit string. If the pattern is detected, then cause output Q to be active-high. If an 011 bit pattern occurs within the same serial data string, cause output P to be active high. If the 011 pattern occurs, cause the state machine to initialize and start over looking for the 01010 pattern. Overlapping 01010 patterns can occur.

For example,

if X = 0 0 1 1 0 1 0 1 0 1 1 1 0 1 0 1 0 1 0 1, then outputs
Q = 0 0 0 0 0 0 0 0 1 0 0 0 0 0 0 0 1 0 1 0
P = 0 0 0 1 0 0 0 0 0 0 1 0 0 0 0 0 0 0 0 0

SOLUTION **1.** Start with an initial state.

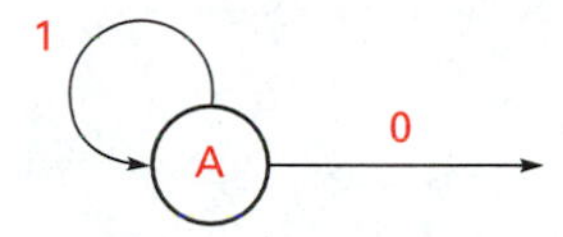

A 0 input is the first correct bit for both patterns and causes a state transition to a new state. A 1 input is incorrect; therefore, the state machine remains in the initial state.

2. State B indicates that the first bit in both input sequences is correct. If $x = 0$, the machine stays in state B; if $x = 1$, a transition is made to a new state.

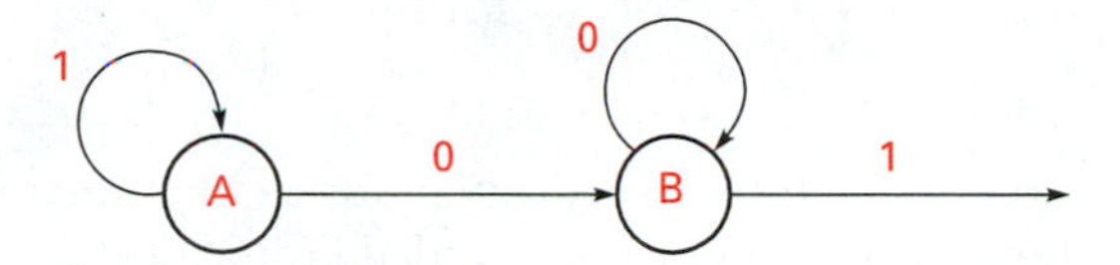

3. State C indicates that the second input bit is correct (for both patterns). If $x = 0$, the correct 01010 pattern is in progress, and the machine changes to a new state. If $x = 1$, then pattern 011 has been detected; therefore, $P = 1$ and the machine resets to state A.

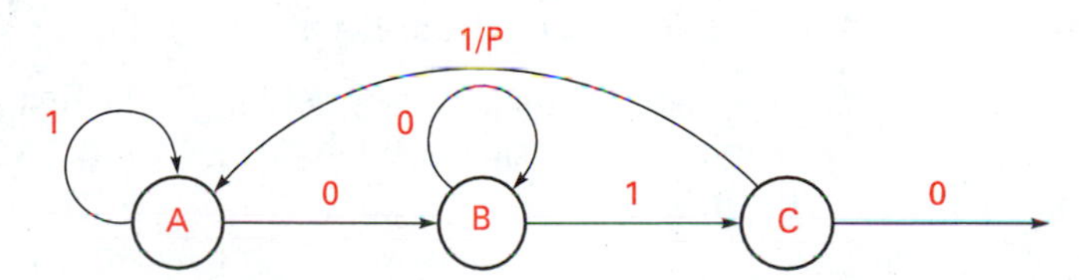

4. State D indicates the third correct bit of pattern 01010 or the first correct bit of the 011 pattern. No output is required from P or Q. Present state D changes to next state B if $x = 0$ and to E if $x = 1$.

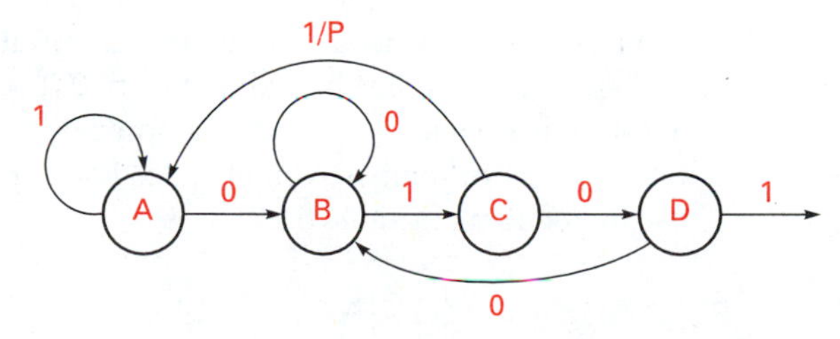

5. State E indicates the fourth correct bit of pattern 01010 or the second correct bit of pattern 011 has been received. If, while in the present state E, $x = 0$, then the fifth bit of pattern 01010 has been detected and output $Q = 1$, and the machine changes to state D. From present state E, the next state will be D if $x = 0$. This is done to accommodate overlapping patterns. A state transition from E to A occurs if $x = 1$, indicating the third and final bit of pattern 011 has been detected; the system resets to state A and output $P = 1$.

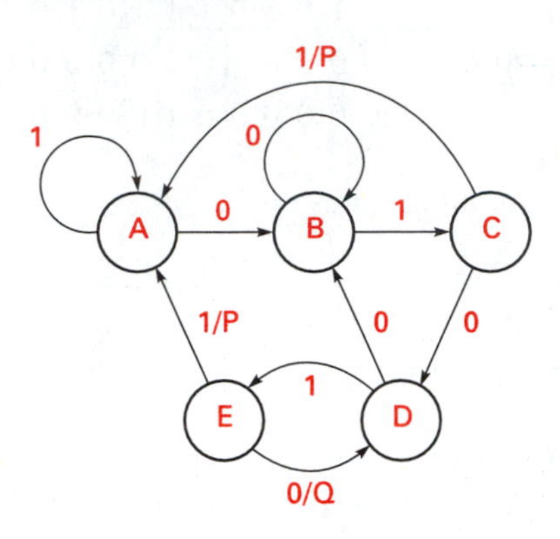

❖

6.4.3 Serial EX-3 to BCD Code Converter

❖ **EXAMPLE 6.7**

Design the Mealy state diagram for a circuit that will convert serial excess-3 code data (input x) into serial binary coded decimal (BCD) data. Because both excess-3 and BCD are four-bit codes the machine is to return to the beginning after four inputs. Input the least significant excess-3 bit first. The serial BCD data output is labeled B. If any non-excess-3 input sequence occurs, cause $z = 1$. The data are serial, so a sequential circuit is necessary to accomplish the conversion.

SOLUTION The ex-3 and BCD codes are as in Table 6.17.

1. Start with an initial state S_0 and input the first ex-3 serial bit. From the table, we see that, if the least significant ex-3 bit is a 1, the BCD bit is a 0. If the ex-3 bit is a 0, then the BCD bit is a 1.

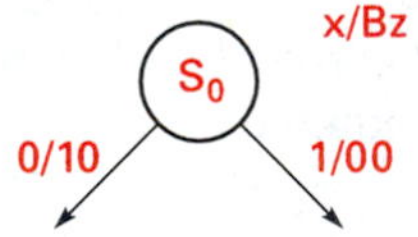

TABLE 6.17
Ex-3 to BCD codes

Ex-3	BCD
0011	0000
0100	0001
0101	0010
0110	0011
0111	0100
1000	0101
1001	0110
1010	0111
1011	1000
1100	1001

2. Two next states are needed, one for each input value of x. States S_1 and S_2 represent the first excess-3 input bit value; if $x = 0$, the system changes to state S_1; if $x = 1$, then the next state is S_2. The second ex-3 input bit causes a transition from either state S_1 or S_2 to some new next state. The correct BCD output is generated for each case by referring to Table 6.17.

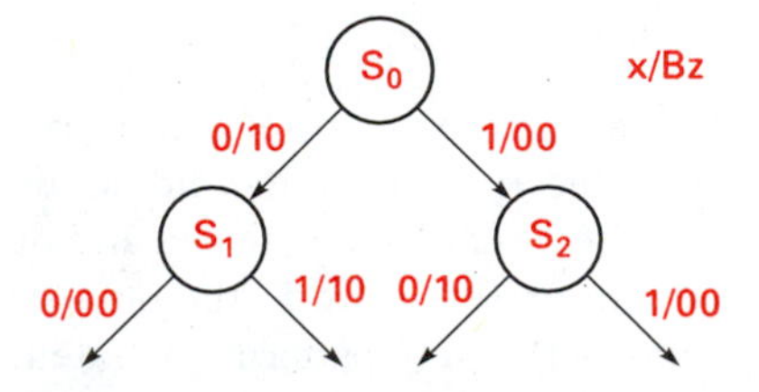

3. Two next-state transition paths exist for both states S_1 and S_2. For now we will assume that a new next state is needed for each transition path. The outputs from each new next state indicate that three inputs have been made. The correct BCD output bits for each transition path is determined from Table 6.17.

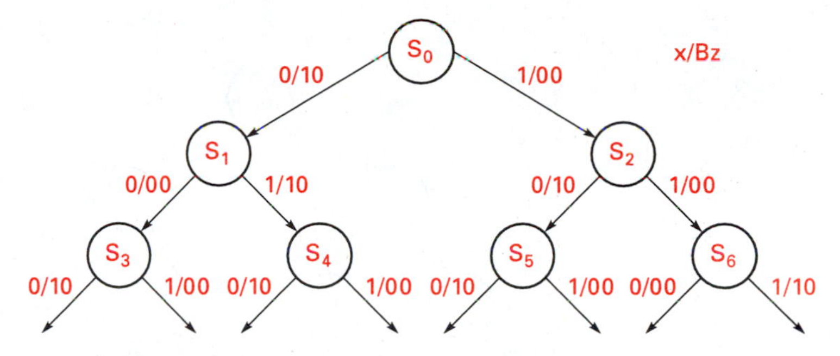

4. Look at the input-output values for states S_4 and S_5 in step 3. Notice that the output transition 0/10 path from S_4 and S_5 are identical, as are the 1/00 transition paths from both states. This means that the two states S_4 and S_5 are redundant and can be combined. We will study state reduction in a more formal manner in Chapter 7.

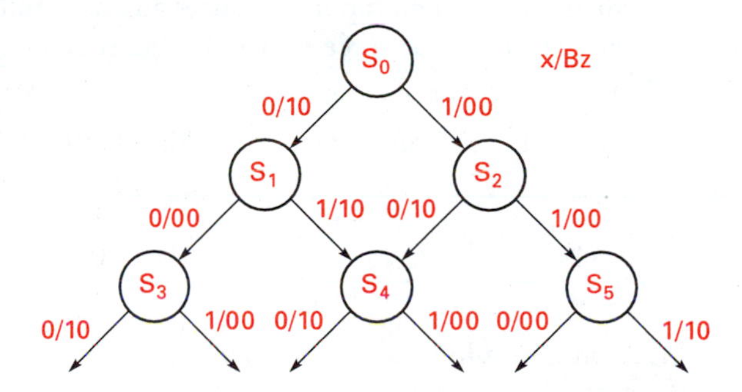

5. Three states S_3, S_4, and S_5 each have two transition paths. A new state is needed for each.

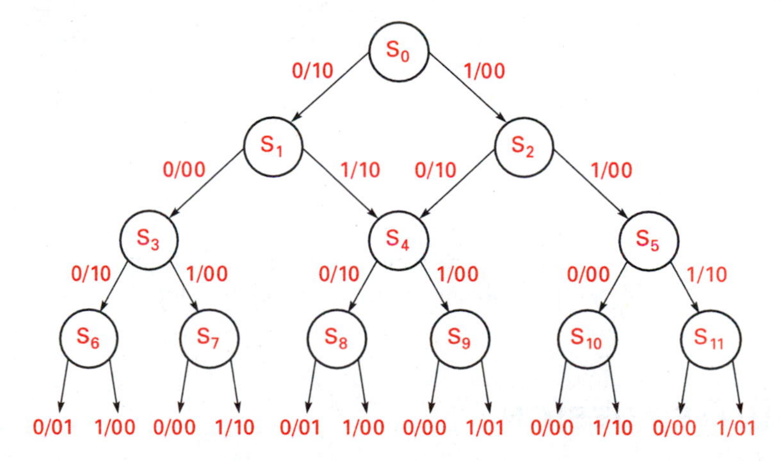

6. States S_6 through S_{11} indicate the reception of the third BCD input bit. The fourth input bit completes the ex-3 sequence; therefore, the next states of present states S_6 through S_{11} must all return to state A. The output from the final state transitions back to state S_0 depends on whether a correct excess-3 value has been received. If

the fourth excess-3 bit is correct, then $B = 0$ or 1 depending on the required BCD value. If the fourth excess-3 input bit is incorrect then $z = 1$.

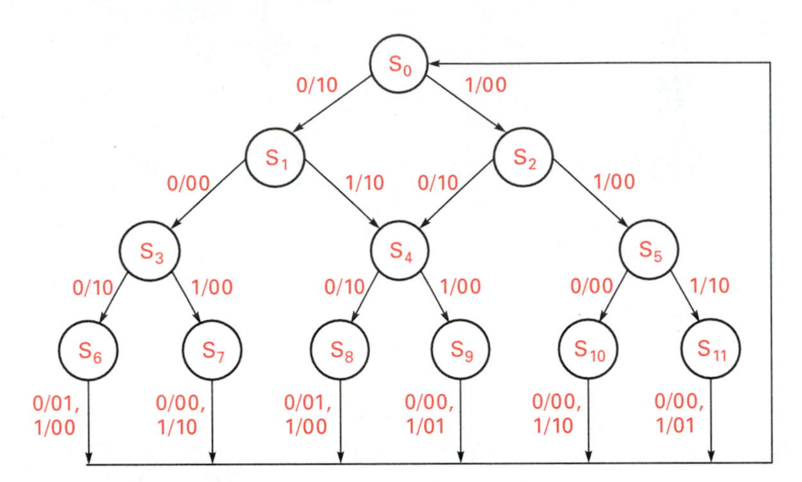

To check the accuracy of the serial excess-3-to-BCD code conversion state diagram, we can assume several input sequences and determine the BCD output sequence.

Let the input sequence be 0101. The sequence of states and resulting BCD output are

Input bit	State transition	BCD output	z output
1	$S_0 \rightarrow S_2$	0	0
0	$S_2 \rightarrow S_4$	1	0
1	$S_4 \rightarrow S_9$	0	0
0	$S_9 \rightarrow S_0$	0	0

Error bit z should be set when a non-excess-3 four-bit input sequence is attempted. Let the input sequence be 1111:

Input bit	State transition	BCD output	z output
1	$S_0 \rightarrow S_2$	0	0
1	$S_2 \rightarrow S_5$	0	0
1	$S_5 \rightarrow S_{11}$	1	0
1	$S_{11} \rightarrow S_0$	0	1

❖

Construction of state diagrams from verbal or written problem statements is not a highly developed systematic process. Insight into the problem being solved is essential, and the only way to develop that is through experience. The basic idea is to understand the problem and then analyze the steps involved, converting each step into state(s).

6.5 COUNTER DESIGN

We will design several sequential counters to illustrate the steps involved from a problem statement through the state diagram and the various tables to a completed logic diagram.

6.5.1 Modulo-8 Synchronous Counter

EXAMPLE 6.8 Design a cyclic modulo-8 synchronous binary counter using *J–K* flip-flops that will count the number of occurrences of an input; that is, the number of times it is a 1. The input variable *x* must be coincident with the clock to be counted. The counter is to count in binary.

SOLUTION Eight states are needed, and the state assignment is implied by the binary count requirement.

1. Construct the state diagram for the modulo-8 counter (Figure 6.29).
2. Next construct a state table (Table 6.18).
3. Eight states are needed, which requires three state variables. Because the state assignment is binary, the transition table is as shown in Table 6.19.

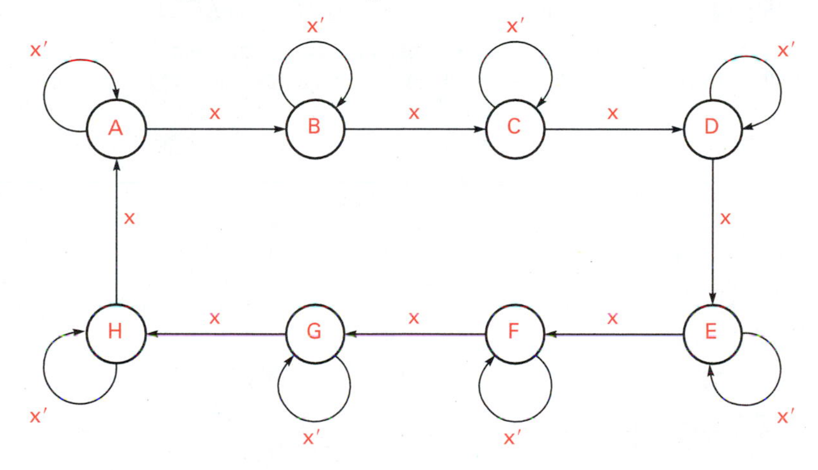

Figure 6.29
Modulo-8 binary counter state diagram

TABLE 6.18
State table for a modulo-8 counter

Present state	Next state $x = 0$	Next state $x = 1$
A	*A*	*B*
B	*B*	*C*
C	*C*	*D*
D	*D*	*E*
E	*E*	*F*
F	*F*	*G*
G	*G*	*H*
H	*H*	*A*

TABLE 6.19
Transition table

Present state			Next state $x = 0$			Next state $x = 1$		
F_3	F_2	F_1	F_3	F_2	F_1	F_3	F_2	F_1
0	0	0	0	0	0	0	0	1
0	0	1	0	0	1	0	1	0
0	1	0	0	1	0	0	1	1
0	1	1	0	1	1	1	0	0
1	0	0	1	0	0	1	0	1
1	0	1	1	0	1	1	1	0
1	1	0	1	1	0	1	1	1
1	1	1	1	1	1	0	0	0

F_3, F_2, and F_1 are the state variable names.

TABLE 6.20
Excitation table, modulo-8 counter using J–K flip-flops

Present state			Next state					
			$x = 0$			$x = 1$		
F_3	F_2	F_1	J_3K_3	J_2K_2	J_1K_1	J_3K_3	J_2K_2	J_1K_1
0	0	0	0d	0d	0d	0d	0d	1d
0	0	1	0d	0d	d0	0d	1d	d1
0	1	0	0d	d0	0d	0d	d0	1d
0	1	1	0d	d0	d0	1d	d1	d1
1	0	0	d0	0d	0d	d0	0d	1d
1	0	1	d0	0d	d0	d0	1d	d1
1	1	0	d0	d0	0d	d0	d0	1d
1	1	1	d0	d0	d0	d1	d1	d1

4. The next step involves creating an excitation table. To do this, we must decide which flip-flop type to use. The problem definition requires the use of JK, resulting in Table 6.20. The counter excitation table uses the data in the transition table and that of the J–K flip-flop excitation table. Refer to the J–K excitation Table 6.12.
5. Data from the excitation table are now loaded into K-maps (Figures 6.30–6.32) to be simplified. All eight possible states for the three state variables are used; therefore,

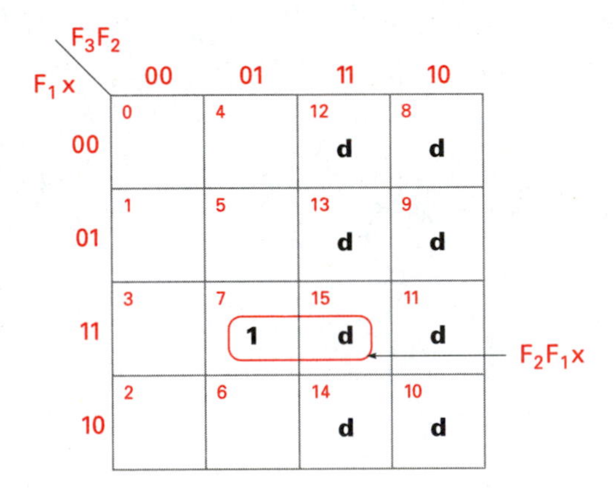

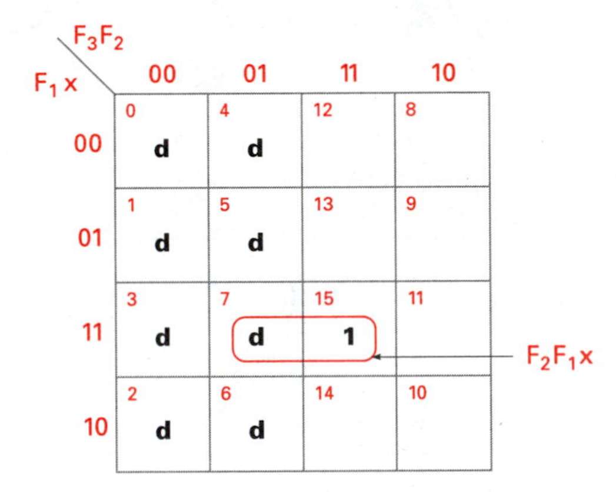

Figure 6.30
J_3 and K_3 K-maps for state variable F_3

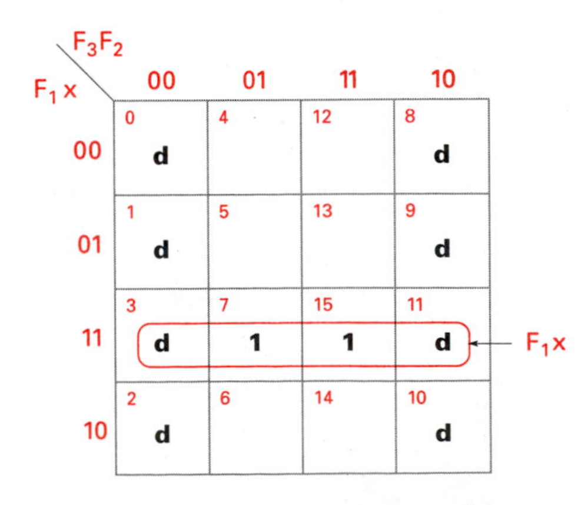

Figure 6.31
J_2 and K_2 K-maps for state variable F_2

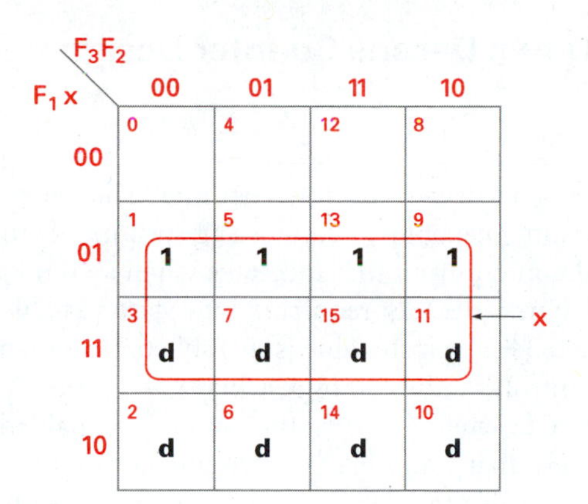

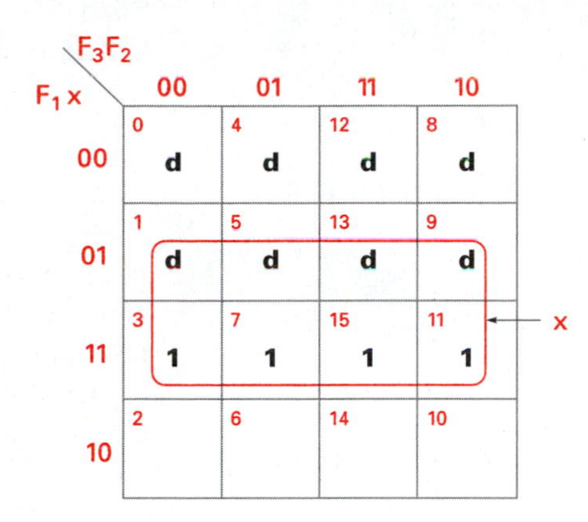

Figure 6.32
J_1 and K_1 K-maps for state variable F_1

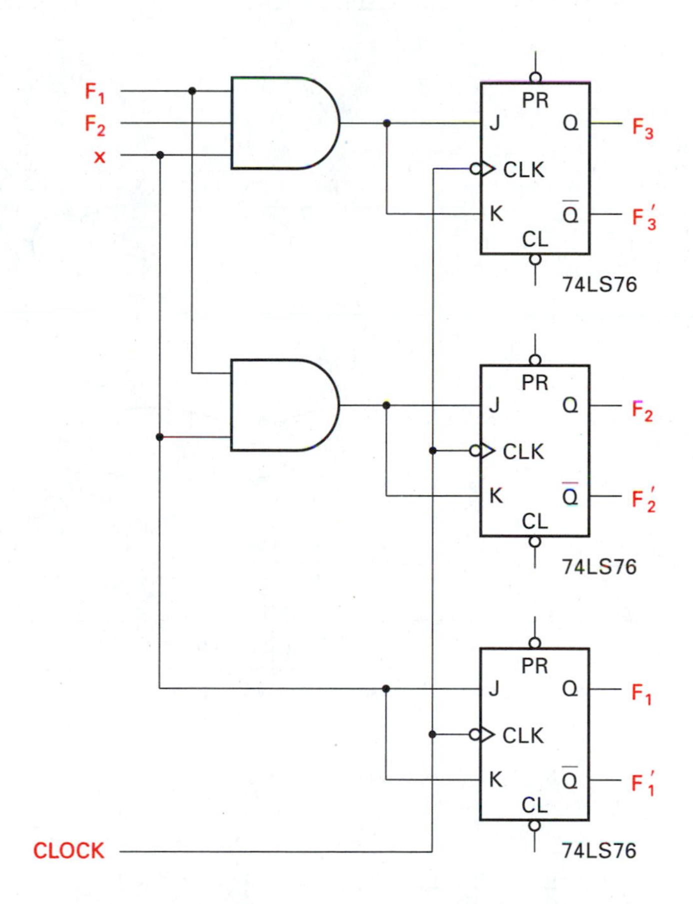

Figure 6.33
Modulo-8 counter logic diagram

no don't-care states exist. Notice that $J = K$ for each of the three sets of excitation variables. The resulting logic diagram is shown in Figure 6.33. ❖

6.5.2 Up-Down Decade Counter Design

❖ **EXAMPLE 6.9**

Design a synchronous decade up-down counter. The counter must have an enable input to provide cascading capability (connecting multiple counters in series). It must also have a terminal count output that indicates when a count of 0 or 10 has occurred. The counter has 10 states (decade) requiring four state variables. The state diagram is illustrated in Figure 6.34. The state table is in Table 6.21, which uses the following symbols.

Up-down control is indicated by x: when $x = 1$, count up; when $x = 0$, count down. Count enable is CE: when CE = 1, the counter is enabled. The terminal count is TC: when TC = 1, terminal count has been reached (count value = 0 or count value = 9).

If we were to realize the decade counter using D flip-flops, we could derive the excitation equations directly from the transition table (Table 6.22). However, the use of J–K flip-flops typically results in fewer gates than a D flip-flop realization. The construc-

Figure 6.34
State diagram for a decade up-down counter

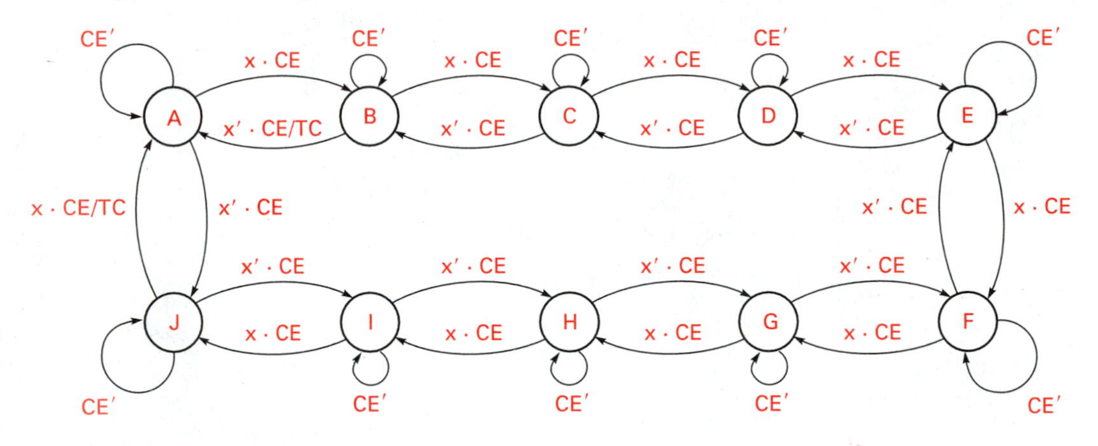

TABLE 6.21
Decade counter state table

Present state	Next state xCE							
	00	TC	01	TC	10	TC	11	TC
A	A	0	J	0	A	0	B	0
B	B	0	A	1	B	0	C	0
C	C	0	B	0	C	0	D	0
D	D	0	C	0	D	0	E	0
E	E	0	D	0	E	0	F	0
F	F	0	E	0	F	0	G	0
G	G	0	F	0	G	0	H	0
H	H	0	G	0	H	0	I	0
I	I	0	H	0	I	0	J	0
J	J	0	I	0	J	0	A	1

TABLE 6.22
Decade counter transition table

Present state				Next state xCE																			
				0 0					0 1					1 0					1 1				
F_4	F_3	F_2	F_1	F_4	F_3	F_2	F_1	TC	F_4	F_3	F_2	F_1	TC	F_4	F_3	F_2	F_1	TC	F_4	F_3	F_2	F_1	TC
0	0	0	0	0	0	0	0	0	1	0	0	1	0	0	0	0	0	0	0	0	0	1	0
0	0	0	1	0	0	0	1	0	0	0	0	0	1	0	0	0	1	0	0	0	1	0	0
0	0	1	0	0	0	1	0	0	0	0	0	1	0	0	0	1	0	0	0	0	1	1	0
0	0	1	1	0	0	1	1	0	0	0	1	0	0	0	0	1	1	0	0	1	0	0	0
0	1	0	0	0	1	0	0	0	0	0	1	1	0	0	1	0	0	0	0	1	0	1	0
0	1	0	1	0	1	0	1	0	0	1	0	0	0	0	1	0	1	0	0	1	1	0	0
0	1	1	0	0	1	1	0	0	0	1	0	1	0	0	1	1	0	0	0	1	1	1	0
0	1	1	1	0	1	1	1	0	0	1	1	0	0	0	1	1	1	0	1	0	0	0	0
1	0	0	0	1	0	0	0	0	0	1	1	1	0	1	0	0	0	0	1	0	0	1	0
1	0	0	1	1	0	0	1	0	1	0	0	0	0	1	0	0	1	0	0	0	0	0	1

tion of the excitation table for the J–K flip-flop can become tedious when problems become very large. The counter we are designing has four state variables and two input variables, requiring six variable K-maps for equation simplification. We can derive the J–K equations from the transition table if we look for changes in the state variables. The J input must be a 1 for a state variable to change from a 0 to a 1. The K input must be a 1 for the state variable to change from a 1 to a 0.

If we find the places where a state variable changes, then we can determine the J and K minterms. For example, state variable F_4 changes from a 0 to a 1 in two places in Table 6.22. In the first row the present state value for state variable F_4 changes from a 0 to a 1 in the 01 next state column. In the eighth row (present state $F_4F_3F_2F_1 = 0111$) state variable F_4 changes from a 0 to a 1, in the 11 next state column.

The J and K minterm numbers are found by decoding the binary value of the present state and input variables. Minterms for J are generated on state variable 0 to 1 transitions, and minterms for K are generated for state variable 1 to 0 transitions.

Any term that is not a minterm is either a 0 or a don't'-care term. Don't-care terms are loaded directly into the K-map. J_4 is a 1 in present state $F_4F_3F_2F_1 = 0000$, and the input $x\text{CE} = 01$. The minterm is found when we combine the present state and input variables, giving $F_4F_3F_2F_1x\text{CE} = 000001_2 = 1_{10}$. Another J_4 minterm exists when $F_4F_3F_2F_1x\text{CE} = 011111_2 = 31_{10}$.

The two minterms for J_4 are

$$J_4 = f(F_4, F_3, F_2, F_1, x, \text{CE}) = \Sigma\,(1, 31)$$

The K_4 minterms are found where the present to next-state change is 1 to 0, which occurs in two places:

$$(F_4,\ F_3,\ F_2,\ F_1,\ x,\ \text{CE}) = 100001_2 = 33_{10}$$
$$(F_4,\ F_3,\ F_2,\ F_1,\ x,\ \text{CE}) = 100111_2 = 39_{10}$$
$$K_4 = f(F_4, F_3, F_2, F_1, x, \text{CE}) = \Sigma\,(33, 39)$$

The J_3 minterms are found for those transitions where $F_3 = 0$ in a present state and 1 in the next state. The J_3 equation is

$$J_3 = f(F_4, F_3, F_2, F_1, x, CE) = \Sigma\ (15, 33)$$

The K_3 minterms are found when the F_3 present to next state transition is 1 to 0. The K_3 equation is

$$K_3 = f(F_4, F_3, F_2, F_1, x, CE) = \Sigma\ (17, 31)$$

The J_2 and K_2 equations are

$$J_2 = f(F_4, F_3, F_2, F_1, x, CE) = \Sigma\ (7, 17, 23, 33)$$
$$K_2 = f(F_4, F_3, F_2, F_1, x, CE) = \Sigma\ (9, 15, 25, 31)$$

The J_1 and K_1 equations are

$$J_1 = f(F_4, F_3, F_2, F_1, x, CE) = \Sigma\ (1, 3, 9, 11, 17, 19, 25, 27, 33, 35)$$
$$K_1 = f(F_4, F_3, F_2, F_1, x, CE) = \Sigma\ (5, 7, 13, 15, 21, 23, 29, 31, 37, 39)$$

Of the 16 states possible when using four state variables, 10 were specified, and the remainder are don't-care states. The don't-care terms that result are

32	16	8	4	2	1	Variable weight	
F_4	F_3	F_2	F_1	x	CE	variable	
1	0	1	0	x	x^*	Don't-care	(40, 41, 42, 43)
1	0	1	1	x	x^*	Don't-care	(44, 45, 46, 47)
1	1	0	0	x	x^*	Don't-care	(48, 49, 50, 51)
1	1	0	1	x	x^*	Don't-care	(52, 53, 54, 55)
1	1	1	0	x	x^*	Don't-care	(56, 57, 58, 59)
1	1	1	1	x	x^*	Don't-care	(60, 61, 62, 63)

where $x\,x^*$ is 00, 01, 10 11, (40, 41, 42, 43, 44, 45, 46, 47, 48, 49, 50, 51, 52, 53, 54, 55, 56, 57, 58, 59, 60, 61, 62, 63).

The don't-care terms resulting from incompletely specified states apply to the simplification of the J–K excitation variables and the output variable, TC. These don't-care terms are in addition to those generated by the individual J and K excitation variables.

The output function, TC, equation is

$$TC = f(F_4, F_3, F_2, F_1, x, CE) = \Sigma\ (5, 39) + \Sigma\ d(40 \text{ to } 63)$$

The four J–K flip-flops require eight six-variable K-maps to simplify the excitation equations and one six-variable map for the output function. The excitation and output K-maps are shown in Figures 6.35–6.39. The complete logic diagram using low-power Schottky ICs is shown in Figure 6.40.

Figure 6.35
J_4 and K_4 K-maps

F_1x CE \ F_4, F_3, F_2	000	001	011	010	100	101	111	110
000	0 d	8 d	24 d	16 d	32	40 d	56 d	48 d
001	1 d	9 d	25 d	17 d	33 1	41 d	57 d	49 d
011	3 d	11 d	27 d	19 d	35	43 d	59 d	51 d
010	2 d	10 d	26 d	18 d	34	42 d	58 d	50 d
100	4 d	12 d	28 d	20 d	36	44 d	60 d	52 d
101	5 d	13 d	29 d	21 d	37	45 d	61 d	53 d
111	7 d	15 d	31 d	23 d	39 1	47 d	63 d	55 d
110	6 d	14 d	30 d	22 d	38	46 d	62 d	54 d

$K_4 = F_1'x'\ CE + F_1 x\ CE$

F_1x CE \ F_4, F_3, F_2	000	001	011	010	100	101	111	110
000	0	8	24	16	32 d	40 d	56 d	48 d
001	1 1	9	25	17	33 d	41 d	57 d	49 d
011	3	11	27	19	35 d	43 d	59 d	51 d
010	2	10	26	18	34 d	42 d	58 d	50 d
100	4	12	28	20	36 d	44 d	60 d	52 d
101	5	13	29	21	37 d	45 d	61 d	53 d
111	7	15	31 1	23	39 d	47 d	63 d	55 d
110	6	14	30	22	38 d	46 d	62 d	54 d

$J_4 = F_3'F_2'F_1'x'\ CE + F_3F_2F_1 x\ CE$

Figure 6.36
J_3 and K_3 K-maps

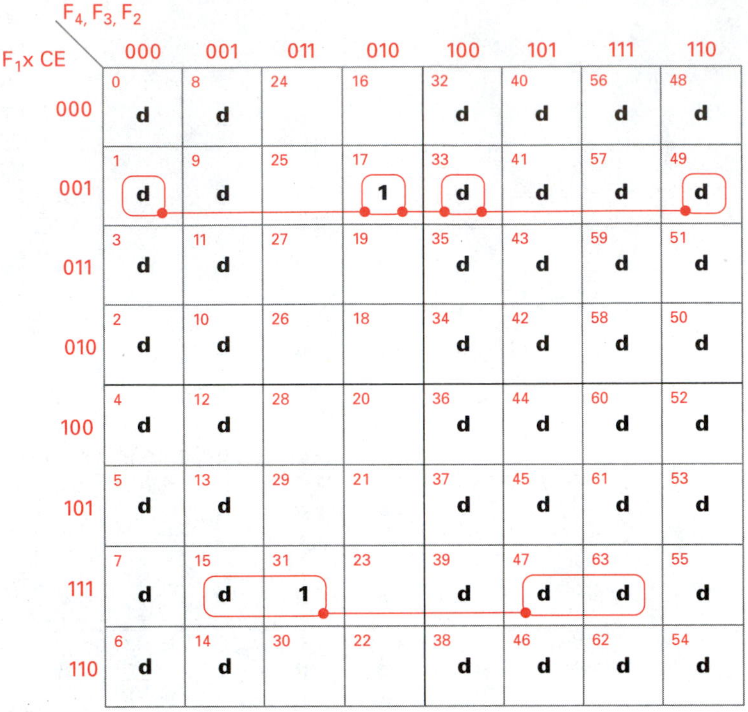

F_1 x CE \ F_4, F_3, F_2	000	001	011	010	100	101	111	110
000	0 d	8 d	24	16	32 d	40 d	56 d	48 d
001	1 d	9 d	25	17 1	33 d	41 d	57 d	49 d
011	3 d	11 d	27	19	35 d	43 d	59 d	51 d
010	2 d	10 d	26	18	34 d	42 d	58 d	50 d
100	4 d	12 d	28	20	36 d	44 d	60 d	52 d
101	5 d	13 d	29	21	37 d	45 d	61 d	53 d
111	7 d	15 d	31 1	23	39 d	47 d	63 d	55 d
110	6 d	14 d	30	22	38 d	46 d	62 d	54 d

$K_3 = F_2' F_1' x' \, CE + F_2 F_1 x \, CE$

F_1 x CE \ F_4, F_3, F_2	000	001	011	010	100	101	111	110
000	0	8	24 d	16 d	32	40 d	56 d	48 d
001	1	9	25 d	17 d	33 1	41 d	57 d	49 d
011	3	11	27 d	19 d	35	43 d	59 d	51 d
010	2	10	26 d	18 d	34	42 d	58 d	50 d
100	4	12	28 d	20 d	36	44 d	60 d	52 d
101	5	13	29 d	21 d	37	45 d	61 d	53 d
111	7	15 1	31 d	23 d	39	47 d	63 d	55 d
110	6	14	30 d	22 d	38	46 d	62 d	54 d

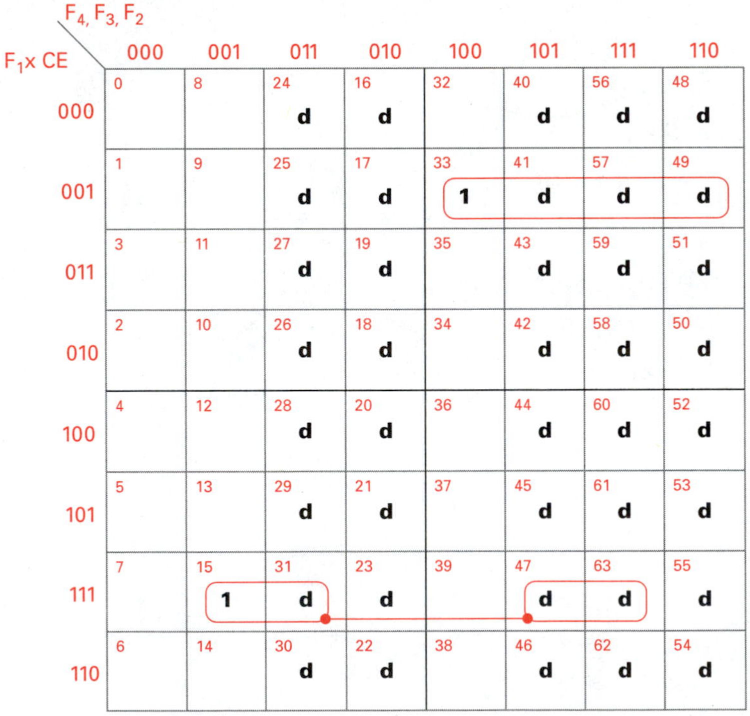

$J_3 = F_4 F_1' x' \, CE + F_2 F_1 x \, CE$

Figure 6.37
J_2 and K_2 K-maps

$F_1 x\ CE$ \ F_4, F_3, F_2	000	001	011	010	100	101	111	110
000	0 d	8	24	16 d	32 d	40 d	56 d	48 d
001	1 d	9 1	25 1	17 d	33 d	41 d	57 d	49 d
011	3 d	11	27	19 d	35 d	43 d	59 d	51 d
010	2 d	10	26	18 d	34 d	42 d	58 d	50 d
100	4 d	12	28	20 d	36 d	44 d	60 d	52 d
101	5 d	13	29	21 d	37 d	45 d	61 d	53 d
111	7 d	15 1	31 1	23 d	39 d	47 d	63 d	55 d
110	6 d	14	30	22 d	38 d	46 d	62 d	54 d

$$K_2 = F_1' x' \, CE + F_1 x \, CE$$

$F_1 x\ CE$ \ F_4, F_3, F_2	000	001	011	010	100	101	111	110
000	0	8 d	24 d	16	32	40 d	56 d	48 d
001	1	9 d	25 d	17 1	33 1	41 d	57 d	49 d
011	3	11 d	27 d	19	35	43 d	59 d	51 d
010	2	10 d	26 d	18	34	42 d	58 d	50 d
100	4	12 d	28 d	20	36	44 d	60 d	52 d
101	5	13 d	29 d	21	37	45 d	61 d	53 d
111	7 1	15 d	31 d	23 1	39	47 d	63 d	55 d
110	6	14 d	30 d	22	38	46 d	62 d	54 d

$$J_2 = F_3 F_1' x' \, CE + F_4 F_1' x' \, CE + F_4' F_1 x \, CE$$

Figure 6.38
J_1 and K_1 K-maps

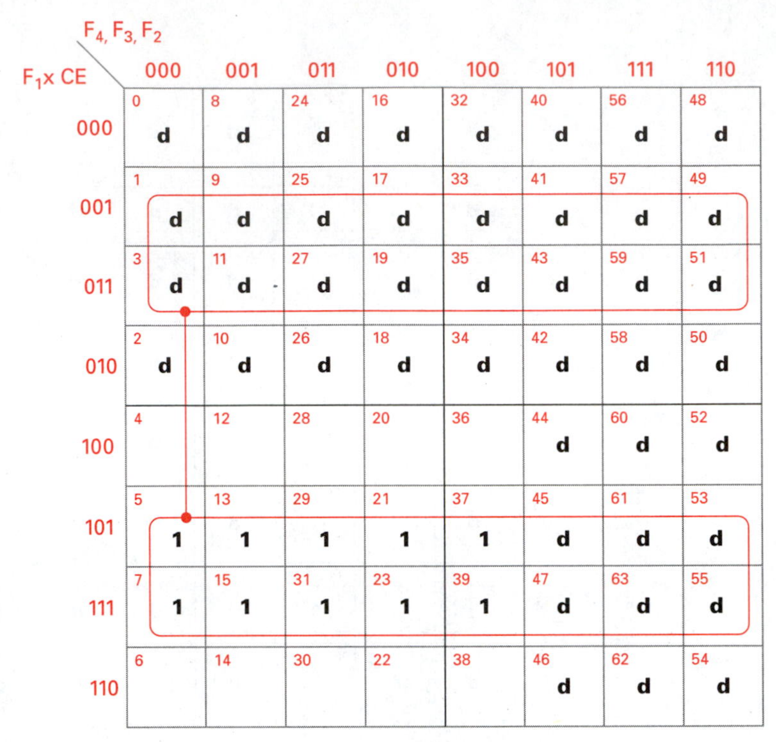

$K_1 = CE$

F_1x CE \ F_4, F_3, F_2	000	001	011	010	100	101	111	110
000	0	8	24	16	32	40: d	56: d	48: d
001	1: 1	9: 1	25: 1	17: 1	33: 1	41: d	57: d	49: d
011	3: 1	11: 1	27: 1	19: 1	35: 1	43: d	59: d	51: d
010	2	10	26	18	34	42: d	58: d	50: d
100	4: d	12: d	28: d	20: d	36: d	44: d	60: d	52: d
101	5: d	13: d	29: d	21: d	37: d	45: d	61: d	53: d
111	7: d	15: d	31: d	23: d	39: d	47: d	63: d	55: d
110	6: d	14: d	30: d	22: d	38: d	46: d	62: d	54: d

$J_1 = CE$

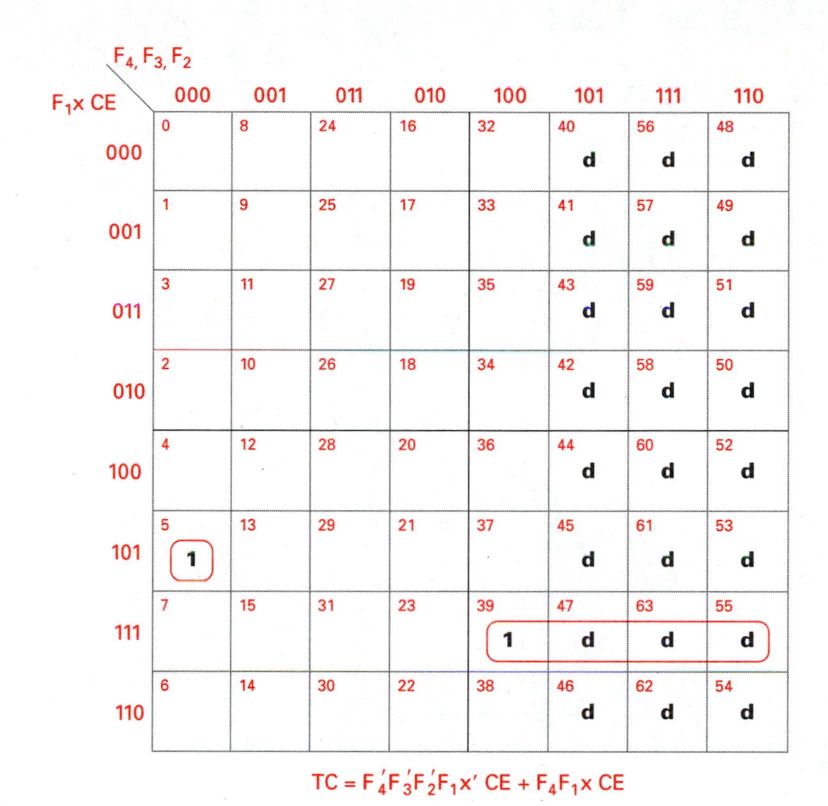

Figure 6.39
K-map for TC

The timing diagram, shown in Figure 6.41, shows how the counter changes from one state to the next when the count enable, CE, is high. When $x = 1$ (count up) and after being reset, the counter starts in state *A* (0000) and progresses through *B* (0001) to state *J* (1001), at which time the terminal count TC goes high. When $x = 0$ (count down), the opposite process occurs. Note that *x* changed from 1 to 0 about midway through the timing diagram (state *E*). Prior to the change in *x* the counter was counting up. When *x* changed to 0, the counter started counting down from state *F* to state *E* and on down to state *A*.

Figure 6.40
Decade up-down counter logic diagram

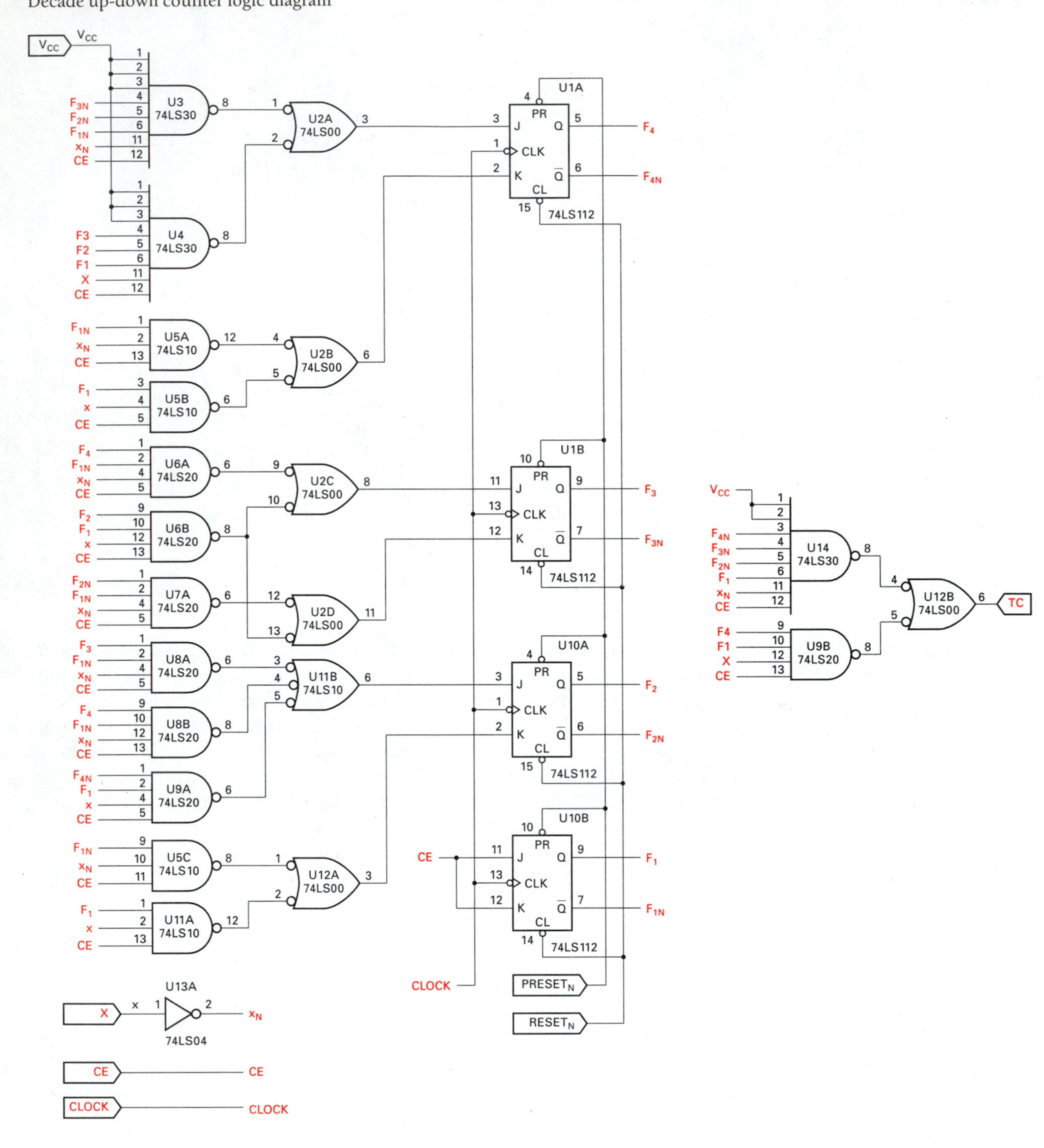

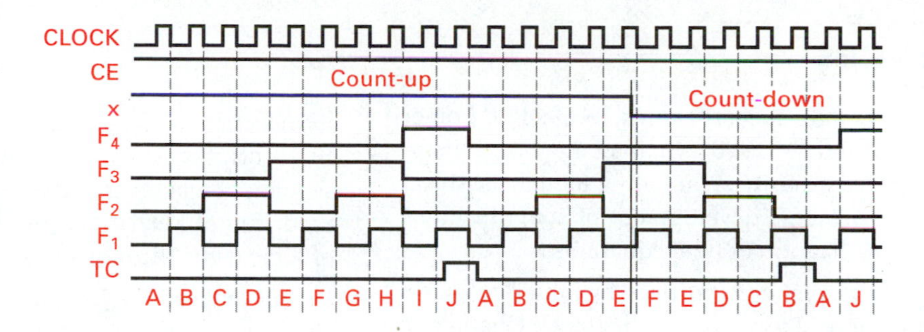

Figure 6.41
Decade counter timing diagram

SUMMARY

A. The chapter started with a review of Mealy and Moore state machine models and then went on to develop synchronous sequential circuit notation.

1. State diagram notation was presented.
   - **a.** Definitions of the variables input, output, state, and excitation.
   - **b.** Definition of states and their relationship to state variables was given, including present state and next state.
   - **c.** State diagrams were derived, showing state symbols, state to state transition symbols, and input and output variables in a state diagram.
2. State tables were developed from state diagrams.
3. Transition tables were developed from state tables. Transition tables illustrate the changes in state variables as the machine changes from a present state to a next state under different input conditions.
4. Excitation tables were developed using the transition tables and the state equations of the four types of flip-flops. An example was realized using each of the four flip-flop types to illustrate the difference in implementation gate cost.
   - **a.** Excitation tables for four flip-flop types were realized.
   - **b.** Excitation equations were derived for each flip-flop type.
   - **c.** Logic diagrams were realized and gate costs compared.

B. Analysis of synchronous sequential circuits was presented.

1. The analysis followed these steps:
   - **a.** Determine state, input, and output variables.
   - **b.** Determine flip-flop type and write the characteristic equation.
   - **c.** Write the excitation equations from inspection of the logic diagram.
   - **d.** Write the next state equations for the state variables.
   - **e.** Write the output equations.
   - **f.** Construct a transition table from the set of next state equations.
   - **g.** Assign names to each state and construct a state table.
   - **h.** construct a state diagram or a timing diagram.
2. The analysis used these examples:
   - **a.** Simple four-state sequential machine.
   - **b.** Chronograph controller sequential circuit.

C. State diagrams were constructed from problem statements, to produce the following:

1. Up-down decade counter.
2. Sequence detectors to detect a serial 10110 pattern, produce an electronic combination lock, and detect a serial pattern of 01010 only if the pattern 011 had not occurred first.
3. Serial ex-3 to BCD code converter.

D. Counters were designed:

1. Modulo-8 synchronous counter.
2. Decade up-down counter.

REFERENCES

Digital Logic and State Machine Design, by David J. Comer (Philadelphia: Saunders, 1990). Chapter 6 presents an introduction to state machines, including model classification. Chapter 7 presents development of state diagrams through to the realization of final equations.

Digital Design Principles and Practices, by John F. Wakerly (Englewood Cliffs, N.J.: Prentice Hall, 1994). Chapters 7 and 8 contain state machine material from Mealy and Moore models to state tables and logic realization.

Fundamentals of Logic Design, by Charles H. Roth, Jr. (Minneapolis–St. Paul: West, 1992). Sequential counters are covered in Unit 12. Analysis of synchronous machines are covered in Unit 13. State tables are developed in Unit 14.

Digital Design, by Morris Mano (Englewood Cliffs, N.J.: Prentice-Hall, 1991). Introduction to sequential machines is presented in Chapter 6.

Introduction to Logic Design, by Sajjan G. Shiva (Glencoe, Ill.: Scott, Foresman and Company, 1988). Synchronous sequential machines are covered in Chapter 8, including modeling of Mealy and Moore machines and realization of logic using flip-flops.

Digital Engineering Design, A Modern Approach, by Richard F. Tinder (Englewood Cliffs, N.J.: Prentice-Hall, 1991). Chapter 5 gives a comprehensive coverage of synchronous sequential circuits.

Digital Circuits, by Kenneth Muchow, Anthony Zeppa, and Bill Deem (Englewood Cliffs, N.J.: Prentice-Hall, 1987). Chapter 11 analyzes MSI counters.

Digital Design, A Practical Course, by Peter Burger (New York: John Wiley and Sons, 1987). Burger introduces sequential machines in Chapter 5. Chapter 6 includes the design of counters.

GLOSSARY

Analysis, of state machines The process of working from a logic diagram to develop a state table, state diagram or timing diagram. The process includes deriving the excitation and output equations, determining the next state equations, constructing the next state table, and timing or state diagrams.

Excitation table A table that specifies the excitation values for each state variable under each state transition condition. A state table is translated to an excitation table for a specific flip-flop type.

Next state The new value of a set of state variables that occurs on the arrival of the next clock pulse. A present state condition points to a next state, determined by the excitation inputs, that will occur on the next clock pulse.

Present state The binary value of a set of state variables prior to changing on the arrival of a clock.

State diagram A graphical representation of the interaction of state transitions under different input conditions. Each state is defined by a set of state variables. States are represented in a state diagram by a circle with an identifying name written inside. State transitions are indicated by directed arcs from one state to another.

State machine Another name for a sequential circuit. A state machine can be either synchronous or asynchronous.

State table A tabular representation of the state changes that occur in a sequential circuit. The state table and state diagram show the same information in different forms.

Synthesis, of state machines To synthesize means to create a whole from parts. The terms *design* and *synthesis* are used interchangeably.

Transition table A transition table describes the state variable changes that occur as a sequential circuit changes states. A transition table is constructed from a state table or a state diagram after the state assignment is known.

QUESTIONS AND PROBLEMS

Section 6.1 to 6.2

1. Draw Mealy and Moore synchronous machine models. Label the excitation variables, state variables, input variables, and output variables in both diagrams.
2. Describe, in words, the meaning of the partial state diagrams in the figure for this problem; identify the appropriate machine model that each partial state diagram illustrates.
3. Given the state diagrams in the figure for this problem, construct the state tables.
4. Construct the transition table for each state table in the figure for problem 3 (use a binary state assignment; that is, 00, 01, 10, 11 or 000, 001, 010, 011, 100, . . .).
5. Using the transition tables constructed in parts c and d of the figure for Problem 3, construct excitation tables for
   a. *R–S* flip-flop.
   b. *J–K* flip-flop.
   c. *D* flip-flop.
   d. *T* flip-flop.
6. Derive and simplify the excitation equations for the excitation tables developed in Problem 5.
7. Determine the gate input cost, for each of the four flip-flop types, for the excitation equations developed in part c of the figure for Problem 3. Which flip-flop type produced the simplest logic results?
8. Identify the input logic, memory, and output logic for the *J–K* realization of Problem 7.
9. Explain why unused states generate don't-care terms when translating a state table to a transition table. Illustrate your response with a sample state table.
10. Explain why *J–K* flip-flops produce more don't-care terms than the other flip-flop types, even when all of the states are specified.

Sections 6.1 and 6.2
Problem 2

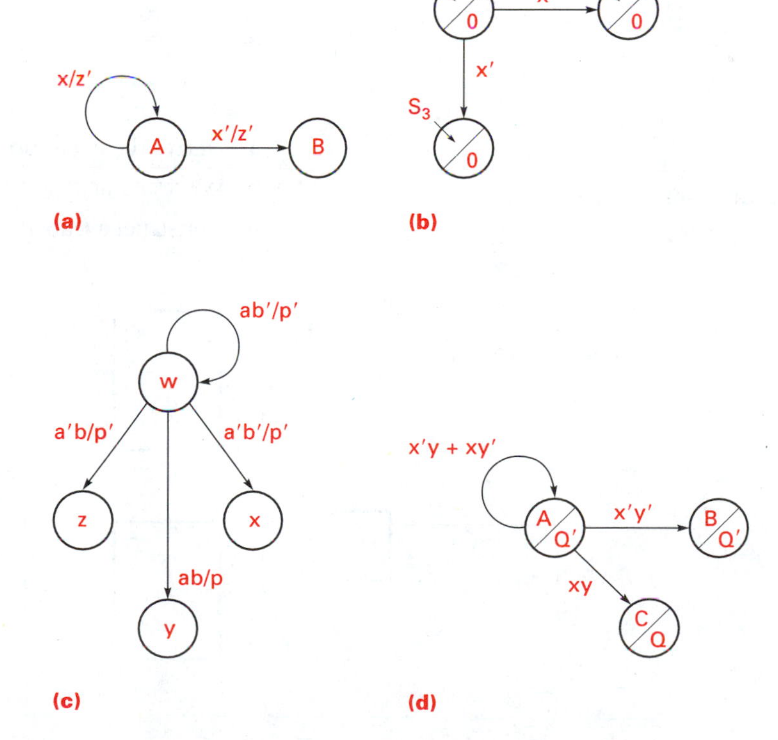

Sections 6.1 and 6.2

Problem 3

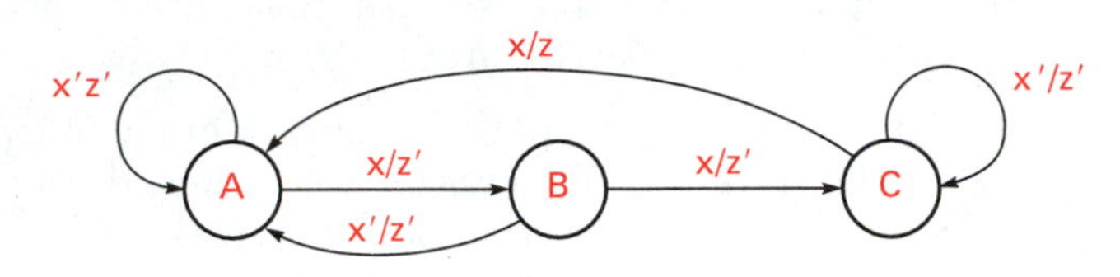

(a)

(b)

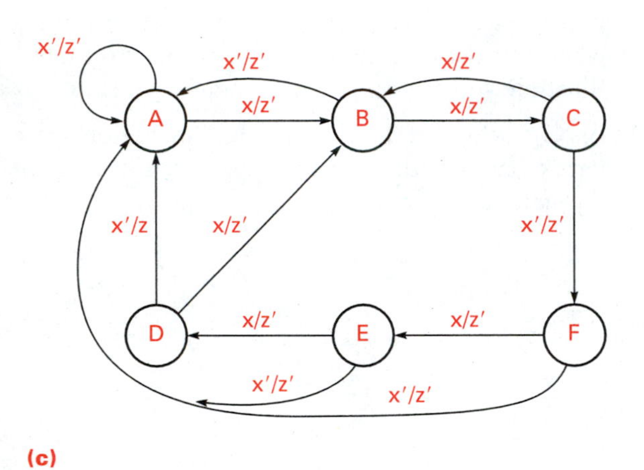

(c)

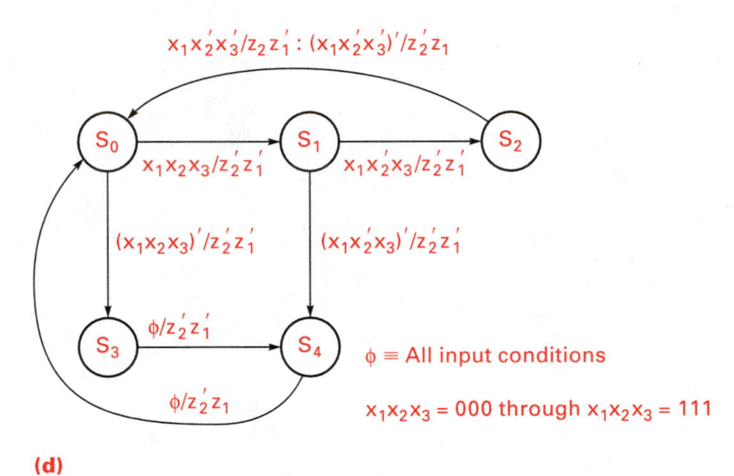

(d)

Section 6.3

11. Given the logic diagram shown in the figure for Problem 11,

a. Derive the excitation and output equations.

b. Write the next state equations.

c. Construct a transition table.

Section 6.3

Problem 11

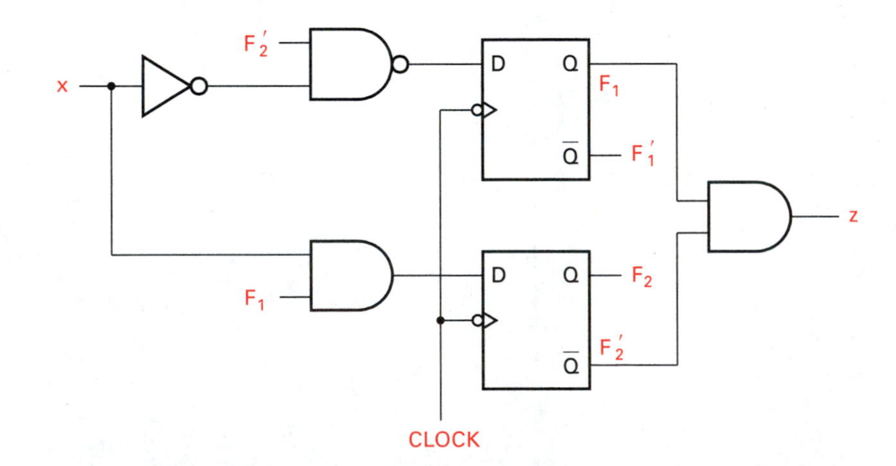

Section 6.3
Problem 11e

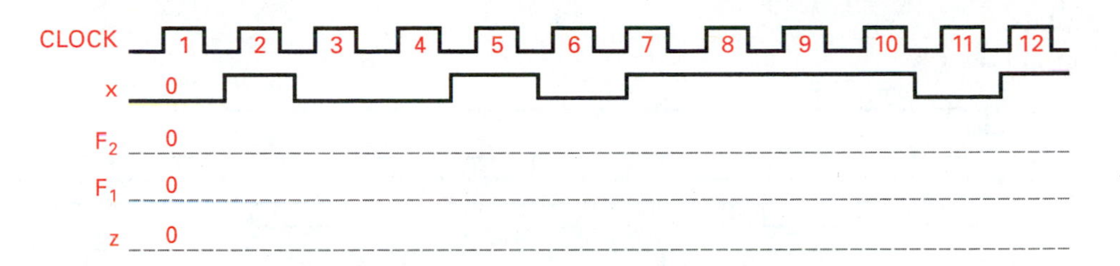

d. Draw the state diagram.

e. Complete the timing diagram in the figure for Problem 11e.

12. Given the logic diagram in the figure for Problem 12,

a. Write the excitation and output equations.

b. Write the next-state equations.

c. Create an excitation table.

d. Construct a state table and a state diagram.

e. Complete the timing diagram in the figure for Problem 12e.

Section 6.3
Problem 12

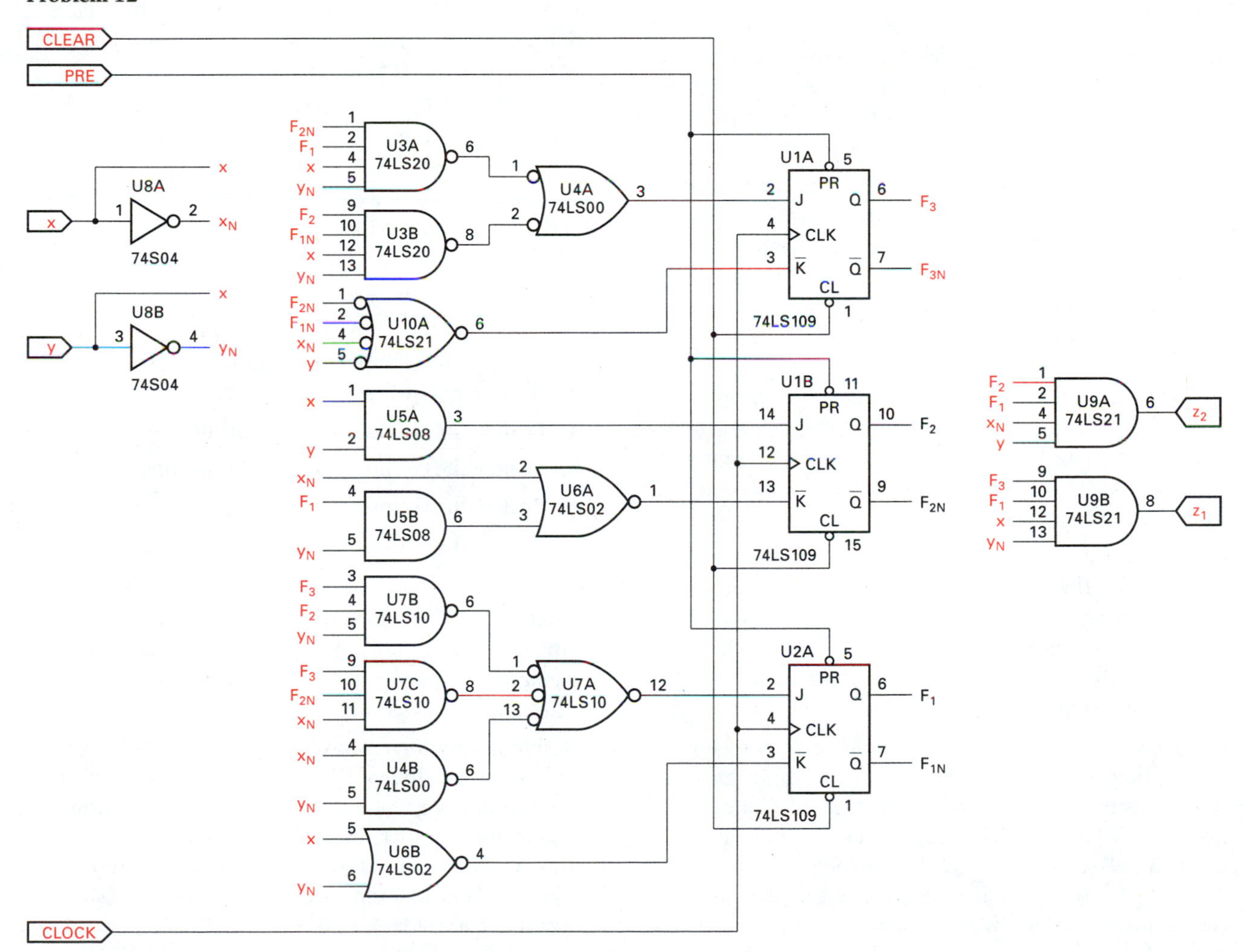

Section 6.3

Problem 12e

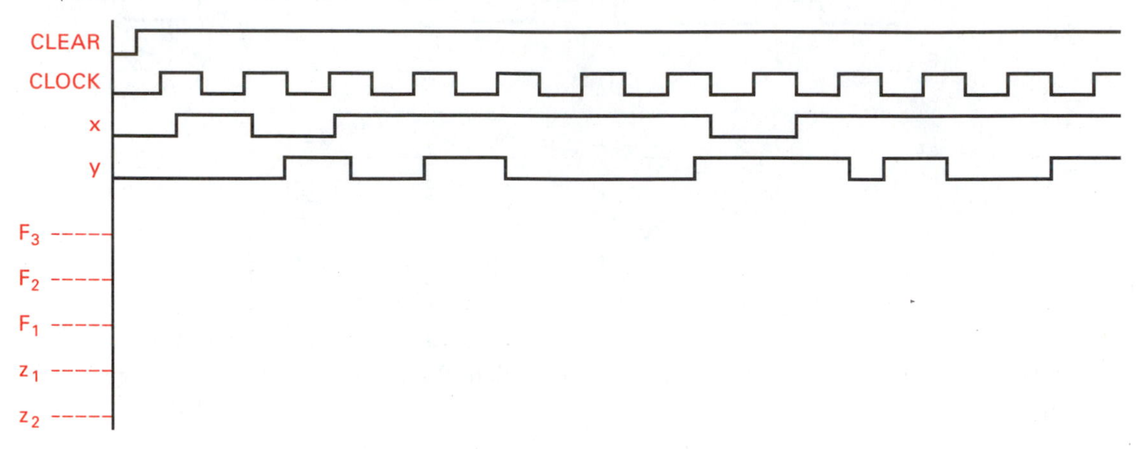

Section 6.4

13. Construct the state diagram for a Mealy sequential circuit that will detect the serial input sequence $x = 010110$. When the complete sequence has been detected, then cause output z to go high.

14. Construct a Moore sequential circuit state diagram that will detect either of the following serial input sequences: $x = 101011$ or 01110. If the first sequence is detected, cause output z_1 to go high; if the second sequence is detected, cause output z_2 to go high. The two input sequences may overlap one another.

However, once an input sequence is in process any x input must continue on the same sequence. Only when a started input sequence is broken may detection shift to the other. Which serial input code "locks" out the other?

15. Construct the state diagram for a Mealy sequential machine that will detect the following input sequences: $x = 01101$ or 01111. If input sequence $x = 01101$ is met, cause $z_1 = 1$. If $x = 01111$, cause $z_2 = 1$. Each input sequence may overlap with itself or the other sequence.

16. Design a Mealy sequential machine state diagram for a code detecting system. Three four-bit input code sequences must be detected and identified. Code sequence 1 = 1010, 0110, 1010. Code sequence 2 = 1101, 1111, 0101; and code sequence 3 = 0010, 0000, 1101. If the first code sequence is detected, then cause $z_1 = 1$, if the second sequence is detected, then $z_2 = 1$, and if the third sequence is detected, then $z_3 = 1$. If an incorrect sequence is generated, no output is active.

17. Design a Moore sequential machine state diagram that will determine whether a four-bit serial input sequence is a legal 8-4-2-1 BCD code. If a legal BCD code sequence is detected, then $z = 1$. If an incorrect code is entered, then $z = 0$.

Section 6.5

18. Design a synchronous modulo-8 binary counter. Use a binary state assignment.

a. Realize the counter using *D* flip-flops.

b. Realize the counter using *J–K* flip-flops.

c. Realize the counter using *T* flip-flops.

d. Realize the counter using *R–S* flip-flops.

19. Design a decade counter using a binary state assignment and *J–K* flip flops such that when two external inputs (x and y) are coincident the counter will increment. A separate clock input provides state transition synchronization.

20. Design an up-down Modulo-16 counter. Use a gray code state assignment. When $x = 1$ the counter is to count up; when $x = 0$ the counter is to count down. When the terminal count of 15 is detected and the unit is counting up, or when the terminal count of 0 is reached and the unit is counting down, cause the terminal count $(TC)' = 0$.

CHAPTER

7 SEQUENTIAL CIRCUIT DESIGN

INTRODUCTION

Chapter 7 continues the synchronous sequential design topics started in Chapter 6. The fundamentals of state diagram notation and development were presented in Chapter 6. In Chapter 7 we develop techniques for the determination of state equivalence and elimination of redundant states. We continue with the development of guidelines for making a state assignment. Finally we develop a notation for describing sequential circuits called *algorithm state machines*.

7.1 STATE EQUIVALENCE

Designers usually construct state diagrams from written or verbal problem statements, without giving consideration to the creation of redundant states. This permits the designer to concentrate on the problem constraints. Once the initial state machine diagram is completed, redundant states can be identified and eliminated.

One state is equivalent to another, and therefore redundant, if the state functions are indistinguishable. The number of states needed to realize a sequential machine directly affects circuit cost, operating speed and design complexity. Therefore, the reduction of equivalent states is desirable (known as **state reduction**). The relationship between states and state variables ($2^x = y$, where x = number of state variables and y = number of states) indicates that if we can reduce the number of states we may also reduce the number of flip-flops.

State Equivalence Theorem

Two states, S_A and S_B, are equivalent if and only if for every possible input X sequence, the outputs are the same and the next states are equivalent.

Let $n(S, X)$ be the next state of present state S if input X is present. If output Y_A = output Y_B and if $n(S_A, X) = n(S_B, X)$, then $S_A = S_B$. If states S_A and S_B occupy rows of a state table, and if $S_A = S_B$, then one row can be eliminated.

Three properties of equivalent states are

1. Symmetry: If $S_A = S_B$, then $S_B = S_A$.
2. Reflexivity: $S_A = S_A$, for any state.
3. Transitivity: If $S_A = S_B$ and $S_B = S_C$, then $S_A = S_C$.

7.2 STATE REDUCTION

Two approaches to state equivalency identification and reduction are presented: equivalence classes and the implication chart. The **equivalence-class** algorithm partitions a set of states into groups or classes based on the next state and output results for a given input sequence. If a group or class contains a single state, no redundancy within that class has occurred. The implication chart identifies classes of states using a graphical technique. Each class is identified by a box in the implication chart. When all of the states that have different next states or outputs have been identified, those remaining are considered to be redundant.

7.2.1 Equivalence Classes

A procedure for state classification in order to determine ***n*-equivalence** is given by the following example. Consider the state table in Table 7.1.

1. Partition the states so that those that share both input and output values will be in the same class. Two classes occur when we do this. States (S_0, S_2, S_4) are grouped because $z = 0$ when $x = 0$ and $z = 1$ when $x = 1$. States (S_1, S_3, S_5) are grouped since $z = 0$ when $x = 0$ and $z = 0$ when $x = 1$. This partition is said to be 1-equivalent.

   $C_1 = (S_0, S_2, S_4)$
   $C_2 = (S_1, S_3, S_5)$

   A next-class table is then constructed, which shows the partition of Table 7.1 into the two classes. Table 7.2 was constructed from Table 7.1 by noting the class in which the next state was found. For example, present state S_0 has a next state of S_4 (when $x = 0$), which belongs in class C_1, so C_1 is written in the $x = 0$ next-

TABLE 7.1
State table

Present state	Next state $x = 0$	z	$x = 1$	z
S_0	S_4	0	S_3	1
S_1	S_5	0	S_3	0
S_2	S_4	0	S_1	1
S_3	S_5	0	S_1	0
S_4	S_2	0	S_5	1
S_5	S_1	0	S_2	0

class column. State S_0 has a next state of S_3 when $x = 1$. S_3 belongs in class C_2, so C_2 is written in the $x = 1$ next-class column.

2. The next partition determines 2-equivalency (an input sequence of 2). We test each state within the classifications found in step 1.

a. Test C_1 for each value of input x. We observe that for $x = 0$ the next class is C_1, and when $x = 1$ the next class is C_2 for all the states contained in C_1. This means that no new partition of class C_1 is possible.

b. Test C_2 for each value of input x. We observe that for $x = 0$ the next class is C_2. For $x = 1$ we observe that C_2 is the next class for states S_1 and S_3, but C_1 is the next class for S_5.

c. The 2-equivalent classifications are

$$C_{11} = (S_0, S_2, S_4)$$
$$C_{21} = (S_1, S_3)$$
$$C_{22} = (S_5)$$

A new next class table is constructed from Table 7.1 to show the 2-equivalent classifications. The 2-equivalent next-class table (Table 7.3) is constructed using Table 7.1 and the 2-equivalent classifications. For example, the $x = 0$ next state, for present state S_3, is S_5, which is in C_{22}; so C_{22} is written in the $x = 0$ next-class column. Likewise, the $x = 1$ next state, for present state S_3, is S_1, which is in C_{21}; so C_{21} is written in the $x = 1$ next class column.

3. Determine 3-equivalency (input sequence of 3).

a. Test C_{11} for each value of input x. A partition can be made so that $C_{111} = (S_0, S_2)$ and $C_{112} = (S_4)$.

b. Test C_{21} for each value of input x. No further partitioning is possible because the next classes for $x = 0$ and $x = 1$ are the same.

c. Test C_{22} for each value of input x. No additional partition is necessary because the class contains only one state.

d. The 3-equivalent classes are

$$C_{111} = (S_0, S_2)$$
$$C_{112} = (S_4)$$
$$C_{211} = (S_1, S_3)$$
$$C_{221} = (S_5)$$

A 3-equivalent next-class table is constructed, as in Table 7.4.

4. Determine 4-equivalency (input sequence of 4).

a. Test $C_{111} = (S_0, S_2)$. Because the next classes are the same for S_0 and S_2 (when $x = 0$ and $x = 1$), no further partitioning is possible.

b. Test $C_{211} = (S_1, S_3)$. Because the next classes are the same for S_1 and S_3 (when $x = 0$ and $x = 1$), no further partitioning is possible.

c. The 4-equivalent classes are the same as the 3-equivalent classes, so no further testing is necessary.

5. The states within a given partition class are redundant. Therefore, state $S_0 = S_2$ and $S_1 = S_3$.

TABLE 7.2
1-equivalency next class table

Present class		Next class	
		$x = 0$	$x = 1$
C_1	S_0	C_1	C_2
	S_2	C_1	C_2
	S_4	C_1	C_2
C_2	S_1	C_2	C_2
	S_3	C_2	C_2
	S_5	C_2	C_1

TABLE 7.3
2-equivalency next-class table

Present class		Next class	
		$x = 0$	$x = 1$
C_{11}	S_0	C_{11}	C_{21}
	S_2	C_{11}	C_{21}
	S_4	C_{11}	C_{22}
C_{21}	S_1	C_{22}	C_{21}
	S_3	C_{22}	C_{21}
C_{22}	S_5	C_{21}	C_{11}

TABLE 7.4
3-equivalency next-class table

Present state		Next class	
		$x = 0$	$x = 1$
C_{111}	S_0	C_{112}	C_{211}
	S_2	C_{112}	C_{211}
C_{112}	S_4	C_{111}	C_{221}
C_{211}	S_1	C_{221}	C_{211}
	S_3	C_{221}	C_{211}
C_{221}	S_5	C_{211}	C_{111}

TABLE 7.5
Modified state table

Present state	Next state $x = 0$	z	Next state $x = 1$	z
S_0	S_4	0	~~S_3~~ S_1	1
S_1	S_5	0	~~S_3~~ S_1	0
~~S_2~~ S_0	S_4	0	S_1	1
~~S_3~~ S_1	S_5	0	S_1	0
S_4	~~S_2~~ S_0	0	S_5	1
S_5	S_1	0	~~S_2~~ S_0	0

TABLE 7.6
Simplified state table

Present state	Next state $x = 0$	z	Next state $x = 1$	z
S_0	S_4	0	S_1	1
S_1	S_5	0	S_1	0
S_4	S_0	0	S_5	1
S_5	S_1	0	S_0	0

Once the redundant states have been identified, a new, simplified state table is constructed. Table 7.5 shows the modifications to Table 7.1. Table 7.6 shows the reduced state table.

States S_2 and S_4 no longer exist. Renaming the states may be desirable at this point. State Table 7.6, with four states, requires two state variables (flip-flops) compared to the six states and three flip-flops needed to realize the original sequential machine.

7.2.2 Implication Charts

Implication charts provide a graphic method of identifying redundant states. The implication chart uses a grid array to list the possible state equivalencies for different inputs. The chart is constructed by listing all of the states except the last along the horizontal axis and all the states except the first along the vertical axis. The intersection of the horizontal and vertical state spaces indicates a possible state equivalency. The unfilled implication chart for Table 7.1 is illustrated in Figure 7.1.

Consider Table 7.1. The top square in Figure 7.1 is defined by the vertical and horizontal intersection of states S_0 and S_1. It represents the possibility of states S_0 and S_1 being equivalent. In a like manner the remaining squares represent state pair equivalency possibilities.

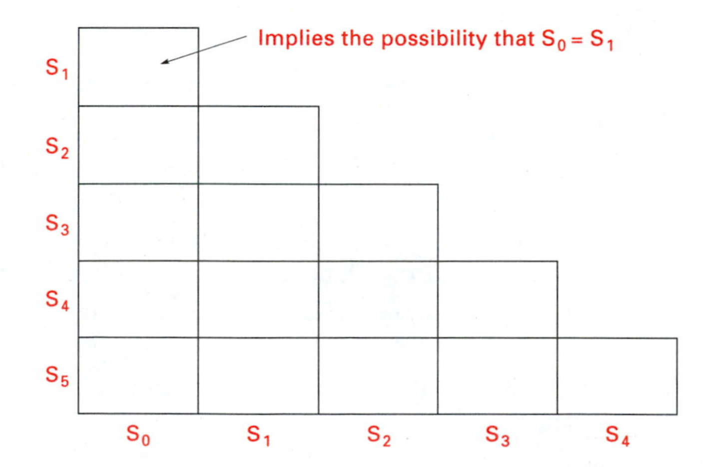

Figure 7.1
Unfilled implication chart for Table 7.1

The implication chart squares are filled in as follows:

1. Place an *X* in the squares where the outputs are different. These states cannot be equivalent.
2. Consider the remaining squares. We need look at only the state equivalency pair(s) because the outputs were taken care of in step 1.
   a. $S_0 \equiv S_2$ iff (if and only if) $S_1 \equiv S_3$. We write the state-equivalency pair S_1, S_3 in the S_0, S_2 square.
   b. $S_0 \equiv S_4$ iff $S_2 \equiv S_4$ and $S_3 \equiv S_5$.
   c. $S_1 \equiv S_5$ iff $S_2 \equiv S_3$.
   d. $S_2 \equiv S_4$ iff $S_1 \equiv S_5$.
   e. $S_3 \equiv S_5$ iff $S_1 \equiv S_5$ and $S_1 \equiv S_2$.
3. Test the equivalencies of each square by testing the implied equivalent state pair(s) written in each square. When the implied equivalent state pair(s) are not equivalent for a given square, then write an *X* in the square.
   a. Implied square $\{S_0, S_2\}$ requires that state pair $\{S_1, S_3\}$ be equivalent. When we look at implied square $\{S_1, S_3\}$ we find that state pair $\{S_1, S_3\}$ is needed. Because state pair $\{S_1, S_3\}$ is dependent on itself for equivalency, it is equivalent (reflexivity, $S_A = S_A$). This means that state pair $\{S_0, S_2\}$ is equivalent.
   b. Implied square $\{S_0, S_4\}$ requires that state pairs $\{S_2, S_4\}$ and $\{S_3, S_5\}$ are necessary for equivalency. Upon investigating the required state pair we find that $\{S_2, S_4\}$ depends on $\{S_1, S_5\}$ and that $\{S_3, S_5\}$ depends on $\{S_1, S_5\}$ and $\{S_1, S_2\}$. Further investigation reveals that $\{S_1, S_2\}$ is not equivalent. Therefore $\{S_3, S_5\}$ is not equivalent. Because $\{S_3, S_5\}$ is not equivalent, we place an *X* in the $\{S_3, S_5\}$ and $\{S_0, S_4\}$ squares.
   c. Implied square $\{S_1, S_5\}$ is dependent on state pair $\{S_2, S_3\}$. Because $\{S_2, S_3\}$ is not equivalent we place an *X* in the $\{S_1, S_5\}$ square and in the $\{S_2, S_4\}$ square.
   d. The remaining state pair possible equivalencies $\{S_0, S_2\}$ and $\{S_1, S_3\}$ are not crossed off; they are equivalent.

Figure 7.2 shows the implication chart with each square filled in with the state equivalency conditions. The resulting simplified state table is shown in Figure 7.3, which is the same result as obtained using the equivalency class approach.

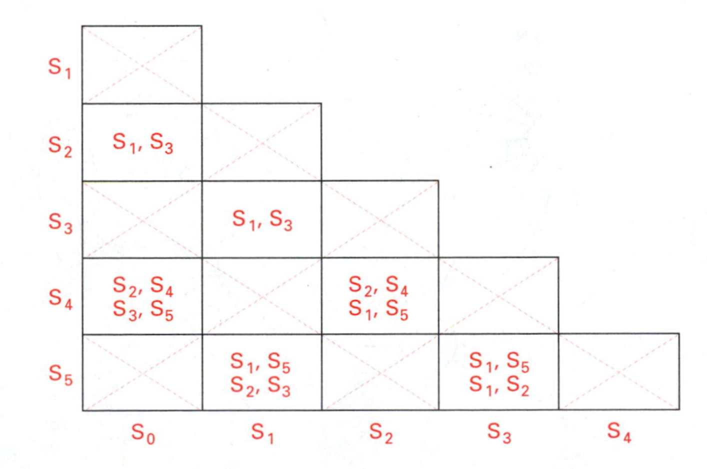

Figure 7.2
Implication chart for state Table 7.1

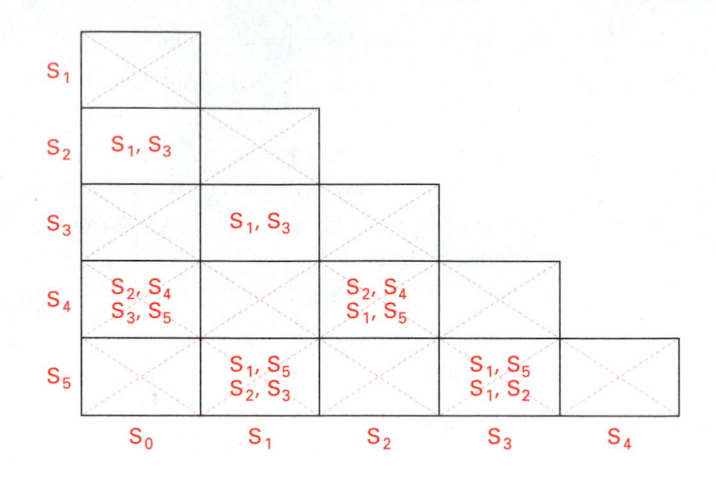

Figure 7.3
Implication chart showing $\{S_0, S_2\}$ and $\{S_1, S_3\}$

❖ **EXAMPLE 7.1** Consider the state diagram in Figure 7.4. Determine any redundant states in the Mealy state diagram shown in that figure.

SOLUTION The state table is shown in Table 7.7. An implication chart, Figure 7.5, is constructed from the state table.

We evaluate state Table 7.7.

1. *X*s are placed in the squares where outputs differ.
2. $A \equiv D$ iff $B \equiv C$, $A \equiv B$, and $D \equiv C$.
3. $A \equiv E$ iff $B \equiv C$.
4. $B \equiv C$ iff $A \equiv E$.
5. $D \equiv E$ iff $C \equiv D$, $B \equiv E$, $A \equiv C$.

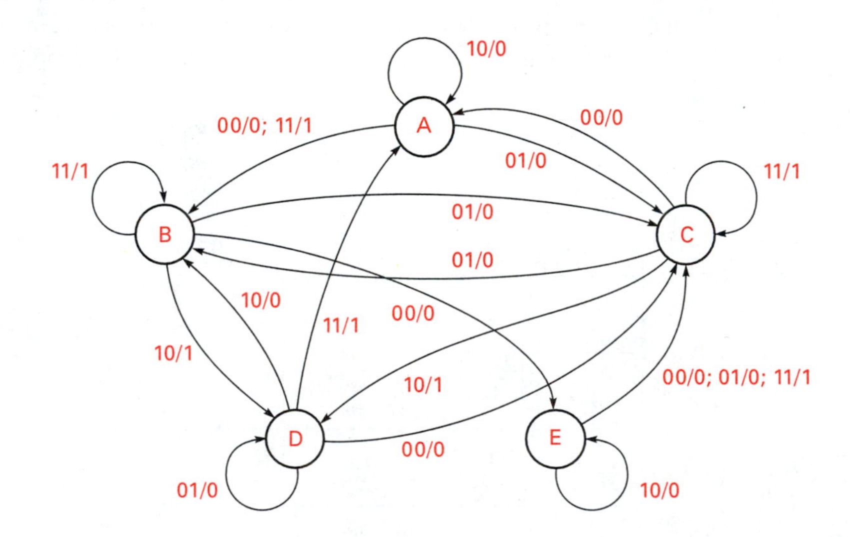

Figure 7.4
State diagram of a Mealy machine

TABLE 7.7
State table for Mealy machine from Figure 7.4

Present state	Next state, xy/z							
	00	z	01	z	11	z	10	z
A	B	0	C	0	B	1	A	0
B	E	0	C	0	B	1	D	1
C	A	0	B	0	C	1	D	1
D	C	0	D	0	A	1	B	0
E	C	0	C	0	C	1	E	0

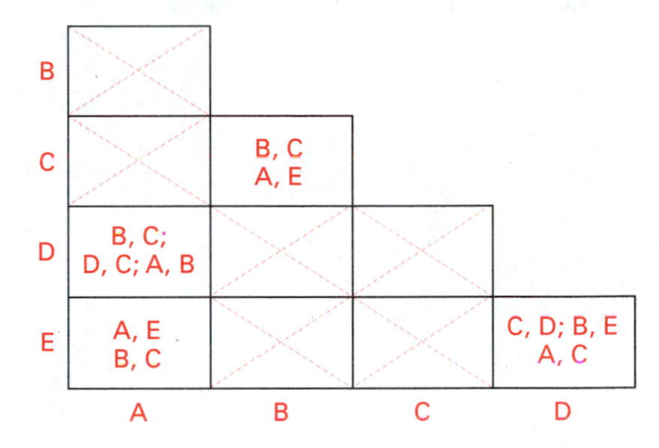

Figure 7.5
Implication chart for Table 7.7

Evaluation of the squares that contain implied state equivalencies gives the following.

1. $A \not\equiv D$, because $A \not\equiv B$ and $D \not\equiv C$.
2. $D \not\equiv E$, because $A \not\equiv C$, $B \not\equiv E$ and $C \not\equiv D$.
3. $A \equiv E$ iff $B \equiv C$ and $B \equiv C$ iff $A \equiv E$.

The result is that $A = E$ and $B = C$ (symmetry if $S_A = S_B$, then $S_B = S_A$). The completed implication chart is shown in Figure 7.6. Table 7.8 is the modified state table.

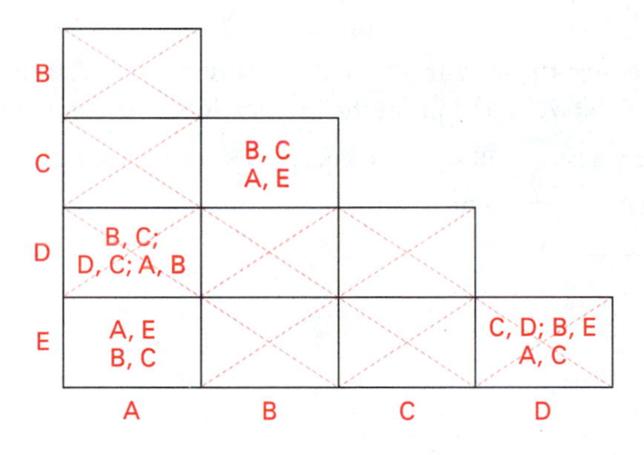

Figure 7.6
Filled-in implication chart for Table 7.7

TABLE 7.8
Modified state table

Present state	Next state, xy/z							
	00	z	01	z	11	z	10	z
A	B	0	~~C~~ B	0	B	1	A	0
B	~~E~~ A	0	~~C~~ B	0	B	1	D	1
~~C~~ B	A	0	B	0	~~C~~ B	1	D	1
D	~~C~~ B	0	D	0	A	1	B	0
~~E~~ A	~~C~~ B	0	~~C~~ B	0	~~C~~ B	1	~~E~~ A	0

TABLE 7.9
Simplified state table with reassigned state names

Present state	Next state, xy/z 00	z	01	z	11	z	10	z
S_0	S_1	0	S_1	0	S_1	1	S_0	0
S_1	S_0	0	S_1	0	S_1	1	S_2	1
S_2	S_1	0	S_2	0	S_0	1	S_1	0

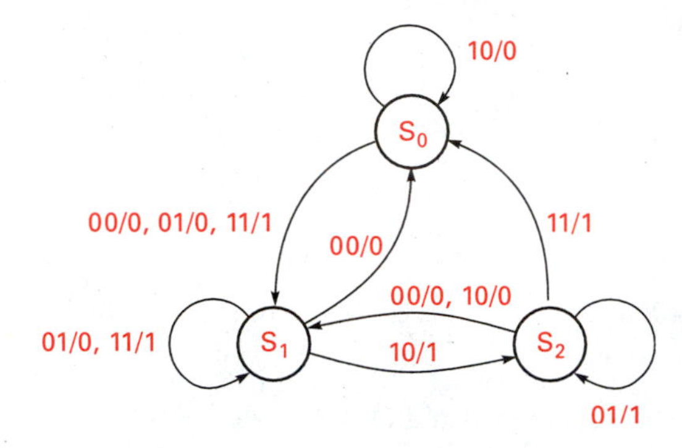

Figure 7.7
Simplified Mealy state diagram

We reassign the state variable names: $A = S_0$, $B = S_1$, $D = S_2$. The new state table, using the reassigned state names, is shown in Table 7.9. The reduced state diagram is shown in Figure 7.7. ❖

❖ EXAMPLE 7.2

Reduce the state table shown in Table 7.10. It has nine states, a single input, and two output variables.

SOLUTION We can reduce the state machine described in Table 7.10 by using an implication chart. Evaluate Table 7.10 and fill in the implication chart shown in Figure 7.8.

1. *X*s are placed in all squares where the outputs differ.
2. Evaluate the remaining squares.
   a. $0 \equiv 1$ iff $0 \equiv 4$ and $1 \equiv 2$.
   b. $0 \equiv 2$ iff $0 \equiv 7$.

TABLE 7.10
State table

Present state	Next state $x = 0$	Z_2	Z_1	$x = 1$	Z_2	Z_1
0	0	0	0	1	0	0
1	4	0	0	2	0	0
2	7	0	0	1	0	0
3	2	0	1	6	1	0
4	6	1	0	5	0	0
5	5	0	1	4	1	1
6	1	0	1	6	1	0
7	3	1	0	8	0	0
8	8	0	1	7	1	1

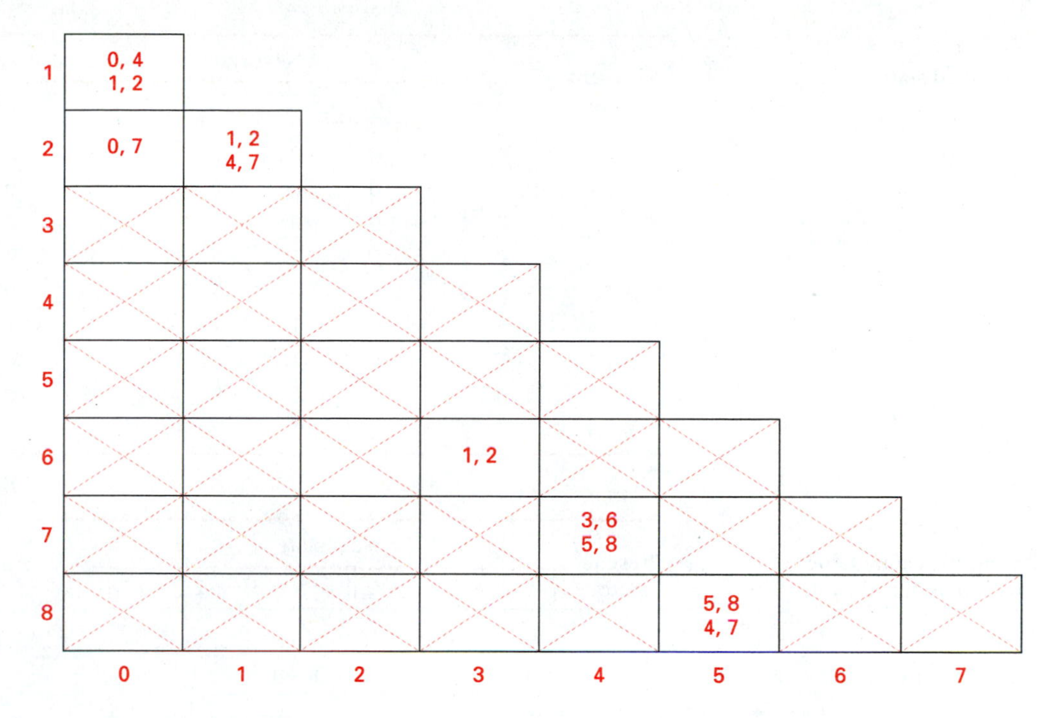

Figure 7.8
Filled-in implication chart for Table 7.10

c. $1 \equiv 2$ iff $4 \equiv 7$.

d. $3 \equiv 6$ iff $1 \equiv 2$.

e. $4 \equiv 7$ iff $3 \equiv 6$ and $5 \equiv 8$.

f. $5 \equiv 7$ iff $4 \equiv 7$.

The implication chart is completed as shown in Figure 7.9.

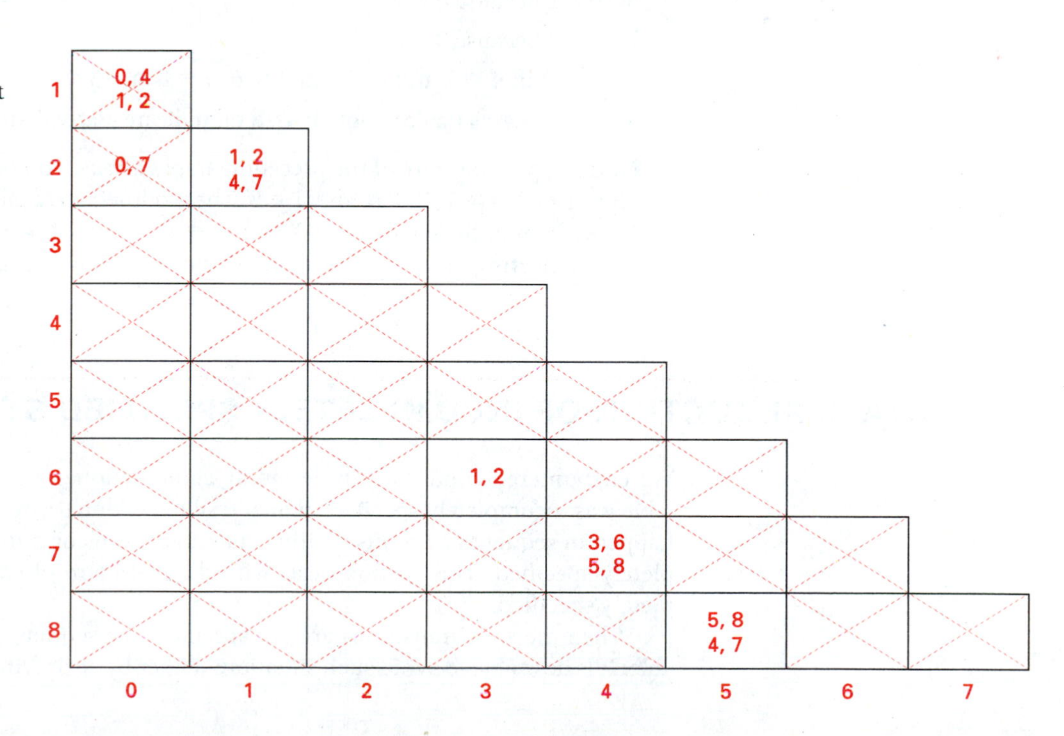

Figure 7.9
Completed implication chart for Table 7.10

TABLE 7.11
Modified state table

Present state	Next state $x=0$	Z_2	Z_1	$x=1$	Z_2	Z_1
0	0	0	0	1	0	0
1	4	0	0	~~2~~ 1	0	0
~~2~~ 1	~~7~~ 4	0	0	1	0	0
3	~~2~~ 1	0	1	~~6~~ 3	1	0
4	~~6~~ 3	1	0	5	0	0
5	5	0	1	4	1	1
~~6~~ 3	1	0	1	~~6~~ 3	1	0
~~7~~ 4	3	1	0	~~8~~ 5	0	0
~~8~~ 5	~~8~~ 5	0	1	~~7~~ 4	1	1

TABLE 7.12
Simplified state table with reassigned names

Present state	Next state $x=0$	Z_2	Z_1	$x=1$	Z_2	Z_1
A	A	0	0	B	0	0
B	D	0	0	B	0	0
C	B	0	1	C	1	0
D	C	1	0	E	0	0
E	E	0	1	D	1	1

3. Evaluation of the remaining state pair squares gives the following:
   a. $0 \neq 1$ because $0 \neq 4$.
   b. $0 \neq 2$ because $0 \neq 7$.
   c. $1 \equiv 2$ iff $4 \equiv 7$ and $4 \equiv 7$ iff $3 \equiv 6$, $5 \equiv 8$ and $3 \equiv 6$ iff $1 \equiv 2$ and $5 \equiv 8$ iff $4 \equiv 7$.
   d. The squares that are not crossed off indicate equivalent state pairs.

   Nothing prevents any of the preceding implications, combining equivalent state pair gives $1 \equiv 2$, $4 \equiv 7$, $3 \equiv 6$, and $5 \equiv 8$. The modified state table is shown in Table 7.11.

   Renaming the states gives $0 \equiv A$, $1 \equiv B$, $3 \equiv C$, $4 \equiv D$, and $5 \equiv E$. A new state table reflecting the new state names is shown in Table 7.12. ❖

7.3 STATE REDUCTION OF INCOMPLETELY SPECIFIED STATE TABLES

We encountered conditions in designing combinational logic circuits where the truth table was incompletely specified, which resulted in don't-care terms. The same thing can happen in sequential circuits if either state transitions or output variables are not completely specified. This section deals with the reduction of state tables that are incompletely specified.

When the state transitions are not specified, the sequential machine is not predictable. It is desirable to avoid such situations, either by specifying inputs sequences so that no next state is unspecified or by assigning next states that are not contrary to the desired result. When next states are assigned to cover those initially unspecified, the machine is no longer incompletely specified.

Unspecified outputs can be assigned without any impact on the state sequence or the final state table. It is usually advantageous to leave don't-care outputs unspecified as long as possible during the state reduction process. This provides for additional flexibility in reducing the state table. Two states are equivalent if they satisfy the properties of equivalent states, given previously. This requires that the outputs be the same for equivalency to exist. *State compatibility* is the term used instead of equivalence when reducing incompletely specified state tables.

Compatibility Definition

States S_A and S_B are **compatible states** if, for every input sequence that affects the two states, the same output sequence occurs when the incompletely specified outputs are specified, regardless of whether S_A or S_B is the starting state.

We will first evaluate state tables with only incompletely specified outputs and then look at those with unspecified next states. Consider Tables 7.13 and 7.14 and their implication charts in Figures 7.10 and 7.11.

TABLE 7.13
Incompletely specified state table

Present state	Next state $X = 0$	Z	Next state $X = 1$	Z
S_0	S_0	0	S_1	0
S_1	S_0	—	S_2	0
S_2	S_3	—	S_3	1
S_3	S_4	—	S_2	0
S_4	S_0	—	S_1	—

TABLE 7.14
Modified incompletely specified state table

Present state	Next state $X = 0$	Z	Next state $X = 1$	Z
S_0	S_0	0	S_1	0
S_1	S_0	— 0	S_2	0
S_2	~~S_3~~ S_1	— 0	~~S_3~~ S_1	1
~~S_3~~ S_1	~~S_4~~ S_0	— 0	S_2	0
~~S_4~~ S_0	S_0	— 0	S_1	— 0

Figure 7.10
Implication chart for Table 7.13

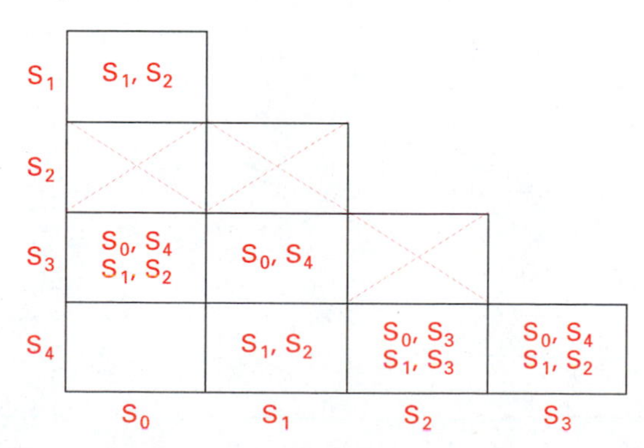

Figure 7.11
Completed implication chart for Table 7.14

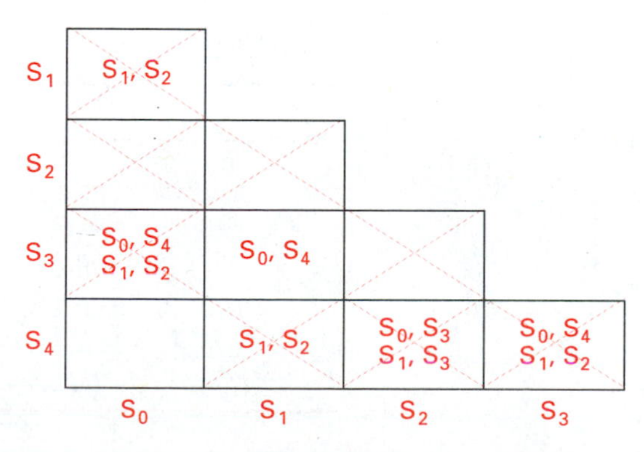

TABLE 7.15
Simplified state table

Present state	Next state $X = 0$	Z	$X = 1$	Z
a	a	0	b	0
b	a	0	c	0
c	b	0	b	1

The following are compatible states:

$S_0 = S_4$, if the output Z, for S_4 when $x = 0$ and $x = 1$, is changed from a don't-care to a 0.

$S_1 = S_3$, if the output Z, for S_1 and S_3 when $x = 0$, is changed from a don't-care to a either a 0 or a 1 (both need to be the same). In this case we assigned both to a 0.

Therefore we assign $Z = 0$ for present state S_4 when $X = 0$ and $X = 1$. The other don't-care outputs are left alone to aid in later simplification of the output function using K-maps.

The states are renamed and a simplified state table (Table 7.15) constructed.

Two basic choices face the logic designer dealing with state reduction of an incompletely specified state table containing don't-care next states. The first is to specify the machine and then apply state reduction techniques. The second is to leave the don't-care states alone as long as possible to retain maximum flexibility in the state reduction process. The second choice requires the application of a reduction tool called *merger graphs*.

7.3.1 Merger Graphs

The **merger graph** contains the same number of vertices as the state table contains states. Each vertex corresponds to a state in the state table. Each compatible state pair is indicated by a line drawn between the two state vertices. Every potentially compatible state pair, with outputs not in conflict but whose next states are different, is connected by a broken line. The implied states are written in the line break between the two potentially compatible states. If two states are incompatible no connecting line is drawn.

Consider Table 7.16 and the merger diagram of Figure 7.12. The merger diagram has a labeled vertex for each state in the machine.

TABLE 7.16
Incompletely specified state table

Present state	Next state 00	01	10	11
A	C/0	—/—	A/0	—/—
B	—/—	E/0	B/0	D/1
C	D/0	B/1	—/—	—/—
D	C/0	A/1	E/0	—/—
E	B/0	—/—	A/0	E/1

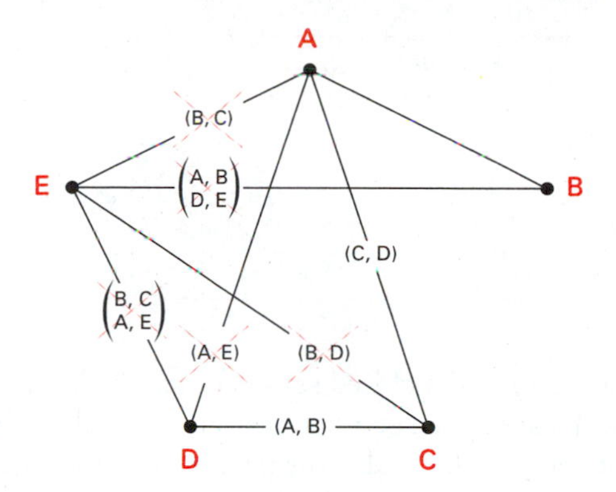

Figure 7.12
Merger graph for state Table 7.16

The merger graph shown in Figure 7.12 illustrates the compatible and potentially compatible states of Table 7.16. States A and B are compatible because no incompatible conflict exists. State pair A, C is implied compatible by state pair C, D. State pair A, E is compatible if states B and C are compatible. State pair A, D is compatible if state pair A, E is compatible. State pair B and C are not compatible because the outputs differ in input column 01. State pair B, D is not compatible due to the different outputs in input column 01. States B and E are compatible if implied compatible pairs A, B and D, E are also compatible. States C and D are implied compatible by state pair A, B. States C and E are implied compatible by state pair B, D. Last, states D and E are implied compatible by state pairs B, C and A, E.

The implied compatible states are evaluated by crossing out all incompatible state pairs, leaving only those that are compatible. An implied compatibility is crossed out when the implied states are found incompatible. State pair A, E is crossed out because the implied compatible state pair B, C is incompatible. Likewise implied compatible state pairs B, E; C, E; A, D; and D, E are crossed off as incompatible.

The compatible state pairs from Figure 7.12 are {A, B}; {A, C}; and {C, D}. The set of compatible state pairs that produces the smallest reduced state table is called the *minimal set of compatibles.* To find the minimal set of compatibles for state reduction it is useful to find what are called the *maximal compatibles.*

Maximal Compatibility

A set of compatible-state pairs is said to be maximal if it is not completely covered by any other set of compatible state pairs. For instance, a single state that is not compatible with any other state has **maximal compatibility.**

The compatible pairs that resulted from Figure 7.12 are not maximal. It is desirable to find the maximal compatibles (a set of states or state pairs that contain states not covered by other compatibles) to develop the minimal set of compatibles needed to construct a reduced state table; that is, partition the states into as few groups as possible, where all the states in a group are compatible with one another. The maximal compatibles can be found by constructing a **compatibility class graph.** This graph has vertices and connecting lines like the merger graph. However, it shows only compatible states. Figure 7.13 is a compatibility class graph for Table 7.16.

Figure 7.13
Compatibility class graph for Table 7.16

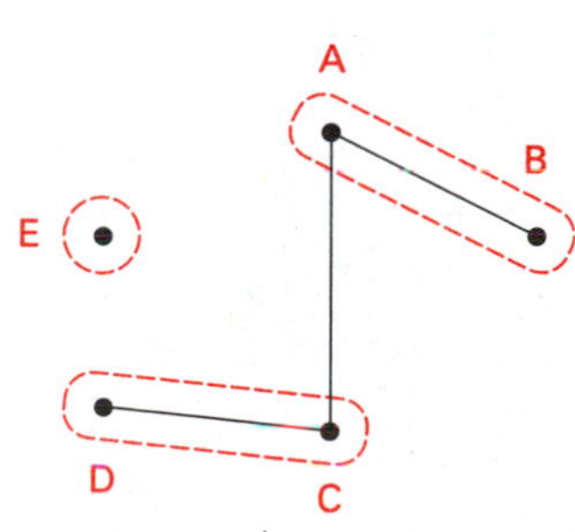

The maximal compatibles are found from the compatibilies class graph by identifying the largest groups of connected compatible state pairs. In this example, maximal compatibles of {A, B}; {C, D}; and {E} exist.

TABLE 7.17
Reduced state table from incompletely specified Table 7.16

Present state	Next state 00	01	10	11
S_0	$S_1/0$	$S_2/0$	$S_0/0$	$S_1/1$
S_1	$S_1/0$	$S_0/1$	$S_2/0$	—/—
S_2	$S_0/0$	—/—	$S_0/0$	$S_2/1$

The next task is to find a closed set of compatibles. The closed compatible set must cover all compatibles, implied compatibles, and all of the original states. In this example, the closed set of compatibles is also the maximal compatibles. Maximal compatible state pairs {A, B}; {C, D}; and {E} cover all of the original states and contain any implied compatibles within a maximal compatible. The closed set of maximal compatibles is

$$[\{A, B\}; \quad \{C, D\}; \quad \{E\}]$$

Each maximal compatible is an element in the closed set and becomes a new state in the reduced state table.

$$\{A, B\} \rightarrow S_0$$
$$\{C, D\} \rightarrow S_1$$
$$\{E\} \rightarrow S_2$$

The reduced state table is shown in Table 7.17. The table is constructed by replacing the original state names with the reduced state names. For example, in the first row of the original state table the present state is A. State A belongs to a compatible class that is now renamed as state S_0. So we write S_0 in place of the present state A. Likewise state/output C/0, in input column 00, is replaced by S_1 in the new state table. The "–/–" don't-care state/output in input column 01 for the present state A row is replaced by $S_2/0$ for original state E/0 in the present state B row. Notice that present state rows A and B combine in the new state table as S_0. The unspecified states are merged with the specified states.

❖ **EXAMPLE 7.3** Simplify the incompletely specified state machine shown in Table 7.18.

SOLUTION The merger diagram indicating compatible and implied compatible pairs is shown in Figure 7.14.

TABLE 7.18
Incompletely specified state machine

Present state	Next state 00	01	10	11
A	—/—	D/0	F/0	C/1
B	A/0	C/0	—/—	D/1
C	E/0	A/0	E/0	—/—
D	—/—	—/—	C/0	G/1
E	C/0	—/—	—/—	F/1
F	D/1	B/0	—/—	E/1
G	E/1	—/—	A/0	C/1

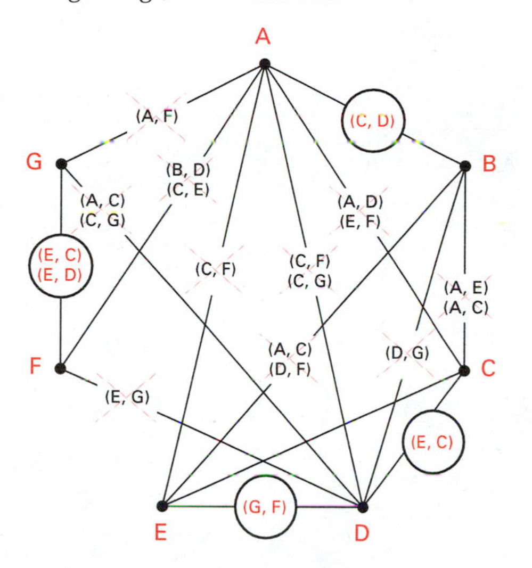

Figure 7.14
Merger diagram for Table 7.18

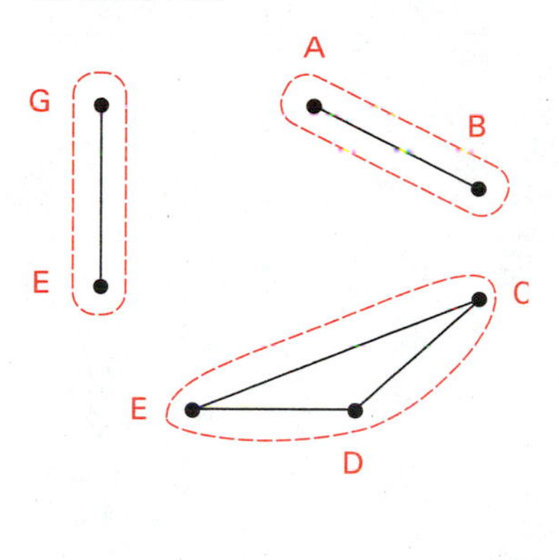

Figure 7.15
Compatibility class graph for Table 7.18

Each of the incompatible state pairs is crossed off, leaving only those that are compatible or implied compatible. The maximal compatible set of state pairs is found in the compatibility class graph shown in Figure 7.15.

A closed set of state pairs is identified for each maximal compatible. In this example, three maximal compatibles exist:

$$\{A, B\}; \quad \{C, D, E\}; \quad \{F, G\}$$

The maximal compatibles are also a closed set.

Assigning new states for the closed set of maximal compatibles gives

$$\{A, B\} \rightarrow S_0$$
$$\{C, D, E\} \rightarrow S_1$$
$$\{F, G\} \rightarrow S_2$$

The simplified state table is shown in Table 7.19.

TABLE 7.19
Simplified state table

Present state	Next state 00	01	10	11
S_0	$S_0/0$	$S_1/0$	$S_2/0$	$S_1/1$
S_1	$S_1/0$	$S_0/0$	$S_1/0$	$S_2/1$
S_2	$S_1/1$	$S_0/0$	$S_0/0$	$S_1/1$

7.4 STATE ASSIGNMENT TECHNIQUES

The assignment of unique binary codes to the state variables of a sequential circuit is called making the **state assignment**. In various sequential circuits designed in Chap-

TABLE 7.20
State table

Present state	Next state	
	$x = 0$	$x = 1$
A	A/0	B/0
B	A/0	C/0
C	D/0	B/0
D	A/1	B/0

TABLE 7.21
State assignments

Present state			Next state			
			$X = 0$		$X = 1$	
	F_2	F_1	D_2	D_1	D_2	D_1
State assignment 1						
A	0	0	0	0/0	0	1/0
B	0	1	0	0/0	1	0/0
C	1	0	1	1/0	0	1/0
D	1	1	0	0/1	0	1/0
State assignment 2						
A	0	0	0	0/0	0	1/0
B	0	1	0	0/0	1	1/0
C	1	1	1	0/0	0	1/0
D	1	0	0	0/1	0	1/0

ter 6, the state assignment was given. Now we wish to investigate a more systematic technique used to assign binary codes to states. The actual binary codes assigned to represent the states have a significant impact on the final realization cost. Therefore, the impact of state assignments on the logic required to realize a circuit is worth investigating.

Consider the state machine of Table 7.20. In Table 7.21, we will realize the circuit using two state assignments to illustrate the effect the assignment has on the final logic.

If D flip-flops are used to realize the state machine then the derived excitation equations for state assignments 1 and 2 are as given. The excitation equations using state assignment 1 are

$$D_2 = F_2F_1'X' + F_2'F_1X$$
$$D_1 = F_2F_1' + F_2X + F_1'X$$

The excitation equations using state assignment 2 are

$$D_2 = F_2F_1X' + F_2'F_1X$$
$$D_1 = X$$

where F_2 and F_1 are state variables and D_2 and D_1 are excitation variables. We can easily see that the equations using state assignment 2 require fewer gates than assignment 1. This simple illustration shows that the state assignment can effect the realization cost as measured in gates or gate inputs for the same state machine.

7.4.1 State Assignment Permutations

The number of possible state assignments, S, for N states realized by M state variables is given by the equation

$$S = (2^M)!/(2^M - N)!$$

Consider a state machine with four states and two state variables. The number of possible state assignments is

$$S = (2^2)!/(2^M - N)! = 4!/1 = 24$$

Let F_2 and F_1 represent the state variables and A, B, C, and D, the states. The 24 possible state assignments are

	1	2	3	4	5	6	7	8	9	10	11	12
	F_2F_1	F_2F_1	F_2F_1	F_2F_1	F_2F_1	F_2F_1	F_2F_1	F_2F_1	F_2F_1	F_2F_1	F_2F_1	F_2F_1
A	00	00	00	00	00	00	01	01	01	01	01	01
B	01	01	10	10	11	11	10	10	11	11	00	00
C	10	11	01	11	01	10	11	00	10	00	10	11
D	11	10	11	01	10	01	00	11	00	10	11	10

	13	14	15	16	17	18	19	20	21	22	23	24
	F_2F_1	F_2F_1	F_2F_1	F_2F_1	F_2F_1	F_2F_1	F_2F_1	F_2F_1	F_2F_1	F_2F_1	F_2F_1	F_2F_1
A	10	10	10	10	10	10	11	11	11	11	11	11
B	11	11	00	00	01	01	00	00	01	01	10	10
C	00	01	11	01	11	00	01	10	00	10	00	01
D	01	00	01	11	00	11	10	01	10	00	01	00

Not all of the possible assignments need to be considered when making the state assignment. Two state assignments are considered equivalent if they produce the same results in realizing the logic. This happens when one state assignment is the complement of another assignment. The first assignment is the complement of the 24th assignment. Both would realize logic of equivalent complexity.

Likewise the second assignment is the complement of the 23rd assignment, and so on for all of the possible state assignments.

Equivalent state assignments also result when the columns are interchanged. That is, if F_2 and F_1 are exchanged, then an equivalent assignment would result. For example, consider state assignment 1 and 3.

State Assignment 1	State Assignment 3	
F_2F_1	F_2F_1	Columns Exchanged
00	00	
01	10	
10	01	
11	11	

State assignment 1 and 3 are identical except that the state variables F_2 and F_1 are swapped.

The following equation gives the number of nonequivalent state assignments, where N = the number of states and M = the number of state variables:

$$S = (2^M - 1)!/[(2^M - N)!\,(M)!]$$

The number of nonequivalent state assignments increases dramatically as the number of states and state variables increase. Table 7.22 illustrates the number of nonequivalent states for several state and state variable values.

It is immediately obvious that trial and error attempts at making state assignments are not practical for five or more states. For this reason a more systematic approach to

TABLE 7.22
Nonequivalent state assignment possibilities for example state and state variable values

No. of states (N)	No. of state variables (M)	No. of nonequivalent states (S)
2	1	1
3	2	3
4	2	3
5	3	140
6	3	420
7	3	840
8	3	840
9	4	10.81×10^{6}
10	4	75.67×10^{6}
16	4	54.48×10^{9}
20	5	143.14×10^{21}

making a state assignment is given. A general solution to the state assignment problem is not available.

To evaluate a state assignment's effectiveness in reducing the excitation logic needed for state transitions of all the states in a sequential circuit would require a computer search to reach an optimum solution. Manual approaches generally evaluate the contribution each state makes to the next-state transition and the contribution that each state makes to generating the output functions. For practical reasons, no manual "best" solution is likely; however, a "reasonable" solution can be found.

7.4.2 State Assignment Algorithm

The excitation logic required to cause state transitions within a sequential circuit is in large part derived from the present-state to next-state transition requirements of the state variables. This logic is combinational and is minimized by using a Karnaugh map or the Quine-McClusky approach. The point is that logic reduction requires the generation of essential prime implicants that are as large as possible. This applies to the reduction of excitation and output combinational logic in a sequential circuit as well. Therefore, a state assignment criteria is to assign state variables in such a way as to maximize the prime implicants' size.

State Assignment Rule 1

States having the same next state, for a given input value, should have state assignments that can be formed into a prime implicant.

Recall the general model for a sequential circuit. Two types of variables are used to generate the excitation inputs to the state variables, inputs, and present-state outputs. If we hold the inputs constant when evaluating the state assignments, then only the state variables affect the size of the potential prime implicants. Consider the partial state diagram in Figure 7.16.

States A, B, C, and D are all present states to next-state E. Each present- to next-state transition has the same input. The K-map in Figure 7.16(b) illustrates the state assignment for rule 1.

Figure 7.16
Illustration of state assignment rule 1

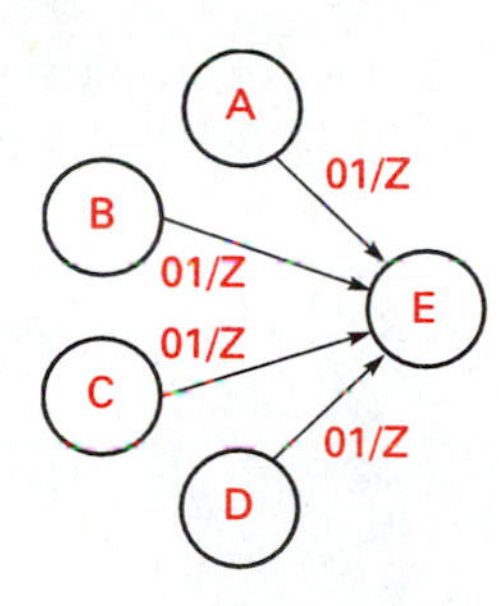

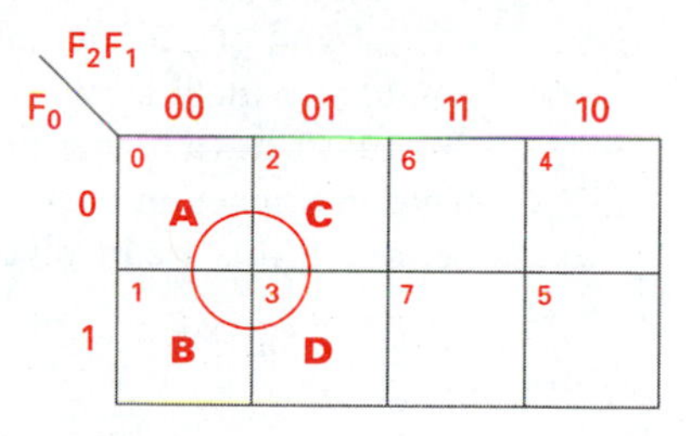

(b) K-Map

Figure 7.17
Illustration of state assignment rule 2

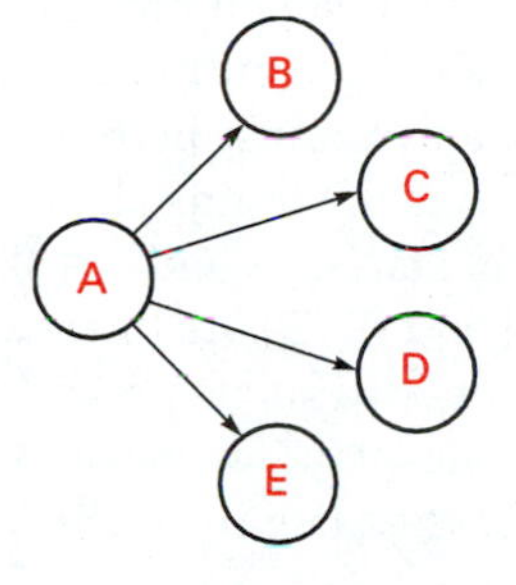

(a) Partial State Diagram

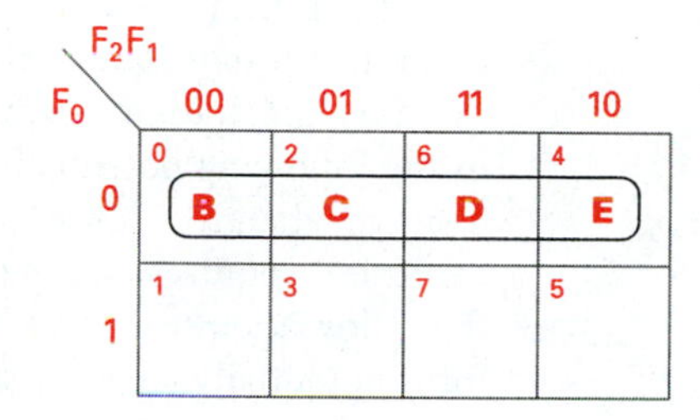

(b) K-Map

State Assignment Rule 2

The next states of a single present state should have state assignments that can be formed into a prime implicant.

Consider the partial state diagram in Figure 7.17.

Rule 2 implies the grouping of each row in a state table. By giving states in the same row adjacent state assignments, the logic needed to realize the state variable excitation equations is reduced.

State Assignment Rule 3

Adjacent assignments should be given to states that have the same outputs.

Consider Table 7.23, which is the same as Table 7.24 except that outputs have been added.

TABLE 7.23
State Table 7.24 with outputs added

State	$X = 0$	$X = 1$
A	A/0	C/0
B	C/1	D/0
C	A/1	F/1
D	B/0	E/1
E	C/0	C/1
F	D/0	E/1

TABLE 7.24
State table

Present state	Next state, X	
	0	1
A	A	C
B	C	D
C	A	F
D	B	E
E	C	C
F	D	E

Based on the application of rule 3, the following state adjacencies should occur: {A, D} and {E, F}. A state assignment that included possible output logic reduction would also include rule 3.

Rule 1 has a higher priority than rule 2 when determining the state adjacencies that have an impact on the excitation logic. Rule 3 has the least priority unless a large number of outputs are involved.

❖ EXAMPLE 7.4

Consider the state table shown in Table 7.24.

Rule 1 requires that present states having the same next states should (if possible) be formed into a prime implicant. The first column ($X = 0$) has two such conditions: {A, C} and {B, E}. States A and C are the present states for next state A and states B and E are the present states for next state C. Because both occur in the $X = 0$ input column, the "for a given input" requirement is met. Finding all of the conditions for rule 1 results in the following potentially adjacent states. When $X = 0$, groups {A, C} and {B, E} are formed; when $X = 1$ groups {A, E} and {D, F} are formed.

Rule 2 requires that the next states of a single present state should have assignments that allow them to form prime implicants. This means that the rows of the state table form potentially adjacent states. Present state rows result in the following potential adjacent states:

{A, C}; {C, D}; {A, F}; {B, E}; {D, E}

Combining the redundancies from the rule 1 and 2 groups gives the following set:

{A, C} 2X; {B, E} 2X; {A, E}; {D, F}; {C, D}; {A, F}; {D, E}

Groups {A, C} and {B, E} occur twice, giving them higher priority than the remaining groups. Now we combine the adjacent state groups from rules 1 and 2 into a single state assignment. We do this by identifying the largest groups and multiple occurring groups and give them the higher priority. The goal now is to find a state assignment that uses as many of the state groups as possible.

Several possible state assignments result from application of the two rules. The K-maps in Figure 7.18 and the assignments in Table 7.25 illustrate three possibilities.

State assignment 1 satisfies the following state adjacency groups: {A, C} 2X; {B, E} 2X; {A, E}; {D, F}; {C, D}; and {A, F}. Group {D, E} was not used in generating the state assignment. State assignment 2 used all of the groups except {D, F}. State assignment 3 used all of the groups except {D, F}, but in a different arrangement than in state assignment 2.

The excitation equations for the three state assignments, after simplification, are as follows. For state assignment 1,

$$D_3 = F_3'F_2X' + F_3'F_2'X + F_1$$

The gate input count is 9.

$$D_2 = F_2F_1'X' + F_2F_1X + F_3F_2'X$$

The gate input count is 12.

$$D_1 = F_3F_2 + F_2F_1'X$$

The gate input count is 7.

Figure 7.18

State assignment K-maps for state Table 7.24

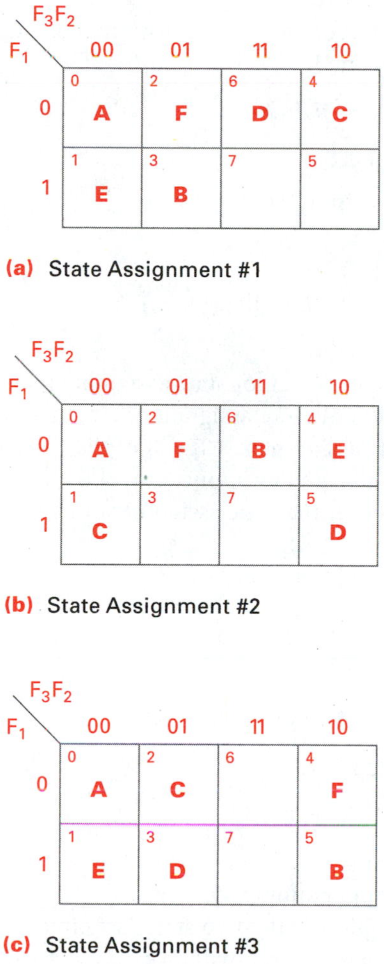

F_1 \ F_3F_2	00	01	11	10
0	0 A	2 F	6 D	4 C
1	1 E	3 B	7	5

(a) State Assignment #1

F_1 \ F_3F_2	00	01	11	10
0	0 A	2 F	6 B	4 E
1	1 C	3	7	5 D

(b) State Assignment #2

F_1 \ F_3F_2	00	01	11	10
0	0 A	2 C	6	4 F
1	1 E	3 D	7	5 B

(c) State Assignment #3

TABLE 7.25

State assignments for state Table 7.24

State	F_3	F_2	F_1
State assignment 1			
A	0	0	0
B	0	1	1
C	1	0	0
D	1	1	0
E	0	0	1
F	0	1	0
d*	1	0	1
d*	1	1	1
State assignment 2			
A	0	0	0
B	1	1	0
C	0	0	1
D	1	0	1
E	1	0	0
F	0	1	0
d*	0	1	1
d*	1	1	1
State assignment 3			
A	0	0	0
B	1	0	1
C	0	1	0
D	0	1	1
E	0	0	1
F	1	0	0
d*	1	1	0
d*	1	1	1

* Don't-care states.

The input gate count is reached by counting the number of gate inputs in each equation. The gate input count gives a relative cost (in gate inputs) of different assignments. The lower the gate input count, the better. The total gate input count of state assignment 1 is 28.

For state assignment 2,

$$D_3 = F_3'F_2 + F_2X + F_3F_1$$

The gate input count is 9.

$$D_2 = F_3'F_1X + F_3F_1X'$$

The gate input count is 8.

$$D_1 = F_3F_1' + F_2'F_1'X + F_2X'$$

The gate input count is 10.

The total gate input count for state assignment 2 is 27.

For state assignment 3,

$$D_3 = F_2F_1'X + F_2F_1X'$$

The gate input count is 8.

$$D_2 = F_3X' + F_2'F_1 + F_3'F_2'X$$

The gate input count is 10.

$$D_1 = F_3F_1' + F_3X + F_2F_1$$

The gate input count is 9.

The total gate input count for state assignment 3 is 27.

State assignment 3 or state assignment 2 realizes the state machine with one fewer gate input than state assignment 1. Each of the three state assignments used six of the seven state groups indicated by application of the two state assignment rules.

By contrast, consider the state assignment in Table 7.26, which did not use the state assignment rules.

TABLE 7.26
Binary state assignment for state Table 7.24

State	F_3	F_2	F_1
A	0	0	0
B	0	0	1
C	0	1	0
D	0	1	1
E	1	0	0
F	1	0	1
d	1	1	0
d	1	1	1

The excitation equations are

$$D_3 = F_2X + F_3F_1X$$

The gate input count is 7.

$$D_2 = F_3F_1' + F_3'F_2'X + F_2'F_1X'$$

The gate input count is 11.

$$D_1 = F_3'F_2'F_1X + F_2F_1'X + F_2F_1X' + F_3F_1X'$$

The gate input count is 17.

The total gate input count is 35.

Although application of the two state assignment rules will not necessarily result in an optimum state assignment, it will give a good assignment.

The state assignment thus far has been concerned with only the excitation functions, ignoring the output functions. Rule 3 includes output functions. ❖

7.4.3 Implication Graph

Another technique for determining state pair adjacencies is the use of a directed-flow graph called an **implication graph**. An implication graph is a flow graph with directed arcs indicating the state transitions of a sequential machine. Each node in the graph represents a state adjacency pair. The directed arc, starting at a node and ending at another node, indicates a state transition. A complete implication graph contains all of the possible state pair transitions of a sequential machine.

An implication subgraph is a part of the implication graph. A subgraph is said to be closed when all of the directed arcs starting in the subgraph also end in the same subgraph. Directed arcs may enter the subgraph but they cannot leave it.

Rules for Using Implication Graphs

Determine the state adjacencies by applying the three state assignment rules. Construct an implication graph and determine the closed subgraphs, if any. Closed subgraphs are composed of a set of state adjacency pairs. Maximize the use of these state pairs in determining the state assignment.

❖ EXAMPLE 7.5

Consider the sequential machine specified in Table 7.27.

The application of the three state assignment rules gives:

Rule 1: {V, W} 2*X*; {U, Z}

Rule 2: {U, V}; {T, Y}; {X, Y} 3X; {W, Z}; {U, W}

Rule 3: {X, T}; {U, W}; {U, Y}; {W, Y}; {V, Z}

The implication graph is drawn by finding a starting state pair and then asking the question, what is the next-state pair for every input combination? Start with state pair {V, W}. From Table 7.27, we find that no next-state pair exists for state pair {V, W}. When A = 0, the next-state for state pair {V, W} is Y and when A = 1 the next-state for state pair {V, W} is *X*. Because no next-state pair exists for {V, W}, no directed arc leaves state pair {V, W}.

TABLE 7.27
State table

Present state	Next state A = 0	Next state A = 1
T	U/1	V/1
U	T/0	Y/0
V	Y/0	X/1
W	Y/0	X/0
X	Z/1	W/1
Y	W/0	U/0
Z	X/0	Y/1

The pair {X, T} is the next-state pair for present-state pair {U, Z} when A = 0 (see Figure 7.19). When A = 1 the present-state pair {U, Z} has a next state of Y. A directed arc between state pair {U, Z} and {X, T} is drawn indicating a possible adjacent state assignment.

State pair {X, T} returns to state pair {U, Z} when A = 0 and to {V, W} when A = 1 (see Figure 7.20). Two directed arcs are drawn to indicate the transitions.

State pair {X, Y} has two next-state pair. When A = 0, the next-state pair is {W, Z}; and when A = 1, the next-state pair is {U, W} (see Figure 7.21). Directed arcs are drawn from state pair {X, Y} to each of the next-state pair.

The process is continued until all of the next-state pair transitions are indicated by directed arcs. Once all of the state pair transition directed arcs are drawn, we look for closed subsets. A closed subset occurs when the directed arcs connecting state pair transitions form a complete loop.

Figure 7.19
Starting implication graph for Table 7.27

Figure 7.20
Continuing construction of implication graph for Table 7.27

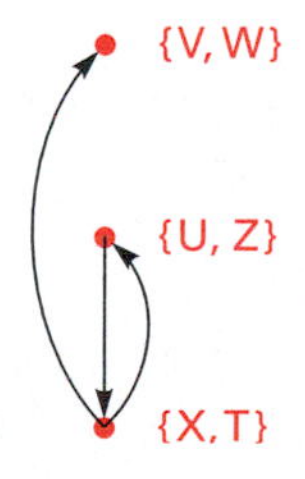

Figure 7.21
Partial implication graph for Table 7.27

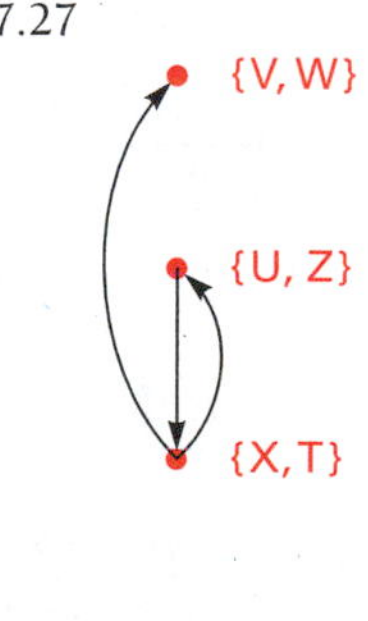

The state pairs produced by rule 3 help simplify the output logic and have no bearing on the excitation logic, so they are not included in the implication graph state pair transitions. The implication graph shown in Figure 7.22 produces two closed subsets. The state pairs that result from these are

Closed subgraph 1: {U, Z} and {T, X}

Closed subgraph 2: {U, V}, {T, Y}, {U, W}, {X, Y}, and {W, Z}

Next we must assign priorities to the state pairs. The larger subgraph takes priority over the smaller one because a greater number of state transitions are involved. State pairs that occur several times are given priority. One possible priority list (several are possible) follows:

1. {X, Y} 3X Subgraph 2
2. {W, Z} Subgraph 2
3. {V, U} Subgraph 2
4. {T, Y} Subgraph 2
5. {U, W} Subgraph 2
6. {T, X} Subgraph 1
7. {U, Z} Subgraph 1
8. {V, W} 2*X* Not part of any subgraph

The starting state is usually assigned all zeros, so that the sequential circuit can be easily initialized by an asynchronous clear or reset input. For that reason we will assign state T to 000, because it is the first state in Table 7.27. The state assignment K-map is shown in Figure 7.23.

Figure 7.22
Implication graph for state Table 7.27

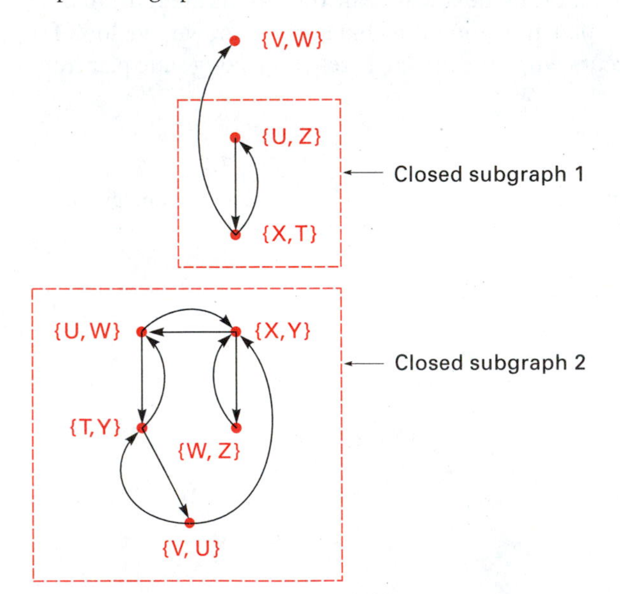

Figure 7.23
State assignment K-map for state Table 7.27

F_1 \ F_3F_2	00	01	11	10
0	0 **T**	2 **Y**	6 **W**	4 **Z**
1	1	3 **X**	7 **U**	5 **V**

7.5 ALGORITHM STATE MACHINES

An *algorithm* is defined as a systematic description of a process or operation. Any process, from a recipe for baking a cake to a complex industrial control problem, can be described in an algorithm. Computer programs are written based on some algorithm for the problem being solved. A tool used by computer programmers to express algorithms is a flowchart. Flowcharts, using accepted symbols, express the operational steps necessary to solve a particular problem.

Algorithm state machine (ASM) charts borrow the flowchart symbols from the computer programmer and use them to express the operation of a sequential circuit. The appearance of an ASM chart is like that of the programmer's flowchart, but the purpose is different. Instead of realizing a problem solution in software, the ASM chart permits the logic designer a hardware solution.

7.5.1 ASM Symbols

ASM charts use three primary symbols: a state symbol, an input condition symbol, and a conditional output symbol. The state symbol is a rectangle with an arrow entering the top and leaving the bottom. A state name is written to the side or on top of the box. Inside the box is written a list of unconditional output(s). If no unconditional output is required, the box interior is left blank. Figure 7.24 illustrates the ASM state symbol.

The condition or decision symbol is a diamond or modified diamond. One arrow enters the top and two leave the sides or a side and the bottom. Written into the decision symbol are the logical conditions for the decision to be either true or false. The condition(s) can be a single variable or a Boolean expression. Figure 7.25 shows the ASM symbols used to indicate decisions.

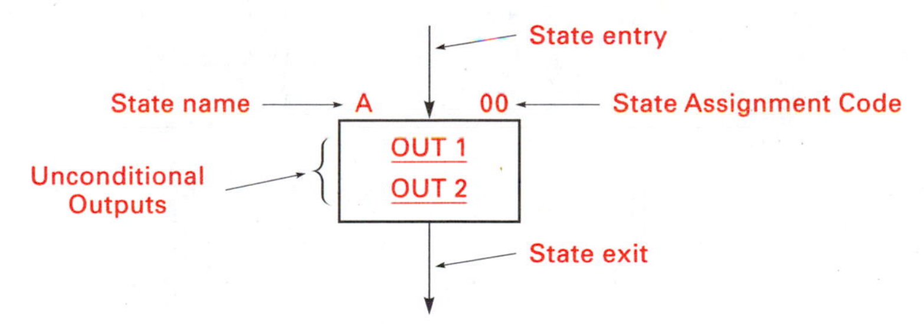

Figure 7.24
ASM state symbol

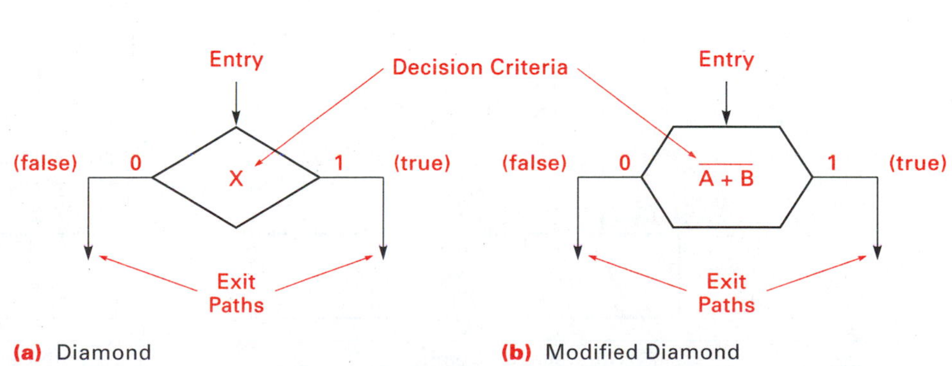

Figure 7.25
ASM decision symbols

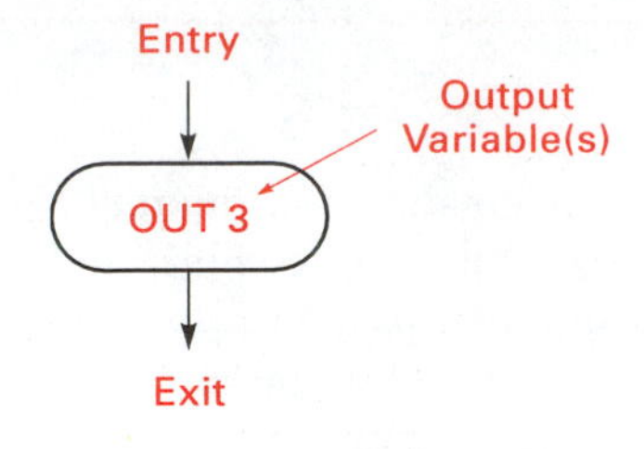

Figure 7.26
ASM conditional output symbol

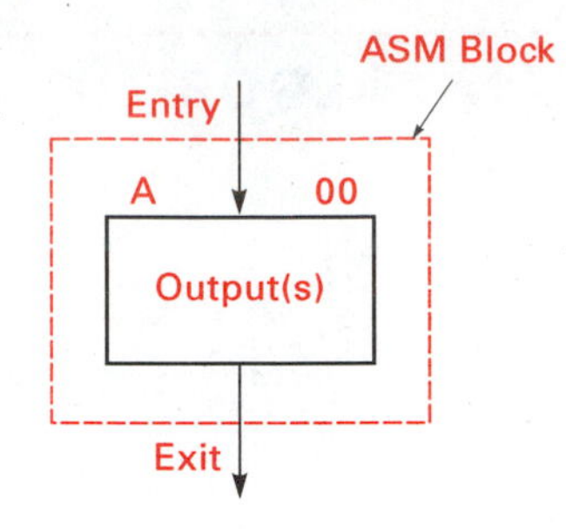

Figure 7.27
Simple ASM block

The conditional output symbol is a modified elipse, as indicated in Figure 7.26. Ouput variables are listed internally to the symbol. There may be a single variable or multiple variables.

An ASM block consists of a state symbol, and/or possibly a state symbol with decision block(s) and/or conditional output symbol(s). The block symbols indicate a process involving a present state, the decisions, and conditional outputs and the next state. The Moore sequential circuit model is represented without conditional output symbols, and the Mealy model includes conditional output symbols.

Figure 7.27 shows an ASM block consisting of a single state with an unconditional output. Figure 7.28 shows more complex ASM blocks. Figure 7.28(a) illustrates a Moore model with an unconditional output in state A; Figure 7.28(b) shows a Moore/Mealy mix with both unconditional (output variable O_1 in state A) and conditional (output variable O_2) outputs.

All of the functions contained within the ASM block are considered to be part of the entry state. For example, Figure 7.28(b) shows state A, which consists of the state symbol, the decision symbol, and the conditional output symbol. If input variables $xy' = 1$, then output O_2 is active until next-state B is reached. If input variables $xy' = 0$, the

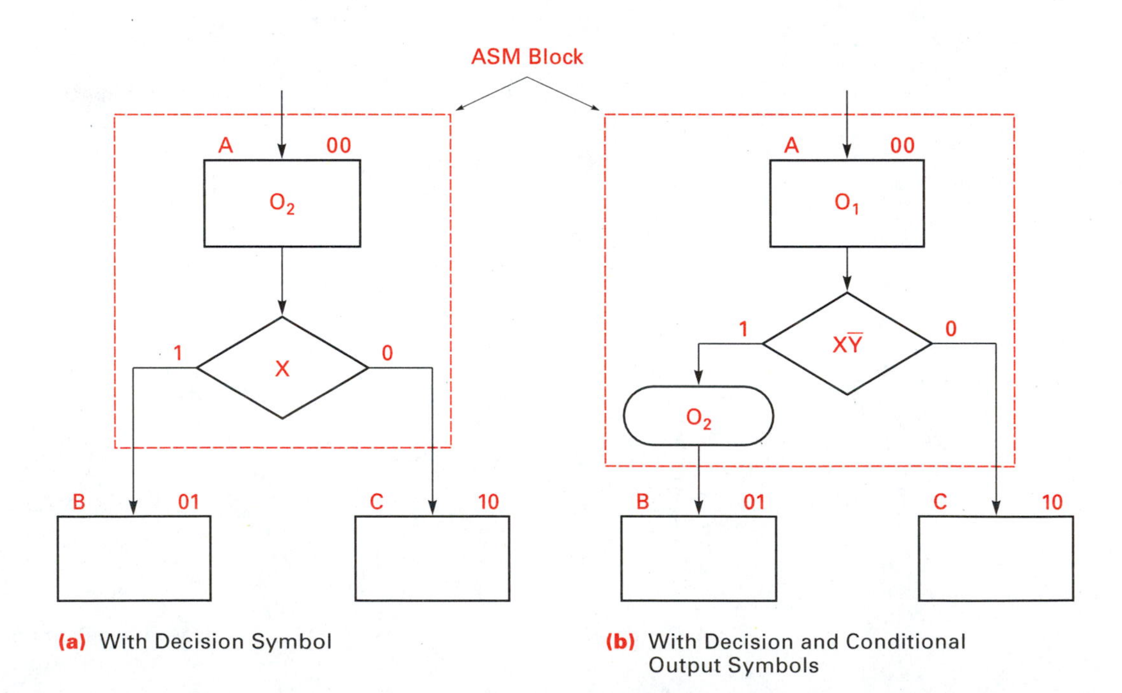

Figure 7.28
ASM blocks

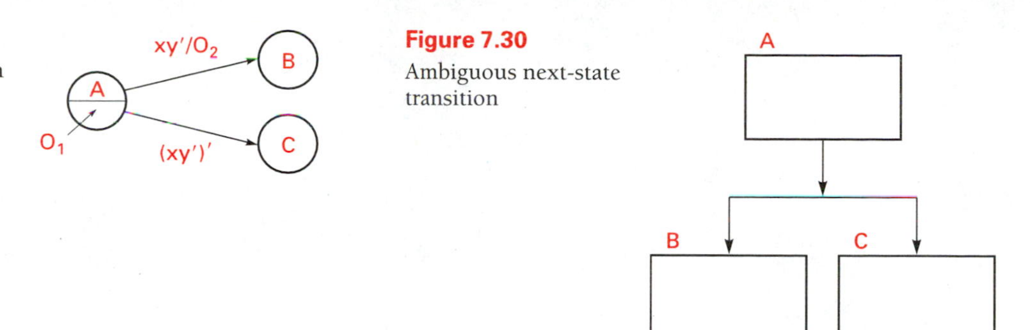

Figure 7.29
Standard state diagram notation for the ASM block in Figure 7.28(b)

Figure 7.30
Ambiguous next-state transition

conditional output is inactive and C is the next state. Output O_1 remains active as long as the machine remains in state A. The standard state diagram notation equivalent of the ASM diagram in Figure 7.28 is shown in Figure 7.29.

Several restrictions are placed on the construction of ASM diagrams. The restrictions are necessary to eliminate any ambiguity. For example, Figure 7.30 shows a present state with ambiguous next-state transitions. Figure 7.31 illustrates an ambiguous decision path. The undefined decision path for the ASM diagram illustrated in Figure 7.31 is due to the uncertainty in the next-state transition from present state A to either state B or state C. Let $xy = 00$. The next state is B, but was the A to B transition through the conditional output O_1 or not? We simply do not know. Let $xy = 11$; in this case, the next state may be B or it may be C—the transition is ambiguous. Figure 7.32 shows the same process in an unambiguous ASM diagram.

ASM charts can depict series or parallel decision paths, as illustrated in Figure 7.33 and Figure 7.34. The parallel decision path depicted in Figure 7.33 indicates a direct

Figure 7.31
Undefined decision path

Figure 7.32
Unambiguous ASM diagram

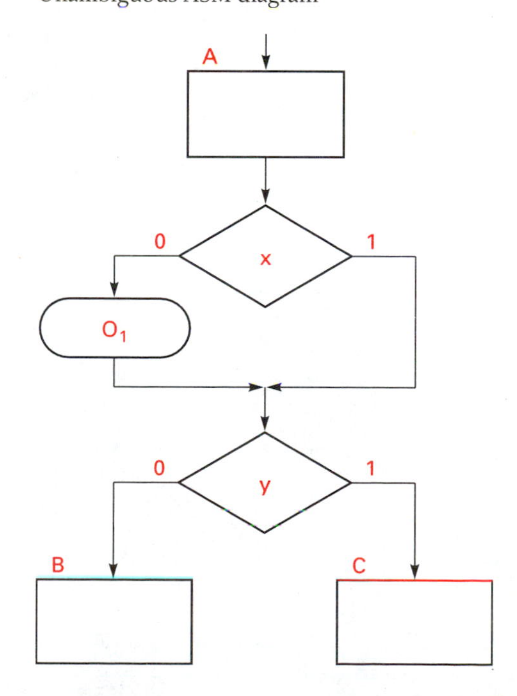

Figure 7.33
Ambiguous parallel decision paths

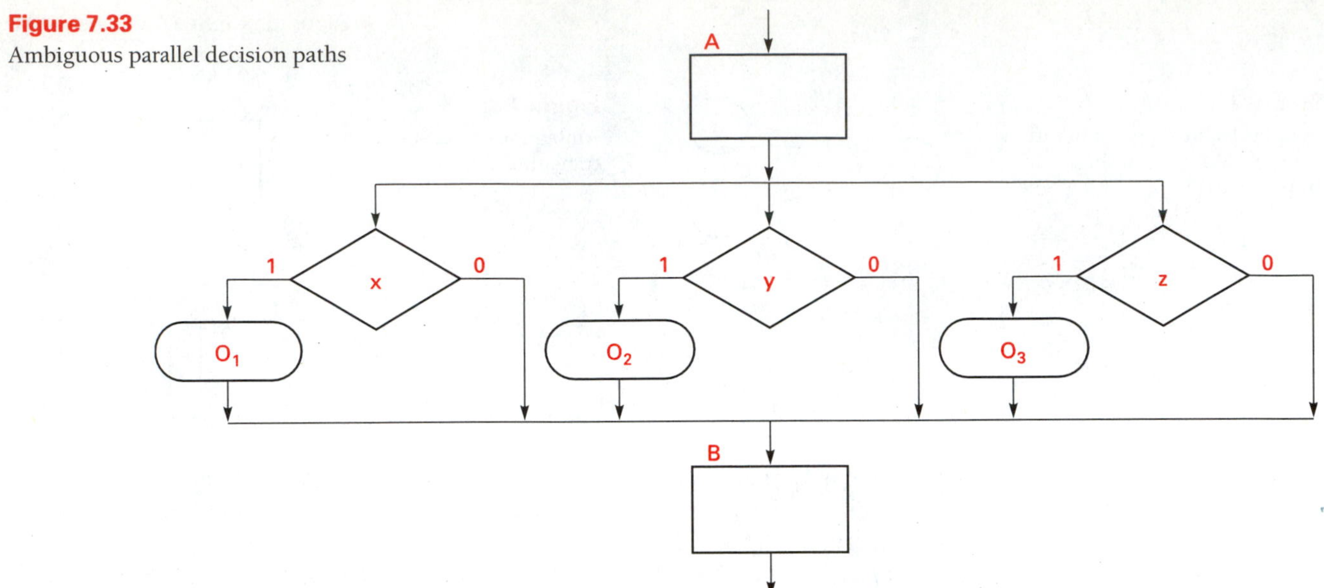

Figure 7.34
Series decision path

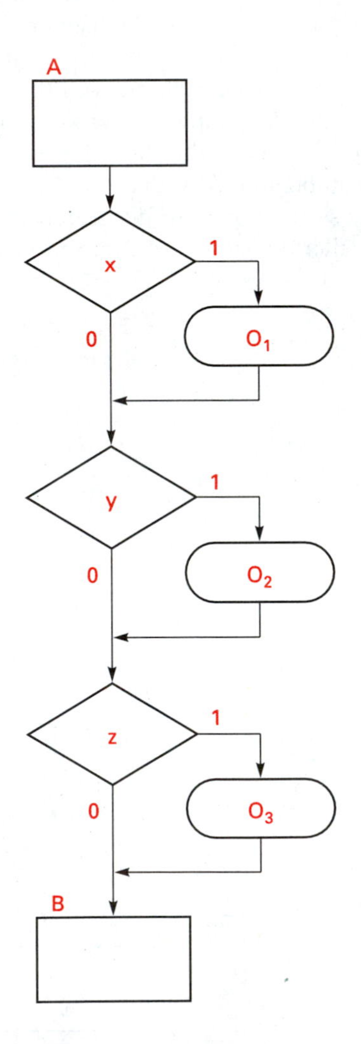

Figure 7.35
Shared ASM decision diamond

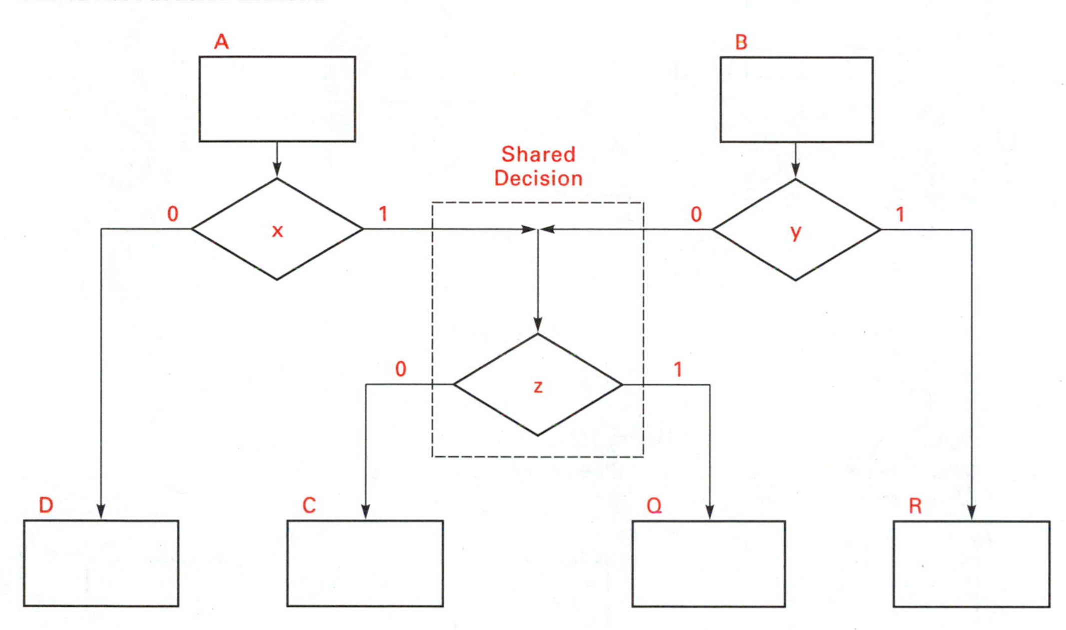

state A to B transition, with no outputs active, when $x + y + z = 0$. Conditional output O_1 is true when $x = 1$, O_2 is true when $y = 1$, and O_3 is true when $z = 1$.

A problem exists with the ASM diagram in Figure 7.33. What happens when $xy = 11$ or $xz = 11$ or $xyz = 111$? Which output or combination of outputs will be active? The diagram is ambiguous. To create an unambiguous parallel decision diagram from Figure 7.33 would require specification of the undefined xyz input conditions.

ASM charts can depict a series decision path as shown in Figure 7.34. In this ASM diagram, the transition from present-state A to next-state B goes through a series of decisions. The direct present-state A to next-state B transition, without generating any active conditional outputs, is made when $x'y'z'$ is true. Conditional output O_1 is true when $x = 1$. If $xyz = 010$, for instance, then the present-state A to next-state B transition would occur and conditional output O_2 would be true.

Decision boxes may be shared between ASM blocks as indicated in Figure 7.35. ASM blocks must contain, at a minimum, a state symbol. They may also contain decision and conditional output symbols. Entry to one ASM block from another must occur at the top of a state symbol. Looping on a state requires a state box and a decision symbol. One decision box exit output loops back to the top of the state symbol. The correct and incorrect methods of looping back on a state are illustrated in Figure 7.36.

One of the simplest state machines is a counter. Consider the modulo-4 counter shown in Figure 7.37. No decisions or conditional outputs are needed. The ASM chart illustrates a simple sequential circuit up-down modulo-4 counter as shown in Figure 7.38. This circuit is a Moore machine with conditional state transitions and no conditional outputs.

A modulo-4 up-down counter requires decision boxes in addition to the state symbols. An external input, *UP*, is the mode control. When $UP = 1$, the counter increments;

Figure 7.36
ASM chart state looping diagrams

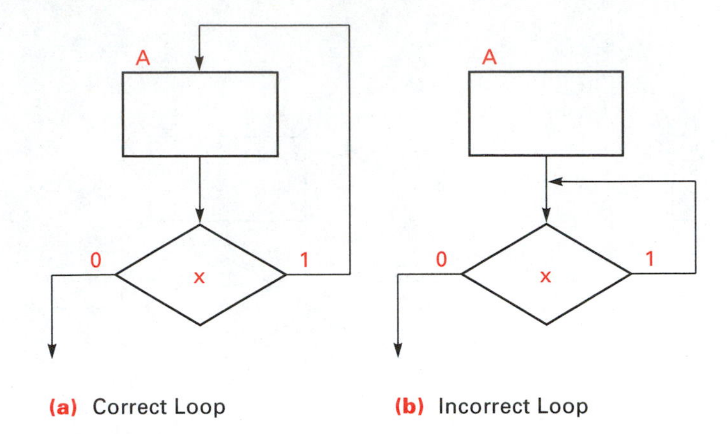

(a) Correct Loop **(b)** Incorrect Loop

Figure 7.37
Modulo-4 counter ASM chart

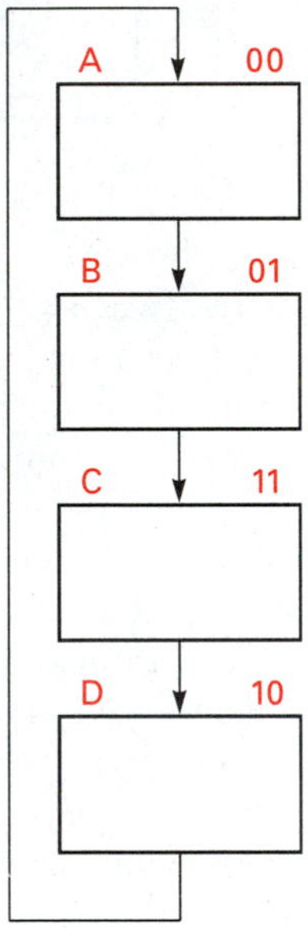

Figure 7.38
Modulo-4 up-down counter ASM chart

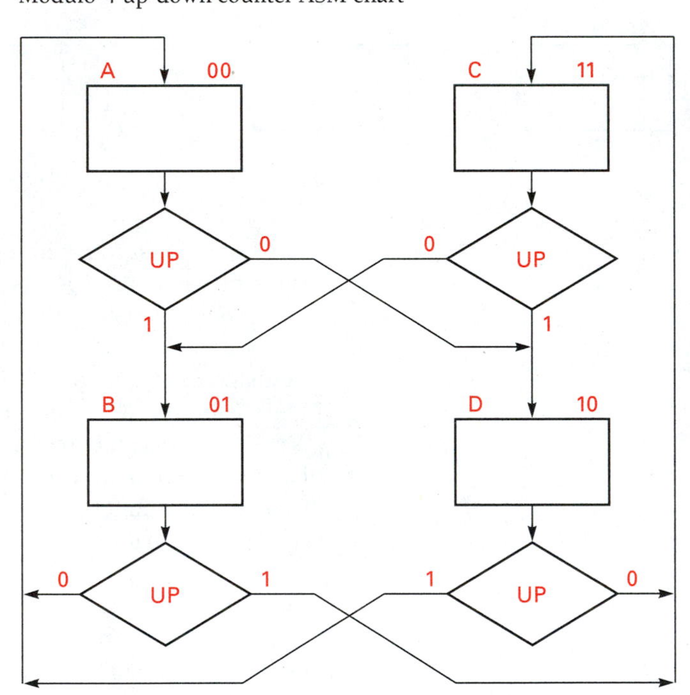

and when $UP = 0$, the counter decrements. The clock input is implied when using ASM charts as it is with state diagram notation.

7.5.2 Elapsed Time Measurement, an ASM Design Example

The best way to illustrate the application of ASM charts in sequential circuit design is by considering the following example.

❖ EXAMPLE 7.6

The measurement of elapsed time between the transmission of an ultrasonic pulse and the resulting return echo is necessary to determine the distance from the transducer to a target. A sequential circuit is needed to control the functions of a 16-bit counter that stores the elapsed time data. The 16-bit elapsed time data word is then transmitted to a microprocessor for processing. Our problem is to design the sequential circuit controller.

SOLUTION We will use a crystal controlled oscillator to produce a 10.000 MHz square wave that is divided by 10, using a decade counter to produce a 1.000 MHz square wave signal. Each period of the 1.000 MHz square wave represents a 1 microsecond (μs) time duration. By counting the number of 1 μs pulses between the transmitted ultrasonic pulse and the return echo, we can determine elapsed time.

The 16-bit elapsed time counter can be realized by using two 74LS393 dual four-bit counter integrated circuits. The ICs are connected in series to produce one 16-bit counter. Two control signals, count-reset and count-enable, are used to control the count operation. Count-reset causes the counter data to be 0 (all 16-bits). The count-enable control signal opens a gate (internal to the IC) to allow counting the 1.000 MHz input clock pulses. When count-enable goes inactive the counting stops and the counter stores the elapsed time between the transmission and receipt of the ultrasonic echo.

We want to send the elapsed time information to a microprocessor for additional processing to determine actual distance. The 16-bit counter output is to be connected to a large-scale integrated circuit that provides an input port to the microprocessor. This integrated circuit, an 8155, was designed to work with Intel's 8085 microprocessor. It has two eight-bit storage registers and additional control logic that enables it to communicate with the microprocessor. Two data input enable signals (active-low) are used to load data into the 8155 internal registers. The sequential circuit controller that we are designing must generate the enable signals to facilitate transfer of the elapsed time data from the counters to the 8155 integrated circuit. We will not be concerned with what the 8155 does to transfer data to the microprocessor or what the microprocessor will do with the information. Figure 7.39 shows the block diagram of the elapsed time measurement system.

Two 74HC04 inverters are used with a 10.000 MHz crystal to form a 10.000 MHz clock that provides the clock input to the controller. The 10.000 MHz clock is divided by 10 using a 74LS196 decade counter. The sequential machine receives the transmit and echo pulses from the ultrasonic transmitter-receiver circuit. The transmit pulse initiates the count operation, and the echo terminates counting. An echo reset is an input to the sequential circuit. The echo reset is necessary to terminate the count process in the event that a transmit occurs and a return echo does not. The timing relationship between the ultrasonic transmit and return echo pulses is shown in Figure 7.40.

The following account is not necessary to know in order to design the controller; however, it is presented to give the reader a sense of perspective in why the controller was needed.

Two 74LS393 integrated circuits are used to count the number of 1.000 MHz pulses that occur during time interval t, as shown in Figure 7.40. This count represents the elapsed time between the transmit and echo pulses. For example, if the roundtrip time from transmit to echo were 10.00 ms, the count value would be 10,000, because each count represents 1 μs.

Once the count-enable gate has been closed, indicating reception of the echo pulse, the active-low AB strobe (enable) is generated, loading the 16-bit count data into the

Figure 7.39
Elapsed time measurement system block diagram

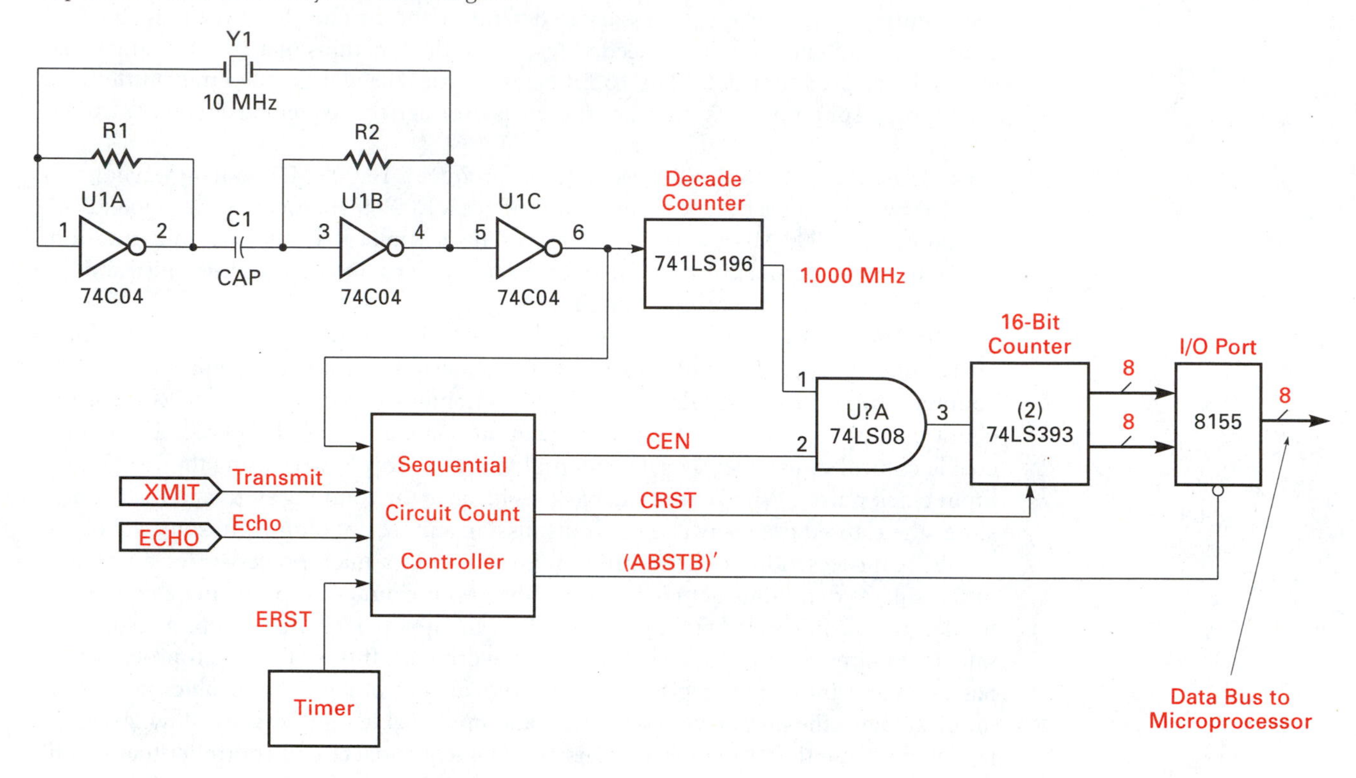

Figure 7.40
Ultrasonic transmit and echo pulse timing

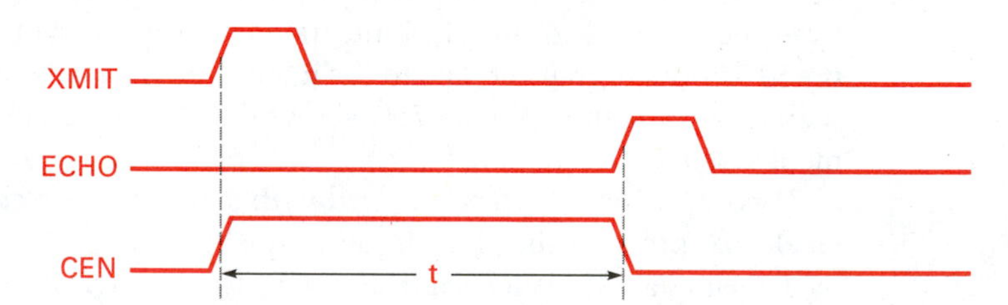

8155 I/O port IC. The 8155 has two eight-bit registers; each has a strobe input to initiate loading data. When the 8155 receives a strobe input, it loads the data into the register and then produces an interrupt signal that is sent to the microprocessor. The interrupt signal informs the microprocessor that data are available to be read (input). The microprocessor in turn generates a RD′ signal that initiates the transmission of the 8155 count data to the microprocessor. This operation takes two read (data input) cycles because the 16-bit count data are sent to the microprocessor eight bits at a time.

Figure 7.41 illustrates the relationship between the count enable signal, ABSTB′ (active low input data strobe), AINTR (A register interrupt), and the microprocessor read (RD′).

The ASM chart for the elapsed time measurement controller is shown in Figure 7.42. Six states are needed for the sequential machine, requiring three flip-flops. The initial or idle state (state A) makes a transition to state B on the first clock pulse. Unconditional

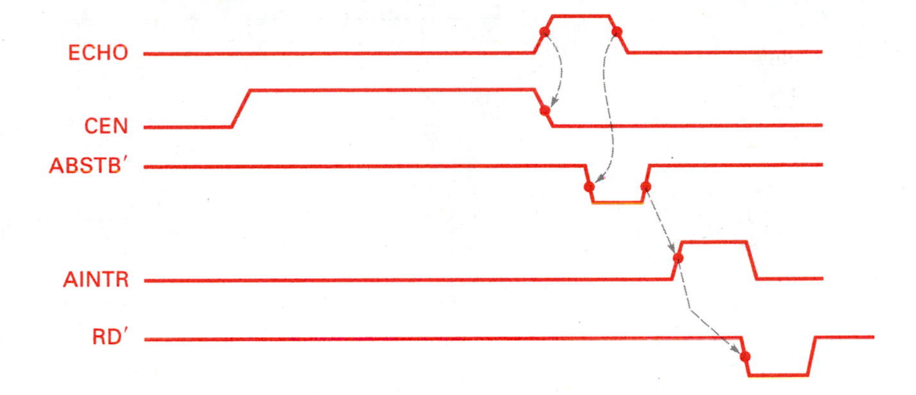

Figure 7.41
8155 I/O device timing

output CRST (count-reset) is active while the machine is in state B, clearing the 16-bit counter. The next clock pulse causes a state B to state C transition. The machine remains in state C until a transmit pulse occurs, initiating a transition to state D. A transition from state C to state D produces unconditional output CEN (count-enable). The active high CEN signal enables the counting of the 1.000 MHz square wave clock pulses. While in state D, several decisions are made. If the echo reset signal (ERST) goes active before an ECHO is returned, the system will go back to state A and start over. If an ECHO is returned before the echo reset signal goes active and if the XMIT (transmit) is not active, then a transition from state D to state E occurs. A diagram illustrating the timing relationship between transmit (XMIT), echo (ECHO), and echo reset (ERST) is given in Figure 7.43.

The reason for including XMIT in the decision process is to ensure that a spurious ECHO is not detected during the transmit pulse. The sequential machine remains in state E until an ECHO pulse becomes inactive. The count-enable (CEN) is active during state D, permitting continued counting of 1.000 MHz clock pulses during the elapsed time between the XMIT and ECHO pulses. Receipt of the ECHO signal causes a state transition from D to E, causing CEN to go inactive (low). At least a clock pulse separates CEN going inactive and ABSTB′ (AB strobe) going active (low). The high-to-low transition of ABSTB′ causes the 16-bit count value to be transferred from the counter to the 8155 I/O integrated circuit. On the clock pulse, after it enters state F, the machine resets back to state A and is ready for the next transmit-echo sequence. Table 7.28 is the state table for the sequential circuit shown in Figure 7.42.

Three input variables produce eight input combinations. Each combination must be accommodated in the state table. The ASM chart indicates only those inputs that cause transitions to occur. The other input variables are considered don't-care. Input combination 111 is very unlikely, for example. No state reduction is possible as indicated by the implication chart shown in Figure 7.44.

Applying the state assignment rules presented in the last section we find

Rule 1: {B, C} 4X; {D, F} 4X; {D, E}; {C, D} 2X;

Rule 2: {D, C}; {A, D, E}; {E, F};

Rule 3: {A, C, E}

Following the rules exactly produces a good state assignment; however, assigning states D and E adjacent assignments gives a better state assignment. The resulting state assignment K-map and table are shown in Figure 7.45.

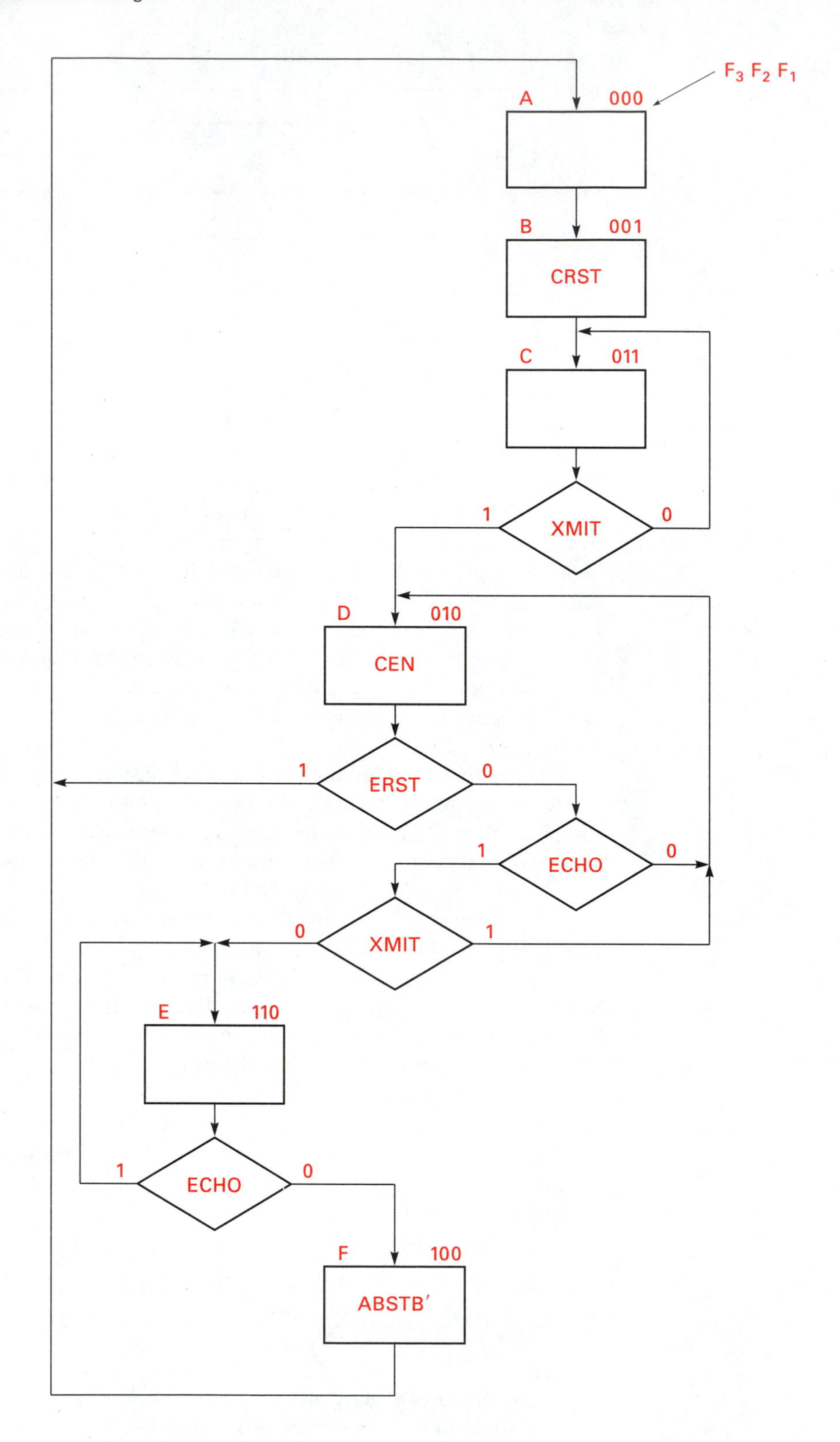

Figure 7.42
Echo reset (ERST) timing

Figure 7.43
Sequential controller ASM chart

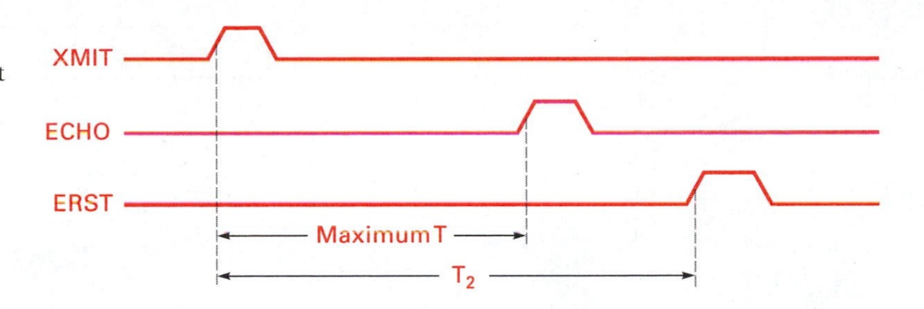

TABLE 7.28
State table for ASM sequential controller

Present state	Next state								Outputs		
	XMIT, ECHO ERST (Inputs)										
	000	001	010	011	100	101	110	111	CRST	CEN	ABSTB′
A	B	B	B	B	B	B	B	B	0	0	1
B	C	C	C	C	C	C	C	C	1	0	1
C	C	C	C	C	D	D	D	D	0	0	1
D	D	A	E	A	D	A	D	A	0	1	1
E	F	F	E	E	F	F	E	E	0	0	1
F	A	A	A	A	A	A	A	A	0	0	0

Figure 7.44
Sequential controller implication chart

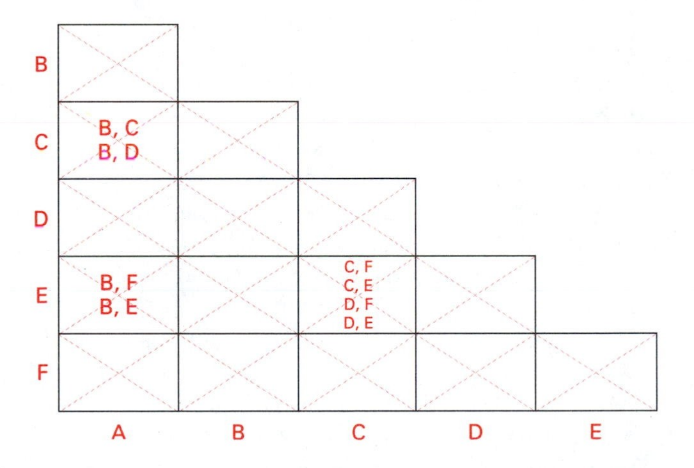

Figure 7.45
State assignment map

F_1 \ F_3F_2	00	01	11	10
0	**A**	**D**	**E**	**F**
1	**B**	**C**	**d**	**d**

TABLE 7.29
Elapsed time sequential machine transition table

Present state	Next state, XMIT, ECHO ERST (Inputs)							
	000	001	010	011	100	101	110	111
$F_3F_2F_1$	$F_3F_2F_1$	$F_3F_2F_1$	$F_3F_2F_1$	$F_3F_2F_1$	$F_3F_2F_1$	$F_3F_2F_1$	$F_3F_2F_1$	$F_3F_2F_1$
000	001	001	001	001	001	001	001	001
001	011	011	011	011	011	011	011	011
011	011	011	011	011	010	010	010	010
010	010	000	110	000	010	000	010	000
110	100	100	110	110	100	100	110	110
100	000	000	000	000	000	000	000	000
101	d	d	d	d	d	d	d	d
111	d	d	d	d	d	d	d	d

From the state table and state assignment, we construct the transition table shown in Table 7.29. The output functions were not included in Table 7.29; the equations can be derived from the state table.

We could decide which flip-flop type to use and then construct an excitation table. However, the use of MEVs and reading state changes directly from the ASM chart in Figure 7.42 eliminates the need for the excitation table. We must recall the excitation relationships for the flip-flop we choose, however. For this example, we shall implement the design using *J–K* flip-flops. The application of MEV terms permits the use of smaller K-maps. For this example, the state variables are used to establish the K-map size and the input variables are entered as MEVs.

We know, based on the *J–K* flip-flop excitation characteristics, that when Q changes $0 \rightarrow 1$, $J = 1$, $K = d$; when Q changes $1 \rightarrow 0$, $J = d$ and $K = 1$. Using this information we can determine the J_x and K_x values from the ASM chart.

The Figure 7.42 ASM chart has the state assignment codes for each state written at the top right corner of each state box. This enables us to determine the present to next-state transitions necessary for finding J and K excitation values. Two K-maps are necessary for each state variable, one for J and one for K. The excitation K-maps are shown in Figure 7.46.

Consider state variable F_1. Find the transitions where F_1 changes from a 0 to a 1. Unconditional state transition A → B produces a 1 in the J_1 K-map square 000 (state A). Therefore, a 1 is written into the J_1 map state A square (000). This is the only transition for F_1. Don't-care terms are written into those squares representing unused state assign ments and *J–K* characteristic don't-cares. The K_1 map is loaded by finding state transitions that produce a 1 to 0 change in state variable F_1. This occurs between states C and D. In this case, the transition is dependent on input variable XMIT being true. The MEV is XMIT and is written in the state-C square (011) of the K_1 map.

Excitation variable J_2 is true where flip-flop F_2 makes 0 to 1 transitions. This occurs when state B changes to state C and requires a 1 in square 001 (state B). The K_2 map is loaded with MEV ECHO′ in square 110 (state E) and MEV ERST in square 010 (state D).

The K-map for J_3 includes a three-variable MEV term (ERST′ ECHO XMIT′) in square 010 (state D). K_3 requires a 1 in square 110 (state F).

The three output function maps are given in Figure 7.47. The simplified excitation and output equations are

$$\begin{aligned}
J_1 &= F_3'F_2' \\
K_1 &= F_2\ \text{XMIT} \\
J_2 &= F_1 \\
K_2 &= F_3\ \text{ECHO}' + F_3'F_1'\ \text{ERST} \\
J_3 &= F_2F_1'\ (\text{ERST}'\ \text{ECHO XMIT}') \\
&= (F_2' + F_1 + \text{ERST} + \text{ECHO}' + \text{XMIT})' \\
K_3 &= F_2' \\
\text{CRST} &= F_2'F_1 \\
\text{CEN} &= F_3'F_2F_1' \\
\text{ABSTB}' &= F_3' + F_2 = (\text{F3F2}')'
\end{aligned}$$

Realization of these equations results in the logic diagram shown in Figure 7.48.

Figure 7.46
Sequential controller excitation maps

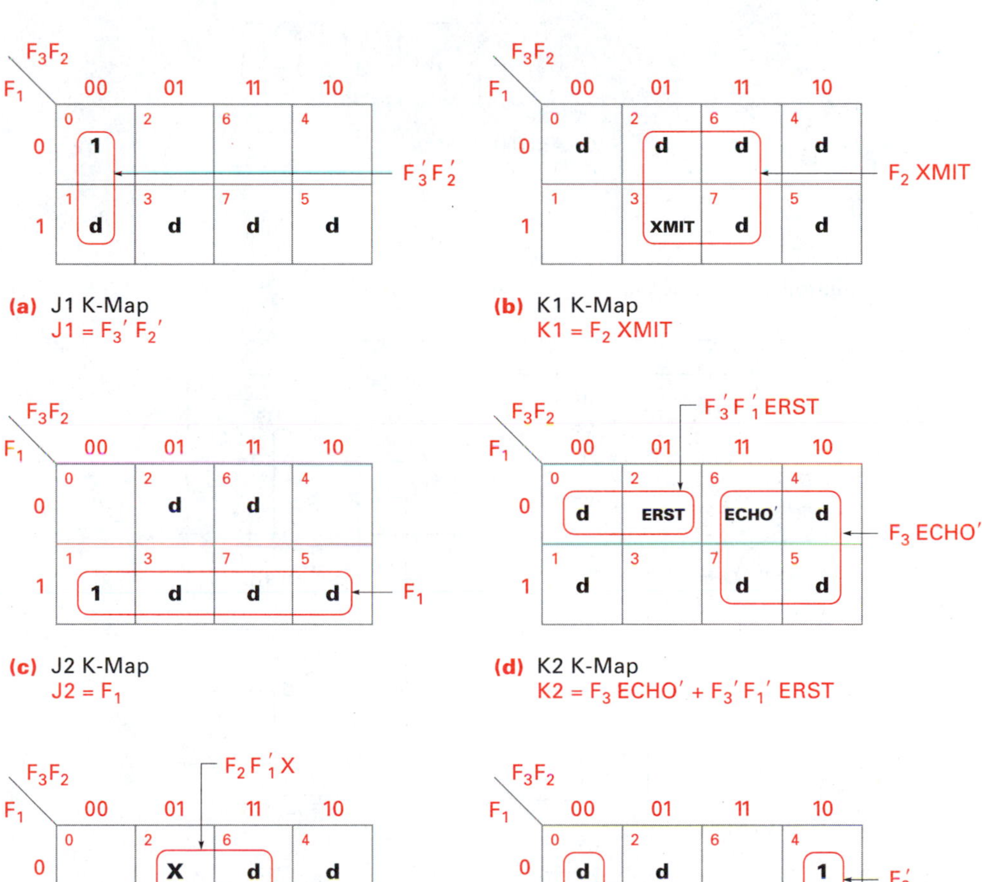

(a) J1 K-Map
$J1 = F_3'\ F_2'$

(b) K1 K-Map
$K1 = F_2$ XMIT

(c) J2 K-Map
$J2 = F_1$

(d) K2 K-Map
$K2 = F_3\ \text{ECHO}' + F_3'\ F_1'\ \text{ERST}$

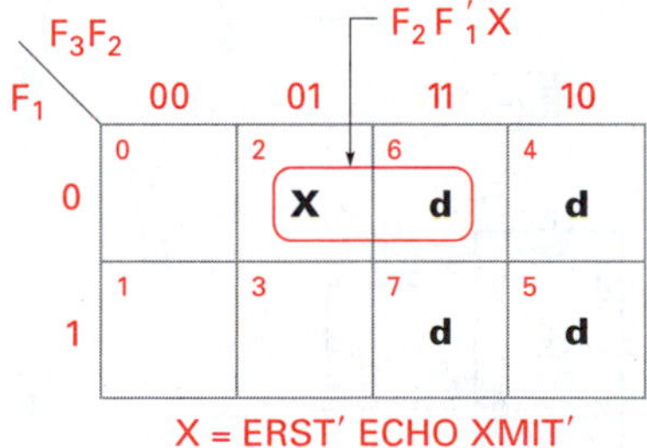

(e) J3 K-Map
$J3 = F_2\ F_1'\ (\text{ERST}'\ \text{ECHO XMIT}')$

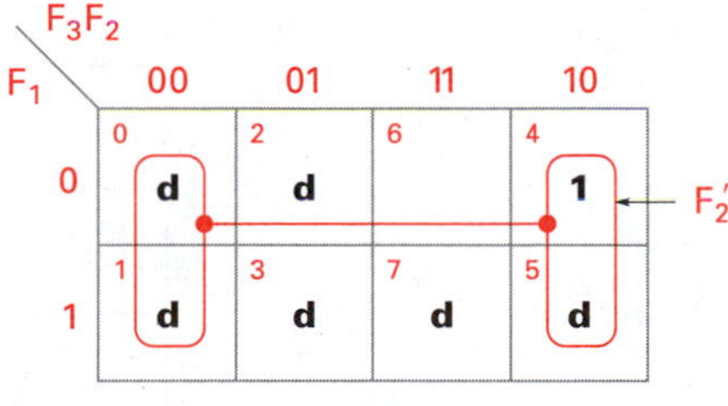

(f) K3 K-Map
$K3 = F_2'$

Figure 7.47
Output functions K-maps

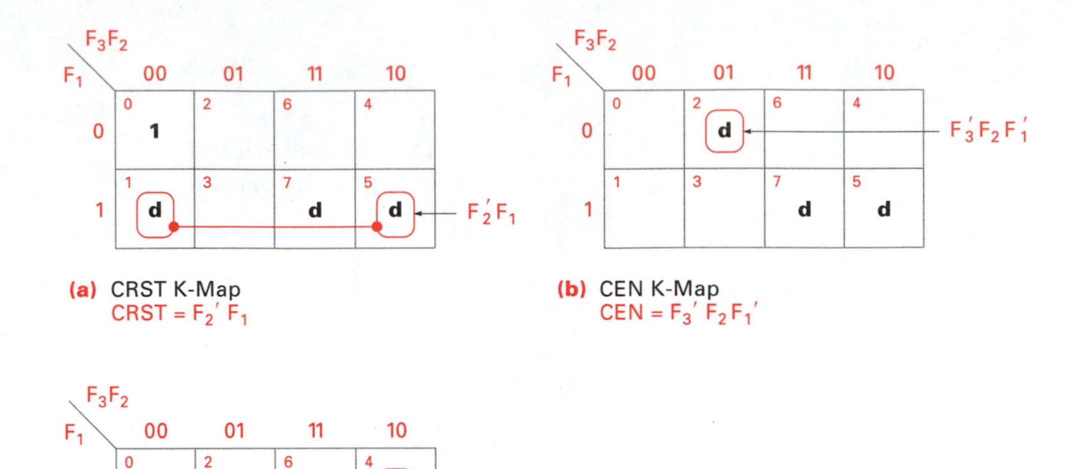

(a) CRST K-Map
$CRST = F_2' F_1$

(b) CEN K-Map
$CEN = F_3' F_2 F_1'$

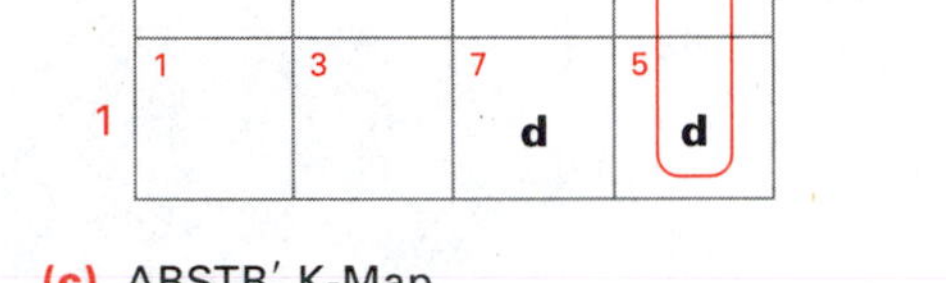

(c) ABSTB′ K-Map
$ABSTB' = (F_3 F_2')'$

Figure 7.48
Sequential controller logic diagram

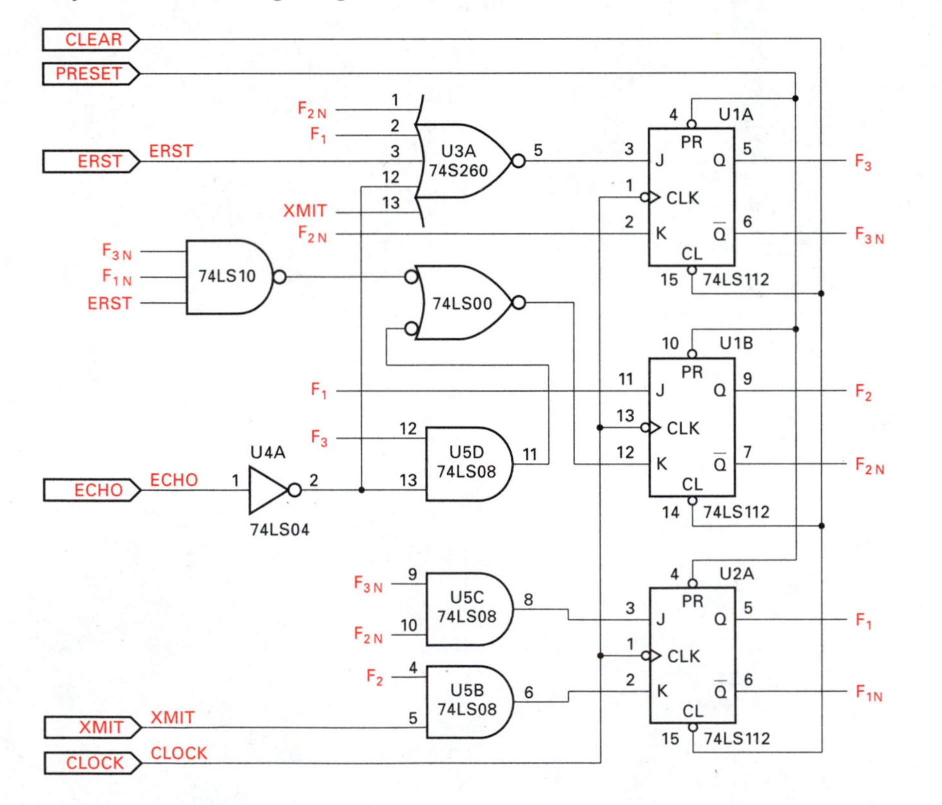

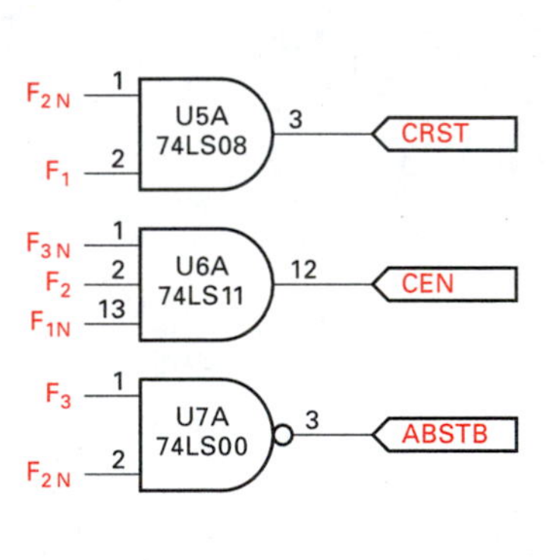

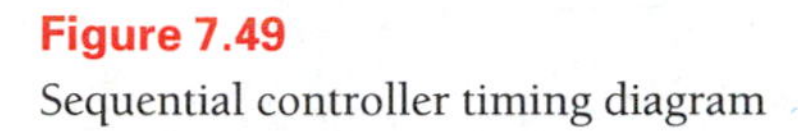

Figure 7.49
Sequential controller timing diagram

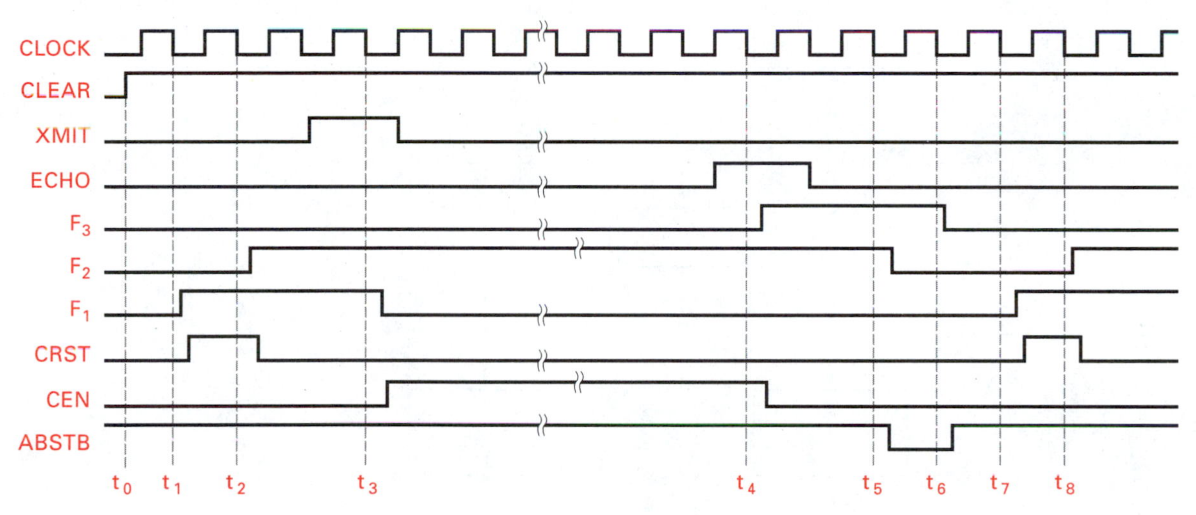

The timing diagram for the sequential circuit is shown in Figure 7.49. The timing diagram was obtained by simulating the circuit using a software applications program called an ORCAD VST logic simulator. The clock frequency in the timing diagram is 10.000 MHz. A break in the timing indicates a time compression between the two portions of the diagram.

The actual simulation time between the transmit and echo pulses for this example was 10^5 ns, representing a count of 100_{10}. At time t_0 the active-low clear signal reset the flip-flops to 000 (state A). The circuit is in state A until the first negative edge of the clock after being reset, then it goes to state B (001) at time t_1. The time from the negative clock edge until F_1 goes high is the delay time of the low-power Schottky (LS) flip-flop.

The clear count signal, CRST, becomes active during state B. At time t_2 the machine changes to state C (011) and CRST goes inactive. The circuit remains in state C (011) until the XMIT input is active. The next clock negative edge after XMIT becomes active, t_3, causes a transition to state D (010) and output CEN (count enable) goes active. CEN gates the 1.000 MHz square wave into 16-bit counter. ABSTB′ goes low at t_5 to transfer the 16-bit count value into the 8155 I/O device in preparation for transfer to the system microprocessor. After ABSTB′ goes low the sequential machine goes back to state A to begin a new transmit-echo time measurement. ❖

7.6 LINKED SEQUENTIAL MACHINES

Sequential machines become **linked sequential machines** when the outputs of one circuit become the inputs to a second circuit, and the outputs of the second circuit become the inputs to the first circuit. Several types of circuit linking have been investigated by researchers (interface, interactive, interpretive, array, and software).

Interface linking occurs when some, but not necessarily all, of the states in one ma-

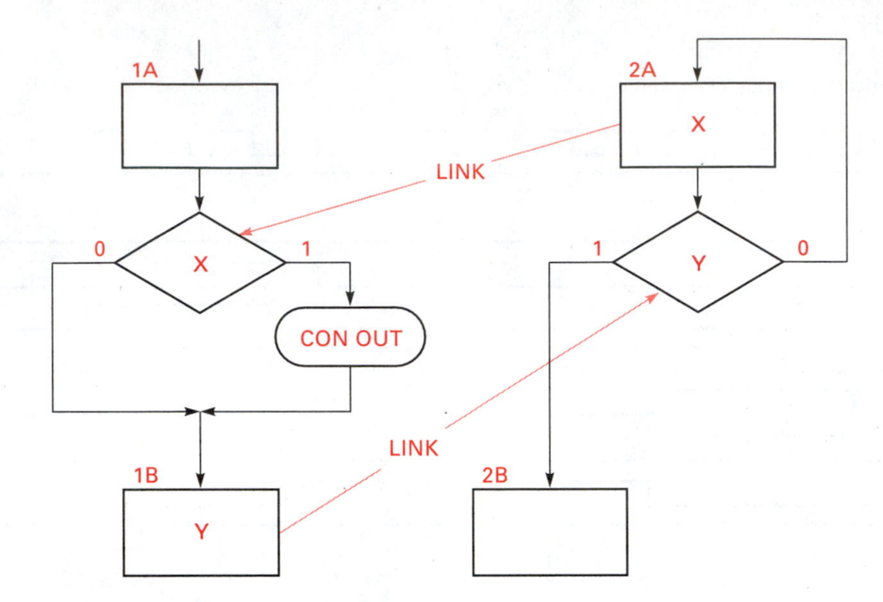

Figure 7.50
Serial linked ASM charts

chine are linked to states of another machine. Sequential circuits may be interfaced linked either serially or in parallel. Figure 7.50 and Figure 7.51 illustrate serial and parallel linked interface sequential machine ASM charts.

Serial interface linked sequential machines are so named for the serial nature of the state changes. Output *X* from state 2A provides the input for conditional state transition from 1A to 1B. State 1B produces output *Y*, which provides the input for the conditional state transition in machine 2. The parallel interface link has interconnected inputs and outputs as illustrated in Figure 7.51. Notice that each conditional state transition is dependent on the output from the other sequential machine. Two linked sequential machines may have some states that are serial and some that are parallel linked.

Iteratively linked sequential machine examples are found in serial adders, subtractors, and comparators. Recall the iterative nature of the comparator design of Chapter 4. A logic problem that can be partitioned into identical cells, with the outputs of one cell providing inputs to the next cell, is said to be *iterative*. If the cells contain sequential machines, then the machines are linked iteratively. Another example of an iterative sequential machine is a shift register.

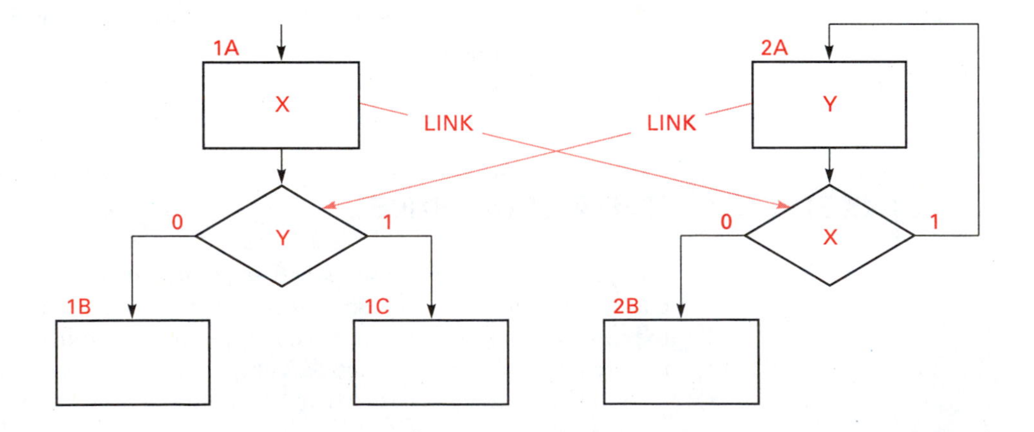

Figure 7.51
Parallel interface linked ASM charts

Interpretative and software-linked sequential machines are beyond the scope of this text (see Clare, *Designing Logic Systems Using State Machines*). Basically, interpretative linked machines are complex systems in which one machine interprets the operation of the other. This can be done by assigning levels of operation to a complex process algorithm so that the results of one level are passed on to another. The higher levels interpret the actions of the lower levels.

Software encodes the solution of a problem in a computer language. Most problem statements can be realized in software or hardware. The use of ASM charts allows the designer to describe an algorithm and then seek either a hardware or a software solution. Many times complex systems requires the linking of modules, where part of an algorithm is performed in hardware and part in software. The hardware may generate inputs needed by the software, and the software generates variable values needed by the hardware. In this fashion they are linked. This type of linking occurs quite often in microprocessor-based control systems.

7.6.1 Computer Simulator and Graphic Plotter Interface, a Linked Sequential Machine Design Example

❖ EXAMPLE 7.7

An interface is needed to connect a computer with a graphic plotter. The plotter draws diagrams based on digital information sent to it by the computer, using 16-bit characters to generate plot vectors (pen up, pen down, direction of carriage movement, and how far the carriage is to move). The plotter is only one of several input-output devices to be connected to the computer; therefore, some means is needed to distinguish the plotter from all the other devices. Other "housekeeping" signals are also necessary to control the plotter and computer activities.

A sequential machine must be designed to provide the necessary interface control activities. A second sequential machine will be designed to simulate the plotter data control activities of the computer so that the interface can be debugged and tested offline from the computer. After the interface is working properly, it will be connected to the computer. The computer simulation sequential machine must provide the same control functions and timing as the computer. A block diagram of the two linked sequential machines is shown in Figure 7.52.

SOLUTION The 16-bit data bus transfers data from the computer to the plotter interface, where it is stored in a buffer memory. The memory design will not be considered in this example problem. Some means, however, must be used to keep track of how many data words are sent to the buffer memory. This means that the interface must have some means of counting data transfers from the computer. When the buffer memory is full, the computer is free to disconnect from the plotter interface and service other input-output units. After the plotter interface has transferred the contents of its buffer memory to the plotter, it is ready to receive more data from the computer. To do this, the interface generates an interrupt signal, informing the computer it is free to receive more plot data. A complete list of terms and the mnemonics used for control signal variable names will be given shortly.

Figure 7.52
Computer simulator-plotter interface system block diagram

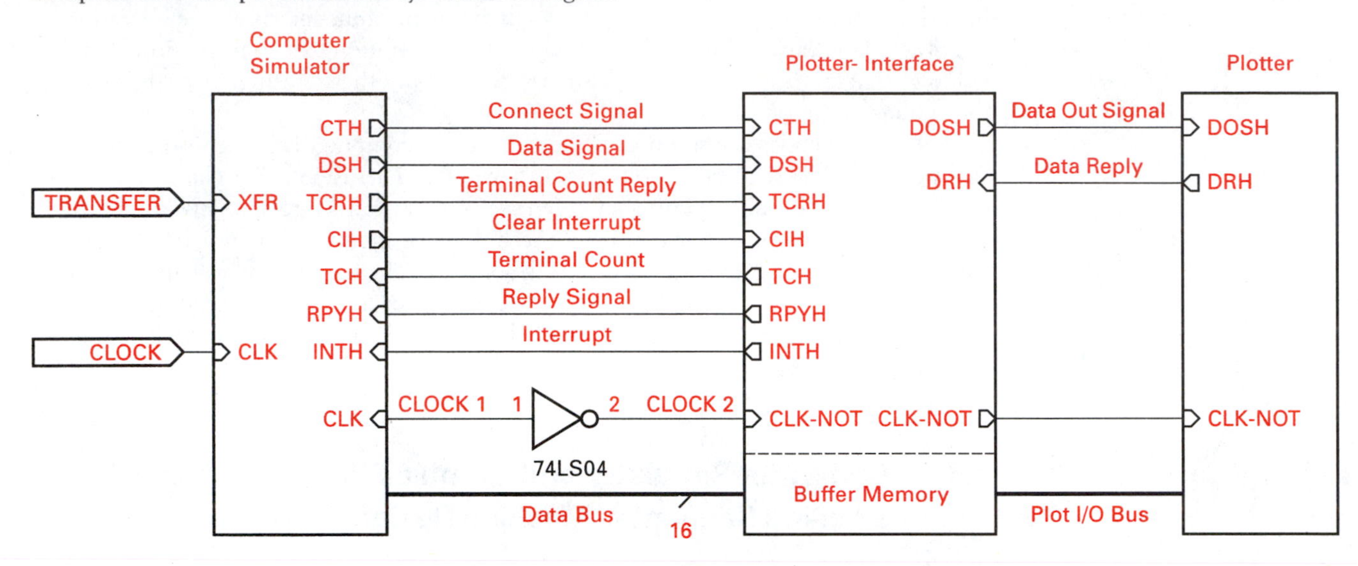

Initiation of data transfer is made by assertion of the transfer (XFR) signal. When XFR goes high, the computer simulator sequential machine must generate a connect signal (CTH). The connect signal (CTH) is received by the plotter interface, which generates a reply signal (RPYH). The interaction between two signals as indicated in Figure 7.53 is called *handshaking*. The term *handshaking* applies when two sequential machines are linked so that the output of one is the input to the second and the output of the second is the input to the first. Notice that the positive edge of the CTH signal causes the low-to-high transition of RPYH; the positive edge of RPYH causes the high-to-low transition of CTH; and the negative edge of CTH causes the negative edge of RPYH. These two signals provide part of the linking between the computer simulator and the plotter interface sequential machines.

A "connect code" must be presented to the plotter interface over the data bus coincidentally with the connect signal to identify the plotter, and not some other I/O device, as the recipient of the data transfer. The actual connect code value is not important for our discussion. Figure 7.54 shows the logic for identifying a proper connect code that is coincident with the connect signal.

The definitions for the control signals indicated in Figure 7.52 are as follows.

Transfer (XFR) Transfer is an active high input to initiate the transmission of data from the computer simulator to the plotter interface. This signal initiates calling the plotter driver subroutine software that begins transfer of data stored in the computer memory to the plotter memory.

Figure 7.53
Connect-reply signal handshaking

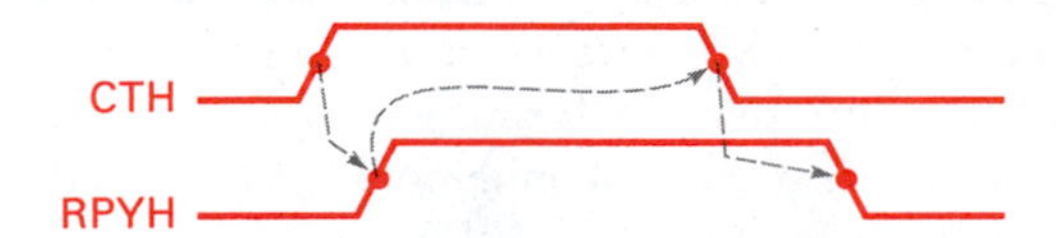

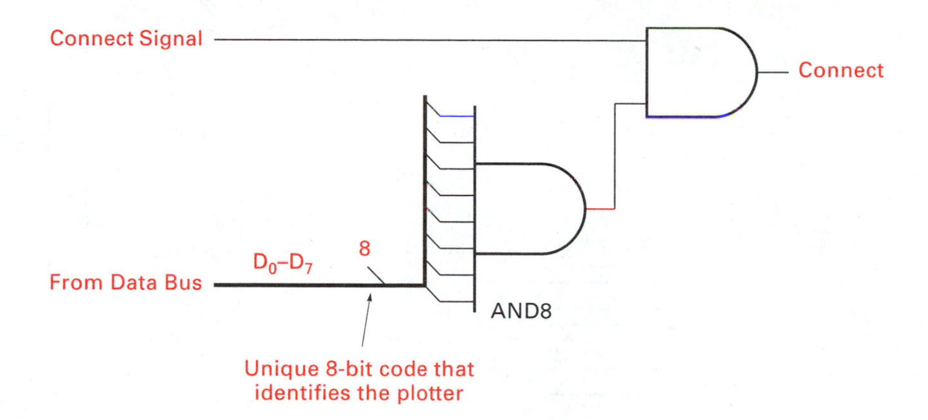

Figure 7.54
Connect code logic

Clock This is a system clock that is common to both sequential machines. The complement of the computer simulator clock is presented to the plotter interface and plotter. This provides a time delay between the two linked sequential machines.

Data signal (DSH) This is an active high output from the computer simulator informing the plotter interface that the data contain information to be written into the buffer memory. Each data signal corresponds to the transmission of a data word.

Reply (RPYH) This is an active high output from the plotter interface indicating reception of the connect signal (CTH) or the data signal (DSH). The reply signal handshakes with the connect signal (CTH) and the data signal (DSH).

Terminal count (TCH) This is an active high output from the plotter interface that indicates that its buffer memory is full.

Terminal count replay (TCRH) This is an active high output from the computer interface that handshakes with the terminal count (TCH).

Interrupt (INTH) This is an active high signal from the plotter interface informing the computer simulator that the buffer memory is empty and a new block of data can be transmitted.

Clear interrupt (CIH) This is an active high handshake response to the interrupt signal (INTH). The clear interrupt signal (CIH) ends a complete data transfer operation. The system repeats a data transfer operation when a new transfer signal (XFR) goes active.

Data out signal (DOSH) This is an active high output from the plotter interface to the plotter, indicating that a 16-bit data word is present on the plot I/O bus.

Data Reply (DRH) This is an active high handshake response by the plotter to the data out signal (DOSH). Each plotter interface to plotter data transfer requires a DOSH to DRH handshake operation.

Data Word Counter

A data word counter requirement is implied by the buffer memory feature. The length of the buffer memory establishes the counter length. The counter must be able to count up and down (increment and decrement) to fill and empty the buffer memory. We will use a 74LS191 4-bit up-down counter integrated circuit for this function. The plotter

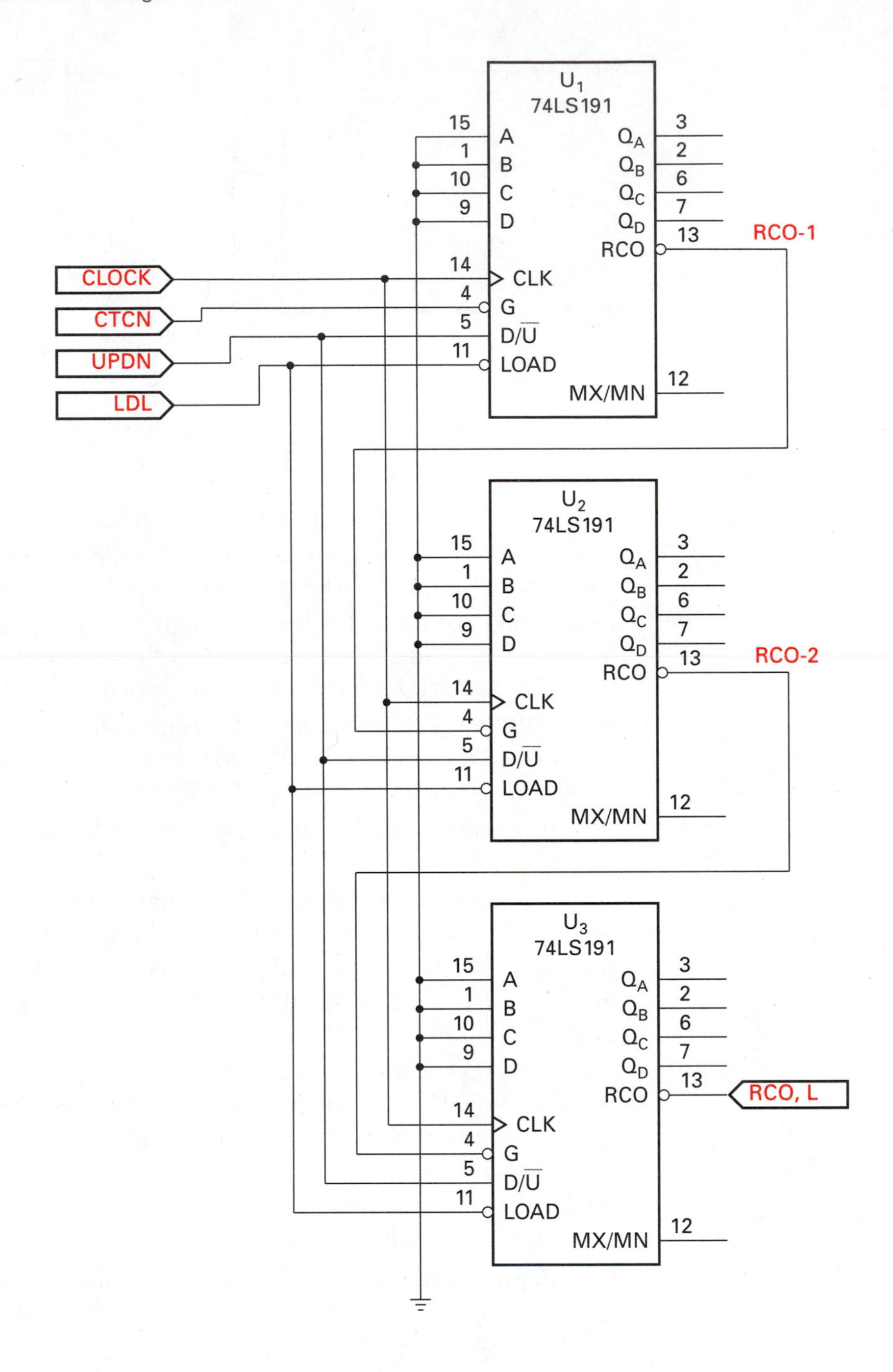

Figure 7.55
A 12-bit up-down binary modulo-4,096 counter

interface sequential machine must link with the counter. Figure 7.55 shows three 74LS191 counters connected to produce a modulo-4,096 counter. A counter of this size would permit a buffer memory of 4,096 or 4K words. Figure 7.56 shows the timing diagram for the 12-bit counter.

The 74LS191 IC counter uses a positive edge trigger clock. The device can load parallel data and then count up or down from the preset input, depending on the status of the mode control input D/U′. An active-low load signal, LOAD′, presets the count

Figure 7.56
12-bit counter timing diagram

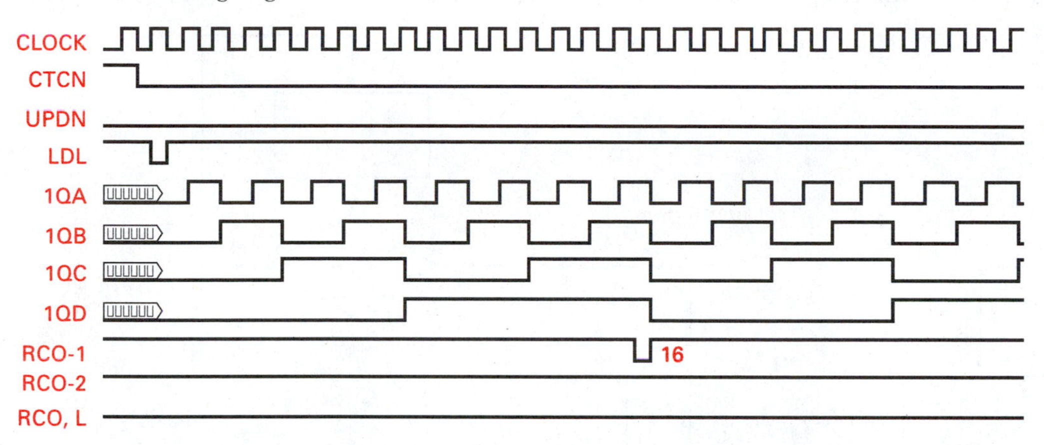

value. Notice that all of the parallel data inputs in Figure 7.56 are connected to ground. When LDL′ (LOAD′ signal for the counter) goes low the counter is preset to 0.

An active-low count-enable input, G′, permits counting. Each flip-flop output is available and a terminal count output, RCO, indicates when the modulo-*N* number is reached. The devices can be cascaded by connecting RCO of one device to the G input of another.

To simulate the design and produce a timing diagram of system signals, a single 74LS191 was used, which limited the buffer memory size to 15 words. This was done for simplicity; expansion to any reasonable count size is possible.

Linking the interface sequential machine and the counter is necessary. The following are interface-to-counter control signal definitions:

Load counter (LDL) This is an active low parallel load signal. It is used here to clear the counter by loading the 0s on the parallel data input lines, because all the parallel input data lines are grounded.

Up-Down mode control (UPDN) This is a counter input signal to control the count direction. When UPDN = 0, the counter counts up; when UPDN = 1, it counts down.

Count-enable (CTCN) This is an active-low counter input to enable counting.

Terminal count (RCO, L) This is an active-low counter output, indicating that the terminal count has been reached. The terminal count when counting up is 16, and when counting down it is 0.

The computer simulator and the plotter interface must operate together. If one sequential machine gets out of sync with the other, the system will hang up. Figure 7.57 illustrates a partial ASM chart for the XFR initiating a transfer operation and the connect-reply handshake.

The ASM charts for the computer simulator and the plotter interface are shown in Figures 7.58 and 7.59. Both ASM charts are shown in their final forms, including the state assignments for the various states. ASM charts are not born full grown, they take time to develop from the initial ideas and requirements to a finished diagram. In fact,

Figure 7.57
Partial ASM chart showing linking between the computer simulator and the plotter interface

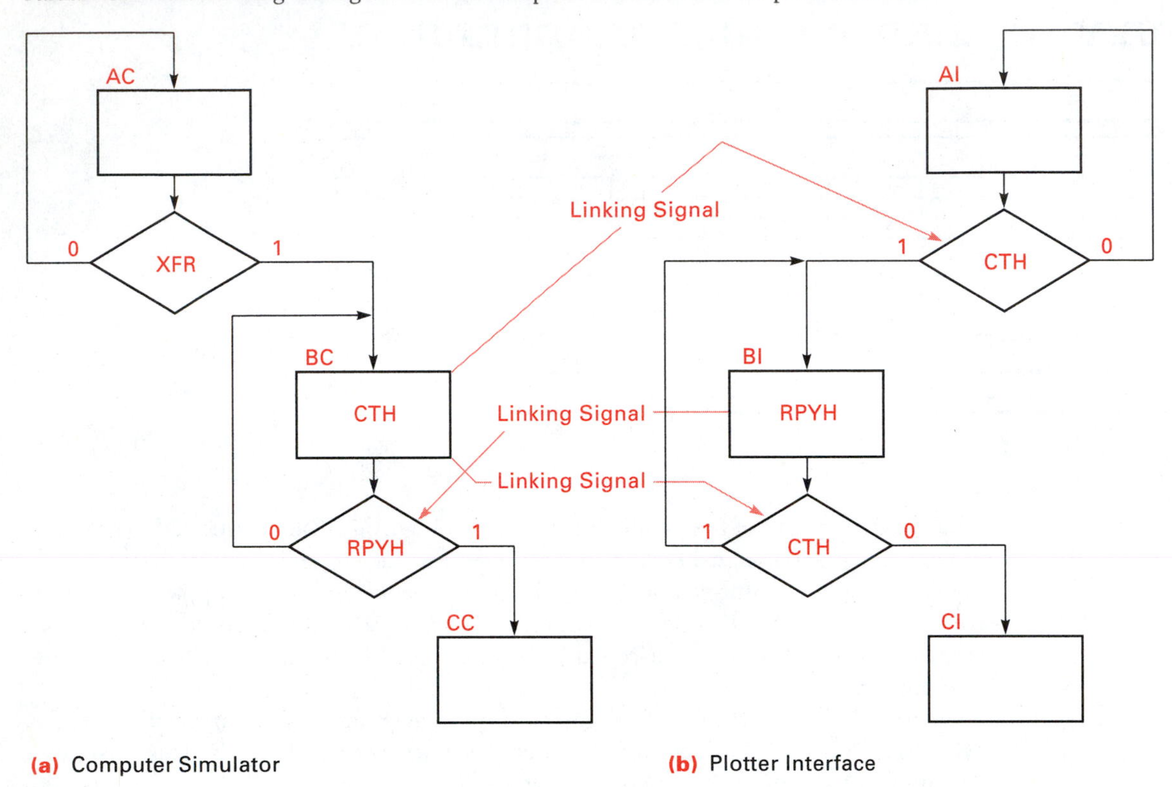

many times the ASM chart is not finished until the sequential circuit is fully tested and functional. Designing sequential machines is an iterative process.

We start the design process with the computer simulator sequential machine. Redundant states are not an issue in this example, because the two linked circuits must operate together. In fact, some linked designs require that states be added to synchronize the timing.

The computer simulator ASM diagram has seven states requiring three state variables. It also has four input variables and four unconditional output variables. A state table can be constructed to determine adjacent states; such a table requires 16 next-state columns. We can also apply the state assignment rules to the ASM chart directly and bypass the need for a state table.

Applying rule 1, we get

{A, G} 4X; {C, D} 6X; {F, G} 4X; {E, F} 4X; {B, D} 4X; {C, E} 2X;
{A, B} 4X; {B, C} 4X

Applying rule 2, we get

{A, B} 8X; {B, C} 8X; {D, E} 8X; {C, D} 8X; {E, F} 8X; {F, G} 8X;
{A, G} 8X; {C, E} 8X

Figure 7.58
Computer simulator ASM chart

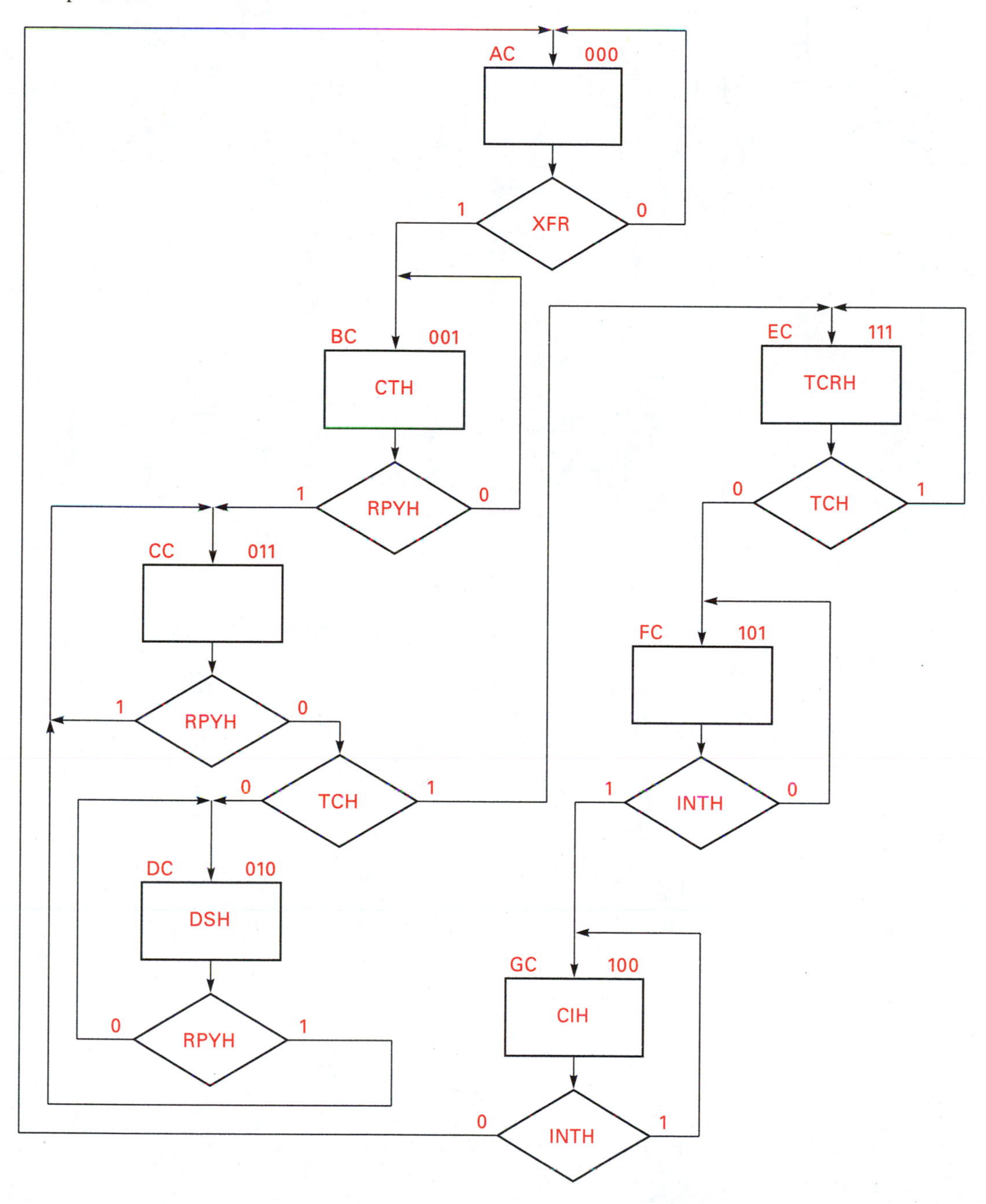

Several state assignments are possible from the adjacency list. Figure 7.60 illustrates one possible state assignment for the computer simulator sequential machine.

The state variables for both machines will be realized using *J–K* flip flops. Map-entered variables will also be used to reduce the K-map size. For both sequential ma-

Figure 7.59
Plotter interface ASM chart

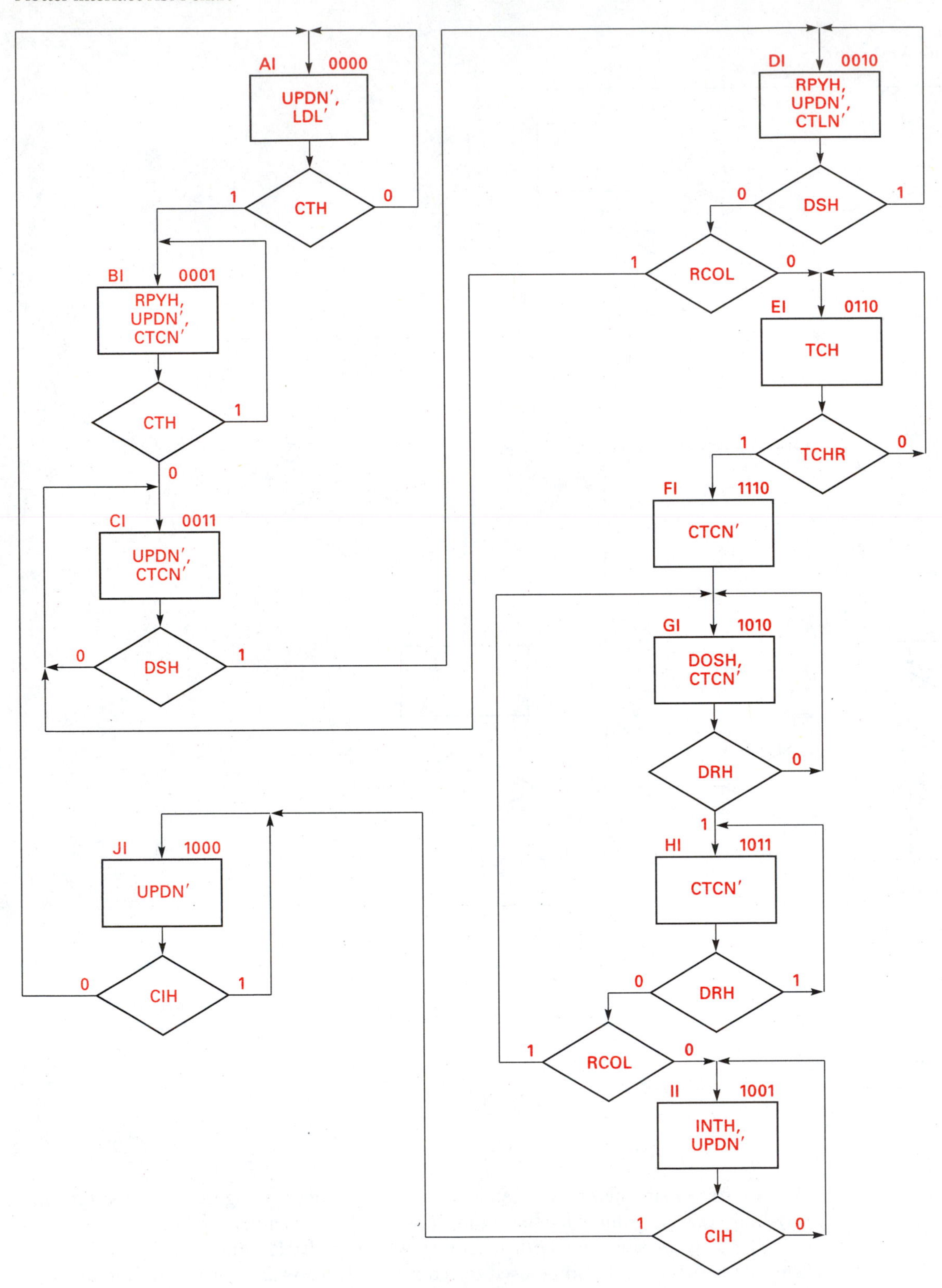

Figure 7.60
Computer simulator sequential machine state assignment and K-map

State	F_3A	F_2A	F_1A
AC	0	0	0
BC	0	0	1
CC	0	1	1
DC	0	1	0
EC	1	1	1
FC	1	0	1
GC	1	0	0

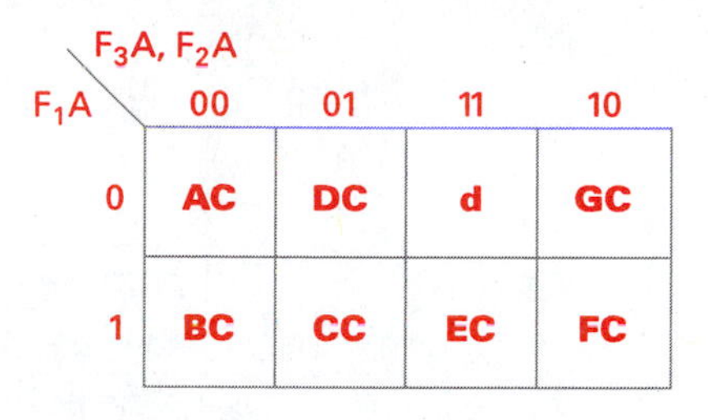

chines we will use a K-map size determined by the number of state variables and apply the input variables as MEVs. For the computer simulator this means three-variable K-maps. Each excitation K-map is loaded directly from the ASM chart. We do this by carefully finding all of the 0-to-1 state variable transitions when loading the J K-map and all of the 1-to-0 state variable transitions when loading the K-map. The excitation K-maps are shown in Figures 7.61–7.65.

Figure 7.61
Computer simulator F_3A K-maps

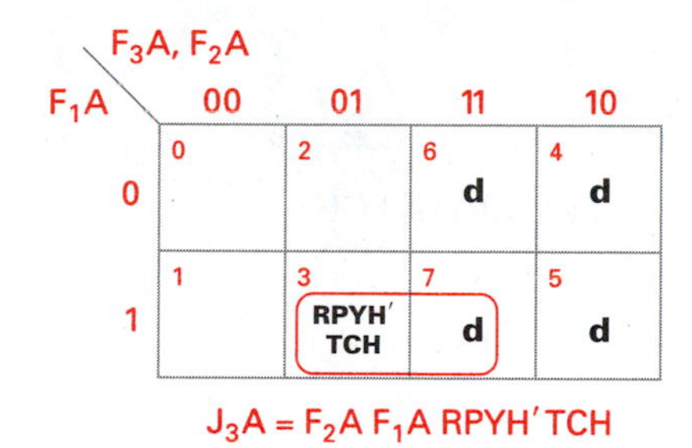

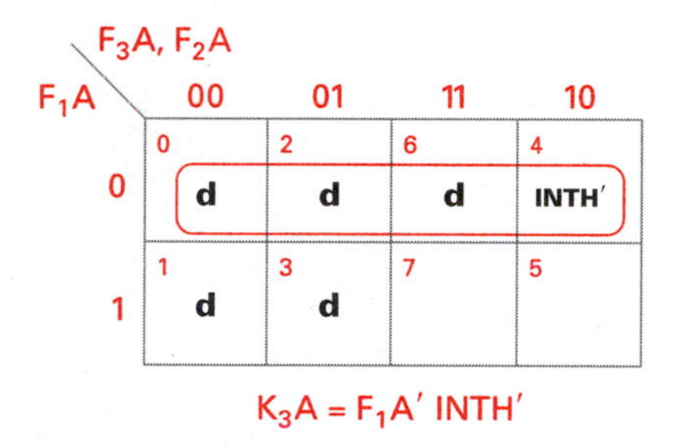

Figure 7.62
Computer simulator F_2A K-maps

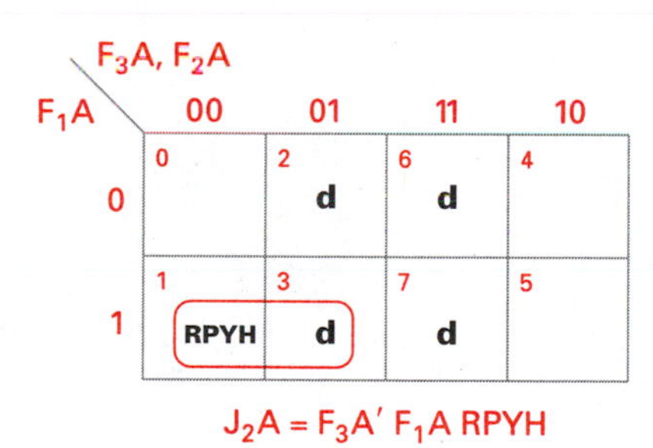

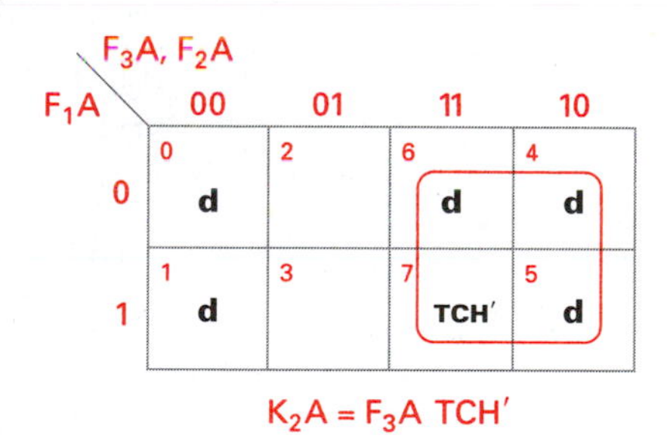

Figure 7.63
Computer simulator F_1A K-maps

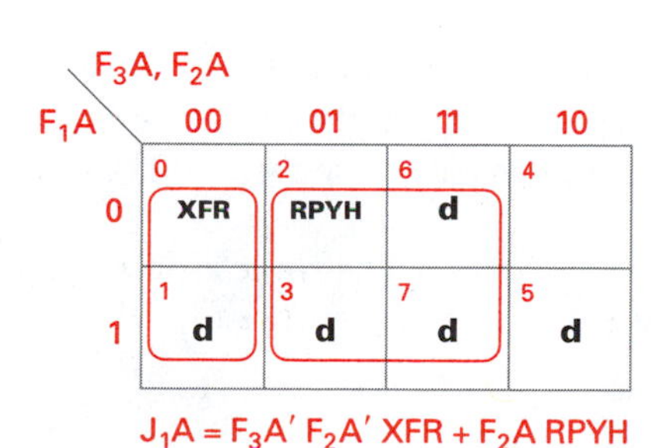

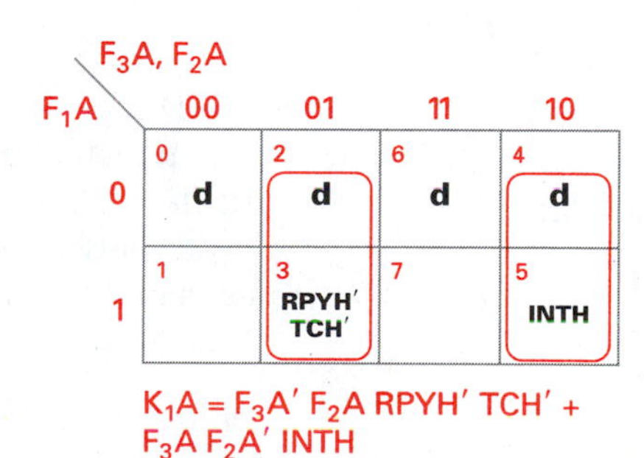

Figure 7.64
Computer simulator CTH and DSH K-maps

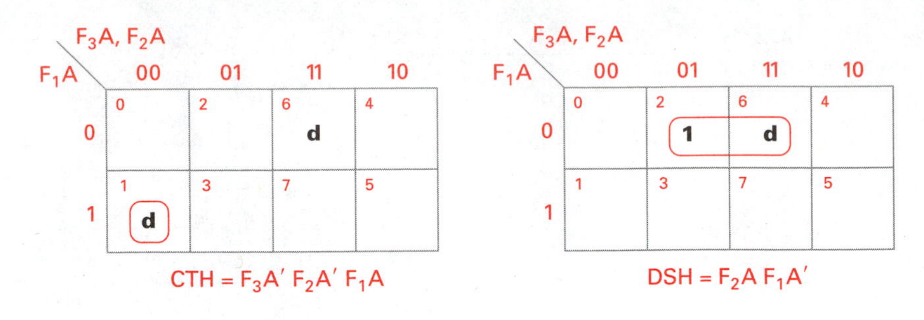

Figure 7.65
Computer simulator TCRH and CIH K-maps

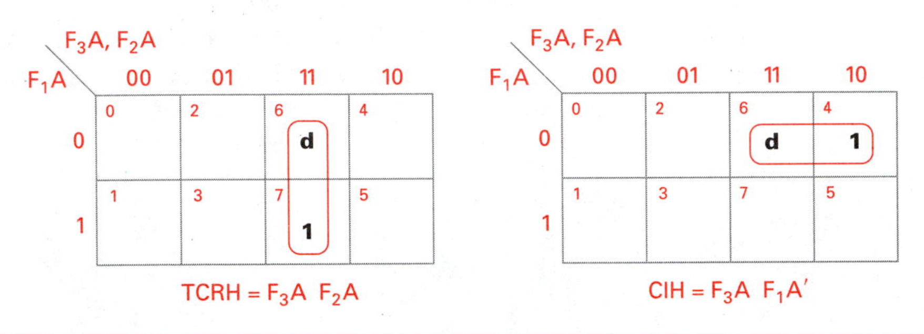

The excitation and output function equations are

$$J_3A = F_2A\ F_1A\ RPYH'TCH$$
$$K_3A = F_1A'\ INTH'$$
$$J_2A = F_3A'\ F_1A\ RPYH$$
$$K_2A = F_3A\ TCH'$$
$$J_1A = F_3A'\ F_2A'\ XFR + F_2A\ RPYH$$
$$K_1A = F_3A'\ F_2A\ RPYH'\ TCH' + F_3A\ F_2A'\ INTH$$
$$CTH = F_3A'\ F_2A'\ F_1A$$
$$DSH = F_2A\ F_1A'$$
$$TCHR = F_3A\ F_2A$$
$$CIH = F_3A\ F_1A'$$

The resulting logic diagram is shown in Figure 7.66. The complemented variables on the logic diagram are indicated with an *N* suffix; for example, the excitation equation for K3A is written

$$K3A = F_1AN\ INTHN$$

The reason for this is that the logic diagram was drawn and then simulated using ORCAD schematic capture and simulation software. The simulation software does not accept punctuation marks in a label, so the *N* suffix is used to indicate a complemented variable.

The plotter interface sequential machine is designed in the same fashion as the computer simulator. The state adjacency groups for rules 1 and 2 are

{A, B}; {B, C}; {C, D}; {D, E}; {E, F}; {F, G}; {G, H}; {H, I}; {I, J}; {A, J}

The resulting state assignment map and table are shown in Figure 7.67 and Table 7.30.

Figure 7.66
Computer simulator logic diagram

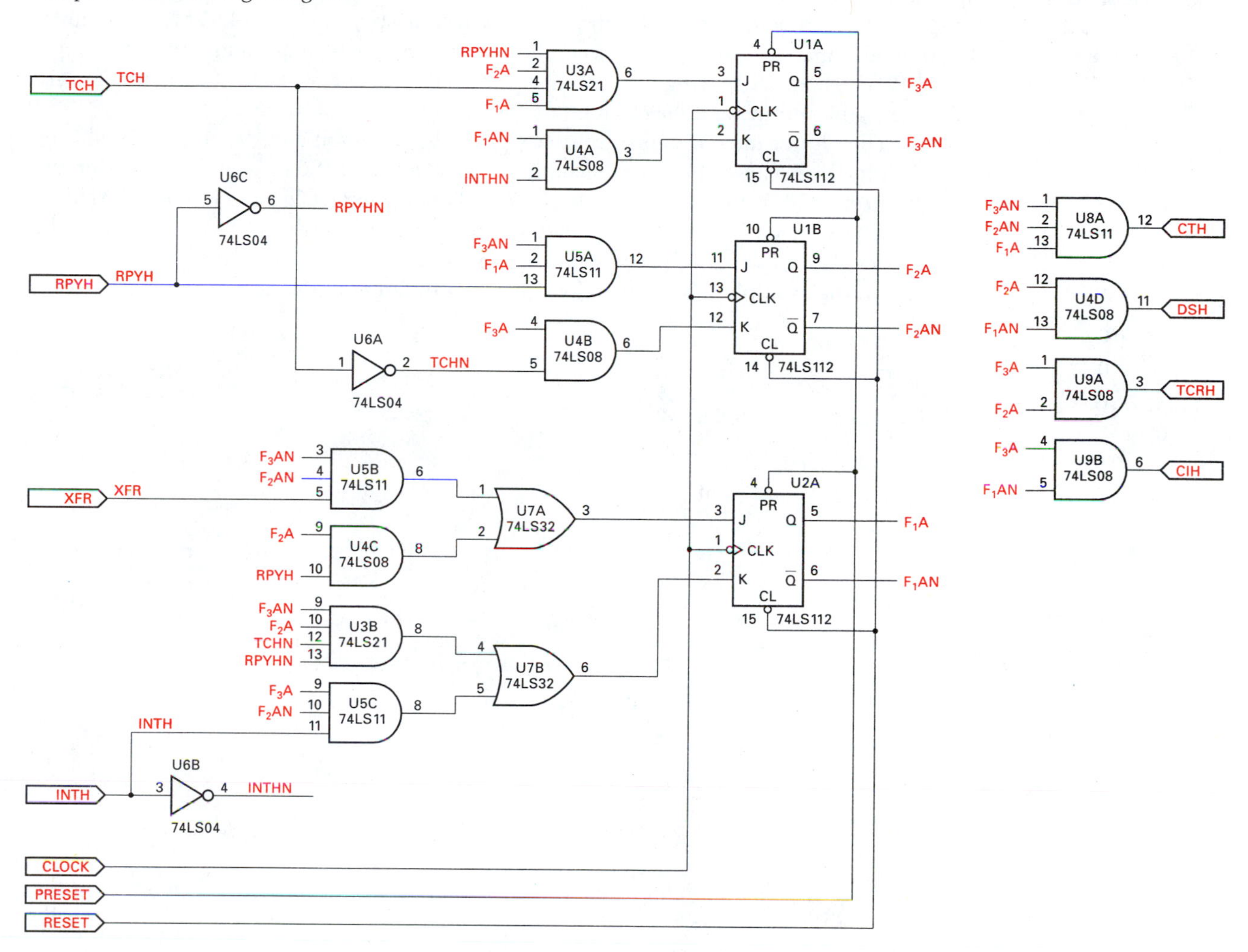

Figure 7.67
Interface state assignment map

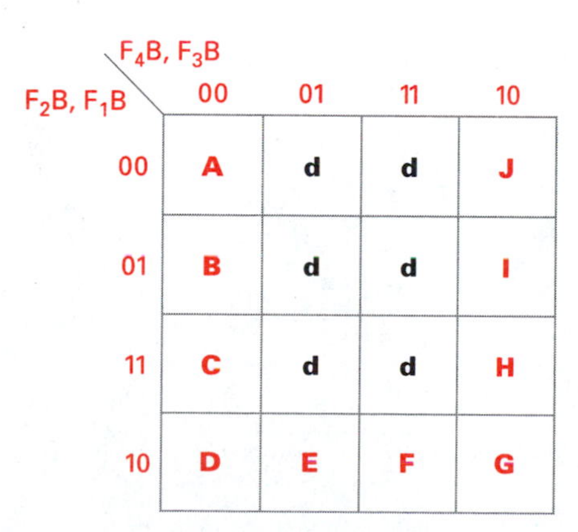

F2B, F1B \ F4B, F3B	00	01	11	10
00	A	d	d	J
01	B	d	d	I
11	C	d	d	H
10	D	E	F	G

TABLE 7.30
Interface state state assignment

State	F_4B	F_3B	F_2B	F_1B
A	0	0	0	0
B	0	0	0	1
C	0	0	1	1
D	0	0	1	0
E	0	1	1	0
F	1	1	1	0
G	1	0	1	0
H	1	0	1	1
I	1	0	0	1
J	1	0	0	0
d	0	1	0	0
d	0	1	0	1
d	1	1	0	0
d	0	1	1	1
d	1	1	0	1
d	1	1	1	1

From the ASM chart the excitation and output K-maps in Figures 7.68–7.75 were derived.

The data word counter for the system will consist of a single 74LS191. The counter will count data signals (DSH) when loading the buffer memory and count data reply (DRH) when transferring the contents of the buffer memory to the plotter. Figure 7.76 shows the counter and the clock input OR gate.

The plotter interface must transfer data to the plotter at a rate that is acceptable by the plotter. Because of the handshaking of the DOSH and DRH signals, the transfer rate is determined by the plotter's response. The plotter interface logic in Figure 7.77 shows the flip-flop (U19A) used to simulate the handshaking between the interface and the plotter. The clock to the plotter-control simulator D flip-flop is the same as to the interface in the example. The plotter clock may be a different frequency; connecting it to the

Figure 7.68
Plotter interface F_4B K-maps

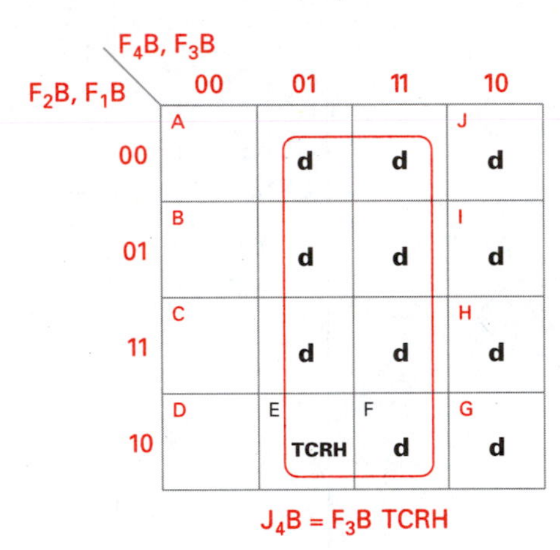

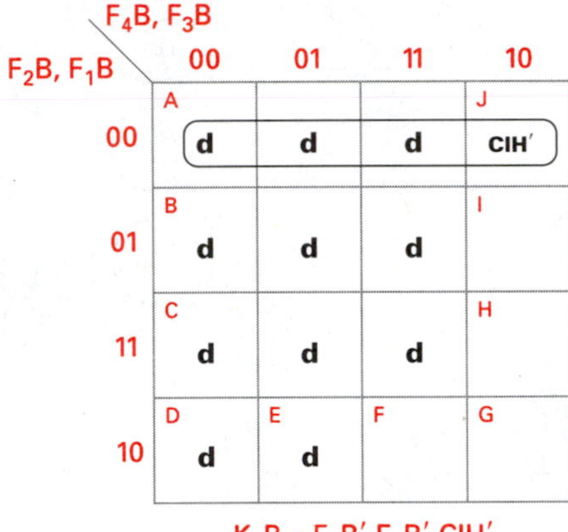

Figure 7.69
Plotter interface F_3B K-maps

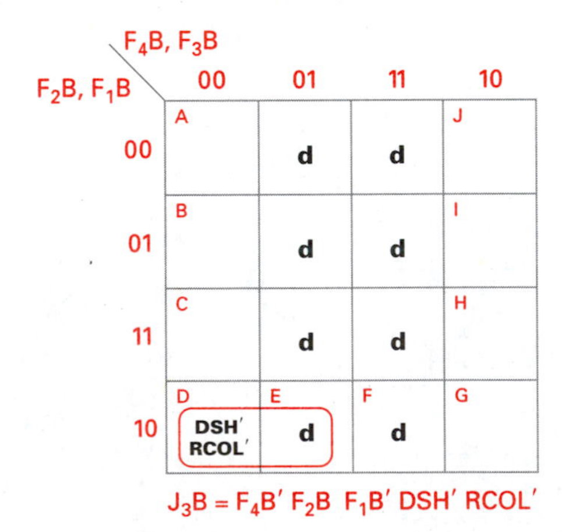

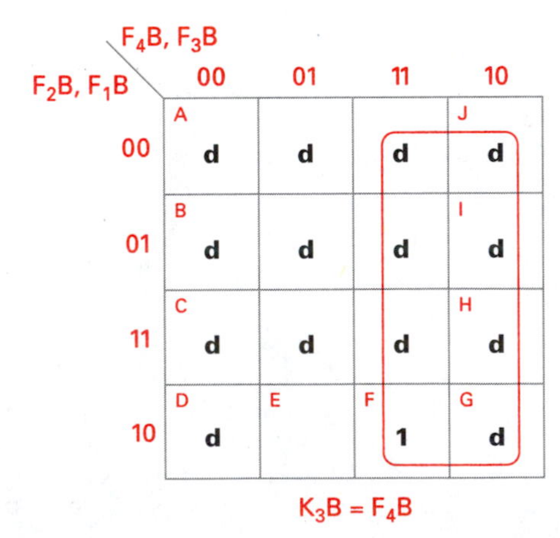

Figure 7.70

Plotter interface F_2B K-maps

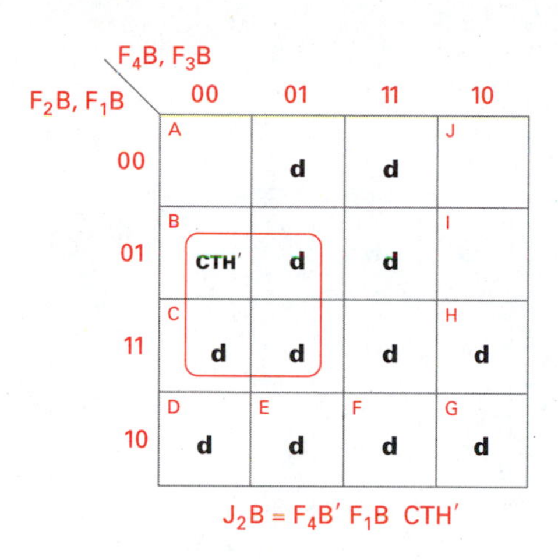

$J_2B = F_4B'\ F_1B\ CTH'$

Table text (the right-hand K_2B K-map; blank cells omitted):

F_2B, F_1B \ F_4B, F_3B	00	01	11	10
00	A d	d	d	J d
01	B d	d	d	I d
11	C	d	d	H DRH′, RCOL′
10	D	E	F	G

$K_2B = F_4B\ F_1B\ DRH'\ RCOL'$

Figure 7.71

Plotter interface F_1B K-maps

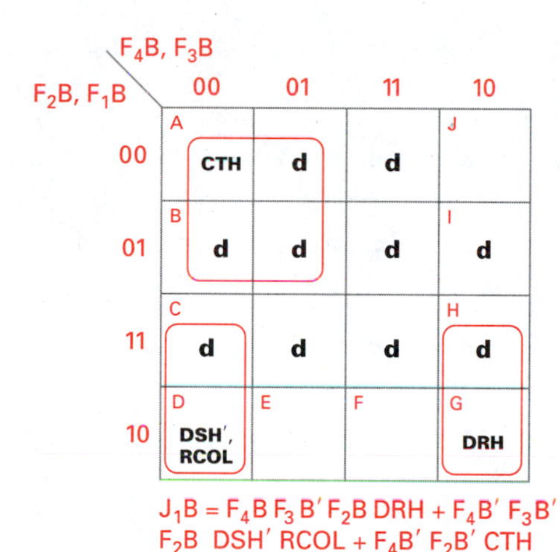

$J_1B = F_4B\ F_3B'\ F_2B\ DRH + F_4B'\ F_3B'\ F_2B\ DSH'\ RCOL + F_4B'\ F_2B'\ CTH$

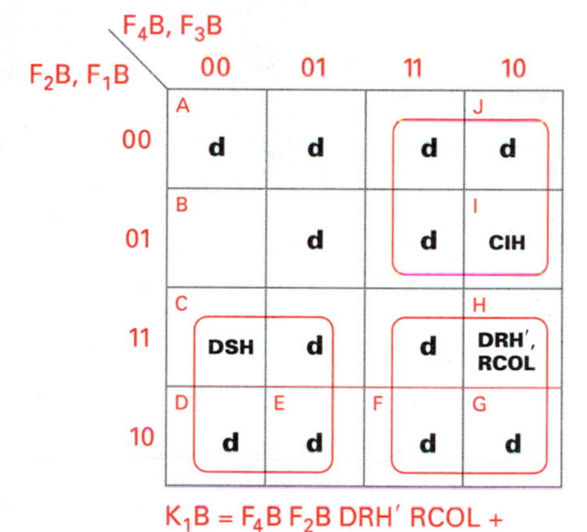

$K_1B = F_4B\ F_2B\ DRH'\ RCOL + F_4B\ F_2B'\ CIH + F_4B'\ F_2B\ DSH$

Figure 7.72

Plotter interface LDL and UPDN output K-maps

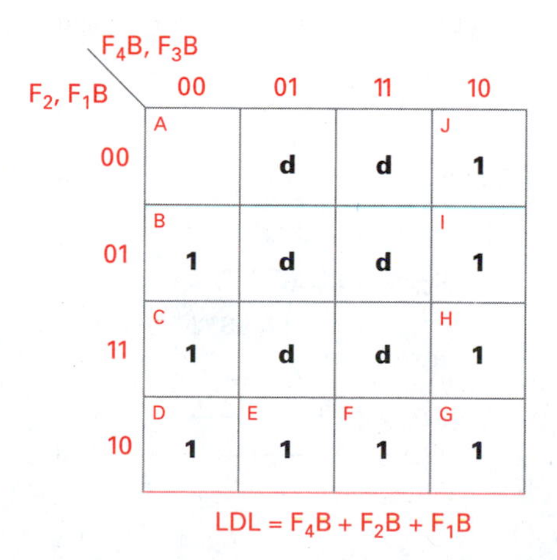

$LDL = F_4B + F_2B + F_1B$

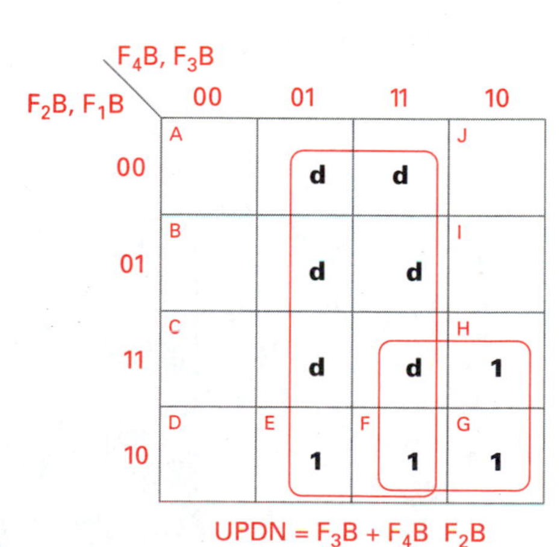

$UPDN = F_3B + F_4B\ F_2B$

Figure 7.73
Plotter interface DOSH and TCH output K-maps

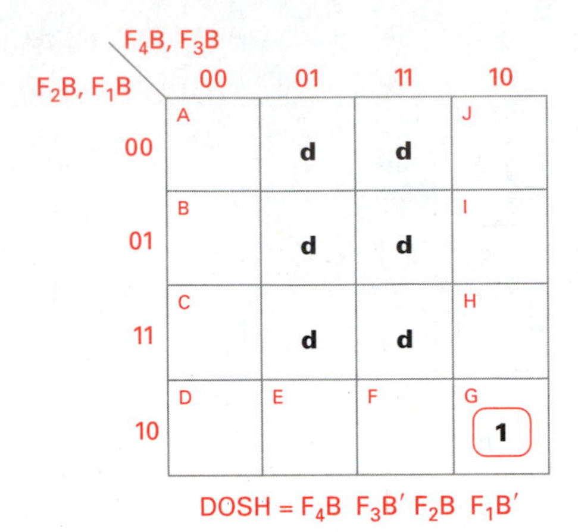

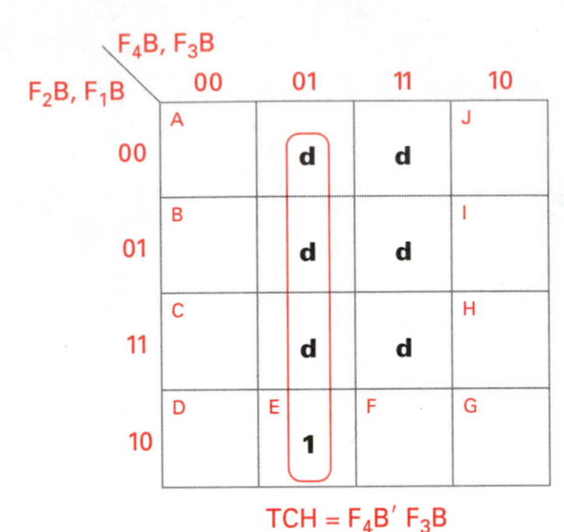

Figure 7.74
Plotter interface CTCH and INTH output K-map

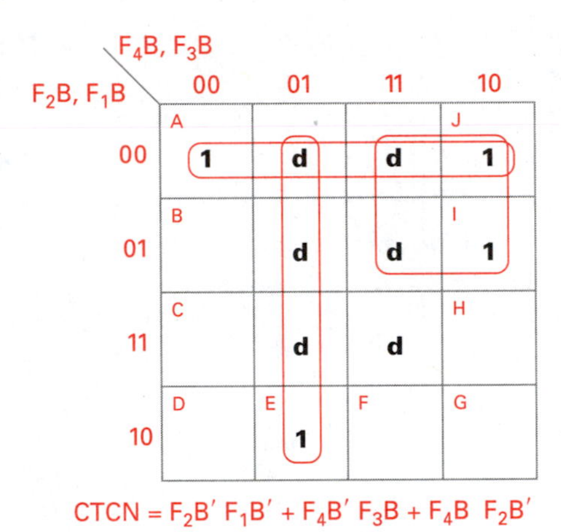

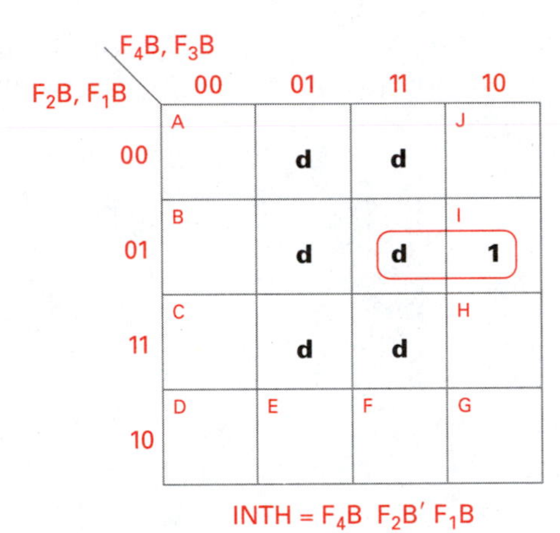

Figure 7.75
Plotter interface RPYH output K-map

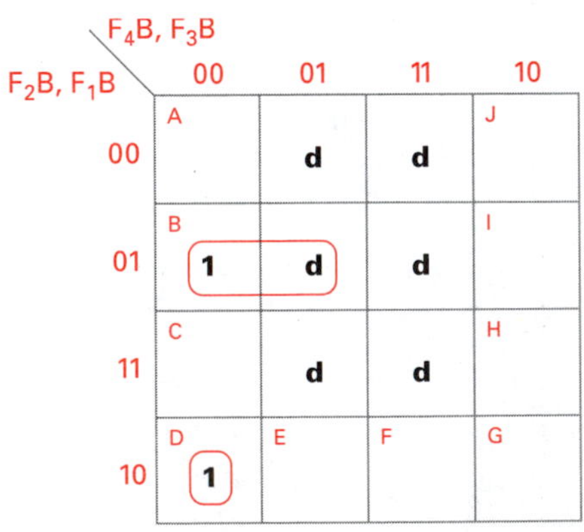

Figure 7.76
Data word counter for loading and unloading buffer memory

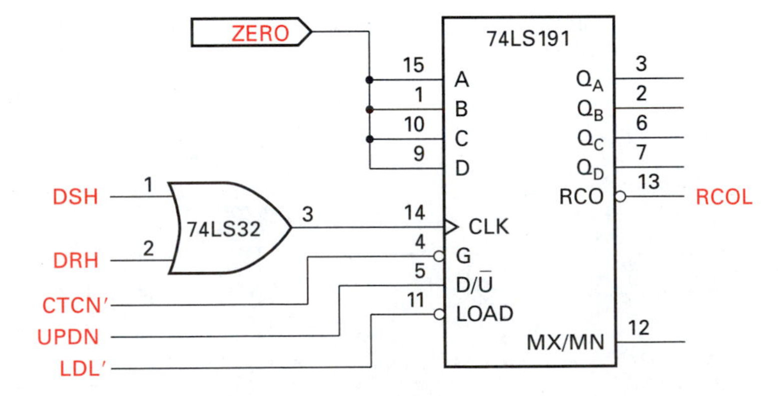

Figure 7.77

Plotter interface logic diagram

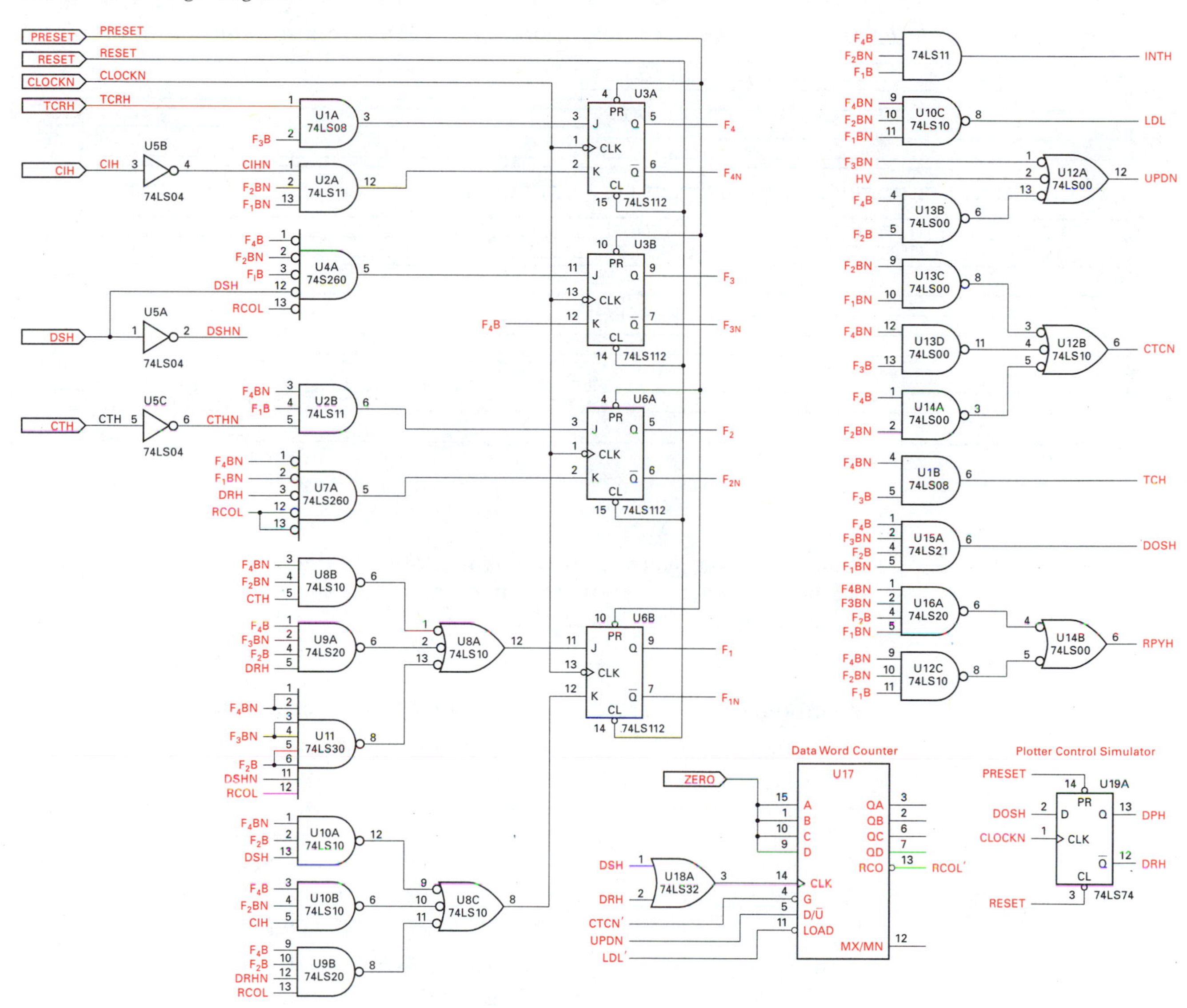

plotter interface clock was done simply for convenience, to illustrate the linking of the state machines.

The complete logic diagram for the plotter interface sequential circuit along with the data word counter and plot-control simulator is shown in Figure 7.77. The timing diagram for the complete operation is illustrated in Figure 7.78. The clock pulsewidth for the timing diagram is 200 ns. The plotter interface clock is inverted from the one shown in the timing diagram. This permits a 200 ns separation between a signal going active in one sequential machine and its reception by the other machine.

Note that 16 data signals are counted when loading the buffer memory and 16 data replies are counted when transfering data to the plotter. State variables for the computer

Figure 7.78
Computer simulator–plotter interface system timing diagram

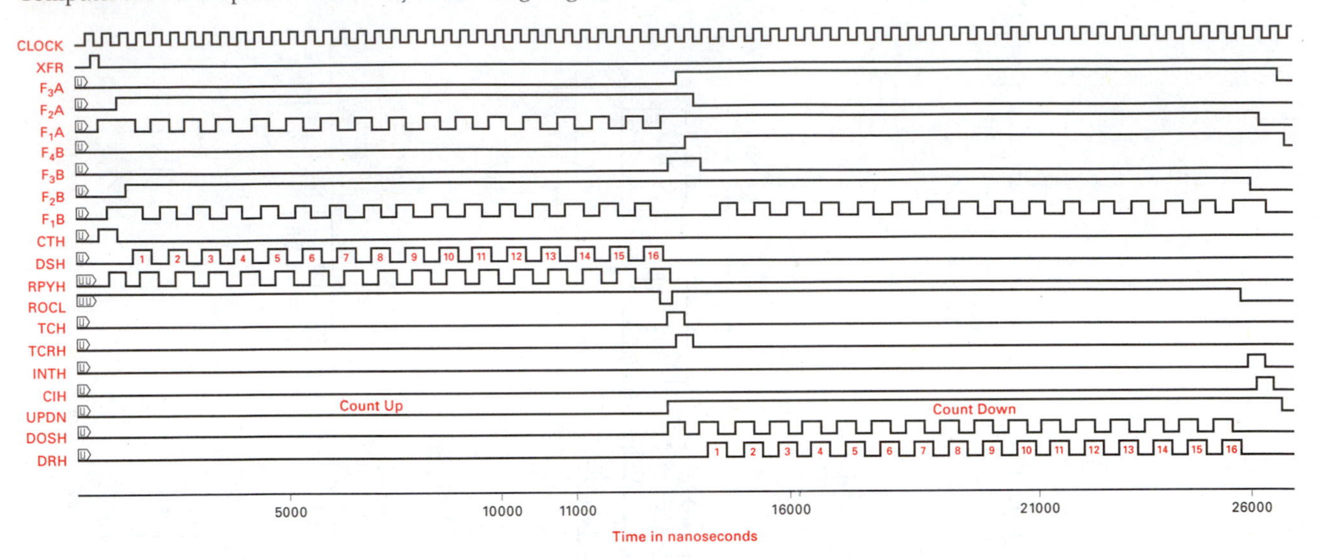

simulator and the plotter interface are included in the timing diagram. From the state variable information, the state transitions can be identified. Compare the ASM charts and the timing diagram; both provide information, in different formats, about the system behavior.

Maximum Clock Speed Calculations for Plotter Interface System

The maximum clock speed is limited by the longest propagation delay paths. By analyzing the logic diagram in Figure 7.66 (computer simulator) and Figure 7.77 (plotter interface), we can determine the longest delay path. Consider the computer simulator logic diagram. The longest delay path is from the RPYH input through U6C (LS04) to U3B (LS21) to U7B (LS32) to the K1A input of U2A (LS112). Output DSH requires an additional delay through U4D to produce output DSH. The total RPYH to DSH propagation delay is the sum of gate delays, flip-flop setup, and propagation delay.

U6C (LS04)	= 22 ns	Maximum propagation delay
U3B (LS21)	= 22 ns	Maximum propagation delay
U7B (LS32)	= 22 ns	Maximum propagation delay
U2A (LS112)	= 20 ns	Setup time
U2A (LS112)	= 20 ns	Excitation to Q output delay
U4D (LS08)	= 27 ns	Maximum propagation delay
	= 133 ns	TOTAL

The longest interface delay path is from the DSH input through U5A (LS04) to U11 (LS30) to U8A (LS10) to the J1B input at U6B (LS112). From U6B, the Q or Q' output is sent to U16A (LS20) or U12C (LS10) to U14B (LS00) where the RPYH output is produced. The total DSH to RPYH delay is the sum of the gate propagation, and the flip-flop setup, and propagation delays.

U5A (LS04)	= 22 ns	Gate delay
U11 (LS30)	= 22 ns	Gate delay
U8A (LS10)	= 22 ns	Gate delay
U6B (LS112)	= 22 ns	Flip-flop setup time
U6B (LS112)	= 20 ns	Flop-flop propagation delay
U16A (LS20)	= 22 ns	Gate delay
U14B (LS00)	= 15 ns	Gate delay
	= 143 ns	TOTAL

The maximum clock speed is determined by finding the minimum time between the negative clock edges. The two sequential circuits are both negative edge triggered, with the interface circuit using an inverted clock. This allows the computer sequential circuit to trigger on the negative and the interface to trigger on the positive clock edges. The linked nature of the two sequential machines, with the output of one becoming the input to the other, necessitates using twice the longest delay path to determine the time interval between clock edges. The reason for multiplying the slowest delay path by 2 is because the same clock frequency is used by both the computer simulator and the plotter interface. If separate clocks were used, the two delay values would simply add. For this example, and using worst case delay values, the minimum clock period time is 286 ns, resulting in a maximum clock frequency of 3.496 MHz.

Figure 7.79 shows a computer simulation of the delays produced through the logic discussed. In this case the 400 ns clock period allows ample time for circuit delay.

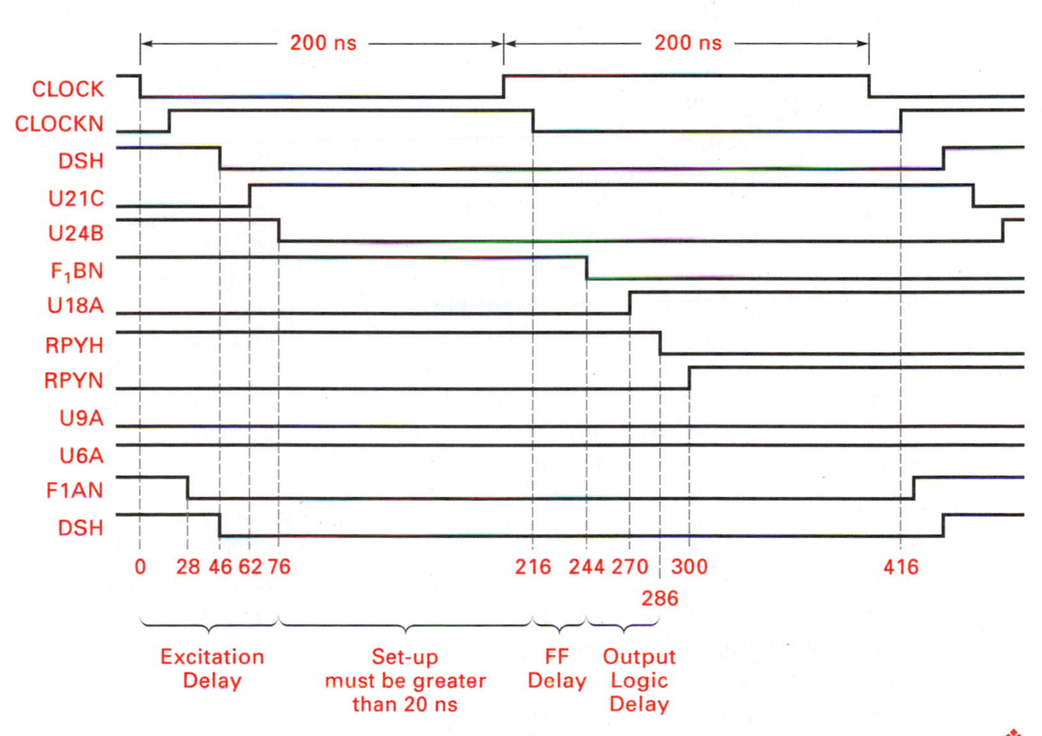

Figure 7.79
Computer simulator–plotter interface propagation delay timing diagram for determination of clock speed

❖

SUMMARY

Chapter 7 continues the discussion of sequential circuits that was started in Chapters 5 and 6.

A. State reduction techniques are presented.

1. The definition of equivalent states is introduced in preparation for reducing redundant states. Two states, S_A and S_B, are said to be equivalent if and only if, for every input sequence, the sequential circuit produces the same output sequence regardless of whether S_A or S_B is the starting state.
2. Two approaches to state reduction are presented.
   - **a.** A procedure for finding redundant states iteratively partitions the state table into equivalency classes of states whose output is the same for a given input. Each partition represents the classes of states whose outputs have been identical: 1-equivalency represents the state partition for a one input; 2-equivalency represents the state partition for two input changes; and n-equivalency represents the state partition for n input changes. The final partition classes occur when further input changes cause no further partitions. The states within the classes are said to be equivalent.
   - **b.** A graphical procedure for determination of redundant states uses an implication chart. The chart is constructed so that each state pair is represented by a square in the chart. The square represents a possible equivalency. State pair equivalency may be dependent on other state pairs also being equivalent. Conditions for state pair equivalency are written in the square. State pairs that cannot be equivalent are crossed off. After all nonequivalent state pairs have been eliminated from consideration the remaining state pairs are considered to be equivalent.
3. A state reduction technique for incompletely specified state tables is presented.
   - **a.** Compatible states are found by using an implication chart.
   - **b.** A merger graph is used as a graphical technique for finding state equivalencies in an incompletely specified state table: all maximal compatible state pairs are found using a compatibility class graph; unspecified next-states and outputs are merged with specified states as an aid in reducing the state table.

B. State assignment is explored.

1. The state assignment problem is presented:

   $S = (2^M - 1)!/[(2^M - N)!\,(M)!]$

   where S = the number of state assignments, M = the number of state variables, and N = the number of states.
2. A method of finding a "good" state assignment using three guidelines is given. States having the same next state for a given input should have adjacent state assignments. The next states of a single present state should have adjacent state assignments. Adjacent assignments would be given to states that have the same output.
3. A technique called an *implication graph* is given for finding the state assignment. The pairing of potential adjacent states is done using the guidelines method. Each pair of potentially adjacent states form a vertex in a diagram. If the input causes a change from one state pair to another, a line is drawn connecting the vertices. The graph is completed after all state pairs have been evaluated for determination of the next-state pair transitions. The implication graph is evaluated for "closed subgraphs," which consist of groups of state pairs that are contained within the connecting lines. Closed subgraphs contain state pairs that should be given adjacent state assignments.

C. Another state machine notation called *algorithm state Machines,* or ASM, is presented.

1. ASM notation uses three basic symbols. The ASM *state symbol* consists of a rectangle with an arrow entering from the top and leaving from the bottom. Written across the top of the rectangle are the state name and state assignment code. Unconditional or Moore output variables' names may be

written inside the rectangle. A *decision diamond* is used to facilitate branches. The branch conditions may be single variables or a Boolean expression written inside the diamond. A single arrow entering from the top is the entry point to the decision. Two exit paths exist, one from each side. One exit path represents the "true" and the other the "false" condition of the statement written inside the diamond. An *ellipse* or oblong is used to represent *conditional* or Mealy *output variables.* The output variable name(s) are written inside the ellipse.

2. Example ASM diagrams are presented.

D. Two design problems using ASM notation are presented.

1. Elapsed time measurement sequential controller.
2. A linked sequential machine consisting of a "computer" and a graphic plotter interface is designed.

REFERENCES

Introduction to Logic Design, by Sajjan G. Shiva (Glencoe, Ill.: Scott, Foresman and Company, 1988). Chapter 8 covers sequential state machines, including state reduction and state assignment.

Digital Logic Design, by Gideon Langholz, Abraham Kandel, and Joe L. Mott (Dubuque, Iowa: Wm. C. Brown, 1988). Chapter 6 covers reduction of incompletely specified state tables.

Fundamentals of Logic Design, 4th ed., by Charles H. Roth (Minneapolis–St. Paul: West, 1992). Unit 15 covers state reduction and state assignment. Unit 22 covers ASM notation.

Computer Aided Logic Design with Emphasis on VLSI, by Frederick J. Hill and Gerald R. Peterson (New York: Wiley, 1993). Chapter 11 covers ASM chart notation, state reduction, and state assignment problems.

Digital Logic and Computer Design, by Richard McCulla (Columbus, Ohio: Merrill, 1992). Chapter 8 introduces synchronous sequential circuit design.

An Engineering Approach to Digital Design, by William F. Fletcher (Englewood Cliffs, N.J.: Prentice Hall, 1980). Chapter 6 covers state reduction and state assignment techniques.

Designing Logic Systems Using State Machines, by Christopher R. Clare (New York: McGraw-Hill, 1973). The entire book is centered around the design of state machines using ASM notation.

Digital Logic Circuit Analysis and Design, by V. Nelson, H. Nagle, B. Carroll, and J. Irwin (Englewood Cliffs, N.J.: Prentice Hall, 1995). Chapter 9 covers state reduction and state assignment.

GLOSSARY

Algorithm state machine (ASM) A notation for describing a sequential state machine that resembles the notation used in flowcharting.

Compatibility class graph A graph constructed from the results of a merger graph. The compatibility class graph indicates the possible grouping of compatible states.

Compatible states Two states, S_A and S_B, are compatible if and only if, for every input sequence that effects the two states, the same output sequence occurs when the incompletely specified outputs are specified, regardless of which state is the starting state. A compatible state is very similar to an equivalent state except for the introduction of incompletely specified outputs.

Equivalence class A group of states that have the same output sequence for a given input sequence. By grouping or classifying states according to their output sequence for a given input sequence, we can eventually identify redundant states.

Implication chart The implication chart is a graphical tool used in eliminating redundant states from a state machine. The graph contains a grid array that lists the possible state equivalencies. A square at the horizontal and vertical intersection represents the possibility that a state pair may be equivalent. If a state pair cannot be equivalent due to differing outputs, an X is written in the array square that represents that state pair. If a state pair equivalency is dependent on other state pair equivalencies, then the dependent state pair(s) are written in the square.

Implication graph A flow graph used to aid in determining the assignment of adjacent binary codes for states. An implication graph is a flow graph with directed arcs indicating the state transitions of a sequential machine. Each node (i) on the graph represents a state adjacency pair. The directed arcs connect the nodes. Subgraphs consist of completely connected groups of nodes. All state pairs within a subgraph should be given adjacent state assignment codes.

Linked sequential machines Two sequential machines that share information in some way. State machines can be linked serially, in parallel, iteratively, or via software. Determination of the type of linking is made by how the two or more sequential machines share information.

Maximal compatibility A set of compatible state pairs is said to be maximally compatible if it is not completely covered by any other set of compatible state pairs. For example, a single state that is not compatible with any other state is a maximal compatible.

Merger graph The merger graph is a tool used to identify compatible state pairs in an incompletely specified state machine for state reduction. Each state is represented by a vertex. Compatible states have a connecting line drawn between them. Incompatible states have no connecting line. States whose compatibility is dependent on other compatible states have a broken line drawn with the dependent compatible state pair(s) written in the broken line.

***n*-equivalency** An equivalence classification is made by how many inputs have been made to a group of states whose outputs have been the same for a given input. When a single input has occurred, a 1-equivalency classification is made. An n-equivalency is when n inputs have occurred.

State assignment Sequential machine states are assigned binary codes. The process of making the assignment of the codes to the states is called making the *state assignment*. Usually the codes are assigned with purpose in mind. For synchronous sequential circuits that usually means finding a code assignment to minimize the logic required to realize the machine.

State reduction When the initial state diagram is made from a verbal problem statement it is easy to introduce redundant states. Sometimes these redundant states are obvious and can be eliminated with trial and error methods. Other times they are not obvious, and other techniques are needed. By eliminating redundant states the machine can accomplish its task with less logic and usually in less time. The process of eliminating redundant states is called *state reduction*. Often the implication chart is used as a tool in state reduction.

QUESTIONS AND PROBLEMS

Section 7.1 to 7.3

1. Reduce the following state tables using the equivalence class state reduction technique.

a.

Present state	Next state $x = 0$	Next state $x = 1$	Output $x = 0$	Output $x = 1$
S_0	S_2	S_1	0	1
S_1	S_3	S_1	0	1
S_2	S_0	S_3	1	0
S_3	S_1	S_2	1	0

b.

Present state	Next state $x = 0$	Next state $x = 1$	Output $x = 0$	Output $x = 1$
S_0	S_3	S_1	0	0
S_1	S_4	S_0	0	1
S_2	S_6	S_5	0	1
S_3	S_0	S_3	1	0
S_4	S_0	S_3	1	0
S_5	S_2	S_1	0	0
S_6	S_0	S_4	1	0

c.

Present state	Next state $X = 0$	Next state $X = 1$	Output $X = 0$	Output $X = 1$
S_0	S_4	S_2	0	0
S_1	S_2	S_0	0	0
S_2	S_1	S_6	0	0
S_3	S_6	S_0	0	0
S_4	S_5	S_1	1	0
S_5	S_4	S_3	0	0
S_6	S_3	S_6	0	0

d.

Present state	Next state/Output $xy = 00$	$xy = 01$	$xy = 10$	$xy = 11$
a	a/0	a/0	b/1	c/0
b	a/0	b/0	d/0	f/1
c	c/0	b/0	b/1	a/0
d	d/0	c/0	e/1	c/0
e	a/0	e/0	b/1	c/0
f	e/0	e/0	f/0	f/0

e.

Present state	Next state $XY = 00$	$XY = 01$	$XY = 10$	$XY = 11$	Output $(Z_2 Z_1)$ $XY = 00$	$XY = 01$	$XY = 10$	$XY = 11$
A	A	B	B	A	00	01	00	00
B	A	C	B	C	00	10	10	00
C	B	C	A	D	00	01	00	00
D	A	C	D	C	00	10	10	00
E	B	A	B	E	00	01	00	00
F	A	B	F	A	10	00	11	00
G	A	D	F	E	10	00	11	00

2. What are the conditions for determining equivalence of two or more states?

3. Reduce the state tables given in Problem 1 using implication charts.

4. Reduce the following incompletely specified state tables using implication charts. Identify the compatible state pair. Specify the value of output don't-cares after reduction.

a.

Present state	Next state $x = 0$	Next state $x = 1$	Output $x = 0$	Output $x = 1$
a	a	b	0	—
b	c	b	0	1
c	d	a	—	1
d	a	e	0	—
e	e	a	—	0

b.

Present state	Next state 00	01	10	11	Output 00	01	10	11
A	B	A	C	B	0	—	0	1
B	B	A	C	A	—	1	0	—
C	A	D	D	B	1	—	—	0
D	E	C	D	C	0	—	1	—
E	C	F	D	C	0	1	0	0
F	A	A	F	E	0	0	—	1

5. Use a merger diagram to reduce the following table.

Present state	Next state $XY = 00$	$XY = 01$	$XY = 10$	$XY = 11$
a	a/—	c/1	e/1	b/1
b	e/0	c/—	—/—	—/—
c	f/0	f/1	e/0	—/—
d	—/—	—/—	a/0	d/1
e	—/—	—/—	a/0	d/1
f	c/0	—/—	b/0	c/1

6. Explain the difference between equivalent and compatible states.

7. Reduce the following state tables using a merger diagram and compatibility graph.

a.

Present state	Next state/Output			
	$xy = 00$	$xy = 01$	$xy = 10$	$xy = 11$
a	a/—	c/1	e/1	b/1
b	e/0	c/—	—/—	—/—
c	f/0	f/1	e/0	—/—
d	—/—	e/0	a/—	—/—
e	—/—	—/—	a/0	d/1
f	c/0	—/—	b/0	c/1

b.

Present state	Next state/ Output	
	0	1
a	b/1	h/1
b	a/0	g/0
c	—/0	f/—
d	d/0	—/1
e	c/0	d/—
f	a/—	c/0
g	—/—	b/—
h	g/0	e/—

8. Why is the transitive postulate (states $A = B$ and $B = A$) applicable for completely specified state machines but not necessarily true for incompletely specified state machines? Illustrate this by showing two example state machine state tables, one specified and the other incompletely specified.

9. Explain how a compatible state pair can be found using an implication chart. Illustrate by an example.

10. What is maximal compatibility? What is the difference between compatible states and maximal compatibility. How does a compatibility class graph determine maximal compatibility, and why is that important in reducing a state machine?

Section 7.4

11. Explain why making the state assignment has an impact on the final realization of a state machine. Illustrate this by creating a simple three-state machine and realizing the design using two different state assignments.

12. How many permutations are possible with

a. two state variables and three states,

b. three state variables and six states,

c. four state variables and 14 states.

13. State, in your own words, the three guidelines for making a state assignment.

14. Create the adjacency groups for the following state tables.

a.

Present state	Next state/ Output	
	$x = 0$	$x = 1$
a	a/0	b/0
b	c/0	d/0
c	a/1	d/0
d	e/0	d/1
e	c/0	d/1

b.

Present state	Next state/ Output	
	$x = 0$	$x = 1$
A	A/0	B/0
B	C/0	E/0
C	B/1	F/0
D	A/0	B/0
E	A/1	E/0
F	G/0	E/0
G	A/0	E/1

c.

Present state	Next state/Output			
	$xy = 00$	$xy = 01$	$xy = 10$	$xy = 11$
a	a/0	a/0	b/0	c/1
b	a/0	b/0	b/0	c/0
c	a/1	b/1	d/0	e/0
d	a/1	b/1	d/0	e/0
e	e/0	e/0	f/0	f/1
f	b/0	b/0	f/0	a/0

15. Plot the state assignment maps for the state tables in Problem 14.

16. Realize the state machine in Problem 15 (table c) using *J–K* flip-flops. Realize the same state table using a binary state assignment (i.e., a = 000, b = 001, c = 010, etc.) and compare the gate input costs.

17. Realize the state machine in Problem 15 (table c) using D flip-flops and compare the costs with the results from Problem 16.

18. Determine the state assignment for the following state tables using the implication graph technique.

a.

Present state	Next state/Output $x = 0$	Next state/Output $x = 1$
a	c/0	b/1
b	e/0	c/0
c	d/0	e/1
d	d/0	b/1
e	e/0	a/1

b.

Present state	Next state/Output $x = 0$	Next state/Output $x = 1$
a	b/0	c/1
b	b/0	d/0
c	e/1	c/1
d	b/0	a/0
e	a/1	d/1

c.

Present state	Next state/Output $xy = 00$	$xy = 01$	$xy = 10$	$xy = 11$
a	b/0	b/0	a/1	a/0
b	b/0	b/0	c/0	b/0
c	a/0	c/0	c/1	d/1
d	a/0	b/1	d/0	e/1
e	a/1	b/0	e/0	e/1

19. Explain, in your own words, the advantages of using the implication graph as an aid in making state assignments. What advantage does the implication graph add to the guideline's algorithm? Explain when the implication graph may not be practical.

Problem 20

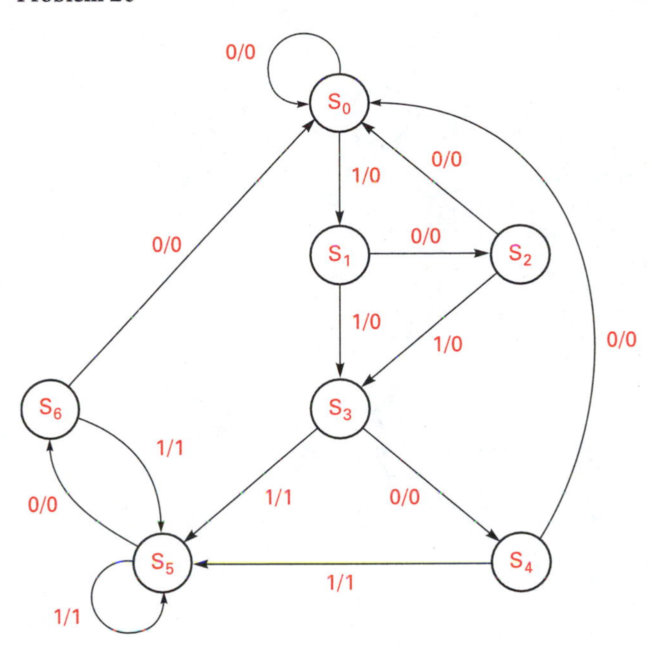

20. Given the state diagram in the illustration for this problem, create a state table, reduce it using a merger diagram and compatibility chart, then create a state assignment using the guidelines and implication graph.

Section 7.5 to 7.6

21. Convert the state diagrams for this problem to an ASM diagram.

22. Design a synchronous circuit that has a single input variable and single output variable. The input data are received serially. Cause the first output bit to be the same value as the first input bit in the serial string (i.e., if $x = 0$, then $z = 0$; if $x = 1$, then $z = 1$). Output z is to change thereafter only when three consecutive input bits have the same value. For example,

$x = 00100111011000$

$z = 00000001111110$

a. Construct an initial ASM diagram describing the system; decide whether a Mealy or Moore machine produces the fewest number of states.

Problem 21

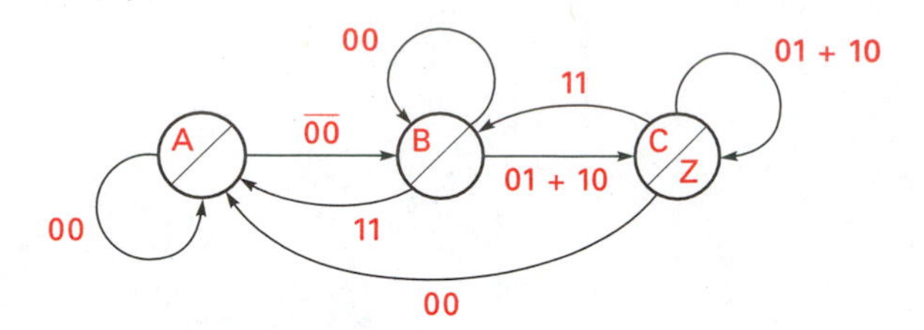

(a)

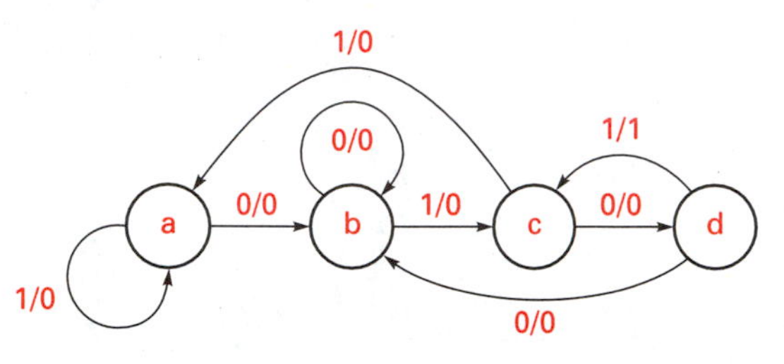

(b)

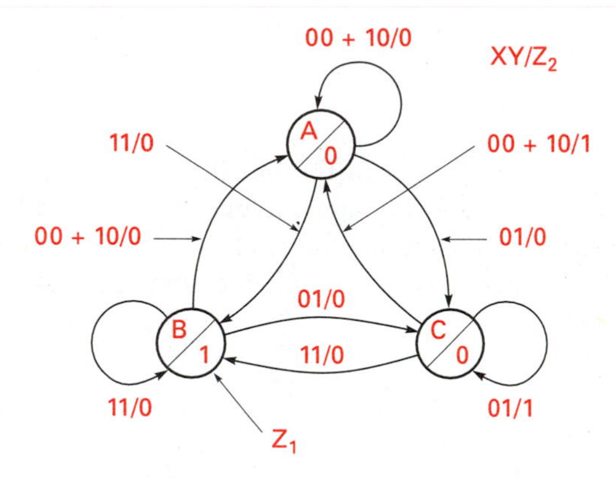

(c)

b. Create a state table, and reduce the table if possible.

c. Determine the number of state variables needed, and create the state assignment to minimize the logic.

d. Construct a new ASM diagram reflecting the minimized state machine and the state assignment.

e. Realize the design using *J–K* flip-flops. Employ the use of MEVs to create the excitation and output equations.

23. Design a synchronous sequential machine that simulates the blackjack card game. Assume that card input and display outputs are available. The operational flowchart of the blackjack game is given in the illustration for this figure.

a. Determine the input and output variables needed.

b. Create an ASM diagram from the operational flowchart.

c. Create a state table, and reduce it, if possible.

d. Determine the state variables needed and construct a simplified ASM diagram that includes the state assignment.

e. Realize the design using *J–K* flip-flops. Use MEVs to create the excitation and output equations.

24. As a project, design a sequential controller to provide switching for a traffic intersection between Main and Pine Streets, as indicated in the figure for this problem.

Sensors are buried in the roadbed to determine traffic flow demand. All light changes go through the following sequence.

A. The light changing from green to yellow to red remains yellow for 15 seconds, before changing to red.

B. Both changing traffic path lights are to remain red for 5 seconds.

C. The red-to-green traffic path light now turns green.

D. Main Street (straight) green is on for 45 seconds before checking sensors. All other traffic path green lights are on for 30 seconds before checking sensors. The traffic flow priorities are as follows:

1. Main Street (straight) light green is the default.
2. Pine Street (straight) is green if no main street traffic is present and the Pine Street (straight) sensor is tripped.
3. Main Street (turn) is green if no Main Street or Pine Street traffic is present and the Main Street (straight) sensor is tripped.
4. If two sensors are tripped then the priority is (a) Main Street (straight) green, then the sec-

Problem 23

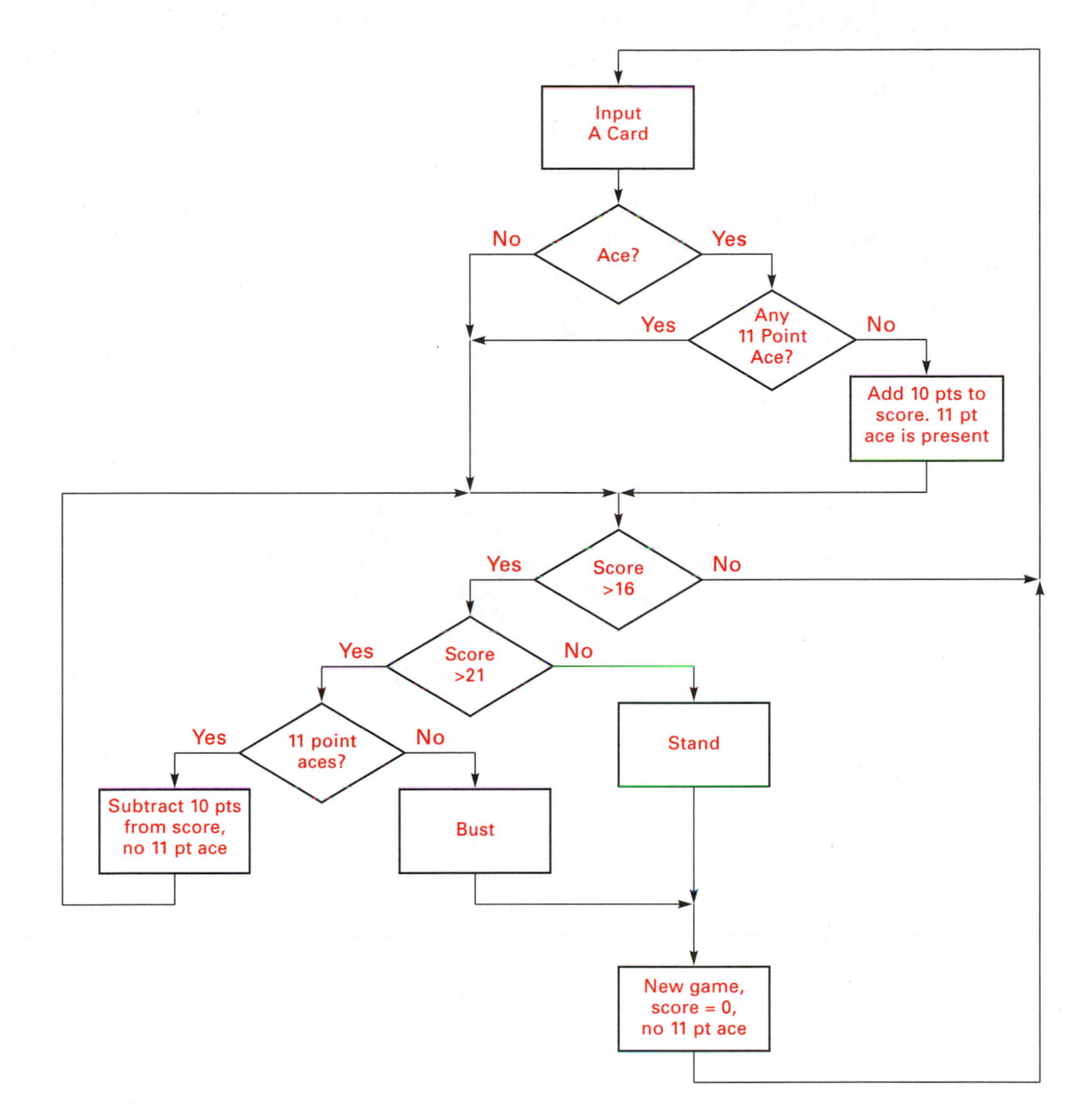

Problem 24

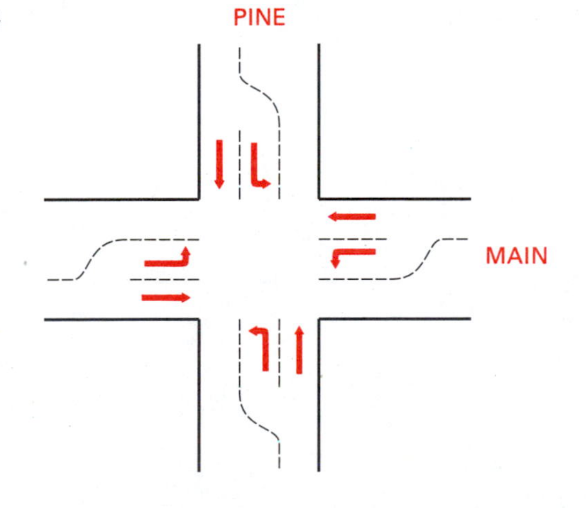

ondary light turns green (Main Street turn, Pine Street straight or Pine Street turn, depending on which sensor is tripped); (b) if Main Street straight sensors are not tripped, then Pine Street straight has priority over the turn paths; (c) if Main Street and Pine Street (both straight) sensors are not tripped, then Main Street turn has priority over Pine Street turn.

5. If three or more sensors are tripped the priority is Main straight, Pine straight, Main turn, Pine turn.

   a. Determine input and output variables and assign mnemonics. A three-light (red, yellow, and green) system is used for each traffic path.

b. Construct a preliminary ASM diagram.

c. Construct a state table and reduce it if possible.

d. Determine the state assignment.

e. Redraw the ASM diagram to include any possible state reductions and state assignment.

f. Realize the design using *J–K* flip-flops.

g. Build the logic and test for functionality.

CHAPTER

8 ASYNCHRONOUS SEQUENTIAL CIRCUITS

INTRODUCTION TO ASYNCHRONOUS SEQUENTIAL MACHINES

Asynchronous means without synchronization; in the case of sequential machines, it means without a synchronizing clock. Synchronous sequential machines change state in response to a synchronizing clock pulse, and asynchronous sequential machines do not use a clock. Synchronous machines require that the internal time delays caused by the excitation logic, flip-flops, and output logic will have transpired prior to the arrival of the next clock pulse. In fact, the synchronous machine logic designer must consider the time delays involved to determine the maximum operating speed of the clock. Asynchronous sequential circuits have no clock; the **internal** or **secondary states** change according to the delay times of the logic. Because of the lack of a synchronizing clock asynchronous sequential circuits are more difficult to design. Care must be taken when selecting the proper state assignment so that delays are predictable and unwanted state transitions are not possible. Because of the lack of a clock, asynchronous sequential circuits are typically faster than their synchronous cousins.

Asynchronous sequential machines have the same state variable output-to-input feedback path as synchronous machines. Synchronous circuit design techniques assume that the input variables are synchronized with the clock. In cases where inputs are not synchronized with the clock, asynchronous design techniques must be considered.

Digital systems may have several subsystems or modules that require synchronizing clock pulses, distributed without clock skew, to each module. In very fast systems, the length of the conductors carrying clock signals must be considered. In such cases, the subunits may operate best asynchronously with respect to one another. For example, consider the transmission of data between one computer system and another. The distances may be too great to send parallel data and a clock. In such cases, the data transmission must be done asynchronously. LSI devices used to accomplish this task are called *universal asynchronous receiver transmitters* (UART). UART's internal control is synchronous but the data transmission is asynchronous. A third reason for the consideration of asynchronous over synchronous is speed; asynchronous systems are faster.

8.1 FUNDAMENTAL AND PULSE MODE ASYNCHRONOUS SEQUENTIAL MACHINES

Figure 8.1 shows a model for an asynchronous sequential machine. The model covers both fundamental and pulse mode circuits. Asynchronous sequential circuit design is divided into either fundamental mode or pulse mode. The difference between fundamental and pulse modes has to do with how the input variables are to be considered.

Consider the diagram in Figure 8.1. The input variables are labeled I, the excitation inputs to the memory are labeled E, the output from memory is labeled S, and the output variables are labeled O. The memory is typically composed of latches, not clocked flip-flops.

Fundamental mode circuit designs assume that the inputs (I) to the asynchronous circuit will change only when the circuit is stable (the state variables (S) are not in the process of transition). Fundamental mode design rules also assume that the inputs are levels, not pulses.

The state variables in fundamental mode circuits can be characterized as delay elements (recall the class-1 sequential machine model of Chapter 5). Delay may be introduced by a latch or simply the propagation delay inherent in the logic devices used to realize the asynchronous circuit.

The behavior of an asynchronous machine is determined by defining what is called a *total* state. The total state is divided into two parts, the input state (the value of I) and the internal or secondary state (the value of S). The total state is defined by the value of (I, S). Total states can be stable or unstable. A stable total state is one that produces no additional state transitions; if I does not change then (I, S) remains stable.

Total state transitions occur as a given input state (I) causes the excitation of the memory to produce a secondary state (S). The secondary state outputs are part of the input to the logic that creates the memory excitation inputs. When the inputs to the combinational excitation logic produce no new value for the secondary states (S), the machine is said to be stable. The time delay to reach a stable state depends not only on the excitation and memory logic delays but also on the state assignment. For example, several secondary state changes could occur before reaching a stable total state.

Fundamental mode models do not permit input state (I) changes until the circuit has reached a stable total state. One can visualize that if changes in (I) were permitted before a stable total state were reached, the circuit might never reach a stable condition.

Another requirement of a fundamental mode machine is that only one input variable can change at a given time. This requirement is imposed because the external input

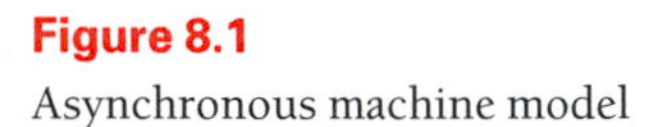

Figure 8.1
Asynchronous machine model

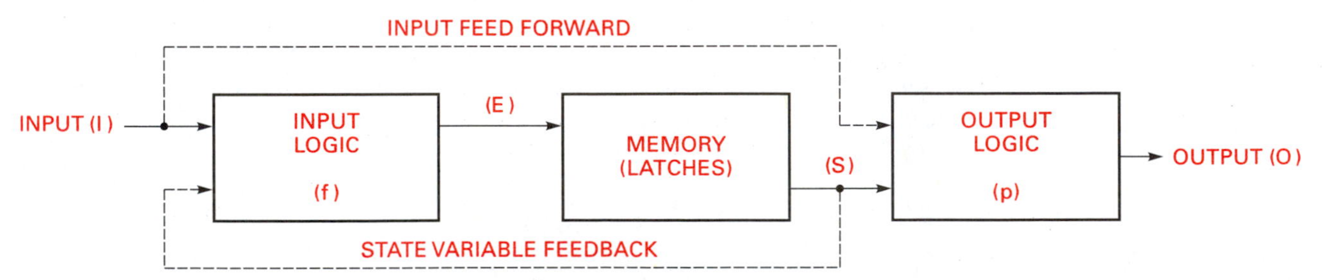

signals are free to change at any time, greatly limiting the possibility that two or more variables will change simultaneously. In practical applications the variations in device delay and length of signal wires (or circuit board traces) prevent simultaneously changing more than one variable. When multiple input variables occur, the input state can make several transitions due to different signal delay paths, which can cause the asynchronous circuit to make erroneous total state transitions.

Pulse mode asynchronous circuit models permit the application of input variables as pulses instead of levels. The requirement limiting input changes to only one variable also applies to pulse mode circuits. The width of the input pulses is critical to the circuit operation. First, the pulses must be long enough for the circuit to respond to the input. Second, the pulse must not be so long that it is still present after the new secondary state is reached.

The minimum input pulsewidth is calculated based on the propagation delay of the excitation logic. The maximum pulsewidth is determined by the total propagation delay through the excitation logic and the memory. For example, assume that the excitation logic were two levels of 5 ns each and the latch delay was 10 ns. The minimum pulsewidth is 10 ns and the maximum is 20 ns. The input pulsewidth requirement restricts the application of pulse mode asynchronous sequential circuits. Circuitry must be added to ensure the input variables are pulses whose widths fall within the required boundaries.

Both fundamental and pulse mode asynchronous circuits use unclocked flip-flops or latches. The common RS latch is available as the 74xx279, a quad RS latch. The 74xx279 uses complemented S and R inputs as shown in Figure 8.2.

The design process for both fundamental and pulse mode asynchronous circuits is similar to that of synchronous circuits. Even though the process steps are similar, differences do exist, mainly timing and input variable restrictions.

The general design steps are as follows:

1. Create a state table or state diagram from the initial verbal problem statement.
2. Remove any redundant states and make a new reduced state table.
3. Create a transition table.
4. Write and simplify the excitation and output Boolean equations.
5. Draw the logic diagram and build or test simulate the circuit.

Figure 8.2
SN74xx279, quad RS latch

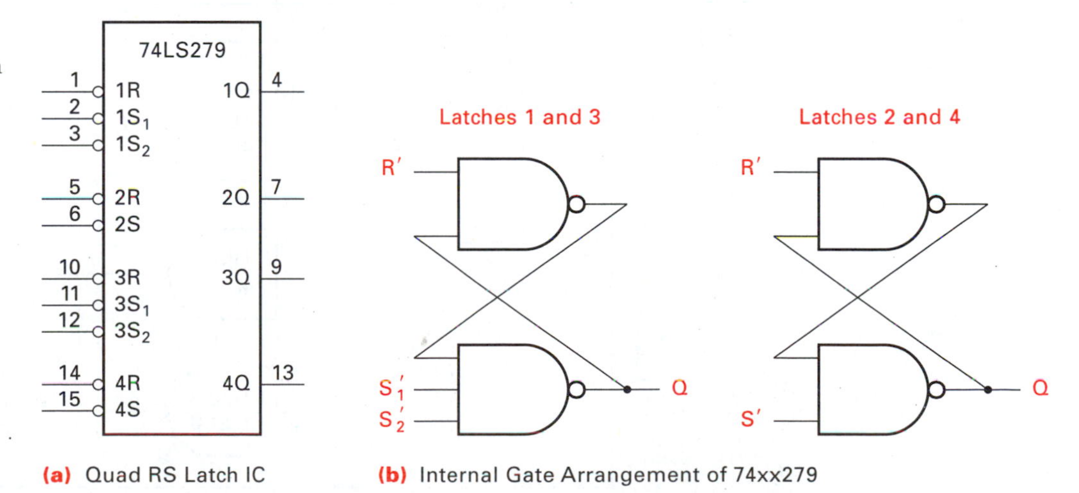

(a) Quad RS Latch IC **(b)** Internal Gate Arrangement of 74xx279

Fundamental mode design techniques modify the familiar state table and create what is called a *flow table* that represents the input, secondary, and total states. Often designers replace the initial state diagram with the flow table to determine total state transitions.

8.2 ANALYSIS OF ASYNCHRONOUS SEQUENTIAL MACHINES

Analysis of pulse mode and fundamental mode circuits is a useful approach to developing insight into asynchronous circuits. Consider the fundamental mode circuit shown in Figure 8.3. No flip-flops are used in this circuit. Two state variable feedback paths exist where X_1 and X_0 provide inputs to the gates that generate the state variables. The feedback essentially creates the latching operation necessary to produce a sequential circuit.

X_1 and X_0 are state variables, I_1 and I_0 are input variables, and Z is an output variable. The input state is determined by the logical value of I_1, I_0; the secondary state is determined by the logical value of X_1 and X_0; and the total state is determined by both the input state and the secondary state (X_1, X_0, I_1, I_0).

The next-secondary state and output equations are derived from the logic:

$$X_1^+ = X_0 I_1 + X_1 I_0$$
$$X_0^+ = X_1' I_1' I_0 + X_1' X_0 I_0 + X_0 I_1$$
$$Z = (X_0 I_1)'$$

Analysis of the example circuit requires an assignment of an input sequence, because the total states are defined by both the state variables and the input variables. The fundamental mode model requires that only one input variable change at a time. A table can be constructed that shows the next-total states, by assuming a present-total state and applying the next-secondary state equations. Table 8.1 illustrates the present-total

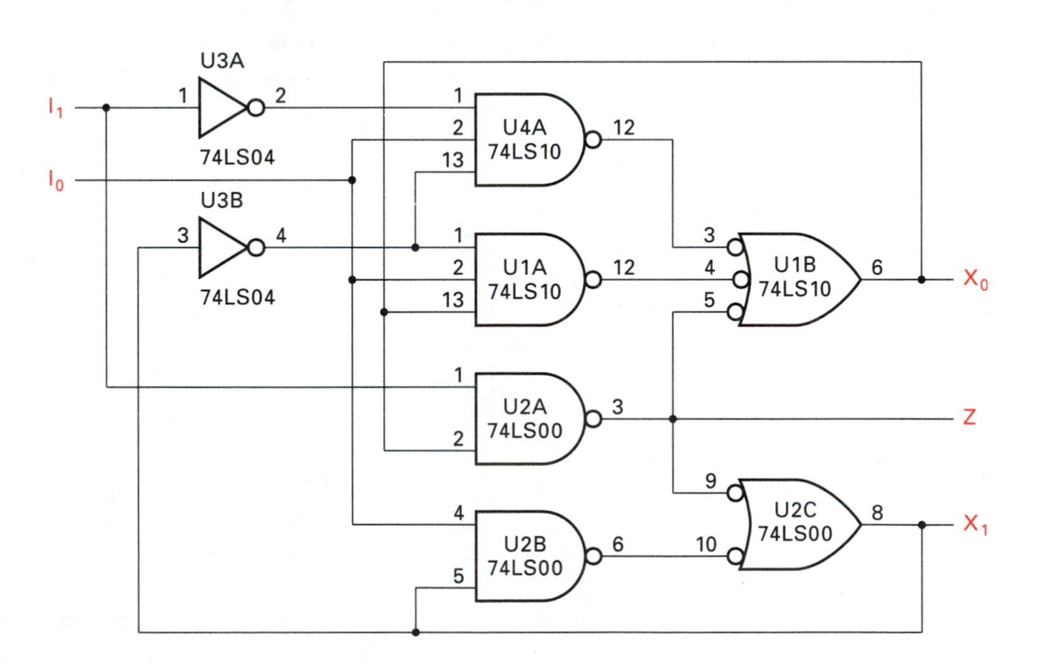

Figure 8.3
Fundamental mode asynchronous circuit

TABLE 8.1
Present-state, next-state table for analysis of an asynchronous sequential circuit

Present total state				Next total state				Stable total state	
X_1	X_0	I_1	I_0	X_1	X_0	I_1	I_0	Yes/No	Z
0	0	0	0	0	0	0	0	Yes	1
0	0	0	1	0	1	0	1	No	1
0	0	1	1	0	0	1	1	Yes	1
0	0	1	0	0	0	1	0	Yes	1
0	1	0	0	0	0	0	0	No	1
0	1	0	1	0	1	0	1	Yes	1
0	1	1	1	1	1	1	1	No	0
0	1	1	0	1	1	1	0	No	0
1	1	0	0	0	0	0	0	No	1
1	1	0	1	1	0	0	1	No	1
1	1	1	1	1	1	1	1	Yes	0
1	1	1	0	1	1	1	0	Yes	0
1	0	0	0	0	0	0	0	No	1
1	0	0	1	1	0	0	1	Yes	1
1	0	1	1	1	0	1	1	Yes	1
1	0	1	0	0	0	1	0	No	1

state, the next-total state, the stability of the next-total state, and the output variable, Z. Next-total state values are found by assigning present-total state values to the Boolean variables in the next-secondary state equations to determine X_1^+ and X_0^+. The next-input state is the same as the present-input state in this table.

When an input change (I_1, I_0) causes a secondary state change (X_1, X_0) and the next-secondary state does not change, the total state is said to be *stable*.

The information in Table 8.1 can be rearranged into a transition table as shown in Figure 8.4. Columns represent input states and rows represent secondary states. The next-secondary state values are written into the squares, each indicating a total state.

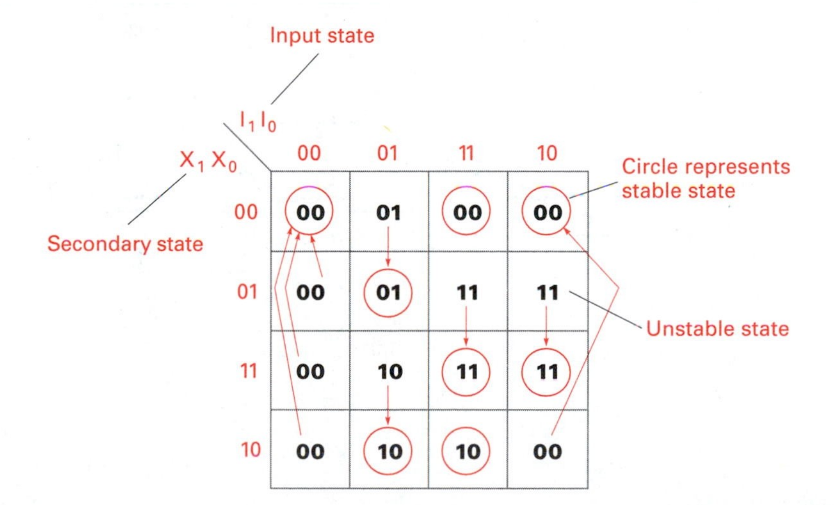

Figure 8.4
Analysis example transition table

Circled secondary states are stable. The directed arcs indicate transitions from unstable states to stable states. For example, if the total state (X_1, X_0, I_1, I_0) were 0001, which is unstable, the next total stable state would be 0101.

Let an input state sequence be I_1, I_0 = 00, 01, 11, 01, 00. Assume the secondary state starts at $X_1, X_0 = 00$. The total state transitions can be determined from the table. Starting at total state $X_1, X_0, I_1, I_0 = 0000$, let the input state change from 00 to 01, the circuit makes a transition to total state 0001, which is unstable, then to 0101, which is stable. The next input change, from 01 to 11, causes a transition from stable-total state 0101 to total state 0111 (unstable), then to total state 1111 (stable). A change in input state from 11 to 01 causes stable-total state 1111 to change to total state 1101 (unstable). Unstable total state 1101 then changes to stable total state 1001. A change in input state from 01 to 00 causes a transition to unstable total state of 1000 and then to stable state 0000, completing the state transitions for the input change.

The timing diagram illustrated in Figure 8.5 shows an input state sequence of I_1, I_0 = 00, 10, 11, 01, 11, 01, 00, 10. Note that only one input variable changes at a time. Stable states remain so until the input states change, at which time an unstable total state occurs. The unstable total state results from the delay time in switching from one stable state to another. The gate delay for 74LS TTL devices is 15 ns, which would produce unstable states whose duration would range from 30 ns (two levels), to 60 ns (four levels).

Figure 8.5 illustrates the timing diagram for the example asynchronous sequential circuit. Input variable I_1 toggles every 75 ns. Input variable I_0 goes high at 135 ns and back low at 335 ns and then repeats 135 ns low and 200 ns high. Notice that state variables X_1 and X_0 as well as active-low output Z operate without any glitches. Output Z goes low for 75 ns once each time input variable I_0 is high.

Study the timing diagram in Figure 8.5 and compare it with the transition table in Figure 8.4. Start at the beginning of the timing diagram. Note that both input and state variables are low, resulting in a stable total state of $X_1, X_0, I_1, I_0 = 0000$. Input variable I_1 goes high at t_a, creating stable total state 0010. At t_b, in the timing diagram, input variable I_0 goes high, causing a transition from stable state 0010 to stable total state

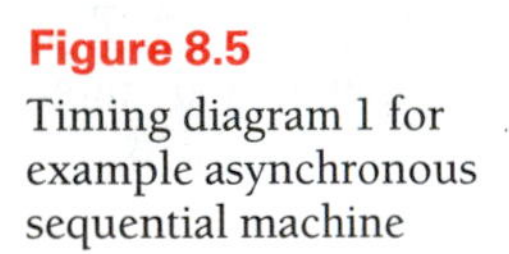
Figure 8.5
Timing diagram 1 for example asynchronous sequential machine

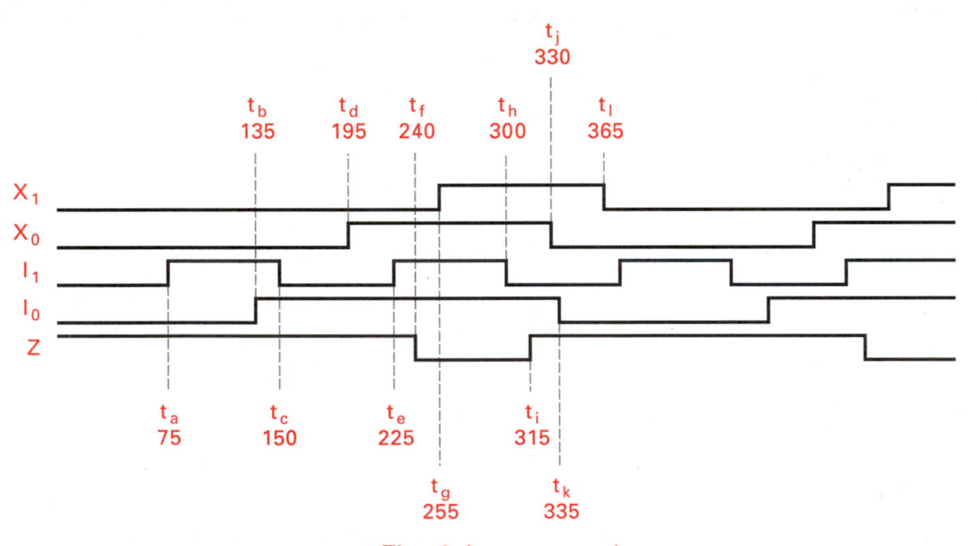

0011. At t_c, I_1 goes low, creating input state 01, causing in turn a transition to unstable total state 0001. The machine remains in this condition until stable total state 0101 is reached at t_d.

The time delay from t_c to t_d was determined by logic gates U3A (I_1 input), U4A, and U1B. The three levels of LS TTL gates have a propagation delay of 45 ns, which is the time $t_d - t_c$. At t_e, the input changes to 11, causing a total state change from stable state 0101 to unstable state 0111 (t_e to t_f) and then to stable total state 1111 at time t_g.

Active-low output Z' goes low at t_f and remains low until t_i. Notice the delay between t_e and t_f and between t_h and t_i. The delay of U2A created these time differences. At time t_h input I_1 went low, causing a transition from total state 1111 to unstable total state 1101. Before the logic could respond to unstable state 1101 input I_0 went low at time t_k, causing a transition to stable total state 1001 and then to unstable total state 1000 at t_k, and finally to stable total state 0000 at t_l. Notice the time difference between t_k and t_j of approximately 5 ns. Because this time is substantially less than the logic propagation delay no secondary state change occurred.

Figure 8.6 shows a timing diagram, where the timing of input variable I_0 is changed. At t_1, 275 ns, I_0 goes high and remains high until t_2, until 325 ns.

Notice the glitch in state variable X_0 at time t_3, 345 ns into the timing diagram. Something happened that is not wanted. The time between the high to low transition of input variable I_0 and the glitch in X_0 at t_3 gives us a clue. Refer to the logic diagram in Figure 8.3, with the understanding that low power Schottky TTL devices are used. The TTL data book lists the propagation delay of a LS10 as a maximum of 15 ns. Consider the status of the circuit prior to t_2 of the timing sequence shown in Figure 8.6. The circuit is in unstable total state, $X_1, X_0, I_1, I_0 = 0001$. Normally state variable X_0 would switch from a 0 to a 1, creating a stable total state of 0101. Before stability is reached, however, input I_0 changes to a 0 at t_2.

The glitch in state variable X_0, at t_3, occurs because the logic propagation delay was greater than the time between t_3 and t_2. Input I_1 must pass through three levels of logic before the output at X_0 can become stable. The time difference between t_3 and t_2 is 20 ns and the logic propagation delay is 45 ns. Therefore the state variable X_0 started to change beginning the glitch, but did not become stable because input I_1 changed at t_4, ending the glitch.

The other glitches shown on the timing diagram indicate similar timing problems. The point of this discussion is to show how important timing is when designing asynchronous sequential circuits. For the given example, faster operation would require using a faster logic family.

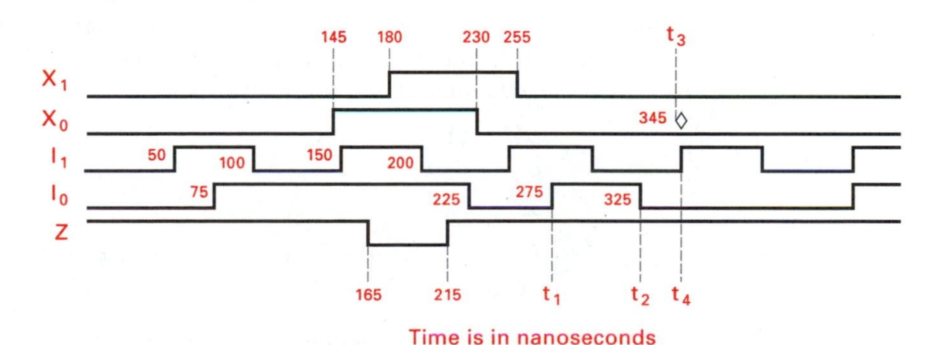

Figure 8.6
Timing diagram 2 for example asynchronous circuit

Figure 8.7
Pulse mode input changes

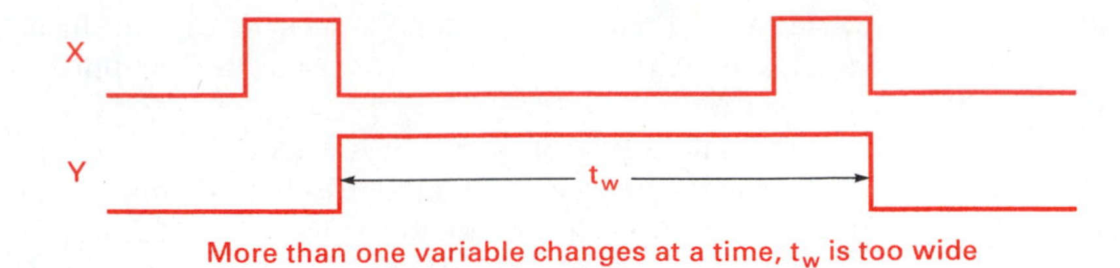

(a) Unacceptable Pulse Mode Input Changes

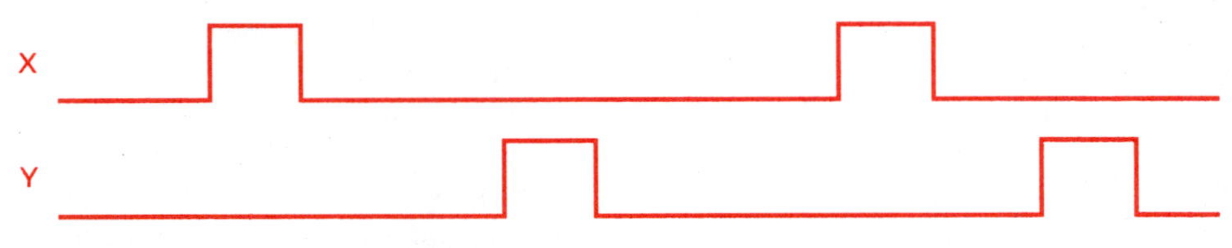

(b) Acceptable Pulse Mode Input Changes

Pulse mode asynchronous sequential circuits rely on input pulses rather than levels. For example, in a coin-operated vending machine, when coins are dropped into the slot, they produce pulses as the coins enter and leave the sensor area. The pulse inputs in pulse mode circuits must be

1. Asserted long enough to cause a change in a state variable(s).
2. Shorter in width than the minimum time delay in the excitation logic.

The first requirement is true because the state variable logic must have time to respond to its excitation or no output change could occur. The second requirement stems from the result if the pulse were too wide. An excessively wide pulse would cause a change in the secondary state, which would be fed back into the excitation logic along with the still present pulse input, causing an error in state transition. Pulse mode circuits allow only one input variable to change at a time.

Because of the input variable pulsewidth restrictions, pulse mode circuits are difficult to design; not because of conceptual problems but because of the timing requirements.

Figure 8.7 shows an example of acceptable and unacceptable input changes for a pulse mode circuit.

Consider the pulse mode circuit logic diagram shown in Figure 8.8. The state variables are realized by two NAND latches. Four input variables X, Y, Z, and EC, one output variable ERR, and two state variables A and B are identified.

The excitation and output equations are

$$S'_A = (Y + Z)' \quad \text{or} \quad S_A = Y + Z$$
$$R'_A = EC' \quad \text{or} \quad R_A = EC$$
$$S'_B = (A'\ X)' \quad \text{or} \quad S_B = A'X$$
$$R'_B = (EC + A'Z)' \quad \text{or} \quad R_B = (EC + A'Z)$$
$$ERR = EC\ B'$$

The characteristic equation for the S′R′ latch is

$$Q_x^+ = S + R'\ Q_x$$

Figure 8.8
Pulse mode asynchronous circuit

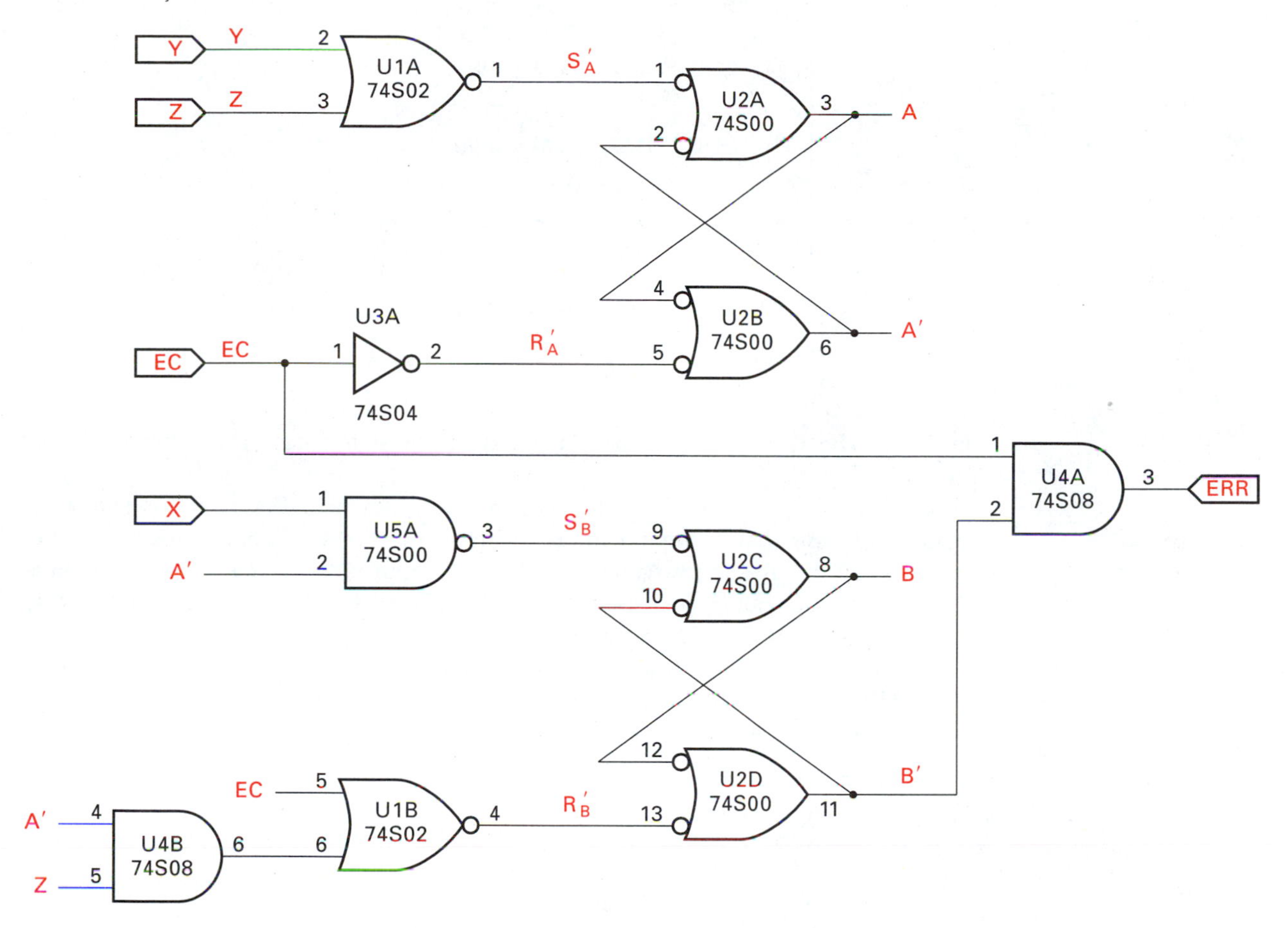

Using the excitation and characteristic equations we can derive the state variable next-state equations, which are

$$Q_A^+ = Y + Z + A\,(EC)'$$
$$Q_B^+ = A'\,X + AB(EC)' + B(EC)'Z'$$

Pulse mode circuits permit only one input change at a time, leading to a simplified transition table. The transition table for the analysis example pulse mode circuit (Figure 8.9) is constructed so that each input variable has a column and the state variables are covered by four rows. For example, if input variable X is true, then no other variable is permitted to be true.

The transition table specifies the next-state and output value for each present-state and input value. The entry in each map square is found by evaluating the next-state and output equations. For example, the top left square represents the present state of Q_A and Q_B = 00, when input X is true. Evaluation of the Q_A next-state equation produces a 0 for Q_A^+ and a 1 for Q_B^+. Therefore a 01 is written in the square as the next-state value. Evaluation of the output equation produces a 0. The total square entry is written 01/0.

Now we can construct the flow table from the transition table as shown in Fig-

Figure 8.9
State variable transition table

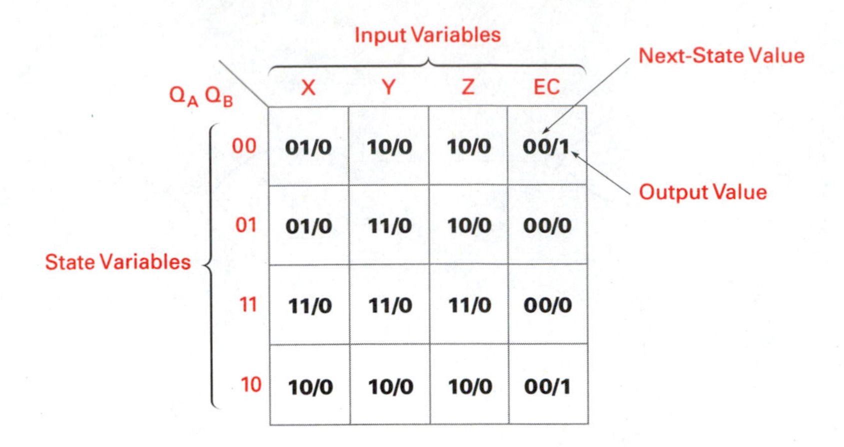

$Q_A Q_B$	X	Y	Z	EC
00	01/0	10/0	10/0	00/1
01	01/0	11/0	10/0	00/0
11	11/0	11/0	11/0	00/0
10	10/0	10/0	10/0	00/1

ure 8.10. The flow table can be converted into a state diagram as shown in Figure 8.11.

A timing diagram for the pulse mode circuit shown in Figure 8.8 is shown in Figure 8.12. Note that the input variables are pulses and that only one changes at a time. Note the changes in state variables A and B. The input pulse in this timing diagram is 25 ns wide, wide enough to cause the Schottky devices to change state but not so wide as to generate errors in state transitions.

Consider the pulse mode asynchronous circuit shown in Figure 8.13. In this circuit a 74LS279 quad R′S′ latch is used to realize the state variables.

The first analysis step is to write the excitation equations:

$$R_1' = Z',\ R_1 = Z$$
$$S_1' = (X + Y)',\ S_1 = X + Y$$
$$R_2' = (Z + YQ_1')',\ R_2 = Z + YQ_1'$$
$$S_2' = (WQ_1')',\ S_2 = WQ_1'$$
$$P = Q_2'Z$$

Figure 8.10
Flow table

	X	Y	Z	EC
S_0	S_1/0	S_3/0	S_3/0	S_0/1
S_1	S_1/0	S_2/0	S_3/0	S_0/0
S_2	S_2/0	S_2/0	S_2/0	S_0/0
S_3	S_3/0	S_3/0	S_3/0	S_0/1

Figure 8.11
Pulse mode asynchronous machine state diagram

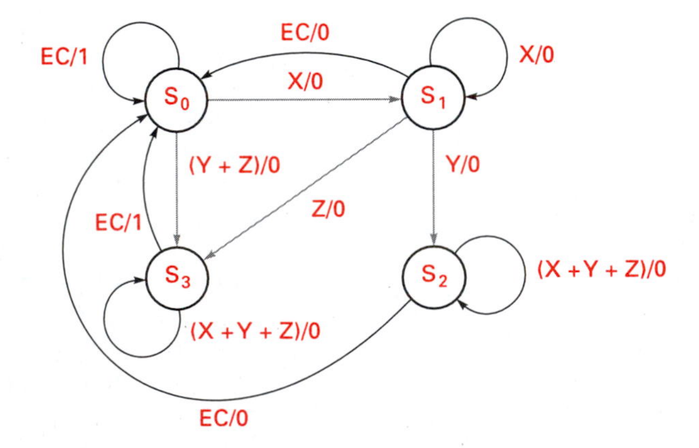

Figure 8.12
Partial timing diagram for the circuit in Figure 8.8

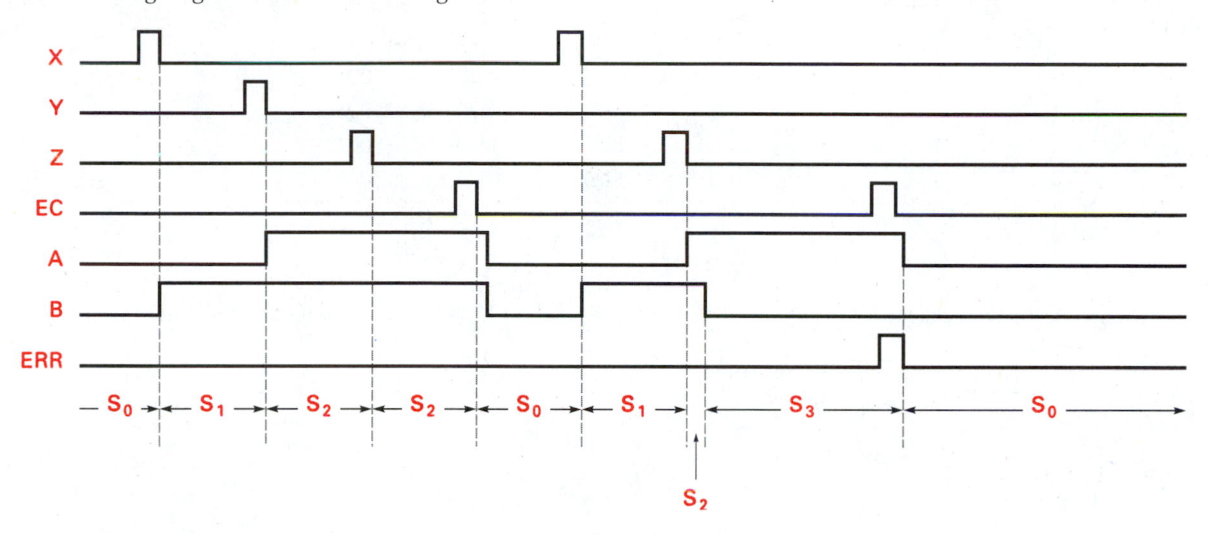

Figure 8.13
Pulse mode asynchronous circuit using a 74LS279 quad S′ R′ latch IC

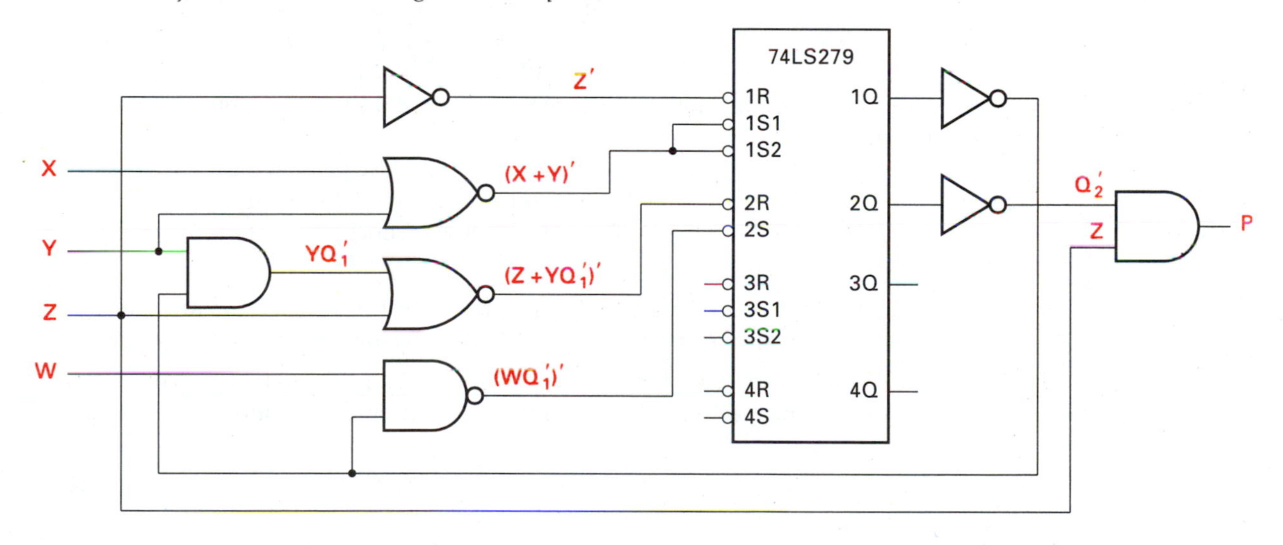

The characteristic equation of an R.L, S.L (active-low excitation inputs) latch is

$$Q_x^+ = S + R' \, Q_x$$

The excitation equations above are all active-low. If any of the terms in the equation is high (true) then the excitation variable is low (true).

We do not know the state assignment or how many don't-care states, if any, may exist. Using the R.L, S.L latch characteristic equations and the excitation equations, the transition table shown in Table 8.2 is constructed.

A state diagram can be constructed from the transition table by assigning a state

TABLE 8.2
Transition table for pulse mode circuit

Present state		Next state/output			
Q_2	Q_1	W	X	Y	Z
0	0	10/0	01/0	01/0	00/1
0	1	01/0	01/0	01/0	00/1
1	1	11/0	11/0	11/0	00/0
1	0	10/0	11/0	01/0	00/0

Figure 8.14
Pulse mode asynchronous circuit state diagram

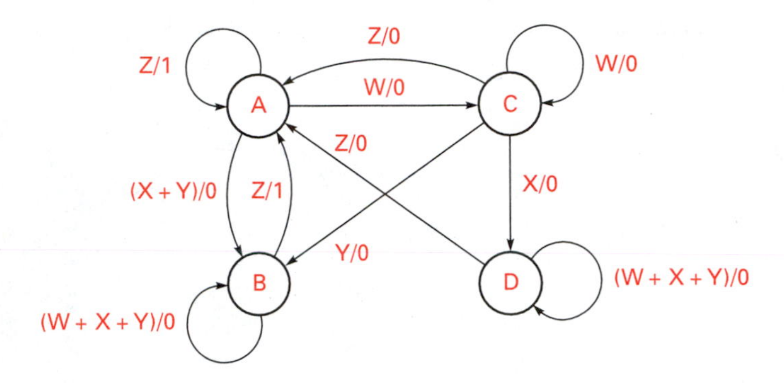

TABLE 8.3
Assignment of names to states

State name	State variables	
	Q_2	Q_1
A	0	0
B	0	1
C	1	0
D	1	1

name to each state as illustrated in Figure 8.14. Table 8.3 shows the assignment of state names to states.

Analysis involves steps to determine how an existing sequential machine works. Design is the process of developing the sequential machine from scratch. The two activities work in opposition to each other. The asynchronous sequential machine design process has the following steps:

1. Construction of a primitive flow table from a set of verbal or written specifications. An intermediate step may include the development of a state diagram.
2. Redundant states are eliminated by state reduction. The next-class, implication chart, and merger graph state reduction techniques were introduced in Chapter 7.
3. The state assignment is made. Asynchronous state machines require a different approach to the state assignment problem than was used for synchronous machines. We will discuss asynchronous machine state assignments in a later section in this chapter.
4. A set of excitation and output equations are derived from the reduced flow tables. These equations must be checked for static and dynamic hazards to ensure correct circuit operation.

8.3 DERIVING FLOW TABLES

Flow tables are used in asynchronous sequential circuit design in the same way state tables are used in synchronous circuit design. The reason for the name *flow table* stems from the behavior of asynchronous state machines. The state changes occur independent of a clock, based on the logic propagation delay, and cause the states to "flow" from one to another.

A primitive flow table is a special case flow table. It is a flow table that has only one stable state for each row in the table. The synthesis process usually starts by the construction of a primitive flow table based on the problem statement. The previously mentioned restriction of only one input variable change applies to the construction of the primitive flow table.

❖ EXAMPLE 8.1

Develop the state diagram and primitive row flow table for a logic system that has two inputs, S and R, and a single output Q^+.

The device is to be an edge triggered SR flip-flop, but without a clock. The device changes state on the rising edges of the two inputs. Static input values are not to have any effect in changing the Q^+ output.

Solution The construction of the state diagram in Figure 8.15 is accomplished in the same fashion as we used in synchronous machines. The initial state A is stable when no input changes have been detected. An SR input change from 00 to 01 (reset) causes a state transition from A to C, and an SR input change from 00 to 10 causes a state transition to state B.

State B is stable with a 10 static SR input and Q^+ is a 1. State C is stable with a 01 static SR input and Q^+ remains a 0. A state transition from B to D occurs when the SR input changes from 10 to 00; the Q^+ output remains a 1. A B to E state change occurs when SR changes from 10 to 11. Once state E is reached the Q^+ output changes from a 1 to a 0.

State D was reached by the SR input transitions of 00 → 10 → 00 (see Figure 8.16). A D to C state transition occurs when SR changes from 00 to 01. Once state C is reached the Q^+ output changes from a 1 to a 0. An SR input sequence of 00 → 01 → 11 causes an A → C → F state transition. The Q^+ output must remain a 0 until the 01 → 11 SR

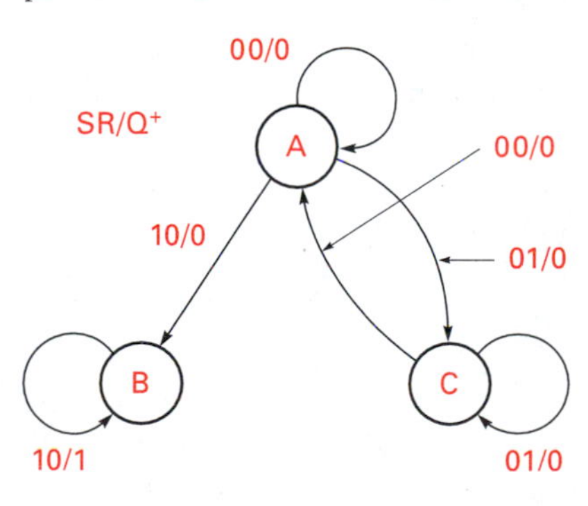

Figure 8.15
Partial state diagram for example problem

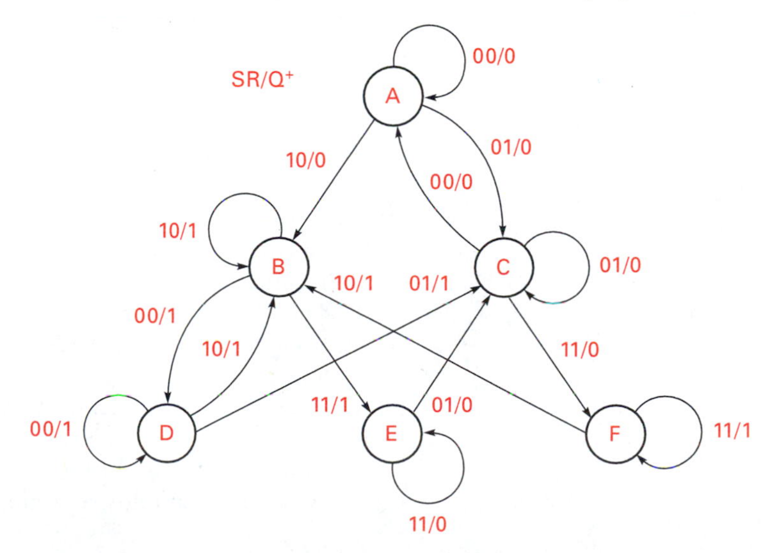

Figure 8.16
Continuation of state diagram construction for the example problem

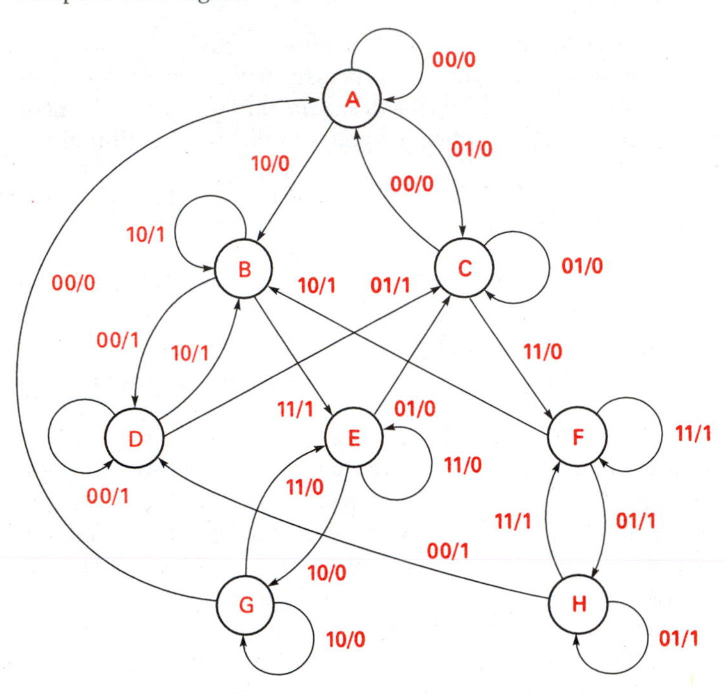

Figure 8.17
Complete state diagram

TABLE 8.4
Primitive row flow table for example problem

Present state	SR 00	Inputs 01	Q 11	Output 10
A	(A)/0	C/0	—	B/0
B	D/1	—	E/1	(B)/1
C	A/0	(C)/0	F/0	—
D	(D)/1	C/1	—	B/1
E	—	C/0	(E)/0	G/0
F	—	H/1	(F)/1	B/1
G	A/0	—	E/0	(G)/0
H	D/1	(H)/1	F/1	—

change occurs. The 11 static input is permitted because only input rising edges cause Q changes. (This is different from the regular SR flip-flop). A state F to B change occurs when SR changes from 11 to 10.

State E represents an SR input sequence of 10 → 11, and the Q^+ output is a 1 for the transition. A transition from state E to state G occurs when SR changes from 11 to 10 and Q^+ is 0. An SR input change of 10 → 11 causes a state G to E transition. Once in state G, a 00 input on SR returns the state machine back to state A.

State H was added to the diagram to permit SR input changes (while in state F) of 11 → 01 → 11. A 10 SR input causes an F to B transition. Figure 8.17 shows the complete state diagram for the edge triggered input SR flip-flop.

Now a primitive flow table (Table 8.4) is constructed from the state diagram shown in Figure 8.17.

Only one stable state exists for each row in the table. The stable state is where no further state transitions occur if the inputs are constant. Note that the stable states are circled. Redundant states are likely to be present in a primitive row flow table. The next task is to construct a merged flow table by application of the merger diagram state reduction technique, first discussed in Chapter 7.

The **merger diagram** contains the same number of vertices as the flow table has states. Each vertex is labeled with the state name. Then lines, which represent possible compatible states, are drawn between states' vertices. If two states are incompatible, then no line is drawn. Recall that compatibility between two states exists if for every input sequence the same output occurs.

The merger diagram of Figure 8.18 is evaluated to find the maximal compatible classes with a compatibility class graph shown in Figure 8.19.

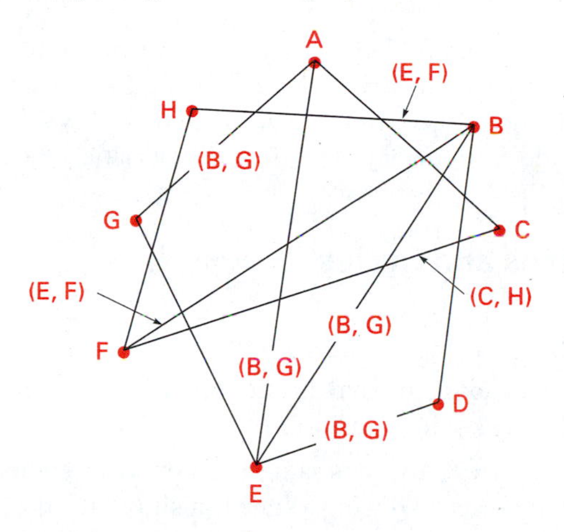

Figure 8.18
Merger diagram for example problem

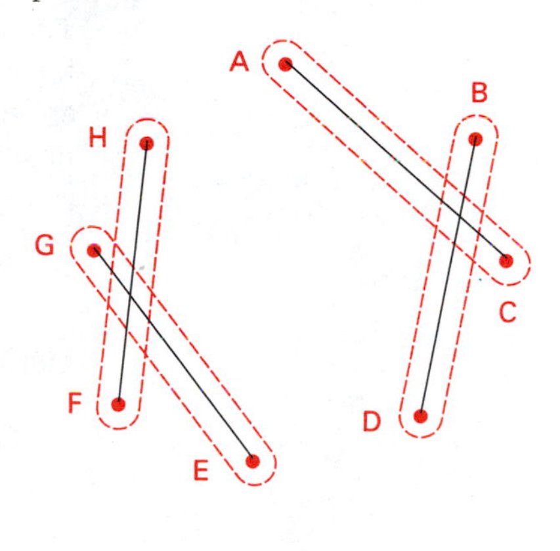

Figure 8.19
Compatibility class graph for example problem

The compatibility class graph shows the following maximally compatible state groups:

{A–C}; {B–D}; {E–G}; {F, H}

This set of maximal compatibles covers all of the original states resulting in the reduced flow table given in Table 8.5. Let

$\{A-C\} \rightarrow S_0$

$\{B-D\} \rightarrow S_1$

$\{E-G\} \rightarrow S_2$

$\{F, H\} \rightarrow S_3$

The reduced state diagram for Table 8.5 is shown in Figure 8.20.

TABLE 8.5
Merged flow table for example problem

Present state	Inputs 00	01	11	10
S_0	(S_0)/0	(S_0)/0	S_3/0	S_1/0
S_1	(S_1)/1	S_0/1	S_2/1	(S_1)/1
S_2	S_0/0	S_0/0	(S_2)/0	(S_2)/0
S_3	S_1/1	(S_3)/1	(S_3)/1	S_1/1

Figure 8.20
Reduced state diagram for example problems shown in Table 8.5

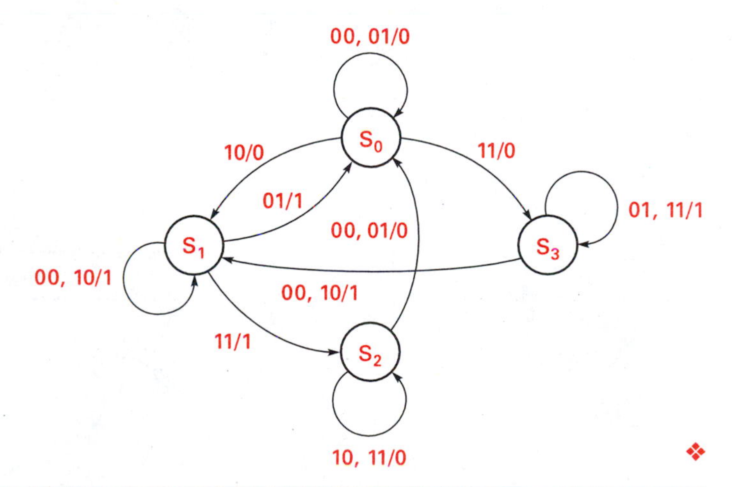

❖

8.4 STATE ASSIGNMENT

The state assignment problem for asynchronous machines is essentially the same as it is for synchronous machines, with one difference. Synchronous machine state assignments are made with the objective of logic reduction. Asynchronous machine state assignments are made with the objective of avoiding critical races.

8.4.1 Races and Cycles

Definitions

Race Races exist in asynchronous machines when two or more state variables change during a state transition.

Noncritical race A race is said to be noncritical when the correct stable next state is eventually reached during a state transition. If a machine, whose partial flow table is shown in Table 8.6, starts in stable total state 0000 and then the input changes to $XY = 01$, the transition path depends on whether F_2 or F_1 changes first. If F_2 changes first the race has a transition path of (F_2, F_1, X, Y) 0000 → 0001 → 1001 → 1101. If F_1 changes first the transition path is (F_2, F_1, X, Y) 0000 → 0001 → 0101 → 1101. Notice that both transition paths lead to total stable state 1101. For this reason the race is noncritical.

Critical race A race becomes critical if the correct next state is not reached during a state transition. Table 8.7 shows a critical race condition for an asynchronous state machine. Assume the machine is in total stable state 0000 and the inputs XY change from 00 to 01. Three possible results can occur depending on which state variable changes first.

TABLE 8.6
Partial flow table showing a noncritical race

		Inputs XY			
F_2	F_1	00	01	11	10
0	0	(00)	11		
0	1		11		
1	0		11		
1	1		(11)		

TABLE 8.7
Partial flow table showing a critical race

		Inputs XY			
F_2	F_1	00	01	11	10
0	0	(00)	11	00	11
0	1		(01)		
1	0		(10)		
1	1		(11)		

In case 1 (desired result, both F_2 and F_1 change at the same time),

F_2 F_1 X Y		F_2 F_1 X Y		F_2 F_1 X Y
0 0 0 0	→	0 0 0 1	→	1 1 0 1

In case 2 (F_2 changes before F_1),

F_2 F_1 X Y		F_2 F_1 X Y		F_2 F_1 X Y
0 0 0 0	→	0 0 0 1	→	1 0 0 1

In case 3 (F_1 changes before F_2),

F_2 F_1 X Y		F_2 F_1 X Y		F_2 F_1 X Y
0 0 0 0	→	0 0 0 1	→	0 1 0 1

The final total stable state is dependent on which state variable changes first or if both change at the same time. The result is not predictable and causes a critical race. Critical races are to be avoided.

Cycle A cycle occurs when an asynchronous machine makes a transition through a series of unstable states. A cycle may exist that eventually reaches a stable state. Table 8.8 shows a partial flow table with a cycle.

Assume that the machine, whose partial flow table is shown in Table 8.8, starts in total stable state (F_2, F_1, X, Y) 0000 and that the input changes from $XY = 00$ to $XY = 01$. The transition path is as follows: the internal state changes from 00 to 10, then to 01 and finally to 11. The transition of state variables through several unstable internal states is a cycle. The cycle introduced in the input column 01 finally reaches a total stable state of 1101.

Now consider what happens when the machine is in total stable state 1101 and the input changes from 01 to 11. A transition from total stable state 1101 to total state 1111 and then to 1011 and to 0011 and to 0111 and to 1011, where it repeats the cycle indefinitely. Notice the arrows in Table 8.8 indicating the repeating cycle.

Three techniques are commonly used for making a critical-race free-state assignment:

1. Shared row state assignment
2. Multiple row state assignment
3. One hot state assignment

TABLE 8.8
Partial flow table showing a cycle

		Inputs XY			
F_2	F_1	00	01	11	10
0	0	(00)	10	01	
0	1		11	10	
1	0		01	00	
1	1		(11)	10	

TABLE 8.9
Asynchronous machine flow table

Present state	Next state			
	00	01	11	10
S_0	(S_0)/0	S_2/0	(S_0)/0	S_1/1
S_1	S_0/0	S_2/1	S_0/0	(S_1)/1
S_2	S_0/0	(S_2)/0	S_0/—	S_1/0

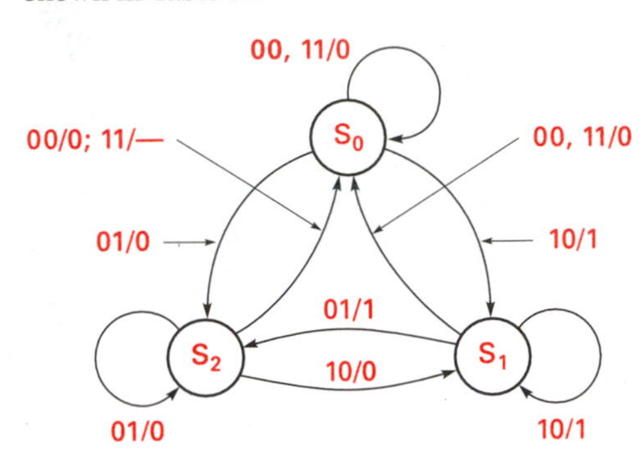

Figure 8.21
State diagram for the asynchronous machine shown in Table 8.9

8.4.2 Shared Row State Assignment

The term **shared row** derives its name from an extra row introduced in a flow table that is "shared" between two stable states. The two original stable states usually have more than one state variable change existing between them. The added "shared" row eliminates the multiple state variable change that may introduce race conditions. Adding a new row to the flow table permits an unstable state transition between the two stable states. Consider the asynchronous machine shown in the flow table of Table 8.9.

The state diagram for the asynchronous machine shown in Table 8.9 is shown in Figure 8.21.

Critical races can be determined by construction of a "transition diagram." A transition diagram contains as many vertices as the machine has states. Figure 8.22 shows the transition diagram for the asynchronous machine of Table 8.9.

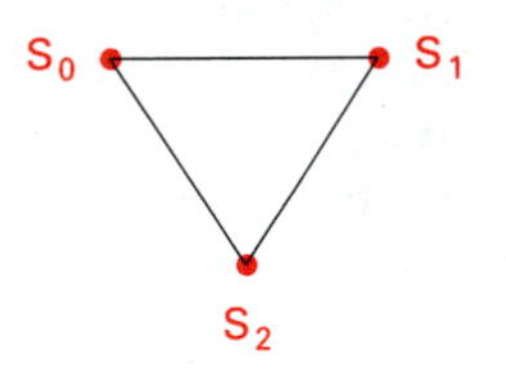

Figure 8.22
Transition diagram for the state machine shown in Table 8.9

Each state in the original flow table is indicated by a vertex in the transition diagram. From the asynchronous machine state diagram or the flow table, we can determine the state transitions that exist. We then indicate the transitions by drawing lines between the states. We find that state transitions between S_0 and S_1 exist so a line is drawn between those two state vertices. Likewise the remaining state transitions result in lines connecting the states.

Transitions exist among three states; each transition requires that only one state variable change to prevent race conditions. Starting with state S_0 and ending with S_2 we attempt to make a state assignment so that only one state variable change exists between each state transition. An attempt is shown in Table 8.10.

TABLE 8.10
Random state assignment for Table 8.9

State	State variable	
	F_2	F_1
S_0	0	0
S_1	0	1
S_2	1	0

If we were to assign state variables as indicated in Table 8.10, then the transition from state S_1 to S_2 would result in a possible critical race, because both state variables change at the same time. Any other possible state assignment for the asynchronous machine would also result in possible critical races for at least one state transition.

The addition of a new state, S_3, positioned between S_1 and S_2, eliminates the potential critical race condition. Figure 8.23 shows the transition diagram with the addition of S_3.

The transition from S_1 to S_2 must now pass through S_3, so a state assignment can be made that contains no race conditions (see Table 8.11). S_3 is a dummy state inserted

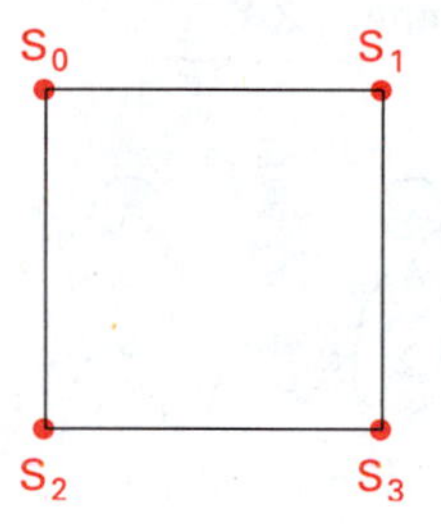

Figure 8.23
Transition diagram with S_3 added

TABLE 8.11
Transition table for flow table with S_3 added

Present state	Next state	
	F_2	F_1
S_0	0	0
S_1	0	1
S_2	1	0
S_3	1	1

TABLE 8.12
Modified flow table for initial flow Table 8.8

Present state	Next state			
	00	01	11	10
S_0	(S_0)/0	S_2/0	(S_0)/0	S_1/1
S_1	S_0/0	S_3/0	S_0/0	(S_1)/1
S_2	S_0/0	(S_2)/0	S_0/1	S_3/0
S_2	—	S_2/0	—	S_1/1

between S_1 and S_2 to avoid a critical race. The addition of new states increases the complexity of the logic and adds delay time to the circuit. But without the added state the asynchronous machine would not work.

The original flow table is modified in Table 8.12 to include state S_3.

Consider the flow table shown in Table 8.13. Each column has two different stable states for a given input, creating a potential critical race for each column. The transition diagram for the flow table is shown in Figure 8.24. Figure 8.25, the expanded transition diagram, shows the addition of new states, inserted between those states where critical races occur.

From the flow table in Table 8.13, we can determine that critical races exist between states S_1 and S_2, S_0 and S_2, S_1 and S_3, and S_0 and S_3.

A state assignment map is useful in identifying adjacent states. Figure 8.26 shows the state assignment map for the expanded transition diagram. The state assignment map is created starting with S_0 being assigned state $F_3F_2F_1 = 000$. States S_1, S_2, and S_3 must be adjacent or have an intermediate state to prevent critical races. We located S_1, S_2, and S_3 logically adjacent to S_0 as shown in Figure 8.26. Other state arrangements could also have been made. An internal state transition from S_1 to S_2 changes two state variables so a dummy state A was inserted between S_1 and S_2. Likewise the remainder of the assignment map is filled in.

The expanded flow table, shown in Table 8.14, includes the original states and the dummy states added to prevent critical races. Notice that dummy state A replaces original state S_2 in the 01 input column S_1. A transition from S_1 to S_2 now goes through state

TABLE 8.13
Flow table with multiple critical races

State	Inputs *X Y*			
	00	01	11	10
S_0	S_2	(S_0)	S_1	(S_0)
S_1	(S_1)	S_2	(S_1)	S_0
S_2	(S_2)	(S_2)	S_3	S_3
S_3	S_1	S_0	(S_3)	(S_3)

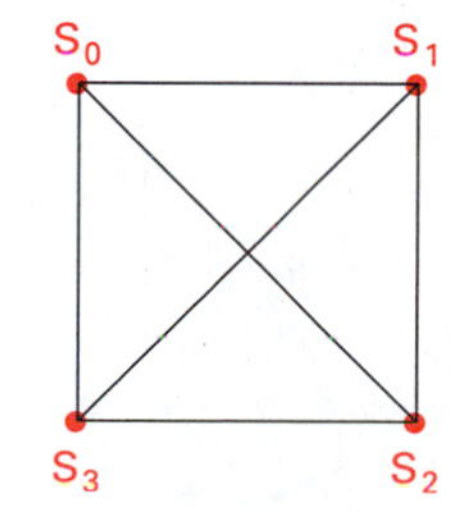

Figure 8.24
Transition diagram for flow Table 8.13

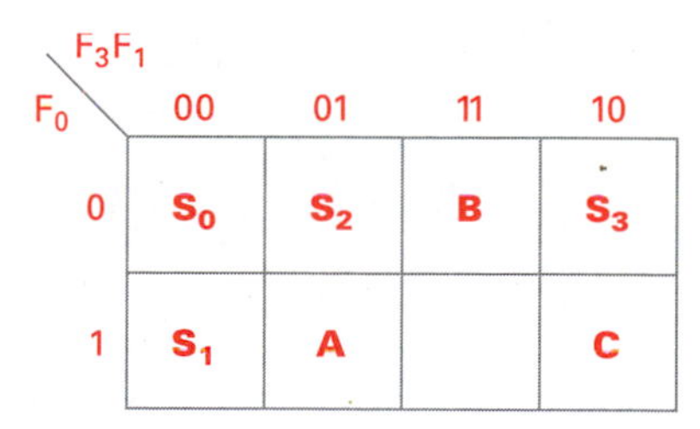

Figure 8.25
Expanded transition diagram

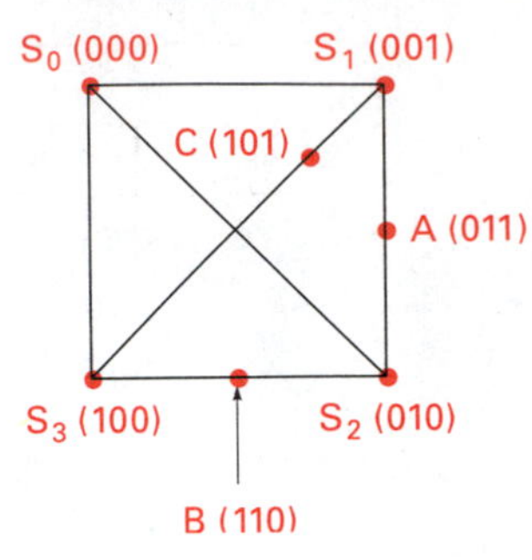

Figure 8.26
State assignment map

TABLE 8.14
Expanded flow table

F_3	F_2	F_1	State	Inputs XY 00	01	11	10
0	0	0	S_0	S_2	(S_0)	S_1	(S_0)
0	0	1	S_1	(S_1)	A	(S_1)	S_0
0	1	0	S_2	(S_2)	(S_2)	B	B
1	0	0	S_3	C	S_0	(S_3)	(S_3)
0	1	1	A	—	S_2	—	—
1	1	0	B	—	—	S_3	S_3
1	0	1	C	S_1	—	—	—

A. Likewise dummy state B replaces original state S_3 in the 11 and 10 input columns of S_2. Transitions from S_2 to S_3 now go through state B.

8.4.3 Multiple Row State Assignment

The multiple row state assignment technique specifies that each row in the original flow table be replaced with two rows. Each new row or total state is equivalent to the original state before splitting; for instance, state $a = a_1 = a_2$. If row a existed in a flow table it would be replaced by rows a_1 and a_2 in an expanded flow table. The assignment for total state a_1 is the complement of the assignment for state a_2. By splitting each original internal state into two parts and by making the state assignments of the split states complements of one another we can create a universal state assignment.

Figure 8.27 and Table 8.15 show the universal state assignment for a four-row flow table, realized with three state variables. Note that each original state (a, b, c, d) is adjacent to every other original state. State a_1 is adjacent to b_2, c_1, and d_1. Likewise state a_2 is adjacent to states b_1, c_2, and d_2.

Figure 8.27
Universal four-row state assignment map

F_1 \ F_3F_2	00	01	11	10
0	a_1	d_1	b_1	c_1
1	b_2	c_2	a_2	d_2

TABLE 8.15
Four-row state assignment

State	State variable F_3	F_2	F_1
a_1	0	0	0
a_2	1	1	1
b_1	1	1	0
b_2	0	0	1
c_1	1	0	0
c_2	0	1	1
d_1	0	1	0
d_2	1	0	1

TABLE 8.16
Universal four-row flow table for Table 8.13

Original state	Split state	State variables F_3	F_2	F_1	Inputs XY 00	01	11	10
S_0	S_{01}	0	0	0	S_{21}	S_{01}	S_{12}	S_{01}
S_0	S_{02}	1	1	1	S_{22}	S_{02}	S_{11}	S_{02}
S_1	S_{11}	1	1	0	S_{11}	S_{21}	S_{11}	S_{02}
S_1	S_{12}	0	0	1	S_{12}	S_{22}	S_{12}	S_{01}
S_2	S_{21}	1	0	0	S_{21}	S_{21}	S_{32}	S_{32}
S_2	S_{22}	0	1	1	S_{22}	S_{22}	S_{31}	S_{31}
S_3	S_{31}	0	1	0	S_{11}	S_{01}	S_{31}	S_{31}
S_3	S_{32}	1	0	1	S_{12}	S_{02}	S_{32}	S_{32}

Figure 8.28
Universal eight-row state assignment map

	00	01	11	10
00	a_1	b_1	c_1	d_1
01	e_1	f_1	g_1	h_1
11	c_2	d_2	a_2	b_2
10	g_2	h_2	e_2	f_2

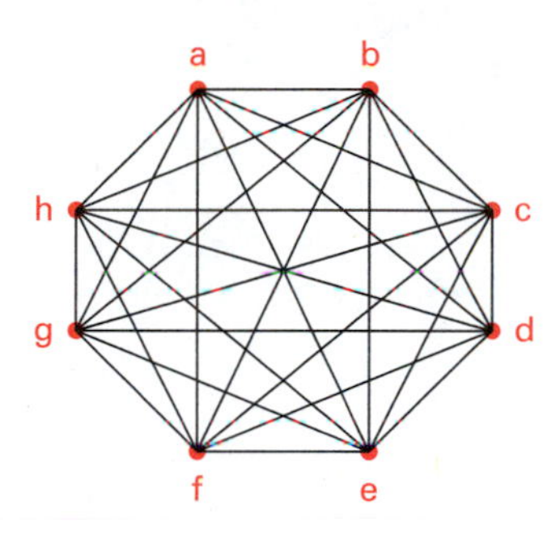

Figure 8.29
Universal eight-row transition diagram

Using the four-row universal state assignment we expand Table 8.13 to what is shown in Table 8.16.

The universal four-row state assignment will work for any four-row flow table. All possible state transitions are covered without any multiple state variable changes, therefore the universal four-row state assignment is critical race free.

An eight-row (four state variables) flow table can be realized with the race free state assignment given in Figure 8.28. Notice, as in a four-row flow table, that the original state variables are split and that the partitioned variable assignments are complements of one another. The universal eight-row transition diagram is shown in Figure 8.29.

8.4.4 One Hot State Assignment

One hot state assignments are made so that only one variable is active or "hot" for each row in the original flow table. This technique requires as many state variables as there are rows in a flow table. Additional rows are introduced to provide single variable changes between internal state transitions.

TABLE 8.17
Flow table

Next state					Outputs			
F_4	F_3	F_2	F_1	State	00	01	11	10
0	0	0	1	a	(a)	b	c	c
0	0	1	0	b	a	(b)	c	d
0	1	0	0	c	a	b	(c)	(c)
1	0	0	0	d	(d)	b	c	(d)

TABLE 8.18
One hot state assignment flow table

State variables					Inputs $X\ Y$			
F_4	F_3	F_2	F_1	State	00	01	11	10
0	0	0	1	a	(a)	~~b~~ Q	~~c~~ R	~~c~~ R
0	0	1	0	b	~~a~~ Q	(b)	~~c~~ S	~~d~~ T
0	1	0	0	c	~~a~~ R	~~b~~ S	(c)	(c)
1	0	0	0	d	(d)	~~b~~ T	~~c~~ U	(d)
0	0	1	1	Q	a	b	—	—
0	1	0	1	R	a	—	c	c
0	1	1	0	S	—	b	c	—
1	0	1	0	T	—	b	—	d
1	1	0	0	U	—	—	c	—

Consider the flow table given in Table 8.17. Four state variables are used to represent the four rows in the table. Each row is represented by a case where only one of the four state variables is a 1. A transition from state a to state b requires two state variable changes. By introducing a new row, one whose state assignment contains 1s where both states a and b have 1s, permits the transition between a and b without a race.

The complete one-hot state assignment flow table is shown in Table 8.18. Notice the additional rows, one for each transition in the original flow table.

The transition from state b to state a is accomplished by passing through dummy state Q. If the machine is in state b and a 00 input is received then the machine will change state to Q and then to a, without any further change in the inputs. State a is replaced by Q in the b row. Row c with a state assignment of 0100 makes a change to state a when the input is 00. State a has an assignment of 0001. By passing through state R, on the way from c to a, races are eliminated. The state transition path is c (0100) $\rightarrow$ R (0101) $\rightarrow$ a (0001).

8.5 ASYNCHRONOUS DESIGN PROBLEMS

Several complete design problems are presented here to illustrate the complete process of synthesizing an asynchronous circuit.

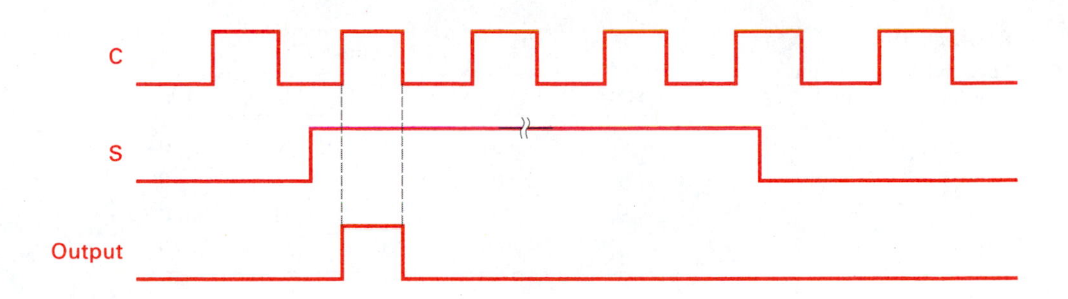

Figure 8.30
Timing diagram for design problem 1

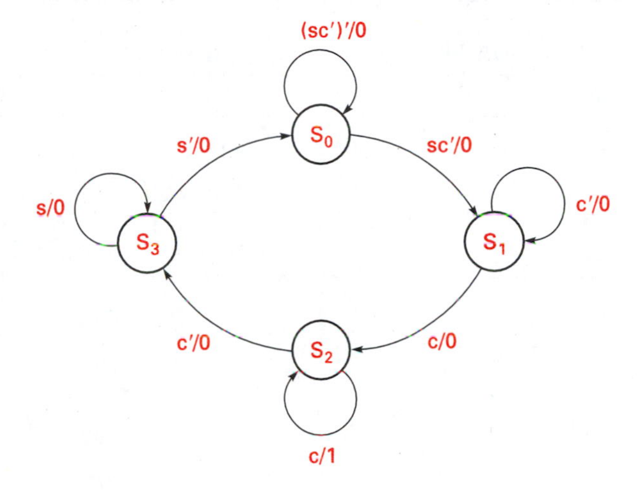

Figure 8.31
State diagram for design problem 1

8.5.1 Asynchronous Design Problem 1

Design an asynchronous sequential machine that will permit passage of a complete, single clock pulse from a continuous stream of input clock pulses, when an external input signal is high. The machine is to ignore the case where both the input clock and the control signal go high at the same time. The resulting output timing would appear as shown in Figure 8.30. A state diagram and flow table are constructed from the problem statement and are shown in Figure 8.31 and Table 8.19.

The state assignment is straightforward for this problem. The transition diagram and transition table are shown in Figure 8.32 and Table 8.20.

The Boolean equations for F_2, F_1, and output Z are derived from the transition table:

$$F_2^+ = f(F_2, F_1, S, C) = \sum(5, 7, 10, 11, 12, 13, 14, 15)$$
$$F_1^+ = f(F_2, F_1, S, C) = \sum(2, 4, 5, 6, 7, 13, 15)$$
$$Z = f(F_2, F_1, S, C) = \sum(13, 15)$$

The equations for F_2, F_1, and Z are loaded into their respective K-maps in Figure 8.33.

The final equations are realized by circling each prime implicant so that static hazards are eliminated. The final equations are

$$F_2^+ = F_2F_1 + F_1C + F_2S$$
$$F_1^+ = F_2'F_1 + F_1C + F_2'SC'$$
$$Z = F_2F_1C$$

Figure 8.34 shows the logic diagram for the circuit.

Figure 8.32
Transition diagram

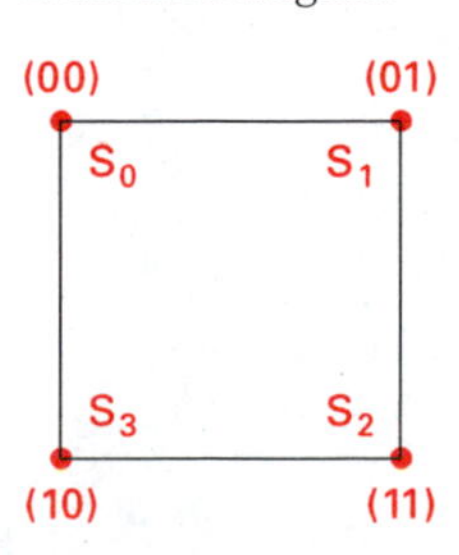

TABLE 8.19
Asynchronous design problem 1 flow table

Internal state	Inputs S C 00	01	11	10
S_0	(S_0)/0	(S_0)/0	(S_0)/0	S_1/0
S_1	(S_1)/0	S_2/0	S_2/0	(S_1)/0
S_2	S_3/0	(S_2)/1	(S_2)/1	S_3/0
S_3	S_0/0	S_0/0	(S_3)/0	(S_3)/0

TABLE 8.20
Transition table

State	State variables F_2	F_1	Inputs S C 00	01	11	10
S_0	0	0	00/0	00/0	00/0	01/0
S_1	0	1	01/0	11/0	11/0	01/0
S_2	1	1	10/0	11/1	11/1	10/0
S_3	1	0	00/0	00/0	10/0	10/0

Figure 8.33
K-maps for asynchronous design problem 1

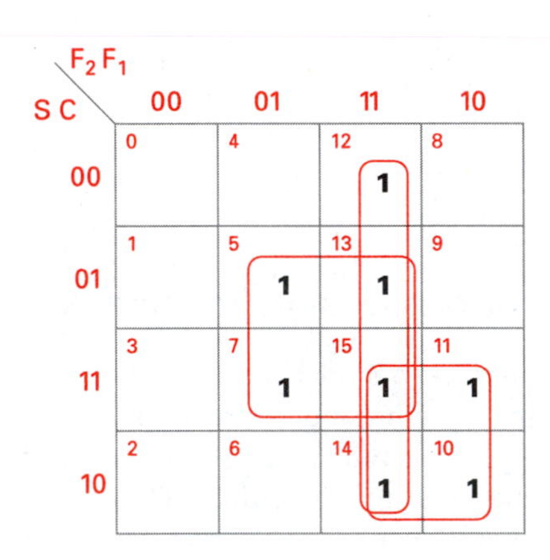

(a) F_2^+ K-Map

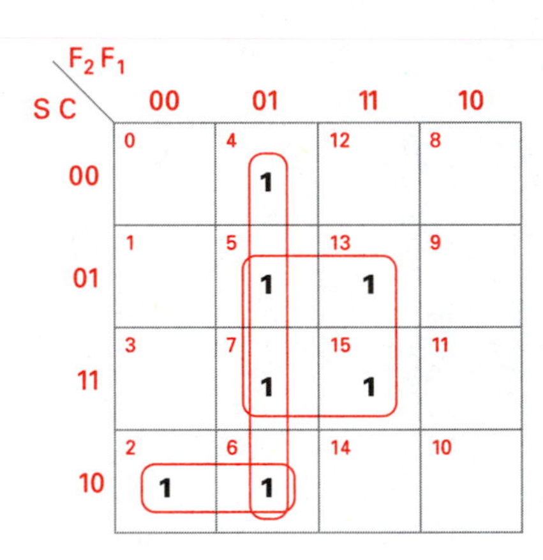

(b) F_1^+ K-Map

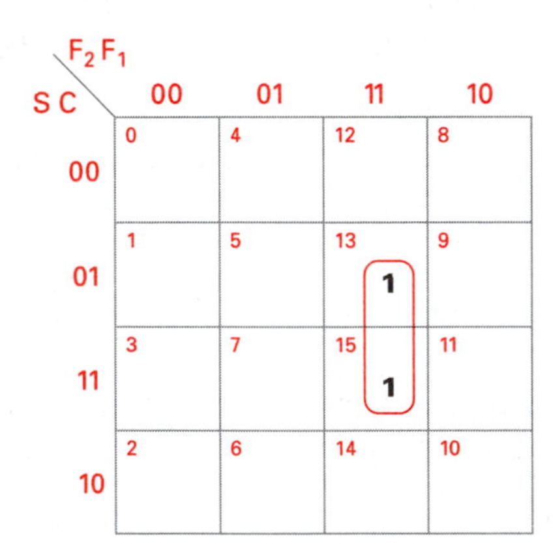

(c) Z K-Map

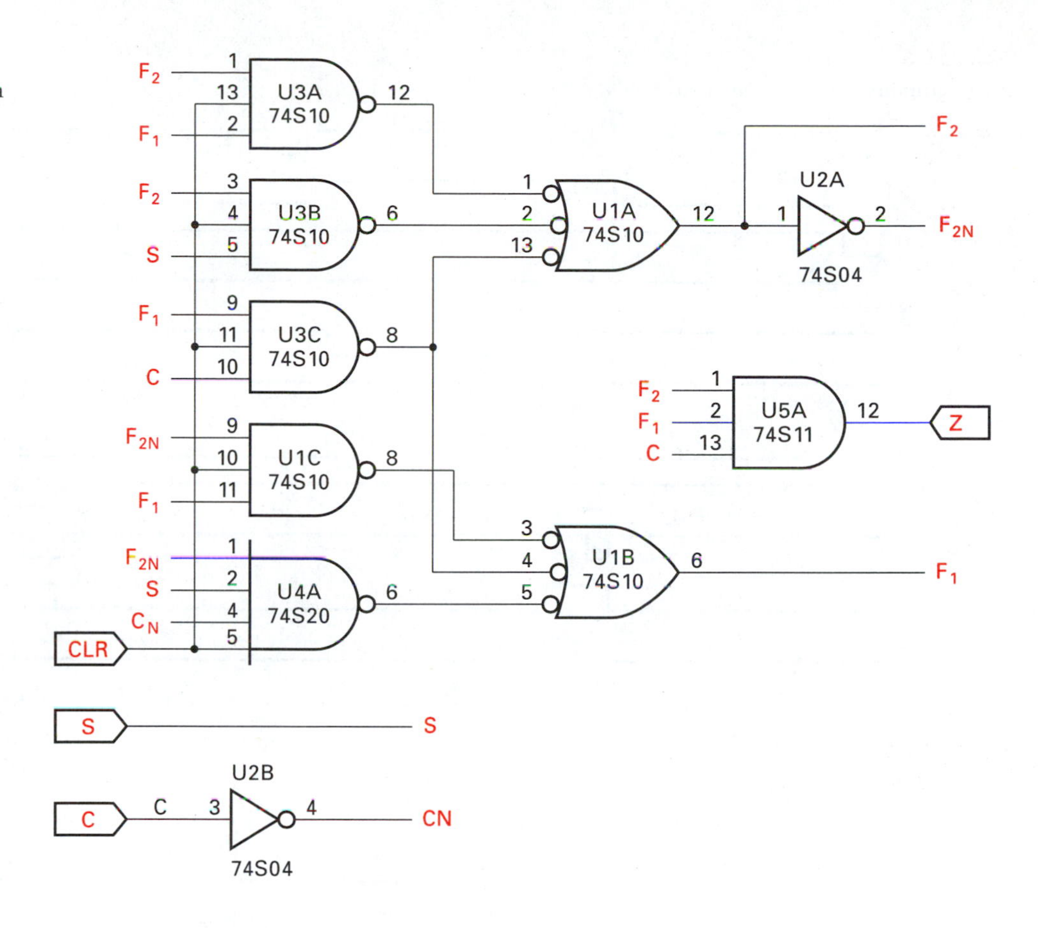

Figure 8.34
Asynchronous design problem 1 logic diagram

The timing diagram for the circuit in Figure 8.34 is shown in Figure 8.35. The input C changes state every 75 ns. An extra input, CLR, was used to force the state machine into a known state. Each gate output timing is also given to help visualize the circuit operation.

8.5.2 Asynchronous Design Problem 2

Design an asynchronous sequential circuit that has two inputs X_2 and X_1 and one output Z. The output is to remain a 0 as long as X_1 is a 0. The first change in X_2 that occurs while X_1 is a 1 will cause Z to be a 1. Z is to remain a 1 until X_1 returns to 0.

1. Construct a timing diagram to represent input and output changes.
2. Construct a state diagram and flow table.
3. Reduce the flow table, if possible, using a merger diagram.
4. Make a critical race free state assignment.
5. Determine the excitation and output equations.
6. Draw the logic diagram and test the circuit.

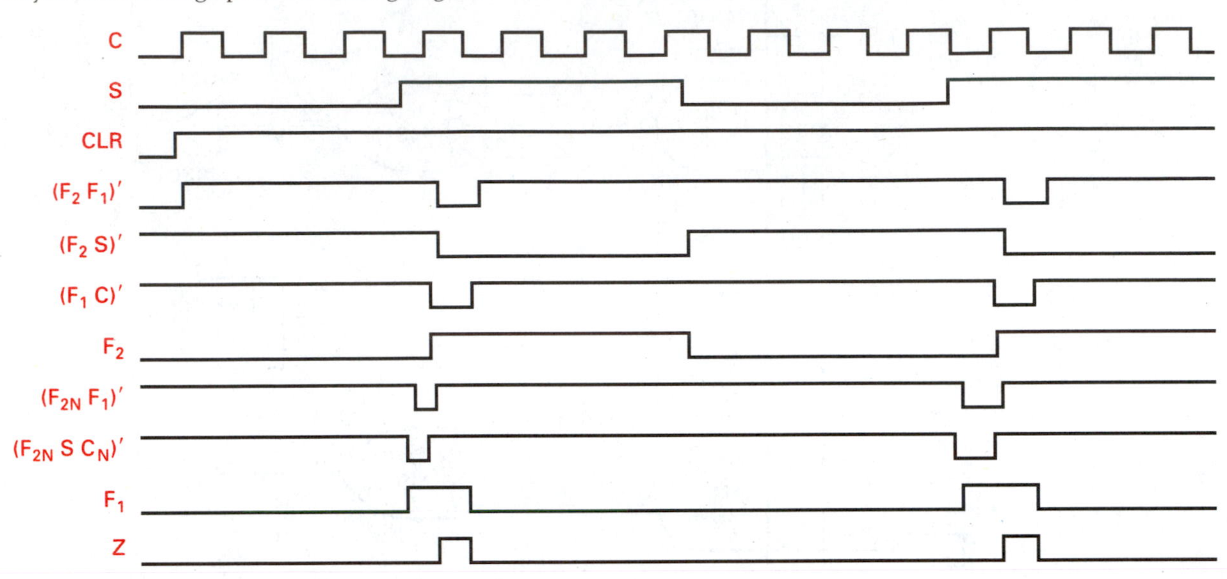

Figure 8.35
Asynchronous design problem 1 timing diagram

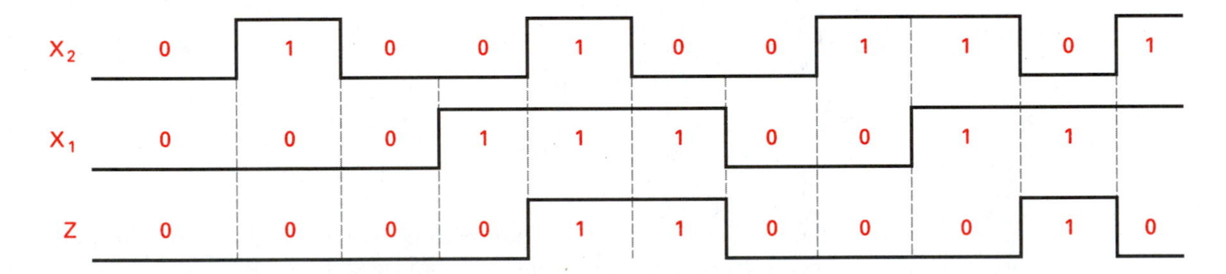

Figure 8.36
Predicted timing diagram

An illustrative, predicted timing diagram is shown in Figure 8.36. A primitive state diagram is shown in Figure 8.37.

A primitive flow table can be constructed from the state diagram or directly from the verbal description of the problem. The purpose of constructing a predicted timing diagram was to help understand the problem. The flow table is shown in Table 8.21.

Present Next-State Outputs

A merger diagram is now constructed to produce a reduced flow table. Figure 8.38 shows the merger diagram. The reduced flow table is shown in Table 8.22. The reduced state diagram is shown in Figure 8.39. A transition diagram, Figure 8.40, is drawn to help determine the state assignment.

A transition between S_0 and S_2 requires that dummy state S_3 be added to prevent critical race. A new state table is constructed reflecting the addition of S_3.

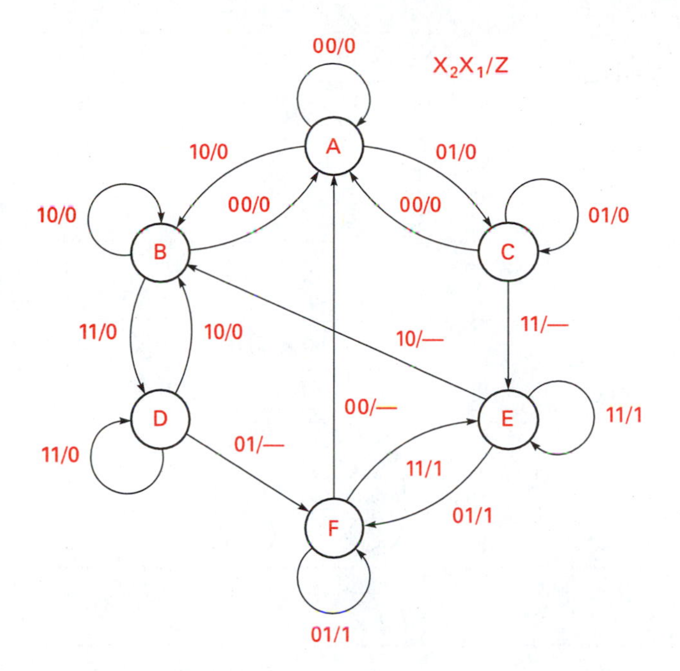

Figure 8.37
Asynchronous design problem 2 state diagram

TABLE 8.21
Primitive flow table

Present state	Next state 00	01	11	10	x_2x_1	Outputs 00	01	11	10
A	(A)	C	—	B		0	0	—	0
B	A	—	D	(B)		0	—	0	0
C	A	(C)	E	—		0	0	—	—
D	—	F	(D)	B		—	—	0	0
E	—	F	(E)	B		—	1	1	—
F	A	(F)	E	—		—	1	1	—

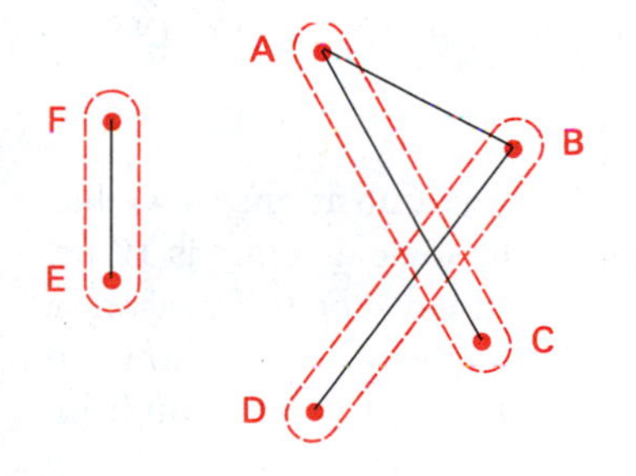

$A \sim C = S_0$

$B \sim D = S_1$

$E \sim F = S_2$

Figure 8.38
Merger diagram

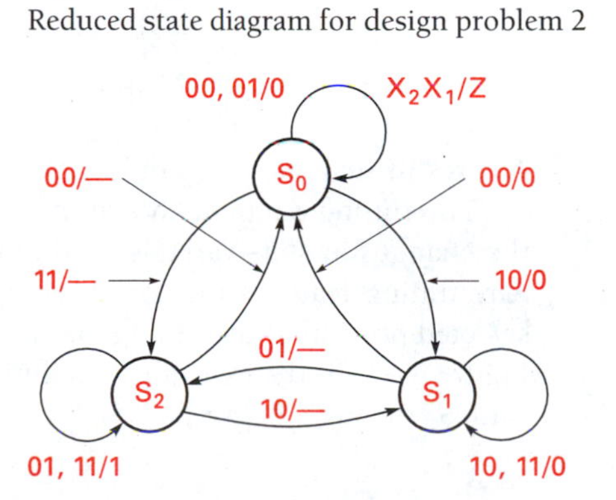

Figure 8.39
Reduced state diagram for design problem 2

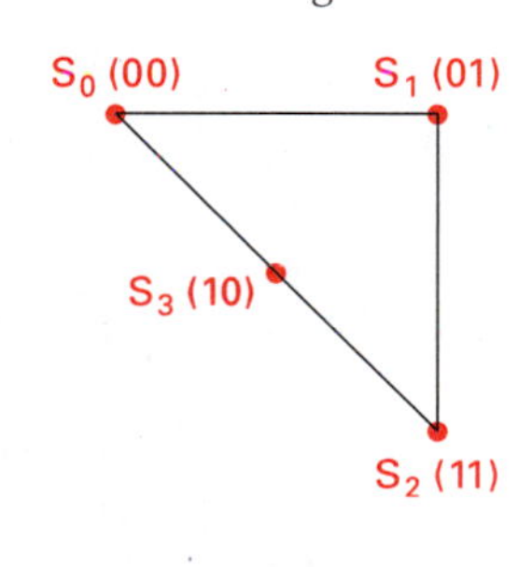

Figure 8.40
Transition diagram

TABLE 8.22
Reduced flow table for design problem 2

Present state	Next state x_2x_1 00	01	11	10	Output 00	01	11	10
S_0	(S_0)	(S_0)	S_2	S_1	0	0	—	0
S_1	S_0	S_2	(S_1)	(S_1)	0	—	0	0
S_2	S_0	(S_2)	(S_2)	S_1	—	1	1	—

TABLE 8.23
Final flow table reflecting the state assignment

Present state	F_2	F_1	Next state x_2x_1 00	01	11	10	Output 00	01	11	10
S_0	0	0	(S_0)	(S_0)	S_3	S_1	0	0	—	0
S_1	0	1	S_0	S_2	(S_1)	(S_1)	0	—	0	0
S_2	1	1	S_3	(S_2)	(S_2)	S_1	—	1	1	—
S_3	1	0	S_0	—	S_2	—	—	—	—	—

TABLE 8.24
Transition table for design problem 2

Present			Next state x_2x_1 00		01		11		10		Output			
	F_2	F_1	F_2^+	F_1^+	F_2^+	F_1^+	F_2^+	F_1^+	F_2^+	F_1^+	00	01	11	10
S_0	0	0	0	0	0	0	1	0	0	1	0	0	—	0
S_1	0	1	0	0	1	1	0	1	0	1	0	—	0	0
S_2	1	1	1	0	1	1	1	1	0	1	—	1	1	—
S_3	1	0	0	0	—	—	1	1	—	—	—	—	—	—

The flow table of Table 8.23 is now converted to a transition table as shown in Table 8.24. From the transition table we can derive the state variable next-state and output equations.

The next-state and output equations are

$$F_2^+ = f(F_2, F_1, X_2, X_1) = \sum(3, 5, 11, 12, 13, 15) + \sum d(9, 10)$$
$$F_1^+ = f(F_2, F_1, X_2, X_1) = \sum(2, 5, 6, 7, 11, 13, 14, 15) + \sum d(9, 10)$$
$$Z = f(F_2, F_1, X_2, X_1) = \sum(13, 15) + \sum d(3, 5, 8, 9, 10, 11, 12, 14)$$

The resulting logic diagram is shown in Figure 8.41.

The timing diagram shown in Figure 8.42 indicates the state transitions as well as the changes in state variables and output. The resolution of the logic diagram is 10 ns. Any timing features that occur in less than 10 ns are not shown. For this reason, a selected portion of the timing diagram was expanded to a resolution of 1 ns as shown in Figure 8.43. In the expanded resolution timing diagram, we can see the transition from state S_2 through S_3 to S_0.

Figure 8.41
Logic diagram for design problem 2

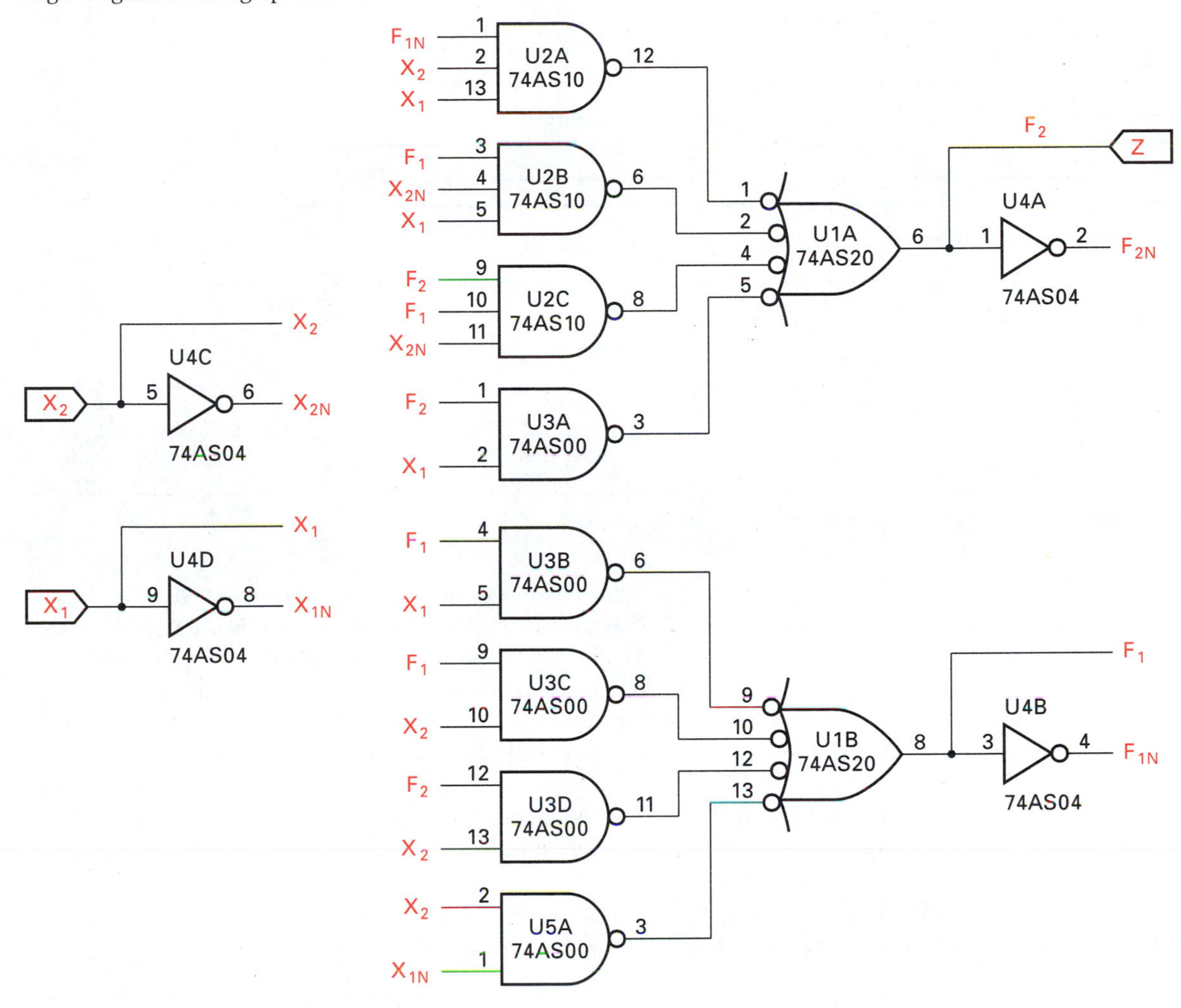

Figure 8.42
Timing diagram with 10 ns resolution

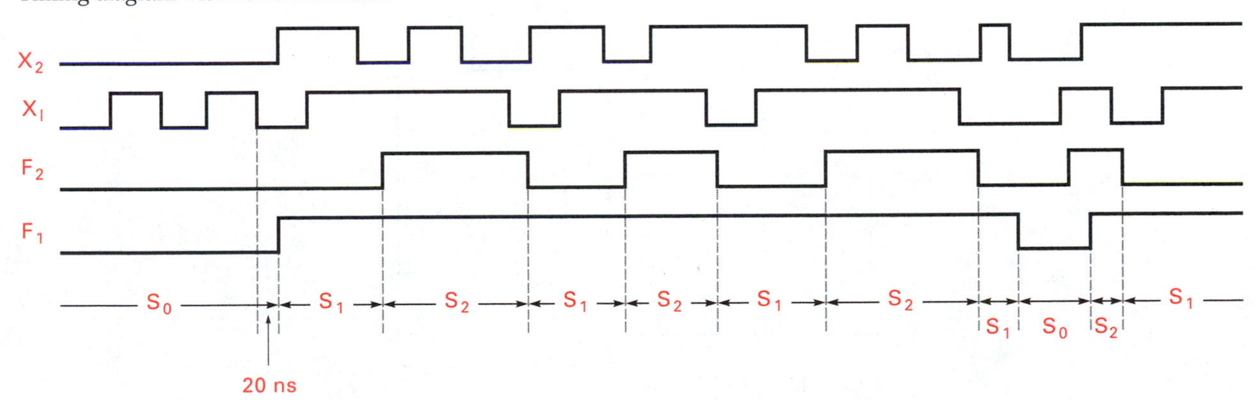

Figure 8.43

Expanded portion timing diagram with 1 ns resolution

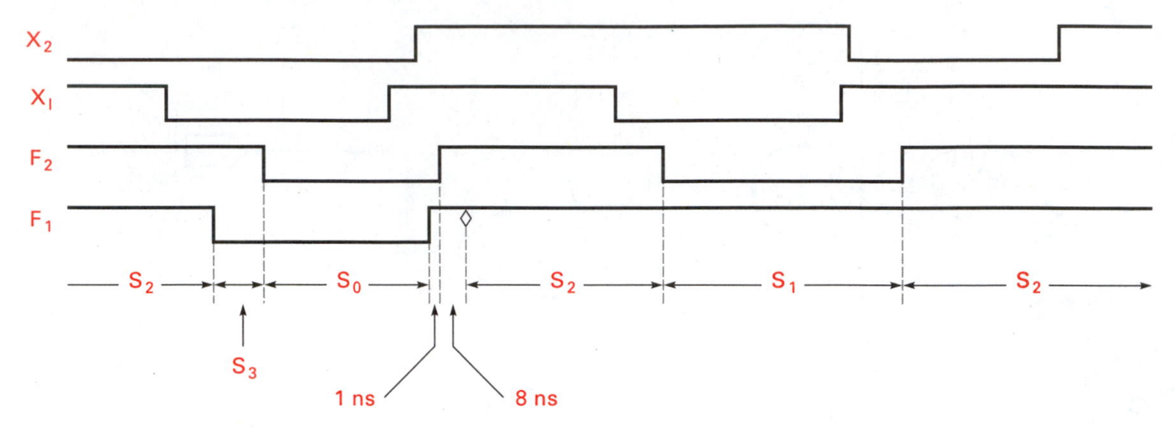

8.6 DATA SYNCHRONIZERS

In order to "synchronize" a system where several modules are using their own clocks, but no common system clock is available, a circuit called a *data synchronizer* is necessary. The task of the circuit is to take what amounts to asynchronous inputs and "synchronize" them; that is, make them consistent with the module clock. Consider the logic in Figure 8.44. A clocked flip-flop is used to control when the input signal is presented to the synchronous logic within the module. The clock is a module clock, not a system clock. The input signal occurs asynchronously with respect to the module clock. We do not know when the input will change. Because the input can change at any time if we do not synchronize the input to the module clock, violation of the setup, hold, and delay timing rules may occur.

The flip-flop shown in the circuit is clocked with the same clock used by the synchronous module system. The asynchronous input is presented to the input flip-flop

Figure 8.44

Simple synchronizing circuit for asynchronous data inputs to a synchronous system

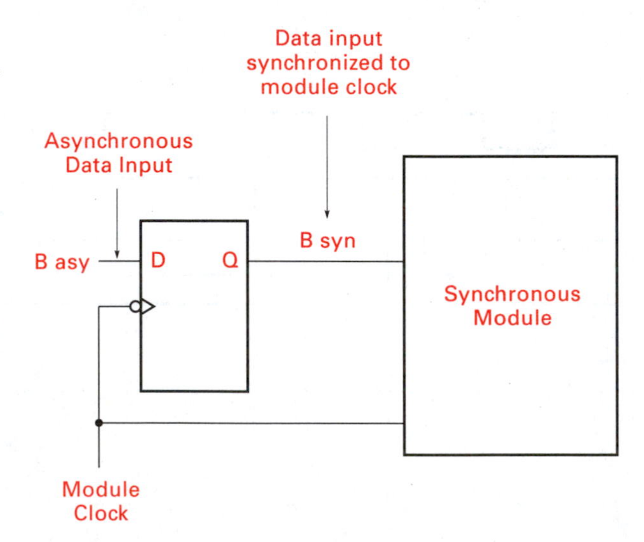

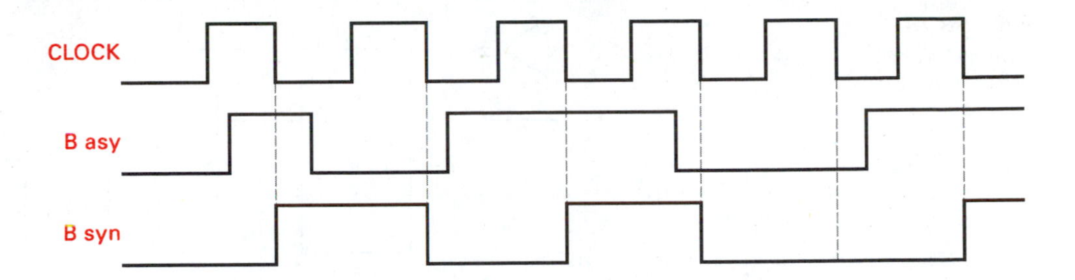

Figure 8.45
Synchronizer timing

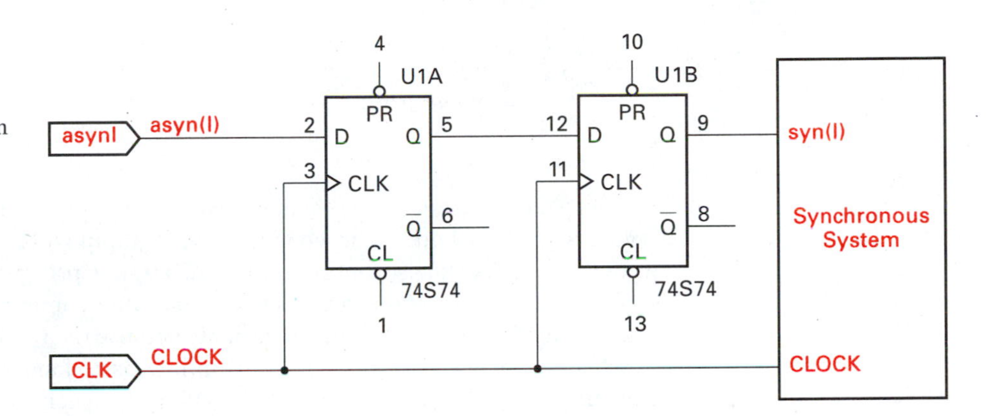

Figure 8.46
Cascaded flip-flops used to synchronize asynchronous inputs to a synchronous system

and presented to the synchronous system delayed by the clock to Q output propagation time. The asynchronous input is now synchronized at the Q output. Figure 8.45 shows the timing of the operation.

When the setup and hold timing requirements are not met, metastability problems occur in the simple synchronizer circuit. By adding two flip-flops in series, with the output of the first providing the input to the second, we can further reduce the possibility of instability. Consider the circuit diagram in Figure 8.46.

Because the asynchronous input can occur at any time, it may violate the timing requirements of the first flip-flop, causing the device to become metastable. Output Q_A is presented to D_B, and Q_B is presented to the synchronous system. Output Q_B is guaranteed to be stable if the Q_A output has come out of its metastable state before the next clock edge that loads Q_B arrives. The probability of metastable states occurring are further reduced by delaying the transfer of the asynchronous input to the synchronous system by one clock period. Figure 8.47 shows a possible timing diagram for the synchronizer circuit.

Figure 8.47
Synchronizer circuit timing diagram

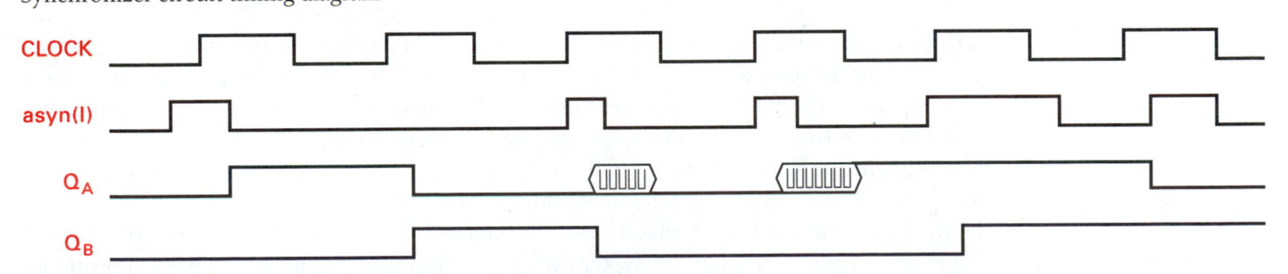

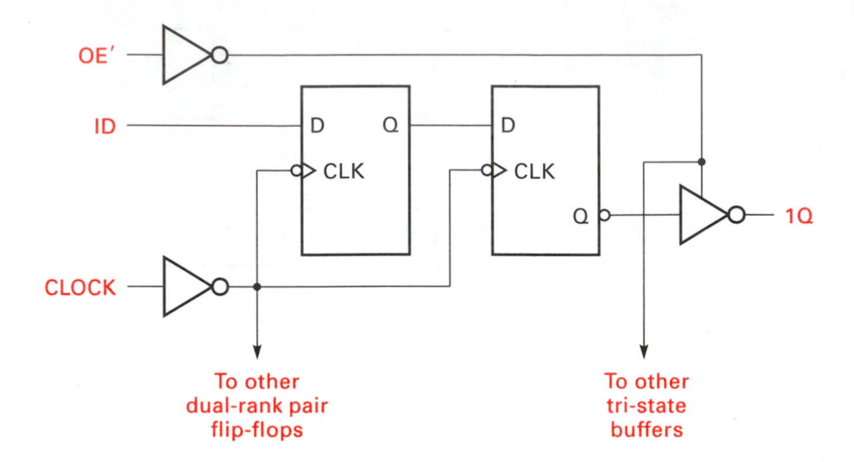

Figure 8.48
74AS4374 octal dual-rank D flip-flop

Metastability states are encountered in two places in the timing diagram of Figure 8.47. They both occur when the asyn(I) input violates the setup time of flip-flop A. In the first case, the Q_A output returns to a 0 state prior to the next clock edge. In the second case, the Q_A output returns to a 1 state after coming out of the metastable state. The Q_B output does not become metastable because the Q_A output is not in a metastable state on the next clock edge. By waiting until Q_A is no longer in a metastable state before clocking, flip-flop B permits synchronization of asynchronous input asyn(I) with the synchronous system and reduces the probability of an indeterminate input. The clock frequency limit occurs when the Q_A output is still in a metastable state at next clock edge. An IC containing "dual-rank" flip-flops suitable for synchronizing asynchronous inputs is shown in Figure 8.48.

The faster speeds permitted by using advanced Schottky technology coupled with the dual-rank D flip-flops reduces the synchronizing problems associated with metastability. The dual-rank flip-flops are used to decouple the metastable state of the first flip-flop that may occur due to an asynchronous input. By the time the second flip-flop is clocked, the first has usually come out of the metastable state. This approach permits asynchronous inputs to be coupled into a synchronous system.

8.6.1 Interface Protocol Asynchronous Cell

Circuits used to synchronize data signals to system module clocks prior to the introduction of the **interface protocol asynchronous cell** (IPAC) required MSI devices like the 74AS4374. The IPAC enables the synchronization of asynchronous inputs to system modules.

Consider a typical "handshake" communication protocol between two modules in a system. One module "requests" data transmission and the other module "grants" the request. The request-grant sequence is necessary to make sure both the sending and receiving units are ready. This was the approach used in the interface design presented in Chapter 7. The plotter interface requested data from the computer and the computer granted the request; both were operating from a common clock.

Asynchronous conditions often exist in digital systems, where a common clock is not practical. Computer systems have information buses over which data, address, and control information is exchanged between memory, input-output (I/O), and the central processing unit. Consider the system diagram shown in Figure 8.49. Each module in

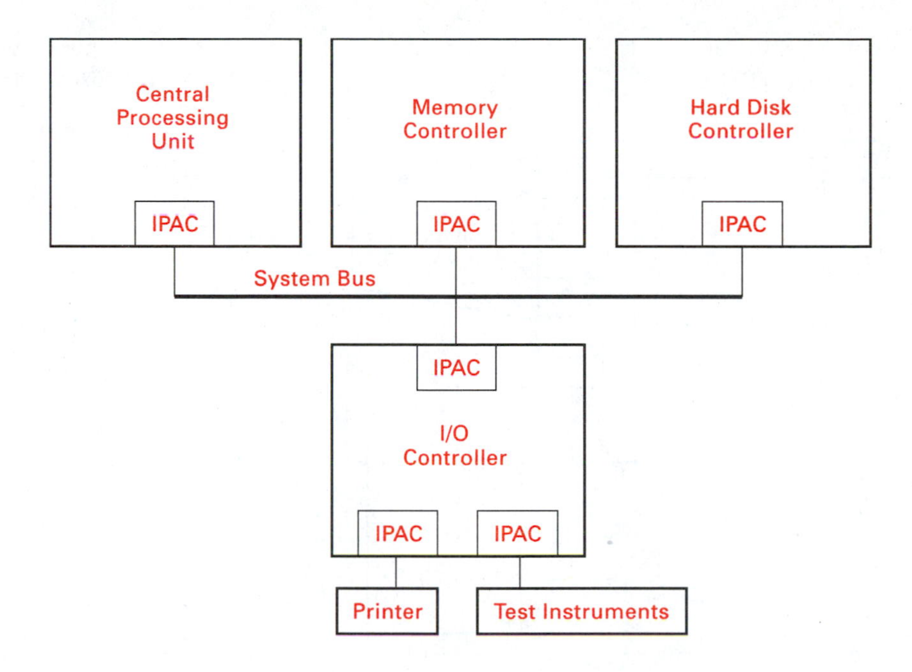

Figure 8.49
Simple computer system block diagram

the system is required to communicate with the other modules over the system bus. For instance, when the central processing unit (CPU) reads data from memory it must place an address on the address bus and then generate the proper control signals necessary to effect a data transfer from the memory to the CPU.

A common clock required for synchronous operation of the module sequential circuit controllers is not practical if the system is to operate at high speeds. Problems of clock skew, extra wires or circuit board traces, fan-out, and reduced system speed limit the use of a shared clocking scheme.

The IPAC blocks associated with each of the system components, shown in Figure 8.49, permits the asynchronous transfer of information between system modules without the need for a common clock. The IPACs are edge activitated registers that need no separate clock to store data. Instead data are stored in the registers in response to changing signal levels of the data itself. The registers can be configured to store data on the rising or falling edge of the data signals. Because data are stored on their own rising or falling edge, there is no need for a clock along with the setup, hold, and delay times associated with synchronous devices. Because no clock setup and hold timing requirements exist the metastability problem is eliminated. The input data signals provide both data inputs and clocks.

Advanced Micro Devices, which created the IPAC, has made the approach available to system designers in the form of a six-cell programmable logic device (PLD), the PAL22IP6. Two types of cells are on the chip: a set-reset latch and a dual input T latch. The latch outputs and external data inputs provide inputs to a combinational logic array, permitting the generation of state variable excitation inputs. Figure 8.50 shows a typical IPAC cell within the 22IP6 integrated circuit.

The collection of AND gates on the left of the diagram in Figure 8.50 form an array producing product terms that can be ORed together to produce state variable inputs. Notice that the flip-flop is shown as either an S–R latch or a dual input T latch. The 22IP6 device contains six such cells, three cells contain R–S and three contain T latches. Exclusive OR gates permit complementing excitation equations. The latch output or a

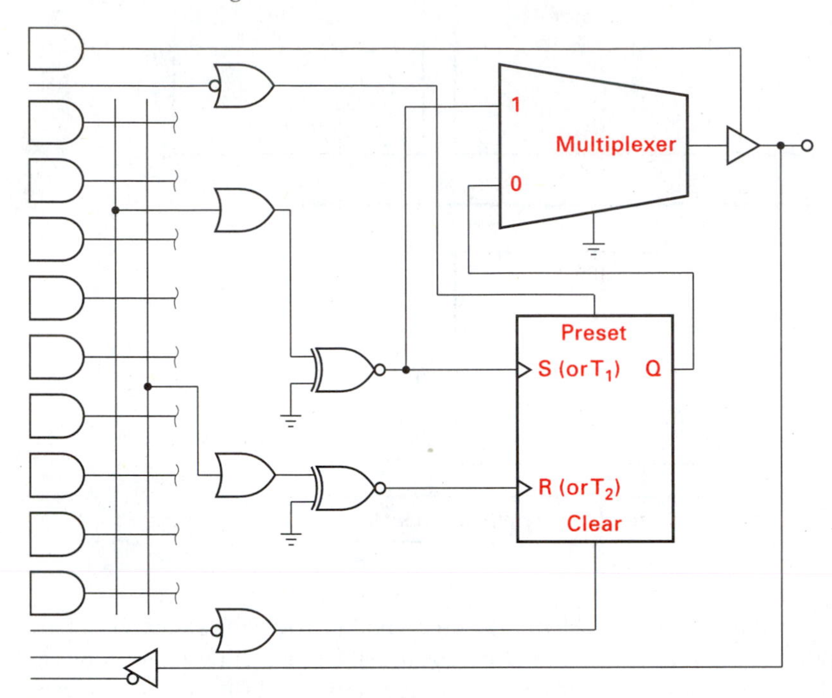

Figure 8.50
AMD 22IPC IPAC logic cell

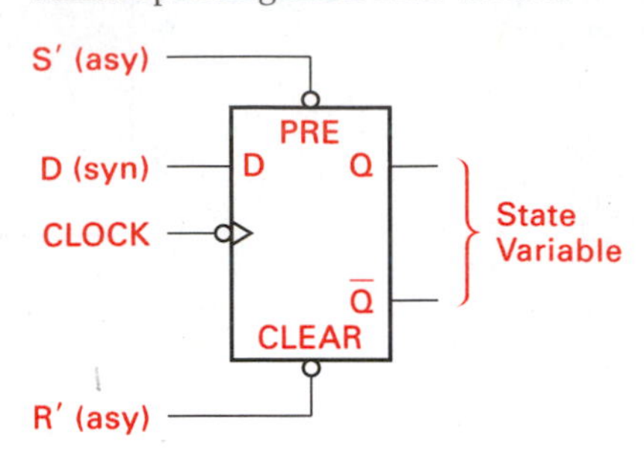

Figure 8.51
Mixed operating mode state variable

combinational output is sent to a multiplexer, so that either can be made available as an output. The output is controlled by a tristate buffer, whose control input is generated from a *P* term. The *Q* output of the flip-flip can also be fed back into the AND array for the realization of asynchronous circuits.

8.7 MIXED OPERATING MODE ASYNCHRONOUS CIRCUITS

Significant problems arise in asynchronous circuit design due to critical races and hazards. Critical races and hazards result from unequal propagation delay paths in the combinational logic that realizes the state variable excitation equations. Static and dynamic hazards can be eliminated by providing holding terms in the excitation logic. Such holding terms increase the logic complexity.

A technique called **mixed operating mode** (MOM) or *self-synchronization* has been proposed by a number of researchers as a solution to asynchronous timing problems. This approach partitions each state variable so that it has both synchronous and asynchronous inputs (many flip-flops have both synchronous and asynchronous inputs on the same chip and standard cell flip-flops can be created with both input types). Figure 8.51 shows a typical mixed mode flip-flop.

By using both synchronous and asynchronous inputs of the same state variable, a particularly difficult problem can be solved. The problem is associated with what are called **essential hazards**. If an essential hazard exists for a given state transition, then the synchronous inputs are used. If an essential hazard does not exist, then the asynchronous inputs are used. This implies that a clock must be generated whenever essential hazards occur between state transitions. And that is exactly what the MOM approach proposes. Extra logic is used to sense the existence of an essential hazard. If one exists, then a clock pulse is generated and distributed to the specific state variable(s) that needs it. By providing such an **internal clock** pulse to only those state variables where essential hazards may occur eliminates the critical races without slowing down all of the state transitions.

Figure 8.52 shows the model diagram for a mixed operating mode system. Two sets

Figure 8.52
Mixed operating mode sequential circuit model

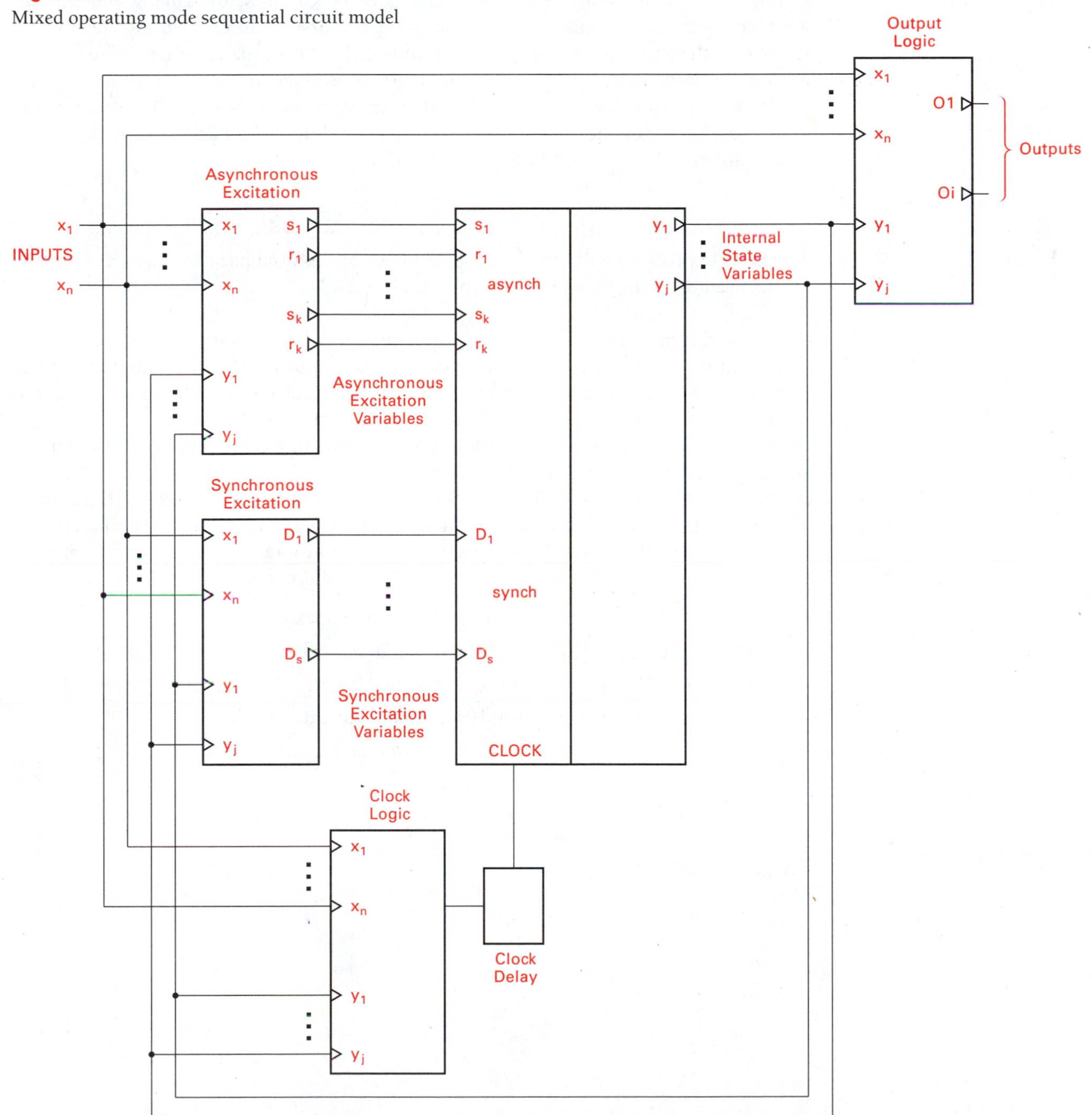

of excitation functions are produced, one for asynchronous and one for the internally clocked synchronous state variable transitions. The asynchronous excitation inputs, $r_1 s_1$ through $r_k s_k$, and the synchronous excitation inputs D_1 through D_s drive state variables y_1 through y_j. A third combinational logic function, derived from the system inputs x_1 through x_n and the state variables y_1 through y_j, produces the clock pulses to the synchronous state variables.

An algorithm for synthesizing MOM circuits is given in a paper, "Hazard-Free Design of Mixed-Operating-Mode Asynchronous Sequential Circuits" by Chiang and Radhakrishnan (*International Journal of Electronics* vol. 68, 1990). The entire design technique implied by the paper is beyond the scope of this text. The approach is directed to finding essential hazards and then designing the logic to produce aperiodic clock pulses and applying them to those state variables when essential hazards exist. All of the other asynchronous techniques of elimination of static and dynamic hazards have first been accomplished. The key is finding essential hazards.

❖ EXAMPLE 8.2

An example problem will illustrate the existence of essential hazards. Consider the essential hazards in the flow table shown in Table 8.25.

SOLUTION An essential hazard exists in an asynchronous circuit when a present total state, S_A, has a different next total state, after one transition of an input variable x_i, than it does after three transitions of x_i. Essential hazards result from uneven propagation delays as an input variable is distributed to the various gates in an asynchronous circuit.

Determination of essential hazards is done by evaluating each stable total state's transitions as a given input variable, x_i, changes.

The essential hazards for Table 8.25 are found by evaluating each stable total state. The initial state is listed followed by the input value listed in parenthesis. The arrows indicate the state transitions. If the new state is the same after one transition as it is after three transitions, no essential hazard exists; if it is different, an essential hazard exists:

1. C(00) → F(01) → F(00) → F(01), no essential hazard.
2. C(00) → B(10) → D(00) → E(10), essential hazard.
3. D(00) → A(01) → C(00) → F(01), essential hazard.
4. D(00) → E(10) → F(00) → A(10), essential hazard.
5. F(00) → F(01) → F(00) → F(01), no essential hazard.

TABLE 8.25
Flow table with essential hazards

Present state	Next state/Output			
	00	01	10	11
A	C/0	(A)/0	(A)/0	E/0
B	D/0	E/1	(B)/0	C/0
C	(C)/0	F/1	B/0	(C)/0
D	(D)/1	A/0	E/0	(D)/0
E	F/0	(E)/1	(E)/1	(E)/0
F	(F)/1	(F)/0	A/1	D/0

6. $F(01) \rightarrow A(10) \rightarrow A(01) \rightarrow A(10)$, no essential hazard.
7. $A(01) \rightarrow C(00) \rightarrow F(01) \rightarrow F(00)$, essential hazard.
8. $E(01) \rightarrow F(00) \rightarrow F(01) \rightarrow F(00)$, no essential hazard.
9. $E(01) \rightarrow E(11) \rightarrow E(01) \rightarrow E(11)$, no essential hazard.
10. $C(11) \rightarrow B(10) \rightarrow C(11) \rightarrow B(10)$, no essential hazard.
11. $C(11) \rightarrow F(01) \rightarrow D(11) \rightarrow A(01)$, essential hazard.
12. $D(11) \rightarrow F(01) \rightarrow D(11) \rightarrow A(01)$, essential hazard.
13. $D(11) \rightarrow A(01) \rightarrow E(11) \rightarrow E(01)$, essential hazard.

The essential hazards are

1. $C(00) \rightarrow B(10); C_0, B_2$.
2. $D(00) \rightarrow A(01); D_0, A_1$.
3. $D(00) \rightarrow E(10); D_0, E_2$.
4. $A(01) \rightarrow C(00); A_1, C_0$.
5. $C(11) \rightarrow F(01); C_3, F_1$.
6. $D(11) \rightarrow A(01); D_3, A_1$.

Finding essential hazard state transitions permits designing the self-synchronous clock logic.

Chiang and Radhakrishnan state and prove the following theorem: Asynchronous flow table input transitions are essential hazard free if self-synchronous clock pulses are applied for the first- or second-state transition where essential hazards exist.

Once the essential hazards are identified, the self-synchronous state transitions are reduced to minimize logic. After minimization of the number of self-synchronous transitions, the state assignment is made and the logic determined for generation of clock, self-synchronous and asynchronous excitation, and output functions. A number of design approaches are under investigation for synthesizing self-synchronous and mixed operating mode logic systems. Mixed operating mode and self-synchronous design techniques may become the methods of choice for logic designers in the future. ❖

SUMMARY

A. Asynchronous sequential machines are introduced.

B. Asynchronous machine modes are specified as the fundamental mode or the pulse mode.

1. In the fundamental mode, only one input is allowed to change at a time and the inputs are considered to be levels.
2. In the pulse mode, only one input is allowed to change at a time, and inputs are considered to pulse (false-true-false) and the input pulses must be long enough to initiate a state change but not so wide that the input is still true after a new state is reached.

C. Methods of analyzing asynchronous machines are presented:

1. Derivation of next-state and output equations.
2. Creation of transition table.
3. Creation of a timing diagram or state diagram.
4. Pulse mode timing.

D. Flow tables are derived.

E. Flow tables are reduced.

F. Different techniques for making a state assignment for asynchronous machines are presented. The problem here is not logic reduction but elimination of races and hazards. Hence, we discussed the following:

1. Races and cycles.
2. Shared row state assignment.
3. Multiple row state assignment.
4. Universal row state assignment.
5. One hot state assignment.

G. Two design examples were presented, from a verbal description to a final logic diagram and timing diagram.

H. Data synchronizers were used to synchronize asynchronous data inputs to subsystem clock. Attempts were made to minimize metastablity problems associated with asynchronous inputs to a synchronous subsystem.

I. Interface protocol asynchronous cell (IPAC), a programmable logic device (PLD) integrated circuit approach to data synchronization, was introduced.

J. The mixed operating mode (MOM) design approach blends both asynchronous and synchronous design techniques by determination of state transitions that produce essential hazards and then generation of logic that produces an aperiodic clock pulse to clock those state variables for transitions that produce essential hazards.

REFERENCES

Computer Aided Logic Design, 4th ed., by Frederick J. Hill and Gerald R. Peterson (New York: Wiley, 1993). Chapter 14 covers level mode asynchronous sequential circuits.

Digital Engineering Design, by Richard F. Tinder (Englewood Cliffs, N.J.: Prentice-Hall, 1991). Chapter 6 has an extensive treatment of asynchronous sequential circuit analysis.

Fundamentals of Logic Design, 4th ed., by Charles H. Roth. (Minneapolis–St. Paul: West, 1992). Unit 23 introduces asynchronous sequential circuit analysis. Unit 24 covers the derivation and reduction of flow tables. Unit 25 covers the state assignment problem. Unit 26 discusses hazards. Unit 27 treats the design of asynchronous circuits.

Digital Design Principles and Practices, by John F. Wakerly (Englewood Cliffs, N.J.: Prentice-Hall, 1994). Asynchronous treatment is limited to coverage of inputs to a synchronous system. Timing hazards are covered in Chapter 3.

Digital Logic and Computer Design, by Thomas R. McCalla (Columbus, Ohio: Merrill, 1992). Fundamental mode circuits are introduced in Chapter 9.

Introduction to Logic Design, by Sajjan G. Shiva (Glencoe, Ill.: Scott, Foresman and Company, 1985). A fairly complete coverage of asynchronous circuits is given in Chapter 11.

Digital Logic Design, by Gideon Langholz, Abraham Kandel, and Joe L. Mott (Dubuque, Iowa: Wm. C. Brown, 1988). Chapter 6 contains a complete coverage of asynchronous circuit analysis and design.

Recent Developments in the Design of Asynchronous Circuits, by J. A. Brzozowski and J. C. Ebergen (technical report from the Computer Science department of the University of Waterloo, Ontario, Canada). The report was presented to the Seventh International Conference, Fundamentals of Computation Theory FCT 1989. It is a development of an abstract algebra description of transition paths in asynchronous circuits along with a presentation of delay models.

Hazard Free Design of Mixed Operating Mode Asynchronous Sequential Circuits, by Jen-Shiun Chiang and Damu Radhakrishnan, *International Journal of Electronics* 68, no. 1 (1990): 23–37. This paper presents a hazard-free design methodology for asynchronous circuits.

"Synthesis of Asynchronous Machines Using Mixed-Operating Mode," by O. Yenersoy, *IEEE Transactions on Computers,* C-28, no. 4 (April 1979). This paper introduces the concept of mixed operating mode sequential circuits.

GLOSSARY

Asynchronous Asynchronous means without a synchronizing clock to control the state transitions of a sequential circuit. The inputs and present state change after the internal delay of the circuit elements. Because no synchronizing clock is used, asynchronous circuits are usually faster than synchronous circuits.

Critical race A critical race occurs in asynchronous circuits when two or more internal state variables change at the same time. The internal delay of the state variable logic paths are almost always different, so state variables can practically never change simultaneously. Because the delay paths are different it is very difficult to predict which state variable will change first. If the changes in state variables result in the possibility of reaching more than one stable state, the race condition is said to be critical.

Cycle A cycle occurs in an asychronous circuit when multiple state transitions occur without changing the inputs.

Dynamic hazard A dynamic hazard occurs in a combinational logic circuit when multiple changes in either a normally 0 or 1 output occurs. Dynamic hazards require at least three levels of logic. Dynamic hazards can be eliminated by a two-level realization of a combinational function.

Essential hazard An essential hazard is caused by unequal propagation delay paths that originate from the same input. Essential hazards are eliminated only by matching the propagation delay paths of a signal. Because the delay paths depend on the final logic design and the circuit layout, the elimination of essential hazards is an application dependent problem.

Flow table A flow table is a tabular method for showing asynchronous circuit state transitions. A row exists for every internal state and columns for each input permutation. Each cell in the flow table shows the next internal state. The output values can be included in the next-state cell or separated into an output section. A four-row flow table follows. The internal states are indicated by the letters a, b, c, d. The input permutations are listed in columns. The next internal state is contained in the row-column intersection points. A separate output section is included in this example flow table.

Present	Next state				Outputs			
state	00	01	11	10	00	01	11	10
a	(a)	c	—	(a)	0	—	0	1
b	a	c	(b)	c	—	0	1	1
c	b	(c)	b	a	—	0	0	0
d	c	(d)	a	c	0	1	1	0

Fundamental mode Asynchronous circuits are typically broken into two main groups: fundamental mode and pulse mode. Other modes such as MOM (mixed operating modes) have been suggested by researchers. A fundamental mode asynchronous circuit is limited to inputs that are levels (0 or 1) and only one input can change at a time. Input level changes can occur no faster than allowed by the slowest propagation path in the circuit.

Hazard Several types of timing problems exist in logic circuits due to propagation delay paths. These timing problems are called *hazards*. Hazards are divided into the following categories: static, dynamic, and essential.

Internal clock In a mixed operating mode circuit some of internal state transitions are clocked. The clock is generated internally, not the externally. The logic that generates the clock is derived from the input variables and state variables. The clock is generated only when needed, therefore, it occurs aperiodically.

Internal or secondary states Asynchronous circuit analysis technique separates the variables used to specify the "state" of the machine into internal state variables (latch outputs) and input variables. The *total state* is defined as the combination of the internal state and the input state.

Interface protocol asynchronous cell (IPAC) A macrocell structure designed by Advanced Micro Devices Company for use in an asynchronous programmable logic device (PLD). The IPAC cells are edge-activated registers that need no separate clock input to latch data. Instead the data are loaded into the registers in response to changes in the data signals themselves.

Merger diagram A graphical technique for identifying possible redundant states. A diagram is drawn with a set of nodes where each node represents a state. A line is drawn between the nodes if, for every input, the same next state is reached and the output is the same. If equivalency between two states is dependent on another state pair equivalency, then those state symbols are written adjacent to the line connecting the two state nodes.

Mixed operating mode (MOM) A design technique that mixes both synchronous and asynchronous inputs to state variables. Those state transitions where essential hazards exist are clocked, and state transitions where an essential hazard does not exist are not clocked. This implies an aperiodic clock. MOM clock pulses are generated by a separate logic that determines when essential hazards occur (by a set of Boolean terms, consisting of input and state variables, that occur for only those state transitions containing an essential hazard). The advantage of MOM is that essential hazard timing problems are solved (which are very difficult with pure asynchronous) with a speed increase over a pure synchronous approach. MOM and other self-clocked design techniques are under investigation by various researchers to find a systematic and reliable design synthesis procedure.

Static hazard A static 0 hazard occurs in a combinational logic circuit when the normally 1 output momentarily goes to a 0. A static 1 hazard occurs when a normally 0 output momentarily goes to a 1. Static hazards are eliminated by using all the prime implicants to cover a function instead of just the essential prime implicants. Static hazards are not normally a problem in combinational circuits. Asynchronous circuits are combinational logic with feedback, and hazards present a real problem.

QUESTIONS AND PROBLEMS

Introduction and Section 8.1

1. Describe in your own words the distinction between synchronous and asynchronous sequential circuits. State the advantages and disadvantages of each design approach.
2. Describe in your own words the distinction between fundamental and pulse mode asynchronous sequential circuits.
3. Define the following (sketch a model diagram and assign variable symbols):
   - **a.** Total stable state.
   - **b.** Input state.
   - **c.** Secondary state.
4. Determine the minimum and maximum input pulse widths for a pulse mode asynchronous sequential circuit that uses
   - **a.** Low-power Schottky TTL devices and has three levels of logic around the state variable to input loop.
   - **b.** Schottky TTL devices and has four levels of logic around the feedback loop.

Section 8.2

5. Analyze the asynchronous state machines in the figures for this problem. For each, determine the state variables and assign state variable, input variable,

Section 8.2
Problem 5(a)

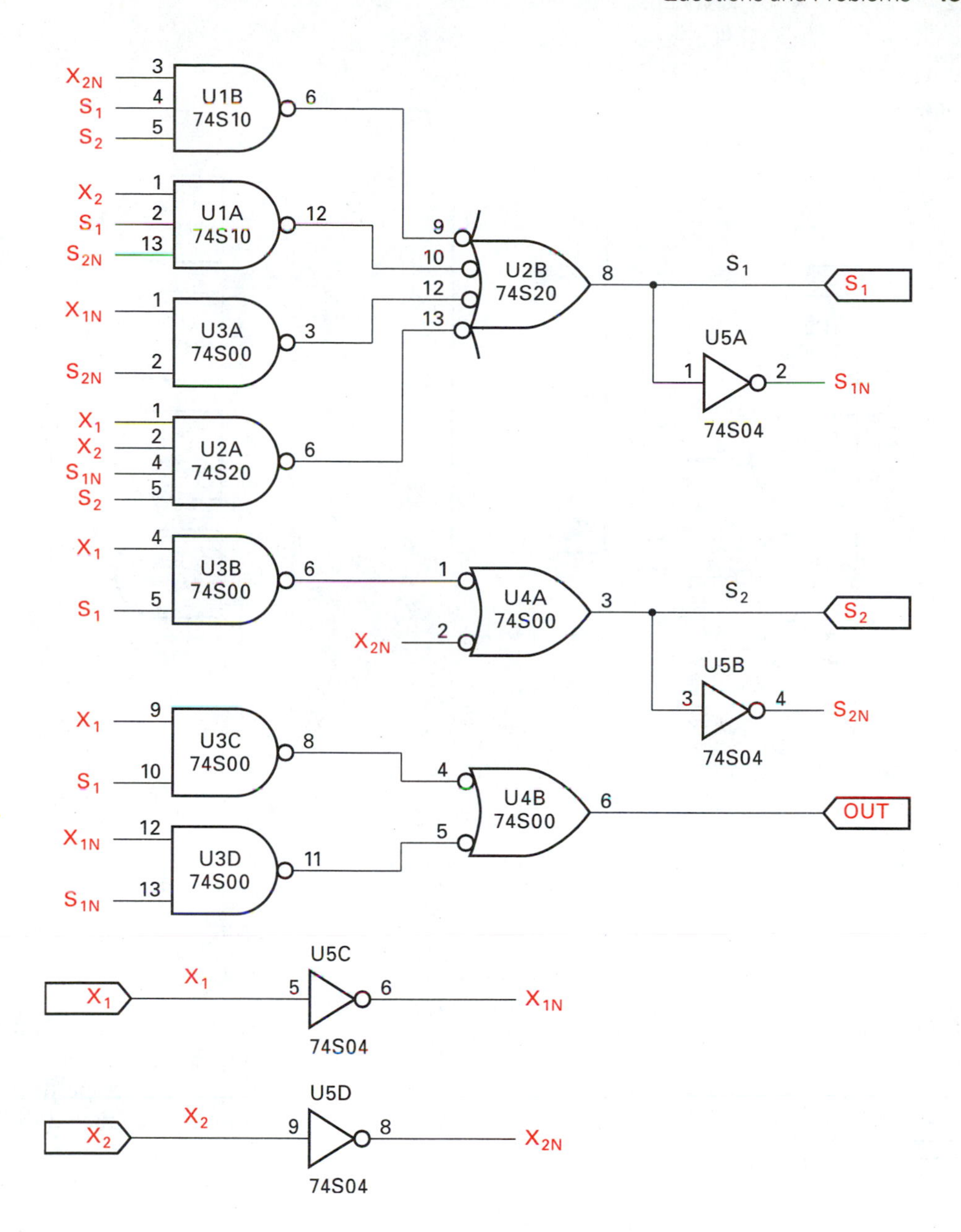

and output variable names. Write the next-state, secondary, and output equations. Construct a transition table and timing diagram for each circuit.

Section 8.3

6. Derive flow tables for the following problem. A storm sewer has two intake pipes, each controlled by separate pumps and a single outlet pipe that regulates the water level in a lift station. The intake pipes fill the lift station well, and the water is removed by gravity flow through the outlet pipe. Three sensors determine the water level in the lift station well. The input pipes, pumps, outlet pipe, and level sensors are indicated in the figure for this problem.

Section 8.2

Problem 5(b)

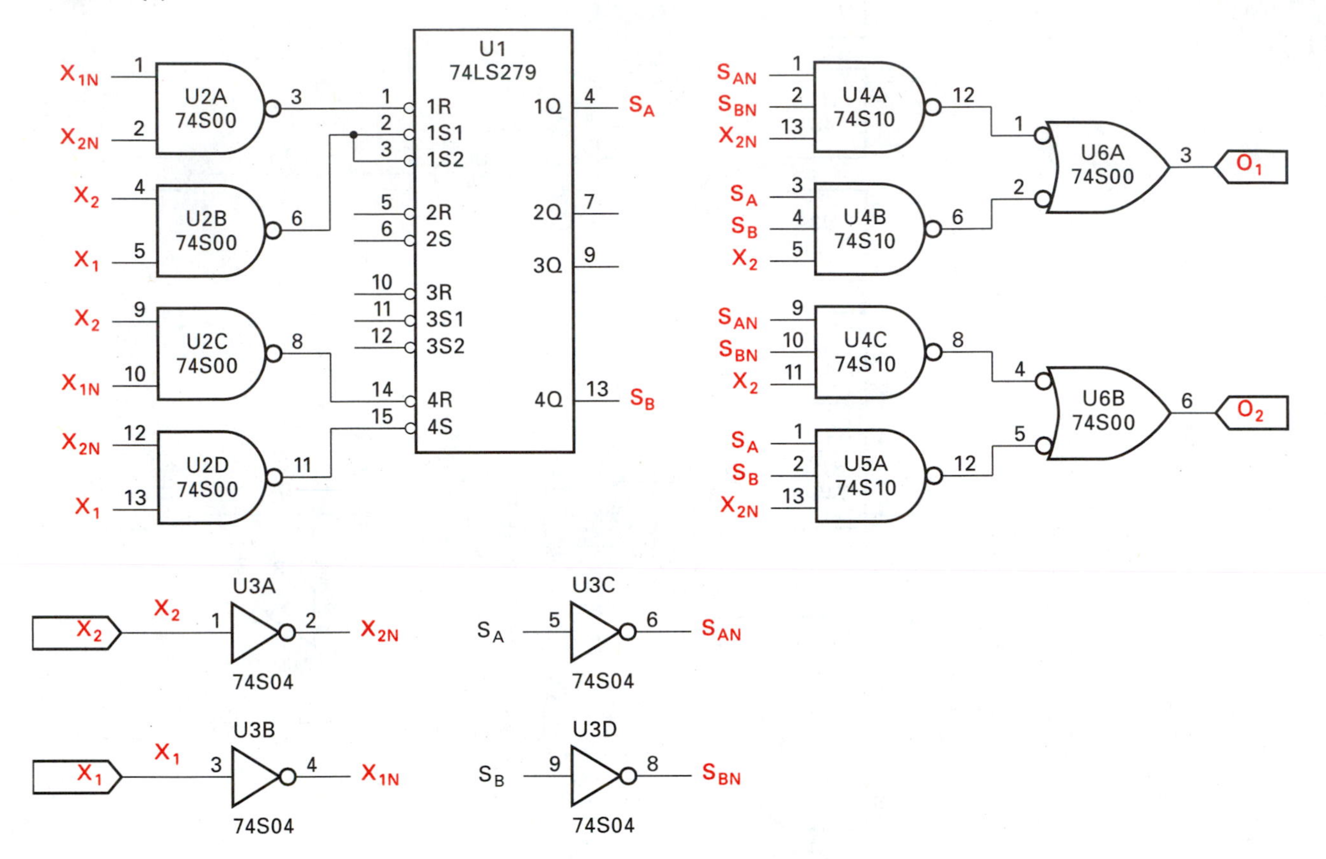

Section 8.3

Problem 6

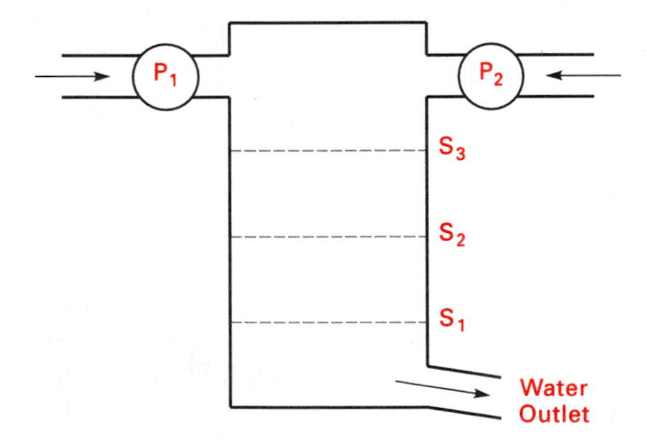

Design the flow table for an asynchronous circuit pump controller. The inputs to the controller are S_3, S_2, and S_1 (binary level sensors). When the water is over the sensor element, a logical 1 is produced. Two outputs exist, one for each pump, P_2 and P_1. Both pumps are to turn on when water is below level S_1. Both pumps are to remain on until the water reaches sensor S_2. At level S_2, P_1 is turned off and P_2 remains on. P_1 remains off and P_2 remains on until the S_3 level is reached, when both pumps are turned off.

7. Derive the flow table for a positive edge triggered clocked T flip-flop. The circuit has two inputs, clock and T, and one output, Q. Cause the flow table to conform to the characteristic equation for a T-flip-flop:

$$Q^{+} = Q \oplus T$$

8. Reduce the following primitive flow table given using a merger diagram.

Present state	Next state/Output			
	00	01	10	11
a	(a)/0	b/—	e/—	—/—
b	a/—	(b)/1	—/—	h/—
c	g/—	—/—	(c)/0	i/—
d	(d)/1	f/—	c/—	—/—
e	a/—	—/—	(e)/0	i/—
f	d/—	(f)/—	—/—	h/—
g	(g)/0	b/—	c/—	—/—
h	—/—	f/—	c/—	(h)/0
i	—/—	b/—	e/—	(i)/0

Section 8.4

9. Define the following and create a simple partial state flow table to illustrate each condition.

a. Race.

b. Noncritical race.

c. Critical race.

d. Cycle.

10. Create a shared row state assignment for the following flow table.

Present state	Next state			
	00	01	10	11
a	(a)	b	(a)	b
b	c	(b)	(b)	(b)
c	(c)	d	b	d
d	(d)	(d)	a	(d)

11. Create a one hot state assignment for the following flow-table and construct a state transition diagram.

Present state	Next state			
	00	01	10	11
S_0	S_2	(S_0)	S_1	(S_0)
S_1	(S_1)	S_2	(S_1)	S_0
S_2	(S_2)	(S_2)	S_3	S_3
S_3	(S_3)	S_0	(S_3)	(S_3)

12. Create a state assignment for Problem 10 using the four-row universal state assignment.

Section 8.5

13. Design a negative edge triggered J–K flip-flop. An edge triggered flip-flop is an asynchronous sequential machine with three inputs (clock, J, and K) and a single output.

a. Create a state diagram.

b. Create a flow table.

c. Reduce the flow table by using a merger diagram.

d. Create a transition diagram and make a race free state assignment.

e. Create a transition table and write the state variable and output Boolean equations.

f. Derive the state variable and output functions that are free from dynamic and static hazards.

g. Draw the resulting logic diagram.

h. Predict the circuit timing by constructing a diagram that shows the input and state variable transitions. (Remember only one input changes at a time.)

i. If you have the opportunity, built the circuit from SSI ICs and verify the circuit's operation.

14. Design an asynchronous control system that is to sort objects of two different widths. The objects are moving down a conveyor. Two photo sensors are positioned to determine the object widths. The outputs from the photo sensors produce inputs to the asynchronous machine. Figure (a) for this problem shows the placement of the photo sensors with respect to the object. The distance between the two detectors is l. Objects come in two different widths, $l + A$ and $l - A$. If the object is narrow ($l - A$) the sensor output timing in figure (b) for this problem is generated. If the object is wide ($l + A$) the sensor output timing in figure (c) for this problem is generated. Design the circuit following the same steps as in Problem 13.

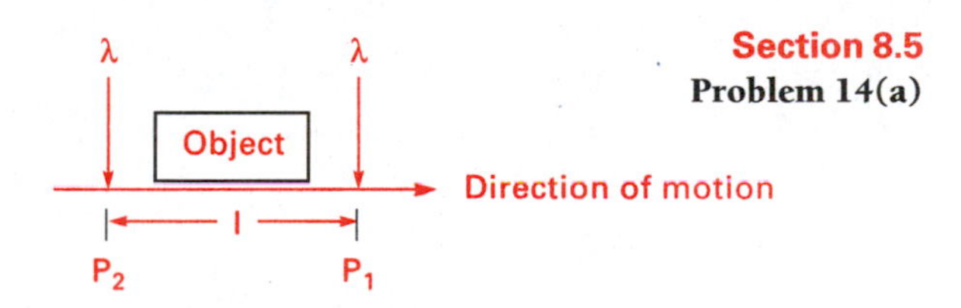

Section 8.5
Problem 14(a)

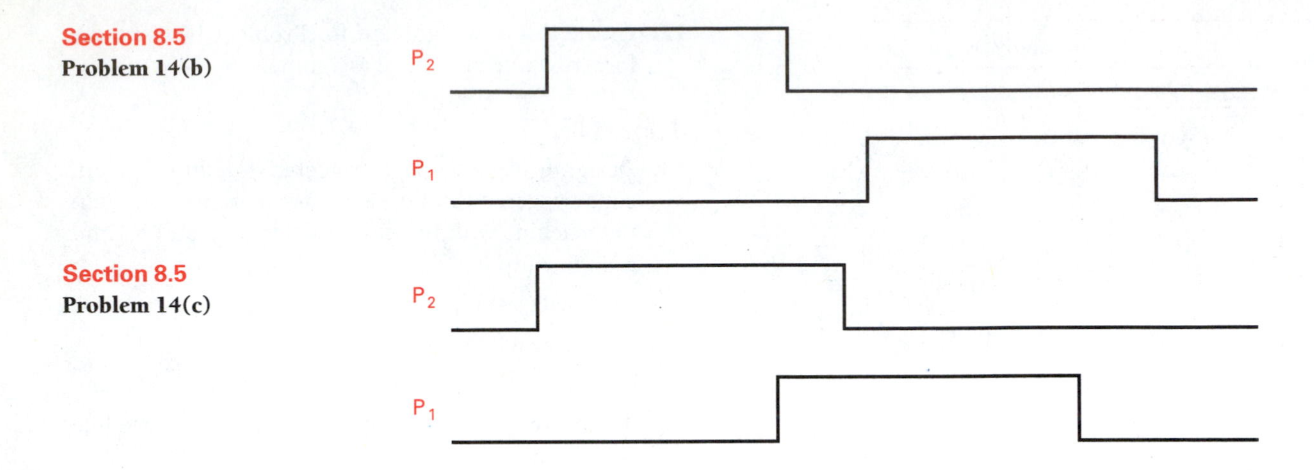

15. Design an asynchronous controller for a vending machine. A single slot is provided that will accept nickels, dimes, and quarters. Each coin value is to be considered as a separate input variable. Products are available at two prices, $.45 and $.55. A separate output is to be produced for each price. When either output is reached a product select button is pushed and the machine resets. Design the circuit following the same steps as in Problem 13.

Sections 8.6 and 8.7

16. Describe the purpose of and draw a simple block diagram for a "data synchronizer."

17. Describe the purpose of an IPAC and describe its operation.

18. Find any essential hazards in the following flow tables.

a.

Present state	Next state 00	01	10	11
S_0	S_2	(S_0)	S_1	(S_0)
S_1	(S_1)	S_2	(S_1)	S_0
S_2	(S_2)	(S_2)	S_3	S_3
S_3	(S_3)	S_0	(S_3)	(S_3)

b.

Present state	Next state 00	01	10	11
A	(A)	(A)	C	B
B	(B)	E	E	(B)
C	A	D	(C)	F
D	B	(D)	F	B
E	B	E	(E)	B
F	A	A	(F)	(F)

19. Sketch the block diagram for a MOM system. Explain advantages and disadvantages of using MOM.

CHAPTER

9

PROGRAMMABLE LOGIC AND MEMORY

INTRODUCTION

Up to now we have been designing digital circuits using small- and medium-scale integrated circuits. These devices are often called *random logic* ICs to separate them from memory and programmable logic devices. Random logic consists of gates, flip-flops, counters, registers, and other functions found in SSI and MSI integrated circuits. Large-scale integrated circuits (LSI) include memory, microprocessor and microprocessor support devices, and a class of devices called *programmable logic devices* or PLDs. PLDs come in a variety of types, but they all have a common characteristic; that is, they all can be custom configured by the user to perform specific functions.

Programmable logic devices consist of an array of identical functional cells. The cell array usually contains an AND-OR network and often includes a flip-flop. Some programmable devices can perform only combinational logic functions; others can perform combinational and sequential functions.

Consider a range of digital integrated circuits. The simplest contain random logic (SSI and MSI), the middle contains programmable logic, and the most complex consists of custom very large-scale integrated circuits (VLSI). Custom designed ICs are the most efficient users of the available silicon area; they are also the most expensive to develop. The PLDs occupy the middle ground; they give improved performance over random logic and a significant cost savings over custom designs. PLD designs typically consume less power, take fewer ICs, and are more reliable than random logic designs. The trade-off is in speed performance. Random logic still has the edge at very high speeds in the ECL family. However, PLD manufacturers are continually developing more efficient and faster PLDs. The real advantage of PLD designs over random logic is the ease of design and the resulting design time savings. We will compare several digital circuits designed with random logic to the programmable logic realization.

We will consider several PLD types:

1. Memory, specifically ROM and EPROM
2. Programmable logic arrays (PLA)
3. Programmable array logic (PAL)
4. Programmable logic devices (such as Altera's PLD)
5. Field programmable gate arrays (such as Xilinx and Actel)

9.1 MEMORY

Our interest in this section is to develop a conceptual understanding of memory devices so that we may apply them in combinational and sequential logic design. A comprehensive coverage of memory is beyond the scope of this book. The reader may want to refer to texts on microprocessor systems and computer architecture for more comprehensive coverage of memory systems and devices.

Integrated circuit memories fall into two broad categories: random access and sequential access. By *random access,* we mean that the address may be presented to the memory in any order (random) without affecting the access time to the stored data. Sequential access memories store and retrieve data sequentially. The access time is not constant; it depends where the data are located. Serial register files, floppy disks, hard disks, and tape systems are examples of sequential access memories. We wish to focus our attention on random access memory and will not cover sequential access memory. These topics are covered in computer organization texts.

Random access memory is further broken down into three categories: read only, read mostly, and read/write. Read only memory (ROM and PROM) stores its data permanently. It was written to only once, either when manufactured or by the user, and then can be read as often as desired. If the device was programmed (data written) by the manufacturer, it is called a **ROM**; if the user programs the device, it is called a **PROM** (programmable read only memory).

Read mostly memory can be written to or programmed more than once. The programming operation is very different from the read operation. In the case of **EPROM** (erasable programmable read only memory) the device must be physically removed from the circuit, erased under an ultraviolet light and then reprogrammed prior to reinsertion back into the circuit. This type of device is most often used in developing circuit prototypes. Another type of read mostly memory is the **EEPROM** (electrically erasable programmable read only memory). It can be rewritten while remaining in the circuit. A special writing process is necessary that requires activating a write pin while presenting address and data information. The writing operation requires several milliseconds and reading requires tens or hundreds of nanoseconds. EEPROM devices are replacing EPROMS in many applications, especially if remote reprogramming is needed.

Read/write memory reads and writes with equal ease. Two basic types of read/write memory are available; static and dynamic. Static is the easiest to use and often the fastest and the most expensive. Dynamic memory gives more bits for the same amount of silicon, but requires some means to refresh the stored data. Without refresh cycles to read and rewrite data the dynamic memory would soon loose information. Most memory in personal computers is dynamic.

Read/write memory is volatile; that is, the power must be continually applied for the data contents to remain valid. If the power is removed, the stored data are lost. Read mostly memory is nonvolatile; that is, the data remain intact even when power is removed. Memory devices used in the design of logic circuits must be of nonvolatile memory.

9.1.1 ROM, PROM, and EPROM

Our concern in this section is with the characteristics of nonvolatile read only and read mostly memories. The random access nature of data selection is important. We wish to

Figure 9.1
ROM (PROM and EPROM) block

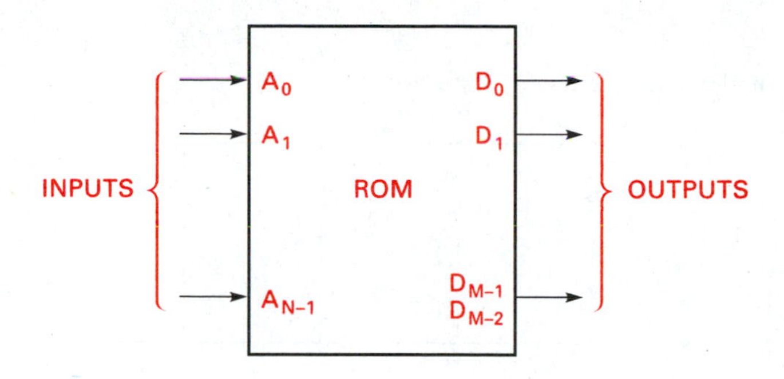

present any random address, within the range of the device, and read the data without concern to address dependent data access time. Access time is a device parameter that tells us how long it takes to find the stored data after presenting an address.

We can consider ROM (also PROM and EPROM) devices as combinational logic circuits, with inputs (address) and outputs (data). The general combinational logic model applies to ROM devices. Figure 9.1 shows a $2^n \times M$ ROM functional block that looks much the same as the model for a combinational logic function. We can also think of ROMs as banks (or locations) of stored data that we can access by presenting an address.

A ROM can be modeled as a combinational logic function; therefore, it can be used to realize any truth table. The address lines act as input variables and the data outputs act as output variables.

A ROM was modeled as a combinational logic circuit in Figure 9.1, where an input vector produced an output vector. A simple ROM can be constructed from a MSI decoder IC, such as the 74LS138 (3×8), a number of diodes and output line buffers. Consider the combinational functions illustrated in Table 9.1. Inputs A_3 to A_1 become addresses to a simple ROM circuit, and D_4 to D_1 are outputs.

A simple ROM constructed from random logic ICs and diodes, shown in Figure 9.2, is used to realize the logic of Table 9.1. Diodes are used to connect the output lines of the decoder to the data output lines that correspond to logic 1s in the truth table. For example, when the input address lines are 000, decoder output 0 is selected. According to the truth table, the required output is 0001, so data line D_1 must be a 1. Note that the output buffers are inverters. The buffer inputs are pulled up to +5 V through current limiting resistors, placing logic 1s on the inverter inputs. To cause an output line to be a

TABLE 9.1
Truth table realized with a ROM

Inputs			Outputs			
A_3	A_2	A_1	D_4	D_3	D_2	D_1
0	0	0	0	0	0	1
0	0	1	0	1	0	1
0	1	0	1	0	0	0
0	1	1	1	1	0	1
1	0	0	0	1	1	0
1	0	1	0	0	0	0
1	1	0	1	0	1	0
1	1	1	1	0	0	1

Figure 9.2
ROM made of a 3 × 8 decoder, diodes, and inverters

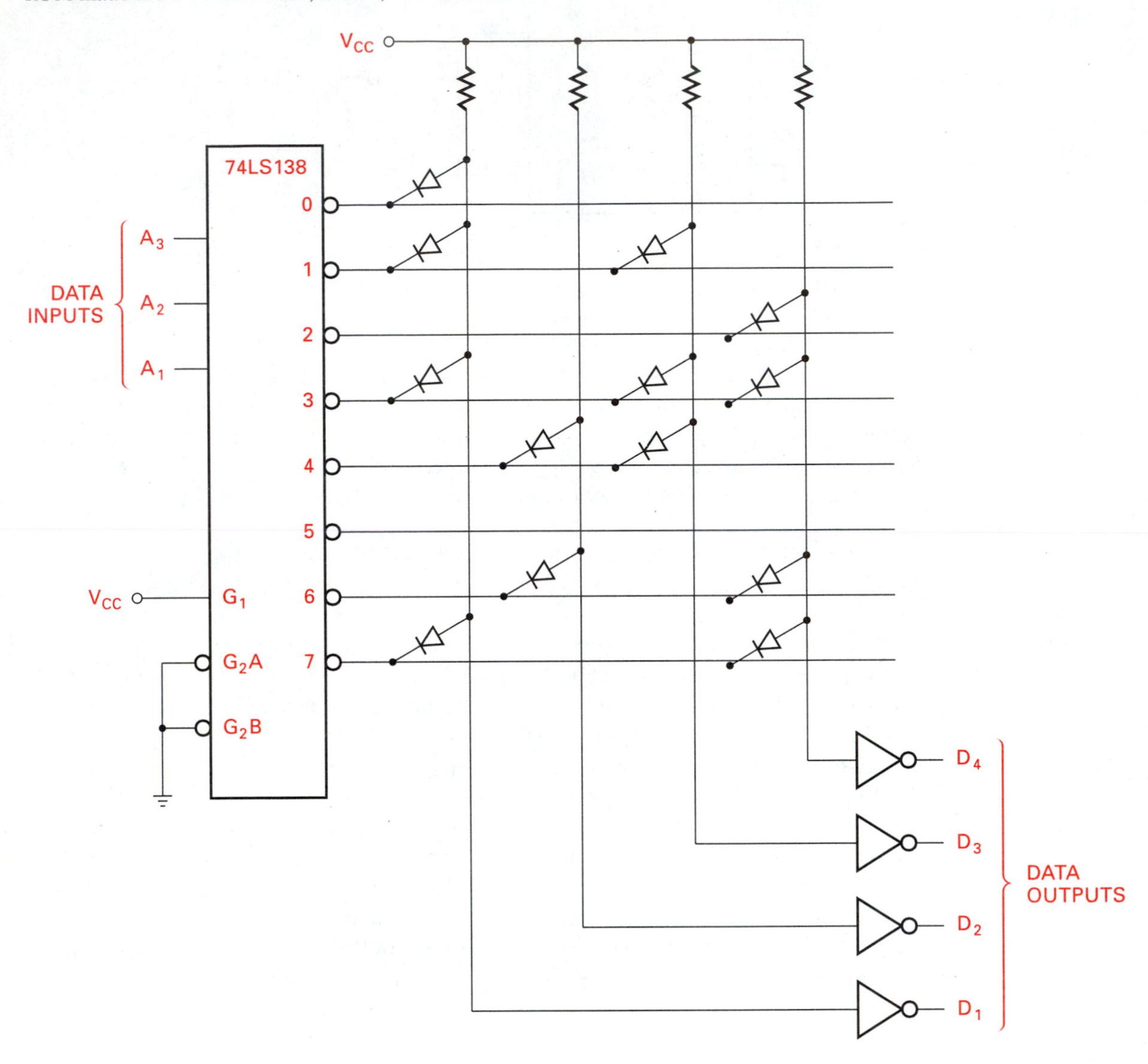

1, the inverter input is clamped to a logical 0 by a conducting diode. A 000 on the decoder address inputs forces output 0 low, which turns on the diode placing a 0 on the input of the D_1 inverter and a 1 on D_1. Input 011 forces decoder output 3 low which turns on the clamp diodes on that decoder output line, forcing the outputs to a 1101.

Field effect or bipolar transistors are used in place of diodes in integrated circuit PROMs. PROMs (programmed by the user) have each address select line connected to a MOSFET (metal oxide silicon field effect transistor) gate or bipolar transistor base, with the source (emitter) grounded and the drain (collector) connected to the vertical data lines. A fuse link exists between ground and the MOSFET source (or bipolar transistor emitter). Because PROMs are programmed by the user, each address select and data line

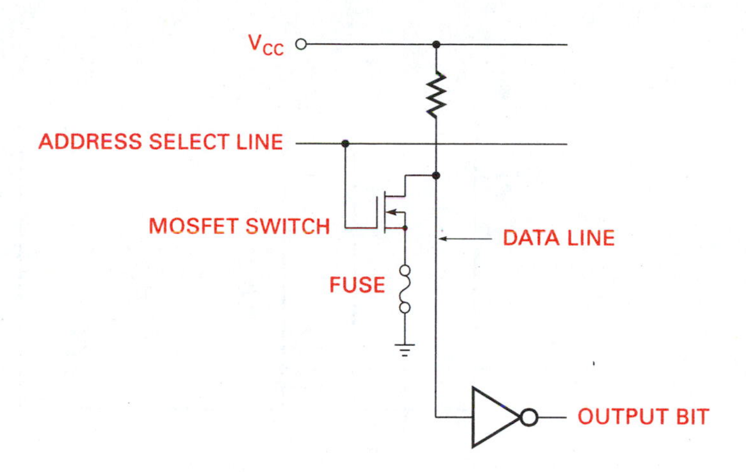

Figure 9.3
Single fused PROM cell

intersection has its own fused MOSFET or transistor. When the fuse is intact the memory cell is configured as a logical 1, and when the fuse is blown (open circuit) the memory cell is a logical 0. Logical 0s are programmed by selecting the appropriate select line and then driving the vertical data line with a pulse of high current. Figure 9.3 illustrates a PROM fused MOSFET memory cell.

EPROMS (erasable programmable read only memory) replace the fused bipolar or FET transistor with a MOSFET that has two gates. The memory cell is configured like a PROM cell, however, no fuse is present. The first gate operates as a conventional MOSFET control terminal. The second gate is called a *floating gate*. Both gates must be activated before the MOSFET can conduct current. The second or floating gate is the key to erasability so the device can be reprogrammed. A MOSFET has a very high input impedance (10^{12} ohms) between the gate and the drain-source current channel. The floating gate acts like one plate on a capacitor with the drain-source channel acting as the other plate. A correct polarity charge, placed on the gate-to-channel capacitor, enables drain-source channel current to flow; no charge inhibits channel current.

When an EPROM is programmed, a charge is placed on the floating gate, permitting channel current to flow and thus establishing the logic level of the memory cell. The floating gate-to-channel voltage is discharged when the device is erased by placing it under ultraviolet light. The ultraviolet (UV) light passes through a transparent quartz lid on the EPROM IC to the exposed MOSFET floating gates. It takes approximately 5 to 20 min of exposure to the UV light to erase the device, which can then be reprogrammed and inserted back into the circuit board.

The internal structure of ROM (PROM, EPROM) consists of an address decoder, a memory array, and an output buffer. Figure 9.4 illustrates the internal structural arrangement of a typical ROM device.

The memory array data are programmed in the ROM when the device is manufactured. In a PROM the data are loaded into the array by the user. The two address decoders are used to select a specific data location. Active-low control inputs CS′ and DE′ enable the row and column address decoders and output tristate buffers so the selected data can be read.

The TBP28S42, a 256 × 8 bit PROM made by Texas Instruments, is a typical PROM. ROM and PROM devices have two access times: one is from the address-to-data outputs, the other is from the control inputs to data outputs. When considering the address-to-

Figure 9.4
Typical ROM internal organization

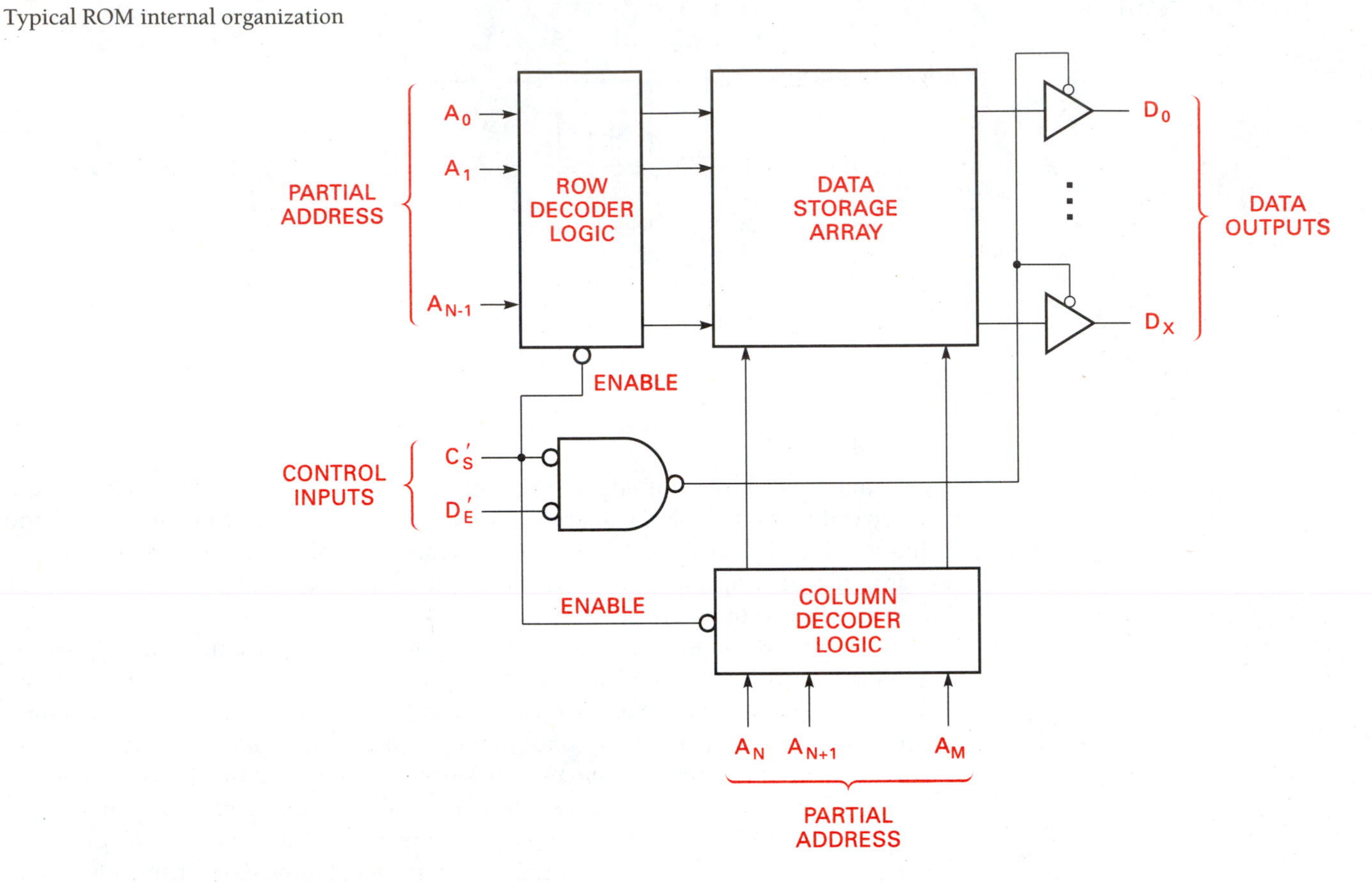

output access time it is assumed that the device control inputs are active. Likewise, when considering the control-to-data output access time, it is assumed that the address inputs are active. The TBP28S42 has a maximum address to data output access time of 70 ns and a chip select to data output access time of 45 ns maximum. Other devices are faster. The Texas Instruments TBP34R162 has address to data and enable to data access times of only 8 ns.

EPROM ICs come in a variety of sizes and access time speeds. Intel, a manufacturer of integrated circuits, makes EPROMS ranging from 2716 (2K × 8 bits) to the 27C040 (512K × 8 bits). Access times range from slow (450 ns for the slowest version of the 2716) to fast (70 ns for the fastest version of the 27C202C (16K × 16 bits)). The EPROMs come in 8- and 16-bit versions to accommodate 8- and 16-bit microprocessor memory systems. Table 9.2 illustrates Intel's part numbers and bit arrangement for their EPROMs.

The EPROMs listed in Table 9.2 are generally used in the development of computer memory systems. The design of such memory systems is not our objective here. Texts on microprocessors and computer architecture as well as manufacturer's data books should be consulted for specifics on devices, programming, and applications in memory systems. Our objective is to illustrate how to use EPROMS to design combinational and sequential logic.

TABLE 9.2
Intel EPROM memory array size and organization

Intel part number	Array size (bits)	Memory organization (bits)
2716	16K	2K × 8
2732	32K	4K × 8
2764	64K	8K × 8
27128	128K	16K × 8
27256	256K	32K × 8
27512	512K	64K × 8
27C010	1M	128K × 8
27C020	2M	256K × 8
27210	1M	64K × 16
27C210	1M	64K × 16
27C220	2M	128K × 16
27C040	4M	512K × 8
27C240	4M	256K × 16

We have shown how to realize combinational logic using ROMs; the same approach works with EPROMs. The advantages of ROMs over EPROMs are faster speed and lower overall cost per bit when large numbers of devices are to be used. The advantage of EPROMs over ROMs is the ease of use during the development stages of design, because erasure and reprogramming are possible.

9.2 USING AN EPROM TO REALIZE A SEQUENTIAL CIRCUIT

We have seen how ROMs, PROMs, and EPROMs can be used to perform a combinational logic function in the previous section. Now we wish to use them in the realization of a sequential logic circuit. Recall the general sequential circuit model presented in Chapter 5. The model lumps the input and output combinational logic together. Some of the combinational logic inputs are input variables (external to the sequential circuit) and some are state variables. The outputs of the combinational logic function are also divided into excitation and output variables. By partitioning the sequential circuit into combinational and state variables, we can realize the combinational logic portion using a ROM or EPROM. Figure 9.5 illustrates the idea.

The address inputs to the ROM/EPROM are input variables (external and state) and the data outputs are either state variables or output variables. The actual data stored in the ROM/EPROM is determined by constructing a data table based on the excitation and output variables values for a given state and input combination.

Consider the elapsed time measurement problem from Chapter 7. The ASM chart is shown in Figure 7.42. Once the state equivalency and state assignment are made we can proceed with the construction of a ROM/EPROM data table. A note of interest in designing sequential machines using an EPROM: The state assignment and state reduction problems are not as important when realizing a sequential circuit using ROM/EPROM unless a variable may be eliminated entirely or the number of state variables reduced.

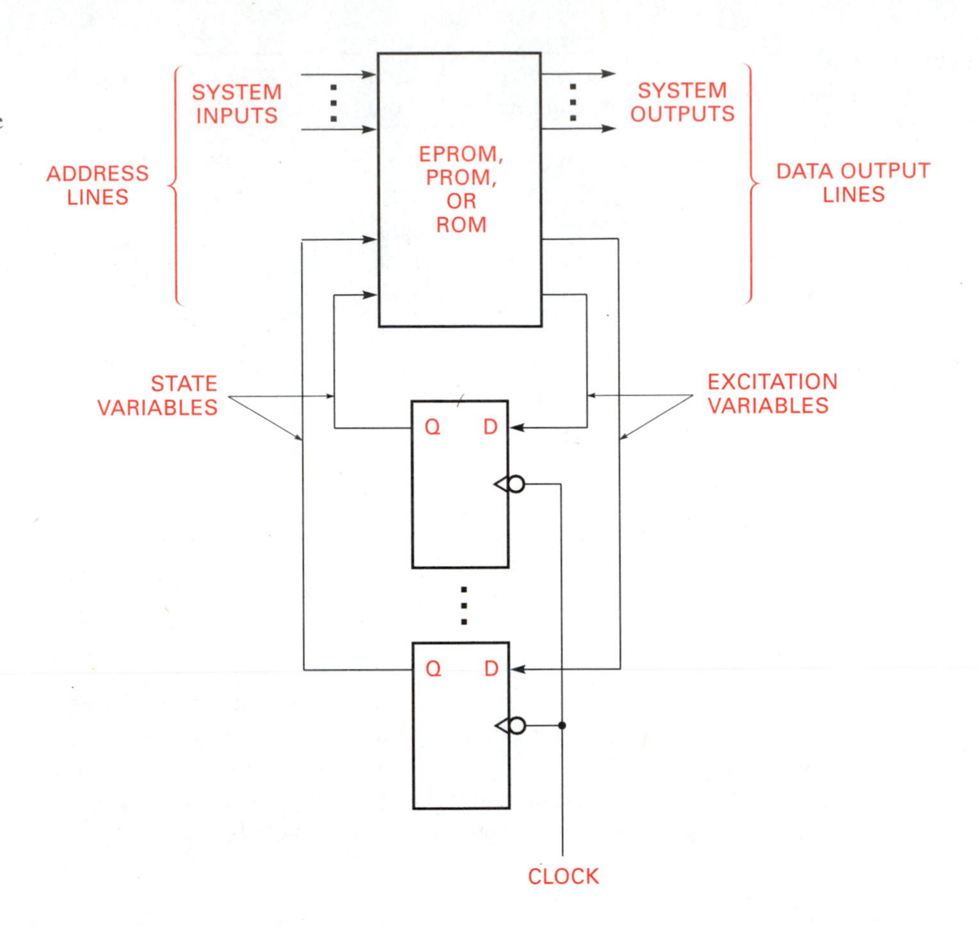

Figure 9.5
Sequential circuit partitioned into an EPROM/ROM and state variable flip-flops

The reason that state reduction should be investigated is that it may reduce the number of state variables and thereby reduce the number of address or data lines to the ROM/EPROM. State reduction or an improved state assignment is not necessary to reduce the number of gates because the internal logic of the ROM/EPROM for a given device is fixed.

The ROM/EPROM data table is constructed from the state table illustrated in Table 7.28. The state variables and input variables are listed as shown in Table 9.3, with F_3 the most and ERST the least significant data bits. The address inputs are connected to the state and input variables. A given state and input condition establishes the next address presented to the memory device. As an address is presented, the stored data are read as excitation and output variables. The excitation variables are presented to the flip-flops and the output variables sent to their destinations. Often it is necessary to buffer the output variables to account for fan-out.

Table 9.3 illustrates the ROM/EPROM data table in binary and Table 9.4 shows the same information in hexadecimal. The information in the tables must be programmed (stored) in the ROM/EPROM. Notice that the state variables are realized using D flip-flops in this example. The reason for using D flip-flops is that fewer excitation variables are needed, thereby reducing the number of output data bits from the memory.

The state variables are realized using a 74LS175 quad D flip-flop IC. A Texas Instru-

TABLE 9.3

Binary data table to be programmed into a TBP28S42 PROM to realize the elapsed time measurement sequential machine

	Address						Data					
Present state	A_5 F_3	A_4 F_2	A_3 F_1	A_2 XMIT	A_1 ECHO	A_0 ERST	D_5 D_3	D_4 D_2	D_3 D_1	D_2 CRST	D_1 CEN	D_0 ABSTB′
A	0	0	0	0	0	0	0	0	1	0	0	1
A	0	0	0	0	0	1	0	0	1	0	0	1
A	0	0	0	0	1	0	0	0	1	0	0	1
A	0	0	0	0	1	1	0	0	1	0	0	1
A	0	0	0	1	0	0	0	0	1	0	0	1
A	0	0	0	1	0	1	0	0	1	0	0	1
A	0	0	0	1	1	0	0	0	1	0	0	1
A	0	0	0	1	1	1	0	0	1	0	0	1
B	0	0	1	0	0	0	0	1	1	1	0	1
B	0	0	1	0	0	1	0	1	1	1	0	1
B	0	0	1	0	1	0	0	1	1	1	0	1
B	0	0	1	0	1	1	0	1	1	1	0	1
B	0	0	1	1	0	0	0	1	1	1	0	1
B	0	0	1	1	0	1	0	1	1	1	0	1
B	0	0	1	1	1	0	0	1	1	1	0	1
B	0	0	1	1	1	1	0	1	1	1	0	1
D	0	1	0	0	0	0	0	1	0	0	1	1
D	0	1	0	0	0	1	0	0	0	0	1	1
D	0	1	0	0	1	0	1	1	0	0	1	1
D	0	1	0	0	1	1	0	0	0	0	1	1
D	0	1	0	1	0	0	0	1	0	0	1	1
D	0	1	0	1	0	1	0	0	0	0	1	1
D	0	1	0	1	1	0	0	1	0	0	1	1
D	0	1	0	1	1	1	0	0	0	0	1	1
C	0	1	1	0	0	0	0	1	1	0	0	1
C	0	1	1	0	0	1	0	1	1	0	0	1
C	0	1	1	0	1	0	0	1	1	0	0	1
C	0	1	1	0	1	1	0	1	1	0	0	1
C	0	1	1	1	0	0	0	1	0	0	0	1
C	0	1	1	1	0	1	0	1	0	0	0	1
C	0	1	1	1	1	0	0	1	0	0	0	1
C	0	1	1	1	1	1	0	1	0	0	0	1
F	1	0	0	0	0	0	0	0	0	0	0	0
F	1	0	0	0	0	1	0	0	0	0	0	0
F	1	0	0	0	1	0	0	0	0	0	0	0
F	1	0	0	0	1	1	0	0	0	0	0	0

continues

TABLE 9.3
Continued

	Address						Data					
Present state	A_5 F_3	A_4 F_2	A_3 F_1	A_2 XMIT	A_1 ECHO	A_0 ERST	D_5 D_3	D_4 D_2	D_3 D_1	D_2 CRST	D_1 CEN	D_0 ABSTB′
F	1	0	0	1	0	0	0	0	0	0	0	0
F	1	0	0	1	0	1	0	0	0	0	0	0
F	1	0	0	1	1	0	0	0	0	0	0	0
F	1	0	0	1	1	1	0	0	0	0	0	0
dc	1	0	1	0	0	0	0	0	0	0	0	1
dc	1	0	1	0	0	1	0	0	0	0	0	1
dc	1	0	1	0	1	0	0	0	0	0	0	1
dc	1	0	1	0	1	1	0	0	0	0	0	1
dc	1	0	1	1	0	0	0	0	0	0	0	1
dc	1	0	1	1	0	1	0	0	0	0	0	1
dc	1	0	1	1	1	0	0	0	0	0	0	1
dc	1	0	1	1	1	1	0	0	0	0	0	1
E	1	1	0	0	0	0	1	0	0	0	0	1
E	1	1	0	0	0	1	1	0	0	0	0	1
E	1	1	0	0	1	0	1	1	0	0	0	1
E	1	1	0	0	1	1	1	1	0	0	0	1
E	1	1	0	1	0	0	1	0	0	0	0	1
E	1	1	0	1	0	1	1	0	0	0	0	1
E	1	1	0	1	1	0	1	1	0	0	0	1
E	1	1	0	1	1	1	1	1	0	0	0	1
dc	1	1	1	0	0	0	0	0	0	0	0	1
dc	1	1	1	0	0	1	0	0	0	0	0	1
dc	1	1	1	0	1	0	0	0	0	0	0	1
dc	1	1	1	0	1	1	0	0	0	0	0	1
dc	1	1	1	1	0	0	0	0	0	0	0	1
dc	1	1	1	1	0	1	0	0	0	0	0	1
dc	1	1	1	1	1	0	0	0	0	0	0	1
dc	1	1	1	1	1	1	0	0	0	0	0	1

ments TBP28S42, 512 × 8, PROM is used for the combinational function in this design. The device is a fusible link user programmable read only memory. A programming unit is necessary to blow the titanium-tungsten fuses for programming the device. The PROM IC has a read access time of 35 ns. It also has tristate outputs controlled by the active-low EN input. Figure 9.6 illustrates the elapsed time measurement controller realized with two integrated circuits.

TABLE 9.4
Hexadecimal data table for elapsed time measurement sequential machine

Address	Data	Address	Data
00	09	20	00
01	09	21	00
02	09	22	00
03	09	23	00
04	09	24	00
05	09	25	00
06	09	26	00
07	09	27	00
08	1D	28	01
09	1D	29	01
0A	1D	2A	01
0B	1D	2B	01
0C	1D	2C	01
0D	1D	2D	01
0E	1D	2E	01
0F	1D	2F	01
10	13	30	21
11	03	31	21
12	33	32	31
13	03	33	31
14	13	34	21
15	03	35	21
16	13	36	31
17	03	37	31
18	19	38	01
19	19	39	01
1A	19	3A	01
1B	19	3B	01
1C	11	3C	01
1D	11	3D	01
1E	11	3E	01
1F	11	3F	01

9.3 PROGRAMMABLE LOGIC DEVICES

Programmable logic devices (PLD) is the generic name given to a group of integrated circuits that can be configured to perform logic functions. The internal circuitry of PLDs are structured during manufacture and configured by the user. ROMS, PROMS, and EPROMS are programmable but are not normally classified as PLDs. We will investigate several programmable logic devices in this section including programmable logic arrays,

Figure 9.6
Two IC realizations of the elapsed time measurement sequential controller

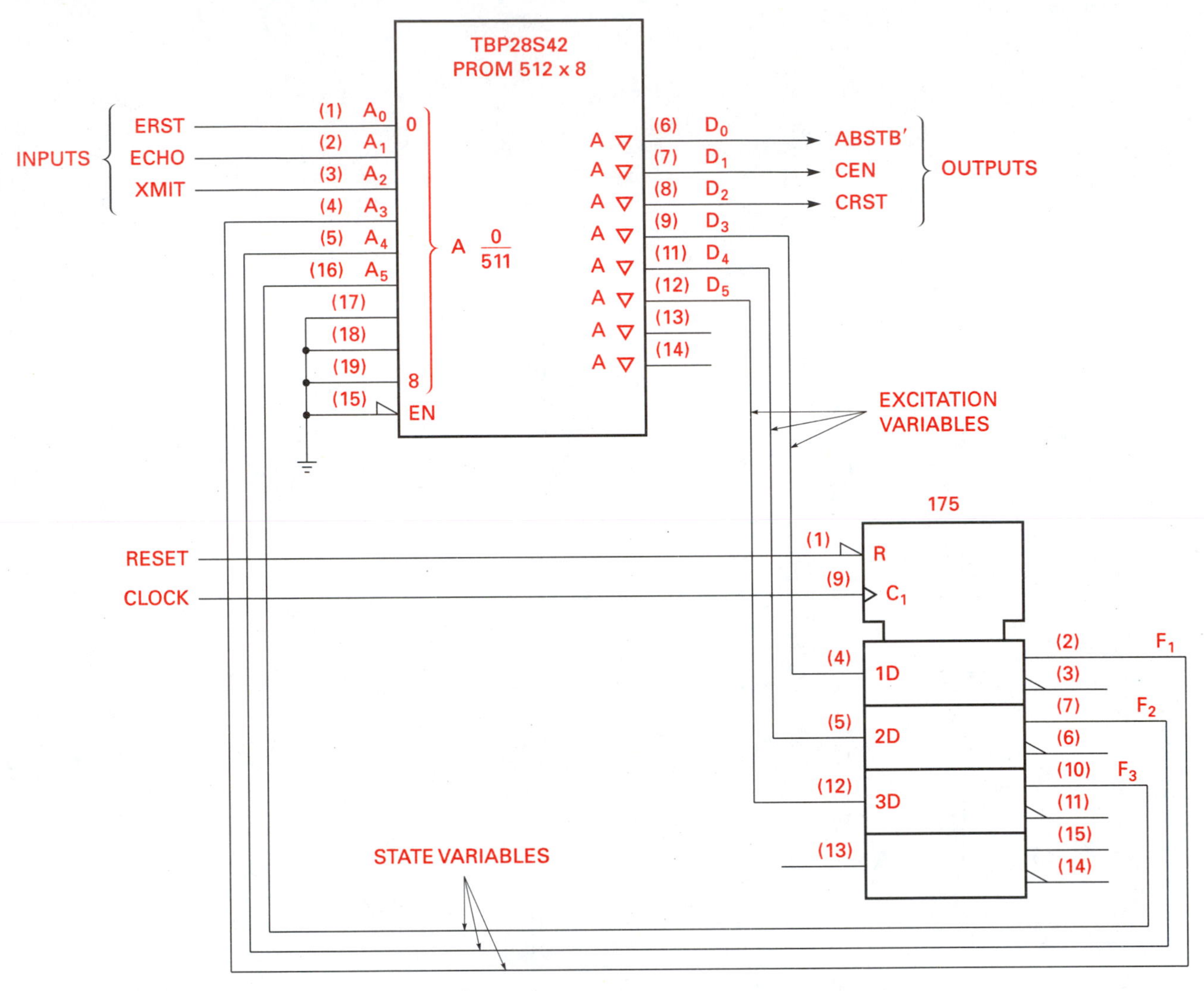

programmable array logic, and generic array logic. Another class of devices called *erasable programmable logic* will be covered in Section 9.4.

9.3.1 Programmable Logic Array (PLA)

A programmable logic array (**PLA**) device contains an array of AND-OR functions whose configuration is determined by the user. AND-OR functions are used to realize sum-of-products functions, therefore, PLAs are combinational logic devices. Any minterm expression within the limits of the number of inputs, array size, and number of outputs can be realized using a PLA.

PLAs also must have an interconnect scheme so that product terms (p-terms) can be ORed together to form output functions. Both the AND and OR terms have programmable interconnections. Most devices are programmed by the designer using an applications program that allows the designer to enter the Boolean functions from which the internal interconnections are made. The programming unit interprets the Boolean function equations and makes the internal connections as we saw in fusible link PROMS. Figure 9.7 illustrates the AND-OR gate arrays and the fused interconnections that are used to create output functions. Note that both active-high and -low input variables are

Figure 9.7
PLA showing configurable AND and OR interconnections

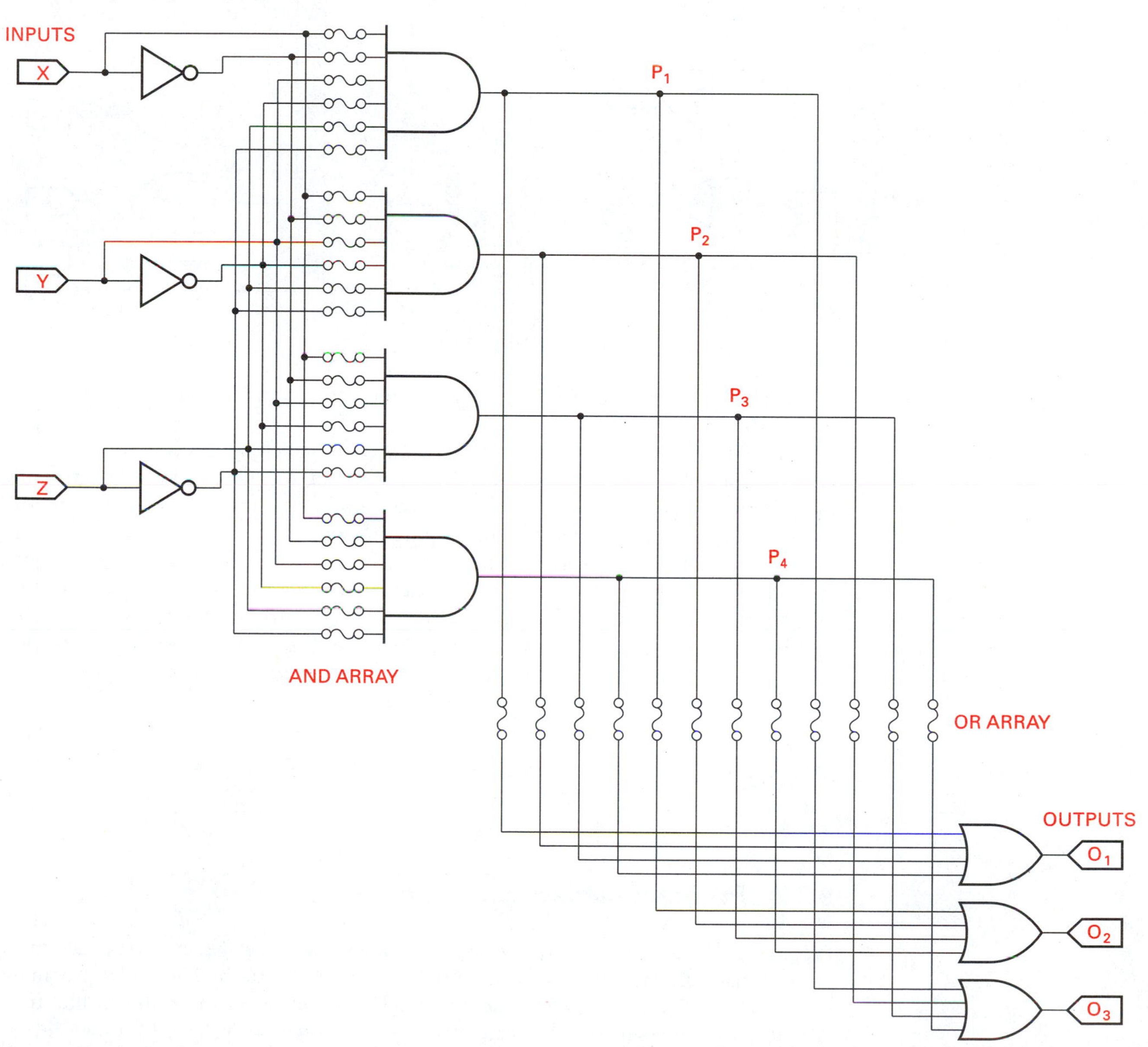

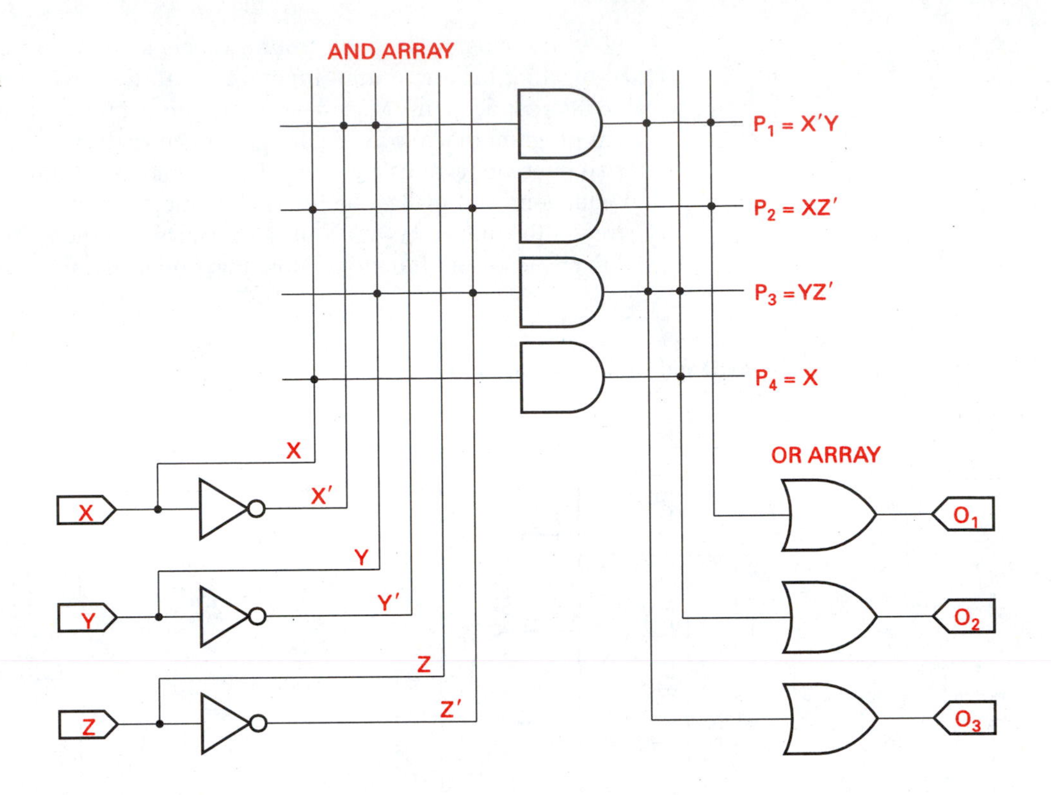

Figure 9.8
PLA realization of the Boolean functions O_3, O_2, and O_1

fused. This permits either active state of an input variable to be used in the generation of a product term.

Each AND gate produces a product term based on which fuses are left intact. Fuses are blown for input variables not used in the generation of a p-term. Product terms are fused inputs to the OR gates. The designer has control over the generation of a p-term and which p-terms are to make up the output function.

For reasonably large PLAs, drawing all of the wires and fuses becomes cumbersome. Instead, each input variable and its complement is shown converging on a single line to an AND gate. If a connection is to be made, an X is drawn at the intersection of the variable or its complement with the p-term input line. The p-term contains the variables where Xs are drawn. Figure 9.8 illustrates the notation for realization of the Boolean functions

$$O_1 = X'Y + XZ'$$
$$O_2 = YZ' + X$$
$$O_3 = X'Y + YZ'$$

9.3.2 Programmable Array Logic

Programmable array logic (PAL) devices are similar to PLA devices except the OR array is fixed instead of programmable. This reduces the flexibility but simplifies programming. Figures 9.9 and 9.10 illustrate the differences between combinational PLA and PAL devices. The Xs mark the interconnection points. Note that the AND array is iden-

Figure 9.9
PLA logic

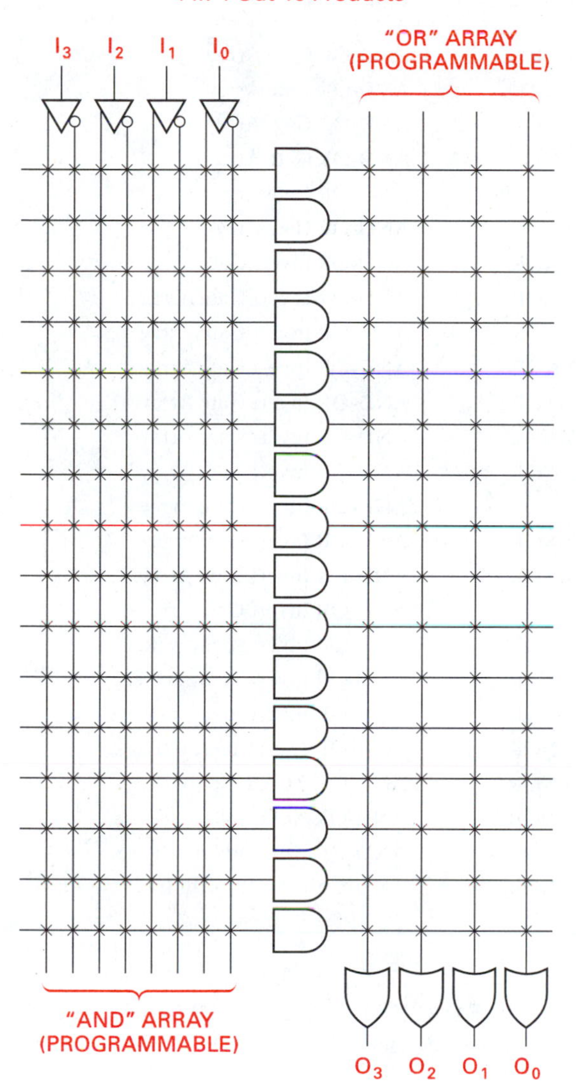

Figure 9.10
PAL logic

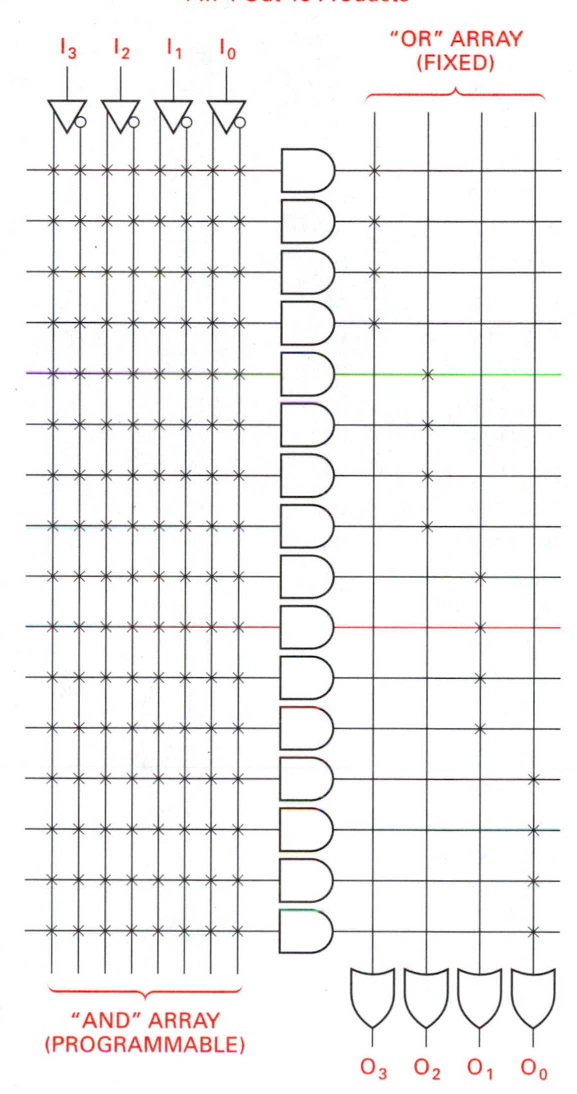

tical in both types of devices. The OR array in the PLA has Xs for each intersection between the p-terms and the OR gate inputs. Each of these intersections is programmable. In the PAL, the OR array is fixed. For example, in Figure 9.10, the O_3 output is created from the first four p-terms, O_2 is created from the next four p-terms, and so on.

A wide range of PAL devices are available to the logic designer, some are strictly combinational, and others contain flip-flops. Table 9.5 presents a partial list of PAL devices and their functions.

Note the part numbers of the PALs listed in Table 9.5. The prefix PAL identifies the device type and the next number specifies the number of inputs. The last number speci-

TABLE 9.5
Partial list of PAL devices

Part number	Input	Output	Programmable I/Os	Feedback Register	Output Polarity	Functions
PAL10H8	10	8			AND-OR	AND-OR Gate Array
PAL12H6	12	6			AND-OR	AND-OR Gate Array
PAL14H4	14	4			AND-OR	AND-OR Gate Array
PAL16H2	16	2			AND-OR	AND-OR Gate Array
PAL16C2	16	2			BOTH	AND-OR Gate Array
PAL20C2	20	2			BOTH	AND-OR Gate Array
PAL10L8	10	8			AND-NOR	AND-OR Invert Gate Array
PAL12L6	12	6			AND-NOR	AND-OR Invert Gate Array
PAL14L4	14	4			AND-NOR	AND-OR Invert Gate Array
PAL16L2	16	2			AND-NOR	AND-OR Invert Gate Array
PAL12L10	12	10			AND-NOR	AND-OR Invert Gate Array
PAL14L8	14	8			AND-NOR	AND-OR Invert Gate Array
PAL16L6	16	6			AND-NOR	AND-OR Invert Gate Array
PAL18L4	18	4			AND-NOR	AND-OR Invert Gate Array
PAL20L2	20	2			AND-NOR	AND-OR Invert Gate Array
PAL16L8	10	2	6		AND-NOR	AND-OR Invert Gate Array
PAL20L10	12	2	8		AND-NOR	AND-OR Invert Gate Array
PAL16R8	8	8		8	AND-NOR	AND-OR Invert Array w/Reg's
PAL16R6	8	6	2	6	AND-NOR	AND-OR Invert Array w/Reg's
PAL16R4	8	4	4	4	AND-NOR	AND-OR Invert Array w/Reg's
PAL20X10	10	10		10	AND-NOR	AND-OR-XOR Invert w/Reg's
PAL20X8	10	8	2	8	AND-NOR	AND-OR-XOR Invert w/Reg's
PAL20X4	10	4	6	4	AND-NOR	AND-OR-XOR Invert w/Reg's
PAL16X4	8	4	4	4	AND-NOR	AND-OR-XOR Invert w/Reg's
PAL16A4	8	4	4	4	AND-NOR	AND-CARRY-OR-XOR Invert w/Reg's

fies the number of output pins. For example, PAL10H8 is a PAL with 10 input pins and 8 output pins. The R in part PAL16R8 indicates registers. The X in PAL 20X8 indicates an AND-OR-XOR logic arrangement.

Programmable I/O PALs allow the logic designer to feed an output function back as an input variable to create a new function. Figure 9.11 illustrates a simple PAL AND-OR array. Figure 9.12 illustrates a programmable I/O arrangement, where the output is brought back into the AND array for use as an input.

Registered output PALS have flip-flops that store the output. Most also have tristate buffers so that several flip-flops can form a data register whose output can be controlled. Figure 9.13 illustrates the logic for a registered PAL. Notice that the flip-flop output is made available as an input to be used in the generation of additional p-terms. This feedback feature allows PALs to be used to design counters, registers, and sequential controllers.

Figure 9.11
Standard I/O PAL

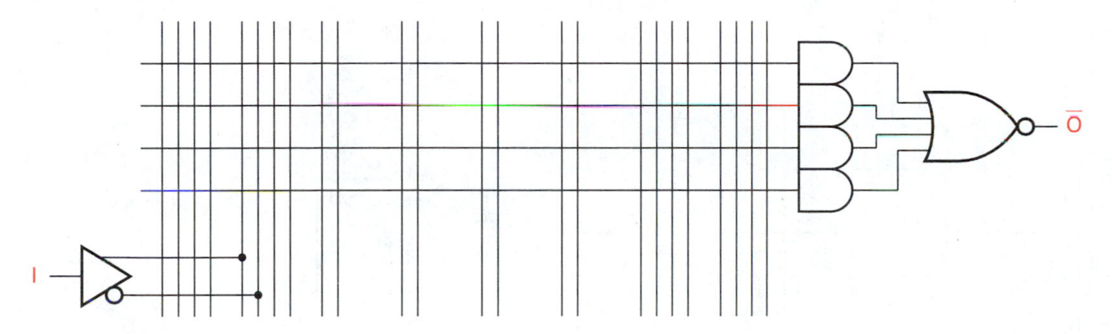

Figure 9.12
Programmable I/O PAL

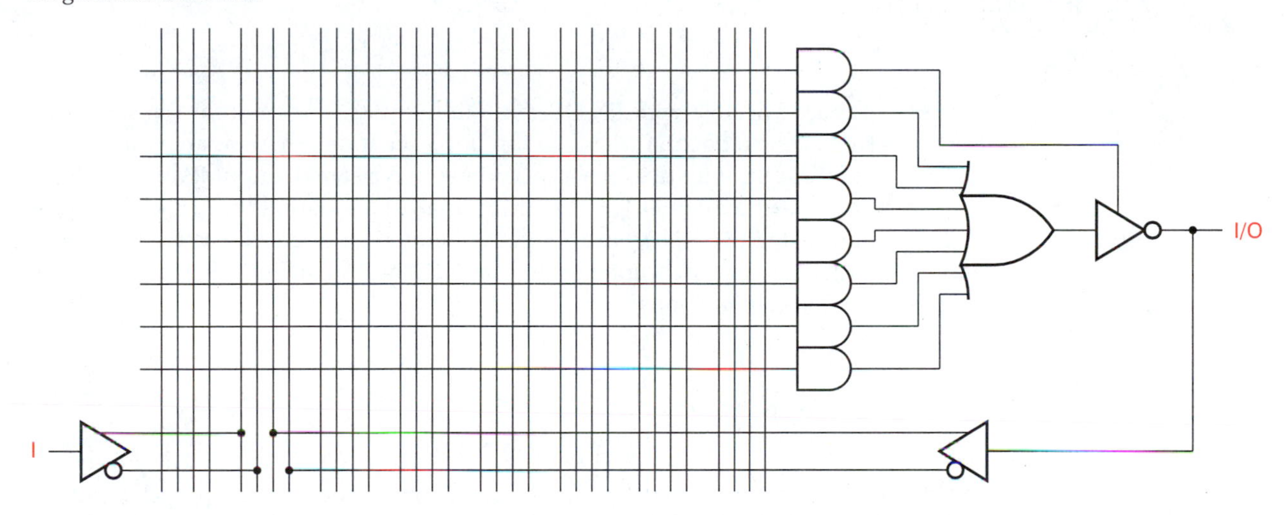

Figure 9.13
Registered output PAL

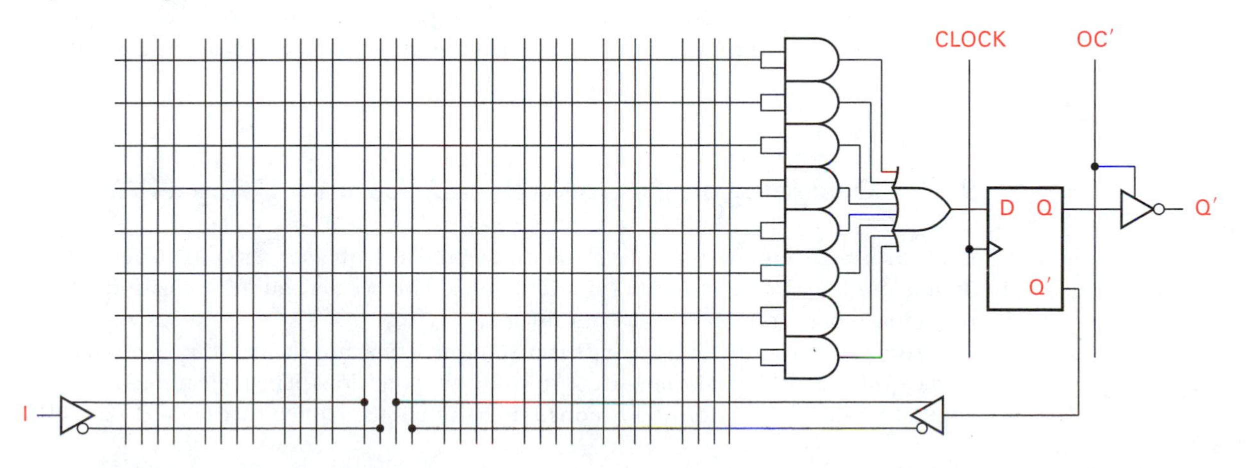

Figure 9.14
EX-OR registered output PAL

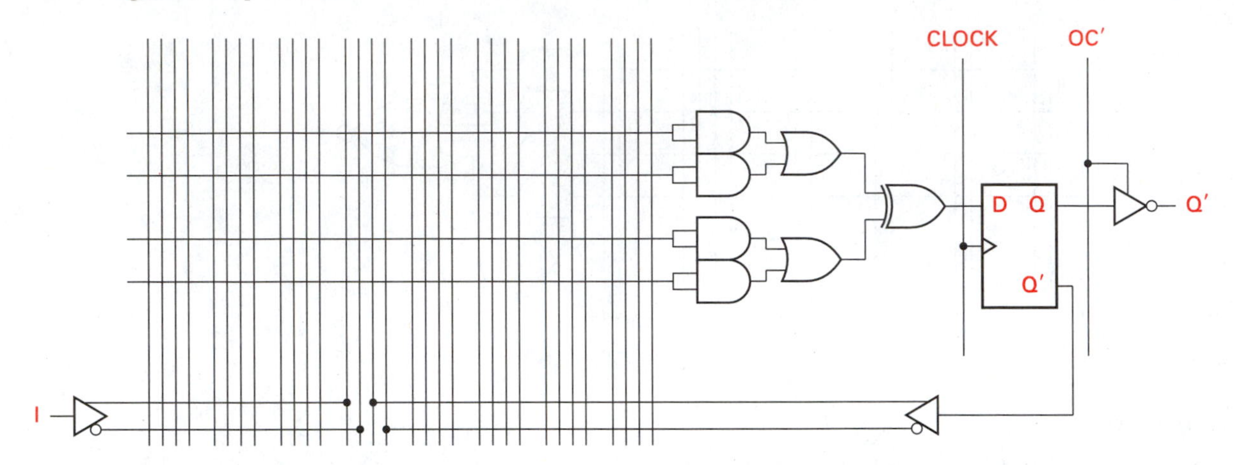

PALs also are available with EX-OR excitation inputs to the output flip-flop. This feature permits logic reduction when the flip-flop excitation functions are more simply realized using EX-ORs. Figure 9.14 shows an EX-OR registered output PAL.

Figure 9.15 illustrates a PAL10H8, with 10 input and 8 output lines. Each output is directly connected to two AND gates. Larger functions can be created by connecting an output line back into one of the inputs. For example, the PAL in Figure 9.15 is connected to realize the Boolean functions

$$O_1 = I_1\, I_2' + I_2'\, I_3$$
$$O_2 = I_1' I_2' I_3 I_4$$
$$O_3 = I_1' I_2 I_3' + I_2\, I_3'\, I_4 + I_4' I_5 + I_3'\, I_4\, I_5'$$

The first two output functions can be realized by connecting the proper inputs to the available p-terms. Output function O_3, however, requires four p-terms and only two are directly available. This means that two intermediate functions X and Y are generated and connected back into two unused input pins to create output O_3. This approach obviously consumes additional input and output pins. It also adds to the output propagation delay time. If either of these restrictions is critical, then another PAL would be chosen.

9.3.3 Designing an Up-Down Decade Counter Using a PAL

PALs with registered outputs can be used to realize sequential circuits. Consider the decade up/down counter design from Chapter 6. The decade counter transition table is reproduced here as Table 9.6 for convenience.

From the transition table we see that two inputs, one output, and four state variables are needed. The PAL16R4 in Figure 9.16 has four D flip-flops. The logic not used for the counter could be used for other, noncounter functions. Notice that eight p-terms are

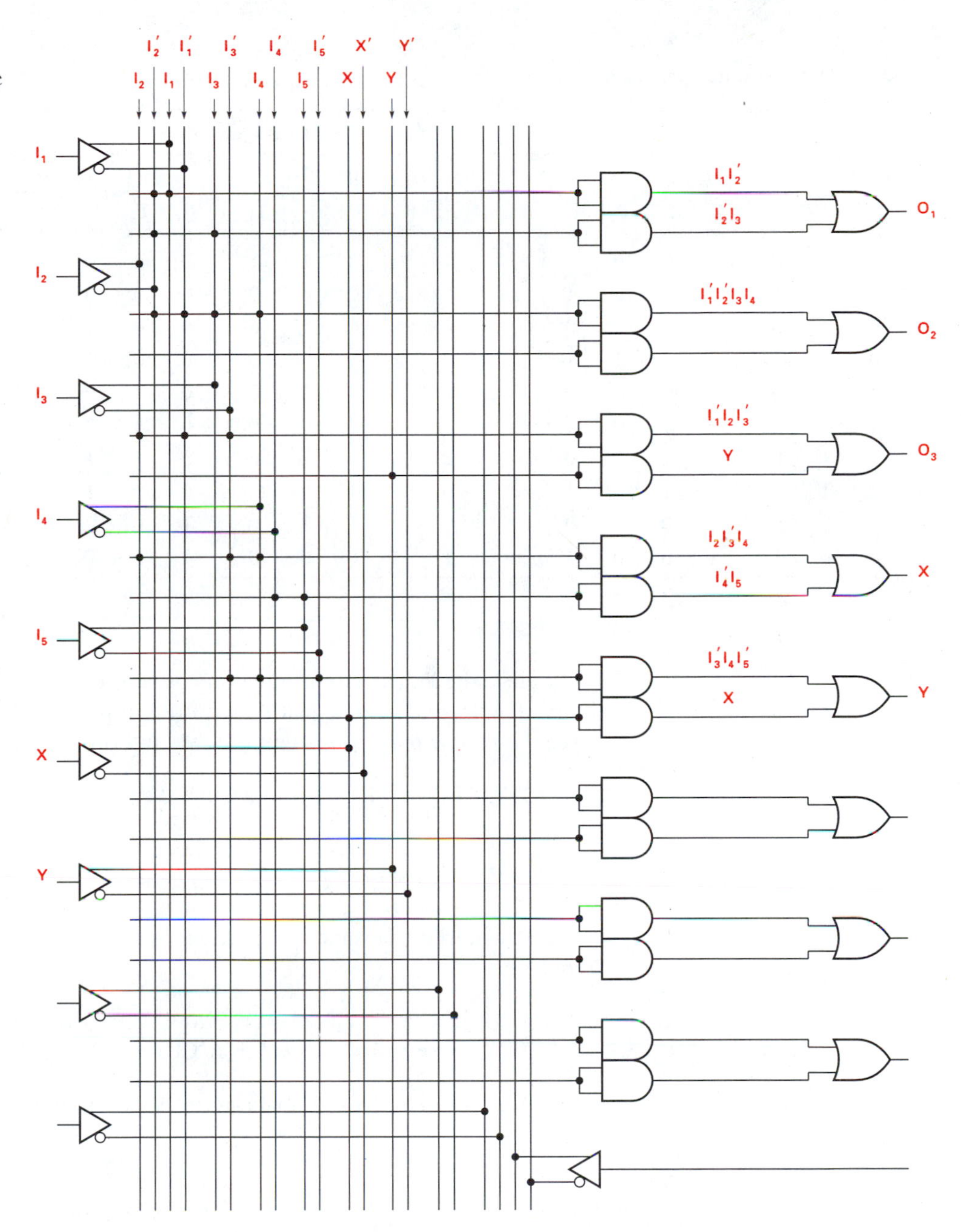

Figure 9.15

A PAL10H8 connected to realize O_1, O_2, and O_3

available for each D excitation input. Simplification of the counter excitation equations would be necessary only if the number of p-terms for any D input exceeds eight p-terms. The following D excitation equations result from the transition table:

$$\begin{aligned} D_4 = {} & F_4F_3'F_2'F_1'X'CE' + F_4F_3'F_2'F_1X'CE' + F_4'F_3'F_2'F_1'X'CE \\ & + F_4F_3'F_2F_1X'CE + F_4F_3'F_2'F_1'XCE' + F_4F_3'F_2'F_1XCE' \\ & + F_4'F_3F_2F_1XCE + F_4F_3'F_2'F_1'XCE \end{aligned}$$

TABLE 9.6
Decade up/down counter transition table

Present state				Next state: X CE =																			
				0 0					0 1					1 0					1 1				
F_4	F_3	F_2	F_1	F_4	F_3	F_2	F_1	TC	F_4	F_3	F_2	F_1	TC	F_4	F_3	F_2	F_1	TC	F_4	F_3	F_2	F_1	TC
0	0	0	0	0	0	0	0	0	1	0	0	1	1	0	0	0	0	0	0	0	0	1	0
0	0	0	1	0	0	0	1	0	0	0	0	0	0	0	0	0	1	0	0	0	1	0	0
0	0	1	0	0	0	1	0	0	0	0	0	1	0	0	0	1	0	0	0	0	1	1	0
0	0	1	1	0	0	1	1	0	0	0	1	0	0	0	0	1	1	0	0	1	0	0	0
0	1	0	0	0	1	0	0	0	0	0	1	1	0	0	1	0	0	0	0	1	0	1	0
0	1	0	1	0	1	0	1	0	0	1	0	0	0	0	1	0	1	0	0	1	1	0	0
0	1	1	0	0	1	1	0	0	0	1	0	1	0	0	1	1	0	0	0	1	1	1	0
0	1	1	1	0	1	1	1	0	0	1	1	0	0	0	1	1	1	0	1	0	0	0	0
1	0	0	0	1	0	0	0	0	0	1	1	1	0	1	0	0	0	0	1	0	0	1	0
1	0	0	1	1	0	0	1	0	1	0	0	0	0	1	0	0	1	0	0	0	0	0	1

Because the D_4 excitation equation contains eight p-terms, no simplification is necessary. The D_3 excitation equation requires 16 p-terms. Simplification is now required to see if it can be reduced to eight or fewer p-terms. After simplification D_3 becomes

$$D_3 = F_4F_1'X'CE + F_3'F_2F_1XCE + F_3F_2'F_1 + F_3F_2F_1' + \begin{matrix} F_3F_2X' \\ \text{or} \\ F_3F_1X' \end{matrix} + F_3CE' + \begin{matrix} F_3F_1X \\ \text{or} \\ F_3F_2'X \end{matrix}$$

The unsimplified excitation equation for D_2 requires 16 p-terms. After simplification D_2 becomes

$$D_2 = F_2CE' + F_2F_1X' + F_2F_1'X + F_4'F_2'F_1XCE + F_3F_2'F_1'X'CE + F_4F_1'X'CE$$

Simplification of the 20 initial p-terms for excitation D_1 results in

$$D_1 = F_1CE' + F_1'CE$$

The output function, TC, occurs under two conditions:

$$TC = F_4F_1XCE + F_4'F_3'F_2'F_1'X'CE$$

The CE input enables the TC tristate output.

Programming the PAL requires that the logic functions be described by Boolean equations. These are converted into a map by a special compiler so that the appropriate fuses can be blown. The fuse map can also be created by marking up the logic diagram indicating the vertical and horizontal coordinates for a connection.

Each *X–Y* coordinate in the array is represented by a location in the PAL fuse map.

Figure 9.16
Up/down counter realized using a PAL16R4

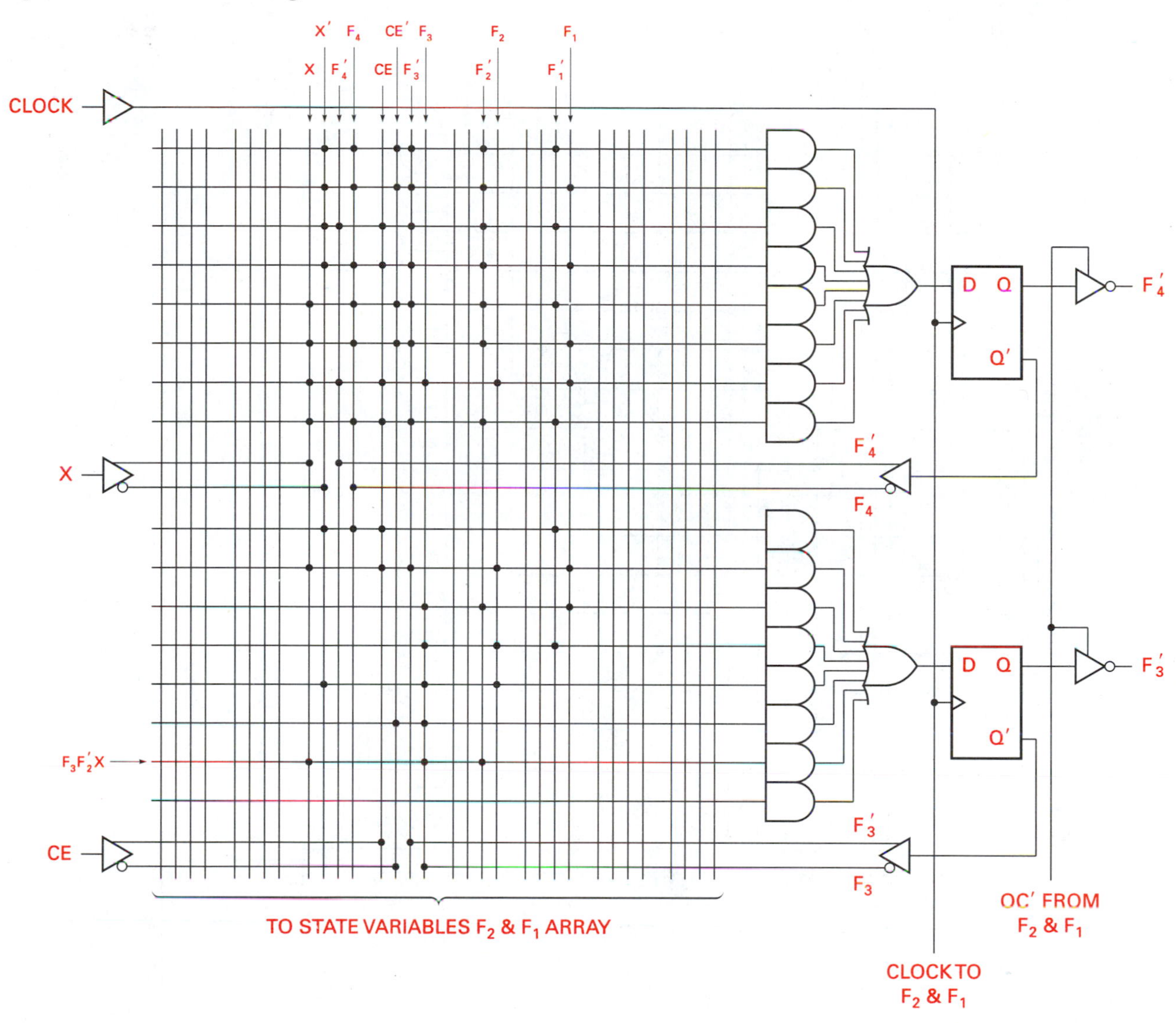

(a) Up/Down Counter State Variables F4 & F3 Realized Using a PAL16R4

The fuse map is called a **JEDEC** (Joint Electron Device Engineering Council) file. The file format establishes a standard that is used by device programmer software for blowing PAL fuses. Some programming units also have simulation software, and the JEDEC file may also contain simulation information, which allows a design to be checked for proper programming.

Most semiconductor companies that manufacture programmable logic also provide programming software to support their chips. For example, the Data I/O Cor-

Figure 9.16
Continued

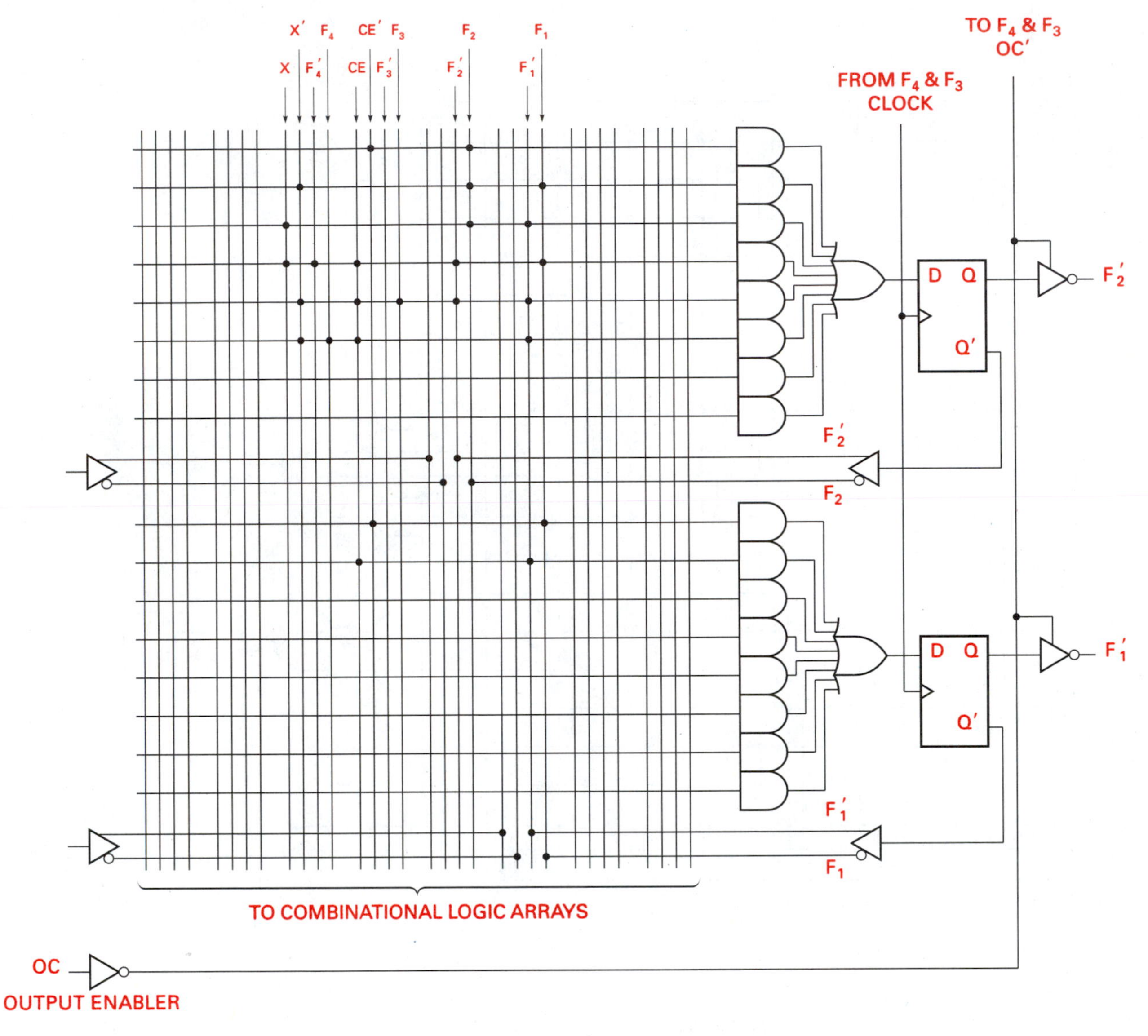

(b) Up/Down Counter State Variables F2 & F1 Realized Using a PAL16R4

poration offers ABEL, Signetics offers AMAZE, and Altera sells MAX+Plus support software. **PALASM** is a programming software compiler that is available in FORTRAN source code.

PAL programmers are also sold by the device makers as well as second party sources. These units vary in their capability and price. They may sell for as little as $300 to $400 or as much as $5000. Some units will program only specific devices and others will

Figure 9.16
Continued

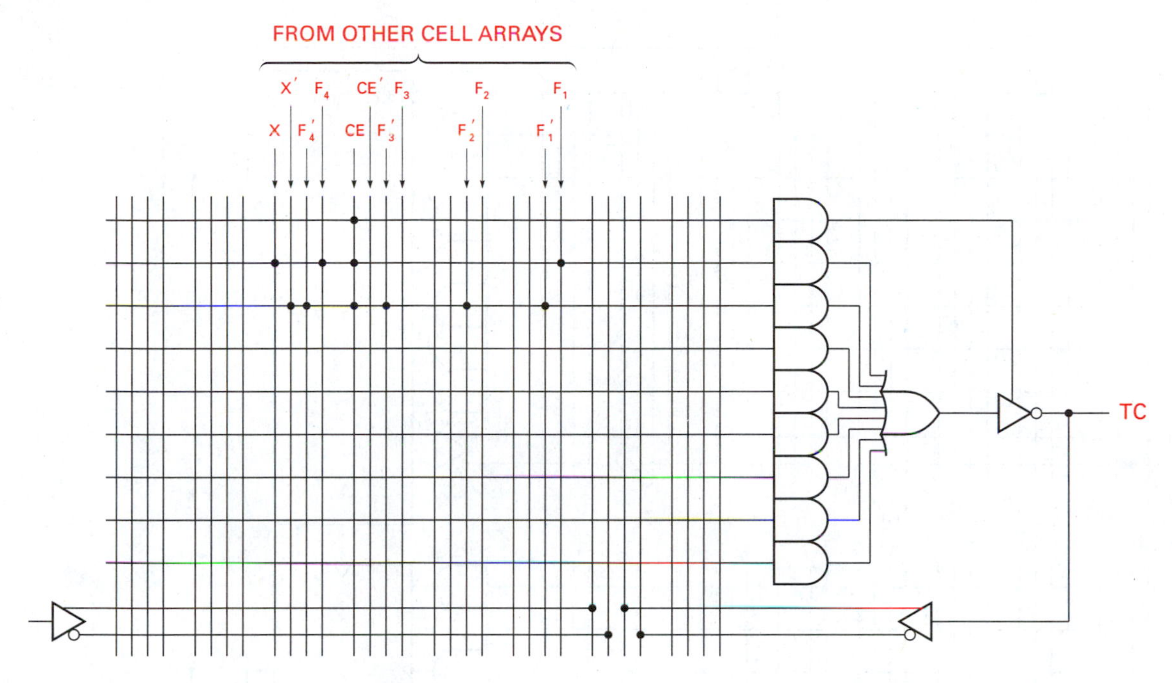

(c) Up/Down Counter TC Output Realized Using a PAL16R4

program almost any PROMs, PLAs, PALs, or PLDs. Some units are stand-alone and others connect to a host personal computer, using JEDEC files to blow fuses.

9.3.4 Generic Array Logic

Another classification of programmable logic devices is the generic array logic or **GAL**. GALs, introduced by Lattice Semiconductor Company, contain a programmable AND array, a fixed OR array, and an output stage. The output state, called an *output logic* **macro cell** (OLMC) by Lattice, contains multiplexers, flip-flops, and tristate output buffers. GAL devices are EEPLDs, which means that they can be reprogrammed while still in the circuit board. The GAL16V8A can have up to 20 inputs (some pins can be either input or registered feedback) and 8 outputs.

The logic diagram for the GAL16V8A is shown in Figure 9.17. Note the OLMC block that receives the p-terms from the inputs or from flip-flops. Also each OLMC output has tristate buffers.

The OLMC unit has three input and one output multiplexers, one tristate flip-flop, and a fixed OR array. Figure 9.18 shows the block diagram of the cell.

Inputs from the programmable AND array are brought into the fixed OR array, which provides one input to a XOR gate. One input to the OR array comes from the

Figure 9.17
GAL16V8A logic diagram

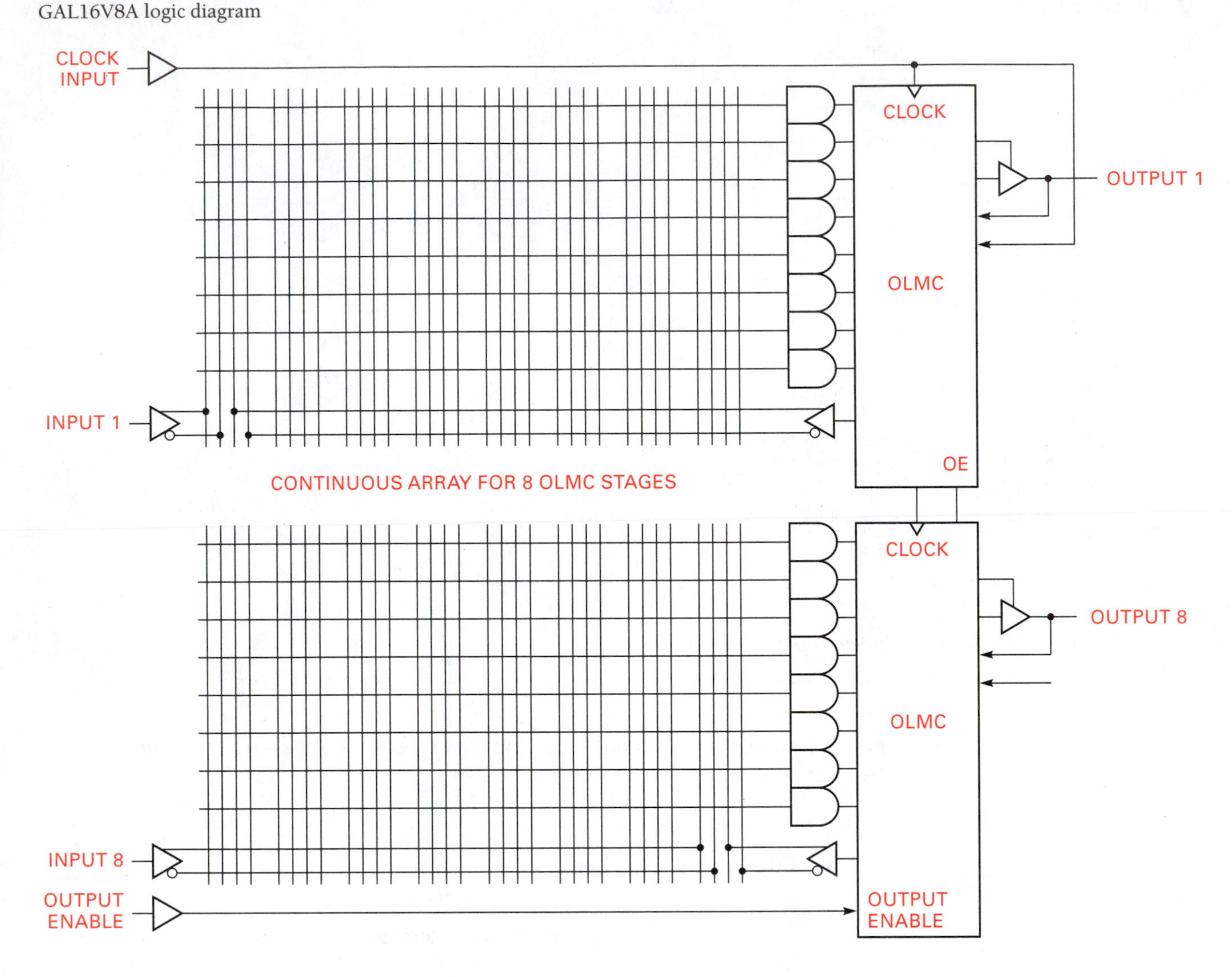

PTMUX output. Two control inputs AC_0 and AC_1 control the PTMUX and TSMUX. The XOR output provides the D input to the cell flip-flop. The excitation level can be active-high or -low, depending on the XOR(n) signal. The flip-flop output and the XOR output are inputs to the OMUX. This means that the cell can be configured as a combinational function or a registered function, depending on the values of AC_0 and $AC_1(n)$.

Registered feedback to the AND array is provided by the flip-flop Q′, the output of tristate output of the OMUX or from an adjacent cell, under the control of AC_0 and $AC_1(n)$. A common positive edge clock provides flip-flop triggering. The tristate output can be controlled by a single output enable (OE) or by a p-term output steered through the TSMUX to the output tristate enable.

Figure 9.18
GAL16V8A output logic macro cell

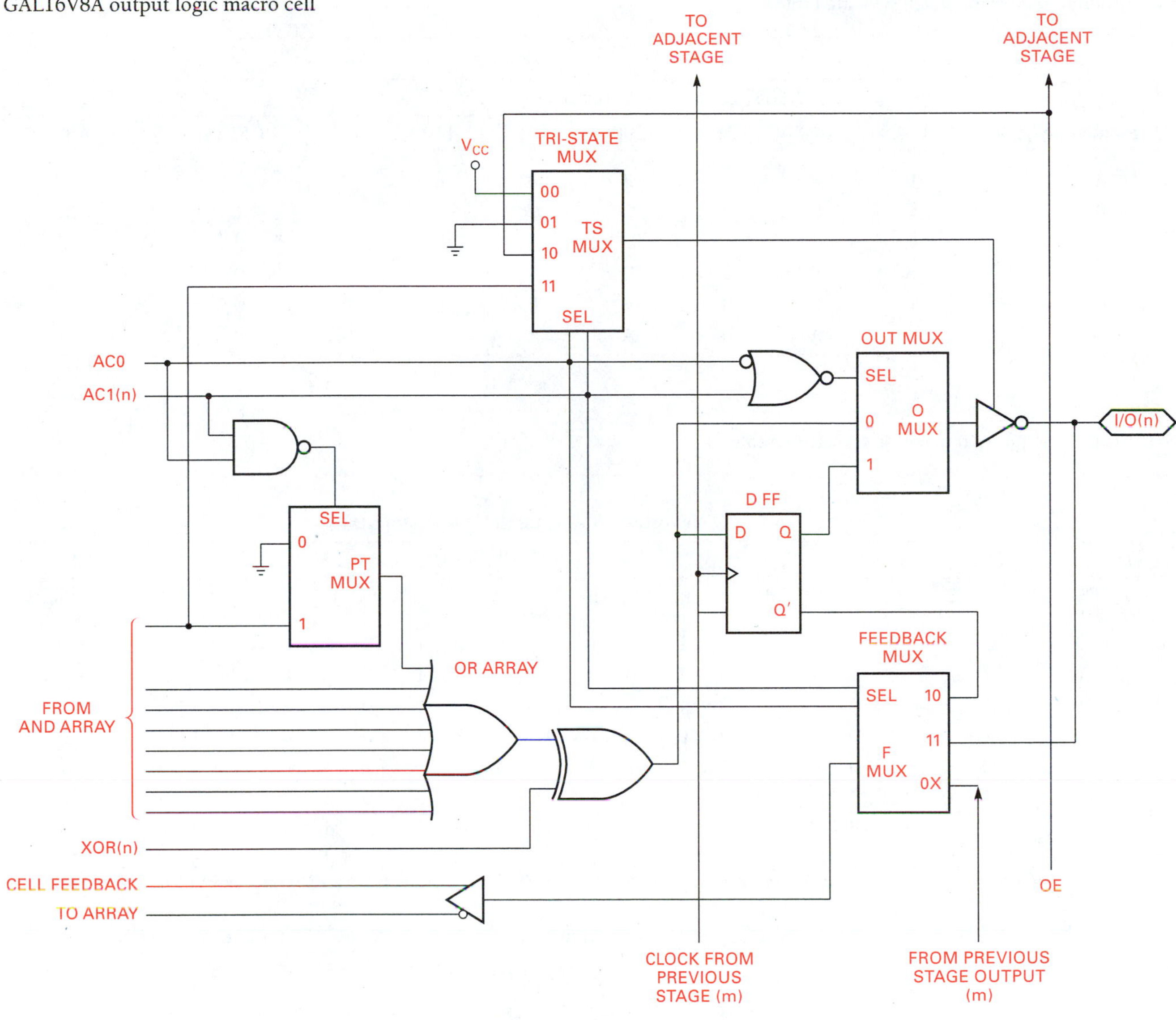

9.3.5 Designing a Synchronous Sequential Circuit Using a GAL

Consider the synchronous sequential circuit first shown in state Table 7.28. The state and transition tables are reproduced here as Tables 9.7 and 9.8. From Tables 9.7 and 9.8 we derive the output equations

$$\mathrm{CRST} = F_3'F_2'F_1$$
$$\mathrm{CEN} = F_3'F_2F_1'$$
$$\mathrm{ABSTB}' = F_3F_2'F_1'$$

TABLE 9.7
Synchronous sequential machine state table

	Next state										
	XMIT, ECHO, ERST (Inputs)								OUTPUTS		
Present state	000	001	010	011	100	101	110	111	CRST	CEN	ABSTB′
A	B	B	B	B	B	B	B	B	0	0	1
B	C	C	C	C	C	C	C	C	1	0	1
C	C	C	C	C	D	D	D	D	0	0	1
D	D	A	E	A	D	A	D	A	0	1	1
E	F	F	E	E	F	F	E	E	0	0	1
F	A	A	A	A	A	A	A	A	0	0	0

TABLE 9.8
Synchronous sequential machine transition table

Present state	Next state, XMIT, ECHO, ERST (Inputs)							
	000	001	010	011	100	101	110	111
F_3 F_2 F_1	F_3 F_2 F_1	F_3 F_2 F_1	F_3 F_2 F_1	F_3 F_2 F_1	F_3 F_2 F_1	F_3 F_2 F_1	F_3 F_2 F_1	F_3 F_2 F_1
0 0 0	0 0 1	0 0 1	0 0 1	0 0 1	0 0 1	0 0 1	0 0 1	0 0 1
0 0 1	0 1 1	0 1 1	0 1 1	0 1 1	0 1 1	0 1 1	0 1 1	0 1 1
0 1 1	0 1 1	0 1 1	0 1 1	0 1 1	0 1 0	0 1 0	0 1 0	0 1 0
0 1 0	0 1 0	0 0 0	1 1 0	0 0 0	0 1 0	0 0 0	0 1 0	0 0 0
1 1 0	1 0 0	1 0 0	1 1 0	1 1 0	1 0 0	1 0 0	1 1 0	1 1 0
1 0 0	0 0 0	0 0 0	0 0 0	0 0 0	0 0 0	0 0 0	0 0 0	0 0 0
1 0 1	d	d	d	d	d	d	d	d
1 1 1	d	d	d	d	d	d	d	d

The D flip-flop excitation equations, derived from Table 9.8 and simplified are

$$D_3 = F_3F_2 + F_2F_1'\text{XMIT}'\ \text{ECHO ERST}'$$
$$D_2 = F_1 + F_3'F_2\ \text{ERST}' + F_3F_2\ \text{ECHO}$$
$$D_1 = F_3'F_2' + F_1\ \text{XMIT}'$$

We will use the GAL16V8A to realize the elapsed time measurement sequential controller. Only three of the eight flip-flops are needed. Three input and three output variables are also needed. Refer to Figures 9.17 and 9.18. We will assign input variables XMIT, ECHO, and ERST to input pins 2, 3, and 4, respectively. Let OLMCs 19, 18, and 17 be assigned to state variables F_3, F_2, and F_1, respectively. The p-terms that generate the excitation inputs to state variables F_3, F_2, and F_1 are brought into the OR array and presented to the D inputs of the flip-flops.

The cells shown in Figure 9.19 represent input and output functions of the output logic macro cell (OLMC) shown in Figure 9.18. The diagram in Figure 9.19 shows a functional subset of the capability of the OLMC; the cell is configured as an input-output function. In this configuration the multiplexer (FMUX) steers the state variables back

Figure 9.19

Output logic macro cell with programmable output enable and functional polarity

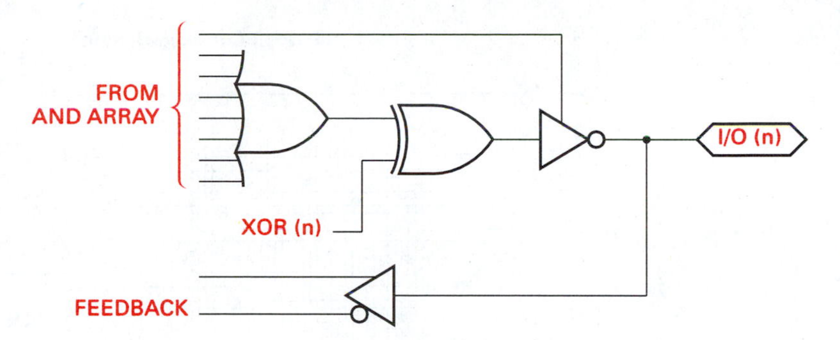

(a) Combinational Input/Output with Programmable Output Enable and Functional Polarity

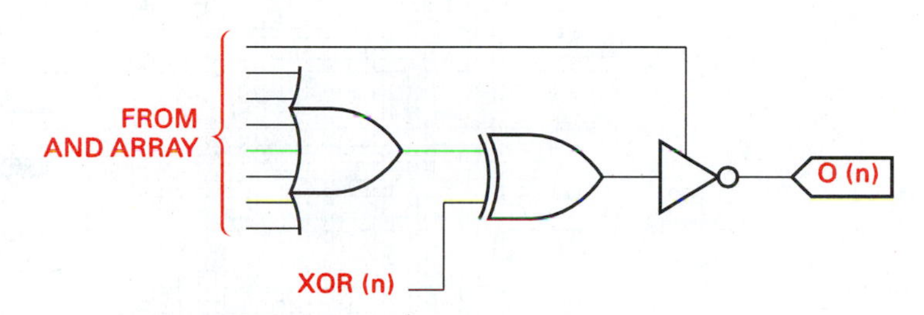

(b) Combinational Output with Programmable Output Enable and Functional Polarity

into the AND array to be used to generate the excitation and output function equations of the elapsed time measurement sequential machine.

Figure 9.19 illustrates a single output cell configured in a complex mode, where each output cell can be programmed as either an input or an output, depending on the value of AC_0 and $AC_1(n)$. Each macro cell has seven p-terms in its AND array. An additional p-term is used as an enable control driving the output tristate buffer.

Figure 9.20 illustrates a single OLMC cell configured in register mode. All of the flip-flops in the register configured cells share a common clock and tristate output buffer enable.

Figure 9.21 illustrates the complete GAL16V8 configured in the complex mode and Figure 9.22 shows the registered mode configuration.

Figure 9.20

Registered output with programmable output function polarity

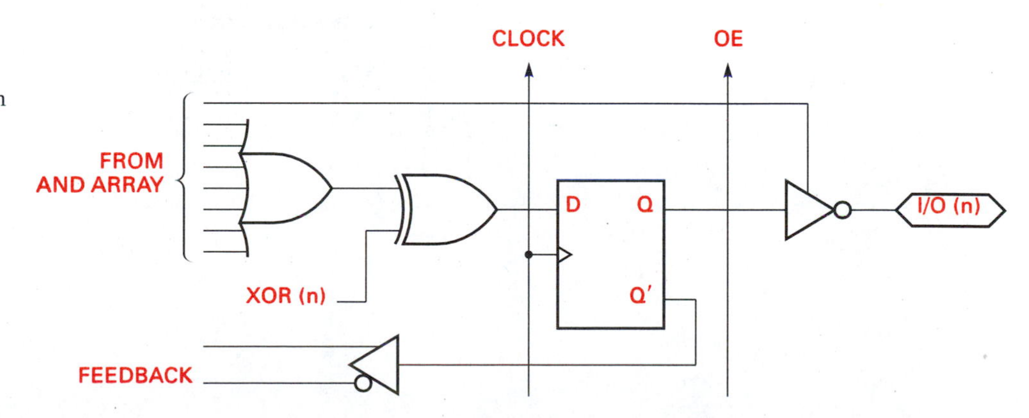

Figure 9.21
GAL16V8 configured in complex mode (courtesy Lattice Semiconductor Corporation)

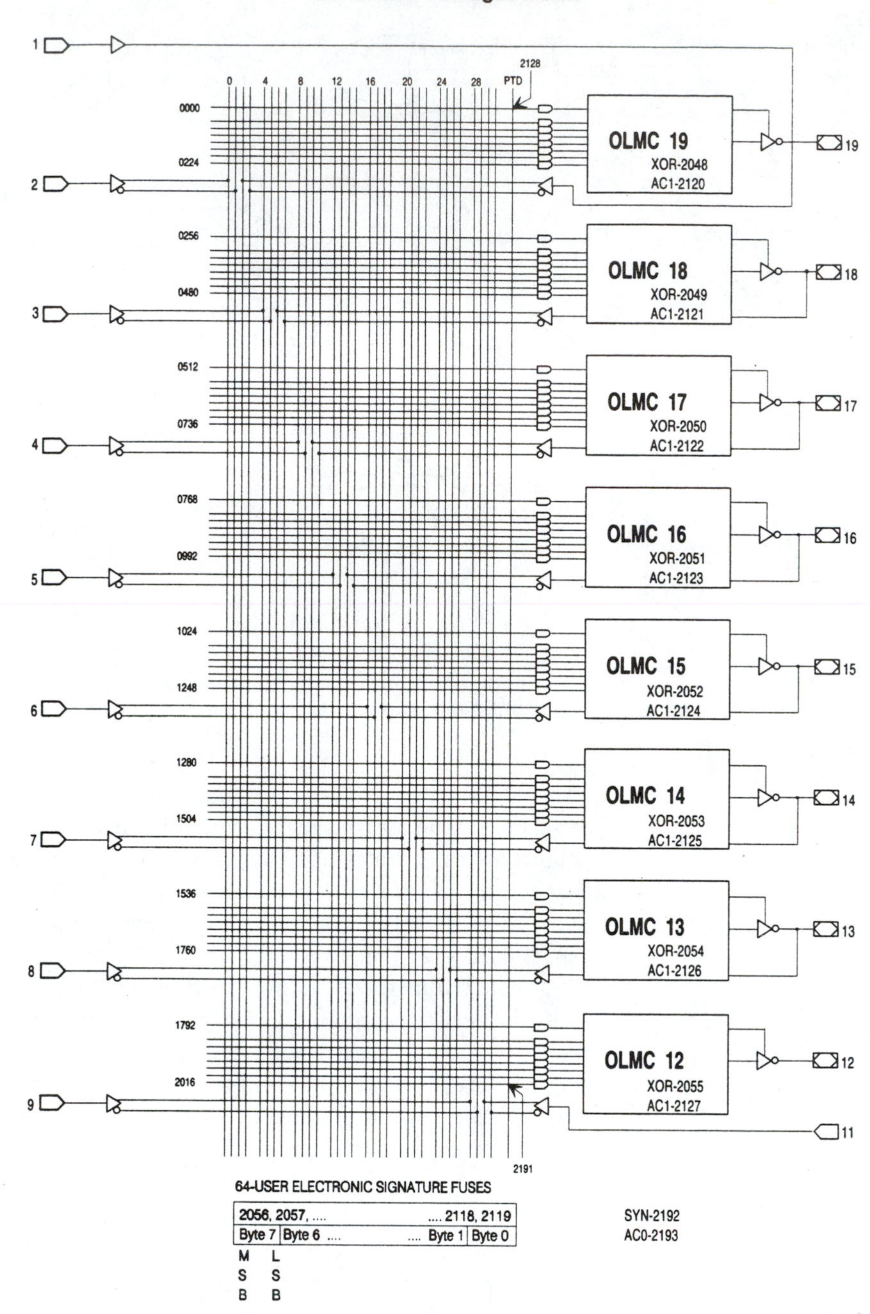

Figure 9.22
GAL16V8 configured in registered mode (courtesy Lattice Semiconductor Corporation)

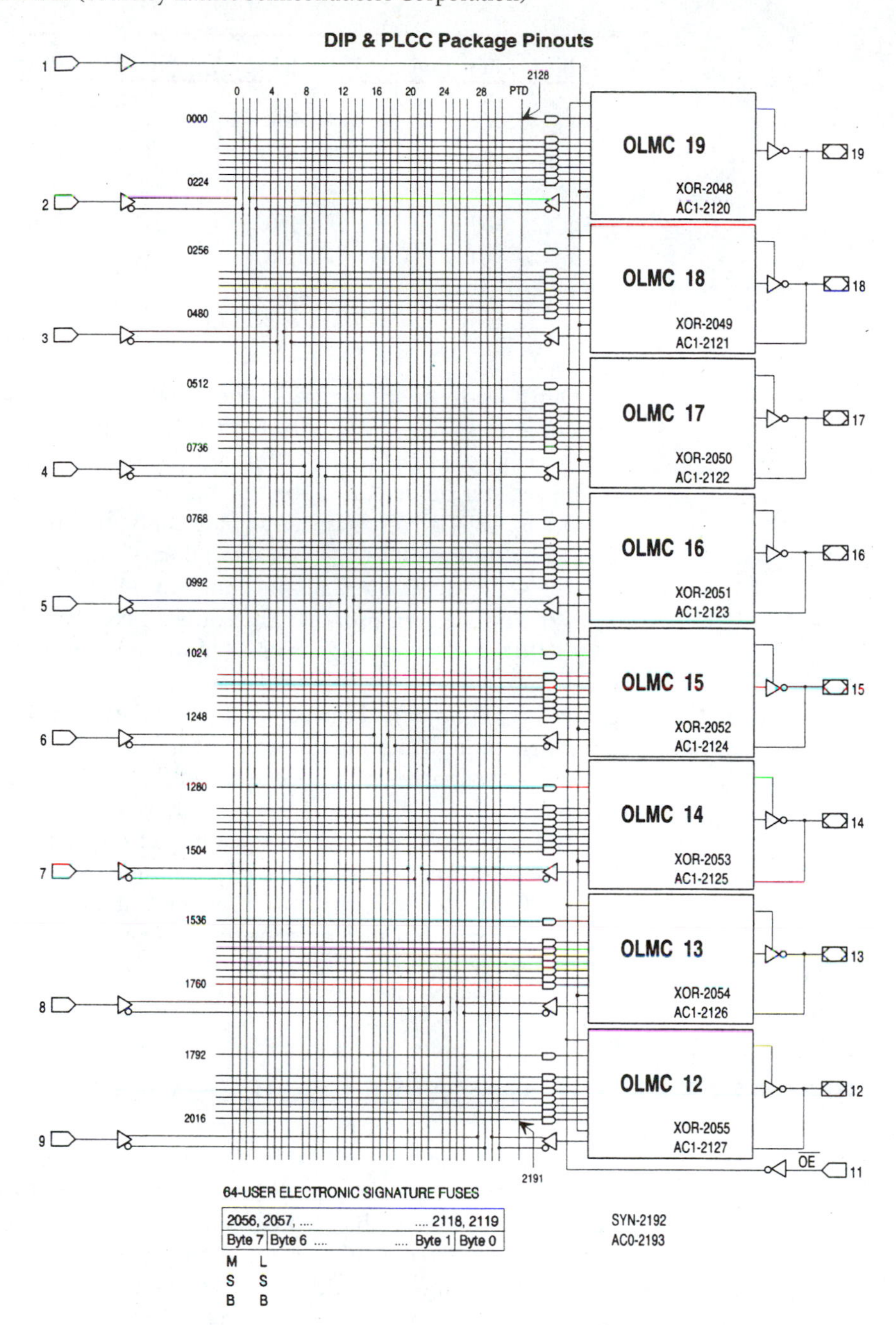

TABLE 9.9

GAL16V8 configuration control bits

Cell	Function	(Mode)	AC_0	ACE(n)	XOR
19	F_3	(Registered)	0	0	1
18	F_2	(Registered)	0	0	1
17	F_1	(Registered)	0	0	1
16	CRST	(Output)	0	1	1
15	CEN	(Output)	0	1	1
14	ABSTB′	(Output)	0	1	0

Our purpose in the design of the elapsed time measurement sequential machine is to configure some OLMC cells in a registered mode and some in an output mode. Because we have assigned cells 19 through 17 to state variables, they need to be configured in the register mode. Cells 16, 15, and 14 will be assigned to output variables CRST, CEN, and ABSTB′, respectively, and are configured in the output mode. Table 9.9 shows the configuration bit values for each cell in our design.

The state variable and output function p-terms are realized in the GAL AND array as in the PAL design process. The software compilers translate the Boolean equations into a JEDEC file for programming connections in the AND array. A document file is created to provide a record of the design. The JEDEC file is then downloaded into a device programmer and the device is programmed. Several companies make device programmers; one such programmer is the LC9000, by Programmable Logic Technologies. The LC9000 consists of a printed circuit board that plugs in to a PC and the software needed to create files and program GAL devices.

The excitation and output equations of the elapsed time measurement sequential machine would be entered into the PC running the programming software. A file is then created by the software that produces the commands to configure the OLMC cells and the GAL device internal interconnections would be made, resulting in a single GAL integrated circuit for the sequential machine.

9.4 ERASABLE PROGRAMMABLE LOGIC DEVICES

The first erasable programmable logic device (**EPLD**) was developed by Altera in 1984. Erasable programmable logic devices combine the features of PALs and EPROMs to create an ultraviolet erasable, reprogrammable logic chip. The Altera EPLDs contain a programmable-AND, fixed-OR array structure and a user configurable I/O cell that contains flip-flops, multiplexers, and tristate I/O buffers. The EPROM type cells are in the AND array and the control portions of the I/O cell.

The EP600 EPLD contains the equivalent of 600 gates. It has 16 macro cells, containing 20 inputs, 16 outputs, and up to 16 flip-flops. The flip-flops can be configured as D, T, R–S, or J–K. Individual resets are available to each macro cell flip-flop. The EP1800 contains 2100 equivalent two-input gates.

Intel also makes EPLDs that range from the eight macro cell, 10 ns t_{pd} D85C508 to the 24 macro cell 25 ns t_{pd} D85C220 device. Intel devices also use the programmable AND, fixed OR array with programmable I/O cells. Figure 9.23 illustrates the general structure of an EPLD.

Figure 9.23
General structure of an EPLD

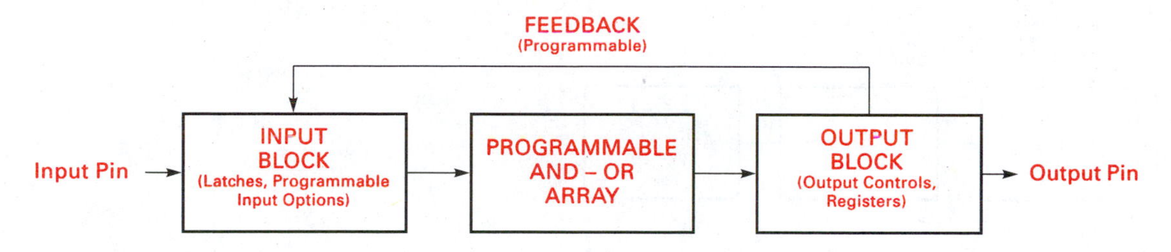

The EPLD can be contrasted with the fusable link PLDs. Table 9.10 illustrates some relative advantages and disadvantages of the EPLD compared to the PLD.

Altera and Intel both have complete hardware and software systems for developing EPLD designs. Altera's MAX+Plus software system block diagram is shown in Figure 9.24.

Logic designs are started by using the design entry module. The module supports both graphical (schematic capture) and text design entry. The graphic editor permits designs to be entered by selecting graphical logic symbols from a library. Standard SSI and MSI logic functions are supported. A comprehensive set of logic function symbols such as AND, OR, NOT, NOR, NAND, XOR, XNOR, counters, decoders, multiplexers, registers, flip-flops, and latches are part of the library. For example, if you wished to include a four-bit counter in your design you could select a 74191 TTL part number and make it part of your design. In addition, third party schematic capture designs (OrCAD, Viewlogic, and Futurenet) can be used as design entry points.

Altera's MAX+Plus text editor supports design entry using AHDL (Altera hardware description language). Hardware description languages provide a textual method for designing logic. A later section in this chapter will go into HDL in more detail.

To implement a design into an EPLD the MAX+Plus compiler converts the design entry file information into a netlist file. Any errors are flagged and the software sends the user back to the design entry module, where the Error Report window highlights the error location.

After an error-free design is completed, it is optimized for realization by a logic synthesizer software module. Redundant logic is minimized and removed. Once optimized, a simulator netlist file is created. This file permits the logic to be simulated prior to programming the actual EPLD device. The simulation software includes the ability to

TABLE 9.10
Comparison of fusable and erasable PLDs

Feature	Fusable PLD	Erasable PLD
Power consumption	Higher (bipolar)	Lower (CMOS)
Programmability	Once (fuse links)	UV-erasable, reprogram
Density	Lower	Higher
Testability	Difficult	Easier
Speed	Fastest	Slower
Security	No, designs can be reconstructed	Yes, protection bits are buried in silicon

Figure 9.24
Altera EPLD design system

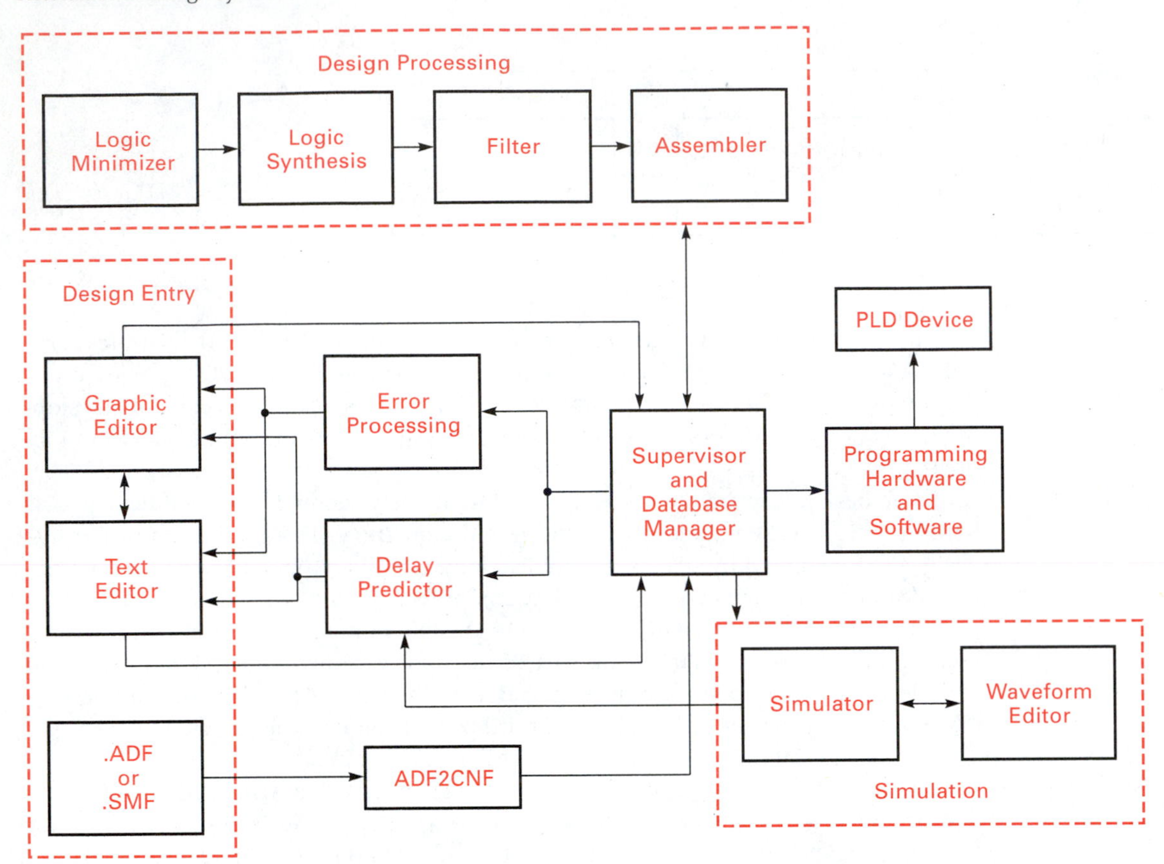

predict timing delays and to place a "probe" at desired points to "see" the circuit's behavior prior to actually programming the EPLD. Waveforms can be viewed using the waveform editor. Once the designer is satisified with the results, the logic functions can be programmed into the EPLD.

Altera's original EPLD integrated circuit family was the EP series. The EP300, 600, 900, 1200, and 1800 devices still offer excellent choices for many college undergraduate laboratories. The MAX series offers higher levels of integration and improved performance characteristics. For example, the EP1800 contains 48 macro cells and up to 64 inputs and 48 outputs. Each macro cell can be configured as D, T, *S–R*, or *J–K* flip-flops. The EP1830 has a propagation delay of 20 ns. The MAX series includes the EPM5016, 5032, through the EPM5192. The EPM5016 contains 16 macro cells, the EPM5032 contains 32 macro cells, and the EPM5192 contains 192 macro cells. A maximum of 252 flip-flops can be used in the EPM5192. Each of these devices has a propagation delay of 15 ns.

Logic designs that include combinational logic, state machines, counters, and registers (including tristate) can be realized using EPLDs. A single EPLD can replace a whole printed circuit board full of SSI and MSI integrated circuits.

Figure 9.25
EPM5000 series EPLD macro cell

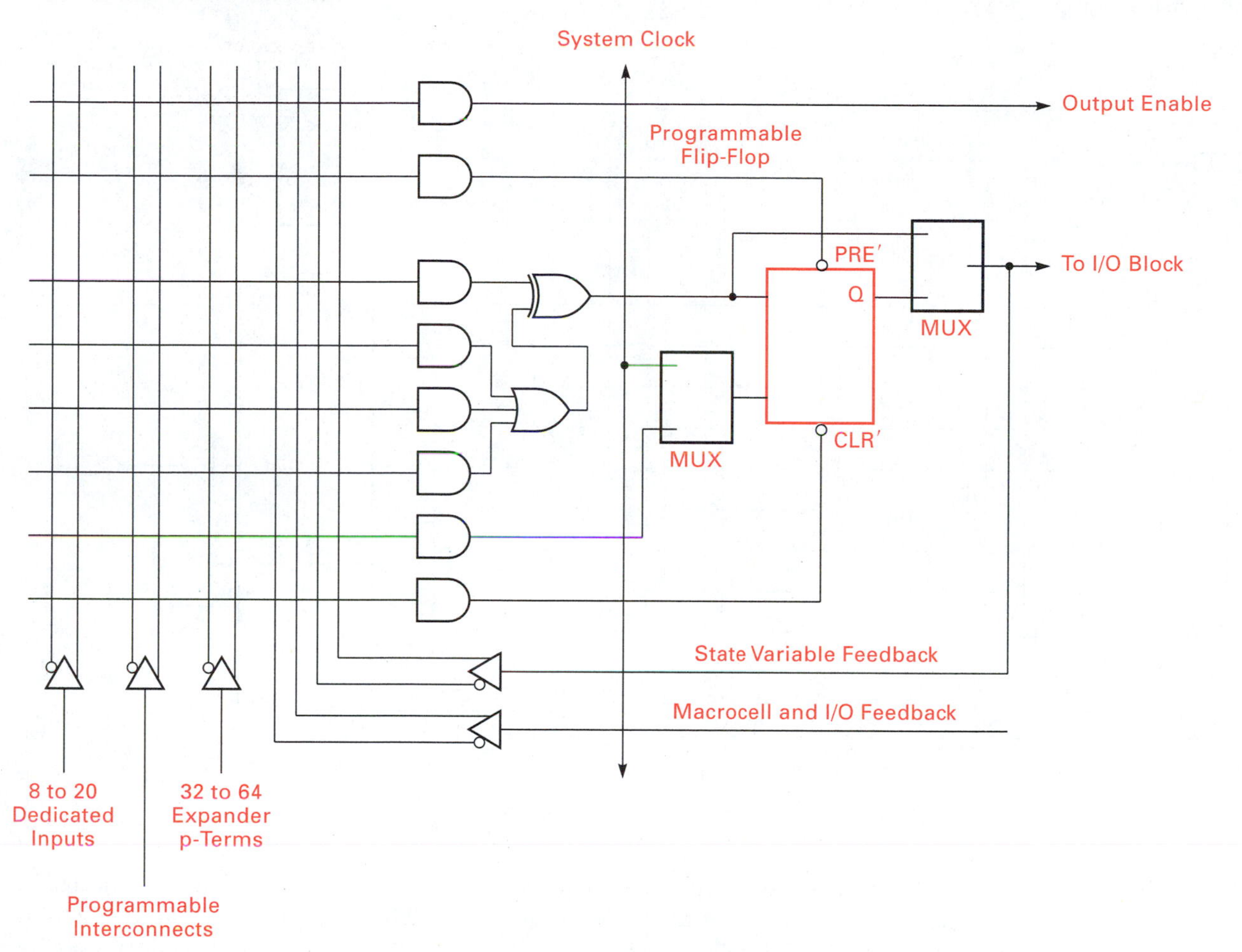

The macro cell for the EPM5000 series EPLD contains a programmable AND array that provides inputs to a fixed OR function. The OR output can be sent to an output pin as a combinational logic function or it can provide the excitation inputs to a flip-flop. Additional product terms can be programmed to provide asynchronous inputs (preset and clear) to the flip-flop. Flip-flop outputs can be fed back to the AND array and/or brought to an output pin. Figure 9.25 illustrates the EPM5000 series macro cell.

9.4.1 Altera EP600 EPLD

The EP600 uses CMOS EPROM technology to make the interconnections between the programmable AND array and the control functions in the I/O macro cell. The device has 4 dedicated external data inputs, 2 synchronous clock inputs, and 16 I/0 pins that can be configured as input, output, or bidirectional.

Figure 9.26
EP600 EPLD macro cell

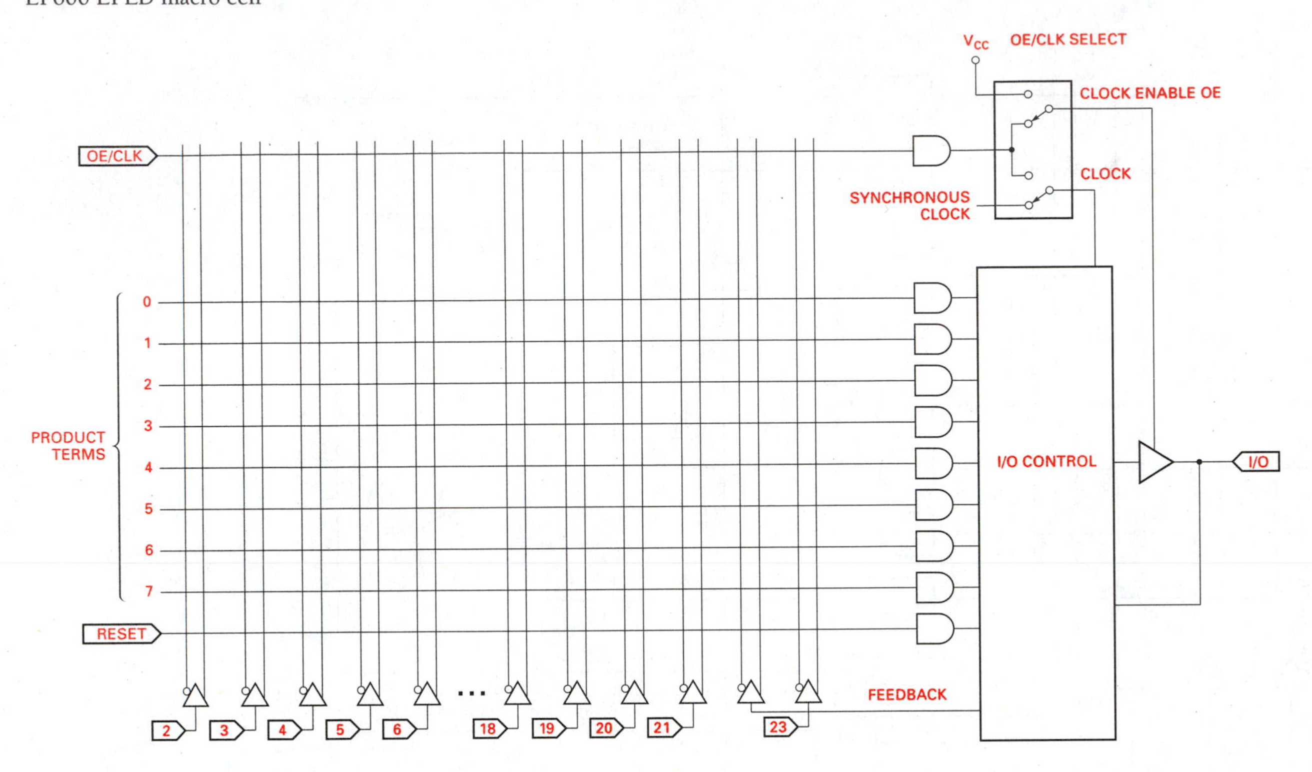

Each of the 16 macro cells contain an AND array with an I/O block as indicated in Figure 9.26. Product terms (p-terms) are generated by the AND array that feeds a fixed OR gate with a controlled complement EX-OR gate. The EX-OR gate provides the ability to invert the function if required. The I/O cell contains a flip-flop that can be configured as a D, T, R–S, or J–K. Each macro cell contains 10 product terms, 8 of which are used for logic realization, 1 is used for a flip-flop reset, and the last for an output enable control.

An EPROM cell array technology permits interconnection of the product terms. When the EPROM cell is erased a connection is made. No connection is realized by programming the floating gate MOSFET EPROM switch.

Two external clocks are presented to the EP600: one for macro-cells 1–8 and the other for macro cells 9–16. This permits two separate positive edge clocking schemes to be realized.

Several multiplexers are used to steer signals to the desired place within the I/O block in each macro cell. EPROM MOSFET switches are used to interconnect the control paths as determined by the JEDEC file when the device is programmed. A programmable design security bit is available that controls access to the data programmed into the device. When this bit is set by the program the internal design is protected. When the device is erased the design protect bit is also erased and the device can be reprogrammed.

9.4.2 Sequential Circuit Realization Using an EP600

The graphic plotter interface/computer simulator linked sequential machine design problem from Chapter 7 can be realized using EPLDs. Consider the two state machines for the design problem. The plotter interface controller requires four state variables plus an additional four for the data word counter. An additional flip-flop is needed by the plotter simulator to simulate the handshaking between the interface and the plotter. Three more flip-flops are needed to realize the computer simulator sequential circuit requiring a total of 12 flip-flops for the entire system. The EP600 has 16 macro cells, each containing a flip-flop, so the entire system could be contained in a single EP600.

Such a system design would not be realized with a single device because the computer simulator and the plotter simulator were designed and built to exercise the plotter interface. Also, the original design limit of only 16 data words is not practical; more would be necessary. Consider, then, the design of the interface with a larger data word counter. Each macro cell can be used as either a combinational or a sequential function, but not both. Because the plotter interface has three outputs (TCH, RPYH, and INTH) to send to the computer simulator and one output (DOSH) to send to the plotter, four macro cells will be occupied as combinational functions. This leaves 12 macro cells to use as sequential functions. Four are needed to realize the sequential machine, leaving eight for the data word counter. Such a design would allow for a 256-data word transfer and still fit in a single EP600.

Implementing the interface using random logic for both the sequential machine and an eight-bit data word counter would take 16 SSI and MSI integrated circuits. Using low-power Schottky devices requires a power supply current of approximately 125 mA. Contrast that with the single EP600 IC and a maximum current requirement of 150 mA.

The plotter interface design developed in Chapter 7 had propagation delay times of 121 ns. These delays were calculated using low-power Schottky TTL ICs, and certainly using Schottky or advanced Schottky would increase the maximum clock frequency. However, compare the low-power Schottky delay times with the propagation delay of the EP600 EPLD. The 121 ns time for the low-power Schottky is slower than the slowest of the EP600 EPLD family, which has a delay of 65 ns.

Inputs and outputs for the plotter interface are as follows:

Inputs (mnemonic)	Definition
1. CTH	Connect (active-high)
2. DSH	Data signal (active-high)
3. TCRH	Terminal count reply (high)
4. CIH	Clear interrupt (high)
5. Clock	Interface synchronous clock
6. Reset	Initialize state machine
7. DRH	Data reply (high) to plotter
8. Zero	Initialize data word count

Outputs (mnemonic)	Definition
1. TCH	Terminal count (high)
2. RPYH	Reply (high)
3. INTH	Interrupt (high)
4. DOSH	Data output signal (high) to plotter

The data word counter inputs and outputs are

1. LDL — Active-low counter load signal used to load all 0s or clear.
2. UPDN — Up-down signal for counting up or down.
3. CTENL — An active-low terminal count signal, used to indicate when buffer memory is full.
4. RCOL — An active-low counter output indicating the terminal count.

The EPLD development software made the assignments of input variables and I/O pins and generated the results in the utilization report. The sequential machine clock is different than the word counter clock. Recall that the EP600 is divided into two banks with separate clocks. We will assign the sequential machine clock to one bank and the data word counter clock to the other bank. Macro cells 9–16 are clocked by the clock input 1 on pin 1. We will assign that bank to the sequential machine state variables and the combinational output variables. This leaves macro cells 1–8 to be clocked by clock 2 on pin 13 or by a p-term generated from the AND array. The sequential machine clock is brought in externally. The word counter clock is derived from logical conditions generated by the sequential machine, requiring that OE/CLK mode 1 be used. Mode 1 generates the clock input to the word counter flip-flops from the AND array. *J–K* flip-flops were used in the random logic realization of the interface sequential machine, so we will use them in the EPLD as well. This means that the eight p-term AND array is split, providing four p-terms for J and four p-terms for K. The state variable *J–K* excitation equations are reproduced here:

$$J_4B = F_3B\ TCRH$$
$$K_4B = F_2B'F_1B'\ CIH'$$
$$J_3B = F_4B'F_2BF_1B'DSH'RCOL'$$
$$K_3B = F_4B$$
$$J_2B = F_4B'F_1B\ CTH'$$
$$K_2B = F_4BF_1BDRH'RCOL'$$
$$J_1B = F_4BF_3B'F_2BDRH + F_4B'F_3B'F_2BDSH'RCOL + F_4B'F_2B'\ CTH$$
$$K_1B = F_4BF_2BDRH'RCOL + F_4BF_2B'CIH + F_4B'F_2BDSH$$

None of the excitation variables requires more than four p-terms, so we will use J–K flip-flops. The output variable equations are

$$TCH = F_4B'F_3B$$
$$INTH = F_4BF_2B'F_1B$$
$$RPYH = F_4B'F_3B'F_2BF_1B' + F_4B'F_2B'F_1B$$
$$DOSH = F_4BF_3B'F_2BF_1B'$$

Designing the data word counter requires the development of internal excitation equations for the eight counter flip-flops as well as the intermediate equations needed to connect the data word counter and the sequential machine. Instead of designing the eight-bit counter from scratch we can use the design of the 74xx191 as a starting point and modify it to suit our purposes. The logic diagram for the 74xx191 is given in Figure 9.27. Not all of the features are needed for the data word counter. We need the

Figure 9.27
74xx191 Up-down counter

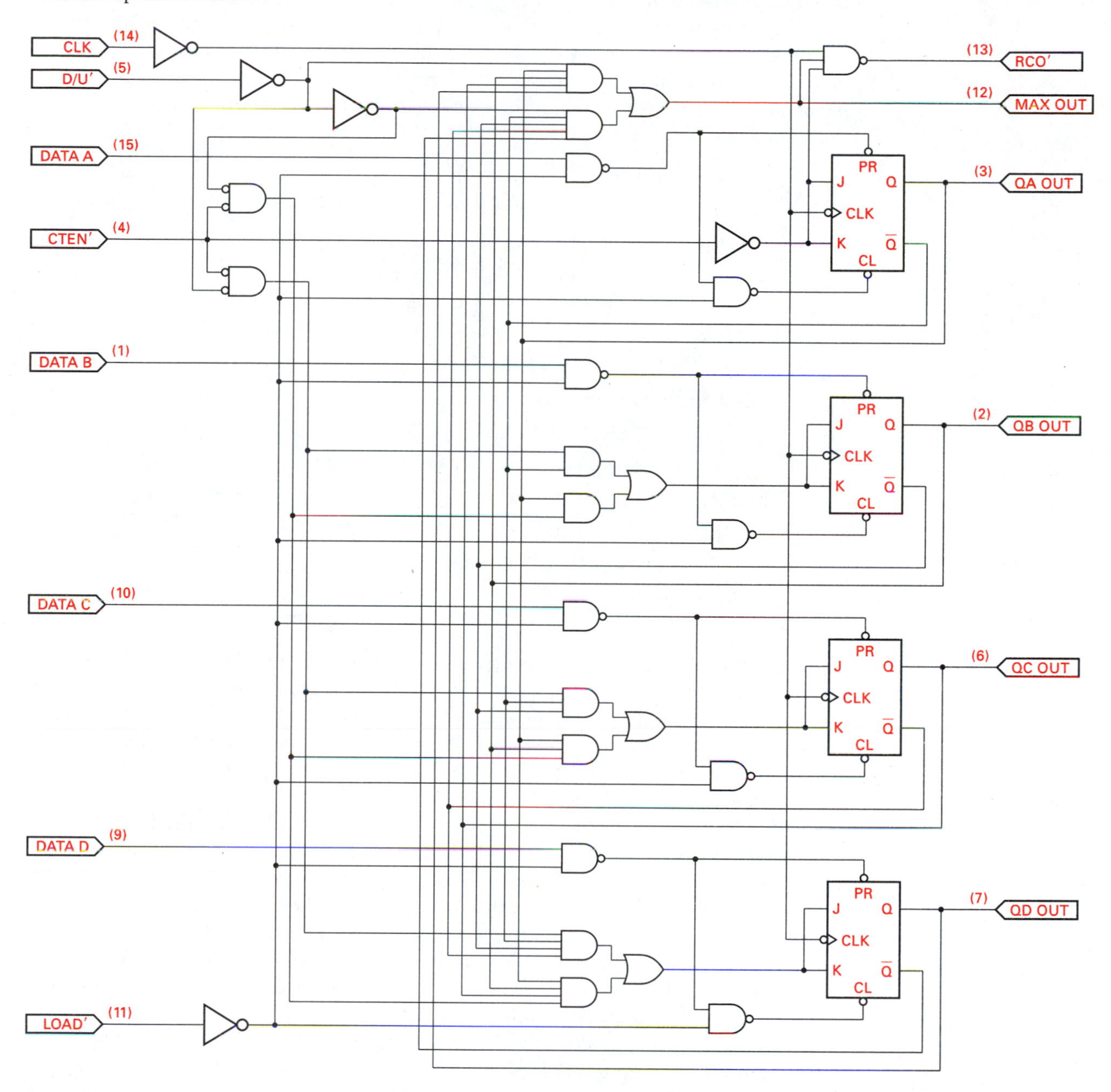

up-down, count enable (CTEN′), reset and terminal count (RCO′) capability. For our purposes we do not need the parallel load feature and we would like to add a common reset to clear the counter. Notice that the *J–K* synchronous inputs to the 74xx191 IC flip-flops are connected together, forming a T flip-flop. Notice as well that the parallel load feature is asynchronous, using the S' and R' inputs to the flip-flop.

The NOTF Altera primitive (no output, T feedback) will work for counter flip-flops. Based on these requirements and the logic diagram, we develop the equations for the Altera realization of the data word counter. The OrCAD schematic capture software will not transfer punctuation marks to the Altera file so all complemented variables must be indicated using an alpha character. In this case we used the *N* suffix to indicate a complemented variable. For instance, the expression for T_{C1} has a term UPDNN, which is the complement of UPDN.

$$
\begin{aligned}
T_{A1} &= \text{CTEN} \\
T_{B1} &= \text{CTEN UPDN } QA_1N + \text{CTEN UPDNN } QA_1 \\
T_{C1} &= \text{CTEN UPDN } QA_1N\ QB_1N + \text{CTEN UPDNN } QA_1\ QB_1 \\
T_{D1} &= \text{CTEN UPDN } QA_1N\ QB_1N\ QC_1N \\
&\quad + \text{CTEN UPDNN } QA_1\ QB_1\ QC_1 \\
RCOL_1N &= \text{CCLK CTEN UPDNN } QA_1\ QB_1\ QC_1\ QD_1 \\
&\quad + \text{CCLK CTEN UPDN } QA_1N\ QB_1N\ QC_1N\ QD_1N \\
\text{CCLK} &= \text{DSH} + \text{DRH} \\
T_{A2} &= CTEN_2 \text{ and } CTEN_2 = RCOL_1 \\
T_{B2} &= CTEN_2 \text{ UPDN } QA_2N + CTEN_2 \text{ UPDNN } QA_2 \\
T_{C2} &= CTEN_2 \text{ UPDN } QA_2N\ QB_2N + CTEN_2 \text{ UPDNN } QA_2\ QB_2 \\
T_{D2} &= CTEN_2 \text{ UPDN } QA_2N\ QB_2N\ QC_2N \\
&\quad + CTEN_2 \text{ UPDNN } QA_2\ QB_2\ QC_2 \\
RCOL_2N &= \text{CCLK } CTEN_2 \text{ UPDN } QA_2N\ QB_2N\ QC_2N\ QD_2N \\
&\quad + \text{CCLK } CTEN_2 \text{ UPDNN } QA_2\ QB_2\ QC_2\ QD_2
\end{aligned}
$$

The eight-bit counter results from cascading two four-bit counters. CTENL (count enable low) provides the count enable input to the least significant four-bit counter. The second four-bit counter (most significant) count enable is made by gating the RCOL from the first four-bit counter into the CTEN input of the second four-bit counter. This means that we can repeat the four-bit counter design twice to form an eight-bit counter.

Figure 9.27 shows the logic diagram for the 74xx191 four-bit up-down counter IC. Figure 9.28 illustrates the logic diagram, drawn with ORCAD, using the EP600 primitives, which will be realized in the EPLD. Notice the INP input primitives shown in Figure 9.28. These are used by the system software to denote input signals that will be assigned to a device pin. Also notice the logic symbols and names; for example, the gates are named AND_3 or OR_2 instead of an IC number. The toggle flip-flop primitive (TOTF) uses a tristate buffer for the output. This primitive was included in the design to facilitate testing purposes. Even though the counter state variables do not need to be brought over to output pins, doing so permits easier testing.

Input modules required by the ORCAD schematic capture software are also indicated for input and output variables. A common reset and preset is included in the EPLD counter design. The terminal count output, $RCOL_2$, is brought out to a device pin for testing. The tristate buffer primitive, CONF (combinational output no feedback), is necessary for that reason (see Figure 9.29). The EP600 output tristate buffer controls defaults to V_{cc} enabling the outputs. The parallel data inputs and the load control signal of the four-bit counter IC are not needed for our data word counter application.

Once the complete logic design has been entered into a schematic capture program (in this case OrCAD) and checked for errors, the design is synthesized by the design processing software and sent to the design verification software module for simulation

Figure 9.28
Logic diagram for plotter interface–data word counter using EPLD primitives

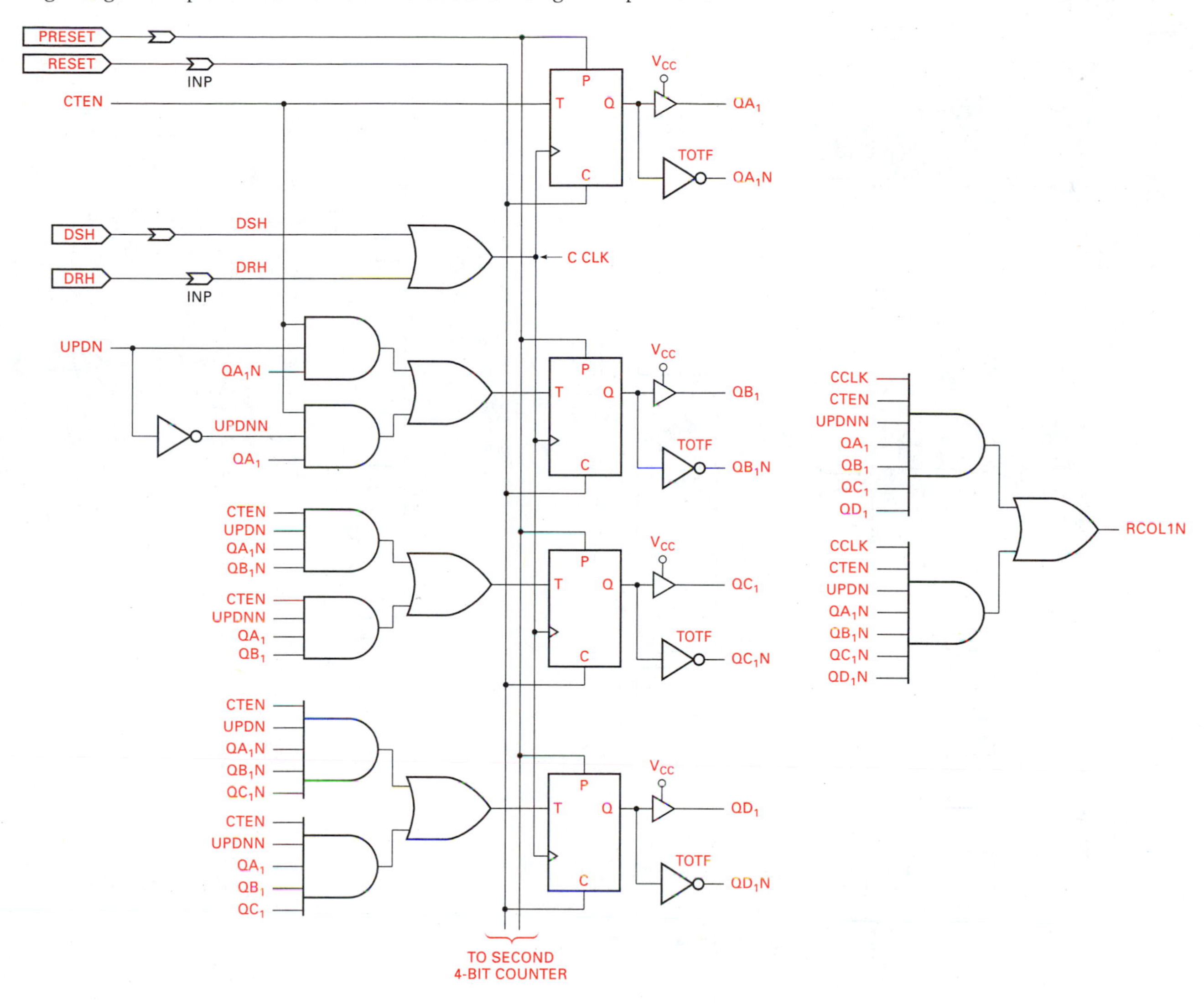

and finally programming. Many times specific logic function designs can be structured as macros. The macros can then be used without having to redesign the entire function from scratch.

9.5 PLD COMPUTER-AIDED DESIGN

Programmable logic devices (PLD) can be configured using a range of approaches. High level languages have been developed that permit design entry using Boolean equations, truth tables, state diagrams, and state tables. Development software for realizing PLD

Figure 9.29

State machine logic diagram for plotter interface using EPLD primitives

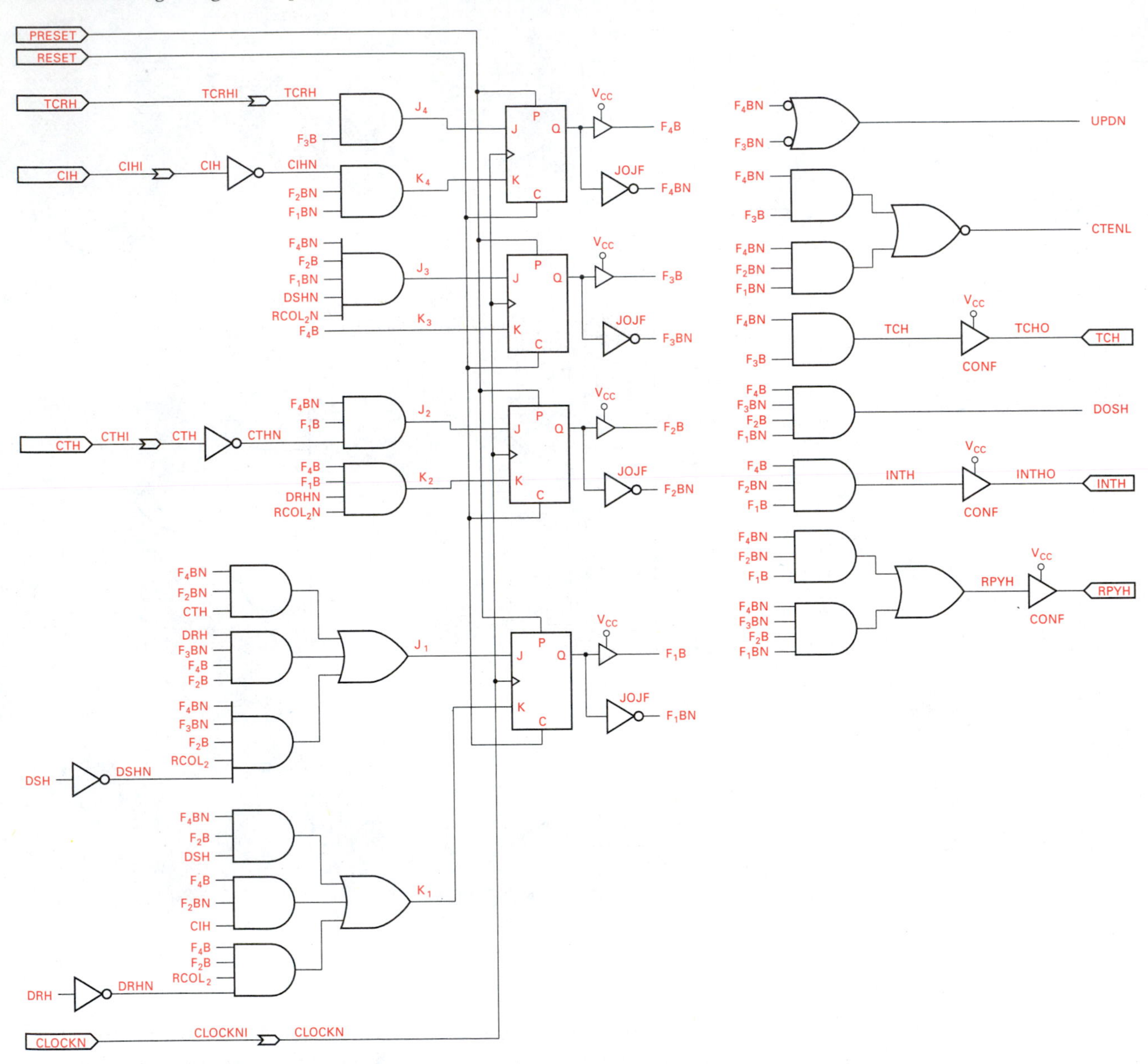

designs are available such as MAX+Plus II from Altera and AMAZE from Signetics. Other generic tools such as PALASM, CUPL, and ABEL are available to logic designers. The general procedure for developing PLD realizations of logic circuits is illustrated in Figure 9.30.

Many different PLD design languages exist; each has its own syntax. The version described in this section was originally developed by OrCAD. We will start with combinational logic and progress to techniques for describing flip-flops and finally deal with counters and registers.

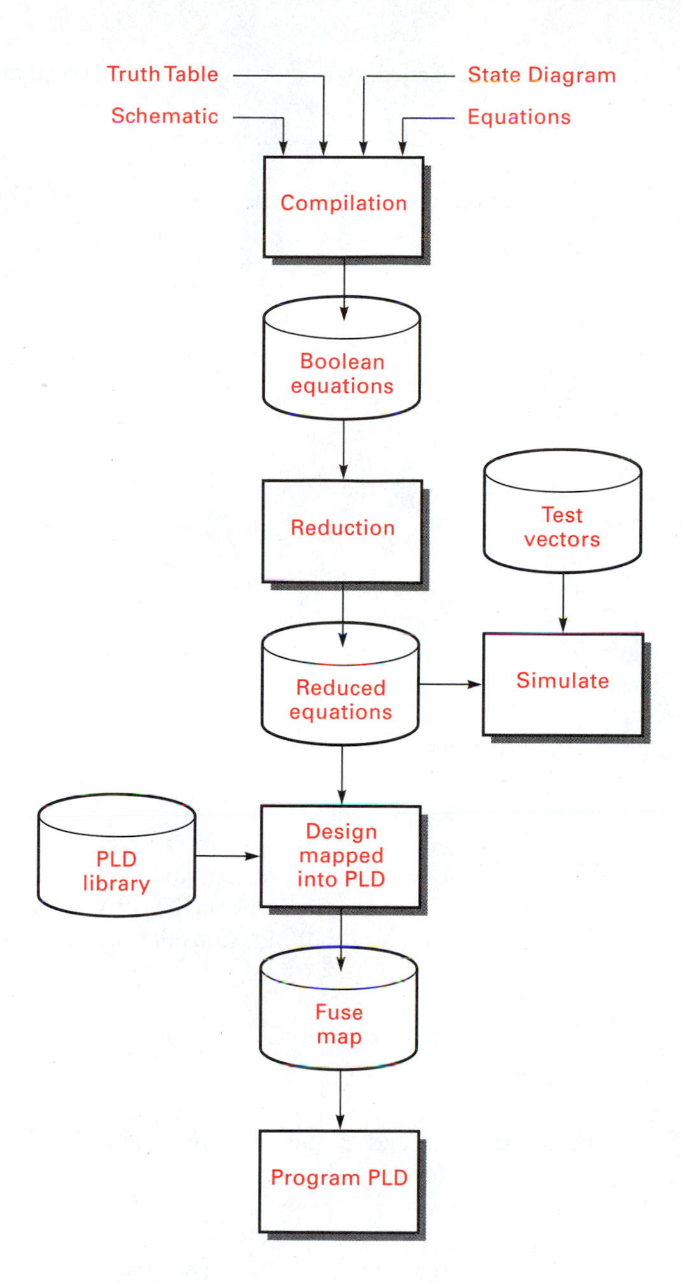

Figure 9.30
PLD design process flowchart

9.5.1 PLD Realization of Combinational Logic

Combinational logic operations can be expressed as a set of Boolean equations such as

$$X_1 = A_1 \& B_1$$
$$X_2 = A_2 \& B_2$$
$$X_3 = A_3 \& B_3$$
$$X_4 = A_4 \& B_4$$

where

$\& = \text{AND}$

The same set of equations can be written more economically as

$$X[1 \ldots 4] = A[1 \ldots 4] \ \& \ B[1 \ldots 4]$$

OR operations use the # symbol. A similar set of OR equations would be written:

$$X_1 = A_1 \ \# \ B_1$$
$$X_2 = A_2 \ \# \ B_2$$
$$X_3 = A_3 \ \# \ B_3$$
$$X_4 = A_4 \ \# \ B_4$$
$$X[1 \ldots 4] = A[1 \ldots 4] \ \# \ B[1 \ldots 4]$$

NOT operations use the prime symbol, ′.

$$X_1 = A_1'$$

NAND, NOR use a combination of &, #, and ′ symbols. NAND is indicated as &′ and NOR as #′. Exclusive OR uses ## and exclusive NOR ##′. A four-bit parity checker sion would be written as

$$X = (a\#\#b\#\#c\#\#d)$$

Tristate enable functions use ?? symbols. For example, X = A ?? (B & C)′ indicates a tristate NAND gate, with A as the enable input.

Inputs and outputs must be defined and declared so that the PLD compiler knows which signal to connect to which I/O pin. I/O can be declared either an input, an output, or a bidirectional input-output pin. Consider the design of a full-adder. The equations are

$$C_o = (X \ \& \ Y) \ \# \ (Y \ \& \ C_i) \ \# \ (X \ \& \ C_i)$$
$$E_o = X \ \#\# \ Y \ \#\# \ C_i$$

Inputs are X, Y, and C_i, outputs are C_o and E_o. The inputs and outputs must be declared. This is done as follows.

in:(X, Y, C_i), ⟵ This statement ends with a comma because more signals are being declared.

out:(C_o, E_o) ⟵ This statement does not have a comma because it is the last one.

Note that the in and out statements are followed by a colon (:), that each input or output variable is separated by a comma (,), and that the in statement ends with a comma. The last statement in a string of input and output declarations does not need a comma.

In and out statements can list as many variables as needed. The only catch is that the PLD device must have enough pins.

Now consider the design of a four-bit adder to be realized using a PAL device. Figure 9.31 shows the full-adder block diagram.

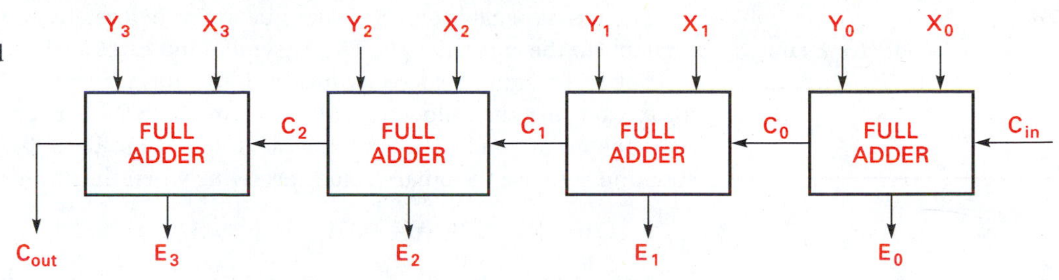

Figure 9.31 Four-bit adder to be realized using a PAL

PAL12H6 in:(X_0, X_1, X_2, X_3, Y_0, Y_1, Y_2, Y_3, C_{IN}),
out:(E_0, E_1, E_2, E_3, C_{OUT})

C_0 = (X_0 & Y_0) # (Y_0 & C_{IN}) # (X_0 & C_{IN})
E_0 = X_0 ## Y_0 ## C_{IN}
C_1 = (X_1 & Y_1) # (Y_1 & C_0) # (X_1 & C_0)
E_1 = X_1 ## Y_1 ## C_0
C_2 = (X_2 & Y_2) # (Y_2 & C_1) # (X_2 & C_1)
E_2 = X_2 ## Y_2 ## C_1
C_{OUT} = (X_3 & Y_3) # (Y_3 & C_2) # (X_3 & C_2)
E_3 = X_3 ## Y_3 ## C_2

The | symbol is used to start each line in the program. There are nine inputs and five outputs for our four-bit adder. The PAL12H6 has 12 inputs and 6 outputs. When the design has been compiled for the specified device (PAL12H6), any equation reductions will be made. Then the reduced design equations will be mapped into the PLD and a fuse map created to blow the appropriate fuses to program the device.

The input and output strings for the adder could also be written as

in: X(0, 1, 2, 3), in: Y(0, 1, 2, 3), in: (C_{IN}),
out: E(0, 1, 2, 3), out: (C_{OUT})

or as

in: X(0 . . . 3), in: Y(0 . . . 3), in: (C_{IN}),
out: E(0 . . . 3), out: (C_{OUT})

This method illustrates how strings of inputs or outputs can be indexed.

If we wished tristate adder outputs we would write the following:

PAL12H6 in: X(0 . . . 3), in:(C_{IN}), out: E(0 . . . 3), out: (C_{OUT})

C_0 = (X_0 & Y_0) # (Y_0 & C_{IN}) # (X_0 & C_{IN})
E_0 = T ?? X_0 ## Y_0 ## C_{IN}
C_1 = (X_1 & Y_1) # (Y_1 & C_0) # (X_1 & C_0)
E_1 = T ?? X_1 ## Y_1 ## C_0
C_2 = (X_2 & Y_2) # (Y_2 & C_1) # (X_2 & C_1)
E_2 = T ?? X_2 ## Y_2 ## C_1
C_{OUT} = T ?? (X_3 & Y_3) # (Y_3 & C_2) # (X_3 & C_2)
E_3 = T ?? (X_3 ## Y_3 ## C_2)

TABLE 9.11
Three-bit decoder truth table

Inputs			Output
I_2	I_1	I_0	O
0	0	0	O_0
0	0	1	O_1
0	1	0	O_2
0	1	1	O_3
1	0	0	O_4
1	0	1	O_5
1	1	0	O_6
1	1	1	O_7

The tristate enable signal T must be active before any of the outputs will be made available to the external pins. The ?? symbol indicates a tristate.

The realization of a decoder further illustrates using a PLD to realize combinational logic. Consider the following truth table in Table 9.11. Each output $O(n)$ is active for only one input condition. For example, $O1 = I_2' \,\&\, I_1' \,\&\, I0$. Another way to write the equation is to use the binary value specifying when the output is true:

$$O1 = I[2 \ldots 0] == 001b$$

The equation, for O1, just written is read as "Output O1 is true when the input variable set of I2, I1, and I0 (I[2 . . . 0]) is equal to 001 binary." We can continue the indexing idea to include the output variable as well:

$$O[n] = I[2 \ldots 0] == n$$

This means that output variable On is dependent on the set of input variables I2, I1, I0 and that it will be active (true) when the binary value of the input variables is equal to n. For example let $n = 5$; then $O5 = I[2 \ldots 0] = 101b$.

We need to establish range for n; and we do this by writing

$$n = 0 \ldots 7: O[n] = I[2 \ldots 0] == n$$

The complete program statement for the 3 × 8 decoder includes the input and output declaration and is written:

|(PLD Device #)
|Title: "3 × 8 Decoder"
|in: I[2 . . . 0], out: O[7 . . . 0]
|$n = 0 \ldots 7: O[n] = I[2 \ldots 0] == n$

Most decoders have one or more enable inputs. To add enable input E we would modify the program given above to:

|(PLD Device #)
|Title "3 × 8 Decoder with Enable"
|in: (I[2 . . . 0], E), out: O[0 . . . 7]
|$n = 0 \ldots 7: O[n] = I[2 \ldots 0] == n \,\&\, E$

The inclusion of E in an AND statement along with n means that when E is true then the output specified by n is also true.

A multiplexer has multiple inputs and only one output. The input data, determined by select inputs, is sent to the output. Consider the truth table in Table 9.12.

TABLE 9.12
3 × 8 multiplexer truth table

Select			Output (O)
S_2	S_1	S_0	
0	0	0	I_0
0	0	1	I_1
0	1	0	I_2
0	1	1	I_3
1	0	0	I_4
1	0	1	I_5
1	1	0	I_6
1	1	1	I_7

Inputs S_2, S_1, and S_0 select which data input $I(n)$ value will be sent to the output O. $I(n)$ can be either a 0 or a 1. To declare the input-output statement, we write

in: (I[7 . . . 0]), S[2 . . . 0], out: O

The value (0 or 1) of output O is determined by the value of the selected (by S_2,S_1,S_0) input I.

$$n = 0 \ldots 7: O = I[n] \,\&\, S[2 \ldots 0]$$

Because both input $I[n]$ and S[2 . . . 0] are necessary we AND them together.

The final program for a 3 × 8 multiplexer is

(PLD Device #)
|Title "3 × 8 Mux"
|in: (I[7 . . . 0], S[2 . . . 0]), out: O
|n = 0 . . . 7: O = I[n] & S[2 . . . 0]

9.5.2 Realizing Truth Tables Using a PLD Language

Truth tables are realized using the PLD language through the key word *Table*. Inputs and outputs of the truth table are written as follows:

|Table: x,y,z ⟶ P
|{ . . . .
| . . . .}

Variables x, y, and z are inputs that map to output variable P. The argument of the table is contained between the two curly brackets. The most common method used to implement the truth table is to use a SOP or POS format. For example,

|Table: a,b,c,d ⟶ T
|{0,1,4,9,12 ⟶ 1
|2,3,5,6,7,8,10,11,13,14,15 ⟶ 0}

All the input combinations are assigned a value of either 0 or 1. Multiple output truth tables are also realized using the PLD language.

❖ **EXAMPLE 9.1** Realize the following expressions:

$$R = f(a, b, c, d) = \sum (0, 1, 3, 9, 12)$$
$$S = f(a, b, c, d) = \sum (0, 2, 3, 4, 7)$$
$$T = f(a, b, c, d) = \sum (1, 2, 3, 5, 9, 13, 15)$$

SOLUTION |(PLD Device #) in: [a,b,c,d], out: [R,S,T]
|Table a,b,c,d ⟶ R
|{0,1,3,9,12 ⟶ 1
|2,4,5,6,7,8,10,11,13,14,15 ⟶ 0}
|Table a,b,c,d ⟶ S
|{0,2,3,4,7 ⟶ 1
|1,5,6,8,9-15 ⟶ 0}
|Table a,b,c,d ⟶ T
|{1,2,3,5,9,13,15 ⟶ 1
|0,4,6,7,8,10,11,12,14} ❖

9.5.3 Realizing Flip-Flops Using a PLD Language

The basic S–R latch is realized by implementing the characteristic equation for the device:

$$Q+ = S + QR'$$

Rewriting the characteristic equation in OrCAD PLD format gives

Q = S # (Q & R′)

The D and T flip-flops are directly supported by the PLD language. The general form for both types is written as

Output = (dff or tff)(data input, clock, reset, set)

Consider the D flip-flop shown in Figures 9.32 and 9.33,

Q = dff((X & Y) # Z, C)

A D flip-flop is specified by the dff term. The synchronous input is specified by the expression (X & Y) # Z. The clock input is specified by C. No asynchronous inputs are indicated.

In the example in Figure 9.33 a tristate output is indicated along with asynchronous inputs S′ and R′:

Q = T ?? dff(X ## Y, C, R′, S′)

T is the tristate enable input. The ?? separates T from the flip-flop-type designation. The synchronous input is specified by the exclusive-OR statement, X ## Y. C indicates the clock, R′ and S′ indicate the active-low asynchrounous reset and set inputs.

9.5.4 Realizing State Machines Using a PLD Language

In PLD languages, state machines are programmed with a procedural language that is much like regular software. The PLD language has its own syntax and restrictions that permit it to work with the models of the actual devices it is to program. PLD languages have IF-THEN and GO-TO statements and LABELS that are used in much the same way as similar statements in C or PASCAL.

Figure 9.32
D flip-flop example

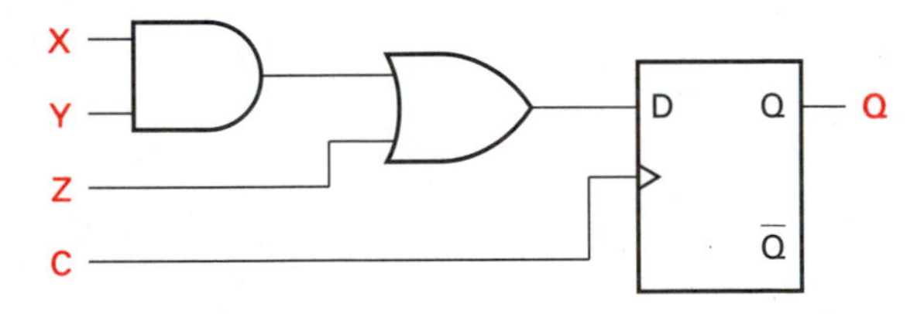

Figure 9.33
D flip-flop example with tristate output

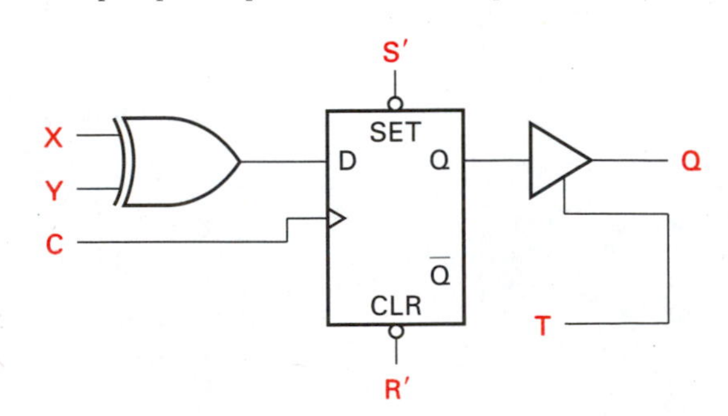

Consider the design of a simple binary counter. The input and output statements appear as we have seen previously.

|(PLD Device #) in: RESET, out: Q[3 . . . 0], clock: CLK
|Title: "Binary Counter"
|conditioning: clock // Q[3 . . . 0]

We have added several key words: *clock:* establishes the name of the clock signal; *title:* permits giving the design a name; *conditioning:* sets the edge (rising as indicated by //, a negative edge clock would have \\) condition for the clock and the flip-flops to receive the clock (Q[3 . . . 0]).

Continuing the counter design we set the state transitions by a procedure statement.

|procedure: RESET, Q[3 . . . 0]
|{0. ⟶ 1
|1. ⟶ 2   : The present state is represented by
|2. ⟶ 3   : the state number # followed
|3. ⟶ 4   : by a period. The arrow ⟶
|4. ⟶ 5   : indicates a "GO-TO" statement.
|5. ⟶ 6   : The next state is represented by
|6. ⟶ 7   : a state #.
|7. ⟶ 8
|8. ⟶ 9
|9. ⟶ 10
|10. ⟶ 11
|11. ⟶ 12
|12. ⟶ 13
|13. ⟶ 14
|14. ⟶ 15

The procedural statement starts by establishing that the reset input is applied to all four state variables by "RESET, Q[3 . . . 0]. The state transitions are, in this case, independent of any external inputs and will occur on the positive edge of the clock. The numbers used to represent states are decimal, but we could have written them in binary, octal, or hexadecimal.

0000b. ⟶ 0001b

or

00h. ⟶ 01h.

Our binary counter consists of four flip-flops, all positive edge triggered, a single input RESET, and four outputs Q3 . . . Q0.

We can also create state machines with input variables effecting the state transitions (Mealy and Moore). Consider the simple state diagram shown in Figure 9.34.

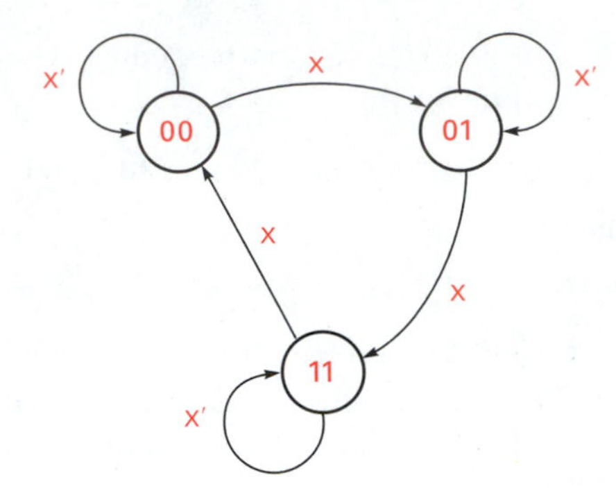

Figure 9.34
Simple state machine

The state transitions are described by the PLD language as

0. X ? ⟶ 1
   X′? ⟶ 0
1. X ? ⟶ 3
   X′ ? ⟶ 1
3. X ? ⟶ 0
   X′? ⟶ 3

If X = 1 then a transition from state "0" to state "1" occurs; else if X = 0, then a loop from state 0 to state 0 occurs.
X′? ⟶ 0

States in this example are indicated by their decimal value state assignment. The ? symbol indicates dependence on the input variable X for state transitions. State machine descriptions do not require that the state assignment be given. Labels can be used. For example, let 0 = A, 1 = B, and 3 = C. Then the same transitions could be written

A. X ? ⟶ B
   X′? ⟶ A
B. X ? ⟶ C
   X′? ⟶ B
C. X ? ⟶ A
   X′? ⟶ C

Unconditional outputs (Moore model) are written as follows:

A. X ? ⟶ A
   X′? ⟶ B
   Z = 1

In this case, a loop from A back to A occurs if X = 1. If X = 0 then a state transition from A to B occurs. In either case, output variable Z = 1 as long as the machine remains in state A.

Conditional outputs (Mealy model) are written as follows:

A. X ? (Z=1 ⟶ A)
   X′? ⟶ B

In this example, a loop back to state A is dependent on input variable X = 1, which produces a 1 for output variable Z. If X = 0 then a transition from state A to B occurs but Z = 0. The statement included inside the ( ) establishes the conditional output and the next-state transition.

Consider the simple state diagram in Figure 9.35.

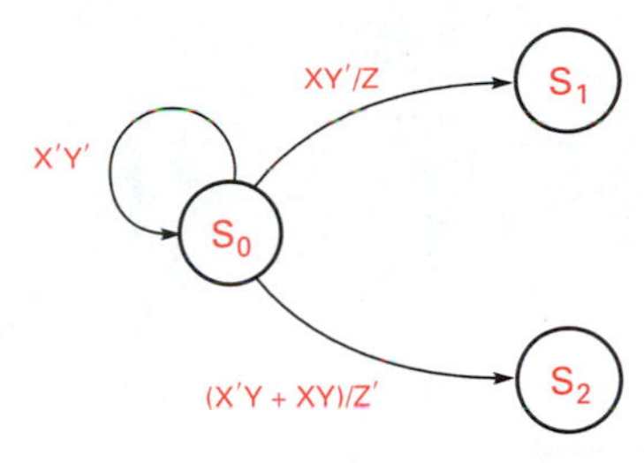

Figure 9.35
State machine with conditional state transitions and outputs

The procedural statements needed to implement the state machine are as follows:

|procedure: Q[1, 0]
|{S_0. (X′ & Y′) ? ⟶ S_0
|(X & Y′) ? (Z = 1 ⟶ S_1)
|(X′ & Y) # (X & Y) ? (Z = 0 ⟶ S_2)}

State assignments do not need to be made before you can enter a design. PLD languages support the use of labels for states. Any alphanumeric string, starting with a letter (strings starting with numbers are not permitted for labels) can be used as a label.

A_1, Idle, Start, Stop, and X are all permitted state labels. A state assignment can be made by the designer by using the key word "states." State assignments may be made in binary, decimal, or hexadecimal. For example,

{states: start = 0000b, A_1 = 0001b, A_2 = 0011b, . . . }
{states: start = 0, A_2 = 1, A_2 = 3, . . . }
{states: start = 0h, A_2 = 1h, A_3 = 3h, . . . }

Consider the design of a simple up-down modulo-6 counter, with conditional outputs as shown in the state diagram of Figure 9.36.

The PLD language description for the counter is

|(PLD Device #) in:[M], out:[Z1,Z2], clock: CLK
|Title: Mod-6 Counter
|Active-high: CLK
|Procedure: Q[2 . . . 0]
|States: A = 0, B = 1, C = 3, D = 2, E = 6, F = 4
|A. M ? ⟶ B
|M′? ⟶ F
|B. M ? ⟶ C
|M′? (Z_2 = 1 ⟶ A)
|C. M ? ⟶ D

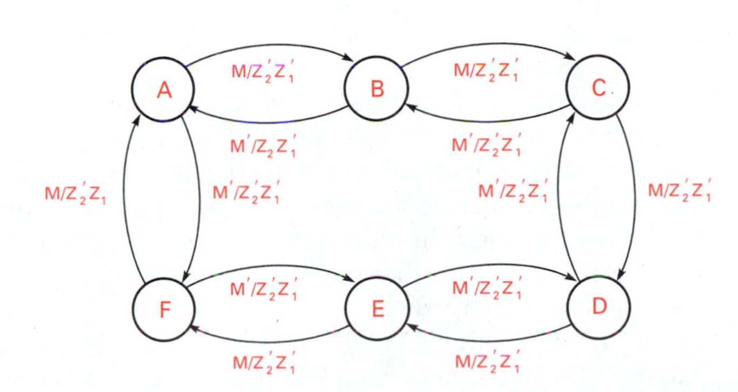

Figure 9.36
Modulo-6 up-down counter state diagram

|M'? ⟶ B
|D. M ? ⟶ E
|M'? ⟶ C
|E. M ? ⟶ F
|M'? ⟶ D
|F. M ? (Z_1 = 1 ⟶ A)
|M'? ⟶ E

9.6 FIELD PROGRAMMABLE GATE ARRAYS

Field programmable gate arrays provide the next step in the programmable logic device hierarchy. The word *field* in the name refers to the ability of the gate array to be programmed for a particular function by the user instead of by the manufacturer of the device. The word *array* is used to denote a series of columns and rows of gates that can be configured by the end user. Field programmable gate arrays (**FPGA**) contain more equivalent gates than EPLDs. The Xilinx XC3090, for example, has 9000 equivalent gates (an equivalent gate is a two-input AND). Standard random logic (SSI and MSI ICs) has given way to the PLA and PAL devices, which in turn, for some applications, are being supplanted by the EPLD. For example, the Altera EP600 has 600 and the EP1800 has 2100 equivalent gates. Field programmable gate arrays differ from standard gate arrays in that the end user can program the device. Standard gate arrays are configured by the manufacturer and require extensive lead time to produce. The basic cell of standard gate array is usually simpler than the FPGA, and the cell interconnection is done by a masking process. Standard gate arrays produce higher density chips than FPGAs and have the most efficient gate-to-end-function utilization. Standard gate arrays are most efficiently used when product numbers are very high, permitting manufacturing costs to be spread over a large number of devices.

Logic design involving PLA and PAL devices required translation software (to convert inputs to blown fuses) and programming hardware (to blow the fuses). The EPLDs needed more extensive software development tools for design implementation than the PLA or PAL devices. The transition to FPGAs involves even more complex design automation tools than the EPLDs.

Two basic FPGA architectures are offered by the product lines of two companies: Xilinx and Actel. Both approaches have cells or modules that can be configured by the end user. Both require a switching matrix to interconnect the cells or modules. The similarity ends there.

The Xilinx architecture uses configurable logic blocks, I/O blocks, and an external memory chip to realize a logic function. The configurable logic block (**CLB**) resources can be used as desired and interconnected with other CLBs to form the completed design. Xilinx uses an external memory chip to store the switch matrix configuration data. By placing the interconnect information in a memory (EPROM or RAM) changes are permitted. This means that the device can be reprogrammed by simply changing the configuration memory data.

Actel, on the other hand, has a simpler basic cell, called a logic module (**LM**), than Xilinx's CLB and a more complex switching matrix. The Actel architecture permits a potentially greater utilization of gate resources, but requires a more complex switching matrix. The architecture also does not allow for reprogramming; once the design is configured it cannot be changed.

The 4000 series FPGAs by Xilinx contain as many as 20,000 equivalent gates. Plessey, another manufacturer of gate arrays, has produced its EP60100 that has 10,000 cells or 40,000 equivalent gates. Equivalent gates or cells alone do not tell the full story. To simply compare one device against another based on gate count would be incomplete. System speed differs from one architecture to another and is application dependent. Flip-flop intensive designs may be better realized using Xilinx devices but large product term Boolean functions may be realized more economically by Actel's approach.

Any FPGA device has switching matrix overhead that does not translate into usable function. To obtain 60% or more utilization from available equivalent gates requires careful evaluation of the application and picking the right architecture.

9.6.1 Xilinx FPGA

Xilinx uses a high-density, high-speed CMOS process to implement its FPGA. Three configurable logic modules make up the FPGA: CLBs, I/O blocks, and switching matrices for interconnections. Figure 9.37 shows a block diagram of the programmable gate array structure.

The CLB consists of a combinational logic array, data multiplexers, and flip-flops. The combinational array function is performed by a 32×1 look-up table that realizes any five-variable Boolean logic function. It can be partitioned to produce two four-variable functions. The program controlled multiplexers are used to route data internally in the CLB.

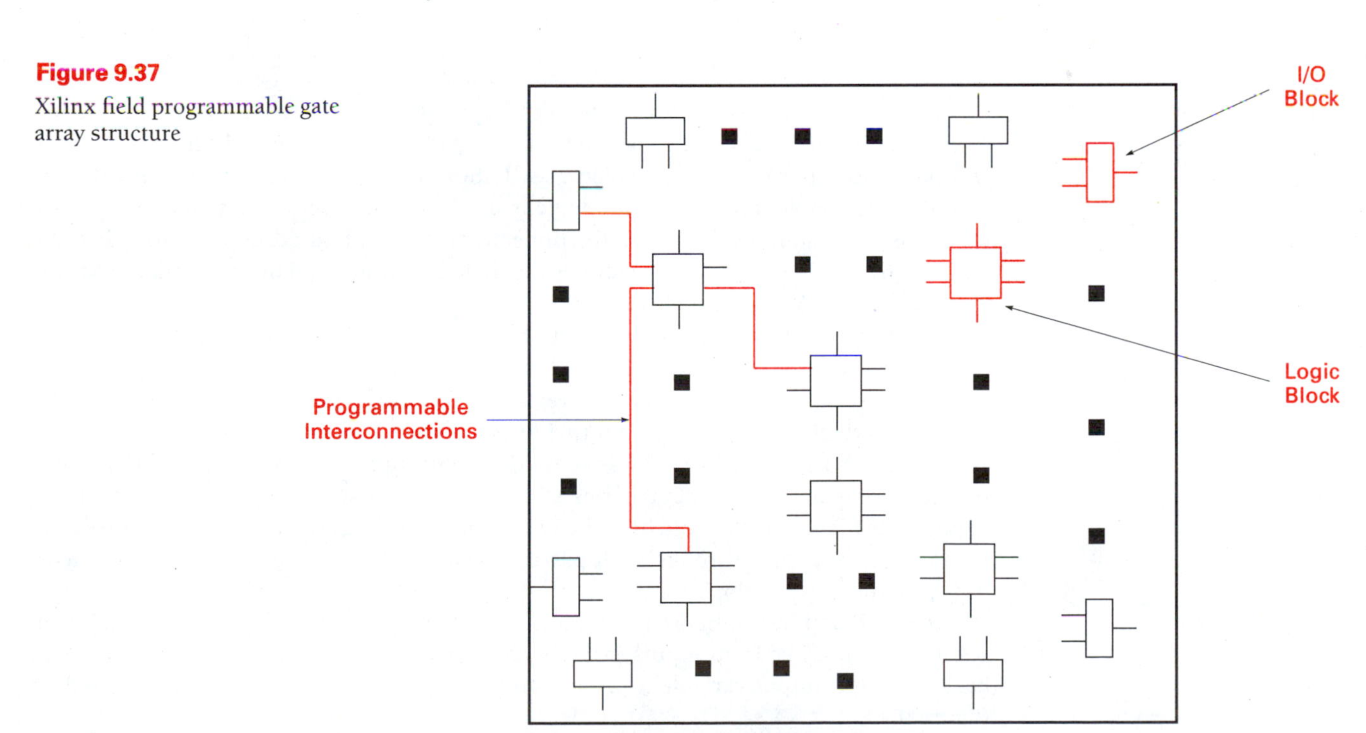

Figure 9.37
Xilinx field programmable gate array structure

Figure 9.38
Xilinx 2000 series FPGA logic block

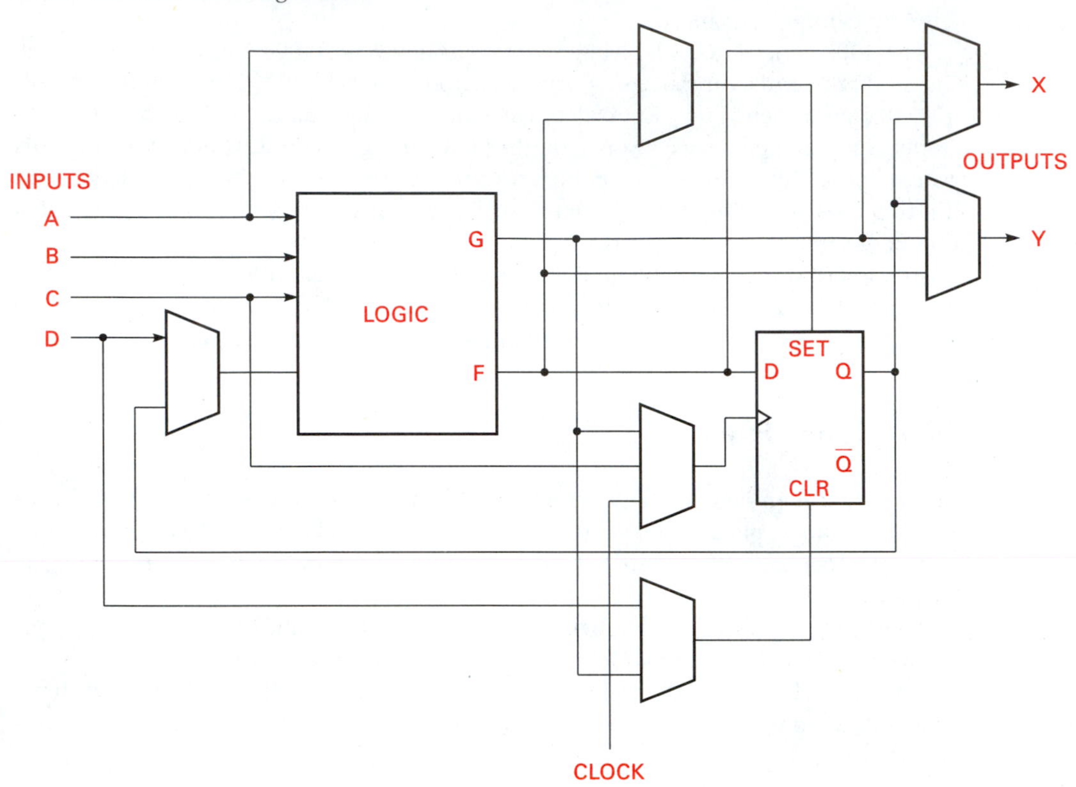

Input-output modules contain two flip-flops, one for registered input and one for registered output, connected to a common I/O pad. Both passive and active pull up of I/O pins and input-output data steering switching is available, providing bidirectional I/O. Programmable interconnection logic is used to configure the CLB and I/O blocks to the user's requirements. The interconnections are configured by data stored in an external memory. Figure 9.38 illustrates the logic functions contained within the Xilinx 2000 series FPGA configurable logic block.

The Xilinx FPGA programmable I/O block consists of two flip-flops, an I/O pad, passive or active pull-up capability, and control switches. One of the flip-flops is used for registered input and the other for registered output, with a single I/O pad. Figure 9.39 shows the I/O block for the Xilinx FPGA.

The 3000 series Xilinx CLB has more data routing multiplexers and an additional flip-flop compared to the 2000 series CLB. Figure 9.40 shows the 3000 series FPGA configurable logic block. Note how the two outputs of the combinational logic function F and G can be routed through multiplexers to either of the two output pins or to the D inputs of the two flip-flops.

The CLB can be configured to perform any logic function of up to a maximum of seven variables, five from inputs external to the CLB and two from CLB flip-flop feedback. A single output variable is present on both array outputs F and G as indicated in Figure 9.41.

Figure 9.39
Xilinx FPGA I/O block

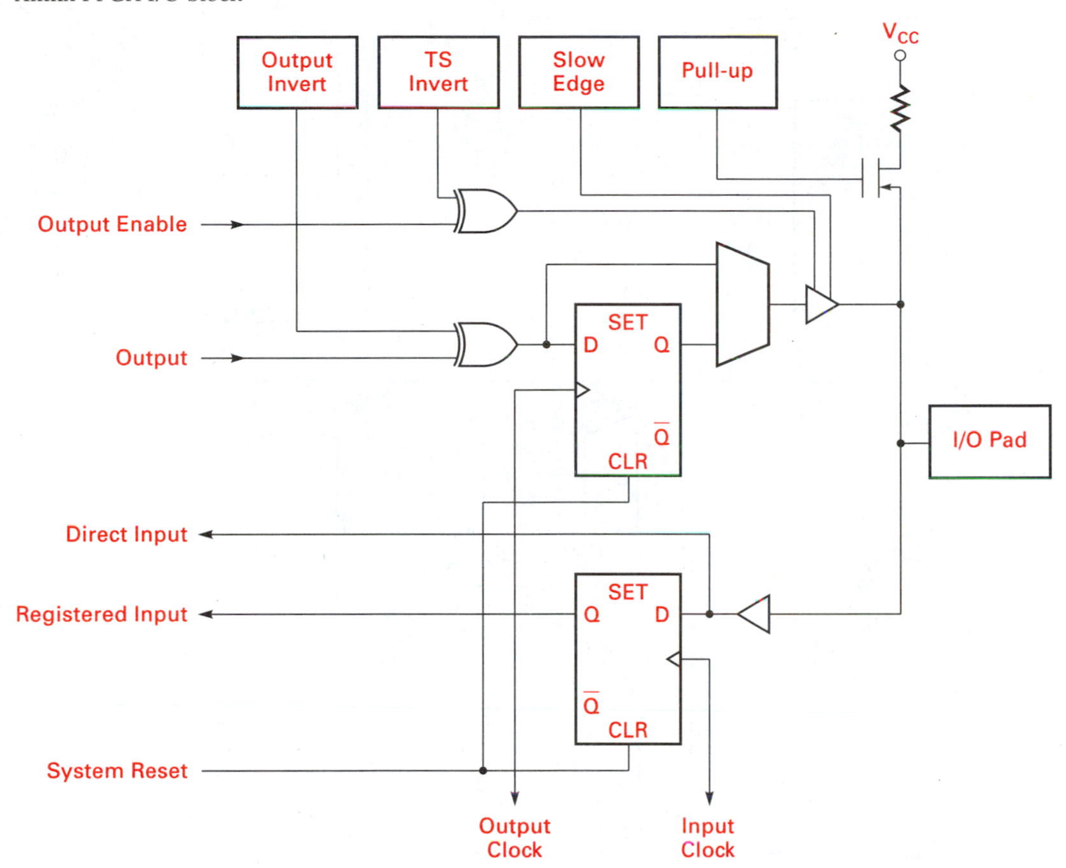

Two combinational logic functions, each consisting of four variables, can also be configured using a single CLB. Combinational array outputs F and G are separated as indicated in Figure 9.42.

A third option permits the two independent four-variable combinational logic function outputs to be multiplexed, under the control of one of the input variables. Figure 9.43 illustrates the multiplexed combinational configured CLB.

Two positive edge trigger D-type flip-flops are also contained in the configurable logic block. The flip-flop D input is driven by a multiplexer as indicated in Figure 9.44. The F and G outputs from the combinational array provide two and DIN a third multiplexer data input. DIN is independent of the combinational array. A clock enable and reset input is common to both flip-flops. Two output multiplexers are available that can select either a flip-flop (QX and QY) or a combinational (F and G) logic output.

Boolean functions requiring more than five variables can be created by cascading CLBs. For example, two CLBs could be used to create two intermediate functions of five variables each, then a third CLB could be used to gate the two intermediate outputs plus an additional three input variables. Figure 9.45 shows three CLBs connected to produce a 13-variable Boolean function. When a CLB is used for a five-variable combinational logic function the flip-flops cannot be used. Resources can be mixed, however; a single

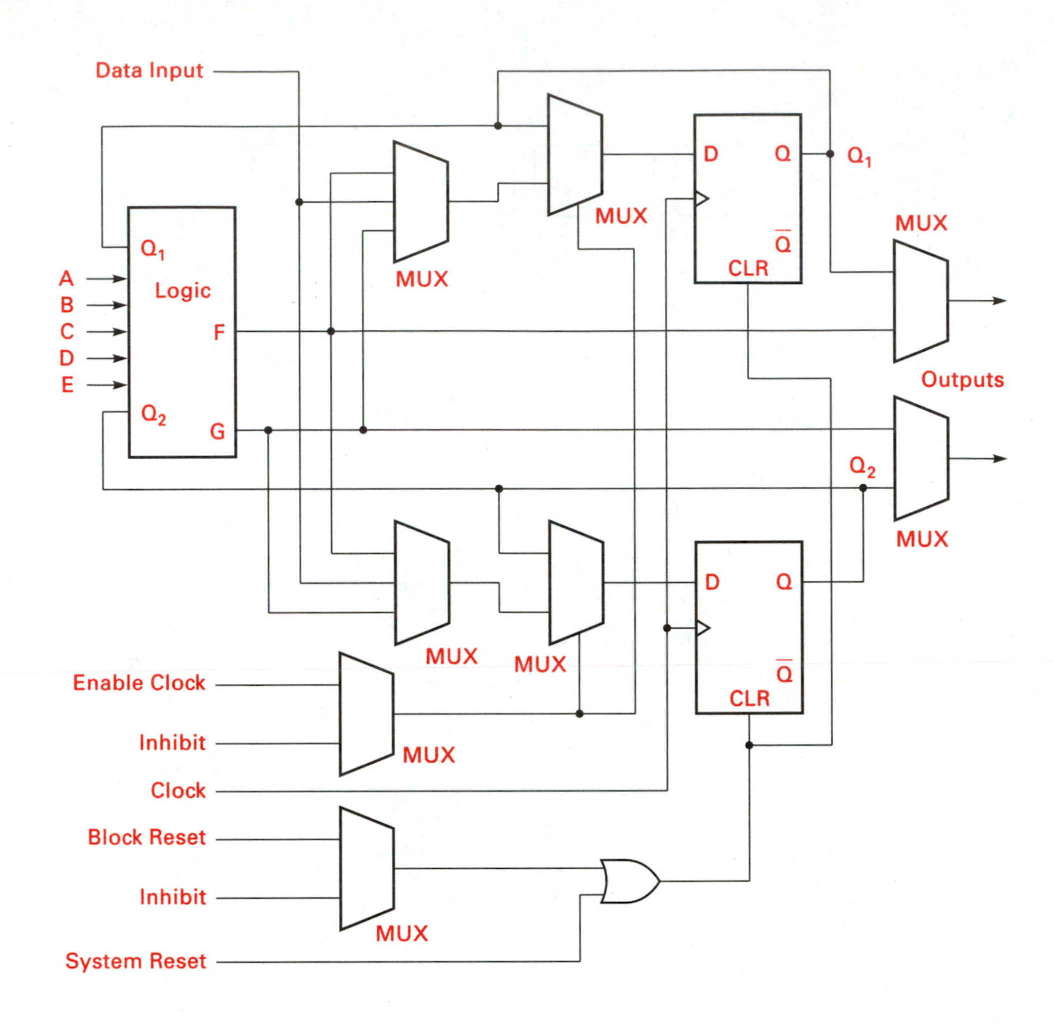

Figure 9.40
Xilinx 3000 series logic block

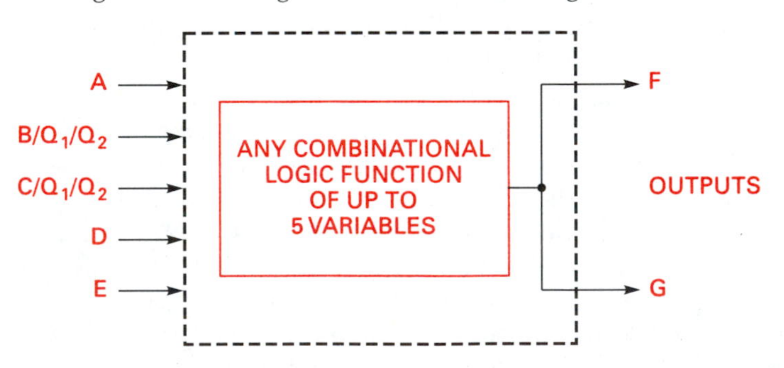

Figure 9.41
A logic block used to generate a combinational logic function

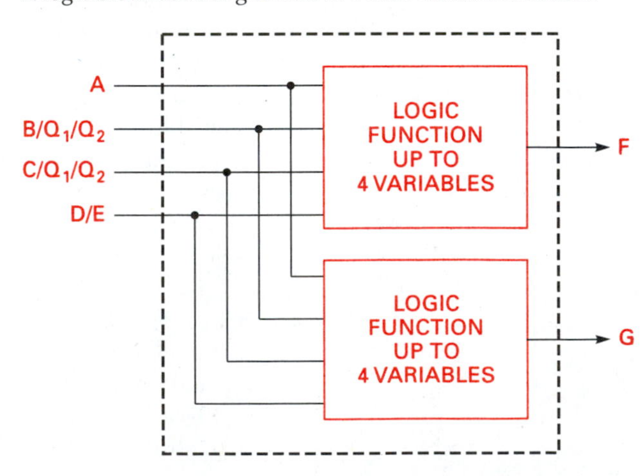

Figure 9.42
A logic block used to generate two four-variable functions

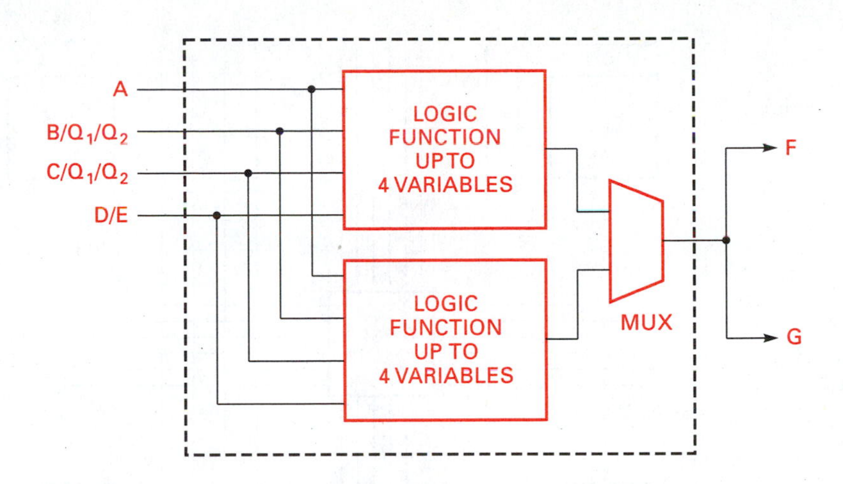

Figure 9.43
Multiplexed combinational logic functions

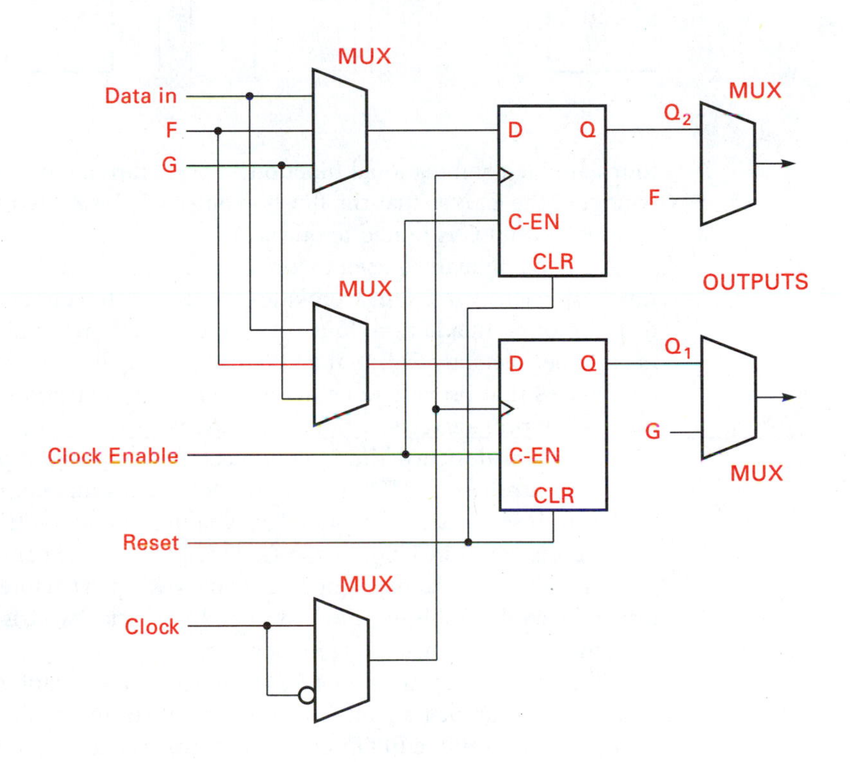

Figure 9.44
Logic block configured to multiplex flip-flop outputs

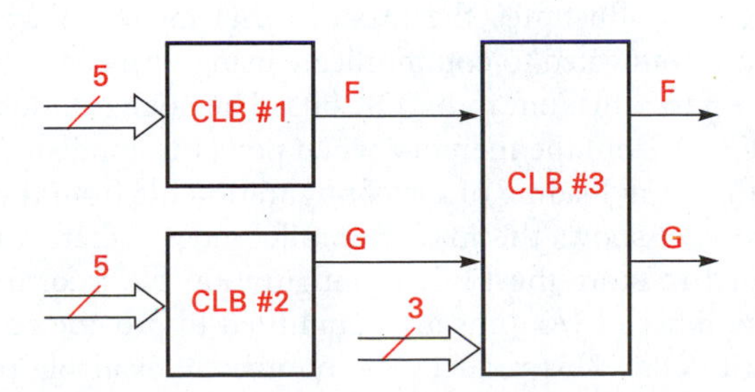

Figure 9.45
Three CLBs interconnected to generate up to a 13–variable Boolean function

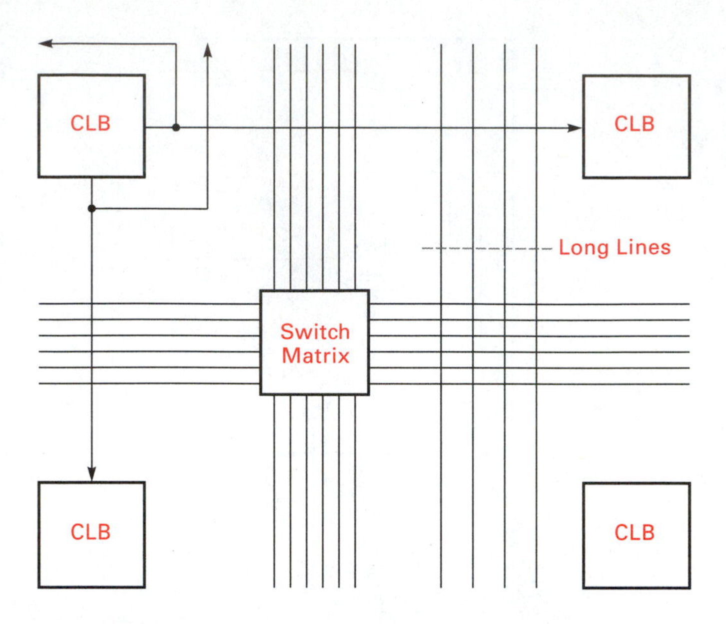

Figure 9.46
Xilinx switch matrix interconnection structure

four-variable combinational function and one flip-flop can be used in a CLB. We could configure the CLB so that the flip-flop output QX is routed to output X and the combinational output G is routed to output Y.

A switching matrix, used to interconnect CLBs and I/O blocks is illustrated in Figure 9.46. Each switch block provides for direct interconnection between adjacent CLBs or I/O blocks. In addition, four vertical and two horizontal long lines permit interconnection between any CLB in the FPGA chip. Long lines are the name given by Xilinx for conductors that provide nonadjacent CLB interconnections. They can be used as data busses that provide register interconnections.

Xilinx has designed the interconnect matrix using a two-layer grid of metal segments. Controllable CMOS pass transistors act as interconnection points in a manner similar to fuse transistors in a PLA or floating gate MOSFETs in an EPROM. Each pass transistor is controlled by a single data bit stored in the configuration file memory.

Vertical and horizontal long lines bypass the internal interconnect matrix and can be driven by configurable logic blocks or an I/O block. Two tristate buffers are located next to each CLB that provide long line drivers.

Xilinx uses an external memory chip to store the configuration interconnect information. Data can be transmitted either serially or in parallel from the external memory to the FPGA. Multiple FPGA chips can be cascaded, with one device acting as a master and others as slaves.

Figure 9.47 illustrates the master serial mode, in which the FPGA and external memory are connected to communicate using serial data. Several memory chips can be cascaded up to a maximum of 32K bits. The configuration file is automatically loaded into the FPGA from the memory when power is applied. A clock internal to the FPGA controls the serial loading of the configuration file from the memory.

Figure 9.48 shows the master parallel mode, where an external PROM or EPROM can be used to store the FPGA configuration file information. In the master parallel mode, a number of I/O pins are committed to provide address and data busses to the external EPROM. This reduces the number of available pins for application use. The

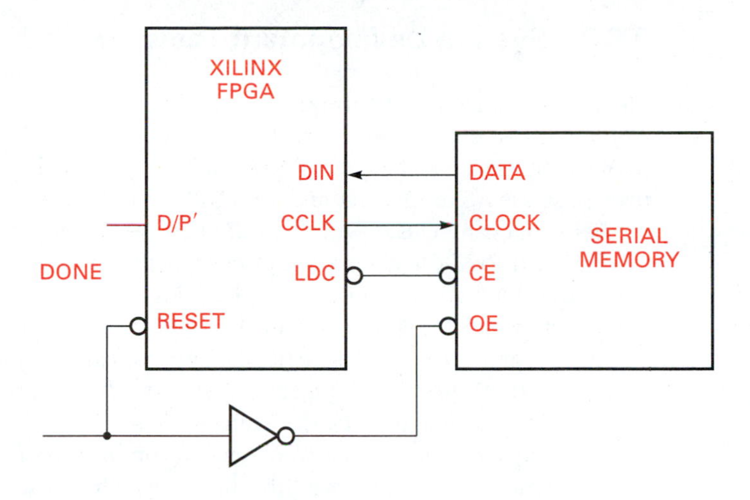

Figure 9.47
Xilinx FPGA serial mode using an external serial memory chip

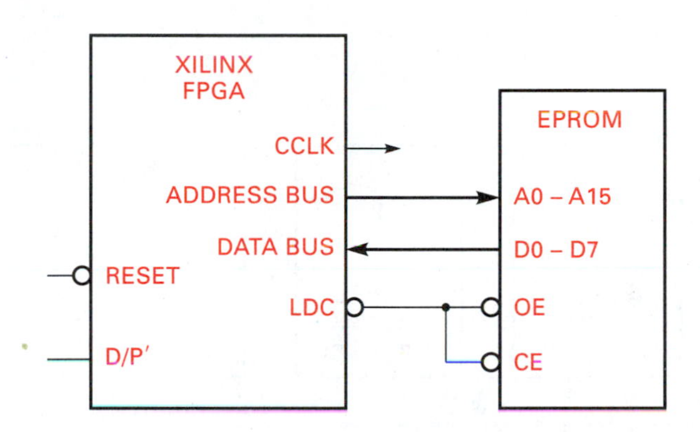

Figure 9.48
Xilinx FPGA parallel mode using an external EPROM

FPGA automatically provides the address sequence to the EPROM upon start-up. Two master parallel modes are possible: one starts the address sequence at 0000_{16} and counts up; the other starts the address at $FFFF_{16}$ and counts down. The reason for both of these address schemes is to provide compatibility with different microprocessor systems that use different initial address starting points.

Table 9.13 shows some of the off-the-shelf Xilinx FPGA devices and lists the number of CLBs, I/O blocks, and other features. Note the number of equivalent gates listed for the devices.

TABLE 9.13
Xilinx FPGA device families (courtesy Xilinx Corporation)

Part	XC2064	XC2018	XC3020	XC3030	XC3042	XC3064	XC3090
Equivalent gates	1200	1800	2000	3000	4200	6400	9000
CLBs	64	100	64	100	144	224	320
IOBs	58	74	64	80	96	120	144
Logic flip-flops	64	100	128	200	288	448	640
On-chip Bus (horizontal long lines)	None	None	16-bit	20-bit	24-bit	32-bit	40-bit

9.6.2 System Development Tools for the Xilinx FPGA

Field programmable gate arrays require a sophisticated development software system, running on an engineering workstation or a high-end PC. The Xilinx Design Manager permits the design to be entered using a variety of methods, including **schematic capture** programs like OrCAD and FUTURENET. Figure 9.49 shows the design entry methods that can be used with the Xilinx development system. ABEL, PALASM, CUPL, Log/IC, and PL Designer are proprietary software packages that exist outside of the Xilinx development system. Any of these logic design packages can be used as a data entry method for the Xilinx system. The schematic editor permits the use of schematic capture programs to draw the logic diagrams, based on a primitive function library and then enter the Xilinx development system. Translators are needed to convert the data entry files to a format used by the Xilinx system.

The schematic capture software depends on libraries to produce primitives that the logic designer can use. The primitive libraries include gates, flip-flops, counters, multiplexers, registers, and other functions. Once the design entry method is selected and a

Figure 9.49
Xilinx design entry system

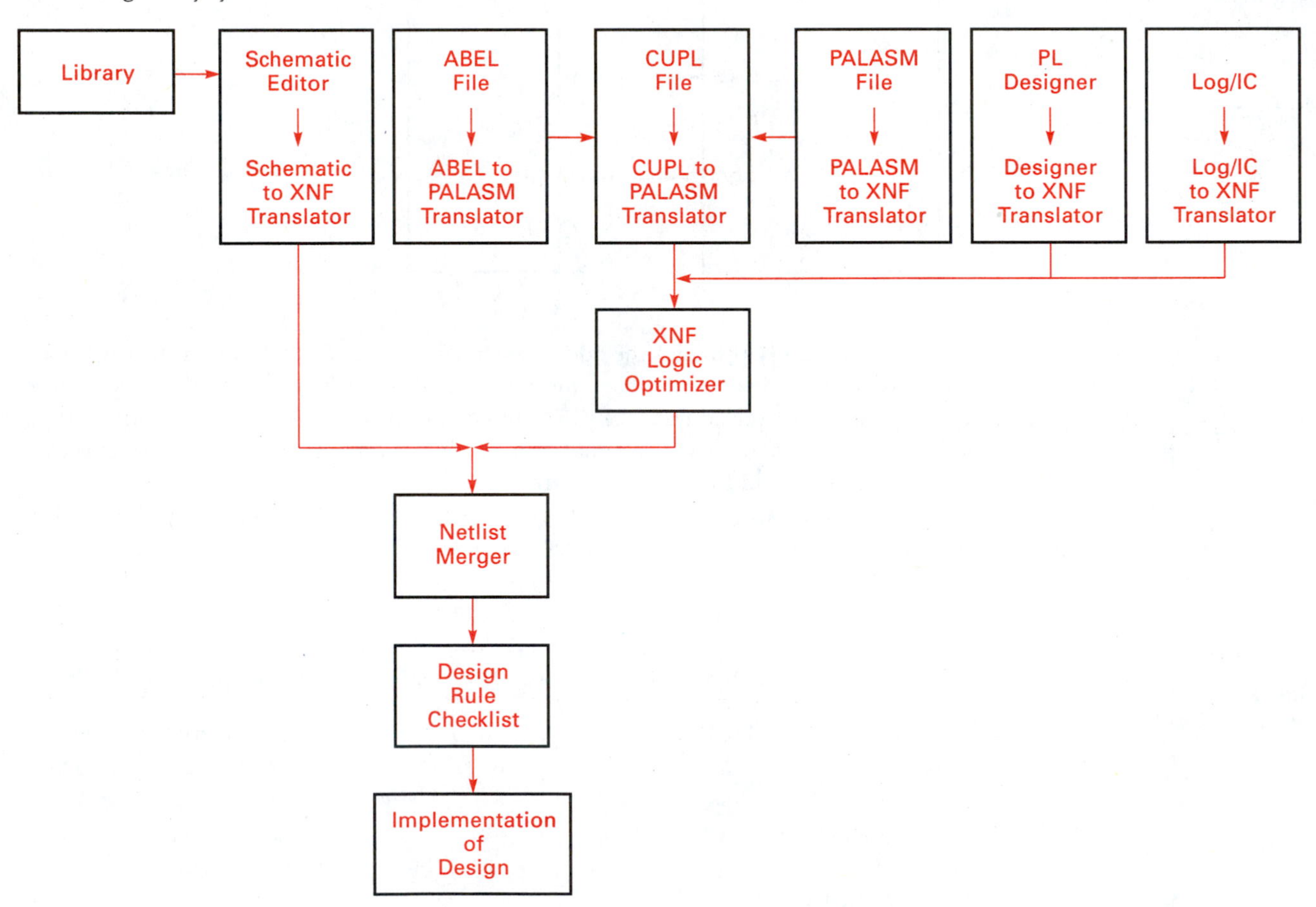

design entered, the information must be translated into a special Xilinx file called an XNF file. After checking for proper primitives and other design rules, the XNF netlist file is passed off to the design implementation software.

Many design entry approaches have simulation capability. Companies that produce schematic capture software such as OrCAD and FUTURENET also have design simulation software as well. Whichever design entry tools are used, they must be compatible with Xilinx. Simulation verifies the functional and preliminary timing of a design. The Xilinx software module provides for in-circuit verification. Final timing verification can be done before or after placing and routing the CLBs and I/O blocks in the FPGA.

The design file netlist, FILENAME.XNF, is passed to the automatic logic and partitioning module for translating the XNF file to an LCA file. The LCA file is used by the automatic placing and routing software to transfer the design file to the FPGA by interconnecting CLBs and I/O blocks.

Four major functional blocks make up the Xilinx development system: schematic entry and simulation, Xilinx netlist translators, Xilinx netlist file (filename.XNF) to logic cell array (filename.LCA) translators, and automatic placing and routing. The automatic placement and routing includes an in-circuit debugger and the XACT editor software. A flowchart of the complete system is shown in Figure 9.50.

9.6.3 Xilinx Macro Library

A macro library of logic functions is provided by Xilinx so the designer does not have to redesign basic functions. Table 9.14 shows representative examples of a number of macro functions along with the number of configurable logic blocks required for each function.

The logic designer can call the macros from the Xilinx library during the schematic entry phase of the design. The development software translates the macro function into the correct number of CLBs or I/O blocks and places them in the FPGA.

Several example function logic diagrams illustrate how a logic diagram would appear when being drawn by a schematic capture program. Figure 9.51 illustrates several macro function logic symbols.

9.6.4 Actel FPGA

Actel is another company that manufacturers FPGA chips. Actel uses a low-power CMOS technology and a company proprietary interconnection fusing technique called *PLICE* (programmable low-impedance circuit element) to produce their version of a field programmable logic array. The companies' first venture into FPGAs in 1988 resulted in the ACT 1 series. The ACT 1020 device has 2000 equivalent gates. Later, the company developed the ACT 2 series with up to 8000 equivalent gates. The basic cell is called a logic module (LM) and consists of an eight-input, one-output combinational logic block that can perform any Boolean function within its I/O limits. Figure 9.52 illustrates Actel's FPGA basic cell logic.

Any logic function can be derived from the basic cell, including multiplexers, adders, flip-flops, and counters. The combinational logic cell design leads to a very high utili-

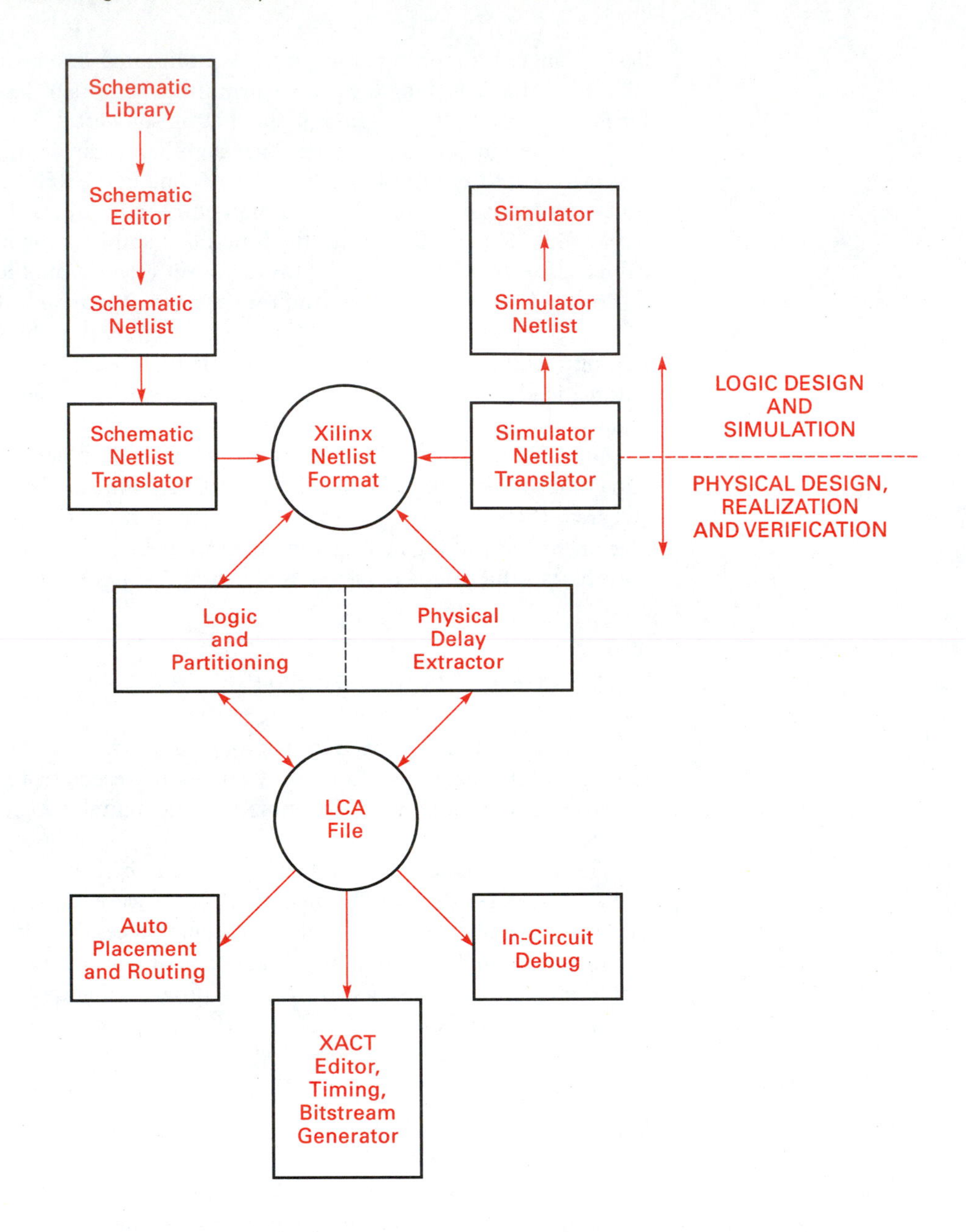

Figure 9.50
Xilinx system development flowchart

zation of available gates, up to 95% in some applications. Actel has a macro library from which the designer can select logic functions to realize a design. In addition to a wide range of combinational functions, macros exist for all flip-flop types. Several multiplexer macros are illustrated in Figure 9.53. Multiplexer functions are used to create a latch macro as shown in Figure 9.54. The R–S latch forms the basis for all the Actel flip-flop macros.

The ACT 1020 FPGA contains 547 logic modules and 69 I/O modules on a single 84-pin chip carrier or grid array device. Several Actel macros and the number of logic modules required for each is illustrated in Table 9.15.

TABLE 9.14
Partial list of Xilinx macro functions

Xilinx mnemonic	Function	#CLBs
GADD	Full adder	1
GEQGT	Equal or greater	1
PIN	Input pad	1 I/O
LD	Data latch	1
FD	D flip-flop	1
FDMS	D FF w/2-in mux and set	1
FJK	J–K flip-flop	1
FT2	2-input toggle FF	1
D2-4	1-of-4 decoder	2
D3-8	1-of-8 decoder	4
74-42	1-of-10 decoder	7
M3-1	3-to-1 multiplexer	2
74-151	8-to-1 multiplexer	7
RD4	4-bit data register	4
RD8	8-bit data register	8
RS4	4-bit shift register	4
74-194	4-bit bidirection SR	12
C4BCP	2-bit binary counter	3
C8JCR	4-bit Johnson counter	4
74-160	4-bit BCD counter	8
74-161	4-bit binary counter	8

Designs can be entered into the Actel development system from several schematic capture programs, including OrCAD, and from Actel's own component libraries. Once the device is programmed it cannot be erased and reprogrammed. If design errors exist, a new device must be programmed.

The Actel FPGA has a different logic cell structure than the Xilinx approach. The Xilinx CLB is more complicated than the LM used by Actel but tends to be more wasteful of gate resources when a combinational function is to be realized. One advantage of the Xilinx approach over that taken by Actel is that the FPGA can be reconfigured by changing the contents of an external memory chip. The disadvantage is that a memory chip must be added, which increases design cost. Actel FPGAs are more efficient in using internal gate resources but once programmed cannot be changed. Both approaches have strong points permitting the logic designer to choose the most appropriate FPGA for a given application.

Figure 9.51
Example macro logic symbols

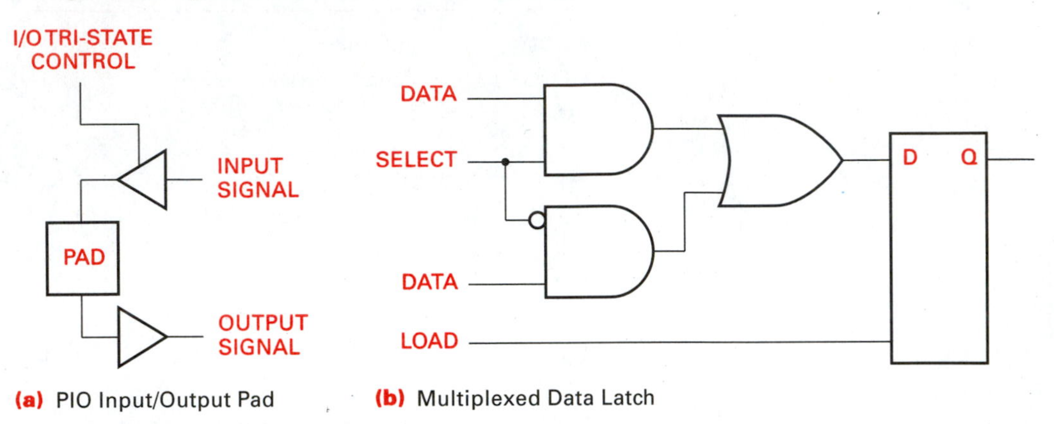

(a) PIO Input/Output Pad **(b)** Multiplexed Data Latch

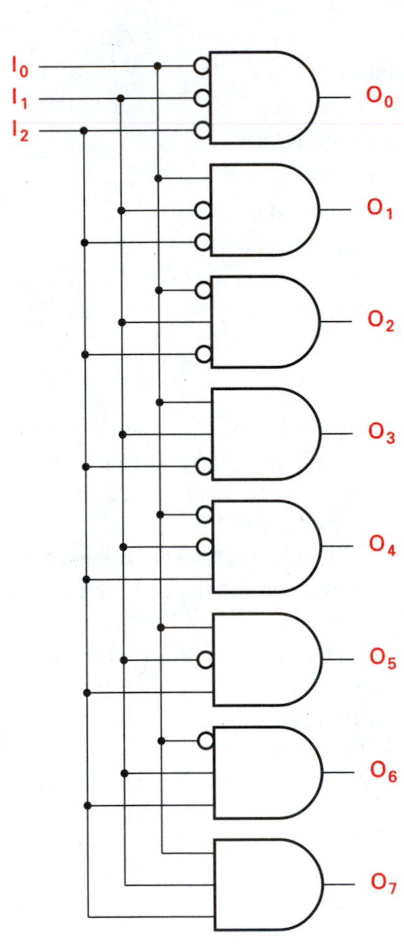

(c) 1-of-8 Decoder

Figure 9.51
Continued

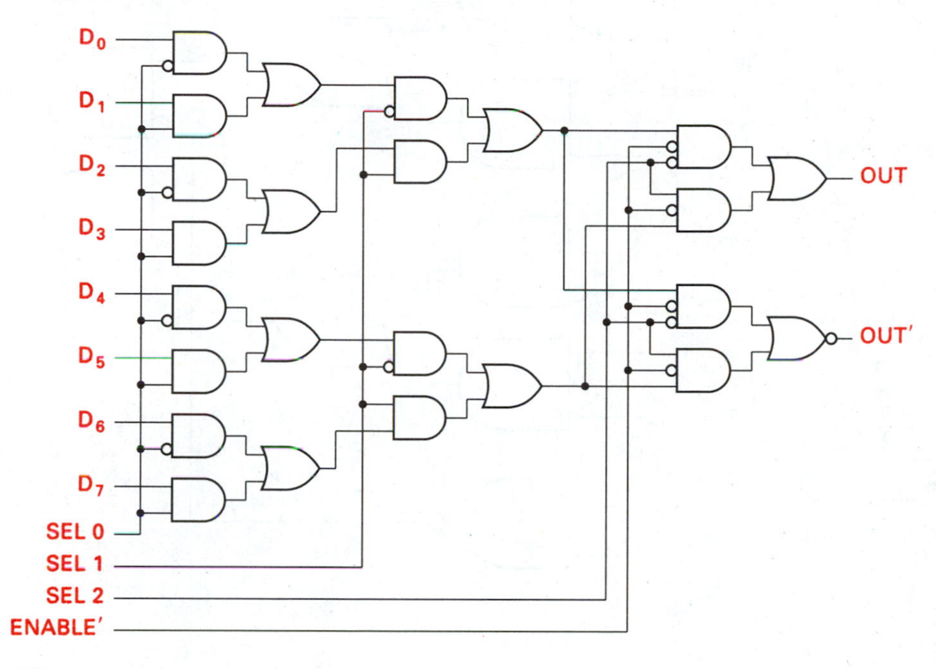

(d) 74xx151 8-to-1 Multiplexer

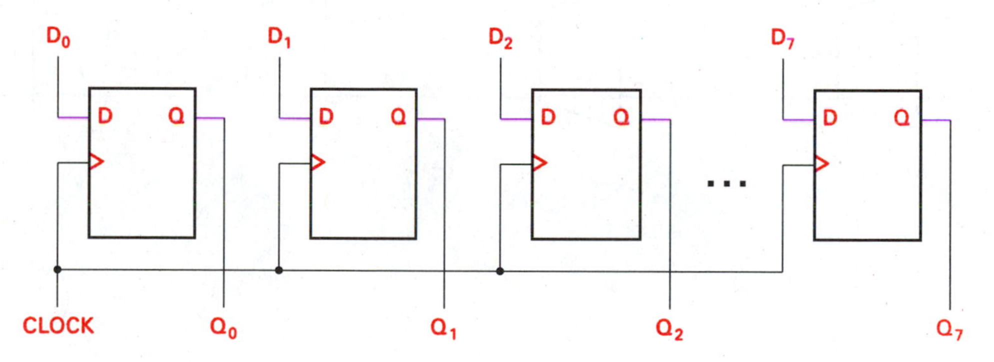

(e) RD8 8-Bit Register

Figure 9.52
Actel basic cell logic

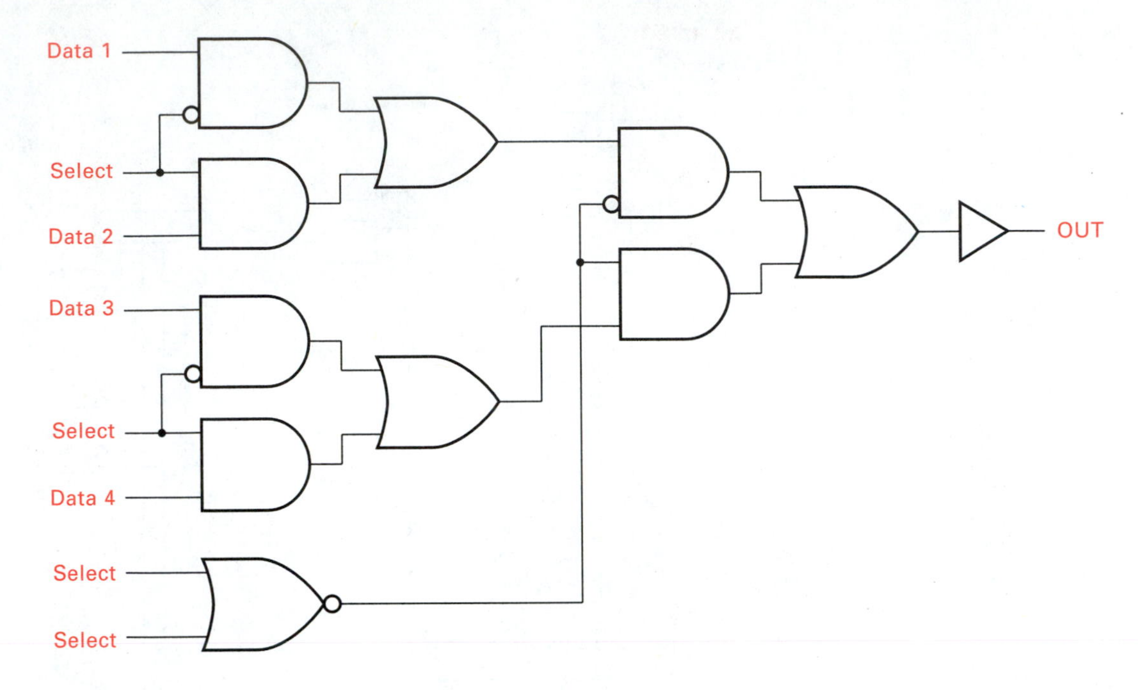

Figure 9.53
Multiplexer macros

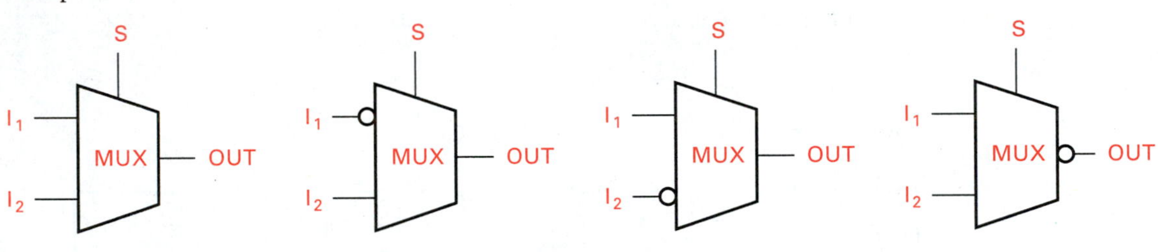

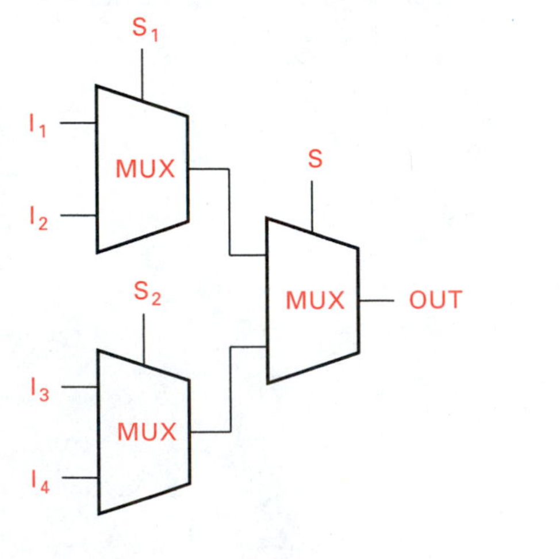

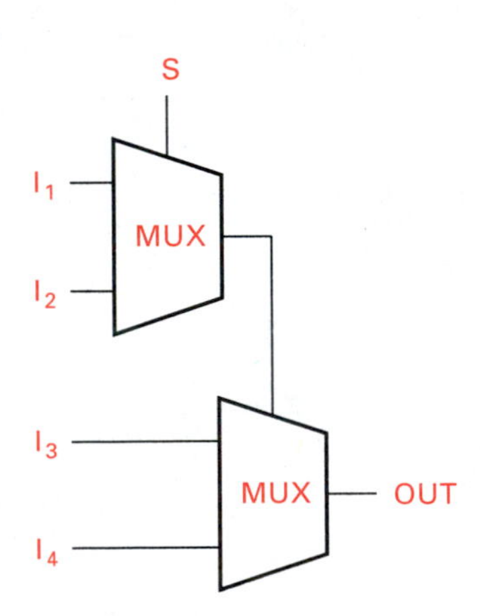

Figure 9.54
Realizing a R–S latch using multiplexers to create an actel macro

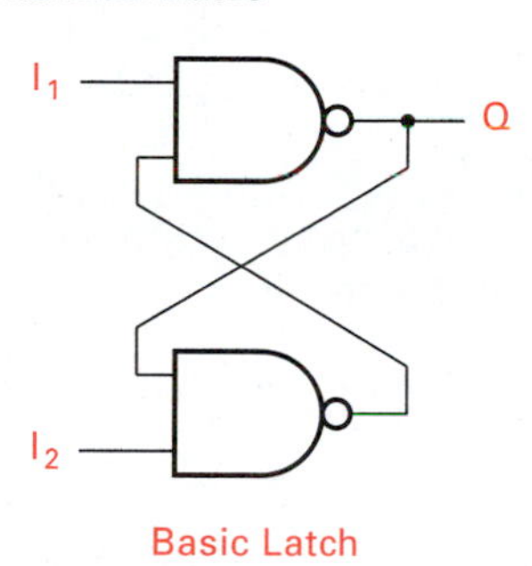

Basic Latch

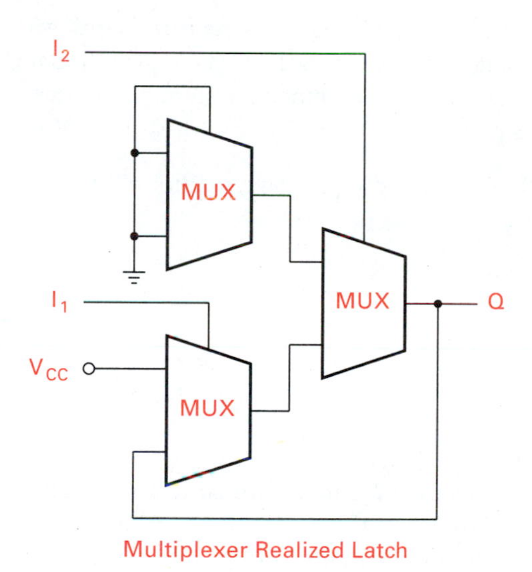

Multiplexer Realized Latch

TABLE 9.15
Actel macros

Macro function	Logic modules
2-input gate	1
3-input gate	1
4-input gate	2
XOR gate	1
XOR OR gate	1
XOR AND gate	1
AND XOR gate	1
AND OR gate	1
OR AND gate	1
Buffer/Inverter	1
2- or 4-input Multiplexer	1
Adder	2
D-latch	1
D-latch with clear	1
D-latch with enable	1
Multiplexed latch	1
D flip-flops	2
Multiplexed D flip-flops	2
D flip-flop with enable	2
JK flip-flop	2

SUMMARY

A. Programmable logic is introduced, including

1. The basic structure of read-only, programmable read-only, and erasable programmable read-only memory.
2. Using ROM. PROM and EPROM to realize combinational logic functions.
3. Using EPROM to realize a sequential function, with examples of using an EPROM to realize the combinational functions and an elapsed time measurement controller realization using a PROM.

B. Programmable logic devices are discussed, focusing on

1. Programmable logic array (PLA) structure.
2. Programmable array logic (PAL): in comparison with PLA, example devices, combinational and sequential devices, realization of combinational logic functions using a PAL, and designing an up-down counter using a PAL.
3. Generic array logic (GAL): its structure, an example device, the macro cell architecture of an ex-

ample device, and sequential machine design using a GAL.

C. Erasable programmable logic devices (EPLD)are introduced, attention centering on

1. General structure of an EPLD.
2. Altera EPLD family: MAX+Plus development system, Altera EPLD primitives, and EP600 EPLD structure (macro cell architecture, operational modes, I/O selection).
3. Sequential machine design using an Altera EP600, including definition of problem and I/O mnemonics, derivation of excitation and output equations, OrCAD schematic capture of sequential machine, and translating OrCAD schematic to Altera design files.

D. Computer-aided design using PLDs is broken down into

1. Combinational logic, with a full adder, a four-bit adder, a three-bit decoder, and a 3×8 multiplexer.
2. Flip-Flops, with synchronous inputs, clocking, and asynchronous inputs.
3. State machines, both Mealy and Moore and counter design

E. Field programmable logic array is introduced:

1. Xilinx FPGA, including its structure (configurable logic block, or CLB, I/O block, switch matrix structure), development system tools, and macros library.
2. Actel FPGA, including its logic module (LM).

REFERENCES

Altera Data Book (Altera Corporation, 1995). The book contains specific specifications of FPGA components manufactured by Altera, including the EP600, EP1800, EPM5000, and EPM7000 series PLDs, plus application notes.

ACT Family Field Programmable Gate Array Data Book (Actel Corporation, 1992). The book contains Actel FPGA component specifications and application notes.

The Programmable Gate Array Data Book (San Jose: Xilinx Corporation, 1992). This book contains Xilinx FPGA component specifications and application notes.

XACT, Logic Cell Array Macro Library (San Jose: Xilinx Corporation, 1992). This book is part of the documentation that comes with Xilinx FPGA development systems. It contains macro library information for logic design realization using Xilinx FPGAs.

Programmable Logic Data Book (Mt. Prospect: Intel Corporation, 1994). This book contains EPLD component specifications and application notes.

Programmable Logic Designer's Guide, by Roger C. Alford (Indianapolis: Howard W. Sams Company, 1989). This book discusses programmable logic in general and PAL, PLA, and EPLD design realization specifically. It also covers some development system software in detail.

Programmable Logic Devices, Technology and Applications, by Geoff Bostock (New York: McGraw-Hill Company, 1988). This is a straightforward treatment of basic programmable device circuits and the technology employed to manufacture programmable devices. It discusses PAL, PLA, EPLD, and custom gate arrays.

Programmable Array Logic Handbook (Sunnyvale: Advanced Micro Devices, 1984). This is a dated but useful treatment of PAL devices and design using AMD PALs.

"Getting Started with PALS," by Robert A. Freedman, *Byte* (January 1987). An introduction to PALs, the article presents basic, introductory ideas justifying designing with PAL devices, including comments on programming units.

"Introduction to Programmable Array Logic," by Vincent J. Coli, *Byte* (January 1987). This is an introduction to PAL devices, design notation and applications.

"Overview of Programmable Hardware," by Phillip Robinson, *Byte* (January 1987). This is an introductory discussion of ROMs, PROMs, EPROMs, PLAs, and EPLDs and where each fits into a logic realization hierarchy. It is a brief comparison of user programmable devices with custom gate arrays.

"Programmable Logic IC Tackles Many Tasks," by Dave Bursky, *Electronic Design* (June 8, 1989). This article discusses Signetics Corporations PLC42VA12, an entry into the EPLD market. The article continues to illustrate the power of EPLD devices in replacing random SSI and MSI logic devices.

"Replace SSI/MSI Glue Chips with a Programmable Array," by David P. Lautzenheiser, *Electronic Design* (August 25, 1988). This article discusses using Xilinx field programmable logic array devices in systems design to replace random SSI-MSI devices. It gives a brief introduction into the architecture of the Xilinx device.

"PLD Architectures Require Scrutiny," by Doug Conner, *Electronic Design News* (September 28, 1989). This very informative article compares Altera's EPM5128, Xilinx's XC3090, Signetics's PLHS502, Intel's 5AC324, and Actel's ACT1 series devices.

"PLD State Machines Cut Controller Design Time," by Michael Treseler, *Electronic Design* (October 27, 1988). This applications example shows how EPLDs can be used in state machine design.

"Programmable Logic Devices," by Charles H. Small, *Electronic Design News* (November 10, 1988). This is an overview of programmable devices, including company names and devices.

"State Machines Solve Control Sequence Problems," by Stan Kopec, *Electronic Design News* (May 26, 1988). This is the first of a two-part series on using Altera EPLDs in designing state machines. The article reviews state machines and proceeds to use an Altera EPLD in realizing a state machine controller.

"Asynchronous State Machines Challenge Digital Designers," by Stan Kopec, *Electronic Design News* (June 9, 1988). This is the second in a two-part series on using Altera EPLDs in the design of sequential machines. This article discusses the IBM microchannel bus controller and applies an EPS448 (stand-alone micro sequencer) to the controller design.

GLOSSARY

CLB (configurable logic block) The basic logic macro cell used by Xilinx in field programmable logic arrays. It contains an AND array, two flip-flops, and control logic.

EEPROM (electrically erasable programmable read only memory) A PROM device that can be reprogrammed while still plugged into the circuit board. Rewriting data takes considerably longer (milliseconds) than reading (nanoseconds). The part is designed to replace EPROMs in applications where in-place reprogramming is desirable.

EPLD (erasable programmable logic device) A UV (ultraviolet) erasable programmable logic device. The same floating-gate MOS technology used for EPROMS applies to the EPLD. The device consists of an AND array (the interconnect links are floating gate MOSFETs), a set of multiplexers, and a flip-flop in each macro cell. Floating gate MOSFET interconnect points are used to switch data paths for different configurations.

EPROM (erasable programmable read only memory) A read only memory that is UV (ultraviolet) erasable and can be reprogrammed. The UV erasable feature is accomplished by using a floating gate MOSFET switch to store bits. Exposure to UV light dissipates the charge on the floating gate, changing the FET source to drain resistance.

FPGA (field programmable gate array) A gate array is a large number of logic gates that can be interconnected to perform some logic function. A field programmable gate array can be configured by the user in the field. Different architectures are available to the user. Xilinx makes a reconfigurable FPGA that uses an external memory to store the gate configuration file. Actel uses a different approach and is not reprogrammable.

GAL (generic array logic) A GAL consists of an AND array with an output logic macro cell that consists of a flip-flop, multiplexers, and an output tristate buffer. Both the AND array and the macro cell resources are programmable.

JEDEC (Joint Electron Device Engineering Council) **file** A set of standards governing data format files for programming programmable logic devices. The JEDEC file is the output of programmable logic design software. It is used to actually program (blow fuses, charge floating gates, or set other logic interconnection points) the logic devices.

LM (logic module) The basic cell used by Actel Corporation in their field programmable logic array. It consists of an eight-input, one-output combinational logic function. Feedback paths are provided to use several LMs in the creation of flip-flops.

Macro cell The basic logic function provided in GAL, EPLD, and FPGA devices. All interconnection done by the user connects together macro cells to form the desired logic functions.

PAL (programmable array logic) A programmable AND array providing p-terms that feed fixed OR gates. PALs also include flip-flops whose excitation functions are derived from the programmable AND array/fixed OR array functions. PALs have fused interconnection points that cannot be reprogrammed once blown.

PALASM A logic design development program that converts Boolean equations into the proper JEDEC file for blowing PAL fuses.

PLA (programmable logic array) A programmable AND array whose product term outputs feed a programmable OR array. The PLA is more flexible than the fixed OR array PAL but requires a more complicated programming device.

PLD (programmable logic device) A generic name given to the family of logic devices whose function is determined by programming interconnection points that configure internal logic functions. PALs, PLAs, EPLDs, and FPGAs are all PLDs.

PROM (programmable read only memory) PROMS are user programmed read only memory devices. They can be programmed only once, because programming blows internal fuses to provide data paths for the desired bit patterns. Most PROMS are bipolar devices (TTL and ECL).

Schematic capture This is the name given to a software package or the process that allows the logic designer to select logic symbols from a library to draw logic diagrams. The software permits the designer to place logic symbols and interconnect them with lines, in a logic diagram drawing, to form a function . The schematic capture software contains a library that has a selection of available functions (TTL SSI/MSI and some LSI parts, for example).

QUESTIONS AND PROBLEMS

Section 9.1 to 9.2

1. Design a four-bit binary full adder using a PROM. Draw the logic symbol of the adder showing all inputs and outputs. Determine the minimum size ($M \times N$ bits) of the PROM. Construct a partial hexadecimal data table, with input values starting at $0A0_{16}$ and ending at $0B0_{16}$. The data table must indicate address and data output values for the given input range.
2. Design a single-digit BCD adder using an EPROM. Determine the EPROM minimum size ($M \times N$ bits). Construct a partial hexadecimal data table showing inputs 70_{10} to 80_{10}. Draw a simple diagram of the adder with all inputs and outputs connected to the EPROM.

3. Design a single-digit hexadecimal-to-gray code converter using a PROM. Construct the truth table (PROM data table). Identify all inputs and outputs in a simple PROM diagram.

4. Design a synchronous modulo-8 binary counter using an EPROM to realize the combinational portion of the circuit. Your design should require two ICs, one being the EPROM and the other a D flip-flop register; include the state table; state assignment; excitation equations; and EPROM data table.

5. Design a synchronous sequential machine that will detect the input sequence 000, 010, 101, 001. When the correct sequence has been detected it causes an output Z_2 to go high. Any incorrect input sequence of four three-bit code words causes output Z_1 to go high. After an input sequence of four three-bit code words, cause the state machine to initialize in preparation for a new input. Determine the minimum EPROM size. Identify inputs and output for the EPROM. Construct the EPROM data table and draw the logic diagram including EPROM used and the state variable ICs used.

Section 9.3

6. Describe the differences between a PLA and a PAL. Use logic diagrams to illustrate your response.

7. Realize the following Boolean equations with a PLA. Draw the equivalent logic diagram using PLA notation.

   a. $A = xy'z + x'yz'$, $B = x + yz'$

   b. $x = a'b'c + a'bc'$, $y = abc + a'b'c + ab$, $z = abc + a'b'c'$.

   c. $P = rst + uv + ut$, $Q = r's + t'$.

8. Draw the PAL equivalent for the following Boolean equations.

   a. $x = abcd' + a'b'c + a$, $y = ab' + c'd$, $z = c + ab'$.

   b. $a = x'y + x'yz$, $b = z' + xy$.

9. Explain how a PAL circuit that has only four product terms available can be used to realize a five-product term function. Draw the logic diagram using PAL notation.

10. A PAL20X8 has

    a ______ input variables.

    b. ______ output variables.

    c. ______ flip-flops.

11. Explain the distinction between a standard I/O, a programmable I/O, and a registered I/O PAL.

12. Explain how a registered I/O PAL can be used to realize a sequential circuit.

13. Realize the modulo-8 counter of Problem 4 with a registered I/O PAL. Use the PAL16R4 PAL illustrated in Figure 9.16.

14. Realize the three-bit sequence detector of Problem 5 with the PAL16R4 PAL illustrated in Figure 9.16. Express the excitation equations using horizontal/vertical intersection numbers as in Problem 13.

15. Explain the distinction between a registered PAL and a GAL. Use logic diagrams to illustrate your response.

16. Realize the equivalent of the TTL 74xx195 shift register using the GAL16V8A. Determine the excitation equations from the logic diagram illustrated in the figure for this problem and how many OLMCs (output logic macro cells) are needed. Identify the configuration control bit values for each of the used OLMCs.

Section 9.4 to 9.5

17. What is the primary distinction between a GAL and an EPLD? What technology is used in the manufacture of an EPLD? Describe the advantages and disadvantages of the EPLD approach versus a registered PAL/GAL approach.

18. Write the PLD description, using the language described in the text, for the following Boolean expressions.

    a. $X = a'b + c' + d$

    b. $Y = (a + b)'(a'b') + c$

    c. $Z = (a + b)' (c' + d')$

19. Write the PLD expressions, including I/O declarations, for the logic functions in parts a and b of the figure for this problem.

Problem 16

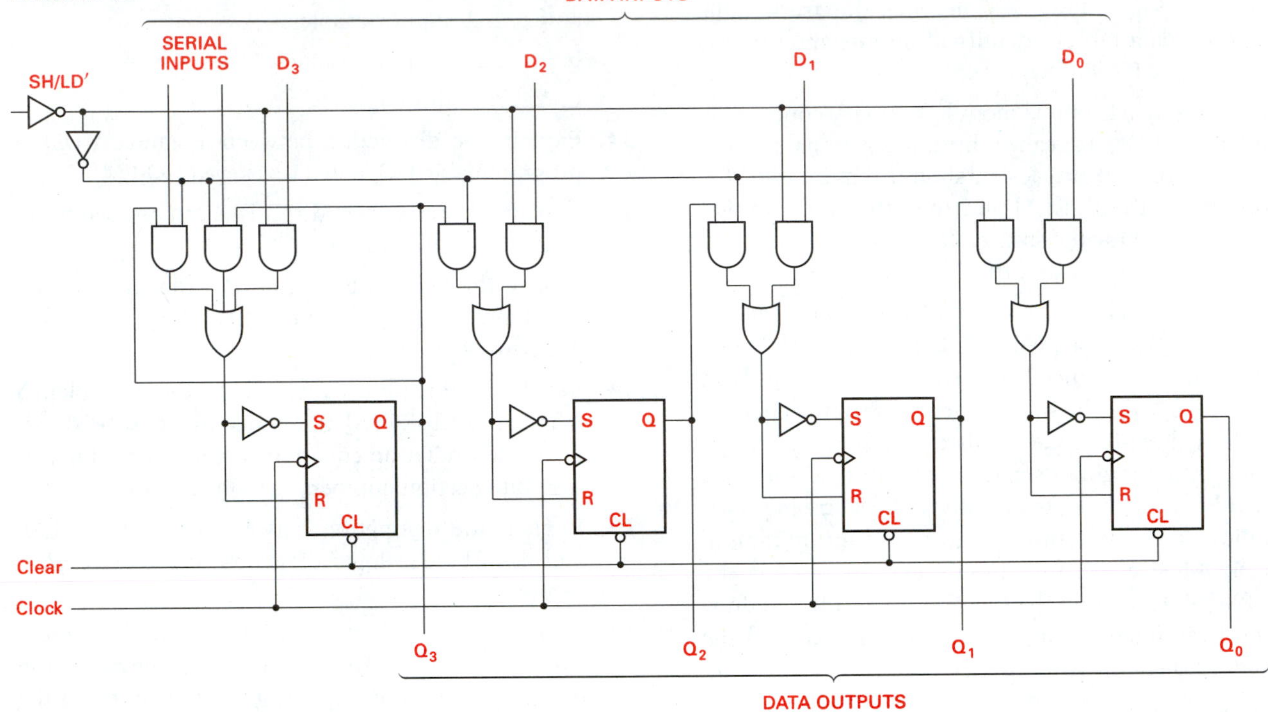

Problem 19

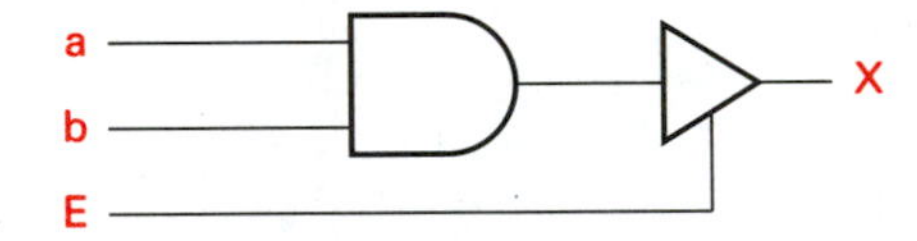

(a)

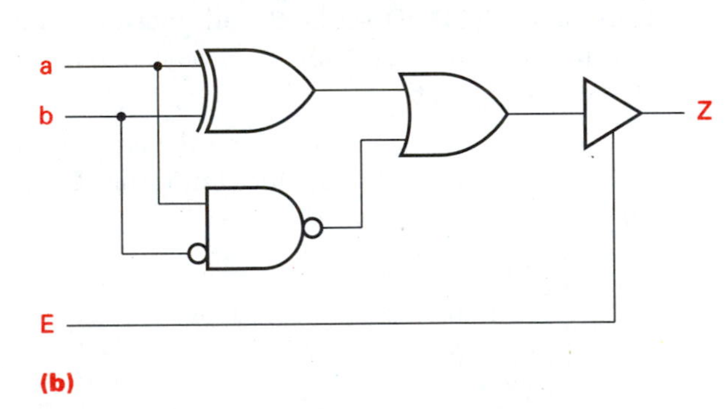

(b)

20. Create the PLD description for a full subtractor. Declare I/O and write the logic functions.

21. Write the PLD descriptions for the following:

a. A 2 × 4 multiplexer (use indexed I/O statements).

b. A four-bit odd parity checker.

c. A tristate D flip-flop with a negatiave edge trigger and active low asynchronous preset and clear inputs.

22. Use the PLD language described in the text to realize the state machine indicated in the state diagram in the figure for this problem.

Problem 22

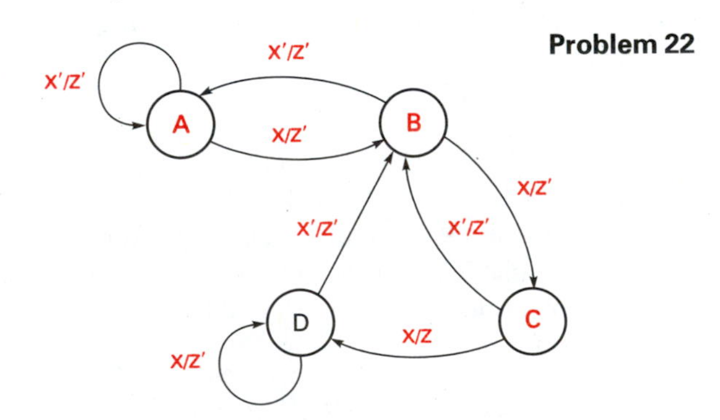

Section 9.6

23. Compare the two different FPGA architectures discussed in this chapter. Illustrate your response using logic diagrams as needed.

24. Explain how the Xilinx FPGA can be reprogrammed.

25. A large combinational logic function requires 24 p-terms. How many Xilinx FPGA configurable logic blocks are needed?

26. A sequential machine uses J–K flip-flops as state variables. One of the state variables excitation equations requires six p-terms to realize. Draw a block diagram, where each block represents a CLB, illustrating how the state variable function can be realized.

27. Both Xilinx and Actel FPGA devices have extensive macro libraries to assist logic designers. Consider a logic design that has three 16-bit data registers and a 16-bit adder. Determine the number of Xilinx CLBs and Actel LMs needed to realize the circuit. Why can direct comparisons of CLBs and LMs be made when comparing designs?

28. Which FPGA approach is likely to give more efficient use of available resources when large combinational functions are needed? Which is likely when large numbers of flip-flops are needed?

CHAPTER

10 DIGITAL INTEGRATED CIRCUITS

INTRODUCTION

Chapter 10 connects the mathematics developed by Boole, DeMorgan, and Shannon with electronic circuits needed to realize the logic presented in previous chapters. Boolean algebra, DeMorgan's theorem, binary codes, and logic symbols, although very important, are insufficient, by themselves, to realize any digital system. Real systems have practical limitations brought about by the characteristics of the electronic circuits used to build them. The design of digital systems requires a knowledge of the characteristics of electronic integrated circuits.

The intent of this chapter is to develop an understanding of the function and characteristics of the three main integrated circuit logic families: transistor–transistor logic (TTL), emitter coupled logic (ECL), and complementary metal oxide semiconductor (CMOS). Each logic family has several subfamilies. For example, TTL logic is further divided into low-power, high-speed, low-power Schottky, Schottky, advanced low-power Schottky, and advanced Schottky logic. Emitter coupled logic also has subfamilies (MECL 10,000; MECL 100,000), as does CMOS logic (4000 series, C, HCT, AC, and ACT series).

All digital integrated circuits are designed using diodes, bipolar transistors, and field effect transistors (FETs). To develop a working understanding of digital integrated circuits requires a knowledge of these devices. We will begin with some simple diodes and bipolar transistor circuits, then cover TTL and ECL integrated circuits before discussing field effect transistors and analyzing CMOS integrated circuits.

10.1 DIODES AS SWITCHES

Diodes and transistors are the primary electronic components used for switching. Switching between two voltage levels, one defined as a logic 1 and the other as the logic 0, is the basis of how electronics circuits implement switching algebra. Digital circuits either conduct or they do not, depending on the level of the input. Two discrete output states exist: one is defined a logical 0 and the other as a logical 1.

Diodes make functional electronic switches, they can be biased so they conduct or into cutoff (not conducting); each state represents a logical level. A typical diode I-V characteristic **curve** is illustrated in Figure 10.1.

It is convenient to model the diode as an ideal device, because it conducts perfectly when on, and no current flows when it is off. The diode can be modeled (thought of) as a mechanical switch. Figure 10.2 illustrates the "ideal" diode model.

The ideal diode model assumes a forward resistance (resistance when conducting), R_f, to be 0 ohms, and the reverse resistance (resistance when not conducting), R_r, to be infinite ohms.

Refining our model into a what is sometimes called a *piecewise-linear* approximate diode model gives the characteristics indicated in Figure 10.3. In Figure 10.3(a), a forward voltage drop exists that offsets the forward transfer curve. The curve illustrated in Figure 10.3(b) adds the fact that the forward resistance is not really 0, giving a slope to the forward transfer curve.

Figure 10.1
Typical diode voltage/current transfer curve

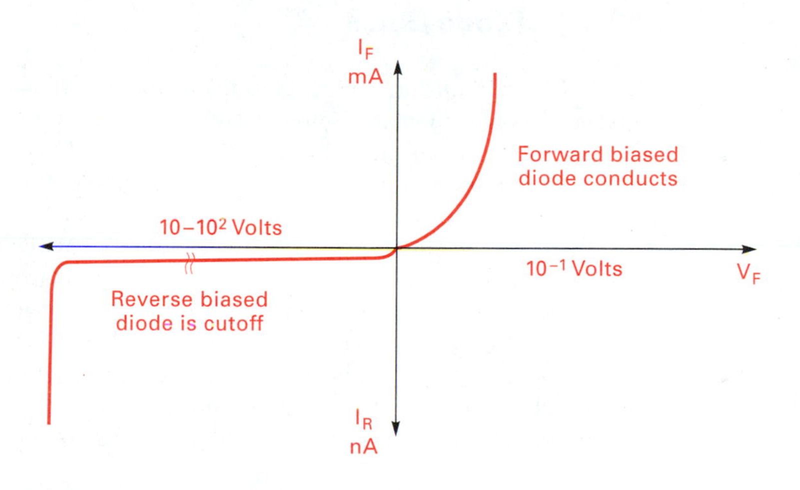

Figure 10.2
Ideal diode model

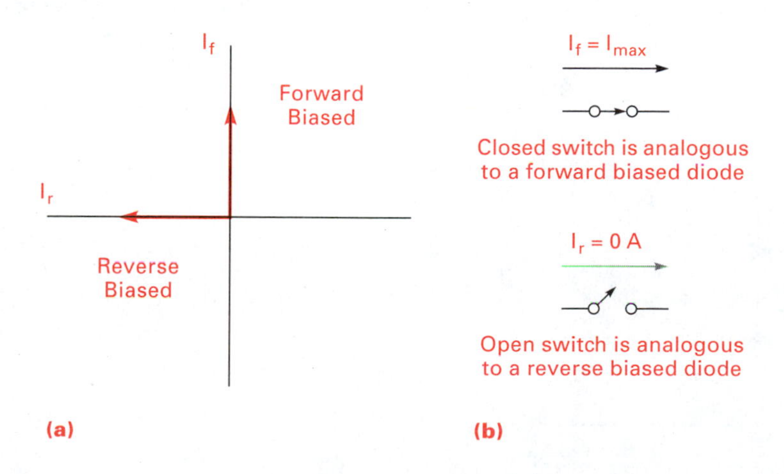

Figure 10.3
Ideal diodes

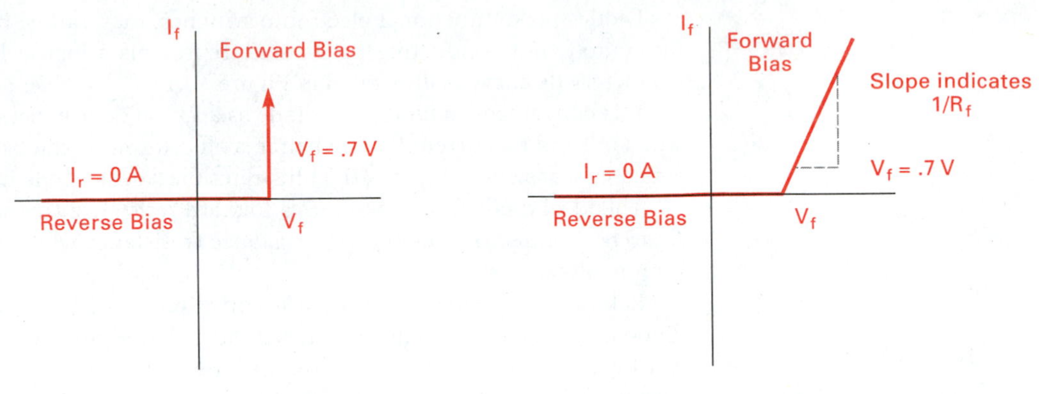

(a) Ideal Diode with Forward Voltage Drop and Zero Forward Resistance

(b) Ideal Diode with Forward Voltage Drop and Non-zero Forward Resistance

10.1.1 Diode Gates

Diode logic circuits are the most fundamental of all logic circuits. A two-input diode AND gate is illustrated in Figure 10.4.

For the circuit illustrated in Figure 10.5, let *A*, *B*, and *C* be logic variables that assume binary values (0, 1). In this example, assign a 1 to +10 V, and a 0 to 0 V or ground. We will model the diodes as ideal switches for our analysis of the circuit operation.

If either of the inputs is 0 V (logical 0), then the switch is closed, providing a current path from $+V_{CC}$, through R and the closed switch, to ground. In the "ideal" model, the closed switch has no "on" resistance, so the output voltage is 0 volts. Both switches must be "open" to generate a logic 1 at the output. That happens when both *A* and *B* are 1. Four combinations of two binary variables are possible. Using the "ideal" switch model, we can illustrate each of the four conditions. These are illustrated in Figure 10.6. Figure 10.6(a) shows the condition where the inputs are both logical 0. S_1 and S_2 are both

Figure 10.4
A two-input diode AND gate

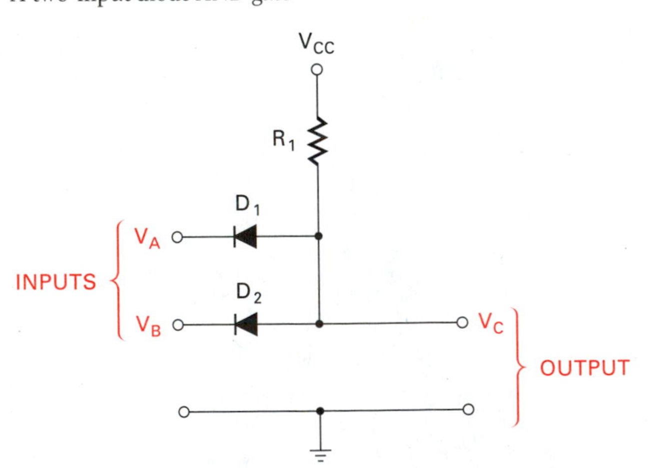

Figure 10.5
Ideal diode model of a two-input AND gate

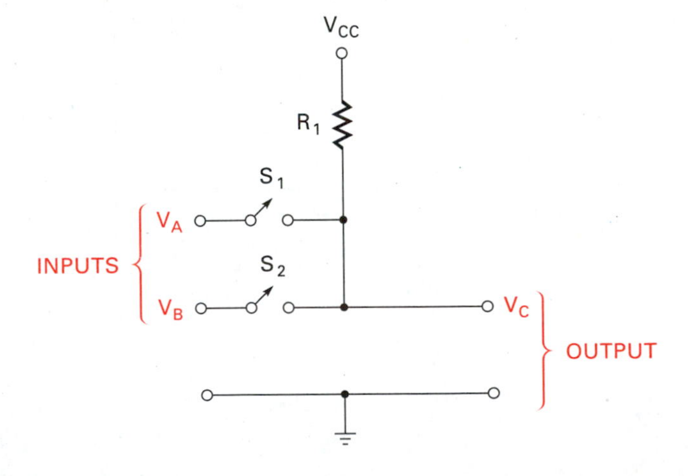

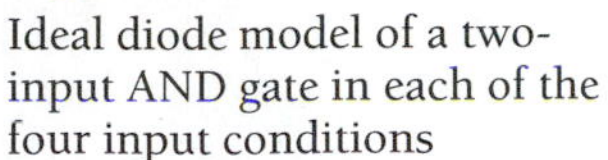

Figure 10.6
Ideal diode model of a two-input AND gate in each of the four input conditions

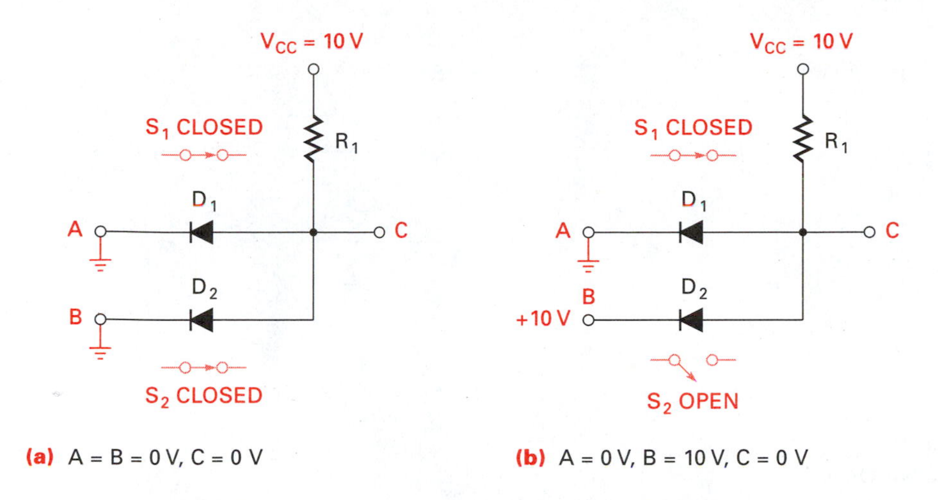

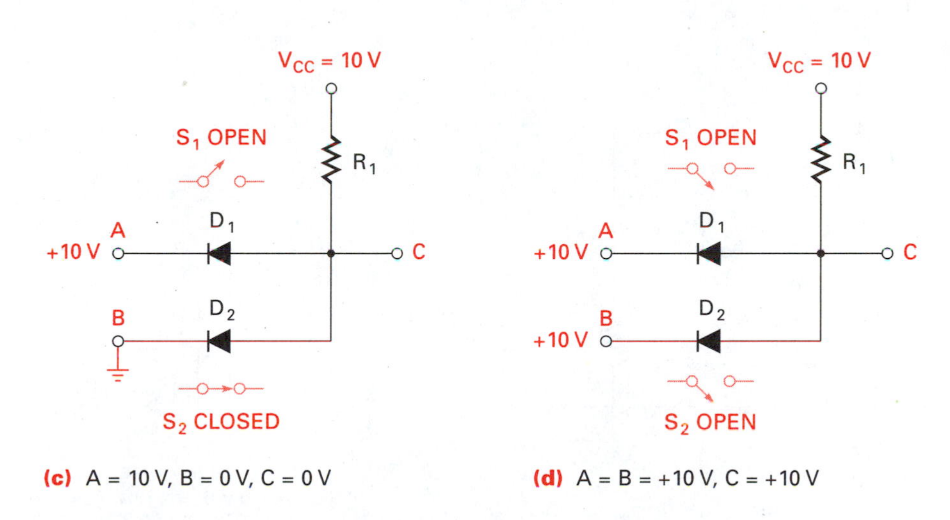

closed, because both diodes are forward biased. This forces the output variable, C, to a logical 0 (ground). Figure 10.6(b) shows the condition where input variables $A = 0$ and $B = 1$, diode D_1 is forward biased, and diode D_2 is reverse biased. The closed switch is parallel to the open switch, clamping the output to ground, forcing the output variable, C, to 0.

In Figure 10.6(c), $A = 1$ and $B = 0$. Switch S_2 is closed, clamping the output to ground, forcing C to be a logical 0. Figure 10.6(d) shows both switches open; therefore, output variable C is free to be pulled up to $+V_{CC}$, resulting in a logical 1 output.

Diodes and a resistor can also be connected to form an OR gate as shown in Figure 10.7. Assume an ideal diode model for D_1 and D_2. Two binary variables have four different combinations of values. Figure 10.8 illustrates each of the four input combinations for the OR gate. In Figure 10.8(a), with both A and B at a logical 0 (ground), no potential exists to produce a voltage drop across R. Therefore, $C = 0$ V, which is a logical 0. In Figure 10.8(b), $A = 0$ and D_1 is off; therefore, no current can flow. $B = 1$ and D_2 is forward biased, allowing current to flow from the +10 V input through D_2 and R, which develops a voltage drop across R, producing a logical 1 at the output.

Figure 10.7
A two-input diode OR gate

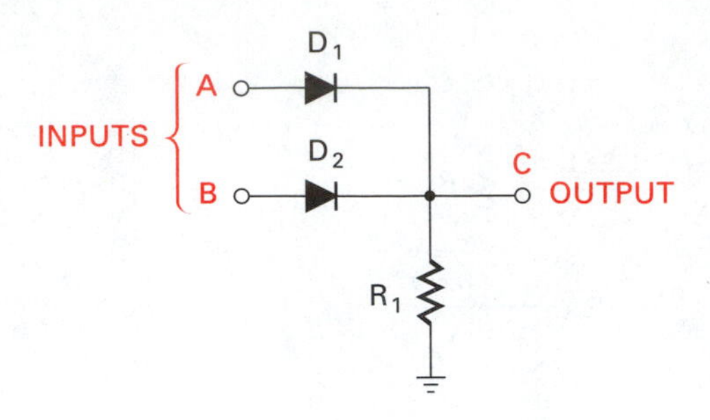

Figure 10.8
Ideal diode OR gate

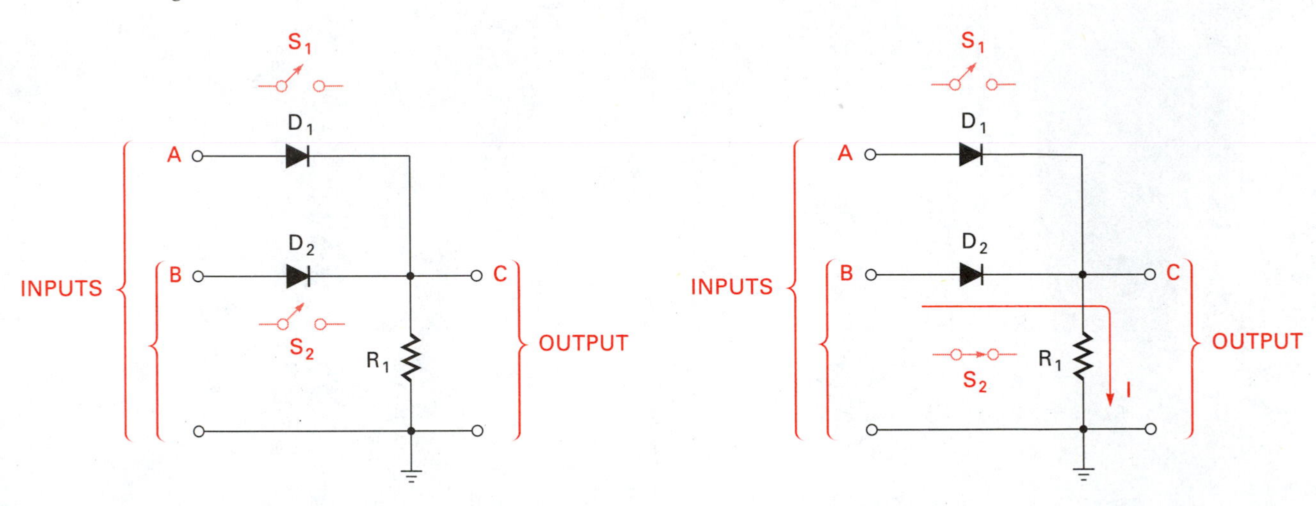

(a) A = B = 0 V, C = 0 V

(b) A = 0 V, B = +10 V, C = +10 V

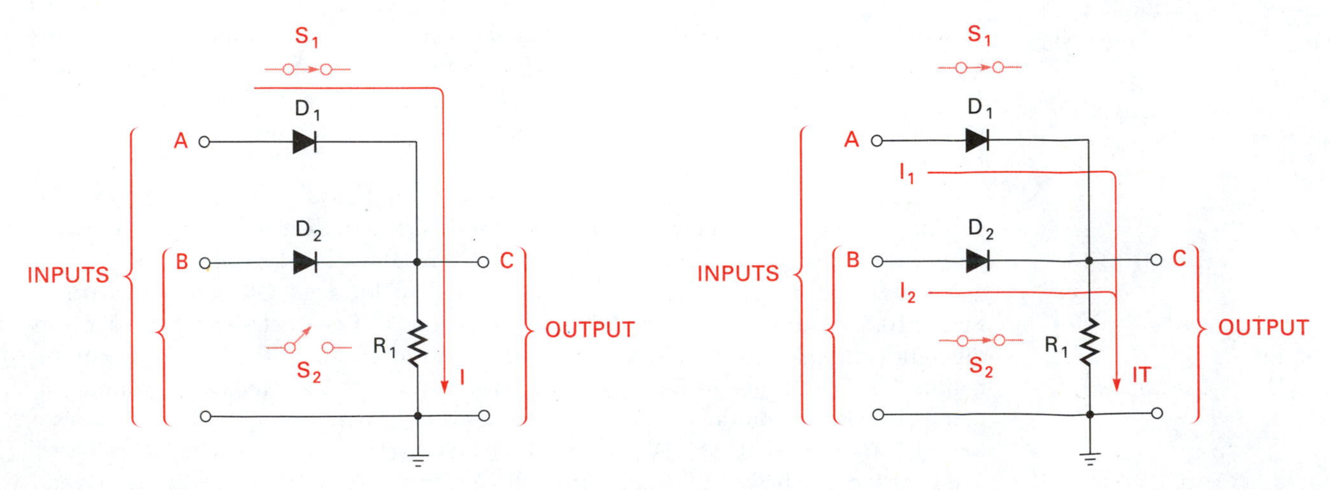

(c) A = +10 V, B = 0 V, C = +10 V

(d) A = +10 V, B = +10 V, C = +10 V

In Figure 10.8(c), $A = 1$ and diode D_1 is forward biased, allowing a current path through R. $B = 0$ and diode D_2 is cut off, so no current path exists through D_2. The two diodes are in parallel, so either one, when conducting, provides a current path to R, and the resulting voltage drop generates a logical 1 output. Finally, in Figure 10.8(d), $A = B = 1$ and both D_1 and D_2 are forward biased, providing current paths through the diodes, converging through R, generating a logical 1 output.

10.2 BIPOLAR TRANSISTOR SWITCH

Inverters (NOT gates) can be easily constructed from a single bipolar transistor operating as a switch. Assume that +5 V is assigned to represent a logical 1 and 0 V is assigned to a logical 0. An inverter circuit performs a logical complement. Because the inputs are restricted to the two voltage levels that represent logical values, the transistor can operate in only two modes, saturation (closed switch) and cutoff (open switch). Consider the circuit shown in Figure 10.9. When the input voltage, $V_{IN} = 0$ V, no base current (input) flows, $I_b = 0$, $V_{be} = 0$ V, and no collector current (output) flows,

$$I_c = 0 \text{ and } V_{CE} = V_{CC} = 5 \text{ V}$$

which is a logical 1. When $v_{IN} = +5$ V, $I_b = (V_{IN} - V_{be})/R_B$, causing the transistor to saturate (closed switch):

$$I_{c(SAT)} = (V_{CC} - V_{CE(SAT)})/R_c$$

Figure 10.10 illustrates the input and output voltage waveforms of the circuit in Figure 10.9.

We can model the switching action for an ideal transistor. That is, when $V_{IN} = 0$ V (logical 0), $I_b = 0$ mA, and $V_{be} = 0$ V, then $I_c = 0$ mA and $V_{CE} = V_{CC}$; therefore, the transistor is cutoff and the output is a logical 1. When $V_{IN} = +5$ V (logical 1), $I_b = I_{b(SAT)}$, and $V_{be} = V_{be(SAT)}$, then $I_c = I_{c(SAT)}$, and $V_{CE} = V_{CE(SAT)}$; therefore, the transistor is in

Figure 10.9
Simple transistor switch

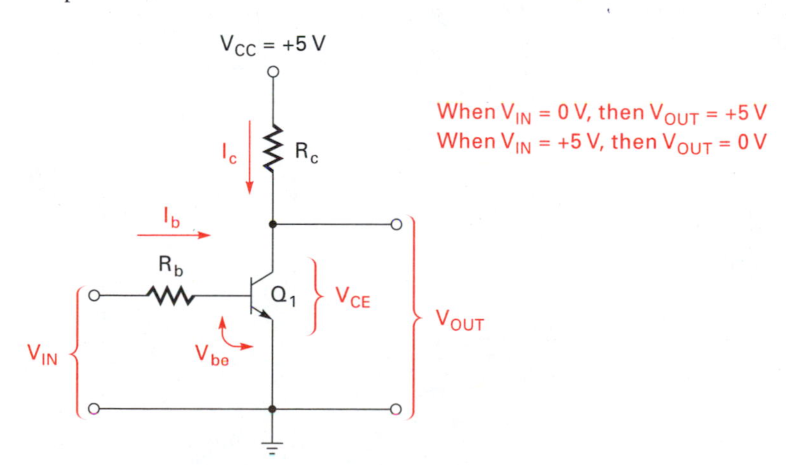

Figure 10.10
Input and output waveforms of the transistor switch

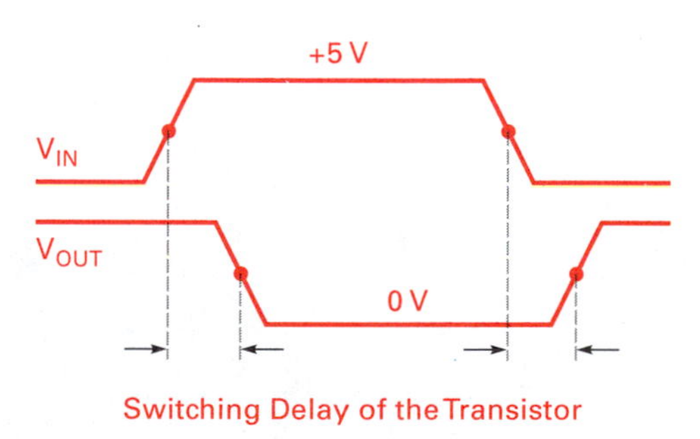

Figure 10.11
Ideal transistor as a switch

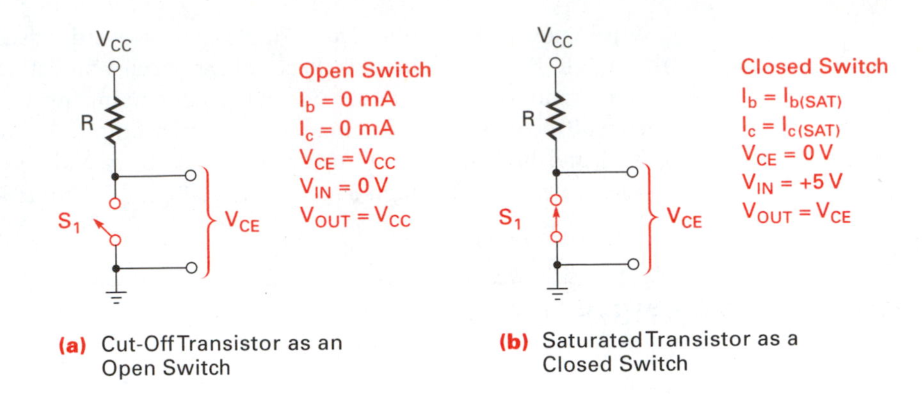

(a) Cut-Off Transistor as an Open Switch

(b) Saturated Transistor as a Closed Switch

saturation and the output is a logical 0. The ideal transistor is a closed switch in saturation and an open switch in cutoff. Whether the switch is open or closed is determined by the logical value of the input. Figure 10.11 illustrates the ideal transistor switch model.

10.3 DIODE TRANSISTOR LOGIC

Diode transistor logic (**DTL**) was the first circuit configuration designed into an integrated circuit (IC). The DTL circuit combined the diode AND gate and the bipolar transistor inverter into a NAND gate. The AND function is performed by the input diodes with a pull-up resistor and the NOT function by the inverting action of the transistor switch. Figure 10.12 illustrates a two-input DTL NAND gate.

Diodes D_1 and D_2 along with R_1 make up the AND function; transistor Q_1 with resistors R_2, R_3, and R_c form an inverter. Together they make up a complete NAND gate.

Figure 10.12
DTL two-input NAND gate, circuit, symbol, and truth table

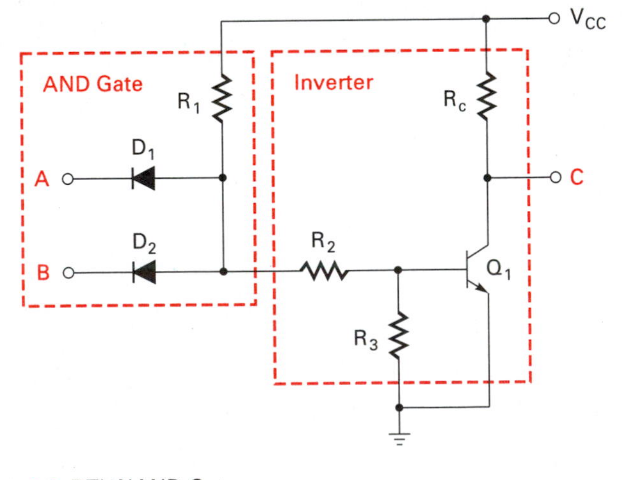

(a) DTL NAND Gate

A 1 — NAND 2 — 3 C
B 2

(b) Logic Symbol

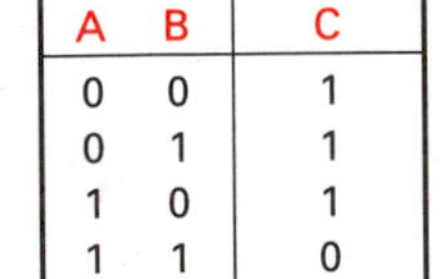

Inputs		Output
A	B	C
0	0	1
0	1	1
1	0	1
1	1	0

(c) Truth Table

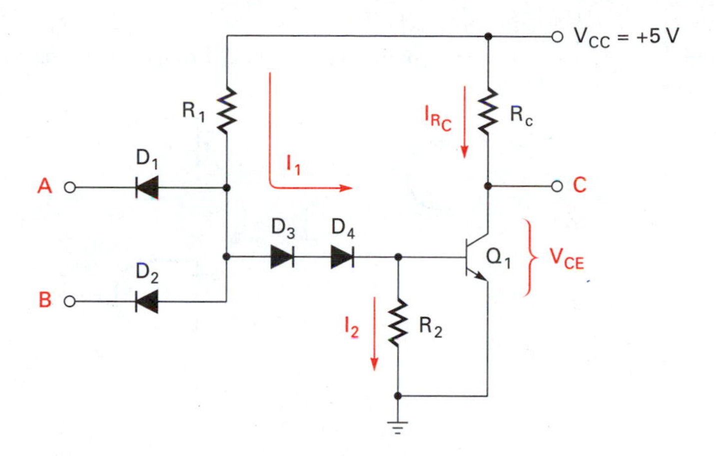

Figure 10.13
DTL two-input NAND gate

When $A = B = 0$, both diodes are forward biased and provide a current path from the +5 V supply through resistor R_1 and the two inputs diodes to ground (logical 0). The forward biased diodes clamp the junction between D_1 and D_2 at a diode drop (.7 V) above ground. The .7 V potential, divided between R_2 and R_3, is insufficient to turn on transistor Q_1. A cutoff transistor is like an open switch, so the output is pulled up to V_{CC}, resulting in a logical 1 output.

Figure 10.13 illustrates the replacement of R_2, Figure 10.12, with two series diodes. The switching characteristics of the diodes (low resistance when on and high resistance when off) permit Q_1 to saturate and cutoff based on the logic inputs.

Consider the case where Q_1 is saturated (both inputs are 1s). Both input diodes, D_1 and D_2, are reverse biased, so we can ignore them (open switch). Diodes D_3 and D_4 are forward biased by the +5 V supply. When forward biased, the diodes present a low resistance to current I_1, which is limited only by **pull-up** resistor R_1. We can see that as I_1 increases the transistor becomes saturated, resulting in a logical 0 output.

The case where transistor Q_1 is cutoff occurs when either or both of the input diodes is forward biased, Figure 10.13, to an approximately .7 V, which is insufficient to turn on D_3, D_4, and Q_1, resulting in a logical 1 output. The cutoff switch diodes (D_3 and D_4) exhibit a very high resistance (10^7 ohms), starving the output transistor of base current.

10.4 EVOLUTION FROM DTL TO TTL

Several problems of a practical circuit nature encountered with the DTL logic IC family ultimately led to the design and development of the transistor–transistor logic (TTL) IC families. One of the main difficulties encountered with DTL circuits occurs because the output stage does not exhibit a low, constant output impedance in both logic states. Lowering the output impedance of the logic gate reduces the amplitude of any signals coupled from one signal wire to another by stray capacitance.

When the DTL output transistor is saturated, the output impedance is low. When the output transistor is cut off, the output impedance is the parallel equivalent of the passive pull-up resistor and the cutoff transistor, which can be approximated as the

Figure 10.14
Output impedance looking into the collector of a DTL NAND gate

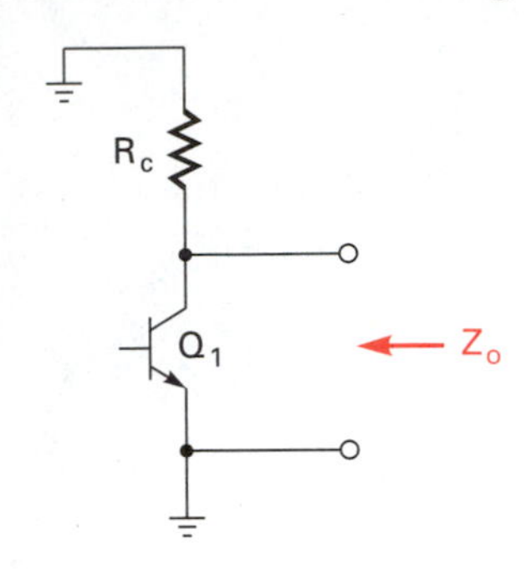

Z_o is the output impedance of Q_1 in parallel with R_c.

Z_o is low when Q_1 is on.
Z_o is equal to R_c when Q_1 is off.

Figure 10.15
Coupling of unwanted signals through stray capacitance

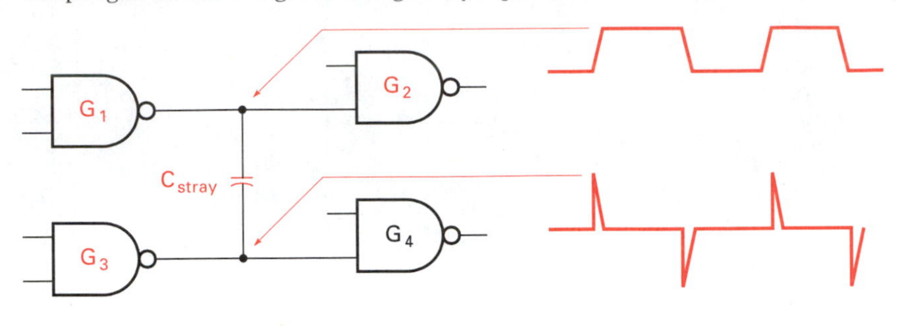

value of the pull-up resistor. In the "on" case, the output impedance was typically less than 100 ohms, and in the "off" condition, the output impedance was typically several K ohms. This variance in impedance, and especially the high logical 1 output impedance led to crosstalk glitches occurring between wires or printed circuit traces adjacent to one another. Figure 10.14 illustrates the output impedance for a DTL gate.

The ac ground at the top of the passive pull-up resistor results from the low impedance of a constant voltage source. Remember that a Thevenin equivalent voltage source's internal resistance is ideally 0. The impedance of a saturated transistor looking in at the collector is very low (both junctions are forward biased). The output impedance for a cutoff transistor is very high (base to collector and base to emitter junctions are reversed biased).

In practice, logic circuits switch states quite rapidly. These high frequency changes can be coupled from one wire or printed circuit trace to another by stray capacitance. The coupling of signals between wires can cause changes in logic levels called **glitches** that may produce errors. Figure 10.15 illustrates the coupling of a "glitch" from one wire to another through stray capacitance.

Assume that the output of G_1 is changing as indicated in Figure 10.15. Assume also that the output of G_3 is a static 1 level. If these gates were DTL circuits, then the output impedance of G_3 (R_{OUT} in Figure 10.16) could be kilo-ohms. The equivalent circuit illustrating the crosstalk due to stray capacitance and the output impedance is shown in Figure 10.16.

The amplitude of the glitches riding on the logic 1 level at the output of G_3 depends on the pulsewidth and rise time of the output of G_1, the value of the stray capacitance, and the output impedance (R_{OUT}) of G_3. The negative spike can go low enough, under the right conditions, that G_4 interprets the level as a 0 instead of the steady state 1, which can produce an error downstream in a digital system. The glitches at the G_4 input are caused by the coupling of the pulses through the stray capacitance into the high output impedance of G_3. The RC network differentiates the legitimate pulses at G_1, producing the spiked waveforms shown in Figure 10.16, as the input of G_4.

A solution to the glitch problem was to design a logic gate with as low an output impedance as possible, because it is impossible to reduce all stray capacitance from any practical logic system. Lowering the output impedance reduces the amplitude of the glitches because the resistance across which a voltage can be developed is lower. If the glitch level changes do not cause a temporary 1 to 0 logic change, then no harm is done.

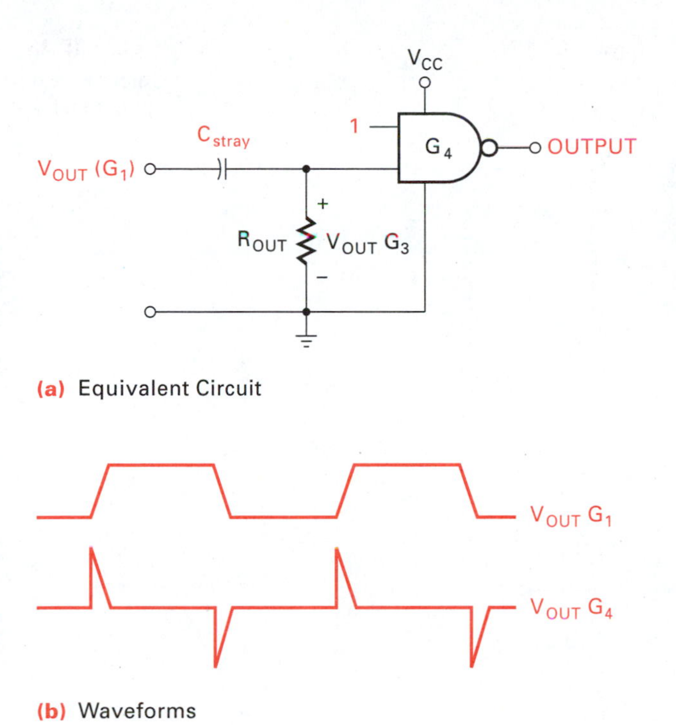

Figure 10.16
Equivalent circuit illustrating crosstalk between logic lines due to stray capacitance and high gate output impedance

The solution to the output impedance problem was met by the creation of a totem-pole output stage, which led to the development of a class of logic devices called *transistor–transistor logic* (TTL).

10.5 TRANSISTOR–TRANSISTOR LOGIC

The **totem-pole** output stage was so named because it resembles the stacked totem pole prominent in Northwest American Indian culture. The constant impedance reduced signal glitches and provided increased ability to drive capacitance loads. The totem-pole transistor configuration is illustrated in Figure 10.17. Q_1 and Q_2 form the output stage. The output is a logic 0 when Q_2 is saturated and Q_1 is cut off. The output impedance is low independent of the logic state because one transistor (either Q_1 or Q_2) is on independent of the logic state.

Because the two transistors of the totem-pole output stage must switch oppositely one from the other, the base input signals of each must be 180 degrees out of phase. This is accomplished by using a phase-splitter circuit. The phase-splitter consists of a single transistor with output signals taken from the emitter and the collector. Figure 10.18 shows the phase splitter driving the totem-pole output stage.

Assume a square wave input to the base of Q_3 in Figure 10.18. When the input voltage is low, Q_3 is cut off, producing a high voltage at the collector and a low voltage at the emitter. The high voltage at the collector of Q_3 is directly coupled to the base of

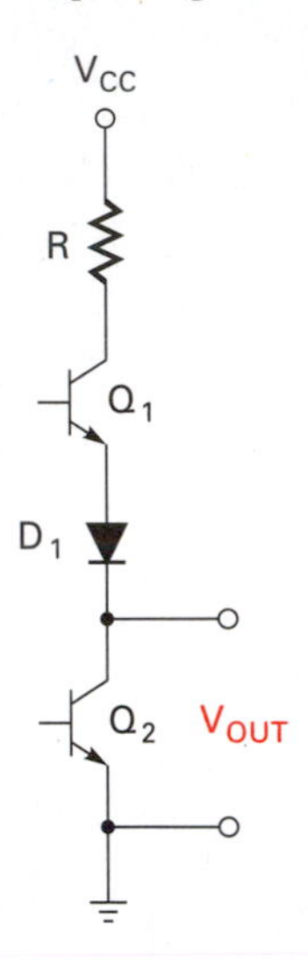

Figure 10.17
TTL totem-pole output stage

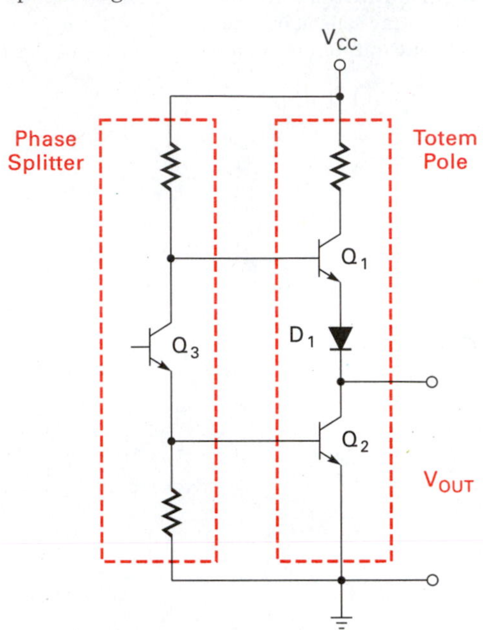

Figure 10.18
Totem-pole output being driven by a phase splitter stage

Q_1, turning on that transistor. The low voltage at the emitter of Q_3 is directly coupled to the base of Q_2, turning off that transistor. The result is a logic 1 at the circuit output. When the input to the phase splitter is a 1, the opposite conditions exist.

The totem-pole output stage was designed to present a constant, low-output impedance to the load. The phase-splitter circuit was necessary to drive the totem pole. If the design development stopped here, the resulting logic circuit would still be classified as DTL, because the input AND circuit would consist of diodes and a pull-up resistor. All the phase splitter and the totem pole did was to solve the output stage problems.

The back-to-back diodes (D_2 and D_1) shown in Figure 10.19(a), can be implemented with an NPN bipolar transistor as shown in Figure 10.19(b). The anode of the gate input diode (D_1) is pulled up to V_{CC} through a resistor. Diode D_2 forms part of the current switch of the DTL circuit shown in Figure 10.13. Figure 10.20 shows multiple input diodes, all connected to a common pull-up resistor. These components form the AND

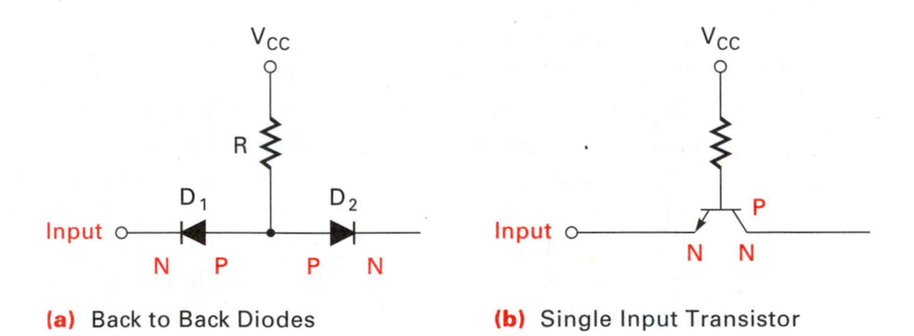

Figure 10.19
Discrete diode to transistor evolution in TTL logic

Figure 10.20
Multiple-emitter transistor replacing discrete input diodes

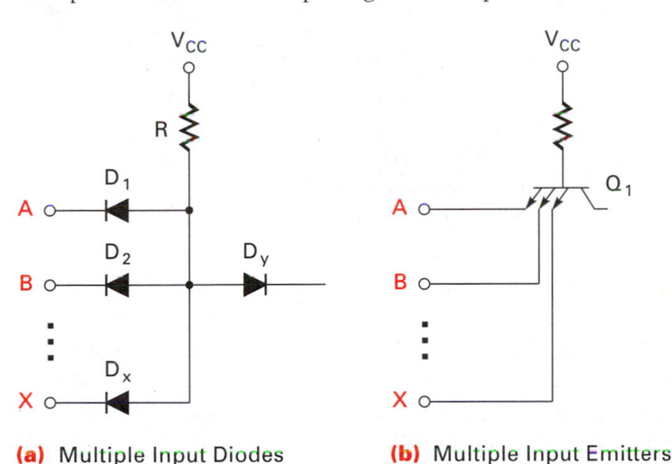

(a) Multiple Input Diodes **(b)** Multiple Input Emitters

Figure 10.21
Two-input TTL NAND gate

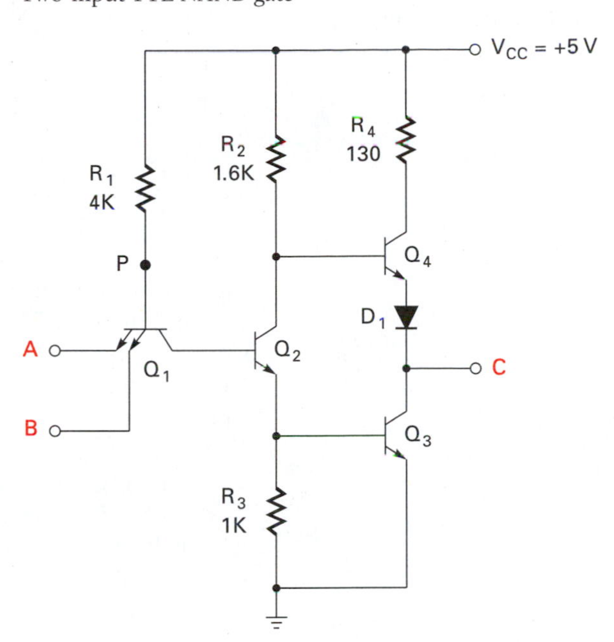

function of the circuit. Diode D_y of Figure 10.20 is the functional equivalent of the base–collector junction of the multiple emitter NPN transistor.

The circuit schematics in Figure 10.19 suggest that a transistor could replace the two diodes. When the early digital integrated circuit designers figured out how to make multiple-emitter transistors, as shown in Figure 10.20(b), the diodes of Figure 10.20(a) were replaced by a single multiple-emitter transistor. Until the development of integrated circuits no significant need existed for multiple-emitter transistors. The lessons learned in designing the DTL integrated circuits led engineers to replace the input diodes with multiple-emitter NPN transistors. The result was an all transistor switching logic circuit called *transistor–transistor logic* or **TTL.**

The connection of the multiple-emitter input transistor to the phase-splitter and the totem-pole output stage completed the TTL NAND gate design. The complete circuit for a two-input TTL NAND gate is shown in Figure 10.21.

10.5.1 TTL Circuit Operation

The circuit operation for various input logic combinations, as shown in Table 10.1, is as follows. Output C = 1, when either one or both emitter–base junctions of Figure 10.21 are forward biased. Under these conditions a current path exists from V_{CC} through R_1 and the forward-biased base–emitter junctions of Q_1 to the 0 input levels. The two emitter–base junctions are in parallel, so either one or both, when forward biased, limit the base of Q_1 to a voltage drop equal to the junction diode potential above the input logic 0 voltage level. The maximum 0 logic level voltage allowed for TTL is .4 V, so the voltage limit at point P, in Figure 10.21, is .4 V + .7 V = 1.1 V. This clamp voltage is

TABLE 10.1
Truth table and transistor junction state for a two-input TTL NAND gate

Inputs		Q_1			Q_2	Q_3	Q_4	
A	B	E_1–B	E_2–B	B–C	B–E	B–E	B–E	Output
0	0	On	On	Off	Off	Off	On	1
0	1	On	Off	Off	Off	Off	On	1
1	0	Off	On	Off	Off	Off	On	1
1	1	Off	Off	On	On	On	Off	0

insufficient to turn on the Q_1 base–collector, the Q_2 base–emitter, and the Q_3 base–emitter junctions. Therefore, all of these transistors are cut off. The cutoff state of transistor Q_2 allows Q_4 to conduct. Resistor R_2 is both the collector pull-up for Q_2 and the base bias resistor for Q_4. When Q_2 is cut off, Q_4 acts as an emitter–follower and conducts. The "on" state of Q_4 and the cutoff state of Q_3 generate a logic 1 at the output terminal (see Table 10.1). When no external load is connected to the logical 1 NAND gate output at terminal C, the emitter current seen by the conducting emitter–follower, Q_4, results from the reverse leakage current of Q_3. When the output is connected to other TTL input terminals, the load current seen by the emitter–follower is the combined input currents, plus the leakage current.

Let inputs A = 1 and B = 1, then output C = 0. Under these conditions both input emitter–base junctions of Q_1 are reverse biased and act as open switches. The base of Q_1, pulled up to a +5 V V_{CC}, is allowed to rise to a potential high enough to forward bias the base–collector junction of Q_1, the base–emitter junction of Q_2, and the base–emitter junction of Q_3. As Q_3 turns on, the output C is driven to a logic 0 voltage level. When Q_2 turns on the base current available to drive Q_4 into conduction is diverted, causing Q_4 to be cut off. Figure 10.22 shows the equivalent circuit of the TTL NAND gate when the output is a 0. Notice that each conducting transistor junction is modeled as a diode.

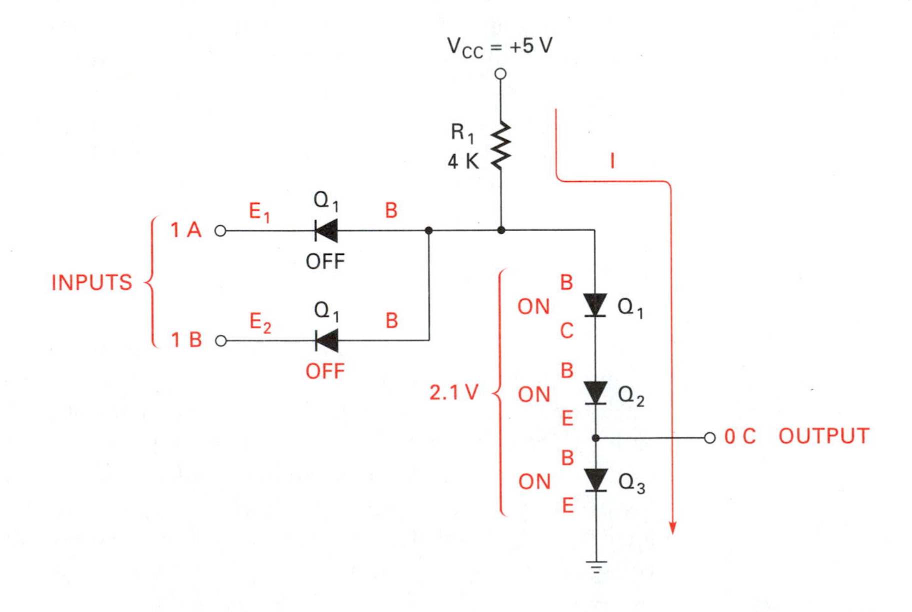

Figure 10.22
Model of the TTL NAND gate when A = B = 1 and C = 0

Figure 10.23
Transistor–transistor logic voltage transfer curve

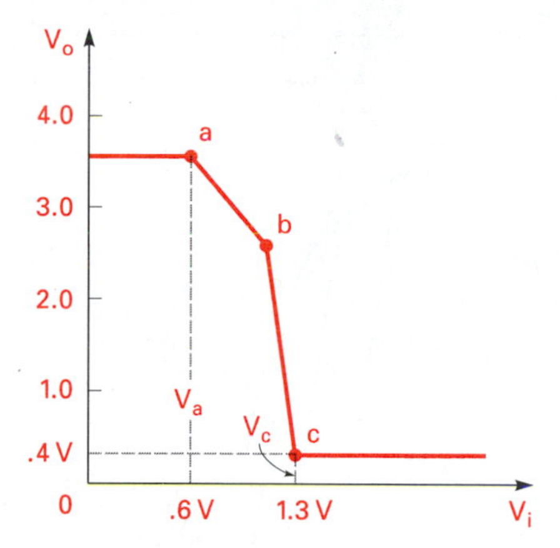

The voltage transfer curve for a standard TTL gate is illustrated in Figure 10.23. The curve depicts the input-output relationship. As the input voltage level increases from 0 V toward +5 V, a logical 0 to a logical 1, the output changes from a logical 1 to a logical 0.

Let inputs A and B be connected together to form a single input, V_i. Then, as V_i increases from 0 V toward +5 V, V_o tracks V_i as indicated by the transfer curve. As long as V_i remains below the maximum voltage assigned to a logical 0, V_o will remain above the voltage assigned to a logical 1. When V_i reaches a threshold voltage, V_t, the base–emitter junction of Q_1 will begin to cut off and the base–collector junction will start to conduct. The base–emitter junctions of Q_1 conduct as long as V_i remains below the threshold voltage, V_t.

From points *a* to *b* on the transfer curve the base–collector junction of Q_1, the base–emitter junction of Q_2, and the base–emitter junction of Q_3 are in the active region; that is, they are conducting but the transistor is not yet at saturation. The voltage gain of transistor Q_2 is determined by the R_2/R_3 ratio during this time. As the value of V_i continues to increase past point *b*, the transfer curve becomes even steeper. This is due to the increased gain of transistor Q_2, resulting from a decrease in emitter impedance. The impedance seen by the emitter of Q_2 consists of resistor R_3, in parallel with the base–emitter junction of Q_3. When Q_3 is cut off, its base–emitter junction appears as a high impedance, so the effective resistance seen by the emitter of Q_2 is R_3.

As transistor Q_3 begins to conduct, the base–emitter junction impedance of Q_3 decreases, further decreasing the emitter impedance seen by transistor Q_2, thereby increasing the Q_2 stage gain and causing the faster dropoff of the voltage transfer curve from point *b* to *c*. At point *c* the curve flattens to a level of the $V_{CE(SAT)}$ of transistor Q_3.

10.5.2 TTL Specifications

The definitions of TTL circuit parameters were given in Section 4.2 of Chapter 4. This section presents additional TTL parameter information. The TTL families are guaranteed to provide the logical 0 and 1 output voltages and currents specified, over the temperature and power supply voltage operating ranges. For example, the voltage transfer curve of Figure 10.23, when subjected to temperature changes, becomes a family of curves shown in Figure 10.24.

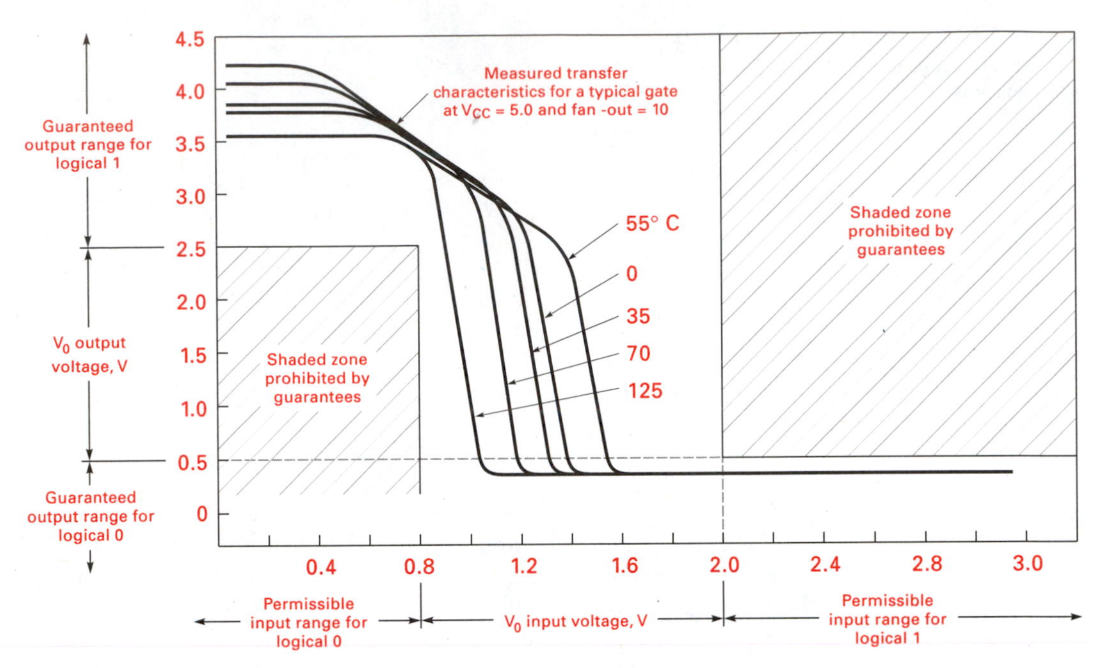

Figure 10.24
TTL voltage transfer curves as a function of temperature (reprinted by permission of Texas Instruments and McGraw-Hill)

A voltage transfer curve exists for each temperature. Notice that the established boundaries for the logical values are never crossed by any of the transfer curves. Also notice that each transfer curve occurs with the output under a full fan-out load. The specified voltages for logical 0 and logical 1 values are defined for a fan-out of 10. This means that a gate output can drive 10 identical gate inputs.

Other circuit parameters, input and output currents, and voltages must also remain intact as temperature changes. Table 10.2 illustrates the voltage levels at different temperatures for both the commercial (74xx) and the military grade (54xx) devices.

From the data in Table 10.2, we can see that the input and output voltage levels defining logical values are not violated for either IC grade over their respective operating temperature ranges.

Table 10.3 shows the output and threshold voltage values as a function of temperature and as a function of power supply voltages.

Each of the IC parameters must remain within the allowable tolerances under the **worst-case** (most extreme) conditions to which the device will be subjected. It would do no good to specify an IC to operate at 70° C, for example, if the logical 0 output voltage were to increase above .4 V or if the fan-out had to be reduced from 10 to 5.

Table 10.2
Temperature and power supply ranges for TTL circuits (reprinted by permission of McGraw-Hill and Texas Instruments)

	Ambient temperature range	Supply voltage range	Fan-out
SN5400	$-55°C \leq T_A \leq 125°C$	$4.5\ V \leq V_{CC} \leq 5.5\ V$	$N = 10$
SN7400	$0°C \leq T_A \leq 70°C$	$4.75\ V \leq V_{CC} \leq 5.25\ V$	
Guaranteed output parameters . . .		$V_{OH} \geq 2.4$ V	$V_{OL} \leq 0.4$ V
Guaranteed input parameters . . .		$V_{IH} \geq 2.4$ V	$V_{IL} \leq 0.8$ V

Table 10.3

TTL parameters as a function of temperature and power supply voltages (reprinted by permission of McGraw-Hill and Texas Instruments)

As function of ambient temperature $T_A \cdot V_{CC} = 5$ V, $N = 10$					
	−55°C	0°C	25°C	70°C	125°C
V_{OH}	3.0	3.1	3.25	3.3	3.5
V_{OL}	0.25	0.29	0.30	0.31	0.32
V_T	1.5	1.4	1.3	1.2	1.0
As function of supply voltage $V_{CC} \cdot T_A = 25°C, N = 10$					
	4.5 V	4.75 V	5.0 V	5.25 V	5.5 V
V_{OH}	2.6	2.85	3.25	3.35	3.55
V_{OL}	0.33	0.32	0.30	0.30	0.30
V_T	1.28	1.29	1.3	1.32	1.35

Each device parameter value must be measured under defined conditions. The following diagrams define the operating conditions for a standard TTL NAND gate and show a partial circuit diagram to illustrate operation.

Figure 10.25 shows the worst-case conditions for input current, I_{IL}. I_{IL} is measured when V_{OL} is maximum (V_{OL} = .4 V); and V_{CC} is maximum for the IC grade (74XX = 5.25 V, 54XX = 5.5 V). The calculated value for input low current is found as

$$I_{IL} = (V_{CC} - V_{be} - V_{IL})/R_1$$

(refer to Figure 10.21).

Figure 10.26 demonstrates the worst-case conditions for measuring the input high current, I_{IH}. These conditions are V_{IH} = 2.4 V (minimum), V_{CC} is maximum for its respective IC grade, and the unused inputs are grounded. The input high current is the leakage current of the input junction.

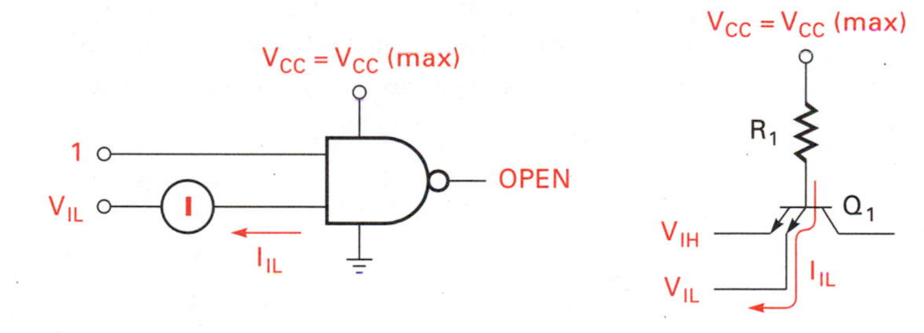

Figure 10.25

I_{IL} worst-case conditions and measurement

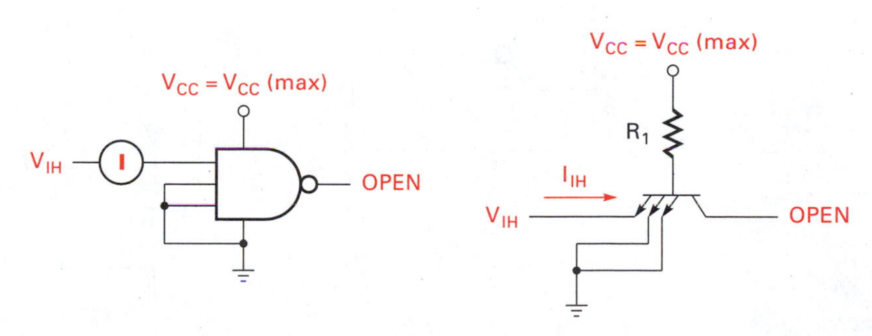

Figure 10.26

Worst-case conditions for measuring I_{IH}

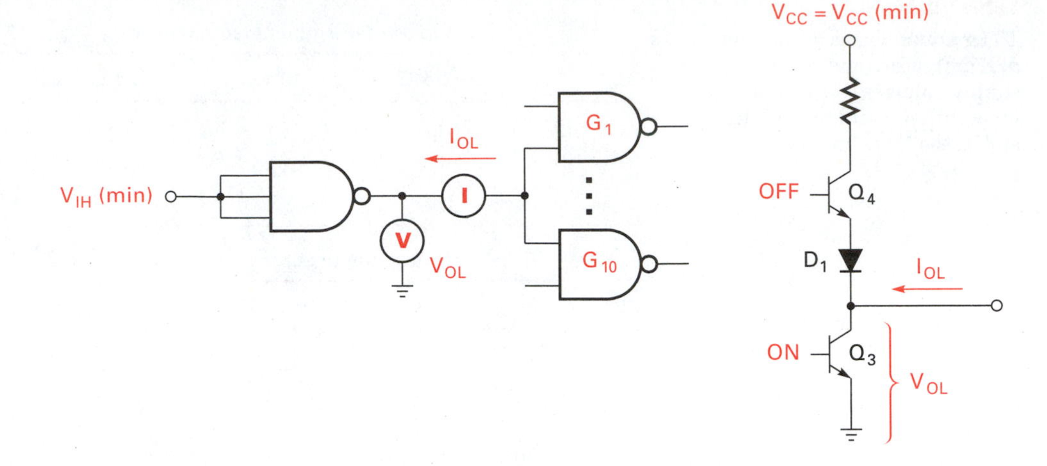

Figure 10.27
Worst-case conditions for measuring I_{OL}

Figure 10.27 illustrates the worst-case conditions for measuring I_{OL}. The output voltage V_{OL} is the $V_{CE(SAT)}$ of transistor Q_3. To ensure that the transistor has sufficient base current to keep it in saturation under full load conditions, the maximum sink current, I_{OL} must also be specified. To establish the conditions under which the transistor is at saturation, full load current I_{OL}, and the low-output voltage (V_{OL}) is specified.

The short-circuit output current I_{OS}, is measured when both inputs are at a logical 0, the power supply voltage V_{CC} is maximum, and the output terminal is shorted to ground. The resulting source output current is specified as I_{OS}. The logic symbol and partial circuit schematic is illustrated in Figure 10.28.

Additional TTL device specifications exist for switching speed. The definitions for delay time was given in Chapter 4, Section 4.2. The test conditions for making propagation delay measurements are described in Figure 10.29.

The diodes in Figure 10.29 are switching diodes that model the switching characteristics of the transistors internal to the TTL gate. The load capacitor models the equivalent capacitance that the output of the gate would see when driving a fan-out of 10 identical gates. The pull-up resistor is given a value of the $R_1/10$, representing a fan-out of 10 for a specific TTL subfamily. Noise margin is another integrated circuit parameter

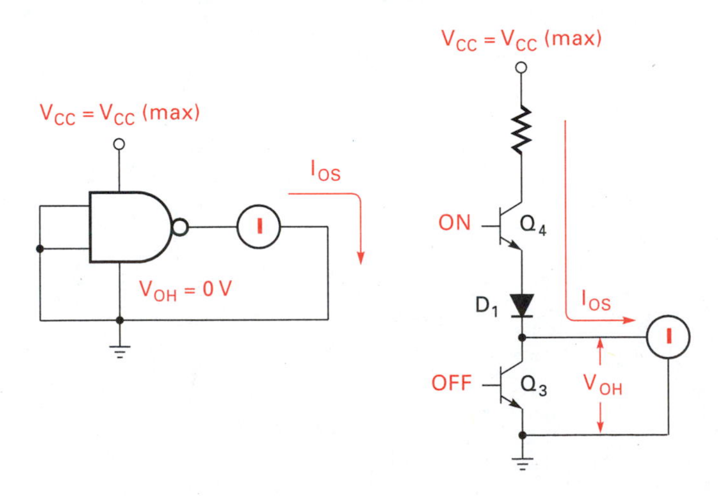

Figure 10.28
Circuit test conditions for measuring I_{OS}

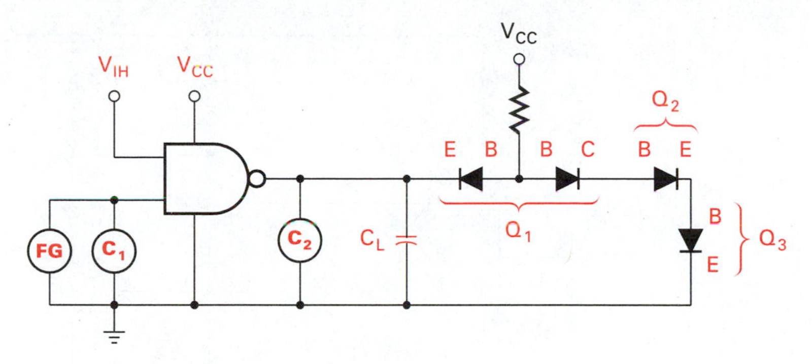

C_1 Channel 1 of Oscilloscope
C_2 Channel 2 of Oscilloscope
FG Function Generator
C_L Stray Capacitance

Figure 10.29
Test setup for measuring propagation delay

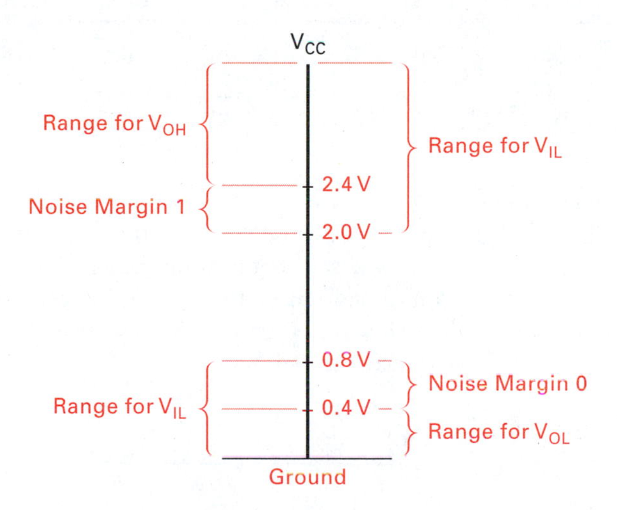

Figure 10.30
Logic level ranges and dc noise margins For TTL

of interest. Noise margin is the difference between the specified output of one logic gate and the specified input of another gate for the same logic level. It is the difference between V_{OL} and V_{IL} for a logical 0 and between V_{OH} and V_{IH} for a logical 1. Figure 10.30 illustrates the logical level ranges for 1s and 0s as well as the dc noise margins for all TTL families.

10.5.3 TTL Subfamilies

Seven TTL sub-families presently are used:

Standard TTL This was the first commercially available TTL subfamily. See the circuit schematic in Figure 10.31. The resistor values for the standard TTL IC are listed in Table 10.4.

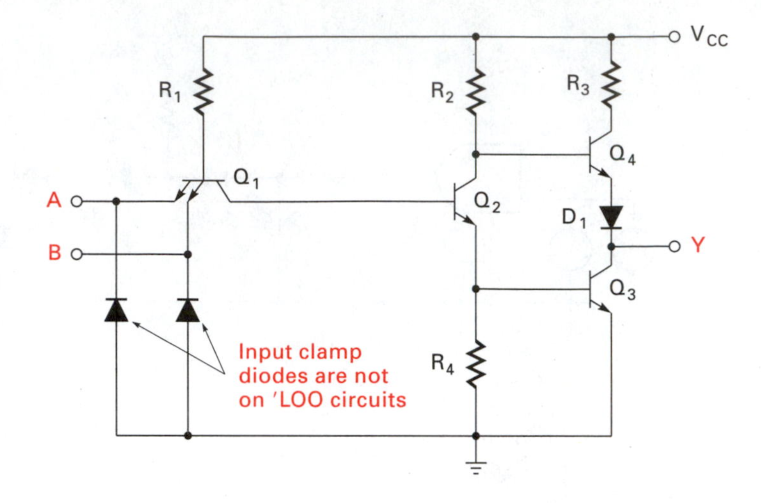

Figure 10.31
Standard and low-power TTL IC schematic

TABLE 10.4
Resistor values for standard and low-power TTL ICs

Circuit	R_1	R_2	R_3	R_4
'00	4K	1.6K	130	1K
'L00	40K	16K	500	12K

Low-power TTL Low-power TTL was one of the early TTL subfamilies. It is no longer widely used, as low-power Schottky has taken over in new designs. It was used to conserve power in applications where speed was not a requirement. The power reduction came from increasing the values of the internal resistors. The circuit schematic is shown in Figure 10.31. The resistor values for the low-power TTL device are listed in Table 10.4.

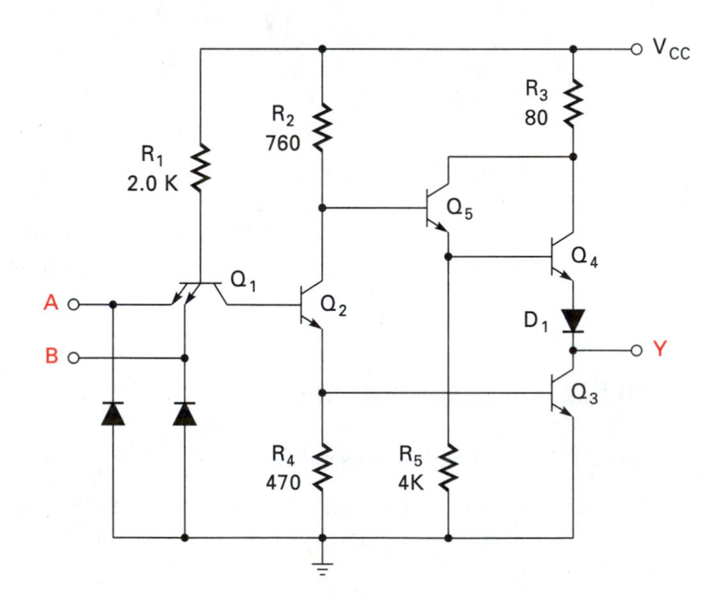

Figure 10.32
High-speed TTL circuit schematic

High-speed TTL The high-speed subfamily was also one of the early TTL ICs. It, like the low-power subfamily, is no longer used in new designs. The high-speed subfamily was originally used in applications requiring higher speeds than the standard TTL subfamily. The higher speed was achieved at the expense of increased power consumption. The speed was gained by decreasing the internal resistance values and by adding a Darlington stage as the top portion of the totem-pole output stage. Figure 10.32 shows the circuit schematic for the high speed (H) TTL device. Compare the resistor values of the high-speed TTL IC with those of the standard and low-power devices.

Low-power Schottky A gain in speed, without giving up low power consumption, was the objective of the low-power Schottky IC designers. Figure 10.33 shows the circuit schematic of the low-power Schottky device. The use of Schottky transistors enabled the designers to greatly reduce stored charge and therefore achieve much higher switching speeds than conventional bipolar transistors allowed. The internal resistor values were kept reasonably high so as not to increase the power consumption. Additional modifications were also made to the basic circuit configuration to improve speed. A dynamic emitter-impedance stage was added to the phase-splitter circuit to improve the active gain during switching. Also a diode resistor network was added to the Darlington active pull-up of the totem-pole output stage. This network provides current switching to remove the stored charge of the non-Schottky transistor of the totem pole.

Schottky The main objective in the design of the first Schottky TTL ICs was switching speed. Figure 10.34 shows the circuit schematic for the Schottky (S) subfamily devices. The Schottky subfamily uses Schottky transistors to reduce junction stored charge and improve switching speed. In this case, however, the resistor values

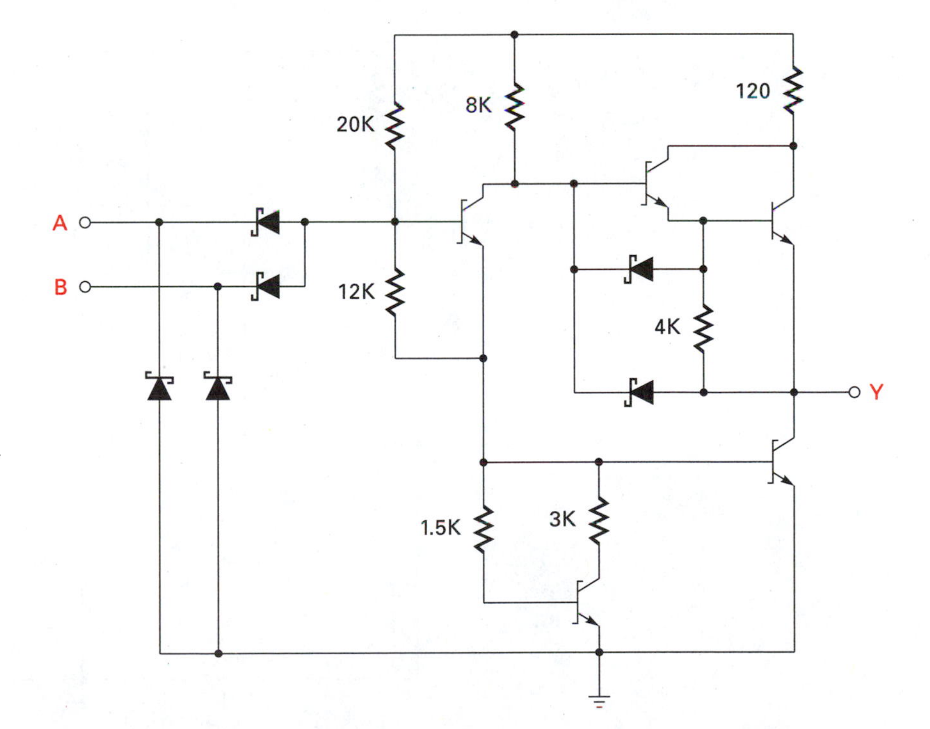

Figure 10.33
Low-power Schottky TTL IC circuit schematic

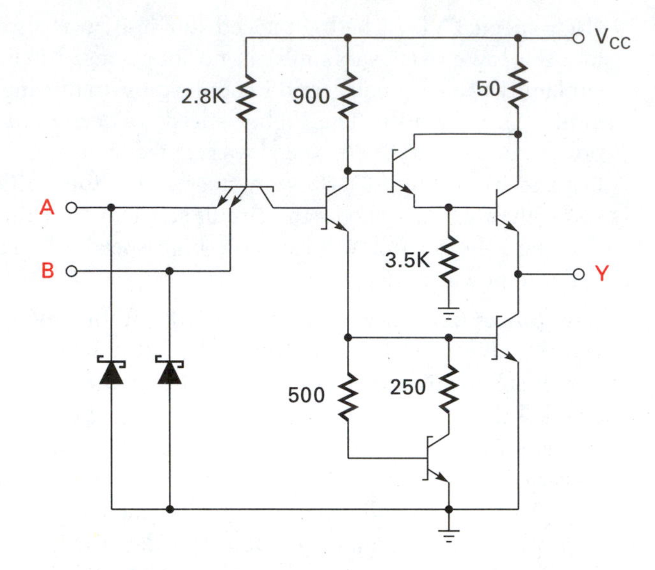

Figure 10.34
Schottky TTL IC circuit schematic

were also reduced, thereby increasing speed at the expense of power consumption when compared to the low-power Schottky devices. Additional circuit modifications were also used. A high-speed current switching Darlington active pull-up in the totem pole, as well as the dynamic emitter impedance seen by the phase-splitter circuit were used.

Advanced low-power Schottky (ALS) Advanced Schottky devices, whose schematic is shown in Figure 10.35, required improved manufacturing processes to de-

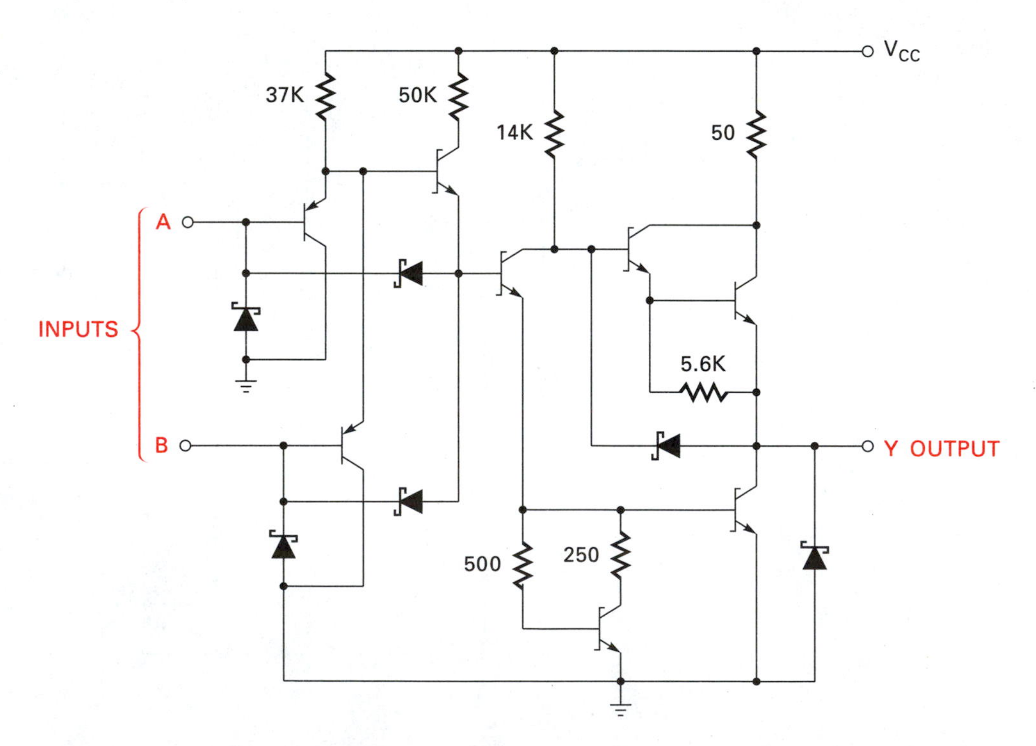

Figure 10.35
Advanced low-power Schottky IC schematic

velop the newest TTL subfamilies. Without going into the details of the processes, the results were smaller device geometries that reduced the interelement capacitances thereby increasing switching speeds. Schottky transistors were used in conjunction with input and output Schottky clamping diodes, which reduced waveform excursions and therefore improved the switching slew rate (*dv/dt*) and reduced propagation delays.

Advanced Schottky Advanced Schottky is the fastest of the TTL families. The speed enhancing techniques used in the design of ALS were improved, permitting even better level clamping, stored charge reduction, and active gain circuits. These are shown in the circuit schematic in Figure 10.36. The circuit is fairly complicated and after a bit of study it is still recognizable as a TTL subfamily. The advanced Schottky multiemitter inputs are protected by Schottky diodes positioned between the input pins and ground. This is to shunt any possible negative input voltage excursions away from the input transistor to ground. A Darlington pair replaces the top totem-pole transistor and an active-emitter circuit is used to drive the bottom totem-pole transistor. Both circuits are used to improve switching speed.

Figure 10.36
Advanced Schottky TTL IC schematic

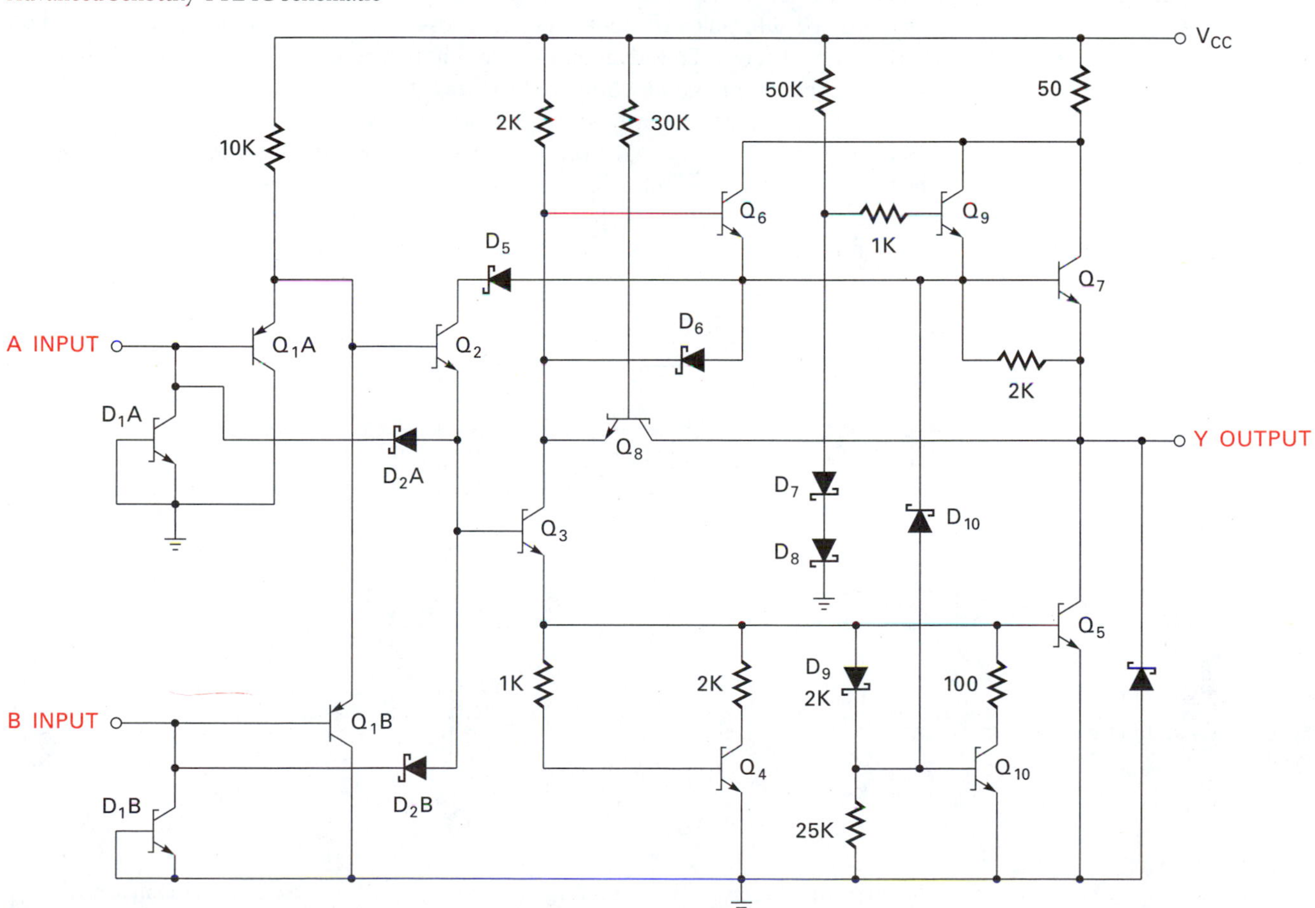

10.5.4 Schottky Junctions

Transistor–transistor logic includes four subfamilies that use **Schottky** diodes and transistors. A Schottky transistor differs from a bipolar transistor in that the Schottky junction consists of a doped semiconductor and metal instead of two differently doped semiconductors.

When the Schottky junction is formed during manufacture, the free electrons from the metal and the semiconductor cross over the physical junction to form a depletion region much like that in a normal semiconductor junction. The electron flow builds a depletion region and thus a barrier potential across the junction. When the barrier potential is high enough to oppose any further electron migration, the electron flow stops. The metal side of the junction forms the anode and the semiconductor side becomes the cathode of the junction. When the junction is forward biased, by applying a positive potential on the metal and a negative potential on the semiconductor, electrons excited by thermal energy cross over the depletion region and become current carriers in the metal. Figure 10.37 illustrates the metal–semiconductor junction of a Schottky barrier junction.

Schottky diodes are sometimes called *hot-carrier* diodes due to the thermal energy production of current carriers from the *n*-doped semiconductor crossing the barrier to the metal side of the junction. When the junction is reverse biased, electrons on the semiconductor side lack sufficient energy to cross the widened depletion region and no majority current flows. The advantage of the Schottky junction is a significantly reduced junction stored charge when compared to a bipolar junction. The reduction in stored charge accounts for significantly decreased switching speeds.

Schottky transistors are created by providing a Schottky clamp diode across the collector-to-base junction of a bipolar transistor. This diode serves to reduce stored charge by clamping the base–collector junction to prevent it going into saturation. The symbols for the Schottky diode and transistor are shown in Figure 10.38.

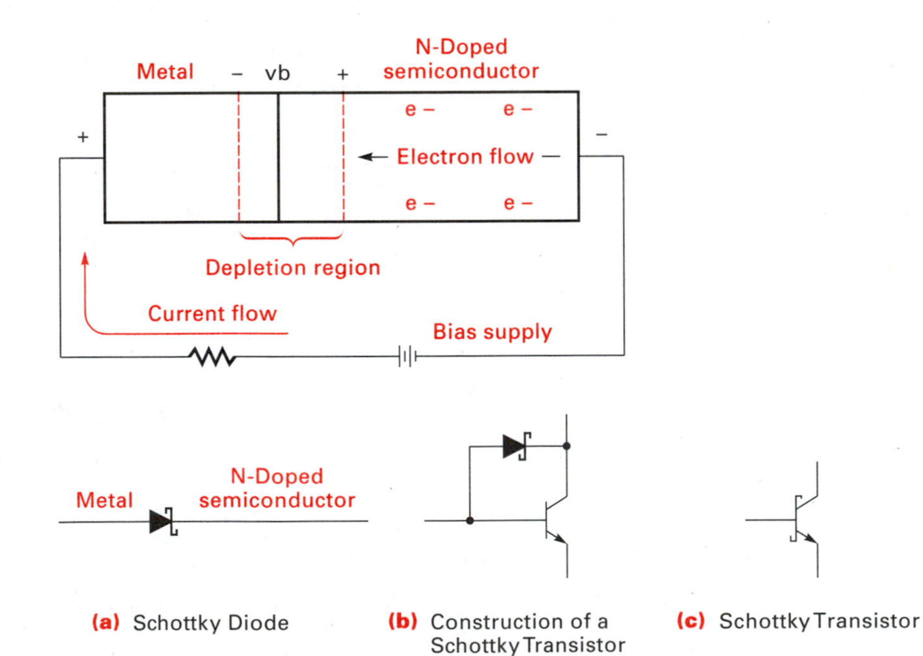

Figure 10.37
Schottky barrier junction

Figure 10.38
Schottky diode and transistor symbols

10.5.5 Comparison of TTL Subfamily Specifications

Each TTL subfamily must conform to the logic voltage level definitions. However, considerable difference exists between subfamilies regarding currents and propagation delays. Table 10.5 lists some of the values for the functional specifications. The number of TTL families and the devices offered within subfamilies continue to grow. A complete set of device specifications is given in data manuals by the companies that make the devices.

Table 10.6 summarizes the TTL subfamilies according to propagation delay and power consumption (ICC average × V_{CC}). A figure of merit associated with digital integrated circuits is called the *speed–power product*. It is the product of the propagation

TABLE 10.5
TTL subfamily specification comparison

Specification		TTL subfamily 7400	74LS00	74S00	74ALS00	74AS00
V_{CC}	4.75 V–5.0 V	→	→	→	→	→
$I_{CC}H$	(max)	8 mA	1.6 mA	16 mA	.85 mA	3.2 mA
$I_{CC}L$	(max)	22 mA	4.4 mA	36 mA	3 mA	17.4 mA
IOS	(max)	55 mA	100 mA	100 mA	70 mA	112 mA
IOL	(max)	16 mA	8 mA	20 mA	8 mA	20 mA
IOH	(max)	.4 mA	.4 mA	1 mA	.4 mA	2 mA
IIH	(max)	50 μA	20 μA	50 μA	20 μA	20 μA
IIL	(max)	1.6 mA	.4 mA	2 mA	.1 mA	.5 mA
VOL	(max)	.4 V	.5 V	.5 V	.5 V	.5 V
VOH	(min)	2.4 V	2.4 V	2.7 V	V_{CC}−2 V	V_{CC}−2 V
VIH	(min)	2.0 V	2.0 V	2.0 V	2.0 V	2.0 V
VIL	(max)	.8 V	.8 V	.8 V	.8 V	.8 V
tpLH	(max)	22 ns	15 ns	4.5 ns	11 ns	4.5 ns
tpHL	(max)	15 ns	15 ns	5 ns	8 ns	4 ns

TABLE 10.6
TTL subfamily speed–power comparison

	Minimizing power					Minimizing delay time				
Circuit technology	Family	Prop. Delay (ns)	PWR Diss (mW)	SPD/PWR product (pJ)	Maximum flip-flop freq. (MHz)	Family	Prop. Delay (ns)	PWR diss (mW)	SPD/PWR product (pJ)	Maximum flip-flop freq. (MHz)
Gold doped	TTL	10	10	100	35	TTL	10	10	100	36
	L TTL	33	1	33	3	H TTL	6	22	132	60
Schottky clamped	LS TTL	9	2	18	45	S TTL	3	19	67	126
	ALS	4	1.2	4.8	70	AS	1.7	8	13.6	200

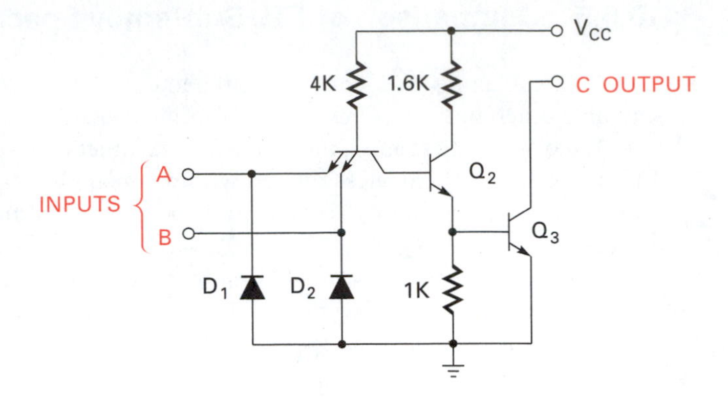

Figure 10.39
Standard TTL open-collector circuit for a two-input NAND gate

delay and the power dissipated by a logic circuit. The lower is the speed–power product, the more efficient the logic circuit. The designer of digital systems is always looking to gain more speed without having to pay a price in increased power dissipation.

10.5.6 Open Collector TTL Circuits

Each of the TTL subfamilies provides integrated circuits with standard totem-pole outputs. In addition, selected circuits within each family have an output stage that requires an external pull-up resistor. These devices are called *open collectors,* because the output drive transistor's collector is left open. It is up to the user to connect the output pin of the gate to an external load that is pulled up to an external power source. Figure 10.39 shows a circuit for the open-collector version of the standard TTL subfamily.

The purpose of the open collector is to provide logic level shifting, drivers to relays or displays, or wired OR connections. We consider the logic level shifting first. Assume a situation where conversion from TTL logic levels to some other logic levels is necessary. The input to the open-collector gate would be the normal TTL voltage levels. The output logic levels must conform to requirements other than TTL. Then the open-collector pull-up resistor and the external supply can be specified to produce the desired new logic level values.

EXAMPLE 10.1 Design a level-shifting circuit that takes standard TTL inputs and drives a new circuit where the logic levels must be a +5 V minimum for a 1 and .6 V or less for a 0.

SOLUTION Choose a TTL open collector circuit that performs the desired function (NAND, NOR, etc.). Choose an external power supply of +6 V to correspond with the new logic 1 level. Determine the load requirements for the new logic levels in terms of logic level currents: I_{OL}, I_{OH}, I_{IL}, and I_{IH}. Assume that the load, for this example, requires the following:

I_{IL} = 1 mA (from load to be sunk into the output transistor of the open-collector gate)

I_{IH} = 100 μA (leakage current from the load to be sourced by the open-collector gate)

Fan-out = 5

V_{IL} = .6 V or less, V_{IH} = 5 V, V_{OL} = .4 V, V_{OH} = 5 V

Use a 7403, quad two-input open-collector NAND IC. With a fan-out of five required, the sinking and sourcing currents for the open-collector output transistor are

I_{OL} = 16 mA (from the 7403 data sheet)

I_{OH} = 250 μA (from the 7403 data sheet)

Calculation of the range of pull-up resistors that will accommodate the required logic and current levels is done next. Two conditions exist: one where the output is a logical 0 and a second where the logical output is a 1.

In the case where the output is at a logical 1,

$R_L(\text{max}) = (V_{CC} - V_{OH})/(nI_{OH} + mI_{IH})$, where n = number of output gates connected to a common pull-up resistor

$R_L(\text{max}) = (6\text{ V} - 5\text{ V})/[1(250\ \mu\text{A}) + 5(100\ \mu\text{A})]$ pull-up

$R_L(\text{max}) = 1\text{ V}/750\ \mu\text{A} = 1.33\text{K ohms}$, m = fan-out, number of inputs

In the case where the output is at a logical 0,

$R_L(\text{min}) = (V_{CC} - V_{OL})/(I_{OL} - mI_{IL})$, where V_{OL} is required by load logic.

$R_L(\text{min}) = (6\text{ V} - .6\text{ V})/[16\text{ mA} - 5(1\text{ mA})]$

$R_L(\text{min}) = 5.4\text{ V}/11\text{ mA} = 491\text{ ohms}$

Any pull-up between the maximum and minimum values for R_L will accommodate the requirements of the load logic and the driving open-collector TTL NAND gate. Figure 10.40 shows the open-collector NAND gate with the external pull-up driving the non-TTL logic load.

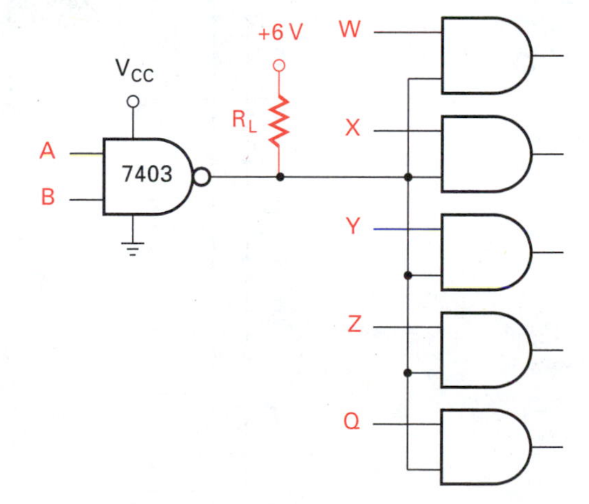

Figure 10.40
TTL open-collector NAND gate driving a non-TTL load

❖ **EXAMPLE 10.2** Design an open-collector to interface with another logic family that requires different logic levels. Also perform a logical OR at the output of the open collector. Calculate the range of values for R_L. A fan-out to eight gates is required.

SOLUTION Let the other logic family logical values be $V_{IL} = .8$ V, $V_{IH} = 5$ V, $I_{IH} = 50\ \mu A$, $I_{IL} = 1$ mA.

Let the logical function be

$$P = (x + y + z)'$$

where

$$x = (AB)'$$
$$y = (CD)'$$
$$z = (EF)'$$

then

$$P = [(AB)' + (CD)' + (EF)']' = (AB)(CD)(EF)$$

Let the external pull-up power supply $V_{CC} = 6$ V to accommodate the required logic 1 input for the load logic. Use the two-input open-collector TTL 7403 device. The values for I_{OH} and I_{OL} are given on the 7403 data sheet as 250 μA and 16 mA, respectively. The logic function, *P*, requires three two-input NAND gates. The NANDs generate the terms to be NORed. The NOR operation is to be accomplished by each of the three open-collector NAND gates sharing a common pull-up resistor. The logic diagram for the problem is shown in Figure 10.41.

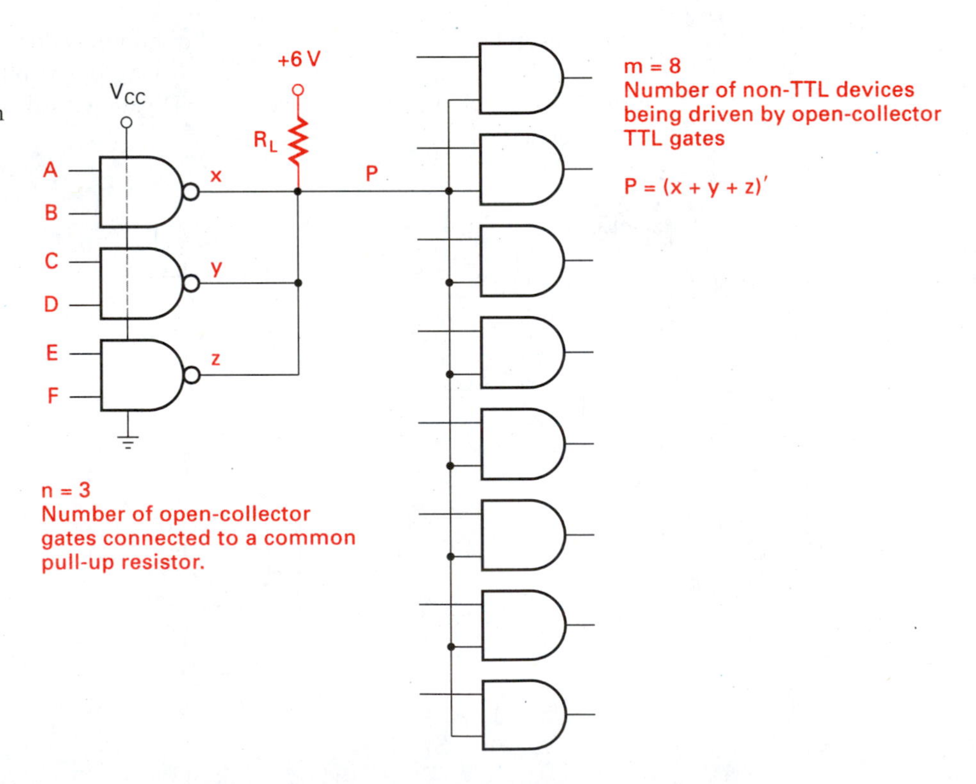

Figure 10.41
Using TTL open-collector gates to perform an OR logic function and drive a non-TTL load

Two conditions exist. In the first, the output is a logical 1:

$$R_L(\max) = (V_{CC} - V_{OH})/(mI_{OH} + nI_{IH})$$
$$R_L(\max) = (6\text{ V} - 5\text{ V})/[(3)(250\ \mu\text{A}) + 8(50\ \mu\text{A})]$$
$$R_L(\max) = 1\text{ V}/1.15\text{ mA} = 870\text{ ohms}$$

In the second case, the output is a logical 0:

$$R_L(\min) = (V_{CC} - V_{OL})/(I_{OL} - m(I_{IL})$$
$$R_L(\min) = (6\text{ V} - 0.6\text{ V})/[16\text{ mA} - 8(1\text{ mA})] = 5.4\text{ V}/8\text{ mA} = 675\text{ ohms}$$

The wired-OR function is created by the common junction of each two-input NAND gate and the external pull-up resistor, R_L. Let each open-collector gate output be labeled as follows:

$$x' = AB$$
$$y' = CD$$
$$z' = EF$$

TABLE 10.7
Truth table for logic function P

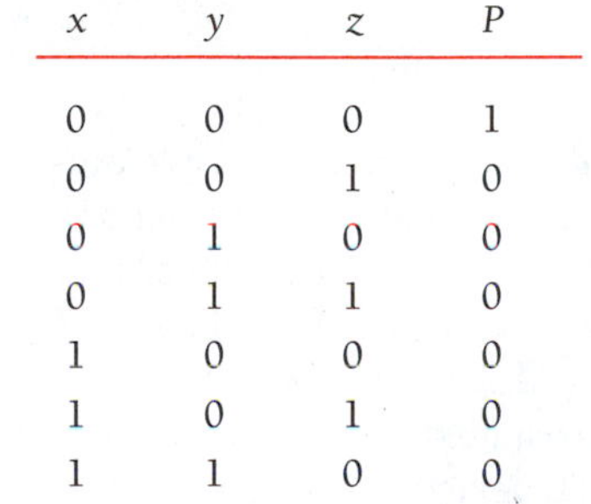

Inputs			Output
x	y	z	P
0	0	0	1
0	0	1	0
0	1	0	0
0	1	1	0
1	0	0	0
1	0	1	0
1	1	0	0
1	1	1	0

The resulting logic function P becomes

$$P = x'y'z'$$

and by DeMorgan, $P = (x + y + z)'$. Also by substitution, $P = [(AB)' + (CD)' + (EF)']'$. A truth table showing all of the combinations of variables x, y, and z is given in Table 10.7.

Consider Figures 10.39 and 10.41. The saturated output transistor shown in Figure 10.39 (logic 0 level output) will pull the output node shown in Figure 10.41 down to a logic 0, because it is a short-circuit path to ground. Any other gate outputs in Figure 10.41 that are at a logic 1 will be "overridden" by the saturated output transistor, clamping the output voltage to the V_{ce} level of the saturated transistor. The only time the output node can be a logical 1 is when all of the transistors connected to the common pull-up resistor are cut off, allowing the output to rise toward the external V_{CC} supply. ❖

Open-collector logic functions are called both *wired-AND* and *wired-OR* because of the principle of duality, so either term is appropriate. A significant application of open-collector devices is to drive small low voltage and low current relays or LED output displays.

❖ **EXAMPLE 10.3** Design a logic circuit to interface to a relay that requires a 24 V supply and 20 mA to operate.

SOLUTION In this example, high-voltage open-collector devices such as the 7416 (inverting) or the 7417 (noninverting) provide the voltage levels needed to interface with the relay. If the relay closure is to occur when the driver input is a 1, an inverting driver is needed. If the relay is to be activated when the driver input is a 0, a noninverting driver is needed. These drivers are shown in Figure 10.42.

The maximum allowable V_{OH} for the 7416 is 30 V, so our 24 V requirement is within the V_{CC} limitation. The I_{OL}, or sinking current capability of the 7416, is 40 mA, so our requirement of 20 mA is within that limit also. The diode in parallel with the relay coil is to prevent high-voltage kickback due to induced EMF resulting from the collapsing magnetic field when the relay switches.

Figure 10.42
TTL open-collector relay driver

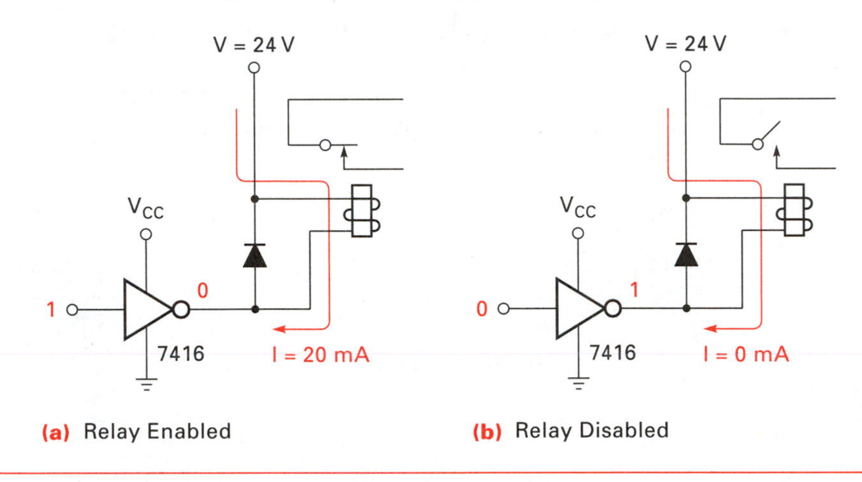

Another application of open-collector devices is the ability to connect multiple output devices to a common data line. Only one such device can be enabled at a time or a bus contention would result. Bus contention occurs when more than one data source has control of a data bus. A data bus is a collection of wires over which digital data are transmitted. The open-collector structure ensures that only one data source is connected to the bus by virtue of the logic realization. More recently, tristate drivers have replaced open collector bus drivers. Section 10.5.7 covers tristate TTL devices.

10.5.7 Tristate TTL Devices

A discussion of the need for tristate TTL ICs was given in Chapter 4, Section 4.10. Our purpose in this section is to describe the TTL tristate circuit in more detail. The problem that tristate circuits solve is to allow multiple data sources to be connected to one data line. **Tristate** circuits permit data multiplexing to occur by switching one desired data source on and the rest off. The high impedance needed to switch a data source off is achieved by turning off both TTL totem-pole output transistors at the same time. Consider the totem-pole TTL output stage shown in Figure 10.43. If both transistors Q_3 and Q_4 are turned off at the same time, then a high-impedance would be "seen" at the output. The question is, then, how to turn off both transistors in the totem pole at the same time. The answer to that question is found by analyzing the circuit shown in Figure 10.43.

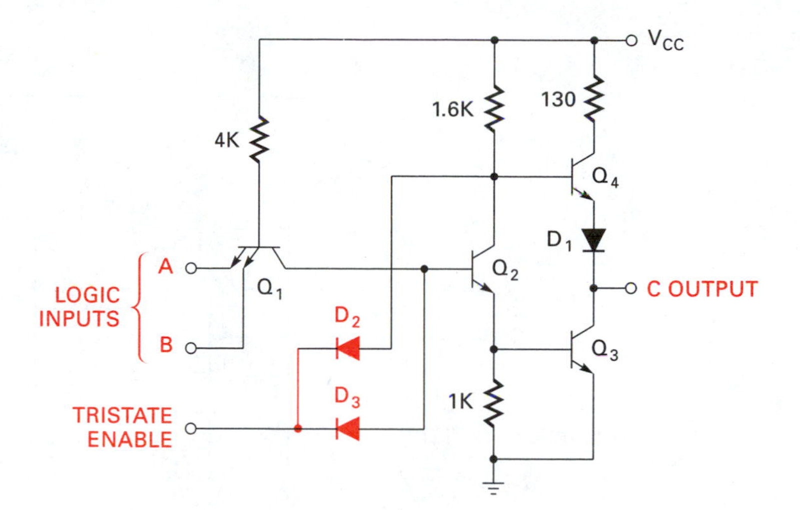

Figure 10.43
Tristate TTL NAND circuit

We see the familiar two-input emitter, the phase-splitter second stage, and the totem-pole output stage of the standard TTL circuit. In addition, a separate enable input is present that drives two diodes, which in turn are connected to the base and collector of Q_2. If D_2 and D_3 are reverse biased, that is, the enable input is high, then the circuit operates like a normal TTL NAND gate and the output impedance is low. This condition exists when the circuit is "enabled" by the tristate enable input. *Enable* in this case means that the device is in its normal mode (not in a high-impedance mode).

If the enable input were a logical 0, then D_2 and D_3 would be forward biased, clamping the base and collector of Q_2 at a diode drop above the logic 0 level. This is sufficient to cause Q_2 to cutoff, in turn cutting off Q_3. Transistor Q_4, normally conducting when Q_3 is off, is also held in cutoff by the clamping action of D_2, because the base current normally available to allow Q_4 to conduct is "steered" away through the forward biased diode. In actual tristate TTL integrated circuits, a buffer, either inverting or noninverting, is present between the enable input and the two tristate steering diodes. Figure 10.44 shows the complete circuit diagram for a two-input TTL NAND tristate circuit. Notice that a separate single-emitter input transistor, Q_5; a second stage transistor, Q_6; and a driver transistor, Q_7; are used to form an inverting TTL buffer. Note also that the collector of Q_7 is directly connected to an extra emitter on Q_1 and a steering diode connected to the collector of Q_2. The extra emitter on Q_1 replaces D_3 in Figure 10.43. The inverting buffer isolates the internal circuit of the NAND gate from the outside world and provides a constant load to the enable input that is consistent with other input loads. Other TTL gates are needed to provide the logic levels to the tristate enable input to provide the control logic necessary to enable or disable the device.

10.5.8 Mixed TTL Subfamily Fan-Out

Each TTL subfamily can support a fan-out of 10 when connected to members of the same subfamily. If, for example, low-power Schottky devices are connected to other low-power Schottky gates the maximum fan-out is 10. If, however, a low-power Schottky

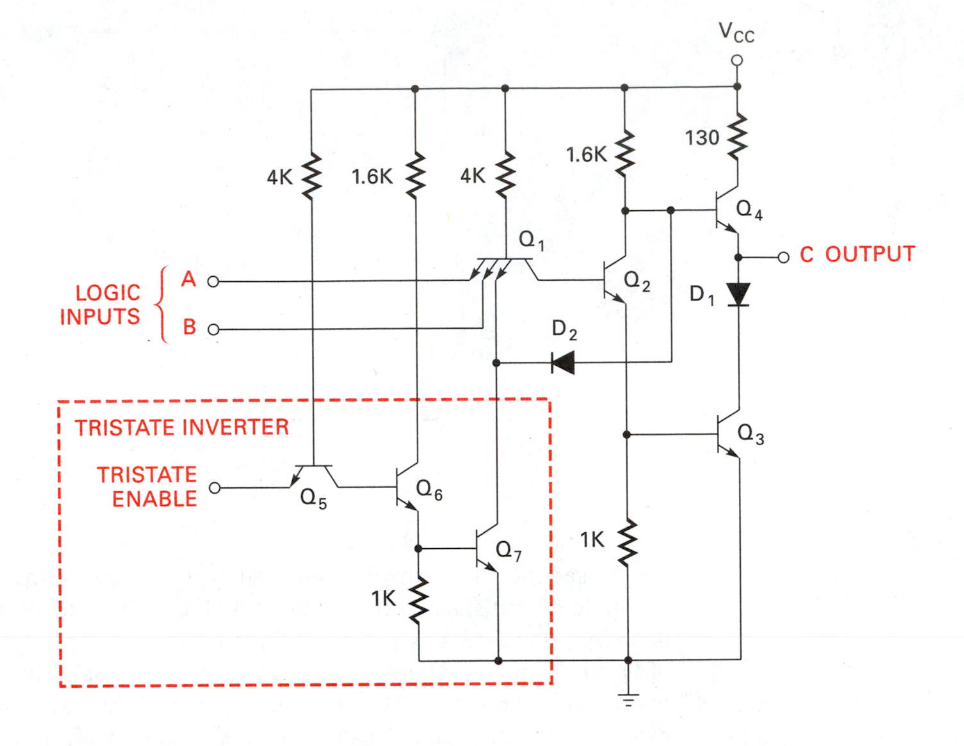

Figure 10.44
Complete tristate TTL two-input NAND circuit

output is connected to high-speed device inputs, then the fan-out is not 10. We determine the appropriate fan-out when dealing with mixed (one TTL subfamily driving the inputs to a different TTL subfamily) logic families by comparing the input and output current levels for both logic states.

❖ EXAMPLE 10.4

A standard TTL IC is to drive high-speed TTL inputs. How many high-speed devices can a single standard IC output accommodate? A comparison of the input and output current levels from a TTL data book gives the information in Table 10.8.

SOLUTION The fan-out is found by determining how many inputs of the high-speed IC can be connected to a single standard driver IC. Two cases exist. In the first, we have a logic 0 output from the driver:

$$\text{Fan-out} = [I_{OL}\ (\text{driver})]/[I_{IL}\ (\text{receiver})] = 16\ \text{mA}/2\ \text{mA} = 8$$

TABLE 10.8
Some TTL input and output currents

Standard	High speed	Low power Schottky	Advanced Schottky
I_{OL} = 16 mA	I_{OL} = 20 mA	I_{OL} = 8 mA	I_{OL} = 20 mA
I_{OH} = 400 μA	I_{OH} = 500 μA	I_{OH} = 400 μA	I_{OH} = 2 mA
I_{IL} = 1.6 mA	I_{IL} = 2 mA	I_{IL} = 400 μA	I_{IL} = .5 mA
I_{IH} = 40 μA	I_{IH} = 50 μA	I_{IH} = 20 μA	I_{IH} = 20 μA

In the second case, we have a logic 1 output from the driver:

$$\text{Fan-out} = [I_{OH}\,(\text{driver})]/[I_{IH}\,(\text{receiver})] = 400\ \mu A/50\ \mu A = 8$$

The fan-out for a standard TTL connected to a high-speed TTL gate is eight, as just computed. Both logic states must be calculated and then the lowest number taken for the fan-out limit. If the computation results in a fractional value, round down. ❖

❖ **EXAMPLE 10.5** Determine the fan-out when an advanced Schottky driver is connected to low-power Schottky receivers.

SOLUTION If we have a logic 0 output from the driver:

$$\text{Fan-out} = I_{OL(AS)}/I_{IL(LS)} = 20\ \text{mA}/.4\ \text{mA} = 50$$

If we have a logic 1 output from the driver:

$$\text{Fan-out} = I_{OH(AS)}/I_{IH(LS)} = 2\ \text{mA}/20\ \mu A = 100$$

In this example, the fan-out limit is set by the 0 case. Any combination of TTL subfamilies may be interconnected. In fact, TTL devices may be connected to other logic families as long as the logic voltage levels (refer to Figure 10.24) are not violated and the output current limits are kept intact. ❖

10.5.9 Other TTL Circuits

The basic TTL gate is the NAND gate. However, for the logic family to be very useful, other functions are also necessary. Figure 10.45 illustrates a two-input AND gate, Fig-

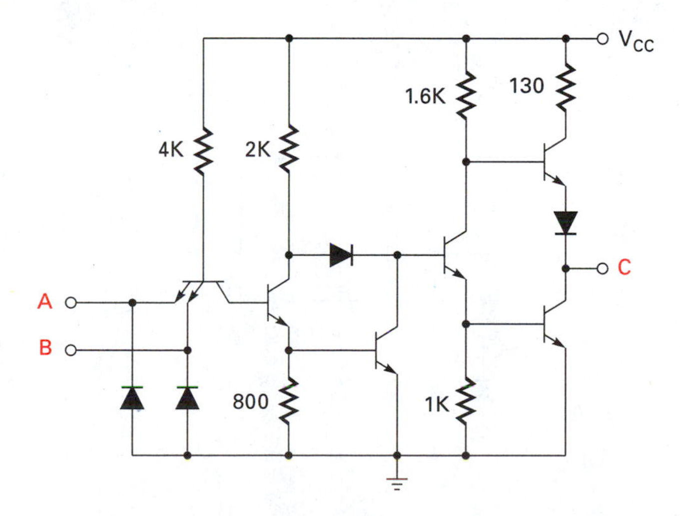

Figure 10.45
TTL two-input AND gate (74x08)

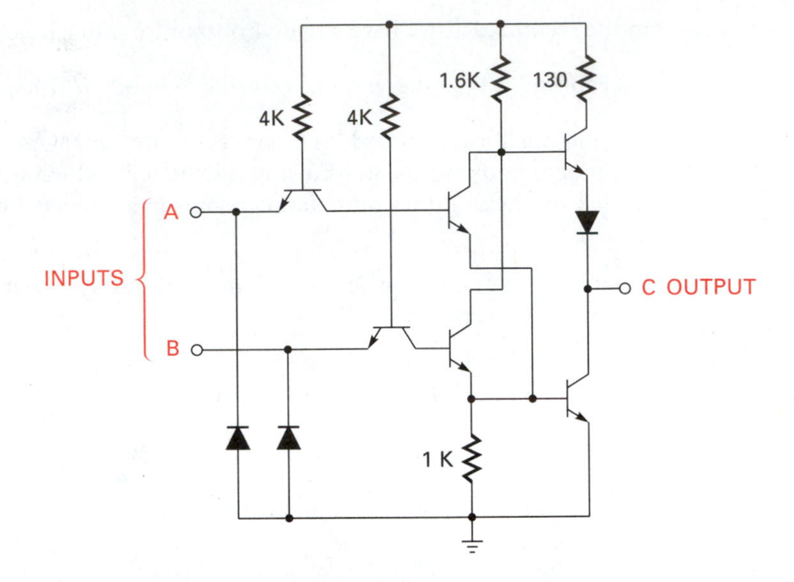

Figure 10.46
TTL two-input NOR gate (74x02)

ure 10.46 shows a two-input NOR gate, and Figure 10.47 shows a quad-input Schmitt trigger NAND gate. Many other logic functions are integrated into the TTL families, including 12-input gates, AND-OR combinations, and a variety of arithmetic functions.

The fan-out rules, totem-pole outputs, logic voltage levels and currents, and other specifications are as discussed previously in this chapter. The power consumption, switching speed, and chip pin-outs must be found by consulting the appropriate data book.

The Schmitt trigger input gate shown in Figure 10.47 was designed to provide definite switching in a slowly changing or "bouncy" input signal. The circuit for the Schmitt

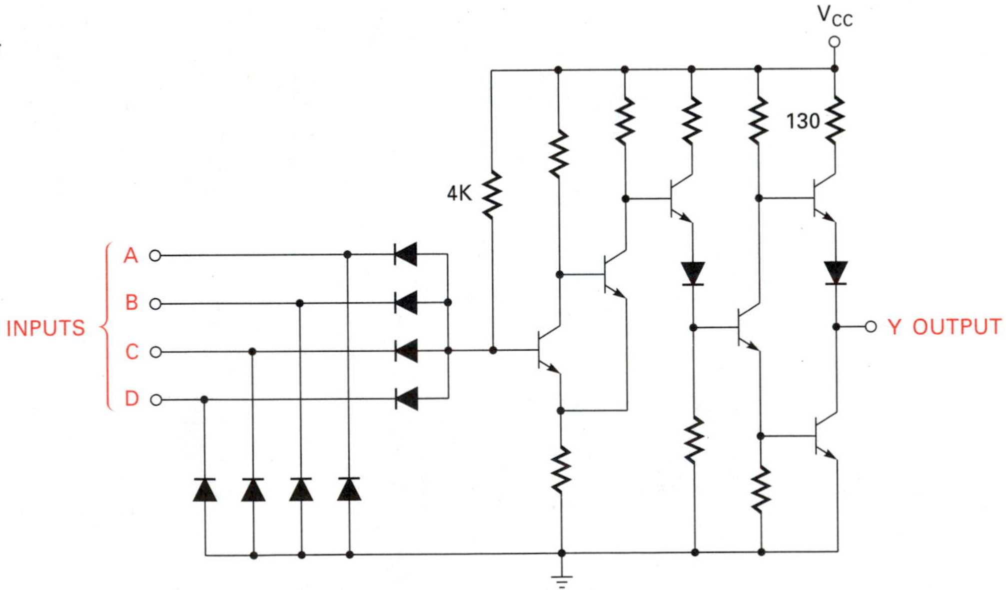

Figure 10.47
TTL four-input Schmitt trigger NAND gate

trigger includes provisions for high gain so that switching can occur rapidly once the input voltage has exceeded a certain threshold. It also has provisions so that, once the output has switched, the input can go back below the input threshold voltage without the output changing state. This feature provides the ability for positive output logic level switching on "bouncy" inputs.

Normal TTL gates would produce output level switching each time the input voltage passed through the input threshold voltage, not at all what the logic designer wants. Consider interfacing your logic design to an input transducer that produces a signal like that shown in Figure 10.48. The normal TTL gate would not give a clean output that could be used by the digital system. The Schmitt trigger gate will produce the output as shown at the bottom of Figure 10.48. The voltage transfer curve for the standard TTL logic families is shown in Figure 10.23. A voltage transfer curve for the Schmitt trigger is illustrated in Figure 10.49. The "box"-like characteristic curve results from what is called **hysteresis.** Hysteresis results from different thresholds used to switch from a 1 to a 0. This property allows the circuit to not switch on the input signal "noise." The circuit will not discount all input signal noise. If the input signal transitions exceed the hysteresis levels (V_{LT} and V_{UT}) then the output will indeed change state. V_{LT} is the lower threshold voltage and V_{UT} is the upper threshold voltage as indicated in Figure 10.48. The circuit "hysteresis" is the difference between V_{LT} and V_{UT}. Although a circuit with hysteresis will not eliminate all false triggering, it will greatly reduce it on noisy inputs.

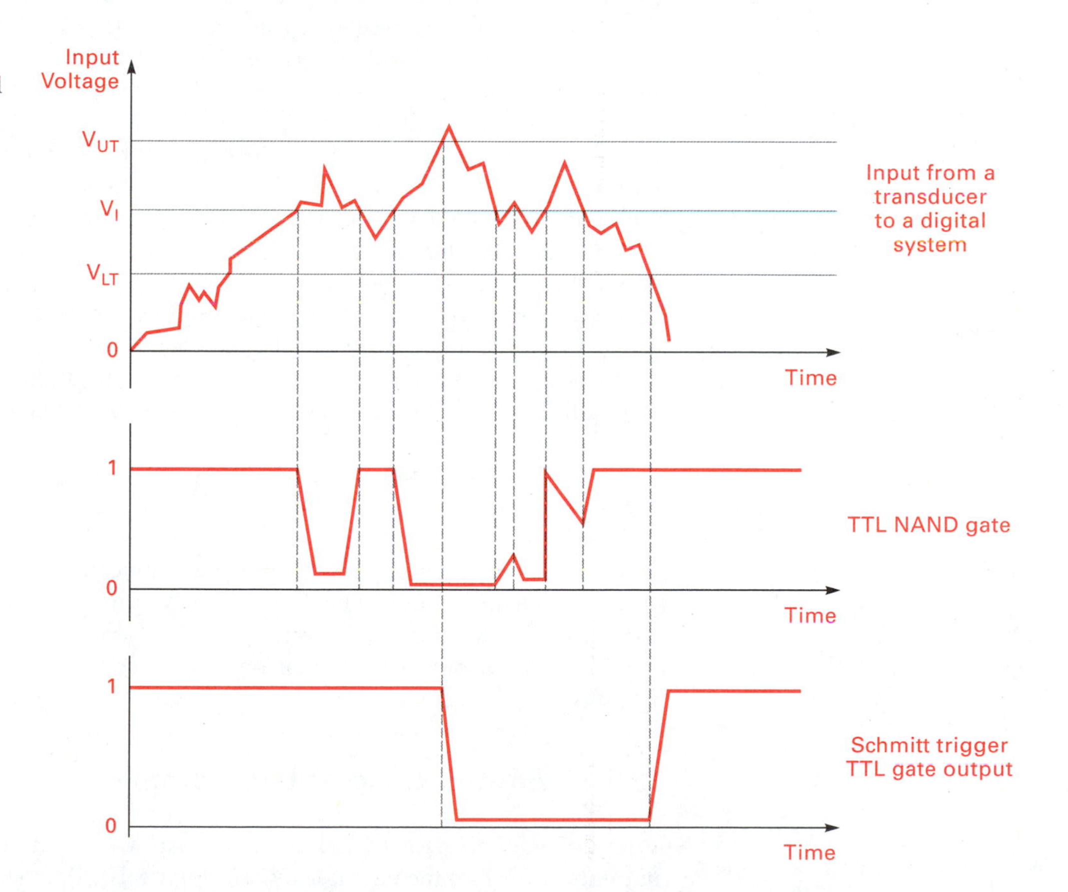

Figure 10.48
Bouncy input and the normal TTL and Schmitt trigger gate output responses

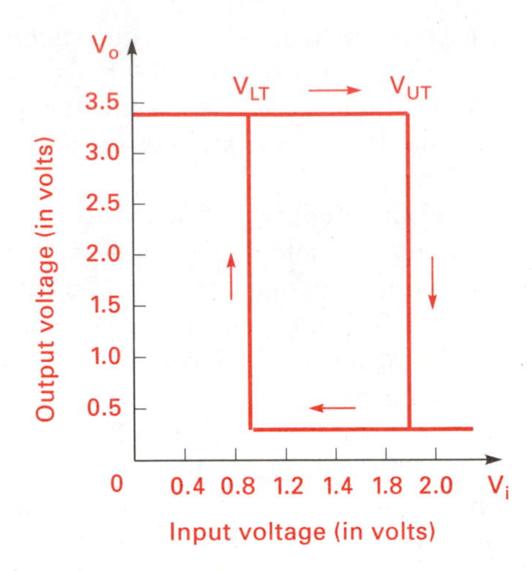

Figure 10.49
Schmitt trigger gate voltage transfer curve

10.6 EMITTER-COUPLED LOGIC

Emitter-coupled logic, **ECL**, is another logic family that uses bipolar transistors. The logic designer's desire for faster switching logic circuits was evident in the development of the various TTL subfamilies. **Speed-up** techniques have been evident in the design and manufacture of digital integrated circuits from the beginning. One of the main speed-up techniques discussed earlier in this chapter was that of keeping the bipolar transistor from going into saturation, which reduced the stored charge problem and improved speed. If we depart from the common-emitter transistor configuration, used in the design of both the DTL and TTL logic families, and develop a circuit configuration around the common-collector circuit, then the saturation problem can be eliminated.

A requirement for saturation in a bipolar transistor is the forward biasing of the base–emitter and the base–collector junctions. The base-to-collector Schottky clamp diode (see Figure 10.38) improved switching speed by keeping the transistor from saturating. The common-collector transistor configuration cannot saturate. Because the output of the ECL switching transistor is taken from the emitter and coupled to a reference transistor's emitter, the new logic family came to be known as *emitter-coupled logic*, or ECL.

Motorola pioneered the development of ECL integrated circuits as far back as 1962, when they introduced the MECL I ECL subfamily. That particular ECL family is still in production.

ECL has not enjoyed as wide a range of application as the TTL families, primarily because it is a much more difficult logic family around which to design a digital system. The reason for this is the speed with which ECL operates and its much narrower noise margins. More about that later; for now let us consider the basic ECL circuit and its operation.

10.6.1 Emitter-Coupled Logic Circuit

A discussion of the basic ECL circuit must start with the development of the differential amplifier. Consider the circuit shown in Figure 10.50, where the two transistors with

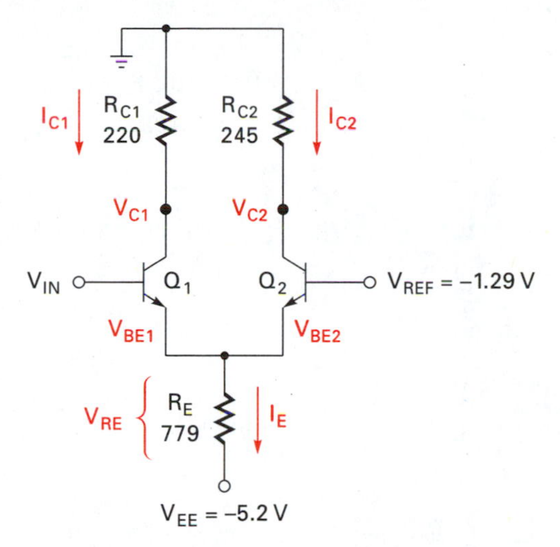

Figure 10.50
Emitter-coupled logic (ECL) differential amplifier

connected emitters form a differential amplifier. In analog applications, the output signal is the voltage gain multiplied by the difference between the two input signals. In switching applications, the inputs are restricted to a defined value for a logical 1 or 0. The differential amplifier configuration provides a circuit that can compare the logic input level to a reference and generate an output level.

Notice that the input is provided at the base of Q_1 and the reference is applied at the base of Q_2. Both transistor emitters form a junction with a common pull-down resistor to V_{EE}.

ECL circuits operate using a negative 5.2 V power supply. The collector's resistors are pulled up to ground. The logic levels for both 0 and 1 are also negative voltages. Each logic family, DTL, TTL, ECL and CMOS, uses different voltage levels to define 0 and 1 logic.

The differential amplifier circuit shown in Figure 10.50 is the heart of the ECL 10K logic family. The reference voltage at the base of Q_2, $V_{REF} = -1.29$ V, provides the voltage point around which Q_1 will switch. Motorola, in its *MECL Device Data* catalog, defines $V_{BE} = .8$ V for the transistors in the ECL 10K family:

$$V_{REF} = -1.29 \text{ V}$$
$$V_{BE} = .8 \text{ V}$$

then

$$V_{E2} = +V_{REF} - (V_{BE2}) = (-1.29 \text{ V} - 0.8 \text{ V}) = -2.09 \text{ V}$$

Because both emitters are at the same node,

$$V_{E2} = V_{E1} = -2.09 \text{ V}$$

V_{B1} must be .8 V greater than V_{E1} for Q_1 to conduct; any voltage more negative than -1.29 V will put Q_1 into cutoff. Q_1 begins to conduct when $V_{IN} = -1.29$ V or greater. Q_1 is cut off when V_{IN} is less than -1.29 V.

Case 1

$$V_{IN} = \text{logical } 1 = -.9 \text{ V (ECL 10K high-input voltage)}$$
$$V_{E1} = V_{IN} - V_{BE1} = -.9 \text{ V} - (.8 \text{ V}) = -1.7 \text{ V}$$

Then

$$-I_E = (V_{EE} - V_{E1})/R_E = [-5.2 \text{ V} - (-1.7 \text{ V})]/779 = -3.5 \text{ V}/779$$
$$I_E = 4.49 \text{ mA}$$

Let $I_E = I_{C1}$, then

$$I_{C1} = 4.49 \text{ mA}$$
$$V_{C1} = -I_{C1}R_{C1} = -(4.49 \text{ mA})(220 \text{ ohms}) = -.988 \text{ V}$$

Case 2

$$V_{IN} = \text{logical } 0 = -1.75 \text{ V (ECL 10K low-input voltage)}$$
$$V_{E1} = V_{E2} = V_{REF} - V_{be2} = -1.29 \text{ V} - .8 \text{ V} = -2.09 \text{ V}$$

With $V_{IN} = -1.75$ V, then Q_1's $V_{BE} = -1.75 \text{ V} - (-2.09 \text{ V}) = +.34$ V.

A base–emitter voltage of .34 V is not enough to turn on Q_1, so it is off. With Q_1 cut off, no majority collector current flows, so V_{C1} is 0 V.

The two logical input voltages at the base of Q_1 switch the collector of Q_1 between 0 V (when V_{IN} is a logical 0) and −.99 V (when V_{IN} is a logical 1). Q_2, the other half of the differential amplifier circuit, switches oppositely from Q_1. When Q_1 is conducting, then Q_2 is cut off; and when Q_1 is cut off, then Q_2 is conducting.

Let Q_1 be conducting, then $V_{IN} = -.9$ V (logical 1 input):

$$V_{E1} = V_{IN} - V_{BE1} = -.9 \text{ V} - .8 \text{ V} = -1.7 \text{ V}$$
$$V_{BE2} = -V_{E1} + V_{B2} = -(-1.7 \text{ V}) + (-1.29 \text{ V}) = +.41 \text{ V}$$

which is not enough to turn on Q_2.

So far we have been discussing the switching operation of the differential amplifier for the ECL gate. We have not yet discussed how the gate performs any useful logical operation. By placing additional transistors in parallel with Q_1 we can obtain additional logical inputs as indicated in Figure 10.51, where Q_1, Q_2, and Q_3 are in parallel.

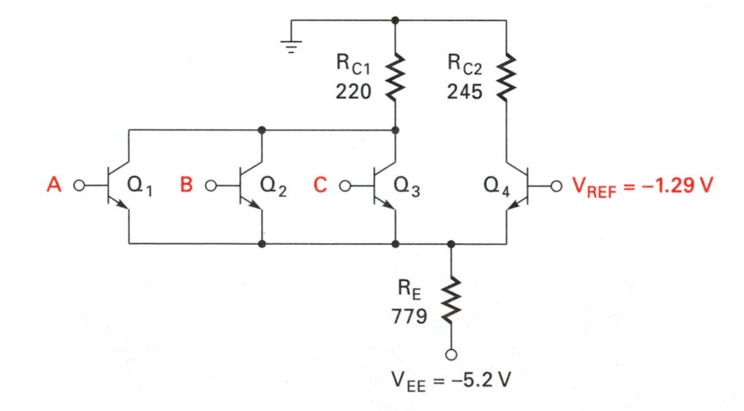

Figure 10.51
A three-input ECL circuit

Figure 10.52
ECL 10K three-input OR-NOR gate

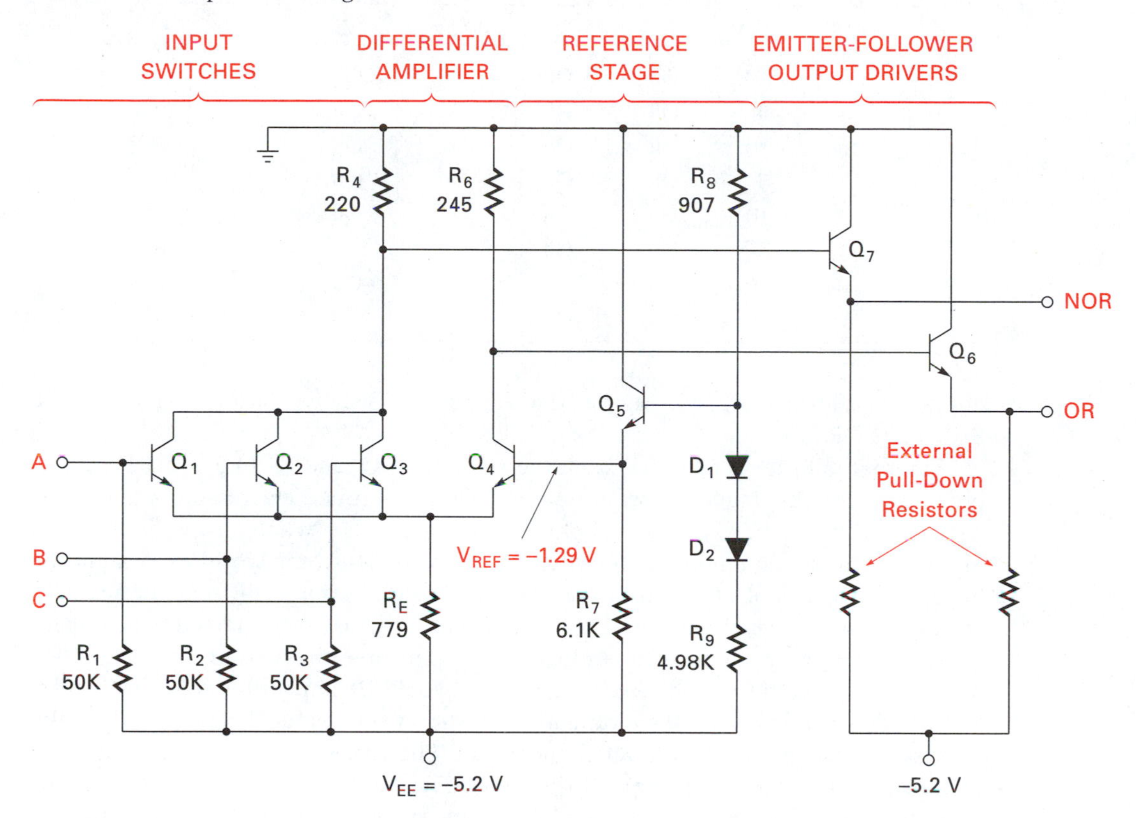

Q_1, Q_2, and Q_3 in Figure 10.51 are switches controlled by inputs V_A, V_B, and V_C. A logical 0 on any input (-1.75 V) turns off that transistor and a logical 1 ($-.9$ V) turns the transistor on. Consider the transistors as "ideal" switches for a moment. Any time the transistor is "on" it is a closed switch, providing a current path from ground (the most positive potential) to -5.2 V. Any closed switch ("on" transistor) provides the current path from ground to $-V_{EE}$.

The "switching" half of the differential amplifier is formed by the three parallel transistors. Any "on" input transistor (Q_1, Q_2, Q_3) forces the reference transistor (Q_4) off. All of the input transistors must be off before Q_4 can conduct. Note that the emitters and collectors of all switching input transistors are connected in common. The complete three-input ECL 10K gate circuit is shown in Figure 10.52.

Consider Figure 10.52; transistors Q_1, Q_2, and Q_3 provide input switching, and transistor Q_4 is the reference half of the differential amplifier. Transistor Q_5 along with R_7, R_8, R_9, D_1, and D_2 produce the -1.29 V reference at the emitter of Q_5. R_1, R_2, and R_3 provide a pull-down resistor for the inputs so that an unused input will be interpreted as a logical 0. Transistors Q_6 and Q_7 are emitter–follower output drivers, providing the ECL circuit with considerable output current drive, and the output impedance is very low. Note that two outputs are available, an OR (Q_6 output) and a NOR (Q_7 output), as shown in Figure 10.52. The designer can use either or both at his or her discretion. Table 10.9 shows the status of each input, the transistor conditions (on or off), and the

TABLE 10.9
Three-input ECL OR/NOR gate: 0 = −1.75 V; 1 = −.9 V

Input variables			Switching transistors			Difference transistor	Output level	
V_a	V_b	V_c	Q_1	Q_2	Q_3	Q_4	OR	NOR
0	0	0	Off	Off	Off	On	0	1
0	0	1	Off	Off	On	Off	1	0
0	1	0	Off	On	Off	Off	1	0
0	1	1	Off	On	On	Off	1	0
1	0	0	On	Off	Off	Off	1	0
1	0	1	On	Off	On	Off	1	0
1	1	0	On	On	Off	Off	1	0
1	1	1	On	On	On	Off	1	0

output levels for the circuit in Figure 10.52. Figure 10.53 shows the logic symbol for the ECL OR/NOR function

Figure 10.53
ECL logic symbol for an OR/NOR

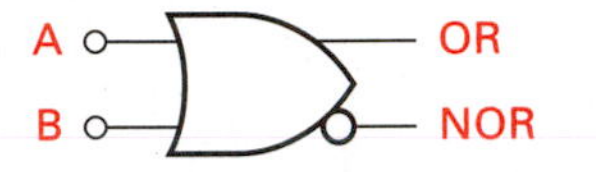

External pull-down resistors, from the emitters of Q_6 and Q_7 to a −5.2 V supply, provide the terminating load for the emitter–follower outputs. The reason for the external pull-down resistors is the speed of the ECL logic family, where propagation delays of 2 ns or less are common. With switching speeds this fast, the world external to the integrated circuit looks like a transmission line. Circuit outputs will generate standing waves that create signal ringing, if not terminated in the correct characteristic impedance. If ringing were to occur in an ECL circuit (left unterminated or improperly terminated), then distinction between logic levels would be impossible. For this reason, the pull-down resistors are the terminating resistors whose value depends on the characteristic impedance of the output environment. The design of proper termination for ECL circuits is beyond the scope of this book. The serious ECL logic designer needs rather extensive background in transmission line theory. A help in this matter is the manufacturer's ECL data catalogs, such as Motorola's *MECL Device Data Book* published in 1989.

10.6.2 ECL Specifications

The definitions for input and output currents and voltages are the same as given in Section 10.5.2. The voltage and current values are different, however. Figure 10.54 and

Figure 10.54
ECL logic levels

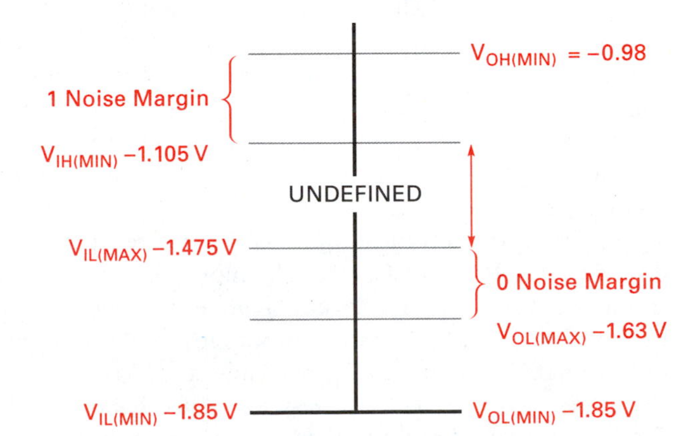

TABLE 10.10
ECL noise margins

Family	Guaranteed worst-case dc noise margin (volts)	Typical dc noise margin (volts)
MECL 10KH	0.150	0.270
MECL 10K	0.125	0.210
MECL III	0.115	0.200

Table 10.10 show the ECL logic levels and noise margins. Figure 10.55 illustrates the ECL voltage transfer curve. Notice that the noise margins for any of the ECL subfamilies are considerably less than TTL.

The Motorola MC10H104 specification sheet shown in Table 10.11 lists the voltages and currents for logical levels, the propagation speeds, and power supply voltages and currents. Notice that three two-input AND gates and a single two-input AND/NAND are available in this particular integrated circuit. Compare the propagation delays of the ECL gates to the advanced Schottky TTL NAND gate. The ECL gate has a 1.9 ns maximum propagation delay at 75° C, compared to the advanced Schottky TTL gate with a maximum delay of 4.5 ns. A comparison of power consumption reveals a dissipation of 25 mW per gate for ECL and a typical 54 mW per gate for the advanced Schottky TTL gate.

An even faster ECL subfamily is the ECL 100K series. The 10K series has a 10 prefix as in the MC*10*H104; the 100K subfamily uses a prefix of 100, as in MC100xxx. The 100K uses a lower power supply voltage of −4.5 V instead of the −5.2 V used by the rest of the ECL subfamilies. The 100K is also faster with a propagation delay less than a nanosecond compared to the 1.9 ns delay of the 10K devices. Power dissipation is almost twice that of the 10K series at 40 mW per gate.

10.6.3 ECL to TTL and TTL to ECL Interfacing

Mixing logic families is sometimes necessary, especially when interfacing between two printed circuit boards or submodules of a digital system. Suppose that a digital system

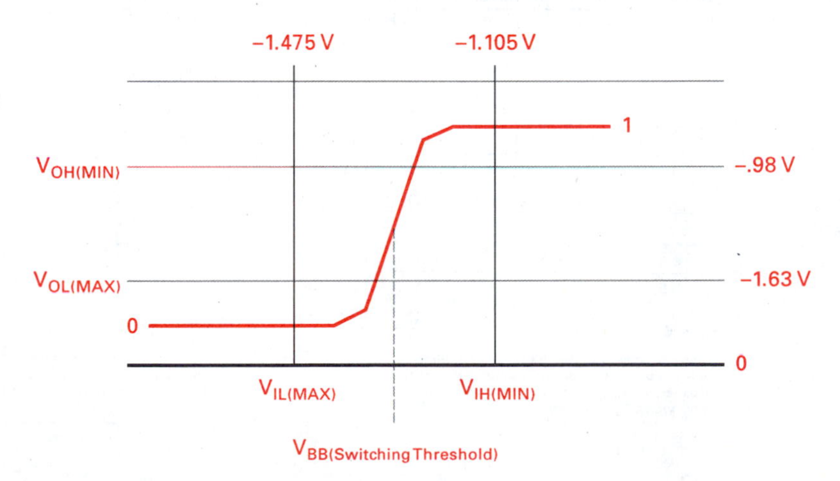

Figure 10.55
ECL voltage transfer curve showing minimum voltage logic levels and threshold

TABLE 10.11
Typical ECL specification data sheet (reprinted by permission from Motorola Corp.)

MC10H104

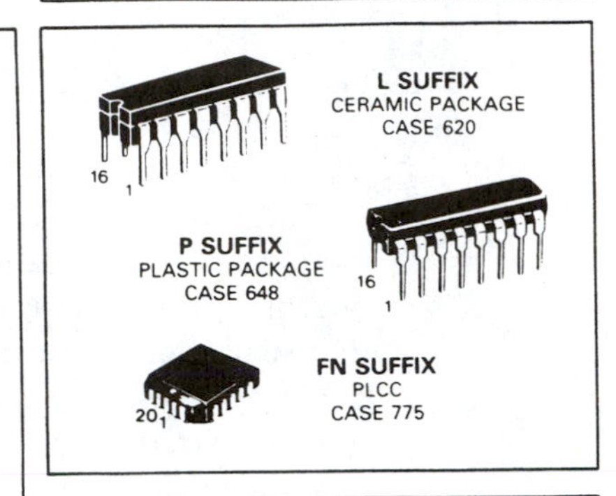

QUAD 2-INPUT AND GATE

The MC10H104 is a quad 2-input AND gate. One of the gates has both AND/NAND outputs available. This MECL 10KH part is a functional/pinout duplication of the standard MECL 10K family part, with 100% improvement in propagation delay, and no increase in power-supply current.

- Propagation Delay, 1.0 ns Typical
- Power Dissipation 25 mW/Gate (same as MECL 10K)
- Improved Noise Margin 150 mV (Over Operating Voltage and Temperature Range)
- Voltage Compensated
- MECL 10K-Compatible

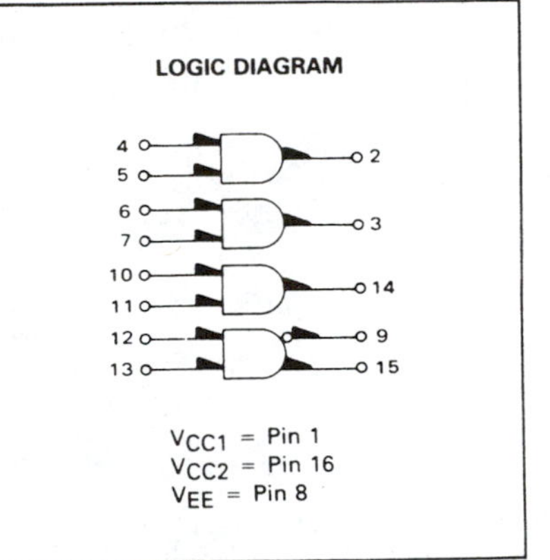

MAXIMUM RATINGS

Characteristic	Symbol	Rating	Unit
Power Supply (V_{CC} = 0)	V_{EE}	−8.0 to 0	Vdc
Input Voltage (V_{CC} = 0)	V_I	0 to V_{EE}	Vdc
Output Current — Continuous — Surge	I_{out}	50 100	mA
Operating Temperature Range	T_A	0–75	°C
Storage Temperature Range — Plastic — Ceramic	T_{stg}	−55 to 150 −55 to 165	°C °C

ELECTRICAL CHARACTERISTICS (V_{EE} = −5.2 V ±5%) (See Note)

Characteristic	Symbol	0° Min	0° Max	25° Min	25° Max	75° Min	75° Max	Unit
Power Supply Current	I_E	—	39	—	35	—	39	mA
Input Current High	I_{inH}	—	425	—	265	—	265	μA
Input Current Low	I_{inL}	0.5	—	0.5	—	0.3	—	μA
High Output Voltage	V_{OH}	−1.02	−0.84	−0.98	−0.81	−0.92	−0.735	Vdc
Low Output Voltage	V_{OL}	−1.95	−1.63	−1.95	−1.63	−1.95	−1.60	Vdc
High Input Voltage	V_{IH}	−1.17	−0.84	−1.13	−0.81	−1.07	−0.735	Vdc
Low Input Voltage	V_{IL}	−1.95	−1.48	−1.95	−1.48	−1.95	−1.45	Vdc

AC PARAMETERS

Characteristic	Symbol	0° Min	0° Max	25° Min	25° Max	75° Min	75° Max	Unit
Propagation Delay	t_{pd}	0.4	1.6	0.45	1.75	0.45	1.9	ns
Rise Time	t_r	0.5	1.6	0.5	1.7	0.5	1.8	ns
Fall Time	t_f	0.5	1.6	0.5	1.7	0.5	1.8	ns

NOTE:
Each MECL 10KH series circuit has been designed to meet the dc specifications shown in the test table, after thermal equilibrium has been established. The circuit is in a test socket or mounted on a printed circuit board and transverse air flow greater then 500 linear fpm is maintained. Outputs are terminated through a 50-ohm resistor to −2.0 volts.

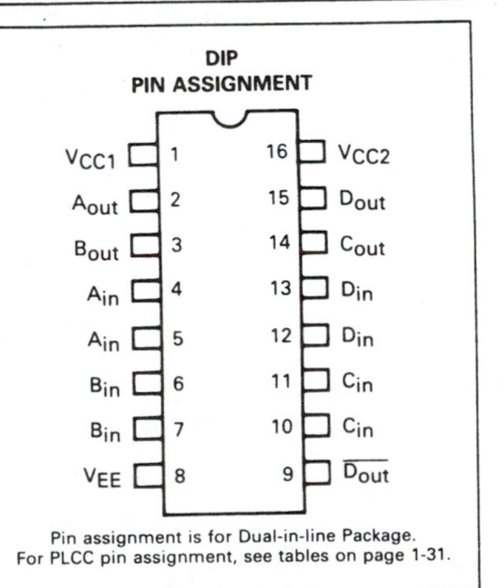

Pin assignment is for Dual-in-line Package.
For PLCC pin assignment, see tables on page 1-31.

was designed using TTL and that a new high-speed ECL module was to be added to the system. Suppose also that signals from the TTL unit were used to control certain activities necessary in the ECL unit and that the ECL unit had to send results back to the TTL part of the system. How would you interconnect the two? The logic levels defining 1 and 0 are different. Fortunately, TTL to ECL and ECL to TTL translator devices are available.

The MC10H124 is a quad TTL-to-ECL translator integrated circuit. The input logic levels are compatible with TTL levels and the output logic levels are compatible with ECL. Table 10.12 is the data sheet for the device. Notice that the values for V_{IH} and V_{IL} are the TTL definitions for 1 and 0. The voltage values for V_{OL} and V_{OH} are those for ECL. Also notice that two power supply voltages, +5 V and −5.2 V, are needed to generate the logic voltage levels for TTL and ECL.

The MC10H125 is a quad ECL to TTL translator for converting ECL logic levels to TTL. Table 10.13 lists the device data. Notice that this chip also uses two power supply voltages and that the inputs are ECL compatible but the outputs are TTL compatible. These devices provide logic designers with a tool that allows them to partition their designs to realize speed and power requirements by choosing either TTL or ECL.

10.7 COMPLEMENTARY METAL OXIDE SEMICONDUCTOR

The third major integrated circuit logic family that we will cover in this chapter is complementary metal oxide semiconductor or **CMOS**. This logic family uses metal oxide semiconductor, MOS, and field effect transistors, FETs, to realize logic functions.

10.7.1 Field Effect Transistors

Before going into CMOS circuits actually used in digital integrated circuits, we need to develop an understanding of field effect transistors and see how they are different from bipolar transistors. Field effect transistors perform the same function in digital circuits as bipolar transistors; that is, they operate as switches. However, bipolar transistors, both NPN and PNP, use holes and electrons as current carriers, and field effect transistors use either holes (P-channel) or electrons (N-channel) as current carriers, but never both in the same device.

If we doped silicon with pentavalent (five valence electrons) material, the result would be a conducting semiconductor using electrons as a current carrier. Trivalent (three electrons in the atom's outer shell) doping of silicon would result in a conductor using holes as the majority current carriers. A FET doped with N-material forms an N-channel transistor and one doped with P-material forms a P-channel device. The doped silicon forms a conductor whose resistance depends on the dopant levels (how many atoms with extra holes or extra electrons were infused into the silicon crystal structure). The doped silicon becomes a "channel" for current. Figure 10.56 illustrates a silicon channel doped with N- or P-material to form a conductor whose resistance depends on dopant levels.

Dynamic control of current in the semiconductor channel can be accomplished by adding an element called a *gate* to the device. The gate is a region diffused into the drain–source channel, which forms a junction field effect transistor (**JFET**). This is

TABLE 10.12
TTL to ECL translator (reprinted with permission from Motorola Corp.)

MC10H124

QUAD TTL-TO-MECL TRANSLATOR

The MC10H124 is a quad translator for interfacing data and control signals between a saturated logic section and the MECL section of digital systems. The 10KH part is a functional/pinout duplication of the standard MECL 10K family part, with 100% improvement in propagation delay, and no increase in power-supply current.

- Propagation Delay, 1.5 ns Typical
- Improved Noise Margin 150 mV (Over Operating Voltage and Temperature Range)
- Voltage Compensated
- MECL 10K-Compatible

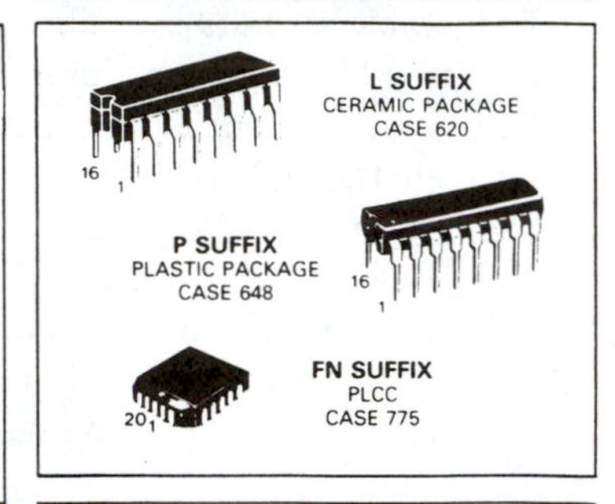

MAXIMUM RATINGS

Characteristic	Symbol	Rating	Unit
Power Supply (V_{CC} = 5.0 V)	V_{EE}	−8.0 to 0	Vdc
Power Supply (V_{EE} = −5.2 V)	V_{CC}	0 to +7.0	Vdc
Input Voltage (V_{CC} = 5.0 V) TTL	V_I	0 to V_{CC}	Vdc
Output Current — Continuous — Surge	I_{out}	50 100	mA
Operating Temperature Range	T_A	0 to +75	°C
Storage Temperature Range — Plastic — Ceramic	T_{stg}	−55 to +150 −55 to +165	°C

ELECTRICAL CHARACTERISTICS (V_{EE} = −5.2 V ±5%, V_{CC} = 5.0 V ± 5.0%)

Characteristic	Symbol	0°		25°		75°		Unit
		Min	Max	Min	Max	Min	Max	
Negative Power Supply Drain Current	I_E	—	72	—	66	—	72	mA
Positive Power Supply Drain Current	I_{CCH}	—	16	—	16	—	18	mA
	I_{CCL}	—	25	—	25	—	25	mA
Reverse Current	I_R							μA
Pin 6		—	200	—	200	—	200	
Pin 7		—	50	—	50	—	50	
Forward Current	I_F							mA
Pin 6		—	−12.8	—	−12.8	—	−12.8	
Pin 7		—	−3.2	—	−3.2	—	−3.2	
Input Breakdown Voltage	$V_{(BR)in}$	5.5	—	5.5	—	5.5	—	Vdc
Input Clamp Voltage	V_I	—	−1.5	—	−1.5	—	−1.5	Vdc
High Output Voltage	V_{OH}	−1.02	−0.84	−0.98	−0.81	−0.92	−0.735	Vdc
Low Output Voltage	V_{OL}	−1.95	−1.63	−1.95	−1.63	−1.95	−1.60	Vdc
High Input Voltage	V_{IH}	2.0	—	2.0	—	2.0	—	Vdc
Low Input Voltage	V_{IL}	—	0.8	—	0.8	—	0.8	Vdc

AC PARAMETERS

Characteristic	Symbol	0° Min	0° Max	25° Min	25° Max	75° Min	75° Max	Unit
Propagation Delay	t_{pd}	0.55	2.25	0.55	2.4	0.85	2.95	ns
Rise Time	t_r	0.5	1.5	0.5	1.6	0.5	1.7	ns
Fall Time	t_f	0.5	1.5	0.5	1.6	0.5	1.7	ns

NOTE:
Each MECL 10KH series circuit has been designed to meet the dc specifications shown in the test table, after thermal equilibrium has been established. The circuit is in a test socket or mounted on a printed circuit board and transverse air flow greater than 500 lfpm is maintained. Outputs are terminated through a 50-ohm resistor to −2.0 volts.

LOGIC DIAGRAM

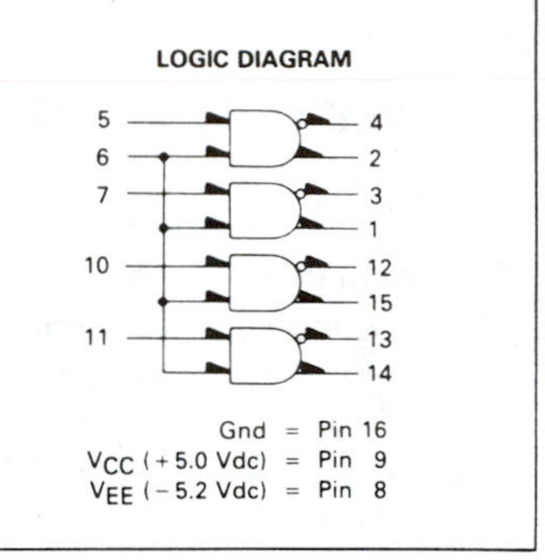

DIP
PIN ASSIGNMENT

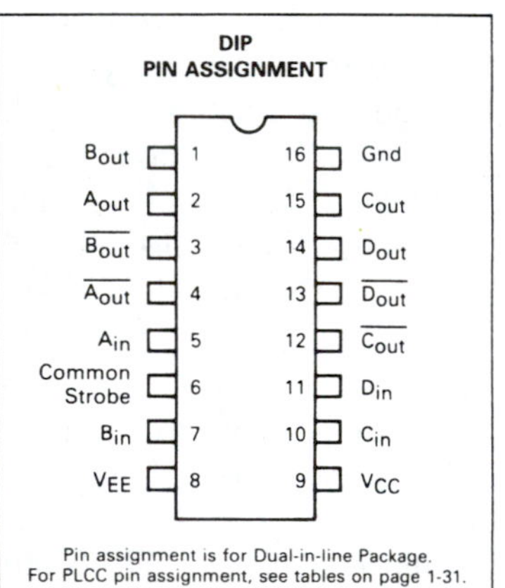

Pin assignment is for Dual-in-line Package.
For PLCC pin assignment, see tables on page 1-31.

TABLE 10.13

ECL to TTL translator (reprinted with permission from Motorola Corp.)

MC10H125

QUAD MECL-TO-TTL TRANSLATOR

The MC10H125 is a quad translator for interfacing data and control signals between the MECL section and saturated logic section of digital systems. The 10KH part is a functional/pinout duplication of the standard MECL 10K family part, with 100% improvement in propagation delay, and no increase in power-supply current.

- Propagation Delay, 2.5 ns Typical
- Voltage Compensated
- Improved Noise Margin 150 mV (Over Operating Voltage and Temperature Range)
- MECL 10K-Compatible

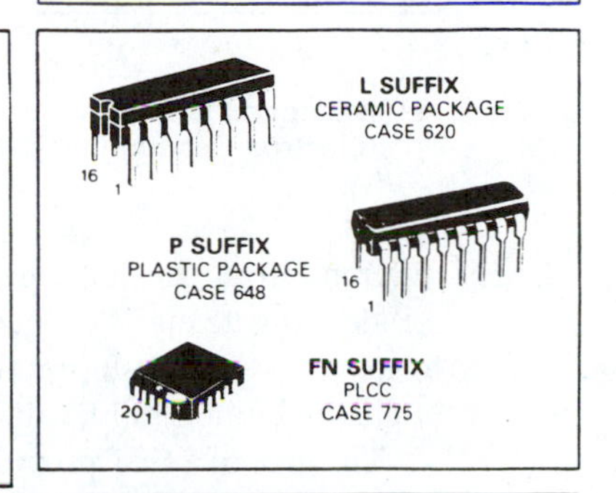

MAXIMUM RATINGS

Characteristic	Symbol	Rating	Unit
Power Supply (V_{CC} = 5.0 V)	V_{EE}	−8.0 to 0	Vdc
Power Supply (V_{EE} = −5.2 V)	V_{CC}	0 to +7.0	Vdc
Input Voltage (V_{CC} = 5.0 V)	V_I	0 to V_{EE}	Vdc
Operating Temperature Range	T_A	0 to +75	°C
Storage Temperature Range — Plastic — Ceramic	T_{stg}	−55 to 150 −55 to 165	°C °C

ELECTRICAL CHARACTERISTICS (V_{EE} = −5.2 V ±5%; V_{CC} = 5.0 V ± 5.0%)
(See Note)

Characteristic	Symbol	0°		25°		75°		Unit
		Min	Max	Min	Max	Min	Max	
Negative Power Supply Drain Current	I_E	—	44	—	40	—	44	mA
Positive Power Supply Drain Current	I_{CCH}	—	63	—	63	—	63	mA
	I_{CCL}	—	40	—	40	—	40	mA
Input Current	I_{inH}	—	225	—	145	—	145	μA
Input Leakage Current	I_{CBO}	—	1.5	—	1.0	—	1.0	μA
High Output Voltage I_{OH} = −1.0 mA	V_{OH}	2.5	—	2.5	—	2.5	—	Vdc
Low Output Voltage I_{OL} = +20 mA	V_{OL}	—	0.5	—	0.5	—	0.5	Vdc
High Input Voltage	V_{IH}	−1.17	−0.84	−1.13	−0.81	−1.07	−0.735	Vdc
Low Input Voltage	V_{IL}	−1.95	−1.48	−1.95	−1.48	−1.95	−1.45	Vdc
Short Circuit Current	I_{OS}	60	150	60	150	50	150	mA
Reference Voltage	V_{BB}	−1.38	−1.27	−1.35	−1.25	−1.31	−1.19	Vdc

AC PARAMETERS

Characteristic	Symbol	0° Min	0° Max	25° Min	25° Max	75° Min	75° Max	Unit
Propagation Delay	t_{pd}	0.8	3.3	0.85	3.35	0.9	3.4	ns
Rise Time	t_r	0.3	1.2	0.3	1.2	0.3	1.2	ns
Fall Time	t_f	0.3	1.2	0.3	1.2	0.3	1.2	ns

NOTE: Each MECL 10KH series circuit has been designed to meet the dc specifications shown in the test table, after thermal equilibrium has been established. The circuit is in a test socket or mounted on a printed circuit board and transverse air flow greater than 500 lfpm is maintained.

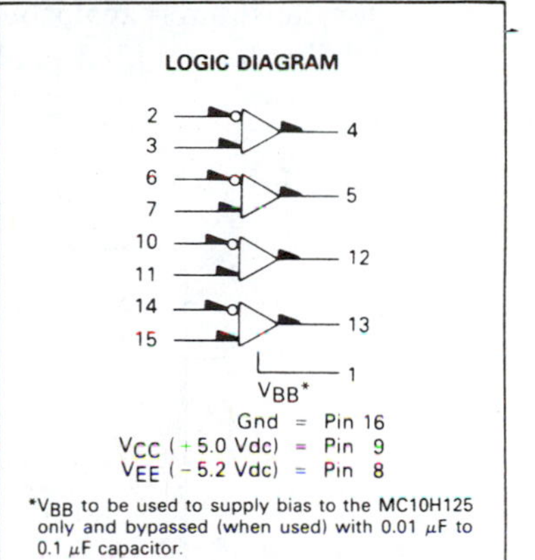

*V_{BB} to be used to supply bias to the MC10H125 only and bypassed (when used) with 0.01 μF to 0.1 μF capacitor.

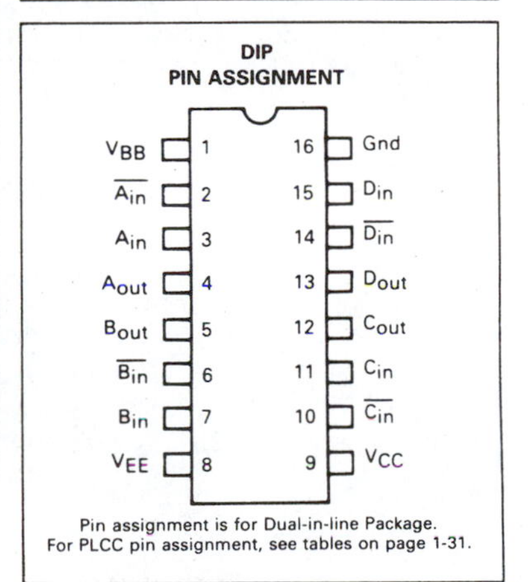

Pin assignment is for Dual-in-line Package.
For PLCC pin assignment, see tables on page 1-31.

Figure 10.56
Doped silicon conductors

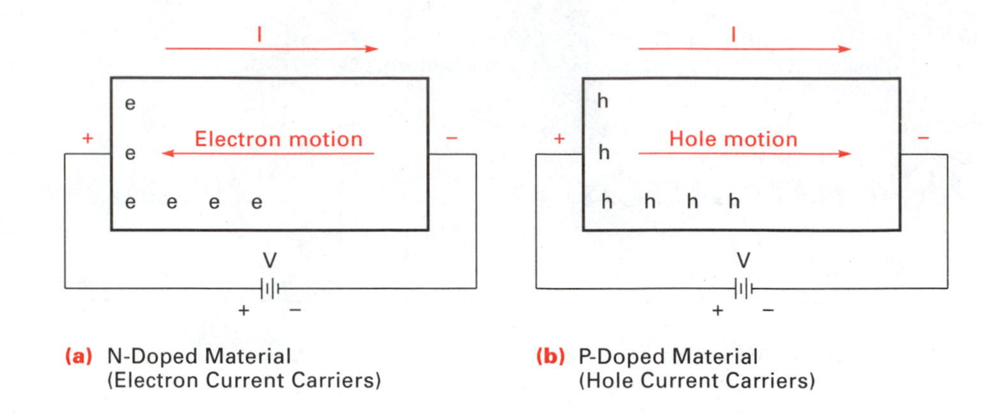

(a) N-Doped Material (Electron Current Carriers)

(b) P-Doped Material (Hole Current Carriers)

different from the logic gate previously discussed. An N-channel junction FET has a P-doped gate and a P-channel JFET has an N-doped gate. The gate diffusion actually forms a P–N or N–P junction with the main current channel. Figure 10.57 shows an N-channel junction FET diagram.

The gate of a FET provides the same control function as the base of a bipolar transistor: the drain is analogous to the collector and the source to the emitter. The schematic symbol for the JFET is compared to that of an NPN bipolar transistor in Figure 10.58.

Figure 10.57
An N-channel junction field effect transistor

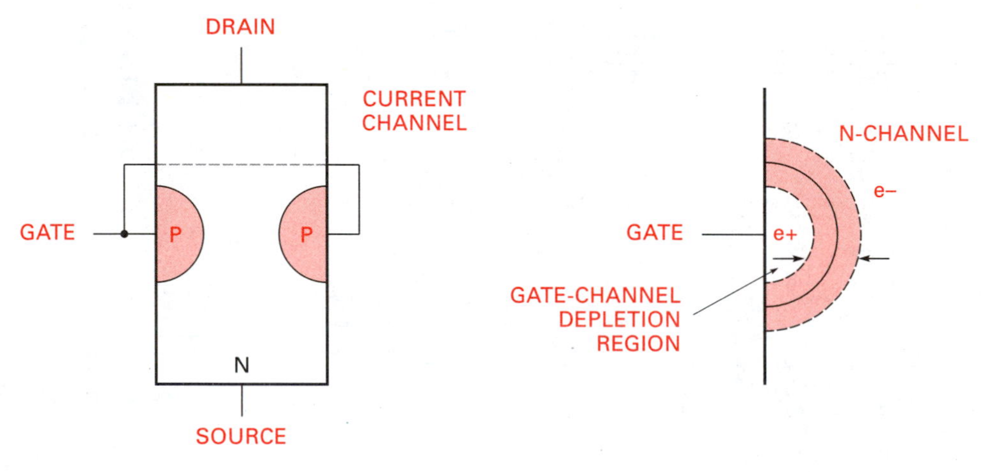

(a) N-Channel Junction FET Diagram

(b) Enlarged Gate-Channel Junction Region

Figure 10.58
JFET and NPN bipolar transistors

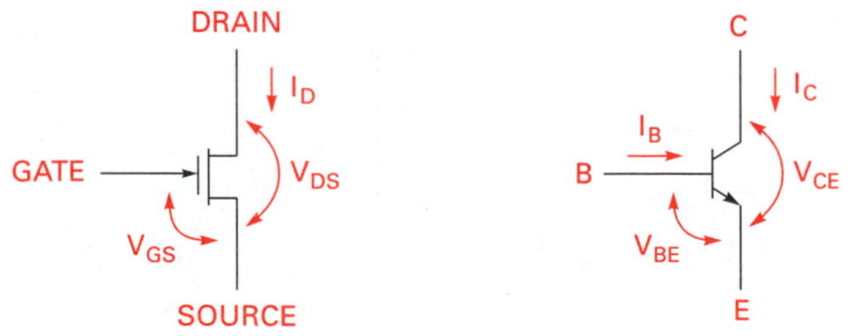

(a) N-Channel JFET Symbol

(b) NPN Bipolar Transistor

The N-channel JFET uses electrons as the drain–source majority current carriers. The gate-channel P–N junction is always reverse biased under normal operating conditions, so only leakage current flows from the gate to the drain–source current channel. This gives the JFET transistor a very high input impedance, on the order of 10^7 ohms. FETs have a forward transfer characteristic, g_m, that expresses the ratio of input voltage, V_{GS}, to output current, I_D:

$$g_m = I_d/V_{GS}$$

where g_m is a conductance parameter, the reciprocal of resistance, 1/R, and describes the relationship between the input voltage, V_{GS}, and output current, I_d. A bipolar transistor's forward transfer parameter is H_{FE} or beta (B). Beta defines the ratio between input current, I_b, and output current, I_c. The bipolar transistor is a current device and the FET is a voltage device. FETs have a characteristic set of curves that illustrates the relationship between input voltage and output current and voltage as shown in Figure 10.59.

The conduction behavior of the P–N gate–channel junction is determined by the junction voltage, V_{GS}. As in any P–N junction a depletion region exists; the depletion region width is driven by the amount of reverse bias across the junction. When $V_{GS} =$ 0 V, for example, the minimum reverse bias is applied and the depletion region is small, allowing maximum area for channel current. This is the equivalent of a bipolar transistor's $I_b = I_{b(SAT)}$. Note that, on the curves shown in Figure 10.59, the maximum drain–source current, I_{DSS}, occurs when $V_{GS} = 0$ V. When the value of V_{GS} increases in a negative direction, the amount of reverse bias on the gate–channel junction increases, causing an increase in the size of the depletion region. Input voltage, V_{GS}, determines the relative area of the drain–source current channel as indicated in Figure 10.60. This, in effect, is seen as an increase in channel resistance that reduces the drain–source current, I_D.

The point at which V_{GS} is sufficiently negative that no majority current flows in the drain–source channel is called *pinch off*. The load line, $1/R_L$, intersects the vertical axis at I_{DSS} and the horizontal axis at V_{DD}. Figure 10.60 is a pictorial representation of the gate depletion region, with increasingly negative values of V_{GS}, and shows the resulting reduction in channel area causing a decrease in drain current.

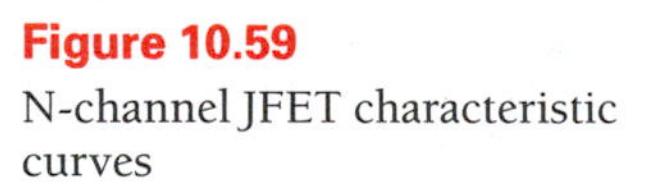

Figure 10.59
N-channel JFET characteristic curves

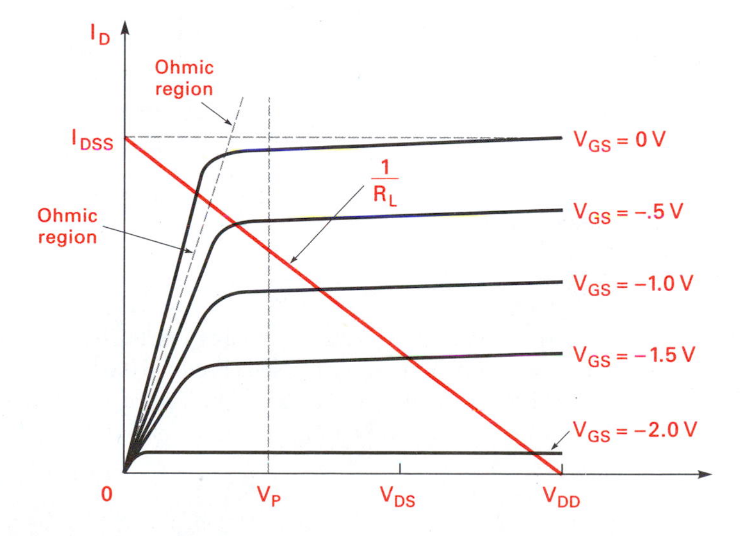

Figure 10.60
Gate channel junction depletion region size increases as V_{GS} goes negative for an N-channel JFET transistor

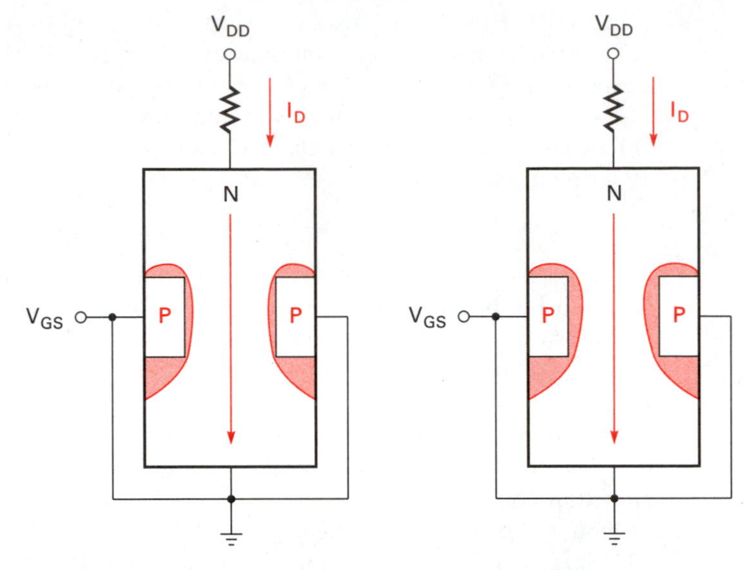

(a) Maximum Channel Area, $V_{GS} = 0$ V, $I_D = I_{DSS}$

(b) Reduced Channel Area, $V_{GS} = -1$ V

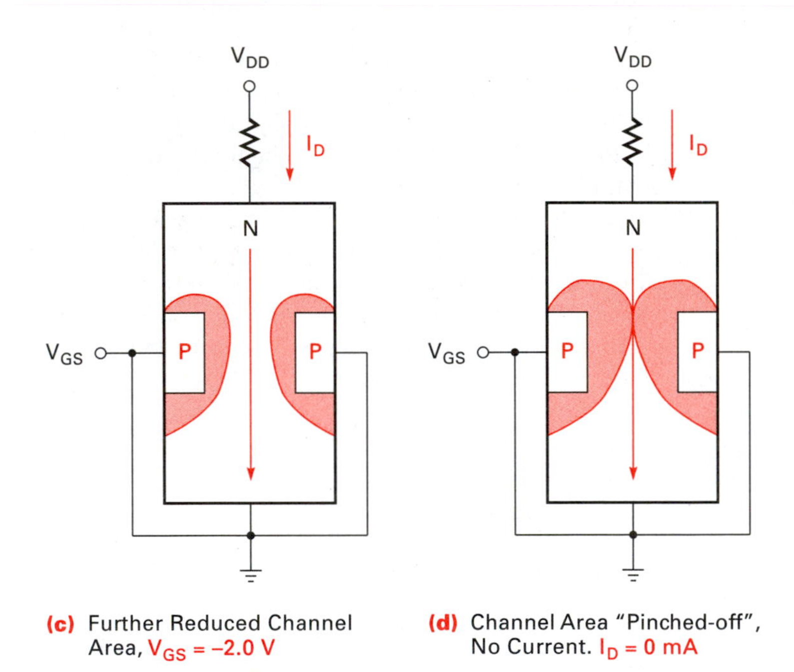

(c) Further Reduced Channel Area, $V_{GS} = -2.0$ V

(d) Channel Area "Pinched-off", No Current. $I_D = 0$ mA

The JFET can be used as a switching circuit in much the same manner as the **BJT** (bipolar junction transistor) is used. Figure 10.61 shows a JFET inverter circuit. The N-channel JFET in the circuit shown in Figure 10.61 is a 2N5458, which is a general purpose FET for switching applications. A partial list of device parameters is given here for convenience:

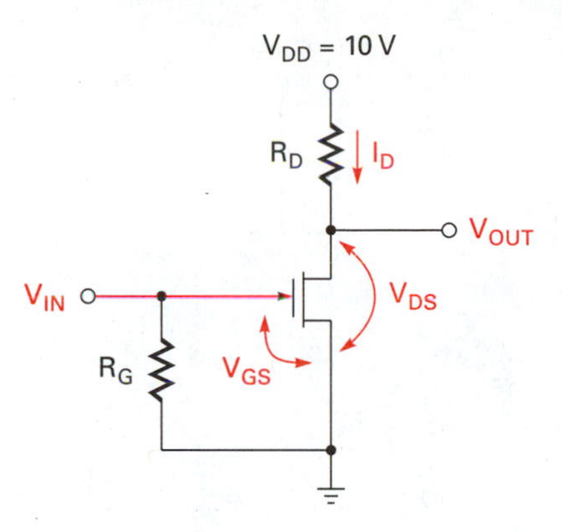

Figure 10.61
An N-channel JFET inverter

$V_{GS(off)} = -1$ V minimum to -7 V maximum

$g_M = 1500$ minimum to 5500 maximum (millisiemens)

$I_{D(off)} = 10$ nA dc maximum

$I_{GSS} = 1$ nA (reverse gate leakage current)

Assume $V_{DD} = 10$ V and assume $I_{D(SAT)} = 2$ mA

The N-channel JFET will be cut off when $V_{IN} = V_{GS(off)} = -3.5$ V (in the middle of the range just given). The FET will be in the ohmic region (on) when $V_{IN} = V_{GS(on)} = 0$ V. By restricting V_{IN} to operate between 0 V and -5 V, we ensure that the transistor will be on (ohmic region) or off.

Although JFETs can be used as logic switches, the MOSFET is the field effect transistor of choice in designing logic integrated circuits.

10.7.2 MOSFETs

The construction of a **MOSFET** differs somewhat from that of a JFET. The drain–source current channel still exists; however, no gate–channel junction is used. Instead the gate is insulated from the channel by a thin layer of silicon dioxide (glass) that prevents any gate current. For this reason the input impedance of the MOSFET is extremely high, on the order of 10^{10} to 10^{12} ohms. The gate-to-source voltage, V_{GS}, provides control of the drain–source channel current, I_D, through an electric field emanating from the gate into the channel, instead of creating a depletion region as in the JFET. The channel electric field strength depends upon the gate voltage amplitude. As V_{GS} changes, the electric field strength changes, which in turn controls the drain–source current.

Two main types of MOSFETs exist, enhancement mode and depletion mode. Each mode can be used to create an N-channel or P-channel device. An N-channel enhancement-mode MOSFET, NMOS, configuration is illustrated in Figure 10.62.

Notice the layer of SiO_2 (glass) insulating the gate from the rest of the MOSFET. This layer of glass is what gives the device the oxide part of its name. Notice that the N-doped regions at both the drain and the source protrude only part way into the substrate. There is no direct current channel between the drain and source. That is, negative current carriers (electrons) are not doped into the entire length of the channel. Current flows

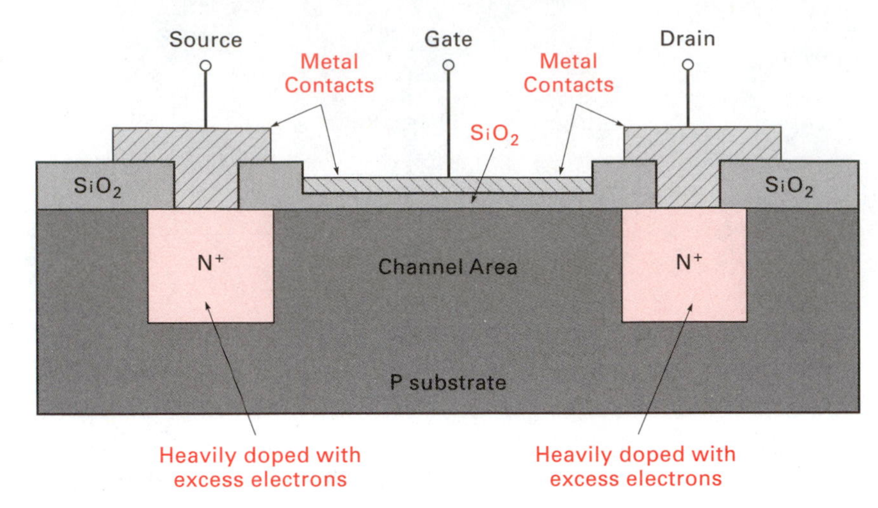

Figure 10.62
NMOS enhancement-mode FET configuration

when a positive electric field, created from a positive gate-to-source potential, V_{GS}, is present in the area between the drain and source terminals. The positive current carriers (holes) are pushed (like charges repel) out of the region. Free electrons in the P-type substrate are attracted to the positive gate potential, inducing a current channel between the drain and the source.

When the drain terminal is at a more positive potential than the source and the gate is at a positive potential with respect to the source, an "enhanced" channel is created and current flow results. The **enhancement-mode** MOSFET causes a current channel to occur by biasing the gate to create an electric field that repels the holes (in the case of an N-channel device) and attracts free electrons. Figure 10.63 illustrates the "enhanced" current channel. Notice that the drain voltage is pulled up to V_{DD} through a resistor and is positive with respect to the source. An electric field is generated by the V_{GS} potential between the gate and source.

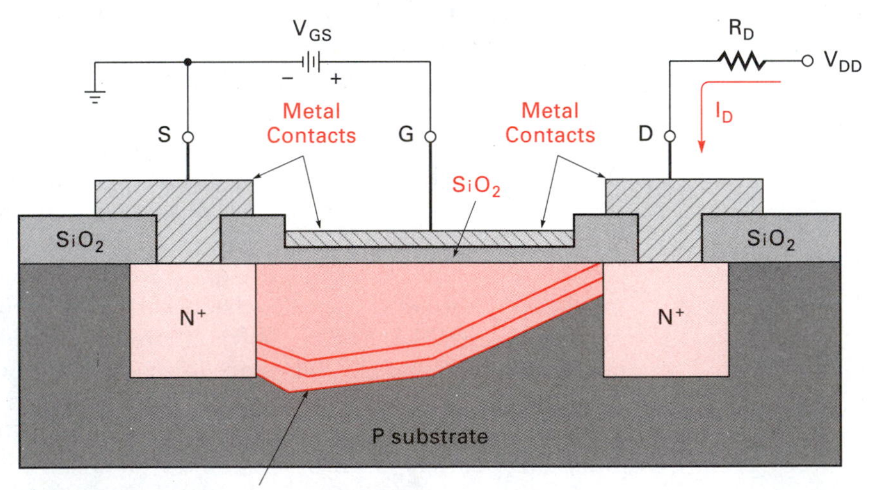

Figure 10.63
NMOS enhancement-mode FET showing drain–source "enhanced" current channel

As the electric field "pushes" the positive holes from the area between the source and drain, the free electrons are attracted to form a current channel. The electric field forms a wedgelike channel from the source to the drain. The N^+ at the drain and source indicate heavily doped N-material. The metal contacts in Figure 10.63 provide electrical connections for the source, drain, and gates.

The same type of operation is possible with a P-channel MOSFET, except that the substrate is N-type and the drain and source areas doped into the substrate are P-type. The electric field polarity needed to "enhance" a current channel is negative instead of positive.

An N-channel MOSFET, NMOS, requires a positive gate voltage to "enhance" a current channel and generate drain current, I_D, flow. The greater is the positive gate-to-source voltage, V_{GS}, the greater is the positive field and therefore the greater the "enhanced" channel area, producing a greater drain current, I_D. Any negative V_{GS} cuts off the drain source current, so $I_D = 0$.

P-channel MOSFETs, PMOS, require a negative V_{GS} to "enhance" a drain-to-source channel; any positive V_{GS} causes the device to be in cutoff. Typical characteristic curves for enhancement-mode NMOS and PMOS field effect transistors are shown in Figure 10.64.

Notice that all values of V_{GS} are positive for the NMOS and negative for the PMOS. The symbols for MOSFETs are different than those used for JFETs. Figure 10.65 shows the symbols for NMOS and PMOS.

Depletion-mode devices are the second major type of MOSFET. Figure 10.66 shows an N-channel depletion-mode MOSFET configuration. Note the lightly doped channel between the more heavily doped drain and source regions. This is different from the enhancement-mode MOSFET, where the channel had to be "created" by the electric field of V_{GS}. In the depletion mode MOSFET a channel exists when $V_{GS} = 0$ V.

NMOS depletion-mode FET operation requires that the majority current carriers (electrons) be pushed out of the lightly doped drain and source channel by the electric

Figure 10.64

Typical enhancement-mode MOSFET characteristic curves

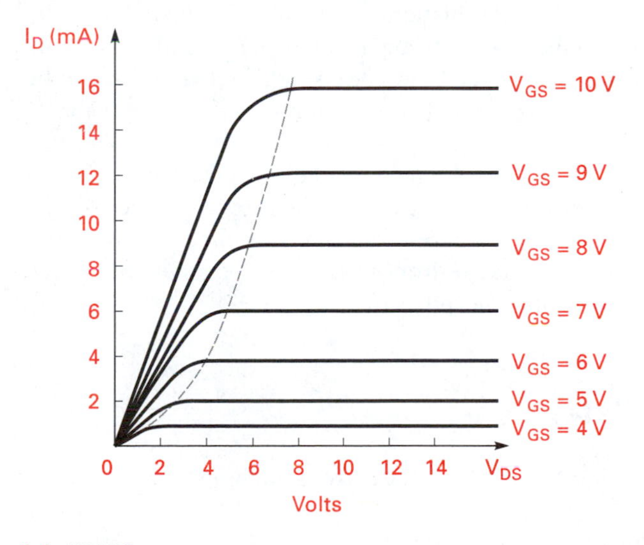

(a) NMOS

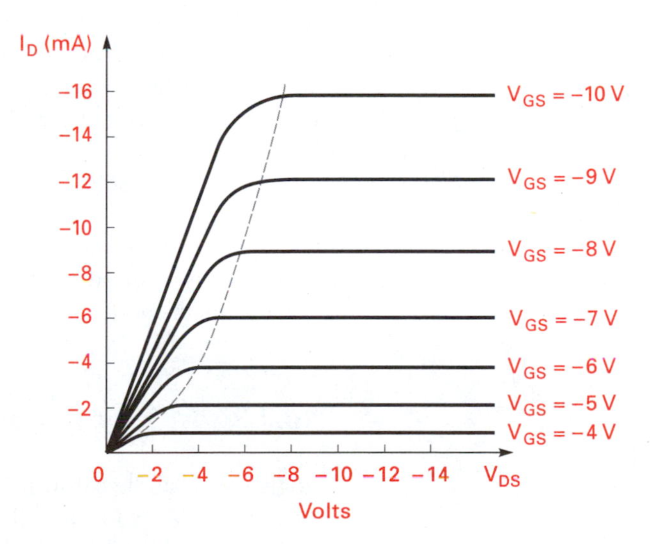

(b) PMOS

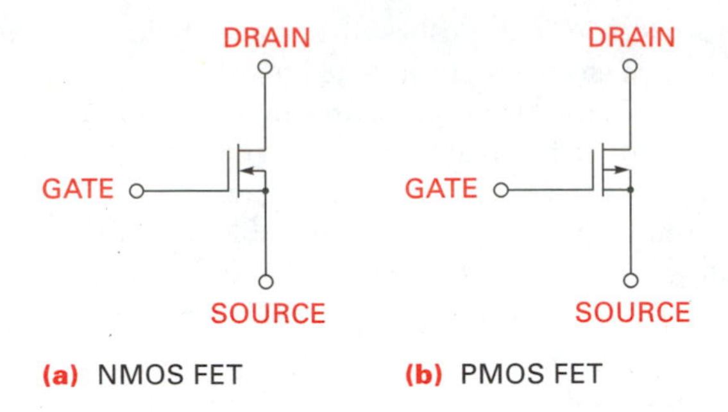

Figure 10.65
NMOS and PMOS FET circuit symbols

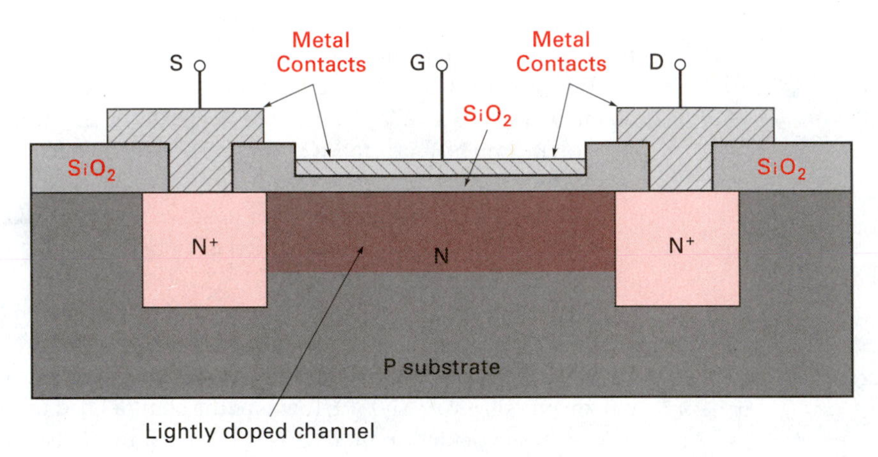

Figure 10.66
NMOS depletion-mode FET configuration

field generated by the gate-to-source voltage, V_{GS} (see Figure 10.66). The enhancement-mode MOSFET "created" the channel; the depletion-mode MOSFET takes away the channel. When the negative potential between gate and source, $-V_{GS}$, is increased, the electrons in the lightly doped region are "pushed" away causing the channel area to be reduced, thus reducing drain current. By varying the amount of $-V_{GS}$, the available current carriers are controlled and therefore I_D is regulated.

Figure 10.67 shows the lightly doped N region channel between the heavily doped N^+ regions being "depleted" by the electric field created by the gate-to-source voltage.

Figure 10.68 shows the characteristic curves for depletion-mode NMOS and PMOS FETs. Note that the values of V_{GS} range from negative to positive, unlike the enhancement-mode devices that allow only one polarity.

10.7.3 MOSFET Logic Gates

Logic gates can be constructed using MOSFETs as the active switching elements. Consider the NMOS inverter shown in Figure 10.69. When the input is a logical 0, the NMOS FET is cut off, causing the output to be a logical 1.

Figure 10.67
Depletion-mode NMOS FET drain–source channel

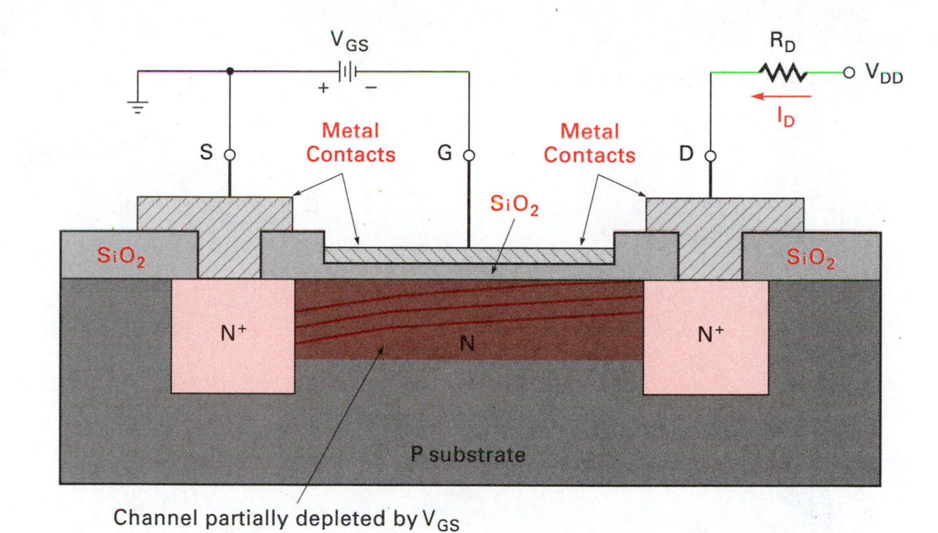

Figure 10.68
Depletion-mode NMOS and PMOS characteristic curves

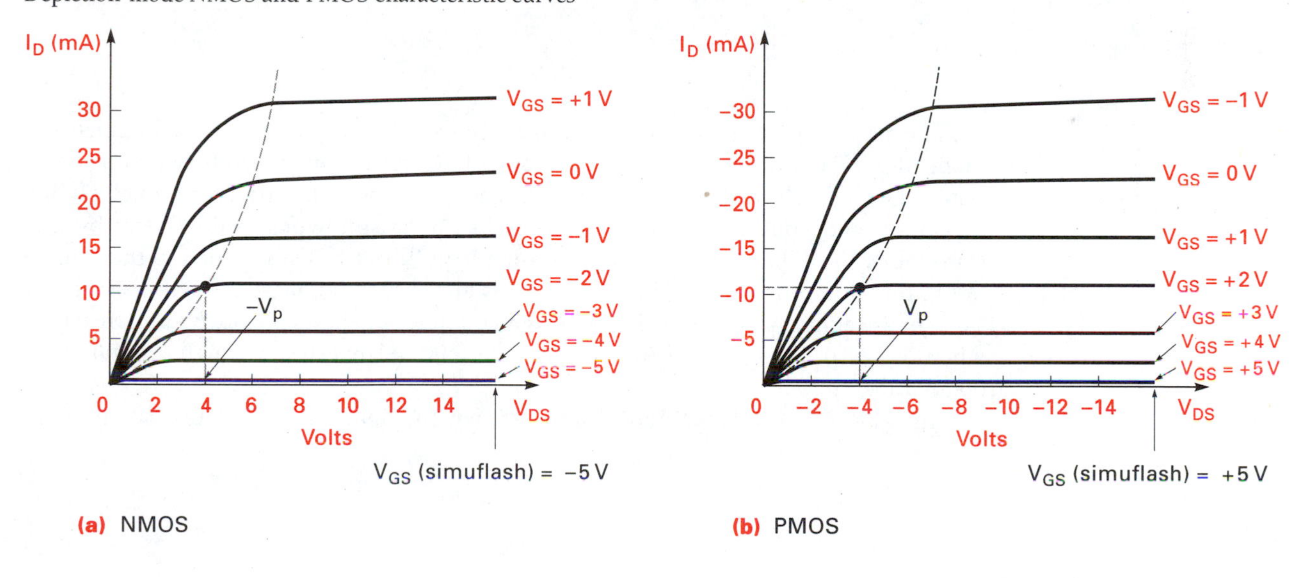

Figure 10.69
Simple NMOS FET inverter using a passive pull-up resistor

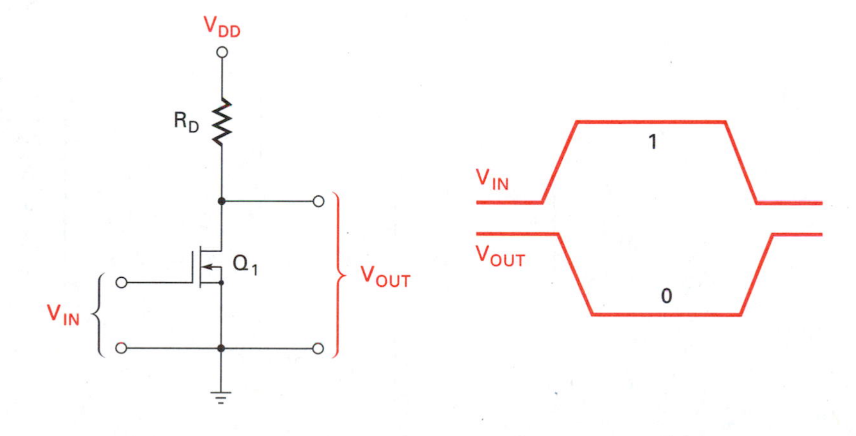

Figure 10.70
NMOS inverter with active pull-up transistor

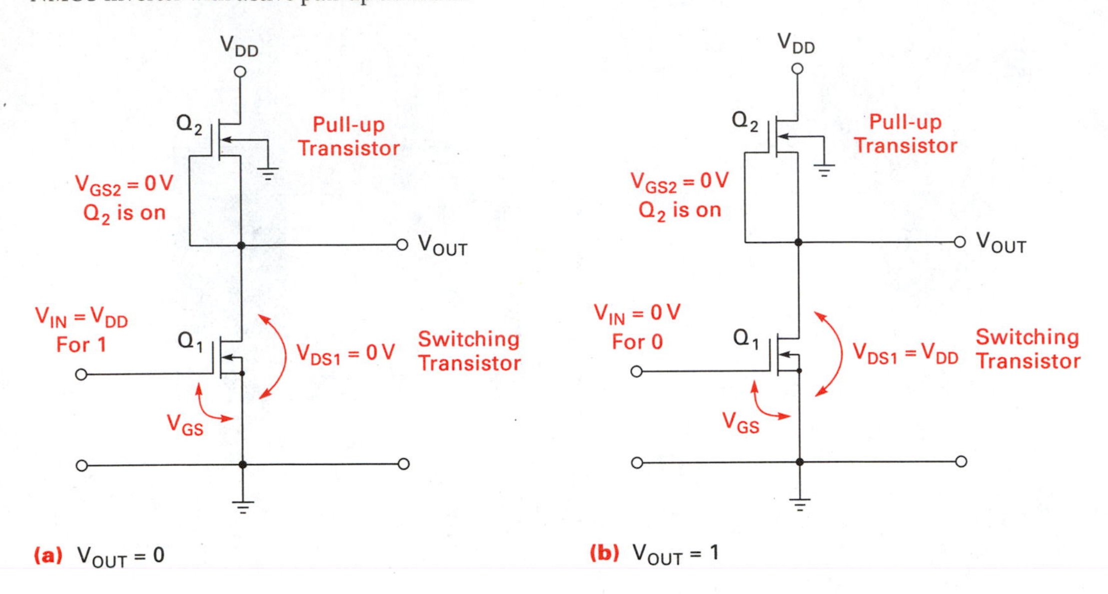

The passive pull-up resistor, R_D, can be replaced by a properly biased depletion-mode MOSFET connected in the drain circuit of the enhancement-mode switching transistor, Q_1, as indicated in Figure 10.70. The drain–source current channel of the depletion-mode pull-up transistor behaves like a resistor whose value is set by the transistor's V_{GS}. The gate of the depletion-mode pull-up transistor is tied to the source of Q_1; therefore, the pull-up transistor's V_{GS} is always 0 volts, biasing the transistor on. The enhancement-mode NMOS FET switch conducts when V_{IN} is positive and turns off when V_{IN} is 0 V. Figure 10.70 illustrates the NMOS inverter when the output is a 0 and a 1.

An NMOS two-input NAND gate is shown in Figure 10.71. Note that the two

Figure 10.71
NMOS NAND gate

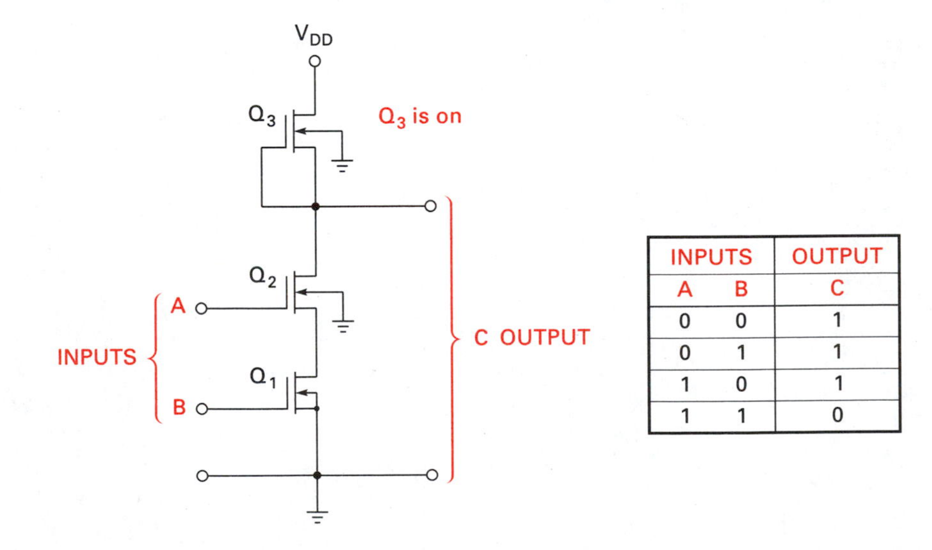

INPUTS		OUTPUT
A	B	C
0	0	1
0	1	1
1	0	1
1	1	0

enhancement-mode switching transistors are connected in series, with a depletion-mode pull-up transistor acting as the active load.

A logical 1 (positive V_{IN}) turns on the respective input transistor and a logical 0 ($V_{IN} = 0$ V) turns the transistor off. The pull-up transistor is biased on by the 0 volt V_{GS}. Both switching transistors must be on to generate a logical 0 at the output. If either or both are off the output is a logical 1. This behavior generates the truth table for a NAND.

Figure 10.72 shows a two-input NMOS NOR gate, with the switching transistors connected in parallel. The same logical input levels apply as with the NAND gate. If either or both of the transistors is on (a logical 1 input) the output is a logical 0. Both switching transistors must be off (logical 0 input) to generate a logical 1 output.

NMOS technology is still used in contemporary devices, particularly in large- and very large-scale integrated circuits, like microprocessors. PMOS was the first of the MOS technologies to be integrated, offering significant power savings and easier to fabricate than the early NMOS circuits. But, as IC fabrication capability improved, the problems associated with manufacturing NMOS were solved. Because of the speed advantage it has over PMOS, NMOS became the technology of choice for logic designers. NMOS transistors are inherently faster than PMOS due to the greater mobility of electrons than holes. PMOS logic circuits still occupy a niche in applications where speed is not important but power dissipation is critical.

10.7.4 CMOS Logic Gates

Complementary metal oxide semiconductors, CMOS, make use of both NMOS and PMOS transistors in the same logic gate. The *C* in CMOS stands for complementary; NMOS and PMOS are complementary devices. NMOS devices have electrons as the majority current carriers, and PMOS transistors use holes as the majority current carriers. The advantage of using both transistors in the same logic gate comes from the value of V_{GS} needed to enable the drain–source current channel. A logic 1 (positive V_{GS}) turns on an NMOS and turns off a PMOS transistor; a logic 0 does the opposite. By connecting the complementary transistors as shown in Figure 10.73 we can create a CMOS inverter.

The NMOS enhancement-mode transistor is on the bottom and the PMOS

Figure 10.72
NMOS two-input NOR gate

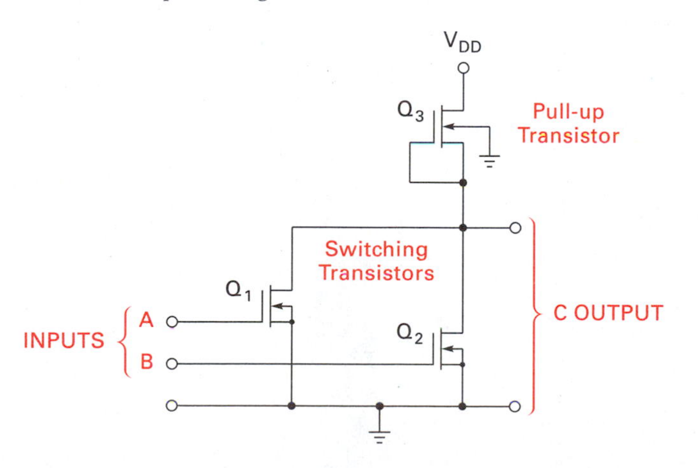

Figure 10.73
CMOS inverter

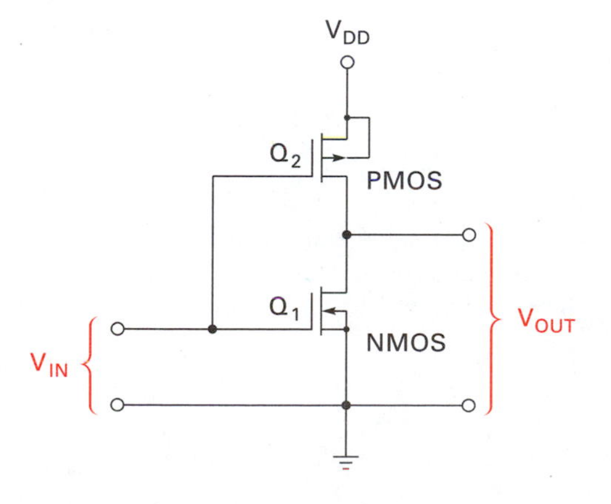

enhancement-mode transistor is at the top of the diagram. Notice that the PMOS transistor substrate arrow is connected to V_{DD}. The gates of both switching transistors are connected to one another, providing the input node to the circuit.

When $V_{IN} = 0$ V (logical 0), then Q_1 is cut off (NMOS needs a $+V_{GS}$). $V_{IN} = 0$ V is also present at the gate of Q_2. Because the substrate of Q_2 is connected to V_{DD} and the gate is at 0 V, then $V_{GS2} = -V_{DD}$, which turns on Q_2. When $V_{IN} = V_{DD}$ (logical 1), then Q_1 is conducting ($+V_{GS}$).

The logical 1 voltage is also at the gate of Q_2, causing that transistor to be cut off. The action of Q_1 and Q_2 is such that when one is conducting the other is cut off, so that the only time both transistors are conducting is during switching. This feature means that the current path from the power supply to ground is through a cut off transistor, resulting in very low power supply current. Both transistors are conducting for a brief time, during switching. Figure 10.74 illustrates the switching action of the two transistors in the CMOS inverter shown in Figure 10.73. The overlap areas in the diagram are expanded to illustrate the time when both transistors are conducting. The early CMOS logic gates had rather slow rise and fall times, causing the switching current calculations to be rather lengthy and dependent on a host of external conditions. The new high-speed CMOS devices have considerably better switching characteristics.

The low power requirement of the CMOS is one of its major advantages, and as we will see a bit later the speed is being improved continually. The voltage transfer curve for a CMOS inverter is shown in Figure 10.75.

As V_{IN} increases from 0 V toward V_{DD}, V_{OUT} remains at or near V_{DD} (logical 1) until the point where V_{IN} reaches the threshold voltage, V_T. Then V_{OUT} starts to "roll off," the slope becoming steeper as V_{IN} continues to increase. The NMOS transistor, Q_1, begins conducting when V_{IN} reaches V_T (point *b* on the curve). From points *b* to *c*, Q_1 goes from cutoff to saturation and Q_2 remains on but not saturated. At point *c* both transistors become saturated. During the steepest part of the transfer curve, both transistors are saturated; this occurs when the input changes state from a logical 0 to a logical 1. From points *d* to *e*, Q_1 comes out of saturation but remains conducting while Q_2 remains in saturation. The curve levels off from points *e* to *f*; Q_1 is on but not saturated while Q_2 is cut off and the output is a 0. Figure 10.76 illustrates a two-input CMOS NAND gate. Note that the two NMOS transistors on the bottom of the diagram are connected in series. Also, the two PMOS transistors on top are connected in parallel. The four transistors form two pairs; each pair contains both N and P MOSFETS with connected gates.

Let V_A and $V_B = 0$. Transistor pair Q_3 and Q_4 (NMOS) are cut off. Transistor pair Q_1 and Q_2 (PMOS) are conducting, creating a logical 1 at the output. Remember that a cutoff MOSFET has a very high drain-to-source resistance.

Let $V_A = 0$ and $V_B = 1$; therefore, Q_3 is cut off, Q_4 is conducting, Q_1 is on, and Q_2 is

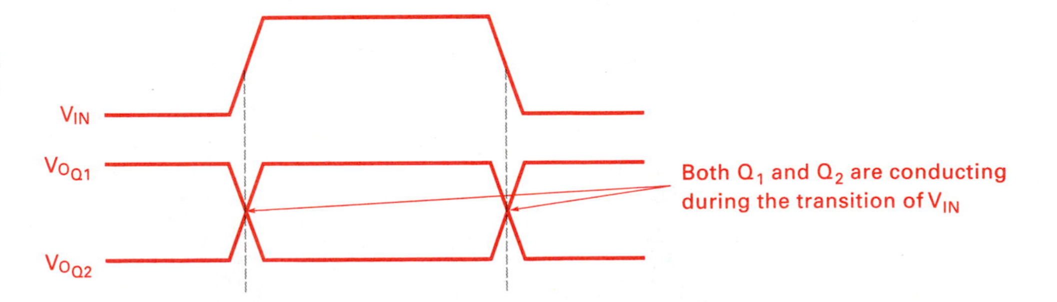

Figure 10.74
Switching action of NMOS and PMOS transistors of the circuit shown in Figure 10.73

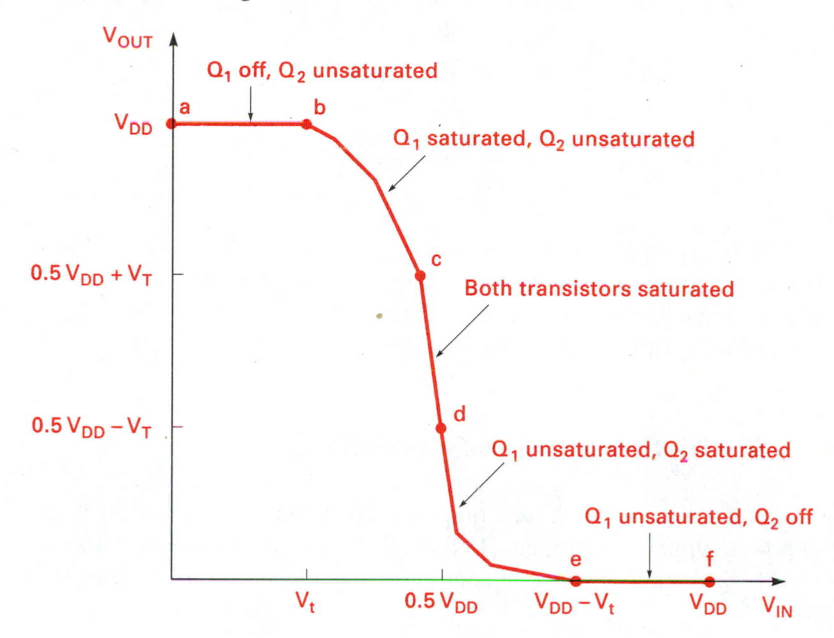

Figure 10.75
CMOS inverter voltage transfer curve

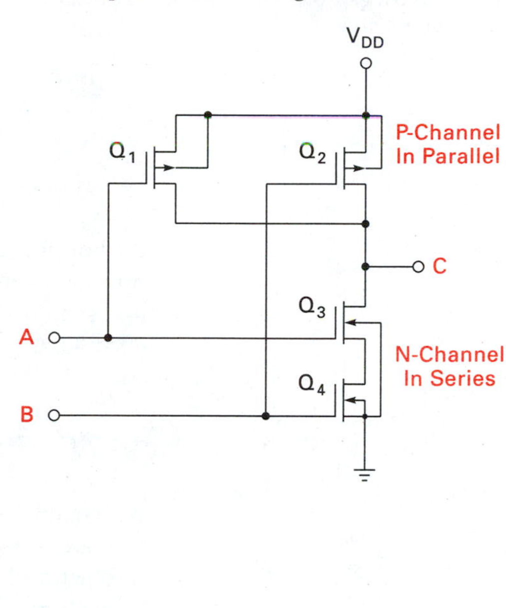

Figure 10.76
A two-input CMOS NAND gate circuit

off. Because Q_1 and Q_2 are in parallel, it takes only one transistor conducting to provide a direct connection between V_{DD} and V_{OUT}; therefore, $V_{OUT} = 1$.

Let $V_A = 1$ and $V_B = 0$. The opposite condition exists in this case. Reverse the on–off conditions of the transistors; therefore, $V_{OUT} = 1$.

Let V_A and $V_B = 1$. Both Q_3 and Q_4 are conducting and Q_1 and Q_2 are cut off. The two off transistors present a high impedance between V_{DD} and the two series conducting transistors. Transistors Q_3 and Q_4 present a low impedance from V_{OUT} to ground. As a result, V_{OUT} is held at or close to ground by the low impedance of the conducting transistors; therefore, the output is a 0.

Figure 10.77 shows a two-input NOR CMOS gate. The two NMOS transistors are on

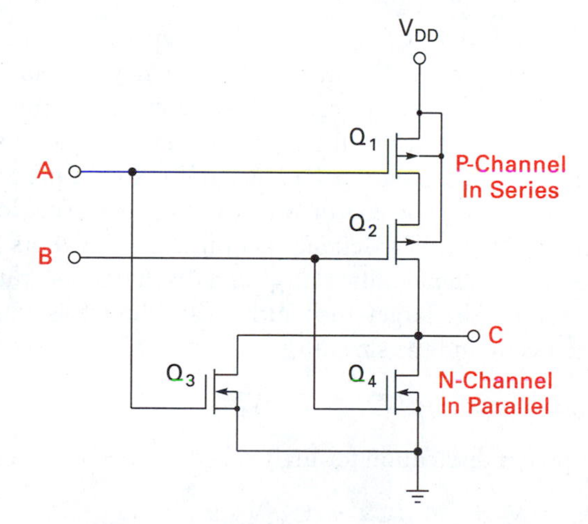

Figure 10.77
CMOS two-input NOR gate circuit

TABLE 10.14
CMOS NOR gate showing transistor status

V_A	V_B	Q_1	Q_2	Q_3	Q_4	V_{OUT}
0	0	On	On	Off	Off	1
0	1	On	Off	Off	On	0
1	0	Off	On	On	Off	0
1	1	Off	Off	On	On	0

the bottom and in parallel (compared to being in series with the NAND). The two PMOS transistors are on top and in series (they were in parallel with the NAND). The transistors switch at the same logic levels as before. The logical operation of the NOR and the effect on the individual transistors is given in Table 10.14.

10.7.5 Power Dissipation for High-Speed CMOS

In general CMOS logic families, all are misers on power consumption and operate from a greater range of power supply voltages than TTL and ECL logic device families. Power dissipation for CMOS gates in a steady state condition (not switching) is given by the following equation:

$$P_D = V_{DD}I_{DD} \tag{10.1}$$

I_{DD} is typically quite low, on the order of microamps, because the current path from V_{DD} to ground is through a cutoff CMOS transistor. The HCT CMOS device family input and output voltage levels are identical to TTL (hence the *T* in HCT). The HCT steady state power dissipation is given by the equation:

$$P_D = I_{DD}V_{DD} + nI_{DD}V_{DD} \tag{10.2}$$

Variable n is the number of inputs at the TTL V_{IH} level.

Dynamic (switching) power dissipation is increased due to capacitive loads being driven by an HC or HCT output:

$$P_D = C_L V_{DD}^2 F \tag{10.3}$$

In equation (10.3), C_L is the total load capacitance seen by the gate output and F is the switching frequency. The switching frequency determines power dissipation because both the N- and P-channel transistors are on briefly during the switching transition. The more transitions that occur, the higher I_{DD} is, because as frequency increases the transient time becomes a greater percentage of the total time.

Load capacitance consists of two parts: C_{ex} (external load) and C_{pd} (internal output capacitance). The load capacitance is of interest due to its impact on signal propagation delay. The output signal must charge and discharge the output capacitance, which takes time. Therefore, the larger the output capacitance is, the longer it takes for voltage changes to occur and the slower the circuit's propagation delay:

$$P_D(\text{dynamic}) = (C_{ex} + C_{pd})V_{DD}^2 F \tag{10.4}$$

The total power dissipation for high-speed CMOS becomes

$$P_D = V_{DD}I_{DD} + (C_{ex} + C_{pd})V_{DD}^2 F \tag{10.5}$$

10.7.6 Propagation Delay for High-Speed CMOS

Propagation delay in HC and HCT CMOS devices is also affected by load capacitance. The output short circuit current, I_s, given as a specification for HC and HCT gates, is 18.5 mA at 25° C, with a V_{DD} of 4.5 V:

$$I_s = C\ V/t \qquad (10.6)$$

where V is the output voltage range.

$V = .5V_{DD}$ because propagation delay is measured at the 50% point and the total voltage swing is approximately V_{DD} to ground. Then equation (10.6) becomes

$$I_s = C\ .5V_{DD}/t \qquad (10.7)$$

The propagation delay equation separates the value C_L, which is a variable depending on the fan-out, circuit board construction, and other factors that affect external load capacitance and the actual output capacitance of the CMOS device. The propagation delay parameters given in the high-speed data books refer to a C_L of 50 pF. More capacitance adds to the delay; less subtracts from the delay:

$$T_{pd} = T_p + [.5V_{DD}(C_L - 50\ \text{pF})]/I_s \qquad (10.8)$$

where

T_p = specified propagation delay with 50 pF C_L

C_L = actual load capacitance (external to device)

T_{pd} = total propagation delay

For example, consider a 74HC00 gate driving a 150 pF load capacitance. Let V_{DD} = 5 V, T_p = 18 ns (with a 50 pF load), C_L = 150 pF, I_s = 19 mA. Then propagation delay is found:

$$t_{pd} = 18\ \text{nS} + 2.5\ \text{V}(150\ \text{pF} - 50\ \text{pF})/19\ \text{mA}$$

$$t_{pd} = 31.1\ \text{ns}$$

10.7.7 CMOS Noise Margins

CMOS integrated circuits have the best noise margins of any of the three basic digital logic families. A logical 0 noise margin (the difference between the output 0 and the required input 0) is between 0 and 30% of the supply voltage. The logical 1 noise margin is between 70% of V_{DD} and 100% of V_{DD}. This gives the CMOS logic family a superior position in electrically "noisy" environments compared to other logic families.

10.7.8 CMOS Subfamilies

The first available CMOS family was the 4000 series, introduced by RCA in 1968. The 4000 series chips were much slower than the newer CMOS devices and for that reason are seldom used in new designs. The 4000 subfamily devices have a high-capacitance metal gate that introduces considerable capacitance to the input pin, causing delays of 50 ns or greater. The series also dissipates approximately 1 μW of static power and .1 mW at a frequency of 100K Hz. This compares to the HC and HCT single digit delay times and power consumptions in the order of 2 nW static and .17 mW dynamic (100K Hz). The 4000 series ICs typically operates with 15 V power supplies.

For power-conscious designers, the 4000 series CMOS was the answer, but for versatility of component choices and low power supply voltages, designers had to wait for the HC and HCT series CMOS devices. The HC series allows digital designers to take advantage of the low power requirements of CMOS and still operate the ICs on +5 V. This permits the systems designer to mix CMOS and TTL integrated circuits on the same printed circuit board without the requirement of another power supply.

The **HC** (high-speed CMOS) series devices retain the voltage transfer curve typical of MOSFET transistors as illustrated in Figure 10.75. This produced excellent noise margins but limited the interfacing of the CMOS part directly to TTL. Enter the **HCT** technology (high-speed CMOS TTL compatible), which has the same logical 0 and 1 voltage levels as TTL devices. This made it possible to selectively interchange CMOS for TTL. The HCT series reached the TTL logic voltage levels when the integrated circuit designers were able to reliably make MOSFET transistors with lower threshold voltages, V_T, than before.

The rules for mixing TTL and HCT CMOS devices are the same as we developed earlier for mixing subfamilies of TTL circuits. The voltage levels are compatible, but current levels are different. **AC** (advanced CMOS) and **ACT** (advanced CMOS TTL compatible) are the latest additions to the CMOS subfamily group. They offer improved switching speed and lower power consumption (speed–power product) than the HC and HCT subfamilies. All are still susceptible to increased power consumption with increasing frequency.

Table 10.15 shows a relative comparison of specifications for the various TTL com-

TABLE 10.15

Contemporary CMOS IC specifications

Parameter	74HC	74HCT	74AC	74ACT
V_{CC} (V)	5	5	5	5
V_{OL} (V) (max)	0.1	0.1	0.1	0.1
V_{OH} (V) (min)	4.9	4.9	4.9	4.9
V_{IL} (V) (max)	1.35	.8	1.35	.8
V_{IH} (V) (min)	3.85	2.0	3.35	2.0
I_{OL} (I) (max) (mA)	4	4	24	24
I_{OH} (I) (mA)	−4	−4	−24	−24
I_{IL} (I) +/−(μA)	1	1	1	1
t_{pd} (ns)	18	18	11.1	12.3
P_d (steady) (μW)	2.5	2.5	5	5
P_d (dynamic) (mW)	.6	.6	.75	.75
P_d (total) (mW)				
at 100K Hz	.0625	.0625	.08	.08
at 1M Hz	.6025	.6025	.755	.755
at 10M Hz	6.0	6.0	7.5	7.5
Speed–power product (pj)				
at 100K Hz	1.1	1.1	.4	.4
at 1M Hz	10.8	10.8	3.9	3.9
at 10M Hz	108	108	39	39

patible CMOS subfamilies. We can see from Table 10.15 that the speed–power product, measured in picojoules, is considerably lower for the AC and ACT CMOS families, making them more desirable for new designs than the older TTL compatible versions of CMOS.

SUMMARY

A. Digital circuit fundamentals are explored:

1. Diode switches and logic gates (AND and OR)
2. Transistor switches (a simple inverter design is presented)
3. Diode transistor logic (DTL NAND gate circuit
4. OTL to TTL evolution is presented

B. TTL circuits are analyzed, including

1. Totem-pole output circuit
2. Phase-splitter to drive the totem-pole output circuit
3. Multiple-emitter input circuit
4. Voltage transfer curve description
5. TTL specifications: temperature effects on the voltage transfer curve, worst-case measurements of voltage and current specifications, and noise margins
6. TTL subfamilies: standard, low-power, and high-speed circuits; low-power Schottky and Schottky circuits; advanced low-power Schottky and advanced Schottky circuits
7. Open-collector TTL circuits: level shifting, wired-OR, and relay drivers
8. Tristate TTL circuit
9. Mixed TTL subfamily fan-out considerations.
10. Other TTL circuits: AND gate and Schmitt trigger TTL

C. ECL (emitter-coupled logic) circuits are discussed:

1. ECL differential amplifier switches
2. Parallel input transistor switches
3. Reference voltage circuit
4. Output drivers (NOR and OR)
5. ECL specifications (logic levels for 0 and 1, noise margins, and voltage transfer curve)
6. ECL to TTL and TTL to ECL level shifting

D. MOSFET logic circuits are explored, including

1. JFET description
2. JFET inverter circuit
3. MOSFET description (enhancement mode and depletion mode)
4. MOSFET inverter circuit
5. CMOS circuit description (NAND and NOR gates)
6. CMOS IC specifications (power dissipation, propagation delay, and noise margin); CMOS subfamilies (HC, HCT, AC, and ACT)

REFERENCES

Pulse and Digital Switching Circuits by Jacob Millman and Herbert Taub (New York: McGraw-Hill, 1965). A detailed development of discrete switching circuits, including the specifics of transistor characteristics and behavior in switching circuit applications is found in this text. Although much of the material has been superseded by the continued development of integrated circuits, this book still remains the "bible" for discrete pulse and switching circuits.

Micro-Electronics, Digital and Analog Circuits and Systems by Jacob Millman (of Millman and Taub fame) (New York: McGraw-Hill, 1979). This is an updated version of *Pulse and Switching Circuits* with extra material in analog as well as digital circuits. This text develops the discrete transistor characteristics as they apply to integrated circuit manufacture. Some discussion of the various process technologies involved in IC development is included.

Digital Electronic Circuits by Glen Glassford (Englewood Cliffs, N.J.: Prentice-Hall, 1988). This work covers the details of device characteristics as applied to switching circuit design used in the construction of digital integrated circuits. It covers TTL, ECL, and CMOS design considerations. Strictly for switching circuit IC design, it does not cover any logic functions or Boolean algebra.

Solid State Pulse Circuits by David Bell (Reston, Va: Reston, 1988). This book provides a conceptual and simple algebraic, trigonometry level analysis of discrete and integrated pulse and switching circuits. The descriptive applications approach provides a foundation of circuit function without becoming enmeshed in the details needed to design actual switching circuits for an IC.

Designing With TTL Integrated Circuits by Texas Instruments (New York: McGraw-Hill, 1972). This is an old but very informative book on TTL circuit behavior. Certain graphs and charts of TTL characteristics are found in this text that are not common in other texts. It was published as an aid to practicing engineers, not as a text.

TTL Data Books. All digital integrated circuit makers have data catalogs that portray their product line. Several Texas Instruments data books are included in this reference section. *Volume I Indexes* has a product guide and general information. *Volume II TTL Data Book* covers standard, low-power, high-speed, low-power Schottky, and Schottky digital small- and medium-scale integrated circuits manufactured by Texas Instruments. *AS/ALS Logic Data Book* is a separate volume that covers advanced and low-power advanced Schottky devices.

The *Emitter-Coupled Logic, ECL, Data Books* included here are published by Motorola. *MECL Systems Design Handbook* (Austin: Motorola, 1971). This book is neither a data book nor a text, but an engineering handbook or reference manual for engineers using MECL devices. It explains, but does not go into detail, the concerns about proper termination, transmission line effects, and other topics specific to using MECL. *MECL Device Data Book* (Austin: Motorola, 1989). This data book covers the device specifications for MECL 10K, MECL III, and other MECL products.

CMOS Data Books are also available from the device manufacturers. *Advanced CMOS, Designers Handbook* (Dallas: Texas Instruments, 1995). This handbook provides a detailed development of CMOS characteristics and behavior without going into a detailed development of device theory. *Advanced CMOS, Data Book* (Dallas: Texas Instruments, 1987). This data book covers the specifications for the AC and ACT CMOS logic devices. *High-Speed CMOS Logic Data Book* (Austin: Motorola, 1995). This data book covers the specifications for the HC and HCT CMOS subfamilies.

Intuitive IC CMOS Evolution by Thomas M. Frederiksen (Santa Clara National Semiconductor Corporation, 1984). This short paperback treatment of CMOS provides an excellent coverage of the development of CMOS. It is written in an engineering handbook style, providing useful information.

Computer-Aided Circuit Analysis Using Spice by Walter Banzhaf (Englewood Cliffs, N.J.: Prentice-Hall, 1989). This book provides an introductory treatment of Spice with an abundance of examples illustrating specific circuits.

Spice, A Guide to Circuit Simulation and Analysis Using PSpice by Paul W. Tuinenga (Englewood Cliffs, N.J.: Prentice-Hall, 1988). Tuinenga works for Microsim Corporation, the developer of PSpice. The book covers the development of circuit component models and how to input specific device parameters into the model for analysis.

GLOSSARY

AC (advanced CMOS) A fast CMOS logic subfamily. It has the best speed–power product of any of the CMOS families.

ACT An advanced TTL-compatible CMOS logic subfamily. The logic voltage levels are the same as TTL.

Active pull-up A "load" connected between a transistor and a positive voltage power supply that is an active device such as a BJT or FET.

Advanced low-power Schottky A subfamily of TTL. Advanced low-power Schottky devices have the best speed–power product of all the TTL subfamilies.

Advanced Schottky The fastest of all the TTL subfamilies.

BJT (bipolar junction transistor) A BJT transistor has two junctions (B–E and B–C), therefore both positive and negative current carriers exist (electrons in N-material and holes in P-material).

CMOS (complementary metal oxide semiconductor) A circuit consisting of both N- and P-channel field effect transistors.

Depletion mode A particular type of field effect transistor, that has a lightly doped drain–source current channel. The input voltage, V_{GS}, modulates the output current, I_d, by depleting the channel to cause less current to flow.

DTL (diode transistor logic) DTL logic was the first digital integrated circuit. It is now obsolete.

ECL (emitter-coupled logic) A bipolar technology integrated circuit. The fastest of the digital integrated circuits, it also consumes the most power.

Enhancement mode A particular type of FET in which the drain–source channel is created by the gate–source bias voltage.

Glitch An unwanted signal, usually a short pulse.

HC A high-speed CMOS digital logic subfamily.

HCT A high-speed TTL compatible CMOS digital logic subfamily.

Hysteresis The lagging of an output response with respect to an input stimulus. Hysteresis applies to the difference in input threshold voltages of a Schmidt trigger circuit: $V_{hys} = V_{UT} - V_{LT}$.

JFET (junction field effect transistor) A reverse biased P–N junction exists between the gate and the drain–source current channel.

MOSFET (metal oxide semiconductor field effect transistor) A thin layer of SiO_2 (glass) exists between the gate and the drain–source channel. MOSFETs can be either depletion or enhancement mode devices.

Schottky The name of a pioneer in electronics who discovered the metal–semiconductor junction. The junction, also called *hot carrier,* is the basis for the speed-up techniques used in the Schottky subfamilies of TTL.

Speed up Techniques used to increase the switching speed of transistors.

Tristate A high-impedance state of a logic gate or flip-flop. The two normal operating states are 0 and 1, which exhibit a low impedance. The third "state," a high-impedance condition, is entered under the control of an additional signal. When the device is in a high-impedance state it is in effect logically an "open circuit."

Totem pole The output stage of a TTL gate. The name *totem pole* stems from the apparent stacking of one transistor on top of another in the schematic diagram, in a fashion resembling the totem poles of Northwest Indian tribes.

TTL (transistor–transistor logic) TTL is a bipolar logic family that evolved from DTL. All of the switching elements are transistors, hence the name TTL.

Voltage transfer curve The voltage input-output relationship of a gate. As v_{IN} increases from 0 V to some maximum, the output v_o changes over a specified range. A graph of that change is called the *voltage transfer curve.*

Worst case The extreme operating conditions for a logic circuit. It is usually specified in terms of temperature, power supply voltage, number of inputs that an output can drive, and switching speed. The circuit must still work under the extreme operating conditions.

QUESTIONS AND PROBLEMS

Introduction and Section 10.1 to 10.2

1. Sketch a diode gate for

a. A three-input AND

b. A three-input OR

2. Design a simple diode two-input AND gate. Assume the following: $V_{CC} = 5$ V, V_d (forward voltage drop) = .65 V (1N914 diode) at a current of 1 mA.

a. Calculate the value of the pull-up resistor.

b. What is the logical 0 voltage level output when the logical 0 input is .2 V?

3. Sketch the ideal diode equivalent circuit for a three-input diode AND gate when the input is 010.
4. Complete the following truth table for a three-input diode OR gate by writing *on* and *off* in the appropriate columns, for each input diode. Write the appropriate 0 or 1 in the output columns.

x	*y*	*z*	D_1	D_2	D_3	Output
0	0	0				
0	0	1				
0	1	0				
0	1	1				
1	0	0				
1	0	1				
1	1	0				
1	1	1				

5. Describe the three regions of operation for a bipolar transistor. Construct a characteristic family of curves showing I_c, V_{CE}, I_b, the load line, and each operation region.
6. Sketch the circuit for a simple bipolar transistor inverter. Sketch the phase relationship between the input and output pulse waveforms.
7. Design a simple inverter circuit like the one in Figure 10.9, with $V_{CC} = 12$ V, $I_{C(SAT)} = 2$ mA, $HFE_{(SAT)} = 10$, $V_{be(SAT)} = .65$ V, and $V_{CE(SAT)} = .15$ V. Assume $V_{IN} = 0$ V for a logical 0 and 12 V for a logical 1.
   a. Calculate R_b and R_c.
   b. Sketch the input and output waveforms. Label the output waveform voltage levels.
8. Given the circuit shown in the figure for this problem, where $V_{be(SAT)} = .72$ V, $V_{CE(SAT)} = .25$ V, $HFE_{(SAT)} = 5$, let $V_{IN} = 12$ V for a logical 1.

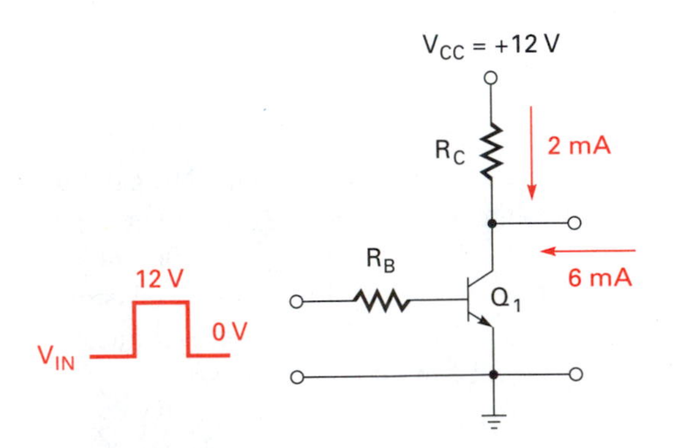

   a. Calculate R_c.
   b. Calculate R_b.
   c. If $HFE_{(SAT)}$ were 10, what is the minimum V_{IN} that ensures that the transistor is in saturation (logical 0 output)?
9. Sketch the input and output waveform of a transistor inverter. Indicate the following on the sketch.
   a. delay time (t_d)
   b. input pulsewidth (t_{pw})

Section 10.3 to 10.4

10. Sketch the circuit diagram for a two-input DTL NAND gate.
11. Assume that the two-input NAND gate shown in Figure 10.12 has the following inputs: A = 0, B = 1. Construct a table showing the on or off status of each diode and the output transistor.
12. Using the circuit configuration illustrated in Figure 10.13, calculate R_1, R_3, and R_c. Let $V_A = V_B = .4$ V (0) or 3 V (1). Find the fan-out. Assume $I_1 = 1$ mA.
13. What happens to the fan-out in Problem 12 if R_1 is increased? Why? Show calculations to support your response.
14. Explain why the high resistance (several K ohms) of R_c in a DTL circuit may cause glitches in adjacent gate inputs.
15. Sketch the equivalent circuit of a DTL gate that is coupling an unwanted glitch to another DTL gate input. Sketch the approximate waveforms at both the switching gate output and the resulting glitch at the "static" gate.
16. PSpice problems (refer to the Appendix):
   a. Do a PSpice analysis of Problem 7. Assign nodes to the circuit diagram. Generate an input waveform that switches between a logic 0 and a logic 1. Generate the PSpice commands and perform a transient analysis. Plot the input and output waveforms. Print the output file showing the voltage and current levels at the 0 and 1 levels.
   b. Do a PSpice analysis of the circuit in Figure 10.13. Perform a transient analysis of the circuit. Plot and print the input-output waveforms. Perform a worst-case analysis by letting the temperature range from 0° to 70°C.

Section 10.5

17. The totem-pole output stage of the TTL circuit solved what problem that plagued DTL logic circuits?

18. Explain how the back-to-back diode circuit shown in Figure 10.19 can be replaced with a single NPN transistor. Describe the status of each PN junction.

19. Sketch a two-input TTL NAND gate circuit. List the status of each transistor in the circuit when A = B = 1.

20. Consider the TTL two-input NAND circuit shown in Figure 10.21. Calculate the current through R_1 when $V_A = .4$ V. Assume $V_{be(SAT)} = .68$ V.

21. In Figure 10.21, $R_4 = 130$ ohms as the pull-up resistor for the totem-pole output stage. Assume that the collector to emitter resistance of both Q_3 and Q_4 is 0 when the transistors are "on." Also assume V_{D1} = .6 V. Calculate I_4 when both Q_3 and Q_4 are active during switching transitions. What would happen if R_4 were 0? Explain the purpose of R_4.

22. Define the following:

    a. I_{OL}

    b. I_{IL}

    c. I_{OS}

    d. t_{pHL}

    e. V_{IH}

    f. V_{OL}

    g. V_{IL}

23. The temperature ranges for TTL are

    a. 54xx

    b. 74xx

24. Sketch the test connection for measuring I_{IL}. What are the test conditions under which the measurement is made?

25. If I_{OL} for a standard TTL NAND gate were increased from 16 mA to 20 mA, what would happen to V_{OL}? Explain why.

26. What is the noise margin for TTL? Explain its purpose.

27. List the TTL subfamilies in order of increasing switching speed. List them in order of increasing power dissipation.

28. Explain the purpose of the Darlington transistor pair found in the top transistor position of the TTL totem pole for the high-speed and Schottky subfamilies.

29. What is a Schottky junction? Draw a sketch of the semiconductor–metal junction and indicate the cathode and anode. Draw the symbol for a Schottky diode and transistor.

30. Design a level shifting circuit, using a TTL open-collector high-voltage buffer integrated circuit (7407), that must drive 10 identical input circuits. The 7407 specifications are $V_{OL} = .7$ V (maximum at $I_{OL} = 40$ mA), $I_{OL} = 40$ mA max., $I_{OH} = .25$ mA maximum, $V_{OH} = 30$ V max. The specifications of the gates to be driven are $V_{IL} = .5$ V, $I_{IL} = 2$ mA, $I_{IH} = 100\ \mu$A, $V_{IH} = 25$ V.

31. Design a wired-AND circuit where standard TTL voltage levels exist, so that the outputs of five NAND gates are wired-ANDed to the inputs of six TTL OR gates.

    a. What TTL part numbers did you choose?

    b. Calculate the pull-up resistor maximum and minimum values when the external power supply is 9 V.

32. Construct a truth table for the circuit you designed in Problem 31. Write the Boolean switching equation.

33. Interface an open-collector inverting buffer to a small relay that requires +12 V at 12 mA to operate.

34. Sketch the circuit diagram for a TTL two-input tristate NAND gate. Identify the logic inputs and the control input. Explain the condition (on or off) of each transistor in the circuit when the tristate enable is active.

35. Explain how tristate circuit outputs can be connected together to form a data bus, so that each output can be switched onto the bus wire.

36. Calculate the maximum fan-out for the following TTL combinations of mixed subfamilies:

    a. Standard TTL to low-power Schottky

    b. Low-power Schottky TTL to standard TTL

    c. Schottky TTL to low-power Schottky TTL

    d. Advanced Schottky to advanced low-power TTL

37. Explain the purpose of a Schmitt trigger input logic gate. For what types of applications is such a gate useful?

Section 10.6

38. Draw the circuit diagram for a two-input ECL OR/NOR gate. Identify each of the functional areas of the circuit: input switches, differential circuit, reference circuit, and output circuit.
39. What are the logic voltage levels for an ECL circuit?
40. ECL is said to be nonsaturating. Explain why this is true and the reason for wanting a nonsaturating logic family.
41. Refer to Figure 10.50. Assume that a reference voltage of −1.5 V were used on the base of transistor Q_2. If $V_{be} = .65$ V, what would be the new switching threshold voltage at the base of Q_1?
42. What are the noise margin levels for ECL?
43. What is the propagation delay for the MC10H104? Is it faster or slower than the advanced Schottky TTL AND gate?
44. Construct a table that shows the on or off status and the logic level output for the ECL OR/NOR gate in Figure 10.52.
45. Let a −6 V power supply rather than the normal −5.2 V be used for the circuit shown in Figure 10.52. Calculate the new output logic levels. Assume $V_{be} = .72$ V and that $V_D = .7$ V.
46. The MC10H125 and MC10H124 are ECL to TTL and TTL to ECL logic level converters. Sketch a logic diagram for a Schottky NAND gate output to be interfaced to an ECL NOR gate input. Determine the propagation delay and power consumption for the three logic levels.

Section 10.7

47. For what is *CMOS* an abbreviation?
48. What is the major difference between a bipolar transistor and a field effect transistor?
49. Draw a diagram illustrating the basic JFET transistor construction. Identify the drain, source, and gate. Explain how the gate can control the drain–source current.
50. Draw the schematic symbols for N-channel JFET, P-channel JFET, N-channel MOSFET, and P-channel MOSFET.
51. Explain the difference between a JFET and a MOSFET.
52. Describe how the enhancement and depletion MOSFET operation modes create the current channel.
53. Draw the schematic diagram for a simple N-channel, enhancement-mode MOS inverter.
54. What is meant by *active pull-up*?
55. Draw the schematic for a simple CMOS inverter. Explain its logical operation.
56. Sketch a waveform diagram showing the input-output relationship of a CMOS inverter.
57. Draw the circuit schematic for CMOS two-input gates:
    - **a.** NAND
    - **b.** NOR
58. Describe the condition of each transistor in the circuit for the different input logic levels for the circuits in Problem 57.
59. Assume that 12 74HC integrated circuits are used to realize a certain function. Assume that each gate output is connected to four inputs, that the load capacitance seen by each output is 50 pF, $V_{DD} = 5$ V, and the operating frequency is 8 MHz. Calculate the total unit power consumption.
60. Assume that a 74HCT logic circuit is connected to a 100 pF load and has a nominal t_{pd} of 16 ns when operated by a 6 V supply voltage. Calculate the circuit propagation delay.
61. A CMOS logic circuit is operating with a +10 V supply. Determine the logic 0 and 1 maximum and minimum levels.
62. List the CMOS subfamilies from slow to fast for T_{pd}. List the same subfamilies from low to high for P_d.
63. Identify the logic IC family that
    - **a.** Has the best noise margin.
    - **b.** Is the fastest.
    - **c.** Has the greatest range of propagation delays in the same basic family.
    - **d.** Has a $V_{OL} = .4$ V.
    - **e.** Has a $V_{IH} = -1.75$ V.
    - **f.** Is used in an open collector.
    - **g.** Has the best speed–power product.
64. Create a table showing the relative speed and power consumption for the following: 74AS, 74LS, 74HCT, ECL 10K, ECL 100K, 74AC, and 74LS.

PSpice Problems (refer to the Appendix)

65. Analyze the diode AND and OR gates using PSpice. Draw the circuit diagram, indicate the nodes, and write the commands for each component. Allow the input to vary between the specified logic levels and use Probe to view the resulting output waveforms. Add a load to the circuit and run PSpice to evaluate the output logic levels.

66. Analyze a simple bipolar transistor switch. Draw the circuit, calculate the resistance values and choose default transistor parameters, identify the nodes, and write the PSpice commands for each component. Choose the V_{CC} and let the input levels change between V_{CC} and ground. Perform a transient analysis of the circuit. Note the propagation delay between the input and output. Place a capacitor across the base resistance (chose a value between 100 and 500 pF) and run the transient analysis again; note the propagation delay. Did it improve? Explain why or why not. Perform the same analysis, but this time place a clamp diode across the base to collector junction of the transistor. Set the V_D of the diode to .25 V. Note and explain the results.

67. Analyze the DTL gate of Figure 10.13. Let the load current be 1 mA from each input connected to the output circuit. Let the fan-out start at 3. Calculate the resistor values using the following transistor assumptions: $V_{be} = .65$ V, $V_{CE(SAT)} = .2$ V, $h_{FE(SAT)} = 10$. Perform a transient analysis using PSpice. Increase the fan-out to six and rerun the analysis. Note and explain any changes.

68. Rerun the TTL PSpice example shown at the beginning of the Appendix. Note the dc voltages at each node in the circuit. Change the values to reflect the low-power TTL circuit. Rerun the analysis note and explain the node voltages and currents.

69. Perform a PSpice analysis for the open-collector circuit example (Example 10.4). Note the logic levels at the open-collector junction.

70. Perform a PSpice analysis of the ECL circuit in Figure 10.52. Identify the nodes and write the commands for each component. Create input logic levels that correspond to the ECL requirements and allow them to change so as to exercise the entire input truth table. Run a transient analysis of the circuit and observe the outputs (OR and NOR).

71. Perform a PSpice analysis for the CMOS circuit of Figure 10.77. Allow the input voltages to change so as to exercise the input truth table for the device. Observe the output waveform. Note the switching action of the MOS transistors. Expand the horizontal scale on the Probe display until you can see the rise and fall times adequately. Explain why it is that CMOS power consumption increases as a function of frequency.

APPENDIX

1 TTL Analysis Spice Exercise

Use PSpice to simulate the voltage transfer curves for a standard TTL inverter. Figure 1.1 shows the circuit configuration with the PSpice nodes.

The PSpice commands connect the various components, establish the power supply voltage level, establish the input voltage (in this example the input voltage will be swept from 0 V to +5 V in .1 V increments). The commands also establish the parameters for the transistors and the diodes, set the resistor values, and call Probe. Probe keeps track of all the values resulting from running the analysis software and plots the result.

TTLINV;	Title of the file
VCC 9 0 DC 5;	Sets +5 V V_{CC}
VIN 1 0 DC 5;	Sets V_{IN} Range
.DC VIN 0 5 .1;	Sets sweep for V_{IN}
R1 2 9 4K;	R_1 = 4K
R2 5 9 1.6K;	R_2 = 1.6K
R3 4 0 1K;	R_3 = 1K
R4 8 9 130;	R_4 = 130 ohms
RL 9 10 400;	RL = 400 ohms
Q1 3 2 1 TRAN;	Establishes Q_1, Q_2, Q_3, Q_4 positions
Q2 5 3 4 TRAN;	
Q3 6 4 0 TRAN;	
Q4 8 5 7 TRAN;	
.MODEL TRAN NPN(BF=50);	All Tran B = 50
D1 7 6 DIO;	Sets D_1 position
D2 10 6 DIO;	Sets D_2 position
.Model DIO D;	Default parameter
.PROBE;	Call Probe
.END;	Done

PSPICE ANALYSIS EXAMPLE

Using PSpice as an analysis tool, develop a voltage transfer curve similar to that in Figure 10.75 for the CMOS inverter in Figure 10.73. The results of the PSpice run are given in Figure 1.2.

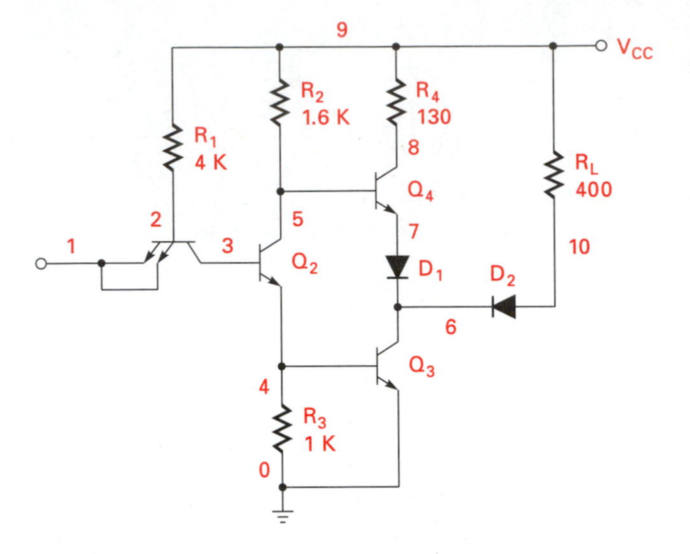

Figure A1.1
TTL inverter with PSpice nodes identified

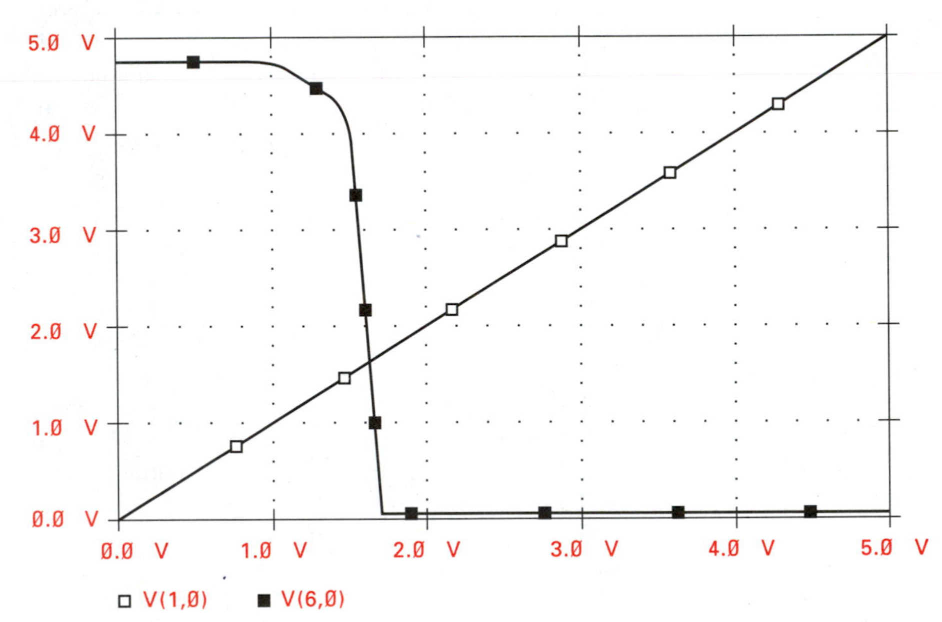

Figure A1.2
PSpice analysis of TTL inverter voltage transfer curve

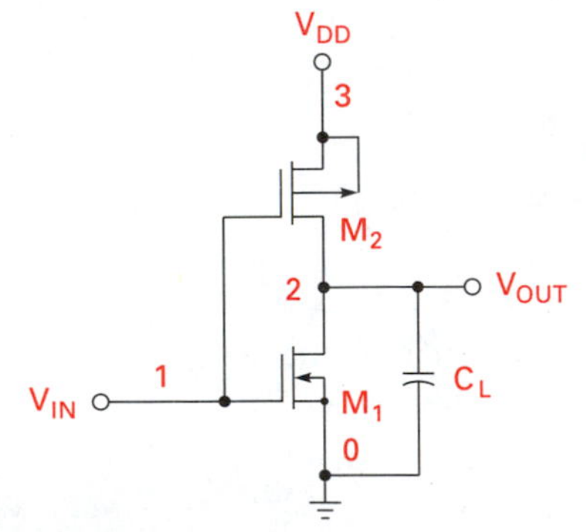

Figure A1.3
CMOS inverter with PSpice nodes assigned

Step 1. Assign node numbers to all of the connections in the circuit. Ground is always assigned 0 as its node number. The order of other node assignments is not critical.

Step 2. Write the Spice commands to simulate the circuit shown in Figure 1.3.

CMOSI;	Title
VDD 3 0 5;	Sets +5 V between nodes 3 and 0
M1 2 1 0 0 N;	Defines connections for both NMOS and
M2 2 1 3 3 P;	PMOS
CL 2 0 20p;	Defines connections and value for C_L
VIN 1 0 DC 5;	Sets connections for dc input voltage
.Model N NMOS (VTO=+2);	Model for NMOS

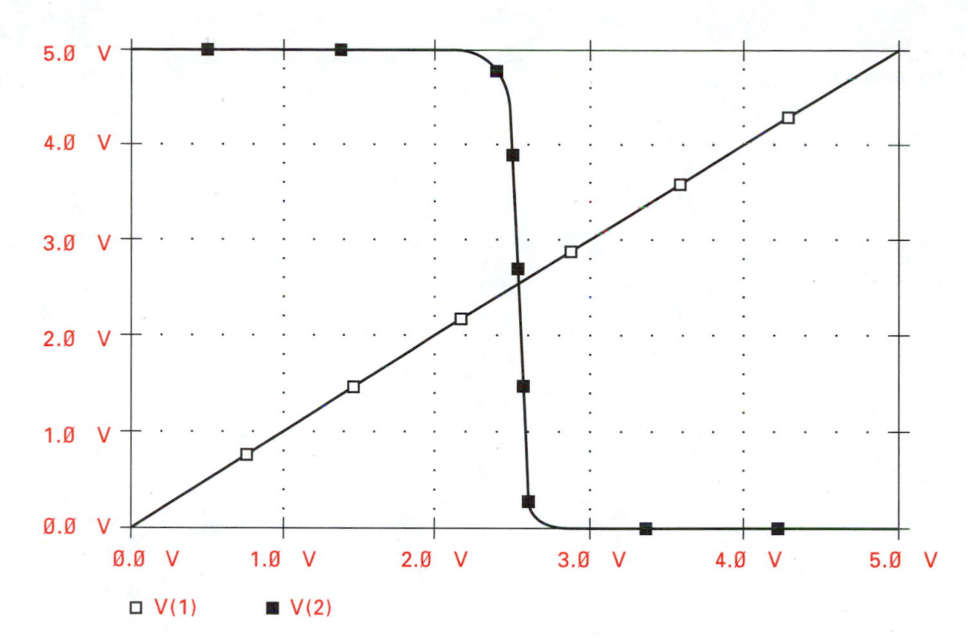

Figure A1.4
CMOS inverter voltage transfer curve run on PSpice

.MODEL P PMOS (VTO+ −2);	Model for PMOS
.DC VIN 0 5 .1;	Establishes V_1 as a variable dc supply starting
;	at 0 V and increasing to +5 V in .1 V steps
.PROBE;	Allows user to investigate current and voltage
;	values in the circuit
.END;	Ends the program

Step 3. When the program has been written under a file called CMOSI.CIR then the PSpice program has access to it and can perform its analysis.

Step 4. Your results should look like Figure 1.4.

The ramp starting at 0 V and continuing to +5 V represents the value of V_{IN} as it is being swept. Notice the point where V_{OUT} starts to "roll off." It appears to be close to V_T of the NMOS transistor. An expanded scale would help us see it more clearly.

Change the PSpice program to read:

```
CMOSI
VDD 3 0 5
M1 2 1 0 0 N
M2 2 1 3 3 P
CL 2 0 20p
VIN 1 0 DC 5
.MODEL N NMOS(VTO=+2)
.MODEL P PMOS(VTO=−2)
.DC VIN 1.75 2.75 .01
.PROBE
.END
```

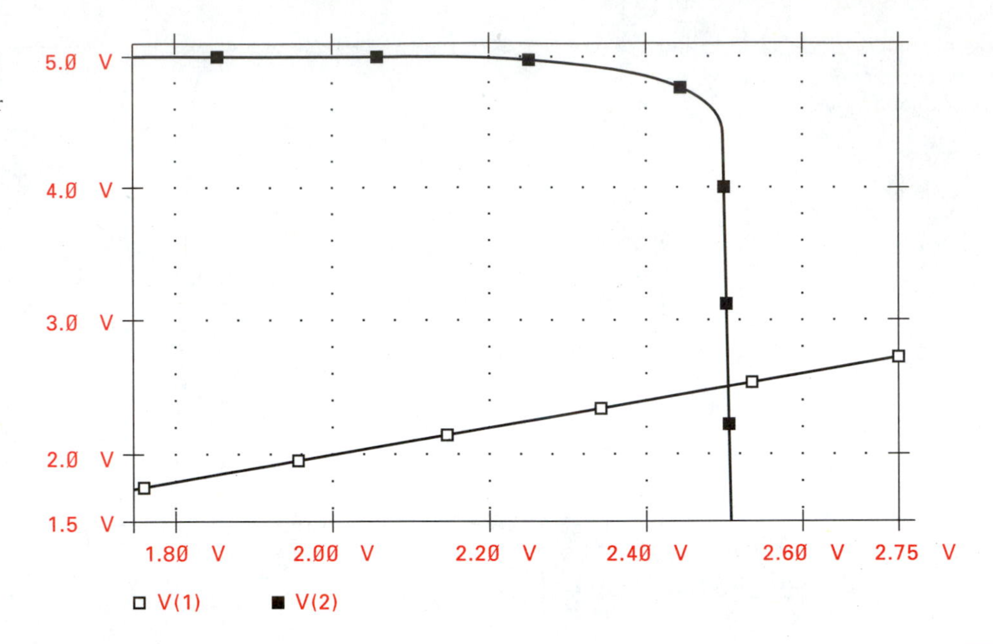

Figure A1.5
Expanded view of V_T point on CMOS inverter voltage transfer curve

The changes in the .DC VIN statement cause the input to be swept from 1.75 V to 2.75 V in 10 mV increments (see Figure 1.5).

In the expanded view we can get a closer look at the threshold point, V_T, where the NMOS transistor begins to conduct. Try an experiment. Change the V_{TO} for the NMOS transistor to 1 V and rerun the voltage transfer curve.

APPENDIX 2

ANSWERS TO ODD-NUMBERED QUESTIONS AND PROBLEMS

CHAPTER 1

1. **a.** Discrete: Broken down into pieces; opposite of continuous; a single part or point that can unambiguously be defined.
   b. Continuous: Without breaks, smooth, no interruptions; a range of values forming a line or curve without gaps or discontinuities.
   c. Digital: Refers to digits or numbers in any radix; digital electronics usually involves binary or base-2 number systems. Digital signals are discrete, not continuous.
   d. Analog: Refers to something that is analogous, has a likeness, or is similar to something else. Analog electronic signals are continuous, whereas voltage, current, or some other electrical parameter is used to represent a physical parameter. For instance, voltage changes may represent changes in temperature.
3. The first "computer on a chip" was developed by INTEL in the early 1970s.
5. An algorithm is a step-by-step method or approach to solving a problem, somewhat like a recipe. Algorithms are very useful in digital design, from both a hardware and a software standpoint, because they help the designer to properly define the problem to be solved and permit the solution to occur in steps.
7. A hierarchy is a structured system where levels represent some feature or property of the idea, thing, or system under consideration. The hierarchy generally describes the system in which the "whole" system consists of subsystems. Each subsystem performs some function of its own. In digital design, use of a hierarchical structure permits different levels to be discussed and understood without the necessity for overwhelming detail of the entire system.
9. **a.** octal (8) **b.** binary (2) **c.** decimal (10) **d.** base-5 (5) **e.** base-3 **f.** hexadecimal (16) **g.** base-12 (12)
11. **a.** 00000, 00001, 00010, 00011, 00100, 00101, 00110, 00111, 01000, 01001, 01010, 01011, 01100, 01101, 01110, 01111, 10000, 10001, 10010, 10011, 10100, 10101, 10110, 10111, 11000, 11001, 11010, 11011, 11100, 11101, 11110
    b. 0, 1, 2, 3, 4, 5, 6, 7, 10, 11, 12, 13, 15, 15, 16, 17, 20, 21, 22, 23, 24, 25, 26, 27, 30, 31, 32, 33, 34, 35, 36
    c. 0, 1, 2, 3, 4, 10, 11, 12, 13, 14, 20, 21, 22, 23, 24, 30, 31, 32, 33, 34, 40, 41, 42, 43, 44, 50, 51, 52, 53, 54, 60
    d. 0, 1, 2, 3, 4, 5, 6, 7, 8, 9, A, B, C, D, E, F, 10, 11, 12, 13, 14, 15, 16, 17, 18, 19, 1A, 1B, 1C, 1D, 1E
13. Convert to octal:
    a. $100, 110_2 = 46_8$
    b. $100, 101, 101.110_2 = 455.6_8$
    c. $010, 000, 111, 001.100, 101_2 = 2071.45_8$
15. Convert to binary:
    a. $12_{10} = 01\ 100_2$
    b. $34.25_{10} = 100\ 010.01_2$

c. $1{,}024.5_{10} = 10\ 000\ 000\ 000.1_2$

d. $255.75_{10} = 11\ 111\ 111.11_2$

17. Convert to decimal:

a. $11001.1_2 = 25.5_{10}$

b. $100111.11_2 = 39.75_{10}$

c. $11001101.111_2 = 205.875_{10}$

19. Convert to binary:

a. $FAC.B_{16} = 1111, 1010, 1100.1011_2$

b. $27AD.9B_{16} = 0010, 0111, 1010, 1101.1001, 1011_2$

c. $CDE2.F5_{16} = 1100, 1101, 1110, 0010.111, 0101_2$

21.

Decimal	BCD
0	0000
1	0001
2	0010
3	0011
4	0100
5	0101
6	0110
7	0111
8	1000
9	1001

23. Convert BCD to decimal:

a. $1001\ 0011\ 1000.0111_{BCD} = 938.7_{10}$

b. $1000\ 0110\ 0010\ 0011.1001\ 0111_{BCD} = 8623.97_{10}$

25. A "self-complementing" code is one where the arithmetic and logic complements are the same. Examples of self-complementing codes are the Excess-3, 631-1, and 2421 BCD codes. These codes are used to reduce the amount of logic required for some arithmetic circuits. The negative of a number can be found by complementing each bit. This permits subtraction by adding the logical complement of the subtrahend to the minuend.

27. A "unit distance" code permits only one bit change between values. Normal binary codes often change several bits. For example, in 0111 (7) to 1000 (8), all four bits change. In a Gray code (unit distance), a 7 is 0100 and an 8 is 1100.

29. Gray code wheel.

31. Find radix and radix-1 complements:

a. 1011.11_2 R-1 (1's) Complement → 0100.00
R (2's) Complement → 0100.01

b. 23.412_5 R-1 (4's) Complement → 21.032
R (3's) Complement → 21.033

c. 92.1_{10} R-1 (9's) Complement → 07.8
R (10's) Complement → 07.9

d. 327.4_8 R-1 (7's) Complement → 450.3
R (8's) Complement → 450.4

e. $AF2.4_{16}$ R-1 (15's) Complement → 50D.9
R (16's) Complement → 50D.A

f. 21.3_4 R-1 (3's) Complement → 12.0
R (4's) Complement → 12.1

33. Perform the indicated operations:

a. $FDC_{16} + A29_{16} = 1A05_{16}$

b. $FE6_{16} - EFC_{16} = 0EA_{16}$

c. $34_5 + 12_5 = 101_5$

d. $24_6 - 15_6 = 05_6$

35. Perform the indicated operations:

a.
$$\begin{array}{r} 234_8 \\ \times\ 24_8 \\ \hline 1160 \\ 470 \\ \hline 6060_8 \end{array}$$

b.
$$\begin{array}{r} FC2_{16} \\ \times\ DE_{16} \\ \hline DC9C \\ CCDA \\ \hline DAA3C \end{array}$$

c.
$$\begin{array}{r} FDE_{16} \\ \times\ F_{16} \\ \hline EE02_{16} \end{array}$$

d.
$$\begin{array}{r} CD.35_{16} \\ \times\ 4.C_{16} \\ \hline 99E7C \\ 334D4 \\ \hline 3CE.BBC_{16} \end{array}$$

37. Perform the indicated operations:

a.
$$\begin{array}{r} 23.4_{10} \\ -19.8_{10} \\ \hline 3.6_{10} \end{array}$$

b.
$$\begin{array}{r} 135.7_8 \\ -\ 67.7_8 \\ \hline 46.0_8 \end{array}$$

c.
$$\begin{array}{r} 321.2_4 \\ -\ 33.3_4 \\ \hline 221.3_4 \end{array}$$

d.
$$\begin{array}{r} FA.3_{16} \\ -0F.F_{16} \\ \hline EA.4_{16} \end{array}$$

e.
$$\begin{array}{r} 10011.11_2 \\ -01111.01_2 \\ \hline 00100.10_2 \end{array}$$

39. Construct an addition table for:

a. Base-4 numbers

		Addend				
		0	1	2	3	10
A	0	0	1	2	3	10
U	1	1	2	3	10	11
G	2	2	3	10	11	12
E	3	3	10	11	12	13
N	10	10	11	12	13	20
D						

b. Base-9 numbers

		Addend									
		0	1	2	3	4	5	6	7	8	10
A	0	0	1	2	3	4	5	6	7	8	10
U	1	1	2	3	4	5	6	7	8	10	11
G	2	2	3	4	5	6	7	8	10	11	12
E	3	3	4	5	6	7	8	10	11	12	13
N	4	4	5	6	7	8	10	11	12	13	14
D	5	5	6	7	8	10	11	12	13	14	15
	6	6	7	8	10	11	12	13	14	15	16
	7	7	8	10	11	12	13	14	15	16	17
	8	8	10	11	12	13	14	15	16	17	18
	10	10	11	12	13	14	15	16	17	18	20

41. The process of complementing the subtrahend and then adding the results to the minuend produces the same result as subtracting the subtrahend from the minuend. The reason for this is because the radix complement of a number is written as $X^* = r_n - X$, where X^* is the radix complement, r is the radix, n is the number of digits used to express x^*, and X is the original number. For example, the 10's complement of 8 is 2 or $2 = 10_1 - 8$. The 10's complement of 68 is 100 − 32. So subtracting from r_n was done when the radix complement was found. Adding the radix complement to the minuend produces the difference.

CHAPTER 2

1. Boolean algebra is called switching algebra when it deals with 2-valued or binary variables.

3. a. AND $s = xy$

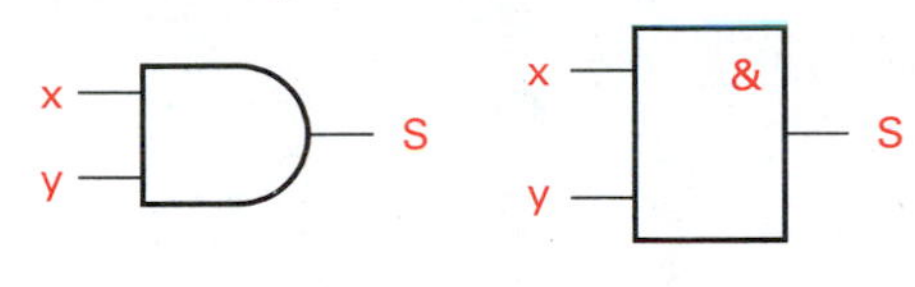

x	y	s
0	0	0
0	1	0
1	0	0
1	1	1

b. OR $s = x + y$

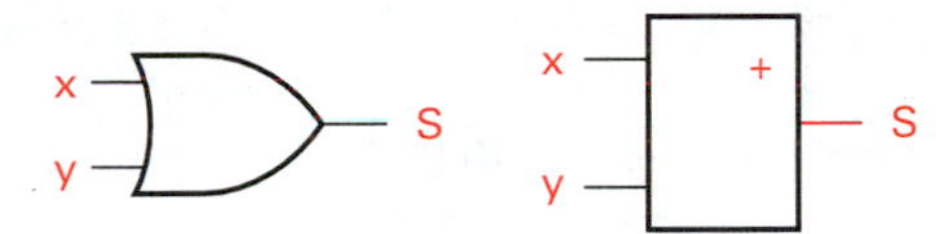

x	y	s
0	0	0
0	1	1
1	0	1
1	1	1

c. NOT $s = x'$

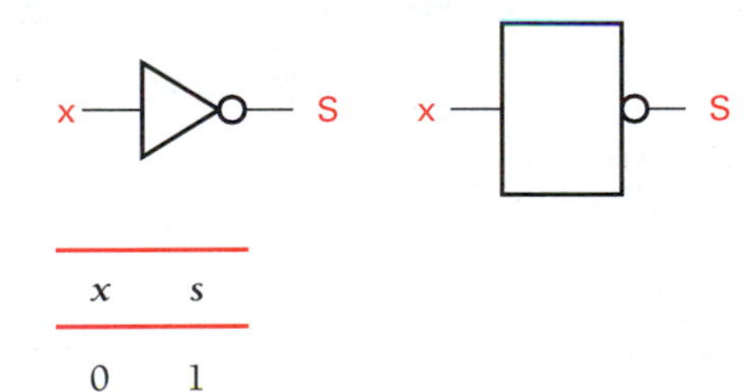

x	s
0	1
1	0

5. a. EX-OR $s = x \oplus y$

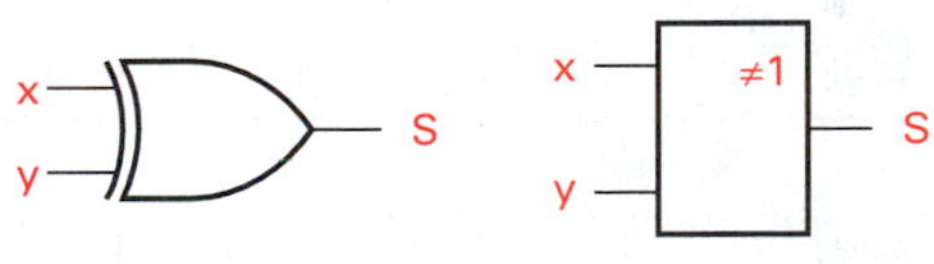

(continued)

x	y	s
0	0	0
0	1	1
1	0	1
1	1	0

b. EX-NOR $s = (x \oplus y)'$

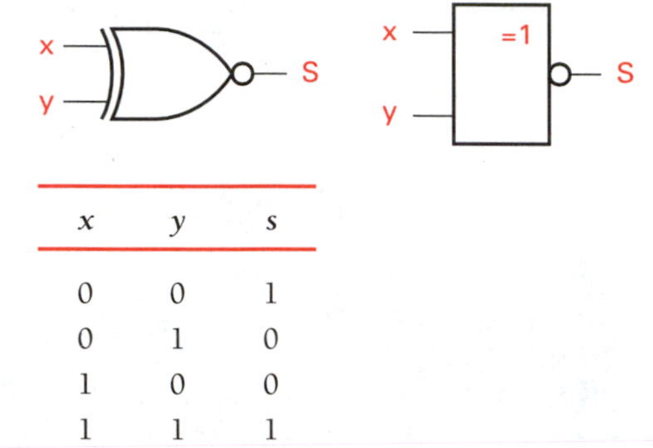

x	y	s
0	0	1
0	1	0
1	0	0
1	1	1

7. Boolean expressions, truth tables, and logic diagrams

9. The number of elements equals the number of people taking the class, and may or may not be the same ones taking an English or math class.

11. An operation is closed with respect to a binary operator if, when the operation is applied to members of the set, the result is also a member of the set.

13. The equivalence operation results in a "true" or "1" output when all the inputs are equal.

15.

			AND		OR	
x	y	z	$(xy)z$	$x(yz)$	$(x+y)+z$	$x+(y+z)$
0	0	0	0	0	0	0
0	0	1	0	0	1	1
0	1	0	0	0	1	1
0	1	1	0	0	1	1
1	0	0	0	0	1	1
1	0	1	0	0	1	1
1	1	0	0	0	1	1
1	1	1	1	1	1	1

17. Binary operations (AND and OR) are commutative on set B, if $xy = yx$ and $x + y = y + x$.

19. Yes. Commutative property.

21.

x	y	xy	$x + xy$
0	0	0	0
0	1	0	0
1	0	0	1
1	1	1	1

$x = x + xy$

23. The complement of a switching function can be found by replacing each variable or constant (0 or 1) with its complement, each AND operator with an OR, and each OR operator with an AND.

25. A functionally complete operation set is one that will allow any combinational logic function to be realized.

27. **a.** $xy'z + xyz' = x(y'z + yz') = x(y \oplus z)$

b. $AB + A'BC' + BC = B(A + A'C' + C) = B(A + C' + C)$
$B(A + C' + C) = B(A + 1) = B(1) = B$

c. $A'B'C' + AB'C' + AB'C = A'B'C' + AB' = B'(A + A'C')$
$B'(A + A'C') = B'(A + C') = AB' + AC'$

d. $ab(cd + c'd) = abd(c + c') = abd$

e. $(x(xy)')(y(xy')) = x(x' + y')(y(x' + y')) = (xx' + xy')(x'y + yy')$
$(xx' + xy')(x'y + yy') = xy' + x'y = x + y$

f. $x'(y + z)' + xy = x'y'z' + xy$

g. $(A + B)'(A' + B')' = A'B'AB = 0$

h. $a'b'c' + a'b'c + ab'c' + abc = a'b'(c + c') + ab'c' + abc = a'b' + a(b'c' + bc) = a'b' + a(b + c)$ or $a'b' + b'c' + abc$

29.

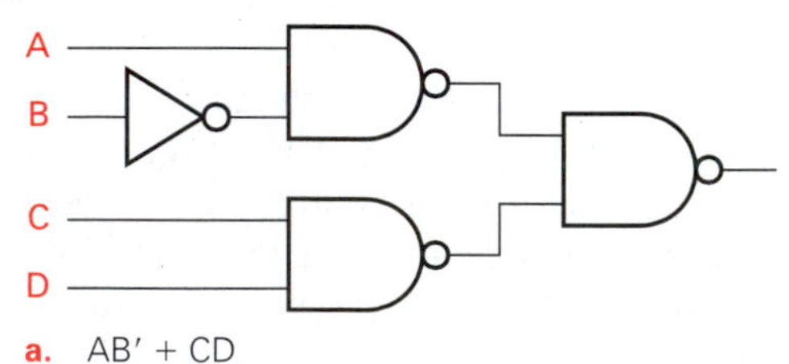

a. AB′ + CD

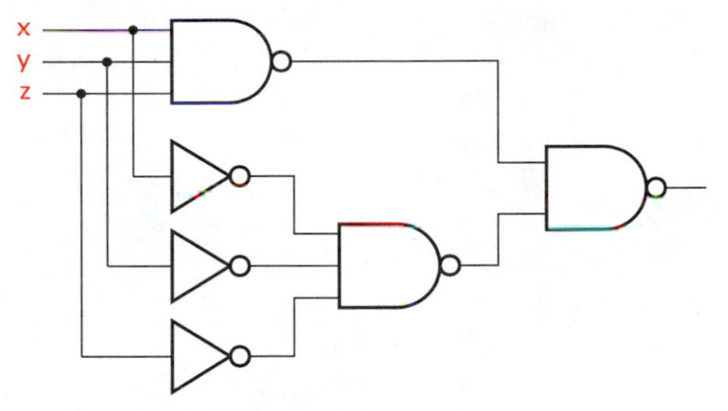

b. xyz + x′y′z′

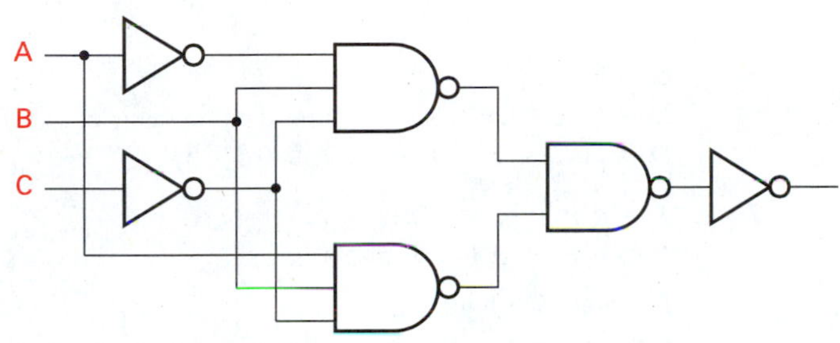

c. (A + B′ + C)(A′ + B′ + C)

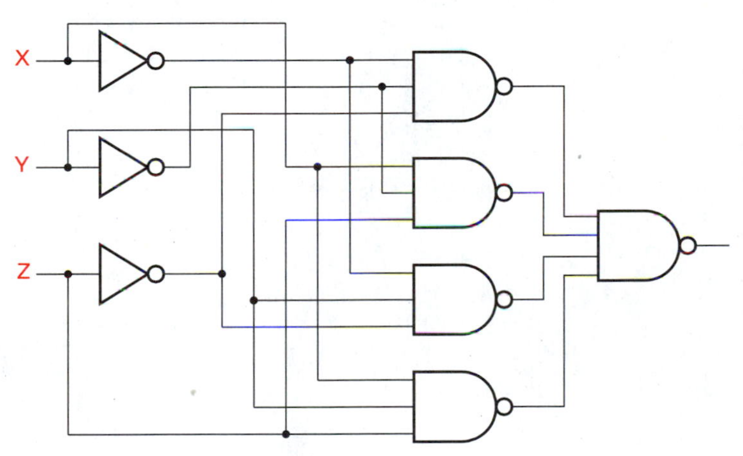

d. X′Y′Z′ + XY′Z + X′YZ′ + XYZ

31. a. $((A + B)(C + D))' = A'B' + C'D'$

A	B	C	D	F
0	0	0	0	0
0	0	0	1	0
0	0	1	0	0
0	0	1	1	1
0	1	0	0	0

continues

A	B	C	D	F
0	1	0	1	0
0	1	1	0	0
0	1	1	1	1
1	0	0	0	0
1	0	0	1	0
1	0	1	0	0
1	0	1	1	1
1	1	0	0	1
1	1	0	1	1
1	1	1	0	1
1	1	1	1	0

b. $(AB + CD) = AB(CD)' + (AB)'CD$
$+ AB(C' + D') + CD(A' + B')$
$ABC' + ABD' + A'CD + B'CD$

A	B	C	D	F
0	0	0	0	1
0	0	0	1	1
0	0	1	0	1
0	0	1	1	1
0	1	0	0	1
0	1	0	1	0
0	1	1	0	0
0	1	1	1	0
1	0	0	0	1
1	0	0	1	0
1	0	1	0	0
1	0	1	1	0
1	1	0	0	1
1	1	0	1	0
1	1	1	0	0
1	1	1	1	0

c. $A'BC + B'D = F$

A	B	C	D	F
0	0	0	0	0
0	0	0	1	1
0	0	1	0	0
0	0	1	1	1

continues

A	B	C	D	F
0	1	0	0	0
0	1	0	1	0
0	1	1	0	1
0	1	1	1	1
1	0	0	0	0
1	0	0	1	1
1	0	1	0	0
1	0	1	1	1
1	1	0	0	0
1	1	0	1	0
1	1	1	0	0
1	1	1	1	0

d. $((a \oplus b'))'((c \oplus d'))' = ab + a'b' + cd' + c'd$

A	B	C	D	F
0	0	0	0	1
0	0	0	1	1
0	0	1	0	1
0	0	1	1	1
0	1	0	0	0
0	1	0	1	1
0	1	1	0	1
0	1	1	1	0
1	0	0	0	0
1	0	0	1	1
1	0	1	0	1
1	0	1	1	0
1	1	0	0	1
1	1	0	1	1
1	1	1	0	1
1	1	1	1	1

CHAPTER 3

1.

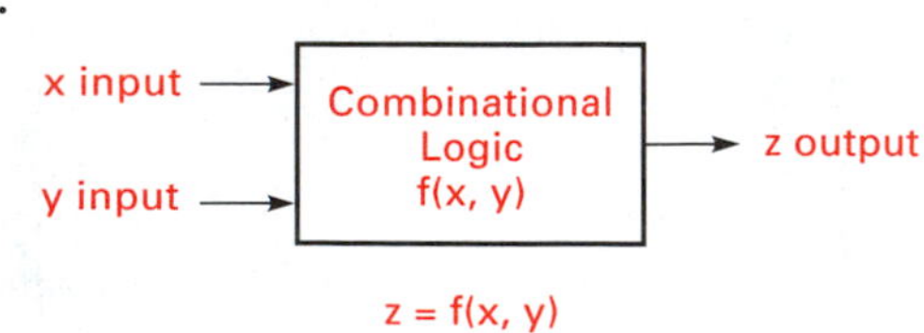

3. The truth table expresses a combinational logic circuit by describing the circuit output(s) logic values as a function of the input variables.

5. a. $W = a'b'c + a'bc + ab'c' + abc$
$W = f(a, b, c) = \Sigma(1, 3, 4, 7)$
$X = a'b'c' + a'bc' + a'bc + ab'c + abc$
$X = f(a, b, c) = \Sigma(0, 2, 3, 5, 7)$

b. $A = wxyz$
$A = f(w, x, y, z) = \Sigma(15)$
$B = wx'yz' + wx'yz + wxyz'$
$B = f(w, x, y, z) = \Sigma(10, 11, 14)$
$C = w'xyz' + w'xyz + wx'y'z + wx'yz + wxy'z + wxyz'$
$C = f(w, x, y, z) = \Sigma(6, 7, 9, 11, 13, 14)$
$D = w'xy'z + w'xyz + wxy'z + wxyz$
$D = f(w, x, y, z) = \Sigma(5, 7, 9, 13, 15)$

7. a. $X = a'b + bc = a'b(c + c') + bc(a + a')$
$X = a'bc + a'bc' + abc + a'bc$

b. $P = (w' + x)(y + z)$
$P = (w' + x + yy' + zz')(y + z + ww' + xx')$
$P = (w' + x + y + z)(w' + x + y + z')$
$\times (w' + x + y' + z)(w' + x + y' + z')$
$\times (w + x + y + z)(w + x' + y + z)$
$\times (w' + x + y + z)(w' + x' + y + z)$

c. $T = p(q' + s)$
$T = pq' + ps$
$T = pq'(s + s') + ps(q + q')$
$T = pq's + pq's' + pqs + pq's$

d. $R = L + M'(N'M + M'L)$
$R = L + M'N'M + M'M'L$
$R = L + M'L$
$R = L(M + M') + M'L$
$R = ML + M'L$

e. $U = r' + s(t + r) + st'$
$U = r' + st + sr + st'$
$U = (r + s' + t')(r + s + t')(r + s' + t)$
$\times (r + s + t)(r' + s' + t')(r' + s + t')$
$\times (r' + s' + t)$

9. a. $Z = f(a, b, c) = \pi(0, 1, 2, 3, 4)$

b. $F = f(w, x, y, z) = \pi(0, 1, 2, 3, 4, 5, 8, 9)$

c. $M_1 = f(S_3, S_2, S_1) = \pi(0, 3, 4, 5, 6, 7)$
$M_2 = f(S_3, S_2, S_1) = \pi(0, 1, 2, 3, 4, 5, 6)$

d. $Z = f(a, b, c, d) = \pi(0, 1, 2, 3, 4, 5, 6, 7, 8, 9)$

e. $A = f(S, E, L, K, D) = \pi(0, 2, 3, 6, 7, 16, 18, 19, 22, 23, 25, 26, 27, 30, 31)$

11. a. $V = f(a, b, c, d) = \Sigma(2, 3, 4, 5, 13, 15) + \Sigma d(8, 9, 10, 11)$

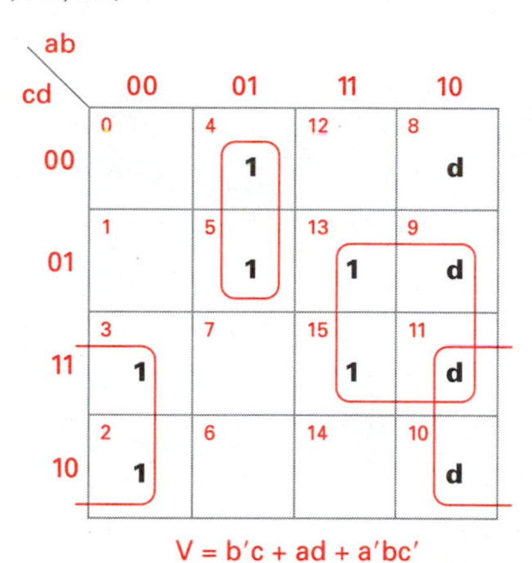

V = b′c + ad + a′bc′

b. $Y = f(u, v, w, x) = \Sigma(1, 5, 7, 9, 13, 15) + \Sigma d(8, 10, 11, 14)$

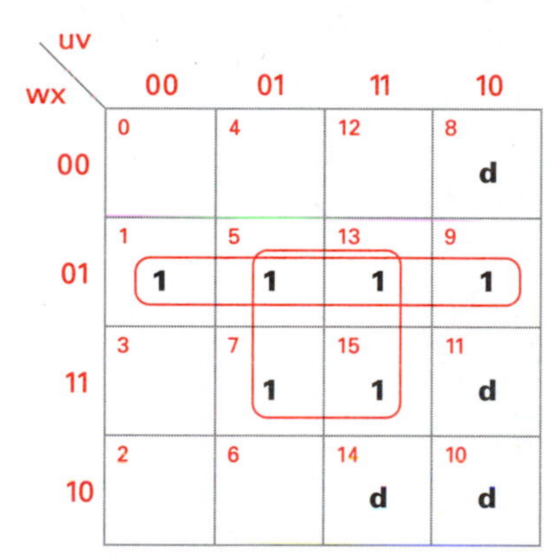

Y = w′x + vx

c. $P = f(r, s, t, u) = \Sigma(0, 2, 4, 8, 10, 14) + \Sigma d(5, 6, 7, 12)$

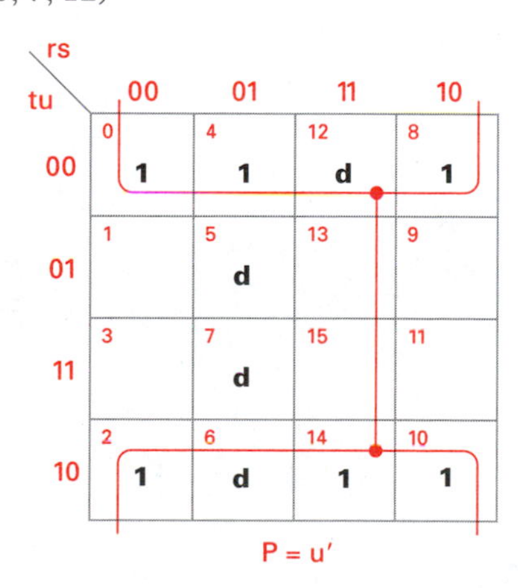

P = u′

d. $F = f(u, v, w, x, y) = \Sigma(0, 2, 8, 10, 16, 18, 24, 26)$

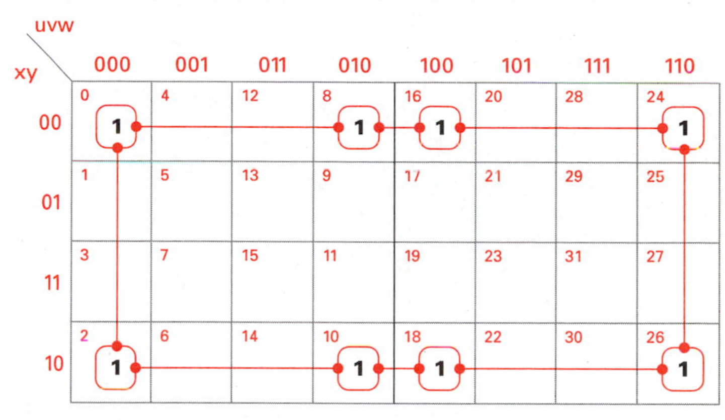

F = w′y′

e. $H = f(a, b, c, d, e) = \Sigma(5, 7, 9, 12, 13, 14, 15, 20, 21, 22, 23, 25, 29, 31)$

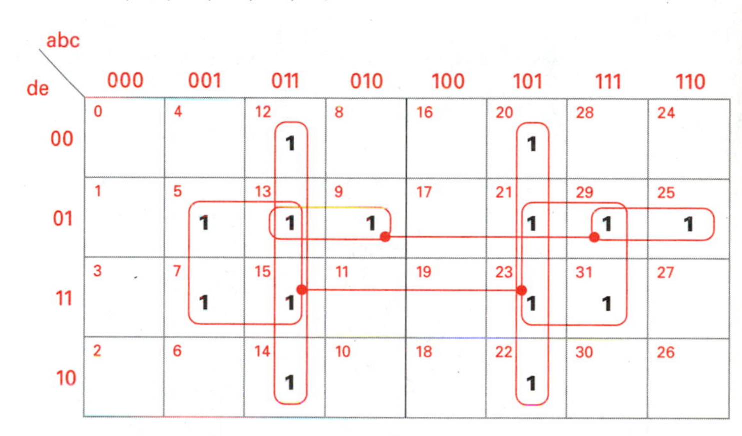

H = a′bc + ab′c + ce + bd′e

f. $M = f(v, w, x, y, z) = \Sigma(1, 3, 4, 6, 9, 11, 12, 14, 17, 19, 20, 22, 25, 27, 28, 30) + \Sigma d(8, 10, 24, 26)$

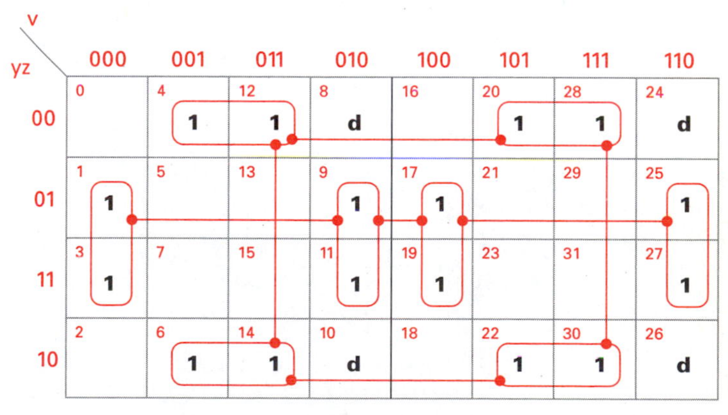

M = x′z + xz′

g. $J = f(a, b, c, d, e, f) = \Sigma(7, 12, 22, 23, 28, 34, 37, 38, 40, 42, 44, 46, 56, 58, 60, 62)$

def \ abc	000	001	011	010	100	101	111	110
000	0	8	24	16	32	40 1	56 1	48
001	1	9	25	17	33	41	57	49
011	3	11	27	19	35	43	59	51
010	2	10	26	18	34 1	42 1	58 1	50
100	4	12 1	28 1	20	36	44 1	60 1	52
101	5	13	29	21	37 1	45	61	53
111	7 1	15	31	23 1	39	47	63	55
110	6	14	30	22 1	38 1	46 1	62 1	54

J = acf′ + ab′ef′ + cde′f′ + a′bc′de + a′c′def + ab′c′de′f

h. $K = f(r, s, t, u, v, w) = \Sigma(9, 11, 13, 15, 25, 27, 29, 31, 41, 45, 47, 57, 59, 61, 63)$

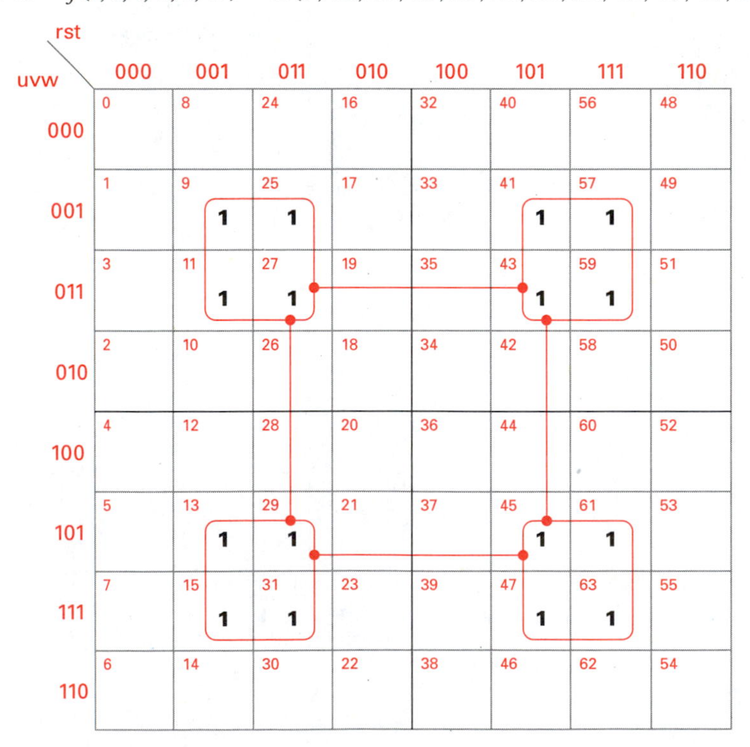

K = tw

13. a. $S = f(a, b, c, d) = \Sigma(1, 5, 7, 8, 9, 10, 11, 13, 15)$

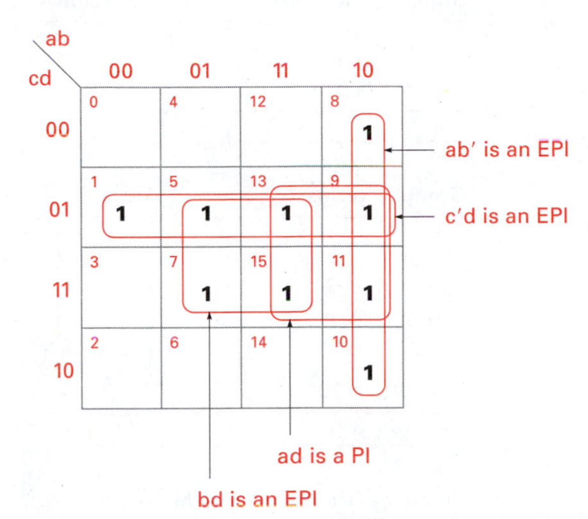

b. $T = f(a, b, c, d, e) = \Sigma(0, 4, 8, 9, 10, 11, 12, 13, 14, 15, 16, 20, 24, 28)$

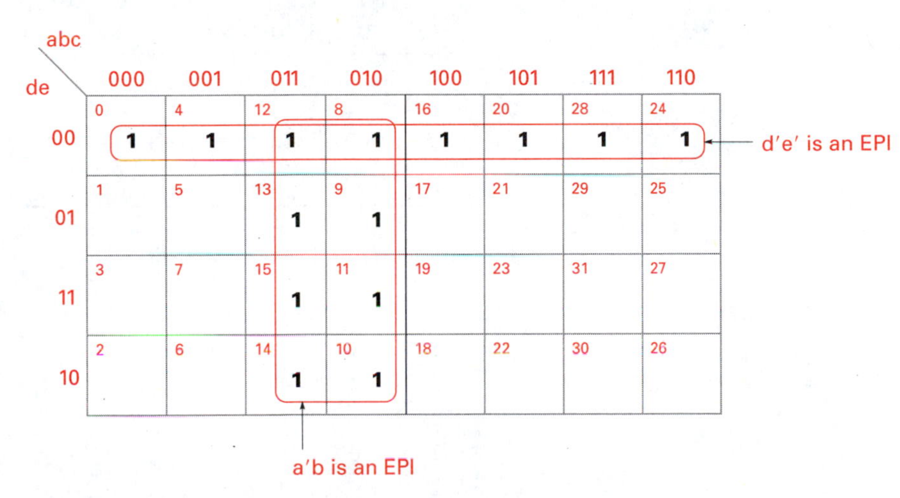

15. $G = f(a, b, c, d) = \Sigma(0, 2, 5, 7, 8, 10, 13, 15)$ and
$G = f(a, b, c, d) = \pi(1, 3, 4, 6, 9, 11, 12, 14)$

cd \ ab	00	01	11	10
00	0: **1**	4: **0**	12: **0**	8: **1**
01	1: **0**	5: **1**	13: **1**	9: **0**
11	3: **0**	7: **1**	15: **1**	11: **0**
10	2: **1**	6: **0**	14: **0**	10: **1**

17. a. $y = f(a, b, c) = \Sigma(1, 3, 5, 6, 7)$

Group	Min-terms	Variables		
		a	b	c
1	1	0	0	1 √
2	3	0	1	1 √
2	5	1	0	1 √
2	6	1	1	0 √
3	7	1	1	1 √

Group	Minterms	Variables		
		a	b	c
1	1, 3, 5, 7	—	—	1

Group	Min-terms	Variables		
		a	b	c
1	1, 3	0	—	1 √
1	1, 5	—	0	1 √
2	3, 7	—	1	1 √
2	5, 7	1	—	1 √
2	6, 7	1	1	—

PI Terms	Decimal	Minterms				
		1	3	5	6	7
c	1, 3, 5, 7	⊗	⊗	⊗		x
ab	6, 7				⊗	x

Both PIs are essential. Therefore, the final expression is $y = c + ab$.

b. $P = f(w, x, y, z) = \Sigma(0, 2, 8, 10)$

Group	Min-terms	Variables			
		w	x	y	z
0	0	0	0	0	0 √
1	2	0	0	1	0 √
1	8	1	0	0	0 √
2	10	1	0	1	0 √

Group	Min-terms	Variables			
		w	x	y	z
0	0, 2	0	0	—	0 √
0	0, 8	—	0	0	0 √
1	2, 10	—	0	1	0 √
1	8,10	1	0	—	0 √

Group	Minterms	Variables			
		w	x	y	z
0	0, 2, 8, 10	—	0	—	0

PI Terms	Decimal	Minterms			
		0	2	8	10
$x'z'$	0, 2, 8, 10	x	x	x	x

$P = x'z'$

c. $R = f(w, x, y, z) = \Sigma(1, 3, 4, 5, 6, 9, 11, 12, 13, 14)$

Group	Min-terms	Variables			
		w	x	y	z
1	1	0	0	0	1 √
1	4	0	1	0	0 √
2	3	0	0	1	1 √
2	6	0	1	1	0 √
2	9	1	0	0	1 √
2	12	1	1	0	0 √
2	5	0	1	0	1 √
3	11	1	0	1	1 √
3	13	1	1	0	1 √
3	14	1	1	1	0 √

Group	Minterms	Variables			
		w	x	y	z
1	1, 3, 9, 11	—	0	—	1
1	1, 5, 9, 13	—	—	0	1
1	4, 6, 12, 14	—	1	—	0
1	4, 5, 12, 13	—	1	0	—

Group	Min-terms	Variables			
		w	x	y	z
1	1, 3	0	0	—	1 √
1	1, 9	—	0	0	1 √
1	1, 5	0	—	0	1 √
1	4, 6	0	1	—	0 √
1	4, 12	—	1	0	0 √
1	4, 5	0	1	0	— √
2	3, 11	—	0	1	1 √
2	6, 14	—	1	1	0 √
2	9, 13	1	—	0	1 √
2	12, 13	1	1	—	0 √
2	12, 14	1	1	—	0 √
2	5, 13	—	1	0	1 √

$R = x'z + xz' + xy'$ or
$R = x'z + xz' + y'z$

PI	Decimal	Minterms									
		1	3	4	5	6	9	11	12	13	14
$x'z$	1, 3, 9, 11	x	⊗				x	⊗			
$y'z$	1, 5, 9, 13	x			x		x			x	
xz'	4, 6, 12, 14			x		⊗			x		⊗
xy'	4, 5, 12, 13			x	x				x	x	

19. a. $V = f(a, b, c, d) = \Sigma(2, 3, 4, 5, 13, 15) + \Sigma d(8, 9, 10, 11)$

Minterm	a	b	c	d	V
0	0	0	0	0	0
1	0	0	0	1	0
2	0	0	1	0	1
3	0	0	1	1	1
4	0	1	0	0	1
5	0	1	0	1	1
6	0	1	1	0	0
7	0	1	1	1	0
8	1	0	0	0	x
9	1	0	0	1	x
10	1	0	1	0	x
11	1	0	1	1	x
12	1	1	0	0	0

continues

Minterm	a	b	c	d	V
13	1	1	0	1	1
14	1	1	1	0	0
15	1	1	1	1	1

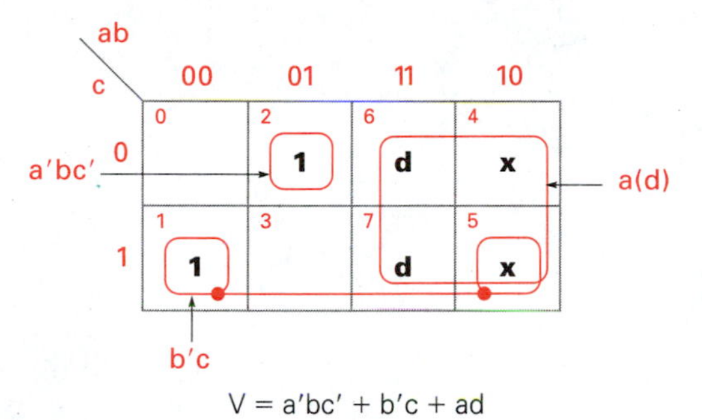

V = a'bc' + b'c + ad

b. $Y = f(u, v, w, x) = \Sigma(1, 5, 7, 13, 15) + \Sigma d(8, 10, 11, 14)$

Minterm	u	y	w	x	Y
0	0	0	0	0	0
1	0	0	0	1	1
2	0	0	1	0	0
3	0	0	1	1	0
4	0	1	0	0	0
5	0	1	0	1	1
6	0	1	1	0	0
7	0	1	1	1	1
8	1	0	0	0	d
9	1	0	0	1	1
10	1	0	1	0	d
11	1	0	1	1	d
12	1	1	0	0	0
13	1	1	0	1	1
14	1	1	1	0	d
15	1	1	1	1	1

x = don't care

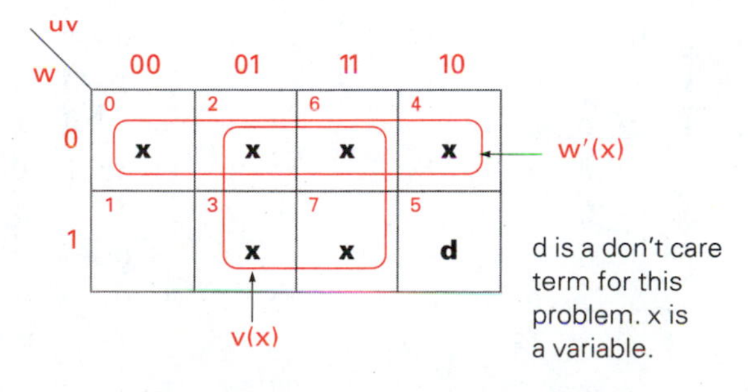

d is a don't care term for this problem. x is a variable.

c. $P = f(r, s, t, u) = \Sigma(0, 2, 4, 8, 10, 14) + \Sigma d(5, 6, 7, 12)$

Minterm	r	s	t	u	P
0	0	0	0	0	1
1	0	0	0	1	0
2	0	0	1	0	1
3	0	0	1	1	0
4	0	1	0	0	1
5	0	1	0	1	x
6	0	1	1	0	x
7	0	1	1	1	x
8	1	0	0	0	1
9	1	0	0	1	0
10	1	0	1	0	1
11	1	0	1	1	0
12	1	1	0	0	x
13	1	1	0	1	0
14	1	1	1	0	1
15	1	1	1	1	0

x = don't care

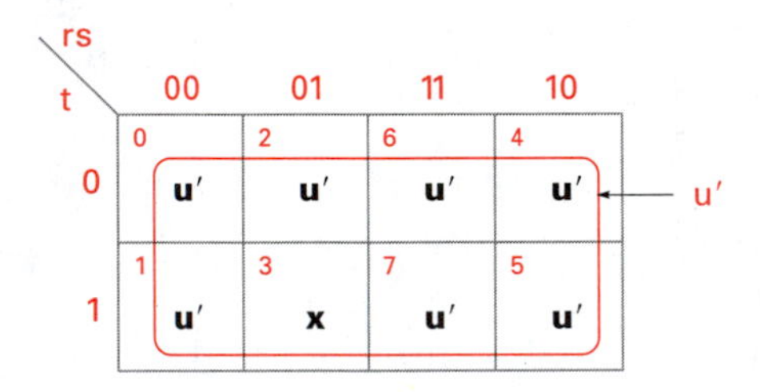

P = u'

d. $F = f(u, v, w, x, y) = \Sigma(0, 2, 8, 10, 16, 18, 24, 26)$

	u	v	w	x	y	F		u	v	w	x	y	F
0	0	0	0	0	0	1	9	0	1	0	0	1	0
1	0	0	0	0	1	0	10	0	1	0	1	0	1
2	0	0	0	1	0	1	11	0	1	0	1	1	0
3	0	0	0	1	1	0	12	0	1	1	0	0	0
4	0	0	1	0	0	0	13	0	1	1	0	1	0
5	0	0	1	0	1	0	14	0	1	1	1	0	0
6	0	0	1	1	0	0	15	0	1	1	1	1	0
7	0	0	1	1	1	0	16	1	0	0	0	0	1
8	0	1	0	0	0	1	17	1	0	0	0	1	0

continues

	u	v	w	x	y	F		u	v	w	x	y	F
18	1	0	0	1	0	1	25	1	1	0	0	1	0
19	1	0	0	1	1	0	26	1	1	0	1	0	1
20	1	0	1	0	0	0	27	1	1	0	1	1	0
21	1	0	1	0	1	0	28	1	1	1	0	0	0
22	1	0	1	1	0	0	29	1	1	1	0	1	0
23	1	1	1	1	1	0	30	1	1	1	1	0	0
24	1	1	0	0	0	1	31	1	1	1	1	1	0

wx \ uv	00	01	11	10
00	0: y'	4: y'	12: y'	8: y'
01	1: y'	5: y'	13: y'	9: y'
11	3	7	15	11
10	2	6	14	10

← w'(y')

F = w'v'

e. $H = f(a, b, c, d, e) = \Sigma(5, 7, 9, 12, 13, 14, 15, 20, 21, 22, 23, 25, 29, 31)$

	a	b	c	d	e	H		a	b	c	d	e	H
0	0	0	0	0	0	0	16	1	0	0	0	0	0
1	0	0	0	0	1	0	17	1	0	0	0	1	0
2	0	0	0	1	0	0	18	1	0	0	1	0	0
3	0	0	0	1	1	0	19	1	0	0	1	1	0
4	0	0	1	0	0	0	20	1	0	1	0	0	1
5	0	0	1	0	1	1	21	1	0	1	0	1	1
6	0	0	1	1	0	0	22	1	0	1	1	0	1
7	0	0	1	1	1	1	23	1	1	1	1	1	1
8	0	1	0	0	0	0	24	1	1	0	0	0	0
9	0	1	0	0	1	1	25	1	1	0	0	1	1
10	0	1	0	1	0	0	26	1	1	0	1	0	0
11	0	1	0	1	1	0	27	1	1	0	1	1	0
12	0	1	1	0	0	1	28	1	1	1	0	0	0
13	0	1	1	0	1	1	29	1	1	1	0	1	1
14	0	1	1	1	0	1	30	1	1	1	1	0	0
15	0	1	1	1	1	1	31	1	1	1	1	1	1

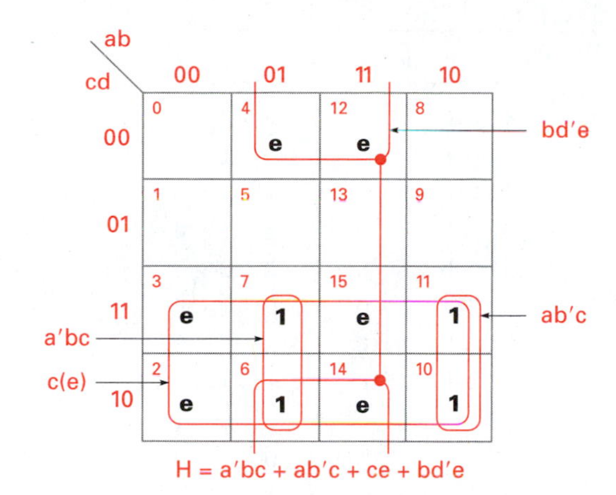

f. $M = f(vwxyz) = \Sigma(1, 3, 4, 6, 9, 11, 12, 14, 17, 19, 20, 22, 25, 27, 28, 30) + \Sigma\, d\,(8, 10, 24, 26)$

	u	*v*	*w*	*x*	*y*	*M*		*u*	*v*	*w*	*x*	*y*	*M*
0	0	0	0	0	0	0	16	1	0	0	0	0	0
1	0	0	0	0	1	1	17	1	0	0	0	1	1
2	0	0	0	1	0	0	18	1	0	0	1	0	0
3	0	0	0	1	1	1	19	1	0	0	1	1	1
4	0	0	1	0	0	1	20	1	0	1	0	0	1
5	0	0	1	0	1	0	21	1	0	1	0	1	0
6	0	0	1	1	0	1	22	1	0	1	1	0	1
7	0	0	1	1	1	0	23	1	1	1	1	1	0
8	0	1	0	0	0	*d*	24	1	1	0	0	0	*d*
9	0	1	0	0	1	1	25	1	1	0	0	1	1
10	0	1	0	1	0	*d*	26	1	1	0	1	0	*d*
11	0	1	0	1	1	1	27	1	1	0	1	1	1
12	0	1	1	0	0	1	28	1	1	1	0	0	1
13	0	1	1	0	1	0	29	1	1	1	0	1	0
14	0	1	1	1	0	1	30	1	1	1	1	0	1
15	0	1	1	1	1	0	31	1	1	1	1	1	0

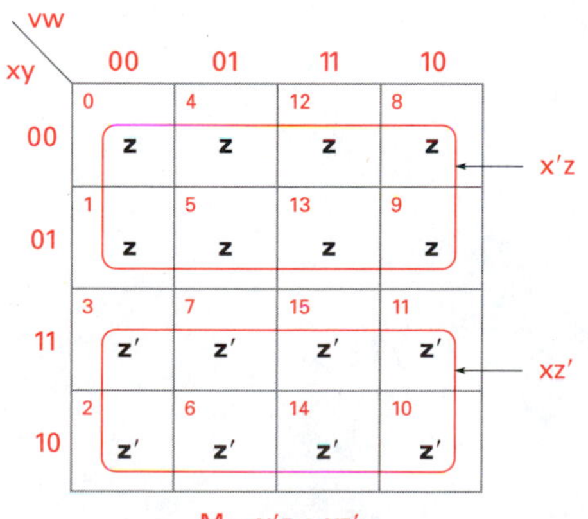

g. $J = f(a, b, c, d, e, f) = \Sigma(7, 12, 22, 23, 28, 34, 37, 38, 40, 42, 44, 46, 56, 58, 60, 62)$

	a	*b*	*c*	*d*	*e*	*f*	*J*
0	0	0	0	0	0	0	0
1	0	0	0	0	0	1	0
2	0	0	0	0	1	0	0
3	0	0	0	0	1	1	0
4	0	0	0	1	0	0	0
5	0	0	0	1	0	1	0
6	0	0	0	1	1	0	0
7	0	0	0	1	1	1	1
8	0	0	1	0	0	0	0
9	0	0	1	0	0	1	0
10	0	0	1	0	1	0	0
11	0	0	1	0	1	1	0
12	0	0	1	1	0	0	1
13	0	0	1	1	0	1	0
14	0	0	1	1	1	0	0
15	0	0	1	1	1	1	0

	a	*b*	*c*	*d*	*e*	*f*	*J*
16	0	1	0	0	0	0	0
17	0	1	0	0	0	1	0
18	0	1	0	0	1	0	0
19	0	1	0	0	1	1	0
20	0	1	0	1	0	0	0
21	0	1	0	1	0	1	0
22	0	1	0	1	1	0	1
23	0	1	0	1	1	1	1
24	0	1	1	0	0	0	0
25	0	1	1	0	0	1	0
26	0	1	1	0	1	0	0
27	0	1	1	0	1	1	0
28	0	1	1	1	0	0	1
29	0	1	1	1	0	1	0
30	0	1	1	1	1	0	0
31	0	1	1	1	1	1	0

	a	*b*	*c*	*d*	*e*	*f*	*J*
32	1	0	0	0	0	0	0
33	1	0	0	0	0	1	0
34	1	0	0	0	1	0	1
35	1	0	0	0	1	1	0
36	1	0	0	1	0	0	0
37	1	0	0	1	0	1	1
38	1	0	0	1	1	0	1
39	1	0	0	1	1	1	0
40	1	0	1	0	0	0	1
41	1	0	1	0	0	1	0
42	1	0	1	0	1	0	1
43	1	0	1	0	1	1	0
44	1	0	1	1	0	0	1
45	1	0	1	1	0	1	0
46	1	0	1	1	1	0	1
47	1	0	1	1	1	1	0

	a	*b*	*c*	*d*	*e*	*f*	*J*
48	1	1	0	0	0	0	0
49	1	1	0	0	0	1	0
50	1	1	0	0	1	0	0
51	1	1	0	0	1	1	0
52	1	1	0	1	0	0	0
53	1	1	0	1	0	1	0
54	1	1	0	1	1	0	0
55	1	1	0	1	1	1	0
56	1	1	1	0	0	0	1
57	1	1	1	0	0	1	0
58	1	1	1	0	1	0	1
59	1	1	1	0	1	1	0
60	1	1	1	1	0	0	1
61	1	1	1	1	0	1	0
62	1	1	1	1	1	0	1
63	1	1	1	1	1	1	0

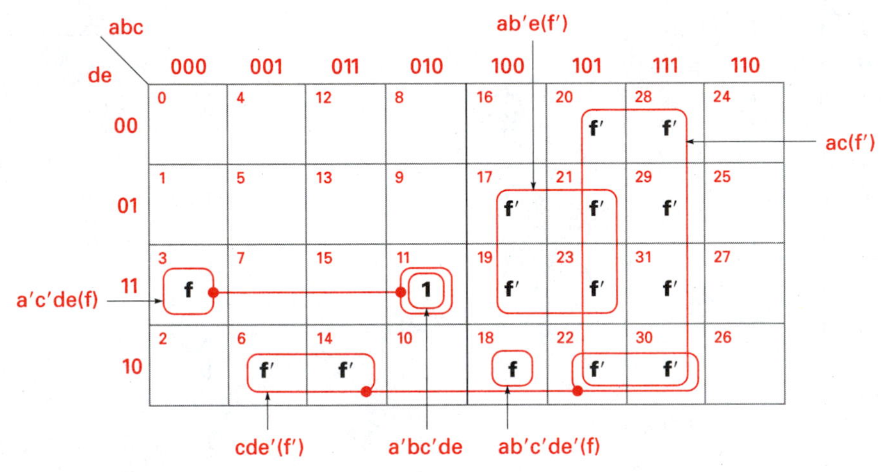

J = a'bc'de + a'c'def + ab'c'de'f + ab'ef' + cde'f' + acf'

h. $K = f(r, s, t, u, v, w) = \Sigma(9, 11, 13, 15, 25, 27, 29, 31, 41, 43, 45, 47, 57, 59, 61, 63)$

	r	s	t	u	v	w	K
0	0	0	0	0	0	0	0
1	0	0	0	0	0	1	0
2	0	0	0	0	1	0	0
3	0	0	0	0	1	1	0
4	0	0	0	1	0	0	0
5	0	0	0	1	0	1	0
6	0	0	0	1	1	0	0
7	0	0	0	1	1	1	0
8	0	0	1	0	0	0	0
9	0	0	1	0	0	1	1
10	0	0	1	0	1	0	0
11	0	0	1	0	1	1	1
12	0	0	1	1	0	0	0
13	0	0	1	1	0	1	1
14	0	0	1	1	1	0	0
15	0	0	1	1	1	1	1

	r	s	t	u	v	w	K
16	0	1	0	0	0	0	0
17	0	1	0	0	0	1	0
18	0	1	0	0	1	0	0
19	0	1	0	0	1	1	0
20	0	1	0	1	0	0	0
21	0	1	0	1	0	1	0
22	0	1	0	1	1	0	0
23	0	1	0	1	1	1	0
24	0	1	1	0	0	0	0
25	0	1	1	0	0	1	1
26	0	1	1	0	1	0	0
27	0	1	1	0	1	1	1
28	0	1	1	1	0	0	0
29	0	1	1	1	0	1	1
30	0	1	1	1	1	0	0
31	0	1	1	1	1	1	1

	r	s	t	u	v	w	K
32	1	0	0	0	0	0	0
33	1	0	0	0	0	1	0
34	1	0	0	0	1	0	0
35	1	0	0	0	1	1	0
36	1	0	0	1	0	0	0
37	1	0	0	1	0	1	0
38	1	0	0	1	1	0	0
39	1	0	0	1	1	1	0
40	1	0	1	0	0	0	0
41	1	0	1	0	0	1	1
42	1	0	1	0	1	0	0
43	1	0	1	0	1	1	1
44	1	0	1	1	0	0	0
45	1	0	1	1	0	1	1
46	1	0	1	1	1	0	0
47	1	0	1	1	1	1	1

	r	s	t	u	v	w	K
48	1	1	0	0	0	0	0
49	1	1	0	0	0	1	0
50	1	1	0	0	1	0	0
51	1	1	0	0	1	1	0
52	1	1	0	1	0	0	0
53	1	1	0	1	0	1	0
54	1	1	0	1	1	0	0
55	1	1	0	1	1	1	0
56	1	1	1	0	0	0	0
57	1	1	1	0	0	1	1
58	1	1	1	0	1	0	0
59	1	1	1	0	1	1	1
60	1	1	1	1	0	0	0
61	1	1	1	1	0	1	1
62	1	1	1	1	1	0	0
63	1	1	1	1	1	1	1

uv \ rst	000	001	011	010	100	101	111	110
00	0	4 w	12 w	8	16	20 w	28 w	24
01	1	5 w	13 w	9	17	21 w	29 w	25
11	3	7 w	15 w	11	19	23 w	31 w	27
10	2	6 w	14 w	10	18	22 w	30 w	26

t(w)

K = tw

21. **a.** Asserted: The "true" condition of a variable.

b. Active high: The asserted or true condition is a "1" or a positive voltage level.

c. Bubble logic: Mixed logic. Both asserted high and asserted low variables are present in the same circuit.

d. Mixed logic: Same as Bubble logic.

23.

X.H, Y.H → O.H

X.H, Y.L → O.L

X.H, Y.H → O.L

X.L, Y.H → O.L

X.H, Y.L → O.H

X.L, Y.L → O.H

X.L, Y.H → O.H

X.L, Y.L → O.L

25. **a.** $X = f(a, b, c) = \Sigma(1, 3, 7)$
$Y = f(a, b, c) = \Sigma(2, 6, 7)$
$X = a'c + bc$
$Y = ab + bc'$
Creating a common term of abc gives: $X = a'c + abc$ and $Y = abc + bc'$.

b. $X = f(a, b, c) = \Sigma(3, 4, 5, 7)$
$Y = f(a, b, c) = \Sigma(3, 4, 6, 7)$
$X = ab' + bc$ and $Y = ac' + bc$. Term bc is shared.

CHAPTER 4

1. **a.** Two input variables and two input constants are needed.

b. Four output variables are necessary to contain the largest value ($3 \times 3 = 9$).

c.

Inputs				Outputs			
M_2	M_1	C_2	C_1	O_4	O_3	O_2	O_1
0	0	1	1	0	0	0	0
0	1	1	1	0	0	1	1
1	0	1	1	0	1	1	0
1	1	1	1	1	0	0	1

$O_4 = f(M_2, M_1) = \Sigma(3)$:

$O_3 = f(M_2, M_1) = \Sigma(2)$:

$O_2 = f(M_2, M_1) = \Sigma(1, 2)$:

$O_1 = f(M_2, M_1) = \Sigma(1, 3)$

$O_4 = M_2M_1$ A K-map isn't needed since only one minterm exists

$O_3 = M_2M_1'$

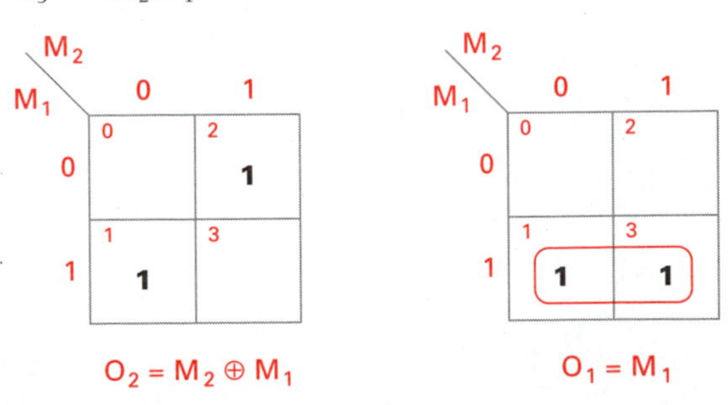

$O_2 = M_2 \oplus M_1$ $O_1 = M_1$

3. a.

BCD				EXCESS-3			
A	B	C	D	W	X	Y	Z
0	0	0	0	0	0	1	1
0	0	0	1	0	1	0	0
0	0	1	0	0	1	0	1
0	0	1	1	0	1	1	0
0	0	0	1	0	1	0	0
0	1	0	0	0	1	1	1
0	1	0	1	1	0	0	0
0	1	1	0	1	0	0	1
0	1	1	1	1	0	1	0
1	0	0	0	1	0	1	1
1	0	0	1	1	1	0	0
1	0	1	0	Don't-care			
1	0	1	1	"			
1	1	0	0	"			
1	1	0	1	"			
1	1	1	0	"			
1	1	1	1	"			

b. $W = f(A, B, C, D) = \Sigma(5, 6, 7, 8, 9) + \Sigma d(10, 11, 12, 13, 14, 15)$

$X = f(A, B, C, D) = \Sigma(1, 2, 3, 4, 9) + \Sigma d(10 - 15)$

$Y = f(A, B, C, D) = \Sigma(0, 3, 4, 7, 8) + \Sigma d(10 - 15)$

$Z = f(A, B, C, D) = \Sigma(0, 2, 4, 6, 8) + \Sigma d(10 - 15)$

c.

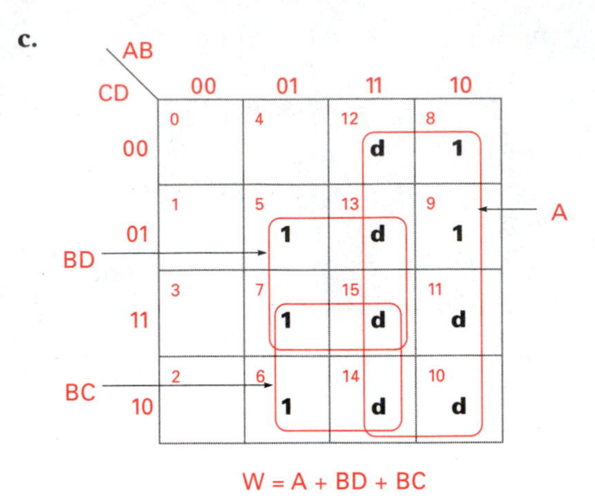

W = A + BD + BC

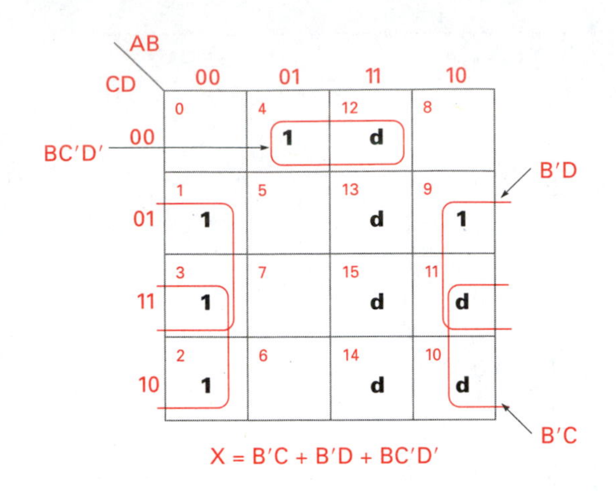

X = B′C + B′D + BC′D′

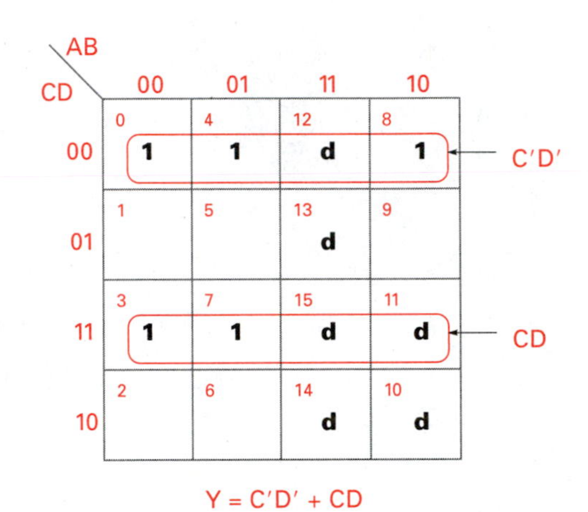

Y = C′D′ + CD

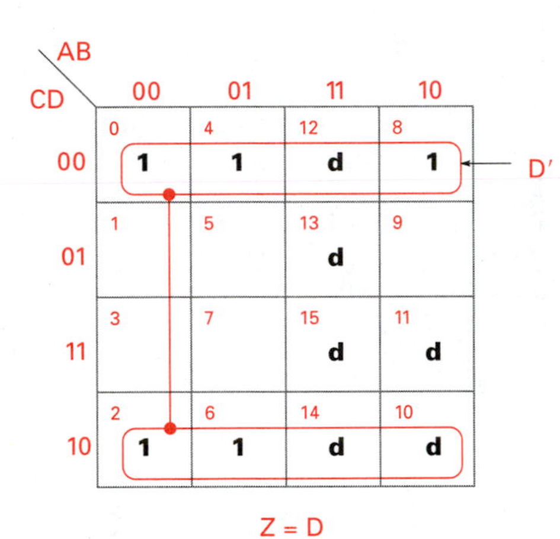

Z = D

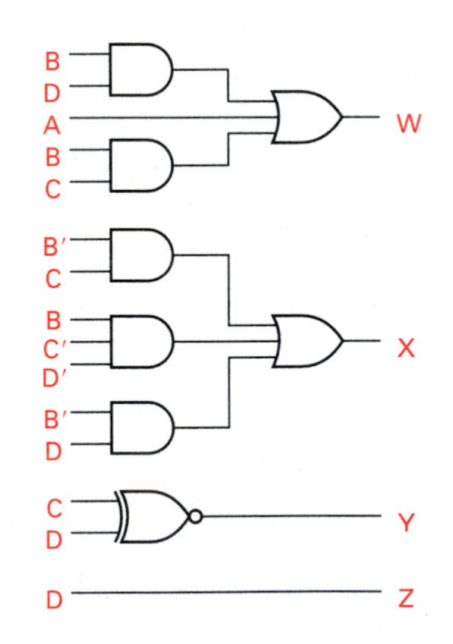

5. a. Truth Table

Excess-3				7-Segment						
M	*N*	*O*	*P*	*a*	*b*	*c*	*d*	*e*	*f*	*g*
0	0	0	0	*x*	*x*	*x*	*x*	*x*	*x*	*x*
0	0	0	1	*x*	*x*	*x*	*x*	*x*	*x*	*x*
0	0	1	0	*x*	*x*	*x*	*x*	*x*	*x*	*x*
0	0	1	1	0	0	0	0	0	0	1
0	1	0	0	1	0	0	1	1	1	1
0	1	0	1	0	0	1	0	0	1	0
0	1	1	0	0	0	0	0	1	1	0
0	1	1	1	1	0	0	1	1	0	0
1	0	0	0	0	1	0	0	1	0	0
1	0	0	1	1	1	0	0	0	0	0
1	0	1	0	0	0	0	1	1	1	1
1	0	1	1	0	0	0	0	0	0	0
1	1	0	0	0	0	0	1	1	0	0
1	1	0	1	*x*	*x*	*x*	*x*	*x*	*x*	*x*
1	1	1	0	*x*	*x*	*x*	*x*	*x*	*x*	*x*
1	1	1	1	*x*	*x*	*x*	*x*	*x*	*x*	*x*

x indicates don't-care terms.

b. $a = f(M, N, O, P)$
$= \Sigma(4, 7, 9) + \Sigma d(0, 1, 2, 13, 14, 15)$

$b = f(M, N, O, P)$
$= \Sigma(8, 9) + \Sigma d(0, 1, 2, 13, 14, 15)$

$c = f(M, N, O, P)$
$= \Sigma(5) + \Sigma d(0, 1, 13, 14, 15)$

$d = f(M, N, O, P)$
$= \Sigma(4, 7, 10, 12) + \Sigma d(0, 1, 2, 13, 14, 15)$

$e = f(M, N, O, P)$
$= \Sigma(4, 6, 7, 8, 10, 12)$
$+ \Sigma d(0, 1, 2, 13, 14, 15)$

$f = f(M, N, O, P)$
$= \Sigma(4, 5, 6, 10) + \Sigma d(0, 1, 2, 13, 14, 15)$

$g = f(M, N, O, P)$
$= \Sigma(3, 4, 10) + \Sigma d(0, 1, 2, 13, 14, 15)$

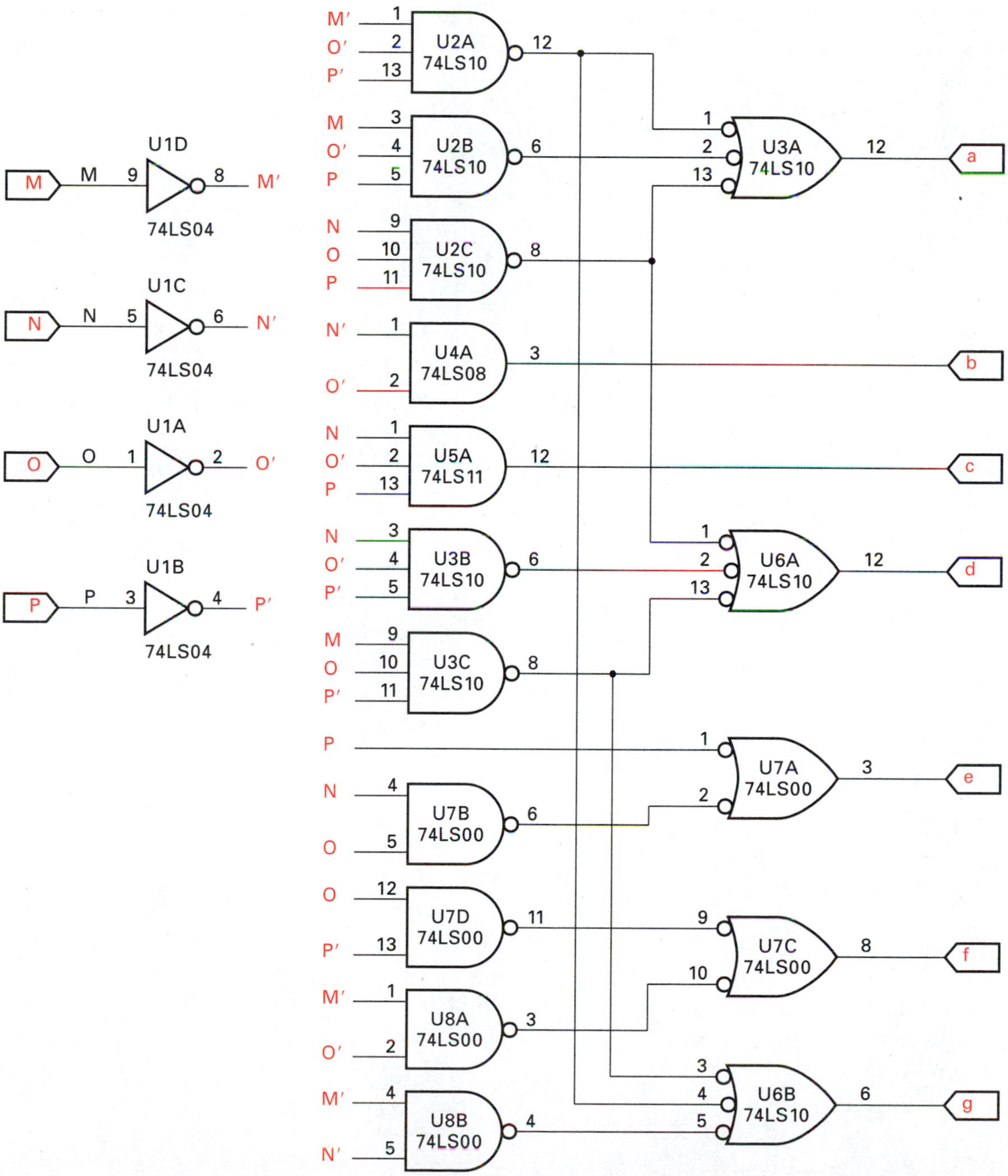

7.

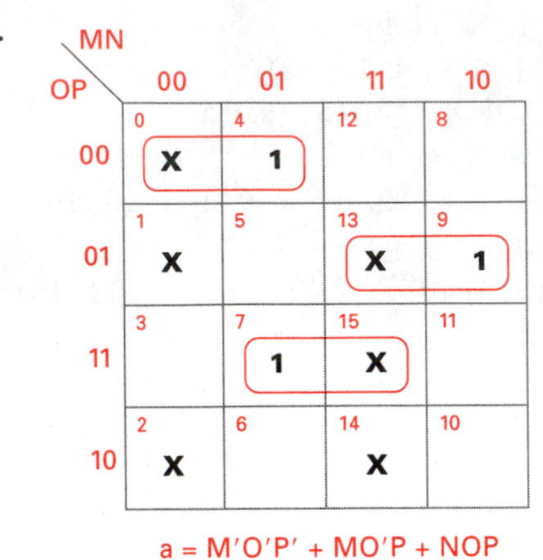

a = M′O′P′ + MO′P + NOP

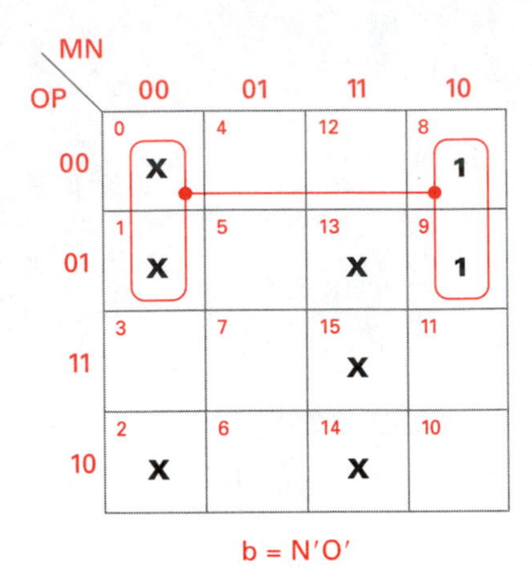

b = N′O′

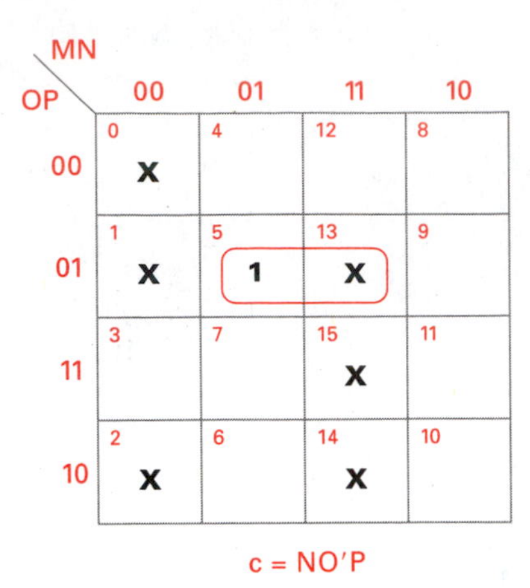

c = NO′P

OP \ MN	00	01	11	10
00	X	1	1	
01	X		X	
11		1	X	
10	X		X	1

d = NO′P′ + NOP + MOP′

OP \ MN	00	01	11	10
00	X	1	1	1
01	X		X	
11		1	X	
10	X	1	X	1

e = P′ + NO

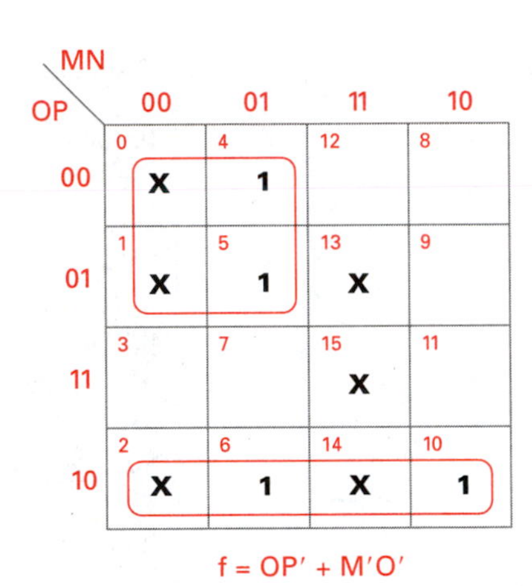

f = OP′ + M′O′

OP \ MN	00	01	11	10
00	X	1		
01	X		X	
11	1		X	
10	X		X	1

g = M′O′P′ + M′N′ + MOP′

The completed simplified equations are:

$a = M'O'P' + MO'P + NOP$

$b = N'O'$

$c = M'O'P$

$d = NO'P' + NOP + MOP'$

$e = P' + NO$

$f = OP' + M'O'$

$g = M'O'P + M'N' + MOP'$

9. a. Two levels are in the longest path:
AND (7408, $t_{pLH} = 27$ nS, $t_{pHL} = 19$ nS)
OR (7432, $t_{pLH} = 15$ nS, $t_{pHL} = 22$ nS).
27 nS + 22 nS = 48nS. Longest path, worst case propagation delay. Five AND gates, 2 OR gates, and 1 Exclusive-OR gate are used. However, (2) 7408 IC packages, (1) OR IC, and (1) EX-OR IC are needed. The power dissipated is:

7408 ($I_{CCH} = 21$ mA, $I_{CCL} = 33$ mA);
$I_{CC(AVE)} = 27$ mA

7432 ($I_{CCH} = 22$ mA, $I_{CCL} = 38$ mA);
$I_{CC(AVE)} = 30$ mA

7486 ($I_{CC} = 50$ mA)

Total average power = $I_{CC(AVE)} \times V_{CC}$
= (2 × 27 mA + 30 mA + 50 mA)(5 V)

Power = 670 mW

11. a. Fan-out: The number of identical gate inputs that can be reliably driven by the output of a gate.

b. Noise margin: The difference between the output and input voltage levels for logical "1" and "0." For example, $V_{OH} = 2.0$ V and $V_{IH} = 2.4$ V. $V_{IH} - V_{OH} = 2.4 \text{ V} - 2.0 \text{ V} = .4$ V. The logical "1" noise margin is 4 V.

c. I_{CCH}: The supply current for an IC package, when all of the outputs are "1."

d. V_{OH}: Voltage output high, or the voltage for a "1" output.

e. T_{PLH}: Time of propagation delay from a logical "0" to a logical "1."

f. Current sink: A condition where the current direction is into the output of a gate. The output transistor "sinks" the current or provides a current path to ground.

g. Current source: A condition where the current direction is from the gate output.

13.

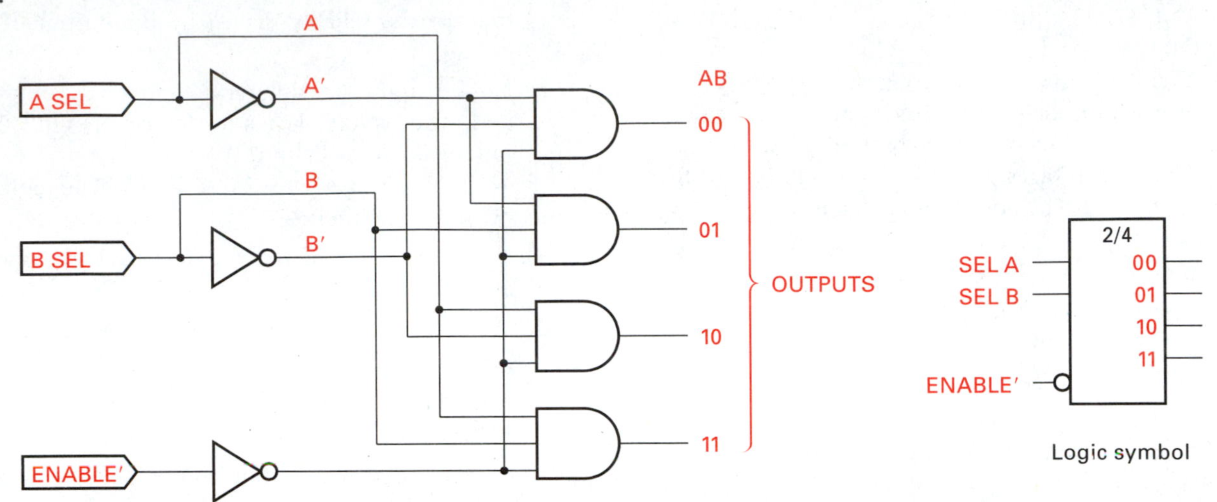

Decoder logic diagram

Inputs			Outputs			
A	*B*	Enable′	00	01	10	11
0	0	0	1	0	0	0
0	1	0	0	1	0	0
1	0	0	0	0	1	0
1	1	0	0	0	0	1
x	*x*	1	0	0	0	0

x is don't-care.
Decoder Truth Table

15.

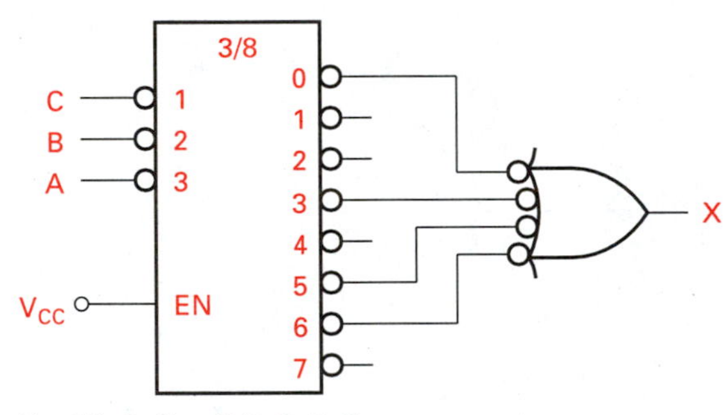

X = f(A, B, C) = Σ (0, 3, 5, 6)

17. a.

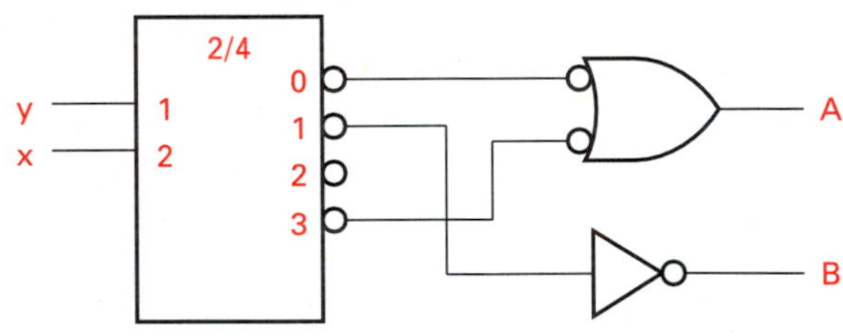

A = f(X, Y) = Σ (0, 3); B = f(X, Y) = (1).

b.

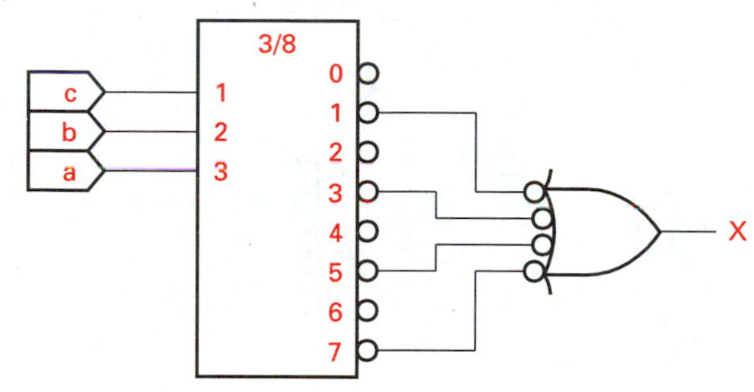

$X = f(A, B, C) = \Sigma\,(1, 3, 5, 7)$

c.

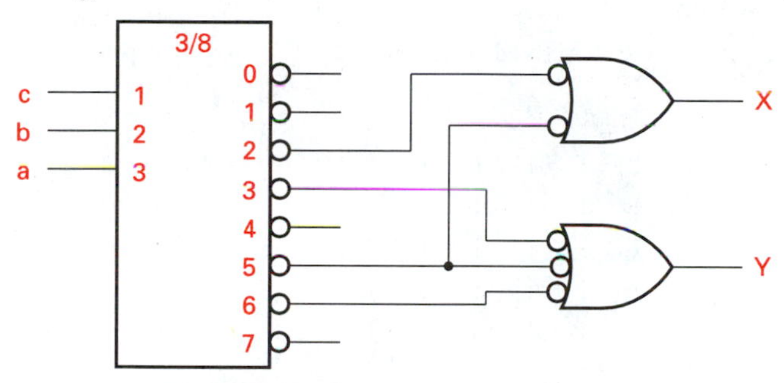

$X = f(a, b, c) = \Sigma\,(2, 5)$;
$Y = f(a, b, c) = \Sigma\,(3, 5, 6)$

d.

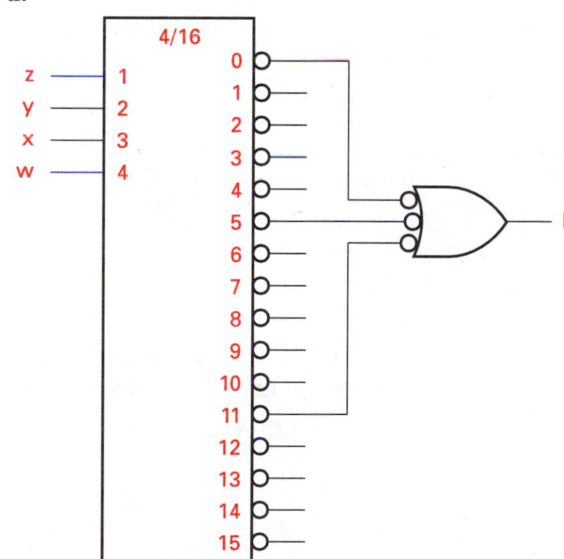

$P = f(w, x, y, z) = \Sigma\,(0, 5, 11)$

e.

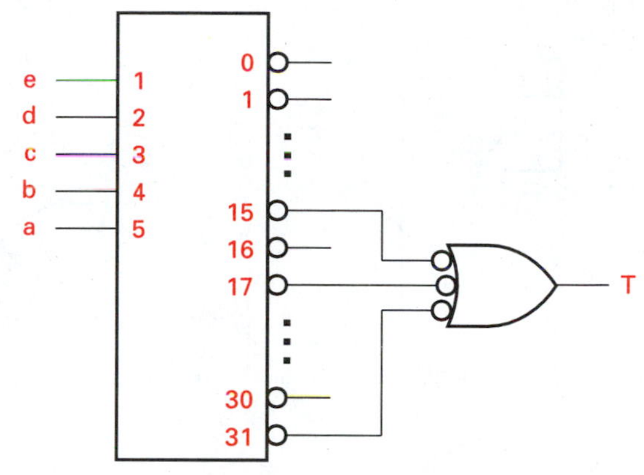

$T = f(a, b, c, d, e) = (15, 17, 31)$

f. Without reduction, this function can be realized using a 5/32 decoder, represented by the 5/32 block below.

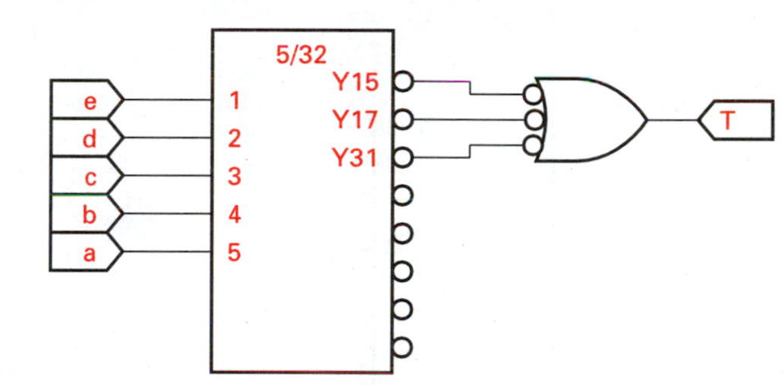

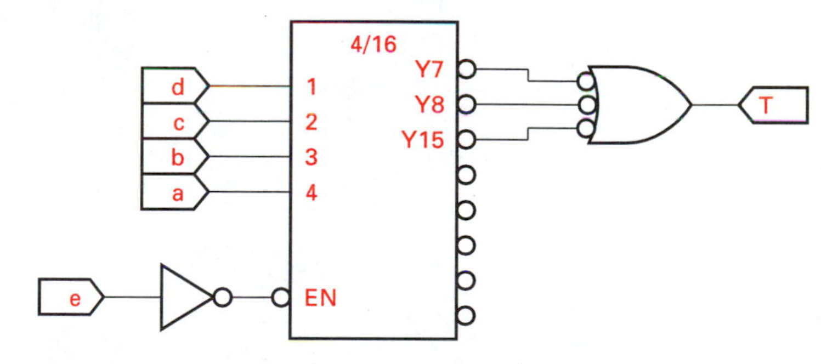

With an examination of the function, it is seen that $e = 1$ in all of the minterms. The function can then be realized in a reduced form using a 4/16 decoder that is enabled by $e = 1$.
$T = a'bcde + ab'c'd'e + abcde$

g.

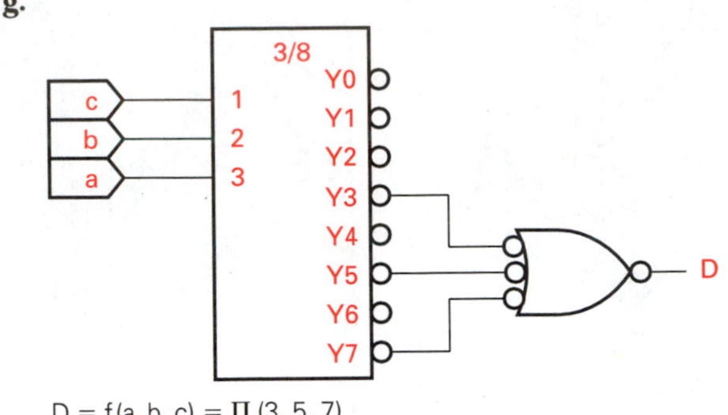

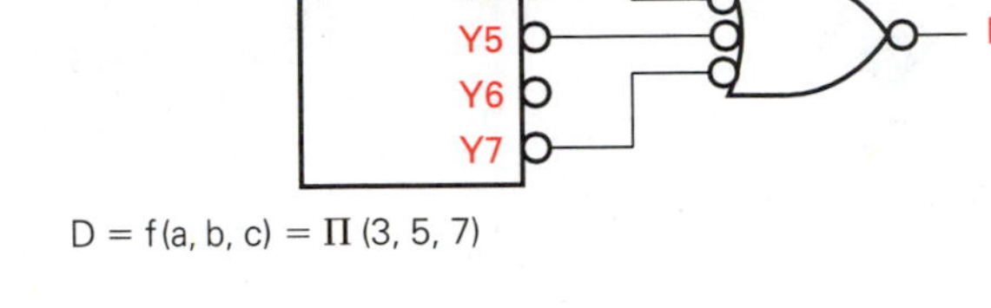

D = f(a, b, c) = Π (3, 5, 7)

h.

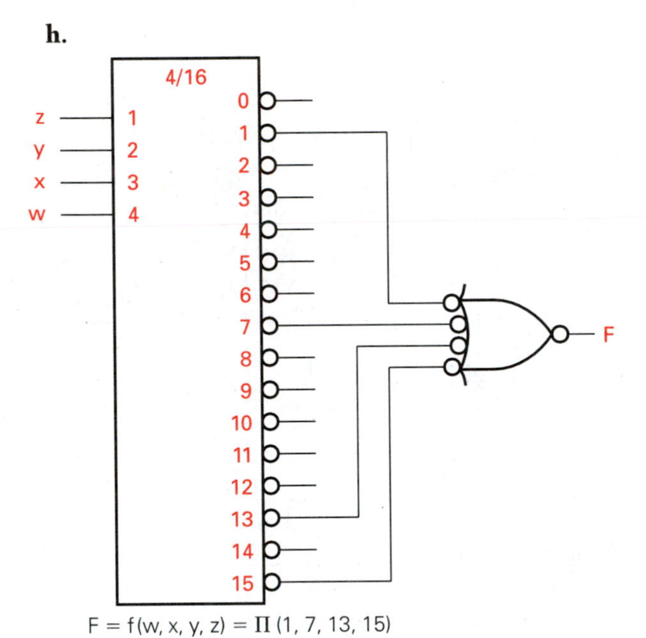

F = f(w, x, y, z) = Π (1, 7, 13, 15)

19.

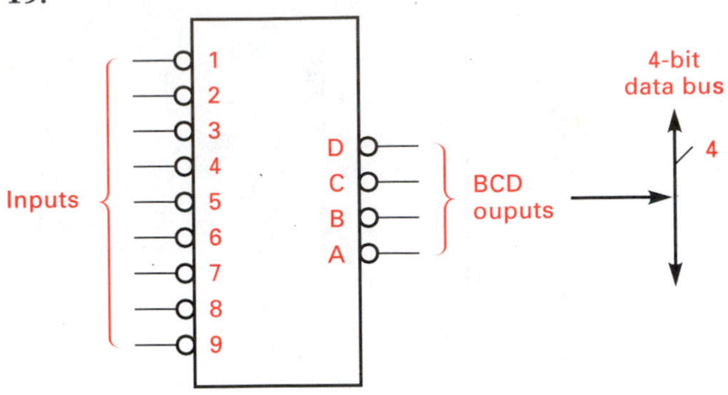

Active low inputs									Active low outputs			
1	2	3	4	5	6	7	8	9	D	C	B	A
1	1	1	1	1	1	1	1	0	0	1	1	0
1	1	1	1	1	1	1	0	1	0	1	1	1
1	1	1	1	1	1	0	1	1	1	0	0	0
1	1	1	1	1	0	1	1	1	1	0	0	1
1	1	1	1	0	1	1	1	1	1	0	1	0
1	1	1	0	1	1	1	1	1	1	0	1	1
1	1	0	1	1	1	1	1	1	1	1	0	0
1	0	1	1	1	1	1	1	1	1	1	0	1
0	1	1	1	1	1	1	1	1	1	1	1	0

When all of the active low inputs are high (1), then active low BCD outputs are all high (1 1 1 1), which is an active low BCD "zero."

21.

Inputs { CH1, CH2, CH3, CH4, CH5 } → Output

Multiplexing permits selection of a single data source from several input data sources for transmittal to an output.

23.

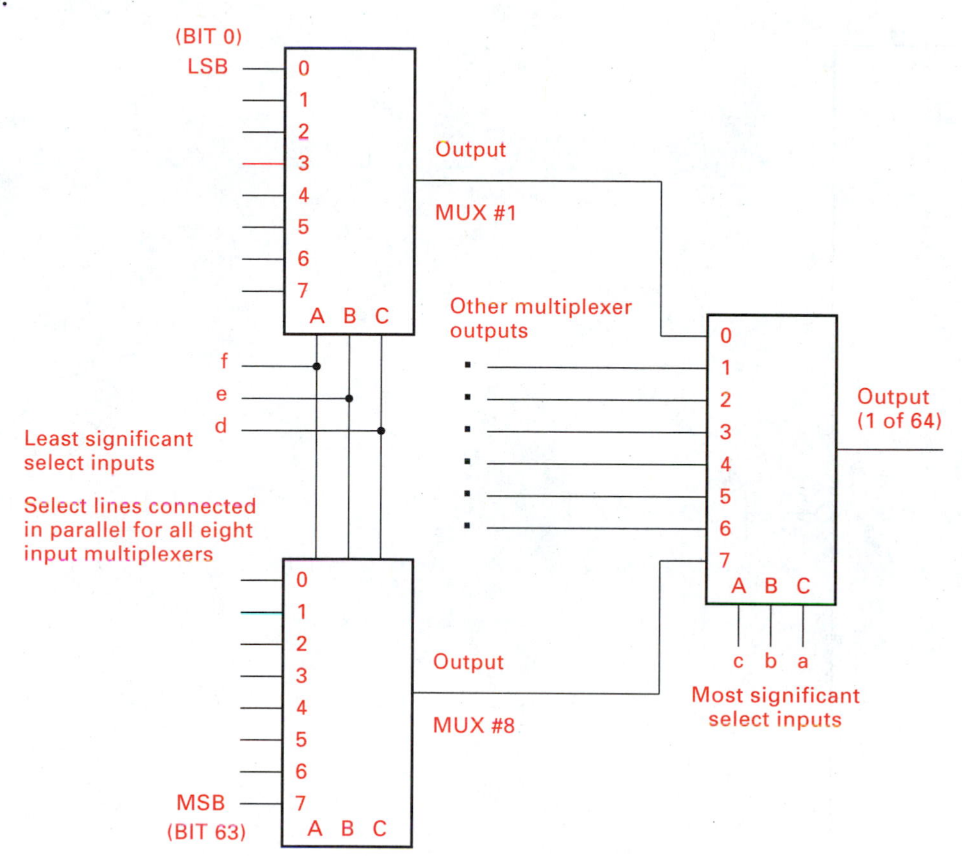

25. a.

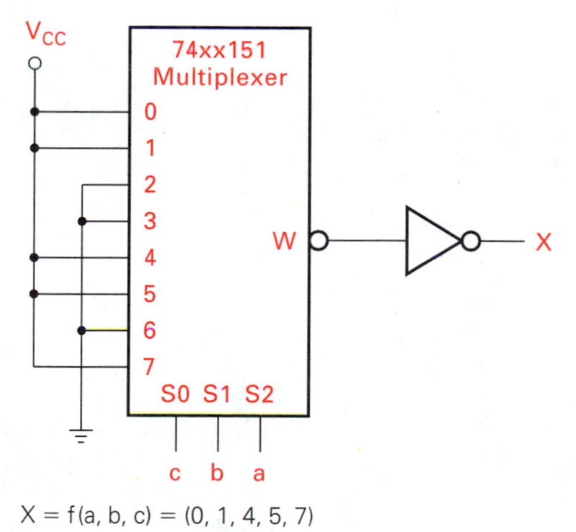

X = f(a, b, c) = (0, 1, 4, 5, 7)

b.

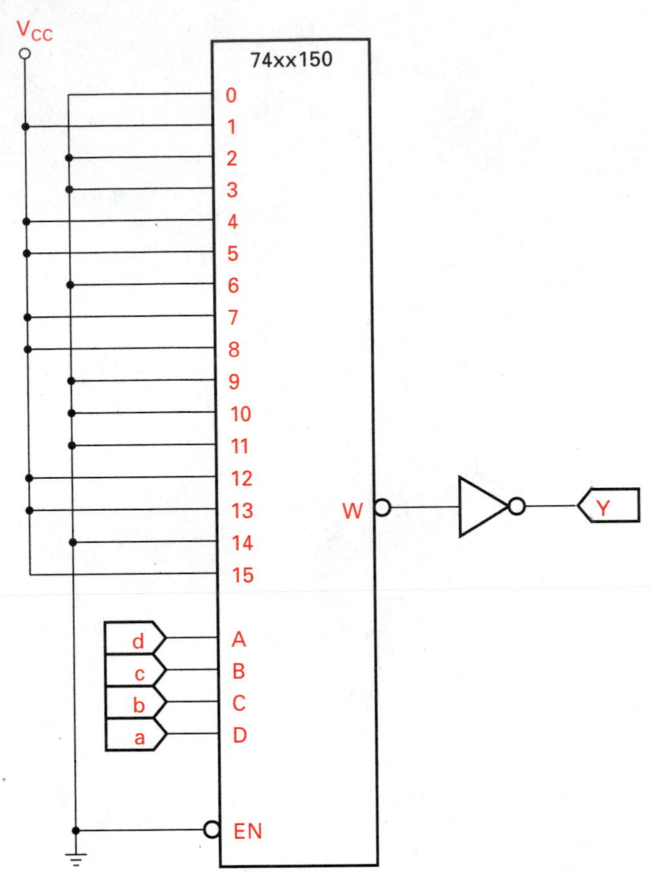

$Y = f(a, b, c, d) = \Sigma\,(1, 4, 5, 7, 8, 12, 13, 15)$
Using 16-1 MUX, 74xx150

c.

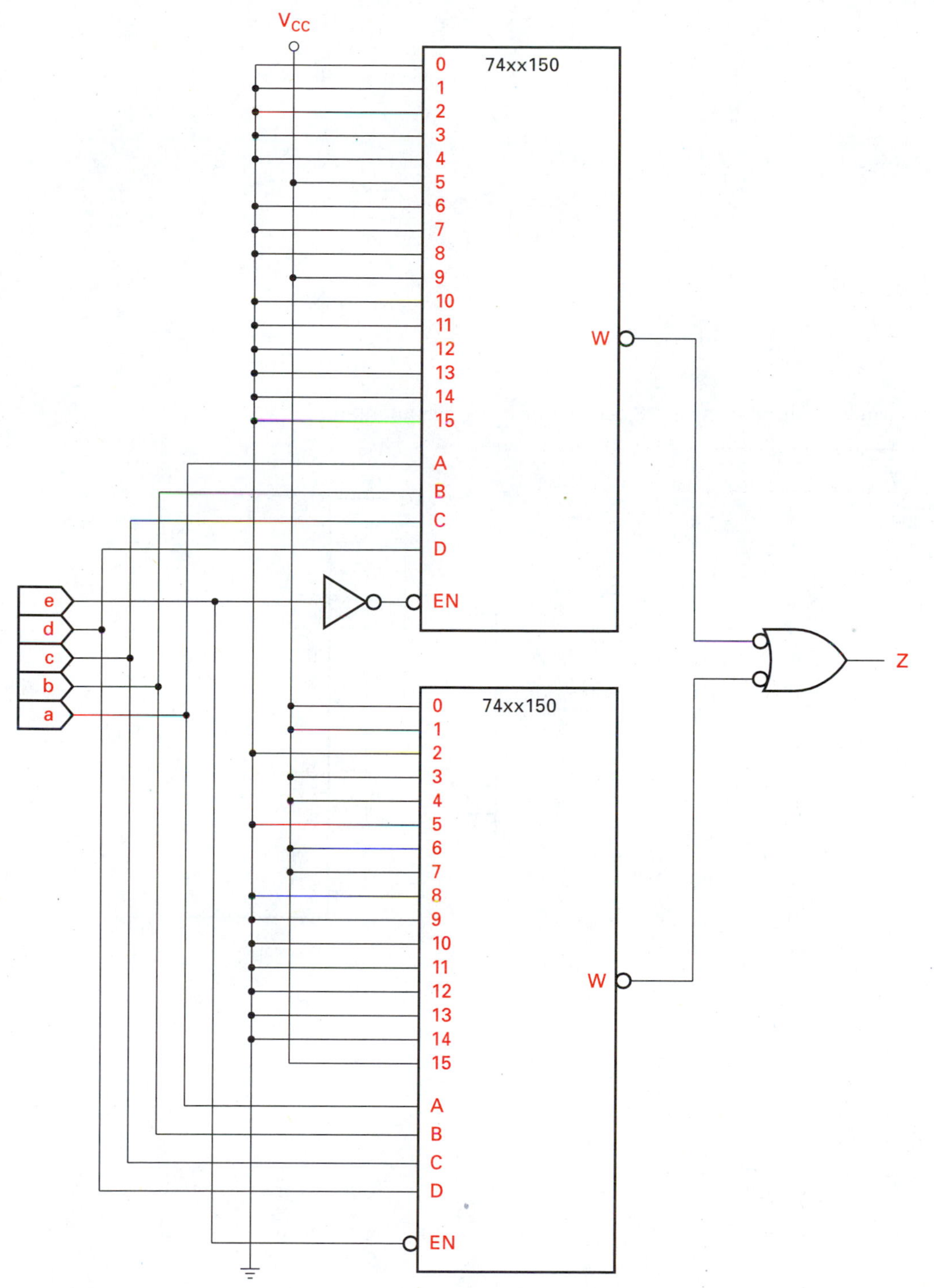

VCC
74xx150
0
1
2
3
4
5
6
7
8
9
10
11
12
13
14
15
A
B
C
D
EN
W
e
d
c
b
a
74xx150
Z

27. a.

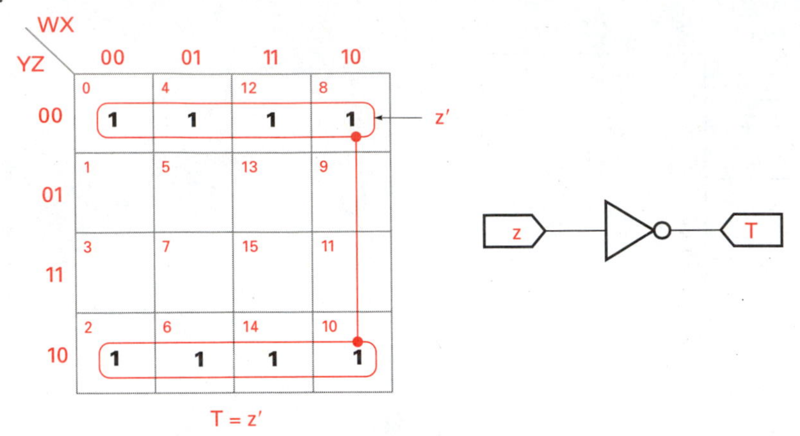

b. This function can be realized using a 3/8 decoder by using z as an input.

	wxy	*z*	*S*	MEV
0	000	0	1	1
	000	1	1	
1	001	0	0	z
	001	1	1	
2	010	0	0	0
	010	1	0	
3	011	0	0	0
	011	1	0	
4	100	0	0	z
	100	1	1	
5	101	0	1	z'
	101	1	0	
6	110	0	1	z'
	110	1	0	
7	111	0	1	z'
	111	1	0	

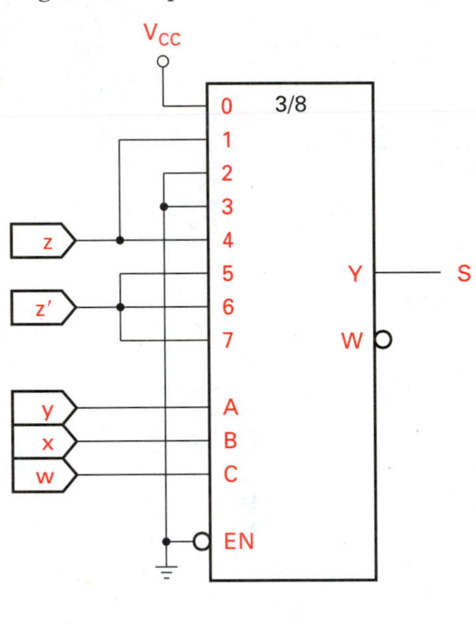

c.

dec	a	b	c	d	e	U		dec	a	b	c	d	e	U	MEV
0	0	0	0	0	0	1		0	0	0	0	0	0	1	
1	0	0	0	0	1	0	0	1	0	0	0	0	1	0	e'
2	0	0	0	1	0	1		2	0	0	0	1	0	1	
3	0	0	0	1	1	0		3	0	0	0	1	1	0	
4	0	0	1	0	0	0		4	0	0	1	0	0	0	
5	0	0	1	0	1	1	1	5	0	0	1	0	1	1	e'
6	0	0	1	1	0	0		6	0	0	1	1	0	0	
7	0	0	1	1	1	1		7	0	0	1	1	1	1	
8	0	1	0	0	0	1		8	0	1	0	0	0	1	
9	0	1	0	0	1	0	2	9	0	1	0	0	1	0	e'
10	0	1	0	1	0	1		10	0	1	0	1	0	1	
11	0	1	0	1	1	0		11	0	1	0	1	1	0	
12	0	1	1	0	0	0		12	0	1	1	0	0	0	
13	0	1	1	0	1	1	3	13	0	1	1	0	1	1	e
14	0	1	1	1	0	0		14	0	1	1	1	0	0	
15	0	1	1	1	1	1		15	0	1	1	1	1	1	
16	1	0	0	0	0	1		16	1	0	0	0	0	1	
17	1	0	0	0	1	0	4	17	1	0	0	0	1	0	e'
18	1	0	0	1	0	1		18	1	0	0	1	0	1	
19	1	0	0	1	1	0		19	1	0	0	1	1	0	
20	1	0	1	0	0	0		20	1	0	1	0	0	0	
21	1	0	1	0	1	1	5	21	1	0	1	0	1	1	e'
22	1	0	1	1	0	0		22	1	0	1	1	0	0	
23	1	0	1	1	1	1		23	1	0	1	1	1	1	
24	1	1	0	0	0	1		24	1	1	0	0	0	1	
25	1	1	0	0	1	0	6	25	1	1	0	0	1	0	e'
26	1	1	0	1	0	1		26	1	1	0	1	0	1	
27	1	1	0	1	1	0		27	1	1	0	1	1	0	
28	1	1	1	0	0	0		28	1	1	1	0	0	0	
29	1	1	1	0	1	1	7	29	1	1	1	0	1	1	e'
30	1	1	1	1	0	0		30	1	1	1	1	0	0	
31	1	1	1	1	1	1		31	1	1	1	1	1	1	

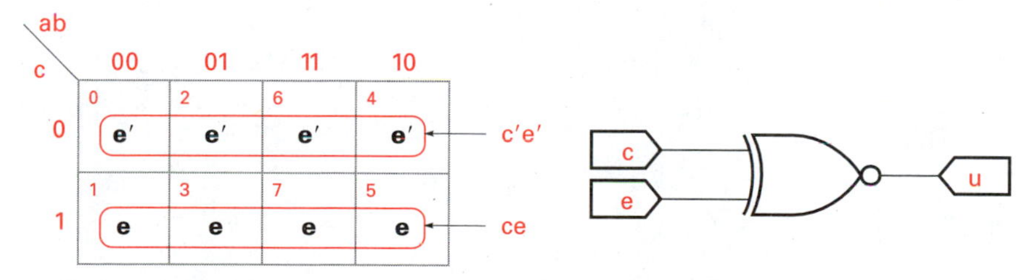

u = f(a, b, c, d, e) = Σ (0, 2, 5, 7, 8, 10, 13, 15, 16, 18, 21, 23, 24, 26, 29, 31)

d.

dec	v	w	x	y	z	A
0	0	0	0	0	0	0
1	0	0	0	0	1	1
2	0	0	0	1	0	0
3	0	0	0	1	1	1
4	0	0	1	0	0	0
5	0	0	1	0	1	1
6	0	0	1	1	0	0
7	0	0	1	1	1	1
8	0	1	0	0	0	0
9	0	1	0	0	1	0
10	0	1	0	1	0	0
11	0	1	0	1	1	0
12	0	1	1	0	0	1
13	0	1	1	0	1	1
14	0	1	1	1	0	1
15	0	1	1	1	1	1
16	1	0	0	0	0	0
17	1	0	0	0	1	1
18	1	0	0	1	0	0
19	1	0	0	1	1	1
20	1	0	1	0	0	0
21	1	0	1	0	1	1
22	1	0	1	1	0	0
23	1	0	1	1	1	1
24	1	1	0	0	0	0
25	1	1	0	0	1	0
26	1	1	0	1	0	0
27	1	1	0	1	1	0
28	1	1	1	0	0	0
29	1	1	1	0	1	0
30	1	1	1	1	0	0
31	1	1	1	1	1	0

	dec	v	w	x	y	z	A	MEV
	0	0	0	0	0	0	0	
0	1	0	0	0	0	1	1	z
	2	0	0	0	1	0	0	
	3	0	0	0	1	1	1	
	4	0	0	1	0	0	0	
1	5	0	0	1	0	1	1	z
	6	0	0	1	1	0	0	
	7	0	0	1	1	1	1	
	8	0	1	0	0	0	0	
2	9	0	1	0	0	1	0	0
	10	0	1	0	1	0	0	
	11	0	1	0	1	1	0	
	12	0	1	1	0	0	1	
3	13	0	1	1	0	1	1	1
	14	0	1	1	1	0	1	
	15	0	1	1	1	1	1	
	16	1	0	0	0	0	0	
4	17	1	0	0	0	1	1	z
	18	1	0	0	1	0	0	
	19	1	0	0	1	1	1	
	20	1	0	1	0	0	0	
5	21	1	0	1	0	1	1	z
	22	1	0	1	1	0	0	
	23	1	0	1	1	1	1	
	24	1	1	0	0	0	0	
6	25	1	1	0	0	1	0	0
	26	1	1	0	1	0	0	
	27	1	1	0	1	1	0	
	28	1	1	1	0	0	0	
7	29	1	1	1	0	1	0	0
	30	1	1	1	1	0	0	
	31	1	1	1	1	1	0	

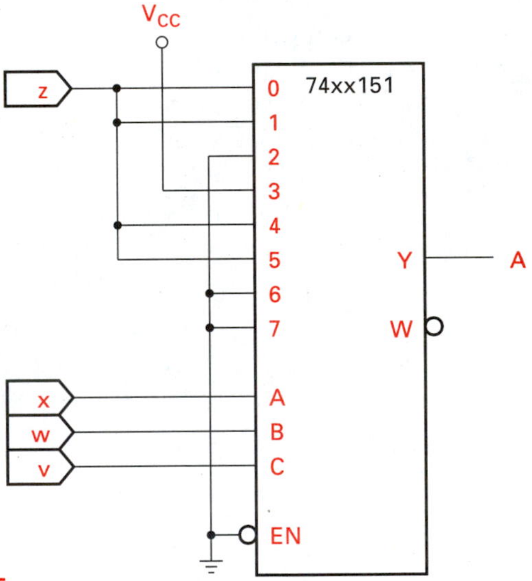

29. a.

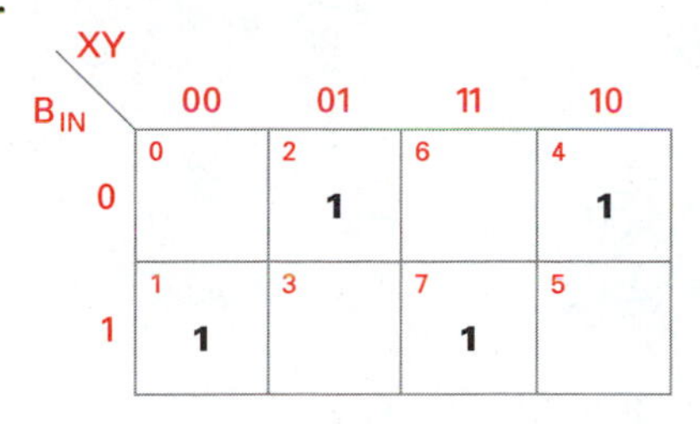

XY
B_{IN}
00 01 11 10
0
1
$X'Y$
$X'B_{IN}$
YB_{IN}
1
1
1
1
B_{OUT}

X	Y	B_{IN}	D	B_{OUT}
0	0	0	0	0
0	0	1	1	1
0	1	0	1	1
0	1	1	0	1
1	0	0	1	0
1	0	1	0	0
1	1	0	0	0
1	1	1	1	1

$D = X'Y'B_{IN} + X'Y'B'_{IN} + XY'B'_{IN} + XYB_{IN}$
$D = X + Y + B_{IN}$
$B_{OUT} = X'Y + X'B_{IN} + YB_{IN}$

b.

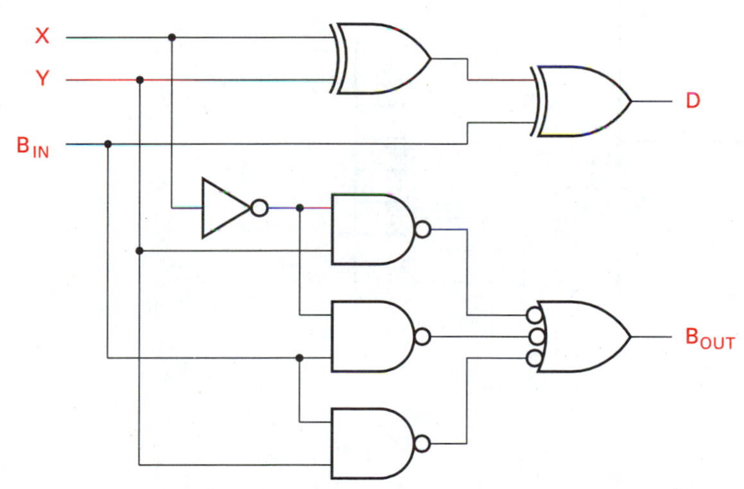

c.

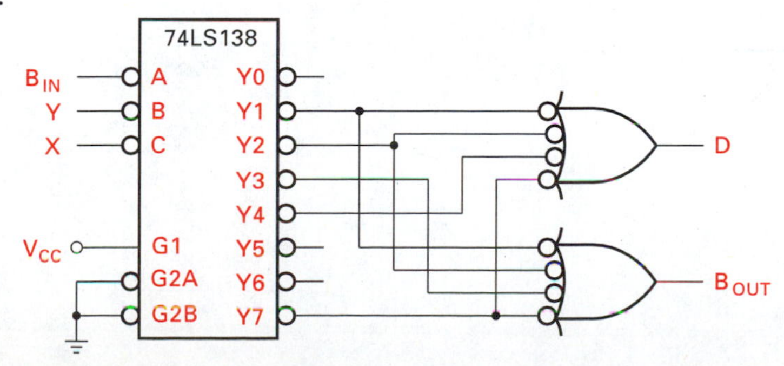

33.

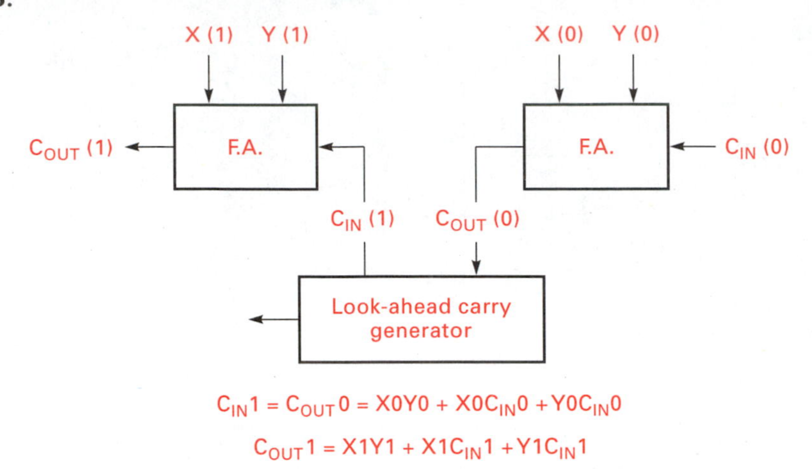

35.

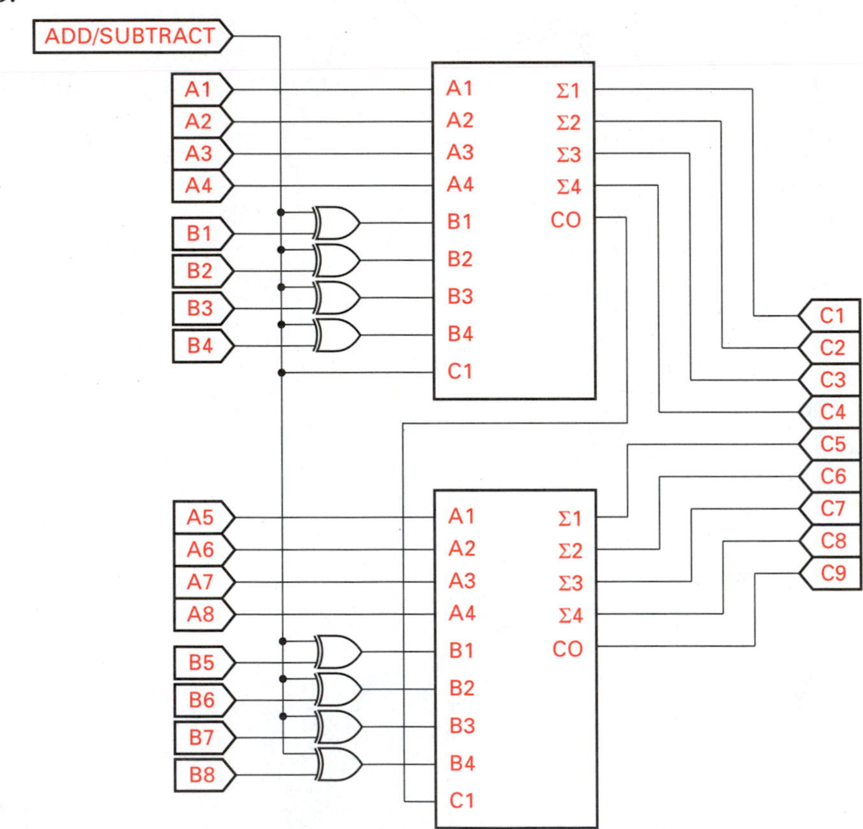

C = A + B when A/S is low.
C = A – B when A/S is high.

37.

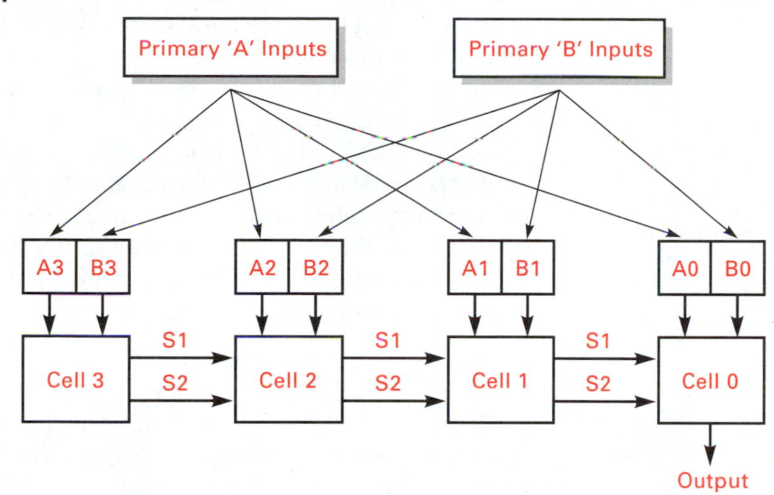

For this solution, SO_n refers to the secondary output to the next cell, and SI_n is the secondary input from the previous cell.

Present condition	SI_2	SI_1	Next condition: Primary inputs A_jB_j 00 SO_2SO_1	01 SO_2SO_1	10 SO_2SO_1	11 SO_2SO_1
A = B	0	0	00	11	01	00
A > B	0	1	01	01	01	01
A < B	1	1	11	11	11	11
Don't care	1	0	*xx*	*xx*	*xx*	*xx*

$$SO_2 = f(SI_2, SI_1, A_jB_j) = \Sigma(1, 12, 13, 14, 15) + \Sigma d(8, 9, 10, 11)$$

$$SO_1 = f(SI_2, SI_1, A_jB_j) = \Sigma(1, 2, 4, 5, 6, 7, 12, 13, 14, 15) + \Sigma d(8, 9, 10, 11)$$

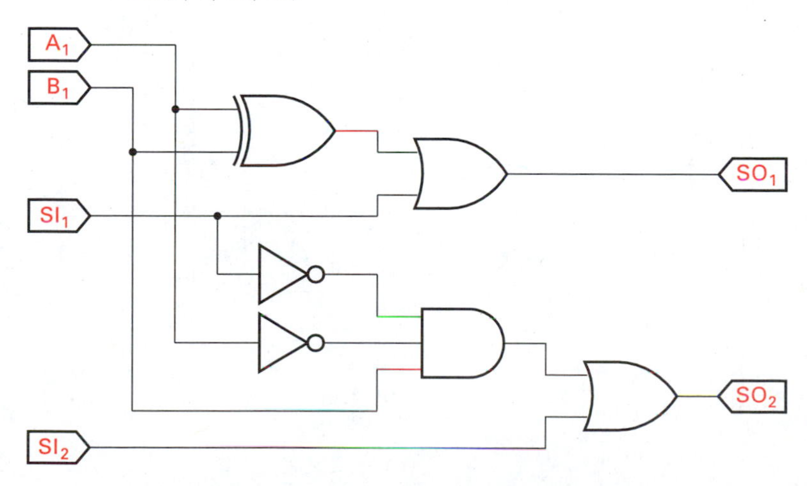

$$SO_2 = SI_1'A_j'B_j + SI_2$$
$$SO_1 = SI_1 + A_j'B_j + A_jB_j'$$

Leftmost cell (cell 3)

Inputs A_jB_j	Outputs SO_2SO_1
00	00
01	11
10	01
11	00

$$SO_2(A_jB_j) = \Sigma(1)$$
$$SO_1(A_jB_j) = \Sigma(1, 2)$$

$$SO_1 = A'B$$
$$SO_2 = A'B + B'A = A \oplus B$$

Leftmost cell (cell 3)

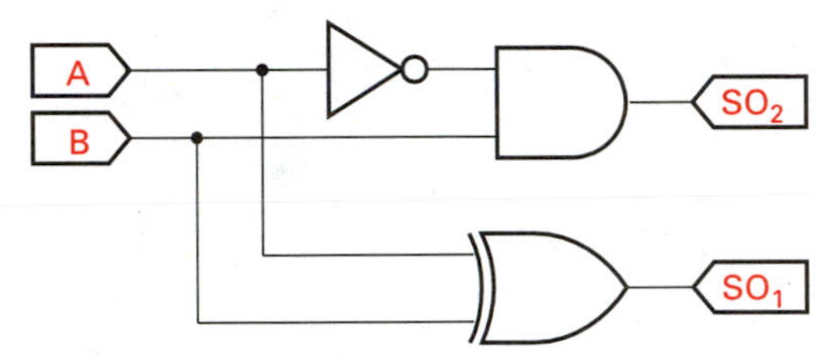

Boundary equations:

$$A = B: f(SO_2, SO_1) = \Sigma(0) + \Sigma d(2) = SO_1'$$
$$A > B: f(SO_2, SO_1) = \Sigma(1) + \Sigma d(2)$$
$$= SO_1'SO_2 + SO_1SO_2'$$
$$= SO_1 \oplus SO_2$$
$$A < B: f(SO_2, SO_1) = \Sigma(3) + \Sigma d(2) = SO_2$$

Output logic

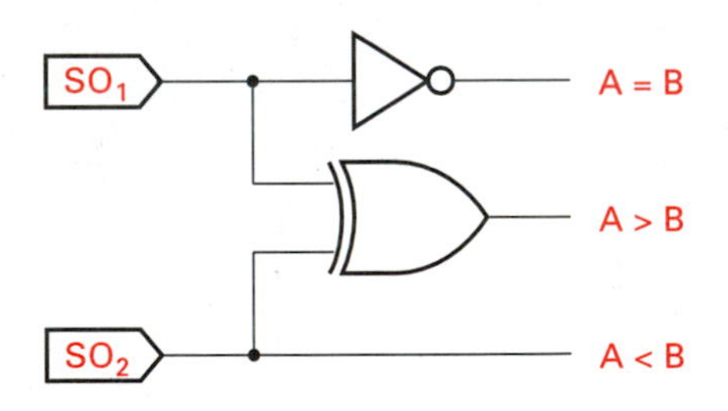

39.

	M	S_3	S_2	S_1	S_0	C_n'
a. AND	1	1	0	1	1	0
b. ADD w/carry	0	0	0	1	1	1
c. SUB w/borrow	0	0	1	0	0	0
d. NOR	1	0	0	1	1	1
e. Transfer A	1	1	1	1	1	0
f. DEC A	0	1	1	1	1	1

45. A tristate logic device has three output states: HIGH, LOW, and HI-Z. HIGH and LOW carry information, while HI-Z carries none.

The purpose of tristate devices is to allow outputs to be connected to the same conductor without interacting. Enable circuitry is used so that one device presents a HIGH or LOW output, and the others are in the high-impedance state. If one device was HIGH, and another LOW, current would flow from the HIGH output into the LOW output, invalidating the data and possibly damaging the devices.

49. A maxterm circuit cannot produce a static-1 hazard because a "true" output is a logic "0."

Hazards occur when an input variable changes, causing the "true" output to be generated by a different EPI. Due to propagation delays, the new EPI may not generate a "true" before "false" has appeared at the output.

51. a.

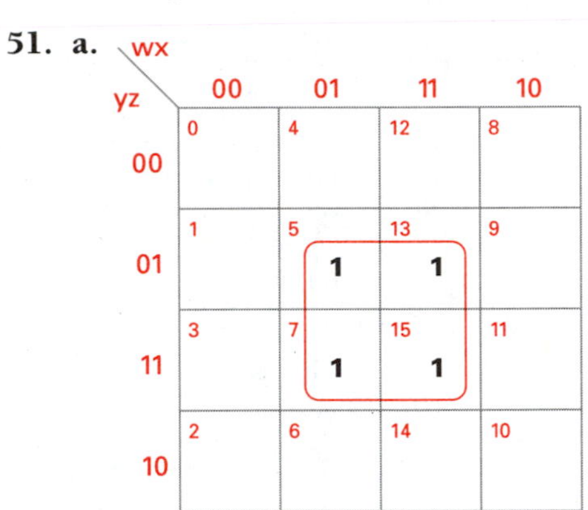

$A = xz$

No static-1 hazards exist.

b.

$X = bd + ab'$

A static-1 hazard exists: a transition of variable "b" could cause a false output. The addition of the group shown by the broken line (ad) will eliminate the hazard.

$X = bd + ab' + ad$

c.

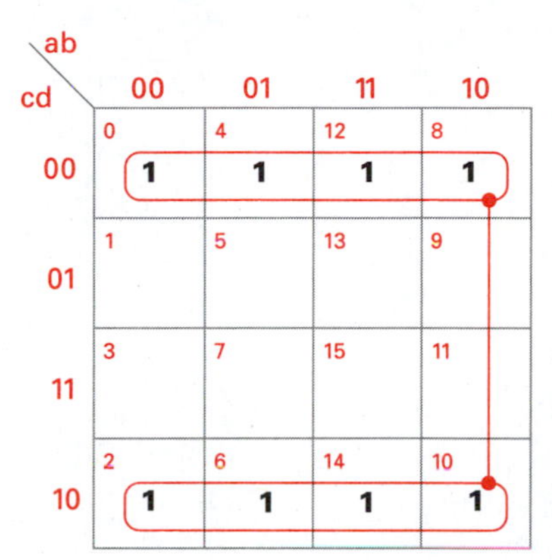

$T = d'$

No static-1 hazard exists.

d.

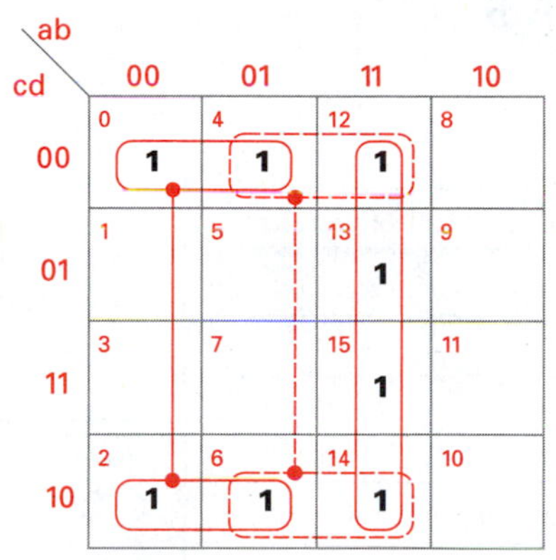

$T = a'd' + ab$

Static-1 hazard on "a" exists that can be eliminated by the addition of group "bd′"

$T = a'd + ab + bd'$

53.

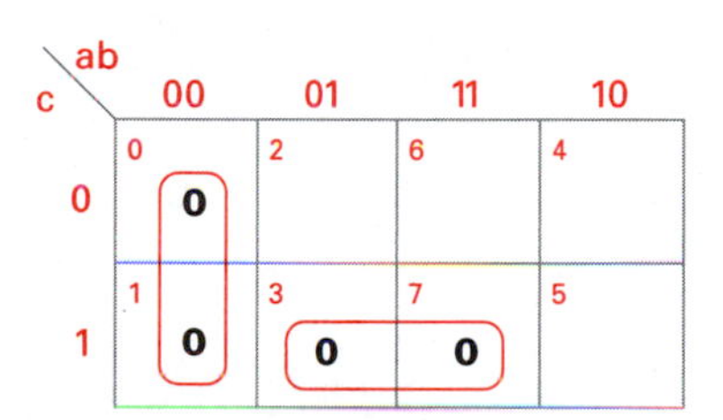

A static-0 hazard on "b" exists.

a

b

c

x

Hazard

CHAPTER 5

1.

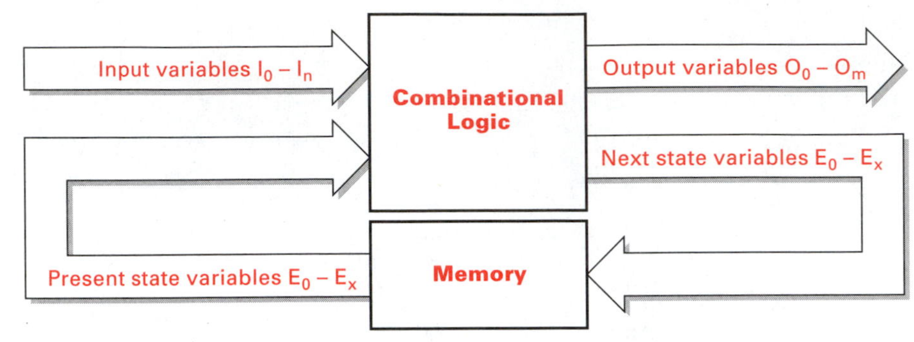

$O = f(I,S)$ $S^+ = f(S,I)$ $E = f(S,I)$

3. a. *R-S* latch

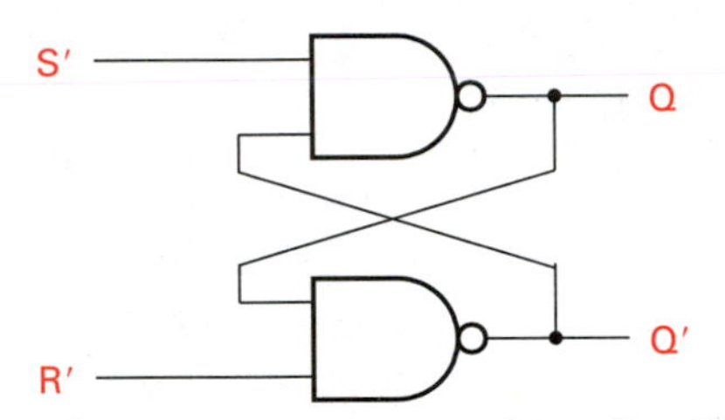

Characteristic equation $Q^+ = S + R'Q$

Truth Table			
S'	R'	Q	Q'
0	0	1 Not allowed	1 Not allowed
0	1	1	0
1	0	0	1
1	1	Last Q	Last Q'

b. Gated *R-S* latch

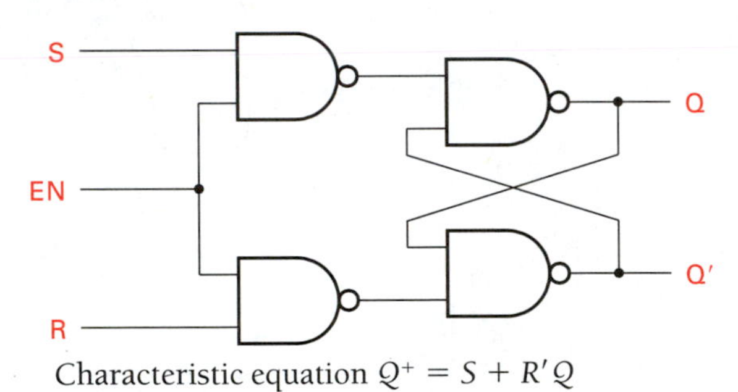

Characteristic equation $Q^+ = S + R'Q$

Truth Table			
Q	S	R	Q^+
0	0	0	0
0	0	1	0
0	1	0	1
0	1	1	Indeterminate
1	0	0	1
1	0	1	0
1	1	0	1
1	1	1	Indeterminate

EN must be “1” for any change to occur.

c. *J-K* flip-flop

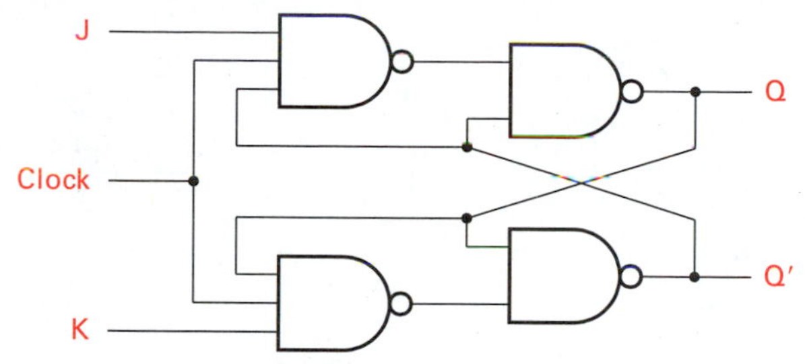

Characteristic equation $Q^+ = JQ' + K'Q$

Truth Table

Q	J	K	Q^+
0	0	0	0
0	0	1	0
0	1	0	1
0	1	1	1
1	0	0	1
1	0	1	0
1	1	0	1
1	1	1	0

d. *T* flip-flop

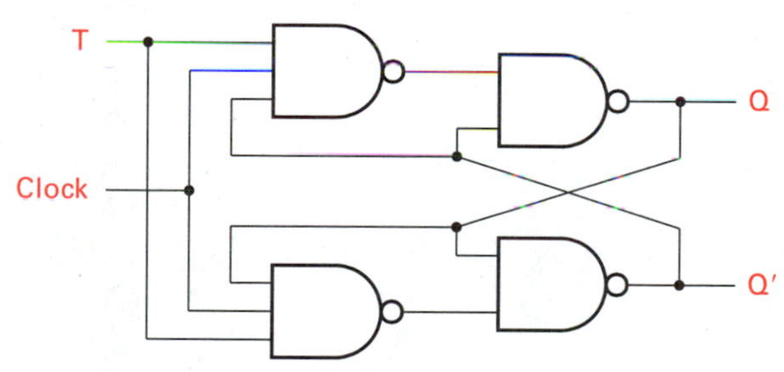

Characteristic equation $Q^+ = TQ' + T'Q$

Truth Table

Q	T	Q^+
0	0	0
0	1	1
1	0	1
1	1	0

e. *D* flip-flop

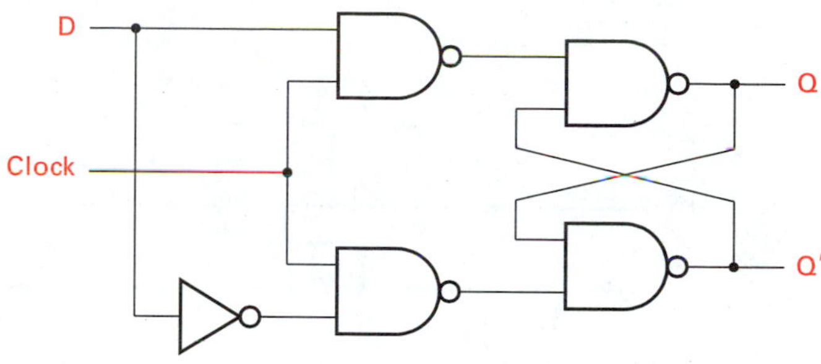

Characteristic equation $Q^+ = D$

Truth Table

Q	D	Q^+
0	0	0
0	1	1
1	0	0
1	1	1

5.

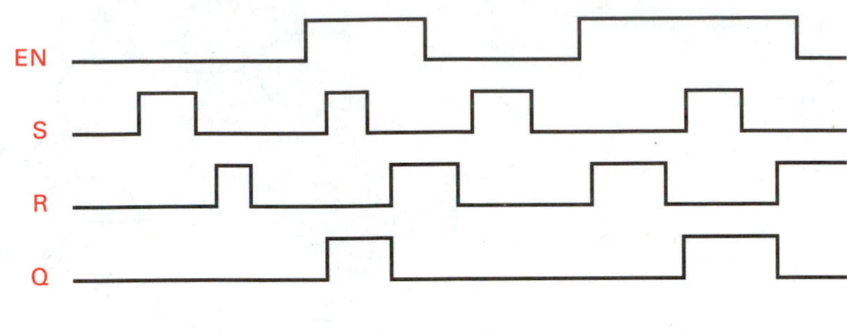

7.

9. The output of a level-triggered flip-flop may change more than once during a clock cycle, whenever the enable is true. An edge-triggered device can transition only once per clock cycle; any change of the inputs between triggering edges will not affect the outputs.

11.

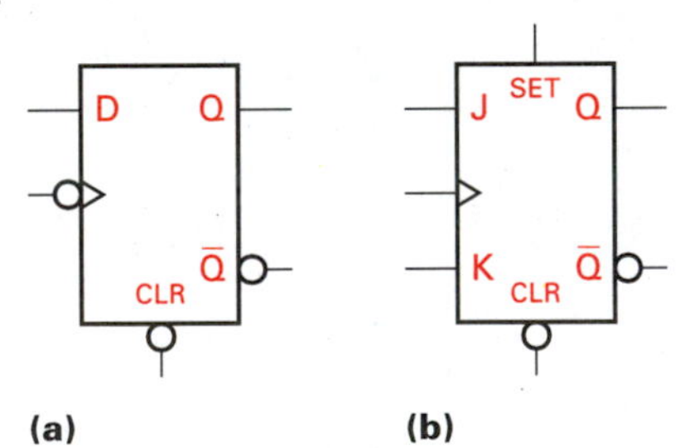

(a) (b)

13.

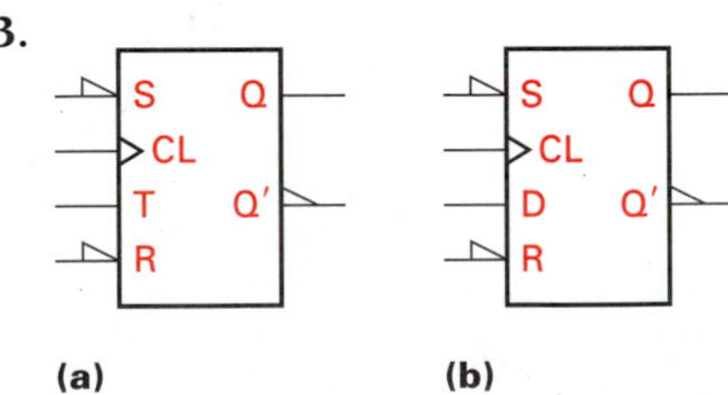

(a) (b)

15. A level-triggered master-slave flip-flop can be set or reset by glitches under certain conditions.

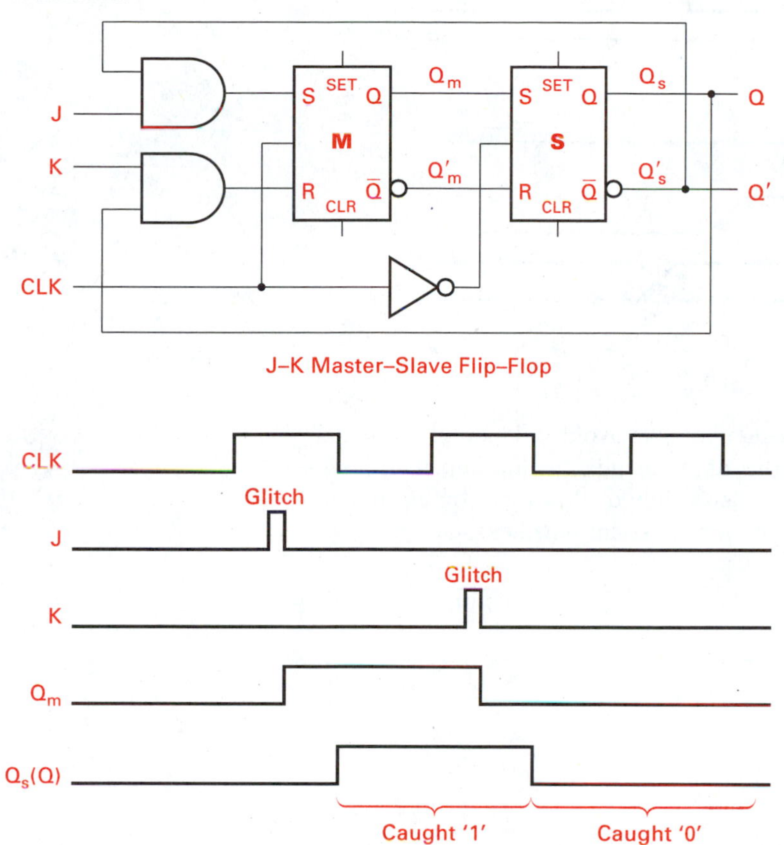

In the master-slave flip-flop shown, when $Q = 0$ the J input is passed through the J AND gate to the master S input. A "1" at the J input will immediately set the master latch. When the clock transitions to low, the slave latch is set by the erroneously set master.

17.

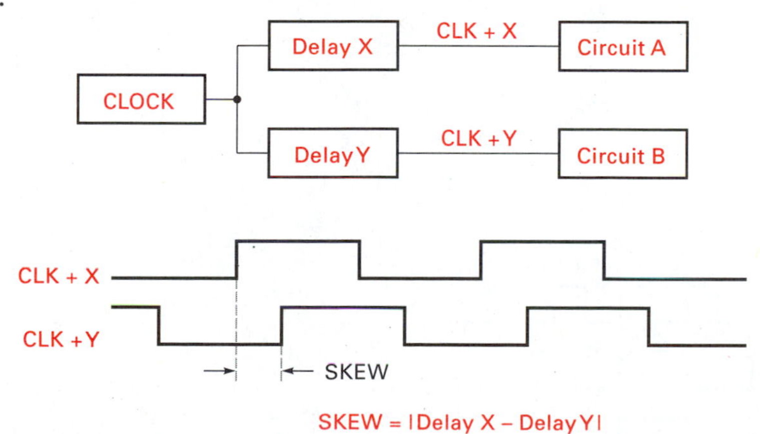

Example of skew problem: assume that CLK + X and CLK + Y trigger two different subcircuits that use the same inputs. If the clocks are skewed, the inputs may change after Circuit A has latched the input, so Circuit B latches a different value.

19.

21.

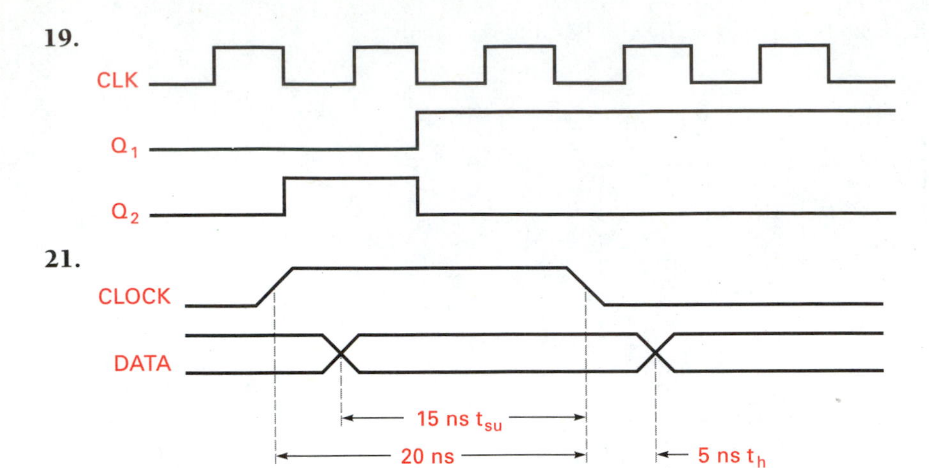

23. The general approach for avoiding metastability is to avoid violation of setup and hold times. Another method is to clock the input into an extra flip-flop, which will delay the input one clock cycle. The extra delay should allow the extra flip-flop to recover from metastability before its output is clocked into the next stage.

25.

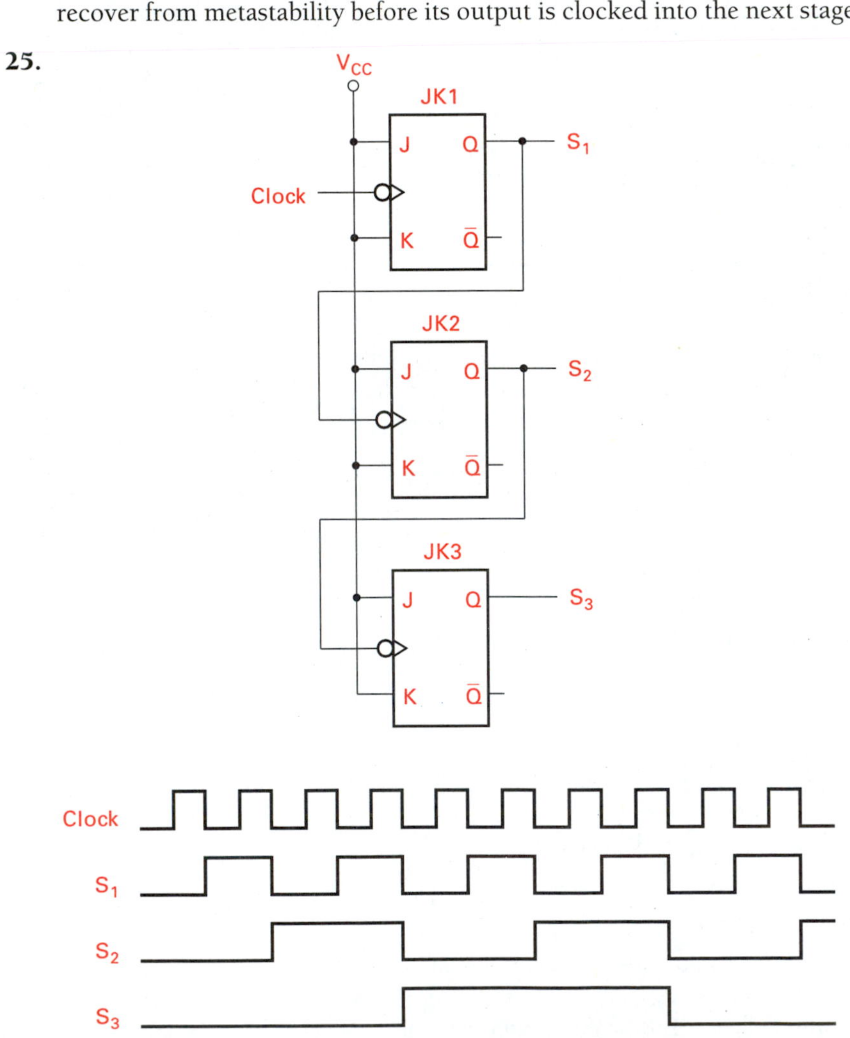

27.

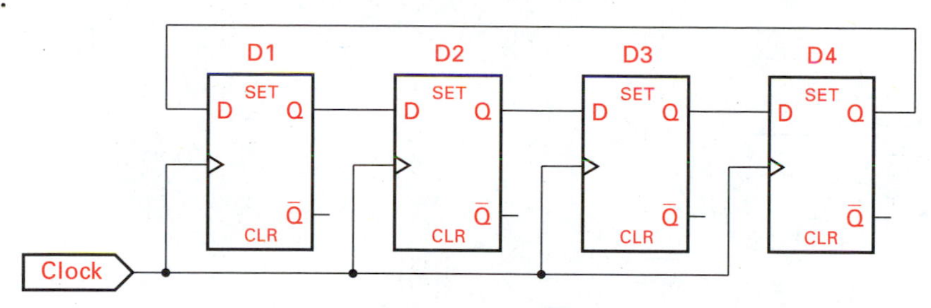

29.

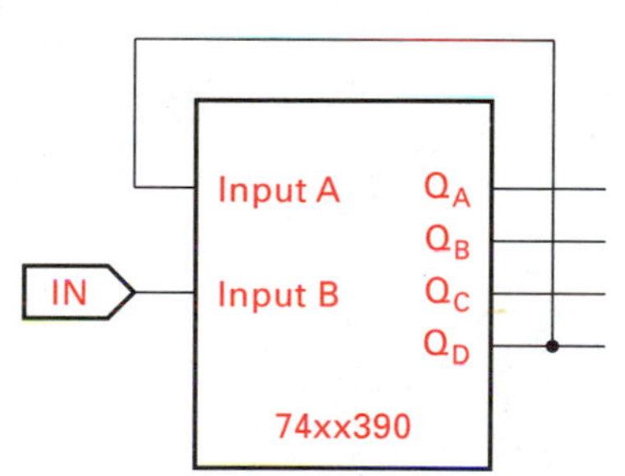

31. $f_{MAX} = 25$ MHz

33. a.

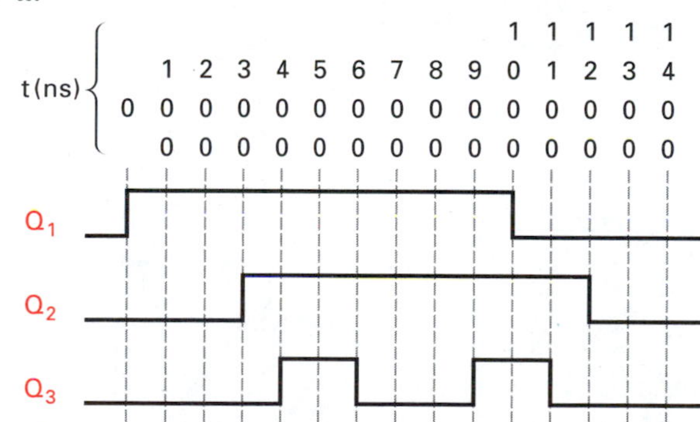

b. 5 MHz

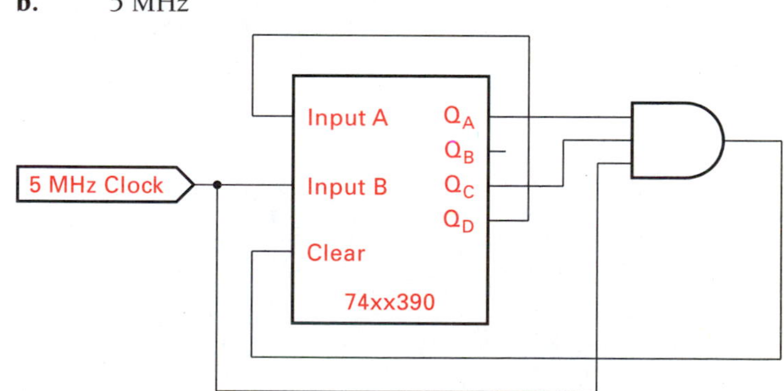

c. Counter logic

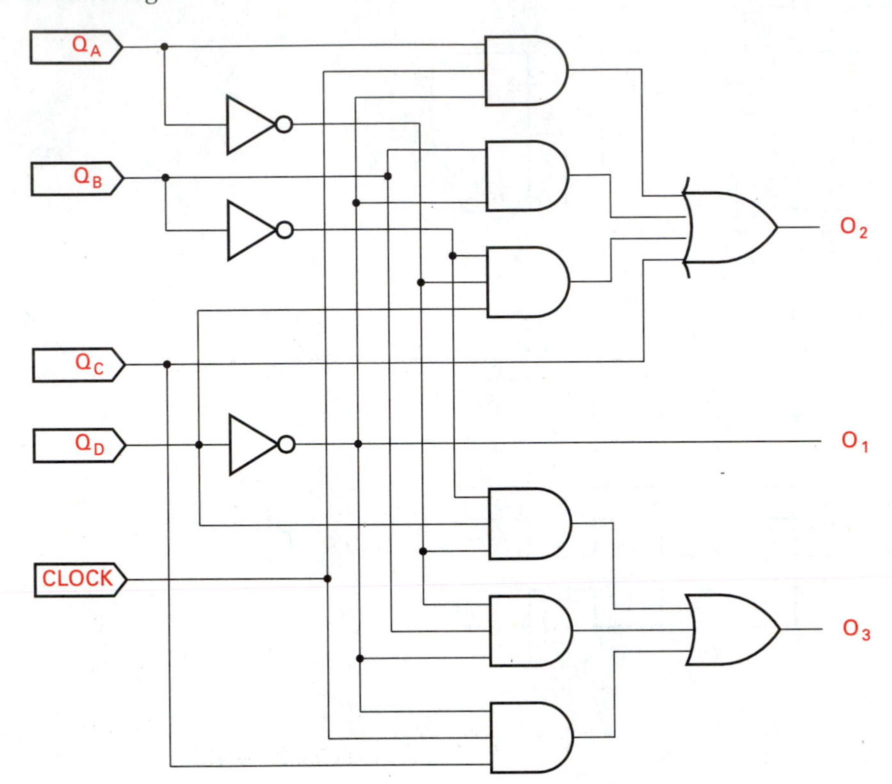

35.

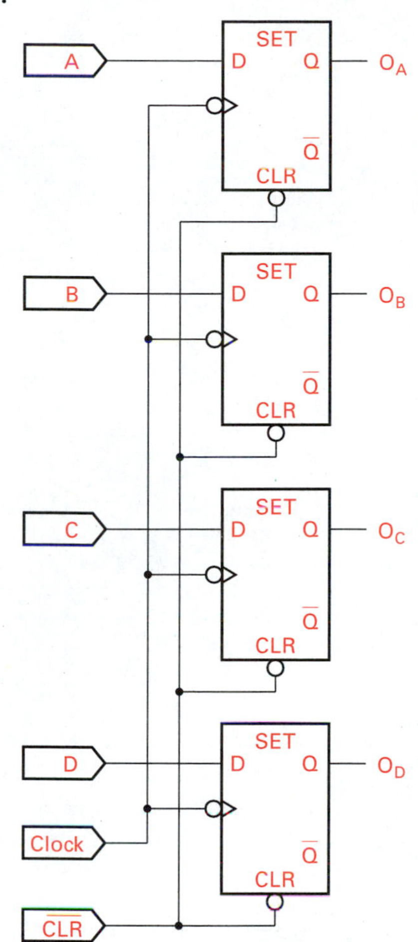

37. 00000110

39. 1001

CHAPTER 6

1.

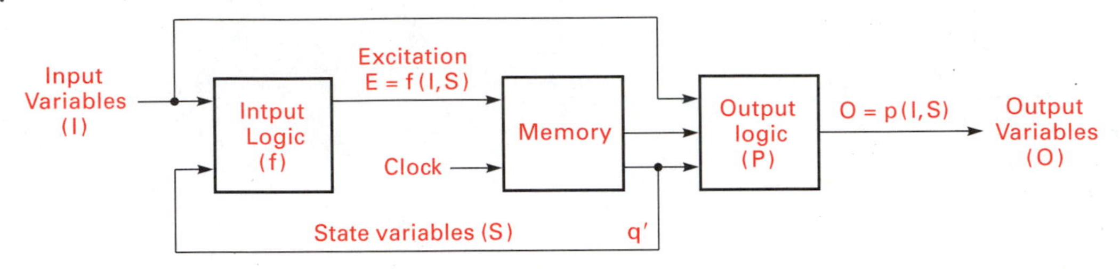

Mealy Machine

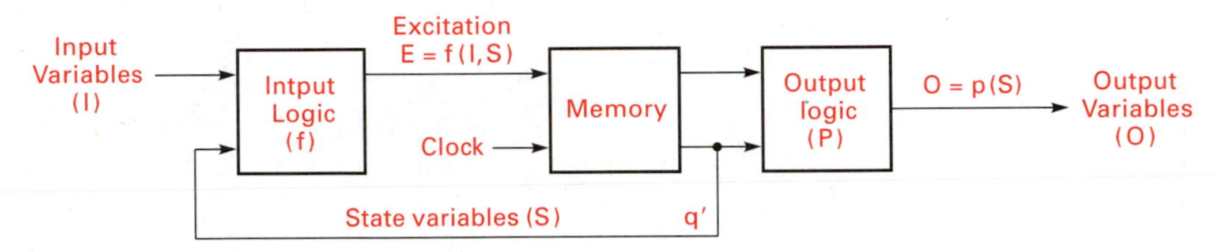

Moore Machine

3. a.

State table

Present state	Next state X'	Next state X	Output (Z) X'	Output (Z) X
A	A	B	0	0
B	A	C	0	0
C	C	A	0	1

b.

State table

Present state	Next state $X'Y'$	Next state $X'Y$	Next state XY'	Next state XY
A	A	C	B	C
B	A	B	C	B
C	B	C	C	A

c.

State table

Present state	Next state X'	Next state X	Output (Z) X'	Output (Z) X
A	A	B	0	0
B	A	C	0	0
C	F	B	0	0
D	A	B	1	0
E	A	D	0	0
F	A	E	0	0

d.

State table

Present state	Next state $X_1 X_2 X_3$								Output $Z_2 Z_1$ $X_1 X_2 X_3$							
	000	001	010	011	100	101	110	111	000	001	010	011	100	101	110	111
S_0	S_3	S_3	S_3	S_3	S_3	S_3	S_3	S_1	00	00	00	00	00	00	00	00
S_1	S_4	S_4	S_4	S_4	S_4	S_2	S_4	S_4	00	00	00	00	00	00	00	00
S_2	S_0	S_0	S_0	S_0	S_0	S_0	S_0	S_0	01	01	01	01	10	01	01	01
S_3	S_4	S_4	S_4	S_4	S_4	S_4	S_4	S_4	00	00	00	00	00	00	00	00
S_4	S_0	S_0	S_0	S_0	S_0	S_0	S_0	S_0	01	01	01	01	01	01	01	01

5. a.

S-R excitation table (6-4c)

Present state	$X = 0$			$X = 1$		
$F_3 F_2 F_1$	$S_3 R_3$	$S_2 R_2$	$S_1 R_1$	$S_3 R_3$	$S_2 R_2$	$S_1 R_1$
000	0d	0d	0d	0d	0d	10
001	0d	0d	01	0d	10	01
010	10	01	10	0d	01	10
011	0d	01	01	0d	10	d0
100	01	0d	0d	01	10	10
101	01	0d	01	d0	0d	10

SR excitation table (6-4d)

Present state	$S_3 R_3$, $X_1 X_2 X_3 =$							
$F_3 F_2 F_1$	000	001	010	011	100	101	110	111
000	0d	0d	0d	0d	0d	0d	0d	0d
001	10	10	10	10	10	0d	10	10
010	0d	0d	0d	0d	0d	0d	0d	0d
011	10	10	10	10	10	10	10	10
100	01	01	01	01	01	01	01	01

$S_2 R_2$								
$F_3 F_2 F_1$	000	001	010	011	100	101	110	111
000	10	10	10	10	10	10	10	0d
001	0d	0d	0d	0d	0d	10	0d	0d
010	01	01	01	01	01	01	01	01
011	01	01	01	01	01	01	01	01
100	0d	0d	0d	0d	0d	0d	0d	0d

$S_1 R_1$								
$F_3 F_2 F_1$	000	001	010	011	100	101	110	111
000	10	10	10	10	10	10	10	10
001	01	01	01	01	01	01	01	01
010	0d	0d	0d	0d	0d	0d	0d	0d
011	01	01	01	01	01	01	01	01
100	0d	0d	0d	0d	0d	0d	0d	0d

b.

J-K excitation table (6-4c)

Present state	$X = 0$			$X = 1$		
$F_3 F_2 F_1$	$J_3 K_3$	$J_2 K_2$	$J_1 K_1$	$J_3 K_3$	$J_2 K_2$	$J_1 K_1$
000	0d	0d	0d	0d	0d	1d
001	0d	0d	d1	0d	1d	d1
010	1d	d1	1d	0d	d1	1d
011	0d	d1	d1	0d	d1	d0
100	d1	0d	0d	d1	1d	1d
101	d1	0d	d1	d0	0d	d1

JK excitation table (6-4d)

Present state	$J_3 K_3$, $X_1 X_2 X_3 =$							
$F_3 F_2 F_1$	000	001	010	011	100	101	110	111
000	0d	0d	0d	0d	0d	0d	0d	0d
001	1d	1d	1d	1d	1d	0d	1d	1d
010	0d	0d	0d	0d	0d	0d	0d	0d
011	1d	1d	1d	1d	1d	1d	1d	1d
100	d1	d1	d1	d1	d1	d1	d1	d1

	J_2K_2							
$F_3F_2F_1$	000	001	010	011	100	101	110	111
000	d0	d0	d0	d0	d0	d0	d0	0d
001	0d	0d	0d	0d	0d	1d	0d	0d
010	d1	d1	d1	d1	d1	d1	d1	d1
011	d1	d1	d1	d1	d1	d1	d1	d1
100	0d	0d	0d	0d	0d	0d	0d	0d

	J_1K_1							
$F_3F_2F_1$	000	001	010	011	100	101	110	111
000	1d	1d	1d	1d	1d	1d	1d	1d
001	d1	d1	d1	d1	d1	d1	d1	d1
010	0d	0d	0d	0d	0d	0d	0d	0d
011	d1	d1	d1	d1	d1	d1	d1	d1
100	0d	0d	0d	0d	0d	0d	0d	0d

	D_2							
$F_3F_2F_1$	000	001	010	011	100	101	110	111
000	1	1	1	1	1	1	1	0
001	0	0	0	0	0	1	0	0
010	0	0	0	0	0	0	0	0
011	0	0	0	0	0	0	0	0
100	0	0	0	0	0	0	0	0

	D_1							
$F_3F_2F_1$	000	001	010	011	100	101	110	111
000	1	1	1	1	1	1	1	1
001	0	0	0	0	0	0	0	0
010	0	0	0	0	0	0	0	0
011	0	0	0	0	0	0	0	0
100	0	0	0	0	0	0	0	0

c.

***D* excitation table (6-4c)**

Present state	$X = 0$			$X = 1$		
$F_3F_2F_1$	D_3	D_2	D_1	D_3	D_2	D_1
000	0	0	0	0	0	1
001	0	0	0	0	1	0
010	1	0	1	0	0	1
011	0	0	0	0	0	1
100	0	0	0	0	1	1
101	0	0	0	1	0	0

***D* excitation table (6-4d)**

Present state	D_3							
	$X_1X_2X_3 =$							
$F_3F_2F_1$	000	001	010	011	100	101	110	111
000	0	0	0	0	0	0	0	0
001	1	1	1	1	1	0	1	0
010	0	0	0	0	0	0	0	0
011	1	1	1	1	1	1	1	1
100	0	0	0	0	0	0	0	0

d.

***T* excitation table (6-4c)**

Present state	$X = 0$			$X = 1$		
$F_3F_2F_1$	T_3	T_2	T_1	T_3	T_2	T_1
000	0	0	0	0	0	1
001	0	0	1	0	1	1
010	1	1	1	0	1	1
011	0	1	1	0	1	0
100	1	0	0	1	1	1
101	1	0	1	0	0	1

***T* excitation table (6-4d)**

Present state	T_3							
	$X_1X_2X_3 =$							
$F_3F_2F_1$	000	001	010	011	100	101	110	111
000	0	0	0	0	0	0	0	0
001	1	1	1	1	1	0	1	0
010	0	0	0	0	0	0	0	0
011	1	1	1	1	1	1	1	1
100	1	1	1	1	1	1	1	1

$F_3F_2F_1$	T_2 000	001	010	011	100	101	110	111
000	1	1	1	1	1	1	1	0
001	0	0	0	0	0	1	0	0
010	1	1	1	1	1	1	1	1
011	1	1	1	1	1	1	1	1
100	0	0	0	0	0	0	0	0

$F_3F_2F_1$	T_1 000	001	010	011	100	101	110	111
000	1	1	1	1	1	1	1	1
001	1	1	1	1	1	1	1	1
010	0	0	0	0	0	0	0	0
011	1	1	1	1	1	1	1	1
100	0	0	0	0	0	0	0	0

9. Unused states generate "don't-care" terms when translating from a state table to a transition table. The number of states required is a function of the problem. For example, a modulo-3 counter requires 3 states. To realize the counter, the states are assigned to combinations of binary variables, with the possible number of states for n variables being 2^n. For the example of the modulo-3 counter, 2 variables will be required, with 4 possible combinations. One combination will be unused, and no state will have that combination as its next state.

Example: modulo-3 counter

State	F_1	F_2
0	0	0
1	0	1
2	1	0
Unused	1	1

11. a. Excitation: $D_1 = (F_2'X')' = F_2 + X$

$D_2 = F_1X$

Output: $Z = F_1F_2'$

b.
$$F_1^+ = (F_2'X)' = F_2 + X$$
$$F_2^+ = F_1X$$
$$Z^+ = F_1^+F_2^{+\prime} = (F_2 + X)(F_1X)$$
$$= (F_2 + X)(F_1' + X')$$
$$= F_1F_2 + F_2X' + F_1'X$$

Transition table

Present state F_1	F_2	Next state $X = 0$: F_1	F_2	$X = 1$: F_1	F_2	Present output Z
0	0	0	0	1	0	0
0	1	1	0	1	0	0
1	0	0	0	1	1	1
1	1	1	0	1	1	0

d.

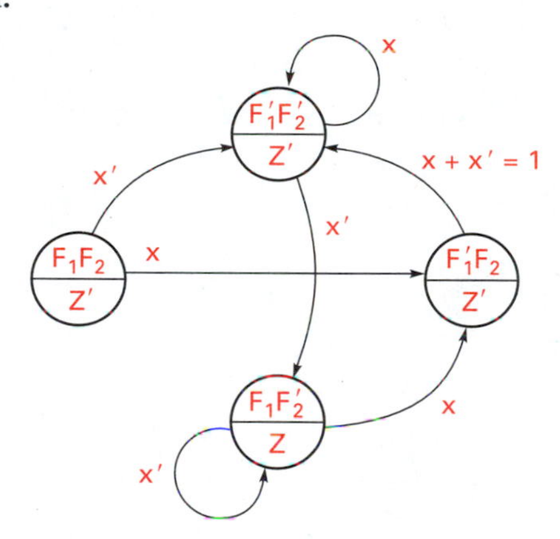

e.

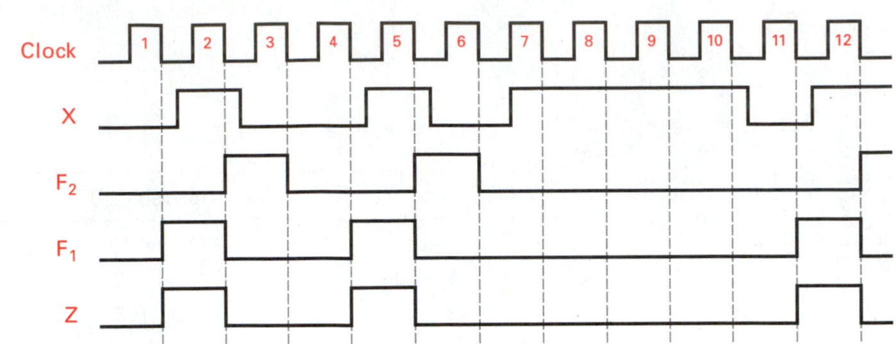
Clock
1
2
3
4
5
6
7
8
9
10
11
12
X
F_2
F_1
Z

13.

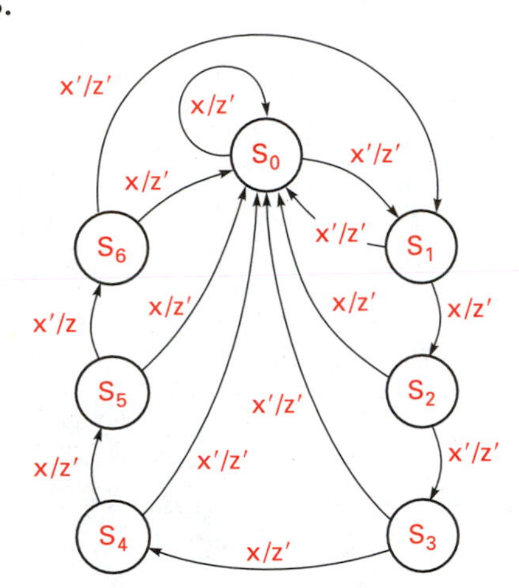
S_0
S_1
S_2
S_3
S_4
S_5
S_6
x/z'
x'/z'
x'/z

15.

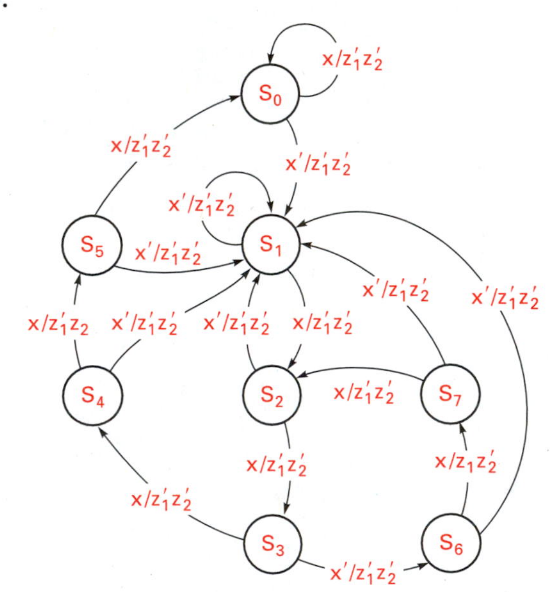
S_0
S_1
S_2
S_3
S_4
S_5
S_6
S_7
$x/z_1'z_2'$
$x'/z_1'z_2'$
$x/z_1'z_2$
x/z_1z_2'

17.

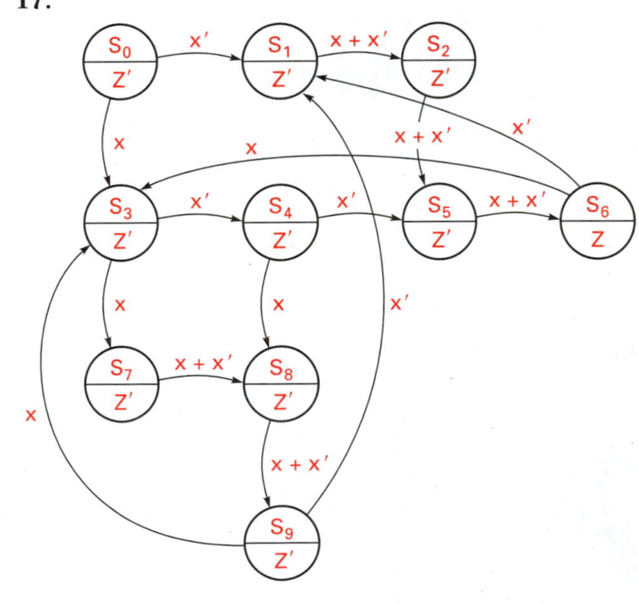
S_0
S_1
S_2
S_3
S_4
S_5
S_6
S_7
S_8
S_9
Z'
Z
x
x'
x + x'

19.

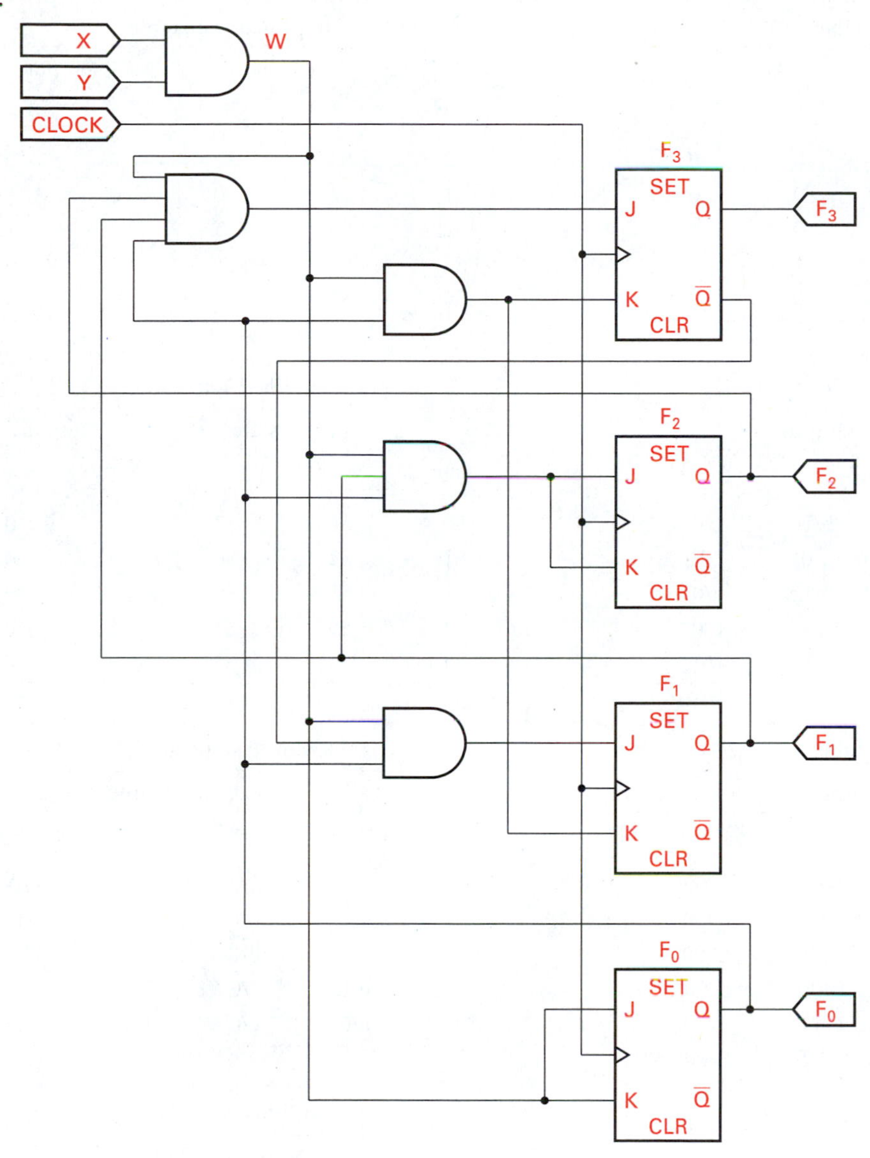

CHAPTER 7

1. a.

Present state	Next state		Output	
	$X = 0$	$X = 1$	$X = 0$	$X = 1$
S0	S2	S1	0	1
S2	S0	S3	1	0

Simplified state table

b.

Simplified state table

Present state	Next state		Output	
	$X = 0$	$X = 1$	$X = 0$	$X = 1$
S0	S3	S1	0	0
S1	S4	S0	0	1
S2	S6	S5	0	1
S3	S0	S3	1	0
S5	S2	S1	0	0

c.

Simplified state table

Present state	Next state $X = 0$	Next state $X = 1$	Output $X = 0$	Output $X = 1$
S0	S4	S2	0	0
S1	S2	S0	0	0
S2	S1	S6	0	0
S4	S5	S1	1	0
S5	S4	S3	0	0

d.

Reduced state table

Present state	Next state/Output $XY = 00$	$XY = 01$	$XY = 10$	$XY = 11$
a	a/0	a/0	b/1	c/0
b	a/0	b/0	d/0	f/1
c	c/0	b/0	b/1	a/0
d	d/0	c/0	e/1	c/0
f	e/0	e/0	f/0	f/0

e.

Simplified state table

Present state	Next state/Output Z_2Z_1 $XY = 00$	$XY = 01$	$XY = 10$	$XY = 11$
A	A/00	B/01	B/00	A/00
B	A/00	C/10	B/10	C/00
C	B/00	C/01	A/00	D/00
E	B/00	A/01	B/00	E/00
F	A/10	B/00	F/11	A/00
G	A/10	D/00	F/11	E/00

3. a.

Simplified state table

Present state	Next state $X = 0$	Next state $X = 1$	Output $X = 0$	Output $X = 1$
S0	S2	S1	0	1
S2	S0	S3	1	0

b.

Simplified state table

Present state	Next state $X = 0$	Next state $X = 1$	Output $X = 0$	Output $X = 1$
S0	S3	S1	0	0
S1	S4	S0	0	1
S2	S6	S5	0	1
S3	S0	S3	1	0
S5	S2	S1	0	0

c.

Simplified state table

Present state	Next state $X = 0$	Next state $X = 1$	Output $X = 0$	Output $X = 1$
S0	S4	S2	0	0
S1	S2	S0	0	0
S2	S1	S6	0	0
S4	S5	S1	1	0
S5	S4	S3	0	0

d.

Reduced state table

Present state	Next state/Output $XY = 00$	$XY = 01$	$XY = 10$	$XY = 11$
a	a/0	a/0	b/1	c/0
b	a/0	b/0	d/0	f/1
c	c/0	b/0	b/1	a/0
d	d/0	c/0	e/1	c/0
f	e/0	e/0	f/0	f/0

e.

Simplified state table

Present state	Next state/Output Z_2Z_1 $XY = 00$	$XY = 01$	$XY = 10$	$XY = 11$
A	A/00	B/01	B/00	A/00
B	A/00	C/10	B/10	C/00
C	B/00	C/01	A/00	D/00
E	B/00	A/01	B/00	E/00
F	A/10	B/00	F/11	A/00
G	A/10	D/00	F/11	E/00

5.

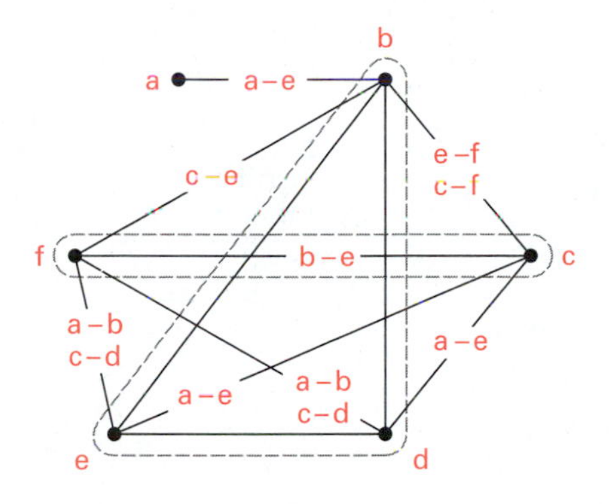

Simplified state table				
Present state	Next state			
	$XY = 00$	$XY = 01$	$XY = 10$	$XY = 11$
a	a/—	c/1	b/1	b/1
b	b/0	c/—	a/0	b/1
c	c/0	c/1	b/0	c/1

7. a. Compatibility chart

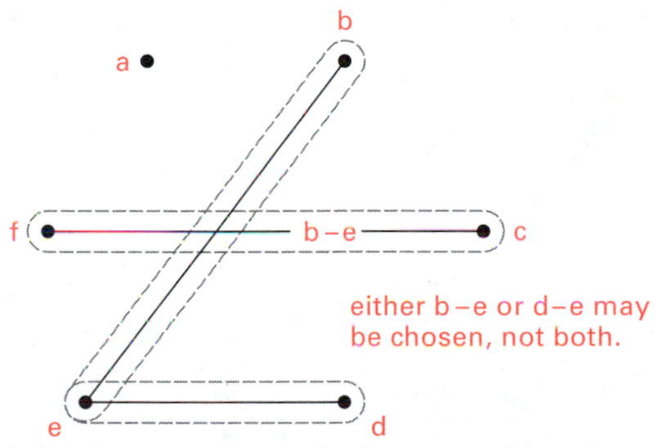

Solution using b-e and c-f as merged groups:

Simplified state table				
Present state	Next state			
	$XY = 00$	$XY = 01$	$XY = 10$	$XY = 11$
a	a/—	c/1	b/1	b/1
b	b/0	c/—	a/0	d/1
c	c/0	c/1	b/0	c/1
d	—/—	—/—	a/0	d/1

Solution using c-f and e-d as merged groups:

Simplified state table				
Present state	Next state			
	$XY = 00$	$XY = 01$	$XY = 10$	$XY = 11$
a	a/—	c/1	d/1	b/1
b	d/0	c/—	—/—	—/—
c	c/0	c/1	d/0	c/1
d	—/—	—/—	a/0	d/1

b. Compatibility graph

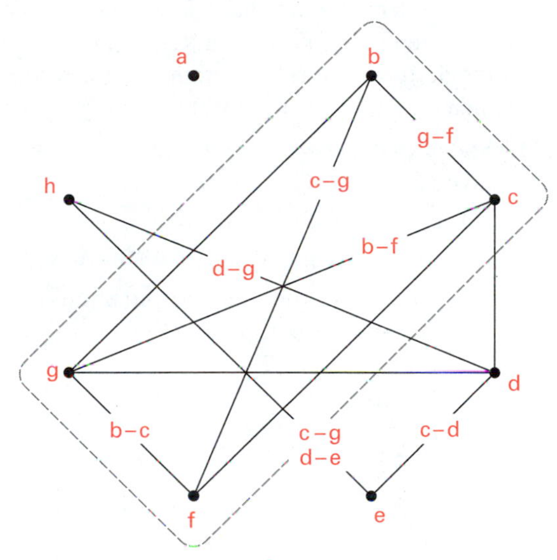

Solution using group b-c-f-g:

Simplified state table		
Present state	Next state/Output	
	0	1
a	b/1	h/1
b	a/0	b/0
d	d/0	—/1
e	b/0	d/—
h	b/0	e/—

9. Compatible state pairs are found with an implication chart by:

1. Placing a dot on the implication chart for each state in the table.

2. Connecting the states that have the same outputs for the same input sequences.
3. Labeling the implied pairs on the lines connecting the states that have different next states.
4. Crossing out the listed implied pairs that are not connected by lines.
5. Crossing out any lines that imply incompatible pairs.

11. The state assignment using adjacent states impacts the realization of a state machine by reducing excitation logic, and possibly output logic. This occurs because adjacent states have fewer variables changing between them. Fewer changing variables require less logic to create the excitation signals that cycle the machine into the next state.

13. 1. States that have the same next state should be given adjacent state assignments.
2. All of the states that are next states for a single state should be given adjacent state assignments.
3. States that have the same output should be given adjacent state assignments.

15. a.

F_1 \ F_3F_2	00	01	11	10
0	c	b	a	Unused
1	d	e	Unused	Unused

b.

F_1 \ F_3F_2	00	01	11	10
0	A	E	G	B
1	D	C	Unused	F

c.

F_1 \ F_3F_2	00	01	11	10
0	a	c	Unused	Unused
1	b	d	e	f

17.

	State assignment		
State	F_3	F_2	F_1
a	0	0	0
b	0	0	1
c	0	1	0
d	0	1	1
e	1	1	1
f	1	0	1

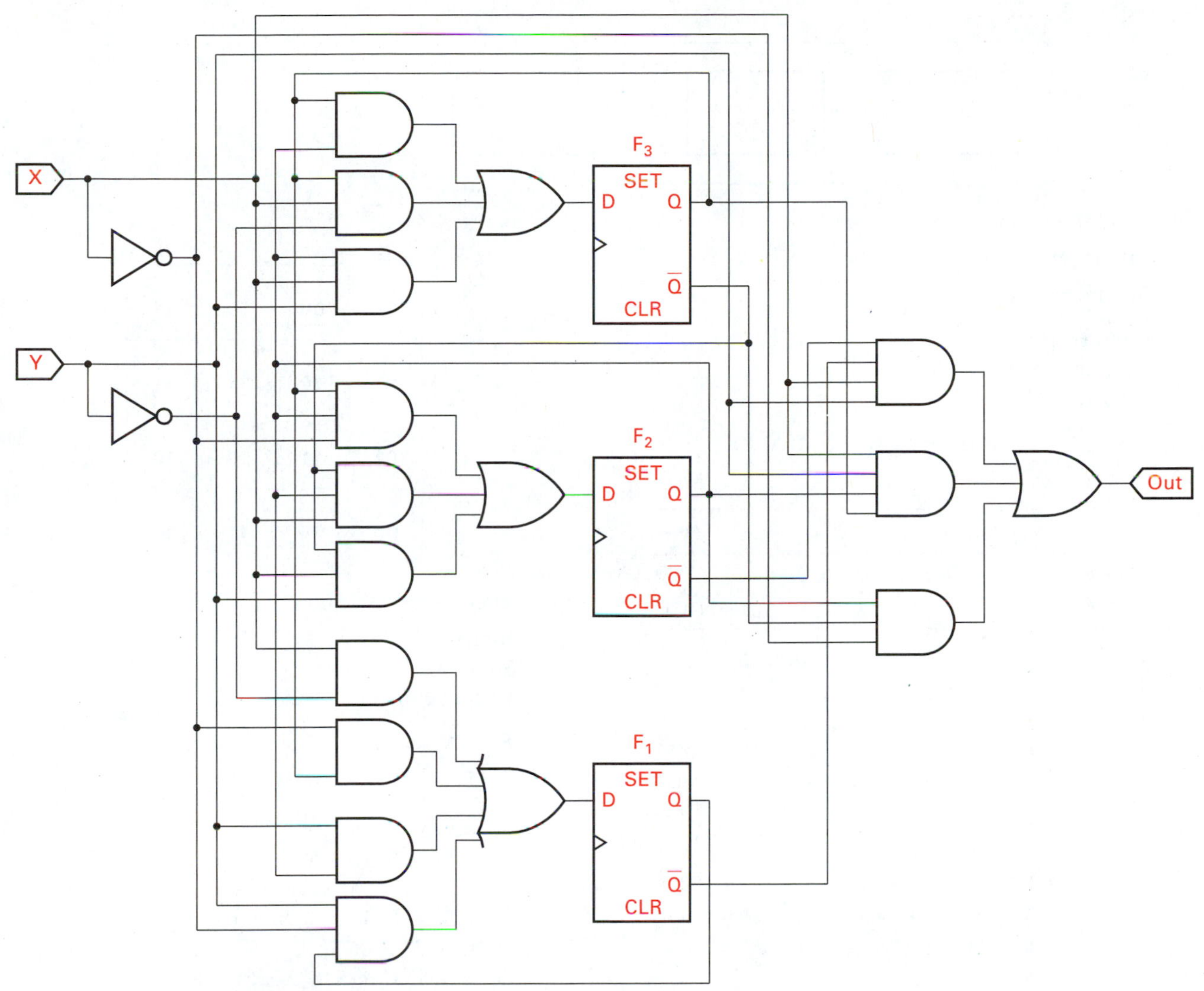

D flip-flop realization

CHAPTER 8

1. An asynchronous sequential circuit has no clock, compared to a synchronous circuit in which all of the state transitions are synchronized with a clock. The synchronous machine has the advantage of ease of design, predictability, and virtual independence of device speed irregularities. The asynchronous sequential machine has the advantage of speed—all of the state transitions occur as quickly as the signals can propagate through the gates. The disadvantages of this are that speed irregularities of gates can cause erratic operation if certain design considerations are not followed. Designing a reliable asynchronous sequential network can take much longer than an equivalent synchronous circuit.

3.

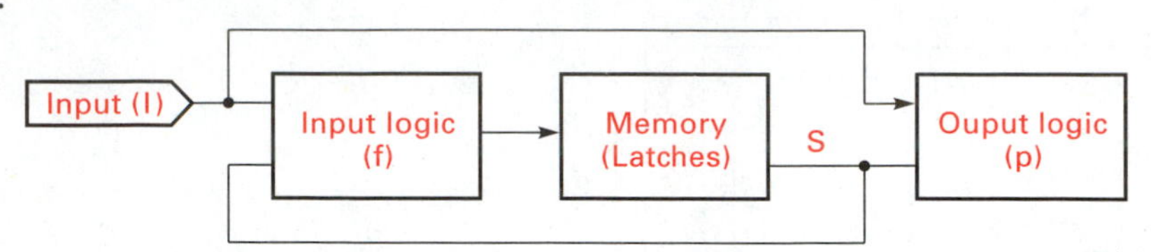

a. Total stable state = S,I

b. Input state = I

c. Secondary state = S

5. a. Input variables: X_2X_1
Output variable: OUT
State variables: S_2S_1

$$S_1 = S_2S_1X_2' + S_2'S_1X_2 + S_2'X_1' + S_2S_1'X_2X_1$$
$$S_2 = S_1X_1 + S_1'X_1'$$

Next state table

Present total state				Next total state					
S_2	S_1	X_2	X_1	S_2	S_1	X_2	X_1	Stable?	OUT
0	0	0	0	0	1	0	0	N	1
0	0	0	1	0	0	0	1	Y	0
0	0	1	0	1	1	1	0	N	1
0	0	1	1	1	0	1	1	N	0
0	1	0	0	0	1	0	0	Y	0
0	1	0	1	1	0	0	1	N	1
0	1	1	0	1	1	1	0	N	0
0	1	1	1	1	1	1	1	Y	1
1	0	0	0	0	0	0	0	N	1
1	0	0	1	0	0	0	1	N	0
1	0	1	0	1	0	1	0	Y	1
1	0	1	1	1	1	1	1	N	0
1	1	0	0	0	1	0	0	N	0
1	1	0	1	1	1	0	1	Y	1
1	1	1	0	1	0	1	0	N	0
1	1	1	1	1	0	1	1	N	1

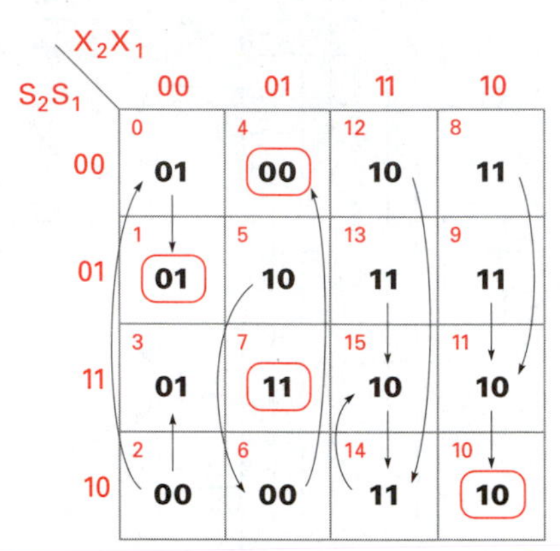

There is no stable total state for input $X_2X_1 = 11$.

b. Input variables: X_2, X_1
State variables: S_B, S_A
Output variables: $O_2\,O_1$

$R_1 = X_2'X_1'$
$S_1 = X_2X_1$
$R_4 = X_2X_1'$
$S_4 = X_2'X_1$
$S_A^+ = S_1 + R_1'S_A = X_2X_1 + (X_2'X_1')'S_A$
$\quad = X_2X_1 + X_2S_A + X_1S_A$
$S_B^+ = S_4 + R_4'S_B = X_2'X_1 + (X_2X_1')'S_B$
$\quad = X_2'X_1 + X_2'S_B + X_1S_B$
$O_1 = S_A'S_B'X_2' + S_AS_BX_2$
$O_1 = S_A'S_B'X_2 + S_AS_BX_2'$

Next state table										
Present total state				Next total state						
S_2	S_1	X_2	X_1	S_2	S_1	X_2	X_1	Stable?	O_2	O_1
0	0	0	0	0	0	0	0	Y	0	1
0	0	0	1	1	0	0	1	N	0	1
0	0	1	0	0	0	1	0	Y	1	0
0	0	1	1	0	1	1	1	N	1	0
0	1	0	0	0	0	0	0	N	0	0
0	1	0	1	1	1	0	1	N	0	0
0	1	1	0	0	1	1	0	Y	0	0
0	1	1	1	0	1	1	1	Y	0	0
1	0	0	0	1	0	0	0	Y	0	0
1	0	0	1	1	0	0	1	Y	0	0
1	0	1	0	0	0	1	0	N	0	0
1	0	1	1	1	1	1	1	N	0	0
1	1	0	0	1	0	0	0	N	1	0
1	1	0	1	1	1	0	1	Y	1	0
1	1	1	0	0	1	1	0	N	0	1
1	1	1	1	1	1	1	1	Y	0	1

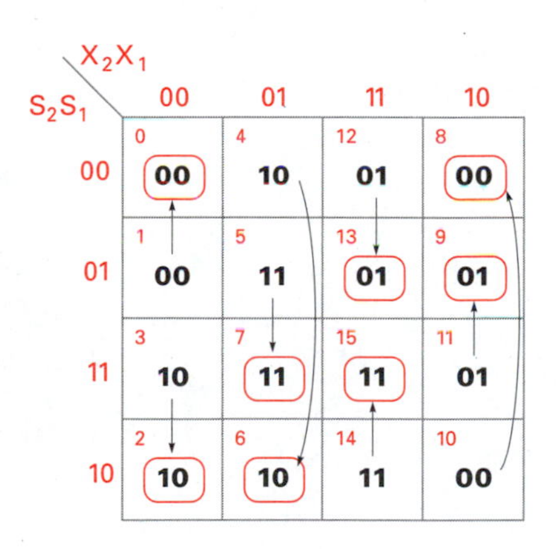

7.

Flow Table				
	Next state/Q			
	CT=			
Present state	00	01	10	11
A	(A/0)	B/0	H/0	—
B	A/0	(B/0)	—	C/1
C	—	D/1	F/1	(C/1)
D	G/1	(D/1)	—	E/0
E	—	B/0	H/0	(E/0)
F	G/1	—	(F/1)	C/1
G	(G/1)	D/1	F/1	—
H	A/0	—	(H/0)	E/0

9. **a.** Race: Two or more state variables change during a state transition.

b. Noncritical race: Race that eventually reaches the correct state.

c. Critical race: Race whose transitions are dependent on which variable changes first.

d. Cycle: Transitions through a series of unstable states.

11.

					Next state			
F_4	F_3	F_2	F_1	State	00	01	10	11
0	0	0	1	S0	a	S0	b	S0
0	0	1	0	S1	S1	c	S1	b
0	1	0	0	S2	S2	S2	d	d
1	0	0	0	S3	S3	e	S3	S3
0	1	0	1	a	S2	—	—	—
0	0	1	1	b	—	—	S1	S0
0	1	1	0	c	—	S2	—	—
1	1	0	0	d	—	—	S3	S3
1	0	0	1	e	—	S0	—	—

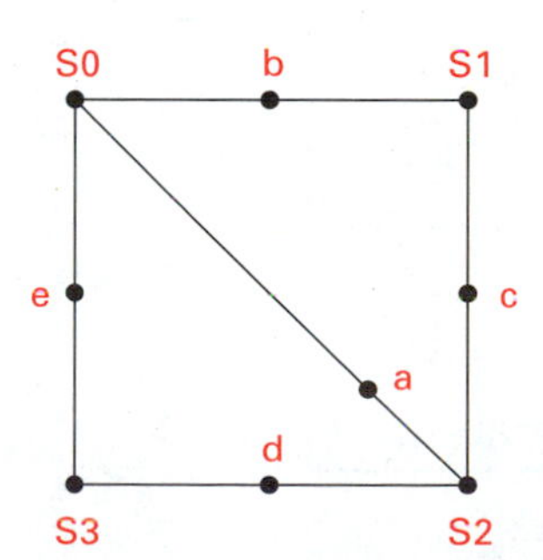

17. An IPAC is used to exchange data between synchronous circuits that do not share a common clock. The IPAC operates by triggering on the edges of the data inputs. Data can be latched to the outputs by the clock of the receiving module, eliminating setup and hold-time violations.

19.

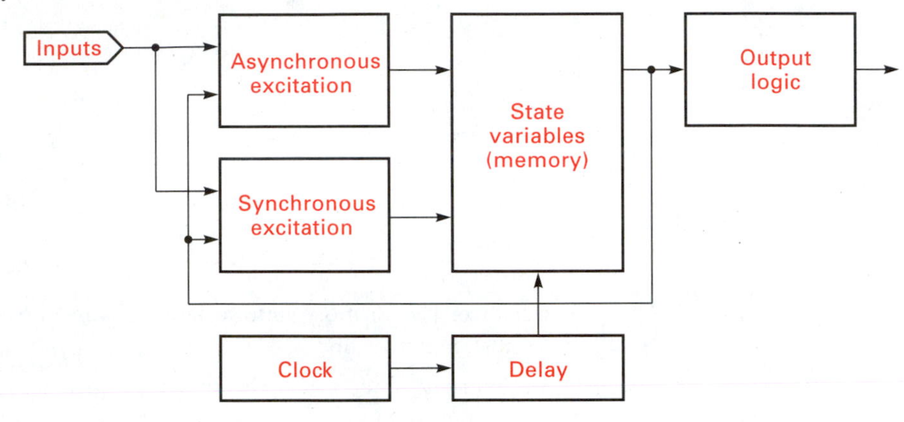

The MOM system has a speed advantage over a purely synchronous sequential circuit. Many of the transitions occur asynchronously. Logic is added to eliminate hazards from different propagation delays. The MOM system has the disadvantage of circuit complexity.

CHAPTER 9

1. The minimum size ROM required is 512 bytes of 5 bits each. The solution shown uses a standard 8-bit wordsize.

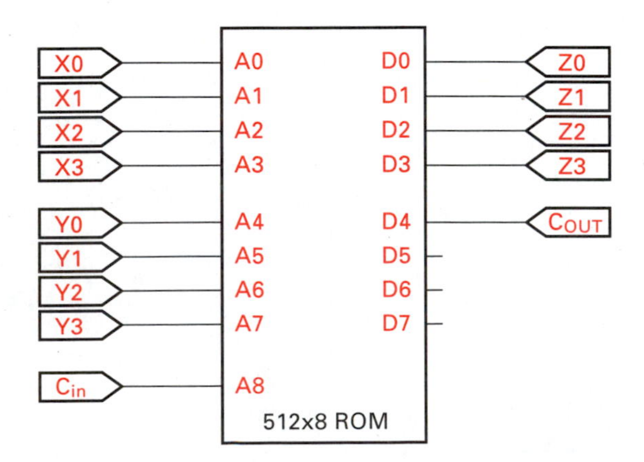

Address	Data
0A0	0A
0A1	0B
0A2	0C
0A3	0D
0A4	0E
0A5	0F
0A6	10
0A7	11
0A8	12
0A9	13
0AA	14
0AB	15
0AC	16
0AD	17
0AE	18
0AF	19
0B0	0B

9. If enough product terms are not available, intermediate functions can be generated that are fed back as input variables to other functions.

15. A GAL has more versatile output logic than a registered PAL. An XOR gate allows the output of the OR array to be selectively inverted. An output multiplexer allows the output to be chosen between a register, or the OR array. A feedback multiplexer allows the selection of various signals to be fed back into the array, including the flip-flop output, the pin output, or the output from an adjacent stage.

17. An EPLD is ultraviolet erasable. It can be reprogrammed. ELPDs are manufactured using CMOS EPROM technology. The advantages of an EPLD are: reprogrammability, lower power consumption, higher density, easier testability, and security (with programmable protection bits.) The disadvantage of an EPLD is: slower operation.

19. a.

```
| in:(a, b, E),
| out:(X)
|
| X = E ?? (a & b)
```

b.

```
| in:(a, b, E),
| out:(Z)
|
| Z = E ?? (a ## b) # (a & b')'
```

21. a.

```
| 2 × 4 Multiplexer
| in:(I[3..0] S[1..0]), out: O
| n = 0..3: O = I[n] & S[1..0]
```

b.

```
| 4-bit odd parity checker
| in:(a, b, c, d), out: P
| P = (a ## b ## c ## d)'
```

c.

Type-D flip-flop with tristate outputs, negative edge trigger, active low asynchronous preset, and clear

```
| in:(D, S, R, E), out: Q, clock: CLK
| conditioning: clock \\ Q
| Q = E ?? dff(D, CLK, S', R')
```

23. The Xilinx FPGA uses an external ROM or EPROM to store the interconnect information, where the Actel FPGA is programmed by blowing fuses within the switching matrix. The logic functions in the Xilinx CLB are performed by a lookup table, where the Actel LM uses a cell of AND and OR gates.

25. 24.

27. Xilinx

3 16-bit data registers: each 16-bit register will use 2 RD8 macro functions, 8 CLBs each, total 48 CLBs.
A 16-bit adder will use 16 GADD macro functions, 1 CLB each, total 16 CLBs.
Total circuit realization: 64 CLBs.

Actel

3 16-bit data registers: each 16-bit register will use 2 DLC8A, DLE8, or DLM8 macro functions, 8 LMs each, total 48 LMs. A 16-bit fast adder will use 1 FADD16 macro function, taking 78 LMs.
Total circuit realization: 126 LMs.

Direct comparison of designs is possible because of the macro function libraries. The libraries provided by Xilinx and Actel provide the block requirements of each macro function.

29. The Actel FPGA allows for higher gate utilization for large combinational logic functions, because of the more complex interconnection logic. Large numbers of flip-flops are more efficiently realized with the Xilinx FPGA because of the CLB's capability of realization of flip-flops and combinational logic within the same logic cell. The Xilinx 3000 series increases efficiency with the addition of a second flip-flop within the cell.

CHAPTER 10

1. a.

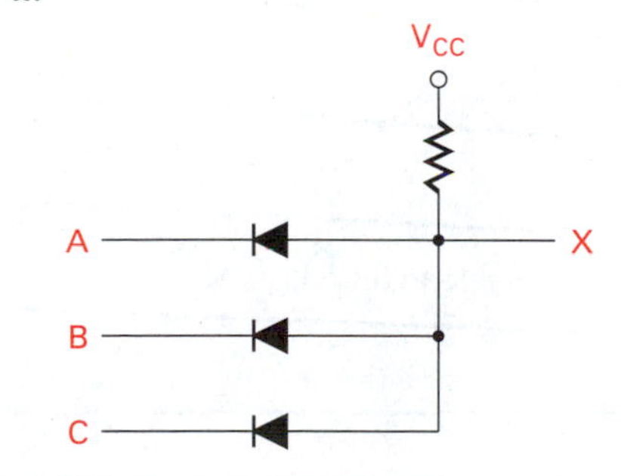

b.

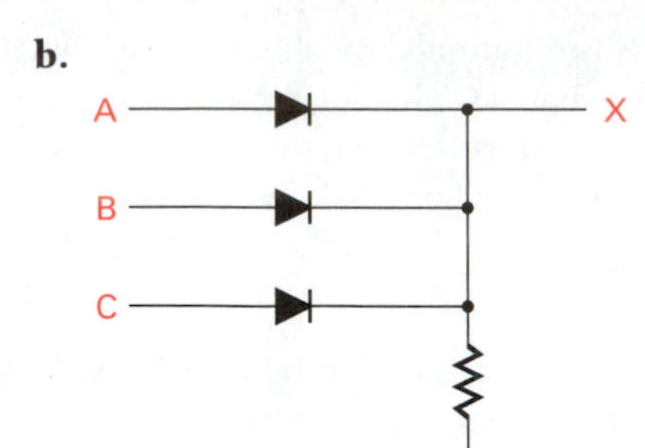

3.

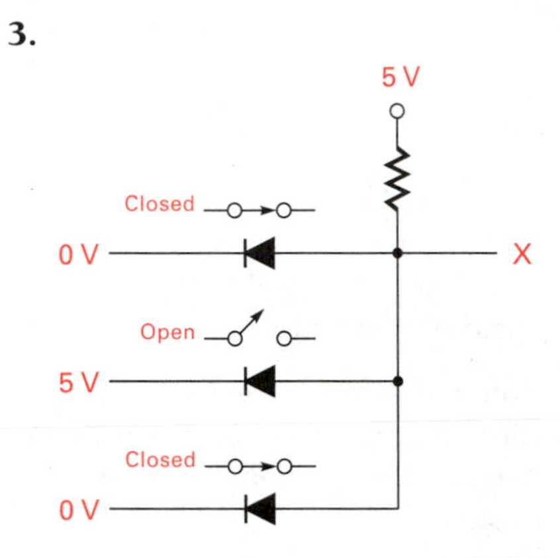

5. Saturation. The transistor is fully ON. From the collector to emitter terminals, the transistor appears as a closed switch (ideal). When a transistor is saturated, applying additional base current will not increase the collector current.

Cut-off. The transistor is fully OFF. From the collector to emitter terminals, the transistor appears as an open switch.

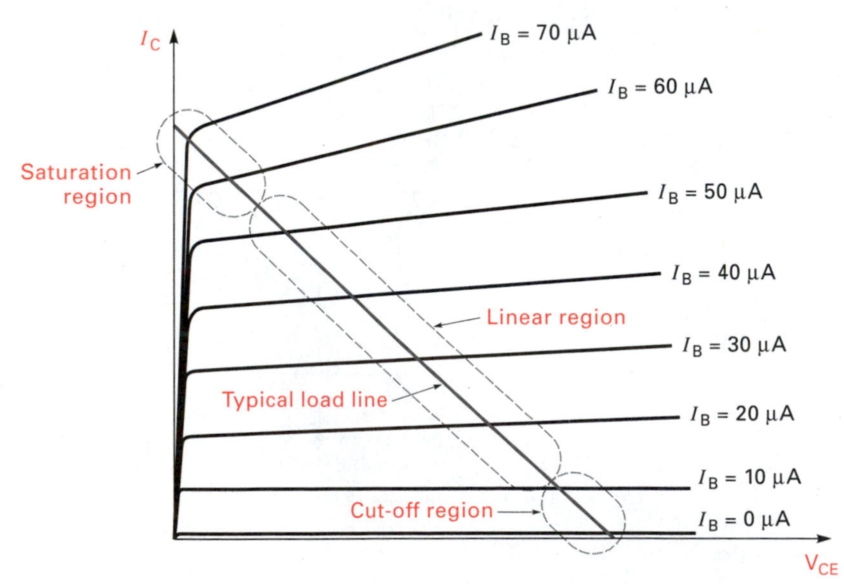

The I_b values shown are representative of relative magnitude.

7.

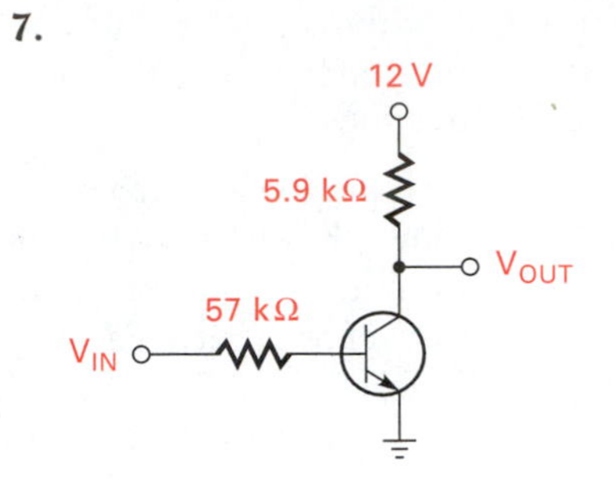

9.

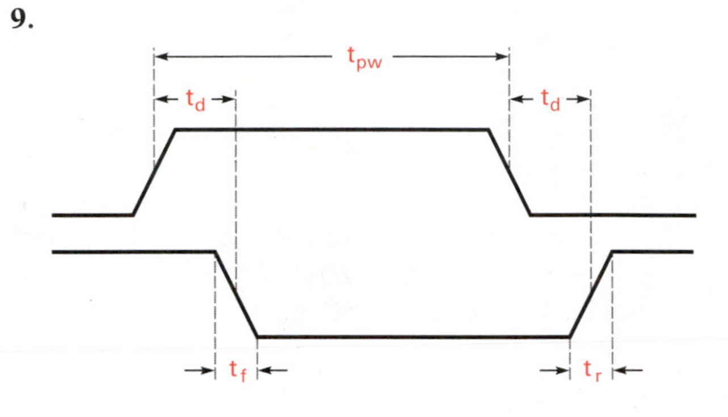

11.

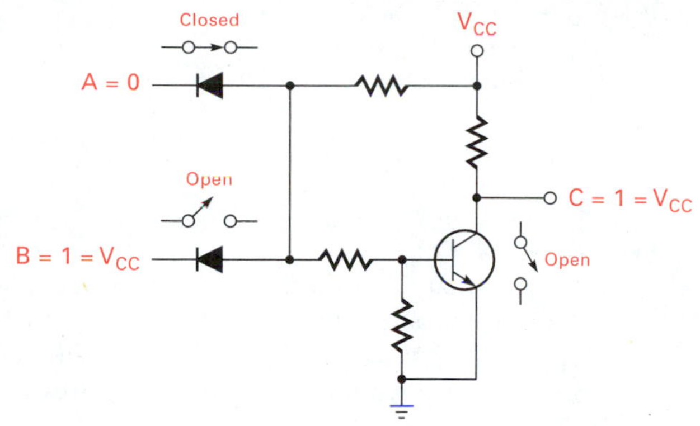

13. The fan-out will increase if R_1 is increased, because less current will be required for the drop across R_1 to be sufficient to cut off the transistor.

15.

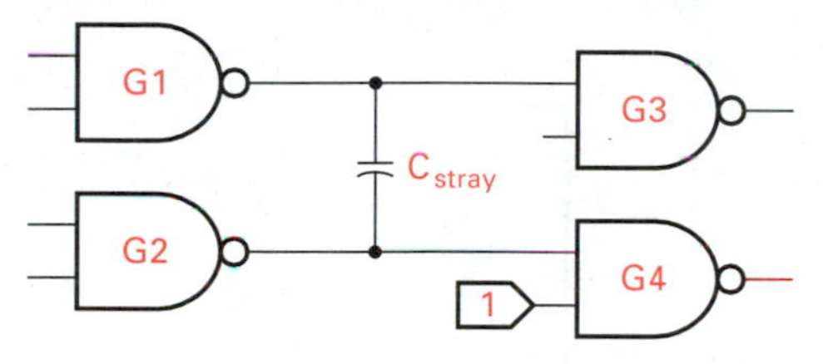

17. The totem-pole output of the TTL circuit solved the noise problem caused by the high-impedance output of the DTL gate by reducing the output impedance.

19.

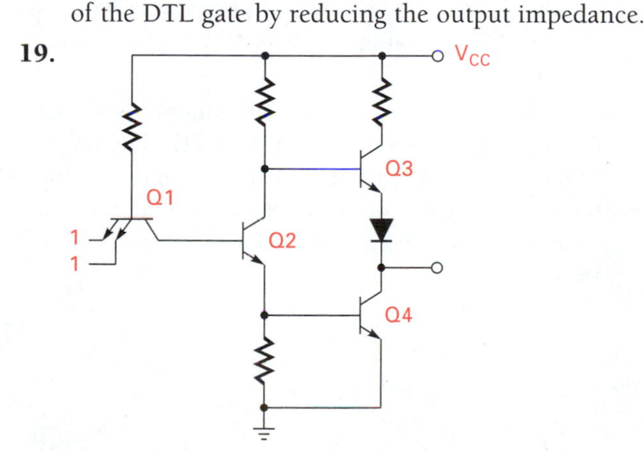

Q1: Cutoff, but the base-collector junction is forward biased.
Q2: ON
Q3: OFF
Q4: ON

21. Ignoring V_{CE}, I_4 will be 33.8 mA when Q_3 and Q_4 are on. If R_4 were zero, I_4 would be unlimited. R_4 limits the current through the totem pole output transistors.

23. **a.** $-55°C \leq T_A \leq 125°C$
b. $0°C \leq T_A \leq 70°C$

25. V_{OL} would increase. The figure I_{OL} is the maximum current that the output transistors can sink without the output voltage rising above 0.4 V.

27. Switching speed, slowest to fastest.
TTL
LSTTL
ALSTTL
STTL
ASTTL

Power consumption, highest to lowest.
STTL
TTL
ASTTL
LSTTL
ALSTTL

29. A Schottky junction is formed from a doped semiconductor and metal.

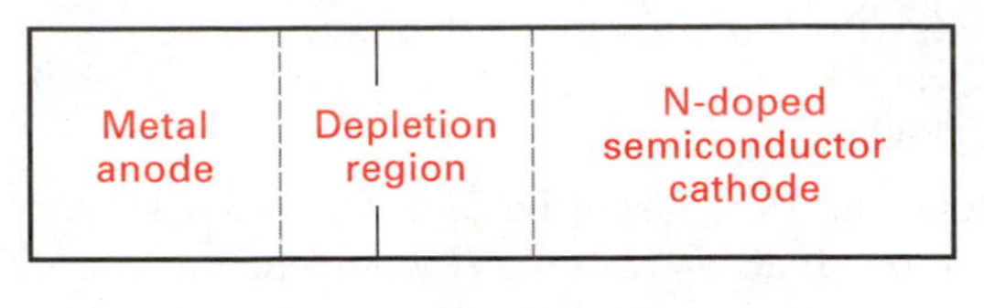

33.

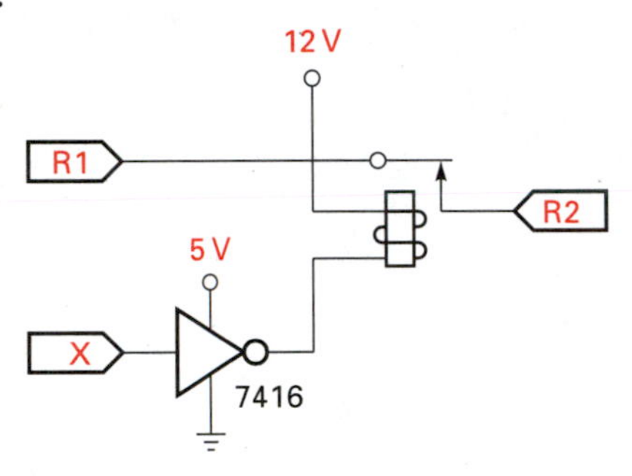

The contacts between R1 and R2 are closed when $X = 1$, and open when $X = 0$.

35. To connect multiple tristate devices to the same bus, the output of only one device at a time can be enabled. Enable circuitry must be connected to each tristate device that keeps all but one device in the high-impedance state.

37. A Schmitt trigger input gate is used to trigger on noisy signals where the input voltage level may not remain at the "true" level long enough for normal TTL circuits, or may fall between the "true" and "false" levels. Schmitt trigger input gates are useful for digital circuits that take their inputs from noisy sources, or from devices that do not have fast, discrete, TTL-level switching characteristics.

39. $V_{IL} = -1.48$ V (MAX)
$V_{IH} = -1.13$ V (MIN)
Both values are for 25°C.

41. -1.5 V.

43. 1.75 ns (max, 25°C)
Faster.

45. $V_{REF} = -1.43$ V
Assuming that all of the voltage drop in the differential stage is across the resistors, $V_{OL} = -1.81$ V. $V_{OH} = -0.72$ V maximum. No further determination is possible without transistor specs or resistor values.

47. Complementary Metal-Oxide Semiconductor.

49.

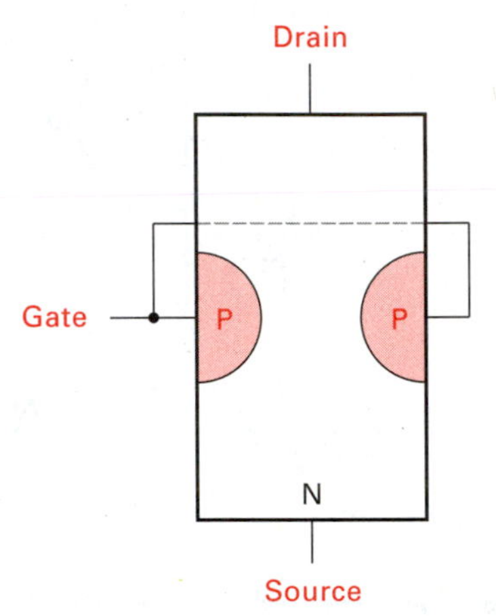

The gate controls the drain-source current by creating depletion regions that extend from the gate material into the substrate. As the gate-source voltage increases, the depletion regions grow larger, decreasing the number of charge carriers available for carrying current from the drain to the gate. At pinch-off, the depletion regions extend fully across the substrate, leaving no carriers for drain-source current.

51. A JFET has a semiconductor PN junction gate that is normally reverse-biased. A MOSFET has no PN junction for a gate. The P and N semiconductor materials are separated by an insulator that prevents current from flowing, regardless of forward- or reverse-biasing.

53.

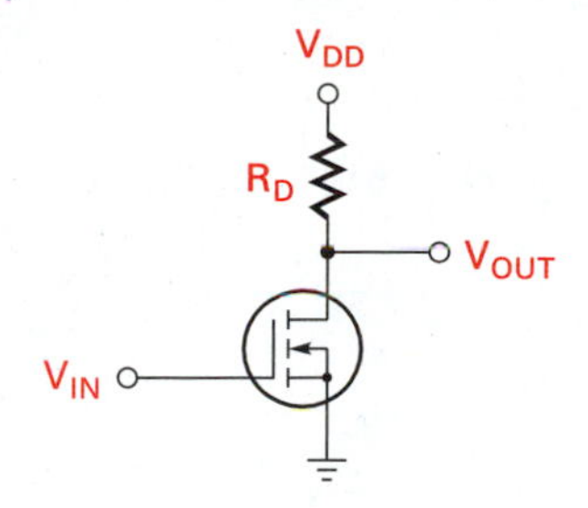

55.

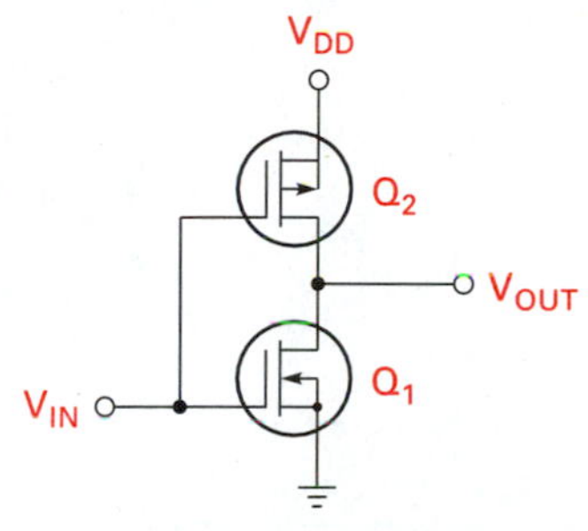

When $V_{IN} = 0$ V, Q_1 is cutoff and Q_2 is on. $V_{OUT} = V_{DD}$.
When $V_{IN} = V_{DD}$, Q_1 is on and Q_2 is cutoff. $V_{OUT} = 0$ V.

57. a.

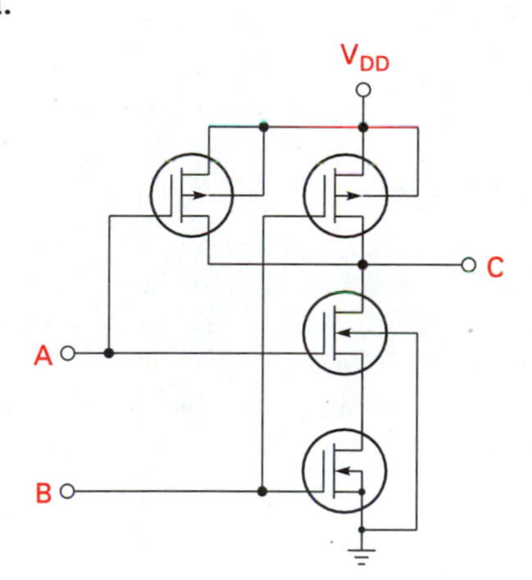

b.

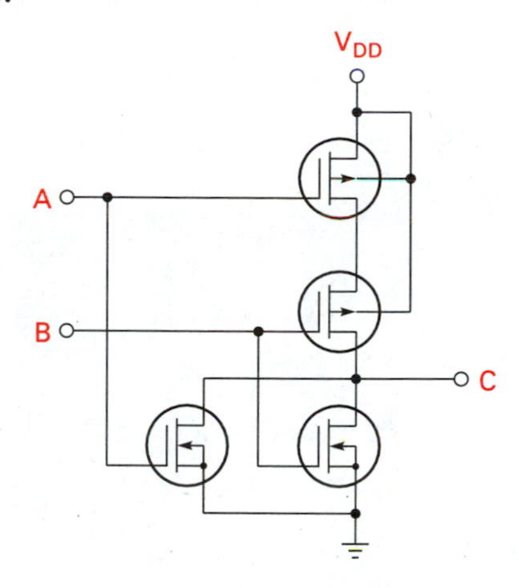

59. 120 mW

61. Logic "0": Min 0 V, Max 3 V
Logic "1": Min 7 V, Max 10 V

63. **a.** CMOS
b. ECL
c. TTL
d. TTL
e. ECL
f. TTL
g. CMOS (74AC, 74ACT)

INDEX